V.S. Beskin • A. Balogh • M. Falanga •
M. Lyutikov • S. Mereghetti • T. Piran •
R.A. Treumann
Editors

The Strongest Magnetic Fields in the Universe

Previously published in *Space Science Reviews* Volume 191,
Issues 1–4, 2015

Springer

Editors
V.S. Beskin
Russian Academy of Sciences
Moscow, Russia

A. Balogh
The Blackett Laboratory
Imperial College London
London, UK

M. Falanga
International Space Science Institute
Bern, Switzerland

M. Lyutikov
Purdue University
West Lafayette, IN, USA

S. Mereghetti
INAF-IASF Milano
Milano, Italy

T. Piran
Racah Institute for Physics
The Hebrew University
Jerusalem, Israel

R.A. Treumann
International Space Science Institute
Bern, Switzerland

ISSN 1385-7525 Space Sciences Series of ISSI
ISBN 978-1-4939-3549-9 ISBN 978-1-4939-3550-5 (eBook)
DOI 10.1007/978-1-4939-3550-5

Library of Congress Control Number: 2016930405

Springer New York Heidelberg Dordrecht London

Cover Image: Numerical simulation of accretion from a disk in strong magnetic fields of a compact object. Taken from Romanowa and Owocki (this issue) Fig. 12.

Printed on acid-free paper

Springer is part of Springer Science+Business Media (www.springer.com)

Foreword: A conclusion to the ISSI series on astrophysical magnetic fields

Magnetic fields are a fundamental component of the physical world on all scales—as constituents of the electromagnetic environment of all matter. The interaction between the magnetic fields and matter is at the heart of many dynamic processes that shape astrophysical objects and their environments. Regions of space threaded by magnetic fields control or at least influence the interactions that take place between them. Observing and understanding magnetic fields and the role they play in physical processes in the solar system and beyond have been the subject of a series of Workshops and publications by the International Space Science Institute since 2008.

The first three volumes covered the origin and dynamics of solar magnetism, the magnetic fields of the planets, and the magnetic field of the Earth. These were meant to constitute a trilogy of magnetic fields within the confines of the solar system. Then came the fourth volume which extended the coverage to the whole Universe, but concentrating on large-scale magnetic fields, comparable to, or smaller in magnitude than those directly measured within the solar system. The four volumes are:

The Origin and Dynamics of Solar Magnetism, M.J. Thompson, A. Balogh, J.L. Culhane, Å. Nordlund, S.K. Solanki, J.-P. Zahn (Eds.), Space Science Series of ISSI, Volume 32, ISBN 978-1-4419-0238-2, published in May 2009
Reprinted from Space Science Reviews Volume 144, No. 1–4, 2009

Planetary Magnetism, U.R. Christensen, A. Balogh, D. Breuer, K.-H. Glaßmeier (Eds.), Space Science Series of ISSI, Volume 33, ISBN 978-1-4419-5900-3, published in September 2010
Reprinted from Space Science Reviews Volume 152, No. 1–4, 2010

Terrestrial Magnetism, G. Hulot, A. Balogh, U.R. Christensen, C. Constable, M. Mandea, N. Olsen (Eds.), Space Science Series of ISSI, Volume 36, ISBN 978-1-4419-7954-4, published in January 2011
Reprinted from Space Science Reviews Volume 155, 1–4, 2010

Large-Scale Magnetic Fields in the Universe, R. Beck, A. Balogh, A. Bykov, R.A. Treumann, L.M. Widrow (Eds.), Space Science Series of ISSI, Volume 39, ISBN 978-1-4614-5727-5, published in October 2012
Reprinted from Space Science Reviews Volume 166, 1–4 and Volume 169, 2012

The topic of the current volume contains reviews of completely new aspects of magnetic fields in the astrophysical Universe—and in many cases of aspects of physical processes and phenomena fundamentally different from those addressed in the first four volumes. The strength of the magnetic fields of the compact objects reviewed in this volume is up to 8 to 10 orders of magnitude higher than that of typical sunspots which are the strongest fields in the solar system. Large-scale astrophysical magnetic fields, such as interstellar and galactic fields are weaker still than the fields experienced in the solar system. The gap between field strengths covered in the previous volumes and those in astrophysical object and environments addressed in the present volume is illustrated in Fig. 1.

The first four volumes in the ISSI series on magnetism have left open the questions related to the generation and effects of the very strong magnetic fields found near the most compact ob-

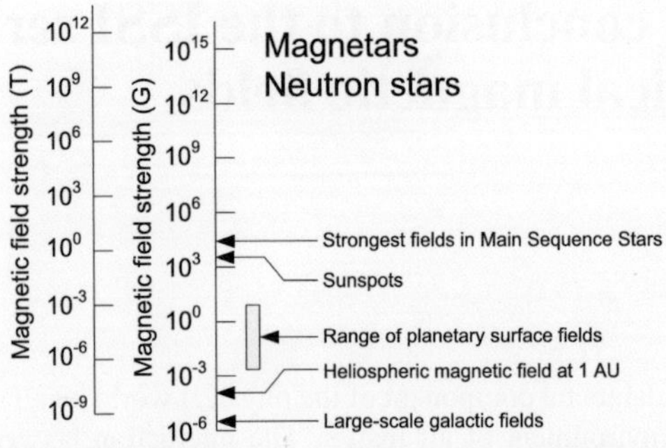

Fig. 1 The range of astrophysical magnetic field strengths, covering 21 orders of magnitude, effectively representative of the whole spectrum of astrophysical phenomena in the Universe. *The lower part of the graph* shows the topics covered by four previous volumes in the Space Science Series of ISSI; the objects with extremely high magnetic fields, neutron stars, magnetars, are the subject of the current volume

jects in the Universe. The motivation for the current volume has been to complete the review of such magnetic fields, associated with Magnetic Stars, White Dwarfs, Neutron Stars and Active Galactic Nuclei (AGNs) and their environments. The reviews presented here describe the current understanding of how the extremely strong magnetic fields of these objects are generated, how they interact with matter and how they generate the broad range of observed phenomena, among them shock waves and many different types of radiation over a very broad spectrum. Of particular interest are the phenomena leading to the generation of astrophysical jets and their complex physics, the occurrence of Gamma Ray Bursts, the formation and extreme properties of relativistic shocks and the production of high-energy particles by the cosmic engines. Radio pulsars (highly magnetised neutron stars) have been studied for the last half century; these have magnetic fields in the range 10^{11} to 10^{13} G. In such strong fields the basic physics changes for the reason that the electron Larmor (gyro) radius drops below the Bohr (atomic) radius. Magnetars are also neutron stars, but with magnetic fields in the range 10^{13} to 10^{16} G. The two or more orders of magnitude difference in the magnetic fields of these objects leads to differences in their energetic output. In particular, the explosive energy in the observed "giant flares" from magnetars, powered by the dissipation of magnetic energy, is of order 10^{44} ergs, about 12 orders of magnitude more energetic than the largest of solar flares. We are indeed in the presence of extreme physical phenomena, ascribed to the ability of these exceptional astrophysical objects to generate the exceptionally strong magnetic fields.

There are fifteen review papers in this volume, providing a comprehensive and up-to-date coverage of a field that remains in a state of rapid evolution. With this volume, the ISSI series on astrophysical magnetic fields concludes an ambitious review of topics that are central to the progress of astrophysics and our understanding of the Universe.

The ISSI Workshop on the Strongest Magnetic Fields in the Universe was held in ISSI, Bern, Switzerland on 3–7 February 2014. The Convenors of the Workshop (André Balogh, Vasily Beskin, Maurizio Falanga, Maxim Lyutikov, Sandro Mereghetti, Tsvi Piran and Rudolf Treumann) and the Editors of this volume are greatly indebted to all the participants of the Workshop who brought their broad range of expertise and interest in astrophysics to deepen our understanding of the issues related to the extreme magnetic fields in the Universe and their parent objects. The resulting collection of review papers was the outcome of the exchanges and fruitful collaboration among the participants; we thank them for their successful efforts to integrate the lessons learned in the different topics, as the reviews in the volume testify. Thanks are also due to the reviewers of the papers; in all cases the reviews were thorough and constructive and the volume bears witness to their contribution. We thank the staff of ISSI for their dedicated support: Prof. Rafael

Rodrigo, Executive Director, and his colleagues Jennifer Fankhauser, Andrea Fischer, Saliba Saliba and Silvia Wenger. Finally the Editors thank the staff of Space Science Reviews, as well as the production staff for their patience on occasion and for an excellently produced volume.

London, UK
Bern, Switzerland

André Balogh
Maurizio Falanga

Contents

Part I
Introduction

DOI 10.1007/978-1-4939-3550-5_1
Reprinted from *Space Science Reviews* Journal, DOI 10.1007/s11214-015-0188-1

Magnetic Fields at Largest Universal Strengths: Overview

V.S. Beskin[1] · A. Balogh[2] · M. Falanga[3] · R.A. Treumann[3]

Received: 20 May 2015 / Accepted: 15 July 2015 / Published online: 29 July 2015

Abstract A brief review is given about the role strong magnetic fields play in the universe. We list the main observational and theoretical achievements treated in the following chapters including a number of open questions which future research is going to attack. Strong fields in the universe exceed any large scale fields by several orders of magnitude, at first glance suggesting that their generation mechanisms would be different. However, it is believed that gravitational collapse and magnetic flux conservation is responsible for the amplification of fields generated in the progenitors to the observed strengths. In this sense the extremely strong fields are mainly fossil, and their variety confirms the different masses and stages where the collapse comes to rest, at the lightest in white dwarfs and at the strongest in magnetars, which are a particular class of neutron stars with strongly inhomogeneous particularly structured crust. Various effects related to the detection of such fields, radiation generation and consequences for the environment are pointed out and referred to the relevant chapters in this volume.

Keywords Magnetic stars: white dwarfs, magnetars, pulsar · Neutron stars · Accretion, outflow and jets

✉ R.A. Treumann
artt@issibern.ch

V.S. Beskin
beskin@td.lpi.ru

A. Balogh
balogh@issibern.ch

M. Falanga
mfalanga@issibern.ch

[1] Lebedev Physical Institute, Russian Academy of Sciences, Moscow, Russian Federation

[2] Imperial College London, London, UK

[3] International Space Science Institute, Bern, Switzerland

1 Introduction

That magnetic fields in the universe are abundant was unknown until the first half of the past century, when sporadic radio emission was discovered from the sun and later also from more distant objects. Such emissions were attributed to synchrotron radiation requiring the presence of moderately strong magnetic fields. Magnetic fields on the sun were estimated in the range of several in the corona to several thousand gauss (G) in sunspots and below. More remote large scale fields turned out to be substantially weaker, of the order of $\mu G < B <$ mG. It is now established that magnetic fields in the universe can be as weak as μG on the large galactic and intergalactic scales (Beck et al. 2012) and possibly as strong as 10^{17} G in magnetars. Obviously, there is a profound difference in scales between these limits. Interstellar fields range in the somewhat stronger (up to mG) field strengths, earth-like planets have fields from fractions of one to several G, the remainder from ten G up to, say, 10^5 G are reserved for stars, non-degenerate stars in the first place. The strongest fields are found in degenerate stars: white dwarfs, neutron stars, including pulsars (see Trümper et al. 1977 for the first detection of fields the order $\sim 10^{13}$ G in the HerX1 pulsar) and, ultimately, magnetars. Moderately strong magnetic fields exist in parts of accretion disks, while strong fields are believed to be involved in the compact objects relatively close to their horizons as well as in the inner parts of their outflow jets. Though it is not known what the ultimate upper limit is for any magnetic fields in the universe, it has been argued (Duncan 2000; Treumann et al. 2014) that such fields are limited by quantum electrodynamics to strengths quite far above the strongest observed magnetar fields. Field strengths could, theoretically, go up to $\sim 10^{32}$ G or so (see Fig. 1). However, in the absence of magnetic monopoles, no mechanism is known which could produce fields of such strengths. Today it is believed that magnetic fields are generated in magnetic dynamo and also battery processes which feed mechanical energy of a turbulent flow of electrically conducting matter into current filaments the integrated effect of which gives rise to observed magnetic fields. Magnetic dynamos of this kind act in the magnetohydrodynamic fluid within magnetic stars. In earthlike planets and neutron stars they act in the metallic-fluid conducting and turbulent matter flow in their interiors. In magnetars, the strongest magnetised objects known so far, some version of neutron stars, it has been shown (Duncan and Thompson 1992; Thompson and Duncan 1995) that neutron star fields have been amplified a bit by further dynamo action during collapse. Concentration of the field in crustal inhomogeneities cares for locally very strong fields responsible for strong magnetic loops reaching out into the magnetar environment. In the environment of highly compact objects, jets and the close part of accretion disks, it seems that dynamo action and magneto-rotational instability (MRI) as well sign responsible for fields substantially stronger than on large scales. Additionally, strong fields can be produced by general relativistic effects known as the Znajek–Blandford (Blandford and Znajek 1977) mechanism acting near rotating black holes and generating magnetic funnels. In all those cases we are in the range of fields much stronger than on large scales though substantially weaker than the above mentioned theoretical limits. Magnetic fields of these comparably high strengths affect the behaviour of matter exposed to them. Once the electron cyclotron energy $\hbar\omega_{ce} \sim e^2/4\pi\varepsilon_0 a_0$ exceeds the Coulomb energy of an electron on an atomic Bohr radius, which happens for fields $B > B_0 \approx 2 \times 10^5$ T $= 2 \times 10^9$ G, the magnetic field affects the atomic properties and fundamentally changes the behaviour of matter. This will necessarily happen in the interiors of neutron stars, magnetars, pulsars. It also affects the matter in the magnetospheres of those objects. On the other hand, when for weaker fields the magnetic energy becomes comparable to the gravitational binding energy it would affect the orbits of bound objects. Such fundamental properties of strong magnetic fields

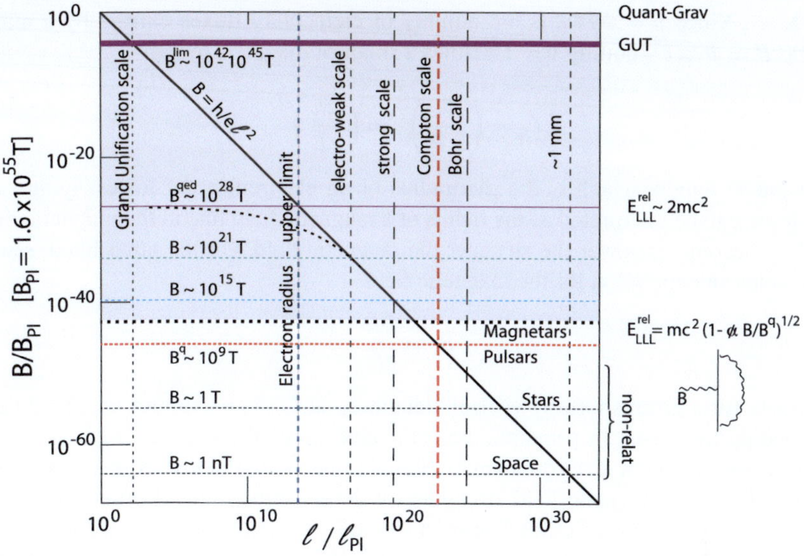

Fig. 1 Log-Log plot scaling of the maximum possible magnetic field strength, B_c, normalized to the (fictitious) Planck-magnetic field, B_{Pl}, as function of fundamental length scales based on Eq. (1). Length scales ℓ on the abscissa are normalized to the Planck length ℓ_{Pl}. The *dotted red cross* indicates the crossing point of the Compton length with the Aharonov–Bohm critical magnetic field line at the so-called quantum limit field $B^q \approx 10^9$ T, the critical field of magnetized neutron stars (pulsars) in agreement with observation of the strongest cyclotron lines. *Horizontal lines* indicate the relation between other length scales and critical magnetic fields under the assumption of validity of the Aharonov–Bohm scaling. Space magnetic fields correspond to scales of $\sim$1 mm. Strongest detected magnetar fields correspond to the first order relativistic correction on the lowest Landau level energy E_{LLL} (shown as graph on the right with $\bar{\alpha} = \alpha/2\pi$ the reduced fine structure constant). Inclusion of higher order corrections would allow for fields of up to $B^{qed} \sim 10^{28}$ T deep in the (*shaded*) relativistic domain which have not been observed. It is interesting that this limit coincides approximately with the measured (Gabrielse et al. 2006) absolute upper limit on the electron radius (*vertical blue dashed line*). The *black dashed curve* indicates a possible deviation of the Aharonov–Bohm scaling near the quantum electrodynamic limit. At GUT scales, fields could theoretically reach values up to $\sim 10^{45}$ T, according to simple Aharonov–Bohm scaling (adapted from Treumann et al. 2014)

are reviewed in the introductory paper Lai (2015) of this volume. The following chapters provide a summary of magnetic fields in various astrophysical candidates of strong fields: magnetic stars, white dwarfs, pulsars, magnetars, their observational evidence, effects and theories.

2 Physics in Very Strong Fields

Quantum mechanics provides an easy way to obtain a first limit on the magnetic field from solution of Schrödinger's equation, first found by Landau (1930), of an electron orbiting in a straight magnetic field. The physical interpretation of this solution was given much later in the Aharonov–Bohm theory (Aharonov and Bohm 1959) of quantization of magnetic flux Φ. From the requirement that the magnetic flux confined in an electron gyration orbit must be single valued, Aharonov and Bohm found that the magnetic flux $\Phi = \nu\Phi_0$ in a magnetic field B is quantized with flux element $\Phi_0 = 2\pi\hbar e$, e the elementary charge, and

$\nu = 1, 2, \ldots$. Since $\nu = \Phi/\Phi_0$ is the number of elementary fluxes carried by a magnetic field, and $B = \Phi/\pi\ell^2$, putting $\nu = 1$ defines a smallest magnetic length

$$\ell_B = \left(\frac{\Phi_0}{\pi B}\right)^{\frac{1}{2}} = \left(\frac{2\hbar}{eB}\right)^{\frac{1}{2}} \tag{1}$$

This magnetic length, which is the gyroradius of an electron in the lowest lying Landau energy level, can be interpreted as the radius of a magnetic field line in the magnetic field B. Field lines become narrower the stronger the magnetic field. On the other hand, rewriting Eq. (1) yields an expression for the magnetic field

$$B_c = \frac{2\hbar}{e\ell_c^2} \tag{2}$$

from which, for a given shortest "critical" length $\ell_B \equiv \ell_c$ the maximum magnetic field B_c corresponding to ℓ_c can, in principle, be estimated. Since the Planck length is the shortest scale where quantum electrodynamics ceases to be valid, the last expression yields the theoretical limit for magnetic field strengths for a very wide range. Putting, for instance, $\ell_c = 2\pi\hbar/mc$ equal to the electron Compton length $\lambda_0 = h/mc$, one obtains the critical pulsar (neutron star) magnetic field strength $B_{ns} \approx 3 \times 10^9$ T $= 3 \times 10^{13}$ G. Approximately this field strength was indeed inferred from observation of the fundamental ($\nu = 1$) electron cyclotron harmonic X-ray line detected from the HerX1 pulsar (Trümper et al. 1977), roughly two decades after Aharonov and Bohm's, and half a century after Landau's theory. However, magnetic fields require a mechanism of generation, i.e. current flow. Since currents are based on lepton (electron, positron) dynamics (including their spin dynamics) current flow is limited once the lepton radius is reached. Its currently best estimate is $r_e \sim 10^{-22}$ m (Gabrielse et al. 2006) which yields the above mentioned theoretical estimate of $B \sim 10^{32}$ G far above any observed field strength.

In addition to current flow, magnetic fields may affect the behaviour of matter. It is known that under classical conditions fields fluid containing oppositely directed magnetic field can annihilate their fields when contacting, a process well known as magnetic reconnection. It causes redistribution of fields and currents, magnetic energy dissipation, plasma jetting and heating as well as violent acceleration of small groups of resonant particles. Such processes, observed and confirmed in near-Earths space also work under astrophysical conditions where in strong field relativistic effects come into play. It has even suggested that this mechanism of magnetic energy annihilation forms the main mechanism of dissipation in strong turbulence of magnetised media. There the injection of mechanical energy at the largest scales causes eddies which themselves cause current flow. Nearly dissipationless cascading of the turbulence down to ever shorter eddy scales leads to the formation of narrow current filaments. Once their width drops below the lepton gyro-radius inertial effects take over, and the current filaments enter into the reconnection regime. The mechanical turbulent energy is then violently dissipated, a mechanism which seems to be important already on the classical scale of moderately strong magnetic fields in stellar environments and coronas and is also expected in stellar interiors.

Much stronger magnetic fields may affect processes on the molecular and atomic scales. This happens when the magnetic energy density becomes comparable to the binding energy density. There is a wealth of phenomena which magnetic fields cause under those conditions, several of them are reviewed in Lai (2015) in the first chapter of this volume. Further aspects of the effects of very strong fields on matter (Ruderman 1974) are discussed in connection with the structure of neutron stars in Pothekhin et al. (2015a, 2015b).

In the following we present a brief overview of the various classes of objects in the universe which support strong magnetic fields. We do not necessarily follow the content of this volume. Instead we just point on the main important and generally interesting facts of interest for the reader who may consult the different chapters for details.

3 Stellar Magnetism

Except for extended and extremely massive objects that will be considered below, strong magnetic fields are a property of a few different classes of stars. From the point of view of generation of such strong fields one may decide between ordinary stars, that is non-degenerate stars and degenerate stars. It is by now known that magnetic fields in either of them result originally from dynamo actions in the stellar interiors even when not acting anymore in which case the fields are 'fossil' remainders of original dynamos.

In non-degenerate stars which are moderately massive objects, the dynamos are believed of acting continuously. Since all information on the fields in those stars (members of the HR-diagram) obtained relies on spectral observations of polarised/unpolarised radiation, the reliability of such measurements is a key problem. Due to limitations in the spatial and temporal resolutions observations provide mostly averages of the real fields which can be polluted by cancellation of oppositely directed magnetic fluxes and distribution of magnetic fields over the stellar surfaces and environments. Average fields can be order s of magnitude smaller than the real spatially resolved magnetic field strengths. Since magnetic fields control also the stellar environment, the various layers of stellar atmospheres, heating rates, chemistry, radiation transport, and even some mass and angular momentum loss, information about magnetic fields is also obtained from observation of such processes.

A critical compilation of the effects of magnetic fields in the different HR-diagram stars, all the various methods inferring about magnetic fields, their strengths and complexity, and the various uncertainties of estimates and interpretation in given here Linsky and Schöller (2015). The paradigm of a 'normal' main sequence star is the Sun. Its average longitudinal surface field is moderate, $\sim \pm 2$ G only, still belonging to strong fields when compared with interstellar values though much weaker than any spatially better resolved fields of $\sim$1.5 kG the surface filling factor of which is $f < 5\ \%$. Pre-main sequence stars generally have substantially stronger average longitudinal fields $<$ 4 kG which channels the disk-accreting matter into small areas on the stellar surface causing hot gas observed in bright He I emission lines. Outside this shocked the field is a multipole field of higher order. Evolution towards the main sequence and higher mass causes weaker fields still of complex structure. Main sequence FM stars cooler than the Sun remain having complex field structures. Zeeman splitting suggests field strength $\sim$3 kG and relatively high filling factors of $f \sim 0.5$. Fields in A- and B-stars on the other hand are lower multipole-like and oblique with respect to their rotation axes, in agreement with dynamo models though chemically peculiar stars seem to be slow rotators only. CP stars among them have rather strong fields $<$ 50 kG which exert strong control of their stellar winds. Weaker fields $<$ 1 kG are found in B stars. Hot main sequence O stars do not have convective zones and therefore, probably, no dynamos. Their fields are believed to be of fossil or just tidal dynamo origin, having strengths $<$ 400 G. Finally post-main sequence stars which ultimately are slow rotators have dynamo generated weaker $\sim$ kG fields which sometimes show rapid (monthly) changes in their magnetic morphology.

4 Generation of Stellar Fields

What concerns the generation mechanisms of stellar magnetism, there is still no complete agreement. Above we stressed that the main mechanism would be the dynamo either ongoing in pre- and main sequence stars or conserved 'fossil' in post-main sequence stars and, as commonly believed, in degenerate stars (white dwarfs, neutron stars, ...). The stellar dynamo remains to be the general paradigm since its first proposal, having been very successful in application to the Sun but substantially less successful in application, for instance, to planetary magnetism. Excluding planets which are not treated in this volume mainly for the reason that observation of their magnetic fields from distance is so far technically impossible, and reference to the magnetic fields of the solar system planets is therefore not necessarily representative, the magnetic dynamo maintains its status of the dominant field generation, mostly assumed to be convection driven though one could also refer to other mechanical sources like tides or general turbulence, accretion driven dynamos and dynamos driven by merging protostellar objects. Indeed, stellar collisions are not as rare as believed. In this case incidence of magnetic fields should increase with stellar mass, which indeed seems to be the case.

The central problem is the question why only a fraction, not even the larger one, of main sequence stars is magnetic. A number of mechanisms, in particular the dynamo and fossil mechanism scenarios are critically reviewed in Ferrario et al. (2015a), mainly in view of their relevance in the generation of the magnetic fields in magnetic white dwarfs and neutron stars, both degenerate stars for which the field must have been either generated in their non-compact progenitors or, as believed, further attributed to accretion from a non-degenerate companion. The latter, however, seems not being feasible for the apparently complete absence of such companions in magnetic white dwarfs. Fossil fields have not just to form during contraction of a star to become a white dwarf (or neutron star), they have also to survive the complex phases of evolution during and after contraction. These processes are indeed nontrivial as they include the decoupling of the convection zone of a contracting star from the core, retardation of the core by magnetic forces and other effects. Thus fields becoming fossil is easy to predict from rough estimates of flux conservation but not easy to demonstrate by taking into account all the physical processes taking place during contraction and collapse, including destruction and dissipation of the progenitor field by magnetic instabilities on all scales including reconnection, and/or reconstruction by magneto-rotational instability effects or turbulence. Extension of dynamo action during contraction of the progenitor and for some time after contraction has therefore been proposed and cannot be discarded as well.

It thus seems that a single universal deterministic process like a convective dynamo is insufficient to stand for all the different magnetic fields attributed to stars from pre-main sequence to neutron stars/magnetars.

5 Degenerate Stars

The two cases of stable degenerate stars which have formed by contraction and collapse are white dwarfs and neutron stars. Among both of them just a fraction is strongly magnetised. Magnetic white dwarfs are reviewed in Ferrario et al. (2015b) which also contains extended tables of essentially all known white dwarfs, their parameters and magnetic fields. The fields cover the range of $10^3 < \langle B \rangle < 10^9$ G. The morphology of the stronger field is fairly complex with presence of higher multipoles. In agreement with estimated ohmic dissipation times of $\sim 10^{11}$ years no relevant decay of their magnetic fields has been found. Indeed, high magnetic fields in white dwarfs correlate with higher masses of the white dwarf thus suggesting that accretion of matter and field is a cause of stronger fields. Weaker fields $\langle B \rangle < 3 \times 10^8$ G are found in binaries (magnetic Cataclysmic Variables) with less detectable multipoles. What concerns the generation of fields in magnetic white dwarfs the fossil hypotheses is paralleled with the assumption of accretion by wind from a hot companion star in the above binaries also explaining their variabilities. At least, the accretion hypothesis of strengthening the field gains increasing theoretical and observational support.

Production of neutron stars, as is well known, proceeds along similar lines as production of white dwarfs with neutron stars being heavier. The gross difference is that they are kept stable not by the degeneration pressure of electrons but of the nuclear matter, neutrons. For reasons of equilibrium and stability of neutrons they consist of a mixture of neutrons, protons and electrons. Neutron stars are heavier than white dwarfs, therefore contraction and collapse provide them with smaller radii, higher densities, and substantially stronger magnetic fields as noted in the introduction. Average fields can become as strong as $\langle B \rangle \sim 10^{14}$ G. Observation of magnetic fields is a clue for gathering information about their internal structure, consistence of their cores and crusts, and their dynamics. Their interiors are superdense, essentially resembling huge atomic nuclei, possibly being superfluid/superconducting, crustal matter being of substantially more complicated structure, their "atmospheres", i.e. their close spatial environments, in highly magnetised relativistic plasma state. Information about any of those regions is gathered solely by observation of radiation in different parts of the electromagnetic spectrum.

Which radiation? At finite surface and atmospheric temperatures $T \sim 10^2$ eV, neutron stars emit thermal radiation in X rays, independent of their magnetisation (Pothekhin et al. 2015a). This is indeed the case for isolated neutron stars as well as for neutron stars in binary systems though the number of such neutron stars with an unambiguously identified thermal component is not large. Once they are magnetised with fields of the order of $B \sim 10^{12}$ G, neutron stars become radio emitters (Beskin et al. 2015), emitting at much lower frequencies in different modes, synchrotron radiation or pulsed radio signals from the obliquely rotating neutron star with beamed radiation. Such radio pulsars are almost precise clocks in the universe; they moreover provide an important and observationally well accessible laboratory for the investigation of high energy classical and quantum problems in electrodynamics. Almost 90 % of the known pulsars are isolated; the remaining 10 % are found in binaries with mostly negligible mass transfer from the companion. Though pulsar physics is by now half a century old, and many aspect of their radiation emission have been understood there remain many unsolved problems, among them the very mechanism of generation of coherent radiation.

Those neutron stars who reside in binaries and accrete matter from their companion star emit in X-rays. Such neutron stars possess magnetic fields up to $B \sim 10^{13}$ G at strengths where the cyclotron resonance line was first detected. Observing this lines provides a simple and precise way of estimating the magnetic field strength though only at the highest fields. At weaker magnetic fields one relies on different and less precise techniques (Revnivtsev and Mereghetti 2015). Being strongly magnetised and in a binary system their magnetospheres are highly distorted by the accretion processes which power their X ray emissions. Geometrical deformation of the magnetosphere sign responsible for their spectral X ray and timing properties. In particular accretion affects the spin of such neutron stars.

6 Magnetars

Strongest magnetic fields, in fact the strongest magnetic fields ever inferred in any natural system, were found in a particular class of neutron stars called magnetars. Such neutron stars have been found to possess surface fields $B \leq 10^{15}$ G and internal magnetic fields up to roughly $B \leq 10^{17}$ G which, originally, occurred to be surprising. Meanwhile it is known that such field strengths are not excluded by fundamental physics (Treumann et al. 2014), and a number of models and mechanisms for generation of such strong fields have been proposed in the recent past (Duncan and Thompson 1992; Duncan 2000; Thompson and Duncan 1995). A critical collection of such models and the properties of magnetars is provided in Mereghetti et al. (2015) for the persistent emission in various regions of the electromagnetic spectrum (spanning the range from radio to X rays) from magnetars and what can be learned from its observation. About thirty magnetars and magnetar candidates are today known in the galactic vicinity, most of them appearing as highly variable X ray emitters suggesting that a substantial amount of matter accretes while the high temporal variability indicates that accretion is either not a permanently stable process or that the magnetic field strength at these high values undergoes some temporal restructuring. Most of this magnetic activity is attributed to processes in the neutron star crust while the core of the star is generally believed to be a type II superconductor which occasionally may expel magnetic flux tubes thus contributing to the temporal variability.

There are many very interesting questions and problems concerning the structure of the crust and core in the presence of the very strong magnetic fields as inferred for magnetars. Such problems have become the subject of most recent investigations beyond those discussed in the present book. Crustal matter seems not to be as simply structured as fluid models including relativistic magnetohydrodynamics assume. There are indications that the strong magnetic fields in such neutron stars cause the matter in the crust to be not only layered but in the layers just beneath the very thin outer crust becoming "pasta-like" vertically (to the surface) smeared out structures of matter that may become mixed with superfluid cells, broken by magnetic flux tubes. Such structures are subject to plastic deformation which may generate magnetic activity in magnetars (Lyutikov 2015). It is not known what effect such a structure might have on the magnetic field distribution in the crust and the neutron star environment. Recent three-dimensional numerical fluid investigations (Wood and Hollerbach 2015)—including Hall effects but ignoring this particular structuring—suggest that small scale crustal magnetic fields may survive for long time and, similar to the butterfly structure of sunspots, occur in bands that, however, move towards the equator where the crustal currents may organise themselves into a strong equatorial electrojet (see Fig. 2).

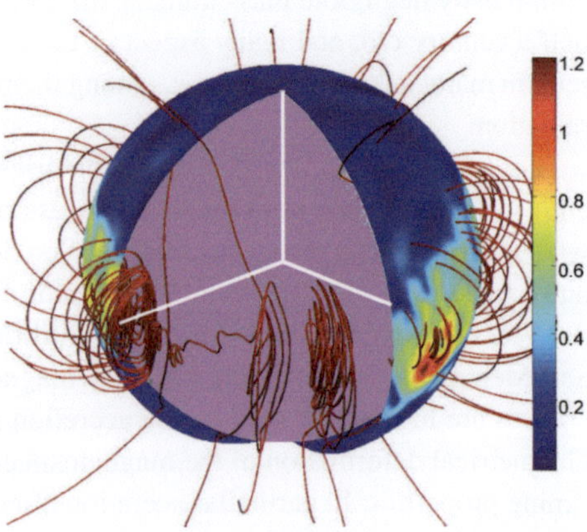

Fig. 2 Magnetic field lines obtained from a fluid-like three-dimensional simulation of a neutron star crust including Hall effect (adapted from Wood and Hollerbach 2015). From its crust sources the field evolves into localised structures which concentrate around the equator indicating an equatorial jet flow

The magnetic stresses of such fields and currents may excite magnetar quakes and related glitches and temporal variations in the magnetic field within estimated up to a million years after formation of the neutron star.

7 The Environments of Strongly Magnetised Objects

Naturally magnetic fields leak out from their body source extending far into the surrounding space and thereby substantially affecting any charged matter present in the stellar environment. Regions where the magnetic field dominates the dynamics of the matter are the magnetospheres. In magnetised white dwarfs, neutron stars, pulsars and magnetars these magnetospheres are strongly controlled by the magnetic field (Beskin et al. 2015; Mereghetti et al. 2015). In rotating neutron stars their spatial extension is limited by the light cylinder, the distance up to that dilute charged matter is believed to corotate with the star as long as it is frozen to the magnetic field, items well known for long time including their problems. The light cyclinder in this case acts as an almost cylindrical surface of injection of matter from the rotating neutron star into the environment. In the opposite case when the fields are not as strong, the star emits a wind which extends radially out to large distances. The best known example is the solar wind which because of the relative weakness of the solar field dominates the latter almost completely. Similar effects happen at magnetised stars causing not only winds but channelled outflows mostly from the polar-dipolar field regions of open field flux tubes. In binary systems the inverse of such outflows is the already mentioned accretion of matter from one star to the other. Most of the relevant questions concerning these latter cases are reviewed in depth in three comprehensive chapters in this volume (Hawley et al. 2015; Kargaltsev et al. 2015; Romanova and Owocki 2015) presenting a wealth of new material based on observation and simulation.

Pulsars, rotating neutron stars, emit most of their energy and, in particular, momentum in ejection of highly relativistic magnetised pair (electron/positron) plasmas in the form of pulsar winds (Kargaltsev et al. 2015) reaching into the region far outside the light cylinder where they become accelerated near the pulsar wind termination shock (Schlickeiser 2002; Bykov and Treumann 2011). The momentum loss causes a spin-down of the pulsar which is a measurable parameter. Spectacular observations of the Crab and Vela pulsars have confirmed this long standing hypothesis including sudden glitches in the rotation period related to flares and eruptive ejections of matter. Otherwise, speed-ups of rotations have been attributed to accretion of matter (Romanova and Owocki 2015) which, however, is not observed in isolated pulsars but restricted to binary systems. The high energy accelerated particles generated radiation outside the termination shock where it is found to produce pulsar wind nebulae which are seen in X rays and Gamma rays. The nebulae populate essentially two morphologies: torus-like formations combined with a jet as in the Crab pulsar, and bow shocks.

In general, magnetised stars affect their environment more or less, depending on whether or not the compact magnetic star possesses an accretion disk, the normal case if it is accompanied by a weakly magnetic of non-magnetic large hot companion which, in the vicinity of a compact object usually fills its Roche-lobe. The strong compact object magnetic field truncates the accretion disk and funnels the disk plasma from along the field to the stars polar region or, in case of magnetically multipolar stars to the multipolar regions. Otherwise plasma may be ejected from the magnetospheres along the open field lines, a process that is the more efficient the faster the compact magnetic star rotates. Fast rotators with accretion may then eject the funnelled accreting mater out along the open fields even before it hits the star's surface. Many such processes can only be treated by numerical simulations of which a representative collection is presented in Romanova and Owocki (2015), while Hawley et al. (2015) covers the knowledge of the properties and instability (mainly to the magneto-rotational instability MRI) of accretion disks that are formed in many those cases under the action of the strong gravity of the central object in combination with its fast rotation that leads to the centrifugal forces. In particular, formation and internal properties of jets and ejection of matter along the open strong magnetic field lines are an important field of research. Break-out, collimation and interaction of jets with the accretion disk and external environment are treated here as well. Some of these processes which have not yet been understood properly for the reason that jets are spatially confined and thus of highly complicated internal structure and that the jet plasma is highly relativistic and thus under some circumstances subject to strong radiation effects in addition to non-negligible interaction between high-energy particles. Processes of this kind are treated to some extent in the last part in this volume on basic processes.

8 Basic Plasma Processes

Three basic plasma processes which lie at the fundament of any understanding of highly magnetised plasma in the vicinity of compact objects in the universe are treated in Granot et al. (2015), Sironi et al. (2015), Kagan et al. (2015). The first deals with the famous gamma ray burst that have been observed since the seventies being about isotropically distributed in the sky. Most of them have been attributed to objects containing strong magnetic fields, thus being compact objects. It is also believed that highly relativistic collisionless shocks play an important role in these gamma ray bursts, in particular in their afterglow. The properties and physics of gamma ray bursts and the different scenarios and models of generation of prompt radiation and afterglow, which have been put forward, are reviewed pointing out their pros and contras. Emphasis is on the role of strong magnetic fields in generation of gamma ray bursts of shortest duration and highest luminosity. It seems that highest luminosities require the presence of very strong magnetic fields like in neutron stars and magnetars.

Collisionless shocks (Balogh and Treumann 2013; Treumann 2009) are one of the important natural tools for heating plasmas and accelerating particles to very high energies (Schlickeiser 2002) as present in cosmic rays. In the astrophysical context they are mostly related to winds and jets emanating from strongly magnetised objects where they become relativistic (Bykov and Treumann 2011). Recently they have also been noted in the context of gamma ray bursts when including general relativistic effects. Understanding their detailed physics is necessary for the correct interpretation of many of the observations of magnetised compact objects, including gamma ray bursts and accretion of the kind discussed in this volume. The current state of the art is summarised in Sironi et al. (2015).

Finally, since magnetic fields imply that they reorganise when contacting each other, the other most important process concerning magnetic fields is collisionless reconnection (Treumann et al. 2014). Under astrophysical conditions, reconnection becomes again relativistic which substantially complicates the process of magnetic rearrangement. Kagan et al. (2015) discusses the relevant processes in collisionless relativistic reconnection in very strong magnetic fields, its effect seen in observations and application to highly magnetic objects. Both, collisionless shocks and reconnection turn out not only to be strong accelerators of particles contributing to cosmic rays but also strong radiation sources in a broad spectrum.

9 Conclusions

The present volume aims on giving a comprehensive review of the state of our knowledge about the strongest magnetic field that have been detected in the universe. These field are many orders of magnitude higher than the average universal magnetic fields (Beck et al. 2012; Widrow et al. 2012; Durrer and Neronov 2013). Their origin, in the progenitors of the most extreme fields, is believed to be mostly dynamical, generated by kind of magnetic dynamos. These fields become substantially increased by magnetic flux conservation when the massive objects contracts to its final compact state. Dynamo action may even prevail during and a short time after collapse (Duncan and Thompson 1992; Thompson and Duncan 1995), a very important point since this prevail not only explains in a few cases the generation of over-strong magnetic field, it also gives an interpretation for the occurrence of strong fields in compact objects in general in particular when they occur in object that are isolated and have not experienced accretion to overcome the gravitational limit for contraction/collapse. The related question is: why do such objects wait until they contract in order to allow the internal dynamo to generate magnetic fields that in contraction become so strong? Such heavy objects should almost immediately collapse according to gravitational theory. Dynamo times are usually long. Hence, dynamos should work as long as the star has not been pushed beyond the limit. The extension of this working time to periods during and after collapse releases this problem. On the other hand, as demonstrated in Ferrario et al. (2015b) the fossil field assumption itself brings its problems with it. Hence the case is not completely settled yet, and more work is expected, mainly numerical and observational, to make progress. Neutron stars, magnetars, pulsars and their environments including magnetospheres, winds, outflow, inflow, accretion, jets and radiation are well treated as far as they concern very strongly magnetised objects. No magnetic strange stars (Chau 1997) have yet been observed. Hence, the processes leading to generation of magnetic fields have probably not to be extended to the inclusion of other new ones which would lead to vastly stronger fields than those observed so far.

References

Y. Aharonov, D. Bohm, Significance of electromagnetic potentials in the quantum theory. Phys. Rev. **115**, 485–491 (1959). doi:10.1103/PhysRev.115.485

A. Balogh, R.A. Treumann, *Physics of Collisionless Shocks: Space Plasma Shock Waves*. ISSI Scientific Report Series, vol. 12 (Springer, New York, 2013). doi:10.1007/978-1-4614-6099-2

R. Beck, A. Balogh, A. Bykov, R.A. Treumann, L.M. Widrow (eds.), *Large-Scale Magnetic Fields in the Universe*. ISSI Space Sci. Ser., vol. 39 (Springer, New York, 2012)

V.S. Beskin, S.V. Chernov, C.R. Gwinn, A.A. Tchekhovskoy, Radio pulsars. Space Sci. Rev. (2015). doi:10.1007/s11214-015-0102-2

R.D. Blandford, R.L. Znajek, Electromagnetic extraction of energy from Kerr black holes. Mon. Not. R. Astron. Soc. **179**, 433–456 (1977)

A.M. Bykov, R.A. Treumann, Fundamentals of collisionless shocks for astrophysical application, 2. Relativistic shocks. Astron. Astrophys. Rev. (2011). doi:10.1007/s00159-011-0042-8

H.F. Chau, On the rotation and magnetic field evolution of superconducting strange stars. Astrophys. J. **479**, 886–901 (1997). doi:10.1086/303898

R.C. Duncan, C. Thompson, Formation of very strongly magnetized neutron stars—implications for gamma-ray bursts. Astrophys. J. Lett. **392**, L9–L13 (1992). doi:10.1086/186413

R.C. Duncan, Physics in ultra-strong magnetic fields, in *Gamma-Ray Bursts, 5th Huntsville Symposium*, vol. 526 (2000), pp. 830–841. doi:10.1063/1.1361651

R. Durrer, A. Neronov, Cosmological magnetic fields: their generation, evolution and observation. Astron. Astrophys. Rev. **21**, 62 (2013). doi:10.1007/s00159-013-0062-7

L. Ferrario, A. Melatos, J. Zrake, Magnetic field generation in stars. Space Sci. Rev. (2015a). doi:10.1007/s11214-015-0138-y

L. Ferrario, D. de Martino, B.T. Gänsicke, Magnetic white dwarfs. Space Sci. Rev. (2015b). doi:10.1007/s11214-015-0152-0

G. Gabrielse, D. Hanneke, T. Kinoshita, M. Nio, B. Odom, New determination of the fine structure constant from the electron g value and QED. Phys. Rev. Lett. **97**, 030802 (2006). doi:10.1103/PhysRevLett.97.03802

J. Granot, T. Piran, O. Bromberg, J.L. Racusin, F. Daigne, Gamma ray bursts as sources of strong magnetic fields. Space Sci. Rev. (2015). doi:10.1007/s11214-015-0102-2

J.F. Hawley, C. Fendt, M. Hardcastle, E. Nokhrina, A. Tchekhovskoy, Disks and jets—gravity, rotation and magnetic fields. Space Sci. Rev. (2015). doi:10.1007/s11214-015-0102-2

D. Kagan, L. Sironi, B. Cerutti, D. Giannios, Relativistic magnetic reconnection in pair plasmas and its astrophysical applications. Space Sci. Rev. (2015). doi:10.1007/s11214-015-0102-2

O. Kargaltsev, B. Cerutti, Y. Lyubarsky, E. Striani, Pulsar-wind nebulae. Space Sci. Rev. (2015). doi:10.1007/s11214-015-0102-2

D. Lai, Physics in very strong magnetic fields. Space Sci. Rev. (2015). doi:10.1007/s11214-015-0137-z

L. Landau, Diamagnetismus der Metalle. Z. Phys. **64**, 629–637 (1930). doi:10.1007/BF01397213

J.L. Linsky, M. Schöller, Observations of strong magnetic fields in nondegenerate stars. Space Sci. Rev. (2015). doi:10.1007/s11214-015-0143-1

M. Lyutikov, Magnetar activity mediated by plastic deformations of neutron star crust. Mon. Not. R. Astron. Soc. **447**, 1407–1417 (2015). doi:10.1093/mnras/stu2413

S. Mereghetti, J.A. Pons, A. Melatos, Magnetars: properties, origin and evolution. Space Sci. Rev. (2015). doi:10.1007/s11214-015-0102-2

A.Y. Pothekhin, A. De Luca, J.A. Pons, Neutron stars—thermal emitters. Space Sci. Rev. (2015a). doi:10.1007/s11214-015-0102-2

A.Y. Pothekhin, J.A. Pons, D. Page, Neutron stars—cooling and transport. Space Sci. Rev. (2015b). doi:10.1007/s11214-015-0102-2

M. Revnivtsev, S. Mereghetti, Magnetic fields of neutron stars in X-ray binaries. Space Sci. Rev. (2015). doi:10.1007/s11214-015-0102-2

M. Romanova, S.P. Owocki, Accretion, outflows and winds from magnetized stars. Space Sci. Rev. (2015). doi:10.1007/s11214-015-0102-2

M. Ruderman, Matter in superstrong magnetic fields, in *Physics of Dense Matter*, ed. by C.J. Hansen. Proc. IAU Symp., vol. 53, Dordrecht, Boston (1974), pp. 117–131

R. Schlickeiser, *Cosmic Ray Astrophysics*. Astron. Astrophys. Library (Springer, Berlin, 2002). doi:10.1007/978-3-662-04814-6

L. Sironi, U. Keshet, M. Lemoine, Relativistic shocks. Space Sci. Rev. (2015). doi:10.1007/s11214-015-0102-2

C. Thompson, R.C. Duncan, The soft gamma repeaters as very strongly magnetized neutron stars—I. Radiative mechanism for outbursts. Mon. Not. R. Astron. Soc. **275**, 255–300 (1995). doi:10.1063/1.1361651

R.A. Treumann, Fundamentals of collisionless shocks for astrophysical application, 1. Non-relativistic shocks. Astron. Astrophys. Rev. **17**, 409–535 (2009). doi:10.1007/s00159-009-0024-2

R.A. Treumann, W. Baumjohann, A. Balogh, The strongest magnetic fields in the universe: how strong can they become? Front. Phys. (2014). doi:10.3389/fphy.2014.00059

J. Trümper, W. Pietsch, C. Reppin, B. Sacco, Evidence for strong cyclotron emission in the hard X-ray spectrum of Her X-1 (eighth Texas symposium on relativistic astrophysics). Ann. N.Y. Acad. Sci. **302**, 538–544 (1977). doi:10.1111/j.1749-6632.1977.tb37072.x

L.M. Widrow, D. Ryu, D.R.G. Schleicher, K. Subramanian, C.G. Tsagas, R.A. Treumann, The first magnetic fields. Space Sci. Rev. **116**, 37–70 (2012). doi:10.1007/s11214-011-9833-5

T.S. Wood, R. Hollerbach, Three dimensional simulation of the magnetic stress in a neutron star crust. Phys. Rev. Lett. **114**, 191101 (2015). arXiv:1501.05149 [astro-ph.SR]. doi:10.1102/PhysRevLett.114.191101

DOI 10.1007/978-1-4939-3550-5_2
Reprinted from *Space Science Reviews* Journal, DOI 10.1007/s11214-015-0137-z

Physics in Very Strong Magnetic Fields

Dong Lai[1]

Received: 3 November 2014 / Accepted: 28 January 2015 / Published online: 5 February 2015

Abstract This paper provides an introduction to a number of astrophysics problems related to strong magnetic fields. The first part deals with issues related to atoms, condensed matter and high-energy processes in very strong magnetic fields, and how these issues influence various aspects of neutron star astrophysics. The second part deals with classical astrophysical effects of magnetic fields: Even relatively "weak" fields can play a strong role in various astrophysical problems, ranging from stars, accretion disks and outflows, to the formation and merger of compact objects.

Keywords Magnetic fields · Stars · Accretion

1 Introduction

The subject "Physics in Very Strong Magnetic Fields" is a very broad one, and the title is also somewhat ambiguous. The first question one may ask is: *How strong a magnetic field is "Strong"?* The answer to this question will depend on the objects one is dealing with, on the issues one is interested in, and on whom one is talking to.

In the following, we will first review issues of strong magnetic fields from a general physics point of view and discuss how these issues may relate to some aspects of neutron star astrophysics. This focus on neutron stars reflects that fact that neutron stars are endowed with the strongest magnetic fields in the universe where fundamental strong-field physics can play an important role. It also reflects the author's own research interest on the subject. For most other astrophysics problems, covering a wide range of sub-fields of astrophysics, magnetic fields are essentially classical, i.e., we are essentially dealing with Maxwell equations. We will discuss why such "weak" magnetic fields can be considered strong, and how such fields play an important role in various astrophysics contexts, ranging from stars and star formation, to disks and outflows, and to stellar mergers.

✉ D. Lai
dong@astro.cornell.edu

[1] Department of Astronomy, Cornell University, Ithaca, NY 14853, USA

2 Atomic and Molecular Physics

When studying matter in magnetic fields, the natural (atomic) unit for the field strength, B_0, is set by equating the electron cyclotron energy $\hbar\omega_{ce}$ to the characteristic atomic energy $e^2/a_0 = 2 \times 13.6$ eV (where a_0 is the Bohr radius), or equivalently by $\hat{R} = a_0$, where $\hat{R} = (\hbar c/eB)^{1/2}$ is the cyclotron radius of the electron. Thus it is convenient to define a dimensionless magnetic field strength b via

$$b \equiv \frac{B}{B_0}; \quad B_0 = \frac{m_e^2 e^3 c}{\hbar^3} = 2.3505 \times 10^9 \text{ G}. \tag{1}$$

For $b \gg 1$, the cyclotron energy $\hbar\omega_{ce}$ is much larger than the typical Coulomb energy, so that the properties of atoms, molecules and condensed matter are qualitatively changed by the magnetic field. In such a strong field regime, the usual perturbative treatment of the magnetic effects (e.g., Zeeman splitting of atomic energy levels) does not apply. Instead, the Coulomb forces act as a perturbation to the magnetic forces, and the electrons in an atom settle into the ground Landau level. Because of the extreme confinement ($\hat{R} \ll a_0$) of the electrons in the transverse direction (perpendicular to the field), the Coulomb force becomes much more effective in binding the electrons along the magnetic field direction. The atom attains a cylindrical structure. Moreover, it is possible for these elongated atoms to form molecular chains by covalent bonding along the field direction. Interactions between the linear chains can then lead to the formation of three-dimensional condensates (see Lai 2001; Harding and Lai 2006 for review).

(i) Atoms: For $b \gg 1$, the H atom is elongated and squeezed, with the transverse size (perpendicular to **B**) $\sim \hat{R} = a_0/b^{1/2} \ll a_0$ and the longitudinal size $\sim a_0/(\ln b)$. Thus the ground-state binding energy $|E| \simeq 0.16\ (\ln b)^2$ (au) (where 1 au = 27.2 eV; the factor 0.16 is an approximate number based on numerical calculations). Thus $|E| = 160540$ eV at $B = 10^{12},\ 10^{14}$ G respectively. In the ground state, the guiding center of the electron's gyro-motion coincides with the proton. The excited states of the atom can be obtained by displacing the guiding center away from the proton; this corresponds to $\hat{R} \to R_s = (2s+1)^{1/2}\hat{R}$ (where $s = 0, 1, 2, \ldots$). Thus $E_s \simeq -0.16\{\ln[b/(2s+1)]\}^2$ (au).

We can imagine constructing a multi-electron atom (with Z electrons) by placing electrons at the lowest available energy levels of a hydrogenic ion. The lowest levels to be filled are the tightly bound states with $\nu = 0$ (zero node in the wavefunction along the field direction). When $a_0/Z \gg \sqrt{2Z-1}\hat{R}$, i.e., $b \gg 2Z^3$, all electrons settle into the tightly bound levels with $s = 0, 1, 2, \ldots, Z-1$. Reliable values for the energy of a multi-electron atom for $b \gg 1$ can be calculated using the Hartree-Fock method or density functional theory, which takes into account the electron-electron direct and exchange interactions in a self-consistent manner.

(ii) Molecules and Chains: In a strong magnetic field, the mechanism of forming molecules is quite different from the zero-field case. The spins of the electrons in the atoms are aligned anti-parallel to the magnetic field, and thus two atoms in their ground states do not bind together according to the exclusion principle. Instead, one H atom has to be excited to the $s = 1$ state before combining (by covalent bond) with another atom in the $s = 0$ state. Since the "activation energy" for exciting an electron in the H atom from s to $(s+1)$ is small, the resulting H_2 molecule is stable. Moreover, in strong magnetic fields, stable H_3, H_4 etc. can be formed in the similar manner. The dissociation energy of the molecule is much greater than the $B = 0$ value: e.g., for H_2, it is 40350 eV at $10^{12}, 10^{14}$ G respectively. A highly magnetized molecule exhibits excitation levels much different from a $B = 0$ molecule.

(iii) Neutron Star Atmospheres and Radiation: An important area of research where the atomic physics in strong magnetic fields plays an important role is the study of neutron star (NS) atmospheres and their radiation (see Potekhin et al. 2014 for more details). Thermal, surface emission from isolated NSs can potentially provide invaluable information on the physical properties and evolution of NS (equation of state at super-nuclear densities, superfluidity, cooling history, magnetic field, surface composition, different NS populations). In recent years, considerable observational resources (e.g. *Chandra* and *XMM-Newton*) have been devoted to such study. For example, the spectra of a number of radio pulsars (e.g., PSR B1055-52, B0656+14, Geminga and Vela) have been observed to possess thermal components that can be attributed to emission from NS surfaces and/or heated polar caps. Phase-resolved spectroscopic observations are becoming possible, revealing the surface magnetic field geometry and emission radius of the pulsar. A number of compact sources in supernova remnants have been observed, with spectra consistent with thermal emission from NSs, and useful constraints on NS cooling physics have been obtained. Surface X-ray emission has also been detected from a number of SGRs and AXPs. Fits to the quiescent magnetar spectra with blackbody or with crude atmosphere models indicate that the thermal X-rays can be attributed to magnetar surface emission at temperatures of $(3\text{–}7) \times 10^6$ K. One of the intriguing puzzles is the absence of spectral features (such as ion cyclotron line around 1 keV for typical magnetar field strengths) in the observed thermal spectra. Clearly, detailed observational and theoretical studies of surface emission can potentially reveal much about the physical conditions and the nature of magnetars.

Of particular interest are the seven isolated, radio-quiet NSs (so-called "dim isolated NSs"; see van Kerkwijk and Kaplan 2007; Haberl 2007). These NSs share the common property that their spectra appear to be entirely thermal, indicating that the emission arises directly from the NS atmospheres, uncontaminated by magnetospheric processes. Thus they offer the best hope for inferring the precise values of the temperature, surface gravity, gravitational redshift and magnetic field strength. The true nature of these sources, however, is unclear at present: they could be young cooling NSs, or NSs kept hot by accretion from the ISM, or magnetars and their descendants. Given their interest, these isolated NSs have been intensively studied by deep *Chandra* and *XMM-Newton* observations. While the brightest of these, RX J1856.5-3754, has a featureless spectrum remarkably well described by a blackbody, absorption lines/features at $E \simeq 0.2\text{–}2$ keV have been detected in six other sources, The identifications of these features, however, remain uncertain, with suggestions ranging from cyclotron lines to atomic transitions of H, He or mid-Z atoms in a strong magnetic field (see Ho and Lai 2004; Ho et al. 2008; Potekhin et al. 2014). Another puzzle concerns the optical emission: For four sources, optical counterparts have been identified, but the optical flux is larger (by a factor of 4–10) than the extrapolation from the black-body fit to the X-ray spectrum. Clearly, a proper understanding/interpretation of these objects requires detailed NS atmosphere modeling which includes careful treatments of atomic physics in strong magnetic fields.

3 Condensed Matter Physics

Several aspects of condensed matter physics in strong magnetic fields play an important role in neutron star astrophysics.

(i) Cohesive Property of Condensed Matter: Continuing our discussion of atoms/molecules in strong magnetic fields, as we add more atoms to a H molecular chain, the energy per atom in a H_n molecule saturates, becoming independent of n. We then have a 1D metal. Chain-chain interactions then lead to 3D condensed matter. The binding energy of magnetized condensed matter at zero pressure can be estimated using the uniform electron gas model. Balancing the electron kinetic (zero-point) energy and the Coulomb energy in a Wigner-Seitz cell (containing one nucleus and Z electrons), we find that the energy per unit cell is of order $E \sim -Z^{9/5}b^{2/5}$. The radius of the cell is $R \sim Z^{1/5}b^{-2/5}$, corresponding to the zero-pressure density $\simeq 10^3 AZ^{3/5}B_{12}^{6/5}$ g cm^{-3} (where A is the mass number of the ion).

Although the simple uniform electron gas model and its Thomas-Fermi type extensions give a reasonable estimate for the binding energy for the condensed state, they are not adequate for determining the cohesive property of the condensed matter. In principle, a three-dimensional electronic band structure calculation is needed to solve this problem. The binding energies of 1D chain for some elements have been obtained using Hartree-Fock method (Neuhauser et al. 1987; Lai et al. 1992). Density functional theory has also been used to calculate the structure of linear chains in strong magnetic fields (Jones 1986; Medin and Lai 2006a, 2006b). Numerical calculations carried out so far indicate that for $B_{12} = 1 - 10$, linear chains are unbound for large atomic numbers $Z \gtrsim 6$. In particular, the Fe chain is unbound relative to the Fe atom; therefore, the chain-chain interaction must play a crucial role in determining whether the 3D zero-pressure Fe condensed matter is bound or not. However, for a sufficiently large B, when $a_0/Z \gg \sqrt{2Z+1}\hat{R}$, or $B_{12} \gg 100(Z/26)^3$, we expect the Fe chain to be bound in a manner similar to the H chain or He chain (Medin and Lai 2006a, 2006b). The cohesive property of magnetized condensed matter is important for understanding the physical condition of the "polar gap" and particle acceleration in pulsars (Medin and Lai 2007).

(ii) Phase Diagram and Equation of State: Given the energies of different bound states of a certain element, one can determine the phase diagram as a function of the field strength B and temperature. This is relevant to the outmost layer of neutron stars (NSs). For a given B, there is a critical temperature below which the phase separation will occur, and the NS surface may be in a condensed state, with negligible gas above it. Some isolated NSs with low surface temperatures may be in such a state (see van Adelsberg et al. 2005; Medin and Lai 2007).

Beyond zero-pressure density, the Coulomb interaction can be neglected, and the effects of magnetic field on the equation of state of matter depend on B, ρ and T. We can define a critical "magnetic density", below which only the ground Landau level is populated (at $T = 0$), given by

$$\rho_B = 0.802 Y_e^{-1} b^{3/2}\ \mathrm{g\,cm^{-3}} = 7.04 \times 10^3 Y_e^{-1} B_{12}^{3/2}\ \mathrm{g\,cm^{-3}}, \tag{2}$$

where $Y_e = Z/A$ is the number of electrons per baryon. We can also define critical "magnetic temperature",

$$T_B \simeq \frac{\hbar\omega_{ce}}{k_B}\left(\frac{m_e}{m_e^*}\right) = 1.34 \times 10^8 B_{12}(1+x_F^2)^{-1/2}\ \mathrm{K}, \tag{3}$$

where $m_e^* = \sqrt{m_e^2 + (p_F/c)^2} = m_e\sqrt{1+x_F^2}$. There are three regimes characterizing the effects of Landau quantization on the thermodynamic properties of the electron gas:

(a) $\rho \lesssim \rho_B$ and $T \lesssim T_B$: In this regime, the electrons populate mostly the ground Landau level, and the magnetic field modifies essentially all the properties of the gas. The field is sometimes termed "strongly quantizing". For example, for degenerate, nonrelativistic electrons ($\rho < \rho_B$ and $T \ll T_F \ll m_e c^2/k_B$, where T_F is the Fermi temperature), the pressure is $P_e = (2/3)n_e E_F \propto B^{-2}\rho^3$. This should be compared with the $B = 0$ expression $P_e \propto \rho^{5/3}$. Note that for nondegenerate electrons ($T \gg T_F$), the classical ideal gas equation of state, $P_e = n_e k_B T$, still holds in this "strongly quantizing" regime.

(b) $\rho \gtrsim \rho_B$ and $T \lesssim T_B$: In this regime, the electrons are degenerate, and populate many Landau levels but the level spacing exceeds $k_B T$. The magnetic field is termed "weakly quantizing". The bulk properties of the gas (e.g., pressure and chemical potential) are only slightly affected by such magnetic fields. However, the quantities determined by thermal electrons near the Fermi surface show large oscillatory features as a function of density or magnetic field strength. These de Haas–van Alphen type oscillations arise as successive Landau levels are occupied with increasing density (or decreasing magnetic field). With increasing T, the oscillations become weaker because of the thermal broadening of the Landau levels; when $T \gtrsim T_B$, the oscillations are entirely smeared out, and the field-free results are recovered.

(c) $T \gtrsim T_B$ or $\rho \gg \rho_B$: In this regime, many Landau levels are populated and the thermal widths of the Landau levels ($\sim k_B T$) are higher than the level spacing. The magnetic field is termed "non-quantizing" and does not affect the thermodynamic properties of the gas.

(iii) Transport Properties: A strong magnetic field can significantly affect the transport properties and thermal structure of a neutron star crust. Even in the regime where the magnetic quantization effects are small ($\rho \gg \rho_B$), the magnetic field can still greatly modify the transport coefficients (e.g., electric conductivity and heat conductivity). This occurs when the effective gyro-frequency of the electron, $\omega_{ce}^* = eB/(m_e^* c)$, where $m_e^* = \sqrt{m_e^2 + (p_F/c)^2}$, is much larger than the electron collision frequency $1/\tau_0$. When $\omega_{ce}^* \tau_0 \gg 1$, the electron heat conductivity perpendicular to the magnetic field, $\kappa_\perp$, is suppressed by a factor $(\omega_{ce}^* \tau_0)^{-2}$. In this classical regime, the heat conductivity along the field, $\kappa_\parallel$, is the same as the $B = 0$ value. In a quantizing magnetic field, the conductivity exhibits oscillatory behavior of the de Haas–van Alphen type. On average, the longitudinal conductivity is enhanced relative to the $B = 0$ value due to quantization. The most detailed calculations of the electron transport coefficients of magnetized neutron star envelopes are due to Potekhin (1999), where earlier references can be found (see Potekhin et al. 2014 for more details).

The thermal structure of a magnetized neutron star envelope has been studied by many authors (see Potekhin et al. 2014 for review). In general, a normal magnetic field reduces the thermal insulation as a result of the (on average) increased $\kappa_\parallel$ due to Landau quantization of electron motion, while a tangential magnetic field (parallel to the stellar surface) increases the thermal insulation of the envelope because the Lamor rotation of the electron significantly reduces the transverse thermal conductivity $\kappa_\perp$. A consequence of the anisotropic heat transport is that for a given internal temperature of the neutron star, the surface temperature is nonuniform, with the magnetic poles hotter and the magnetic equator cooler (see, e.g., Shabaltas and Lai 2012 for a recent application).

4 High Energy Physics: QED in Strong Magnetic Fields

In superstrong magnetic fields, a number of quantum-electrodynamic (QED) processes are important. A well-known one is single-photon pair production, $\gamma \to e^+ + e^-$. This process is forbidden at zero-field, but is allowed for $B \neq 0$, and is one of the dominant channels for pair cascade in pulsar magnetospheres (Sturrock 1971; Medin and Lai 2010). Another process is photon splitting, $\gamma \to \gamma + \gamma$, which can attain appreciable probability for sufficiently strong fields. The critical QED field strength is set by $\hbar\omega_{ce} = m_e c^2$, i.e.,

$$B_Q = \frac{m_e^2 c^3}{e\hbar} = 4.414 \times 10^{13}\ \text{G}. \tag{4}$$

Above B_Q, many of these QED effects become important.

A somewhat surprising strong-field QED effect is vacuum polarization, which makes even an "empty" space birefringent for photons propagating through it. This can significantly affect radiative transfer in neutron star atmospheres and the observed spectral and x-ray polarization signals even for modest field strengths. We discuss this issue below.

The magnetized plasma of a NS atmosphere is birefringent. An X-ray photon, with energy $E \ll E_{ce} = \hbar\omega_{ce} = 1.16 B_{14}$ MeV (where $B_{14} = B/(10^{14}$ G)), propagating in such a plasma can be in one of the two polarization modes: The ordinary mode (O-mode) has its electric field **E** oriented along the **B**-**k** plane (**k** is along direction of propagation), while the extraordinary mode (X-mode) has its **E** perpendicular to the **B**-**k** plane. Since charge particles cannot move freely across the magnetic field, the X-mode photon opacity (e.g., due to free-free absorption or electron scattering) is suppressed compared to the zero-field value, $\kappa_X \sim (E/E_{Be})^2 \kappa_{(B=0)}$, while the O-mode opacity is largely unchanged, $\kappa_O \sim \kappa_{(B=0)}$ (e.g., Meszaros 1992). Vacuum polarization can change this picture in an essential way. In the presence of a strong magnetic field, vacuum itself becomes birefringent due to virtual e^+e^- pairs. Thus in a magnetized NS atmosphere, both the plasma and vacuum polarization contribute to the dielectric tensor of the medium. The vacuum polarization contribution is of order $10^{-4}(B/B_Q)^2 f(B)$ (where $B_Q = m_e^2 c^3/e\hbar = 4.414 \times 10^{13}$ G, and $f \sim 1$ is a slowly varying function of B), and is quite small unless $B \gg B_Q$. However, even for "modest" field strengths, vacuum polarization can have a dramatic effect through a "vacuum resonance" phenomenon. This resonance arises when the effects of vacuum polarization and plasma on the polarization of the photon modes "compensate" each other. For a photon of energy E (in keV), the vacuum resonance occurs at the density $\rho_V \simeq 0.964 Y_e^{-1} B_{14}^2 E^2 f^{-2}\ \mathrm{g\,cm^{-3}}$, where Y_e is the electron fraction (Lai and Ho 2002). Note that ρ_V lies in the range of the typical densities of a NS atmosphere. For $\rho \gtrsim \rho_V$ (where the plasma effect dominates the dielectric tensor) and $\rho \lesssim \rho_V$ (where vacuum polarization dominates), the photon modes are almost linearly polarized—they are the usual O-mode and X-mode described above; at $\rho = \rho_V$, however, both modes become circularly polarized as a result of the "cancellation" of the plasma and vacuum polarization effects When a photon propagates outward in the NS atmosphere, its polarization state will evolve adiabatically if the plasma density variation is sufficiently gentle. Thus the photon can convert from one mode into another as it traverses the vacuum resonance. The conversion probability $P_{\rm conv}$ depends mainly on E and atmosphere density gradient; for a typical atmosphere density scale height (~ 1 cm), adiabatic mode conversion requires $E \gtrsim 1$–2 keV (Lai and Ho 2003a). Because the O-mode and X-mode have vastly different opacities, the vacuum polarization-induced mode conversion can significantly affect radiative transfer in magnetar atmospheres. In particular, the effect tends to deplete the high-energy tail of the thermal spectrum (making it closer to blackbody) and reduce the width of the ion cyclotron line or other spectral lines (Ho and Lai 2003, 2004; Lai and Ho 2003a; van Adelsberg and Lai 2006). It is tempting to suggest that the absence of lines in the observed thermal spectra of several AXPs is a consequence of the vacuum polarization effect at work in these systems.

We also note that even for "ordinary" NSs (with $B \sim 10^{12}$–10^{13} G), vacuum resonance has a profound effect on the polarization signals of the surface emission; this may provide a direct probe of strong-field QED in the regime inaccessible at terrestrial laboratories (Lai and Ho 2003b; Wang and Lai 2009; see Lai 2010 for a review). Such polarization signals will be of interest for future X-ray polarimetry detectors/missions.

Finally, magnetic fields can modify neutrino processes that take place in neutron stars. For example, in proto-neutron stars with sufficiently strong B-fields, the neutrino cross sections and emission rates, as well as their angular dependences, can be affected, and these can contribute to the natal velocity kick imparted to the neutron star (e.g., Arras and Lai 1999a, 1999b; Maruyama et al. 2014).

5 "Classical" Astrophysics

For most areas of astrophysics, magnetic fields are "classical". That is, we are dealing with Maxwell's equations, MHD and classical plasma physics. The quantization, microscopic effects discussed previous sections are not relevant. Nevertheless, these classical magnetic field effects are important, interesting and rich. We will highlight some of these in the following.

5.1 Clouds, Stars and Compact Objects

The first effect of "classical" magnetic fields is that they can influence the equilibrium of bound objects via the so-called magnetic Virial theorem. For a spherical cloud or star of mass M and mean radius R, static equilibrium requires that the ratio of the magnetic energy and gravitational energy be less than unity, i.e.,

$$\frac{E_{\rm mag}}{E_{\rm grav}} \sim \frac{B_{\rm in}^2 R^3/6}{GM^2/R} \sim \frac{1}{6\pi^2 G}\left(\frac{\Phi}{M}\right)^2 \lesssim 1, \tag{5}$$

where the second equality assumes that the dominant internal magnetic field takes form of a large-scale poloidal field, and $\Phi = \pi R^2 B_{\rm in}$ is the magnetic flux threading the cloud.

In the context of star formation, clouds (cores) with $E_{\rm mag}/E_{\rm grav} \gtrsim 1$ cannot collapse on a dynamical timescale, but require ambipolar diffusion to eliminate the magnetic flux. This process is perhaps relevant for the formation of low-mass stars (e.g., Shu et al. 1999), although in recent years the roles of turbulence in the molecular clouds have been recognized (McKee and Ostriker 2007).

For neutron stars (with $M \simeq 1.4 M_\odot$ and $R \simeq 10$ km), equation (5) implies $B_{\rm in} \lesssim 10^{18}$ Gauss. This is the maximum field strength achievable in all astrophysical objects.

What do we know observationally about magnetic fields of isolated neutron stars? For radio pulsars, the dipole magnetic fields are inferred indirectly from the measured P and $\dot{P}$ (rotation period and period derivative), and the assumption that the spindown is due to magnetic dipole radiation. For most "regular" pulsars, the magnetic fields thus derived lie in the range of $10^{12\text{–}13}$ G. A smaller population, so-called "millisecond pulsars", have fields in the range of $10^{8\text{–}9}$ G. How such a "weak" field evolves from the regular field of $10^{12\text{–}13}$ G remains unclear (see Payne and Melatos 2004). In recent years, a number of "High-B Radio Pulsars" have also been found: these have $B \sim 10^{14}$ G, comparable to magnetars.

Magnetars are neutron stars powered by energy dissipation of magnetic fields. They usually have dipole fields (as inferred from P and $\dot{P}$ based on x-ray timing) of $B \gtrsim 10^{14}$ G. Interestingly, a number of low-field ($\sim 10^{13}$ G) magnetars have also been found recently (Rea et al. 2010), although the internal fields could be higher. Indeed, there is growing evidence that there exist hidden magnetic fields inside neutron stars. This is the case for the neutron star in Kes 79 SNR: It has a dipole field of 3×10^{10} G, but the internal field buried inside its crust could be larger than 10^{14} G, based on its observed large x-ray pulse fraction of 60 % (Halpern and Gotthelf 2010; Shabaltas and Lai 2012; Viganò et al. 2013). In the case of SGR 0418+5729, the dipole field is less than a few times 10^{12} G, but internal field could be much stronger (Turolla et al. 2011).

Another way to assess whether a magnetic field is "strong" is to look at the energetics. For magnetars, even in quiescence, the x-ray luminosity is $L \sim 10^{34-36}$ erg s^{-1}, much larger than the spindown luminosity ($I\Omega\dot{\Omega}$). The giant flares of the three SGRs indicate that a much larger internal field is possible. For example, the December 2004 flare of SGR 1806-20 has a total energy of 10^{46} erg, suggesting an internal field of at least a few times 10^{14} G.

What is the origin of such strong magnetic fields? It is intriguing to note that (Reisenegger 2013) for upper main-sequence stars (radius $10^{6.5}$ km), white dwarfs (10^4 km) and neutron stars (10 km), the maximum observed magnetic fields ($10^{4.5}$ G, 10^9 G and 10^{15} G respectively) all correspond to similar maximum magnetic flux $\Phi_{\rm max} = \pi R^2 B_{\rm max} \sim 10^{17.5-18}$ G km^2. This seems to suggest a fossil origin of the strongest magnetic fields. However, recent observations of magnetic white dwarfs (and their populations in binaries) indicate the strong magnetic fields ($\gtrsim$ a few MG) of white dwarfs originate from binary mergers (Wickramasinghe et al. 2014). So perhaps the strongest magnetic fields found in magnetars is the result of dynamo action in the proto-neutron star phase (Thompson and Duncan 1993). In any case, since $E_{\rm mag}/E_{\rm grav} \lesssim 10^{-6}$ (assuming no significant hidden magnetic fields), these magnetic fields have a negligible effect on the global static equilibrium of the star.

5.2 Stellar Envelopes and "Outside"

Although astrophysically observed magnetic fields have a negligible effect on the global equilibrium of a star, they can strongly influence the local "static" equilibrium of stellar envelopes. A notable example is neutron star (NS) crust. Because of the evolution of crustal magnetic fields due to a combination of Hall drift and Ohmic diffusion, the NS crust can break (e.g. Pons and Perna 2011). This occurs when $B^2/(8\pi) \gtrsim \mu\theta_{\rm max}$ (where μ is the shear modulus and $\theta_{\rm max}$ is the maximum strain of the crust), or $B \gtrsim 2 \times 10^{14}(\theta_{\rm max}/10^{-3})^{1/2}$ G. The consequences of the crustal breaking (and its manifestations such as magnetar flares) are not clear. They depend on whether the breaking is fast or slow. The energy release and whether the energy can get out of the NS are also uncertain (see Link 2014; Beloborodov and Levin 2014).

Of course, outside the star, even a "weak" magnetic field can be quite "strong" and dominates the dynamics of the flow. Such magnetically dominated region is relevant to the magnetic braking of stars. In the case of radio pulsars, the electrodynamics and physical processes in the magnetosphere are ultimately responsible for most of the observed phenomena of pulsars. In recent years, there has been significant progress in ab initio calculations of pulsar magnetospheres (e.g. Tchekhovskoy et al. 2013), although it remains unclear whether the current theoretical approach can adequately explain some of the enigmatic pulsar phenomena (such as mode-switching in radiation; e.g. Hermsen et al. 2013). The magnetospheres of magnetars have also been studied: Unlike radio pulsars, the closed field line regions play an important role (e.g. Thompson et al. 2002; Beloborodov 2013).

Finally, further away from pulsars, we have pulsar wind nebulae, where pulsar wind impinges upon a supernova remnant, creating a broad spectrum of non-thermal radiation (from radio to gamma rays). The ultimate source of this radiation is the pulsar's rotational energy, and magnetic field plays an important role in making such a "transfer of energy" possible (e.g. Amato 2014).

5.3 Accretion Disks

Magnetic fields play a number of important roles in accretion disks. First, we have magnetically dominated disks. These occur when $B^2/(8\pi) \gtrsim \rho v_k^2/2$, where v_k is the Keplerian velocity of the disk and ρ is the density. In the last few years, a number of studies have shown that the innermost region of a disk around a black hole may accumulate large magnetic flux, and relativistic jets can be generated through the Blandford-Znajek process (e.g. McKinney et al. 2012). However, a physical understanding of the state transition and jets (both steady and episodic) from black-hole x-ray binaries remains elusive (Fender and Belloni 2012; Yuan and Narayan 2014).

Outflows can be launched from disks with large-scale super-thermal magnetic fields (at the disk surface), $B^2/(8\pi) \gtrsim \rho c_s^2/2$ (where c_s is the sound speed). Such magnetocentrifugal winds/outflows (a la Blandford-Payne) may occur in x-ray binaries (in the thermal state) and in protostars (see Konigl and Pudritz 2000 for a review). Large-scale magnetic fields can affect the oscillations and waves associated with disks (e.g., Tagger and Pellet 1999; Tagger and Varniere 2006; Yu and Lai 2013, 2015).

Such large-scale strong magnetic fields are unlikely to be produced in the disk by dynamo processes, and must be advected inward from large radii. This is an important issue that has received a lot of theoretical attention. The radial inward advection speed is $|u_r| \sim \nu/r$, and the outward Ohmic diffusion speed is $u_{\rm diff} \sim (\eta/H)(B_r/B_z)$, where ν is the disk viscosity, η is the magnetic diffusivity and H is the disk thickness. Clearly, the net outcome depends on the magnetic Prandtl number $P_r = \nu/\eta$, which is typically of order unity (based on local MRI turbulence simulations; Lesur and Longaretti 2009). Recent work has emphasized the importance of proper treatment of vertical structure of the disk (e.g., the electric conductivity is higher at the disk surface, so the field advection is faster than mass advection; Lovelace et al. 2009; Guilet and Ogilvie 2013). Outflows/winds can extract angular momentum from the disk, enhancing the radial advection of magnetic fields (e.g. Cao and Spruit 2013).

Finally, even "weak" sub-thermal magnetic fields can play an important role in accretion disks. It is now well-established that for most astrophysical disks, MRI (magneto-rotational instability) driven turbulence is responsible for generating the anomalous viscosity needed for accretion to proceed (Balbus and Hawley 1998). It is also recognized that the strength of the turbulence depends on the net vertical field threading the disk (Hawley et al. 1995; Simon et al. 2013). Recent works have emphasized the roles of non-ideal MHD effects in suppressing turbulence in proto-planetary disks (Bai and Stone 2013; Bai 2014).

5.4 Disk Accretion onto Magnetic Stars

Disk accretion onto magnetic central objects occurs in a variety of astrophysical contexts, ranging from classical T Tauri stars, and cataclysmic variables (intermediate polars), to accretion-powered X-ray pulsars. The basic picture of disk-magnetosphere interaction is well known: The stellar magnetic field disrupts the accretion flow at the magnetospheric boundary and funnels the plasma onto the polar caps of the star or ejects it to infinity. The magnetosphere boundary is located where the magnetic and plasma stresses balance,

$$r_m = \xi \left(\frac{\mu^4}{G M \dot{M}^2} \right)^{1/7}, \qquad (6)$$

where M and μ are the mass and magnetic moment of the central object, $\dot{M}$ is the mass accretion rate and ξ is a dimensionless constant of order 0.5–1. Roughly speaking, the funnel flow occurs when r_m is less than the corotation radius r_c (where the disk rotates at the same rate as the star). For $r_m \gtrsim r_c$, centrifugal forces may lead to ejection of the accreting matter ("propeller" effect).

Over the years, numerous theoretical studies have been devoted to understanding the interaction between accretion disks and magnetized stars. Many different models have been developed (see Lai 2014 for a review). In parallel to these theoretical studies, there have been many numerical simulations, with increasing sophistication. These simulations are playing an important role in elucidating the physics of magnetosphere-disk interaction in various astrophysical situations (see Romanova et al. 2014 and Zanni 2014 for review).

The problem of magnetosphere-disk interaction has many applications: (i) Rotation rate of protostars: Many protostars are found to have rotation rates about 10 % of breakup. Magnetosphere spin equilibrium (r_m equals the corotation radius) has long been suggested, although magnetosphere/stellar winds may also play a role (Gallet and Bouvier 2013). (ii) Spinup/spindown of accreting x-ray pulsars: Many x-ray pulsars have been observed to exhibit changing spinup and spindown behavior over timescales of years. For example, 4U1626-67 is an accreting pulsar with spin period 7.66 s. The clean spinup before 1990.6 was followed by a clean spindown, and another spinup phase starting 2008.2. The spindown/spinup transition lasted 150 days. Understanding this spindown/spinup behavior and its correlation with the accretion rate remains an outstanding unsolved problem.

When the stellar field lines penetrate some region of the disk, they provide a linkage between the star and the disk. These field lines are twisted by differential rotation between the stellar rotation Ω_s and the disk rotation $\Omega(r)$, generating a toroidal field. However, when the toroidal field becomes comparable to the poloidal field, the flux tube connecting the star and the disk will start expanding. This field inflation is driven by the pressure associated with the toroidal field. As the fields open up, the star-disk linkage is broken. Such field-opening behavior has been well-established through theoretical studies and numerical simulations in the contexts of solar flares and accretion disks (Lovelace et al. 1995). Given this constraint on the toroidal twist, steady-state disk-star linkage is possible only very near corotation. In general, we should expect a quasi-cyclic behavior, involving several stages: (1) The stellar field penetrates the inner region of the disk; (2) The linked field lines are twisted; (3) The resulting toroidal fields drive field inflation; (4) Reconnection of the inflated field restores the linkage. The whole cycle then repeats (see Aly and Kuijpers 1990; Uzdensky et al. 2002). This quasi-cyclic behavior may be relevant to QPOs observed in low-mass X-ray binaries (see van der Klis 2006 for a review; Shirakawa and Lai 2002a, 2002b) and other systems, as well as give rise to episodic outflows and winds (Zanni and Ferreira 2013).

Finally, we note that in the standard picture of magnetic star-disk interaction, it is usually assumed that the stellar spin axis is aligned with the disk axis (the disk normal vector). This seems reasonable since the star may have gained substantial angular momentum from the accreting gas in the disk. However, magnetic interaction between the star and the inner region of the disk may (if not always) change this simple picture (Lai 1999, 2003), giving rise to stellar spin-disk misalignment. This has application to spin-orbit misalignment in exoplanetary systems (Lai et al. 2011; Foucart and Lai 2011).

5.5 Magnetic Fields in the Formation of Compact Objects

In the "standard" picture of core-collapse supernovae leading to the formation of neutron stars, neutrino heating behind the stalled accretion shock, plus various hydrodynamical instabilities, are responsible for the explosion. Magnetic fields play a negligible role in this picture. However, there is a long list of theoretical works exploring the role of magnetic fields in supernovae (LeBlanc and Wilson 1970; Bisnovatyi-Kogan et al. 1976; Moiseenko et al. 2006; Burrows et al. 2007). The key requirement for the magnetic field to play a role is that the pre-SN core must have sufficiently rapid rotation—this is rather uncertain observationally. A technical challenge is that if one starts out with a modest magnetic field, and use MRI dynamo to amplify the field, it is important that the MRI scale is resolved in the numerical code—this is currently not achieved unless the initial field is greater than 10^{15} G (see however, Bisnovatyi-Kogan et al. 2014; Popov and Samokhin 1976).

In general, newly formed magnetars can play two roles in supernovae. (i) They can power the explosion if the initial spin period of the proto-neutron star is less than ~ 3 ms and the magnetic field is 10^{15} G or higher (Bodenheimer and Ostriker 1974; Thompson et al. 2004). (ii) For modest rotation period (~ 10 ms), the released rotational energy does not affect the explosion itself, but can still impact the SN lightcurves (since the spindown timescale, about days to weeks, is comparable to the photon diffusion time through the remnant). Such energy injection may help explain some of the super-luminous SNe with $L \gtrsim 10^{44}$ erg s^{-1} (Kasen and Bildsten 2010; Woosley 2010). In this regard, it is of interest to note that many central compact objects in SNRs have been found to possess rather weak dipole fields ($B \lesssim 10^{12}$ G) and slow rotation (period ~ 0.1 s), although the internal fields may be much stronger.

Magnetic fields play an important role in the central engine of long Gamma-Ray Bursts (GRBs). Two scenarios are often discussed: (1) With rapid rotation, core collapse leads to the formation of a hyper-accreting black hole. Neutrino heating and magnetic fields (via Blandford-Znajek process) then lead to the production of relativistic jets (Zhang et al. 2003). (2) Core collapse leads to the formation of millisecond magnetars, which power the GRB outflows/jets. Recent observations of long-lasting ($\sim 10^4$ s) x-ray emission/flares suggest that long-lasting central engine may be needed for some GRBs (Kumar and Zhang 2014). Also, the observed high polarization in reverse-shock emission indicates that large-scale magnetic fields are present in the GRB jets (Mundell et al. 2013).

5.6 Magnetic Fields in Merging Compact Binaries

There are two types of merging compact binaries that are of great interest: (1) NS/NS and NS/BH binaries: These produce gravitational waves that are detectable by LIGO/VIRGO and generate electromagnetic counterparts in the form of short GRBs and kilo-novae. (2) Compact WD/WD binaries: These produce various exotic outcomes (R CrB stars, AM CVn binaries, and possibly accretion-induced collapse and SN Ia), and generate low-frequency gravitational waves detectable by LISA/NGO.

In recent years, there have been significant progress in simulating (in full General Relativity) the merger of NS/NS binaries (e.g. Shibata and Taniguchi 2006) and NS/BH binaries (Foucart et al. 2013). Simulations with magnetic fields are also becoming possible (Giacomazzo et al. 2011; Palenzuela et al. 2013), although much remains to be understood.

One issue of great interest is the merger of the NS magnetospheres prior to the merger of the stars. The combined binary system can behave as a single unipolar inductor producing radio waves that may be detectable (Hansen and Lyutikov 2001), although this is highly uncertain and detailed calculations are difficult. Nevertheless, a robust upper limit of the energy dissipation power in the magnetosphere that can be generated prior to NS merger can be obtained (Lai 2012). This upper limit indicates that the magnetospheric dissipation will not affect the orbital decay rate (and the gravitational waveform), although the prospect for radio detection remains uncertain.

Another issue of interest is the production of magnetic fields during NS/NS binary merger. Since the binary NSs cannot be spin-synchronized (because of the rapid orbital decay in the last few minutes of the binary lifetime), strong velocity shear is present when the two stars touch each other: Kelvin-Helmholtz instability develops at the interface, which may then lead to the generation of strong magnetic fields (Price and Rosswog 2006). Recent studies, however, suggest the dynamical impact of such magnetic fields may be limited to the shear layer (Obergaulinger et al. 2010), although the situation is not entirely clear (see Giacomazzo et al. 2014). Finally, the magnetic field in the merger remnant is of great importance. This situation is similar to the remnant in core-collapse supernova: Can the initial (weak/modest) magnetic field be amplified by differential rotation and MRI dynamo (Is MRI resolved in the simulation)? How are winds/outflows/jets produced? Is a black-hole or millisecond magnetar formed in the merger remnant?

Acknowledgements I thank ISSI for hosting the workshop on Strongest Magnetic Fields in the Universe. I thank my collaborators and students who have helped me understand various topics discussed in this review. I also thank K. Anderson and J. Zanazzi for reading parts of the first draft. This work has been supported in part by NSF grants AST-1008245, 1211061, and NASA grant NNX12AF85G.

References

J.J. Aly, J. Kuijpers, Astron. Astrophys. **227**, 473 (1990)
E. Amato, Int. J. Mod. Phys. Conf. Ser. **28**, 60160 (2014)
P. Arras, D. Lai, Astrophys. J. **519**, 745 (1999a)
P. Arras, D. Lai, Phys. Rev. D **60**, 043001 (1999b)
X.-N. Bai, arXiv:1409.2511 (2014)
X.-N. Bai, J.M. Stone, Astrophys. J. **769**, 76 (2013)
S.A. Balbus, J.F. Hawley, Rev. Mod. Phys. **70**, 1 (1998)
A.M. Beloborodov, Astrophys. J. **777**, 114 (2013)
A.M. Beloborodov, Y. Levin, Astrophys. J. **794**, L24 (2014)
G.S. Bisnovatyi-Kogan, I.P. Popov, A.A. Samokhin, Astrophys. Space Sci. **41**, 287 (1976)
G.S. Bisnovatyi-Kogan, S.G. Moiseenko, N.V. Ardeljan, arXiv:1408.2395 (2014)
I.P. Popov, A.A. Samokhin, Astrophys. Space Sci. **41**, 287 (1976)
P. Bodenheimer, J.P. Ostriker, Astrophys. J. **191**, 465 (1974)
A. Burrows, L. Dessart, E. Livne, C.D. Ott, J. Murphy, Astrophys. J. **664**, 416 (2007)
X. Cao, H.C. Spruit, Astrophys. J. **765**, 149 (2013)
R. Fender, T. Belloni, Science **337**, 540 (2012)
F. Foucart, D. Lai, Mon. Not. R. Astron. Soc. **412**, 2799 (2011)
F. Foucart, L. Buchman, M.D. Duez et al., Phys. Rev. D **88**, 064017 (2013)
F. Gallet, J. Bouvier, Astron. Astrophys. **556**, A36 (2013)
B. Giacomazzo, L. Rezzolla, L. Baiotti, Phys. Rev. D **83**, 044014 (2011)
B. Giacomazzo, J. Zrake, P. Duffell, A.I. MacFadyen, R. Perna, arXiv:1410.0013 (2014)
J. Guilet, G.I. Ogilvie, Mon. Not. R. Astron. Soc. **430**, 822 (2013)
F. Haberl, Astrophys. Space Sci. **308**, 181 (2007)
J.P. Halpern, E.V. Gotthelf, Astrophys. J. **709**, 436 (2010)
B.M.S. Hansen, M. Lyutikov, Mon. Not. R. Astron. Soc. **322**, 695 (2001)
A.K. Harding, D. Lai, Rep. Prog. Phys. **69**, 2631 (2006)
J.F. Hawley, C.F. Gammie, S.A. Balbus, Astrophys. J. **440**, 742 (1995)
W. Hermsen, J.W.T. Hessels, L. Kuiper et al., Science **339**, 436 (2013)
W.C.G. Ho, D. Lai, Mon. Not. R. Astron. Soc. **338**, 233 (2003)
W.C.G. Ho, D. Lai, Astrophys. J. **607**, 420 (2004)
W.C.G. Ho, A.Y. Potekhin, G. Chabrier, Astrophys. J. Suppl. Ser. **178**, 102 (2008)
P.B. Jones, Mon. Not. R. Astron. Soc. **218**, 477 (1986)
D. Kasen, L. Bildsten, Astrophys. J. **717**, 245 (2010)
A. Konigl, R.E. Pudritz, *Protostars and Planets IV*. V. Mannings et al.(University of Arizona Press, Tucson, 2000), p. 759. arXiv:astro-ph/9903168
P. Kumar, B. Zhang, arXiv:1410.0679 (2014)
D. Lai, Astrophys. J. **524**, 1030 (1999)
D. Lai, Rev. Mod. Phys. **73**, 629 (2001)
D. Lai, Astrophys. J. **591**, L119 (2003)
D. Lai, Astrophys. J. **757**, L3 (2012)
D. Lai, in *X-Ray Polarimetry: A New Window in Astrophysics*, ed. by R. Bellazzini, E. Costa et al.(Cambridge University Press, Cambridge, 2010), p. 157. arXiv:0906.4717
D. Lai, Eur. Phys. J. Web Conf. **64**, 1001 (2014). arXiv:1402.1903

D. Lai, E.E. Salpeter, S.L. Shapiro, Phys. Rev. A **45**, 4832–4847 (1992). doi:10.1103/PhysRevA.45.4832
D. Lai, F. Foucart, D.N.C. Lin, Mon. Not. R. Astron. Soc. **412**, 2790 (2011)
D. Lai, W.C.G. Ho, Astrophys. J. **566**, 373 (2002)
D. Lai, W.C.G. Ho, Astrophys. J. **588**, 962 (2003a)
D. Lai, W.C. Ho, Phys. Rev. Lett. **91**, 071101 (2003b)
J.M. LeBlanc, J.R. Wilson, Astrophys. J. **161**, 541 (1970)
G. Lesur, P.-Y. Longaretti, Astron. Astrophys. **504**, 309 (2009)
B. Link, Mon. Not. R. Astron. Soc. **441**, 2676 (2014)
R.V.E. Lovelace, M.M. Romanova, G.S. Bisnovatyi-Kogan, Mon. Not. R. Astron. Soc. **275**, 244 (1995)
R.V.E. Lovelace, D.M. Rothstein, G.S. Bisnovatyi-Kogan, Astrophys. J. **701**, 885 (2009)
T. Maruyama, M.-K. Cheoun, J. Hidaka et al., Phys. Rev. D **90**, 067302 (2014)
C.F. McKee, E.C. Ostriker, Annu. Rev. Astron. Astrophys. **45**, 565 (2007)
J.C. McKinney, A. Tchekhovskoy, R.D. Blandford, Mon. Not. R. Astron. Soc. **423**, 3083 (2012)
Z. Medin, D. Lai, Phys. Rev. A **74**, 062507 (2006a)
Z. Medin, D. Lai, Phys. Rev. A **74**, 062508 (2006b)
Z. Medin, D. Lai, Mon. Not. R. Astron. Soc. **382**, 1833 (2007)
Z. Medin, D. Lai, Mon. Not. R. Astron. Soc. **406**, 1379 (2010)
P. Meszaros, *High-Energy Radiation from Magnetized Neutron Stars* (University of Chicago Press, Chicago, 1992)
S.G. Moiseenko, G.S. Bisnovatyi-Kogan, N.V. Ardeljan, Mon. Not. R. Astron. Soc. **370**, 501 (2006)
C.G. Mundell, D. Kopač, D.M. Arnold et al., Nature **504**, 119 (2013)
D. Neuhauser, S.E. Koonin, K. Langanke, Phys. Rev. A **36**, 4163 (1987)
M. Obergaulinger, M.A. Aloy, E. Muller, Astron. Astrophys. **515**, A30 (2010)
C. Palenzuela, L. Lehner, S.L. Liebling et al., Phys. Rev. D **88**, 043011 (2013)
D.J.B. Payne, A. Melatos, Mon. Not. R. Astron. Soc. **351**, 569 (2004)
J.A. Pons, R. Perna, Astrophys. J. **741**, 123 (2011)
A.Y. Potekhin, Astron. Astrophys. **351**, 787 (1999)
A.Y. Potekhin, A. De Luca, J.A. Pons, arXiv:1409.7666 (2014)
D.J. Price, S. Rosswog, Science **312**, 719 (2006)
N. Rea, P. Esposito, R. Turolla et al., Science **330**, 944 (2010)
A. Reisenegger, arXiv:1305.2542 (2013)
M.M. Romanova, R.V.E. Lovelace, M. Bachetti et al., Eur. Phys. J. Web Conf. **64**, 5001 (2014)
N. Shabaltas, D. Lai, Astrophys. J. **748**, 148 (2012)
F.H. Shu, A. Allen, H. Shang, E.C. Ostriker, Z.-Y. Li, *The Origin of Stars and Planetary Systems*. NATO ASIC Proc., vol. 540 (1999), p. 193
M. Shibata, K. Taniguchi, Phys. Rev. D **73**, 064027 (2006)
A. Shirakawa, D. Lai, Astrophys. J. **564**, 361 (2002a)
A. Shirakawa, D. Lai, Astrophys. J. **565**, 1134 (2002b)
J.B. Simon, X.-N. Bai, P.J. Armitage, J.M. Stone, K. Beckwith, Astrophys. J. **775**, 73 (2013)
P.A. Sturrock, Astrophys. J. **164**, 529 (1971)
M. Tagger, R. Pellet, Astron. Astrophys. **349**, 1003 (1999)
M. Tagger, P. Varniere, Astrophys. J. **652**, 1457 (2006)
A. Tchekhovskoy, A. Spitkovsky, J.G. Li, Mon. Not. R. Astron. Soc. **435**, L1 (2013)
C. Thompson, R.C. Duncan, Astrophys. J. **408**, 194 (1993)
C. Thompson, M. Lyutikov, S.R. Kulkarni, Astrophys. J. **574**, 332 (2002)
T.A. Thompson, P. Chang, E. Quataert, Astrophys. J. **611**, 380 (2004)
R. Turolla, S. Zane, J.A. Pons, P. Esposito, N. Rea, Astrophys. J. **740**, 105 (2011)
D. Uzdensky, A. Königl, C. Litwin, Astrophys. J. **565**, 1191 (2002)
M. van Adelsberg, D. Lai, A.Y. Potekhin, P. Arras, Astrophys. J. **628**, 902 (2005)
M. van Adelsberg, D. Lai, Mon. Not. R. Astron. Soc. **373**, 1495 (2006)
M. van der Klis, in *Compact Stellar X-ray Sources*, ed. by W. Lewin, M. van der Klis (Cambridge University Press, Cambridge, 2006)
M.H. van Kerkwijk, D.L. Kaplan, Astrophys. Space Sci. **308**, 191 (2007)
D. Viganò, N. Rea, J.A. Pons et al., Mon. Not. R. Astron. Soc. **434**, 123 (2013)
C. Wang, D. Lai, Mon. Not. R. Astron. Soc. **398**, 515 (2009)
D.T. Wickramasinghe, C.A. Tout, L. Ferrario, Mon. Not. R. Astron. Soc. **437**, 675 (2014)
S.E. Woosley, Astrophys. J. **719**, L204 (2010)
C. Yu, D. Lai, Mon. Not. R. Astron. Soc. **429**, 2748 (2013)
C. Yu, D. Lai, arXiv:1501.05727 (2015)
F. Yuan, R. Narayan, Annu. Rev. Astron. Astrophys. **52**, 529 (2014)
C. Zanni, Eur. Phys. J. Web Conf. **64**, 5005 (2014)
C. Zanni, J. Ferreira, Astron. Astrophys. **550**, A99 (2013)
W. Zhang, S.E. Woosley, A.I. MacFadyen, Astrophys. J. **586**, 356 (2003)

Part II
Magnetic Fields in Stars

DOI 10.1007/978-1-4939-3550-5_3
Reprinted from *Space Science Reviews* Journal, DOI 10.1007/s11214-015-0143-1

Observations of Strong Magnetic Fields in Nondegenerate Stars

Jeffrey L. Linsky[1] · Markus Schöller[2]

Received: 22 November 2014 / Accepted: 4 March 2015 / Published online: 14 March 2015

Abstract We review magnetic-field measurements of nondegenerate stars across the Hertzprung–Russell diagram for main sequence, premain sequence, and postmain sequence stars. For stars with complex magnetic-field morphologies, which includes all G–M main sequence stars, the analysis of spectra obtained in polarized vs unpolarized light provides very different magnetic measurements because of the presence or absence of cancellation by oppositely directed magnetic fields within the instrument's spatial resolution. This cancellation can be severe, as indicated by the spatially averaged magnetic field of the Sun viewed as a star. These averaged fields are smaller by a factor of 1000 or more compared to spatially resolved magnetic-field strengths. We explain magnetic-field terms that characterize the fields obtained with different measurement techniques. Magnetic fields typically control the structure of stellar atmospheres in and above the photosphere, the heating rates of stellar chromospheres and coronae, mass and angular momentum loss through stellar winds, chemical peculiarity, and the emission of high energy photons, which is critically important for the evolution of protoplanetary disks and the habitability of exoplanets. Since these effects are governed by the star's magnetic energy, which is proportional to the magnetic-field strength squared and its fractional surface coverage, it is important to measure or reliably infer the true magnetic-field strength and filling factor across a stellar disk. We summarize magnetic-field measurements obtained with the different observing techniques for different types of stars and estimate the highest magnetic-field strengths. We also comment on the different field morphologies observed for stars across the H–R diagram, typically inferred from Zeeman-Doppler imaging and rotational modulation observations.

✉ J.L. Linsky
jlinsky@jila.colorado.edu
Fax: +1-303-492-5235

M. Schöller
mschoell@eso.org
Fax: +49-89-32006-530

1 JILA/University of Colorado and NIST, Boulder, CO, USA

2 European Southern Observatory, Garching, Germany

Keywords Magnetic fields · Pre-main sequence stars · Main sequence stars · Post-main sequence stars

List of Acronyms and Abbreviations

BBLP	Broad-band linear polarization
BOB	B fields in OB stars consortium
CFHT	Canada-France-Hawaii Telescope
CP	Chemically preculiar A- and B-type stars
CRIRES	CRyogenic high-resolution InfraRed Echelle Spectrograph
DI	Doppler imaging
ESO	European Southern Observatory
ESP_aDO_nS	a high-resolution spectrograph on the CFHT telescope
FEROS	Fiber-fed Extended Range Optical Spectrograph
FORS1	FOcal Reducer and low dispersion Spectrograph
FORS2	New version of FORS1
FUV	Far ultraviolet (912–1700 Å)
GTO	Guaranteed time observer
HARPS	High Accuracy Radial Velocity Planet Searcher
HARPSpol	HARPS with polarizing optics
HINODE	Solar-B satellite
IVQU	Components of the Stokes polarimetry vector
LTE	Local thermodynamic equilibrium
LSD	Least-squares deconvolution
MDI	Michelson Doppler imager
MHD	Magnetohydrodynamics
MiMeS	Magnetism in Massive Stars collaboration
MOST	Microvariability and Oscillations of STars satellite
NIM	Near Infrared Magnetograph
NSO	National Solar Observatory
RGB	Red giant branch stars
SB	Spectroscopic binary stars
SOFIN	An optical high-resolution spectrograph on the Nordic Optical Telescope
SOHO	Solar and Heliospheric Observatory
SOT	Solar Optical Telescope
SOT/SP	SOT Spectropolarimeter
SPB	Slowly pulsating B-type stars
SVD	Singular value decomposition method for Stokes profile reconstruction
UV	Ultraviolet (912–3000 Å)
VLT	Very Large Telescope
ZAMS	Zero-age main sequence
ZDI	Zeeman Doppler imaging

1 Introduction and Terminology

A prominent solar physicist once remarked,

> "If the Sun did not have a magnetic field, it would be as uninteresting as most astronomers consider it to be."

Table 1 Magnetic field terminology

Measurement type	Spatially unresolved observations	
	Filling factor included	Filling factor separate
Stokes I unsigned line-of-sight spectra (no cancellation)	Mean unsigned magnetic-field strength (or flux density) (unpolarized Zeeman broadening) $\langle B\rangle = B_{\mathrm{MOD}} f$	Unsigned magnetic-field modulus and filling factor (unpolarized Zeeman splitting) B_{MOD} and f
Stokes V or IVQU signed line-of-sight spectropolarimetry (with cancellation)	Net longitudinal magnetic-field strength (or flux density) (Zeeman broadening) $\langle B_z\rangle = B_{\mathrm{NET}} f$	Net magnetic-field modulus and filling factor (Zeeman splitting) B_{NET} and f

Such statements about the Sun and stars in general ignore the important ways in which magnetic fields typically control the structure of stellar atmospheres in and above the photosphere, the heating rates of stellar chromospheres and coronae, mass and angular momentum loss through stellar winds, chemical peculiarity, and the emission of high energy photons, which is critically important for the evolution of protoplanetary disks and the habitability of exoplanets. Thus, stellar structure models and phenomenological descriptions are woefully incomplete without a clear picture of the star's magnetic-field morphology and variability. Here we will summarize the magnetic-field measurement techniques, observations, and phenomena of nondegenerate stars. For other summaries of this topic, see the review papers by Reiners (2012) and Donati and Landstreet (2009). For a review of theoretical ideas that have been proposed to explain the origins and morphologies of magnetic fields in nondegenerate stars, see the following chapter where Ferrario et al. (2015) provide an overview of turbulent dynamo theories including a description of the inverse cascade of magnetic energy to large scales and the back reaction of strong fields leading to saturation. Also see the review by Donati and Landstreet (2009) and references therein and the recent papers by Nelson et al. (2013, 2014) on solar-like dynamos.

Before proceeding, it is essential to understand magnetic-field terminology (see Table 1). Unfortunately, these terms are often misapplied, which makes it difficult to compare theory with observations and to identify the strongest magnetic fields on the stellar surface.

Vector Magnetic-Field Strength $\mathbf{B}(\mathbf{x}, t)$ is the true strength of the local magnetic-field vector at a given location and time in the stellar photosphere (see Fig. 1). Although this is an essential quantity that must be compared with theory, it is very difficult to measure. There are very few measurements of $\mathbf{B}(\mathbf{x}, t)$ on the Sun (e.g., Rabin 1992) and, except for stars with the simplest magnetic-field morphologies, none for stars because magnetographs have apertures that are not able to resolve magnetic fields that are inhomogeneous on small spatial scales. Many magnetic-field measurements are polarimetric, in which case there is cancellation by oppositely directed fields. When such cancellations cannot be corrected for, we use the term net magnetic-flux density, $\mathbf{H}(\mathbf{x}, t)$ to describe the vector field with cancellation.

Mean Unsigned Magnetic-Field Strength Unpolarized spectra can be used to measure the unsigned magnetic field by studying the excess broadening of high Landé g spectral lines compared to otherwize similar low Landé g lines or by analyzing resolved line splitting when the magnetic field is strong. Since the magnetic-field strength varies across the stellar surface, it is useful to define a "mean-unsigned–magnetic-field strength", $\langle B\rangle$, which is an average value of the magnetic field across the stellar surface, and an "unsigned-magnetic-field modulus" also called the "mean magnetic-field modulus" (e.g., Hubrig et al.

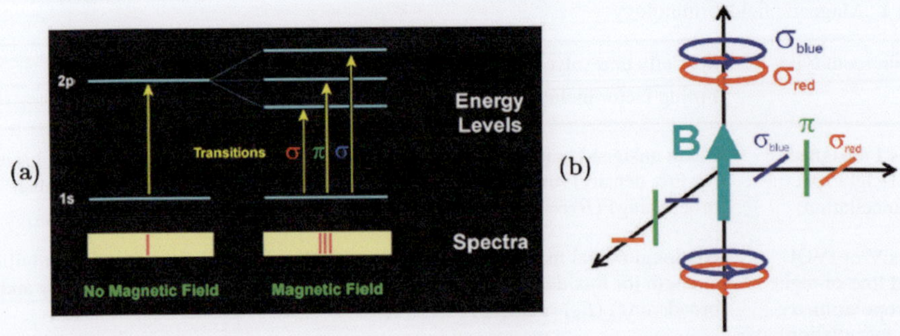

Fig. 1 *Left*: The splitting $\Delta\lambda = 46.67 g\lambda^2 B$ in mÅ of a 2p level in the presence of a magnetic field, where B is in kGauss and g is the Landé splitting factor. Since the splitting is proportional to the magnetic-field strength squared, the best opportunity for actually measuring B is with infrared instrumentation. *Right*: Circular polarization (Stokes vector V) is observed when the magnetic field is along the line of sight, and linear polarization (Stokes vectors Q and U) is observed when the magnetic field is observed perpendicular to the line of sight. Figure from Reiners (2012)

2005), B_{MOD}, which is the peak magnetic-field strength. These two quantities are related by $\langle B\rangle = B_{\mathrm{MOD}} f$, where the filling factor f is the fraction of the stellar surface covered by B_{MOD}. In the simplest case, one can approximate the distribution of magnetic field strengths within an instrument's field of view by a single magnetic-field strength B_{MOD} filling a fractional area f and zero magnetic field filling the remaining fractional area, $1 - f$. When the signal-to-noise ratio and spectral resolution are excellent, especially in the infrared where the λ^2 dependence of the σ and π component splitting resolves the Zeeman pattern, B_{MOD} can be measured from the wavelength difference of the split components and f can be determined by fitting the depth of the Zeeman absorption components. In a few cases, for example the measurements of the classical T Tauri star BP Tau by Johns-Krull et al. (1999), observers have been able to identify a distribution of magnetic-field strengths with corresponding filling factors, such that $B_{\mathrm{MOD}} f = \sum_i B_{\mathrm{MOD,i}} f_i$. When the magnetic field is too weak to resolve the splitting of the Zeeman components, one can measure only $\langle B\rangle = B_{\mathrm{MOD}} f$.

Net Longitudinal Magnetic-Field Strength Spectropolarimeters measure net magnetic fields produced by the cancellation of oppositely oriented magnetic fields subtended within the instrument's field of view—either the observed stellar hemisphere, the resolution element obtained with Zeeman Doppler-imaging observations (see below), or the spatial resolution element on the Sun. We use the term "net longitudinal magnetic-field strength", $\langle B_z\rangle$, to refer to the magnetic field averaged over the spatial resolution element of the detector on the Sun or the entire star, including field cancellation. We refer to the peak signed magnetic field in the resolution element as the "net magnetic-field modulus", B_{NET}, which is the largest signed magnetic field strength seen anywhere on the solar surface away from sunspots or at any rotational phase of a star. The relation between these two quantities is $\langle B_z\rangle = B_{\mathrm{NET}} f$, where f is the filling factor.

Since stellar magnetic fields are often complex on small spatial scales, unpolarized instruments measure $\langle B\rangle$, which for stars with complex magnetic-field morphologies like the Sun is much larger than the net longitudinal field $\langle B_z\rangle$. For stars with simple dipolar fields like many of the Ap stars, these two quantities can be similar. Both instruments with and without polarizing optics have great difficulty identifying the contribution of the strongest magnetic fields, which typically occur in dark starspots, when measuring unresolved stellar emission. The presence of strong magnetic fields in dark starspots has only been measured on the Sun with high-resolution infrared instruments as we will show. For all other observations of the Sun and stars, instruments measure a magnetic quantity that is smaller, and often much smaller, than the highest value of $\mathbf{B}(\mathbf{x}, t)$.

2 Techniques for Measuring Magnetic Fields

In this section, we discuss magnetic field observations obtained by four direct and three indirect techniques. For additional discussion of this topic see Reiners (2012) and Donati and Landstreet (2009).

Net Circular Polarization This is the classical technique used for more than 100 years to measure net longitudinal magnetic-field strengths, $\langle B_z \rangle$. The magnetic signal is provided by the difference between opposite circular- polarization spectra (blue–red in Fig. 1) for a spectral line with a large Landé g factor. The required instrumentation is relatively simple, but oppositely directed fields cancel within a spatial resolution element. Also, since these measurements are only sensitive to the longitudinal component of the magnetic-field vector, they refer to the net magnetic-field strength rather than the vector magnetic-field strength, $\mathbf{B}(\mathbf{x}, t)$. Since circular polarization is not sensitive to nonmagnetic line broadening, this technique can measure weak magnetic fields. However, signal saturation when measuring strong magnetic fields can limit the range of measurable fields. When observing stars with largely organized magnetic fields, such as the chemically peculiar stars, measurements of $\langle B_z \rangle$ are invaluable tools to learn about the net magnetic-field modulus, $B_{\rm NET}$, and the magnetic topology, since the magnetic field changes in a characteristic way while the star rotates.

Zeeman Broadening Spectral lines with large Landé g factors are broadened more than lines with small g factors. The excess broadening of a high g line compared to a low g line provides a measure of the unsigned magnetic-field strength averaged over the stellar hemisphere, $\langle B \rangle$. However, the high and low g lines must have nearly the same broadening, except for the magnetic component, as likely occurs for two spectral lines of similar strength formed in the same region of the atmosphere and preferably from the same element or ion. This technique first developed by Robinson et al. (1980) requires only high-resolution spectra obtained without polarizing optics. Since many effects can broaden a spectral line, this technique works best for measuring strong magnetic fields.

Zeeman Splitting When high-resolution spectroscopy in the infrared makes it feasible, the resolved σ components of an absorption line provide a measurement of the average value of $\langle B \rangle$ over the stellar surface or observed region on the Sun. For the strong magnetic fields of the Ap stars, Zeeman splitting is also observed in the visible.

Zeeman Doppler Imaging The Doppler-imaging (DI) technique provides relative brightness images of the surface of a rotating star in a spectral line. At each rotational phase, the bright and dark regions along each stellar longitude modify the spectral line profile at their corresponding Doppler shifts. Analysis of high-resolution spectra obtained at many rotational phases lead to a synthetic stellar image with the bright and dark regions displayed at their proper locations across the stellar surface. The Zeeman Doppler-imaging (ZDI) technique provides magnetic-field maps of the stellar surface with comparable spatial resolution. Instruments that measure circular polarization (i.e., Stokes vector V, which is proportional to

$\langle B_z \rangle$) provide stellar surface maps of $B_{\mathrm{NET}} f$. Instruments that measure both circular and linear polarization (i.e., Stokes vectors QU, which are proportional to the transverse magnetic field) provide surface maps of the radial, toroidal, and meridional components of $B_{\mathrm{NET}} f$. This technique is described in detail by Donati and Landstreet (2009). Since the spectral resolution set by the rotation rate and the spectral resolution cannot resolve the small scale structure of $\mathbf{B}(\mathbf{x}, t)$, such instruments measure the vector or longitudinal components of the net magnetic-flux density, $\mathbf{H}(\mathbf{x}, t)$.

Nonthermal Radio Emission Active late-type stars are often detected by their gyrosynchrotron radio emission produced by relativistic electrons spiraling in coronal magnetic fields. Observations over a wide spectral range can provide estimates of the coronal magnetic-field strength. Inactive stars like the Sun emit gyroresonance emission produced by thermal electrons spiraling in coronal magnetic fields. The wavelength of third-harmonic gyroresonant emission from solar active regions provides an accurate measure of coronal magnetic-field strengths.

Sunspots and Starspots Optical solar images show dark sunspots and bright faculae in the photosphere. Magnetic-field measurements in sunspots show the strongest magnetic fields on the solar surface and similar measurements in bright faculae show moderate-strength fields. Dark sunspots are generally thought to be produced by magnetic suppression of convective energy transport, and the bright faculae are typically explained by magnetic heating in the low chromosphere. Doppler images and rotational modulation studies provide evidence for very large starspots in active stars, providing an indirect indicator of magnetic fields.

Indicators of Magnetic Heating Stellar chromospheres ($T = 10^4$–10^5 K) and coronae ($T \geq 10^6$ K) are generally thought to be heated by either the dissipation of magnetohydrodynamic (MHD) waves or the direct conversion of magnetic to kinetic energy (microflaring). Both mechanisms involve magnetic fields, and the hotter outer layers produce emission lines of Ca II, hydrogen (H-α and Lyman-α), and many highly ionized species in the UV and X-ray spectra. These emission features provide strong evidence for the presence of magnetic fields but do not measure magnetic-field strengths.

3 What can be Learned from a Spatially Resolved Star, the Sun?

By virtue of its proximity and brightness, the Sun provides the best laboratory for studying magnetic fields on a star, but even for this special case there are important limitations. Figure 2 shows a high-resolution magnetogram of the whole Sun and an expanded portion near disk center obtained with the Michelson Doppler Imager (MDI) on the Solar and Heliospheric Observatory (SOHO) spacecraft. The figure shows the net longitudinal magnetic-field strength, $\langle B_z \rangle$, obtained with the circular polarization- differencing technique by observing the Ni I 6768 Å line with a Landé $g = 1.43$. With $2'' \times 2''$ pixels, MDI could not measure unsigned magnetic-field strengths as the magnetic-field morphology is inhomogeneous on smaller spatial scales. Since the largest $\langle B_z \rangle$ values shown in this figure are ± 250 G, the maximum resolved magnetic-field strengths, B_{MOD}, must be much larger. How much larger? To answer this question, we consider another observing technique.

For Zeeman-sensitive lines with large Landé g factors observed in the infrared, the wavelength-splitting pattern obtained when opposite circular-polarization spectra are subtracted is larger than the linewidth. In this case, the observed splitting is proportional to the

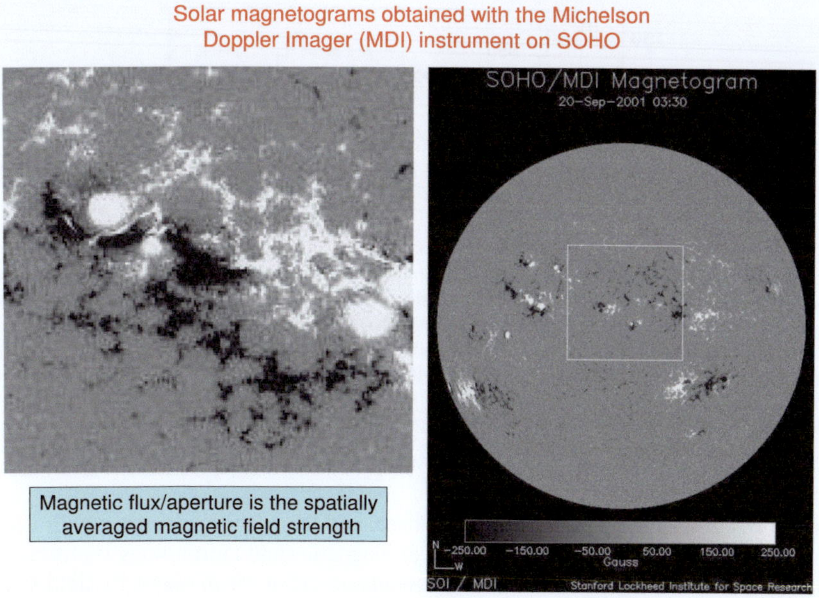

Fig. 2 Magnetic-flux densities of the whole Sun (*right*) and the center of the solar disk (*left*) obtained by the MDI instrument on the Solar and Heliospheric Observatory (SOHO) spacecraft with a resolution of 2″. *White* and *black* refer to opposite orientations of the net longitudinal magnetic-field strength

net magnetic-field strength modulus, $B_{\rm NET}$, and the amplitude of the circular polarization signal indicates the approximate filling factor of the magnetic field in the solar photosphere at this spatial resolution. Rabin (1992) used this Zeeman-splitting technique to measure the distribution of magnetic fields in a solar active region by observing the Fe I 1.565 μm line. The distribution shown in Fig. 3 has a peak near $B_{\rm NET} = 1300$ Gauss with a width of 270 Gauss. Rabin estimated that the filling factor for the 1300 Gauss field is $f = 0.1$–0.2, implying that $\langle B_z \rangle = B_{\rm NET} f$ should be approximently 130–260 Gauss. By comparison, the distribution of net longitudinal magnetic-field strengths measured in the optical spectrum with the National Solar Observatory (NSO) magnetograph in the same active region at the same time ranges between 0 and 500 Gauss with most resolution elements showing $\langle B_z \rangle \leq 250$ Gauss. Since both instruments had similar spatial resolution of 2", the two data sets are consistent. This comparison suggests that magnetic flux measurements without independent estimates of f would underestimate the magnetic-field modulus by a factor of 5–10, even for the high spatial-resolution observations of the Sun. The lack of spatial resolution for stellar observations suggests that there is an even larger discrepancy between $B_{\rm NET}$ and $\langle B_z \rangle$ for stars.

The Solar Optical Telescope/Spectropolarimeter (SOT/SP) on the *HINODE* satellite provides 0.″32 angular resolution and 10^{-3} polarization sensitivity. Viticchié et al. (2011) have inverted a representive sample of these Stokes V data with a code that allows for unresolved magnetic elements smaller than even this extremely high resolution. They found a distribution of net magnetic-field modulus values with the largest fraction of the observed field at $B_{\rm NET} = 1600$ Gauss. They also found that 95.5 % of their observed region in the photosphere had no, or very weak, magnetic fields such that the $\langle B_z \rangle = 66$ G. This analysis provides a picture of the nearest star in which photospheric magnetic fields are concentrated in very thin flux tubes with a distribution of field strengths up to 1600 Gauss. Such stars occupy a

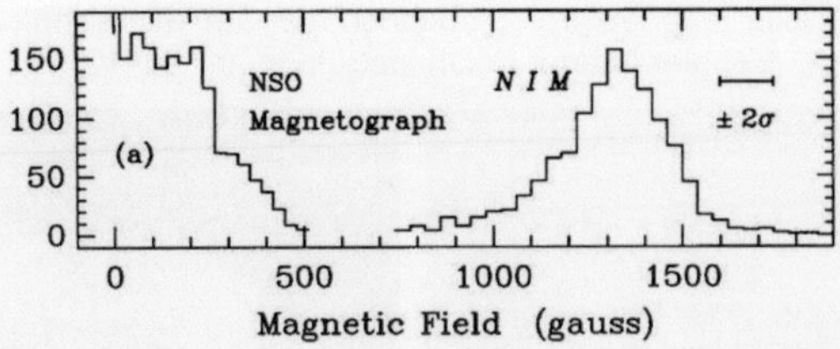

Fig. 3 Plotted on the right is the histogram (y axis) of the net magnetic-field modulus, $B_{\rm NET}$, measured in a solar active region by the Near Infrared Magnetograph (NIM). The λ^2 splitting of the Zeeman σ components permits measurements of true magnetic field strengths in the infrared. Plotted on the left is the histogram (also y axis) of the net-longitudinal magnetic-field strength (also called flux density), $\langle B_z \rangle$, in the same active region observed at the same time by the National Solar Observatory (NSO) magnetograph. The x-axis plots the corresponding magnetic-field strengths measured in the two different ways. The NIM was able to measure $B_{\rm NET}$ values from the complete splitting in Stokes V spectra of the Fe I 1.565 µm line with Landé $g = 3.00$, whereas the NSO magnetograph observing in the optical could not detect the splitting of spectral lines in Stokes V, resulting in field cancellation and averaging of the inhomogeneous field. Figure from Rabin (1992)

very small fraction of the photosphere at its base, and then rapidly decrease with height as the flux tubes expand. If this picture of the solar magnetic-field morphology is representative of other convective stars, then from measurements of stellar mean magnetic-field strengths (signed or unsigned) it will be very difficult to infer magnetic modulus values, filling factors, and magnetic energies.

The strongest solar magnetic fields are found in sunspot umbrae. Moon et al. (2007) analyzed Stokes IVQU spectra of the Fe I 6301.5 Å line observed in a sunspot umbra with the SOT/SP instrument on *HINODE*. In these 0.″6 resolution spectra, they found maximum net field strengths of about 4500 Gauss. These field strengths are about 1000 Gauss stronger than typical ground-based measurements (e.g., Pevtsov et al. 2014) likely because of the absence of spartial smearing due to seeing and minimal scattered light in the SOT/SP data.

Measuring the net longitudinal magnetic field of the Sun viewed as an unresolved point source by a distant observer can be done by studying the solar spectrum reflected by an asteroid or by integrating whole disk magnetograms. An example of the first approach is the study of the solar spectrum reflected by Vesta that Daou et al. (2006) found to have $\langle B_z \rangle = -4 \pm 3$ Gauss. This is a very small net field compared to the active stars described below. An example of the second approach is the measurement by Plachinda et al. (2011) who found $\langle B_z \rangle = \pm 2$ Gauss at the maximum of the solar activity cycle and ± 0.2 Gauss at the minimum. Schrijver and Liu (2008) found total magnetic fluxes of 25×10^{22} Maxwells at times of solar minima and 100×10^{22} Maxwells at the maxima of the sunspot cycle by integrating global magnetograms obtained with the SOHO/MDI instrument. For comparison, individual active regions on the Sun have magnetic fluxes in the range 10^{20} to 3×10^{22} Maxwells. For a more detailed description of the solar magnetic field see, for example, Schrijver and Zwaan (2000).

4 Pre-Main Sequence Stars

Magnetic fields play key roles in essentially all aspects of the star formation process (e.g., McKee and Ostriker 2007). Models of magnetically driven accretion and outflows successfully reproduce many observational properties of low-mass pre-main sequence stars, also called young stellar objects (YSOs). While indirect observational evidence for the presence of magnetic fields in these stars is manifested in strong X-ray, FUV (912–1700 Å), and UV

(912–3000 Å) emission, flaring, and gyrosynchrotron radio emission (e.g., Feigelson and Montmerle 1999), there are now many direct observations of YSO magnetic fields.

4.1 T Tauri Stars

The evolution of YSOs from dense clumps of interstellar gas and dust to main sequence stars proceeds through several stages (Feigelson and Montmerle 1999). The earliest phase of a protostar (referred to as a Class 0 object) consists of a collapsing massive disk detectable only at millimeter and far-infrared wavelengths. After about 10^5 years, the collapsing cloud has formed a Class I object consisting of a heavily obscured star heated by the release of gravitational energy and a large circumstellar disk. After about 10^6 years, the star becomes optically visible above and below the plane of the disk, although the visible spectrum is veiled by emission from the disk. These stars are called Classical T Tauri stars (CTTSs) or Class II objects. Strong UV emission seen in CTTSs results mostly from accretion of gas along magnetic field lines from the disk to the star, producing postshock regions with strongly enhanced emission in the He I $\lambda5876$ and other lines. Strong X-ray emission can be produced in the postshock region, closed-field lines in the stellar corona, and where stellar and disk magnetic fields interact. In the last stage of evolution to the main sequence, which occurs between a few million years and 10^7 years, most of the gas and dust from the disk has condensed into planets or has been accreted onto the star, which is now called a Weak-lined T Tauri star (WTTS) or Class III object. WTTSs are often called naked TTSs because their ultraviolet emission line spectrum is no longer obscured by a disk.

Zeeman broadening of infrared absorption lines provides a powerful tool for measuring the unsigned magnetic-field strengths, $\langle B \rangle$, of CTTSs. Johns-Krull (2007) summarized previous studies and included new measurements for 15 CTTSs (mostly spectral types K7–M2) in the $\sim$2 Myr old Taurus region. Analysis of spectropolarimetry of four Ti I lines at wavelengths near 2.2 μm allowed him to solve for the distribution of field strengths and their sum to obtain $\langle B \rangle$. These field strengths do not correlate with the predictions of several magnetospheric models and are typically twice as large as predicted by pressure equilibrium with the photospheric gas, unlike the case for main sequence stars. He argued that the need for horizontal pressure balance requires that strong fields must cover the entire stellar surface of CTTSs. Since spectropolarimetric measurements (see below) indicate much weaker fields for these stars, the fields needed for horizontal pressure balance must have complex morphologies rather than be simple inclined dipoles. These strong complex fields are dynamo-driven in the stellar convective zone starting from "seed" primordial fields. The role of strong magnetic fields in controlling the accretion flow from disks to small areas at high latitudes on the stellar surface is now generally accepted, see the review by Bouvier et al. (2007). The Reiners (2012) review paper includes an updated list of Zeeman-broadening measurements for pre-main sequence stars.

The Zeeman-broadening technique has permitted Johns-Krull and collaborators to sample the unsigned magnetic fields of pre-main sequence stars with a range of ages younger and older than the CTTSs in Taurus. Johns-Krull et al. (2009) obtained the first measurement of the magnetic-field strength of a Class 1 protostar WL 17. They obtained $\langle B \rangle = 2.9 \pm 0.43$ kG, which corresponds to three times the photospheric gas pressure indicating that the entire photosphere of this embedded $\sim 10^5$ yr protostar is likely dominated by strong fields. Yang and Johns-Krull (2011) studied 14 Class II CTTSs (spectral types K3–M1.5) in the 1 My old Orion Nebular Cluster. These stars have measured $\langle B \rangle$ between 1.3 and 3.45 kG with magnetic pressures about twice that of the photospheric gas pressure. The $\sim$10 Myr old TW Hya association contains stars at the very end of the pre-main sequence phase. Using the same observational technique, Yang et al. (2008) found that magnetic fields of five Class III naked TTSs with spectral types K7–M3 have $\langle B \rangle = 3.1$–4.9 kG. For these stars the photospheric magnetic pressure also dominates the gas pressure. Comparing the magnetic fields of the youngest to the oldest pre-main sequence stars, Yang and Johns-Krull (2011) found no dependence on age for the unsigned magnetic-field strength but a systematic decrease in the magnetic flux ($4\pi R_\star^2 \langle B \rangle$) with age as a result of the systematic decrease of stellar radius with age.

Spectropolarimetry provides the basis for mapping the topological structure of large-scale magnetic fields of pre-main sequence stars. While the magnetic fields of many CTTSs have now been mapped, the 1.5 Myr star BP Tau observed by Johns-Krull et al. (1999), Donati et al. (2008b), and Chen and Johns-Krull (2013) provides an illustrative example. All three studies found that the circular polarization seen in the He I $\lambda 5876$ and several other emission lines, which are produced in postshock accretion regions, occur in very small areas (2–3 % of the stellar surface) generally located at high latitudes. The longitudinal magnetic field strength in BP Tau's accretion region was measured by Johns-Krull et al. (1999) to be $\langle B \rangle = 2460 \pm 120$ G by the Zeeman-broadening technique but as large as 12 kG by Donati et al. (2008b) using the ZDI technique. The reason for this large discrepancy is uncertain. The correspondence of Stokes I and Stokes V images led Donati et al. (2008b) to conclude that the accretion region overlies a large dark starspot in the photosphere near the star's rotational pole. All three studies agree that the dark starspot covers about 25 % of the visible surface and that the remainder of the star is threaded with strong, but tangled, magnetic fields with an unsigned magnetic field strength $\langle B \rangle = 2.6 \pm 0.3$ kG, but only an upper limit to the mean longitudinal magnetic field measured by differencing absorption line profiles in opposite circular polarizations. Donati et al. (2008b) found that a simple approximation to the magnetic-field topology is the sum of tilted dipole and octopole components, but only 90 % of the large scale magnetic energy is poloidal, with the remainder torroidal. Since the measured magnetic-field strength exceeds pressure equilibrium with the photospheric gas, the entire surface must be covered by strong radial or complex magnetic fields. According to Donati et al. (2008b), the strong fields are likely produced by dynamo processes in the star's convective zone rather than being primordial.

Donati and an international team of collaborators have nearly completed the Magnetic Protostars and Planets (MaPP) project to map the surface brightness and magnetic-field topology of 15 CTTSs using the ESPaDOnS spectropolarimeter on the 3.6 m Canada-France-Hawaii Telescope and the NARVAL spectropolarimeter on the 2 m Télescope Bernard Lyot. The observed CTTSs include the very low mass 0.35 $M_\odot$ star V2247 Oph (Donati et al. 2010a) and the 0.7 $M_\odot$ CTTSs DN Tau (Donati et al. 2013) and AA Tau (Donati et al. 2010b), which like BP Tau have fully convective interiors. Their observing program also includes the higher mass stars TW Hya (Donati et al. 2011) and the 1.7–1.8 $M_\odot$ stars CV Cha and CR Cha (Hussain et al. 2009), which have radiative cores. The CTTSs range in age from about 1.5 Myr [AA Tau, BP Tau, and V2247 Oph (Donati et al. 2010a)] to about 8 Myr (TW Hya). Figure 4 plots the luminosities and effective temperatures for these and other CTTSs and identifies their locations along pre-main sequence evolution tracks for their masses. The basic properties of large-scale magnetic fields of CTTSs shown in Fig. 4 suggest that these properties may change as stars develop a radiative core. However, this suggestion is based on only two stars (V4046 Sgr A and B) and there are many competing effects including spin-down with mass loss and spin-up with accretion and decreasing radius with age. More data are needed to address this question.

An important criterion for classification as a CTTS is evidence of accretion of gas from the circumstellar disk to the star. For CTTSs with dipolar magnetic fields, the footpoints of

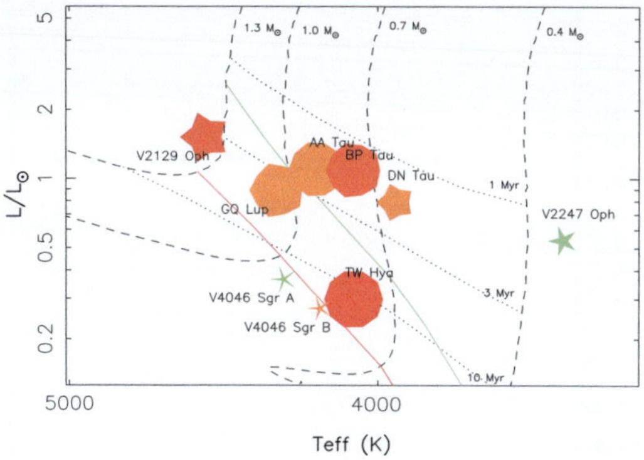

Fig. 4 The basic properties of large-scale magnetic topologies of CTTSs obtained by the ZDI technique. *Symbol size* indicates the relative magnetic intensities. *Symbol color* illustrates the magnetic-field configurations (*red* to *blue* for purely poloidal to purely toroidal fields). Symbol shape depicts the relative axisymmetry of the poloidal field component (*decagon* and *stars* for purely axisymmetric and purely non-axisymmetric poloidal fields, respectively). *Dashed lines* are pre-main sequence evolution tracks and *dotted lines* are the corresponding isochrones. CTTSs begin to develop radiative cores below the *green line* and convective envelopes become thinner than $0.5R_\star$ below the *red line*. Figure and caption from Donati et al. (2013)

accretion columns, as indicated by postshock heated emission in the He I 5876 Å, Ca II, and hydrogen lines, occur in small areas (about 2 % of the observed disk) located in high-latitude regions where strong magnetic fields overlie spots that are optically dark in the photosphere like sunspots. On the other hand, stars with very complex fields, such as the lowest mass CTTS V2247 Oph, the accretion impact area is at a range of latitudes (Johnstone et al. 2014), and the underlying photosphere may not appear to be dark. The longitudinal magnetic fields where the accretion occurs are typically in the range $\langle B \rangle = 2$–3 kG and appear to be mainly radial. These spot fields do not contribute to the average stellar magnetic field obtained from Stokes V measurements of photospheric absorption lines, because the photospheric emission from these starspots is faint.

Outside of the accretion shock regions, the stellar magnetic field can be approximated by a sum of multipoles typically misaligned with respect to the rotational pole. For many of the mapped CTTSs, the octopole component is stronger and often very much stronger than the dipole. AA Tau is an exception. In general, the more complex the geometry of the large-scale field, the weaker the dipole component and the more distributed the magnetic flux across the stellar surface (Johnstone et al. 2014). The percent of magnetic energy in the toroidal field ranges from < 5 % for the oldest star TW Hya to 50 % or more for V2247 Oph and the high mass stars V2129 Oph, CR Cha, and CV Cha. The coupling of stellar magnetic fields with magnetic fields in the inner disk slows stellar rotation, because the Keplerian rotation speed of the inner disk is generally slower than that of the star, while accretion of disk gas adds angular momentum to the star, thereby increasing its rotation rate. Hussain et al. (2009) and Donati et al. (2011) have called attention to the role that stellar radiative cores can play in this angular momentum interchange. As CTTSs age and evolve toward the main sequence, they develop radiative cores and weaker dipole fields. The effect is to weaken the star-disk magnetic coupling, allowing accretion to speed up the stellar rotation unhindered. Johnstone et al. (2014) describe the correlation of simpler magnetic-field geometry with slower rotation and larger size of X-ray-emitting stellar coronae. Vidotto et al. (2014) show that the increase in stellar rotation rates with age is consistent with the weakening of the star-disk deceleration torque with age.

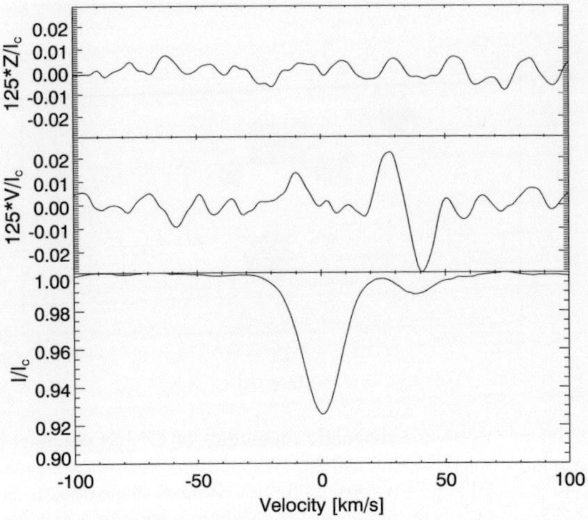

Fig. 5 Analysis of the presence of a magnetic field in both components of the Herbig Ae SB2 system HD 104237 by applying the SVD method to polarimetric spectra with the High Accuracy Radial Velocity (HARPS) instrument. The upper spectrum presents the diagnostic null (or Z) spectrum, which is associated with the Stokes V spectrum by using subexposures with identical waveplate orientation. Below the null spectrum, one finds the Stokes V and Stokes I spectra. The measured-net longitudinal magnetic field for the primary is about 13 G, while for the secondary it is 128 G

4.2 Herbig Ae/Be Stars

Magnetic fields have also been detected in half a dozen Herbig Ae/Be stars (e.g., Hubrig et al. 2009c). Similar to T Tauri stars, Herbig Ae/Be stars show clear signatures of surrounding disks as evidenced by a strong infrared excess and are actively accreting material.

Current theories are neither able to present a consistent scenario of how the magnetic fields in Herbig Ae/Be stars are generated nor how these fields interact with the circumstellar environment consisting of a combination of disk, wind, accretion, and jets. On the other hand, understanding the interaction between the central stars, their magnetic fields, and their protoplanetary disks is crucial for reconstructing the Solar System's history and accounting for the diversity of exo-planetary systems.

Before 2004, the only magnetic field detection of about 50 G had been reported for the optically brightest ($V = 6.5$) Herbig Ae star HD 104237 (Donati et al. 1997), but no further publications confirming this detection existed until recently. Using high-resolution, high signal-to-noise spectropolarimetric observations with the High Accuracy Radial velocity (HARPSpol) instrument, Hubrig et al. (2013b) detected a weak net-mean-longitudinal magnetic field. And only now have Hubrig et al. (in preparation) been able to demonstrate that the secondary component possesses a magnetic field (see Fig. 5). Spectropolarimetric studies from 2004 to 2008 reported the discovery of magnetic fields in seven other Herbig Ae/Be stars (Wade et al. 2005, 2007; Catala et al. 2007; Hubrig et al. 2004, 2006a, 2007). Later on, a study of 21 Herbig Ae/Be stars with the FOcal Reducer and low dispersion Spectrograph (FORS 1) revealed the presence of magnetic fields in six additional stars (Hubrig et al. 2009b). More recent studies involved the outbursting magnetic binary Z CMa (Szeifert et al. 2010), the Herbig Ae star HD 101412 with resolved magnetically split lines, and HD 31648 = MWC 480 (Hubrig et al. 2010, 2011d).

Spectropolarimetric observations of a sample of 21 Herbig Ae/Be stars observed with FORS 1 have been used to search for a link between the presence of a magnetic field and

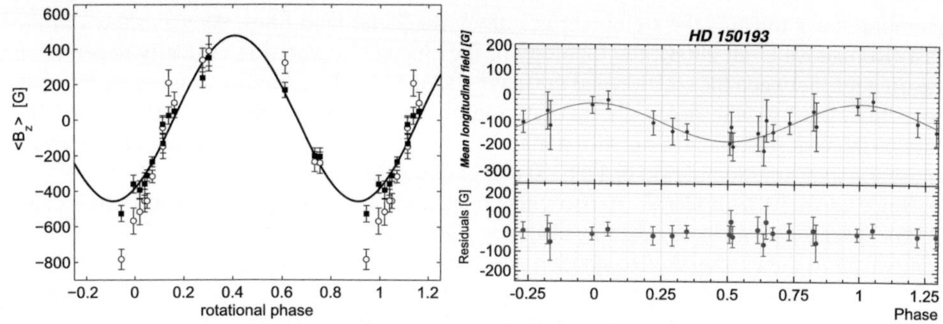

Fig. 6 *Left panel*: Phase diagram of HD 101412 with the best sinusoidal fit for the net longitudinal magnetic-field measurements using all lines (*filled squares*) and hydrogen lines (*open circles*). *Right panel*: Phase diagram of HD 150193 with the best sinusoidal fit for $\langle B_z \rangle$ using all lines. The residuals (observed–calculated) are shown in the lower panel

other stellar properties (Hubrig et al. 2009c). This study did not indicate any correlation of the strength of the net longitudinal magnetic field with disk orientation, disk geometry, or the presence of a companion. No simple dependence on the mass-accretion rate was found, but the range of observed-field values qualitatively supported the expectations from magnetospheric accretion models with dipole-like field geometries. Both the magnetic field strength and the X-ray emission showed hints of a decline with age in the range of $\sim$ 2–14 Myr probed by the sample, supporting a dynamo mechanism that decays with age. Furthermore, the stars seemed to obey the universal power-law relation between magnetic flux and X-ray luminosity established for the Sun and main-sequence active dwarf stars.

A series of net longitudinal magnetic-field measurements was recently obtained at low resolution with the multimode instrument FORS 2 at the Very Large Telescope (VLT) for the Herbig Ae/Be stars HD 97048, HD 101412, HD 150193, and HD 176386 (Hubrig et al. 2011a). Magnetic fields of the order of $\langle B_z \rangle = 120$–250 G were for the first time detected in these stars a few years ago during a visitor run with FORS 1 in May 2008 (Hubrig et al. 2009b). In these observations, Herbig Ae/Be stars exhibit a single-wave variation in $\langle B_z \rangle$ during the stellar rotation cycle. This behavior is usually considered as evidence of a dominant dipolar contribution to the magnetic-field topology. Presently, the Herbig Ae star HD 101412 possesses the strongest net-longitudinal magnetic field ever measured in any Herbig Ae star, with a surface magnetic field $\langle B_z \rangle$ up to 3.5 kG (Wade et al. 2005, 2007). HD 101412 is also the only Herbig Ae/Be star for which the rotational Doppler effect was found to be small in comparison to the magnetic splitting, and several spectral lines observed in unpolarized light at high dispersion are resolved into magnetically split components (Hubrig et al. 2010, 2011a).

To date, magnetic-field geometries have been studied for the two SB2 systems HD 200775 (B3 primary) and V380 Ori (B9 primary) (Alecian et al. 2008, 2009), and presumably single stars HD 101412, HD 97048, HD 150193, and HD 176386 (Hubrig et al. 2010, 2011a). As an example, phase diagrams of the magnetic data for the Herbig Ae/Be stars HD 101412 and HD 150193 folded with the determined magnetic/rotation periods are presented in Fig. 6.

The magnetic-field model for the Herbig Ae star HD 101412 is described by a centered dipole with a polar magnetic-field strength B_d between 1.5 and 2 kG and an inclination of the magnetic axis to the rotation axis β of $84 \pm 13°$ (Hubrig et al. 2011d). The fact that the dipole axis is located close to the stellar equatorial plane is very intriguing in view of the generally assumed magnetospheric accretion scenario that magnetic fields channel the accretion flows towards the stellar surface along magnetic field lines. As was shown in the past (Romanova et al. 2003), the topology of the channeled accretion critically depends on the tilt angle between the rotation and the magnetic axis. For large inclination angles β, many polar field lines would thread the inner region of the disk, while the closed lines cross the path of the disk matter, causing strong magnetic braking, which could explain the observed unusually long rotation period of HD 101412 of about 42 days.

Since about 70 % of Herbig Ae/Be stars appear in binary/multiple systems (Baines et al. 2006), special care must be taken in assigning the measured magnetic field to the particular component in the Herbig Ae/Be system. Alecian et al. (2008) reported on the discovery of a dipolar magnetic field in the Herbig Be star HD 200775, which is a double-lined spectroscopic binary system. However, the magnetic field was discovered not in the component possessing a circumstellar disk and dominating the Hα emission; thus the evolutionary status of the primary B3 component is unclear yet (Benisty et al. 2013). Similar to the case of HD 200775, the frequently mentioned discovery of a magnetic field in the Herbig SB2 system HD 72106 (Alecian et al. 2009) refers to the detection only in the primary component, which is a young main-sequence star, but not in the Herbig Ae secondary (Folsom et al. 2008). The same uncertainty in the evolutionary status applies to the magnetic field detection in the system V380 Ori reported by Alecian et al. (2009). The authors report on the presence of a dipole magnetic field of polar strength 2.12 ± 0.15 kG on the surface of the chemically peculiar primary V380 Ori system. V380 Ori has spectral type around B9 and has been observed in great detail over many wavelength ranges (e.g., Hamann and Persson 1992; Rossi et al. 1999; Stelzer et al. 2006). It has a close infrared companion, with a separation of $0.15''$ at PA 204° (Leinert et al. 1997). Alecian et al. (2009) found that the primary in the V380 Ori system is itself a spectroscopic binary with a period of 104 days, with the secondary being a massive T Tauri star. More recently, Reipurth et al. (2013) report that V380 Ori is a hierarchical quadruple system with a fourth component at a distance of $8.8''$ and position angle 120.4°. Since no periodicity was found in the behavior of the emissions in hydrogen, helium, calcium, and oxygen lines (the lines determining the Herbig Ae/Be nature), it is possible that the primary chemically peculiar component with the detected-dipolar magnetic field is already in advanced age, and the Herbig Be status of the primary is merely based on the appearance of emission in the above-mentioned lines belonging to the secondary T Tauri component.

The presence of a magnetic field on the surface of the Herbig Ae star HD 190073 has been known for several years. The first measurement of a net longitudinal magnetic field in HD 190073 was published by Hubrig et al. (2009c), indicating that $\langle B_z \rangle = 84 \pm 30$ G measured on FORS 1 low-resolution spectra at the 2.8σ level. Catala et al. (2007) then observed this star using the ESPaDOnS spectrograph mounted on the Canada-France-Hawaii Telescope (CFHT) and confirmed the presence of a weak magnetic field, $\langle B_z \rangle = 74 \pm 10$ G, at a higher significance level. A few years later Hubrig et al. (2009c) reported $\langle B_z \rangle = 104 \pm 19$ G using FORS 1 measurements. The measurement of the net longitudinal magnetic field using the available archival HARPS observations from May 2011, $\langle B_z \rangle = 91 \pm 18$ G, presented in this work fully confirms the presence of a rather stable weak field. Surprisingly, new observations of this star during July 2011 and October 2012 by Alecian et al. (2013) detected variations of the Zeeman signature in the LSD spectra on time scales of days to weeks.

The authors suggest that the detected variations of Zeeman signatures are the result of the interaction between the fossil field and the ignition of a dynamo field generated in the newly born convective core. Careful spectropolarimetric monitoring over the next years will be important to confirm the reported variability of the magnetic field. Furthermore, since HD 190073 is very likely a binary system (Baines et al. 2006), special care must be taken in the interpretation of magnetic-field measurements.

As mentioned above, all previously studied Herbig Ae/Be stars exhibit a single-wave variation in the net longitudinal magnetic field during the stellar rotation cycle. These observations are usually considered as evidence for a dominant dipolar contribution to the magnetic field topology. Magnetospheric accretion theories traditionally consider that simple ∼kG dipolar magnetic fields truncate the disk and force in-falling gas to flow along the field lines. The assumption of dominant dipolar fields is usually made for simplicity or because of the lack of available information about the true large-scale magnetic field topology of these stars. Indeed, the recent work of Adams and Gregory (2012) shows that high-order field components may even play a dominant role in the physics of the gas inflow, as the accretion columns approach the star.

The rather new diagnostic He I λ 1.083 μm emission line is considered as probing inflow (accretion) and outflow (winds) in the star-disk interaction region of accreting T Tauri and Herbig Ae/Be stars. The uniqueness of this probe derives from the metastability of this transition and makes it a good indicator of wind and funnel flow geometry (Edwards et al. 2006). Furthermore, according to Edwards et al., the He I line appears in emission for stronger mass-accretion rates and in net absorption for low mass-accretion rates. Modeling of this line allowed Gregory et al. (in preparation) for the first time to study the influence of field topologies on the star-disk interaction. Their models use magnetic fields with an observed degree of complexity, as determined via field extrapolation from stellar magnetic maps.

In Fig. 7, one can find recent high-resolution CRyogenic high-resolution Infrared Echelle Spectrograph (CRIRES) observations of the spectral regions containing the He I λ 1.083 μm line and the hydrogen recombination line Paγ at 1.094 μm over the rotation period of HD 101412 (Hubrig et al. 2012a). The rather strong variation of the line profile of the He I line indicates that the magnetic field of this star is likely more complex than a dipole field. Variable behavior of the He I λ 1.083 μm line was also discovered in a recent X-shooter spectra of the magnetic Herbig Ae stars HD 190073 and PDS 2 (see Fig. 8).

Knowledge of the magnetic-field strength and topology is indispensable for understanding the magnetospheres of Herbig Ae/Be stars and their interaction with the circumstellar environment consisting of a combination of disk, wind, accretion, and jets. Progress in understanding the disk-magnetosphere interaction can, however, only come from studying a sufficient number of targets in detail to look for various patterns encompassing this type of pre-main sequence stars.

Before concluding this topic, we call attention to the work of Bagnulo et al. (2012) and Landstreet et al. (2014), who have questioned the credibility of previous detections of weak magnetic fields at the 3–4σ level with the FORS1 instrument on the ESO VLT. They find that consistency checks among different observations of the same star reveal external errors that are at least 30–40 % larger than can be explained by the cited photon statistics and that alternative reduction procedures provide significantly different (and often smaller) values for the mean longitudinal fields. They suggest that small instrument flexures and velocity shifts and "occasional outliers" likely limit the ability of FORS1 to obtain reliable fields weaker than 100–200 G. Bagnulo et al. (2012) argued that many of the published detections of weaker fields are spurious and not confirmed by measurements with high-resolution spectropolarimeters such as ESPaDOnS (Silvester et al. 2009), but that stronger field measurements by FORS1, for example many Ap/Bp stars, are generally confirmed (Landstreet et al. 2014).

Bagnulo et al. (2012) found that of the five Herbig Be stars that Hubrig et al. (2009c) cites as new detections with errors of 3.4–4.5σ, four (PDS2, HD 100546, HD 135344B, and HD 176386) appear to be spurious and two (HD 97048 and HD 150193) are possible

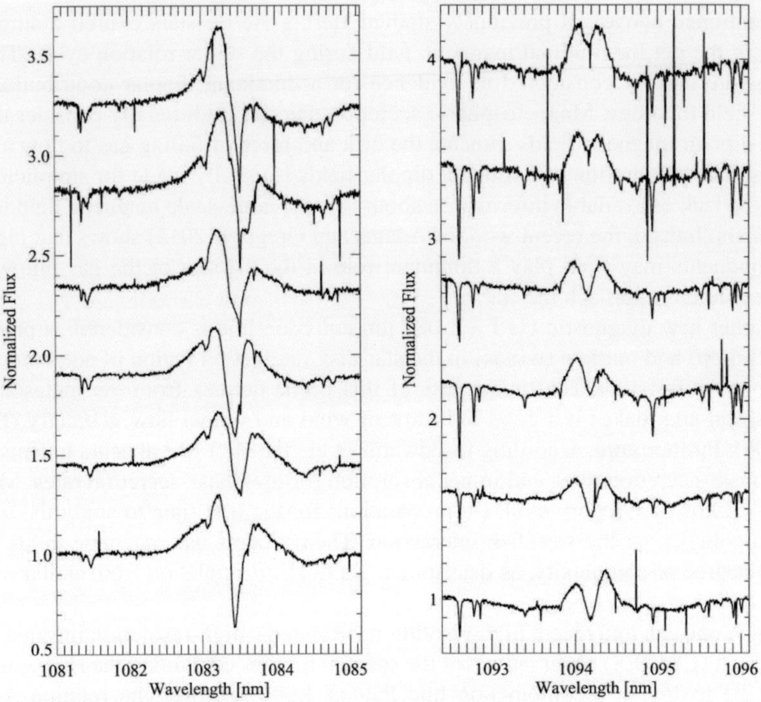

Fig. 7 Recent observations of HD 101412 with the CRyogenic high-resolution Infrared Echelle Spectrograph (CRIRES). *Left panel*: The variability of the He I λ 1.083 μm line profile over the rotation period. Obviously, the field of HD 101412 appears more complex than just a dipole. *Right panel*: Variations of the hydrogen recombination line Paγ at 1.094 μm at the same rotation phases. The Paγ line at 1.094 μm is frequently employed for calculating the mass accretion rate in the way presented by Gatti et al. (2008)

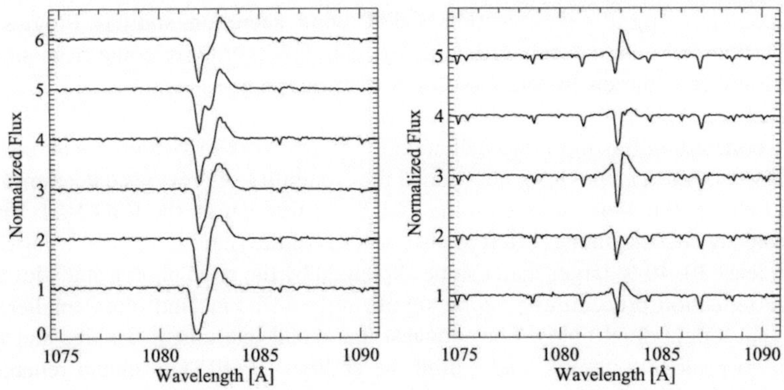

Fig. 8 Recent X-shooter observations of the He I λ 1.083 μm line profile in the magnetic Herbig Ae stars HD 190073 (*left panel*) and PDS 2 (*right panel*) at different epochs. The spectra are shifted vertically for clarity

detections. In subsequent observations with FORS2, Hubrig et al. (2011d) found weak sinusoidally varying fields in HD 97048, HD 150193, and HD 176386, but the cited errors are in the 3–4σ range. At present, the identification of magnetic fields in a few Herbig Be stars (e.g., HD 76106B, HD 101412 and HD 190073) are confirmed, weak fields appear to be present but are not definitely established in several stars, but most Herbig Be stars do not have detected magnetic fields.

5 Main Sequence Stars

5.1 Low Mass Stars

The very small values of the surface-averaged net longitudinal component of the solar magnetic field are a consequence of the almost complete cancellation of the spectropolarimetric signal by oppositely oriented magnetic fields. Similar measurements for nearby bright G and K stars like ξ Boo A and 61 Cyg A also show weak magnetic fields of 10–30 Gauss and 13 Gauss, respectively (Plachinda and Tarasova 2000; Plachinda et al. 2001). One must conclude that the typical morphology of late-type dwarf stars is highly complex, and, therefore, different observing techniques are required to understand the magnetic-field strengths and morphologies of these stars.

5.1.1 Zeeman Broadening Observations

Robinson et al. (1980) first showed that measurements of Zeeman broadening of spectral lines in unpolarized light can be used to measure the mean magnetic-field strength $\langle B \rangle = B_{\rm MOD} f$ averaged over the surface of a late-type star. This technique described in Sect. 2 has been applied to a number of mostly active stars (see tables in Reiners 2012), but the technique has its limitations as described in detail by Anderson et al. (2010) and Saar (1988).

Anderson et al. (2010) analyzed high-resolution optical spectra of four stars with spectral types similar to the Sun, two slowly rotating quiet stars, 61 Vir (G6 V) and the Sun (G2 V) viewed as a star by reflected light from Ganymede, and two rapidly rotating active stars, 59 Vir (G0 V) and HD 68456 (F6 V). They computed spectral line profiles by radiative transfer calculations in local thermodynamic equilibrium (LTE) model atmospheres with a range of broadening parameters to best fit the observed line profiles. These calculations were for both single-component and two-component model atmospheres for which one component could be hotter (plage region), the same, or cooler (starspot) than the mean temperature structure. They found that $\langle B \rangle$ is consistent with zero for the two quiet stars, 600–1200 Gauss for HD 68456, and about 500 Gauss for 59 Vir. However, they identified a number of systematic effects (e.g., line blends, overly simplified treatment of turbulence, and LTE atmospheres) and concluded that a more fruitful approach would be to analyze Zeeman broadening of infrared spectral lines with large Landé g values, because then Zeeman broadening is much larger than the other broadening mechanisms.

An early example of applying the near-infrared Zeeman-broadening technique to measure magnetic fields was the study of the active star ϵ Eri (K2 V) and two other K stars (40 Eri and σ Dra) by Valenti et al. (1995). They observed the Fe I 1.56485 μm line with Landé $g = 3.0$ and 15 other nearby Fe I lines with smaller g factors. By modelling the Fe I line profiles with marginally resolved σ components, they were able to measure separately the unsigned magnetic-field modulus $B_{\rm MOD} = 1.44 \pm 0.2$ kG and filling factor $f = 0.088 \pm 0.025$. The corresponding value of $B_{\rm MOD} f = 0.13$ kG. With the same technique, Johns-Krull and Valenti (1996) modelled the Fe I 8468.4 Å line with Landé $g = 2.5$ in four M dwarfs. Figure 9 shows the significant Zeeman broadening of the Fe I lines of the two active stars, EV Lac (M4.5e) and GJ 729 (M4.5e), relative to less active comparison

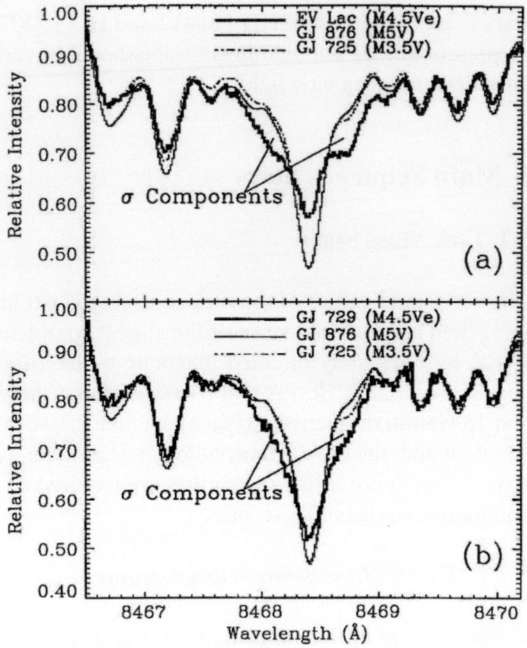

Fig. 9 Comparison of Stokes I spectra of the Fe I 8468.40 Å line of two active M dwarfs (EV Lac and GJ 729) compared to less active stars. This comparison demonstrates that the excess line broadening for the two active stars is magnetic rather than blending by TiO lines, which should be the same for all of these stars. Figure from Johns-Krull and Valenti (1996)

stars that the authors identify as the splitting of the σ components of the stellar magnetic fields. Since this splitting is larger than other broadening mechanisms, their spectral synthesis analysis determined the mean magnetic fields and filling factors separately rather than as the product of $B_{\mathrm{MOD}} f$. They found $B_{\mathrm{MOD}} = 3.8 \pm 0.5$ kG and $f = 0.50 \pm 0.13$ for EV Lac and $B_{\mathrm{MOD}} = 2.6 \pm 0.3$ kG and $f = 0.5 \pm 0.13$ for Gl 729. These magnetic field strengths are larger than the maximum field strengths seen in solar active regions and likely result from the higher pressure in M dwarf photospheres and the balance between magnetic pressure $(B^2/8\pi)$ in magnetic flux tubes and gas pressure in the nonmagnetic photosphere.

Figure 10 shows that for G–M dwarfs, the observed magnetic-field strengths are approximately equal to the equipartition field set by pressure balance between the magnetic pressure in flux tubes and the nonmagnetic-photospheric gas pressure outside of the flux tubes. This pressure balance is also seen in solar active regions (plages) but not in sunspots, where the emergent spectrum is depressed to higher pressure layers below the normal photosphere. T Tauri stars show the same effect, although the reason is not yet clear. This figure also shows that f increases with rotation period according to the relation $f = 79\text{–}64\log(P)$. Cuntz et al. (1999) have used a similar relation in computing a two-component (nonmagnetic and magnetic flux tube) model chromosphere for ϵ Eri. This relation probably does not contain significant bias resulting from the different rotation rates or activity of the magnetic star and the comparison low activity star, because the widths of low Landé g spectral lines in the active and inactive stars have the same shape whereas the shapes of the high Landé g lines differ greatly and show marginally or completely resolved sigma components for the active star.

Reiners and Basri (2007, 2010) and Shulyak et al. (2011) have extended this study of M dwarf magnetic fields to include more than 80 stars in the spectral range M2.0 to M9.5 by fitting the Zeeman-broadened lines of molecular FeH near 1.0 μm. The average magnetic-field strengths of these stars range up to $\langle B \rangle = 4$ kG (see Fig. 11) with a large range at each spectral type. Except for the coolest stars in their sample (spectral types M7 to M9.5),

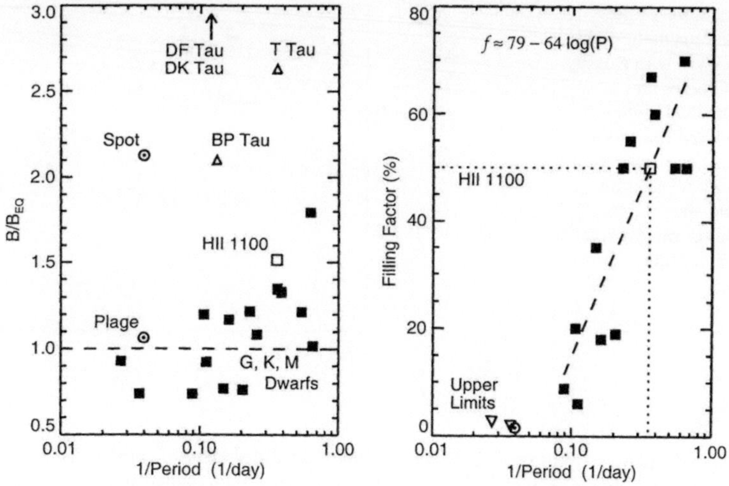

Fig. 10 (*left*) Ratio of the measured-unsigned magnetic-field modulus to the equipartition magnetic field strength when the magnetic pressure ($B^2_{\mathrm{MOD}}/8\pi$) equals the photospheric gas pressure. The *dashed-horizontal line* corresponds to measured-magnetic fields equal to the equipartition field. *Triangles* indicate T Tauri stars and ⊙ symbols refer to solar features (plage and sunspot). (*right*) Filling factors ($100 \times f$) vs. rotational periods for G–M dwarf stars. Figure from Valenti and Johns-Krull (2001)

Fig. 11 A summary of mean-unsigned magnetic-field strengths $\langle B \rangle = B_{\mathrm{MOD}} f$ of M dwarf stars obtained mostly with the Zeeman-broadening technique (Reiners 2012). For stars at each spectral type, there is a wide range in Bf, but a maximum value near 4000 G

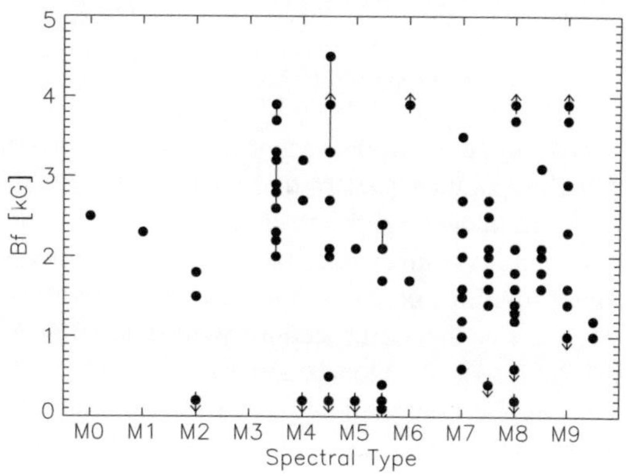

the magnetic field strength increases with rotational velocity, as is seen in G and K stars. Although these measurements could not separate B_{MOD} from f, many of these stars have $B_{\mathrm{MOD}} f \approx 4$ kG, similar to EV Lac, for which $f = 0.50 \pm 0.13$ (Johns-Krull and Valenti 1996). This summary suggests that for the active cool M dwarfs, about half of the stellar photosphere has magnetic fields as strong as $B_{\mathrm{MOD}} \approx 8$ kG.

The Ti I spectral lines near 2.2 μm, which provide even larger Zeeman splittings, are useful for measuring magnetic-field strengths and filling factors of M dwarfs like AD Leo (Saar and Linsky 1985). The combination of large Zeeman splitting and high signal/noise Ti I spectra of the T Tauri star TW Hya (see Sect. 4.1) permitted Valenti and Johns-Krull (2001) to determine the distribution of magnetic-field strengths and corresponding filling factors across the stellar surface.

The measured values of $B_{\mathrm{MOD}} f$ are correlated with the commonly used indicators of magnetic activity and heating. Figure 12 shows the correlation of $B_{\mathrm{MOD}} f$ with the Rossby

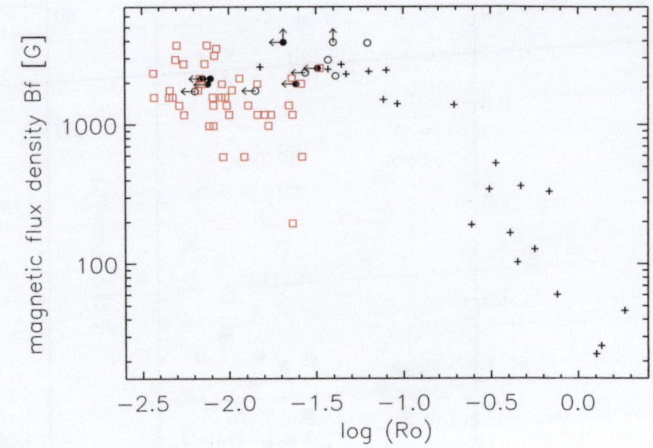

Fig. 12 Correlation of the mean-unsigned magnetic-field strength, $B_{\mathrm{MOD}}f$, with Rossby number R_0. Crosses are solar-type stars, and circles are M0–M6 stars. *Red squares* are M7–M9 stars. These data show a rotation-activity relation for slowly rotating stars (large R_0) and saturation for rapidly rotating stars (small R_0). Figure from Reiners (2012)

number, $R_0 = P_{\mathrm{rot}}/\tau_{\mathrm{conv}}$, where P_{rot} is the stellar rotational period, and τ_{conv} is the convective turnover time, which is not a well-defined theoretical quantity. It is usually computed empirically as the mass-dependent convective turnover time that minimizes scatter in the rotation-activity relation as described by Reiners (2012). "R_0 is an important parameter for the dynamo generation of magnetic fields. Figure 12 shows the log-log rotation-activity relation for slower-rotating stars and saturation for the faster-rotating stars. The ratio of Hα luminosity to bolometric luminosity, log $[L_{\mathrm{H}\alpha}/L_{\mathrm{bol}}]$ vs. log R_0 shows the same type of correlation (Reiners 2012).

5.1.2 Zeeman Doppler Imaging

Doppler imaging produces a brightness map of a stellar surface by inverting a time sequence of high-resolution spectra that contain the brightness and Doppler shift of absorption lines produced in each spatial resolution element on the stellar surface (Vogt and Penrod 1983). Although such inversions have many solutions, several regularization techniques lead to robust solutions of the two-dimensional brightness maps. Zeeman-Doppler imaging is the extension of the DI procedure with polarizing optics, as first proposed by Semel (1989) and developed by Donati and Brown (1997). The first applications of ZDI to stars used only circular polarization data (Stokes V), but recent developments in instrumentation and polarized radiative-transfer calculations (Piskunov and Kochukhov 2002) allow the analysis of both circular- and linear-polarized spectra (Stokes IVQU). Donati and Landstreet (2009) provide a comprehensive review of this topic.

ZDI works because the longitudinal and transverse components of the vector magnetic field produce different Zeeman patterns as a function of velocity when the magnetic vector is viewed from the advancing limb to the star's central meridian and then to the receding limb. The circular polarization Zeeman patterns (net after cancellation) are different for the radial, azimuthal (toroidal), and meridional components of the magnetic field, as shown in Fig. 13. Linear-polarization Zeeman patterns show different center-to-limb effects for the radial, azimuthal, and meridional magnetic fields. Since high signal-to-noise data are required to detect weak circular and especially linear polarization, Donati and Brown (1997) pioneered the use of a very large number of spectral lines to create ZDI images.

Before describing some of the major accomplishments of ZDI, we should mention the limitations of the technique. First, and most important, ZDI measures the large-scale net magnetic-field strength and not the unsigned magnetic-field strength because the circular and linear polarization data are net values after cancellation by oppositely oriented fields in

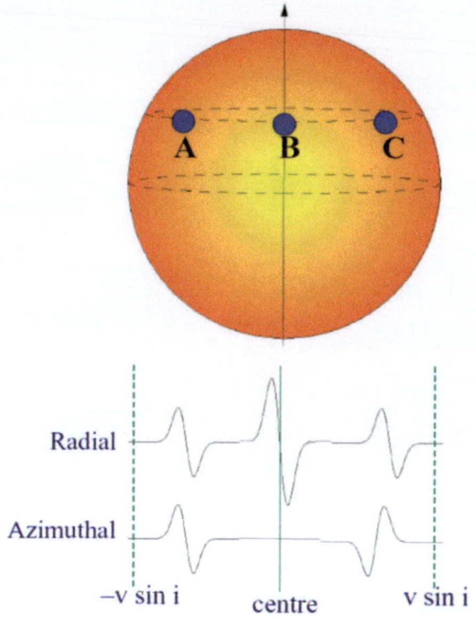

Fig. 13 A schematic representation of the circular-polarization Zeeman spectrum (V/I) of an absorption line formed in a magnetic field. If the magnetic field is radial relative to the star (and therefore longitudinal at disk center as viewed by an observer), the Zeeman pattern is strongest on the stellar meridian and retains the same symmetry, but weakens, as the radial field is observed towards the stellar limbs. If the magnetic field is azimuthal (E–W direction) relative to the star (and therefore transverse at disk center and longitudinal at the limbs), the Zeeman pattern is strongest at the limbs (but opposite symmetry) and zero at disk center. If the magnetic field is in the meridional direction (N–S direction, not shown in the figure), the Zeeman pattern is observed only toward the poles. Figure from Hussain (2004)

each spatial resolution element. Also important are the effects of irregular or large gaps in the time spacing, uncertainty in the inclination of the star's rotation axis relative to the line of sight, possible crosstalk between the Stokes parameters, line blends, different temperatures (and thus continuum brightness) of magnetic compared to nonmagnetic regions, evolution of the stellar magnetic topology during the multiday observing sequence, and poorly understood instrumental and systematic errors. Despite these limitations, ZDI studies have found that the topologies of active G–M stars, which are younger and likely precursers of older stars like the Sun, differ considerably from what we know about the Sun in the following ways:

Importance of Azimuthal Fields ZDI images of active solar-mass stars typically show strong azimuthal fields that are not seen in inactive stars like the Sun. A good example is the study by Morgenthaler et al. (2012) of the magnetic field on ξ Boo A, a G8 V star rotating almost four times faster than the Sun ($P_{\rm rot} = 6.43$ days). As shown in Fig. 14, the reconstructed net magnetic field includes a strong azimuthal component that contains most of the net magnetic field in the 2007 and 2011 data sets when Ca II emission indicated that the star was most active. For solar-mass stars, the magnetic energy in large-scale fields increases with the rotation rate (Vidotto et al. 2014): most of this magnetic energy is in the poloidal field for slow rotators like the Sun, but the fraction stored in azimuthal fields increases with rotation rate and dominates the poloidal fields when $P_{\rm rot} \leq 12$ days (Petit et al. 2008; see Fig. 15). ZDI images of the very rapidly rotating ($P_{\rm rot} = 0.51479$ days)

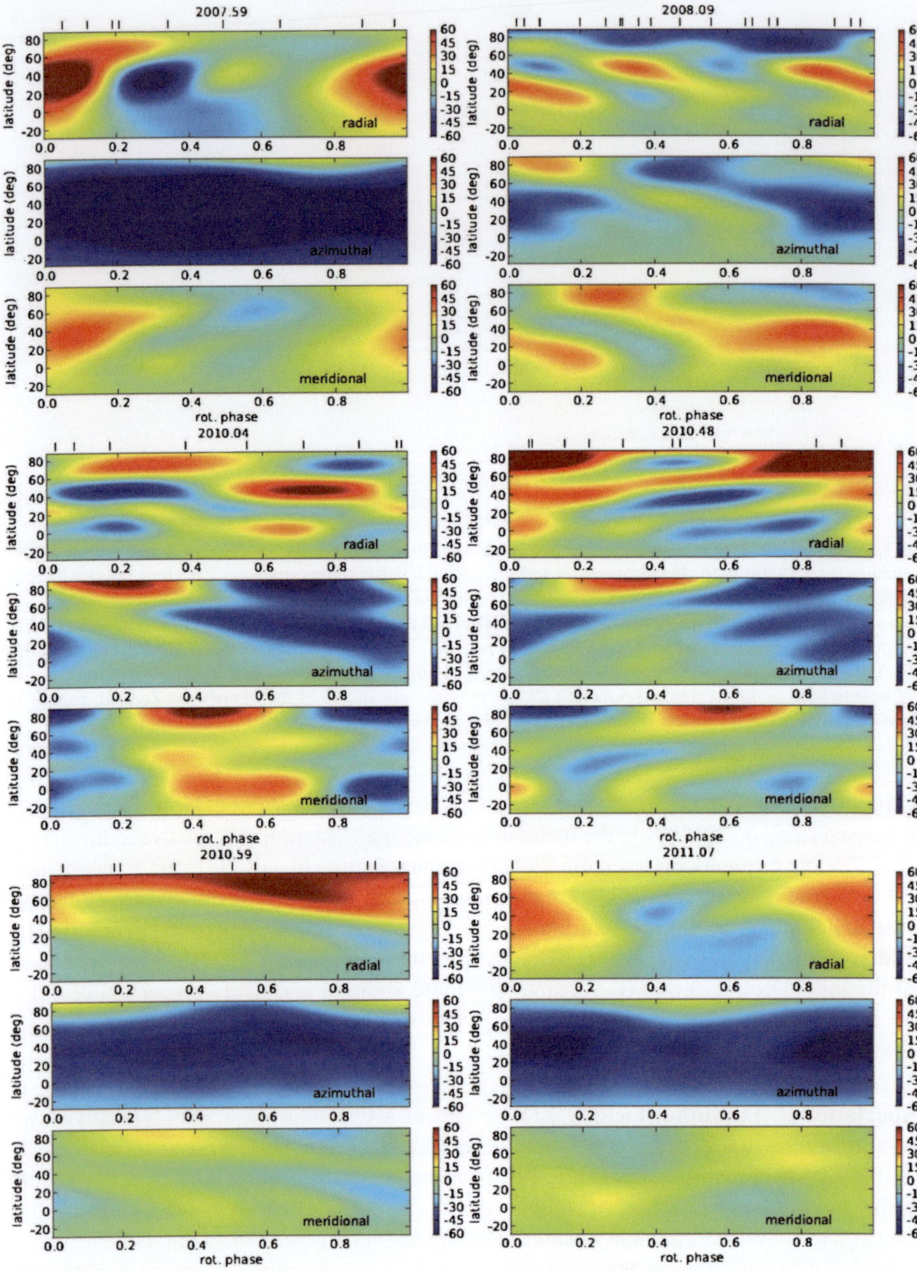

Fig. 14 Reconstructed net magnetic-field maps of ξ Boo A for the radial, azimuthal,and meridional components obtained at five times between 2007 and 2011 by Morgenthaler et al. (2012). The strong azimuthal component contains the most of the magnetic energy when the star is most active (2007 and 2011)

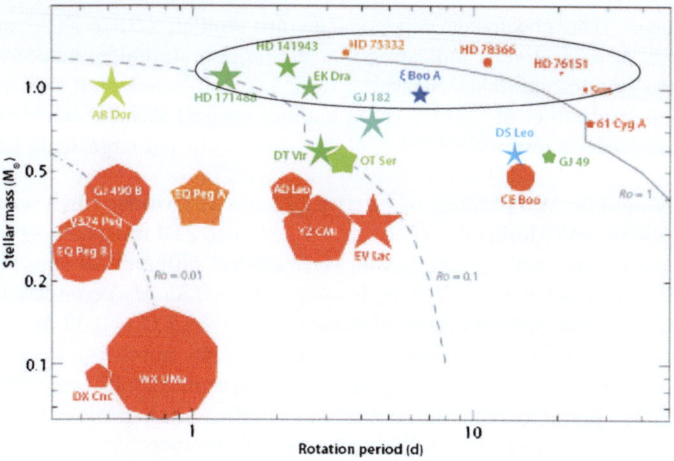

Fig. 15 A summary of the large-scale magnetic topologies and fluxes obtained from ZDI images as a function of stellar mass, rotation period, and Rossby number (R_0). Symbol size indicates the relative magnetic energy density, and symbol color indicates field topology with blue for purely toroidal and red for purely poloidal fields. The symbol shape indicates the degree of asymmetry (decagon for purely axisymmetric poloidal fields and stars for purely nonaxisymmetric poloidal fields). The encircled area contains the Sun and more rapidly rotating solar-mass stars. Figure adapted from Donati and Landstreet (2009)

young K0 V star AB Dor show strong azimuthal fields both at high latitude (70°–80°), which appears to encircle a polar spot, and at intermediate latitudes with opposite sign (Donati et al. 1999; Hussain et al. 2002). The net magnetic field in these azimuthal fields exceeds 1 kG in some places with the high latitude azimuthal flux the largest measured flux on this star.

Relation of Net Magnetic Fields to Unsigned Field Strengths The net magnetic fields for ξ Boo A shown in Fig. 14 would suggest that $B_{\mathrm{NET}} = 30 - 100$ G if one assumes that $f = 1.0$ and no field cancellation. This result should be compared to the measurements of the spatially resolved solar magnetic-field modulus, $B_{\mathrm{MOD}} \approx 1600$ G. The value of B_{MOD} for ξ Boo A should be similar to, or somewhat larger than, solar as the star is slightly cooler than the Sun with higher photospheric gas pressure and, therefore, higher equipartition magnetic-field strength. Zeeman-broadening measurements of ξ Boo A lie in the range $\langle B \rangle = B_{\mathrm{MOD}} f = 0.34 - 0.48$ kG (Reiners 2012). Marcy and Basri (1989) measured $\langle B \rangle = 0.35$ kG with $B_{\mathrm{MOD}} = 1600$ G, similar to the Sun, and $f = 0.22$. If one assumes this value for the filling factor, the observed net magnetic field is 30–100 G corresponds to $B_{\mathrm{NET}} = 140$–450 G, which can be compared to the resolved field strength $B_{\mathrm{MOD}} = 1600$ G. Since the magnetic energy density is proportional to B^2_{MOD}, field cancellation means that the ZDI image only represents 1–10 % of the photospheric magnetic energy of this star. Other ZDI images likely also represent only a small fraction of the magnetic energy in a stellar photosphere. Lang et al. (2014) estimate the amount of small-scale flux not found in ZDIs of M stars and extrapolate the small-scale and large-scale photospheric fields into the stellar coronae.

Dependence on Age and Rotation Vidotto et al. (2014) find that for non-accreting main-sequence stars in the unsaturated regime, that is stars older than 10^7 years, magnetic fields decay with increasing age and rotation period P_{rot}. They find this decay to be the case for both $\langle B \rangle$, measured from Stokes I spectra, and for an unsigned version of $\langle B_z \rangle$, measured from ZDI images. The power-law dependencies are similar, $\langle B \rangle \propto P_{\rm rot}^{-1.7}$ and unsigned $\langle B_z \rangle \propto P_{\rm rot}^{-1.32\pm0.14}$, respectively, indicating that the large-scale fields, measured from ZDI images, and the small-scale fields, measured from Zeeman broadening and splitting spectra, are coupled (Vidotto et al. 2014). These authors suggest that the same dynamo field-generation processes are responsible for both the small-scale and large-scale fields.

Large-Scale Magnetic Morphology of Early and Late M Dwarfs In a series of papers, Donati et al. (2008a) and Morin et al. (2008, 2010) obtained ZDI images of 23 M dwarf stars to study whether the magnetic topologies of M dwarfs are different for stars with radiative cores and presumably solar-like $\alpha\Omega$-type dynamos ($M > 0.35\ M_\odot$) compared to stars that are fully convective with different types of dynamo processes ($M < 0.35\ M_\odot$). One finding is that for stars with similar parameters (i.e., similar Rossby numbers), there are two very different magnetic topologies (strong dipolar and weaker multipolar) that may indicate a bistability with two possible dynamo states or an age effect (cf. Gastine et al. 2013; Vidotto et al. 2014). Despite this topological diversity, many of the more massive M dwarfs ($M > 0.5\ M_\odot$) have strong azimuthal fields and weak poloidal fields, but with decreasing mass, axisymmetric poloidal fields dominate over the toroidal fields and produce stronger large-scale magnetic fluxes, as shown in Fig. 15. The strong azimuthal fields seen in the more massive M dwarfs, which have radiative cores, are also seen in the young classical T Tauri stars (see Sect. 4.1), which are fully convective. Why these two star types with very different interior structures have similar magnetic topologies has not been explained and may provide an important clue concerning magnetic dynamos.

There is evidence for evolution of the reconstructed magnetic topologies for the less massive stars on time scales of less than one year. The fully convective stars likely have a higher degree of large-scale field organization as the ratio of net magnetic-field strength measured in the circular polarization data to the unsigned magnetic-field strength inferred from the magnetic-broadening data is only about 15 % for these stars compared to only a few percent for the more massive stars with radiative cores. The more massive M stars, like the G and K stars, have complex fields on small spatial scales.

Variable Magnetic Topology ZDI images of the solar-type star HD 190771, which rotates nearly three times faster than the Sun ($P_{\rm rot} = 8.8$ days), show large changes in the magnetic morphology in the course of one year (2007 to 2008; Petit et al. 2009). These changes include polarity reversal in the axisymmetric and azimuthal fields and a sharp decrease in the magnetic energy stored in the azimuthal component.

Nonpotential Magnetic Fields and Large Magnetic Loops Hussain et al. (2002) showed that the magnetic field of the very rapidly rotating star AB Dor contains about 20 % non-potential energy in the photosphere and about 14 % nonpotential energy in the corona. This nonpotential energy indicates strong electric currents in both locations, which is very different from the slowly rotating Sun. Also, sequences of high-resolution spectra provide evidence for large slingshot-type prominences anchored at high latitudes that are probably magnetic loops extending outward to several stellar radii that co-rotate with the photosphere.

5.2 The Chemically Peculiar Stars

Spectra of chemically peculiar (CP) main-sequence A- and B-type stars contain abnormally strong or weak absorption lines of certain elements (e.g., Si, Sr, Cr, Eu, or He). These stars generally have magnetic fields that can be detected through circular-polarization observations of spectral lines. Observables, such as the magnitudes in various photometric bands,

Table 2 Different groups of chemically peculiar stars

Peculiarity type	Spectral type	T_{eff} range	Magnetic	Spots
He-strong	B1-B4	17 000–21 000	yes	yes
He-weak	B4-B8	13 000–17 000	yes	yes
Si	B7-A0	9000–14 000	yes	yes
HgMn	B8-A0	10 000–14 000	yes?	yes!
SrCrEu	A0-F0	7000–10 000	yes	yes
Am	A0-F0	7000–10 000	yes?	no

spectral line equivalent widths, and magnetic-field properties, vary with the same period, which can range from half a day to several decades. Abnormal line strengths correspond to element overabundances (by up to 5–6 dex with respect to the Sun) in the stellar outer layers. The CP star class is roughly represented by three subclasses: the magnetic Ap and Bp stars, the metallic-line Am stars, and the HgMn stars. An overview of the different groups of CP stars can be found in Table 2. Indirect evidence for kG magnetic fields in these stars consists of strong gyrosynchrotron emission detected at cm wavelengths (Linsky et al. 1992) and an X-ray flare seen in the A0p star IQ Aur (Robrade and Schmitt 2011). About a few percent of intermediate-mass main-sequence stars are identified as Ap/Bp stars (Landstreet et al. 2007; Donati and Landstreet 2009).

The first detection of the net-longitudinal magnetic field in a star other than the Sun was achieved in CS Vir by Babcock in 1946 (Babcock 1947). Today, net-longitudinal magnetic-field measurements throughout the variation period have been obtained for no more than 100 stars. The resolution of magnetically split lines requires a sufficiently strong magnetic field and slow rotation. Resolved magnetically split lines were first discovered in Babcock's star, HD 215441 (Babcock 1960), for which he measured a net magnetic-field modulus of $B_{MOD} \sim 34$ kG, which is the strongest magnetic-field modulus measured in an Ap star to date. However, Elkin et al. (2010) found that the Ap star HD 75049 with $B_{MOD} \sim 28$ kG (see Fig. 16) is a close rival to Babcock's star, Hubrig et al. (2005) found that the cool Ap star HD 154708 has $B_{MOD} \sim 24.5$ kG, and Kochukhov et al. (2011) showed that the He-strong star HD 37776 may have an even stronger surface magnetic field. In 1987, 12 stars with magnetically resolved lines were known, but only four of those were studied throughout their variation period. By 2001, 44 stars with magnetically resolved lines were known, with 24 of those were studied throughout their variation period (Mathys et al. 1997; in preparation). The first systematic determinations of the crossover effect and the mean quadratic magnetic field were published in 1995 by Mathys (1995a, 1995b). A full phase coverage was achieved for about two dozen stars. The bulk of the published material on broad-band linear polarization (BBLP) was gathered by Leroy between 1990 and 1995 (Leroy 1995, and references therein). Variations in BBLP are well studied for about 15 stars.

As shown in Fig. 17, the strongest magnetic fields tend to be found in more massive stars, but also in fast-rotating stars (Hubrig et al. 2000). All CP stars with rotation periods exceeding 1000 days have magnetic fields below 6.5 kG. However, a great many Ap stars have short periods typically 2–4 days (Hubrig et al. 2000). From the finding that the net-longitudinal magnetic field averaged over the stellar disk is not zero, one can directly conclude that the magnetic field must be organized on a larger scale, either as a dipole or a superposition of a dipole and a quadrupole. The circular polarization from tangled, solar-like magnetic fields mostly cancels out in a disk integration. The magnetic field of Ap stars thus has a significant dipole-like component. For a dipole, the ratio between the longitudinal magnetic field and the magnetic-field modulus $\langle B_z \rangle / B_{NET}$ is 0.3, for a quadrupole it is 0.05. If toroidal or higher-order multipolar components were sufficient to account for the observed net-longitudinal magnetic field, these components would induce strong distortions of the spectral line profiles in Stokes I, i.e., in integral light, which is not seen.

The magnetic field covers the whole CP stellar surface homogeneously, i.e., the distribution of the field strength over the star is fairly narrow. Evidence for this distribution comes from the fact that the magnetic field is observed at all phases, the continuum is reached between the split components of resolved lines, and that the resolved magnetically split components are rather narrow (Mathys et al. 1997).

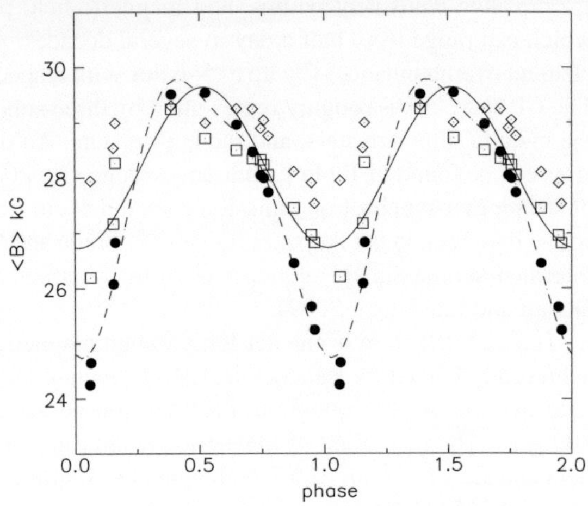

Fig. 16 Variation of the magnetic field modulus of HD 75049 with rotational phase. Different symbols refer to measurements using different ions. The *solid line* is the fit to the data obtained using the hydrogen lines, and the *dashed line* is the fit using the FeII lines. Figure from Elkin et al. (2010)

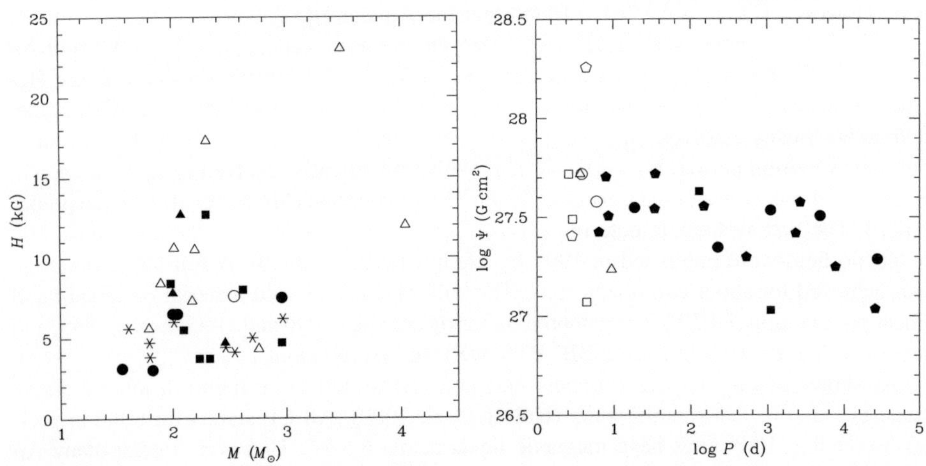

Fig. 17 *Left*: Plot of the mean magnetic field modulus as a function of stellar mass for Ap stars. Triangles are for stars with $P_{rot} \leq 10$ days, squares for periods 10–1000 days, and asterisks for periods longer than 1000 days. *Right*: Plot of magnetic flux as a function of rotational period for Ap stars. Filled symbols are for stars with magnetically-resolved spectral lines. Open symbols are for stars for which mean quadratic fields are available. Figures are from Hubrig et al. (2000)

Magnetic fields have severe effects on the structure of stellar outer layers. They are responsible for magnetically controlled winds and elemental-abundance stratification. Evidence for abnormal atmospheric structure comes from the profiles of hydrogen Balmer lines in cool Ap stars that cannot be fitted by conventional models. This structure has also a potential impact on the net-longitudinal magnetic-field determination by Balmer-line polarimetry. The core-wing anomaly (Cowley et al. 2001) of the hydrogen Balmer lines leads to the impossibility of fitting the Balmer lines with one temperature. For example, to fit the Hβ line in HD 965, one needs to assume $T_{\rm eff} = 5500$ K for the core of the line and $T_{\rm eff} = 7000$ K for the wings.

The magnetic field is not symmetric with respect to the stellar rotation axis. Other surface features, e.g., the abundance distribution, are determined by the magnetic field. Since the observed variations result from changing aspects of the visible hemisphere as the star rotates, the variation period must be the rotation period of the star. No intrinsic variations of the magnetic field have been observed in Ap stars over time scales of decades.

In early models of the magnetic field, a quasi-sinusoidal variation of the longitudinal magnetic field was assumed. In the simplest model, a dipole centered at the star's center and with an axis inclined with respect to the stellar rotation axis, was employed. Stars with magnetically resolved lines show that the magnetic-field modulus generally has one maximum and one minimum per rotation period, even for stars with a reversing-longitudinal magnetic field (Mathys et al. 1997) (see Fig. 16). From these observations, a centered dipole can be ruled out. Alternative models include a dipole that is offset along its axis (parameters: i, β, $B_{\rm d}$, a), or a collinear dipole plus a quadrupole (parameters: i, β, $B_{\rm d}$, $B_{\rm q}$), with i the inclination angle of the star with respect to the line of sight, β the inclination angle of the magnetic field with respect to i, $B_{\rm d}$ the strength of the dipole, $B_{\rm q}$ the strength of the quadrupole, and a the offset of the dipole with respect to the star's center. The models must make a good match with four observables: the maximum and the minimum of both the longitudinal magnetic field and the magnetic-field modulus. Both classes of models are equivalent to first order.

Additional constraints on the magnetic field geometry can come from the crossover and the mean-quadratic magnetic field. A collinear dipole plus a quadrupole and an octupole give good first approximations in many cases (Landstreet and Mathys 2000). The dipole primarily accounts for the longitudinal magnetic field, the quadrupole gives the field strength contrast between the poles, and the octupole is responsible for the equator-to-pole field strength contrast. Asymmetric variation curves can be determined from some magnetic-field moments. They exist when the magnetic field is not symmetric about an axis passing through the center of the star (Mathys 1993) and can be described with a generalized-multipolar model (Bagnulo et al. 2000, and references therein). The input observables for these models are all available observables of the magnetic field: $\langle B_z \rangle$, $\langle x B_z \rangle$, $\sqrt{\langle B^2 \rangle + \langle B_z^2 \rangle}$, $\langle B_{\rm NET} \rangle$, and the BBLP, where x is the stellar equatorial velocity times the sine of the inclination angle. Landolfi et al. (1998) has shown that the inversion of these measured quantities provide information on the strength and orientations of the dipole and quadrupole components of the stellar magnetic field subject to the limitations imposed by measurement errors. In particular, the quantity $\sqrt{\langle B^2 \rangle + \langle B_z^2 \rangle}$ is needed to determine the presence and properties of the quadrupole component of the magnetic field. A χ^2 minimization between the predicted and observed values of the observables at phases distributed throughout the rotation period will determine the final model for the geometric structure of the magnetic field.

Ultimately, a direct inversion of the line profiles recorded in all four Stokes parameters will allow one to derive magnetic-field maps without a-priori assumptions. Since inversion is an ill-posed problem, a regularization condition is needed. This condition is achieved with the Zeeman Doppler-imaging technique (ZDI) (Piskunov and Kochukhov 2002). It is very demanding in terms of the signal-to-noise ratio in the data, spectral resolution, and phase coverage. So far, these inversions are restricted to a few individual stars (e.g., Kochukhov et al. 2004; Lüftinger et al. 2010; Kochukhov and Wade 2010; Silvester et al. 2012; Rusomarov et al. 2013). Inversion of the phase-resolved full Stokes IVQU spectrophotometric data set for the A0pSiEuHg star α^2CVn revealed a global dipolar-like topology with localized spots of higher field intensity (Kochukhov and Wade 2010). Recent magnetic-field maps of α^2 CVn obtained by Silvester et al. (2014b, 2014a) confirm the complex substructure of the magnetic field and show that low-order multipole models fail to match the observed maps. They also find that the magnetic-field topology of this star has been stable for the last 10 years and that maps based on strong absorption lines tend to smooth the finer scale structure seen in maps constructed from weak absorption lines. The chemical abundance enhancements are correlated with the radial magnetic field in patterns that are not predicted by theoretical models. In particular, some elements (e.g., O, Cl, and Eu) are enhanced in areas with negative radial magnetic fields, whereas other elements (e.g., Fe, Cr, Si, Ti, and Nd) are deleted where the magnetic fields are weak but overabundant where $B_{\rm MOD} \sim$2 kG, irrespective of sign.

The magnetic-field modulus much better characterizes the intrinsic-stellar magnetic field than the net-longitudinal magnetic field, which is much more dependent on the geometry of the observation. Most Ap stars with magnetically resolved lines have a magnetic-field modulus (averaged over the stellar rotation period) between 3 and 9 kG, but there is a lower cutoff of the distribution at 2.8 kG, (Landstreet et al. 2007; Aurière et al. 2007), which corresponds to the thermal equipartition fields in the photospheres of Ap and Bp stars (Donati and Landstreet 2009). For Ap stars with masses greater than 3 $M_\odot$, the magnetic-field strenghs decline rapidly with age, but the lower mass Ap stars do not show this behavior (Landstreet et al. 2007). One should be able to resolve lines down to 1.7 kG or lower at some rotation phases of some stars, but only for one target is the resolution sufficient to observe down to 2.2 kG. The lower limit of the magnetic-field distribution is roughly temperature independent; hotter stars may have stronger magnetic fields than cooler stars (Mathys et al. 1997).

Ap star variation periods span five orders of magnitude. Until recently, there appeared to be no systematic differences between short- and long-period stars. A confirmation that very long periods are indeed rotation periods has been brought by BBLP (Leroy et al. 1994). The systematic study of 40 Ap stars with resolved magnetically split lines has doubled the number of known Ap stars with $P > 30$ days (Mathys et al. 1997). The distribution of periods longer than 1 year is compatible with a uniform distribution on a logarithmic scale. No star with $P > 150$ d has a magnetic-field modulus exceeding 7.5 kG. More than 50 % of the Ap stars with resolved lines and shorter periods have a magnetic-field modulus above this value (Mathys et al. 1997). In the collinear dipole plus quadrupole and octupole model, the angle between the magnetic and rotation axes β is generally smaller than 20° for stars with $P > 30$ d, unlike short period magnetic Ap stars for which this angle is usually large (Landstreet and Mathys 2000).

HgMn stars are chemically peculiar stars with spectral type B8 to A0 and $T_{\rm eff} =$ 10 000–14 000 K. They show extreme overabundance of Hg (up to 6 dex) and/or Mn (up to 3 dex). They display the most obvious departures from abundances expected within the context of nucleosynthesis (Cowley and Aikman 1975). More than 150 HgMn stars are known, many of which are found in young associations (Sco-Cen, Orion OB1). They are among the most slowly rotating stars on the upper main sequence and have exceptionally stable atmospheres with an average rotational velocity of $\langle v \sin i \rangle = 29$ km/s, which leads to extremely sharp-lined spectra. They are the best-suited targets to study isotopic and hyperfine structure. More than 2/3 of the HgMn stars belong to spectroscopic binary (SB) systems with a prevalence of $P_{orb} \approx 3$–20 d, and many HgMn stars are in multiple systems. The spectrum variability seen in HgMn stars is due to the presence of chemical spots. These stars do not have strong large-scale organized magnetic fields, but tangled magnetic fields are possible. They do not have enhanced strengths of rare earth elements but rather of the heavy elements W, Re, Os, Ir, Pt, Au, Hg, Tl, Pb, and Bi, which makes these stars natural laboratories for the study of heavy elements. They also show anomalous isotopic abundances for the elements He, Hg, Pt, Tl, Pb, and Ca.

One of the most exciting objects containing a HgMn star is the triple system AR Aur. The inner two stars constitute the only known eclipsing binary including a HgMn star. This binary has an orbital period of 4.13 d and an age of 4 Myr. The two stars are of spectral types B9V and B9.5V, and while the primary HgMn star is exactly on the zero-age main sequence (ZAMS), the secondary is still contracting (e.g., Nordstrom and Johansen 1994). Hubrig et al. (2012b) used observations with SOFIN high-resolution optical spectrograph at the Nordic Optical Telescope to study the distribution of different elements over the surface of the primary HgMn star using the Doppler-mapping technique. From the same data set, they also measured a weak longitudinal magnetic field in the primary star with 3–4σ errors. AR Aur shows a similar behavior to other HgMn systems discussed by Hubrig et al. (2012b). The results suggest a correlation between the proposed magnetic field, the abundance anomalies, and the binary properties. For the synchronously rotating components of the SB2 system AR Aur, the stellar surfaces facing the companion star usually display low-abundance element spots and negative magnetic-field polarity. The surface of the opposite hemisphere, as a rule, is covered by high-abundance element spots and the putative magnetic field is positive at the rotation phases of the best-spot visibility (Hubrig et al. 2012b). Since the reanalysis of the FORS1 data that Hubrig et al. (2006b) used to identify weak longitudinal fields in four HgMn stars did not confirm the presence of fields in these stars (Bagnulo et al. 2012) and Kochukhov et al. (2013) found no evidence for tangled magnetic fields in any HgMn stars, the present conclusion is that these stars do not have measured magnetic fields.

The He-strong early B stars are the most massive and hottest of the chemically peculiar stars. They are very rapid rotators; the most rapidly rotating star in this class, HR 7355, has a rotational period of 0.52 days, about 90 % of critical. Rivinius et al. (2013) showed that this star's magnetic-field topology is more complex than an oblique dipole and that the magnetic poles are the regions with the strongest He overabundances and the lowest metal overabundances. The B2Vp star σ Ori E shares many of the same properties as HR 7355 with a similar magnetic-field topology and chemical abundances at the magnetic poles, but with a slower rotational period (1.19 days) and higher T_{eff} (Oksala et al. 2012). For a third member of the He-strong subgroup, HD 37776, ZDI shows a very complex magnetic field with a magnetic-field modulus, $B_{MOD} = 43$–49 kG, depending on the assumed topology, that could be stronger than for Babcock's star (Kochukhov et al. 2011). The magnetic field appears to contain a toroidal component.

5.3 Pulsating B Stars

β Cep stars are short-period (3–8 h) pulsating variables of spectral type O9 to B3 (corresponding to a mass range of 8–20 $M_{\odot}$) along the main sequence that pulsate in low-order pressure (p) and/or gravity (g) modes. Slowly pulsating B (SPB) stars show variability with periods of the order of 1 d, are less massive (3–9 $M_{\odot}$) main sequence B-type stars, and have multiperiodic high-order low-degree g mode oscillations.

A long-term monitoring project aimed at asteroseismology of a large sample of SPB and β Cep stars was started by researchers of the Institute of Astronomy of the University of

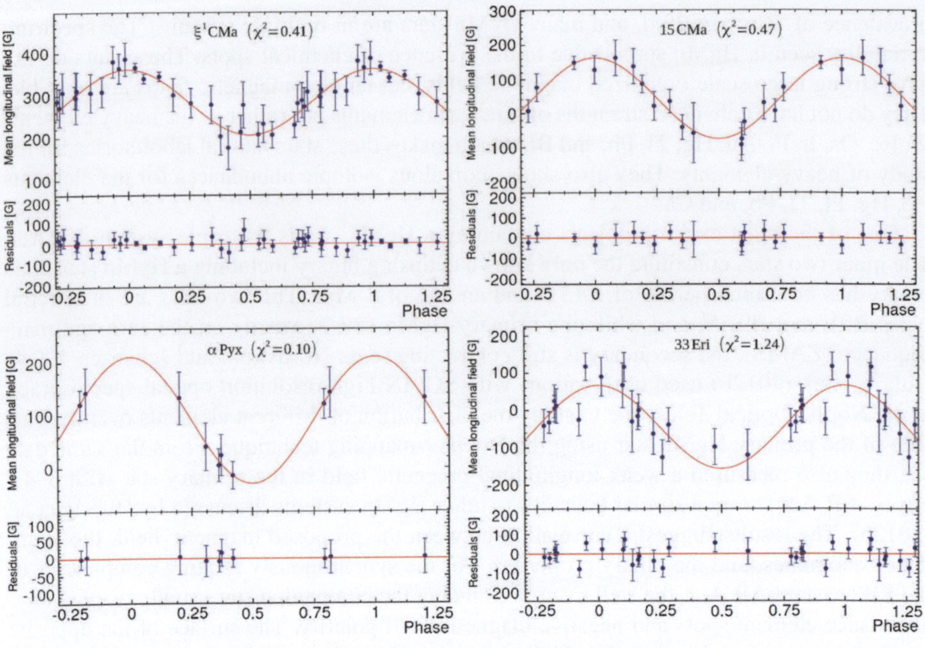

Fig. 18 Phase diagrams with the best sinusoidal fit for the net longitudinal magnetic-field strength measurements. The residuals (observed–calculated) are shown in the lower panels. The deviations are mostly of the same order as the error bars; no systematic trends are obvious, which justifies a single sinusoid as a fit function

Leuven more than 10 years ago. In the first publication on a magnetic survey of pulsating B-type stars with FORS 1, Hubrig et al. (2006a) announced detections of weak net-longitudinal magnetic fields of a few hundred Gauss in a number of SPB stars and in the β Cep star ξ^1 CMa, whose field of the order of 300–400 G is one of the largest among all currently known magnetic β Cep stars. Bagnulo et al. (2012) has confirmed the detection of two of the SPB stars (HD 53921 = V450 Car and HD 208057 = 16 Peg) and listed two others as possible detections (HD 74195 = o Vel and HD 152511 = V847 Ara). Reanalysis of the data for one or a few observations of the other SPD stars did not confirm the earlier detections of weak longitudinal magnetic fields. In the continuation of their FORS1 observing program, Hubrig et al. (2009a) listed six β Cephei stars and 18 SPB stars as having magnetic fields at the $\geq 3\sigma$ level, but Bagnulo et al. (2012) was able to confirm only one of the β Cephei stars (HD136504 = ϵ Lup) and none of the individual observations of the other SPB stars. The β Cephei star ξ^1 CMa (HD 46328), however, is a confirmed detection with a strong magnetic field $\langle B_z \rangle = 338 \pm 11$ G (Silvester et al. 2009, 2012). HD 180642 = V1449 Aql is another β Cephei stars with a strong longitudinal magnetic field and a dipolar field of about 3 kG. The differences between FORS1 detections (Hubrig et al. 2006b, 2009a) and the nondetections based on a alternative reductions of the same data by Bagnulo et al. (2012) are discussed at the end of Sect. 4.2.

In Fig. 18, one can see the results of magnetic-field monitoring of four β Cep and SPB stars. Although many of the individual measurements of the stars shown in the figure are not significant at the 3σ level, the plots of longitudinal magnetic fields phased with the rotational period reveal patterns that can be simply explained by oblique rotator models. From FORS 1/2 and the higher resolution SOFIN observations, Hubrig et al. (2011c) determined

Table 3 Measurements of the mean longitudinal magnetic field using high-resolution HARPS spectra

Object	MJD	S/N	$\langle B_z \rangle$
HD 74195	55605.217	220	-70 ± 21
HD 74195	55606.130	300	-14 ± 18
HD 74560	55605.221	240	56 ± 19
HD 74560	55606.134	280	8 ± 18
HD 74560	55607.177	350	-35 ± 15
HD 85953	55600.305	230	79 ± 20

a rotation period of $P = 2.1795$ d for ξ^1 CMa, which is nearly a factor of two smaller than found by Fourtune-Ravard et al. (2011), who determined $P \sim 4.2680$ d from ESPaDOnS observations. Note that in the latter work, the impact of pulsations on the magnetic-field measurements from high-resolution spectra was not taken into account. The effects of not including pulsations in the analysis of low-resolution FORS1 spectra can be large but not readily quantifiable (Hubrig et al. 2011e), whereas pulsation effects on high-resolution Stokes V data may not be large (Silvester et al. 2012).

Among the sample of SPB stars with detected magnetic fields using FORS 1, three stars, HD 74195, HD 74560, and HD 85953, have been observed in February 2011 with the high-resolution ($R = 115,000$) polarimeter HARPSpol, installed at the European Southern Observatory (ESO) 3.6 m telescope on La Silla, in the framework of the guaranteed-time observer (GTO) program 086.D-0240(A). The star HD 85953 was observed once, whereas HD 74195 was observed on two different nights, and HD 74560 was observed on three different nights. Additional spectra were downloaded from the ESO archive and reduced using the HARPS data reduction software available at the ESO headquarters in Germany. For the measurements of the magnetic fields, the moment technique developed by Mathys (e.g., Mathys 1991) was used. Formally significant detections above the 3σ level were achieved in HD 85953 and in one observation of HD 74195 (see Table 3; Hubrig et al., in preparation). In line with the discoveries of weak magnetic fields in pulsating stars, Briquet et al. (2013) found a magnetic field in the hybrid SPB/β Cep star HD 43317.

The pulsation amplitudes for the three-studied pulsating stars range from 4.5 to 25 mmag. The study of correlations between the strength of magnetic fields and pulsational characteristics (Hubrig et al. 2009b) indicates that it is possible that stronger magnetic fields appear in stars with lower-pulsating frequencies and smaller-pulsating amplitudes. Spectra for all three sources can be found in Fig. 19. Spectral variability is evident for the two objects with more than one observation.

From the FEROS (Fiber-fed Extended Range Optical Spectrograph) time series, one can find line profile variability for V1449 Aql with a pulsating frequency of $f_{\rm puls} = 5.487$ d^{-1}. The variability in the spectra of V1449 Aql and the impact of pulsations on the polarimetric spectra can be seen in Fig. 20. Neglecting the rapid changes in the line profiles shapes due to pulsations in the analysis of spectropolarimetric data can lead to nondetections of magnetic fields in these stars (Hubrig et al. 2011e) or even spurious magnetic detections (Schneer et al. 2006). The need for a careful observing procedure is especially important for large amplitude pulsators like V1449 Aql (Hubrig et al. 2011e).

5.4 Be Stars

Rapidly rotating Be stars lose mass and initially accumulate it in a rotating circumstellar disk. Much of the mass loss is in the form of outbursts, and thus additional mechanisms,

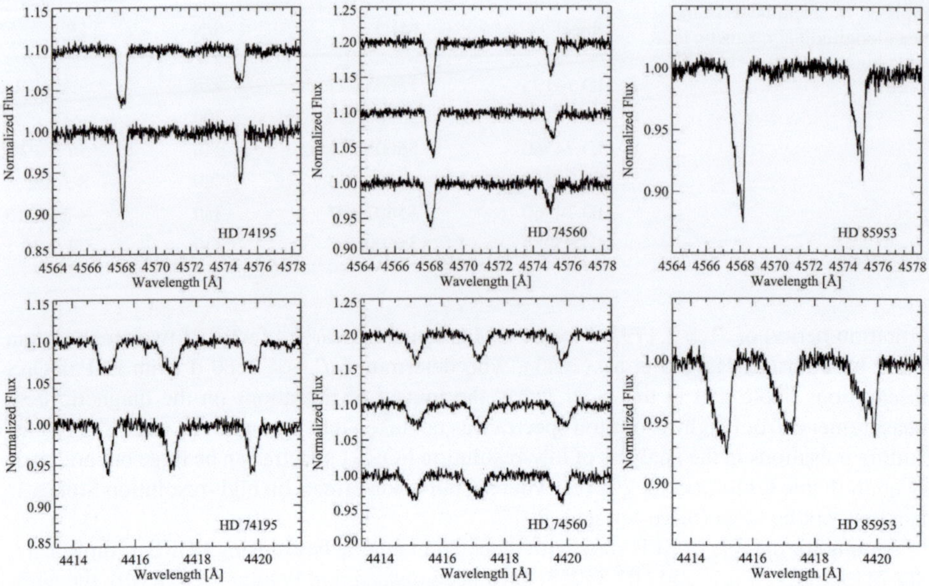

Fig. 19 Spectral variability as seen in HARPS Stokes *I* spectra. *Left*: HD 74195, *middle*: HD 74560, *right*: HD 85953

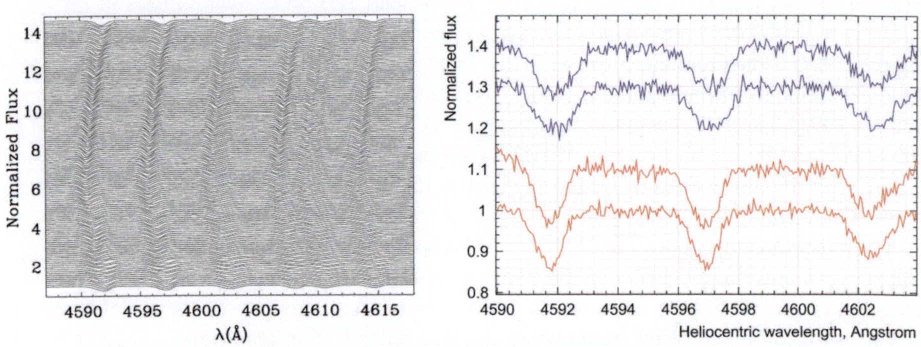

Fig. 20 *Left*: Time series of FEROS (Fiber-fed Extended Range Optical Spectrograph) spectra for V1449 Aql showing pulsational line-profile variability in the spectral region 4590–4615 Å. The pulsation phase zero is at the bottom. *Right*: Variability of the output spectra in two SOFIN subexposures taken with the quarter-wave plate angles separated by 90° taken around HJD 2455398.530. The lower two spectra, $(I+V)_0$ and $(I-V)_0$, correspond to the first subexposure, while the upper spectra, $(I-V)_{90}$ and $(I+V)_{90}$, correspond to the second subexposure. The strong effect of pulsations on the line-profile shapes and the line positions is clearly visible between the spectra of the first subexposure with a duration of 20 min and the spectra of the second subexposure with the same duration

such as the beating of nonradial pulsation modes or magnetic flares, must be at work. Indirect evidence for the presence of a magnetic field are variations of X-ray emission and transient features in absorption line profiles. Angular momentum transfer to a circumstellar disk, channeling stellar wind matter, and accumulation of material in an equatorial disk are more easily explained if magnetic fields can be invoked. Fifteen Be stars have been observed with the hydrogen polarimeter by Barker et al. (1985) using Hβ, but no magnetic fields were detected. One Be star with a reported magnetic field, ω Ori (Neiner et al. 2003), was not confirmed as magnetic by recent observations.

A sample of Be stars in the field and in the cluster NGC 3766 (14.5–25 Myr old) was observed by Hubrig et al. (2009b) in 2006–2008 with FORS 1. A few Be stars show weak magnetic fields with the strongest field detected in HD 62367 ($\langle B_z \rangle = 117 \pm 38$ G, $m_V = 7.1$). Usually, the detected magnetic fields are below 100 G (see Figs. 21 and 22). Since the magnetic fields are weak, it is difficult to determine their large-scale structure. The cluster NGC 3766 appears to be extremely interesting, where evidence can be found for the presence of a magnetic field in seven early-B type stars (among them three Be stars) out of the observed 14 cluster members (Hubrig et al. 2009a).

For nine early-type Be stars, the authors obtained time-resolved magnetic-field measurements over ∼one hour (up to 30 measurements) with FORS 1 at the VLT. For λ Eri, they were able to detect a period of $P = 21.1$ min in the magnetic-field measurements (see Fig. 23). The spectral line profiles of λ Eri exhibit short-time periodic variability (see Fig. 24) because of non-radial pulsations with a period of 0.7 d (Kambe et al. 1993b). Fur-

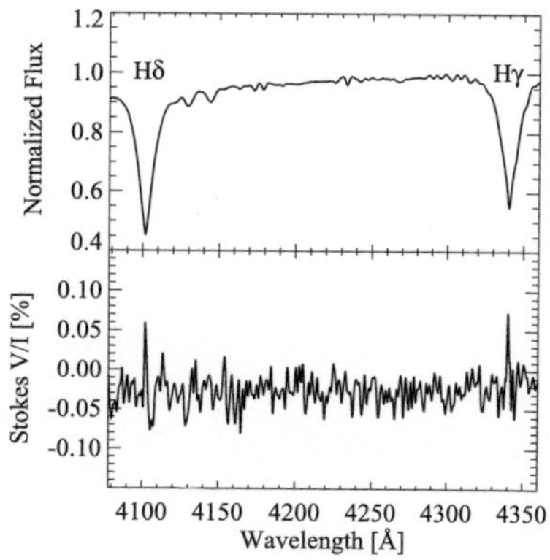

Fig. 21 Stokes I and Stokes V spectra of the Be star o Aqr ($\langle B_z \rangle = 98 \pm 31$ G) in the region including the Hδ and Hγ lines

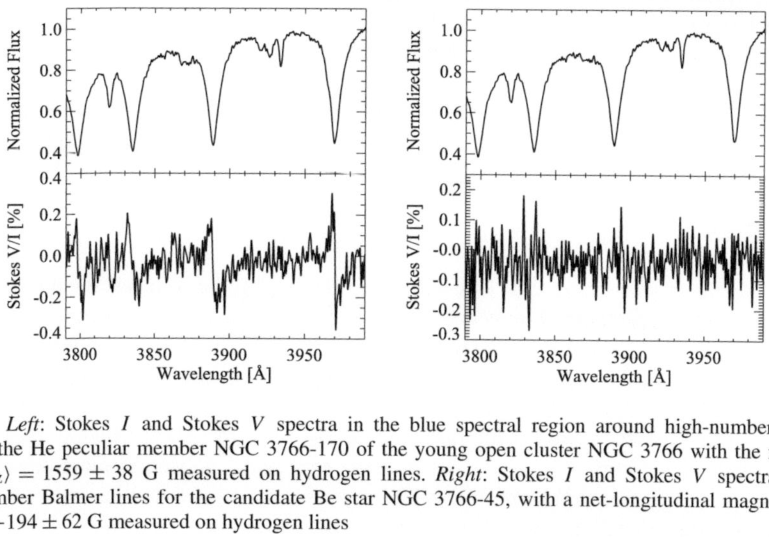

Fig. 22 *Left*: Stokes I and Stokes V spectra in the blue spectral region around high-number Balmer lines of the He peculiar member NGC 3766-170 of the young open cluster NGC 3766 with the magnetic field $\langle B_z \rangle = 1559 \pm 38$ G measured on hydrogen lines. *Right*: Stokes I and Stokes V spectra around high-number Balmer lines for the candidate Be star NGC 3766-45, with a net-longitudinal magnetic field $\langle B_z \rangle = -194 \pm 62$ G measured on hydrogen lines

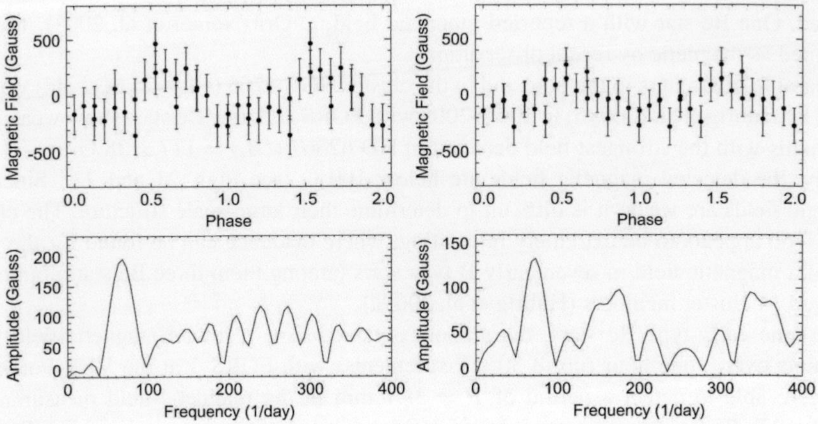

Fig. 23 Phase diagram and amplitude spectrum for the net-longitudinal magnetic-field strengths of λ Eri in 2006 August using hydrogen lines (*left*) and all lines (*right*)

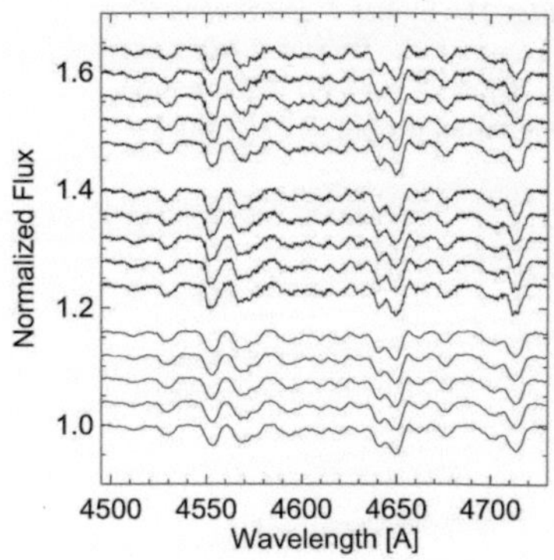

Fig. 24 Spectrum variability of λ Eri on three different nights: on 2006 August 8 (*bottom*), 2007 November 27 (*middle*), and 2007 November 28 (*top*)

thermore, Smith (1994) detected dimples with a duration of 2–4 h. Are these strong local magnetic fields?

Apart from λ Eri, four other stars showed indications of magnetic-cyclic variability on the scales of tens of minutes (Hubrig et al. 2009a). A similar magnetic-field periodicity ($P = 8.8$ min) was detected for the B0 star θ Car (Hubrig et al. 2008b). These stars are good candidates for future time-resolved magnetic-field observations with high-resolution spectropolarimeters.

5.5 OB Stars

The presence of a convective envelope has been assumed to be a necessary condition for significant magnetic activity. Magnetic activity is found all the way from the late A-type stars (e.g., in Altair: Robrade and Schmitt 2009) with very shallow convective envelopes down to the coolest fully convective M-type stars as late as spectral type M9.5 (Reiners and Basri 2007, 2010).

On the other hand, advances in instrumentation over the past decades have led to magnetic-field detections in a small, but gradually growing, subset of massive stars that frequently present cyclic wind variability, Hα emission variations, nonthermal radio/X-ray emission, and transient features in absorption-line profiles.

Magnetic fields have fundamental effects on the evolution of massive stars, their rotation, and on the structure, dynamics, and heating of radiative winds. The origin of the magnetic fields is still under debate: it has been argued that magnetic fields could be "fossil", or magnetic fields may be generated by strong binary interaction, i.e., in stellar mergers, or during a mass transfer or common-envelope evolution.

To identify and model the physical processes responsible for the generation of magnetic fields in massive stars, it is important to understand whether:

- most magnetic stars are slowly rotating,
- magnetic fields appear in stars at a certain age,
- magnetic fields are generated in stars in special environments: Do some clusters contain a larger number of magnetic massive stars, similar to the Ap/Bp content in different clusters (NGC 2516 has the largest number of magnetic Ap stars and X-ray sources)?
- magnetic fields are produced through binary interaction,
- X-ray emission can be used as an indirect indicator for the presence of magnetic fields.

5.5.1 Magnetic Fields in O-type Stars

Early indications of the presence of magnetic fields in O-type stars came from (a) non-thermal radio emission (Beiging et al. 1989; Abbott et al. 1986), indicating gyrosynchrotron emission from energetic electrons in magnetic fields, and (b) X-ray emission in the Fe XXIV and Fe XXV lines (Schulz et al. 2000), indicating plasma as hot as 6.1×10^7 K in the wind of the O7 V star θ^1 Ori C, which is far hotter than predicted by radiation-driven shocks in the stellar wind, The first direct detections of a magnetic field in an O-type star were reported by Donati et al. (2002) for θ^1 Ori C and Donati et al. (2006) for HD 191612. Today, about two dozen magnetic O-type stars are known.

Hubrig et al. (2008a) were the first to determine net-longitudinal magnetic-field strengths for a large sample of O-type stars with an accuracy of a few tens of Gauss, using FORS 1. Very few magnetic fields stronger than 300 G were detected in the studied sample, suggesting that large-scale dipole-like magnetic fields with polar magnetic-field strengths higher than 1 kG are not common among O-type stars. Their studies of massive stars revealed that the presence of a magnetic field can be expected in stars of different classification categories and at different evolutionary stages. No physical properties are known that define particular classes of stars as nonmagnetic. The inability to detect magnetic fields in massive stars in earlier studies could be related to the weakness of these fields, which can, in some stars, be as small as only a few tens of Gauss.

In recent years, two major surveys have been aimed at better understanding the nature and origin of magnetic fields in OB stars and the physics of their atmospheres, winds, and magnetospheres. MiMeS (Wade et al. 2012) focused on high-resolution ($R \approx 65\,000$–$110\,000$) spectropolarimetry with Narval, ESPaDOnS, and HARPS. The BOB consortium (Morel et al. 2014) uses 35.5 nights on FORS 2 and HARPS over a period of 2.5 years. It concentrates on main-sequence OB stars and do not consider, e.g., Be or Wolf-Rayet stars. One half of their targets are early B stars, one third are late O stars, and the rest is made up of several late B stars with only very few early O stars. So far, the BOB consortium has observed of $\sim$100 OB stars, with only very few targets in common with MiMeS. It aims to show consistent detections using different reduction and analysis techniques, independently carried out by two teams.

Overall, the data obtained by various authors seem to confirm that the occurrence of fields above the detectability threshold ($\sim$ 100–200 G) is low in massive stars. An exact estimation of the incidence rate is still pending and may be revised upwards in the future.

While an exact estimate of the incidence rate is still pending, Wade and the MiMeS Collaboration (2014) suggests that about 7 % of Galactic O stars have magnetic fields above this measurement threshold. They strongly suggest that there is a bimodal distribution of O-type stars consisting of a small population of stars with strong (1–3 kG) magnetic fields, generally thought to have an oblique rotator structure, and a much larger population of stars with undetected fields smaller than about 100 G. If confirmed by subsequent measurements, this dicotomy would be an extension to higher mass stars of the "magnetic desert" that Aurière et al. (2007) has proposed for intermediate-mass stars. Citing biases for measuring weak magnetic fields, especially for more rapidly-rotating stars, Fossati et al. (2015) argues that there is no clear evidence for a "magnetic desert" for high mass stars. Recent measurements of magnetic fields on ϵ CMa and β CMa with the high-resolution HARPSpol spectrometer on the VLT now demonstrate the feasibility of obtaining credible longitudinal magnetic fields strengths smaller than 10 G (Fossati et al. 2015) at least for bright stars. New measurements with even better sensitivity are needed to quantify the distribution of magnetic field strengths for OB stars, which is essential for understanding the origin of their magnetic fields and the effects of magnetic fields on the evolution of OB stars. Fortunately, the measurement of weak magnetic fields is helped by magnetic OB stars having, on average, rotation speeds significantly lower than the rest of the OB star population.

A kinematic analysis of known magnetic O-type stars using the best available astrometric, spectroscopic, and photometric data indicates that a magnetic field is more frequently detected in candidate runaway stars than in stars belonging to clusters or associations (Hubrig et al. 2011). The results obtained so far allowed the authors to constrain preliminarily the conditions conducive for the presence of magnetic fields and to derive the first trends for their occurrence rate and field-strength distribution.

To investigate statistically whether magnetic fields in O-type stars are ubiquitous or appear only in stars with a specific spectral classification, certain age, or in a special environment, Hubrig et al. (2011b, 2011) acquired spectro-polarimetric observations with FORS 2. They detected a magnetic field at a significance level of 3σ in eleven stars. The strongest net-longitudinal magnetic fields were measured in two Of?p stars: $\langle B_z \rangle = 381 \pm 122$ G for CPD–28 2561 and $\langle B_z \rangle = 297 \pm 62$ G for HD 148937. Both magnetic fields were detected by them for the first time, the latter in an earlier study (Hubrig et al. 2008a).

Walborn (1973) introduced the class of Of?p stars as the subset of massive O stars that display recurrent spectral variations in certain spectral lines, sharp emission or P Cygni profiles in the He I and Balmer lines, and strong C III emission lines around 4650 Å. Only five Galactic Of?p stars are currently known (HD 108, NGC 1624-2, CPD-28 2561, HD 148937, and HD 191612), and they have all been found to harbor magnetic fields (Martins et al. 2010; Wade et al. 2012a; Hubrig et al. 2011b, 2008a; Donati et al. 2006). Interestingly, a kinematical assessment of space velocities of the three brightest in this class (HD 108, HD 148937, and HD 191612) indicates that all three can be considered as candidate runaway stars (Hubrig et al. 2011b).

The excellent potential of FORS 2 for the detection and investigation of magnetic fields in massive stars is demonstrated in Fig. 25, which shows FORS 2 observations collected between 2008 and May 2011 of the Of?p star HD 148937 together with ESPaDOnS observations obtained at the CFHT (Wade et al. 2012b). The measurement errors for both ESPaDOnS and FORS 1/2 observations are of similar order.

In most magnetic O stars, strong magnetic fields seem to give rise to spectral peculiarities and/or drive periodic line-profile variations (e.g., the Of?p stars or θ^1 Ori C). In contrast, Grunhut et al. (2012) observed a narrow-lined O9.7 V star (HD 54879) hosting an oblique rotator field with a dipole strength of 880 ± 50 G that does not display any evidence for abundance anomalies. Only the broad emission-like Hα profile is variable. This variability may

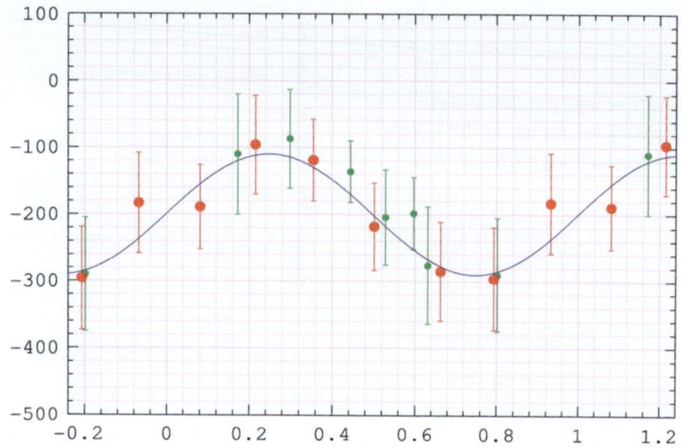

Fig. 25 Net-longitudinal magnetic-field variation of HD 148937 over the 7.032 d period determined by Nazé et al. (2010). *Red* (*large*) symbols correspond to ESPaDOnS observations (Wade et al. 2012b), while *green* (*small*) symbols are FORS 1 and FORS 2 measurements (Hubrig et al. 2008b, 2011b, in preparation)

be related to the presence of a centrifugal magnetosphere where wind material is trapped in closed-magnetic loops and prevented from falling back to the star by centrifugal forces. Petit et al. (2013) classified slowly rotating O-type stars as having dynamical magnetospheres and the more rapidly rotating B-type stars as having centrifugal magnetospheres.

Another interesting discovery is the detection of a magnetic field in a multiple system in the Trifid Nebula (Hubrig et al. 2014), which is a very young and active site of star formation. From observations of the three brightest components (A, C, and D) identified in the central part of this nebula, Kohoutek et al. (1999) clearly detected a circularly polarized signal with FORS2 in component C (HD 164492C). In contrast, no such features were visible for the two other components (an early O star and a Herbig Be star).

The O9.5 V star ζ Ophiuchi is a well-known rapidly rotating runaway star with extremely interesting characteristics. It undergoes episodic mass loss seen as emission in Hα, and it is possible that it rotates at almost break-up speed with $v \sin i = 400$ km s^{-1} (Kambe et al. 1993a). It is probably associated with the pulsar PSR B1929+10. Tetzlaff et al. (2010) suggested that both objects were ejected from Upper Scorpius during the same supernova event. Spectropolarimetric observations of ζ Oph by Hubrig et al. (2011) with FORS 1 in 2008 revealed the presence of a net-longitudinal magnetic field $\langle B_z \rangle_{\mathrm{NET}} = 141 \pm 45$ G. Hubrig et al. (2013a) obtained nine additional spectropolarimetric observations with FORS 2 over the rotation period in 2011. The net-longitudinal magnetic field shows a change of polarity, and its variation over the rotation cycle can be represented by a sinusoidal fit with a semiamplitude of ~ 160 G. FORS 2 measurements using all spectral lines can be phased with a period of 1.3 d (Fig. 26). This period is roughly twice the 0.643 d period found from the variation of the He I λ6678 line. MOST (Microvariability and Oscillations of STars) satellite observations discovered a dozen significant oscillation frequencies between 1 and 10 cycles day^{-1} (Walker et al. 2005). The suggested rotation periods using UV and X-ray data range from 0.77 to 0.98 d.

6 Post-Main Sequence Stars

The magnetic-field strengths and topology of post-main sequence stars are very different between rapidly rotating and slowly rotating stars. Subgiant and giant stars can rotate rapidly when members of tidally locked spectroscopic binary systems have orbital periods less than about 20 days, such as the RS CVn systems, or rapid rotators by some other process, such as the FK Com systems. Zeeman-broadening measurements of the primary stars in the

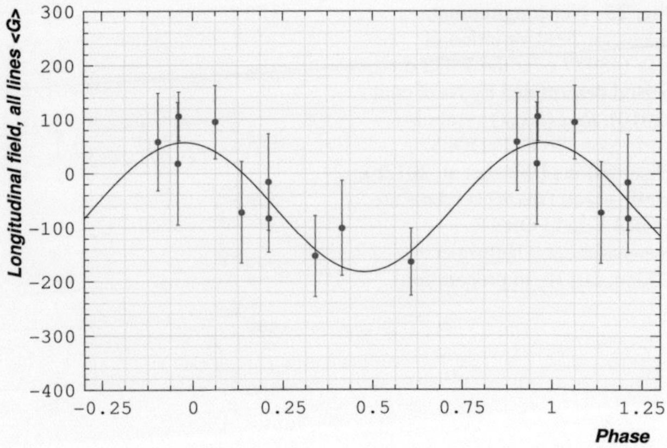

Fig. 26 Phase diagram for the best sinusoidal fit corresponding to the period of 1.3 days for the longitudinal magnetic-field measurements using the whole spectrum for ζ Oph

RS CVn-type systems VY Ari (K3 III–IV) and II Peg (K2–3 IV–V) indicate strong fields with large filling factors. For VY Ari, Bopp et al. (1989) found $B_{\rm MOD} = 2000 \pm 300$ G with $f = 0.66 \pm 0.14$, and for II Peg, Saar (1996) found $B_{\rm MOD} \approx 3000$ G with $f \approx 0.60$

ZDI has provided information on the topology of the magnetic fields of several rapidly rotating giant stars. For example, Petit et al. (2004a) found for HD 199178, an FK Com-type star with $P_{\rm rot} = 3.3$ days, that the radial magnetic flux contains roughly 85 % of the large-scale magnetic energy, but there is also significant azimuthal flux. Comparison with the Zeeman-broadening data for rapidly rotating dwarfs suggests that most of the magnetic energy is in small-scale unresolved magnetic fields. Changes in the magnetic topology are seen on time scales as short as two weeks, and the differential rotation is solar-like, but about 1.5 times faster. Donati et al. (2003) and Petit et al. (2004b) find that ZDI images of the very active primary star (G5 IV) in the HR 1099 system ($P_{\rm rot} = 2.84$ days) show that unlike HD 199178, the magnetic energy in the G5 IV star is mostly toroidal with changes seen in the smaller-scale structure on time scales of 4–6 weeks and has a much smaller differential rotation than the Sun. Since $\langle B_z \rangle$ is only 40–120 Gauss, most of the photospheric energy is at smaller scales than ZDI can resolve. The rapid rotation and deep convective zones of these stars produce the strong dynamos that power the magnetic fields in these stars.

Single giants and supergiants show a diversity of magnetic properties. We list here five examples in order of increasing rotational period that exemplify this diversity as inferred from ZDI measurements. V390 Aur is a G5 III active star (strong Ca II and X-ray emission) with $P_{\rm rot} = 9.825$ days located at the base of the red giant branch (RGB). Konstantinova-Antova et al. (2012) find that this star has a moderately strong magnetic field $\langle B_z \rangle$ that varies between $+2$ and -16 G with a strong toroidal component likely produced by an efficient $\alpha\Omega$ dynamo. β Cet is a somewhat more evolved K0 III star also located at the base of the RGB with $P_{\rm rot} = 215$ days. This star has an axisymmetric poloidal field dominated by a dipole. Tsvetkova et al. (2013) argue that this is a fossil field left over from when β Cet was an Ap star on the main sequence. Continuing to the more slowly rotating stars, we mention Pollux (β Gem), an inactive K0 III giant located in the RGB with $P_{\rm rot} \approx 590$ days. Aurière et al. (2009) find that this star has a weak, but measurable, $\langle B_z \rangle = -0.46 \pm 0.04$ G that they suggest is produced by a weak $\alpha\Omega$ dynamo. EK Boo is an M5 III star with $P_{\rm rot} \approx 846$ days located at the tip of the RGB or on the asymptotic giant branch (AGB). Konstantinova-Antova et al. (2010) measure a variable $\langle B_z \rangle$ between -0.1 and 8 G that could be produced by a weak $\alpha\Omega$ dynamo. Finally, Aurière et al. (2010) measured a magnetic field of about 1 G on the M2 Iab supergiant Betelgeuse (α Ori). With its very slow rotation period of about 17 years, this star's very weak field may be created by a local small-scale dynamo.

7 Conclusions

It is essential both to measure stellar magnetic fields and to understand what physical quantity is actually measured, because magnetic fields play many critical roles concerning the structure and energy balance in stellar atmospheres. These roles include creating the heat input in stellar chromospheres and coronae, controlling the mass and angular momentum loss from the star, facilitating chemical peculiarity, and energizing the emission of high-energy photons, which are critically important for the evolution of proto-planetary disks and the habitability of exoplanets. Since these effects are governed by the star's magnetic energy, which is proportional to the magnetic-field strength squared and its fractional surface coverage, it is essential to measure or credibly infer the true magnetic-field strength and filling factor across a stellar disk. Magnetic-field strengths and filling factors have been measured only for the Sun and then only with extreme care. We have indicated how and when it is feasible to estimate magnetic-field strengths for stars. There are now many examples for which the large-scale magnetic morphologies of stars have been inferred with ZDI techniques, but the small scale morphologies are matters of inference rather than measurement. The following points summarize our main conclusions:

Magnetic-Field Complexity and Cancellation Two critical questions that should be answered to understand any stellar magnetic-field measurement are (a) whether or not oppositely oriented magnetic fields cancel as detected by the observing instrument, and (b) whether or not the magnetic field has a complex structure across the surface of the star or in a spatial-resolution element for ZDI measurements. High-resolution spectra, especially in the infrared, but also in the optical for stars with very strong magnetic fields, measure the component of the magnetic field along the line of sight without cancellation from Zeeman broadening of absorption lines when the magnetic splitting is smaller than the intrinsic line width or Zeeman splitting when the field strength is large enough to separate the Zeeman pattern in wavelength compared to the intrinsic line width. On the other hand, all spectropolarimetric measurements include potential field cancellation. Except for the chemically peculiar A and B stars, most stars have complex magnetic-field structures that are unresolved by existing instruments. Therefore, most spectropolarimetric magnetic-field measurements will not measure true magnetic-field strengths. Field cancellation will be especially important for solar-like main sequence stars for which the Sun provides the test case of extreme complexity, but many other types of stars likely also have complex magnetic fields.

Magnetic Field Terminology We reserve the vector quantity "magnetic-field strength" for measurements in which the field is spatially resolved (or corrected for lack of resolution) without cancellation. We use the term "modulus" to refer to the peak magnetic-field strength. For spectral measurements with no cancellation, $B_{\rm MOD}$ refers to the "unsigned magnetic-field modulus", and for spectropolarimetric measurements in which there is field cancellation, $B_{\rm NET}$ refers to the "net magnetic-field modulus". When observing spectral lines that are Zeeman split, it is possible to separate out the corresponding "modulus" from the filling factor.

What Can Be Learned from the Sun? As the only star that can be spatially resolved, the Sun provides a unique test case for understanding the morphology and strength of a stellar magnetic field and how different observing techniques provide very different measurements of magnetic-field quantities. When observed as a star-like point source either in the light reflected by an asteroid or by co-adding all pixels in solar magnetograms, the net-longitudinal magnetic-field strength $\langle B_z \rangle$ is at most ± 2 G at maximum activity and usually far less. By contract, high-spatial resolution spectropolarimetry of magnetically sensitive infrared lines with Zeeman splitting exceeding the line width reveals that the solar magnetic field is highly filamentary with magnetic-field modulus values in the range 1300–1600 G, in equipartition with the photospheric gas pressure, and very small filling factors of about 4.5 %. The magnetic energy of a 1600 G field covering 4.5 % of the solar surface is 1400 times larger than the energy content of a 2 Gauss field covering 100 % of the surface. Spatial averaging, even with subarcsecond-sized apertures, such as with the Solar Optical Telescope (SOT) instrument on the *HINODE* satellite, require correction for unresolved-magnetic structures to provide sensible estimates of the photospheric magnetic-field modulus and filling factor. Since stars with convective zones like the Sun, likely also have highly complex magnetic-field morphologies, one should expect that spectrophotometric observations will provide net magnetic-field strengths far smaller than the magnetic-field modulus because of cancellation in spatially unresolved observations. Only unpolarized spectra of magnetically split lines offer the possibility of measuring the stellar magnetic-field modulus and filling factor but without information on the magnetic-field morphology.

Pre-Main Sequence Stars Pre-main sequence stars have strong magnetic fields and small areas where accretion from the circumstellar disk along magnetic field lines impact their surface creating postshock hot gas observed as bright emission in the He I $\lambda 5876$ and other lines. For most CTTSs, the accretion shock is co-spatial with strong radial magnetic fields at or near the magnetic poles and overlies dark starspots in the photosphere. The stellar magnetic field away from the accretion impact area can be approximated as a multipole with the octopole component often stronger than the dipole. For some CTTSs, the toroidal component of the field is comparable to, or even larger than, the poloidal component. As the higher mass CTTSs evolve toward the main sequence, they develop radiative cores and complex magnetic-field topologies leading to weaker magnetic dipoles and weaker interactions with their circumstellar disk. This evolution leads to a weakening of the star-disk deceleration torque and an increase in the stellar rotation rate as a result of the positive accretion torque and the decrease in stellar radius with age. Herbig Ae/Be stars are the higher mass analogs of the T Tauri stars with accretion from circumstellar disks and magnetic fields that are mostly dipolar and often inclined by large angles from the rotation axis. HD 101412 has the strongest net-longitudinal magnetic field ($\langle B_z \rangle = 3.5$ kG) yet measured for a Herbig Ae star.

Main-Sequence F–M Stars Main-sequence stars cooler than the Sun have complex magnetic fields with magnetic-field strengths and morphologies that depend systematically on spectral type and rotation period. Zeeman-broadening measurements show a pattern of increasing magnetic-field modulus, $B_{\rm MOD}$, with decreasing stellar mass, effective temperature, and rotation period. Magnetic-field modulus values correlate with commonly used indicators of magnetic heating and activity. For those stars in which Zeeman splitting is observed, the magnetic-field modulus and filling factor can be far larger than for the Sun. For example, the magnetic-field modulus and filling factor in the photosphere of the M3.5e star EV Lac were measured to be $B_{\rm MOD} = 3.8 \pm 0.5$ kG and $f = 0.50 \pm 0.13$. The corresponding magnetic energy per unit surface area is 50 times larger than for the Sun. Many M dwarf stars show similarly large $B_{\rm MOD}$ and f values. Most G–M dwarf stars have magnetic pressures close to equipartition with their nonmagnetic photospheric gas pressures and have filling factors that increase systematically with decreasing rotational period. ZDI provides information on the large-scale magnetic-field morphology, but ZDI images likely represent only a small fraction of the magnetic energy for these stars as a result of cancellation within each resolution element. Unlike the Sun, cooler and more rapidly rotating dwarf stars have strong azimuthal fields that become stronger than poloidal fields for solar mass stars rotating faster than a 12-day period. Fully convective M dwarfs show a higher degree of magnetic field organization than higher mass stars as indicated by a lower degree of field cancellation. ZDI has provided evidence for changes in the large-scale magnetic morphology and even reversals in the poloidal magnetic field on yearly time scales. The very rapidly rotating K dwarf star AB Dor shows a magnetic field with a high percent of nonpotential magnetic energy in the photosphere and corona available for flaring.

Chemically Peculiar and Normal B-Type Stars Unlike cooler main sequence stars with convective zones, the chemically peculiar A- and B-type stars show magnetic-field structures that are organized on large scales–usually dipole or coaligned dipole, quadrupole, and octopole fields that are oriented at an angle with respect to the rotation axis. However, recent ZDI maps of the He-strong star HD 37776 show a magnetic topology that is too complex to be fit by a low-order multipole model. The three types of chemically peculiar stars, the magnetic Ap and Bp stars, metallic-line Am stars, and the HgMn stars, are often slow rotators compared to chemically normal stars with the same effective temperature. The CP stars are typically pre-main sequence stars or have just reached the main sequence with very strong magnetic fields that vary in rotational phase with the chemical peculiarities. The strongest magnetic-field modulus ever measured for a nondegenerate star is either for Babcock's star, $B_{\rm MOD} = 34$ kG, or for HD 37776, $B_{\rm MOD} = 43$–49 kG. Such strong magnetic fields control the winds and elemental abundance stratification of these stars. The magnetic fields of these stars are generally assumed to be primordial rather than dynamo generated and are strongest for the more massive and fastest rotating stars in this class. Unlike the chemically peculiar B-type stars, other B-type stars have either weak or undetected magnetic fields. Pulsating B stars of the β Cep and SPB classes have $\langle B_z \rangle \approx 400$ G or less. Be stars also have weak or undetected magnetic fields with $\langle B_z \rangle \approx 100$ G. The β Cephei star with the strongest longitudinal magnetic field (V1449 Aql) appears to have an oblique rotator magnetic structure.

O Stars The hottest main sequence stars, spectral types O and early-B, are not expected to have dynamo-generated magnetic fields because the stars do not have convective zones, but they could have "fossil" fields left over from the primordial nebulae out of which these stars recently formed or magnetic fields generated by strong tidal forces between binary stars. Spectrophotometric observations have detected magnetic fields for about two dozen O-type stars with net-longitudinal magnetic-field strengths generally less than 400 G. The O9.7 V star HD 54879 may be unusual with a dipole strength of 2 kG, and ζ Oph is a very rapidly rotating runaway star with a detected magnetic field. Some O-type stars show indirect evidence for magnetic fields including very high ionization states in X-ray spectra and very strong radio emission that can be explained by magnetic shocks in their massive winds.

Post-Main Sequence Stars As stars evolve off the main sequence, they expand, rotate more slowly, and, as a consequence, are expected to have weaker magnetic fields generated by their $\alpha\Omega$ dynamos. As stars progress up the red giant branch from subgiants to class-III red giants to M supergiants, their magnetic-flux measurements show a sequence of decreasing magnetic flux consistent with this picture. Their magnetic fields are assumed to be complex like their main-sequence predecessor stars, and poloidal or toroidal large-scale geometries have been detected in a few cases. Prominent exceptions to this decline in magnetic fields with evolution up the red giant branch are those close binaries whose rapid rotation is enforced by tides leading to synchronized orbital and rotational periods. For the G or K-type subgiant or giant RS CVn systems and the FK Com stars, rapid rotation rejuvenates strong $\alpha\Omega$ dynamos leading to kG magnetic-field moduli detected by Zeeman-broadening techniques and toroidal magnetic-field morphologies detected by ZDI for the more rapidly rotating stars. In one case, HD 199178, ZDI measurements have detected remarkably rapid changes in the field morphology on a time scale of 4–6 weeks.

Acknowledgements We thank the referee for a careful reading of the manuscript and many useful suggestions. We thank the International Space Science Institute (ISSI) for their hospitality and the convenors of the ISSI workshop "The Strongest Magnetic Fields in the Universe" for organizing a stimulating workshop. JLL thanks the Space Telescope Science Institute for support of grants GO-12464 and GO-11616 to observe and analyze spectra of exoplanet host stars and premain sequence stars.

References

D.C. Abbott, J.H. Beiging, E. Churchwell, A.V. Torres, Radio emission from galactic Wolf–Rayet stars and the structure of Wolf–Rayet winds. Astrophys. J. **303**, 239–261 (1986). doi:10.1086/164070

F.C. Adams, S.G. Gregory, Magnetically controlled accretion flows onto young stellar objects. Astrophys. J. **744**, 55 (2012). doi:10.1088/0004-637X/744/1/55

E. Alecian, C. Catala, G.A. Wade, J.-F. Donati, P. Petit, J.D. Landstreet, T. Böhm, J.-C. Bouret, S. Bagnulo, C. Folsom, J. Grunhut, J. Silvester, Characterization of the magnetic field of the Herbig Be star HD200775. Mon. Not. R. Astron. Soc. **385**, 391–403 (2008). doi:10.1111/j.1365-2966.2008.12842.x

E. Alecian, G.A. Wade, C. Catala, S. Bagnulo, T. Böhm, J.-C. Bouret, J.-F. Donati, C.P. Folsom, J. Grunhut, J.D. Landstreet, Magnetism and binarity of the Herbig Ae star V380 Ori†. Mon. Not. R. Astron. Soc. **400**, 354–368 (2009). doi:10.1111/j.1365-2966.2009.15460.x

E. Alecian, C. Neiner, S. Mathis, C. Catala, O. Kochukhov, J. Landstreet, The dramatic change of the fossil magnetic field of HD 190073: evidence of the birth of the convective core in a Herbig star? Astron. Astrophys. **549**, 8 (2013). doi:10.1051/0004-6361/201220796

R.I. Anderson, A. Reiners, S.K. Solanki, On detectability of Zeeman broadening in optical spectra of F- and G-dwarfs. Astron. Astrophys. **522**, 81 (2010). doi:10.1051/0004-6361/201014769

M. Aurière, G.A. Wade, J. Silvester, F. Lignières, S. Bagnulo, K. Bale, B. Dintrans, J.-F. Donati, C.P. Folsom, M. Gruberbauer, A. Hui Bon Hoa, S. Jeffers, N. Johnson, J.D. Landstreet, A. Lèbre, T. Lueftinger, S. Marsden, D. Mouillet, S. Naseri, F. Paletou, P. Petit, J. Power, F. Rincon, S. Strasser, N. Toqué, Weak magnetic fields in Ap/Bp stars. Evidence for a dipole field lower limit and a tentative interpretation of the magnetic dichotomy. Astron. Astrophys. **475**, 1053–1065 (2007). doi:10.1051/0004-6361:20078189

M. Aurière, G.A. Wade, R. Konstantinova-Antova, C. Charbonnel, C. Catala, W.W. Weiss, T. Roudier, P. Petit, J.-F. Donati, E. Alecian, R. Cabanac, S. van Eck, C.P. Folsom, J. Power, Discovery of a weak magnetic field in the photosphere of the single giant Pollux. Astron. Astrophys. **504**, 231–237 (2009). doi:10.1051/0004-6361/200912050

M. Aurière, J.-F. Donati, R. Konstantinova-Antova, G. Perrin, P. Petit, T. Roudier, The magnetic field of Betelgeuse: a local dynamo from giant convection cells? Astron. Astrophys. **516**, 2 (2010). doi:10.1051/0004-6361/201014925

H.W. Babcock, Zeeman effect in stellar spectra. Astrophys. J. **105**, 105 (1947). doi:10.1086/144887

H.W. Babcock, The 34-KILOGAUSS magnetic field of HD 215441. Astrophys. J. **132**, 521 (1960). doi:10.1086/146960

S. Bagnulo, M. Landolfi, G. Mathys, M. Landi Degl'Innocenti, Modelling of magnetic fields of CP stars. III. The combined interpretation of five different magnetic observables: theory, and application to beta Coronae Borealis. Astron. Astrophys. **358**, 929–942 (2000)

S. Bagnulo, J.D. Landstreet, L. Fossati, O. Kochukhov, Magnetic field measurements and their uncertainties: the FORS1 legacy. Astron. Astrophys. **538**, 129–150 (2012). doi:10.1051/0004-6361/201118098

D. Baines, R.D. Oudmaijer, J.M. Porter, M. Pozzo, On the binarity of Herbig Ae/Be stars. Mon. Not. R. Astron. Soc. **367**, 737–753 (2006). doi:10.1111/j.1365-2966.2006.10006.x

P.K. Barker, J.M. Marlborough, J.D. Landstreet, I.B. Thompson, A search for magnetic fields in Be stars. Astrophys. J. **288**, 741–745 (1985). doi:10.1086/162841

J.H. Beiging, D.C. Abbott, E. Churchwell, A survey of radio emission from galactic OB stars. Astrophys. J. **340**, 518–536 (1989). doi:10.1086/167414

M. Benisty, K. Perraut, D. Mourard, P. Stee, G.H.R.A. Lima, J.B. Le Bouquin, M. Borges Fernandes, O. Chesneau, N. Nardetto, I. Tallon-Bosc, H. McAlister, T. Ten Brummelaar, S. Ridgway, J. Sturmann, L. Sturmann, N. Turner, C. Farrington, P.J. Goldfinger, Enhanced H activity at periastron in the young and massive spectroscopic binary HD 200775. Astron. Astrophys. **555**, 113 (2013). doi:10.1051/0004-6361/201219893

B.W. Bopp, S.H. Saar, C. Ambruster, P. Feldman, R. Dempsey, M. Allen, S.C. Barden, The active chromosphere binary HD 17433 (VY Arietis). Astrophys. J. **339**, 1059–1072 (1989). doi:10.1086/167360

J. Bouvier, S.H.P. Alencar, T.J. Harries, C.M. Johns-Krull, M.M. Romanova, Magnetospheric accretion in classical T Tauri stars, in *Protostars and Planets V*, ed. by B. Reipurth, D. Jewitt, K. Keil Protostars and Planets, vol. 5, 2007, pp. 479–494

M. Briquet, C. Neiner, B. Leroy, P.I. Pápics, Discovery of a magnetic field in the CoRoT hybrid B-type pulsator HD 43317. Astron. Astrophys. **557**, 16 (2013). doi:10.1051/0004-6361/201321779

C. Catala, E. Alecian, J.-F. Donati, G.A. Wade, J.D. Landstreet, T. Böhm, J.-C. Bouret, S. Bagnulo, C. Folsom, J. Silvester, The magnetic field of the pre-main sequence Herbig Ae star HD 190073. Astron. Astrophys. **462**, 293–301 (2007). doi:10.1051/0004-6361:20066264

W. Chen, C.M. Johns-Krull, Spectropolarimetry of the classical T Tauri star BP Tau. Astrophys. J. **776**, 113–132 (2013). doi:10.1088/0004-637X/776/2/113

C.R. Cowley, G.C.L. Aikman, Nuclear and nonnuclear abundance patterns in the manganese stars. Astrophys. J. **196**, 521–524 (1975). doi:10.1086/153432

C.R. Cowley, S. Hubrig, T.A. Ryabchikova, G. Mathys, N. Piskunov, P. Mittermayer, The core-wing anomaly of cool Ap stars. Abnormal Balmer profiles. Astron. Astrophys. **367**, 939–942 (2001). doi:10.1051/0004-6361:20000539

M. Cuntz, W. Rammacher, P. Ulmschneider, Z.E. Musielak, S.H. Saar, Two-component theoretical chromosphere models for K dwarfs of different magnetic activity: exploring the Ca II emission-stellar rotation relationship. Astrophys. J. **522**, 1053–1068 (1999). doi:10.1086/307689

A.G. Daou, C.M. Johns-Krull, J.A. Valenti, Spectropolarimetry of the classical T Tauri star T Tauri. Astron. J. **131**, 520–526 (2006). doi:10.1086/498306

J.-F. Donati, S.F. Brown, Zeeman–Doppler imaging of active stars. V. Sensitivity of maximum entropy magnetic maps to field orientation. Astron. Astrophys. **326**, 1135–1142 (1997)

J.-F. Donati, J.D. Landstreet, Magnetic fields of nondegenerate stars. Annu. Rev. Astron. Astrophys. **47**, 333–370 (2009). doi:10.1146/annurev-astro-082708-101833

J.-F. Donati, M. Semel, B.D. Carter, D.E. Rees, A. Collier Cameron, Spectropolarimetric observations of active stars. Mon. Not. R. Astron. Soc. **291**, 658 (1997)

J.-F. Donati, A. Collier Cameron, G.A.J. Hussain, M. Semel, Magnetic topology and prominence patterns on AB Doradus. Mon. Not. R. Astron. Soc. **302**, 437–456 (1999). doi:10.1046/j.1365-8711.1999.02095.x

J.-F. Donati, J. Babel, T.J. Harries, I.D. Howarth, P. Petit, M. Semel, The magnetic field and wind confinement of θ^1 Orionis C. Mon. Not. R. Astron. Soc. **333**, 55–70 (2002). doi:10.1046/j.1365-8711.2002.05379.x

J.-F. Donati, A. Collier Cameron, M. Semel, G.A.J. Hussain, P. Petit, B.D. Carter, S.C. Marsden, M. Mengel, A. López Ariste, S.V. Jeffers, D.E. Rees, Dynamo processes and activity cycles of the active stars AB Doradus, LQ Hydrae and HR 1099. Mon. Not. R. Astron. Soc. **345**, 1145–1186 (2003). doi:10.1046/j.1365-2966.2003.07031.x

J.-F. Donati, I.D. Howarth, J.-C. Bouret, P. Petit, C. Catala, J. Landstreet, Discovery of a strong magnetic field on the O star HD 191612: new clues to the future of θ^1 Orionis C*. Mon. Not. R. Astron. Soc. **365**, 6–10 (2006). doi:10.1111/j.1745-3933.2005.00115.x

J.-F. Donati, J. Morin, P. Petit, X. Delfosse, T. Forveille, M. Aurière, R. Cabanac, B. Dintrans, R. Fares, T. Gastine, M.M. Jardine, F. Lignières, F. Paletou, J.C. Ramirez Velez, S. Théado, Large-scale magnetic topologies of early M dwarfs. Mon. Not. R. Astron. Soc. **390**, 545–560 (2008a). doi:10.1111/j.1365-2966.2008.13799.x

J.-F. Donati, M.M. Jardine, S.G. Gregory, P. Petit, F. Paletou, J. Bouvier, C. Dougados, F. Ménard, A. Collier Cameron, T.J. Harries, G.A.J. Hussain, Y. Unruh, J. Morin, S.C. Marsden, N. Manset, M. Aurière, C. Catala, E. Alecian, Magnetospheric accretion on the T Tauri star BP Tauri. Mon. Not. R. Astron. Soc. **386**, 1234–1251 (2008b). doi:10.1111/j.1365-2966.2008.13111.x

J.-F. Donati, M.B. Skelly, J. Bouvier, M.M. Jardine, S.G. Gregory, J. Morin, G.A.J. Hussain, C. Dougados, F. Ménard, Y. Unruh, Complex magnetic topology and strong differential rotation on the low-mass T Tauri star V2247 Oph. Mon. Not. R. Astron. Soc. **402**, 1426–1436 (2010a). doi:10.1111/j.1365-2966.2009.15998.x

J.-F. Donati, M.B. Skelly, J. Bouvier, S.G. Gregory, K.N. Grankin, M.M. Jardine, G.A.J. Hussain, F. Ménard, C. Dougados, Y. Unruh, et al., Magnetospheric accretion and spin-down of the prototypical classical T Tauri star AA Tau. Mon. Not. R. Astron. Soc. **409**, 1347–1361 (2010b). doi:10.1111/j.1365-2966.2010.17409.x

J.-F. Donati, S.G. Gregory, S.H.P. Alencar, J. Bouvier, G.A.J. Hussain, M. Skelly, C. Dougados, M.M. Jardine, F. Ménard, C. Dougados, M.M. Romanova, et al., The large-scale magnetic field and poleward masss accretion of the classical T Tauri star TW Hya. Mon. Not. R. Astron. Soc. **417**, 472–487 (2011). doi:10.1111/j.1365-2966.2010.19288.x

J.-F. Donati, S.G. Gregory, S.H.P. Alencar, G.A.J. Hussain, J. Bouvier, M.M. Jardine, F. Ménard, C. Dougados, M.M. Romanova (the MaPP collaboration), Magnetospheric accretion on the fully convective classical T Tauri star DN Tau. Mon. Not. R. Astron. Soc. **436**, 881–897 (2013). doi:10.1093/mnras.436..881D

S. Edwards, W. Fischer, L. Hillenbrand, J. Kwan, Probing T Tauri accretion and outflow with 1 micron spectroscopy. Astrophys. J. **646**, 319–341 (2006). doi:10.1086/504832

V.G. Elkin, G. Mathys, D.W. Kurtz, S. Hubrig, L.M. Freyhammer, A rival for Babcock's star: 30-kG variable magnetic field in the Ap star HD 75049. Mon. Not. R. Astron. Soc. **402**, 1883–1891 (2010). doi:10.1111/j.1365-2966.2009.16015.x

E.D. Feigelson, T. Montmerle, High-energy processes in young stellar objects. Annu. Rev. Astron. Astrophys. **37**, 363–408 (1999). doi:10.1146/annurev.astro.37.1.363

L. Ferrario, A. Melatos, J. Zrake, Magnetic field generation in stars. Space Science Rev. (2015)

C.P. Folsom, G.A. Wade, O. Kochukhov, E. Alecian, C. Catala, S. Bagnulo, T. Böhm, J.-C. Bouret, J.-F. Donati, J. Grunhut, D.A. Hanes, J.D. Landstreet, Magnetic fields and chemical peculiarities of the very young intermediate-mass binary system HD 72106. Mon. Not. R. Astron. Soc. **391**, 901–914 (2008). doi:10.1111/j.1365-2966.2008.13946.x

L. Fossati, N. Castro, T. Morel, N. Langer, M. Briquet, T.A. Carroll, S. Hubrig, M.F. Nieva, L.M. Oskinova, N. Przybilla, F.R.N. Schneider, M. Schöller, S. Simón-Diaz, I. Ilyin, A. de Koter, A. Reisenegger, H. Sana (the BOB collaboration), B fields in OB stars (BOB): on the detection of weak magnetic fields in the two early B-type stars β CMa and ϵ CMa. Possible lack of a "magnetic desert" in massive stars. Astron. Astrophys. **574**, 20–34 (2015). doi:10.1051/0004-6361/201424986

C. Fourtune-Ravard, G.A. Wade, W.L.F. Marcolino, M. Shultz, J.H. Grunhut, H.F. Henrichs, (MiMeS Collaboration), The constant magnetic field of ξ^1 CMa: geometry or slow rotation? in *IAU Symposium*, ed. by C. Neiner, G. Wade, G. Meynet, G. Peters IAU Symposium, vol. 272, 2011, pp. 180–181. doi:10.1017/S1743921311010234

T. Gastine, J. Morin, L. Duarte, A. Reiners, U.R. Christensen, J. Wicht, What controls the magnetic geometry of M dwarfs? Astron. Astrophys. **549**, 5 (2013). doi:10.1051/0004-6361/201220317

T. Gatti, A. Natta, S. Randich, L. Testi, G. Sacco, Accretion properties of T Tauri stars in σ Orionis. Astron. Astrophys. **481**, 423–432 (2008). doi:10.1051/0004-6361:20078971

J.H. Grunhut, G.A. Wade, J.O. Sundqvist, A. ud-Doula, C. Neiner, R. Ignace, W.L.F. Marcolino, T. Rivinius, A. Fullerton, L. Kaper, B. Mauclaire, et al., Investigating the spectroscopic, magnetic and circumstellar variability of the O9 star HD 57682. Mon. Not. R. Astron. Soc. **426**, 2208–2227 (2012). doi:10.1111/j.1365-2966.2012.21799.x

F. Hamann, S.E. Persson, Emission-line studies of young stars. II. The Herbig Ae/Be stars. Astrophys. J. Suppl. Ser. **82**, 285 (1992). doi:10.1086/191716

S. Hubrig, N.V. Kharchenko, M. Schöller, The kinematic characteristics of magnetic O-type stars. Astron. Nachr. **332**, 65 (2011). doi:10.1002/asna.201011479

S. Hubrig, P. North, G. Mathys, Magnetic AP stars in the Hertzsprung–Russell diagram. Astrophys. J. **539**, 352–363 (2000). doi:10.1086/309189

S. Hubrig, L.M. Oskinova, M. Schöller, First detection of a magnetic field in the fast rotating runaway Oe star ζ Ophiuchi. Astron. Nachr. **332**, 147 (2011). doi:10.1002/asna.201111516

S. Hubrig, M. Schöller, R.V. Yudin, Magnetic fields in Herbig Ae stars. Astron. Astrophys. **428**, 1–4 (2004). doi:10.1051/0004-6361:200400091

S. Hubrig, N. Nesvacil, M. Schöller, P. North, G. Mathys, D.W. Kurtz, B. Wolff, T. Szeifert, M.S. Cunha, V.G. Elkin, Detection of an extraordinarily large magnetic field in the unique ultra-cool Ap star HD 154708. Astron. Astrophys. **440**, 37–40 (2005). doi:10.1051/0004-6361:200500164

S. Hubrig, R.V. Yudin, M. Schöller, M.A. Pogodin, Accurate magnetic field measurements of Vega-like stars and Herbig Ae/Be stars. Astron. Astrophys. **446**, 1089–1094 (2006a). doi:10.1051/0004-6361:20053794

S. Hubrig, M. Briquet, M. Schöller, P. De Cat, G. Mathys, C. Aerts, Discovery of magnetic fields in the βCephei star ξ^1 CMa and in several slowly pulsating B stars*. Mon. Not. R. Astron. Soc. **369**, 61–65 (2006b). doi:10.1111/j.1745-3933.2006.00175.x

S. Hubrig, M.A. Pogodin, R.V. Yudin, M. Schöller, R.S. Schnerr, The magnetic field in the photospheric and circumstellar components of Herbig Ae stars. Astron. Astrophys. **463**, 1039–1046 (2007). doi:10.1051/0004-6361:20066090

S. Hubrig, M. Schöller, R.S. Schnerr, J.F. González, R. Ignace, H.F. Henrichs, Magnetic field measurements of O stars with FORS 1 at the VLT. Astron. Astrophys. **490**, 793–800 (2008a). doi:10.1051/0004-6361:200810171

S. Hubrig, M. Briquet, T. Morel, M. Schöller, J.F. González, P. De Cat, New insights into the nature of the peculiar star θ Carinae. Astron. Astrophys. **488**, 287–296 (2008b). doi:10.1051/0004-6361:200809972

S. Hubrig, M. Schöller, I. Savanov, R.V. Yudin, M.A. Pogodin, S. Štefl, T. Rivinius, M. Curé, Magnetic survey of emission line B-type stars with FORS 1 at the VLT. Astron. Nachr. **330**, 708 (2009a). doi:10.1002/asna.200911236

S. Hubrig, M. Briquet, P. De Cat, M. Schöller, T. Morel, I. Ilyin, New magnetic field measurements of β Cephei stars and slowly pulsating B stars. Astron. Nachr. **330**, 317 (2009b). doi:10.1002/asna.200811187

S. Hubrig, B. Stelzer, M. Schöller, C. Grady, O. Schütz, M.A. Pogodin, M. Curé, K. Hamaguchi, R.V. Yudin, Searching for a link between the magnetic nature and other observed properties of Herbig Ae/Be stars and stars with debris disks. Astron. Astrophys. **502**, 283–301 (2009c). doi:10.1051/0004-6361/200811533

S. Hubrig, M. Schöller, I. Savanov, J.F. González, C.R. Cowley, O. Schütz, R. Arlt, G. Rüdiger, The exceptional Herbig Ae star HD 101412: the first detection of resolved magnetically split lines and the presence of chemical spots in a Herbig star. Astron. Nachr. **331**, 361 (2010). doi:10.1002/asna.201011346

S. Hubrig, M. Schöller, I. Ilyin, C.R. Cowley, Z. Mikulášek, B. Stelzer, M.A. Pogodin, R.V. Yudin, M. Curé, Characterising the magnetic fields of the Herbig Ae/Be stars HD 97048, HD 150193, HD 176386, and MWC 480. Astron. Astrophys. **536**, 45 (2011a). doi:10.1051/0004-6361/201117407

S. Hubrig, M. Schöller, N.V. Kharchenko, N. Langer, W.J. de Wit, I. Ilyin, A.F. Kholtygin, A.E. Piskunov, N. Przybilla (Magori Collaboration), Exploring the origin of magnetic fields in massive stars: a survey of O-type stars in clusters and in the field. Astron. Astrophys. **528**, 151 (2011b). doi:10.1051/0004-6361/201016345

S. Hubrig, I. Ilyin, M. Schöller, M. Briquet, T. Morel, P. De Cat, First magnetic field models for recently discovered magnetic β cephei and slowly pulsating B stars. Astrophys. J. Lett. **726**, 5 (2011c). doi:10.1088/2041-8205/726/1/L5

S. Hubrig, Z. Mikulášek, J.F. González, M. Schöller, I. Ilyin, M. Curé, M. Zejda, C.R. Cowley, V.G. Elkin, M.A. Pogodin, R.V. Yudin, Rotationally modulated variations and the mean longitudinal magnetic field of the Herbig Ae star HD 101412. Astron. Astrophys. **525**, 4 (2011d). doi:10.1051/0004-6361/201015806

S. Hubrig, I. Ilyin, M. Briquet, M. Schöller, J.F. González, N. Nuñez, P. De Cat, T. Morel, The strong magnetic field of the large-amplitude β Cephei pulsator V1449 Aquilae. Astron. Astrophys. **531**, 20 (2011e). doi:10.1051/0004-6361/201117261

S. Hubrig, F. Castelli, J.F. González, V.G. Elkin, G. Mathys, C.R. Cowley, B. Wolff, M. Schöller, Line identification in high-resolution, near-infrared CRIRES spectra of chemically peculiar and Herbig Ae stars. Astron. Astrophys. **542**, 31 (2012a). doi:10.1051/0004-6361/201218968

S. Hubrig, J.F. González, I. Ilyin, H. Korhonen, M. Schöller, I. Savanov, R. Arlt, F. Castelli, G. Lo Curto, M. Briquet, T.H. Dall, Magnetic fields of HgMn stars. Astron. Astrophys. **547**, 90 (2012b). doi:10.1051/0004-6361/201219778

S. Hubrig, M. Schöller, I. Ilyin, N.V. Kharchenko, L.M. Oskinova, N. Langer, J.F. González, A.F. Kholtygin, M. Briquet (Magori Collaboration), Exploring the origin of magnetic fields in massive stars. II. New magnetic field measurements in cluster and field stars. Astron. Astrophys. **551**, 33 (2013a). doi:10.1051/0004-6361/201220721

S. Hubrig, I. Ilyin, M. Schöller, G. Lo Curto, HARPS spectropolarimetry of Herbig Ae/Be stars. Astron. Nachr. **334**, 1093 (2013b). doi:10.1002/asna.201311948

S. Hubrig, L. Fossati, T.A. Carroll, N. Castro, J.F. González, I. Ilyin, N. Przybilla, M. Schöller, L.M. Oskinova, T. Morel, N. Langer, R.D. Scholz, N.V. Kharchenko, M.-F. Nieva, B fields in OB stars (BOB): the discovery of a magnetic field in a multiple system in the Trifid nebula, one of the youngest star forming regions. Astron. Astrophys. **564**, 10 (2014). doi:10.1051/0004-6361/201423490

G.A.J. Hussain, Stellar surface imaging: mapping brightness and magnetic fields. Astron. Nachr. **325**, 216–220 (2004). doi:10.1002/asna.200310219

G.A.J. Hussain, A.A. van Ballegooijen, M. Jardine, A. Collier Cameron, The coronal topology of the rapidly rotating K0 dwarf AB doradus. I. using surface magnetic field maps to model the structure of the stellar corona. Astrophys. J. **575**, 1078–1086 (2002). doi:10.1086/341429

G.A.J. Hussain, A. Collier Cameron, M.M. Jardine, N. Dunstone, J. Ramirez Velez, .C. Stempels, J.-F. Donati, M. Semel, G. Aulanier, T. Harries, et al., Surface magnetic fields on two accreting T Tauri stars: CV Cha and CR Cha. Mon. Not. R. Astron. Soc. **398**, 189–200 (2009). doi:10.1111/j.1365-2966.2009.14881.x

C.M. Johns-Krull, The magnetic fields of classical T Tauri stars. Astrophys. J. **664**, 975–985 (2007). doi:10.1086/519017

C.M. Johns-Krull, J.A. Valenti, Detection of strong magnetic fields on M Dwarfs. Astrophys. J. Lett. **459**, 95 (1996). doi:10.1086/309954

C.M. Johns-Krull, J.A. Valenti, A.P. Hatzes, A. Kanaan, Spectropolarimetry of magnetospheric accretion on the classical T Tauri star BP Tauri. Astrophys. J. **510**, 41–44 (1999). doi:10.1086/311802

C.M. Johns-Krull, T.P. Greene, G.W. Doppmann, K.R. Covey, First magnetic field detection on a class I Protostar. Astrophys. J. **700**, 1440–1448 (2009). doi:10.1088/0004-637X/700/2/1440

C.P. Johnstone, M. Jardine, S.G. Gregory, J.-F. Donati, G. Hussain, Classical T Tauri stars: magnetic fields, coronae and star-disk interactions. Mon. Not. R. Astron. Soc. **437**, 3202–3220 (2014). doi:10.1093/mnras/stt2107

E. Kambe, H. Ando, R. Hirata, Shortterm line-profile variations and episodic mass loss in the Be-Star Zeta-Ophiuchi. Astron. Astrophys. **273**, 435 (1993a)

E. Kambe, H. Ando, R. Hirata, G.A.H. Walker, E.J. Kennelly, J.M. Matthews, Line-profile variations of Lambda Eridani in emission and quiescence. Publ. Astron. Soc. Pac. **105**, 1222–1231 (1993b). doi:10.1086/133299

O. Kochukhov, G.A. Wade, Magnetic Doppler imaging of α^2 Canum Venaticorum in all four Stokes parameters: unveiling the hidden complexity of stellar magnetic fields. Astron. Astrophys. **513**, 13–25 (2010). doi:10.1051/0004-6361/2009113860

O. Kochukhov, S. Bagnulo, G.A. Wade, L. Sangalli, N. Piskunov, J.D. Landstreet, P. Petit, T.A.A. Sigut, Magnetic Doppler imaging of 53 Camelopardalis in all four Stokes parameters. Astron. Astrophys. **414**, 613–632 (2004). doi:10.1051/0004-6361:20031595

O. Kochukhov, A. Lundin, I. Romanyuk, D. Kudryavtsev, The extraordinary complex magnetic fields of the helium-strong star HD 37776. Astrophys. J. **726**, 24–31 (2011). doi:10.1088/0004-637X/726/1/24

O. Kochukhov, V. Makaganiuk, N. Piskunov, S.V. Jeffers, C.M. Johns-Krull, C.U. Keller, M. Rodenhuis, F. Snik, H.C. Stempels, J.A. Valenti, Are there tangled magnetic fields on HgMn stars? Astron. Astrophys. **554**, 61 (2013). doi:10.1051/0004-6361/201321467

L. Kohoutek, P. Mayer, R. Lorenz, Photometry and spectroscopy of the central star of the Trifid nebula. Astron. Astrophys. Suppl. Ser. **134**, 129–133 (1999). doi:10.1051/aas:1999128

R. Konstantinova-Antova, M. Aurière, C. Charbonnel, N.A. Drake, K.-P. Schröder, I. Stateva, E. Alecian, P. Petit, R. Cabanac, Direct detection of a magnetic field in the photosphere of the single M giant EK Bootis. How common is magnetic activity among M giants? Astron. Astrophys. **524**, 57 (2010). doi:10.1051/0004-6361/201014503

R. Konstantinova-Antova, M. Aurière, P. Petit, C. Charbonnel, S. Tsvetkova, A. Lèbre, R. Bogdanovski, Magnetic field structure in single late-type giants: the effectively single giant V390 Aurigae. Astron. Astrophys. **541**, 44 (2012). doi:10.1051/0004-6361/201116690

M. Landolfi, S. Bagnulo, M. Landi Degl'Innocenti, Modelling of magnetic fields of CP stars I. A diagnostic method for dipole and quadrupole fields from Stokes I and V observations. Astron. Astrophys. **338**, 111–121 (1998)

J.D. Landstreet, G. Mathys, Magnetic models of slowly rotating magnetic Ap stars: aligned magnetic and rotation axes. Astron. Astrophys. **359**, 213–226 (2000)

J.D. Landstreet, S. Bagnulo, L. Fossati, On the consistency of magnetic field measurements of Ap stars: lessons learned from the FORS1 archive. Astron. Astrophys. **572**, 113–125 (2014). doi:10.1051/0004-6361/201424749

J.D. Landstreet, S. Bagnulo, V. Andretta, L. Fossati, E. Mason, J. Silaj, G.A. Wade, Searching for links between magnetic fields and stellar evolution II. The evolution of magnetic fields as revealed by observations of Ap stars in open clusters and associations. Astron. Astrophys. **470**, 685–698 (2007). doi:10.1051/0004-6361:20077343

P. Lang, M. Jardine, J. Morin, J.-F. Donati, S. Jeffers, A.A. Vidotto, R. Fares, Modelling the hidden magnetic field of low-mass stars. Mon. Not. R. Astron. Soc. **439**, 2122–2131 (2014). doi:10.1093/mnras/stu091

C. Leinert, A. Richichi, M. Haas, Binaries among Herbig Ae/Be stars. Astron. Astrophys. **318**, 472–484 (1997)

J.L. Leroy, Linear polarimetry of AP stars. V. A general catalogue of measurements. Astron. Astrophys. Suppl. Ser. **114**, 79 (1995)

J.L. Leroy, S. Bagnulo, M. Landolfi, E. Landi Degl'Innocenti, A long period model for the magnetic star gamma Equulei. Astron. Astrophys. **284**, 174–178 (1994)

J.L. Linsky, S.A. Drake, T.S. Bastian, Radio emission from chemically peculiar stars. Astrophys. J. **393**, 341–356 (1992). doi:10.1086/171509

T. Lüftinger, O. Kochukhov, T. Ryabchikova, N. Piskunov, W.W. Weiss, I. Ilyin, Magnetic Doppler imaging of the roAp star HD 24712. Astron. Astrophys. **509**, 71 (2010). doi:10.1051/0004-6361/200811545

G.W. Marcy, G. Basri, Physical realism in the analysis of stellar magnetic fields. II—K dwarfs. Astrophys. J. **345**, 480–488 (1989). doi:10.1086/167921

F. Martins, J.-F. Donati, W.L.F. Marcolino, J.-C. Bouret, G.A. Wade, C. Escolano, I.D. Howarth (Mimes Collaboration), Detection of a magnetic field on HD108: clues to extreme magnetic braking and the Of?p phenomenon. Mon. Not. R. Astron. Soc. **407**, 1423–1432 (2010). doi:10.1111/j.1365-2966.2010.17005.x

G. Mathys, Spectropolarimetry of magnetic stars. II—The mean longitudinal magnetic field. Astron. Astrophys. Suppl. Ser. **89**, 121–157 (1991)

G. Mathys, Magnetic field diagnosis through spectropolarimetry, in *IAU Colloq. 138: Peculiar Versus Normal Phenomena in A-Type and Related Stars*, ed. by M.M. Dworetsky, F. Castelli, R. Faraggiana Astronomical Society of the Pacific Conference Series, vol. 44, 1993, p. 232

G. Mathys, Spectropolarimetry of magnetic stars. IV. The crossover effect. Astron. Astrophys. **293**, 733–745 (1995a)

G. Mathys, Spectropolarimetry of magnetic stars. V. The mean quadratic magnetic field. Astron. Astrophys. **293**, 746–763 (1995b)

G. Mathys, S. Hubrig, J.D. Landstreet, T. Lanz, J. Manfroid, The mean magnetic field modulus of AP stars. Astron. Astrophys. Suppl. Ser. **123**, 353–402 (1997). doi:10.1051/aas:1997103

C.F. McKee, E.C. Ostriker, Theory of star formation. Annu. Rev. Astron. Astrophys. **45**, 565–687 (2007). doi:10.1146/annurev.astro.45.051806.110602

Y.-J. Moon, Y.-H. Kim, Y.-D. Park, K. Ichimoto, T. Sakurai, J. Chae, K.S. Cho, S. Bong, Y. Suematsu, S. Tsuneta, Y. Katsukawa, M. Shimojo, T. Shimizu, R.A. Shine, T.D. Tarbell, A.M. Title, B. Lites, M. Kubo, S. Nagata, T. Yokoyama, Hinode SP vector magnetogram of AR10930 and its cross-comparison with MDI. Publ. Astron. Soc. Jpn. **59**, 625 (2007). doi:10.1093/pasj/59.sp3.S625

T. Morel, N. Castro, L. Fossati, S. Hubrig, N. Langer, N. Przybilla, M. Schöller, T. Carroll, I. Ilyin, A. Irrgang, L. Oskinova, F. Schneider, S. Simon Díaz, M. Briquet, J.-F. González, N. Kharchenko, M.-F. Nieva, R.-D. Scholz, A. de Koter, W.-R. Hamann, A. Herrero, J. Maíz Apellániz, H. Sana, R. Arlt, R. Barbá, P. Dufton, A. Kholtygin, G. Mathys, A. Piskunov, A. Reisenegger, H. Spruit, S.-C. Yoon, The B Fields in OB Stars (BOB) Survey (2014)

A. Morgenthaler, P. Petit, S. Saar, S.K. Solanki, J. Morin, S.C. Marsden, M. Aurière, B. Dintrans, R. Fares, T. Gastine, J. Lanoux, F. Lignières, F. Paletou, J.C. Ramírez Vélez, S. Théado, V. Van Grootel, Long-term magnetic field monitoring of the Sun-like star ξ Bootis A. Astron. Astrophys. **540**, 138 (2012). doi:10.1051/0004-6361/201118139

J. Morin, J.-F. Donati, P. Petit, X. Delfosse, T. Forveille, L. Albert, M. Aurière, R. Cabanac, B. Dintrans, R. Fares, T. Gastine, M.M. Jardine, F. Lignières, F. Paletou, J.C. Ramirez Velez, S. Théado, Large-scale magnetic topologies of mid M dwarfs. Mon. Not. R. Astron. Soc. **390**, 567–581 (2008). doi:10.1111/j.1365-2966.2008.13809.x

J. Morin, J.-F. Donati, P. Petit, X. Delfosse, T. Forveille, M.M. Jardine, Large-scale magnetic topologies of late M dwarfs. Mon. Not. R. Astron. Soc. **407**, 2269–2286 (2010). doi:10.1111/j.1365-2966.2010.17101.x

Y. Nazé, A. Ud-Doula, M. Spano, G. Rauw, M. De Becker, N.R. Walborn, New findings on the prototypical Of?p stars. Astron. Astrophys. **520**, 59 (2010). doi:10.1051/0004-6361/201014333

C. Neiner, A.-M. Hubert, Y. Frémat, M. Floquet, S. Jankov, O. Preuss, H.F. Henrichs, J. Zorec, Rotation and magnetic field in the Be star omega Orionis. Astron. Astrophys. **409**, 275–286 (2003). doi:10.1051/0004-6361:20031086

N.J. Nelson, B.P. Brown, A.S. Brun, M.S. Miesch, J. Toomre, Magnetic wreaths and cycles in convective dynamos. Astrophys. J. **762**, 73–92 (2013). doi:10.1088/0004-637X/762/73

N.J. Nelson, B.P. Brown, A.S. Brun, M.S. Miesch, J. Toomre, Buoyant magnetic loops generated by global convective action. Sol. Phys. **289**, 441–458 (2014). doi:10.1007/s11207-012-0221-4

B. Nordstrom, K.T. Johansen, Radii and masses for young star AR Aurigae. Astron. Astrophys. **282**, 787–800 (1994)

M.E. Oksala, G.A. Wade, R.H.D. Townsend, S.P. Owocki, O. Kochukhov, C. Neiner, E. Alecian, J. Grunhut (the MiMeS Collaboration), Revisiting the rigidly rotating magnetosphere model for σ Ori E—I. Observations and data analysis. Mon. Not. R. Astron. Soc. **419**, 959–970 (2012). doi:10.1111/j.1365-2966.2011.19753.x

P. Petit, J.-F. Donati, G.A. Wade, J.D. Landstreet, S. Bagnulo, T. Lüftinger, T.A.A. Sigut, S.L.S. Shorlin, S. Strasser, M. Aurière, J.M. Oliveira, Magnetic topology and surface differential rotation on the K1 subgiant of the RS CVn system HR 1099. Mon. Not. R. Astron. Soc. **348**, 1175–1190 (2004a). doi:10.1111/j.1365-2966.2004.07420.x

P. Petit, J.-F. Donati, J.M. Oliveira, M. Aurière, S. Bagnulo, J.D. Landstreet, F. Lignières, T. Lüftinger, S. Marsden, D. Mouillet, F. Paletou, S. Strasser, N. Toqué, G.A. Wade, Photospheric magnetic field and surface differential rotation of the FK Com star HD 199178. Mon. Not. R. Astron. Soc. **351**, 826–844 (2004b). doi:10.1111/j.1365-2966.2004.07827.x

P. Petit, B. Dintrans, S.K. Solanki, J.-F. Donati, M. Aurière, F. Lignières, J. Morin, F. Paletou, J. Ramirez Velez, C. Catala, R. Fares, Toroidal versus poloidal magnetic fields in Sun-like stars: a rotation threshold. Mon. Not. R. Astron. Soc. **388**, 80–88 (2008). doi:10.1111/j.1365-2966.2008.13411.x

P. Petit, B. Dintrans, A. Morgenthaler, V. Van Grootel, J. Morin, J. Lanoux, M. Aurière, R. Konstantinova-Antova, A polarity reversal in the large-scale magnetic field of the rapidly rotating sun HD 190771. Astron. Astrophys. **508**, 9–12 (2009). doi:10.1051/0004-6361/200913285

V. Petit, S.P. Owocki, G.A. Wade, D.H. Cohen, J.O. Sundqvist, M. Gagné, J. Maíz Apellániz, M.E. Oksala, D.E. Bohlender, T. Rivinius, H.F. Henrichs, E. Alecian, R.H.D. Townsend, A. ud-Doula (MiMeS Collaboration), A magnetic confinement versus rotation classification of massive-star magnetospheres. Mon. Not. R. Astron. Soc. **429**, 398–422 (2013). doi:10.1093/mnras/sts344

A.A. Pevtsov, L. Bertello, A.G. Tlatov, A. Kilcik, Y.A. Nagovitsyn, E.W. Cliver, Cyclic and long-term variation of sunspot magnetic fields. Sol. Phys. **289**, 593–602 (2014). doi:10.1007/s11207-012-0220-5

N. Piskunov, O. Kochukhov, Doppler imaging of stellar magnetic fields. I. Techniques. Astron. Astrophys. **381**, 736–756 (2002). doi:10.1051/0004-6361:20011517

S. Plachinda, N. Pankov, D. Baklanova, General magnetic field of the Sun as a star (GMF): variability of the frequency spectrum from cycle to cycle. Astron. Nachr. **332**, 918 (2011). doi:10.1002/asna.201111591

S.I. Plachinda, T.N. Tarasova, Magnetic field variations with a rotational period on solar-like star ξ Bootis A. Astrophys. J. **533**, 1016–1022 (2000). doi:10.1086/308694

S.I. Plachinda, C.M. Johns-Krull, T.N. Tarasova, Direct measurements of the general magnetic field on the solar-like stars, in *Odessa Astronomical Publications* vol. 14 (2001), p. 219

D. Rabin, A true-field magnetogram in a solar plage region. Astrophys. J. Lett. **390**, 103–106 (1992). doi:10.1086/186382

A. Reiners, Observations of cool-star magnetic fields. Living Rev. Sol. Phys. **9**, 1 (2012). doi:10.12942/lrsp-2012-1

A. Reiners, G. Basri, The first direct measurements of surface magnetic fields on very low mass stars. Astrophys. J. **656**, 1121–1135 (2007). doi:10.1086/510304

A. Reiners, G. Basri, A volume-limited sample of 63 M7-M9.5 dwarfs. II. Activity, magnetism, and the fade of the rotation-dominated dynamo. Astrophys. J. **710**, 924–935 (2010). doi:10.1088/0004-637X/710/2/924

B. Reipurth, J. Bally, C. Aspin, M.S. Connelley, T.R. Geballe, S. Kraus, I. Appenzeller, A. Burgasser, HH 222: a giant herbig-haro flow from the quadruple system V380 Ori. Astron. J. **146**, 118 (2013). doi:10.1088/0004-6256/146/5/118

T. Rivinius, R.H.D. Townsend, O. Kochukhov, S. Stefl, D. Baade, L. Barrera, T. Szeifert, Basic parameters and properties of the rapidly rotating magnetic helium-strong B star HR 7355. Mon. Not. R. Astron. Soc. **429**, 177–188 (2013). doi:10.1093/mnras/sts323

R.D. Robinson, S.P. Worden, J.W. Harvey, Observations of magnetic fields on two late-type dwarf stars. Astrophys. J. Lett. **236**, 155–158 (1980). doi:10.1086/183217

J. Robrade, J.H.M.M. Schmitt, Altair—the "hottest" magnetically active star in X-rays. Astron. Astrophys. **497**, 511–520 (2009). doi:10.1051/0004-6361/200811348

J. Robrade, J.H.M.M. Schmitt, New X-ray observations of IQ Aurigae and α^2 Canum Venaticorum. Probing the magnetically channeled wind shock model in A0p stars. Astron. Astrophys. **531**, 58–68 (2011). doi:10.1051/0004-6361/201116843

M.M. Romanova, G.V. Ustyugova, A.V. Koldoba, J.V. Wick, R.V.E. Lovelace, Three-dimensional simulations of disk accretion to an inclined dipole. I. Magnetospheric flows at different Θ. Astrophys. J. **595**, 1009–1031 (2003). doi:10.1086/377514

C. Rossi, L. Errico, M. Friedjung, F. Giovannelli, G. Muratorio, R. Viotti, A. Vittone, On the nature of the Herbig Be star V 380 Orionis. Astron. Astrophys. Suppl. Ser. **136**, 95–105 (1999). doi:10.1051/aas:1999200

N. Rusomarov, O. Kochukhov, N. Piskunov, S.V. Jeffers, C.M. Johns-Krull, C.U. Keller, V. Makaganiuk, M. Rodenhuis, F. Snik, H.C. Stempels, J.A. Valenti, Three-dimensional magnetic and abundance mapping of the cool Ap star HD 24712 I. Spectropolarimetric observations in all four Stokes parameters. Astron. Astrophys. **558**, 8–23 (2013). doi:10.1051/0004-6361/201220950

S.H. Saar, Improved methods for the measurement and analysis of stellar magnetic fields. Astrophys. J. **324**, 441–465 (1988). doi:10.1086/165907

S.H. Saar, Recent magnetic fields measurements of stellar, in *Stellar Surface Structure*, ed. by K.G. Strassmeier, J.L. Linsky IAU Symposium, vol. 176, 1996, p. 237

S.H. Saar, J.L. Linsky, The photospheric magnetic field of the dM3.5e flare star AD Leonis. Astrophys. J. Lett. **299**, 47–50 (1985). doi:10.1086/184578

R.S. Schneer, H.F. Henrichs, R.D. Oudmaijer, J.H. Telting, On the Halpha emission from the beta Cephei system. Astron. Astrophys. **459**, 21–24 (2006). doi:10.1051/0004-6361:20066392

C.J. Schrijver, Y. Liu, The global solar magnetic field through a full sunspot cycle: observations and model results. Sol. Phys. **252**, 19–31 (2008). doi:10.1007/s11207-008-9240-6

C.J. Schrijver, C. Zwaan, *Solar and Stellar Magnetic Activity* (Cambridge University Press, Cambridge, 2000)

N.S. Schulz, C.R. Canizares, D. Huenemoerder, J.C. Lee, X-ray line emission from the hot stellar wind of θ^1 Ori C. Astrophys. J. **545**, 135–139 (2000). doi:10.1086/317891

M. Semel, Zeeman–Doppler imaging of active stars. I—Basic principles. Astron. Astrophys. **225**, 456–466 (1989)

D. Shulyak, A. Seifahrt, A. Reiners, O. Kochukhov, N. Piskunov, Rotation, magnetism and metallicity of M dwarf systems. Mon. Not. R. Astron. Soc. **418**, 2548–2557 (2011). doi:10.1111/j.1365-2966.2011.19644.x

J. Silvester, O. Kochukhov, G.A. Wade, Stokes IQUV magnetic Doppler imaging of Ap stars—III. Next generation chemical abundance mapping of α^2 CVn. Mon. Not. R. Astron. Soc. **444**, 1442–1452 (2014a). doi:10.1093/mnras/stu1531

J. Silvester, O. Kochukhov, G.A. Wade, Stokes IQUV magnetic Doppler imaging of Ap stars—II. Next generation magnetic Doppler imaging of α^2 CVn. Mon. Not. R. Astron. Soc. **440**, 182–192 (2014b). doi:10.1093/mnras/stu306

J. Silvester, C. Neiner, H.F. Henrichs, G.A. Wade, V. Petit, E. Alecian, A.-L. Huat, C. Martayan, J. Power, O. Thizy, On the incidence of magnetic fields in slowly pulsating B, Beta Cehpei and B-type emission line stars. Mon. Not. R. Astron. Soc. **398**, 1505–1511 (2009). doi:10.1111/j.1365-2966.2009.15208.x

J. Silvester, G.A. Wade, O. Kochukov, S. Bagnulo, C.P. Folsom, D. Hanes, Stokes IQUV magnetic Doppler imaging of Ap stars—I. ESPaDOnS and NARVAL observations. Mon. Not. R. Astron. Soc. **426**, 1003–1030 (2012). doi:10.1111/j.1365-2966.2012.21587.x

M.A. Smith, Photospheric activity in selected Be stars: lambda ERI and gamma CAS, in *Pulsation; Rotation; and Mass Loss in Early-Type Stars*, ed. by L.A. Balona, H.F. Henrichs, J.M. Le Contel IAU Symposium, vol. 162, 1994, p. 241

B. Stelzer, G. Micela, K. Hamaguchi, J.H.M.M. Schmitt, On the origin of the X-ray emission from Herbig Ae/Be stars. Astron. Astrophys. **457**, 223–235 (2006). doi:10.1051/0004-6361:20065006

T. Szeifert, S. Hubrig, M. Schöller, O. Schütz, B. Stelzer, Z. Mikulášek, The nature of the recent extreme outburst of the Herbig Be/FU Orionis binary Z Canis Majoris. Astron. Astrophys. **509**, 7 (2010). doi:10.1051/0004-6361/200913704

N. Tetzlaff, R. Neuhäuser, M.M. Hohle, G. Maciejewski, Identifying birth places of young isolated neutron stars. Mon. Not. R. Astron. Soc. **402**, 2369–2387 (2010). doi:10.1111/j.1365-2966.2009.16093.x

S. Tsvetkova, P. Petit, M. Aurière, R. Konstantinova-Antova, G.A. Wade, C. Charbonnel, T. Decressin, R. Bogdanovski, Magnetic field structure in single late-type giants: β Ceti in 2010–2012. Astron. Astrophys. **556**, 43 (2013). doi:10.1051/0004-6361/201321051

J.A. Valenti, C. Johns-Krull, Magnetic field measurements for cool stars, in *Magnetic Fields Across the Hertzsprung–Russell Diagram*, ed. by G. Mathys, S.K. Solanki, D.T. Wickramasinghe The Astronomical Society of the Pacific Conference Series, vol. 248, 2001, p. 179

J.A. Valenti, G.W. Marcy, G. Basri, Infrared Zeeman analysis of epsilon Eridani. Astrophys. J. **439**, 939–956 (1995). doi:10.1086/175231

A.A. Vidotto, S.G. Gregory, M. Jardine, J.-F. Donati, P. Petit, J. Morin, C.P. Folsom, J. Bouvier, A.C. Cameron, G. Hussain, S. Marsden, I.A. Waite, R. Fares, S. Jeffers, J.D. do Nascimento Jr, Stellar magnetism: empirical trends with age and rotation. Mon. Not. R. Astron. Soc. **441**, 2361–2374 (2014). doi:10.1093/mnras/stu728

B. Viticchié, J. Sánchez Almeida, D. Del Moro, F. Berrilli, Interpretation of HINODE SOT/SP asymmetric Stokes profiles observed in the quiet Sun network and internetwork. Astron. Astrophys. **526**, 60 (2011). doi:10.1051/0004-6361/201015391

S.S. Vogt, G.D. Penrod, Doppler imaging of spotted stars—application to the RS Canum venaticorum star HR 1099. Publ. Astron. Soc. Pac. **95**, 565–576 (1983). doi:10.1086/131208

G.A. Wade, the MiMeS Collaboration, Review: Magnetic fields of O-type stars (2014). arXiv:1411.3604

G.A. Wade, J.H. Grunhut (MiMeS Collaboration), The MiMeS survey of magnetism in massive stars, in *Circumstellar Dynamics at High Resolution*, ed. by A.C. Carciofi, T. Rivinius Astronomical Society of the Pacific Conference Series, vol. 464, 2012, p. 405

G.A. Wade, D. Drouin, S. Bagnulo, J.D. Landstreet, E. Mason, J. Silvester, E. Alecian, T. Böhm, J.-C. Bouret, C. Catala, J.-F. Donati, Discovery of the pre-main sequence progenitors of the magnetic Ap/Bp stars? Astron. Astrophys. **442**, 31–34 (2005). doi:10.1051/0004-6361:200500184

G.A. Wade, S. Bagnulo, D. Drouin, J.D. Landstreet, D. Monin, A search for strong, ordered magnetic fields in Herbig Ae/Be stars. Mon. Not. R. Astron. Soc. **376**, 1145–1161 (2007). doi:10.1111/j.1365-2966.2007.11495.x

G.A. Wade, J. Maíz Apellániz, F. Martins, V. Petit, J. Grunhut, N.R. Walborn, R.H. Barbá, M. Gagné, E. García-Melendo, J. Jose, A.F.J. Moffat, Y. Nazé, C. Neiner, A. Pellerin, M. Penadés Ordaz, M. Shultz, S. Simón-Díaz, A. Sota, NGC 1624-2: a slowly rotating, X-ray luminous Of?cp star with an extraordinarily strong magnetic field. Mon. Not. R. Astron. Soc. **425**, 1278–1293 (2012a). doi:10.1111/j.1365-2966.2012.21523.x

G.A. Wade, J. Grunhut, G. Gräfener, I.D. Howarth, F. Martins, V. Petit, J.S. Vink, S. Bagnulo, C.P. Folsom, Y. Nazé, N.R. Walborn, R.H.D. Townsend, C.J. Evans, The spectral variability and magnetic field characteristics of the Of?p star HD 148937. Mon. Not. R. Astron. Soc. **419**, 2459–2471 (2012b). doi:10.1111/j.1365-2966.2011.19897.x

N.R. Walborn, The space distribution of the O stars in the solar neighborhood. Astron. J. **78**, 1067–1073 (1973). doi:10.1086/111509

G.A.H. Walker, R. Kuschnig, J.M. Matthews, P. Reegen, T. Kallinger, E. Kambe, H. Saio, P. Harmanec, D.B. Guenther, A.F.J. Moffat, S.M. Rucinski, D. Sasselov, W.W. Weiss, D.A. Bohlender, H. Božić, O. Hashimoto, P. Koubský, R. Mann, D. Ruždjak, P. Škoda, M. Šlechta, D. Sudar, M. Wolf, S. Yang, Pulsations of the Oe star ζ ophiuchi from MOST satellite photometry and ground-based spectroscopy. Astrophys. J. Lett. **623**, 145–148 (2005). doi:10.1086/430254

H. Yang, C.M. Johns-Krull, Magnetic field measurements of T Tauri stars in the Orion Nebula Cluster. Astrophys. J. **729**, 83–90 (2011). doi:10.1088/004-637X/729/2/83

H. Yang, C.M. Johns-Krull, J.A. Valenti, Magnetic properties of young stars in the TW Hydrae Association. Astron. J. **136**, 2286–2294 (2008). doi:10.1088/0004-6256/136/6/2286

DOI 10.1007/978-1-4939-3550-5_4
Reprinted from *Space Science Reviews* Journal, DOI 10.1007/s11214-015-0138-y

Magnetic Field Generation in Stars

Lilia Ferrario[1] · Andrew Melatos[2] · Jonathan Zrake[3]

Received: 5 November 2014 / Accepted: 11 February 2015 / Published online: 4 March 2015

Abstract Enormous progress has been made on observing stellar magnetism in stars from the main sequence (particularly thanks to the MiMeS, MAGORI and BOB surveys) through to compact objects. Recent data have thrown into sharper relief the vexed question of the origin of stellar magnetic fields, which remains one of the main unanswered questions in astrophysics. In this chapter we review recent work in this area of research. In particular, we look at the fossil field hypothesis which links magnetism in compact stars to magnetism in main sequence and pre-main sequence stars and we consider why its feasibility has now been questioned particularly in the context of highly magnetic white dwarfs. We also review the fossil versus dynamo debate in the context of neutron stars and the roles played by key physical processes such as buoyancy, helicity, and superfluid turbulence, in the generation and stability of neutron star fields.

Independent information on the internal magnetic field of neutron stars will come from future gravitational wave detections. Coherent searches for the Crab pulsar with the Laser Interferometer Gravitational Wave Observatory (LIGO) have already constrained its gravitational wave luminosity to be $\lesssim 2\ \%$ of the observed spin-down luminosity, thus placing a limit of $\lesssim 10^{16}$ G on the internal field. Indirect spin-down limits inferred from recycled pulsars also yield interesting gravitational-wave-related constraints. Thus we may be at the dawn of a new era of exciting discoveries in compact star magnetism driven by the opening of a new, non-electromagnetic observational window.

We also review recent advances in the theory and computation of magnetohydrodynamic turbulence as it applies to stellar magnetism and dynamo theory. These advances offer in-

✉ L. Ferrario
Lilia.Ferrario@anu.edu.au

A. Melatos
amelatos@unimelb.edu.au

J. Zrake
zrake@stanford.edu

1 Mathematical Sciences Institute, The Australian National University, ACT 2601, Canberra, Australia

2 School of Physics, University of Melbourne, Melbourne, Australia

3 Kavli Institute of Particle Astrophysics and Cosmology, Stanford University, Stanford, USA

sight into the action of stellar dynamos as well as processes which control the diffusive magnetic flux transport in stars.

Keywords Magnetic fields · Main sequence stars · White dwarfs · Neutron stars

1 Introduction

It was Hale (1908), one of the greatest astrophysicist of the twentieth century, who built the first spectroheliograph and used it to establish that sunspots are magnetic and grouped in pairs of opposite polarities. Further pioneering work on solar magnetism conducted by Hale and collaborators in (1919) revealed an East-West direction of polarity in the sunspots' magnetic fields exhibiting a mirror symmetry with respect to the solar equator. Such polarity was observed to undergo inversion according to the 11 years solar cycle. This is commonly known as the "Hale-Nicholson law".

The first detection of a magnetic field in a star other than our own Sun, 78 Vir, was obtained in (1947) by Babcock. In (1958), Babcock published the first catalogue of magnetic stars and in 1960 he discovered what is still today the most magnetic main sequence star known, HD 215441 ($\sim 3.4 \times 10^4$ G). This became known as "Babcock's star". Surface fields discovered in more recent years that rival in strength that of Babcock's star are those of HD 154708 ($\sim 2.45 \times 10^4$ G; Hubrig et al. 2005), HD 137509 ($\sim 2.9 \times 10^4$ G; Mathys 1995; Kochukhov 2006) and HD 75049 ($\sim 3 \times 10^4$ G; Freyhammer et al. 2008; Elkin et al. 2010).

Following the detections of strong fields in main sequence stars, Blackett (1947) suggested that if magnetic flux is conserved, some white dwarfs should exhibit magnetic fields of up to $10^7 - 10^8$ G. However, spectroscopic surveys aimed at detecting magnetic fields in white dwarfs yielded negative results (Preston 1970). Kemp (1970) argued that electrons spiralling in a magnetic field would emit linearly and circularly polarised radiation that should be detectable in the continuum of strongly magnetic white dwarfs. A spectropolarimetric survey of white dwarfs led to the discovery of strong circular polarisation in the continuum of the white dwarf Grw $+ 70°8247$ (Kemp et al. 1970).

Baade and Zwicky (1934) first suggested that some stars could be made up of neutrons and that a supernova could be the result of a rapid transition of a normal star into a neutron star. Baade (1942) and Minkowski (1942) found unusual emission arising from the central parts of the Crab Nebula. Later on radio pulsations in this nebula were discovered by Staelin and Reifenstein (1968). Such radio emission had been predicted by Shklovsky (1953) as caused by relativistic electrons spiralling around magnetic field lines.

Woltjer (1964) first proposed that under magnetic flux conservation, if a star contracts to the density expected in a neutron star, then the surface field strength could be amplified to values of up to 10^{14}–10^{16} G. The first pulsar, a highly magnetised rapidly spinning neutron star (Pacini 1968), was discovered in 1967 by Jocelyn Bell (Hewish et al. 1968).

In this chapter we review progress made on the origin of magnetic fields in stars. The origin of magnetic fields is still a major unresolved problem in astrophysics. The "fossil field" hypothesis is often invoked to link magnetism in compact stars to magnetism on the main sequence (see Sects. 3.1 and 4.1). In the fossil scenario, some fraction of the magnetic flux of the progenitor star is conserved during the collapse process, because the stellar plasma is highly conducting and hence, by Faraday's law, the magnetic field is 'frozen in' to the fluid (Woltjer 1964; Ruderman 1972; Braithwaite and Spruit 2004a; Ferrario and Wickramasinghe 2006a, 2007, 2008a). Under these circumstances, the field strength B scales with radius R of the star as $B \propto R^{-2}$. A main sequence B star with radius 5 $R_{\odot}$ and field $\sim 1{,}000$ G compresses to give a neutron star with a field of $\sim 10^{14}$ G; a main sequence A star with radius 1.6 $R_{\odot}$ and field $\sim 1{,}000$ G compresses to give a magnetic white dwarf with a field of $\sim 10^{7}$ G. The fossil field scenario is indeed quite attractive and can explain the existence of magnetic fields even in the strongest magnets in the universe, the so-called "magnetars". However, the feasibility of the fossil field hypothesis does have its problems and has been recently questioned, in the context of highly magnetic white dwarfs, on the basis of recent observational results related to their binarity (see Sect. 3.2). The fossil versus dynamo-generated fields debate for neutron stars is analysed in Sect. 4.1.1. Alternative explanations for the origin of fields in stars are presented in Sects. 2, 3.2 and 4.2.

In this chapter we also review recent progress in magnetohydrodynamics (MHD) turbulence theory and computation as it applies to stellar magnetism (see Sect. 5), which underpins the operation of stellar dynamos, and controls diffusive magnetic flux transport in stars.

2 Magnetism in Non-degenerate Stars

Magnetic fields in main sequence stars have been measured mainly via spectropolarimetry. Different techniques measure different field properties, e.g., line-of-sight component, volume-averaged field strength, or dipole moment. A thorough review on magnetic field measurements of non-degenerate stars across the Hertzsprung–Russell diagram can be found in this book in the chapter by Linsky and Schöller.

Direct measurements of magnetic fields in the chemically peculiar main sequence Ap and Bp stars, which form about 10 % of stars in the 1.6–5 $M_{\odot}$ range, have revealed the existence of strong ($\sim 3 \times 10^{2}$–3×10^{4} G) large scale fields (e.g. Aurière et al. 2007; Donati and Landstreet 2009). Landstreet et al. (2007a) found that the majority of magnetic objects have an average longitudinal field (line of sight component of the surface field) $B > 0.25$ kG in every 1 $M_{\odot}$ mass bin between 2 $M_{\odot}$ and 9 $M_{\odot}$ (Bagnulo et al. 2006a; Landstreet et al. 2007a). The lack of fields below 300 G, value that corresponds to the strength at which the magnetic field is in equipartition with the gas pressure in the stellar photosphere, is not a detection threshold effect since Aurière et al. (2010) have set a 3σ upper limit of longitudinal fields down to 1–10 G depending on stellar brightness. The absence of magnetic stars below this cutoff has been referred to as "the Ap/Bp magnetic desert". This indicates that Ap and Bp stars are not simply the high field tail of a continuous field distribution. Landstreet et al. (2008) also find that the field strengths seem to show some decline with stellar age but the field incidence does not. Such large scale fields are observed throughout the main sequence phase and, in more recent years, have also been observed in a small number of stars on the red giant branch (Aurière et al. 2008).

The recent detections of sub-gauss fields in Vega (Lignières et al. 2009), and possibly in the Am stars Sirius A (Petit et al. 2011) and β UMa and θ Leo (Blazère et al. 2014) have unveiled a new class of magnetic stars which is at the 1 gauss end of the magnetic desert, thus potentially suggesting a dichotomy between strong and ultra-weak magnetic fields among intermediate-mass stars (Lignières et al. 2014). This intriguing observational result has prompted Braithwaite and Cantiello (2013) to investigate the origin of these low fields. They argued that such fields could be the remnants of fields already present or formed during or immediately after the star formation stage. Hence, these fields would still be evolving on a timescale that is comparable to the age of the star. According to these studies, all intermediate and high mass stars lacking strong fields should display sub-gauss field strengths that would slowly decline over their main sequence lifetime. There has been some recent effort to test this prediction by Wade et al. (2014a), Neiner et al. (2014b) and Neiner et al. (2014a) but without any clear conclusion.

Alecian et al. (2008) conducted a spectropolarimetric study of Herbig Ae/Be (HAeBe) objects (Herbig 1960). HAeBe stars are pre-main sequence stars of 2–15 $M_{\odot}$ which are still embedded in their protostellar gas-dust envelope and exhibit emission lines in spectra of type A/B. The observations of Alecian et al. (2008), Alecian et al. (2009), Hubrig et al. (2009) and more recently Alecian et al. (2013a) and Hubrig et al. (2013) have revealed that about 7 % of HAeBe objects are magnetic. These studies have also found that they display large scale dipolar fields of strength comparable to that of Ap and Bp stars, under the assumption of conservation of magnetic flux. Interestingly, although some magnetic HAeBe stars are in the totally radiative evolutionary phase while others have already developed convective cores, they seem to share the same magnetic field strength and structure thus indicating that the convective core is not responsible for generation or destruction of fields (Alecian et al. 2013b).

The Magnetism in Massive Stars (MiMeS, Wade et al. 2011) project, the MAGORI (Hubrig et al. 2011; Schöller et al. 2011) project and the B fields in OB star (BOB) Collaboration (Hubrig et al. 2014) have conducted large surveys of bright and massive Galactic stars of spectral types B and O. It is now clear that about 10 % of all stars with radiative envelopes in the mass range 1.5–50 $M_{\odot}$ possess large scale mostly dipolar magnetic fields (Grunhut and Wade 2013). Even more interestingly, the MiMes project has revealed that all of the basic field characteristics do not vary significantly from the coolest spectral types F0 ($\sim 1.5\ M_{\odot}$) to the hottest spectral types O4 ($\sim 50\ M_{\odot}$) stars thus supporting a common formative scenario over a very large range of stellar masses (Wade et al. 2014b). Furthermore, the upper dipolar field limits placed on the undetected magnetic O stars sample studied by the MiMeS collaboration are 40 G at 50 % confidence, and 130 G at 80 % confidence (Wade and the MiMeS Collaboration 2014). This result seems to indicate that the field distribution of massive stars may also have a magnetic desert, similar to that observed in intermediate-mass stars. However, the recent work of Fossati et al. (2014), who detected weak longitudinal fields of $\lesssim 30$ G in the early B-type main sequence stars β CMa and ϵ CMa, appears to support a more continuous distribution of fields in massive stars. They also claim that weak fields in massive stars could be more widespread than currently observed because of the numerous observational biases associated with the detection of weak fields in massive stars using current instrumentation and techniques.

The vexed question on the origin of fields in main sequence stars is still unanswered. The two main hypothesis are the fossil field, according to which magnetism in stars is a relic of the interstellar field from which the star was born (e.g. Borra et al. 1982; Moss 2001), and dynamo action taking place in the rotating cores of main sequence stars. Both theories are unable to explain why only a small subset of main sequence stars is magnetic, although the fossil field theory does assume that differences such as initial density and magnetic field strength in interstellar clouds could be responsible for this. However, it is peculiar that observations have never revealed the presence of magnetism in both components of a binary system. The BinaMIcS (Binarity and Magnetic Interactions in various classes of Stars) project (Alecian et al. 2014; Neiner and Alecian 2013) is a collaborative endeavour that has been recently setup to investigate the phenomenon of magnetism in close (orbital periods shorter than 20 d) binary systems. The study of the magnetic properties of these systems, at a detection limit of 150 G, will give us crucial information on the origin of fields in magnetic main sequence stars which are well known to be rare in binaries (Carrier et al. 2002). Since the components of binaries share the same history, the study of the two stars will help us unravel the importance of their physical parameters and their studies should enable us to distinguish effects caused by initial conditions at formation from those caused later on by evolutionary processes. Preliminary studies of 314 massive stars in close binaries have revealed the presence of fields in 6 systems yielding an incidence of magnetism of $\lesssim 2\ \%$ which is up to 5 times lower than what is observed in single stars. This confirms the results of Carrier et al. (2002) which were based on a smaller sample of A-type stars.

In order to explain the rarity of short period binaries containing magnetic main sequence stars and also the fact that magnetism is only present in a small percentage of single main sequence stars, Ferrario et al. (2009) proposed that magnetic fields in main sequence stars could form when two protostellar objects merge late on their approach to the main sequence and when at least one of them has already acquired a radiative envelope. The N-body simulations of pre-main sequence evolution of Railton et al. (2014) have indicated that in high initial cluster densities the number of collisions between stars is twice as high as that of stars already on the main sequence. Furthermore, they find that massive stars generally form through the merging of lower mass stars. Thus, the expectation is that the incidence of magnetism should increase with mass, which seems to be demonstrated by the observations of Power et al. (2007) for a volume limited sample of A and B type stars with $M \leq 4\ M_{\odot}$. Observations also seem to be in general agreement with the predictions of Ferrario et al. (2009) that the Ap and Bp stars should have pre-main sequence progenitors with similar magnetic field flux and structure.

Another merger scenario has been proposed by Tutukov and Fedorova (2010). They propose that the coalescence of the two components of a close binary system with masses in the range 0.7–1.5 $M_{\odot}$, which are expected to have convective envelopes and strong magnetic fields, would be responsible for the formation of Ap and Bp stars with $M \lesssim 3\ M_{\odot}$. In this picture, the magnetic Ap and Bp stars with radiative envelopes owe their strong magnetic fields to progenitors with convective envelopes. Tutukov and Fedorova (2010) speculate that systems such as the W UMa-type contact binaries could represent the precursory phase of these merging events.

However all of these predictions are so far only of a very qualitative nature and need to be supported by detailed quantitative calculations in the future.

3 Magnetic White Dwarfs

Magnetism in white dwarfs is revealed either through Zeeman splitting and circular polarisation of spectral lines or, at very high field strengths, as continuum polarisation. The magnetic white dwarfs, which represent about 8–13 % of the total white dwarf population (Liebert et al. 2003; Kawka et al. 2007) and exhibit polar field strengths of $\sim 10^3$–10^9 G, have been discovered mainly in optical sky surveys. At spectral resolutions of about 10 Å, the Zeeman triplet in the lower members of the Balmer and Lyman series can be easily recognised at fields of $\sim 10^6$–10^7 G when the splitting in intensity spectra is smaller than other broadening factors, such as pressure broadening. At higher spectral resolutions, it is possible to detect fields down to $\sim 10^5$ G. However, circular spectropolarimetry is a much more sensitive observational technique that is used to measure even lower ($< 10^5$ G) fields, since it can independently measure the σ^+ and σ^- oppositely polarised components of the Zeeman split absorption features. On the other hand, due to the faintness of white dwarfs, fields below $\sim 3 \times 10^4$ G have only been recently detected. Spectropolarimetric observations on the ESO VLT by Aznar Cuadrado et al. (2004), Jordan et al. (2007) revealed fields down to $\sim 10^3$ G in about 10 % of the surveyed white dwarfs. Landstreet et al. (2012) have confirmed this finding by conducting a survey of objects randomly drawn from a list of nearby cool ($T_{\rm eff} \lesssim 14{,}000$ K) WDs.

Observations seem to indicate that there is a paucity of white dwarfs with fields in the intermediate field range 10^5–10^6 G (Kawka and Vennes 2012), reminiscent of the magnetic desert of Ap and Bp stars. However, this finding has not been fully corroborated by observations and future surveys may find objects also in this field range. In this context we need to stress that while magnetic fields in bright main sequence and pre main sequence stars have been mainly found via circular polarisation observations, only a small percentage of magnetic white dwarfs has been discovered using this method. This non-systematic approach has created, over the past 30 years, a sample of stars whose properties are difficult to investigate since the observational biases are not fully understood and thus are difficult to remove.

Another interesting characteristic of strongly magnetic white dwarfs is that their mean mass (0.85 ± 0.04 $M_\odot$) is higher than that of non-magnetic or weakly magnetic white dwarfs (0.593 ± 0.002 $M_\odot$, see the chapter on magnetic white dwarfs in this book). This indicate that their progenitors are more massive than those of ordinary white dwarfs. Interestingly, the observations of Vennes (1999) of the EUVE/Soft X-ray selected ultra-massive white dwarfs ($M > 1.2$ $M_\odot$) have found that ~ 20 % of them are strongly magnetic. Thus it appears that the incidence of magnetism in the high field group increases with white dwarf mass and hence with progenitor mass, unless they result from a merger (e.g. EUVE J0317-853 Barstow et al. 1995; Ferrario et al. 1997).

3.1 Origin of Fields in Highly Magnetic White Dwarfs: The Fossil Hypothesis

According to the "strong" fossil field hypothesis, magnetic stars originate in gas clouds whose fields are at the upper end of the field distribution observed in the interstellar medium (10^{-6} and 10^{-4} G, Heiles 1997). One of the very first articles on star formation and flux freezing is that of Mestel (1966). Briefly, in the simple model proposed by Tout et al. (2004), a star of mass M that collapses from the interstellar medium entraps a magnetic flux $\Phi \propto B_{\rm ISM} M^{2/3}$ Mx where $B_{\rm ISM}$ is the interstellar (primordial) field. The variations expected in the interstellar magnetic field will determine the distribution of magnetic fluxes in the protostars. The magnetic flux lines would freeze in the radiative upper layers of the emerging star that will then evolve towards the main sequence. Assuming perfect magnetic flux conservation, a contraction by a 10^7 factor could in principle create fields of the order of 10^8 G in a main sequence star. However during the cloud's collapse most of the initial magnetic flux would be lost via ohmic dissipation or ambipolar diffusion, so that much lower fields than this upper limit can be realised. The survival of significant large-scale magnetic flux through the pre-main sequence evolution has been addressed by Moss (2003).

According to the "weak" fossil field hypothesis, stellar fields could just be the remnants of those generated by the dynamos of active pre-main sequence stars or by late mergers of protostars with at least one of the two having a radiative envelope (see Sect. 2 and Ferrario et al. 2009). The magnetic field flux would then be conserved during the subsequent evolution from main sequence to the white dwarf stage.

Under the fossil field hypothesis, the magnetic Ap and Bp stars in the mass range 1.5–8 $M_\odot$ would be the progenitors of the highly magnetic white dwarfs ($B \geq 10^6$ G). The paucity of white dwarfs with fields in the range 10^5–10^6 G could be interpreted as being related to the lower field cut off of Ap and Bp stars.

The main issue is whether magnetic fields can survive complex phases of evolution when a star develops convective and radiative zones that contract and expand in size with time.

Tout et al. (2004) point out that a dynamo-generated field in the convective regions of a star is transient and has no large-scale component. A star can conserve its primordial fossil field as long as this star possesses a radiative zone throughout its life, because the field would be pumped out of any developing convective region into an adjacent radiative region. The conclusions of Tout et al. (2004) is that a poloidal fossil field that is present in a main sequence star can appear in the compact star phase with a similar field structure and with its magnetic flux nearly conserved.

There have been many semi-analytical and numerical calculations that have addressed the issues concerning the evolution of a magnetic field during the main sequence phase. Braithwaite and Spruit (2004b) and Braithwaite and Nordlund (2006a) have shown that stable magnetic fields with roughly equal poloidal and toroidal field strengths can exist in the radiative interior of a star and their exterior appearance would be that of a dipole with minor contributions from higher multipoles (see also Mestel and Moss 2010). Interestingly, they find that Ohmic dissipation diffuses the field outward over time so that the field strength at the surface of the star increases while the field structure in the stellar interior would switch from being mainly toroidal to poloidal. Hence they predict an increase in the surface field strength with the age of the star. In order to test these theoretical findings, it is necessary to study a sample of Ap and Bp stars with known age and fraction of the main sequence evolution completed. However, it is quite difficult to estimate age-related quantities for magnetic stars (see Bagnulo et al. 2006b). The evolution of magnetic fields from the zero-age main sequence (ZAMS) to the terminal-age main sequence (TAMS) has been investigated observationally by Bagnulo et al. (2006b) and Landstreet et al. (2007b) through the study of Ap stars in clusters, since the age of a cluster can generally be established to better than about ± 0.2 dex. They find that magnetic fields exist at the ZAMS phase for stellar masses 2–5 $M_\odot$ when stars have fractional ages (that is, fraction of the main sequence evolution completed) below about 0.05 and for fractional ages of less than about $\lesssim 0.10$ for masses up to 9 $M_\odot$. This is consistent with the existence of fields in the late pre-main sequence stars Herbig AeBe (see Sect. 2). However, the study of Landstreet et al. (2007b) also reveals that the evolutionary time of magnetic fields is less straightforward to interpret. For stars with $M \gtrsim 3\ M_\odot$ they find that the field strength decreases on a timescale of about $2\text{–}3 \times 10^7$ yr, in agreement with Kochukhov (2006). However it is not clear whether the magnetic flux really decreases or whether it is the conservation of magnetic flux that causes the surface field to decrease as the star expands. Interestingly, for stars with $M \lesssim 3\ M_\odot$ they find no conclusive evidence of field strength or magnetic flux reduction on a time scale of a few 10^8 yr.

Support for the fossil field hypothesis also comes from spectropolarimetric studies of fifty red giants by Auriere et al. (2013). This work has revealed the existence of four magnetic giants that have been identified as probable descendants of Ap/Bp stars.

Population synthesis calculations carried out by Wickramasinghe and Ferrario (2005) have shown that starting with a distribution of magnetic fields on the main sequence, as observed for the Ap and Bp stars and under the assumption of magnetic flux conservation, the predicted magnetic field and mass distribution in white dwarfs are in general agreement with observations, provided that a modified initial to final mass relation is employed for magnetic white dwarfs (their "scenario A"). That is, in order to fit the mass distribution, one needs to assume that mass loss mechanisms are partly inhibited by the presence of strong fields. However, such models can at most yield an incidence of magnetism of ~ 5 % as compared to the observed $\sim 8\text{–}12$ % (see also Kawka et al. 2007). Wickramasinghe and Ferrario (2005) have shown that it is possible to achieve a better agreement with observations if one assumes that in addition to the actually observed fields in Ap and Bp stars, about 40 % of stars more massive than 4.5 $M_\odot$ (early B-type stars) have fields of 10–100 G and evolve

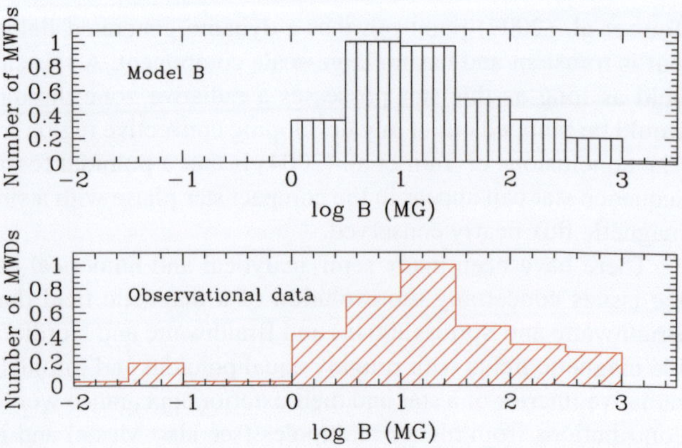

Fig. 1 Population synthesis calculations of Wickramasinghe and Ferrario (2005) (their scenario B) for the field distributions of magnetic white dwarfs based on the fossil field hypothesis compared with observations

into high fields magnetic white dwarfs (see Fig. 1). This is their "scenario B" which would only be viable if magnetic B-type stars do not have a "magnetic desert", which seems to be, albeit not conclusively, supported by the work of Fossati et al. (2014).

Larger and more sensitive spectropolarimetric surveys of white dwarfs may confirm the classification of white dwarfs within either the very low ($\lesssim 10^4$ G) or the very high field group ($\gtrsim 10^6$ G), hence forming a bimodal field distribution. This would point to two different channels for field formation. It is interesting to note that the weak magnetic fields which have recently been detected in Vega and Sirius (Lignières et al. 2009; Petit et al. 2011), would scale up to a surface field of a few $\sim 10^4$ G at the white dwarf stage under magnetic flux conservation. Such weakly magnetised white dwarfs would belong to the low field component of the bimodal white dwarf field distribution.

3.2 Origin of Fields in Highly Magnetic White Dwarfs: The Stellar Merger Hypothesis

We have plotted in Fig. 2 the ratio of poloidal magnetic flux to mass of the highest field magnetic white dwarfs and of the most magnetic main sequence stars. This diagram shows that the two groups of stars share similar characteristics indicating that their magnetic fields may have a common origin. However, the possible evolutionary link between the two populations, and thus the viability of the fossil field hypothesis, has been questioned by Tout et al. (2008) and Wickramasinghe et al. (2014). The main argument hinges on the fact, first highlighted by Liebert et al. (2005), that there is not a single example of a high field magnetic white dwarf ($B \gtrsim$ a few 10^6 G) with a non-degenerate companion star (generally an M dwarf, but see Ferrario 2012) in a non-interacting binary. Further searches on a much larger sample of objects conducted by Liebert et al. (2015) have confirmed the hypothesis that magnetic field in white dwarfs and binarity (with M or K dwarfs) are independent at a 9 σ level. However, we cannot invoke some as yet unknown physical mechanism that could inhibit the formation of a strong magnetic field in a white dwarf when a companion star is present, because such a mechanism would also prevent the formation of Magnetic Cataclysmic Variables (MCVs) which are interacting binaries consisting of a magnetic white dwarfs with a low-mass red dwarf companion. This curious lack of detached white dwarf—non-degenerate star systems indicates that there are no known progenitors of MCVs (for more details see also the chapter on magnetic white dwarfs in this book).

An merger scenario for the generation of fields has also been proposed by Nordhaus et al. (2011). Here a low-mass star would be tidally disrupted by its proto-white dwarf compan-

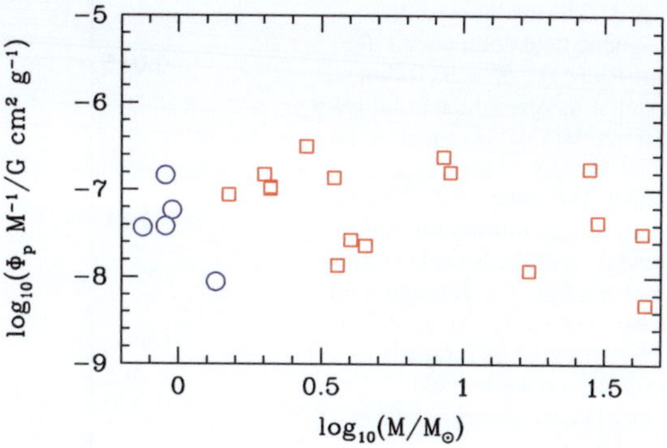

Fig. 2 The ratio of magnetic flux to mass ϕ_P/M for the most magnetic main sequence stars (*squares*) and white dwarfs (*circles*) (Wickramasinghe et al. 2014)

ion during common envelope evolution to form an accretion disc. In this disc seed fields would be amplified through turbulence and shear and then advected on to the object that will become an isolated highly magnetic white dwarf. García-Berro et al. (2012) conducted three-dimensional hydrodynamic simulations of merging double degenerates and argued that a hot and differentially rotating convective corona would form around the more massive star. Their population synthesis calculations of double degenerate mergers are in general agreement with observations. Similar population synthesis calculations conducted for a wide range of merging astrophysical objects have also been carried out by Bogomazov and Tutukov (2009).

Following the work of Tout et al. (2008), Wickramasinghe et al. (2014) proposed that the fields are generated by an Ω dynamo within the common envelope of a binary system where a weak seed poloidal field would wind up by differential rotation. The closer the cores of the two stars are drawn before the envelope is ejected, the stronger the final field of the star emerging from common envelope will be. Therefore the strongest fields occur when the two stars merge, forming an isolated highly magnetic white dwarf. If the two stars do not coalesce but emerge from the common envelope when they are about to transfer mass, they become the MCVs (Tout et al. 2008).

Observations indicate that highly magnetic stars are typified by (see Fig. 2)

$$10^{-8.5} < \frac{\Phi_P/M}{\mathrm{G\,cm^2\,g^{-1}}} < 10^{-6.5}, \tag{1}$$

where $\Phi_p = R^2 B_p$ is poloidal magnetic flux and M and R are the total mass and radius of the star respectively.

If the dynamo mechanism generates a magnetic field from differential rotation $\Delta\Omega$, then we have

$$0 \le \Delta\Omega \le \Omega_{\mathrm{crit}} = \frac{1}{\tau_{\mathrm{dyn}}} = \sqrt{\frac{GM}{R^3}}. \tag{2}$$

Toroidal and poloidal fields are unstable on their own (Braithwaite 2009). The process that gives rise to the decay of toroidal field leads to the generation of poloidal field with one component stabilising the other and thus limiting field growth until a stable configuration is reached. The results of Fig. 2 have been interpreted by Wickramasinghe et al. (2014) as follows. If E_p, E_ϕ and $E = E_p + E_\phi$ are the poloidal, toroidal and total magnetic energies respectively and $\eta = E/U$, $\eta_p = E_p/U$ and $\eta_\phi = E_\phi/U$ are, respectively, the ratios of poloidal and toroidal magnetic to thermal energy U, then by scaling to the observed maximum poloidal flux, Wickramasinghe et al. (2014) find that

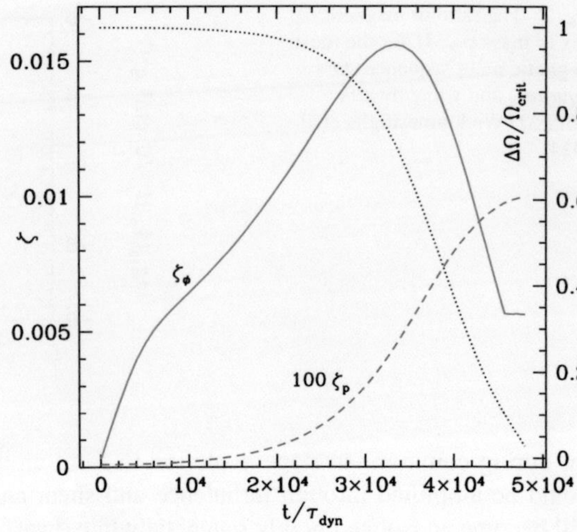

Fig. 3 The evolution of the magnetic field components B_p and B_ϕ ($\zeta = \sqrt{\eta} \approx B$) following a stellar merger which produces a differential rotation equal to the break up spin of the merged object. The decay of differential rotation $\Delta\Omega$ follows the *right hand axis* while the fields follow the *left hand axis*. Toroidal field decays unless $\eta_p > a\eta^2$ and this determines the final ratio of toroidal to poloidal field (Wickramasinghe et al. 2014)

$$\eta_p = \frac{10^{-8}}{\lambda}\left(\frac{\hat{\Phi}_p}{10^{-6.5}\ \mathrm{G\,cm^2\,g^{-1}}}\right)^2, \tag{3}$$

where $\hat{\Phi}_p$ is the ratio of magnetic flux to stellar mass. The observational data in Fig. 2 show that the maximum η_p is independent of the mass and type of star. In non-rotating stars, a stable poloidal-toroidal configuration must satisfy (Braithwaite 2009)

$$a\eta^2 < \eta_p < 0.8\eta, \tag{4}$$

where $a \approx 1$ is a buoyancy factor for main sequence stars. The left hand side inequality is caused by the stabilising effect of a poloidal field on the Taylor instability in purely toroidal fields. A lower limit to the poloidal field is determined by the relative importance of magnetic to gravitational–thermal energy through buoyancy effects. The upper limit is due to the stability of poloidal fields which requires they be not significantly larger than the toroidal field. Braithwaite (2009) argued that the same inequalities are also likely to hold for stable fields in rotating stars.

Figure 3, where $\zeta = \sqrt{\eta} \propto B$, shows how B_p and B_ϕ evolve as $\Delta\Omega$ decreases from its maximum at $\Delta\Omega = \Omega_{\mathrm{crit}}$. Their single model parameter is chosen to give the observed maximum $\eta_p \approx 10^{-8}$, so that $\zeta_p \approx 10^{-4}$. Toroidal field is initially generated by the winding up of the seed poloidal field through differential rotation. As soon as the toroidal field is large enough, the poloidal field starts to grow. As long as the differential rotation is sufficiently large, the toroidal field continues to grow up to a maximum value. Then it decreases until it reaches equilibrium with the poloidal field and $a\eta^2 = \eta_p$. The magnetic torque extinguishes $\Delta\Omega$ to yield a final object that rotates as a solid body. Wickramasinghe et al. (2014) find that the dynamical timescale for a 2 $M_\odot$ star is of the order of 40 min so this evolution is completed in about 3.7 yr which is a lot smaller than the corresponding Kelvin-Helmholtz timescale (2.3×10^6 yr for a 2 $M_\odot$ star). The conclusion of Wickramasinghe et al. (2014) is that the final poloidal field is stable and proportional to the initial quantity of differential rotation, but is independent of the size of small initial seed field and.

More sophisticated α–Ω dynamo models are described in Nordhaus et al. (2007), although these calculations are aimed at understanding the different problem related to the shaping mechanisms in planetary nebulae.

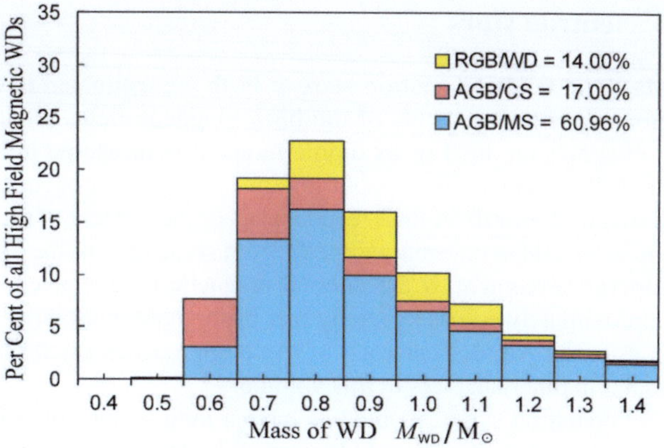

Fig. 4 Theoretical mass distribution of magnetic white dwarfs for $\alpha = 0.1$. The contributions are separated according to their pre-CE progenitors. AGB = Asymptotic Giant Branch; MS = Main Sequence; RGB = Red Giant Branch; CS = Convective Star and WD = White Dwarf. Other paths also contribute but are less important than those shown. The Galactic disc age is chosen to be 9.5 Gyr

In any case, the work of Nordhaus et al. (2011) highlights that for the dynamo mechanism of Wickramasinghe et al. (2014) to be viable, the transport of a strong field generated in the envelope to the proto-white dwarf's surface would require a diffusion coefficient $\geq 10^{21}$–10^{22} $\mathrm{g\,cm^{-3}}$ which is unphysical. Hence Nordhaus et al. (2011) propose an alternative scenario whereby a companion star, of sufficiently low mass to avoid a premature ejection of the envelope, would be tidally destroyed by the gravitational field of the primary star and would form a disc extending all the way to the surface of the proto-white dwarf. Accretion would then transport the strong fields that are formed in the disc towards the proto-white dwarf where they would become anchored to its surface.

The population synthesis calculations carried out by Briggs et al. (2015) have given further support to the stellar merging hypothesis for the origin of fields in the highly magnetic white dwarfs. Briggs et al. (2015) find that the mass distribution and the fraction of stars that merge during a common envelope phase are in good agreement with the observations of magnetic white dwarfs for a wide range of common envelope efficiency parameter α. Figure 4 shows the contributions from the various merging pre-common envelope progenitor pairs for $\alpha = 0.1$. They find a theoretical incidence of magnetism among white dwarfs, for $\alpha = 0.1$–0.3, of about 13–19 %, which is consistent with observational results (e.g. Kawka et al. 2007).

For the sake of completeness, we remark that similar population synthesis calculations, but related to the viability of the core-degenerate scenario for Type Ia supernovae, have been conducted by Ilkov and Soker (2013) (and references therein).

Finally, we would like to bring the attention of the reader to the calculations of Duez et al. (2010a, 2010b), Duez and Mathis (2010) (and references therein) on the stability of non-force-free magnetic equilibria in radiative stellar regions (cores of Sun-like stars, envelopes of intermediate and high mass stars, and compact stars). Through the use of semi-analytic techniques they constructed and tested an axisymmetric non-force-free magnetostatic equilibrium. Their numerical calculations recovered the instabilities which are characteristic to purely poloidal and toroidal magnetic fields and proved computationally, and for the first time, that only a mixed configuration is stable under all types of perturbations. Their important results can be applied to describe the magnetic equilibrium topology in stellar radiative regions (instead of choosing the initial field configuration arbitrarily) and to provide initial conditions for magneto-rotational transport in state-of-the-art stellar evolution codes (e.g., see the models of Maeder and Meynet 2014, for giants and supergiant stars). More general applications of these results to multi-dimensional magnetohydrodynamic computations are also envisaged.

4 Neutron stars

Magnetic fields of neutron stars at birth are estimated to be in the range $\sim 10^{11}$–10^{15} G. However, measurements of the birth magnetic fields of neutron stars are always indirect. Sometimes the field or its dipole moment is measured today, e.g., from neutron star spin-down or Zeeman spectropolarimetry of suitable progenitors, and then extrapolated backwards or forwards in time respectively in the context of an evolutionary model. Otherwise, the birth field is inferred indirectly from some relic of the birth event, e.g., calorimetry of the supernova remnant. If the internal magnetic field evolves slowly, so that its strength today approximately equals its strength at birth, then gravitational wave upper limits also provide constraints. We discuss each of these approaches briefly below. More information can be found in other chapters in this volume.

Population synthesis models have a long history of being used to infer the birth fields of neutron stars (Gunn and Ostriker 1970; Hartman et al. 1997; Faucher-Giguère and Kaspi 2006; Kiel et al. 2008). Radio timing measurements of the spins, fields, and radio luminosities of the current neutron star population are combined with prescriptions for binary evolution, source kinematics in the Galactic gravitational potential, radio emission properties (e.g., beaming), supernova kicks, and observational selection effects to constrain the spin and field distributions at birth. Population synthesis models have benefited recently from the discovery of many new radio pulsars, both isolated and recycled, in large-sale radio multi-beam surveys (Manchester et al. 2001; Morris et al. 2002; Kramer et al. 2003; Faulkner et al. 2004; Hobbs et al. 2004), an updated Galactic electron density map (Cordes and Lazio 2002), and proper motions from very-long-baseline interferometry. The results are that the birth spins are inferred to be normally distributed, with average birth period $\langle P_0 \rangle = 0.30$ s and standard deviation $\sigma_{P_0} = 0.15$ s, while the birth fields B_0 are log-normally distributed, with $\langle \log B_0 \rangle = 12.65$ and $\sigma_{\log B_0} = 0.55$ (B_0 in gauss) (Faucher-Giguère and Kaspi 2006; Kiel et al. 2008). The latter studies find no evidence for magnetic field decay over $\lesssim 0.1$ Gyr, assuming that the radio luminosity scales roughly as the square root of the spin-down luminosity. The no-decay conclusion sits in partial tension with the inference of field decay over $\sim 10^4$ yr in a different sub-population of neutron stars, e.g., the magnetars and the thermal X-ray sources (Pons et al. 2012). Faucher-Giguère and Kaspi (2006) also find no evidence for bimodality in the distributions of P_0 and space velocity.

The energetics of the birth event, which are related to B_0 and P_0, leave their imprint on the supernova remnant. From X-ray measurements of the remnant's radius, temperature, and emission measure, one can infer the total blast energy assuming Sedov expansion (Reynolds and Chevalier 1981; Reynolds 2008). The blast wave is powered by the core-collapse event, which depends weakly on B_0 and P_0, and the relativistic wind emitted by the newly born pulsar, whose luminosity scales $\propto B_0^2 P_0^{-4}$. Drawing on X-ray Multi-Mirror Mission Newton (XMM-Newton) data, Vink and Kuiper (2006) showed that the pulsar wind played an insignificant role powering the blast wave in two anomalous X-ray pulsars (AXPs) and one soft-gamma-ray repeater (SGR), implying $P_0 \gtrsim 5$ ms and hence no significant protoneutron star dynamo in these three objects at least. Steep density gradients in the interstellar medium can modify this conclusion, especially if the gradient correlates with the interstellar magnetic field (Vigelius et al. 2007).

Gravitational wave upper limits and future detections furnish independent information on the *internal* magnetic field of a neutron star. At the time of writing, coherent searches for the Crab pulsar with the Laser Interferometer Gravitational Wave Observatory (LIGO) constrain its gravitational wave luminosity to be $\lesssim 2$ % of the observed spin-down luminosity, thereby limiting the internal field to $B_0 \lesssim 10^{16}$ G (Abbott et al. 2010; Aasi et al. 2014). An analogous result has been obtained for the Vela pulsar (Abadie et al. 2011; Aasi et al. 2014).

The gravitational wave strain scales as the square of the spin frequency, so the best constraints come from rapid rotators. Two classes of object are especially interesting in this regard. First, the large deformation of a newly born, fast-spinning ($P_0 \sim 1$ ms) magnetar caused by a super-strong internal magnetic field ($> 10^{16}$ G) radiates powerful gravitational waves, which should be detectable with Advanced-LIGO-class detectors up to the distance of the Virgo cluster (Stella et al. 2005; Dall'Osso and Stella 2007; Dall'Osso et al. 2009; Mastrano et al. 2011; Melatos and Priymak 2014). A future gravitational-wave detection of a millisecond magnetar would provide a direct measurement of B_0, if the distance is known (e.g., from a counterpart whose redshift is measured electromagnetically). Second, recycled millisecond pulsars have low external magnetic dipole moments, but their internal fields may be much stronger, if the external dipole is reduced by accretion-driven diamagnetic screening (Payne and Melatos 2004), which leaves the birth field in the interior untouched. Combining this scenario with rapid rotation and tight electromagnetic spin-down limits (ellipticity $\epsilon \lesssim 10^{-8}$ in some cases), recycled millisecond pulsars already yield some of the best constraints on the internal B_0 by any method, ruling out $B_0 \gtrsim 10^{13}$ G in some objects (Mastrano and Melatos 2012), depending on the poloidal-toroidal flux ratio and whether the core is superconducting or not.

The above conclusions change, if the internal field evolves significantly in ordinary pulsars (as opposed to magnetars), cf. Faucher-Giguère and Kaspi (2006). Radio pulse evolution over long time-scales may be a signature of magnetic polar wandering or plate tectonics, where the field wanders inside the star (Macy 1974; Ruderman 1991; Melatos 2012). The recent discovery of changes in the flux ratio (0.1 per century) and component separation (0.6 degrees per century) of the Crab's radio pulses (Lyne et al. 2013) is noteworthy in this regard, although other explanations like radiative precession are also possible (Melatos 2000; Barsukov et al. 2013, 2014). Stairs et al. (2000) discovered switching between two discrete magnetospheric states (and hence two pulse shapes and spin-down rates) in PSR B1828-11.

The central, enduring debate regarding the origin of neutron star magnetic fields revolves, as it does in magnetic white dwarfs (see Sect. 3), around whether the field is a fossilised relic of the progenitor's field or is generated afresh by dynamo action in the protoneutron star in the first ~ 10 s after core collapse. Below, we summarise briefly the pros and cons of each scenario, drawing heavily on an excellent review by Spruit (2009). The reader is directed to the latter reference for more discussion.

4.1 Origin of Fields in Neutron Stars: The Fossil Hypothesis

Although the surface magnetic fields in pre-main sequence stars, Ap/Bp, OB stars, magnetic white dwarfs and neutron stars cover a wide range of magnetic field strength values (from a few 10^2 to 10^{15} Gauss), the magnetic fluxes near the upper limit of the observed field ranges are all a few 10^{27} G cm^2. We show in Fig. 5 the magnetic fluxes of the most magnetic compact stars and non-degenerate stars with radiative envelopes. This finding has been used in support of the fossil field hypothesis for the origin of fields.

The fossil field hypothesis for the origin of fields in neutron stars was first proposed by Woltjer (1964). Ruderman (1972) also remarked that the similarity of magnetic fluxes in magnetic white dwarfs and neutron stars could be explained through flux conservation during the evolution from main sequence to compact star stage.

Ferrario and Wickramasinghe (2006b) have investigated the effects of the fossil field hypothesis for the origin of magnetic fields in neutron stars by carrying out population synthesis calculations for different assumptions on the distribution of the magnetic flux of massive

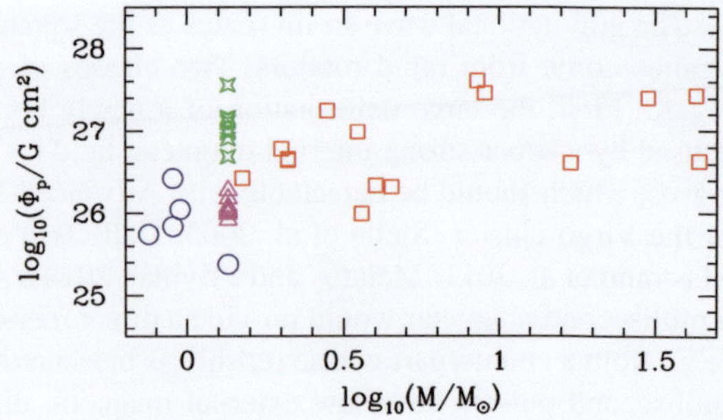

Fig. 5 The magnetic flux ϕ_P for the most magnetic main sequence stars (*squares*), white dwarfs (*circles*), high field radio pulsars (*triangles*) and magnetars (*stars*)

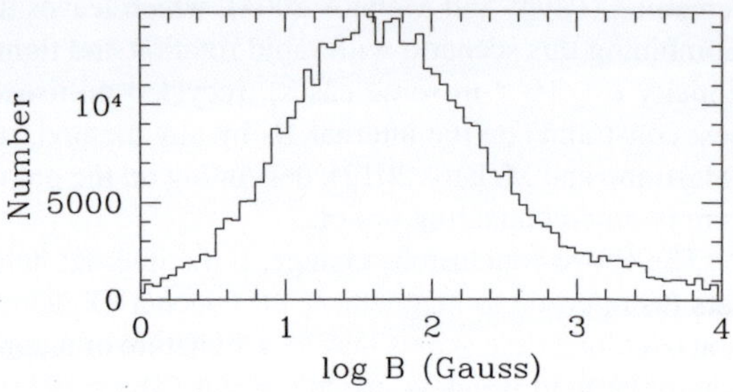

Fig. 6 Predicted magnetic field distribution of massive stars (8–45 $M_\odot$), according to the population synthesis modelling of Ferrario and Wickramasinghe (2006b) under the assumption of fossil fields and magnetic flux conservation

($\geq 8\ M_\odot$) main sequence stars and on the dependence of the initial birth period of neutron stars. They used the observed properties of the population of isolated radio pulsars in the 1374 MHz Parkes Multi-Beam Survey (Manchester et al. 2001) to constrain model parameters. These were then used to deduce the required magnetic properties of their progenitor stars. Their conclusion is that the fossil field hypothesis, which does not allow for magnetic flux loss in the post-main sequence evolution, requires a very specific distribution of magnetic fields for massive main sequence stars as shown in Fig. 6 for an assumed dipolar field structure. In this picture the field distribution in massive stars would be continuous and all massive stars would be magnetic. This distribution is predicted to have a peak near 46 G with low- and high-field wings covering a field range from 1 to 10^4 G. About 8 per cent of main sequence stars would require to have fields in excess of 10^3 G and these would the progenitors of the highest-field neutron stars. This main sequence distribution of B0–O type stars is an observable quantity and thus a prediction of the fossil field model.

Further population synthesis calculations of the observed properties of magnetars were carried out by Ferrario and Wickramasinghe (2008b) under the assumption that magnetars originate from main sequence stars that are much more massive than those giving rise to normal radio pulsars. These studies were prompted by (i) the observation that most magnetars have been linked to progenitors in the mass range ~ 20–45 $M_\odot$ (Gaensler et al. 2005; Muno et al. 2006) and (ii) the discovery of strong fields in massive O-type stars (see Sect. 2). Thus fossil magnetic fluxes similar to those observed in the magnetars may already be present in stellar cores prior to their collapse to neutron star and would explain the presence of fields of up to $\sim 10^{15}$ G in magnetars. According to Ferrario and Wickramasinghe (2008b), the anomalous X-ray emission of magnetars would be caused by the decay of the toroidal field that does not contribute to the spin down of the neutron star. Ferrario and Wickramasinghe (2008b) predict a number of active magnetars (see Fig. 7) that is consistent with the number of sources detected by ROSAT and a Galactic birth rate compatible to that inferred by Kouveliotou et al. (1998).

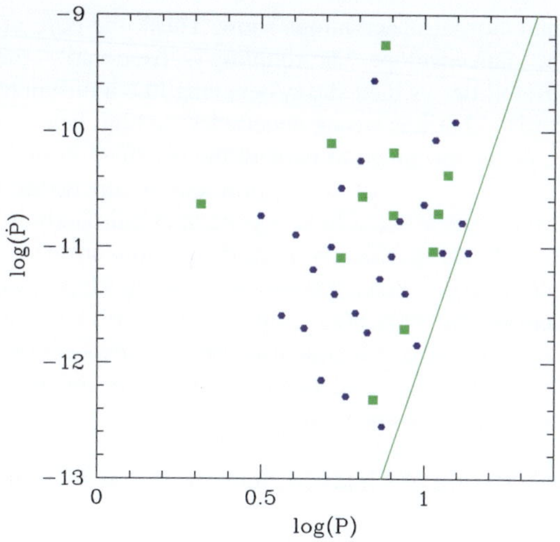

Fig. 7 *Filled squares*: observed magnetars; *filled circles*: magnetars as derived from the modelling of Ferrario and Wickramasinghe (2008b). The *solid line* is an empirically determined boundary (a "magnetar death line") given by $\log(\dot{P}) = 8.4\log(P) - 20$ (Zhang 2003)

Since stars of spectral types B0 to O are the progenitors of neutron stars, it remains to be seen whether there is a sufficiently large range of magnetic fluxes in these stars to explain the entire range of magnetic fluxes in neutron stars. So far, observations of magnetism in massive stars by the MiMeS, MAGORI and BOB collaborations have led to the discovery of a large number of new objects. The paper by Petit et al. (2013) lists the physical, rotational and magnetic properties of about 60 highly magnetic massive OB stars. Most of the detected fields are in the $> 10^3$ G regime with masses $> 8\ M_\odot$ and radii $> 5\ R_\odot$. Under magnetic flux conservation, they would be the progenitors of the highest magnetic field neutron stars ($> 10^{13}$ G).

4.1.1 Problems with the Fossil Hypothesis

Taken at face value, the fossil field arguments seem to be reasonable. Certainly, there is no trouble accounting for the magnetar end of the distribution, since typical surface fields measured in Ap and Bp stars compress to give fields $\gtrsim 10^{13}$ G in neutron stars. If anything, there is a problem at the lower end, where it is hard to produce neutron stars with $B \sim 10^{10}$–10^{11} G without some other accretion-related physics. In fact, a main sequence star with a surface field $B \sim 300$ G and a radius $R \sim 2\ R_\odot$ compresses to give $B \sim 6 \times 10^{12}$ G. There are three problems with the fossil picture. First, the innermost $\sim 1.4\ M_\odot$ of the progenitor, which collapses to form a neutron star, occupies ~ 2 % of the progenitor's cross-subsectional area, reducing the maximum compressed field to $B \sim 10^{14}$ G. One can raise the limit somewhat, if the progenitor's field is centrally concentrated, but the three-dimensional MHD simulations of Braithwaite and Nordlund (2006b) (which assume polytropic, spherically symmetric density and temperature profiles and neglect core convection) do not predict much central concentration. Second, the magnetar birth rate is comparable to the ordinary neutron star birth rate (Keane and Kramer 2008; Woods 2008), yet, as already mentioned in Sect. 4.1, there may be too few known progenitors with $B_0 \gtrsim 10$ kG to create magnetars in the numbers observed. Furthermore, the relatively low incidence of intermediate-strength fields (10–100 G) (Wade and the MiMeS Collaboration 2014) may also represent a problem for the fossil field hypothesis, unless future, more sensitive spectropolarimetric surveys of massive stars discover that intermediate fields (e.g., see Fossati et al. 2014) are indeed much more common than current observations show. Third, the core of the progenitor couples magnetically to the giant envelope. The coupling is 'frictional'—the field lines are disrupted into a tangle by instabilities as they shear, behaving like a turbulent magnetic viscosity rather than 'bungee cords'—but it is strong nonetheless at high fields, scaling as B_0^2. Consequently, the core decelerates too much to explain the distribution of neutron star spin periods observed today. However, note that this is a problem for any neutron star with a magnetised progenitor, even if the neutron star's field is generated ultimately in a dynamo; it is not an argument against fossil fields specifically. Indeed, a protoneutron star dynamo may never take root, if the progenitor core rotates too slowly. Core-envelope coupling has stimulated the suggestion that neutron star spins are imparted by the birth kick during supernova core collapse (Spruit and Phinney 1998). Alignment is therefore expected between the spin axis and proper motion, for which some evidence exists in X-ray images of supernova remnants and pulsar radio polarimetry (Wang et al. 2006).

4.2 Origin of Fields in Neutron Stars: Dynamo Generated Fields

In this model, the seed field is amplified by a dynamo in the proto-neutron star (Ruderman and Sutherland 1973; Duncan and Thompson 1992; Thompson and Duncan 1993; Bonanno et al. 2005). There are two main kinds of protoneutron star dynamos: those driven by convection and those driven by differential rotation. Convection-driven dynamos were first analysed in the neutron star context by Thompson and Duncan (1993). A thermal luminosity L is transported outwards through the star by turbulent convection with characteristic eddy velocity $V \approx (L/4\pi R^2 \rho)^{1/3}$, where ρ is the mean stellar density, and this generates a magnetic field of strength $B_{\rm PNS} \approx (4\pi\rho V^2)^{1/2}$, if the mechanical and magnetic stresses are in equipartition. For a typical photon luminosity, we obtain $B_{\rm PNS} \approx 10^9$ G $(L/10^{38}\ {\rm erg\,s^{-1}})^{1/3}(R/10^8\ {\rm cm})^{7/6}$. For a typical neutrino luminosity, $B_{\rm PNS}$ is $\sim 10^2$ times higher ($L \sim 10^{44}\ {\rm erg\,s^{-1}}$). When the protoneutron star collapses from $R = 10^8$ cm to 10^6 cm, the field is compresses further by flux conservation to give $B \approx 10^4 B_{\rm PNS}$.

A convective dynamo faces several obstacles to explain all neutron star magnetic fields, although it may certainly explain a subset. The most serious obstacle is the assumption of equipartition. Numerical simulations reveal time and again that self-sustaining dynamos do not reach equipartition (e.g., Cook et al. 2003) because then the magnetic stresses would paradoxically self-quench the driving shear. In a variety of contexts, and for various numerical set-ups, the magnetic stresses saturate at $\lesssim 5\ \%$ of the mechanical stresses (Cook et al. 2003; Braithwaite 2006). These theoretical results are consistent with observations of the Sun, whose dipole amounts to $\approx 0.3\ \%$ of equipartition (Charbonneau 2014). A second obstacle arises when stratification quenches convection ~ 10 s after the dynamo begins. Does the field retreat or leave behind a permanent dipole moment? The danger here is that pockets ('domains') of tilted polarity are left behind, if the convective eddies turn over faster than they are quenched, producing a dipole moment lower than one inferred in magnetars (e.g., from spin down Kouveliotou et al. 1998), even if the higher-order multipoles are substantial. The consequences for magnetic structure of forming convective pockets (which 'insulate') and radiative pockets (which 'conduct') has been explored by Tout et al. (2004).

A second type of dynamo is driven by differential rotation. It is sometimes presumed that a sheared, magnetised fluid develops a toroidal field component B_ϕ, which reacts back to shut off the shear completely. In practice, the back reaction can only ever be partial in a continuously driven system; if the fluid velocity gradient were to vanish momentarily, B_ϕ generation would cease, and the gradient would eventually reassert itself in response to the driver (Spruit 2009). Microphysically, quenching is avoided because toroidal wind-up is limited to $B_\phi \lesssim 10 B_p$ by instabilities, where B_p is the poloidal field component (Spruit 2009).

Hence it is necessary to grow B_ϕ and B_p together in a differentially driven dynamo. The dynamo loop can be closed by Tayler (Braithwaite 2006) or magneto-rotational (Moiseenko et al. 2006) instabilities, for example. The growth window is open for ~ 10 s, before the star stratifies stably, and lasts for a time $\approx R/v_A$ or $(\Delta\Omega)^{-1}$ in the latter scenarios respectively, where v_A is the Alfvén speed and $\Delta\Omega$ is the angular velocity shear. Differential rotation can also be driven by stellar mergers; the reader is referred to Sect. 3.2 and Wickramasinghe et al. (2014) for details.

Braithwaite (2006) showed that a shear-driven dynamo can even persist in a stably stratified star. Fluid is displaced parallel to equipotential surfaces in the Tayler instability. Hence, the condition for dynamo operation reduces to requiring that the maximum hydromagnetic mode wavelength allowed by buoyancy exceeds the minimum wavelength set by ohmic dissipation—a condition which is easily satisfied in a highly conducting neutron star. Braithwaite (2006) showed explicitly through simulations that a Tayler dynamo is indeed self-sustaining in the presence of strong stratification and that it saturates with $B_\phi \sim 5B_p$ and $B_\phi^2/8\pi \approx 10^{-2}\rho V^2$, i.e., at 1 % of equipartition. However, Zahn et al. (2007) have strongly contested this result, finding that the Tayler instability cannot sustain a dynamo even at high magnetic Reynolds number.

A mean-field variant of the shear-driven dynamo was analysed in the neutron star context by Mastrano and Melatos (2011). In a mean-field dynamo, the turbulent electromotive force $\langle \mathbf{v} \times \mathbf{B} \rangle$ in Faraday's law is the sum of two terms: the 'α' term proportional to $\mathbf{B}$, which arises from turbulent vorticity, and the so-called '$\boldsymbol{\Omega} \times \mathbf{J}$' term, which is proportional to $\eta_{ij}(\nabla \times \mathbf{B})_j$ and arises from off-diagonal entries in the resistivity tensor η_{ij} (Moffatt 1978; Biskamp 1997). A natural seat for a mean-field dynamo is a tachocline, where the angular velocity jumps sharply (Ott et al. 2006; Burrows et al. 2007); in a protoneutron star, this may occur at the boundary between the convection and neutron finger zones (Bonanno et al. 2003, 2005). Mastrano and Melatos (2012) proved that a tachocline mean-field dynamo operates self-sustainingly in a protoneutron star, if η_{ij} is anisotropic and the jumps in 'α' (vorticity) and angular velocity are coincident, and also (less realistically) if η_{ij} is isotropic but the jumps are displaced.

4.2.1 Generation Physics

The ultimate fate of a dynamo-generated magnetic field depends on buoyancy—does the field float out of the star or remain anchored in the core?—and helicity—does the field maintain its large-scale integrity against the action of hydromagnetic instabilities, or is it shredded?

Buoyancy exists in a neutron star because, in β equilibrium, the ratio of charged to neutral species densities increases with depth (Reisenegger and Goldreich 1992). Magnetic flux tubes ('light fluid') rise to the surface if the Alfvén speed satisfies $v_A \gtrsim HN$, where H is the hydrostatic scale height, and N is the Brunt-Väisälä frequency (Reisenegger and Goldreich 1992; Spruit 2009). The field rises most easily when the protoneutron star is young and neutrinos help to lower N to $\lesssim 10\ \mathrm{rad\,s^{-1}}$. By contrast, in a normal neutron star, theory predicts $N \gg 10^2\ \mathrm{rad\,s^{-1}}$ (Reisenegger and Goldreich 1992). In this way, magnetic loops inside the star penetrate the surface and add up to form a permanent dipole moment, as long as their footpoints remain anchored in the cooling crust or core and do not float away entirely.

Assuming that the field does remain anchored, is it stable? Many analyses have shown that a purely poloidal magnetic field in a highly conducting plasma is always unstable (Markey and Tayler 1973; Wright 1973; Flowers and Ruderman 1977). If we regard the star's eastern and western hemispheres as two bar magnets, it is energetically favourable for the magnets to flip and counteralign, destroying the dipole moment (Flowers and Ruderman 1977; Marchant et al. 2011); in practice, the instability proceeds by forming Rayleigh-Taylor 'mushrooms' in the toroidal direction. Similarly, a purely toroidal magnetic field is always unstable (Tayler 1973; Akgün et al. 2013). However, linked poloidal-toroidal fields appear to be stable under a range of conditions, both in analytic calculations (Prendergast 1956; Tayler 1973; Akgün et al. 2013) and numerical simulations (Braithwaite and Spruit 2004a; Braithwaite and Nordlund 2006b; Braithwaite 2009).

Linked poloidal-toroidal fields are stabilised, because their helicity is conserved approximately in a highly conducting plasma. Helicity is defined as $H = \int d^3\mathbf{x}\mathbf{A}\cdot\mathbf{B}$, where $\mathbf{A}$ is the magnetic vector potential; it is a gauge-dependent quantity which describes the number of times one flux tubes winds around another (Biskamp 1997). In a weakly resistive plasma, H is conserved globally, even when H is scrambled locally on individual flux tubes. Hence, if a dynamo-generated field has $H \neq 0$ initially, the twist can never be undone completely, leading to $B \neq 0$ in the final state. If magnetic energy is minimised during the field's evolution, it tends to a force-free configuration, cf. laboratory spheromaks (Broderick and Narayan 2008; Mastrano and Melatos 2008).

Numerous papers have been written recently reporting stability calculations on poloidal-toroidal fields, e.g. (Braithwaite and Spruit 2004a; Braithwaite and Nordlund 2006b; Akgün et al. 2013; Armaza et al. 2013; Ciolfi and Rezzolla 2013; Gourgouliatos et al. 2013; Lander and Jones 2012; Mitchell et al. 2013; Passamonti and Lander 2013). A comprehensive review lies outside the scope of this chapter; we simply mention a few highlights. In a barotropic star, it has become apparent that stability depends in subtle ways on the boundary conditions and the entropy distribution within the star (Ciolfi et al. 2011; Lander and Jones 2012). In a non-barotropic star, magnetic field configurations with an equatorial 'torus' are generically stable, when the poloidal-toroidal flux ratio is 'modest'. Akgün et al. (2013) obtained the stability condition $0.25 < B_\phi^2/B_p^2 < 0.5(\gamma_{\rm ad}/\gamma - 1)|E_g|/(B_p^2/8\pi)$, where $\gamma_{\rm ad}$ and γ are the adiabatic and non-barotropic specific heat ratios respectively, and $|E_g|$ is the total gravitational potential energy. Non-barotropic solutions also match self-consistently to an external dipole (or any arbitrary multipole), extending the power of future observational tests involving gravitational waves (Mastrano et al. 2011, 2013; Mastrano and Melatos 2012). Finally, in a superconducting star, the stability problem changes character fundamentally for two reasons: the Lorentz force is much stronger ($|\mathbf{F}_B| \propto H_{\rm c1}B \sim 10^3 B^2$, where $H_{\rm c1}$ is the critical field in a type II superconductor), and its vectorial structure is more complicated $[\mathbf{F}_B \propto \mathbf{B}\times(\nabla\times\mathbf{H}_{\rm c1}) + \rho\nabla(|\mathbf{B}|\partial|\mathbf{H}_{\rm c1}|/\partial\rho)]$. The equilibrium magnetic structure and its stability are then controlled by $H_{\rm c1}/\langle B\rangle$ at the crust-core boundary (Glampedakis et al. 2012; Lander et al. 2012; Lander 2013, 2014). In fact, the internal field configuration now depends very strongly on field strength, suggesting significant differences between the interior fields of pulsars and those of magnetars (Lander 2014).

4.2.2 Superfluid Turbulence

Does magnetic field evolution in a neutron star conclude ~ 10 s after the protoneutron star is born and stably stratifies? The traditional consensus has been in the affirmative, except for the slow ohmic and Hall evolution described elsewhere in this volume and in Geppert et al. (2012), Pons et al. (2012), Viganò et al. (2012). However, this picture may need modification in light of recent work suggesting that the neutron superfluid and charged components to which it is coupled are turbulent. In this scenario, internal magnetic activity may be ongoing, even though the magnetic flux thereby produced struggles to rise buoyantly to the surface.

Hints of abrupt magnetospheric changes observed on ~ 1 yr time-scales (see Sect. 4)—much longer than the dynamical time of the magnetosphere, but much shorter than the ohmic and Hall time-scales—are relevant in this context.

In the standard picture, the vorticity field inside a neutron star is uniform and quasi-static. Macroscopically, this means that there is zero meridional circulation. Microscopically, it means that the superfluid is threaded by a rectilinear array of vortices, each carrying a quantum of circulation $\kappa = h/(2m_n)$, where h is Planck's constant and m_n is the mass of the neutron. As the star brakes electromagnetically, an angular velocity lag $\Delta\Omega$ builds up between the crust and superfluid, whose value is set by the balance between the Magnus and pinning forces. One finds $\rho\kappa R\Delta\Omega \approx E_{\rm pin}(\xi_{\rm coh}\xi_{\rm pin})^{-1}$ and hence $\Delta\Omega \approx 1\ {\rm rad\,s^{-1}}$ independent of Ω and $\dot{\Omega}$, where $E_{\rm pin}$ is the pinning energy, $\xi_{\rm coh}$ is the superfluid coherence length, and $\xi_{\rm pin}$ is the pinning site separation (Warszawski and Melatos 2013). Glitches reset $\Delta\Omega$ but only partially, because there is no observed correlation between glitch sizes and waiting times in most pulsars (Melatos et al. 2008).

The lag $\Delta\Omega$ changes the flow structure completely. It is a well-known result of fluid mechanics that a shear flow in a sphere cannot be purely toroidal. The boundary conditions induce Ekman circulation, wherein interior fluid is drawn into a viscous boundary layer, spun down through contact with the crust, then recycled back into the interior (Abney and Epstein 1996; van Eysden and Melatos 2010). This occurs with or without a rigid core. Furthermore, Ekman circulation is known to be unstable at high Reynolds numbers (see Sect. 5.2) Re $\gtrsim 10^3$, forming unsteady, nonaxisymmetric flow patterns like herringbone waves (Re $\approx 6 \times 10^3$) and Taylor–Gortler vortices (Re $\approx 10^4$) (Nakabayashi 1983). In a neutron star, where the effective Reynolds number is much greater (Re $\gtrsim 10^7$ for e^-–e^- shear viscosity modified by Landau damping by transverse plasmons; see Shternin and Yakovlev 2008), wave modes resembling fully developed turbulence are expected (Melatos and Peralta 2007).

At the microscopic level, the vorticity field is disrupted by vortex-line instabilities. One example is the Donnelly–Glaberson instability, familiar from laboratory experiments with liquid helium and Bose-Einstein condensates (Glaberson et al. 1974). If there is a superfluid counterflow directed along the vortices, Kelvin waves are excited and amplified by mutual friction to produce a reconnecting, self-sustaining vortex tangle (Tsubota et al. 2003, 2013). This process has a low threshold (counterflow $\approx$ ${\rm mm\,s^{-1}}$, easily exceeded during Ekman pumping) and a fast growth time ($\lesssim$ one spin period) (Peralta et al. 2006). Recently, a related Kelvin-wave instability has been discovered by Link (2012a; 2012b), which does not rely on a counterflow and arises from imperfect pinning, which creates a lag to amplify the Kelvin waves (cf. Glampedakis et al. 2009). This instability grows over $\sim$ days, which is still fast.

Turbulent, tangled vorticity is an important ingredient in the story of magnetic field generation. The turbulent neutron condensate couples to charged components in the star through mutual friction and entrainment (Mendell 1998; Prix et al. 2002; Haskell et al. 2009, 2012; Chamel 2013), as well as through fluxoid-vortex interactions (Srinivasan et al. 1990; Jahan-Miri 2010). Consequently, the charged components circulate too (Peralta et al. 2005). Simulations that incorporate the magnetic field dynamics explicitly are required to understand the behaviour of the coupled fluids in detail. There is reason to be hopeful that such simulations will be conducted in the next few years, motivated by the prospect of gravitational wave experiments.

Can superfluid turbulence be quenched by stratification, i.e., by the buoyancy force produced by the charge-neutral ratio increasing with depth (see Sect. 4.2.1)? Counterintuitively, perhaps, the situation is borderline. Stratified turbulence is characterised by two dimensionless quantities: the Froude number, Fr, which is related to the buoyancy force, and the Reynolds number, Re (Brethouwer et al. 2007). Even for ${\rm Fr}^{-1} \gtrsim 10^2$, the turbulence is not quenched, if Re is sufficiently large. Turbulence persists for Re $\gtrsim$ Fr^{-2} or equivalently $R^2(\Delta\Omega)^3 \gtrsim \nu N^2$, where ν is the kinematic viscosity (Iida et al. 2009; Lasky et al. 2013). Neutron stars lie near the boundary defined by the above condition in the Fr^{-1}–Re plane. In the regime Re $\gtrsim$ Fr^{-2} and Fr$^{-1} \gtrsim 10^2$, the flow is modified strongly away from Kolmogorov isotropy, even though it remains circulatory; the Earth's atmosphere and oceans also occupy this regime (Brethouwer et al. 2007; Chung and Matheou 2012).

Another possibility is that superfluid turbulence is quenched by magnetic stresses. Again, though, the case is unclear. At the microscopic level, the imperfect pinning instability which drives a vortex tangle (Link 2013) persists even when a magnetic field is present; in fact, Link's (2013) analysis assumes magnetic locking of the viscous proton-electron plasma to the crust. Does the B_ϕ wound up by flux freezing erase the shear? More simulations are needed, but for now it seems unlikely: the magnetic back reaction saturates at $\lesssim 1$ % of the mechanical (shear) stress in simulations attempted to date (Cook et al. 2003; Braithwaite 2006), and the Tayler instability self-organizes such that $\mathbf{B} \cdot \nabla\Omega = 0$, i.e., such that wind-up stops, before B_ϕ quenches the process (Duez et al. 2006; Rüdiger et al. 2009). Magnetised Ekman pumping acts on the rapid time-scale $\approx (\Omega R/v_A)^{2/3}\Omega^{-1}$ (where v_A is the Alfvén speed) in simple magnetic geometries, e.g., uniform magnetisation (Easson 1979b). However, it is known to operate much more slowly in complicated geometries and stratified conditions (Goedbloed and Poedts 2004; Melatos 2012). Finally, Easson showed that many—perhaps most—realistic magnetic topologies (e.g., those with closed loops) cannot enforce corotation (Easson 1979a); see also Melatos (2012). Physically, this occurs because the toroidal component of the magnetic field perturbation created by spin down satisfies a nonconservative equation of motion in a superconductor with Fermi liquid forces (Easson 1979a). Again, large-scale numerical simulations are needed to determine if the class of magnetic topologies that permit corotation is restricted or generic.

5 Status of MHD Turbulence Theory and Computation

The basic physics underlying stellar magnetic fields lies in the theory of magnetised fluids, or magnetohydrodynamics (MHD). Of particular importance is the manifestation of magnetohydrodynamic turbulence because its presence underpins the operation of stellar dynamos, and controls diffusive magnetic flux transport in stars. This section is intended to give a high-level overview of recent progress in MHD turbulence theory and computation as it applies to stellar magnetism.

5.1 Equations of Motion

The equations of compressible MHD can be written as

$$\frac{\partial \rho}{\partial t} = \nabla \cdot (\rho \mathbf{u}) \tag{5}$$

$$\frac{\partial \mathbf{u}}{\partial t} = -\nabla \cdot (\rho \mathbf{u}\mathbf{u}) - \nabla p + \rho\nu\nabla^2\mathbf{u} \tag{6}$$

$$\frac{\partial \mathbf{B}}{\partial t} = \nabla \times (\mathbf{u} \times \mathbf{B}) + \eta\nabla^2\mathbf{B} \tag{7}$$

where in Eq. (6) $\mathbf{u}$ is the fluid velocity, ρ is the mass density, and ν is the kinematic viscosity. Equation (7) is Faraday's law together with the phenomenological Ohm's law $\mathbf{E} = -\mathbf{u} \times \mathbf{B} + \eta\mathbf{J}$. In general an equation for the transport of energy must also be supplied along with an equation of state relating the gas pressure to its temperature and density. However, if incompressibility is assumed, as it is in most analytical treatments of turbulent dynamo, the energy equation is superfluous.

5.2 Dimensionless Numbers

If L and u represent characteristic length and velocity magnitudes of the system in question, then the ratio $\mathrm{Re} = uL/\nu$ is referred to as the Reynolds number. It represents the relative importance of inertial and viscous forces. It can also be seen as the ratio of the viscous time $t_{\mathrm{visc}} = L^2/\nu$ to the advective time $t_{\mathrm{adv}} = L/u$. Flows having large Re are in general susceptible to turbulence as their dynamics are dominated by the non-linear coupling in Eq. (6). Due to the enormous length scales involved in astrophysical flows, it is commonly appropriate to drop the viscous term altogether and just work with the Euler equation.

Analogously, one can define the *magnetic* Reynolds number $\mathrm{Rm} = uL/\eta$ which measures, correspondingly the relative importance of magnetic induction and diffusion. It is also the ratio $t_{\mathrm{res}}/t_{\mathrm{adv}}$ where $t_{\mathrm{res}} = L^2/\eta$ is the damping time scale for magnetic fluctuations of scale L. The limiting case of infinite Rm corresponds to the case of a perfectly conducting fluid. This limit has the special property that the magnetic field is "frozen" into the fluid. Flux-freezing is a conservation law for the which the quantity $\int_S \mathbf{B} \cdot d\mathbf{A}$ is a constant of the motion, when S represents a surface which is deformed along with the flow. The limit of vanishing viscosity and resistivity is referred to as ideal MHD.

One can also define the magnetic Prandtl number $\mathrm{Pm} = \mathrm{Rm}/\mathrm{Re}$ which characterises the relative importance of viscous and resistive dissipation. In turbulent flows, it also corresponds to the ratio of length scales where the velocity and magnetic fields become smooth under the action of their respective dissipation mechanisms. Thus, for flows having large Pm, the velocity field is smoothed out by viscosity over scales where the magnetic field is still frozen into the fluid.

As an example, let us estimate the magnetic Prandtl number for various parts of the sun. The resistively of a fully ionised plasma is given by Priest (2014)

$$\eta = \frac{m_e}{\mu_0 n_e e^2 \tau_e} \tag{8}$$

where m_e is the electron mass, n_e is electron number density, and τ_e is the electron mean free path. In the solar photosphere this gives $\eta \sim 10^4\ \mathrm{m^2\,s^{-1}}$. Thus vortices on the solar surface of size ~ 500 km rotating with typical speed of $1\ \mathrm{km\,s^{-1}}$ (Simon and Weiss 1997) have magnetic Reynolds number $\mathrm{Rm} \sim 5 \times 10^4$. By comparison, the Reynolds number is of order 10^2 making the magnetic Prandtl number of order 10^2. A similar estimate for the solar convection zone gives $\mathrm{Pm} \sim 10^{-2}$. The magnetic Prandtl number of diffuse astrophysical plasmas is vastly larger, for example it is $\sim 10^{15}$ (Maron and Cowley 2001) in the Galaxy and as large as 10^{22} (Schekochihin et al. 2002b) in the intergalactic medium. These values reflect their extraordinarily low collisionality.

Other dimensionless numbers of interest include the Lundquist number $S = Lv_A/\eta$ where $v_A = B/\sqrt{4\pi\rho}$ is the speed of Alfvén wave propagation. The sonic and Alfvén Mach numbers are flow speed measured in units of their corresponding wave propagation speeds. The sonic Mach number $\mathcal{M}$ is also roughly the square-root of the ratio of kinetic to thermal energy densities, and similarly with the Alfvén Mach number $\mathcal{M}_A$ except with the magnetic energy instead of the thermal energy.

5.3 Conserved Quantities

Conserved quantities in MHD belong to one of three types (Bekenstein 1987). The first kind, referred to as type A, is where the closed line integral of a vector field is a constant of the motion, where the path of circulation is carried along with the local flow. The canonical example is that of Kelvin's circulation theorem, which states that in a perfect fluid the circulation $\Gamma = \oint_C \mathbf{u} \cdot \mathbf{ds}$ is a constant of the motion, where the closed path C is advected and deformed along with the flow. Kelvin's circulation theorem has been generalised to relativistic hydrodynamics by Taub (1959), and to non-relativistic and relativistic MHD by Bekenstein and Oron (2000). Another example is the Alfvén law of magnetic flux conservation, which states that $\Phi = \oint_C \mathbf{A} \cdot \mathbf{ds}$ is a constant of the motion. This law can also be stated as the constancy of magnetic flux through the surface whose boundary is the closed contour C. It formally applies in ideal MHD, where the viscosity and resistivity are vanishing. This law, commonly referred to as "flux freezing", is generally considered a good approximation in astrophysical plasmas, where the microphysical diffusion time of magnetic fields is far longer than other dynamical times. However, for turbulent flows in which the velocity field stays chaotic down to scales below those of interest (referred to as "rough" velocity fields) turbulent diffusivity effectively violates the perfect fluid assumptions. The violation of Kelvin's circulation theorem in high Reynolds number hydrodynamic flows has been analysed by Chen et al. (2006), and similarly for the Alfvén theorem by Eyink (2011).

The second type of conservation law (type B) is where a local fluid quantity is a constant of the motion along streamlines. Bernoulli's theorem is the canonical example of type B conservation laws.

Type C conservation laws apply to the global conservation of scalar or vector quantities whose transport is strictly local. These are like Gauss's law in that they relate the conserved quantity's time rate of change in a finite volume to its flux through the volume's bounding surface. They can thus be stated in either differential or integral form, the latter of which forms the basis for modern numerical methods for solving the hydrodynamic and MHD equations. The prototypical type C conservation law is the continuity equation $\dot{\rho} + \nabla \cdot \rho\mathbf{u} = 0$. In the absence of viscous and resistive effects, momentum and magnetic flux are also conserved in this way, where $\rho\mathbf{u}$ is replaced with a generic flux function for that conserved quantity. Similarly, in the absence of heat transport by conduction, the evolution of total energy can also be written as a continuity equation.

5.4 Consequences of Helicity Conservation

A particularly important conserved quantity is the magnetic helicity $H_M = \int d^3x \mathbf{A} \cdot \mathbf{B}$. H_M is a global topological characterisation of the magnetic field, which indicates the degree to which magnetic field loops are mutually linked. It is analogous to the kinetic helicity $H_K = \int d^3x \mathbf{u} \cdot \boldsymbol{\omega}$ (where $\omega = \nabla \times \mathbf{u}$ is the vorticity) which represents the degree to which the vortex lines are mutually linked. In a perfect fluid, both helicity measures are a constant of the motion.

Magnetic helicity is important because it tends to establish magnetic fields well above the turbulence integral scale. This consequence of magnetic helicity conservation was first investigated by Frisch et al. (1975), where it was predicted that the magnetic energy would peak around the scale H_M/E_B (where E_B is the total magnetic energy), even if the injection of magnetic energy took place at much smaller scales. This process is referred to as inverse cascading because the magnetic energy cascades from small scales toward large scales, unlike the direct cascade of energy observed in hydrodynamic turbulence. It is the underlying mechanism for the formation of large-scale dynamos which may be responsible for establishing stellar magnetic fields.

5.5 Inverse Cascade of Magnetic Energy

Magnetohydrodynamic turbulence exhibits an unusual phenomenon where, at least in its freely decaying state, magnetic field correlations arise over distances much longer than the original eddy size. This *inverse cascading* effect is reminiscent of what is seen in two-dimensional (but not three-dimensional) hydrodynamic turbulence. Although this effect may be related to the conservation of magnetic helicity (Frisch et al. 1975), until recently it was not appreciated that it happens even when the field is initially helicity-free. That is, when it is topologically equivalent to the uniform (or zero) magnetic field.

Inverse cascading of magnetic energy was first seen in numerical simulations by Meneguzzi et al. (1981). It has since been confirmed by many others using direct numerical simulations (Christensson et al. 2001; Banerjee and Jedamzik 2004; Kahniashvili et al. 2010; Tevzadze et al. 2012), quasi-analytical techniques such as the "Eddy damped quasi normal mode" approximation (Son 1999) and also turbulence "shell models" (Kalelkar and Pandit 2004). Such inverse cascading is a process of magnetic field self-assembly, whereby magnetic energy provided by random motions at very small scales, may over time establish dynamically important magnetic fields over scales that are much larger. For example, the sun's convective cells, which set the turbulence integral scale, are far smaller than the solar diameter. The ordered magnetic fields that extend over the whole surface of the sun (and many other stars) may come from such a self-assembly process. Dynamos capable of establishing magnetic fields on scales far larger than the turbulence integral scale are generically referred to as large scale dynamos.

The literature to date is still conflicted on whether non-zero magnetic helicity is a necessary condition for inverse cascading to occur. It was shown by Olesen (1997) and Shiromizu (1998) that inverse cascading could be expected even for non-helical magnetic field configurations, as a consequence of rescaling symmetries native to the Navier-Stokes equations. But no inverse cascading was seen in numerical studies based on EDQNM theory (Son 1999) or direct numerical simulations with relatively low resolution (Christensson et al. 2001; Banerjee and Jedamzik 2004).

Very recently, three numerical studies have appeared which confirm the existence of inverse cascading in non-helical three-dimensional MHD turbulence. Brandenburg et al. (2014) carried out high resolution direct numerical simulations of compressible MHD turbulence with magnetic energy injected at very small scales. The results confirm the growth of magnetic energy spectrum $P_M(k,t)$ at low wavenumbers. Zrake (2014) found similar results in high resolution simulations of freely decaying relativistic MHD turbulence. In the latter study, the evolution of $P_M(k,t)$ was parametrised around a self-similar ansatz which may be used to evaluate the time required for magnetic fields to be established over scales arbitrarily larger than the energy injection scale. Berera and Linkmann (2014) also found inverse cascades in decaying MHD turbulence. That study utilised a novel technique, whereby the system was evolved hydrodynamically with a passive magnetic field. The resulting energy spectrum also exhibits growth of magnetic field fluctuations at large scales, suggesting that inverse cascading of passively advected vector fields is a generic property of hydrodynamic turbulence.

5.6 Small-Scale Turbulent Dynamo

In general, the term "dynamo" refers to the conversion of kinetic energy into magnetic energy. In magnetohydrodynamic theory, a number of distinct processes can facilitate this conversion. The only thing they all have in common is the presence of turbulence. Many excellent reviews have been written on this the connection between MHD turbulence and dynamo processes: Brandenburg and Subramanian (2005), Lazarian and Cho (2005), Kulsrud and Zweibel (2008), Tobias et al. (2011). Here we will present a very brief overview of a process known as small-scale turbulent dynamo.

Small-scale turbulent dynamo operates in highly conducting turbulent fluids, even when the turbulence is isotropic and non-helical. Conceptually it is quite simple. Magnetic field lines carried along by the turbulent flow become increasingly chaotic and distorted over time, leading to enhancement of their energy. When the field is very weak, i.e. $v_A \ll u$, its back-reaction on the fluid can be ignored, and the field evolves as a passively advected vector field in otherwise hydrodynamic turbulent flow. This approximation is referred to as *kinematic* small-scale dynamo, and was first studied by Kazantsev (1968), but more modern analytic treatments exist such as that of Vincenzi (2001). The original formulation, known as the Kraichnan–Kazantsev dynamo starts with the assumption of an arbitrary spectrum of velocity fluctuations that are delta-correlated in time. Despite its simplicity, this model makes accurate predictions regarding the time rate of change and spectral energy distribution $P_M(k,t)$ of the magnetic field, where k is the inverse length scale. In particular, solutions for $P_M(k,t)$ that are exponentially growing in time exist when the microphysical diffusion coefficient η is small relative to the turbulent diffusivity, a condition which is met when the Reynolds number of the flow is sufficiently large.

The Kraichnan–Kazantsev dynamo predicts for the unstable growing solutions, (1) that $P_M(k,t) \propto k^{3/2}$ and peaks at the resistive cutoff, and (2) that the magnetic energy exponentiates at the fastest eddy turnover time of the turbulent hydrodynamic cascade. The first numerical experiments to confirm the basic features of the Kraichnan–Kazantsev dynamo were by Meneguzzi et al. (1981). At the time, direct numerical simulations were exceedingly expensive, and those simulations were limited to grid sizes of 32^3 and 64^3. Since then, many groups have continued this investigation utilising high performance parallel computing architectures, and modern numerical schemes for obtaining solutions to the MHD equation.

Although the original work by Kazantsev predicted that the magnetic Reynolds number Rm had to be "sufficiently large" for small scale dynamo action to occur, the exact value of the critical Rm needed to be found through numerical experiments. For the case of $\mathrm{Pm} = 1$, Haugen et al. (2004b) found that the critical magnetic Reynolds number $\mathrm{Rm}_{\mathrm{crit}}$ for dynamo action was about 35. For smaller Pm, it was found that $\mathrm{Rm}_{\mathrm{crit}}$ increases as Pm decreases. This may be partially understood as having to do with the flow properties in the resistive interval. When the Pm is very large, the resistive scale lies deep below the viscous cutoff, so that magnetic energy peaks where the flow is very smooth. However, when Pm is smaller than one, the resistive scale moves into the turbulent inertial range of the flow, and the velocity field is *rough* at the resistive scale. Haugen et al. (2004c) found that $\mathrm{Rm}_{\mathrm{crit}}$ increased to 70 when the flow is supersonic, even when $\mathrm{Pm} = 1$. Federrath et al. (2011) found that for compressible flows the dynamo growth rate depends upon how the turbulence is forced; vortically driven flows produced more efficient dynamo than did dilatational forcing. It was argued analytically by Boldyrev and Cattaneo (2004) that in the small Pm regime, the critical magnetic Reynolds number should be a strong function of the *roughness* exponent α of the velocity field, $v_\ell \propto \ell^\alpha$, such that smaller α (a rougher field) requires a larger Rm to produce small-scale kinematic dynamo.

A central question around small-scale dynamo is the nature of its saturation. The exponentially growing solution of the Kazantsev model holds only as long as the dynamo is in the kinematic regime, so that magnetic back-reaction on the flow is negligible. Once the field attains a strength high enough to compete with the fluid inertia, the kinematic assumption breaks down, and some sort of saturation behaviour is expected. This saturation is a highly nonlinear problem, and thus very difficult to study analytically. Schekochihin et al. (2002a) argued that in the non-linear stage, at least for very large Pm, magnetic energy would remain concentrated in the sub-viscous range, moving toward the inertial range on a resistive rather than dynamical time. Were this to be the case, equipartition of the turbulent kinetic and magnetic energy densities at the largest scales would require an asymptotically long time to be established. However, this scenario has not been observed in the numerical literature. Instead, the emerging consensus (other differences aside) (Cho and Vishniac 2000; Brandenburg et al. 2003; Haugen et al. 2004a; Schleicher et al. 2013; Schekochihin et al. 2004; Maron and Cowley 2001; Zrake and MacFadyen 2011, 2013) is that magnetic energy shifts into the inertial range regardless of the magnetic Prandtl number. Once there, the field attains coherence at increasing scales until finally reaching the energy-containing scale of turbulence, thus establishing scale-by-scale equipartition. Both the kinetic and magnetic energy follow a Kolmogorov spectrum, with the magnetic energy per unit wavelength exceeding that of the kinetic energy by a factor of around 2.

The time-scale for the flow to fully establish scale-by-scale equipartition is still not completely clear. It was argued by Beresnyak (2012) that non-linear small-scale dynamo exhibits universality in the sense that a constant fraction C_E of the work done by turbulent pumping accumulates in magnetic energy. In this scenario, scale-by-scale equipartition should be attained after a time $1/C_E$. Careful modelling of the approach to saturation by Zrake and MacFadyen (2013) finds that indeed scale-by-scale equipartition is reached after a number of dynamical times that is independent of the Reynolds number.

6 Conclusions

In this chapter we have reviewed progress made on the origin of magnetic fields in intermediate and massive main sequence stars and in compact stars.

Observations of magnetic fields in white dwarfs may indicate that the distribution is bimodal, exhibiting a paucity of objects between 10^5–10^6 G, although further sensitive spectropolarimetric surveys conducted on 8 m class telescopes are needed in order to confirm this finding. Should this bimodality be confirmed, then two different channels for field formation and evolution could be at play. In particular, one could identify weakly magnetic main sequence stars such as Vega and Sirius as the progenitors of weakly magnetic white dwarfs belonging to the low field component. This would lend some support to the fossil field hypothesis for the origin of fields in compact stars.

However, the viability of the fossil field hypothesis has been recently questioned because of the absence of magnetic white dwarfs paired with non-degenerate stars. These observations mirror the dearth of magnetic non-degenerate A, B and O stars in short-period binary systems. It is not clear whether there is any connection or analogy to be made between fields on the main sequence and compact star phases, but a merger scenario for field formation has been suggested to overcome some of the problems raised by recent observations.

Here we stress that the pairing properties of magnetic white dwarfs do not necessarily invalidate the viability of the fossil scenario since more than one formation channel could be at work. However, the complete lack, rather than just a dearth, of magnetic white dwarfs paired with non-degenerate companions in non-interacting systems, raises strong doubts on whether the fossil route can be the main channel for the formation of magnetic white dwarfs. If magnetic white dwarfs were the descendants of magnetic main sequence stars, then magnetic white dwarfs should be commonly found in binary systems to reflect the incidence of binarity observed in (magnetic) stars of F to late B spectral types. However, there is *not one single example* of a magnetic white dwarf paired with a non-degenerate companion in a non-interacting binary.

In order to explain this curious finding, it has been proposed that fields in isolated highly magnetic white dwarfs could be generated during common envelope evolution when the two components of a binary system merge.

The observational situation surrounding birth magnetic fields of neutron stars remains frustratingly uncertain, because what measurements exist are indirect. Population synthesis models point, on balance, to moderate birth fields $\sim 10^{12.65\pm0.55}$ G. Zeeman spectropolarimetry has revealed strongly magnetised (i.e., a few thousand gauss) progenitors across a range of spectral classes, but even so it seems that the numbers may be insufficient to account for the magnetar birth rate under the fossil field scenario. At least for magnetars, therefore, a proto-neutron star dynamo is favoured, driven by differential rotation and the Tayler or magneto-rotational instabilities rather than thermal convection (Spruit 2009). Emergence (through buoyancy), stability (through helicity), and ongoing activity (through superfluid turbulence) play roles in determining the strength of the (observable) surface field. Future gravitational wave observations offer the best prospects for direct internal field measurements. Until these observations become available we can only say that, as for magnetic white dwarfs, it is possible that both fossil and dynamo processes are at work in different neutron stars.

We have also reviewed the theory of magnetohydrodynamic turbulence because its presence underpins the operation of stellar dynamos and controls diffusive magnetic flux transport in stars. But MHD turbulence poses a great theoretical challenge due to its inherent complexities, and many features of its operation are yet to be satisfactorily explained. Advances in numerical algorithms and computing resources have yielded considerable gains in the empirical grounding of various phenomenological pictures.

Acknowledgements The authors wish to thank the International Space Science Institute (ISSI) for hospitality and support. This paper was supported by an Australian Research Council Discovery Project grant. Andrew Melatos thanks Alpha Mastrano for extensive assistance with typesetting Sect. 4. We would like to thank the Referee for carefully reading our manuscript and for making many valuable comments that have helped us improve this review paper.

References

J. Aasi, J. Abadie, B.P. Abbott, R. Abbott, T. Abbott, M.R. Abernathy, T. Accadia, F. Acernese, C. Adams, T. Adams et al., Astrophys. J. **785**, 119 (2014). doi:10.1088/0004-637X/785/2/119

J. Abadie, B.P. Abbott, R. Abbott, M. Abernathy, T. Accadia, F. Acernese, C. Adams, R. Adhikari, C. Affeldt, B. Allen et al., Astrophys. J. **737**, 93 (2011). doi:10.1088/0004-637X/737/2/93

B.P. Abbott, R. Abbott, F. Acernese, R. Adhikari, P. Ajith, B. Allen, G. Allen, M. Alshourbagy, R.S. Amin, S.B. Anderson et al., Astrophys. J. **713**, 671 (2010). doi:10.1088/0004-637X/713/1/671

M. Abney, R.I. Epstein, J. Fluid Mech. **312**, 327 (1996). doi:10.1017/S0022112096002030

T. Akgün, A. Reisenegger, A. Mastrano, P. Marchant, Mon. Not. R. Astron. Soc. **433**, 2445 (2013). doi:10.1093/mnras/stt913

E. Alecian, C. Catala, G.A. Wade, J.-F. Donati, P. Petit, J.D. Landstreet, T. Böhm, J.-C. Bouret, S. Bagnulo, C. Folsom, J. Grunhut, J. Silvester, Mon. Not. R. Astron. Soc. **385**, 391 (2008). doi:10.1111/j.1365-2966.2008.12842.x

E. Alecian, G.A. Wade, C. Catala, S. Bagnulo, T. Böhm, J.-C. Bouret, J.-F. Donati, C.P. Folsom, J. Grunhut, J.D. Landstreet, Mon. Not. R. Astron. Soc. **400**, 354 (2009). doi:10.1111/j.1365-2966.2009.15460.x

E. Alecian, G.A. Wade, C. Catala, J.H. Grunhut, J.D. Landstreet, S. Bagnulo, T. Böhm, C.P. Folsom, S. Marsden, I. Waite, Mon. Not. R. Astron. Soc. **429**, 1001 (2013a). doi:10.1093/mnras/sts383

E. Alecian, C. Neiner, S. Mathis, C. Catala, O. Kochukhov, J. Landstreet, Astron. Astrophys. **549**, 8 (2013b). doi:10.1051/0004-6361/201220796

E. Alecian, C. Neiner, G.A. Wade, S. Mathis, D. Bohlender, D. Cébron, C. Folsom, J. Grunhut, J.-B. Le Bouquin, V. Petit, H. Sana, A. Tkachenko, A. ud-Doula, BinaMIcS Collaboration, arXiv e-prints (2014)
C. Armaza, A. Reisenegger, J. Alejandro Valdivia, P. Marchant, arXiv e-prints (2013)
M. Aurière, G.A. Wade, J. Silvester, F. Lignières, S. Bagnulo, K. Bale, B. Dintrans, J.F. Donati, C.P. Folsom, M. Gruberbauer, A. Hui Bon Hoa, S. Jeffers, N. Johnson, J.D. Landstreet, A. Lèbre, T. Lueftinger, S. Marsden, D. Mouillet, S. Naseri, F. Paletou, P. Petit, J. Power, F. Rincon, S. Strasser, N. Toqué, Astron. Astrophys. **475**, 1053 (2007). doi:10.1051/0004-6361:20078189
M. Aurière, R. Konstantinova-Antova, P. Petit, C. Charbonnel, B. Dintrans, F. Lignières, T. Roudier, E. Alecian, J.F. Donati, J.D. Landstreet, G.A. Wade, Astron. Astrophys. **491**, 499 (2008). doi:10.1051/0004-6361:200810502
M. Aurière, G.A. Wade, F. Lignières, A. Hui-Bon-Hoa, J.D. Landstreet, I.K. Iliev, J.-F. Donati, P. Petit, T. Roudier, S. Théado, Astron. Astrophys. **523**, 40 (2010). doi:10.1051/0004-6361/201014848
M. Auriere, F. Lignieres, R. Konstantinova-Antova, C. Charbonnel, P. Petit, S. Tsvetkova, G.A. Wade, arXiv e-prints (2013)
R. Aznar Cuadrado, S. Jordan, R. Napiwotzki, H.M. Schmid, S.K. Solanki, G. Mathys, Astron. Astrophys. **423**, 1081 (2004). doi:10.1051/0004-6361:20040355
W. Baade, Astrophys. J. **96**, 188 (1942). doi:10.1086/144446
W. Baade, F. Zwicky, Proc. Natl. Acad. Sci. **20**, 254 (1934). doi:10.1073/pnas.20.5.254
H.W. Babcock, *Contributions from the Mount Wilson Observatory/Carnegie Institution of Washington*, vol. 727 (1947), pp. 1–15
H.W. Babcock, Astrophys. J. Suppl. Ser. **3**, 141 (1958). doi:10.1086/190035
S. Bagnulo, J.D. Landstreet, E. Mason, V. Andretta, J. Silaj, G.A. Wade, Astron. Astrophys. **450**, 777 (2006a). doi:10.1051/0004-6361:20054223
S. Bagnulo, J.D. Landstreet, E. Mason, V. Andretta, J. Silaj, G.A. Wade, Astron. Astrophys. **450**, 777 (2006b). doi:10.1051/0004-6361:20054223
R. Banerjee, K. Jedamzik, Phys. Rev. D **70**(12), 123003 (2004). doi:10.1103/PhysRevD.70.123003
M.A. Barstow, S. Jordan, D. O'Donoghue, M.R. Burleigh, R. Napiwotzki, M.K. Harrop-Allin, Mon. Not. R. Astron. Soc. **277**, 971 (1995)
D.P. Barsukov, O.A. Goglichidze, A.I. Tsygan, Mon. Not. R. Astron. Soc. **432**, 520 (2013). doi:10.1093/mnras/stt501
D.P. Barsukov, O.A. Goglichidze, A.I. Tsygan, J. Phys. Conf. Ser. **496**(1), 012013 (2014). doi:10.1088/1742-6596/496/1/012013
J. Bekenstein, A. Oron, Phys. Rev. E **62**(4), 5594 (2000). doi:10.1103/PhysRevE.62.5594
J.D. Bekenstein, Astrophys. J. **319**, 207 (1987). doi:10.1086/165447
A. Berera, M. Linkmann, Inverse cascades and the evolution of decaying magnetohydrodynamic turbulence (2014)
A. Beresnyak, Phys. Rev. Lett. **108**, 35002 (2012)
D. Biskamp, *Nonlinear Magnetohydrodynamics* (1997)
P.M.S. Blackett, Nature **159**, 658 (1947). doi:10.1038/159658a0
A. Blazère, P. Petit, F. Lignières, M. Aurière, T. Böhm, G. Wade, in *SF2A-2014: Proceedings of the Annual Meeting of the French Society of Astronomy and Astrophysics*, ed. by J. Ballet, F. Martins, F. Bournaud, R. Monier, C. Reylé (2014), pp. 463
A.I. Bogomazov, A.V. Tutukov, Astron. Rep. **53**, 214 (2009). doi:10.1134/S1063772909030032
S. Boldyrev, F. Cattaneo, Phys. Rev. Lett. **92**, 144501 (2004)
A. Bonanno, L. Rezzolla, V. Urpin, Astron. Astrophys. **410**, 33 (2003). doi:10.1051/0004-6361:20031459
A. Bonanno, V. Urpin, G. Belvedere, Astron. Astrophys. **440**, 199 (2005). doi:10.1051/0004-6361:20042098
E.F. Borra, J.D. Landstreet, L. Mestel, Annu. Rev. Astron. Astrophys. **20**, 191 (1982). doi:10.1146/annurev.aa.20.090182.001203
J. Braithwaite, Astron. Astrophys. **449**, 451 (2006). doi:10.1051/0004-6361:20054241
J. Braithwaite, Mon. Not. R. Astron. Soc. **397**, 763 (2009). doi:10.1111/j.1365-2966.2008.14034.x
J. Braithwaite, M. Cantiello, Mon. Not. R. Astron. Soc. **428**, 2789 (2013). doi:10.1093/mnras/sts109
J. Braithwaite, Å. Nordlund, Astron. Astrophys. **450**, 1077 (2006a). doi:10.1051/0004-6361:20041980
J. Braithwaite, Å. Nordlund, Astron. Astrophys. **450**, 1077 (2006b). doi:10.1051/0004-6361:20041980
J. Braithwaite, H.C. Spruit, Nature **431**, 819 (2004a). doi:10.1038/nature02934
J. Braithwaite, H.C. Spruit, Nature **431**, 819 (2004b). doi:10.1038/nature02934
A. Brandenburg, K. Subramanian, Phys. Rep. **417**(1–4), 1 (2005)
A. Brandenburg, N.E.L. Haugen, W. Dobler, eprint arXiv (2003)
A. Brandenburg, T. Kahniashvili, A.G. Tevzadze, eprint arXiv:1404.2238 (2014)
G. Brethouwer, P. Billant, E. Lindborg, J.-M. Chomaz, J. Fluid Mech. **585**, 343 (2007). doi:10.1017/S0022112007006854

G.P. Briggs, L. Ferrario, C.A. Tout, D.T. Wickramasinghe, J.R. Hurley, Mon. Not. R. Astron. Soc. **447**, 1713 (2015). doi:10.1093/mnras/stu2539
A.E. Broderick, R. Narayan, Mon. Not. R. Astron. Soc. **383**, 943 (2008). doi:10.1111/j.1365-2966.2007.12634.x
A. Burrows, L. Dessart, E. Livne, C.D. Ott, J. Murphy, Astrophys. J. **664**, 416 (2007). doi:10.1086/519161
F. Carrier, P. North, S. Udry, J. Babel, Astron. Astrophys. **394**, 151 (2002). doi:10.1051/0004-6361:20021122
N. Chamel, Phys. Rev. Lett. **110**(1), 011101 (2013). doi:10.1103/PhysRevLett.110.011101
P. Charbonneau, Annu. Rev. Astron. Astrophys. **52**, 251 (2014). doi:10.1146/annurev-astro-081913-040012
S. Chen, G.L. Eyink, M. Wan, Z. Xiao, Phys. Rev. Lett. **97**, 144505 (2006)
J. Cho, E.T. Vishniac, Astrophys. J. **538**, 217 (2000)
M. Christensson, M. Hindmarsh, A. Brandenburg, Phys. Rev. E **64**(5), 056405 (2001). doi:10.1103/PhysRevE.64.056405
D. Chung, G. Matheou, J. Fluid Mech. **696**, 434 (2012). doi:10.1017/jfm.2012.59
R. Ciolfi, L. Rezzolla, Mon. Not. R. Astron. Soc. **435**, 43 (2013). doi:10.1093/mnrasl/slt092
R. Ciolfi, S.K. Lander, G.M. Manca, L. Rezzolla, Astrophys. J. Lett. **736**, 6 (2011). doi:10.1088/2041-8205/736/1/L6
J.N. Cook, S.L. Shapiro, B.C. Stephens, Astrophys. J. **599**, 1272 (2003). doi:10.1086/379283
J.M. Cordes, T.J.W. Lazio, arXiv Astrophysics e-prints (2002)
S. Dall'Osso, L. Stella, Astrophys. Space Sci. **308**, 119 (2007). doi:10.1007/s10509-007-9323-0
S. Dall'Osso, S.N. Shore, L. Stella, Mon. Not. R. Astron. Soc. **398**, 1869 (2009). doi:10.1111/j.1365-2966.2008.14054.x
J.-F. Donati, J.D. Landstreet, Annu. Rev. Astron. Astrophys. **47**, 333 (2009). doi:10.1146/annurev-astro-082708-101833
M.D. Duez, Y.T. Liu, S.L. Shapiro, M. Shibata, B.C. Stephens, Phys. Rev. D **73**(10), 104015 (2006). doi:10.1103/PhysRevD.73.104015
V. Duez, S. Mathis, Astron. Astrophys. **517**, 58 (2010). doi:10.1051/0004-6361/200913496
V. Duez, J. Braithwaite, S. Mathis, Astrophys. J. Lett. **724**, 34 (2010a). doi:10.1088/2041-8205/724/1/L34
V. Duez, S. Mathis, S. Turck-Chièze, Mon. Not. R. Astron. Soc. **402**, 271 (2010b). doi:10.1111/j.1365-2966.2009.15955.x
R.C. Duncan, C. Thompson, Astrophys. J. Lett. **392**, 9 (1992). doi:10.1086/186413
I. Easson, Astrophys. J. **233**, 711 (1979a). doi:10.1086/157432
I. Easson, Astrophys. J. **228**, 257 (1979b). doi:10.1086/156842
V.G. Elkin, G. Mathys, D.W. Kurtz, S. Hubrig, L.M. Freyhammer, Mon. Not. R. Astron. Soc. **402**, 1883 (2010). doi:10.1111/j.1365-2966.2009.16015.x
G.L. Eyink, Phys. Rev. E **83**(5), 056405 (2011). doi:10.1103/PhysRevE.83.056405
C.-A. Faucher-Giguère, V.M. Kaspi, Astrophys. J. **643**, 332 (2006). doi:10.1086/501516
A.J. Faulkner, I.H. Stairs, M. Kramer, A.G. Lyne, G. Hobbs, A. Possenti, D.R. Lorimer, R.N. Manchester, M.A. McLaughlin, N. D'Amico, F. Camilo, M. Burgay, Mon. Not. R. Astron. Soc. **355**, 147 (2004). doi:10.1111/j.1365-2966.2004.08310.x
C. Federrath, G. Chabrier, J. Schober, R. Banerjee, R.S. Klessen, D.R.G. Schleicher, Phys. Rev. Lett. **107**, 114504 (2011)
L. Ferrario, Mon. Not. R. Astron. Soc. **426**, 2500 (2012). doi:10.1111/j.1365-2966.2012.21836.x
L. Ferrario, D. Wickramasinghe, Mon. Not. R. Astron. Soc. **367**, 1323 (2006a). doi:10.1111/j.1365-2966.2006.10058.x
L. Ferrario, D. Wickramasinghe, Mon. Not. R. Astron. Soc. **367**, 1323 (2006b). doi:10.1111/j.1365-2966.2006.10058.x
L. Ferrario, D. Wickramasinghe, Mon. Not. R. Astron. Soc. **375**, 1009 (2007). doi:10.1111/j.1365-2966.2006.11365.x
L. Ferrario, D. Wickramasinghe, Mon. Not. R. Astron. Soc. **389**, 66 (2008a). doi:10.1111/j.1745-3933.2008.00527.x
L. Ferrario, D. Wickramasinghe, Mon. Not. R. Astron. Soc. **389**, 66 (2008b). doi:10.1111/j.1745-3933.2008.00527.x
L. Ferrario, S. Vennes, D.T. Wickramasinghe, J.A. Bailey, D.J. Christian, Mon. Not. R. Astron. Soc. **292**, 205 (1997)
L. Ferrario, J.E. Pringle, C.A. Tout, D.T. Wickramasinghe, Mon. Not. R. Astron. Soc. **400**, 71 (2009). doi:10.1111/j.1745-3933.2009.00765.x
E. Flowers, M.A. Ruderman, Astrophys. J. **215**, 302 (1977). doi:10.1086/155359
L. Fossati, N. Castro, T. Morel, N. Langer, M. Briquet, T.A. Carroll, S. Hubrig, M.F. Nieva, L.M. Oskinova, N. Przybilla, F.R.N. Schneider, M. Scholler, S. Simon-Diaz, I. Ilyin, A. de Koter, A. Reisenegger, H. Sana, BOB Collaboration, arXiv e-prints (2014)

L.M. Freyhammer, V.G. Elkin, D.W. Kurtz, G. Mathys, P. Martinez, Mon. Not. R. Astron. Soc. **389**, 441 (2008). doi:10.1111/j.1365-2966.2008.13595.x
U. Frisch, A. Pouquet, J. Leorat, A. Mazure, J. Fluid Mech. **68**, 769 (1975)
B.M. Gaensler, N.M. McClure-Griffiths, M.S. Oey, M. Haverkorn, J.M. Dickey, A.J. Green, Astrophys. J. Lett. **620**, 95 (2005). doi:10.1086/428725
E. García-Berro, P. Lorén-Aguilar, G. Aznar-Siguán, S. Torres, J. Camacho, L.G. Althaus, A.H. Córsico, B. Külebi, J. Isern, Astrophys. J. **749**, 25 (2012). doi:10.1088/0004-637X/749/1/25
U. Geppert, J. Gil, G. Melikidze, J. Pons, D. Viganò, in *Electromagnetic Radiation from Pulsars and Magnetars*, ed. by W. Lewandowski, O. Maron, J. Kijak. Astronomical Society of the Pacific Conference Series, vol. 466 (2012), p. 187
W.I. Glaberson, W.W. Johnson, R.M. Ostermeier, Phys. Rev. Lett. **33**, 1197 (1974). doi:10.1103/PhysRevLett.33.1197
K. Glampedakis, N. Andersson, D.I. Jones, Mon. Not. R. Astron. Soc. **394**, 1908 (2009). doi:10.1111/j.1365-2966.2008.13995.x
K. Glampedakis, N. Andersson, S.K. Lander, Mon. Not. R. Astron. Soc. **420**, 1263 (2012). doi:10.1111/j.1365-2966.2011.20112.x
J.P.H. Goedbloed, S. Poedts, *Principles of Magnetohydrodynamics* (2004)
K.N. Gourgouliatos, A. Cumming, A. Reisenegger, C. Armaza, M. Lyutikov, J.A. Valdivia, Mon. Not. R. Astron. Soc. **434**, 2480 (2013). doi:10.1093/mnras/stt1195
J.H. Grunhut, G.A. Wade, in *EAS Publications Series*, vol. 64 (2013), pp. 67. doi:10.1051/eas/1364009
J.E. Gunn, J.P. Ostriker, Astrophys. J. **160**, 979 (1970). doi:10.1086/150487
G.E. Hale, Astrophys. J. **28**, 315 (1908). doi:10.1086/141602
G.E. Hale, F. Ellerman, S.B. Nicholson, A.H. Joy, Astrophys. J. **49**, 153 (1919). doi:10.1086/142452
J.W. Hartman, D. Bhattacharya, R. Wijers, F. Verbunt, Astron. Astrophys. **322**, 477 (1997)
B. Haskell, N. Andersson, A. Passamonti, Mon. Not. R. Astron. Soc. **397**, 1464 (2009). doi:10.1111/j.1365-2966.2009.14963.x
B. Haskell, P.M. Pizzochero, T. Sidery, Mon. Not. R. Astron. Soc. **420**, 658 (2012). doi:10.1111/j.1365-2966.2011.20080.x
N.E.L. Haugen, A. Brandenburg, W. Dobler, Astrophys. Space Sci. **292**, 53 (2004a)
N.E. Haugen, A. Brandenburg, W. Dobler, Phys. Rev. E **70**, 16308 (2004b)
N.E.L. Haugen, A. Brandenburg, A.J. Mee, Mon. Not. R. Astron. Soc. **353**, 947 (2004c)
C. Heiles, Astrophys. J. Suppl. Ser. **111**, 245 (1997). doi:10.1086/313010
G.H. Herbig, Astrophys. J. Suppl. Ser. **4**, 337 (1960). doi:10.1086/190050
A. Hewish, S.J. Bell, J.D.H. Pilkington, P.F. Scott, R.A. Collins, Nature **217**, 709 (1968). doi:10.1038/217709a0
G. Hobbs, A. Faulkner, I.H. Stairs, F. Camilo, R.N. Manchester, A.G. Lyne, M. Kramer, N. D'Amico, V.M. Kaspi, A. Possenti, M.A. McLaughlin, D.R. Lorimer, M. Burgay, B.C. Joshi, F. Crawford, Mon. Not. R. Astron. Soc. **352**, 1439 (2004). doi:10.1111/j.1365-2966.2004.08042.x
S. Hubrig, N. Nesvacil, M. Schöller, P. North, G. Mathys, D.W. Kurtz, B. Wolff, T. Szeifert, M.S. Cunha, V.G. Elkin, Astron. Astrophys. **440**, 37 (2005). doi:10.1051/0004-6361:200500164
S. Hubrig, B. Stelzer, M. Schöller, C. Grady, O. Schütz, M.A. Pogodin, M. Curé, K. Hamaguchi, R.V. Yudin, Astron. Astrophys. **502**, 283 (2009). doi:10.1051/0004-6361/200811533
S. Hubrig, M. Schöller, N.V. Kharchenko, N. Langer, W.J. de Wit, I. Ilyin, A.F. Kholtygin, A.E. Piskunov, N. Przybilla, Magori Collaboration, Astron. Astrophys. **528**, 151 (2011). doi:10.1051/0004-6361/201016345
S. Hubrig, I. Ilyin, M. Schöller, G. Lo Curto, Astron. Nachr. **334**, 1093 (2013). doi:10.1002/asna.201311948
S. Hubrig, L. Fossati, T.A. Carroll, N. Castro, J.F. González, I. Ilyin, N. Przybilla, M. Schöller, L.M. Oskinova, T. Morel, N. Langer, R.D. Scholz, N.V. Kharchenko, M.-F. Nieva, Astron. Astrophys. **564**, 10 (2014). doi:10.1051/0004-6361/201423490
O. Iida, N. Tsuzuki, Y. Nagano, Theor. Comput. Fluid Dyn. **23**, 109 (2009). doi:10.1007/s00162-009-0101-1
M. Ilkov, N. Soker, Mon. Not. R. Astron. Soc. **428**, 579 (2013). doi:10.1093/mnras/sts053
M. Jahan-Miri, Astrophys. J. **725**, 29 (2010). doi:10.1088/0004-637X/725/1/29
S. Jordan, R. Aznar Cuadrado, R. Napiwotzki, H.M. Schmid, S.K. Solanki, Astron. Astrophys. **462**, 1097 (2007). doi:10.1051/0004-6361:20066163
T. Kahniashvili, A. Brandenburg, A.G. Tevzadze, B. Ratra, eprint arXiv (2010)
C. Kalelkar, R. Pandit, Phys. Rev. E **69**(4), 046304 (2004). doi:10.1103/PhysRevE.69.046304
A. Kawka, S. Vennes, Mon. Not. R. Astron. Soc. **425**, 1394 (2012). doi:10.1111/j.1365-2966.2012.21574.x
A. Kawka, S. Vennes, G.D. Schmidt, D.T. Wickramasinghe, R. Koch, Astrophys. J. **654**, 499 (2007). doi:10.1086/509072
A.P. Kazantsev, Sov. Phys. JETP **26**, 1031 (1968)
E.F. Keane, M. Kramer, Mon. Not. R. Astron. Soc. **391**, 2009 (2008). doi:10.1111/j.1365-2966.2008.14045.x

J.C. Kemp, Astrophys. J. **162**, 169 (1970). doi:10.1086/150643
J.C. Kemp, J.B. Swedlund, J.D. Landstreet, J.R.P. Angel, Astrophys. J. Lett. **161**, 77 (1970). doi:10.1086/180574
P.D. Kiel, J.R. Hurley, M. Bailes, J.R. Murray, Mon. Not. R. Astron. Soc. **388**, 393 (2008). doi:10.1111/j.1365-2966.2008.13402.x
O. Kochukhov, Astron. Astrophys. **454**, 321 (2006). doi:10.1051/0004-6361:20064932
C. Kouveliotou, S. Dieters, T. Strohmayer, J. van Paradijs, G.J. Fishman, C.A. Meegan, K. Hurley, J. Kommers, I. Smith, D. Frail, T. Murakami, Nature **393**, 235 (1998). doi:10.1038/30410
M. Kramer, J.F. Bell, R.N. Manchester, A.G. Lyne, F. Camilo, I.H. Stairs, N. D'Amico, V.M. Kaspi, G. Hobbs, D.J. Morris, F. Crawford, A. Possenti, B.C. Joshi, M.A. McLaughlin, D.R. Lorimer, A.J. Faulkner, Mon. Not. R. Astron. Soc. **342**, 1299 (2003). doi:10.1046/j.1365-8711.2003.06637.x
R.M. Kulsrud, E.G. Zweibel, Rep. Prog. Phys. **71**, 6901 (2008)
S.K. Lander, Phys. Rev. Lett. **110**(7), 071101 (2013). doi:10.1103/PhysRevLett.110.071101
S.K. Lander, Mon. Not. R. Astron. Soc. **437**, 424 (2014). doi:10.1093/mnras/stt1894
S.K. Lander, D.I. Jones, Mon. Not. R. Astron. Soc. **424**, 482 (2012). doi:10.1111/j.1365-2966.2012.21213.x
S.K. Lander, N. Andersson, K. Glampedakis, Mon. Not. R. Astron. Soc. **419**, 732 (2012). doi:10.1111/j.1365-2966.2011.19720.x
J.D. Landstreet, S. Bagnulo, V. Andretta, L. Fossati, E. Mason, J. Silaj, G.A. Wade, Astron. Astrophys. **470**, 685 (2007a). doi:10.1051/0004-6361:20077343
J.D. Landstreet, S. Bagnulo, V. Andretta, L. Fossati, E. Mason, J. Silaj, G.A. Wade, Astron. Astrophys. **470**, 685 (2007b). doi:10.1051/0004-6361:20077343
J.D. Landstreet, S. Bagnulo, V. Andretta, L. Fossati, E. Mason, J. Silaj, G.A. Wade, Contrib. Astron. Obs. Skaln. Pleso **38**, 391 (2008)
J.D. Landstreet, S. Bagnulo, G.G. Valyavin, L. Fossati, S. Jordan, D. Monin, G.A. Wade, Astron. Astrophys. **545**, 30 (2012). doi:10.1051/0004-6361/201219829
P.D. Lasky, M.F. Bennett, A. Melatos, Phys. Rev. D **87**(6), 063004 (2013). doi:10.1103/PhysRevD.87.063004
A. Lazarian, J. Cho, Phys. Scr. **2005**, 32 (2005). doi:10.1238/Physica.Topical.116a00032
J. Liebert, P. Bergeron, J.B. Holberg, Astron. J. **125**, 348 (2003). doi:10.1086/345573
J. Liebert, D.T. Wickramasinghe, G.D. Schmidt, N.M. Silvestri, S.L. Hawley, P. Szkody, L. Ferrario, R.F. Webbink, T.D. Oswalt, J.A. Smith, M.P. Lemagie, Astron. J. **129**, 2376 (2005). doi:10.1086/429639
J. Liebert, L. Ferrario, D.T. Wickramasinghe, P. Smith, Astrophys. J. **1**, 1 (2015)
F. Lignières, P. Petit, T. Böhm, M. Aurière, Astron. Astrophys. **500**, 41 (2009). doi:10.1051/0004-6361/200911996
F. Lignières, P. Petit, M. Aurière, G.A. Wade, T. Böhm, in *IAU Symposium*, vol. 302 (2014), pp. 338. doi:10.1017/S1743921314002440
B. Link, Mon. Not. R. Astron. Soc. **421**, 2682 (2012a). doi:10.1111/j.1365-2966.2012.20498.x
B. Link, Mon. Not. R. Astron. Soc. **422**, 1640 (2012b). doi:10.1111/j.1365-2966.2012.20740.x
B. Link, arXiv e-prints (2013)
A. Lyne, F. Graham-Smith, P. Weltevrede, C. Jordan, B. Stappers, C. Bassa, M. Kramer, Science **342**, 598 (2013). doi:10.1126/science.1243254
W.W. Macy Jr., Astrophys. J. **190**, 153 (1974). doi:10.1086/152859
A. Maeder, G. Meynet, Astrophys. J. **793**, 123 (2014). doi:10.1088/0004-637X/793/2/123
R.N. Manchester, A.G. Lyne, F. Camilo, J.F. Bell, V.M. Kaspi, N. D'Amico, N.P.F. McKay, F. Crawford, I.H. Stairs, A. Possenti, M. Kramer, D.C. Sheppard, Mon. Not. R. Astron. Soc. **328**, 17 (2001). doi:10.1046/j.1365-8711.2001.04751.x
P. Marchant, A. Reisenegger, T. Akgün, Mon. Not. R. Astron. Soc. **415**, 2426 (2011). doi:10.1111/j.1365-2966.2011.18874.x
P. Markey, R.J. Tayler, Mon. Not. R. Astron. Soc. **163**, 77 (1973)
J. Maron, S. Cowley, eprint arXiv (2001)
A. Mastrano, A. Melatos, Mon. Not. R. Astron. Soc. **387**, 1735 (2008). doi:10.1111/j.1365-2966.2008.13365.x
A. Mastrano, A. Melatos, Mon. Not. R. Astron. Soc. **417**, 508 (2011). doi:10.1111/j.1365-2966.2011.19290.x
A. Mastrano, A. Melatos, Mon. Not. R. Astron. Soc. **421**, 760 (2012). doi:10.1111/j.1365-2966.2011.20350.x
A. Mastrano, P.D. Lasky, A. Melatos, Mon. Not. R. Astron. Soc. **434**, 1658 (2013). doi:10.1093/mnras/stt1131
A. Mastrano, A. Melatos, A. Reisenegger, T. Akgün, Mon. Not. R. Astron. Soc. **417**, 2288 (2011). doi:10.1111/j.1365-2966.2011.19410.x
G. Mathys, Astron. Astrophys. **293**, 746 (1995)
A. Melatos, Mon. Not. R. Astron. Soc. **313**, 217 (2000). doi:10.1046/j.1365-8711.2000.03031.x
A. Melatos, Astrophys. J. **761**, 32 (2012). doi:10.1088/0004-637X/761/1/32
A. Melatos, C. Peralta, Astrophys. J. Lett. **662**, 99 (2007). doi:10.1086/518598

A. Melatos, M. Priymak, Astrophys. J. **794**, 170 (2014). doi:10.1088/0004-637X/794/2/170
A. Melatos, C. Peralta, J.S.B. Wyithe, Astrophys. J. **672**, 1103 (2008). doi:10.1086/523349
G. Mendell, Mon. Not. R. Astron. Soc. **296**, 903 (1998). doi:10.1046/j.1365-8711.1998.01451.x
M. Meneguzzi, U. Frisch, A. Pouquet, Phys. Rev. Lett. **47**, 1060 (1981)
L. Mestel, Mon. Not. R. Astron. Soc. **133**, 265 (1966)
L. Mestel, D. Moss, Mon. Not. R. Astron. Soc. **405**, 1845 (2010). doi:10.1111/j.1365-2966.2010.16558.x
R. Minkowski, Astrophys. J. **96**, 199 (1942). doi:10.1086/144447
J.P. Mitchell, J. Braithwaite, N. Langer, A. Reisenegger, H. Spruit, arXiv e-prints (2013)
H.K. Moffatt, *Magnetic Field Generation in Electrically Conducting Fluids* (1978)
S.G. Moiseenko, G.S. Bisnovatyi-Kogan, N.V. Ardeljan, Mon. Not. R. Astron. Soc. **370**, 501 (2006). doi:10.1111/j.1365-2966.2006.10517.x
D.J. Morris, G. Hobbs, A.G. Lyne, I.H. Stairs, F. Camilo, R.N. Manchester, A. Possenti, J.F. Bell, V.M. Kaspi, N.D. Amico, N.P.F. McKay, F. Crawford, M. Kramer, Mon. Not. R. Astron. Soc. **335**, 275 (2002). doi:10.1046/j.1365-8711.2002.05551.x
D. Moss, in *Magnetic Fields Across the Hertzsprung-Russell Diagram*, ed. by G. Mathys, S.K. Solanki, D.T. Wickramasinghe. Astronomical Society of the Pacific Conference Series, vol. 248 (2001), p. 305
D. Moss, Astron. Astrophys. **403**, 693 (2003). doi:10.1051/0004-6361:20030431
M.P. Muno, J.S. Clark, P.A. Crowther, S.M. Dougherty, R. de Grijs, C. Law, S.L.W. McMillan, M.R. Morris, I. Negueruela, D. Pooley, S. Portegies Zwart, F. Yusef-Zadeh, Astrophys. J. Lett. **636**, 41 (2006). doi:10.1086/499776
K. Nakabayashi, J. Fluid Mech. **132**, 209 (1983). doi:10.1017/S0022112083001561
C. Neiner, E. Alecian, in *EAS Publications Series*, vol. 64, ed. by K. Pavlovski, A. Tkachenko, G. Torres (2013), pp. 75. doi:10.1051/eas/1364010
C. Neiner, C.P. Folsom, A. Blazere, in *SF2A-2014: Proceedings of the Annual Meeting of the French Society of Astronomy and Astrophysics*, ed. by J. Ballet, F. Martins, F. Bournaud, R. Monier, C. Reylé (2014a), pp. 163
C. Neiner, D. Monin, B. Leroy, S. Mathis, D. Bohlender, Astron. Astrophys. **562**, 59 (2014b). doi: 10.1051/0004-6361/201323093
J. Nordhaus, E.G. Blackman, A. Frank, Mon. Not. R. Astron. Soc. **376**, 599 (2007). doi:10.1111/j.1365-2966.2007.11417.x
J. Nordhaus, S. Wellons, D.S. Spiegel, B.D. Metzger, E.G. Blackman, Proc. Nat. Acad. Sci. **108**, 3135 (2011). doi:10.1073/pnas.1015005108
P. Olesen, Phys. Lett. B **398**(3–4), 321 (1997). doi:10.1016/S0370-2693(97)00235-9
C.D. Ott, A. Burrows, T.A. Thompson, E. Livne, R. Walder, Astrophys. J. Suppl. Ser. **164**, 130 (2006). doi:10.1086/500832
F. Pacini, Nature **219**, 145 (1968). doi:10.1038/219145a0
A. Passamonti, S.K. Lander, Mon. Not. R. Astron. Soc. **429**, 767 (2013). doi:10.1093/mnras/sts372
D.J.B. Payne, A. Melatos, Mon. Not. R. Astron. Soc. **351**, 569 (2004). doi:10.1111/j.1365-2966.2004.07798.x
C. Peralta, A. Melatos, M. Giacobello, A. Ooi, Astrophys. J. **635**, 1224 (2005). doi:10.1086/497899
C. Peralta, A. Melatos, M. Giacobello, A. Ooi, Astrophys. J. **651**, 1079 (2006). doi:10.1086/507576
P. Petit, F. Lignières, M. Aurière, G.A. Wade, D. Alina, J. Ballot, T. Böhm, L. Jouve, A. Oza, F. Paletou, S. Théado, Astron. Astrophys. **532**, 13 (2011). doi:10.1051/0004-6361/201117573
V. Petit, S.P. Owocki, G.A. Wade, D.H. Cohen, J.O. Sundqvist, M. Gagné, J. Maíz Apellániz, M.E. Oksala, D.A. Bohlender, T. Rivinius, H.F. Henrichs, E. Alecian, R.H.D. Townsend, A. ud-Doula, MiMeS Collaboration, Mon. Not. R. Astron. Soc. **429**, 398 (2013). doi:10.1093/mnras/sts344
J.A. Pons, D. Viganò, U. Geppert, Astron. Astrophys. **547**, 9 (2012). doi:10.1051/0004-6361/201220091
J. Power, G.A. Wade, D.A. Hanes, M. Aurier, J. Silvester, in *Physics of Magnetic Stars*, ed. by I.I. Romanyuk, D.O. Kudryavtsev, O.M. Neizvestnaya, V.M. Shapoval (2007), pp. 89
K.H. Prendergast, Astrophys. J. **123**, 498 (1956). doi:10.1086/146186
G.W. Preston, Astrophys. J. Lett. **160**, 143 (1970). doi:10.1086/180547
E. Priest, *Magnetohydrodynamics of the Sun* (Cambridge University Press, Cambridge, 2014), 110778266X. http://books.google.com/books?id=RhL7AgAAQBAJ&pgis=1
R. Prix, G.L. Comer, N. Andersson, Astron. Astrophys. **381**, 178 (2002). doi:10.1051/0004-6361:20011499
A.D. Railton, C.A. Tout, S.J. Aarseth, Proc. Astron. Soc. Aust. **31**, 17 (2014). doi:10.1017/pasa.2014.10
A. Reisenegger, P. Goldreich, Astrophys. J. **395**, 240 (1992). doi:10.1086/171645
S.P. Reynolds, Annu. Rev. Astron. Astrophys. **46**, 89 (2008). doi:10.1146/annurev.astro.46.060407.145237
S.P. Reynolds, R.A. Chevalier, Astrophys. J. **245**, 912 (1981). doi:10.1086/158868
M. Ruderman, Annu. Rev. Astron. Astrophys. **10**, 427 (1972). doi:10.1146/annurev.aa.10.090172.002235
M. Ruderman, Astrophys. J. **366**, 261 (1991). doi:10.1086/169558
M.A. Ruderman, P.G. Sutherland, Nat. Phys. Sci. **246**, 93 (1973). doi:10.1038/physci246093a0

G. Rüdiger, D.A. Shalybkov, M. Schultz, M. Mond, Astron. Nachr. **330**, 12 (2009). doi:10.1002/asna.200811042

A.A. Schekochihin, S.C. Cowley, G.W. Hammett, J.L. Maron, J.C. McWilliams, New J. Phys. **4**(1), 84 (2002a)

A.A. Schekochihin, J.L. Maron, S.C. Cowley, J.C. McWilliams, Astrophys. J. **576**(2), 806 (2002b). doi:10.1086/341814

A.A. Schekochihin, S.C. Cowley, J.L. Maron, J.C. McWilliams, Phys. Rev. Lett. **92**, 64501 (2004)

D.R.G. Schleicher, J. Schober, C. Federrath, S. Bovino, W. Schmidt, New J. Phys. **15**(2), 23017 (2013)

M. Schöller, S. Hubrig, I. Ilyin, N.V. Kharchenko, M. Briquet, J.F. González, N. Langer, L.M. Oskinova, MAGORI Collaboration, Astron. Nachr. **332**, 994 (2011). doi:10.1002/asna.201111606

T. Shiromizu, Phys. Lett. B **443**(1–4), 127 (1998). doi:10.1016/S0370-2693(98)01348-3

I.S. Shklovsky, Dokl. Akad. Nauk SSSR **90**, 983 (1953)

P.S. Shternin, D.G. Yakovlev, Phys. Rev. D **78**(6), 063006 (2008). doi:10.1103/PhysRevD.78.063006

G.W. Simon, N.O. Weiss, Astrophys. J. **489**(2), 960 (1997). doi:10.1086/304800

D. Son, Phys. Rev. D **59**(6), 063008 (1999). doi:10.1103/PhysRevD.59.063008

H. Spruit, E.S. Phinney, Nature **393**, 139 (1998). doi:10.1038/30168

H.C. Spruit, in *IAU Symposium*, vol. 259 ed. by K.G. Strassmeier, A.G. Kosovichev, J.E. Beckman (2009), pp. 61. doi:10.1017/S1743921309030075

G. Srinivasan, D. Bhattacharya, A.G. Muslimov, A.J. Tsygan, Curr. Sci. **59**, 31 (1990)

D.H. Staelin, E.C. Reifenstein III, Science **162**, 1481 (1968). doi:10.1126/science.162.3861.1481

I.H. Stairs, A.G. Lyne, S.L. Shemar, Nature **406**, 484 (2000). doi:10.1038/35020010

L. Stella, S. Dall'Osso, G.L. Israel, A. Vecchio, Astrophys. J. Lett. **634**, 165 (2005). doi:10.1086/498685

A.H. Taub, Arch. Ration. Mech. Anal. **3**, 312 (1959)

R.J. Tayler, Mon. Not. R. Astron. Soc. **161**, 365 (1973)

A.G. Tevzadze, L. Kisslinger, A. Brandenburg, T. Kahniashvili, Astrophys. J. **759**(1), 54 (2012). doi:10.1088/0004-637X/759/1/54

C. Thompson, R.C. Duncan, Astrophys. J. **408**, 194 (1993). doi:10.1086/172580

S.M. Tobias, F. Cattaneo, S. Boldyrev, eprint arXiv (2011)

C.A. Tout, D.T. Wickramasinghe, L. Ferrario, Mon. Not. R. Astron. Soc. **355**, 13 (2004). doi:10.1111/j.1365-2966.2004.08482.x

C.A. Tout, D.T. Wickramasinghe, J. Liebert, L. Ferrario, J.E. Pringle, Mon. Not. R. Astron. Soc. **387**, 897 (2008). doi:10.1111/j.1365-2966.2008.13291.x

M. Tsubota, K. Kasamatsu, T. Araki, eprint arXiv:cond-mat/0309364 (2003)

M. Tsubota, M. Kobayashi, H. Takeuchi, Phys. Rep. **522**, 191 (2013). doi:10.1016/j.physrep.2012.09.007

A.V. Tutukov, A.V. Fedorova, Astron. Rep. **54**, 156 (2010). doi:10.1134/S1063772910020083

C.A. van Eysden, A. Melatos, Mon. Not. R. Astron. Soc. **409**, 1253 (2010). doi:10.1111/j.1365-2966.2010.17387.x

S. Vennes, Astrophys. J. **525**, 995 (1999). doi:10.1086/307949

D. Viganò, J.A. Pons, J.A. Miralles, Comput. Phys. Commun. **183**, 2042 (2012). doi:10.1016/j.cpc.2012.04.029

M. Vigelius, A. Melatos, S. Chatterjee, B.M. Gaensler, P. Ghavamian, Mon. Not. R. Astron. Soc. **374**, 793 (2007). doi:10.1111/j.1365-2966.2006.11193.x

D. Vincenzi, eprint arXiv (2001)

J. Vink, L. Kuiper, Mon. Not. R. Astron. Soc. **370**, 14 (2006). doi:10.1111/j.1745-3933.2006.00178.x

G.A. Wade, MiMeS Collaboration, arXiv e-prints (2014)

G.A. Wade, E. Alecian, D.A. Bohlender, J.-C. Bouret, D.H. Cohen, V. Duez, M. Gagné, J.H. Grunhut, H.F. Henrichs, N.R. Hill, O. Kochukhov, S. Mathis, C. Neiner, M.E. Oksala, S. Owocki, V. Petit, M. Shultz, T. Rivinius, R.H.D. Townsend, J.S. Vink, in *IAU Symposium*, vol. 272, ed. by C. Neiner, G. Wade, G. Meynet, G. Peters (2011), pp. 118. doi:10.1017/S1743921311010131

G.A. Wade, C.P. Folsom, P. Petit, V. Petit, F. Lignières, M. Aurière, T. Böhm, Mon. Not. R. Astron. Soc. **444**, 1993 (2014a). doi:10.1093/mnras/stu1541

G.A. Wade, J. Grunhut, E. Alecian, C. Neiner, M. Aurière, D.A. Bohlender, A. David-Uraz, C. Folsom, H.F. Henrichs, O. Kochukhov, S. Mathis, S. Owocki, V. Petit, in *IAU Symposium*, vol. 302, (2014b), pp. 265. doi:10.1017/S1743921314002233

C. Wang, D. Lai, J.L. Han, Astrophys. J. **639**, 1007 (2006). doi:10.1086/499397

L. Warszawski, A. Melatos, Mon. Not. R. Astron. Soc. **428**, 1911 (2013). doi:10.1093/mnras/sts108

D.T. Wickramasinghe, L. Ferrario, Mon. Not. R. Astron. Soc. **356**, 1576 (2005). doi:10.1111/j.1365-2966.2004.08603.x

D.T. Wickramasinghe, C.A. Tout, L. Ferrario, Mon. Not. R. Astron. Soc. **437**, 675 (2014). doi:10.1093/mnras/stt1910

L. Woltjer, Astrophys. J. **140**, 1309 (1964). doi:10.1086/148028

P.M. Woods, in *40 Years of Pulsars: Millisecond Pulsars, Magnetars and More*, ed. by C. Bassa, Z. Wang, A. Cumming, V.M. Kaspi. American Institute of Physics Conference Series, vol. 983 (2008), pp. 227. doi:10.1063/1.2900149
G.A.E. Wright, Mon. Not. R. Astron. Soc. **162**, 339 (1973)
J.-P. Zahn, A.S. Brun, S. Mathis, Astron. Astrophys. **474**, 145 (2007). doi:10.1051/0004-6361:20077653
B. Zhang, in *Astrophysics and Space Science Library*, vol. 298, ed. by K.S. Cheng, K.C. Leung, T.P. Li (2003), pp. 27
J. Zrake, Astrophys. J. **794**(2), 26 (2014). doi:10.1088/2041-8205/794/2/L26
J. Zrake, A.I. MacFadyen, Astrophys. J. **744**(1), 32 (2011)
J. Zrake, A.I. MacFadyen, Astrophys. J. **769**(2), 29 (2013). doi:10.1088/2041-8205/769/2/L29

DOI 10.1007/978-1-4939-3550-5_5
Reprinted from *Space Science Reviews* Journal, DOI 10.1007/s11214-015-0152-0

Magnetic White Dwarfs

Lilia Ferrario[1] · Domitilla de Martino[2] · Boris T. Gänsicke[3]

Received: 7 November 2014 / Accepted: 17 April 2015 / Published online: 12 May 2015

Abstract In this paper we review the current status of research on the observational and theoretical characteristics of isolated and binary magnetic white dwarfs (MWDs).

Magnetic fields of isolated MWDs are observed to lie in the range 10^3–10^9 G. While the upper limit cutoff near 10^9 G appears to be real, the lower limit is more difficult to investigate. The incidence of magnetism below a few 10^3 G still needs to be established by sensitive spectropolarimetric surveys-conducted on 8 m class telescopes.

Highly magnetic WDs tend to exhibit a complex and non-dipolar field structure with some objects showing the presence of higher order multipoles. There is no evidence that fields of highly magnetic WDs decay over time, which is consistent with the estimated Ohmic decay times scales of $\sim 10^{11}$ yrs. The slow rotation periods ($\sim$ 100 yrs) inferred for a large number of isolated MWDs in comparison to those of non-magnetic WDs (a few days) suggest that strong magnetic fields augment the braking of the stellar core.

MWDs, as a class, also appear to be more massive (0.784 ± 0.047 $M_\odot$) than their weakly or non-magnetic counterparts (0.663 ± 0.136 $M_\odot$).

MWDs are also found in binary systems where they accrete matter from a low-mass donor star. These binaries, called magnetic Cataclysmic Variables (MCVs), comprise about 20–25 % of all known CVs. Zeeman and cyclotron spectroscopy of MCVs have revealed the presence of fields in the range $\sim$ 7–230 MG. Complex field geometries have been inferred in the high field MCVs (the polars) whilst magnetic field strength and structure in the lower field group (intermediate polars, IPs) are much harder to establish.

✉ L. Ferrario
Lilia.Ferrario@anu.edu.au

D. de Martino
demartino@oacn.inaf.it

B.T. Gänsicke
Boris.Gaensicke@warwick.ac.uk

1 Mathematical Sciences Institute, The Australian National University, Canberra, ACT 2601, Australia

2 INAF, Osservatorio Astronomico di Capodimonte, Via Moiariello 16, 80131 Naples, Italy

3 Department of Physics, University of Warwick, Coventry CV4 7AL, UK

The MCVs exhibit an orbital period distribution which is similar to that of non magnetic CVs. Polars dominate the distribution at orbital periods $\lesssim 4$ h and IPs at longer periods. It has been argued that IPs above the 2–3 hr CV period gap with magnetic moments $\gtrsim 5 \times 10^{33}$ G cm^3 may eventually evolve into polars. It is vital to enlarge the still incomplete sample of MCVs to understand not only their accretion processes but also their evolution.

The origin of fields in MWDs is still being debated. While the fossil field hypothesis remains an attractive possibility, field generation within the common envelope of a binary system has been gaining momentum, since it would explain the absence of MWDs paired with non-degenerate companions and also the lack of relatively wide pre-MCVs.

Keywords Magnetic fields · Magnetic white dwarfs · Magnetic cataclysmic variables · Binary systems

1 Introduction

The Sloan Digital Sky Survey (SDSS, York et al. 2000) has increased the number of known MWDs with fields B in the range 2–1000 MG from fewer than 70, as listed in Wickramasinghe and Ferrario (2000) to over 600 in 2015 (see Gänsicke et al. 2002; Schmidt et al. 2003; Vanlandingham et al. 2005; Külebi et al. 2009; Kepler et al. 2013, 2015, and this work). However, their space density estimates are still debated. Volume-limited samples suggest that $\sim$ 10–20 % of WDs are magnetic (Kawka et al. 2007; Giammichele et al. 2012; Sion et al. 2014), whereas magnitude-limited samples indicate that only $\sim$ 2–5 % are magnetic (Liebert et al. 2003a; Kepler et al. 2013, 2015). This discrepancy may be partly resolved by correcting for the difference in search volume for the MWDs, since, on average, they are more massive than their non-magnetic counterparts (see Sect. 4 and Liebert et al. 2003a).

A spectropolarimetric survey of a small sample of cool ($\lesssim$ 14000 K) WDs conducted by Landstreet et al. (2012) found that the probability of detecting a kG field in DA[1] WDs is $\sim$ 10 % per decade of field strength but also stress the inability of current precision measures to reveal whether there is a lower cutoff to the field strengths in WDs or there is a field below which all WDs are magnetic.

Furthermore, there seems to be a paucity of young MWDs in the intermediate field range (0.1–1 MG, see Kawka and Vennes 2012). The reason for this dearth of objects is not clear since they should be easily detected in most spectropolarimetric surveys such as that conducted by Aznar Cuadrado et al. (2004).

The ESO SNIa Progenitor Survey (SPY) also pointed to a peculiarly low percentage of MWDs in the field range 0.1–1 MG (Koester et al. 2001). Therefore, the magnetic field distribution of WDs could be bimodal exhibiting a high field (1–1000 MG) population and a low field (< 0.1 MG) one. Intensive searches for magnetic fields in the direct progenitors of MWDs have so far drawn a blank, both among planetary nebulae (Asensio Ramos et al. 2014; Leone et al. 2014), and hot subdwarfs (Mathys et al. 2012; Savanov et al. 2013).

MWDs can also be found in CVs, which are close binary systems where a WD accretes matter from a late-type main sequence companion through Roche-lobe overflow. Orbital periods are less than a day and orbital separations of the order of the solar radius making these binaries relatively compact. Because of their abundance[2] and proximity, CVs are key objects to understand close binary evolution. In particular, those hosting a strongly magnetic

[1] DA type WDs have a hydrogen rich atmosphere.

[2] 1166 CVs are reported in the 7.20 (Dec. 2013) version of the Ritter and Kolb (2003) catalogue.

WD ($B \gtrsim 1$ MG) allow us to study accretion and emission processes in a strong magnetic field environment as well as improving our understanding of the influence of magnetic fields in close binary evolution.

The number of known MCVs has also increased dramatically over the years from about 60 as listed in Wickramasinghe and Ferrario (2000) to about 170. Magnetic field measures are however available only for about half of them most of which in the range $\sim$7–230 MG.

Here we review the current status of research on MCVs and MWDs and outline key aspects and open problems to be investigated in the future. Previous reviews on the topic can be found in Angel (1978), Angel et al. (1981), Cropper (1990), Chanmugam (1992), Patterson (1994), Wickramasinghe and Ferrario (2000).

2 Historical Background

2.1 Isolated Magnetic White Dwarfs

Grw+70°8247, discovered by Kuiper (1934), was the first WD to be classified as magnetic when Kemp et al. (1970) demonstrated that its light was strongly circularly polarised. Although the spectrum of Grw+70°8247 appeared to be nearly featureless, close inspection by Minkowski (1938) and Greenstein and Matthews (1957) revealed the presence of unusual shallow broad absorption bands near 3650 Å, 4135 Å and 4466 Å, which became known as "Minkowski bands". The spectral features of Grw+70°8247 remained unidentified till the mid-80s, when the first computations of the hydrogen transitions in strong magnetic fields became available (see Sect. 3.1). These calculations allowed the spectral features of Grw+70°8247 to be identified as Zeeman shifted hydrogen lines in a magnetic field of 100–320 MG (Angel et al. 1985; Greenstein et al. 1985; Wickramasinghe and Ferrario 1988a). In particular, the famous Minkowski band near 4135 Å was shown to be a Zeeman component of Hβ shifted some 700 Å from its zero field position.

Following the discovery of polarisation in Grw+70°8247, Angel and Landstreet (1971a) detected circular polarisation in another object, G195-19. Follow-up observations conducted by Angel and Landstreet (1971b) revealed that the circular polarisation in G195-19 varies at the spin period of the star under the assumption that the magnetic axis is inclined with respect to the spin axis of the WD (the 'oblique rotator' of Stibbs 1950). In both Grw+70°8247 and G195-19, some linear polarisation was also detected, although at a much lower level than circular polarisation. Since the observations showed that the polarisation data in G195-19 did not follow a sinusoidal behaviour, Landi Degl'Innocenti (1976) raised the possibility that magnetic spots were responsible for the observed asymmetries.

Landstreet and Angel (1971) soon discovered circular polarisation in a third WD, G99-37 and a few years later, polarisation was also detected in GD229 (Swedlund et al. 1974a). The spectral features of GD229 remained a mystery for more than 20 years until the calculations for He I transitions in a strong magnetic field became available (see Sect. 3.1). The absorption structures were thus interpreted as stationary line transitions of helium in a magnetic field of 300–700 MG (Jordan et al. 1998; Wickramasinghe et al. 2002).

The sample of known MWDs has rapidly increased since Kemp's first discovery in 1970. Over the past 40 years, thanks to surveys such as the Hamburg/ESO Quasar Survey (HQS, Wisotzki et al. 1991), the Edinburgh-Cape survey (Kilkenny et al. 1991), and, as already noted, the SDSS, their number has grown to more than 600 (if we also count objects with uncertain or no field determination) thus providing a large enough sample to allow meaningful statistical studies of their characteristics (see Table 1).

2.2 Magnetic Cataclysmic Variables

Known as a Novalike since 1924, AM Her was the first CV discovered to emit soft X-rays by the *UHURU* and *SAS-3* satellites in 1976 (Hearn et al. 1976). Follow-up optical observations revealed variable linear (up to 7 %) and circular (up to 9 %) polarisation at the 3.09 h binary orbital period (Tapia 1977). Systems with similar characteristics were named AM Her-type variables. The name "polar" was introduced later for AM Her and other objects identified as X-ray sources that also showed polarised optical light (see Warner 1995).

The binary system DQ Her was discovered in the mid-50s to display a 71 s periodic variability (Walker 1956) and only about two decades later was also found to be weakly polarised (Swedlund et al. 1974b). In 1978, 33 s optical pulsations were detected in AE Aqr, but not in polarised light (Patterson 1979). In the following years, fast optical periodic variations at periods much shorter than the orbital one ($P_{\rm rot} \ll P_{\rm orb}$) were found in other CVs. These systems were first called DQ Her-type variables and then renamed "intermediate polars" (IPs) (Patterson and Steiner 1983; Patterson 1994; Warner 1995). This led to the division of MCVs into two groups, polars and IPs.

The properties of polarised radiation were first studied by Chanmugam and Dulk (1981) and Meggitt and Wickramasinghe (1982). The magnetic moment of the accreting WD in polars ($\mu \gtrsim 5 \times 10^{33}$ G cm^3) is sufficient to lock the stars into synchronous rotation with the orbital period ($P_{\rm orb} \sim 70$–480 mins). On the other hand, the magnetic moment of the WD in IPs is not high enough to phase-lock the stars into synchronous rotation with the orbit, resulting in WD spin periods $P_{\rm s}$ that are shorter than their orbital periods.

Because of their strong soft X-ray emission, the number of polars increased significantly thanks to soft X-ray surveys, and in particular that conducted in the nineties by the *ROSAT* satellite (Beuermann 1999). To date ~110 systems of this class are known hosting WDs with surface field strengths $B \sim 7$–230 MG (see Sect. 3 and Table 2 for a complete list of known systems as of December 2014). The IPs constituted a minor group of harder X-ray sources and remained elusive objects until the recent hard X-ray surveys conducted by the *INTEGRAL*/IBIS and *Swift*/BAT satellites (Bird et al. 2010; Baumgartner et al. 2013). The flux limits ($\sim 10^{-11}$ erg cm^{-2} s^{-1}) of these surveys can detect sources up to 1 kpc away for X-ray luminosities $L_x \sim 10^{33}$ erg s^{-1}. IPs, being the brightest and hardest X-ray sources among CVs, account for ~ 20 % of the Galactic hard X-ray sources discovered (Barlow et al. 2006). Thus, the number of identified IPs has now increased to ~ 55 systems (see Table 3 and reference therein and also Bernardini et al. 2012, 2013), with ~60 candidates still awaiting confirmation through X-ray follow-ups with sensitive facilities such as XMM-Newton and NuSTAR.[3]

3 Magnetic Field Measures

In this section we shall review the techniques that are routinely adopted to determine the magnetic field strength of isolated MWDs and of accreting MWDs in binary systems.

3.1 Field Determination in Isolated Magnetic White Dwarfs

Similarly to their non-magnetic counterparts, most MWDs are hydrogen-rich (DA). They are classified as DAp or DAH, indicating the method used in their discovery. Thus, "p"

[3]Further details on IP type CVs can be found at http://asd.gsfc.nasa.gov/Koji.Mukai/iphome/iphome.html.

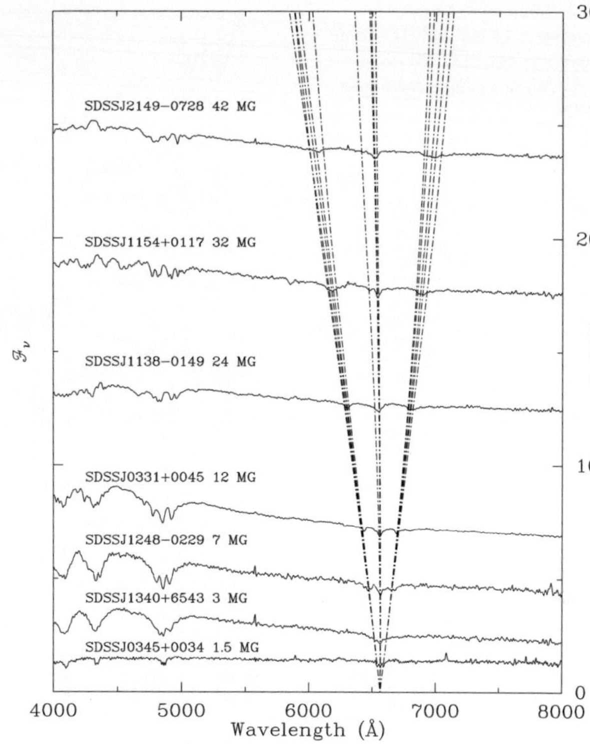

Fig. 1 The Zeeman effect on H_α in the linear and quadratic regimes for fields of 1.5–42 MG. The quadratic effect becomes gradually more important in the higher members of the Balmer series and as the field strength increases (Schmidt et al. 2003)

stands for polarisation measurements and "H" for Zeeman splitting (see Sion et al. 1983, for more information on the WD spectral classification system). An understanding of their properties relies heavily on the theory of the hydrogen atom at strong magnetic fields.

Depending on field strength, different Zeeman regimes become relevant to the understanding of MWD spectra. If (n, l, m_l) are the quantum numbers corresponding to zero field, then the removal of the m_l degeneracy will give rise to the linear Zeeman regime ($\sim$ 1–5 MG for Balmer lines). Here the energy levels are shifted by $\frac{1}{2}m_l h\omega_C$, where $\omega_C = \frac{eB}{m_e c}$ is the cyclotron frequency of a free electron, m_e and e are the mass and charge of the electron respectively and c is the speed of light. A line is split into three components: an unshifted central π component ($\Delta m_l = 0$), a redshifted σ_+ ($\Delta m_l = +1$) component and a blueshifted σ_- ($\Delta m_l = -1$) component. Some Zeeman triplets observed in MWDs are shown in Fig. 1 (Schmidt et al. 2003). Circular spectropolarimetry across lines can be effectively used to detect low fields ($B \lesssim 1$ MG) when Zeeman splits are unresolvable in the flux spectra. When viewed along the magnetic field, the σ_- and the σ_+ components are circularly polarised with opposite signs.

As the field strength increases and/or n increases, the quadratic effect becomes gradually more important and the l degeneracy is also removed (inter-l mixing regime). The energy shifts depend on the excitation of the electron and the π and σ components are shifted by different amounts from their zero field positions. The quadratic shift is comparable to the linear shift for, e.g. H_δ, at $B \sim 4$ MG (see Fig. 1). The first Zeeman calculations in this intermediate field strength regime were conducted in 1974 by Kemic for fields up to $\sim$ 20 MG.

Fig. 2 Stationary Zeeman components of H_α and H_β (from Vanlandingham et al. 2005) in the spectra of strongly magnetic MWDs

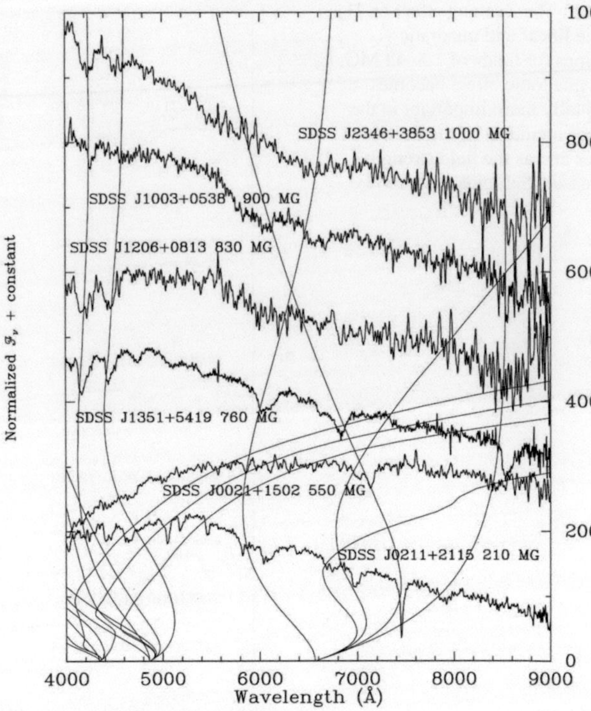

As the field progressively increases, the Coulomb and magnetic field forces become comparable in strength and neighbouring n manifolds overlap (inter-n mixing regime). In the "strong field mixing regime" the magnetic field dominates (see Wickramasinghe and Ferrario 2000). The first published data of wavelengths and oscillator strengths of hydrogen transitions in the infrared to ultraviolet bands in the presence of very strong magnetic fields (up to 10^6 MG) were published in the mid 80s by Roesner et al. (1984), Forster et al. (1984), Henry and O'Connell (1984, 1985), Wunner et al. (1985), and more recently by Schimeczek and Wunner (2014).

An important characteristic which is clearly visible in the field against wavelength curves diagrams is that the σ^+ components become nearly 'stationary'. That is, appreciable changes in B only yield small changes in wavelength. This is very useful in establishing the field of MWDs, since the features corresponding to these turning points will have the smallest amount of magnetic broadening and will have the largest influence on the observed field averaged spectrum. We display in Fig. 2 some spectra of strongly magnetic WDs showing the presence of stationary components (Vanlandingham et al. 2005).

Another interesting effect in the presence of strong magnetic fields ($\gtrsim 100$ MG) and local electric fields in highly ionised plasmas, is an increase of the oscillator strength of the "forbidden" $1s_0 \rightarrow 2s_0$ component at the expense of the π $(1s_0 \rightarrow 2p_0)$ component. This was first detected in *HST* observations of RE J0317–853 (Burleigh et al. 1999), and later seen in the MCV AR UMa (Gänsicke et al. 2001; Schmidt et al. 1996).

3.1.1 Magnetic Field Evolution in Isolated Magnetic White Dwarfs

There is no evidence for field evolution along the cooling curve, that is, the mean field strength and the distribution about this mean appear to be independent of effective tempera-

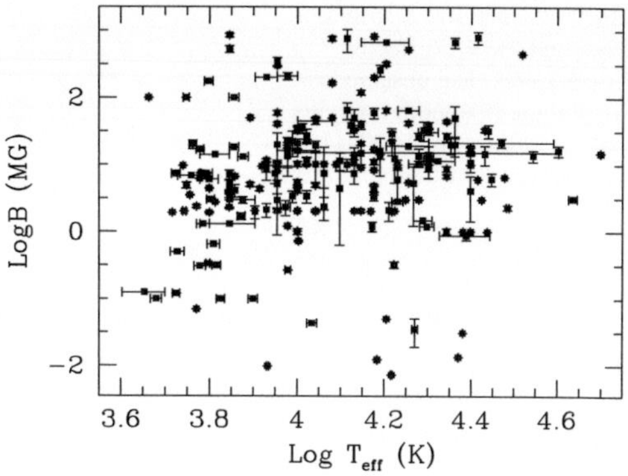

Fig. 3 Magnetic field strength against effective temperature in MWDs showing no indication for field evolution with cooling age (this work)

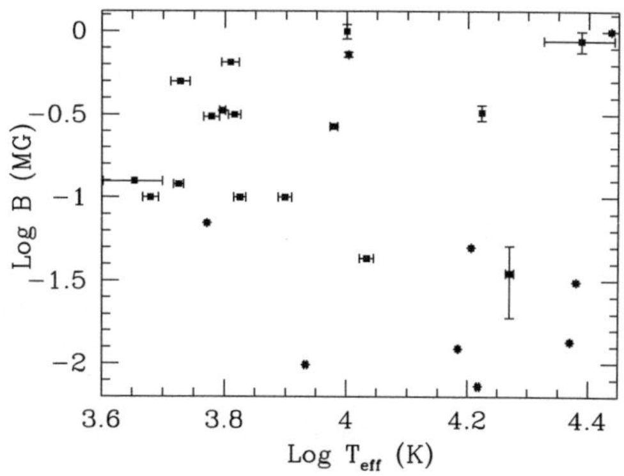

Fig. 4 Magnetic field strength against effective temperature in MWDs with $B \lesssim 1$ MG showing some correlation between field and temperature (this work)

ture (see Fig. 3, this work). The Pearson Product Moment Correlation test gives a correlation coefficient $r = 0.03$, which indicates that these variables are not related.

The free Ohmic decay time can be estimated from

$$t_{\text{ohm}} \sim \frac{4\pi\sigma L^2}{c^2}$$

where L is the length scale over which the magnetic field varies and σ is the electrical conductivity. If we set $L \sim R$ (where R is the stellar radius) and σ equal to the value expected in the degenerate cores of WDs then we have $t_{\text{ohm}} \sim 2\text{–}6 \times 10^{11}$ yr almost independently of mass (Cumming 2002). The lack of evidence for correlation between magnetic field strength and effective temperature is consistent with these long decay time scales.

We note that Kawka and Vennes (2012) find that the distribution of field strengths below 1 MG versus cooling ages may show some selections effects (see their Fig. 10). That is, objects with fields $\lesssim 50$ kG appear to be younger (that is hotter) than those with field $\gtrsim$ 50 kG. We have calculated a correlation coefficient $r = -0.192$ for objects with $B \lesssim 1$ MG, indicating that this effect does exist also in our sample of weakly magnetic WDs listed in Table 1 (see Fig. 4). This trend could be caused by the fact that cool WDs ($\lesssim 7000$ K) do not have narrow and deep lines in their spectra that are good magnetic field tracers unless heavy element lines are present (Kawka and Vennes 2012). Another possibility is that the estimation of parameters such as effective temperature and gravity using models for non-magnetic WDs *may* introduce biases even in the presence of very low fields. However, the latter possibility is more difficult to ascertain at the present time. A possible explanation for

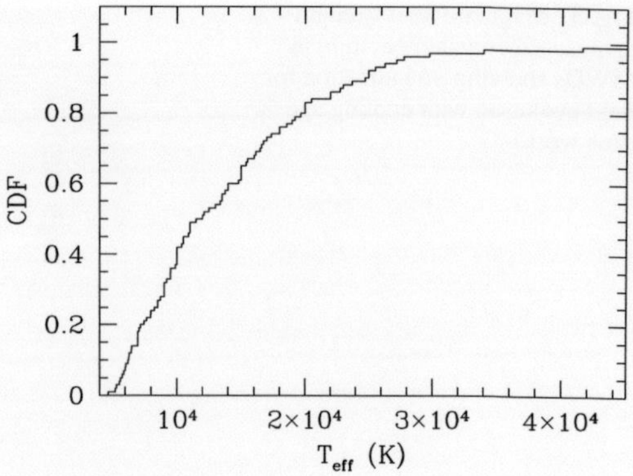

Fig. 5 Cumulative distribution function of MWD effective temperatures. The observed distribution is smooth implying that the birthrate of MWDs has not changed over time (this work)

why this effect is not apparent if we take all MWDs could be because their temperatures are estimated using a wide range of methods so that biases cancel each other out.

Finally, we can state that the claim originally made by Liebert and Sion (1979) and supported by Fabrika and Valyavin (1999) that there is a higher incidence of magnetism among cool WDs than among hot WDs does not appear to be corroborated by the present enlarged sample of MWDs. We show in Fig. 5 the cumulative distribution function of the effective temperatures of the observed sample MWDs (see Table 1). We note that this function is smooth over the entire range of effective temperatures $T_{\rm eff} = 4000$–45000 K thus indicating that the birthrate of MWDs has not significantly changed over the age of the Galactic disk.

3.2 Field Determination of White Dwarfs in Binary Systems

Direct measurements of the WD magnetic field strength in the high field magnetic CVs, the polars, can be obtained either (i) through Zeeman splitting of the photospheric hydrogen absorptions lines when these systems enter low accretion states (as described in Sect. 3.1) or (ii) through the modelling of cyclotron emission features that characterise the optical to IR spectra during intermediate and high accretion states (see Wickramasinghe and Ferrario 2000) or (iii) via the study of Zeeman features arising from the halo of matter surrounding the accretion shock.

When accreting at low rates, polars reveal the WD photosphere and thus can allow the detection of the Zeeman σ_+, σ_- and π components of Balmer line absorptions. Thus some polars have their field determined through Zeeman spectroscopy of photospheric lines in the optical wavelength range (Wickramasinghe and Martin 1985; Ferrario et al. 1992; Schwope et al. 1995a). The highest magnetised polars, AR UMa (230 MG, Schmidt et al. 1996) and AP CrB (144 MG, Gänsicke et al. 2004), were instead detected through Zeeman split absorption features in the UV range. We show in Fig. 6 an ultraviolet spectrum of AR UMa covering the range 917–1182 Å when the system was in its typical low-accretion state. The spectral absorption features are caused by Lyα–Lyγ Zeeman transitions. The modelling indicates a dipolar field strength of about 235 MG offset along its axis by a 0.15 of the stellar radius (Hoard et al. 2004).

The optical to IR spectra of polars may also reveal the typical undulation of cyclotron humps. At low temperatures, the position of the nth harmonic for a magnetic field B and viewing angle near 90° can be estimated using

$$\lambda_{\rm n} = \frac{10710}{n}\left(\frac{10^8\ {\rm G}}{B}\right)\ \text{Å}.$$

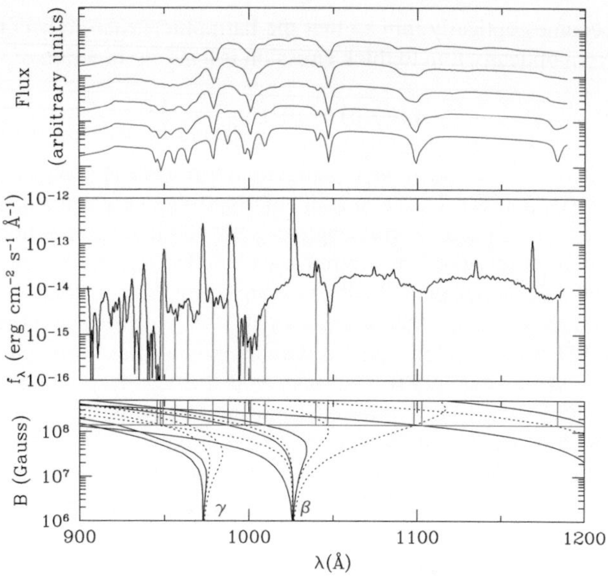

Fig. 6 FUSE spectrum of the high field MCV AR UMa (*middle panel*) compared with hydrogen Lyman transitions for $1\ \text{MG} \le B \le 500\ \text{MG}$ (*bottom panel*) and with photospheric model spectra (*top panel*). The model correspond (from *top* to *bottom*) to dipole field strengths and fractional offsets of 200 MG, 0.0; 215 MG, 0.10; 235 MG, 0.15; 260 MG, 0.20; and 280 MG, 0.25. Negative offsets imply that we view the weaker field hemisphere, where the field distribution is more uniform. Dashed lines represent normally forbidden components that are enabled by the strong electric fields present in highly magnetic WDs (Hoard et al. 2004)

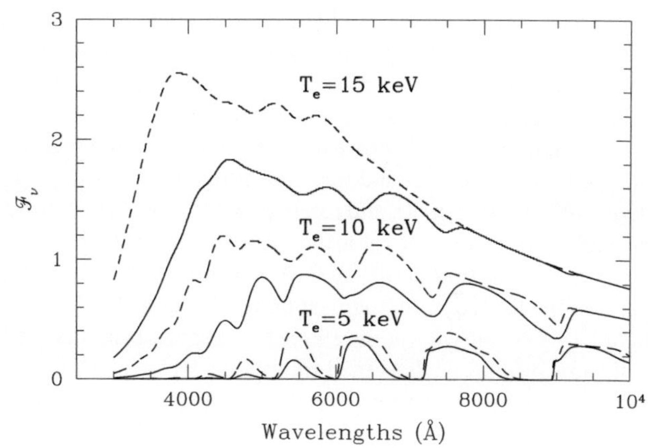

Fig. 7 Theoretical cyclotron spectra for a field $B = 30$ MG as a function of electron temperature T_e for a viewing angle $\theta = 90°$. The *solid* and *dashed curves* are for optical depth parameters $\Lambda = 2 \times 10^5$ and $\Lambda = 10^6$ respectively (from Wickramasinghe and Ferrario 2000)

Cyclotron lines in polars are only seen when the shocks are viewed at large angles θ to the field direction. The intensity is at its maximum at the fundamental and rapidly decreases as n increases, with the rate of decline depending on temperature.

We show in Fig. 7 a number of theoretical cyclotron emission spectra. These have been obtained using the Wickramasinghe and Meggitt (1985) constant Λ (or "point source") models which assume uniform conditions in the shock but allow for optical depth effects. The parameters of these models have been chosen to yield the characteristic cyclotron undulation in the optical band.

At long wavelengths these spectra display an optically thick Rayleigh-Jeans tail while at shorter wavelengths they are characterised by a power law spectrum modulated by cyclotron lines. The flat-topped profiles of the harmonic peaks at low harmonic numbers implies optically thick emission. As the harmonic number increases the opacity drops and the shock becomes optically thin so that the harmonic structure becomes clearly visible. The switch from optically thin to thick emission is a strong function of the optical depth parameter Λ

$$\Lambda = 2.01 \times 10^6 \left(\frac{s}{10^6\ \mathrm{cm}}\right)\left(\frac{N_\mathrm{e}}{10^{16}\ \mathrm{cm}^{-3}}\right)\left(\frac{3 \times 10^7\ \mathrm{G}}{B}\right)$$

where s is a characteristic path length through the post-shock region and N_e is the electron density number. The parameter Λ is approximately equal to the optical depth at the cyclotron fundamental at a viewing angle $\theta = 90°$ to the field direction. Cyclotron emission in MCVs has been observed at infrared wavelengths (e.g. Bailey et al. 1991; Ferrario et al. 1993a, 1996; Campbell et al. 2008b,a), optical (e.g. Visvanathan and Wickramasinghe 1979; Wickramasinghe et al. 1989; Ferrario et al. 1994; Schwope and Beuermann 1990a; Schwope et al. 1995b), and in a few systems with the highest field strengths also at ultraviolet wavelengths (e.g. Rosen et al. 2001; Gänsicke et al. 2001; Ferrario et al. 2003).

The values of Λ inferred from the modelling of cyclotron emission ($\sim 10^5$–10^6) are much lower than those expected for a bremsstrahlung-dominated shock ($\sim 10^8$–10^9, e.g., Lamb and Masters 1979; Chanmugam and Dulk 1981), thus indicating that the cyclotron radiation mainly comes from strongly cyclotron cooled shock regions characterised by low specific accretion rates.

The lack of harmonic features seen in the intensity spectra of most polars also supports the hypothesis that the emission regions of these systems are extended and structured. This flat energy distribution has been attributed to magnetic field spread and to the optical depth parameter varying across a wide region characterised by different specific mass flow rates. More realistic models that take into consideration the effects of extension and temperature and density distribution across the emission region were constructed by, e.g., Wickramasinghe and Ferrario (1988b), Ferrario and Wickramasinghe (1990), Rousseau et al. (1996).

Through the careful fitting of cyclotron harmonic features it is possible to determine the magnetic field strength and physical parameters of the post-shock region (see Wickramasinghe and Ferrario 2000, and references therein). Many polars have measured field strengths through time-resolved cyclotron spectroscopy. Information on the accretion geometry can also be gained through the study of harmonics that shift with orbital phase due to different parts of the accretion region becoming visible as the WD rotates (e.g. Schwope and Beuermann 1990b; Burwitz et al. 1996b). The cyclotron study of the phase resolved spectra of eclipsing systems such as the bright polar HU Aquarii by Schwope et al. (2003) has proven to be particularly important to establish the accretion geometry of these objects.

In many cases two independent sets of cyclotron lines arising from regions with different magnetic field strengths have been found, with the main accreting pole possessing a weaker field (Ferrario et al. 1993a, 1996; Schwope et al. 1995a; Schwope 1996; Campbell et al. 2008b). These studies indicate that the magnetic field distribution is not that of a centred dipole and offsets as high as 10–20 % are often inferred (Wickramasinghe and Ferrario 2000). Evidence of non-centred dipole field distribution also comes from Zeeman components which are seen against strong cyclotron emission and are formed in the free fall material surrounding the WD pole, often named "halo" (Achilleos et al. 1992a; Schwope et al. 1995a). The study of all these different components have shown that the field strength obtained from the modelling of photospheric Zeeman lines, $B_{\mathrm{Zeem,phot}}$, is different from the field strength obtained from the study of halo Zeeman features, $B_{\mathrm{Zeem,halo}}$, with the latter comparable to the field strength B_{cyc} obtained through the modelling of cyclotron humps. This is because the field strength measured from photospheric Zeeman split lines is averaged over the entire visible hemisphere of the WD while the fields derived from cyclotron modelling or from the study of halo Zeeman features arise from regions close to the visible accreting pole.

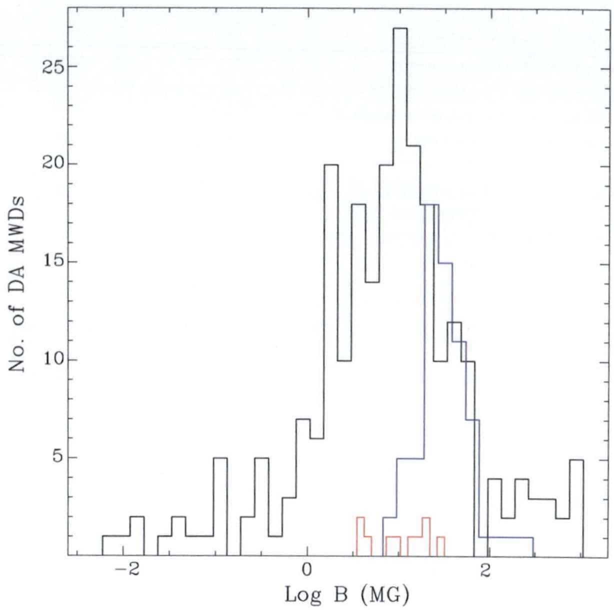

Fig. 8 Distributions of magnetic field strength in polars (*blue line*) and IPs (*red line*) compared to that of single magnetic WDs (*black line*). This figure has been prepared using the data in Tables 1, 2, and 3 (this work)

From time-resolved polarimetry (e.g. Potter et al. 2004) and spectropolarimetry (e.g. Beuermann et al. 2007) detailed information on the complexity (quadrupole or even multipoles) of magnetic field topology in these systems can be obtained (see also Sect. 3.3). However, in systems with ages $\gtrsim$ 1 Gyr a substantial decay of multipole components could be expected and thus short period MCVs may not have complex fields (Beuermann et al. 2007). Magnetic field strengths have been measured or estimated for ~86 WDs in binaries (see Tables 2 and 3 for a complete list of known systems as of December 2014). Using the main pole magnetic field strength, Fig. 8 depicts the magnetic field distribution of polars compared to that of single MWDs listed in Table 1 with the latter having fields in the range 0.1–1000 MG. The polars clearly populate a more restricted range of field strengths, 7–230 MG, with a mean value of 38 MG.

Fields strengths above 230 MG, which are detected in single magnetic WDs, are not found in polars and there is no clear explanation for this yet. High magnetic field polars could be difficult to identify due to selection effects because these systems would be highly intermittent soft X-ray sources such as AR UMa. Hameury et al. (1989) explained the paucity of very high field polars in terms of their very short lifetimes due to efficient loss of angular momentum via magnetic braking mechanism. However, it appears more likely that the strong fields in polars may *decrease* the efficiency of magnetic braking, which would result in a slower evolution of their orbital periods and in lower accretion luminosities (Li and Wickramasinghe 1998; Webbink and Wickramasinghe 2002; Araujo-Betancor et al. 2005a). On the other hand, if the magnetic field is generated during the CE phase, then the highest fields could only be produced when the two stars merge to give rise to an isolated MWD (see Sect. 6 and the chapter on the origin of magnetic fields in this book).

The lowest surface averaged magnetic field strength measured in a polar is 7 MG in V2301 Oph which was modelled by Ferrario et al. (1995) with a dipolar field of 12 MG offset by 20 % from the centre of the WD. The lack of lower field synchronous systems could be explained if the asynchronous IPs represent the low field tail of the magnetic field strength distribution in MCVs. However this is difficult to prove, because the absence of

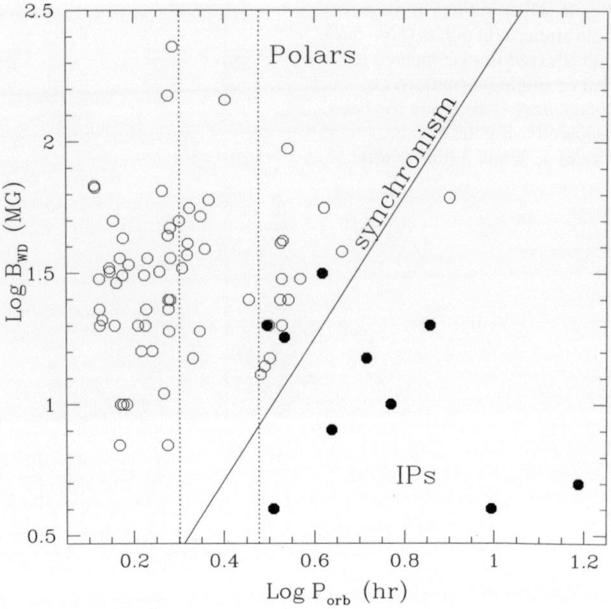

Fig. 9 The magnetic field strength and orbital period diagram for polars (*empty circles*) and IPs (*filled circles*). The *line* at which synchronism is expected for a mean mass accretion rate as a function of $P_{\rm orb}$ is reported as a *solid line* (adapted from Beuermann 1999). The *dotted vertical lines* mark the orbital period gap

low accretion states in IPs prevents the WD photosphere to become visible, thus precluding the detection of photospheric Zeeman split lines. Most of these systems do not show polarised optical/IR emission or cyclotron features, which also prevents the determination of the WD magnetic field. Whether the lack of polarisation in the optical/IR is caused by weaker magnetic fields or is due to efficient depolarisation mechanisms is difficult to ascertain. So far only ten IPs are known to be circularly polarised at a level $\lesssim 1$–$3\ \%$ (Penning et al. 1986; Piirola et al. 1993, 2008; Buckley et al. 1995; Potter et al. 1997, 2012; Katajainen et al. 2007; Butters et al. 2009). In these IPs, the field strengths are estimated to be in the range ~ 5–20 MG, with V405 Aur (Piirola et al. 2008) possessing the highest ($\sim$30 MG) field (see Table 3). Seven are also found to show a soft X-ray blackbody component. The fields of these IPs, shown in Fig. 8, are at the low-field end of the distribution and partially overlap with the low field polars. Their orbital periods are all above the CV period gap (see Fig. 9) and, given the large uncertainties in the estimates, they are below or close to the line at which synchronism is expected to occur (see Fig. 9, Beuermann 1999). These systems could in fact be the progenitors of the low-field polars, as suggested by Norton et al. (2004). As the number of polarised IPs has increased by a factor of three in the last few years, further polarimetric and spectropolarimetric studies of all known IPs are crucial to establish their field strengths and accretion properties.

3.3 Field Topology

When discussing the magnetic field strength in isolated MWDs, some attention has to be given to the definition of B, as the field strength measured from observations is usually an average value over the visible hemisphere of the WD. In the absence of any other information, it is common practice to assume a dipolar field configuration, in which case the field varies by a factor two between the magnetic pole and the equator. When averaging over the visible hemisphere, the relative weighting of regions with different fields strengths depends

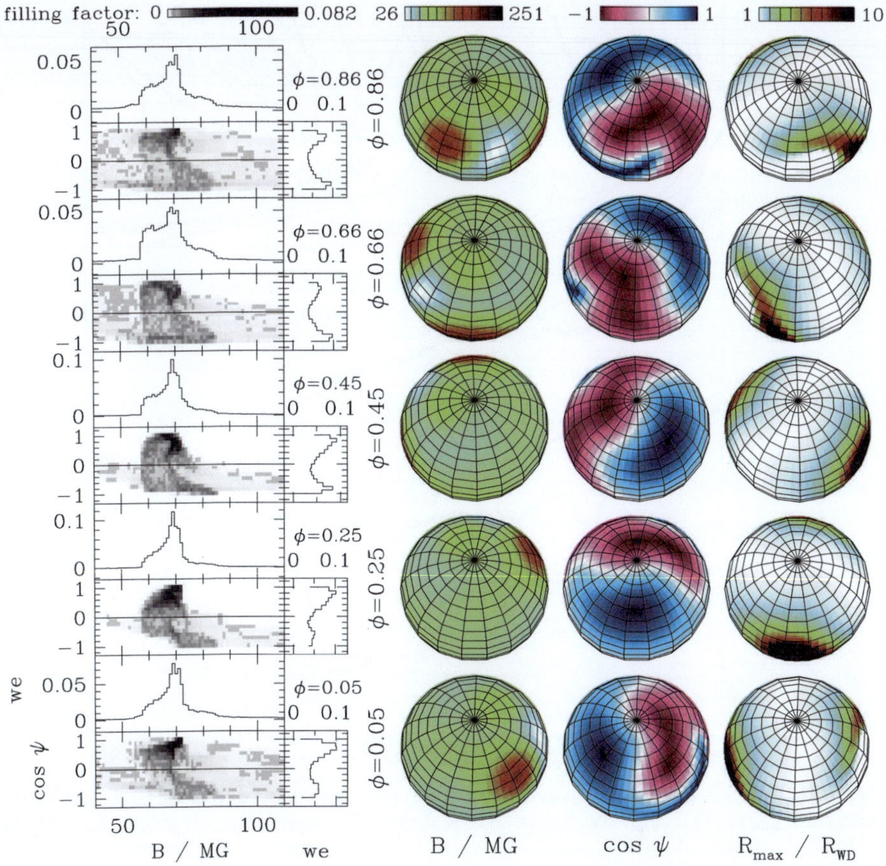

Fig. 10 Field topology of the MWD PG 1015+014, derived from a tomographic analysis of time-resolved spectropolarimetry (Euchner et al. 2005b). Fields in the range 50–80 MG are detected (*left panel*) with a highly non-dipolar configuration (*middle panels*: surface field and angle between the line-of-sight and the magnetic field direction). The maximum radial distance reached by field lines in units of the WD radius is shown in the *right-most panel*

on the angle between the observer's line of sight and the magnetic field axis (see Figure 2 of Achilleos et al. 1992b).

However, detailed spectroscopy and spectropolarimetry have demonstrated in a number of cases that the field topologies can be very complex (e.g. Putney and Jordan 1995). Several sophisticated tomographic reconstruction methods have been developed to map the field topology (Donati et al. 1994; Euchner et al. 2002). The application of these methods to both single MWDs (Euchner et al. 2005b, 2006) and MWDs in polars (Beuermann et al. 2007) reveals a startling complexity of the field topologies (Fig. 10).

3.4 Beyond Hydrogen

The bulk of all WDs have hydrogen-dominated atmospheres, both in magnitude-limited (Kleinman et al. 2013) and volume-limited (Giammichele et al. 2012) samples, and the same is true for MWDs.

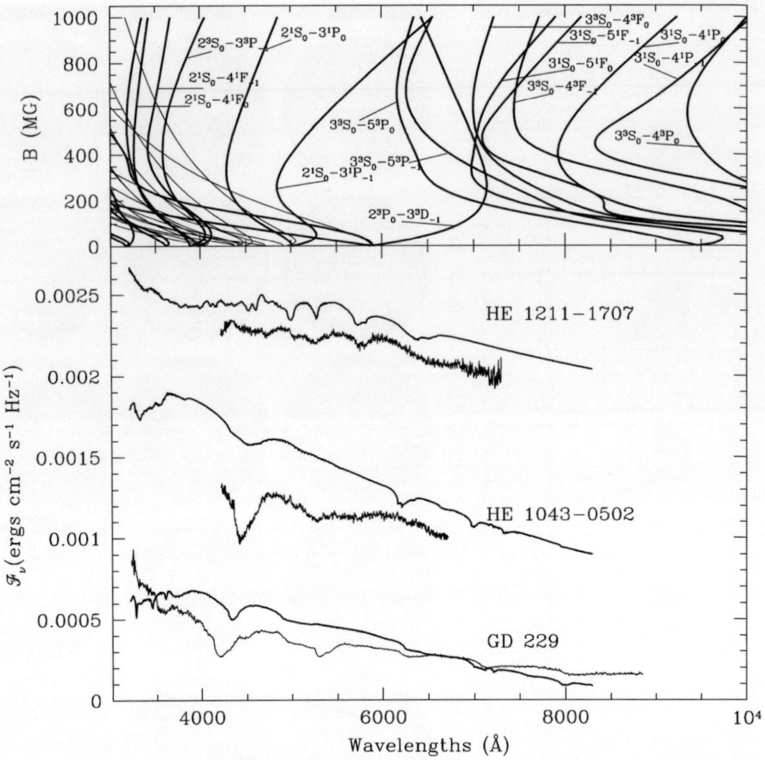

Fig. 11 A comparison of centred dipole models for helium rich WDs with observations of GD229, HE 1043-0502, and HE 1211-1707 (Wickramasinghe et al. 2002). The models have polar fields $B_d = 520$ MG (GD 229), 820 MG (HE 1043-0502) and 50 MG (HE 1211-1707). The observation and theory mismatch in GD229 could be due to resonances in the HeI bound-free opacities for which there is at present no adequate theory

At temperatures above ~ 10000 K, MWDs can also exhibit Zeeman split HeI lines in their spectra (DBp WDs). The two electron problem is much more difficult to treat and calculations for $n \leq 5$ singlet and triplet states for $m = 0, \pm 1, \pm 2, \pm 3$ only became available in the late 90s (Jordan et al. 1998; Jones et al. 1999; Becken et al. 1999; Becken and Schmelcher 2000b,a, 2001; Al-Hujaj and Schmelcher 2003). We show in Fig. 11 a comparison between centred dipole models and observations for three helium-rich MWDs.

Cool helium-dominated atmospheres develop deep convection zones that may dredge-up core material into the photosphere, resulting in atomic or molecular carbon features. The maximum carbon contamination in these DQ WDs is expected around $\simeq 12000$ K. A number of very cool DQs show broad absorption troughs reminiscent of C_2 Swan bands, without however matching the laboratory wavelengths of the Swan bands. Spectropolarimetry has revealed magnetism in some of these DQp stars (see Fig. 12; Liebert et al. 1978; Schmidt et al. 1999b), with suggested (but rather uncertain) field strengths of ~ 100 MG. However, not all DQp WDs show polarisation, and the nature of the peculiar absorption bands is not fully settled—suggestions are absorption by C_2H (Schmidt et al. 1995a) or pressure-shifted Swan bands (Liebert and Dahn 1983; Hall and Maxwell 2008; Kowalski 2010). A number of cool WDs exhibit polarisation in the absorption band of CH (Angel and Landstreet

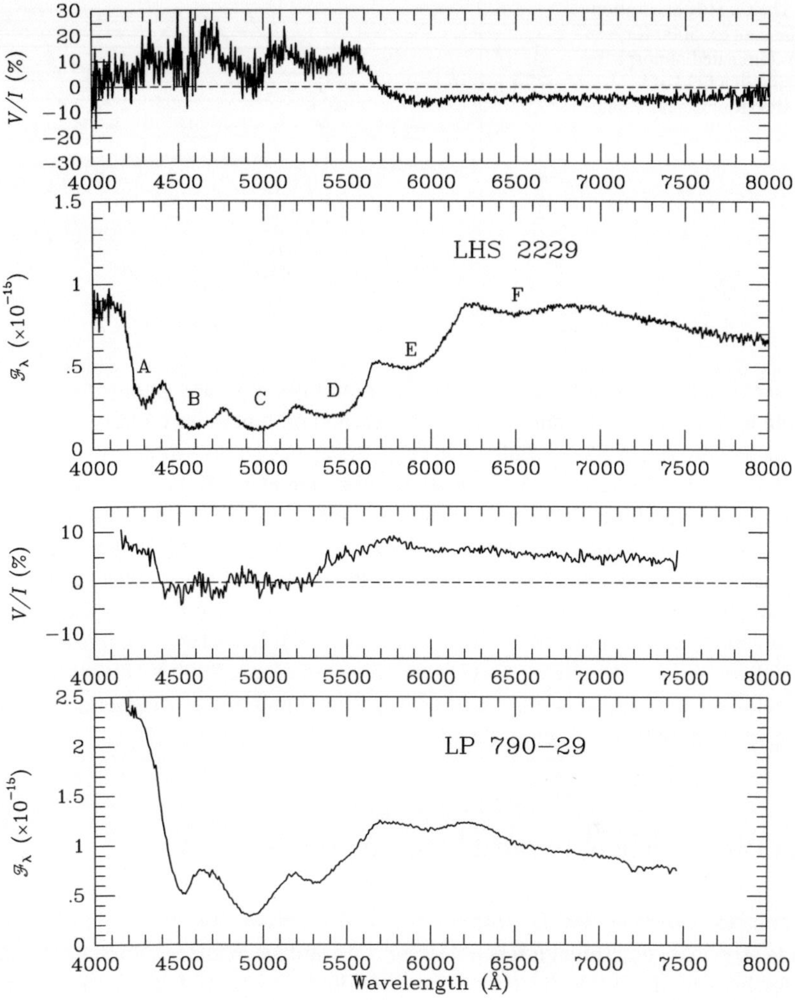

Fig. 12 Flux and polarisation spectra of two confirmed magnetic DQp stars, with crude B-field estimates of ~ 100 MG (Schmidt et al. 1999b)

1974; Vornanen et al. 2010). Berdyugina et al. (2007) applied new calculations of molecular magnetic dichroism to observations of G 99-37, a cool helium-rich MWD showing strongly polarised molecular bands, and estimated a field of 7.5 ± 0.5 MG.

Recently, Dufour et al. (2007) discovered a new class of WDs with carbon-dominated atmospheres, that are substantially hotter than the "classic" cool DQ stars. Short-periodic photometric variability detected in several of the hot DQs (Montgomery et al. 2008; Dufour et al. 2008b; Dunlap et al. 2010) was initially interpreted as non-radial pulsations, but the discovery of a 2.11 d modulation in SDSS J000555.90-100213.5 (Lawrie et al. 2013) casts some doubt on this hypothesis. The additional discovery of magnetic fields among the hot DQs (Dufour et al. 2008a,b; Williams et al. 2013a) suggests that they may represent a peculiar and rare path in WD evolution. Currently, 5 out of 14 known hot DQs are magnetic (Vornanen et al. 2013).

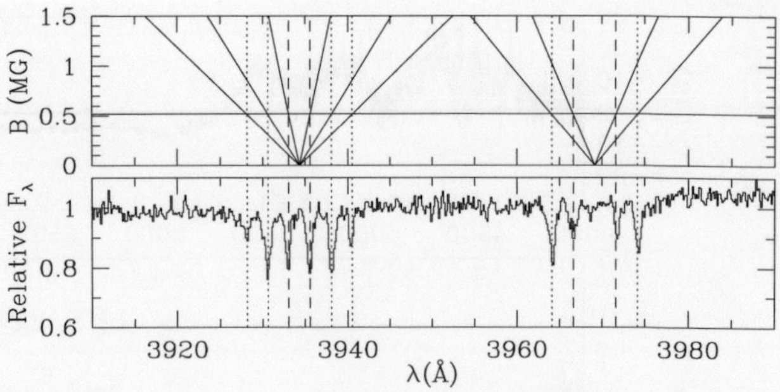

Fig. 13 The Ca H/K absorption lines in the cool ($\simeq 5000$ K) hydrogen-dominated atmosphere Zeeman-split in a field of $B = 0.5$ MG (Kawka and Vennes 2011)

Because of their high surface gravity, WDs should not have any photospheric elements apart from hydrogen, helium, and dredge-up of carbon or, much more rarely, oxygen (Gänsicke et al. 2010). Yet, ~ 25 % of all WDs show traces of metals—most commonly Ca and Si, but also Mg, Fe, Na and other elements (Zuckerman et al. 2003; Koester et al. 2014). This photospheric pollution requires recent or ongoing accretion and the widely accepted hypothesis for the origin of the material is planetary debris (Jura 2003). This hypothesis is corroborated by the detection of close-in circumstellar dust and gas (Gänsicke et al. 2006a; Farihi et al. 2009). A very small fraction of these debris-polluted WDs show magnetic fields (Reid et al. 2001; Kawka and Vennes 2011; Farihi et al. 2011, see Fig. 13). It is interesting to note that all known metal-polluted MWDs are very cool ($T_{\rm eff} < 7000$ K, Kawka and Vennes 2014), including the sample of strongly metal-polluted and highly magnetic DZ WDs recently discovered by Hollands et al. (2015).

4 Mass Distribution of Isolated Magnetic White Dwarfs

Mass estimates of isolated MWDs with fields $\gtrsim 1$ MG are available only for a small number of objects. The mass determination is not straightforward since there is no Stark broadening theory for high field MWDs. Thus, even state-of-the art models fail to fully account for the magnetic effects in MWD atmospheres. As a consequence, the effective temperatures derived often remain inherently uncertain (Fig. 15), and the same is true for the implied MWD masses (e.g. Külebi et al. 2010). The standard procedure of fitting the Balmer lines for $T_{\rm eff}$ and $\log g$ can only be applied to MWDs with fields below a few MG and even in these cases the results have to be treated with some caution (e.g. Ferrario et al. 1998; Dupuis et al. 2003). For higher field strengths, mass estimates are derived from the combination of effective temperatures, parallaxes, and a mass-radius relation. While only a small number of MWDs have accurate parallaxes, an estimate of the distance can be determined for MWDs that have non-degenerate WD companions (e.g. Girven et al. 2010; Dobbie et al. 2012, 2013), or for MWDs in open clusters (Külebi et al. 2013b). The situation will dramatically improve in the next few years, when parallaxes for practically all known MWDs will become available from the ESA satellite Gaia.

Taking into account the caveats mentioned above, the mean mass of high field isolated MWDs ($B \gtrsim 1$ MG) is 0.784 ± 0.047 M$_\odot$. High field MWDs also exhibit a strong tail that extends to the Chandrasekhar limit. The most recent estimate for the mean mass of non-magnetic DA WDs is 0.663 ± 0.136 M$_\odot$ (Tremblay et al. 2013). That the mean mass of

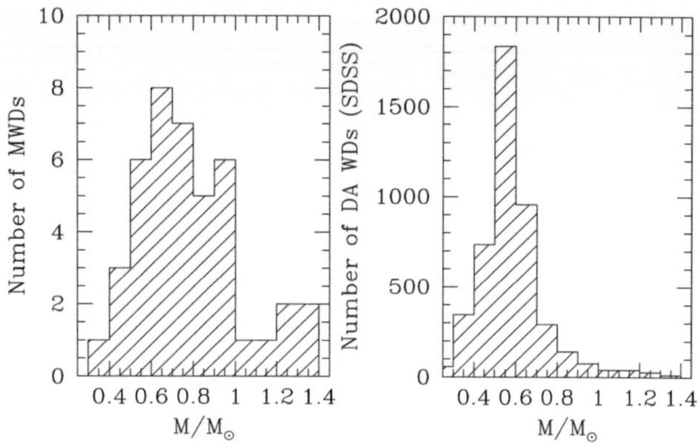

Fig. 14 *Left panel*: The mass distribution of MWDs (Ferrario and Wickramasinghe 2010). *Right panel*: The mass distribution of non-magnetic DA WDs from SDSS (Kepler et al. 2007)

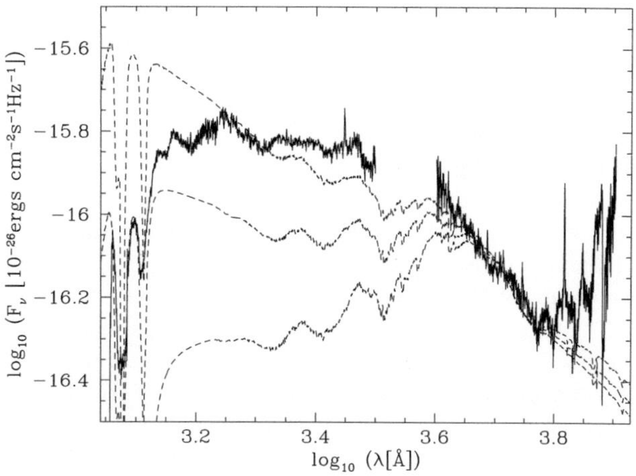

Fig. 15 The combined ultraviolet/optical spectral energy distribution of the high-field (230 MG) MCV AR UMa (*solid line*), along with model spectra for $T_{\rm eff} = 15000$, 20000, 25000 K (*dashed lines*, *top* to *bottom*). Magnetic effects cannot be fully described by current atmosphere models, and consequently the effective temperature of high-field MWDs remains poorly constrained even in the presence of excellent data (Gänsicke et al. 2001)

MWDs is higher than that of their non-magnetic counterparts was first noted by Liebert (1988). The mass distribution of all magnetic and non-magnetic WDs is shown in Fig. 14.

5 Spin Periods of Isolated Magnetic White Dwarfs

The majority of non-magnetic WDs are slow rotators, with even high-resolution spectroscopy usually only providing lower limits on $v \sin i$ (Karl et al. 2005; Berger et al. 2005). Asteroseismology shows that the spin periods are typically a few days (Fontaine and Brassard 2008; Greiss et al. 2014) and that the angular momentum of the stellar core is lost before the WD stage (Charpinet et al. 2009).

In MWDs, the magnetic effects in their atmospheres can give rise to noticeable spectroscopic (Liebert et al. 1977; Latter et al. 1987), photometric (Brinkworth et al. 2004, 2005), and polarimetric variability (Schmidt and Norsworthy 1991; Piirola and Reiz 1992).

The observed rotation periods of MWDs span a wide range, from 725 s (RE J0317–853, Barstow et al. 1995; Ferrario et al. 1997a) to lower limits of decades, if not centuries (Berdyugin and Piirola 1999; Beuermann and Reinsch 2002; Jordan and Friedrich 2002) (see Table 1).

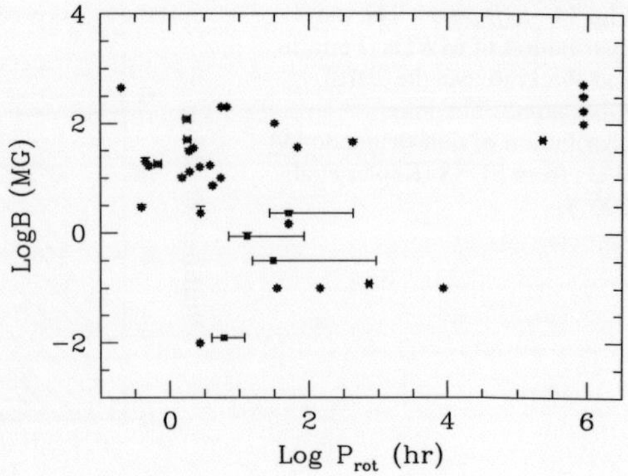

Fig. 16 Rotation periods of isolated MWDs against their magnetic field strength (this work)

We show in Fig. 16 (this work) the MWDs with known magnetic fields and rotation (spin) periods, the latter determined from polarimetry and photometry (e.g. Brinkworth et al. 2013). Inspection of the magnetic field versus spin period distribution may suggest the existence of two groups. One of them characterised by "short" periods of hours to weeks, and the other by rotation periods of decades to centuries as estimated for objects that have been monitored over many decades (e.g., Grw+70°8247, discovered over 80 years ago; Kuiper 1934; Brinkworth et al. 2013). It seems that very long period MWDs tend to possess high fields while short period ones do not show any preferred field strength. Clearly further observations to measure rotation periods of MWDs are needed to ascertain the existence of two groups of MWDs with different rotational properties.

The rotation rate of MWDs holds potentially some crucial clues on their nature and origin. For instance, slow rotators could be the descendants of the magnetic main sequence Ap/Bp stars (and their fields would thus be of fossil origin), whereas fast rotators could be the products of binary interaction, though Külebi et al. (2013a) have argued that in the case of a merger magnetospheric interactions of the MWD with the debris disk may slow down the rotation rather quickly. Unfortunately, the statistics of rotation periods have only slowly improved, however Brinkworth et al. (2013) have demonstrated that the search for photometric variability is relatively cheap in terms of observational requirements, and could extensively be used for future work. An interesting hypothesis is that rapidly rotating MWDs may be detectable as sources of gravitational wave radiation (Heyl 2000).

For the sake of completeness, we recall that the spin period of the MWD in MCVs is dictated by the interaction between the magnetic field of the MWD and that of the secondary and/or by the torque of the accretion flow. That is, for strong fields the rotation of the MWD is locked to the orbital period (polars), and for weaker fields the MWD is rotating faster than the orbital period (IPs).

6 Origin of Magnetic Fields in White Dwarfs and Magnetic Cataclysmic Variables

WDs are often found paired with Main-Sequence (MS) companion stars (generally M dwarfs, but see Ferrario 2012). A glaring anomaly is that there are no examples of fully detached MWD-MS pairs, as first noted by Liebert et al. (2005a) through the study of 501 objects with composite WD+MS spectra from the Sloan Digital Sky Survey (SDSS) DR3. Even the most recent work of Rebassa-Mansergas et al. (2013), which has yielded 3419

SDSS DR8 WD-MS binary candidates, does not contain objects consisting of a MWD with a non-degenerate companion. Further searches conducted through visual inspection of all SDSS spectra of WDs with a red excess have confirmed the hypothesis that magnetic field and binarity (with M or K dwarfs) are independent at a 9σ level (Liebert et al. 2015). Such a pairing is also absent from catalogues of high field MWDs (Kawka et al. 2007; Kepler et al. 2013, and this work).

Thus, although all magnitude-limited surveys of WDs have lead to the discovery of at least $\sim 2\ \%$ of strongly magnetised WDs ($B \gtrsim 1$ MG), these objects are never found paired with a non-degenerate companion star. Yet, about 20–25 % of CVs host a MWD, thus raising some serious questions regarding the progenitors, and thus the origin, of MCVs (see Liebert 2009).

In the late 90s, surveys such as the HQS and the SDSS have revealed the existence of a small number of cool MWDs which accrete matter from the wind of their low-mass MS companions (e.g. Reimers et al. 1999; Reimers and Hagen 2000; Schwope et al. 2002b, 2009; Schmidt et al. 2005b, 2007; Vogel et al. 2007, 2011; Parsons et al. 2013). The accretion rate, which is about 10^{-13}–$10^{-14}\ M_{\odot}\ yr^{-1}$, is a few orders of magnitude larger than that observed in detached non-magnetic Post Common Envelope Binaries (PCEBs, e.g. Parsons et al. 2013, and references therein). Comprehensive studies of these systems have unveiled that the secondary is an active late-type main sequence star underfilling its Roche-lobe (see Schwope et al. 2009, and references therein). Accretion onto the cool MWD primary, constant over years (Schwarz et al. 2001), is consistent with what is expected from the wind emanating from the active companion (Schwope et al. 2002b) and captured by the MWD. The spectra of these systems exhibit strong cyclotron harmonics humps superimposed on the WD+M dwarf stellar continuum. Their very peculiar colour is the reason why the first such systems were uncovered in surveys whose science goal was to identify active galactic nuclei. The suggestion is that these systems could be the progenitors of the high field MCVs. Thus their initial class name "Low-Accretion Rate Polars" ("LARPS", Schwope et al. 2002b) is a misnomer, so they were renamed "Pre-Polars" ("PREPS", Schwope et al. 2009).

There are now ten systems that have been classified as PREPS (see Table 2). The WD magnetic fields determined for these systems cluster in the 60–70 MG range with only one system above (108 MG, Schwope et al. 2009) and two below (42 MG and 8 MG, Schmidt et al. 2007; Parsons et al. 2013). This field clustering may be due to selection effects. In any case, the field strengths are certainly consistent with the hypothesis that these wind accreting magnetic systems are the progenitors of the high field MCVs. The low magnetic field of 8 MG found in SDSS J030308.35+0054441.1 could instead suggest that this system may evolve into an IP (Parsons et al. 2013).

There are however a few polars that undergo prolonged low-accretion states that cannot be reconciled with wind accretion and therefore these can be rightfully named LARPS (see e.g. Schmidt et al. 2005a, Breedt et al. 2012, and Table 2).

The PREPS hypothesis is consistent with the scenario first proposed by Tout et al. (2008) and further developed by Wickramasinghe et al. (2014) for the origin of fields in MCVs. They have raised the possibility that the strong differential rotation expected in the CE phase may lead to the generation, by the dynamo mechanism, of a magnetic field that becomes frozen into the degenerate core of the pre-WD in MCVs (see also Nordhaus et al. 2011; García-Berro et al. 2012; Kissin and Thompson 2015, about field generation in MWDs). The dynamo mechanism responsible for magnetic field generation during the CE phase proposed by Wickramasinghe et al. (2014) is presented in the chapter of this book on the origin of magnetic fields in stars.

The binary population synthesis calculations of Briggs et al. (2015) are compatible with the hypothesis that MWDs originate from stars merging during common envelope evolution and are also consistent with the observation that MWDs are on average more massive than their non-magnetic counterparts (see Sect. 4).

However, other formation channels for MWDs may be at work and, for instance, the fossil field hypothesis for the origin of high fields MWDs, as proposed by Wickramasinghe and Ferrario (2005), cannot be dismissed. However, the fact that there is no known MWD paired with a non-degenerate companion of M to K spectral type is a serious challenge to the fossil field hypothesis. Of course it is possible that some MWDs could be hidden in the glare of luminous companions (e.g. Ferrario 2012). Should large enough numbers of these 'Sirius-type systems' hosting a MWD be discovered, this finding could point to different (or additional) formation channels for MWDs. However here we need to stress that Sirius-type systems could not be the progenitors of the MCVs.

According to the stellar merging hypothesis, the absence of relatively wide pre-MCVs can be explained if magnetic systems are born either in an already semi-detached state or with the two stars close enough for the MWD to capture the wind of its companion as it happens in PREPS. Any other alternative would trigger the question "where are the progenitors of the PREPS?" (Schwope et al. 2009). In the merging star picture, the highest fields are expected to occur when the two stars merge to produce an isolated MWD. The MCVs would then arise when the two stellar cores come close enough to each other to generate a strong field in the WD but fail to merge. Wider core separations after the end of the CE phase would result in non magnetic pre-CVs or in systems where the two stars will never interact. These systems, single stars and binaries that never undergo a CE evolution would account for the populations of low-field WDs in CVs, wide binaries and single isolated WDs. The few known wide binaries consisting of a MWD paired with a non-magnetic WD would result from triple star evolution where two stars merged to produce the MWD and the third one evolved as a single star to become a non-magnetic WD (e.g. LB 11146, RXJ 0317-853, EUVE J1439+750, CBS 229; Glenn et al. 1994a; Barstow et al. 1995; Ferrario et al. 1997a; Vennes et al. 1999; Dobbie et al. 2013).

One would expect that some PREPS should have hotter WDs. However, as Tout et al. (2008) have pointed out, WDs cool down to an effective temperature of 15000 K in only $\sim 10^7$ years and at a period of 2 h, the orbital decay time scale due to gravitational wave radiation is about 3×10^9 years which is sufficient for a WD to cool down considerably and reach the observed effective temperatures of the PREPS before Roche lobe accretion begins. Therefore, although some 20–25 % of CVs are magnetic, their birthrate may be considerably lower than that of non-magnetic CVs and their presence in large numbers could simply be a reflection of their longer lifespans because of the reduction in magnetic braking (see Sect. 7).

Extensive searches for magnetic fields in central stars of planetary nebulae and in hot subdwarfs have until now yielded negative results (Asensio Ramos et al. 2014; Leone et al. 2014; Mathys et al. 2012; Savanov et al. 2013). Should this finding be verified by further spectropolarimetric observations, it would again be consistent with the view that the origin of fields in MWDs is intimately related to their binary nature.

Another possibility that cannot be a priori excluded may involve the screening of the WD magnetic field by an unmagnetised layer of material at the end of the CE phase. Should this be the case, magnetic systems would not distinguish themselves as such in the early post CE phase until their field re-emerges when the stars are in near contact (as in PREPS) or in contact (as in MCVs). Because of the observed low effective temperatures of the WDs in PREPS this phase must last $\gtrsim 10^9$ years. Estimates of the diffusion and advection rates of a dipolar magnetic field would thus suggest the presence of $\sim 0.2\ M_\odot$ of hydrogen-rich material screening the field in order to explain the observations (Cumming 2002). However, we need to stress that the calculations of Cumming (2002) are for accreting WDs under the assumption that they retain the accreted mass. The situation in the context of CE evolution is different, in so far as the screening would be caused by a layer of matter which is retained following an incomplete ejection of the envelope.

7 Evolutionary Aspects of Magnetic White Dwarf Binaries

Close binary evolution theory predicts compact binary systems with low-mass donors to evolve towards short orbital periods through angular momentum losses (King 1988). At long orbital periods ($\gtrsim 3$ h) magnetic braking (Verbunt and Zwaan 1981; Rappaport et al. 1983) acts as main mechanism that ceases at $\sim$3 h when secondaries become fully convective and shrink within their Roche lobe. Mass transfer stops until contact is re-established near $\sim$2 h. This gives rise to the so-called 2–3 h CV orbital period gap (Rappaport et al. 1983; Spruit and Ritter 1983; Davis et al. 2008). Below the period gap gravitational radiation drives the systems towards the orbital period minimum, $\sim$ 70–80 min, as first suggested by Paczynski and Sienkiewicz (1981) (but see also Townsley and Bildsten 2003; Townsley and Gänsicke 2009) until the mass of the donor star becomes too small to sustain nuclear burning and the star becomes degenerate. The present day CV population is expected to densely populate the orbital period distribution close to the period minimum (the "orbital minimum spike"). These systems would then be expected to 'bounce back' and evolve toward longer periods (Kolb and Baraffe 1999). The discrepancy between observations and theory has been a major issue for a long time and only recently mitigated by deep optical surveys, such as the SDSS which has unveiled the long-sought low accretion rate and short orbital period CVs (Gänsicke et al. 2009, see their Fig. 17, top panel). The CV space density also suffers from discrepancies between theoretical predictions (De Kool 1992; Politano 1996) and observations (Schreiber and Gänsicke 2003; Pretorius et al. 2007), although there are recent claims by Reis et al. (2013) that some of these disagreements may now be resolved. The study of Reis et al. (2013) of Swift X-ray spectra of an optically selected sample of nearby CVs has revealed the existence of a population of objects whose X-ray luminosities are an order of magnitude fainter than found in earlier studies indicating that the space density of CVs may be larger than previously forecast and thus in better agreement with population synthesis calculations.

Among the MCVs, the polars dominate the period distribution at short ($\lesssim$4 h) orbital periods while the IPs dominate the distribution at longer periods (see Fig. 17, mid and bottom panels). It is not clear yet whether the IPs, as a class, have generally lower fields than polars or the field strengths of both sub-classes are similar but IPs still need to synchronise (King and Lasota 1991; Norton et al. 2004). Unlike polars, the majority of IPs has not been detected in polarisation searches except for ten systems that may eventually evolve into polars (see Sect. 3). Norton et al. (2004, 2008) argued that IPs above the gap with WD surface magnetic moments $\mu_{\rm WD} \gtrsim 5 \times 10^{33}$ G cm^3 and $P_{\rm orb} \gtrsim$3 h will eventually evolve into polars. IPs below the gap, however, are not expected to evolve into polars. Recently new short period IPs have been discovered both in X-ray (de Martino et al. 2005; Bernardini et al. 2012; Woudt et al. 2012a; Bernardini et al. 2013; Thorstensen and Halpern 2013) and optical bands (Rodríguez-Gil et al. 2004; Araujo-Betancor et al. 2005b; Southworth et al.

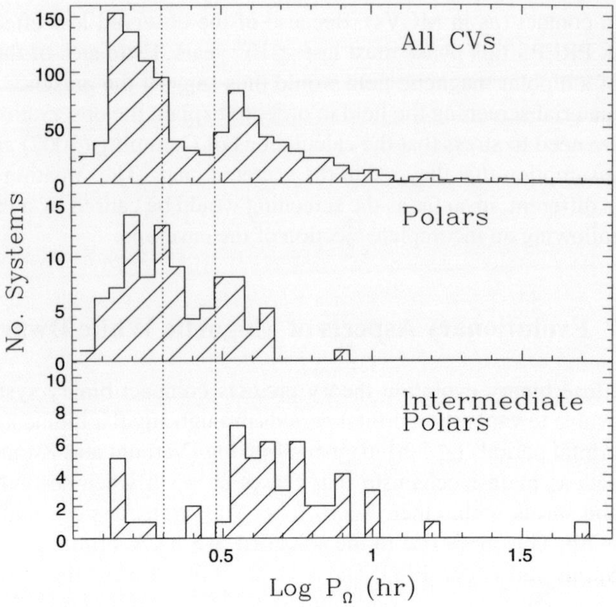

Fig. 17 The orbital period distribution of CVs (*top*) and of the magnetic types Polars (*middle*) and IPs (*bottom*). The latest version (v7.20) of the Ritter and Kolb (2003) CV catalogue is used. A few identifications were corrected. *The vertical lines* mark the 2–3 h orbital period gap

2007). From these findings Pretorius and Mukai (2014) have suggested the existence of a low-luminosity (hence low-accretion rate) population of IPs that still need to be unveiled. To enlarge the sample of short period IPs is clearly crucial to understand MCVs and their evolution.

Although the evolution of MCVs was expected to be similar to that of non-magnetic CVs, Wickramasinghe and Wu (1994) predicted a scenario where the strong magnetic field in polars reduces the efficiency of the magnetic braking mechanism (see also Wickramasinghe and Ferrario 2000; Webbink and Wickramasinghe 2002). As a consequence, the mass transfer rates would be lower requiring longer evolutionary timescales than those of non-magnetic systems. The first strong observational evidence in support of this scenario was presented by Sion (1999) on the basis of simple considerations of compressional heating and structure changes in response to accretion in non-magnetic CVs versus the known effective temperatures of the WDs in MCVs. This finding was later confirmed by Araujo-Betancor et al. (2005a) for a larger sample of MCVs with exposed MWDs. They showed that polars possess systematically cooler WDs than non-magnetic CVs. Since the WD effective temperature is a good proxy of the secular mass transfer rate (Townsley and Gänsicke 2009) it turns out that CVs with highly magnetised WDs accrete at lower rates and thus evolve on longer timescales. This could explain the higher incidence of magnetism observed among CVs ($\sim$ 20–25 %) as compared to that of isolated MWDs (Araujo-Betancor et al. 2005a).

The MCVs, being stronger X-ray emitters than non-magnetic CVs, are claimed to be important constituent of the galactic X-ray source population at low luminosities. The deep *Chandra* survey of the galactic centre has revealed a large number of low-luminosity hard X-ray sources attributed to MCVs of the IP type (Muno et al. 2004a; Hong et al. 2012). The hard X-ray surveys of the Galactic Ridge conducted by the *INTEGRAL* (Revnivtsev et al. 2009, 2011), *Rossi-XTE* (Sazonov et al. 2006) and *Suzaku* (Yuasa et al. 2012) satellites have also resolved a large fraction of the diffuse emission into discrete low luminosity hard X-ray sources largely attributed to coronally active stars and MCVs of the IP type. The true contribution of these systems to the X-ray luminosity function of the Galactic X-ray source population at low luminosities is still under investigation (Muno et al. 2004b; Revnivtsev et al. 2006; Yuasa et al. 2012; Reis et al. 2013; Pretorius et al. 2013; Pretorius and Mukai 2014; Warwick 2014).

8 Magnetic Accretion

The magnetic field of the WD primaries influences the accretion geometry and emission properties of MCVs. It determines the dynamics of accretion when the dynamical timescale is of the order of the timescale of magnetic interaction. This occurs at the magnetospheric boundary where the magnetic pressure balances the ram pressure of the infalling material: $\rho v^2 \simeq B^2/8\pi$ (Frank et al. 1985). The magnetospheric radius, $R_{\rm mag}$ depends on the magnetic field strength of the WD and the system parameters (see the review by Wickramasinghe 2014). Among polars, the WD magnetic moment is strong enough ($\mu_{\rm WD} \gtrsim 10^{34}$ G cm^3) to prevent the formation of an accretion disc, because $R_{\rm mag}$ is of the order of the orbital separation. Thus the accretion stream flowing into the primary Roche lobe along a ballistic trajectory is channelled by the field lines towards the magnetic poles of the WD (Ferrario et al. 1989a; Schwope 1996) forming an accretion "funnel". The radiation in the optical and IR bands is characterised by strongly circularly and linearly polarised cyclotron emission from the stand-off shocks located at the base of the accretion funnels. Because of phase locking, radial velocity and light variations are seen only at the orbital period.

There are currently four confirmed polars that show a small degree (several percent) of asynchronism between the spin and orbital periods and two candidate systems (Campbell and Schwope 1999, and see also Table 2). Doppler tomography of the near-synchronous polar BY Cam (Schwarz et al. 2005) has uncovered that most of the matter flowing towards the magnetic WD is accreted via a funnel that extends by $\sim 180°$ in azimuth. This implies that the stream can travel around the WD, making the accretion pattern in this system resemble that of IPs (see below). Thus, asynchronous polars are particularly important for our understanding of magnetic accretion since they display characteristics common to both Polars and IPs.

The reason for this asynchronism is not clear, but it has been proposed that it could be caused by recent nova eruptions (e.g. Nova V1500 Cyg, Schmidt and Stockman 1991; Schmidt et al. 1995b). Furthermore among newly identified MCVs there are a handful of systems with weakly desynchronised MWDs (see Table 3), such as Paloma (Schwarz et al. 2007), IGR J19552+0044 (Bernardini et al. 2013), and V598 Peg (Southworth et al. 2007) indicating that the distinction between polars as synchronous rotators and IPs is not as sharp as formerly believed.

In most IPs, accretion discs or rings usually form (but see Hellier 2014, for a review on accretion in IPs) and $R_{\rm mag} \lesssim R_{\rm cor}$ where $R_{\rm cor} = (GM_{\rm WD}/\Omega^2_{\rm rot})^{1/3}$ is the co-rotation radius at which the Keplerian angular velocity equals the spin angular velocity of the primary and $R_{\rm mag} = \phi\, 2.7 \times 10^{10} \mu_{33}^{4/7} \dot{M}_{16}^{-2/7}\, M_{\rm WD,\odot}^{-1/7}$ cm, where $\phi \sim 1$ is a parameter that takes into account the departure from the spherically symmetric case, μ_{33} is the WD magnetic moment in units of 10^{33} G cm^3, M_{16} is the mass accretion rate in units of 10^{16} g s^{-1} and $M_{\rm WD,\odot}$ is the WD mass in solar units (Norton and Watson 1989; Hellier 1995). The details of the threading region depend on the magnetic diffusivity of the disc and the toroidal field produced by the shear between the Keplerian disc and the co-rotating disc (see Wickramasinghe 2014). Close to the WD, the flow in IPs consists of two magnetically confined accretion "curtains" fed by the disk or ring (Ferrario et al. 1993; Hellier 1995).

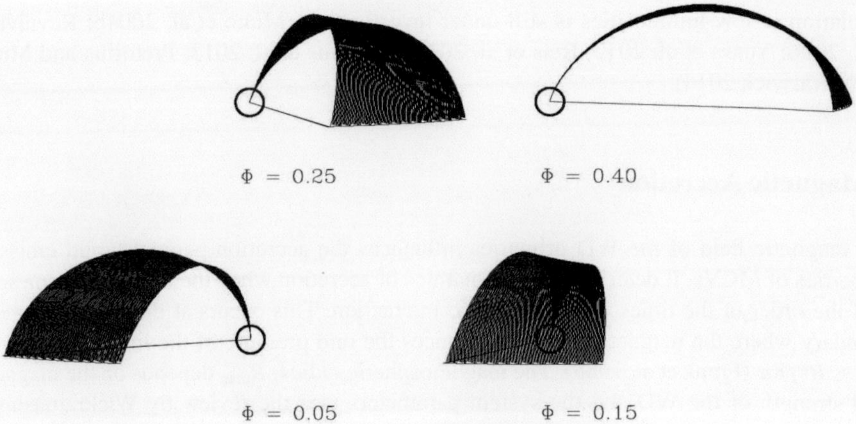

Fig. 18 Accretion funnel in a polar obtained using an orbital inclination, $i = 85°$, and a dipole inclination, $\theta = 15°$. The diagrams for phases $\phi > 0.5$ are mirror images of those shown here (Ferrario et al. 1993)

Since the WD in IPs is not phase-locked into synchronous rotation with the orbit, the emission variations are observed to occur at the spin period of the WD, P_s, at the beat period $P_{beat} = (P_s^{-1} - P_{orb}^{-1})^{-1}$ and often at both. The multi-component model of Ferrario and Wickramasinghe (1999) to calculate the optical and X-ray power spectra of disced and discless MCVs has revealed that as the magnetic field strength of the WD increases, the cyclotron emission from the shocks becomes comparable to the optical radiation from the magnetically confined flow and the dominant power shifts from the beat frequency to the WD spin frequency. Thus those MWDs in IPs with sufficiently large magnetic moments, such as V 2400 Oph (Buckley et al. 1997), do not accrete via a disc while in many other IPs, such as FO Aqr, the stream from the companion star can flow over the disc (Norton et al. 1992; Beardmore et al. 1998; Hellier 2014).

One of the most significant characteristic that distinguishes disc-fed accretion from stream-fed accretion is the amplitude of the radial velocity variations. Disc-fed accretion is characterised by low radial velocity amplitudes ($\sim$ 50–100 km s^{-1}), while stream-fed accretion by high radial velocity amplitudes ($\sim$ 500–1000 km s^{-1}).

Here we need to stress that although curtains are generally associated with asynchronous systems, under certain accretion conditions even polars may exhibit extended coupling regions and a curtain-type structure for the magnetically confined infalling matter. Similarly, some IPs may sometime show funnel-like flows typical of polars. In this context we note that Schwope et al. (2004) have shown how tomographic techniques can be effectively used to infer the accretion geometry of MCVs.

We show schematic diagrams of magnetically channelled accretion flows in typical polars and IPs in Fig. 18 and Fig. 19 respectively. As shown in these figures, the observed modulation arises from projection area effects and viewing aspect. Depending on the orbital phase ϕ, parts of the accretion curtains are either self-occulted or are hidden by the disc or the body of the WD (Ferrario et al. 1993; Ferrario and Wickramasinghe 1993b).

The discovery of Zeeman-split emission lines in the circular polarisation spectra of AR UMa (Schmidt et al. 1999a) and V884 Her (Schmidt et al. 2001b) have enabled the investigation of the magnetic, thermal and dynamical structure of accretion funnels. The

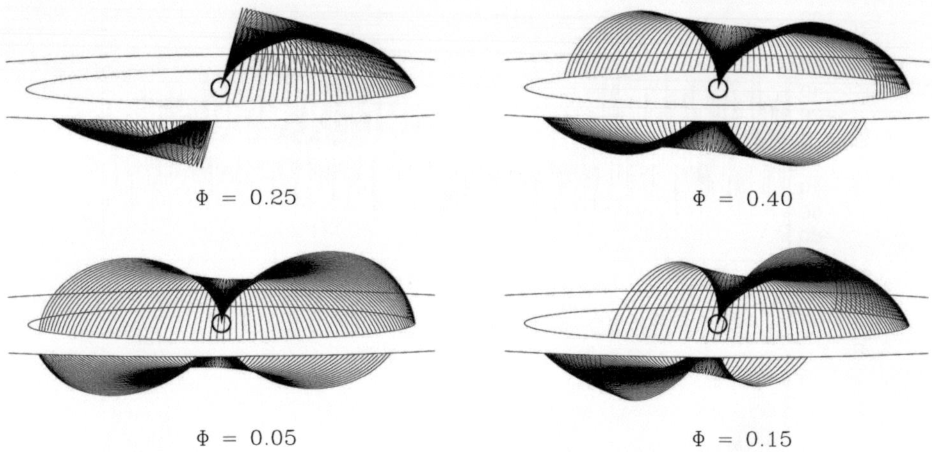

Fig. 19 Visible portions of the two accretion curtains in IPs obtained using a viewing angle $i = 70°$ and a magnetic colatitude of $\theta = 15°$. The diagrams for $\phi > 0.5$ are mirror images of those shown here (Ferrario et al. 1993)

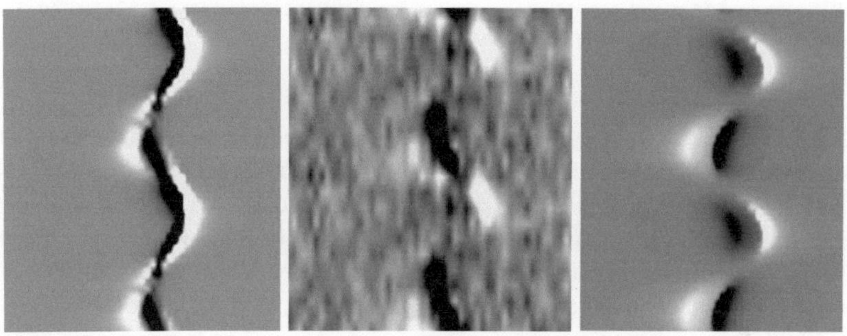

Fig. 20 *Middle panel*: observed phase-dependence of the circularly polarised flux for the region surrounding H_β. Wavelength runs from 4750 Å to 4950 Å, and orbital phase advances upward, with $\phi = 0.25$ at the *bottom* and $\phi = 0.25$ at the *top* of the *figure*, covering two complete orbital cycles. The polarised flux is white for negative circular polarisation and black for positive circular polarisation. *Left panel*: "standard" funnel model with $\theta_d = 30°$, $\phi_d = 90°$ and $i = 60°$. *Right panel*: idealised plane parallel slab model for field aligned flow in the transition region from ballistic stream to funnel flow (Ferrario et al. 2002)

modelling of Ferrario et al. (2002) of these two very high field polars using the funnel structures developed by Ferrario and Wehrse (1999) has revealed that the polarisation spectrum is very sensitive to velocity and field gradients, as shown in Fig. 20. They find that the bulk of the observed emission arises from two components, (i) the material in the magnetically channelled funnel flow and (ii) the threading region at the base of the funnel where the stream changes from ballistic to co-rotational. The latter makes the main contribution in the high field system AR UMa which is viewed at large inclination angles while the former dominates at the low orbital inclinations of V884 Her (Ferrario et al. 2002). Thus, the study of polarised line emission, which is only possible for systems which possess very strong fields, allows one to unravel the contributions from the stream, funnel, and transition region.

Close to the WD surface the densest parts of the supersonic infalling matter produce a strong shock where the gas is heated up to temperatures of ∼ 10–40 keV. The shock tem-

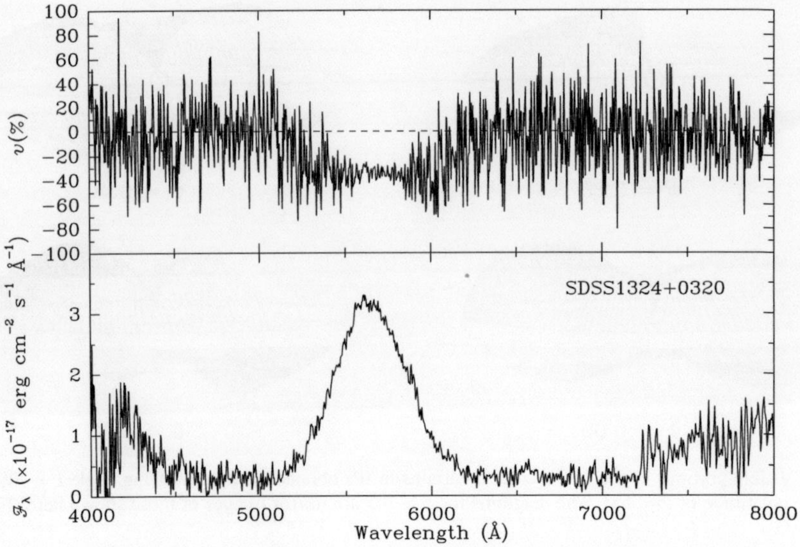

Fig. 21 *Bottom panel*: Phase-averaged flux spectrum of the PREP PZ Vir (=SDSS J1324+0320). *Top panel*: phase-averaged circular polarisation spectrum (Szkody et al. 2003)

perature is $T_s = 3.7 \times 10^8\ M_{WD,\odot}/R_9$ K, where R_9 is the WD radius in units of 10^9 cm. The post-shock flow becomes subsonic and cools via thermal bremsstrahlung (hard X-rays) and cyclotron radiation (optical/IR) (Aizu 1973; King and Lasota 1979; Lamb and Masters 1979). The relative proportion of cyclotron to bremsstrahlung radiations depends on the WD field strength and specific mass accretion rate. Cyclotron dominates at high field strengths and/or low local mass accretion rates, thus, if L_{cyc} and L_{brem} are the cyclotron and bremsstrahlung luminosities respectively, then $L_{cyc} > L_{brem}$. The temperatures of ions and electrons are different in this case (two-temperature fluid). Bremsstrahlung instead dominates at low field and/or high local mass accretion rates and the ion and electron temperatures are about the same (one-temperature fluid). The WD surface also intercepts these primary emissions, which are partially reflected and thermalised. The poles of the WD are then heated at temperatures $\sim$ 20–40 eV and emit a blackbody-like spectrum. However, the prediction that the black body luminosity, L_{BB}, should be $\sim L_{brem} + L_{cyc}$ was not confirmed in AM Her and other polars which showed a prominent soft X-ray component. This was referred to as the "soft X-ray puzzle" (see also Beuermann 1999). Kuijpers and Pringle (1982) suggested that at high local mass accretion rates ($\dot{m} \sim 100\ \mathrm{g\,cm^{-2}\,s^{-1}}$) the pressure of material is so high that the shock can be buried in the WD atmosphere and that the heating of the WD from below produces an intense blackbody-like emission. This component does not enter in the energy balance, thus solving the soft X-ray puzzle. In AM Her most of the reprocessed radiation was found to emerge in the FUV/UV range (Gänsicke et al. 1995). In this context, we note that the recent modelling of the accretion heated spot of the WD using a temperature distribution across the region can well reproduce the soft X-ray spectrum in AM Her (Beuermann et al. 2012).

The structure of the post-shock region has been the subject of several studies. Detailed one-dimensional two-fluid hydrodynamic calculations coupled with radiative transfer equations for cyclotron and bremsstrahlung were performed for different regimes by Woelk and

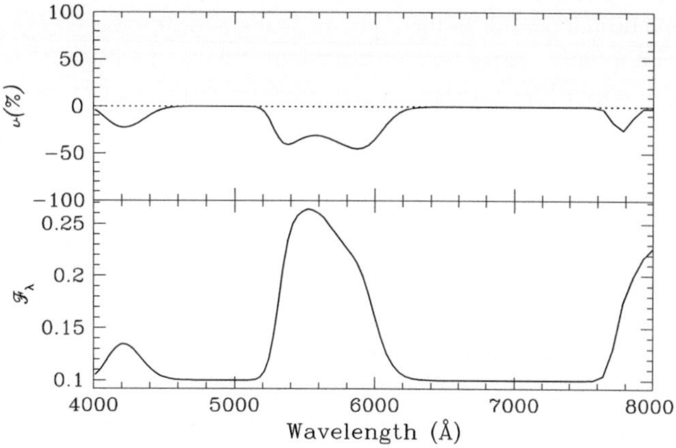

Fig. 22 Phase-averaged cyclotron model for the PREP PZ Vir (=SDSS J1324+0320) corresponding to a magnetic field strength $B = 64$ MG. *Lower panel*: flux. *Upper panel*: circular polarisation

Beuermann (1992, 1996), Thompson and Cawthorne (1987), Kuijpers and Pringle (1982) and Fischer and Beuermann (2001). They also include the so-called bombardment regime, which is associated with very low specific mass accretion rates and high field strengths. These very low accretion rates do not allow the formation of a hydrodynamic shock, and the photosphere is heated directly by particle bombardment. This is the typical accretion regime of PREPS, as indicated by the observations of systems such as PZ Vir (=SDSS J1324+0320, Fig. 21). The very low electron temperatures and specific accretion rates derived by the modelling of these systems (see Fig. 22 and Ferrario et al. 2005) are found to be consistent with the bombardment regime. The absence of Zeeman features from the photosphere of the WD further strengthens the hypothesis that these systems are not polars in a low state of accretion but pre-MCV with very cool, old, WDs.

When fitting the X-ray spectra, early computations of the structure of the post-shock flow using one-temperature calculations (Cropper et al. 1999) were later demonstrated to provide higher WD masses than those obtained using hydrodynamical models that included two-temperature effects (Saxton et al. 2005) and dipolar field geometry (Saxton et al. 2007).

In the IPs, accretion generally occurs via magnetically confined curtains (see Fig. 19) forming arc-shaped shock regions around the WD magnetic poles (Ferrario et al. 1993; Ferrario and Wickramasinghe 1993b; Hellier 1995). Since these systems are strong hard X-ray emitters, the post-shock region mainly cools via thermal bremsstrahlung. Because of the large arc-shaped footprints, the reprocessed radiation was initially expected to emerge in the EUV range. However, the *ROSAT* survey revealed the existence of a few IPs with a soft X-ray component similar to that observed in polars. Observations with the *XMM-Newton* satellite have increased the number of IPs that exhibit a soft blackbody component to ~ 18, although these cover a wider range of temperatures than those in polars (Anzolin et al. 2008, 2009; Bernardini et al. 2012, and reference therein). Since the soft X-ray component is only a small fraction of the hard X-ray luminosity, it is consistent with reprocessing.

Interestingly, recent X-ray observations with *XMM-Newton* of polars in high states of accretion have revealed an increasing number of systems that do not exhibit a distinct soft X-ray component but rather a more 'IP-like' X-ray spectrum (Ramsay and Cropper 2004; Vogel et al. 2008; Ramsay et al. 2009; Bernardini et al. 2014). However, the magnetic fields and orbital periods of these polars do not appear to be very dissimilar from all other polars with a more classic type of behaviour. Hence, the distinction between the two subclasses now appears less marked than ever before, requiring further investigations.

9 Conclusions

To date there are about ~ 250 MWDs with well determined fields (see Table 1) and over ~ 600 if we also count objects with no or uncertain field determination (see Kepler et al. 2013, 2015). These MWDs have been discovered following surveys such as the SDSS, HQS and the Cape Survey. The enlarged sample has shown that the field distribution of MWDs is in the range 10^3–10^9 G. While the high field cut-off appears to be real, the low field one is currently determined solely by the sensitivity of current spectropolarimetric surveys. Observations also indicate that MWDs may be divided into two groups: a high field group (1–1000 MG), where most objects are found, and a low field group (< 0.1 MG), whose importance still needs to be determined by much more sensitive spectropolarimetric surveys conducted on 8 m class telescopes.

The high field group of MWDs differs from the low field group in terms of average mass (see also Kepler et al. 2013). That is, high field MWDs exhibit a higher average mass ($\sim 0.85\ M_\odot$) than weakly magnetic or non-magnetic WDs ($\sim 0.66\ M_\odot$). High field MWDs also have a relatively strong tail that extends to the Chandrasekhar limit.

The significant increase in the number of MWDs has also led to new insights on the nature of magnetism. However, we still need to construct (i) more realistic model atmospheres that allow for the presence of magnetic fields and (ii) stellar evolution tracks of intermediate mass stars that take into consideration both fossil and dynamo generated fields. Such calculations may be able to tell us whether all WDs are magnetic at some level.

The origin of fields in highly magnetic WDs is currently being debated. Although the newly proposed scenario that all high field MWDs (single and in binaries) are the result of close binary evolution and mergers is gaining momentum, the fossil field hypothesis cannot be totally dismissed. The attractiveness of the merger hypothesis lies mostly in its ability to explain why there are no wide binaries consisting of a MWD with a non-degenerate companion star and why MWDs are on average more massive than their non-magnetic or weakly magnetic counterparts.

Observations of magnetic WDs in interacting binaries obtained in the last decade have also opened interesting questions on their evolution, accretion and emission properties. Forthcoming surveys such as SDSS-IV and VPHAS+ (Drew et al. 2014) in the optical and in the X-rays, such as the one expected to be conducted by *eROSITA* (e.g., Schwope 2012) will allow the discovery of new systems providing new exciting challenges.

The study of isolated and binary MWDs is likely to remain at the forefront of research for many years to come.

Acknowledgements The authors gratefully acknowledge the organisers of this workshop for the timely opportunity to review our understanding of the strongest magnetic fields in the Universe, and ISSI for their warm hospitality. The research leading to these results has received funding from the European Research Council under the European Union's Seventh Framework Programme (FP/2007-2013)/ERC Grant Agreement n. 320964 (WDTracer). DDM acknowledges financial support by ASI/INAF under contract I/037/12/0.

Appendix: Tables of Magnetic White Dwarfs and Magnetic Cataclysmic Variables

Table 1 Magnetic white dwarfs

WD	Other names	Comp.	B_p	T_{eff} (MG)	M (K)	P_{rot} ($M_\odot$)	References notes
0003-103	SDSS J000555.91-100213.4	DQ	1.47*	19420 ± 920	$\log g = (8.0)$	2.110 ± 0.045 d	1, 2, 3
0005-148	NLTT 347, SDSS J010319.70032501.0	DA	-0.0046 ± 0.0019	6400 ± 180	0.59	...	4
0009+501	LHS 1038, G 217-37, NLTT 574	DA	0.316*	6540 ± 150	0.74 ± 0.04	hrs–2.5 d	5, 6, 4
0011-134	LHS 1044, G 158-045	DA	16.7 ± 0.6	6010 ± 120	0.71 ± 0.07	30 min-days	7, 8, 9
0015+004	SDSS J001742.44+004137.4	DB	8.3	15000	...	...	10
0018+147	SDSS J002129.00+150223.7	DA	530.69 ± 63.56	7000	...	...	11, 12
0038-084	NLTT 2219	DA	0.307*	6000 ± 180	0.59	...	4
0040+000	SDSS J004248.19+001955.3	DA	2	11000	...	...	DD, 10, 12
0041-102	Feige 7, L 795-7	DBA	35	20000	$\log g = (8.0)$	131.606 min	13, 14, 15
0051+115	HS 0051+1145, PHL 886	DA	0.240 ± 0.010	...	...	...	16
0058-044	GD 9, GR 407, PHL 940	DA	0.325 ± 0.035	16700	$\log g = 8.07$	...	17, 16
0104+149	SDSS J010647.92+151327.8	DQ	not confirmed	23430 ± 1680	$\log g = (8.0)$	...	2
0140+130	SDSS J014245.37+131546.4	DB	4	15000	...	...	10
0155+003	SDSS J015748.15+003315.1	DZ	3.49 ± 0.05*	5700	...	...	10, 18
0159-032	1H 0201-029	DA	6	26000	$\log g = (8.0)$	...	19
0208+002	SDSS J021116.34+003128.5	DA	341.31 ± 54.34	9000	...	...	10, 12
0209+210	SDSS J021148.22+211548.2	DA	166.16 ± 7.41	12000	...	...	11, 12
0231+263	SDSS J023420.63+264801.7	DA	32.82 ± 6.26	13500	...	...	12
0233-083	SDSS J023609.40-080823.9	DQA	5	10000	...	...	11
0236-269	HE 0236-2656	DB	...	6000–7000	...	...	20
0239+109	G 4-34, LTT 10886	DA	0.725 ± 0.025	10060	$\log g = 8.73$	...	DD, 17, 16, 21
0253+508	KPD 0253+508	DA	17	15000	$\log g = (8.0)$	3.79 ± 0.05 hr	22, 23, 24
0257+080	LHS 5064, G 76-48	DA	0.1*	6680 ± 150	0.57 ± 0.09	6 d	25, 16, 4, 9
0301-006	SDSS J030407.40-002541.7	DA	10.95 ± 0.98	15000	...	...	26, 12
0307-428	1H0307-426	DA	10	25000	$\log g = (8.0)$	...	19

Table 1 (*Continued*)

WD	Other names	Comp.	B_p	T_{eff} (MG)	M (K)	P_{rot} ($M_\odot$)	References notes
0315-293	NLTT 10480, LHS 5070, LP 887-66	DAZ	0.5*	5340 ± 190	0.58	…	27, 4, 28
0315+422	SDSS J031824.19+422651.0	DA	10.12 ± 0.10	10500	…	…	12
0321-026	KUV 03217-0240	DA	< 1	27370	$\log g = 8.45$	…	21
0322-019	G 77-50, NLTT 10871, LHS 1547	DAZ	0.120*	5310 ± 100	0.60 ± 0.01	28–33 d	29, 4
0323+051	SDSS J032628.17+052136.3	DA	16.87 ± 2.41	25000	…	…	12
0325-857	RE J0317-853, EUVE J0317-855	DA	185–450	33000	1.34 ± 0.03	725 s	DD, 30, 31, 32, 33
0329+005	KUV 0329+0035, SDSS J033145.69+004517.0	DA	13.13 ± 1.00	15500	…	…	26, 10, 12
0330-000	HE 0330-0002	DB	…	6000–7000	…	…	20, 26, 10
	SDSS J033320.36+000720.6	DA	849.30 ± 51.75	7000	…	…	12
0340-068	SDSS J034308.18-064127.3	DA	9.96 ± 2.06	13000	…	…	10, 12
0342+004	SDSS J034511.11+003444.3	DA	1.96 ± 0.42	8000	…	…	26, 10, 12
0350+098	1RXS J035315.5+095700	…	…	…	…	…	21
0410-114	G 160-51, NLTT 12758	DA	1.7 ± 0.2	7440 ± 150	0.83	…	DD, 34, 4
0413-077	40 Eri B	DA	0.0073*	16490 ± 84	0.497 ± 0.005	…	35, 4
0416-096	NLTT 13015, LP 714-52	DA	6–7.5	5745 ± 405	0.59	variable	4
0446-789	BPM 3523	DA	0.0135*	23450 ± 20	0.49 ± 0.01	…	36, 16
0503-174	LHS 1734, LP 777-001	DA	7.3 ± 0.2	5300 ± 120	0.37 ± 0.07	…	37, 7
0548-001	G 99-37	DQB	7.3 ± 0.3	6200 ± 200	0.69 ± 0.02	4.117 hr	38, 13, 39, 40, 41, 42
0553+053	G 99-47,LTT 17891	DA	20 ± 3	5790 ± 110	0.71 ± 0.03	26.8 ± 0.7 min	13, 43, 44, 39, 37, 9
0616-649	EUVE J0616-649	DA	14.8	50000	$\log g = (8.0)$	…	45
0637+477	GD 77	DA	1.2 ± 0.2	14870 ± 120	0.69	…	1, 46
0728+642	G 234-4	DA	0.125*	4500 ± 500	…	…	8
	SDSS J073549.19+205720.9	DZ	6.12 ± 0.06	6000	…	…	18
0745+302	SDSS J074850.48+301944.8	DA	6.75 ± 0.41	22000	…	…	11, 12
0745+303	SDSS J074853.07+302543.5	DA	11.4	21000 ± 2000	0.81 ± 0.09	…	47
0746+172	SDSS J074924.91+171355.4	DA	13.99 ± 1.30	20000	…	…	12
0749+173	SDSS J075234.96+172525.0	DA	10.30 ± 1.23	9000	…	…	12
0755+358	SDSS J075819.57+354443.7	DA	26.40 ± 3.94	22000	…	…	10, 12

Table 1 (*Continued*)

WD	Other names	Comp.	B_p	$T_{\rm eff}$ (MG)	M (K)	$P_{\rm rot}$ ($M_\odot$)	References notes
0756+437	G 111-49	DA	180–220	8500 ± 500	…	6.68 hr	44, 8, 9
0801+124	SDSS J080359.93+122944.0	DA	40.70 ± 2.13	9000	…	…	12
0801+186	SDSS J080440.35+182731.0	DA	48.47 ± 2.93	11000	…	…	11, 12
0802+220	SDSS J080502.29+215320.5	DA	6.11 ± 1.29	28000	…	…	11, 12
0804+397	SDSS J080743.33+393829.2	DA	65.75 ± 18.52	13000	…	…	10, 12
0806+376	SDSS J080938.10+373053.8	DA	39.74 ± 5.41	14000	…	…	11, 12
0814+043	SDSS J081648.71+041223.5	DA	10.13 ± 8.03	11500	…	…	11, 12
0814+201	SDSS J081716.39+200834.8	DA	3.37 ± 0.44	7000	…	…	12
0816+376	GD 90	DA	9	14000	$\log g = (8.0)$	…	48, 13, 43, 49, 8
0821-252	EUVE J0823-254	DA	2.8–3.5	43200 ± 1000	1.20 ± 0.04	…	50
0825+297	SDSS J082835.82+293448.7	DA	33.40 ± 10.53	19500	…	…	11, 12
	SDSS J083200.38+410937.9	DZ	2.35 ± 0.11*	5900	…	…	18
0825+822	SDSS J083448.63+821059.1	DA	14.44 ± 4.57	27000	…	…	12
0836+201	EG 59 (Mislabeled as EG 61)	DA	2.83 ± 0.19	17000 ± 500	0.82 ± 0.05	…	51, 52
0837+273	SDSS J084008.50+271242.7	DA	10	12250	…	…	11
	SDSS J083945.56+200015.7	DA	3.38 ± 0.49	15000	…	…	12
0839+026	SDSS J084155.74+022350.6	DA	5.00 ± 0.99	7000	…	…	10, 12
0843+488	SDSS J084716.21+484220.4	DA	~ 3	19000	…	…	10
0848+121	SDSS J085106.12+120157.8	DA	2.03 ± 0.10	11000	…	…	12
0853+163	PG 0853+164,LB 8915	DBA	0.75–1.0	21200–27700	$\log g = (8.0)$	2–24 hr	8, 53, 9, 4
0853+169	SDSS J085523.87+164059.0	DA	12.6 ± 1.0	20000 ± 500	1.12 ± 0.11	…	12, 52
	SDSS J085550.67+824905.3	DA	10.82 ± 2.99	25000	…	…	12
0855+416	SDSS J085830.85+412635.1	DA	3.38 ± 0.19	7000	…	…	10, 12
	SDSS J090222.98+362539.6	DZ	1.92 ± 0.05*	6300	…	…	18
0903+083	SDSS J090632.66+080716.0	DA	5.98 ± 3.02	17000	…	…	11, 12
0904+358	SDSS J090746.84+353821.5	DA	22.40 ± 8.80	16500	…	…	11, 12
0907+083	SDSS J091005.44+081512.2	DA	1.01	25000	…	…	12

Table 1 (*Continued*)

WD	Other names	Comp.	B_p	$T_{\rm eff}$ (MG)	M (K)	$P_{\rm rot}$ ($M_\odot$)	References notes
0908+422	SDSS J091124.68+420255.9	DA	35.20 ± 5.83	10250	...	...	11, 12
0911+059	SDSS J091437.40+054453.3	DA	9.16 ± 0.77	17000	...	...	11, 12
0912+536	G195-19	DB	~ 100	7160 ± 190	0.75 ± 0.02	1.3301 d	37, 13, 54
0915+211	SDSS J091833.32+205536.9	DA	2.04 ± 0.10	14000	...	...	12
0922+014	SDSS J092527.47+011328.7	DA	2.04	10000	...	...	10, 12
	SDSS J092646.88+132134.5	DA	210 ± 25.1	9500 ± 500	0.62 ± 0.10	...	DD, 55
0930+010	SDSS J093313.14+005135.4	DQB	...	...	...	...	like LHS 2229, 10
0931+105	SDSS J093356.40+102215.7	DA	2.11 ± 0.49	8500	...	...	11, 12
0931+394	SDSS J093409.90+392759.3	DA	1.01	10000	...	...	12
0931+507	SDSS J093447.90+503312.2	DA	7.35 ± 2.21	8900	...	...	11, 12
0939+211	SDSS J094235.02+205208.3	DA	39.21 ± 4.55	20000	...	...	12
0941+458	SDSS J094458.92+453901.2	DA	15.91 ± 9.10	15500	...	...	11, 12
0945+246	LB 11146a	DA	670	16000 ± 2000	$0.90^{+}_{-0.14}$	...	DD, 56, 57
0952+094	SDSS J095442.91+091354.4	DQ	...	...	...	...	11
0957+022	SDSS J100005.67+015859.2	DA	19.74 ± 10.26	9000	...	...	10, 12
	SDSS J100346.66003123.1	DZ	4.37 ± 0.05	6300	...	...	18
1001+058	SDSS J100356.32+053825.6	DA	672.07 ± 118.63	23000	...	...	11, 12
1004+304	SDSS J100657.51+303338.1	DA	1.0 ± 0.1	10000	...	...	12
1004+128	SDSS J100715.55+123709.5	DA	5.41	18000	...	...	complex B, 11, 12
1005+163	SDSS J100759.80+162349.6	DA	19.18 ± 3.36	11000	...	...	12
1011+371	SDSS J101428.09+365724.3	DA	11.09 ± 1.50	10500	...	...	12
1008+290	LHS 2229	DQB	~ 100	4600	...	...	58, 59
1012+093	SDSS J101529.62+090703.8	DA	4.09 ± 0.86	7200	...	...	11, 12
1013+044	SDSS J101618.37+040920.6	DA	2.01	10000	...	...	10, 12
1015+014	PG 1015+014, SDSS J101805.04+011123.5	DA	120 ± 10	14000	$\log g = (8.0)$	105^{+12}_{-8} min	60, 23, 61, 9
1019+200	SDSS J102239.06+194904.3	DA	2.94 ± 0.71	9000	...	...	12
1019+274	SDSS J102220.69+272539.8	DA	4.91 ± 0.31	11000	...	...	12

Table 1 (*Continued*)

WD	Other names	Comp.	B_p	T_{eff} (MG)	M (K)	P_{rot} ($M_\odot$)	References notes
1017+367	GD 116, Ton 1206	DA	65 ± 5	16000	...	...	62
1018-103	EC 10188-1019	DA	3	17720	$\log g = 8.52$	...	21
1026+117	LHS 2273	DA	18	7160 ± 170	0.59	35–45 min	25, 9
	HS 1031+0343	DA	6.1 ± 0.3	...	...	...	16
1031+234	PG 1031+234, Ton 527	DA	~ 200–1000	~ 15000	...	5.53 ± 0.05 hr	63, 9
1032+214	LP 372-41, NLTT 24770, SDSS J103532.53+212603.5	DA	2.96 ± 0.33	7000–8000	...	...	34, 12
1033+656	SDSS J103655.38+652252.0	DQ	4*	...	...	...	64, 10
1036-204	LP 790-29	DQB	50	7800	$\log g = (8.0)$	24–28 yrs	58, 65, 66, 59
1043-050	HE 1043-0502	DB	~ 820	~ 15000	...	...	20, 67
1045-091	HE 1045-0908	DA	16	10000 ± 1000	$\log g = (8.0)$	2.7 hr	68
1050+598	SDSS J105404.38+593333.3	DA	17.41 ± 7.90	9500	...	...	10, 12
1053+656	SDSS J105628.49+652313.5	DA	29.27 ± 5.78	16500	...	...	10, 12
1054+042	SDSS J105709.81+041130.3	DA	2.03	8000	...	...	12
1105-048	LTT 4099	DA	0.0123*	15280 ± 20	0.52 ± 0.01	>3	36, 69, 16, 4
1107+602	SDSS J111010.50+600141.4	DA	6.5	30000	...	...	10, 12
1111+020	LSPM J1113+0146, SDSS J111341.33+014641.7	DQB	...	...	...	...	like LP 790-29, 10, 59
1115+101	SDSS J111812.67+095241.4	DA	3.38 ± 0.72	10500	...	...	11, 12
1117-113	SDSS J112030.34-115051.1	DA	8.90 ± 1.02	20000	...	...	12
1120+324	SDSS J112257.10+322327.8	DA	11.38 ± 3.42	12500	...	...	12
	SDSS J232538.93+044813.1	DZ	6.56 ± 0.09*	7200	...	...	18
1120+101	SDSS J112328.49+095619.3	DA	1.21	9500	...	...	12
1126-008	SDSS J112852.88-010540.8	DA	2	11000	...	...	10, 12
1126+499	SDSS J112924.74+493931.9	DA	5.31 ± 0.64	10000	...	...	11, 12
1129+284	SDSS J113215.38+280934.3	DA	3.01 ± 0.82	7000	...	...	12
1131+521	SDSS J113357.66+515204.8	DA	8.64 ± 0.78	22000	...	...	10, 12
1135+579	SDSS J113756.50+574022.4	DA	5.00 ± 0.34	7800	...	...	11, 12
1136-015	LBQS 1136-0132, SDSS J113839.51-014903.0	DA	22.71 ± 1.26	10500	...	...	70, 10, 12

Table 1 (*Continued*)

WD	Other names	Comp.	B_p	T_{eff} (MG)	M (K)	P_{rot} ($M_\odot$)	References notes
1137+614	SDSS J114006.37+611008.2	DA	50.19 ± 17.78	13500	...	...	10, 12
1145+487	SDSS J114829.00+482731.2	DA	32.47 ± 7.11	27500	...	...	11, 12
	SDSS J115224.51+160546.7	DZ	2.72 ± 0.04*	6500	...	...	18
1151+015	SDSS J115418.14+011711.4	DA	33.47 ± 2.07	27000	...	...	10, 12
1156+619	SDSS J115917.39+613914.3	DA	20.10 ± 6.70	23000	...	...	10, 12
1159+619	SDSS J120150.10+614257.0	DA	11.35 ± 1.53	10500	...	...	11, 12
1203+085	SDSS J120609.80+081323.7	DA	760.63 ± 281.66	13000	...	...	11, 12
1204+444	SDSS J120728.96+440731.6	DA	2.03	16750	...	...	11, 12
1209+018	SDSS J121209.31+013627.7	DA	10.12 ± 0.93	10000	...	90 min	L5–L8 comp., 71, 12
1211-171	HE 1211-1707	DB	50	~ 12000	...	...	20, 9
1212-022	LHS 2534, SDSS 121456.39023402.7	DZ	1.92*	5200	...	...	72, 10, 18
1214-001	SDSS J121635.37-002656.2	DA	59.70 ± 10.23	15000	...	...	26, 10, 12
1219+005	SDSS J122209.44+001534.0	DA	14.70 ± 4.70	14000	...	...	26, 10, 12
1220+234	PG 1220+234	DA	3	26540	0.81	...	73, 21
1220+484	SDSS J122249.14+481133.1	DA	8.05 ± 2.24	9000	...	...	11, 12
1221+422	SDSS J122401.48+415551.9	DA	22.36 ± 3.02	9500	...	...	11, 12
1231+130	SDSS J123414.11+124829.6	DA	4.32 ± 0.27	8200	...	...	11, 12
1233-052	HE 1233-0519	DA	0.61 ± 0.01	...	...	...	17, 16
1235+422	LHS 5222, NLTT 31347	DQ	...	...	...	...	113
1245+413	SDSS J124806.38+410427.2	DA	7.03 ± 1.19	7000	...	...	11, 12
1246+296	SDSS J124836.31+294231.2	DA	3.95 ± 0.25	7000	...	...	12
1246-022	SDSS J124851.31-022924.7	DA	7.36 ± 2.19	13500	...	...	10, 12
1248+161	SDSS J125044.42+154957.4	DA	20.71 ± 3.66	10000	...	...	11, 12
1252+564	SDSS J125416.01+561204.7	DA	38.86 ± 9.03	13250	...	...	11, 12
	SDSS J125434.65+371000.1	DA	4.10 ± 0.35	10000	...	...	12
1254+345	HS 1254+3440, SDSS J125715.54+341439.3	DA	11.45 ± 0.71	8500	...	...	74, 12
1300+590	SDSS J130033.48+590407.0	DA	~ 6	6300 ± 300	0.54 ± 0.06	...	DD, 75
1309+853	G256-7	DA	4.9 ± 0.5	~ 5600	0.5	...	44, 8, 73

Table 1 (*Continued*)

WD	Other names	Comp.	B_p	$T_{\rm eff}$ (MG)	M (K)	$P_{\rm rot}$ ($M_\odot$)	References notes
1312+098	PG 1312+098	DA	10	~ 20000	...	5.42839 hr	23, 8
1317+135	SDSS J132002.48+131901.6	DA	2.02	14750	...	...	11, 12
1327+594	SDSS J132858.20+590851.0	DA	18	25000	...	...	11
1328+307	G165-7, SDSS J133059.26+302953.2	DZ	0.65*	6440 ± 210	0.57 ± 0.17	...	76, 18
1330+015	G62-46	DA	7.36 ± 0.11	6040	0.25	...	DD, 77
1331+005	SDSS J133359.86+001654.8	DQB	...	...	...	...	like LHS 2229, 10, 59
1332+643	SDSS J133340.34+640627.4	DA	10.71 ± 1.03	13500	...	...	10, 12
1334+486	GD 359, SDSS J170751.91+353239.97	DA	2.7	...	...	...	78
1339+659	SDSS J134043.10+654349.2	DA	4.32 ± 0.76	15000	...	...	10, 12
1346+383	SDSS J134820.79+381017.2	DA	13.65 ± 2.66	35000	...	...	12
1349+545	SBS 1349+5434, SDSS J135141.13+541947.4	DA	761.00 ± 56.42	12000	...	...	79, 11, 12
1350-090	LP 907-037	DA	0.268*	9520 ± 140	0.83 ± 0.03	...	5, 80
1405+501	SDSS J140716.66+495613.7	DA	12.49 ± 6.20	20000	...	...	12
1416+256	SDSS J141906.19+254356.5	DA	2.03 ± 0.10	9000	...	...	12
	SDSS J142625.71+575218.3	DQB	~ 1.2	19830 ± 750	$\log g = 9.0$	...	pulsating DQ, 81
1425+375	SDSS J142703.40+372110.5	DA	27.04 ± 3.20	19000	...	...	11, 12
1428+282	SDSS J143019.05+281100.8	DA	9.34 ± 1.44	9000	...	...	12
1430+432	SDSS J143218.26+430126.7	DA	1.01	24000	...	...	11, 12
1430+460	SDSS J143235.46+454852.5	DA	12.29 ± 6.98	16750	...	...	11, 12
1440+753	EUVE J1439+750	DA	14–16	20000–50000	0.88-1.19	...	DD, 45
1444+592	SDSS J144614.00+590216.7	DA	4.42 ± 3.79	12500	...	...	10, 12
1452+435	SDSS J145415.01+432149.5	DA	2.35 ± 0.88	11500	...	...	11, 12
	LSPM J1459+0851	DA	~ 2	5535 ± 45	$\log g = (8.0)$	...	T4.5 $\pm$ 0.5 comp., 82
1503-070	GD 175	DA	2.3	6990	0.70 ± 0.13	...	DD, 37
	SDSS J150746.80+520958.0	DA	65.2 ± 0.3	18000 ± 1000	0.99 ± 0.05	...	DD, 55
1506+399	CBS 229, SDSS J150813.24+394504.9	DA	18.9	18000 ± 2000	0.81 ± 0.09	...	DD, 11, 12, 47
1509+425	SDSS J151130.20+422023.0	DA	22.40 ± 9.41	9750	...	...	11, 12

Table 1 (*Continued*)

WD	Other names	Comp.	B_p	$T_{\rm eff}$ (MG)	M (K)	$P_{\rm rot}$ ($M_\odot$)	References notes
1511+076	SDSS J151415.65+074446.5	DA	35.34 ± 2.80	10000	…	…	12
1516+612	SDSS J151745.19+610543.6	DA	13.98 ± 7.36	9500	…	…	10, 12
1521+191	SDSS J152401.60+185659.2	DA	11.96 ± 1.85	13500	…	…	12
1531-022	GD 185	DA	0.035 ± 0.016*	18620 ± 285	0.88 ± 0.03	…	uncertain, 83, 80
1533+423	SDSS J153532.25+421305.6	DA	5.27 ± 4.05	18500	…	…	10, 12
	SDSS J153642.53+420519.2	DZ	9.59 ± 0.04*	5500	…	…	18
1533-057	PG 1533-057	DA	31 ± 3	20000 ± 1040	0.94 ± 0.18	1.89 ± 0.001 hr	84, 80, 22, 9
1537+532	SDSS J153829.29+530604.6	DA	13.99 ± 3.82	13500	…	…	10, 12
1536+085	SDSS J153843.10+084238.2	DA	13.20 ± 4.34	9500	…	…	12
1539+039	SDSS J154213.48+034800.4	DA	8.35 ± 2.60	8500	…	…	10, 12
1541+344	SDSS J154305.67+343223.6	DA	4.09 ± 2.67	25000	…	…	12
	SDSS J155708.04+041156.52	DA	41	…	…	…	78
1603+492	SDSS J160437.36+490809.2	DA	59.51 ± 4.64	9000	…	…	10, 12
1610+330	CBS 418	…	…	…	…	…	21
1639+537	GD 356	DA	13	7510 ± 210	0.67 ± 0.07	0.0803 d	H in emission, 85, 86, 87, 88, 114
1641+241	SDSS J164357.02+240201.3	DA	2	16500	…	…	11, 12
1645+372	SDSS J164703.24+370910.3	DA	2.10 ± 0.67	16250	…	…	11, 12
1647+591	G 226-29, V* DN Dra, NLTT 43637	DA	<0.005	…	…	…	pulsating DA, 89, 69
1648+342	SDSS J165029.91+341125.5	DA	3.38 ± 0.67	9750	…	…	11, 12
1650+355	SDSS J165203.68+352815.8	DA	7.37 ± 2.92	11500	…	…	10, 12
1650+334	SDSS J165249.09+333444.9	DA	5.07 ± 4.18	9000	…	…	12
1653+385	NLTT 43806, SDSS J165445.69+382936.5	DAZ	0.07*	5900	…	…	90
1658+440	PG 1658+440	DA	2.3 ± 0.2	30510 ± 200	1.31 ± 0.02	6 hr 4 d	91, 50, 9
1702+322	SDSS J170400.01+321328.7	DA	50.11 ± 25.08	23000	…	…	11, 12
1713+393	NLTT 44447	DA	1.3	7000 ± 1000	0.59	…	92
1715+601	SDSS J171556.29+600643.9	DA	2.03	13500	…	…	11, 12
1719+562	SDSS J172045.37+561214.9	DA	19.79 ± 5.42	22500	…	…	26, 10, 12

Table 1 *(Continued)*

WD	Other names	Comp.	B_p	T_{eff} (MG)	M (K)	P_{rot} ($M_\odot$)	References notes
1722+541	SDSS J172329.14+540755.8	DA	32.85 ± 3.56	10000	...	...	26, 10, 12
1728+565	SDSS J172932.48+563204.1	DA	27.26 ± 7.04	10500	...	...	10, 12
1743-520	BPM 25114	DA	36	~ 20000	$\log g = (8.0)$	2.84 d	93, 43, 94
1748+708	G 240-72	DB	$\gtrsim 100$	5590 ± 90	0.81 ± 0.01	$\gtrsim 100$ yr	95, 96, 97, 37
1814+248	G 183-35	DA	~ 14	6500 ± 500	$\log g = (8.0)$	~ 50 min-yr	44, 8
1818+126	G 141-2	DA	~ 3	6340 ± 130	0.26 ± 0.12	...	DD, 98, 25
1820+609	LP 103-294, G 227-28	DBA	≤ 0.1	4780 ± 140	0.48 ± 0.05	months-years	37, 8, 9
1829+547	G 227-35	DBA	170–180	6280 ± 140	0.90 ± 0.07	$\gtrsim 100$ yr	99, 100, 44, 37
1900+705	Grw +70°8247	DA	320 ± 20	16000	0.95 ± 0.02	$\gtrsim 100$ yr	101, 85, 102, 103, 37
1939+401	NGC 6819-8	DA	10.3 ± 1.1	19000 ± 1000	0.50 ± 0.05	...	104, 52
1953-011	NLTT 48454, G 92-40	DA	0.1–0.5	7920 ± 200	0.74 ± 0.03	1.44176 d	magnetic spot, 37, 105, 106
2010+310	GD 229	DB	520	18000	$\gtrsim 1$	$\gtrsim 100$ yr	101, 107, 97, 67
2022+130	SDSS J202501.10+131025.6	DA	10.10 ± 1.76	17000	...	...	12
2039-682	GJ 2149, LTT 8190	DA	0.05*	16050	...	...	83, 4
2043-073	SDSS J204626.15-071037.0	DA	2.03	8000	...	...	10, 12
2049-004	SDSS J205233.52-001610.7	DA	13.42 ± 3.73	19000	...	...	10, 12
2051-208	HK 22880134	DA	0.22-0.29	...	...	Variable	16
2105-820	L 24-52, LTT 8381, G J820.1	DAZ	0.043*	10800 ± 290	0.75 ± 0.03	...	37, 83, 16, 108
2146+005	SDSS J214900.87+004842.8	DA	10.09 ± 4.71	11000	...	...	11, 12
2146-077	SDSS J214930.74-072812.0	DA	44.71 ± 1.92	22000	...	...	10, 12
2149+002	SDSS J215135.00+003140.5	DA	~ 300	9000	...	...	10
2149+126	SDSS J215148.31+125525.5	DA	20.76 ± 1.39	14000	...	...	10, 12
2153-512	GJ 841B, BPM 27606	DQ	1.3	6100 ± 200	...	...	109, 58, 42
	SDSS J220029.09-074121.5	DQ	Very weak	21240 ± 180	$\log g = (8.0)$	...	2
2202-000	SDSS J220435.05+001242.9	DA	1.02 ± 0.10	22000	...	...	12
2215-002	SDSS J221828.59-000012.2	DA	257.54 ± 48.71	15500	...	...	10, 12
2225+176	NLTT 53908	DAZ	0.334 ± 0.003	6250 ± 70	$\log g = (7.87 \pm 0.12)$	...	28

Table 1 (*Continued*)

WD	Other names	Comp.	B_p	T_{eff} (MG)	M (K)	P_{rot} ($M_\odot$)	References notes
2245+146	SDSS J224741.46+145638.8	DA	42.11 ± 2.83	18000	...	...	10, 12
2254+076	SDSS J225726.05+075541.7	DA	16.17 ± 2.81	40000	...	...	12
2329-291	...	DA	0.031	24000	...	...	83, 4
2316+123	KUV 813-14, KUV 23162+1220	DA	45 ± 5	11000 ± 1000	$\log g = (8.0)$	17.856 d	23, 44
2317+008	SDSS J231951.73+010909.3	DA	9.35	8300	...	...	11, 12
2320+003	SDSS J232248.22+003900.9	DA	21.40 ± 3.36	20000–39000	...	...	26, 10, 12
2321-010	SDSS J232337.55-004628.2	DB	4.8	15000	...	...	10
2329+267	PG 2329+267	DA	2.31 ± 0.59	9400 ± 240	0.61 ± 0.16	2.767 hr	110, 37, 9
2343+386	SDSS J234605.44+385337.7	DA	798.1 ± 163.6	26000	...	...	11, 12
2343-106	SDSS J234623.69-102357.0	DA	9.17 ± 1.58	8500	...	...	11, 12
2359-434	LP 988-088, LTT 9857, LHS 1005	DA	0.0098*	8570 ± 50	0.98 ± 0.04	2.69479	Different B detected, 83, 36, 111, 16, 108, 4, 112

References: (1) Schmidt et al. (1992b); (2) Dufour et al. (2008a); (3) Lawrie et al. (2013); (4) Kawka and Vennes (2012); (5) Schmidt and Smith (1994); (6) Valyavin et al. (2005); (7) Bergeron et al. (1992); (8) Putney (1997); (9) Brinkworth et al. (2013); (10) Schmidt et al. (2003); (11) Vanlandingham et al. (2005); (12) Külebi et al. (2009); (13) Angel (1977); (14) Achilleos et al. (1992b); (15) Liebert et al. (1977); (16) Koester et al. (2009); (17) Koester et al. (2001); (18) Hollands et al. (2015); (19) Achilleos et al. (1991); (20) Schmidt et al. (2001a); (21) Gianninas et al. (2011); (22) Achilleos and Wickramasinghe (1989); (23) Schmidt and Norsworthy (1991); (24) Friedrich et al. (1997); (25) Bergeron et al. (1997); (26) Gänsicke et al. (2002); (27) Kawka and Vennes (2011); (28) Kawka and Vennes (2014); (29) Farihi et al. (2011); (30) Barstow et al. (1995); (31) Ferrario et al. (1997a); (32) Vennes et al. (2003); (33) Burleigh et al. (1999); (34) Arazimova et al. (2009); (35) Fabrika et al. (2000); (36) Aznar Cuadrado et al. (2004); (37) Bergeron et al. (2001); (38) Angel and Landstreet (1974); (39) Bues and Pragal (1989); (40) Dufour et al. (2005); (41) Berdyugina et al. (2007); (42) Vornanen et al. (2010); (43) Wickramasinghe and Martin (1979); (44) Putney and Jordan (1995); (45) Vennes et al. (1999); (46) Giovannini et al. (1998); (47) Dobbie et al. (2013); (48) Angel et al. (1974a); (49) Martin and Wickramasinghe (1984); (50) Ferrario et al. (1998); (51) Claver et al. (2001); (52) Külebi et al. (2013b); (53) Wesemael et al. (2001); (54) Angel et al. (1972); (55) Dobbie et al. (2012); (56) Liebert et al. (1993); (57) Glenn et al. (1994a); (58) Schmidt et al. (1999b); (59) Hall and Maxwell (2008); (60) Wickramasinghe and Cropper (1988); (61) Euchner et al. (2005b); (62) Saffer et al. (1989); (63) Schmidt et al. (1986); (64) Liebert et al. (2003b); (65) Bues (1999); (66) Jordan and Friedrich (2002); (67) Wickramasinghe et al. (2002); (68) Euchner et al. (2005a); (69) Valyavin et al. (2006); (70) Foltz et al. (1989); (71) Schmidt et al. (2005a); (72) Reid et al. (2001); (73) Liebert et al. (2003a); (74) Hagen et al. (1987); (75) Girven et al. (2010); (76) Dufour et al. (2006) (77) Bergeron et al. (1993); (78) Kepler et al. (2015); (79) Liebert et al. (1994); (80) Liebert et al. (2005b); (81) Dufour et al. (2008b); (82) Day-Jones et al. (2011); (83) Koester et al. (1998); (84) Liebert et al. (1985); (85) Greenstein et al. (1985); (86) Ferrario et al. (1997b); (87) Brinkworth et al. (2004); (88) Wickramasinghe et al. (2010); (89) Schmidt and Grauer (1997); (90) Zuckerman et al. (2011); (91) Schmidt et al. (1992a); (92) Kawka and Vennes (2006); (93) Martin and Wickramasinghe (1978); (94) Wegner (1977); (95) Angel et al. (1974b); (96) Angel (1978); (97) Berdyugin and Piirola (1999); (98) Greenstein and Saha (1986); (99) Angel et al. (1975); (100) Cohen et al. (1993); (101) Angel et al. (1985); (102) Wickramasinghe and Ferrario (1988a); (103) Jordan (1992); (104) Kalirai et al. (2008); (105) Maxted et al. (2000); (106) Brinkworth et al. (2005); (108) Landstreet et al. (2012); (109) Wickramasinghe and Bessell (1979); (110) Moran et al. (1998); (111) Kawka et al. (2007); (112) Gary (2014); (113) Vornanen et al. (2013); (114) Li et al. (1998)

Table 2 Magnetic white dwarfs in synchronous cataclysmic variables

Name	$B_{cyc,1}$ (MG)	$B_{cyc,2}$ (MG)	$B_{Zeem,phot}$ (MG)	$B_{Zeem,halo}$ (MG)	T_{eff} (K)	P_{rot} (min)	Mass ($M_\odot$)	References notes
Polars								
EQ Cet (=RX J0128.8-2339)	34	45	...	...	...	92.8	...	1, 2
CV Hyi (=RX J0132.7-6554)	68	...	...	...	...	77.8	...	3
BL Hyi (=H 0139-68)	23	...	21	12	13300	113.6	0.71	4, 5, 6
RX J0154.0-5947	...	...	...	...	...	80	...	7
FL Cet (=SDSS J0155+0028)	29	...	...	...	...	87.1	...	8
AI Tri (=RX J0203.8+2959)	38	...	...	...	...	275.5	...	9
BS Tri (=RX J0209.4+2832)	...	...	...	...	...	96.3	...	10
CW Hyi (=RBS0324)	13	...	...	...	...	181.8	...	11
WW Hor (=EXO 023432-5232)	25	...	...	...	...	114.6	1.1	12
EF Eri (=2A0311-227)	21	17	...	15	9500	81	0.9	13, 14, 15
CSS091109 J035759+102943	...	...	...	...	114.0	...	16	
SDSS J032855.00+052254.2	33	...	...	...	...	122.0	...	17
VY For (=EXO 0329.9-2606)	...	...	...	...	...	228.0	...	18
UZ For (=EXO 0333.3-2554)	56	75	...	...	...	126.5	0.7	19, 20
RX J0425.6-5714	>50	...	...	...	...	85.8	...	21
IW Eri (=RBS0541)	...	...	...	...	...	87.1	...	11
RS Cae (=RX J0453.4-4213)	36	...	...	...	...	102.1	...	22, 23
HY Eri (=RX J0501.7-0359)	25	...	...	...	...	171.3	0.4	24
V1309 Ori (=RX J0515.6+0105)	61	...	...	...	<20000	479.0	0.6–0.7	25, 26, 27
IPHAS J052832.69+283837.6	...	...	...	...	...	...	...	28
UW Pic (=RX J0531.5-4624)	19	...	...	...	...	133.4	...	29, 30
BY Cam (=H 0538+608)	41	...	...	...	...	199.3	1.04	31, 32, 33, $P_{orb} = 201.3$
RX J0649.8-0737	...	...	...	...	...	265	...	34
LW Cam (=RX J0704.2+6203)	20	...	...	...	...	97.3	...	35
CSS081231 J071126+440405	...	...	...	...	...	117.2	...	36
HS Cam (=RX J0719.2+6557)	...	...	...	...	...	98.2	0.75	37
V654 Aur (=SDSS J072910.2+365838)	...	...	...	...	...	150	...	38

Table 2 (*Continued*)

Name	$B_{cyc,1}$ (MG)	$B_{cyc,2}$ (MG)	$B_{Zeem,phot}$ (MG)	$B_{Zeem,halo}$ (MG)	T_{eff} (K)	P_{rot} (min)	Mass ($M_\odot$)	References notes
RX J0749.1-0549	...	...	...	...	...	215	...	34
EU Lyn (=SDSS J075240.45+362823.2)	...	...	...	...	...	164	...	38
V516 Pup (=RX J0803.4-4748)	39	...	...	...	...	136.8	...	39
CSS100108 J081031+002429	...	...	...	...	...	116.2	...	40
VV Pup (=2E 0812.8-1853)	31	56	...	...	12100	100.4	0.73	6, 41
EG Lyn (=RBS0696)	...	...	...	...	...	99.4	...	11
EU Cnc (=G 186)	41	...	...	...	...	125.4	≥0.68	42, 43
FR Lyn (=SDSS J085414.02+390537.2)	44	...	...	...	...	113.3	...	44, 45
SDSS J085909.18+053654.5	...	...	...	...	...	143.8	...	45
SDSS J092122.84+203857.1	32	...	...	...	...	84.2	...	46, 47
HU Leo (=SDSSJ 092444.48+080150.9)	...	...	...	...	...	131.24	...	48
MN Hya (=RX J0929.1-2404)	42	...	...	...	...	203.4	...	49, 50
SDSSJ 093839.25+534403.8	...	...	...	...	...	$\lesssim 120$	51	
RX J0953.1+1458	...	...	...	...	...	102	...	7
1RXS J100211.4-192534	...	...	...	...	...	107	~0.5	52
1RXS J100734.4-201731	94	...	...	...	...	208.6	0.8	52, 53
GG Leo (=RBS0842)	23	...	...	...	...	79.9	~1.0	54, 55
V381 Vel (=RX J1016.9-4103)	52	...	...	...	...	134	...	56
FH UMa (=RBS0904)	>20	...	...	...	...	80	...	57
EK UMa (=1E 1048.5+5421)	47	...	...	...	...	114.5	...	58, 59
AN UMa (=PG 1101+453)	36	...	...	...	...	114.8	...	60, 61
ST LMi (=CW 1103+254)	12	21	19	...	10800	113.9	0.45	6, 62, 63, 64
AR UMa (=1ES 1113+432)	$\gtrsim 160$	240	230	...	~20000	115.9	...	65, 66
DP Leo (=1E 1114+18.2)	31	76	...	...	13500	89.8	...	11, 67, 68
V1033 Cen (=RX J1141.3-6410)	20	...	...	...	...	189.4	...	69, 70
EU UMa (=RX J1149.9+2844)	43	...	...	...	...	90.1	...	71, 72
SDSSJ 120724.69+223529.8	...	...	...	...	...	...	...	73
V379 Vir (=SDSS J121209.31+013627.7)	7	...	...	...	~10000	88.4	...	74, 75, 76, LARP

Table 2 *(Continued)*

Name	$B_{cyc,1}$ (MG)	$B_{cyc,2}$ (MG)	$B_{Zeem,phot}$ (MG)	$B_{Zeem,halo}$ (MG)	T_{eff} (K)	P_{rot} (min)	Mass ($M_\odot$)	References notes
WD 1248+161 (=SDSS J125044.42+154957.4)	20	...	...	...	10000	86.3	...	77, LARP
EV UMa (=RX J1307.8+5351)	30	...	...	...	...	79.7	...	78
2XMM J131223.4+173659	≤10	...	...	...	...	91.9	...	79
V1043 Cen (=RX J1313.2-3259)	56	...	...	...	15000	251.4	...	80, 81
SDSS J133309.19+143706.9	...	...	...	...	...	132	...	46, 47
SDSS J134441.83+204408.4	65	...	...	...	...	110	...	82
V834 Cen (=1E 1405-45.1)	23	...	22	23	14300	101.5	0.66	6, 83, 84, 85, 86
SDSS J142256.31-022108.0	...	...	...	...	...	202	...	87
V895 Cen (=EUVE J1429-38.0)	...	...	...	...	13800	285.9	...	6, 88
IGR J14536-5522	20	...	...	...	...	189.4	...	89
CSS100216 J150354-220711	...	...	...	...	...	133.4	...	40
SDSS J151415.65+074446.5	36	...	...	...	10000	88.7	...	77, LARP
SDSS J153023.64+220646.4	...	...	...	...	...	...	...	90
BM CrB (=SDSS J154104.66+360252.9)	33	...	...	...	...	84	...	44, 45
2XMM J154305.5-522709	...	...	...	...	...	143	...	91
MR Ser (=PG 1550+191)	25	...	28	25	14000	113.5	0.5	6, 64, 92
AP CrB (=RX J1554.2+2721)	110	...	144	...	~20000	151.9	...	93, 94, 95
V519 Ser (=1RXS J161008.0+035222)	15	...	...	...	...	190.5	...	96, 97
V1189 Her (=SDSS J162936.53+263519.5)	...	...	...	...	...	134	...	44
1RXS J170053.7+400354	...	...	...	...	...	116.4	...	38
V1007 Her (=RBS1646)	50	...	...	...	...	119.9	...	98
V2301 Oph (=1H 1752+081)	...	...	...	7	10000	113.0	0.8	99, 100, 101
V884 Her (=RX J1802.1+1804)	150	...	...	150	...	113.3	...	102, 103
AM Her (=3U1809+50)	14	...	13	...	19800	185.6	0.78	104, 105, 106
XGPS-I J183251-100106	...	...	...	...	...	89.0	...	107
V347 Pav (=RX J1844.7-7418)	10	20	...	...	12300	90.1	...	6, 108, 109
1RXS J184542.4+483134	...	...	...	...	...	79.1	...	110
MT Dra (=RX J1846.9+5538)	15	...	...	...	...	128.7	...	111

Table 2 (*Continued*)

Name	$B_{cyc,1}$ (MG)	$B_{cyc,2}$ (MG)	$B_{Zeem,phot}$ (MG)	$B_{Zeem,halo}$ (MG)	T_{eff} (K)	P_{rot} (min)	Mass ($M_\odot$)	References notes
EP Dra (=1H 1903+689)	...	...	...	16	...	104.6	0.43	112
CTCV J1928-5001	20	...	...	...	...	101.0	...	113, 114
QS Tel (=RX J1938.6-4613)	47	75	60	...	17500	139.9	...	115
V1432 Aql (=RX J1940.2-1025)	30	...	...	...	~35000	202.5	1.2	116, 117, 118, $P_{orb} = 201.9$
CSS100805 J194428-420209	...	...	...	...	...	91.9	...	119
V393 Pav (=RX J1957.1-5738)	...	...	...	16	...	98.8	...	120
QQ Vul (=1E 2003+22.5)	30	...	...	...	...	222.5	0.58	64, 121, 122
V349 Pav (=Drissen V211b)	...	...	...	...	...	159.7	...	123
V4738 Sgr (=RX J2022.6-3954)	67	...	...	...	...	78.0	...	3
SDSS J205017.83-053626.7	...	...	...	...	...	94.2	...	124
HU Aqr (=RX J2107.9-0517)	37	...	20	...	14000	125.0	0.9	6, 125, 126, 127
V1500 Cyg (=Nova Cyg 1975)	$\gtrsim 25$	...	...	...	70000	197.5	0.9	128, 129, $P_{orb} = 201.0$
CD Ind (=RX J2115.7-5840)	11	...	...	...	...	109.6	0.79	130, 131, 132, $P_{orb} = 110.9$
CE Gru (=Grus V1)	32	...	...	...	...	108.5	~1.0	133, 134, 135
SDSS J215427.19+155713.0	...	...	...	...	...	96.9	...	82
V388 Peg (=RX J2157.5+0855)	20	...	...	...	...	202.5	...	136
SWIFT J2218.4+1925	...	...	...	...	...	129.5	0.97	137, 138
2XMM J225036.9+573154	...	...	...	...	...	174.2	...	139
CP Tuc (=AX J2315-592)	...	...	10	...	...	89.0	0.68	33, 140, 141
1RXS J231603.9-052713	25	...	...	...	...	209.5	1.0	96
SWIFT J2319.4+2619	...	...	...	...	...	180.6	...	142
Pre-polars								
SDSS J030308.35+005444.1	...	...	8	...	9150	193.6	0.84	143
SDSS J083751.0+383012.5	...	...	...	...	...	178.8	...	87
HS 0922+1333	66	81	...	...	≤8000	242.4	...	144, 145, 146
WX LMi (=HS 1023+3900)	61	70	...	...	...	166.9	...	147, 148, 149
IL Leo (=SDSS J103100.55+202832.2)	42	...	...	...	9500	83.2	...	150

Table 2 (*Continued*)

Name	$B_{cyc,1}$ (MG)	$B_{cyc,2}$ (MG)	$B_{Zeem,phot}$ (MG)	$B_{Zeem,halo}$ (MG)	T_{eff} (K)	P_{rot} (min)	Mass ($M_{\odot}$)	References notes
SDSS J105905.06+272755.4	57	...	...	...	≤ 8500	150	...	150
SDSS J120615.73+510047.0	108	...	...	...	9000	197	...	151
PZ Vir (=J132411.57+032050.4)	63	...	...	...	$\lesssim 7500$	158.7	...	152, 153, 154
MQ Dra (=SDSS J155331.11+551614.4)	58	...	...	...	$\lesssim 10000$	263.5	...	152, 154
SDSS J204827.91+005008.9	62	...	...	...	7500	252	...	154

References: (1) Schwope et al. (1999); (2) Campbell et al. (2008b); (3) Burwitz et al. (1997); (4) Ferrario et al. (1996); (5) Schwope et al. (1995a); (6) Araujo-Betancor et al. (2005a); (7) Beuermann (1999); (8) Szkody et al. (2002); (9) Schwarz et al. (1998); (10) Denisenko et al. (2006); (11) Schwope et al. (2002c); (12) Bailey et al. (1988); (13) Ferrario et al. (1996); (14) Achilleos et al. (1992a); (15) Schwope and Christensen (2010); (16) Schwope and Thinius (2012) (17) Szkody et al. (2007); (18) Beuermann et al. (1989); (19) Ferrario et al. (1989b); (20) Schwope et al. (1990); (21) Halpern et al. (1998); (22) Burwitz et al. (1996b); (23) Traulsen et al. (2014); (24) Burwitz et al. (1999); (25) Staude et al. (2001); (26) Shafter et al. (1995); (27) de Martino et al. (1998); (28) Skinner et al. (2014); (29) Reinsch et al. (1994); (30) Romero-Colmenero et al. (2003); (31) Cropper et al. (1989); (32) Mason et al. (1998); (33) Ramsay (2000); (34) Motch et al. (1998); (35) Tovmassian et al. (2001); (36) Schwope et al. (2015); (37) Tovmassian et al. (1997); (38) Homer et al. (2005); (39) Schwarz and Greiner (1999); (40) Woudt et al. (2012b); (41) Wickramasinghe et al. (1989); (42) Pasquini et al. (1994); (43) Williams et al. (2013b); (44) Gänsicke et al. (2009); (45) Szkody et al. (2005); (46) Schmidt et al. (2008); (47) Southworth et al. (2015); (48) Southworth et al. (2010); (49) Ramsay and Wheatley (1998); (50) Buckley et al. (1998a); (51) Szkody et al. (2009); (52) Ramsay and Cropper (2003); (53) Thomas et al. (2012); (54) Burwitz et al. (1998); (55) Ramsay and Cropper (2004); (56) Greiner and Schwarz (1998); (57) Singh et al. (1995); (58) Beuermann et al. (2009); (59) Cropper et al. (1990); (60) Schneider and Young (1980); (61) Cropper et al. (1989); (62) Ferrario et al. (1993a); (63) Schmidt et al. (1983); (64) Mukai and Charles (1987); (65) Schmidt et al. (1996); (66) Gänsicke et al. (2001); (67) Cropper and Wickramasinghe (1993); (68) Schwope et al. (2002a); (69) Cieslinski and Steiner (1997); (70) Buckley et al. (2000); (71) Howell et al. (1995); (72) Schwope (1995); (73) Szkody et al. (2011); (74) Schmidt et al. (2005b); (75) Burleigh et al. (2006); (76) Farihi et al. (2008); (77) Breedt et al. (2012); (78) Osborne et al. (1994); (79) Vogel et al. (2008); (80) Thomas et al. (2000); (81) Gänsicke et al. (2000); (82) Szkody et al. (2014); (83) Wickramasinghe et al. (1987); (84) Ferrario et al. (1992); (85) Schwope and Beuermann (1990a); (86) Cropper et al. (1986); (87) Hilton et al. (2009); (88) Howell et al. (1997); (89) Potter et al. (2010); (90) Breedt et al. (2014); (91) Servillat et al. (2012); (92) Schwope et al. (1993b); (93) Thorstensen and Fenton (2002); (94) Gänsicke et al. (2004); (95) Schwope et al. (2006); (96) Rodrigues et al. (2006); (97) Thorstensen et al. (2015); (98) Greiner et al. (1998); (99) Ferrario et al. (1995); (100) Silber et al. (1994); (101) Barwig et al. (1994); (102) Szkody et al. (1995); (103) Schmidt et al. (2001b); (104) Wickramasinghe and Martin (1985); (105) Bailey et al. (1991); (106) Gänsicke et al. (2006b); (107) Hui et al. (2012); (108) Ramsay et al. (1996); (109) Potter et al. (2000); (110) Shchurova et al. (2013); (111) Schwarz et al. (2002); (112) Schwope and Mengel (1997); (113) Tappert et al. (2004); (114) Potter et al. (2005); (115) Schwope et al. (1995b); (116) Friedrich et al. (1996); (117) Rana et al. (2005); (118) Schmidt and Stockman (2001); (119) Coppejans et al. (2014); (120) Thomas et al. (1996); (121) Schwope et al. (2000); (122) Cropper (1998); (123) Wickramasinghe et al. (1993); (124) Potter et al. (2006); (125) Schwope et al. (1993a); (126) Schwope et al. (1997b); (127) Glenn et al. (1994b); (128) Schmidt and Stockman (1991); (129) Schmidt et al. (1995b); (130) Ramsay et al. (1999a); (131) Schwope et al. (1997a); (132) Ramsay et al. (2000); (133) Tuohy et al. (1988); (134) Wickramasinghe et al. (1991); (135) Ramsay and Cropper (2002); (136) Tovmassian et al. (2000); (137) Thorstensen and Halpern (2013); (138) Bernardini et al. (2014); (139) Ramsay et al. (2009); (140) Beuermann et al. (2007); (141) Ramsay et al. (1999b); (142) Shafter et al. (2008); (143) Parsons et al. (2013); (144) Reimers and Hagen (2000); (145) Tovmassian and Zharikov (2007); (146) Vogel et al. (2011); (147) Reimers et al. (1999); (148) Schwarz et al. (2001); (149) Vogel et al. (2007); (150) Schmidt et al. (2007); (151) Schwope et al. (2009); (152) Szkody et al. (2003); (153) Southworth et al. (2015); (154) Schmidt et al. (2005b)

Table 3 Magnetic white dwarfs in asynchronous cataclysmic variables

Name	B_{cyc} (MG)	P_{rot} (min)	P_{orb} (min)	Mass ($M_\odot$)	References notes
V1033 Cas (=IGR J00234+6141)	...	563.5	242.0	0.9	1, 2
V709 Cas (=RX J0028.8+5917)	...	312.8	320.0	0.96	3, 4, 2
V515 And (=XSS J00564+4548)	...	465.5	163.9	0.79	5, 6, 7
XY Ari (=H 0253+193)	...	206.3	363.9	1.0	8, 9, 2
GK Per (=Nova Persei 1901)	...	351.3	2875.4	0.9	10, 11, 2
V1062 Tau (=H 0459+246)	...	3704	598.9	0.7	2, 4, 12
UU Col (=RX J0512.2-3241)	10–30	863.5	207.0	0.6	13, 14, 15
Paloma (=1RXS J052430.2+424449)	...	8175.4	157.2	...	16
TV Col (=2A 0526-328)	...	1909.7	329.2	0.78	17, 18, 2
TX Col (=1 H0542-407)	...	1911	343.2	0.7	19, 2
V405 Aur (=RX J0558.0+5353)	32	545.4	249.6	0.89	20, 21, 22, 2
MU Cam (=1RXS J062518.2+733433)	...	1187.2	283.1	0.74	23, 2
V902 Mon (=IPHAS J062746.41+014811.3)	...	2210	489.6	...	23b
V647 Aur (=1RXS J063631.9+353537)	...	932.9	207.9	0.74	24, 7
V418 Gem (=1RXS J070407.9+262501)	...	480.7	262.8	$\lesssim 0.5$	25, 26
BG CMi (=3 A0729+103)	~4	913.5	194.1	0.7	27, 28, 2
V667 Pup (=Swift J0732.5-1331)	...	512.4	336.2	...	4
PQ Gem (=RX J0751.2+1444)	9–21	833.4	311.6	0.65	29, 30, 31, 2
HT Cam (=RX J0757.0+6306)	...	515.1	86.0	0.6	32, 33
DW Cnc (=HS 0756+1624)	...	2314.7	86.1	...	34
WX Pyx (=1E 0830.9-2238)	...	1559.2	318	...	35
EI UMa (=1H0832+488)	...	741.6	386.1	...	36, 37
IGR J08390-4833	...	1480.8	480	0.95	7
VZ Sex (=1RXS J094432.1+035738)	...	2450	214.1	...	38, 4
YY Dra (=DO Dra)	...	529.3	238.1	0.8	39

Table 3 (*Continued*)

Name	B_{cyc} (MG)	P_{rot} (min)	P_{orb} (min)	Mass ($M_{\odot}$)	References notes
V1025 Cen (=RX J1238.2-3842)	...	2146.6	84.6	0.5	40, 2
EX Hya (=4U 1249-28)	...	4021.6	98.3	0.79	41, 42
IGR J15094-6649	$\gtrsim$10	809.4	353.4	0.89	43, 44, 7
NY Lup (=1RXS J154814.5-452845)	>4	693.0	591.8	1.09	44, 45, 2
IGR J16500-3307	...	571.9	217.0	0.92	7
1RXS J165443.5-191620	...	546.7	222.9	...	46
V2400 Oph (=RX J1712.6-2414)	9–27	927.7	205.8	0.8	47, 48, 2
IGR J17195-4100	...	1053.7	240.3	0.86	49, 50, 7
V2731 Oph (=1RXS J173021.5-055933)	5	128.0	925.3	0.96	51, 52, 53
AX J1740.1-2847	...	730	125	...	54, 55
AX J1740.2-2903	...	628.6	343.3	...	56
V1323 Her (=1RXS J180340.0+401214)	...	1520.5	264.1	0.69	51, 26
1RXS J180431.1-273932	...	494	...	0.8	57
DQ Her (=Nova Her 1934)	...	70.8	278.8	0.60	58, 59,
IGR J18173-2509	...	1663.4	91.9	0.96	7, 56
IGR J18308-1232	...	1820	322.4	0.85	7, 56
AX J1832.3-0840	...	1552.3	...	...	54
AX J1853.3-0128	...	477.6	87.2	...	56
V1223 Sgr (=4 U1851-31)	...	745.5	201.9	0.65	60, 61, 2
IGR J19267+1325	...	935.1	206.9	...	56
V2306 Cyg (=WGA J1958.2+3232)	8	1466.7	261.0	0.8	62, 63, 64, 2
IGR J19552+0044	...	4960	101.7	0.77	56, 65
AE Aqr (=1E 2037.5-0102)	...	33.1	592.8	0.63	66, 67, 68
V2069 Cyg (=RX J2123.7+4217)	...	743.1	448.8	0.82	7, 69
1RXS J213344.1+510725	$\gtrsim$20	570.8	431.6	0.93	70, 71, 72

Table 3 (*Continued*)

Name	B_{cyc} (MG)	P_{rot} (min)	P_{orb} (min)	Mass ($M_{\odot}$)	References notes
FO Aqr (=H 2215-086)	...	1254.5	290.9	0.61	73, 2
AO Psc (=H 2252-035)	...	805.2	215.5	0.55	74, 75, 2
CC Scl (=1RXS J231532.3-304855)	...	389.5	840.9	...	76
V598 Peg (=SDSS J233325.92+152222.1)	...	2500	83.12	...	77, 78
V455 And (=HS 2331+390)	...	67.6	81.1	...	79

References: (1) Bonnet-Bidaud et al. (2007); (2) Brunschweiger et al. (2009); (3) de Martino et al. (2001); (4) Thorstensen et al. (2010); (5) Kozhevnikov (2012); (6) Bonnet-Bidaud et al. (2009); (7) Bernardini et al. (2012); (8) Allan et al. (1996); (9) Hellier (1997b); (10) Crampton et al. (1986); (11) Mauche (2004); (12) Hellier et al. (2002); (13) Burwitz et al. (1996a); (14) de Martino et al. (2006b); (15) Katajainen et al. (2010); (16) Schwarz et al. (2007); (17) Augusteijn et al. (1994); (18) Rana et al. (2004); (19) Buckley and Tuohy (1989); (20) Harlaftis and Horne (1999); (21) Skillman (1996); (22) Piirola et al. (2008); (23) Staude et al. (2003); (24) Kozhevnikov (2014); (25) Patterson et al. (2011); (26) Anzolin et al. (2008); (27) Kim et al. (2005); (28) Chanmugam et al. (1990); (29) Hellier (1997a); (30) Evans et al. (2006); (31) Potter et al. (1997); (32) Kemp et al. (2002); (33) de Martino et al. (2005); (34) Patterson et al. (2004); (35) Joshi et al. (2011); (36) Baskill et al. (2005); (37) Thorstensen (1986); (38) de Martino (2007); (39) Haswell et al. (1997); (40) Buckley et al. (1998b); (41) Mauche et al. (2009); (42) Beuermann and Reinsch (2008); (43) Pretorius (2009); (44) Potter et al. (2012); (45) de Martino et al. (2006a), (46) Scaringi et al. (2011); (47) Buckley et al. (1995); (48) Vaeth (1997); (49) Girish and Singh (2012); (50) Pretorius (2009); (51) Gänsicke et al. (2005); (52) de Martino et al. (2008); (53) Butters et al. (2009); (54) Kaur et al. (2010); (55) Britt et al. (2013); (56) Thorstensen and Halpern (2013); (57) Masetti et al. (2012); (58) Zhang et al. (1995); (59) Bloemen et al. (2010); (60) Osborne et al. (1985); (61) Jablonski and Steiner (1987); (62) Norton et al. (2002); (63) Zharikov et al. (2002); (64) Uslenghi et al. (2001); (65) Bernardini et al. (2013); (66) Welsh et al. (1993); (67) de Jager et al. (1994); (68) Echevarría et al. (2008); (69) Thorstensen and Taylor (2001); (70) Katajainen et al. (2007); (71) Bonnet-Bidaud et al. (2006); (72) Anzolin et al. (2009); (73) Patterson et al. (1998); (74) Kaluzny and Semeniuk (1988); (75) Hellier et al. (1991); (76) Woudt et al. (2012a); (77) Southworth et al. (2007); (78) Hilton et al. (2009); (79) Araujo-Betancor et al. (2005b)

References

Achilleos, N., Wickramasinghe, D.T.: Astrophys. J. **346**, 444 (1989)
Achilleos, N., Remillard, R.A., Wickramasinghe, D.T.: Mon. Not. R. Astron. Soc. **253**, 522 (1991)
Achilleos, N., Wickramasinghe, D.T., Wu, K.: Mon. Not. R. Astron. Soc. **256**, 80 (1992a)
Achilleos, N., Wickramasinghe, D.T., Liebert, J., Saffer, R.A., Grauer, A.D.: Astrophys. J. **396**, 273 (1992b)
Aizu, K.: Prog. Theor. Phys. **49**, 1184 (1973)
Al-Hujaj, O.-A., Schmelcher, P.: Phys. Rev. A **68**(5), 053403 (2003)
Allan, A., Hellier, C., Ramseyer, T.F.: Mon. Not. R. Astron. Soc. **282**, 699 (1996)
Angel, J.R.P.: Astrophys. J. **216**, 1 (1977)
Angel, J.R.P.: Annu. Rev. Astron. Astrophys. **16**, 487 (1978)
Angel, J.R.P., Landstreet, J.D.: Astrophys. J. Lett. **164**, 15 (1971a)
Angel, J.R.P., Landstreet, J.D.: Astrophys. J. Lett. **165**, 71 (1971b)
Angel, J.R.P., Landstreet, J.D.: Astrophys. J. **191**, 457 (1974)
Angel, J.R.P., Illing, R.M.E., Landstreet, J.D.: Astrophys. J. Lett. **175**, 85 (1972)
Angel, J.R.P., Carswell, R.F., Beaver, E.A., Harms, R., Strittmatter, P.A.: Astrophys. J. Lett. **194**, 47 (1974a)
Angel, J.R.P., Hintzen, P., Strittmatter, P.A., Martin, P.G.: Astrophys. J. Lett. **190**, 71 (1974b)
Angel, J.R.P., Hintzen, P., Landstreet, J.D.: Astrophys. J. Lett. **196**, 27 (1975)
Angel, J.R.P., Borra, E.F., Landstreet, J.D.: Astrophys. J. Suppl. Ser. **45**, 457 (1981)
Angel, J.R.P., Liebert, J., Stockman, H.S.: Astrophys. J. **292**, 260 (1985)
Anzolin, G., de Martino, D., Bonnet-Bidaud, J.-M., Mouchet, M., Gänsicke, B.T., Matt, G., Mukai, K.: Astron. Astrophys. **489**, 1243 (2008)
Anzolin, G., de Martino, D., Falanga, M., Mukai, K., Bonnet-Bidaud, J.-M., Mouchet, M., Terada, Y., Ishida, M.: Astron. Astrophys. **501**, 1047 (2009)
Araujo-Betancor, S., Gänsicke, B.T., Long, K.S., Beuermann, K., de Martino, D., Sion, E.M., Szkody, P.: Astrophys. J. **622**, 589 (2005a)
Araujo-Betancor, S., Gänsicke, B.T., Hagen, H.-J., Marsh, T.R., Harlaftis, E.T., Thorstensen, J., Fried, R.E., Schmeer, P., Engels, D.: Astron. Astrophys. **430**, 629 (2005b)
Arazimova, E., Kawka, A., Vennes, S.: Publ. Obs. Astron. Beograd **86**, 297 (2009)
Asensio Ramos, A., Martínez González, M.J., Manso Sainz, R., Corradi, R.L.M., Leone, F.: Astrophys. J. **787**, 111 (2014)
Augusteijn, T., Heemskerk, M.H.M., Zwarthoed, G.A.A., van Paradijs, J.: Astron. Astrophys. Suppl. Ser. **107**, 219 (1994)
Aznar Cuadrado, R., Jordan, S., Napiwotzki, R., Schmid, H.M., Solanki, S.K., Mathys, G.: Astron. Astrophys. **423**, 1081 (2004)
Bailey, J., Wickramasinghe, D.T., Hough, J.H., Cropper, M.: Mon. Not. R. Astron. Soc. **234**, 19 (1988)
Bailey, J., Ferrario, L., Wickramasinghe, D.T.: Mon. Not. R. Astron. Soc. **251**, 37 (1991)
Barlow, E.J., Knigge, C., Bird, A.J., Dean, A.J., Clark, D.J., Hill, A.B., Molina, M., Sguera, V.: Mon. Not. R. Astron. Soc. **372**, 224 (2006)
Barstow, M.A., Jordan, S., O'Donoghue, D., Burleigh, M.R., Napiwotzki, R., Harrop-Allin, M.K.: Mon. Not. R. Astron. Soc. **277**, 971 (1995)
Barwig, H., Ritter, H., Barnbantner, O.: Astron. Astrophys. **288**, 204 (1994)
Baskill, D.S., Wheatley, P.J., Osborne, J.P.: Mon. Not. R. Astron. Soc. **357**, 626 (2005)
Baumgartner, W.H., Tueller, J., Markwardt, C.B., Skinner, G.K., Barthelmy, S., Mushotzky, R.F., Evans, P., Gehrels, N.: Astrophys. J. Suppl. Ser. **297**, 19 (2013)
Beardmore, A.P., Mukai, K., Norton, A.J., Osborne, J.P., Hellier, C.: Mon. Not. R. Astron. Soc. **297**, 337 (1998). doi:10.1046/j.1365-8711.1998.01382.x
Becken, W., Schmelcher, P., Diakonos, F.K.: J. Phys. B, At. Mol. Phys. **32**, 1557 (1999)
Becken, W., Schmelcher, P.: J. Phys. B, At. Mol. Phys. **33**, 545 (2000a)
Becken, W., Schmelcher, P.: J. Comput. Appl. Math. **126**, 449 (2000b)
Becken, W., Schmelcher, P.: Phys. Rev. A **63**(5), 053412 (2001)
Berdyugin, A.V., Piirola, V.: Astron. Astrophys. **352**, 619 (1999)
Berdyugina, S.V., Berdyugin, A.V., Piirola, V.: Phys. Rev. Lett. **99**(9), 091101 (2007)
Berger, L., Koester, D., Napiwotzki, R., Reid, I.N., Zuckerman, B.: Astron. Astrophys. **444**, 565 (2005)
Bergeron, P., Ruiz, M.-T., Leggett, S.K.: Astrophys. J. **400**, 315 (1992)
Bergeron, P., Ruiz, M.-T., Leggett, S.K.: Astrophys. J. **407**, 733 (1993)
Bergeron, P., Ruiz, M.T., Leggett, S.K.: Astrophys. J. Suppl. Ser. **108**, 339 (1997)
Bergeron, P., Leggett, S.K., Ruiz, M.T.: Astrophys. J. Suppl. Ser. **133**, 413 (2001)
Bernardini, F., de Martino, D., Falanga, M., Mukai, K., Matt, G., Bonnet-Bidaud, J.-M., Masetti, N., Mouchet, M.: Astron. Astrophys. **542**, 22 (2012)

Bernardini, F., de Martino, D., Mukai, K., Falanga, M., Andruchow, I., Bonnet-Bidaud, J.-M., Masetti, N., Buitrago, D.H.G., Mouchet, M., Tovmassian, G.: Mon. Not. R. Astron. Soc. **435**, 2822 (2013)
Bernardini, F., de Martino, D., Mukai, K., Falanga, M.: Mon. Not. R. Astron. Soc. **445**, 1403 (2014)
Beuermann, K.: In: Aschenbach, B., Freyberg, M.J. (eds.) *Highlights in X-Ray Astronomy*, p. 410 (1999)
Beuermann, K., Reinsch, K.: Astron. Astrophys. **381**, 487 (2002)
Beuermann, K., Reinsch, K.: Astron. Astrophys. **480**, 199 (2008)
Beuermann, K., Thomas, H.-C., Giommi, P., Tagliaferri, G., Schwope, A.D.: Astron. Astrophys. **219**, 7 (1989)
Beuermann, K., Euchner, F., Reinsch, K., Jordan, S., Gänsicke, B.T.: Astron. Astrophys. **463**, 647 (2007)
Beuermann, K., Diese, J., Paik, S., Ploch, A., Zachmann, J., Schwope, A.D., Hessman, F.V.: Astron. Astrophys. **507**, 385 (2009)
Beuermann, K., Burwitz, V., Reinsch, K.: Astron. Astrophys. **543**, 41 (2012)
Bird, A.J., Malizia, A., Bazzano, A., Barlow, E.J., Bassani, L., Hill, A.B., Bélanger, G., Capitanio, F., Clark, D.J., Dean, A.J., Fiocchi, M., Götz, D., Lebrun, F., Molina, M., Produit, N., Renaud, M., Sguera, V., Stephen, J.B., Terrier, R., Ubertini, P., Walter, R., Winkler, C., Zurita, J.: Astrophys. J. Suppl. Ser. **186**, 1 (2010)
Bloemen, S., Marsh, T.R., Steeghs, D., Østensen, R.H.: Mon. Not. R. Astron. Soc. **407**, 1903 (2010)
Bonnet-Bidaud, J.M., Mouchet, M., de Martino, D., Silvotti, R., Motch, C.: Astron. Astrophys. **445**, 1037 (2006)
Bonnet-Bidaud, J.M., de Martino, D., Falanga, M., Mouchet, M., Masetti, N.: Astron. Astrophys. **473**, 185 (2007)
Bonnet-Bidaud, J.-M., de Martino, D., Mouchet, M.: ATel **1895** (2009)
Breedt, E., Gänsicke, B.T., Girven, J., Drake, A.J., Copperwheat, C.M., Parsons, S.G., Marsh, T.R.: Mon. Not. R. Astron. Soc. **423**, 1437 (2012)
Breedt, E., Gänsicke, B.T., Drake, A.J., Rodríguez-Gil, P., Parsons, S.G., Marsh, T.R., Szkody, P., Schreiber, M.R., Djorgovski, S.G.: Mon. Not. R. Astron. Soc. **443**, 3174 (2014)
Briggs, G.P., Ferrario, L., Tout, C.A., Wickramasinghe, D.T., Hurley, J.R.: Mon. Not. R. Astron. Soc. **447**, 1713 (2015)
Brinkworth, C.S., Burleigh, M.R., Wynn, G.A., Marsh, T.R.: Mon. Not. R. Astron. Soc. **348**, 33 (2004)
Brinkworth, C.S., Marsh, T.R., Morales-Rueda, L., Maxted, P.F.L., Burleigh, M.R., Good, S.A.: Mon. Not. R. Astron. Soc. **357**, 333 (2005)
Brinkworth, C.S., Burleigh, M.R., Lawrie, K., Marsh, T.R., Knigge, C.: Astrophys. J. **773**, 47 (2013)
Britt, C.T., Torres, M.A.P., Hynes, R.I., Jonker, P.G., Maccarone, T.J., Greiss, S., Steeghs, D., Groot, P., Knigge, C., Dieball, A., Nelemans, G., Mikles, V.J., Gossen, L.: Astrophys. J. **769**, 120 (2013)
Brunschweiger, J., Greiner, J., Ajello, M., Osborne, J.: Astron. Astrophys. **496**, 121 (2009)
Buckley, D.A.H., Tuohy, I.R.: Astrophys. J. **344**, 376 (1989)
Buckley, D.A.H., Sekiguchi, K., Motch, C., O'Donoghue, D., Chen, A.-L., Schwarzenberg-Czerny, A., Pietsch, W., Harrop-Allin, M.K.: Mon. Not. R. Astron. Soc. **275**, 1028 (1995)
Buckley, D.A.H., Haberl, F., Motch, C., Pollard, K., Schwarzenberg-Czerny, A., Sekiguchi, K.: Mon. Not. R. Astron. Soc. **287**, 117 (1997)
Buckley, D.A.H., Ferrario, L., Wickramasinghe, D.T., Bailey, J.A.: Mon. Not. R. Astron. Soc. **295**, 899 (1998a)
Buckley, D.A.H., Cropper, M., Ramsay, G., Wickramasinghe, D.T.: Mon. Not. R. Astron. Soc. **299**, 83 (1998b)
Buckley, D.A.H., Cropper, M., van der Heyden, K., Potter, S.B., Wickramasinghe, D.T.: Mon. Not. R. Astron. Soc. **318**, 187 (2000)
Bues, I.: In: Solheim, S.-E., Meistas, E.G. (eds.) *11th European Workshop on White Dwarfs*. Astronomical Society of the Pacific Conference Series, vol. 169, p. 240 (1999)
Bues, I., Pragal, M.: In: Wegner, G. (ed.) *IAU Colloq. 114: White Dwarfs*. Lecture Notes in Physics, vol. 328, p. 329 (Springer, Berlin, 1989)
Burleigh, M.R., Jordan, S., Schweizer, W.: Astrophys. J. Lett. **510**, 37 (1999)
Burleigh, M.R., Marsh, T.R., Gänsicke, B.T., Goad, M.R., Dhillon, V.S., Littlefair, S.P., Wells, M., Bannister, N.P., Hurkett, C.P., Martindale, A., Dobbie, P.D., Casewell, S.L., Baker, D.E.A., Duke, J., Farihi, J., Irwin, M.J., Hewett, P.C., Roche, P., Lewis, F.: Mon. Not. R. Astron. Soc. **373**, 1416 (2006)
Burwitz, V., Reinsch, K., Beuermann, K., Thomas, H.-C.: Astron. Astrophys. **310**, 25 (1996a)
Burwitz, V., Reinsch, K., Schwope, A.D., Beuermann, K., Thomas, H.-C., Greiner, J.: Astron. Astrophys. **305**, 507 (1996b)
Burwitz, V., Reinsch, K., Beuermann, K., Thomas, H.-C.: Astron. Astrophys. **327**, 183 (1997)
Burwitz, V., Reinsch, K., Schwope, A.D., Hakala, P.J., Beuermann, K., Rousseau, T., Thomas, H.-C., Gänsicke, B.T., Piirola, V., Vilhu, O.: Astron. Astrophys. **331**, 262 (1998)

Burwitz, V., Reinsch, K., Beuermann, K., Thomas, H.-C.: In: Hellier, C., Mukai, K. (eds.) *Annapolis Workshop on Magnetic Cataclysmic Variables*. Astronomical Society of the Pacific Conference Series, vol. 157, p. 127 (1999)
Butters, O.W., Katajainen, S., Norton, A.J., Lehto, H.J., Piirola, V.: Astron. Astrophys. **496**, 891 (2009)
Campbell, C.G., Schwope, A.D.: Astron. Astrophys. **343**, 132 (1999)
Campbell, R.K., Harrison, T.E., Kafka, S.: Astrophys. J. **683**, 409 (2008a)
Campbell, R.K., Harrison, T.E., Mason, E., Howell, S., Schwope, A.D.: Astrophys. J. **678**, 1304 (2008b)
Chanmugam, G.: Annu. Rev. Astron. Astrophys. **30**, 143 (1992)
Chanmugam, G., Dulk, G.A.: Astrophys. J. **244**, 569 (1981)
Chanmugam, G., Frank, J., King, A.R., Lasota, J.-P.: Astrophys. J. Lett. **350**, 13 (1990)
Charpinet, S., Fontaine, G., Brassard, P.: Nature **461**, 501 (2009)
Cieslinski, D., Steiner, J.E.: Mon. Not. R. Astron. Soc. **291**, 321 (1997)
Claver, C.F., Liebert, J., Bergeron, P., Koester, D.: Astrophys. J. **563**, 987 (2001)
Cohen, M.H., Putney, A., Goodrich, R.W.: Astrophys. J. Lett. **405**, 67 (1993)
Coppejans, D.L., Woudt, P.A., Warner, B., Körding, E., Macfarlane, S.A., Schurch, M.P.E., Kotze, M.M., Breytenbach, H.B., Gulbis, A.A.S., Coppejans, R.: Mon. Not. R. Astron. Soc. **437**, 510 (2014)
Crampton, D., Fisher, W.A., Cowley, A.P.: Astrophys. J. **300**, 788 (1986)
Cropper, M.: Space Sci. Rev. **54**, 195 (1990)
Cropper, M.: Mon. Not. R. Astron. Soc. **295**, 353 (1998)
Cropper, M., Wickramasinghe, D.T.: Mon. Not. R. Astron. Soc. **260**, 696 (1993)
Cropper, M., Menzies, J.W., Tapia, S.: Mon. Not. R. Astron. Soc. **218**, 201 (1986)
Cropper, M., Mason, K.O., Allington-Smith, J.R., Branduardi-Raymont, G., Charles, P.A., Mittaz, J.P.D., Mukai, K., Murdin, P.G., Smale, A.P.: Mon. Not. R. Astron. Soc. **236**, 29 (1989)
Cropper, M., Mason, K.O., Mukai, K.: Mon. Not. R. Astron. Soc. **243**, 565 (1990)
Cropper, M., Wu, K., Ramsay, G., Kocabiyik, A.: Mon. Not. R. Astron. Soc. **306**, 684 (1999)
Cumming, A.: Mon. Not. R. Astron. Soc. **333**, 589 (2002)
Davis, P.J., Kolb, U., Willems, B., Gänsicke, B.T.: Mon. Not. R. Astron. Soc. **389**, 1563 (2008)
Day-Jones, A.C., Pinfield, D.J., Ruiz, M.T., Beaumont, H., Burningham, B., Gallardo, J., Gianninas, A., Bergeron, P., Napiwotzki, R., Jenkins, J.S., Zhang, Z.H., Murray, D.N., Catalán, S., Gomes, J.: Mon. Not. R. Astron. Soc. **410**, 705 (2011)
de Jager, O.C., Meintjes, P.J., O'Donoghue, D., Robinson, E.L.: Mon. Not. R. Astron. Soc. **267**, 577 (1994)
De Kool, M.: Astron. Astrophys. **261**, 188 (1992)
de Martino, D.: In: di Salvo, T., Israel, G.L., Piersant, L., Burderi, L., Matt, G., Tornambe, A., Menna, M.T. (eds.) *The Multicolored Landscape of Compact Objects and Their Explosive Origins*. American Institute of Physics Conference Series, vol. 924, p. 524 (2007)
de Martino, D., Barcaroli, R., Matt, G., Mouchet, M., Belloni, T., Beuermann, K., Chiappetti, L., Done, C., Gänsicke, B.T., La Franca, F., Mukai, K.: Astron. Astrophys. **332**, 904 (1998)
de Martino, D., Matt, G., Mukai, K., Belloni, T., Bonnet-Bidaud, J.M., Chiappetti, L., Gänsicke, B.T., Haberl, F., Mouchet, M.: Astron. Astrophys. **377**, 499 (2001)
de Martino, D., Matt, G., Mukai, K., Bonnet-Bidaud, J.-M., Gänsicke, B.T., Gonzalez-Perez, J.M., Haberl, F., Mouchet, M., Solheim, J.-E.: Astron. Astrophys. **437**, 935 (2005)
de Martino, D., Bonnet-Bidaud, J.-M., Mouchet, M., Gänsicke, B.T., Haberl, F., Motch, C.: Astron. Astrophys. **449**, 1151 (2006a)
de Martino, D., Matt, G., Mukai, K., Bonnet-Bidaud, J.-M., Burwitz, V., Gänsicke, B.T., Haberl, F., Mouchet, M.: Astron. Astrophys. **454**, 287 (2006b)
de Martino, D., Matt, G., Mukai, K., Bonnet-Bidaud, J.-M., Falanga, M., Gänsicke, B.T., Haberl, F., Marsh, T.R., Mouchet, M., Littlefair, S.P., Dhillon, V.: Astron. Astrophys. **481**, 149 (2008)
Denisenko, D.V., Pavlinsky, M.N., Sunyaev, R.A., Aslan, Z., Khamitov, I., Parmaksizoglu, M.: Astron. Lett. **32**, 252 (2006)
Dobbie, P.D., Baxter, R., Külebi, B., Parker, Q.A., Koester, D., Jordan, S., Lodieu, N., Euchner, F.: Mon. Not. R. Astron. Soc. **421**, 202 (2012)
Dobbie, P.D., Külebi, B., Casewell, S.L., Burleigh, M.R., Parker, Q.A., Baxter, R., Lawrie, K.A., Jordan, S., Koester, D.: Mon. Not. R. Astron. Soc. **428**, 16 (2013)
Donati, J.F., Achilleos, N., Matthews, J.M., Wesemael, F.: Astron. Astrophys. **285**, 285 (1994)
Drew, J.E., Gonzalez-Solares, E., Greimel, R., Irwin, M.J., Küpcü Yoldas, A., Lewis, J., Barentsen, G., Eislöffel, J., Farnhill, H.J., Martin, W.E., Walsh, J.R., Walton, N.A., Mohr-Smith, M., Raddi, R., Sale, S.E., Wright, N.J., Groot, P., Barlow, M.J., Corradi, R.L.M., Drake, J.J., Fabregat, J., Frew, D.J., Gänsicke, B.T., Knigge, C., Mampaso, A., Morris, R.A.H., Naylor, T., Parker, Q.A., Phillipps, S., Ruhland, C., Steeghs, D., Unruh, Y.C., Vink, J.S., Wesson, R., Zijlstra, A.A.: Mon. Not. R. Astron. Soc. **440**, 2036 (2014)
Dufour, P., Bergeron, P., Fontaine, G.: Astrophys. J. **627**, 404 (2005)

Dufour, P., Bergeron, P., Schmidt, G.D., Liebert, J., Harris, H.C., Knapp, G.R., Anderson, S.F., Schneider, D.P.: Astrophys. J. **651**, 1112 (2006)
Dufour, P., Liebert, J., Fontaine, G., Behara, N.: Nature **450**, 522 (2007)
Dufour, P., Fontaine, G., Liebert, J., Schmidt, G.D., Behara, N.: Astrophys. J. **683**, 978 (2008a)
Dufour, P., Fontaine, G., Liebert, J., Williams, K., Lai, D.K.: Astrophys. J. Lett. **683**, 167 (2008b)
Dunlap, B.H., Barlow, B.N., Clemens, J.C.: Astrophys. J. Lett. **720**, 159 (2010)
Dupuis, J., Chayer, P., Vennes, S., Allard, N.F., Hébrard, G.: Astrophys. J. **598**, 486 (2003)
Echevarría, J., Smith, R.C., Costero, R., Zharikov, S., Michel, R.: Mon. Not. R. Astron. Soc. **387**, 1563 (2008)
Euchner, F., Jordan, S., Beuermann, K., Gänsicke, B.T., Hessman, F.V.: Astron. Astrophys. **390**, 633 (2002)
Euchner, F., Jordan, S., Reinsch, K., Beuermann, K., Gänsicke, B.T.: In: Koester, D., Moehler, S. (eds.) *14th European Workshop on White Dwarfs*. Astronomical Society of the Pacific Conference Series, vol. 334, p. 269 (2005a)
Euchner, F., Reinsch, K., Jordan, S., Beuermann, K., Gänsicke, B.T.: Astron. Astrophys. **442**, 651 (2005b)
Euchner, F., Jordan, S., Beuermann, K., Reinsch, K., Gänsicke, B.T.: Astron. Astrophys. **451**, 671 (2006)
Evans, P.A., Hellier, C., Ramsay, G.: Mon. Not. R. Astron. Soc. **369**, 1229 (2006)
Fabrika, S., Valyavin, G.: In: Solheim, S.-E., Meistas, E.G. (eds.) *11th European Workshop on White Dwarfs*. Astronomical Society of the Pacific Conference Series, vol. 169, p. 214 (1999)
Fabrika, S.N., Valyavin, G.G., Burlakova, T.E., Barsukova, E.A., Monin, D.N.: In: Glagolevskij, Y.V., Romanyuk, I.I. (eds.) *Magnetic Fields of Chemically Peculiar and Related Stars*, p. 218 (2000)
Farihi, J., Burleigh, M.R., Hoard, D.W.: Astrophys. J. **674**, 421 (2008)
Farihi, J., Jura, M., Zuckerman, B.: Astrophys. J. **694**, 805 (2009)
Farihi, J., Dufour, P., Napiwotzki, R., Koester, D.: Mon. Not. R. Astron. Soc. **413**, 2559 (2011)
Ferrario, L.: Mon. Not. R. Astron. Soc. **426**, 2500 (2012)
Ferrario, L., Wehrse, R.: Mon. Not. R. Astron. Soc. **310**, 189 (1999)
Ferrario, L., Wickramasinghe, D.T.: Astrophys. J. **357**, 582 (1990)
Ferrario, L., Wickramasinghe, D.T.: Mon. Not. R. Astron. Soc. **265**, 605 (1993b)
Ferrario, L., Wickramasinghe, D.T.: Mon. Not. R. Astron. Soc. **309**, 517 (1999)
Ferrario, L., Wickramasinghe, D.T.: In: Werner, K., Rauch, T. (eds.) *American Institute of Physics Conference Series*. American Institute of Physics Conference Series, vol. 1273, p. 378 (2010)
Ferrario, L., Tuohy, I.R., Wickramasinghe, D.T.: Astrophys. J. **341**, 327 (1989a)
Ferrario, L., Wickramasinghe, D.T., Bailey, J., Tuohy, I.R., Hough, J.H.: Astrophys. J. **337**, 832 (1989b)
Ferrario, L., Wickramasinghe, D.T., Bailey, J., Hough, J.H., Tuohy, I.R.: Mon. Not. R. Astron. Soc. **256**, 252 (1992)
Ferrario, L., Bailey, J., Wickramasinghe, D.T.: Mon. Not. R. Astron. Soc. **262**, 285 (1993a)
Ferrario, L., Wickramasinghe, D.T., King, A.R.: Mon. Not. R. Astron. Soc. **260**, 149 (1993)
Ferrario, L., Wickramasinghe, D.T., Bailey, J.A., Buckley, D.A.H.: Mon. Not. R. Astron. Soc. **268**, 128 (1994)
Ferrario, L., Wickramasinghe, D., Bailey, J., Buckley, D.: Mon. Not. R. Astron. Soc. **273**, 17 (1995)
Ferrario, L., Bailey, J., Wickramasinghe, D.: Mon. Not. R. Astron. Soc. **282**, 218 (1996)
Ferrario, L., Vennes, S., Wickramasinghe, D.T., Bailey, J.A., Christian, D.J.: Mon. Not. R. Astron. Soc. **292**, 205 (1997a)
Ferrario, L., Wickramasinghe, D.T., Liebert, J., Schmidt, G.D., Bieging, J.H.: Mon. Not. R. Astron. Soc. **289**, 105 (1997b)
Ferrario, L., Vennes, S., Wickramasinghe, D.T.: Mon. Not. R. Astron. Soc. **299**, 1 (1998)
Ferrario, L., Wickramasinghe, D.T., Schmidt, G.: Mon. Not. R. Astron. Soc. **331**, 736 (2002)
Ferrario, L., Wickramasinghe, D.T., Schmidt, G.: Mon. Not. R. Astron. Soc. **338**, 340 (2003)
Ferrario, L., Wickramasinghe, D.T., Schmidt, G.D.: In: Hameury, J.-M., Lasota, J.-P. (eds.) *The Astrophysics of Cataclysmic Variables and Related Objects*. Astronomical Society of the Pacific Conference Series, vol. 330, p. 411 (2005)
Fischer, A., Beuermann, K.: Astron. Astrophys. **373**, 211 (2001)
Foltz, C.B., Latter, W.B., Hewett, P.C., Weymann, R.J., Morris, S.L., Anderson, S.F.: Astron. J. **98**, 665 (1989)
Fontaine, G., Brassard, P.: Publ. Astron. Soc. Pac. **120**, 1043 (2008)
Forster, H., Strupat, W., Rosner, W., Wunner, G., Ruder, H., Herold, H.: J. Phys. B, At. Mol. Phys. **17**, 1301 (1984)
Frank, J., King, A.R., Raine, D.J.: *Accretion Power in Astrophysics*, (1985)
Friedrich, S., Staubert, R., Lamer, G., Koenig, M., Geckeler, R., Baessgen, M., Kollatschny, W., Oestreicher, R., James, S.D., Sood, R.K.: Astron. Astrophys. **306**, 860 (1996)
Friedrich, S., Koenig, M., Schweizer, W.: Astron. Astrophys. **326**, 218 (1997)
Gänsicke, B.T., Beuermann, K., de Martino, D.: Astron. Astrophys. **303**, 127 (1995)
Gänsicke, B.T., Beuermann, K., de Martino, D., Thomas, H.-C.: Astron. Astrophys. **354**, 605 (2000)

Gänsicke, B.T., Schmidt, G.D., Jordan, S., Szkody, P.: Astrophys. J. **555**, 380 (2001)
Gänsicke, B.T., Euchner, F., Jordan, S.: Astron. Astrophys. **394**, 957 (2002)
Gänsicke, B.T., Jordan, S., Beuermann, K., de Martino, D., Szkody, P., Marsh, T.R., Thorstensen, J.: Astrophys. J. Lett. **613**, 141 (2004)
Gänsicke, B.T., Marsh, T.R., Edge, A., Rodríguez-Gil, P., Steeghs, D., Araujo-Betancor, S., Harlaftis, E., Giannakis, O., Pyrzas, S., Morales-Rueda, L., Aungwerojwit, A.: Mon. Not. R. Astron. Soc. **361**, 141 (2005)
Gänsicke, B.T., Marsh, T.R., Southworth, J., Rebassa-Mansergas, A.: Science **314**, 1908 (2006a)
Gänsicke, B.T., Long, K.S., Barstow, M.A., Hubeny, I.: Astrophys. J. **639**, 1039 (2006b)
Gänsicke, B.T., Dillon, M., Southworth, J., Thorstensen, J.R., Rodríguez-Gil, P., Aungwerojwit, A., Marsh, T.R., Szkody, P., Barros, S.C.C., Casares, J., de Martino, D., Groot, P.J., Hakala, P., Kolb, U., Littlefair, S.P., Martínez-Pais, I.G., Nelemans, G., Schreiber, M.R.: Mon. Not. R. Astron. Soc. **397**, 2170 (2009)
Gänsicke, B.T., Koester, D., Girven, J., Marsh, T.R., Steeghs, D.: Science **327**, 188 (2010)
García-Berro, E., Lorén-Aguilar, P., Aznar-Siguán, G., Torres, S., Camacho, J., Althaus, L.G., Córsico, A.H., Külebi, B., Isern, J.: Astrophys. J. **749**, 25 (2012)
Gary, B.: http://brucegary.net/WDE/WD2359-434/WD2359-434.htm (2014)
Giammichele, N., Bergeron, P., Dufour, P.: Astrophys. J. Suppl. Ser. **199**, 29 (2012)
Gianninas, A., Bergeron, P., Ruiz, M.T.: Astrophys. J. **743**, 138 (2011)
Giovannini, O., Kepler, S.O., Kanaan, A., Wood, A., Claver, C.F., Koester, D.: Balt. Astron. **7**, 131 (1998)
Girish, V., Singh, K.P.: Mon. Not. R. Astron. Soc. **427**, 458 (2012)
Girven, J., Gänsicke, B.T., Külebi, B., Steeghs, D., Jordan, S., Marsh, T.R., Koester, D.: Mon. Not. R. Astron. Soc. **404**, 159 (2010)
Glenn, J., Liebert, J., Schmidt, G.D.: Publ. Astron. Soc. Pac. **106**, 722 (1994a)
Glenn, J., Howell, S.B., Schmidt, G.D., Liebert, J., Grauer, A.D., Wagner, R.M.: Astrophys. J. **424**, 967 (1994b)
Greenstein, J.L., Matthews, M.S.: Astrophys. J. **126**, 14 (1957)
Greenstein, J.L., Saha, A.: Astrophys. J. **304**, 721 (1986)
Greenstein, J.L., Henry, R.J.W., Oconnell, R.F.: Astrophys. J. Lett. **289**, 25 (1985)
Greiner, J., Schwarz, R.: Astron. Astrophys. **340**, 129 (1998)
Greiner, J., Schwarz, R., Wenzel, W.: Mon. Not. R. Astron. Soc. **296**, 437 (1998)
Greiss, S., Gänsicke, B.T., Hermes, J.J., Steeghs, D., Koester, D., Ramsay, G., Barclay, T., Townsley, D.M.: Mon. Not. R. Astron. Soc. **438**, 3086 (2014)
Hagen, H.-J., Groote, D., Engels, D., Haug, U., Toussaint, F., Reimers, D.: Astron. Astrophys. **183**, 7 (1987)
Hall, P.B., Maxwell, A.J.: Astrophys. J. **678**, 1292 (2008)
Halpern, J.P., Leighly, K.M., Marshall, H.L., Eracleous, M., Storchi-Bergmann, T.: Publ. Astron. Soc. Pac. **110**, 1394 (1998)
Hameury, J.M., King, A.R., Lasota, J.P.: Mon. Not. R. Astron. Soc. **237**, 845 (1989)
Harlaftis, E.T., Horne, K.: Mon. Not. R. Astron. Soc. **305**, 437 (1999)
Haswell, C.A., Patterson, J., Thorstensen, J.R., Hellier, C., Skillman, D.R.: Astrophys. J. **476**, 847 (1997)
Hearn, D.R., Richardson, J.A., Clark, G.W.: Astrophys. J. Lett. **210**, 23 (1976)
C. Hellier, Eur. Phys. J. Web Conf. **64**, 7001 (2014)
Hellier, C.: In: Buckley, D.A.H., Warner, B. (eds.) *Magnetic Cataclysmic Variables*. ASP Conf. Ser., vol. 85, p. 185 (1995)
Hellier, C.: Mon. Not. R. Astron. Soc. **288**, 817 (1997a)
Hellier, C.: Mon. Not. R. Astron. Soc. **291**, 71 (1997b)
Hellier, C., Cropper, M., Mason, K.O.: Mon. Not. R. Astron. Soc. **248**, 233 (1991)
Hellier, C., Beardmore, A.P., Mukai, K.: Astron. Astrophys. **389**, 904 (2002)
Henry, R.J.W., O'Connell, R.F.: Astrophys. J. Lett. **282**, 97 (1984)
Henry, R.J.W., O'Connell, R.F.: Publ. Astron. Soc. Pac. **97**, 333 (1985)
Heyl, J.S.: Mon. Not. R. Astron. Soc. **317**, 310 (2000)
Hilton, E.J., Szkody, P., Mukadam, A., Henden, A., Dillon, W., Schmidt, G.D.: Astron. J. **137**, 3606 (2009)
Hoard, D.W., Schmidt, G.D., Szkody, P., Ferrario, L., Fraser, O., Wolfe, M.A., Gänsicke, B.T.: Astron. J. **128**, 1894 (2004)
Hollands, M., Gänsicke, B., Köster, D.: Mon. Not. R. Astron. Soc. **450**, 681 (2015)
Homer, L., Szkody, P., Chen, B., Henden, A., Schmidt, G.D., Fraser, O.J., Saloma, K., Silvestri, N.M., Taylor, H., Brinkmann, J.: Astrophys. J. **620**, 929 (2005)
Hong, J., van der Berg, M., Grindlay, J., Servillat, M., Zhao, P.: Astrophys. J. **746**, 165 (2012)
Howell, S.B., Sirk, M.M., Malina, R.F., Mittaz, J.P.D., Mason, K.O.: Astrophys. J. **439**, 991 (1995)
Howell, S.B., Craig, N., Roberts, B., McGee, P., Sirk, M.: Astron. J. **113**, 2231 (1997)
Hui, C.Y., Seo, K.A., Hu, C.P., Lin, L.C.C., Chou, Y.: Astrophys. J. **759**, 109 (2012)
Jablonski, F., Steiner, J.E.: Astrophys. J. **323**, 672 (1987)

Jones, M.D., Ortiz, G., Ceperley, D.M.: Astron. Astrophys. **343**, 91 (1999)
Jordan, S.: Astron. Astrophys. **265**, 570 (1992)
Jordan, S., Friedrich, S.: Astron. Astrophys. **383**, 519 (2002)
Jordan, S., Schmelcher, P., Becken, W., Schweizer, W.: Astron. Astrophys. **336**, 33 (1998)
Joshi, V.H., Ashok, N.M., Banerjee, D.P.K.: Bull. Astron. Soc. India **39**, 259 (2011)
Jura, M.: Astrophys. J. Lett. **584**, 91 (2003)
Kalirai, J.S., Hansen, B.M.S., Kelson, D.D., Reitzel, D.B., Rich, R.M., Richer, H.B.: Astrophys. J. **676**, 594 (2008)
Kaluzny, J., Semeniuk, I.: Inf. Bull. Var. Stars **3145**, 1 (1988)
Karl, C.A., Napiwotzki, R., Heber, U., Dreizler, S., Koester, D., Reid, I.N.: Astron. Astrophys. **434**, 637 (2005)
Katajainen, S., Butters, O.W., Norton, A.J., Lehto, H.J., Piirola, V.: Astron. Astrophys. **475**, 1011 (2007)
Katajainen, S., Butters, O., Norton, A.J., Lehto, H.J., Piirola, V., Berdyugin, A.: Astrophys. J. **724**, 165 (2010)
Kaur, R., Wijnands, R., Paul, B., Patruno, A., Degenaar, N.: Mon. Not. R. Astron. Soc. **402**, 2388 (2010)
Kawka, A., Vennes, S.: Astrophys. J. **643**, 402 (2006)
Kawka, A., Vennes, S.: Astron. Astrophys. **532**, 7 (2011)
Kawka, A., Vennes, S.: Mon. Not. R. Astron. Soc. **425**, 1394 (2012)
Kawka, A., Vennes, S.: Mon. Not. R. Astron. Soc. **439**, 90 (2014)
Kawka, A., Vennes, S., Schmidt, G.D., Wickramasinghe, D.T., Koch, R.: Astrophys. J. **654**, 499 (2007)
S.B. Kemic, Wavelengths and strengths of hydrogen and helium transitions in large magnetic fields. JILA Pub. **1153**(113), 1 (1974)
Kemp, J.C., Swedlund, J.B., Landstreet, J.D., Angel, J.R.P.: Astrophys. J. Lett. **161**, 77 (1970)
Kemp, J., Patterson, J., Thorstensen, J.R., Fried, R.E., Skillman, D.R., Billings, G.: Publ. Astron. Soc. Pac. **114**, 623 (2002)
Kepler, S.O., Kleinman, S.J., Nitta, A., Koester, D., Castanheira, B.G., Giovannini, O., Costa, A.F.M., Althaus, L.: Mon. Not. R. Astron. Soc. **375**, 1315 (2007)
Kepler, S.O., Pelisoli, I., Jordan, S., Kleinman, S.J., Koester, D., Külebi, B., Peçanha, V., Castanheira, B.G., Nitta, A., Costa, J.E.S., Winget, D.E., Kanaan, A., Fraga, L.: Mon. Not. R. Astron. Soc. **429**, 2934 (2013)
Kepler, S.O., Pelisoli, I., Koester, D., Ourique, G., Kleinman, S.J., Romero, A.D., Nitta, A., Eisenstein, D.J., Costa, J.E.S., Külebi, B., Jordan, S., Dufour, P., Giommi, P., Rebassa-Mansergas, A.: Mon. Not. R. Astron. Soc. **446**, 4078 (2015)
Kilkenny, D., O'Donoghue, D., Stobie, R.S.: Mon. Not. R. Astron. Soc. **248**, 664 (1991)
Kim, Y.G., Andronov, I.L., Park, S.S., Jeon, Y.-B.: Astron. Astrophys. **441**, 663 (2005)
King, A.R.: Q. J. R. Astron. Soc. **29**, 1 (1988)
King, A.R., Lasota, J.P.: Mon. Not. R. Astron. Soc. **188**, 653 (1979)
King, A.R., Lasota, J.-P.: Astrophys. J. **378**, 674 (1991)
Kissin, Y., Thompson, C.: ArXiv e-prints arXiv:1501.07197 (2015)
Kleinman, S.J., Kepler, S.O., Koester, D., Pelisoli, I., Peçanha, V., Nitta, A., Costa, J.E.S., Krzesinski, J., Dufour, P., Lachapelle, F.-R., Bergeron, P., Yip, C.-W., Harris, H.C., Eisenstein, D.J., Althaus, L., Córsico, A.: Astrophys. J. Suppl. Ser. **204**, 5 (2013)
Koester, D., Dreizler, S., Weidemann, V., Allard, N.F.: Astron. Astrophys. **338**, 612 (1998)
Koester, D., Napiwotzki, R., Christlieb, N., Drechsel, H., Hagen, H.-J., Heber, U., Homeier, D., Karl, C., Leibundgut, B., Moehler, S., Nelemans, G., Pauli, E.-M., Reimers, D., Renzini, A., Yungelson, L.: Astron. Astrophys. **378**, 556 (2001)
Koester, D., Voss, B., Napiwotzki, R., Christlieb, N., Homeier, D., Lisker, T., Reimers, D., Heber, U.: Astron. Astrophys. **505**, 441 (2009)
Koester, D., Gänsicke, B.T., Farihi, J.: Astron. Astrophys. **566**, 34 (2014)
Kolb, U., Baraffe, I.: Mon. Not. R. Astron. Soc. **309**, 1034 (1999)
Kowalski, P.M.: Astron. Astrophys. **519**, 8 (2010)
Kozhevnikov, V.P.: Mon. Not. R. Astron. Soc. **422**, 1518 (2012)
Kozhevnikov, V.P.: Mon. Not. R. Astron. Soc. **443**, 2444 (2014)
Kuijpers, J., Pringle, J.E.: Astron. Astrophys. **114**, 4 (1982)
Kuiper, G.P.: Publ. Astron. Soc. Pac. **46**, 287 (1934)
Külebi, B., Jordan, S., Euchner, F., Gänsicke, B.T., Hirsch, H.: Astron. Astrophys. **506**, 1341 (2009)
Külebi, B., Jordan, S., Nelan, E., Bastian, U., Altmann, M.: Astron. Astrophys. **524**, 36 (2010)
Külebi, B., Ekşi, K.Y., Lorén-Aguilar, P., Isern, J., García-Berro, E.: Mon. Not. R. Astron. Soc. **431**, 2778 (2013a)
Külebi, B., Kalirai, J., Jordan, S., Euchner, F.: Astron. Astrophys. **554**, 18 (2013b)
Lamb, D.Q., Masters, A.R.: Astrophys. J. Lett. **234**, 117 (1979)
Landi Degl'Innocenti, E.: Astrophys. J. **209**, 208 (1976)

Landstreet, J.D., Angel, J.R.P.: Astrophys. J. Lett. **165**, 67 (1971)
Landstreet, J.D., Bagnulo, S., Valyavin, G.G., Fossati, L., Jordan, S., Monin, D., Wade, G.A.: Astron. Astrophys. **545**, 30 (2012)
Latter, W.B., Schmidt, G.D., Green, R.F.: Astrophys. J. **320**, 308 (1987)
Lawrie, K.A., Burleigh, M.R., Dufour, P., Hodgkin, S.T.: Mon. Not. R. Astron. Soc. **433**, 1599 (2013)
Leone, F., Corradi, R.L.M., Martínez González, M.J., Asensio Ramos, A., Manso Sainz, R.: Astron. Astrophys. **563**, 43 (2014)
Li, J., Wickramasinghe, D.T.: Mon. Not. R. Astron. Soc. **300**, 718 (1998)
Li, J., Ferrario, L., Wickramasinghe, D.T.: Astrophys. J. Lett. **503**, 151 (1998)
Liebert, J.: Publ. Astron. Soc. Pac. **100**, 1302 (1988)
Liebert, J.: J. Phys. Conf. Ser. **172**(1), 012040 (2009)
Liebert, J., Dahn, C.C.: Astrophys. J. **269**, 258 (1983)
Liebert, J., Sion, E.M.: Astrophys. J. Lett. **20**, 53 (1979)
Liebert, J., Angel, J.R.P., Stockman, H.S., Spinrad, H., Beaver, E.A.: Astrophys. J. **214**, 457 (1977)
Liebert, J., Angel, J.R.P., Stockman, H.S., Beaver, E.A.: Astrophys. J. **225**, 181 (1978)
Liebert, J., Schmidt, G.D., Sion, E.M., Starrfield, S.G., Green, R.F., Boroson, T.A.: Publ. Astron. Soc. Pac. **97**, 158 (1985)
Liebert, J., Bergeron, P., Schmidt, G.D., Saffer, R.A.: Astrophys. J. **418**, 426 (1993)
Liebert, J., Schmidt, G.D., Lesser, M., Stepanian, J.A., Lipovetsky, V.A., Chaffe, F.H., Foltz, C.B., Bergeron, P.: Astrophys. J. **421**, 733 (1994)
Liebert, J., Bergeron, P., Holberg, J.B.: Astron. J. **125**, 348 (2003a)
Liebert, J., Harris, H.C., Dahn, C.C., Schmidt, G.D., Kleinman, S.J., Nitta, A., Krzesiński, J., Eisenstein, D., Smith, J.A., Szkody, P., Hawley, S., Anderson, S.F., Brinkmann, J., Collinge, M.J., Fan, X., Hall, P.B., Knapp, G.R., Lamb, D.Q., Margon, B., Schneider, D.P., Silvestri, N.: Astron. J. **126**, 2521 (2003b)
Liebert, J., Wickramasinghe, D.T., Schmidt, G.D., Silvestri, N.M., Hawley, S.L., Szkody, P., Ferrario, L., Webbink, R.F., Oswalt, T.D., Smith, J.A., Lemagie, M.P.: Astron. J. **129**, 2376 (2005a)
Liebert, J., Bergeron, P., Holberg, J.B.: Astrophys. J. Suppl. Ser. **156**, 47 (2005b)
Liebert, J., Ferrario, L., Wickramasinghe, D.T., Smith, P.: Astrophys. J. **804**, 93 (2015)
Martin, B., Wickramasinghe, D.T.: Mon. Not. R. Astron. Soc. **183**, 533 (1978)
Martin, B., Wickramasinghe, D.T.: Mon. Not. R. Astron. Soc. **206**, 407 (1984)
Masetti, N., Nucita, A.A., Parisi, P.: Astron. Astrophys. **544**, 114 (2012)
Mason, P.A., Ramsay, G., Andronov, I., Kolesnikov, S., Shakhovskoy, N., Pavlenko, E.: Mon. Not. R. Astron. Soc. **295**, 511 (1998)
Mathys, G., Hubrig, S., Mason, E., Michaud, G., Schöller, M., Wesemael, F.: Astron. Nachr. **333**, 30 (2012)
Mauche, C.W.: In: Vrielmann, S., Cropper, M. (eds.) *IAU Colloq. 190: Magnetic Cataclysmic Variables*. Astronomical Society of the Pacific Conference Series, vol. 315, p. 120 (2004)
Mauche, C.W., Brickhouse, N.S., Hoogerwerf, R., Luna, G.J.M., Mukai, K., Sterken, C.: Inf. Bull. Var. Stars **5876**, 1 (2009)
Maxted, P.F.L., Ferrario, L., Marsh, T.R., Wickramasinghe, D.T.: Mon. Not. R. Astron. Soc. **315**, 41 (2000)
Meggitt, S.M.A., Wickramasinghe, D.T.: Mon. Not. R. Astron. Soc. **198**, 71 (1982)
Minkowski, R.: Annu. Rep. Dir. Mt. Wilson Obs. **1**, 38 (1938)
Montgomery, M.H., Williams, K.A., Winget, D.E., Dufour, P., De Gennaro, S., Liebert, J.: Astrophys. J. Lett. **678**, 51 (2008)
Moran, C., Marsh, T.R., Dhillon, V.S.: Mon. Not. R. Astron. Soc. **299**, 218 (1998)
Motch, C., Guillout, P., Haberl, F., Krautter, J., Pakull, M.W., Pietsch, W., Reinsch, K., Voges, W., Zickgraf, F.-J.: Astron. Astrophys. Suppl. Ser. **132**, 341 (1998)
Mukai, K., Charles, P.A.: Mon. Not. R. Astron. Soc. **226**, 209 (1987)
Muno, M.P., Baganoff, F.K., Bautz, M.W., Feigelson, E.D., Garmire, G.P., Morris, M.R., Park, S., Ricker, G.R., Townsley, L.K.: Astrophys. J. **613**, 326 (2004a)
Muno, M.P., Arabadjis, J.S., Baganoff, F.K., Bautz, M.W., Brandt, W.N., Broos, P.S., Feigelson, E.D., Garmire, G.P., Morris, M.R., Ricker, G.R.: Astrophys. J. **613**, 1179 (2004b)
Nordhaus, J., Wellons, S., Spiegel, D.S., Metzger, B.D., Blackman, E.G.: Proc. Natl. Acad. Sci. **108**, 3135 (2011)
Norton, A.J., Watson, M.G.: Mon. Not. R. Astron. Soc. **237**, 715 (1989)
Norton, A.J., Watson, M.G., King, A.R., Lehto, H.J., McHardy, I.M.: Mon. Not. R. Astron. Soc. **254**, 705 (1992)
Norton, A.J., Quaintrell, H., Katajainen, S., Lehto, H.J., Mukai, K., Negueruela, I.: Astron. Astrophys. **384**, 195 (2002)
Norton, A.J., Wynn, G.A., Somerscales, R.V.: Astrophys. J. **614**, 349 (2004)
Norton, A.J., Butters, O.W., Parker, T.L., Wynn, G.A.: Astrophys. J. **672**, 524 (2008)
Osborne, J.P., Rosen, R., Mason, K.O., Beuermann, K.: Space Sci. Rev. **40**, 143 (1985)

Osborne, J.P., Beardmore, A.P., Wheatley, P.J., Hakala, P., Watson, M.G., Mason, K.O., Hassall, B.J.M., King, A.R.: Mon. Not. R. Astron. Soc. **270**, 650 (1994)
Paczynski, B., Sienkiewicz, R.: Astrophys. J. Lett. **248**, 27 (1981)
Parsons, S.G., Marsh, T.R., Gänsicke, B.T., Schreiber, M.R., Bours, M.C.P., Dhillon, V.S., Littlefair, S.P.: Mon. Not. R. Astron. Soc. **436**, 241 (2013)
Pasquini, L., Belloni, T., Abbott, T.M.C.: Astron. Astrophys. **290**, 17 (1994)
Patterson, J.P.: Astrophys. J. **234**, 978 (1979)
Patterson, J.: Publ. Astron. Soc. Pac. **106**, 209 (1994)
Patterson, J., Steiner, J.E.: Astrophys. J. Lett. **264**, 61 (1983)
Patterson, J., Kemp, J., Richman, H.R., Skillman, D.R., Vanmunster, T., Jensen, L., Buckley, D.A.H., O'Donoghue, D., Kramer, R.: Publ. Astron. Soc. Pac. **110**, 415 (1998)
Patterson, J., Thorstensen, J.R., Vanmunster, T., Fried, R.E., Martin, B., Campbell, T., Robertson, J., Kemp, J., Messier, D., Armstrong, E.: Publ. Astron. Soc. Pac. **116**, 516 (2004)
Patterson, J., Thorstensen, J.R., Sheets, H.A., Kemp, J., Vican, L., Uthas, H., Boyd, D., Potter, M., Krajci, T., Campbell, T., Roberts, G., Starkey, D., Goff, B.: Publ. Astron. Soc. Pac. **123**, 130 (2011)
Penning, W.R., Schmidt, G.D., Liebert, J.: Astrophys. J. **301**, 881 (1986)
Piirola, V., Reiz, A.: Astron. Astrophys. **259**, 143 (1992)
Piirola, V., Hakala, P., Coyne, G.V.: Astrophys. J. Lett. **410**, 107 (1993)
Piirola, V., Vornanen, T., Berdyugin, A., Coyne, G.V.: Astrophys. J. **684**, 558 (2008)
Politano, M.: Astrophys. J. **465**, 338 (1996)
Potter, S.B., Cropper, M., Mason, K.O., Hough, J.H., Bailey, J.A.: Mon. Not. R. Astron. Soc. **285**, 82 (1997)
Potter, S.B., Cropper, M., Hakala, P.J.: Mon. Not. R. Astron. Soc. **315**, 423 (2000)
Potter, S.B., Romero-Colmenero, E., Watson, C.A., Buckley, D.A.H., Phillips, A.: Mon. Not. R. Astron. Soc. **348**, 316 (2004)
Potter, S.B., Augusteijn, T., Tappert, C.: Mon. Not. R. Astron. Soc. **364**, 565 (2005)
Potter, S.B., O'Donoghue, D., Romero-Colmenero, E., Buckley, D.A.H., Woudt, P.A., Warner, B.: Mon. Not. R. Astron. Soc. **371**, 727 (2006)
Potter, S.B., Buckley, D.A.H., O'Donoghue, D., Romero-Colmenero, E., O'Connor, J., Fourie, P., Evans, G., Sass, C., Crause, L., Still, M., Butters, O.W., Norton, A.J., Mukai, K.: Mon. Not. R. Astron. Soc. **402**, 1161 (2010)
Potter, S.B., Romero-Colmenero, E., Kotze, M., Zietsman, E., Butters, O.W., Pekeur, N., Buckley, D.A.H.: Mon. Not. R. Astron. Soc. **420**, 2596 (2012)
Pretorius, M.L.: Mon. Not. R. Astron. Soc. **395**, 386 (2009)
Pretorius, M.L., Mukai, K.: Mon. Not. R. Astron. Soc. **442**, 2580 (2014)
Pretorius, M.L., Knigge, C., O'Donoghue, D., Henry, J.P., Gioia, I.M., Mullis, C.R.: Mon. Not. R. Astron. Soc. **382**, 1279 (2007)
Pretorius, M.L., Knigge, C., Schwope, A.D.: Mon. Not. R. Astron. Soc. **432**, 570 (2013)
Putney, A.: Astrophys. J. Suppl. Ser. **112**, 527 (1997)
Putney, A., Jordan, S.: Astrophys. J. **449**, 863 (1995)
Ramsay, G.: Mon. Not. R. Astron. Soc. **314**, 403 (2000)
Ramsay, G., Cropper, M.: Mon. Not. R. Astron. Soc. **335**, 918 (2002)
Ramsay, G., Cropper, M.: Mon. Not. R. Astron. Soc. **338**, 219 (2003)
Ramsay, G., Cropper, M.: Mon. Not. R. Astron. Soc. **347**, 497 (2004)
Ramsay, G., Wheatley, P.J.: Mon. Not. R. Astron. Soc. **301**, 95 (1998)
Ramsay, G., Cropper, M., Wu, K., Potter, S.: Mon. Not. R. Astron. Soc. **282**, 726 (1996)
Ramsay, G., Buckley, D.A.H., Cropper, M., Harrop-Allin, M.K.: Mon. Not. R. Astron. Soc. **303**, 96 (1999a)
Ramsay, G., Potter, S.B., Buckley, D.A.H., Wheatley, P.J.: Mon. Not. R. Astron. Soc. **306**, 809 (1999b)
Ramsay, G., Potter, S., Cropper, M., Buckley, D.A.H., Harrop-Allin, M.K.: Mon. Not. R. Astron. Soc. **316**, 225 (2000)
Ramsay, G., Rosen, S.R., Hakala, P., Barclay, T.: Mon. Not. R. Astron. Soc. **395**, 416 (2009)
Rana, V.R., Singh, K.P., Schlegel, E.M., Barrett, P.: Astron. J. **127**, 489 (2004)
Rana, V.R., Singh, K.P., Barrett, P.E., Buckley, D.A.H.: Astrophys. J. **625**, 351 (2005)
Rappaport, S., Verbunt, F., Joss, P.: Astrophys. J. **275**, 713 (1983)
Rebassa-Mansergas, A., Agurto-Gangas, C., Schreiber, M.R., Gänsicke, B.T., Koester, D.: Mon. Not. R. Astron. Soc. **433**, 3398 (2013)
Reid, I.N., Liebert, J., Schmidt, G.D.: Astrophys. J. Lett. **550**, 61 (2001)
Reimers, D., Hagen, H.-J.: Astron. Astrophys. **358**, 45 (2000)
Reimers, D., Hagen, H.-J., Hopp, U.: Astron. Astrophys. **343**, 157 (1999)
Reinsch, K., Burwitz, V., Beuermann, K., Schwope, A.D., Thomas, H.-C.: Astron. Astrophys. **291**, 27 (1994)
Reis, R.C., Wheatley, P.J., Gänsicke, B.T., Osborne, J.P.: Mon. Not. R. Astron. Soc. **430**, 1994 (2013)
Revnivtsev, M., Sazonov, S., Gilfanov, M., Churazov, E., Sunyaev, R.: Astron. Astrophys. **452**, 169 (2006)

Revnivtsev, M., Sazonov, S., Churazov, E., Forman, W., Vikhlinin, A., Sunyaev, R.: Nature **458**, 1142 (2009)
Revnivtsev, M., Sazonov, S., Forman, W., Churazov, E., Sunyaev, R.: Mon. Not. R. Astron. Soc. **414**, 495 (2011)
Ritter, H., Kolb, U.: Astron. Astrophys. **404**, 301 (2003)
Rodrigues, C.V., Jablonski, F.J., D'Amico, F., Cieslinski, D., Steiner, J.E., Diaz, M.P., Hickel, G.R.: Mon. Not. R. Astron. Soc. **369**, 1972 (2006)
Rodríguez-Gil, P., Gänsicke, B.T., Araujo-Betancor, S., Casares, J.: Mon. Not. R. Astron. Soc. **349**, 367 (2004)
Roesner, W., Wunner, G., Herold, H., Ruder, H.: J. Phys. B, At. Mol. Phys. **17**, 29 (1984)
Romero-Colmenero, E., Potter, S.B., Buckley, D.A.H., Barrett, P.E., Vrielmann, S.: Mon. Not. R. Astron. Soc. **339**, 685 (2003)
Rosen, S.R., Rainger, J.F., Burleigh, M.R., Mittaz, J.P.D., Buckley, D.A.H., Sirk, M.M., Lieu, R., Howell, S.B., de Martino, D.: Mon. Not. R. Astron. Soc. **322**, 631 (2001)
Rousseau, T., Fischer, A., Beuermann, K., Woelk, U.: Astron. Astrophys. **310**, 526 (1996)
Saffer, R.A., Liebert, J., Wagner, R.M., Sion, E.M., Starrfield, S.G.: Astron. J. **98**, 668 (1989)
Savanov, I.S., Romaniuk, I.I., Semenko, E.A., Dmitrienko, E.S.: Astron. Rep. **57**, 751 (2013)
Saxton, C.J., Wu, K., Cropper, M., Ramsay, G.: Mon. Not. R. Astron. Soc. **360**, 1091 (2005)
Saxton, C.J., Wu, K., Canalle, J.B.G., Cropper, M., Ramsay, G.: Mon. Not. R. Astron. Soc. **379**, 779 (2007)
Sazonov, S., Revnivtsev, M., Gilfanov, M., Churazov, E., Sunyaev, R.: Astron. Astrophys. **450**, 117 (2006)
Scaringi, S., Connolly, J., Patterson, S., Thorstensen, J.R., Uthas, H., Knigge, C., Vican, L., Monard, B., Rea, R.e.: Astron. Astrophys. **530**, 6 (2011)
Schimeczek, C., Wunner, G.: Astrophys. J. Suppl. Ser. **212**, 26 (2014)
Schmidt, G.D., Grauer, A.D.: Astrophys. J. **488**, 827 (1997)
Schmidt, G.D., Norsworthy, J.E.: Astrophys. J. **366**, 270 (1991)
Schmidt, G.D., Stockman, H.S.: Astrophys. J. **371**, 749 (1991)
Schmidt, G.D., Smith, P.S.: Astrophys. J. Lett. **423**, 63 (1994)
Schmidt, G.D., Stockman, H.S.: Astrophys. J. **548**, 410 (2001)
Schmidt, G.D., Stockman, H.S., Grandi, S.A.: Astrophys. J. **271**, 735 (1983)
Schmidt, G.D., West, S.C., Liebert, J., Green, R.F., Stockman, H.S.: Astrophys. J. **309**, 218 (1986)
Schmidt, G.D., Bergeron, P., Liebert, J., Saffer, R.A.: Astrophys. J. **394**, 603 (1992a)
Schmidt, G.D., Stockman, H.S., Smith, P.S.: Astrophys. J. Lett. **398**, 57 (1992b)
Schmidt, G.D., Bergeron, P., Fegley, B.: Astrophys. J. **443**, 274 (1995a)
Schmidt, G.D., Liebert, J., Stockman, H.S.: Astrophys. J. **441**, 414 (1995b)
Schmidt, G.D., Szkody, P., Smith, P.S., Silber, A., Tovmassian, G., Hoard, D.W., Gänsicke, B.T., de Martino, D.: Astrophys. J. **473**, 483 (1996)
Schmidt, G.D., Hoard, D.W., Szkody, P., Melia, F., Honeycutt, R.K., Wagner, R.M.: Astrophys. J. **525**, 407 (1999a)
Schmidt, G.D., Liebert, J., Harris, H.C., Dahn, C.C., Leggett, S.K.: Astrophys. J. **512**, 916 (1999b)
Schmidt, G.D., Vennes, S., Wickramasinghe, D.T., Ferrario, L.: Mon. Not. R. Astron. Soc. **328**, 203 (2001a)
Schmidt, G.D., Ferrario, L., Wickramasinghe, D.T., Smith, P.S.: Astrophys. J. **553**, 823 (2001b)
Schmidt, G.D., Harris, H.C., Liebert, J., Eisenstein, D.J., Anderson, S.F., Brinkmann, J., Hall, P.B., Harvanek, M., Hawley, S., Kleinman, S.J., Knapp, G.R., Krzesinski, J., Lamb, D.Q., Long, D., Munn, J.A., Neilsen, E.H., Newman, P.R., Nitta, A., Schlegel, D.J., Schneider, D.P., Silvestri, N.M., Smith, J.A., Snedden, S.A., Szkody, P., Vanden Berk, D.: Astrophys. J. **595**, 1101 (2003)
Schmidt, G.D., Szkody, P., Silvestri, N.M., Cushing, M.C., Liebert, J., Smith, P.S.: Astrophys. J. Lett. **630**, 173 (2005a)
Schmidt, G.D., Szkody, P., Vanlandingham, K.M., Anderson, S.F., Barentine, J.C., Brewington, H.J., Hall, P.B., Harvanek, M., Kleinman, S.J., Krzesinski, J., Long, D., Margon, B., Neilsen, E.H. Jr., Newman, P.R., Nitta, A., Schneider, D.P., Snedden, S.A.: Astrophys. J. **630**, 1037 (2005b)
Schmidt, G.D., Szkody, P., Henden, A., Anderson, S.F., Lamb, D.Q., Margon, B., Schneider, D.P.: Astrophys. J. **654**, 521 (2007)
Schmidt, G.D., Smith, P.S., Szkody, P., Anderson, S.F.: Publ. Astron. Soc. Pac. **120**, 160 (2008)
Schneider, D.P., Young, P.: Astrophys. J. **240**, 871 (1980)
Schreiber, M.R., Gänsicke, B.T.: Astron. Astrophys. **406**, 305 (2003)
Schwarz, R., Greiner, J.: In: Hellier, C., Mukai, K. (eds.) *Annapolis Workshop on Magnetic Cataclysmic Variables*. Astronomical Society of the Pacific Conference Series, vol. 157, p. 139 (1999)
Schwarz, R., Schwope, A.D., Beuermann, K., Burwitz, V., Fischer, J.-U., Fried, R., Lehmann, I., Mantel, K.-H., Mengel, S., Metzner, A., Misselt, K., Notni, P., Reinsch, K., Shafter, A., Thomas, H.-C.: Astron. Astrophys. **338**, 465 (1998)
Schwarz, R., Schwope, A.D., Staude, A.: Astron. Astrophys. **374**, 189 (2001)
Schwarz, R., Greiner, J., Tovmassian, G.H., Zharikov, S.V., Wenzel, W.: Astron. Astrophys. **392**, 505 (2002)

Schwarz, R., Schwope, A.D., Staude, A., Remillard, R.A.: Astron. Astrophys. **444**, 213 (2005)
Schwarz, R., Schwope, A., Staude, A., Rau, A., Hasinger, G., Urrutia, T., Motch, C.: Astron. Astrophys. **473**, 511 (2007)
Schwope, A.D.: In: Klare, G. (ed.) *Reviews in Modern Astronomy*. Reviews in Modern Astronomy, vol. 8, p. 125 (1995)
Schwope, A.D.: In: Evans, A., Wood, J.H. (eds.) *IAU Colloq. 158: Cataclysmic Variables and Related Objects*. Astrophysics and Space Science Library, vol. 208, p. 189 (1996)
Schwope, A.: Mem. Soc. Astron. Ital. **83**, 844 (2012)
Schwope, A.D., Beuermann, K.: Astron. Astrophys. **238**, 173 (1990a)
Schwope, A.D., Beuermann, K.: Astron. Astrophys. **238**, 173 (1990b)
Schwope, A.D., Christensen, L.: Astron. Astrophys. **514**, 89 (2010)
Schwope, A.D., Mengel, S.: Astron. Nachr. **318**, 25 (1997)
Schwope, A.D., Thinius, B.: Astron. Nachr. **333**, 717 (2012)
Schwope, A.D., Beuermann, K., Thomas, H.-C.: Astron. Astrophys. **230**, 120 (1990)
Schwope, A.D., Thomas, H.C., Beuermann, K.: Astron. Astrophys. **271**, 25 (1993a)
Schwope, A.D., Beuermann, K., Jordan, S., Thomas, H.-C.: Astron. Astrophys. **278**, 487 (1993b)
Schwope, A.D., Beuermann, K., Jordan, S.: Astron. Astrophys. **301**, 447 (1995a)
Schwope, A.D., Thomas, H.-C., Beuermann, K., Burwitz, V., Jordan, S., Haefner, R.: Astron. Astrophys. **293**, 764 (1995b)
Schwope, A.D., Buckley, D.A.H., O'Donoghue, D., Hasinger, G., Truemper, J., Voges, W.: Astron. Astrophys. **326**, 195 (1997a)
Schwope, A.D., Mantel, K.-H., Horne, K.: Astron. Astrophys. **319**, 894 (1997b)
Schwope, A.D., Schwarz, R., Greiner, J.: Astron. Astrophys. **348**, 861 (1999)
Schwope, A.D., Catalán, M.S., Beuermann, K., Metzner, A., Smith, R.C., Steeghs, D.: Mon. Not. R. Astron. Soc. **313**, 533 (2000)
Schwope, A.D., Brunner, H., Buckley, D., Greiner, J., Heyden, K.v.d., Neizvestny, S., Potter, S., Schwarz, R.: Astron. Astrophys. **396**, 895 (2002c)
Schwope, A.D., Hambaryan, V., Schwarz, R., Kanbach, G., Gänsicke, B.T.: Astron. Astrophys. **392**, 541 (2002a)
Schwope, A.D., Brunner, H., Hambaryan, V., Schwarz, R.: In: Gänsicke, B.T., Beuermann, K., Reinsch, K. (eds.) *The Physics of Cataclysmic Variables and Related Objects*. Astronomical Society of the Pacific Conference Series, vol. 261, p. 102 (2002b)
Schwope, A.D., Thomas, H.-C., Mante, K.-H., Haefner, R., Staude, A.: Astron. Astrophys. **402**, 201 (2003)
Schwope, A.D., Staude, A., Vogel, J., Schwarz, R.: Astron. Nachr. **325**, 197 (2004)
Schwope, A.D., Schreiber, M.R., Szkody, P.: Astron. Astrophys. **452**, 955 (2006)
Schwope, A.D., Nebot Gomez-Moran, A., Schreiber, M.R., Gänsicke, B.T.: Astron. Astrophys. **500**, 867 (2009)
Schwope, A.D., Mackebrandt, F., Thinius, B.D., Littlefield, C., Garnavich, P., Oksanen, A., Granzer, T.: Astron. Nachr. **336**, 115 (2015)
Servillat, M., Grindlay, J., van den Berg, M., Hong, J., Zhao, P., Allen, B.: Astrophys. J. **748**, 32 (2012)
Shafter, A.W., Reinsch, K., Beuermann, K., Misselt, K.A., Buckley, D.A.H., Burwitz, V., Schwope, A.D.: Astrophys. J. **443**, 319 (1995)
Shafter, A.W., Davenport, J.R.A., Güth, T., Kattner, S., Marin, E., Sreenivasamurthy, N.: Publ. Astron. Soc. Pac. **120**, 374 (2008)
Shchurova, A.V., Pavlenko, E.P., Malanushenko, V.P.: Adv. Astron. Space Phys. **3**, 79 (2013)
Silber, A.D., Remillard, R.A., Horne, K., Bradt, H.V.: Astrophys. J. **424**, 955 (1994)
Singh, K.P., Szkody, P., Barrett, P., White, N.E., Fierce, E., Silber, A., Hoard, D.W., Hakala, P.J., Piirola, V., Sohl, K.: Astrophys. J. Lett. **453**, 95 (1995)
Sion, E.M.: Publ. Astron. Soc. Pac. **111**, 532 (1999)
Sion, E.M., Greenstein, J.L., Landstreet, J.D., Liebert, J., Shipman, H.L., Wegner, G.A.: Astrophys. J. **269**, 253 (1983)
Sion, E.M., Holberg, J.B., Oswalt, T.D., McCook, G.P., Wasatonic, R., Myszka, J.: Astron. J. **147**, 129 (2014)
Skillman, D.R.: Publ. Astron. Soc. Pac. **108**, 130 (1996)
Skinner, J.N., Thorstensen, J.R., Lépine, S.: Astron. J. **148**, 115 (2014)
Southworth, J., Gänsicke, B.T., Marsh, T.R., de Martino, D., Aungwerojwit, A.: Mon. Not. R. Astron. Soc. **378**, 635 (2007)
Southworth, J., Copperwheat, C.M., Gänsicke, B.T., Pyrzas, S.: Astron. Astrophys. **510**, 100 (2010)
Southworth, J., Tappert, C., Gänsicke, B.T., Copperwheat, C.M.: Astron. Astrophys. **573**, 61 (2015)
Spruit, H.C., Ritter, H.: Astron. Astrophys. **124**, 267 (1983)
Staude, A., Schwope, A.D., Schwarz, R.: Astron. Astrophys. **374**, 588 (2001)
Staude, A., Schwope, A.D., Krumpe, M., Hambaryan, V., Schwarz, R.: Astron. Astrophys. **406**, 253 (2003)

Stibbs, D.W.N.: Nature **165**, 195 (1950)
Swedlund, J.B., Wolstencroft, R.D., Michalsky, J.J. Jr., Kemp, J.C.: Astrophys. J. Lett. **187**, 121 (1974a)
Swedlund, J.B., Kemp, J.C., Wolstencroft, R.D.: Astrophys. J. Lett. **193**, 11 (1974b)
Szkody, P., Silber, A., Hoard, D.W., Fierce, E., Singh, K.P., Barrett, P., Schlegel, E., Piirola, V.: Astrophys. J. Lett. **455**, 43 (1995)
Szkody, P., Anderson, S.F., Agüeros, M., Covarrubias, R., Bentz, M., Hawley, S., Margon, B., Voges, W., Henden, A., Knapp, G.R., Vanden Berk, D.E., Rest, A., Miknaitis, G., Magnier, E., Brinkmann, J., Csabai, I., Harvanek, M., Hindsley, R., Hennessy, G., Ivezic, Z., Kleinman, S.J., Lamb, D.Q., Long, D., Newman, P.R., Neilsen, E.H., Nichol, R.C., Nitta, A., Schneider, D.P., Snedden, S.A., York, D.G.: Astron. J. **123**, 430 (2002)
Szkody, P., Anderson, S.F., Schmidt, G., Hall, P.B., Margon, B., Miceli, A., SubbaRao, M., Frith, J., Harris, H., Hawley, S., Lawton, B., Covarrubias, R., Covey, K., Fan, X., Murphy, T., Narayanan, V., Raymond, S., Rest, A., Strauss, M.A., Stubbs, C., Turner, E., Voges, W., Bauer, A., Brinkmann, J., Knapp, G.R., Schneider, D.P.: Astrophys. J. **583**, 902 (2003)
Szkody, P., Henden, A., Fraser, O.J., Silvestri, N.M., Schmidt, G.D., Bochanski, J.J., Wolfe, M.A., Agüeros, M., Anderson, S.F., Mannikko, L., Downes, R.A., Schneider, D.P., Brinkmann, J.: Astron. J. **129**, 2386 (2005)
Szkody, P., Henden, A., Mannikko, L., Mukadam, A., Schmidt, G.D., Bochanski, J.J., Agüeros, M., Anderson, S.F., Silvestri, N.M., Dahab, W.E., Oguri, M., Schneider, D.P., Shin, M.-S., Strauss, M.A., Knapp, G.R., West, A.A.: Astron. J. **134**, 185 (2007)
Szkody, P., Anderson, S.F., Hayden, M., Kronberg, M., McGurk, R., Riecken, T., Schmidt, G.D., West, A.A., Gänsicke, B.T., Nebot Gomez-Moran, A., Schneider, D.P., Schreiber, M.R., Schwope, A.D.: Astron. J. **137**, 4011 (2009)
Szkody, P., Anderson, S.F., Brooks, K., Gänsicke, B.T., Kronberg, M., Riecken, T., Ross, N.P., Schmidt, G.D., Schneider, D.P., Agüeros, M.A., Gomez-Moran, A.N., Knapp, G.R., Schreiber, M.R., Schwope, A.D.: Astron. J. **142**, 181 (2011)
Szkody, P., Everett, M.E., Howell, S.B., Landolt, A.U., Bond, H.E., Silva, D.R., Vasquez-Soltero, S.: Astron. J. **148**, 63 (2014)
Tapia, S.: Astrophys. J. Lett. **212**, 125 (1977)
Tappert, C., Augusteijn, T., Maza, J.: Mon. Not. R. Astron. Soc. **354**, 321 (2004)
Thomas, H.-C., Beuermann, K., Schwope, A.D., Burwitz, V.: Astron. Astrophys. **313**, 833 (1996)
Thomas, H.-C., Beuermann, K., Burwitz, V., Reinsch, K., Schwope, A.D.: Astron. Astrophys. **353**, 646 (2000)
Thomas, H.-C., Beuermann, K., Reinsch, K., Schwope, A.D., Burwitz, V.: Astron. Astrophys. **546**, 104 (2012)
Thompson, A.M., Cawthorne, T.V.: Mon. Not. R. Astron. Soc. **224**, 425 (1987)
Thorstensen, J.R.: Astron. J. **91**, 940 (1986)
Thorstensen, J.R., Taylor, C.J.: Mon. Not. R. Astron. Soc. **326**, 1235 (2001)
Thorstensen, J.R., Fenton, W.H.: Publ. Astron. Soc. Pac. **114**, 74 (2002)
Thorstensen, J.R., Halpern, J.: Astron. J. **146**, 107 (2013)
Thorstensen, J.R., Peters, C.S., Skinner, J.N.: Publ. Astron. Soc. Pac. **122**, 1285 (2010)
Thorstensen, J.R., Taylor, C.J., Peters, C.S., Skinner, J.N., Southworth, J., Gaensicke, B.T.: Astron. J. **149**, 128 (2015)
Tout, C.A., Wickramasinghe, D.T., Liebert, J., Ferrario, L., Pringle, J.E.: Mon. Not. R. Astron. Soc. **387**, 897 (2008)
Tovmassian, G.H., Zharikov, S.V.: Astron. Astrophys. **468**, 643 (2007)
Tovmassian, G.H., Greiner, J., Zickgraf, F.-J., Kroll, P., Krautter, J., Thiering, I., Zharykov, S.V., Serrano, A.: Astron. Astrophys. **328**, 571 (1997)
Tovmassian, G.H., Greiner, J., Schwope, A.D., Szkody, P., Schmidt, G., Zickgraf, F.-J., Serrano, A., Krautter, J., Thiering, I., Zharikov, S.V.: Astrophys. J. **537**, 927 (2000)
Tovmassian, G.H., Szkody, P., Greiner, J., Zharikov, S.V., Zickgraf, F.-J., Serrano, A., Krautter, J., Thiering, I., Neustroev, V.: Astron. Astrophys. **379**, 199 (2001)
Townsley, D.M., Bildsten, L.: Astrophys. J. Lett. **596**, 227 (2003)
Townsley, D.M., Gänsicke, B.T.: Astrophys. J. **693**, 1007 (2009)
Traulsen, I., Reinsch, K., Schwope, A.D., Schwarz, R., Walter, F.M., Burwitz, V.: Astron. Astrophys. **562**, 42 (2014)
Tremblay, P.-E., Ludwig, H.-G., Steffen, M., Freytag, B.: Astron. Astrophys. **559**, 104 (2013)
Tuohy, I.R., Ferrario, L., Wickramasinghe, D.T., Hawkins, M.R.S.: Astrophys. J. Lett. **328**, 59 (1988)
Uslenghi, M., Tommasi, L., Treves, A., Piirola, V., Reig, P.: Astron. Astrophys. **372**, 1 (2001)
Vaeth, H.: Astron. Astrophys. **317**, 476 (1997)

Valyavin, G., Bagnulo, S., Fabrika, S., Reisenegger, A., Wade, G.A., Han, I., Monin, D.: Odessa Astron. Publ. **18**, 135 (2005)
Valyavin, G., Bagnulo, S., Fabrika, S., Reisenegger, A., Wade, G.A., Han, I., Monin, D.: Astrophys. J. **648**, 559 (2006)
Vanlandingham, K.M., Schmidt, G.D., Eisenstein, D.J., Harris, H.C., Anderson, S.F., Hall, P.B., Liebert, J., Schneider, D.P., Silvestri, N.M., Stinson, G.S., Wolfe, M.A.: Astron. J. **130**, 734 (2005)
Vennes, S., Ferrario, L., Wickramasinghe, D.T.: Mon. Not. R. Astron. Soc. **302**, 49 (1999)
Vennes, S., Schmidt, G.D., Ferrario, L., Christian, D.J., Wickramasinghe, D.T., Kawka, A.: Astrophys. J. **593**, 1040 (2003)
Verbunt, F., Zwaan, C.: Astron. Astrophys. **100**, 7 (1981)
Visvanathan, N., Wickramasinghe, D.T.: Nature **281**, 47 (1979)
Vogel, J., Schwope, A.D., Gänsicke, B.T.: Astron. Astrophys. **464**, 647 (2007)
Vogel, J., Byckling, K., Schwope, A., Osborne, J.P., Schwarz, R., Watson, M.G.: Astron. Astrophys. **485**, 787 (2008)
Vogel, J., Schwope, A.D., Schwarz, R.: Astron. Astrophys. **530**, 117 (2011)
Vornanen, T., Berdyugina, S.V., Berdyugin, A.V., Piirola, V.: Astrophys. J. Lett. **720**, 52 (2010)
Vornanen, T., Berdyugina, S.V., Berdyugin, A.: Astron. Astrophys. **557**, 38 (2013)
Walker, M.F.: Astrophys. J. **123**, 68 (1956)
Warner, B.: In: Buckley, D.A.H., Warner, B. (eds.) *Magnetic Cataclysmic Variables*. Astronomical Society of the Pacific Conference Series, vol. 85, p. 3 (1995)
Warwick, R.S.: Mon. Not. R. Astron. Soc. **445**, 66 (2014)
Webbink, R.F., Wickramasinghe, D.T.: Mon. Not. R. Astron. Soc. **335**, 1 (2002)
Wegner, G.: Mem. Soc. Astron. Ital. **48**, 27 (1977)
Welsh, W.F., Horne, K., Gomer, R.: Astrophys. J. Lett. **410**, 39 (1993)
Wesemael, F., Liebert, J., Schmidt, G.D., Beauchamp, A., Bergeron, P., Fontaine, G.: Astrophys. J. **554**, 1118 (2001)
D. Wickramasinghe, Eur. Phys. J. Web Conf. **64**, 3001 (2014)
Wickramasinghe, D.T., Bessell, M.S.: Mon. Not. R. Astron. Soc. **188**, 841 (1979)
Wickramasinghe, D.T., Cropper, M.: Mon. Not. R. Astron. Soc. **235**, 1451 (1988)
Wickramasinghe, D.T., Ferrario, L.: Astrophys. J. **327**, 222 (1988a)
Wickramasinghe, D.T., Ferrario, L.: Astrophys. J. **334**, 412 (1988b)
Wickramasinghe, D.T., Ferrario, L.: Publ. Astron. Soc. Pac. **112**, 873 (2000)
Wickramasinghe, D.T., Ferrario, L.: Mon. Not. R. Astron. Soc. **356**, 1576 (2005)
Wickramasinghe, D.T., Martin, B.: Mon. Not. R. Astron. Soc. **188**, 165 (1979)
Wickramasinghe, D.T., Martin, B.: Mon. Not. R. Astron. Soc. **212**, 353 (1985)
Wickramasinghe, D.T., Meggitt, S.M.A.: Mon. Not. R. Astron. Soc. **214**, 605 (1985)
Wickramasinghe, D.T., Wu, K.: Mon. Not. R. Astron. Soc. **266**, 1 (1994)
Wickramasinghe, D.T., Tuohy, I.R., Visvanathan, N.: Astrophys. J. **318**, 326 (1987)
Wickramasinghe, D.T., Ferrario, L., Bailey, J.: Astrophys. J. Lett. **342**, 35 (1989)
Wickramasinghe, D.T., Ferrario, L., Cropper, M., Bailey, J.: Mon. Not. R. Astron. Soc. **251**, 137 (1991)
Wickramasinghe, D.T., Ferrario, L., Bailey, J.A., Drissen, L., Dopita, M.A., Shara, M., Hough, J.H.: Mon. Not. R. Astron. Soc. **265**, 29 (1993)
Wickramasinghe, D.T., Schmidt, G., Ferrario, L., Vennes, S.: Mon. Not. R. Astron. Soc. **332**, 29 (2002)
Wickramasinghe, D.T., Farihi, J., Tout, C.A., Ferrario, L., Stancliffe, R.J.: Mon. Not. R. Astron. Soc. **404**, 1984 (2010)
Wickramasinghe, D.T., Tout, C.A., Ferrario, L.: Mon. Not. R. Astron. Soc. **437**, 675 (2014)
Williams, K.A., Winget, D.E., Montgomery, M.H., Dufour, P., Kepler, S.O., Hermes, J.J., Falcon, R.E., Winget, K.I., Bolte, M., Rubin, K.H.R., Liebert, J.: Astrophys. J. **769**, 123 (2013a)
Williams, K.A., Howell, S.B., Liebert, J., Smith, P.S., Bellini, A., Rubin, K.H.R., Bolte, M.: Astron. J. **145**, 129 (2013b)
Wisotzki, L., Reimers, D., Wamsteker, W.: Astron. Astrophys. **247**, 17 (1991)
Woelk, U., Beuermann, K.: Astron. Astrophys. **256**, 498 (1992)
Woelk, U., Beuermann, K.: Astron. Astrophys. **306**, 232 (1996)
Woudt, P.A., Warner, B., Gulbis, A., Coppejans, R., Hambsch, F.-J., Beardmore, A.P., Evans, P.A., Osborne, J.P., Page, K.L., Wynn, G.A., van der Heyden, K.: Mon. Not. R. Astron. Soc. **427**, 1004 (2012a)
Woudt, P.A., Warner, B., de Budé, D., Macfarlane, S., Schurch, M.P.E., Zietsman, E.: Mon. Not. R. Astron. Soc. **421**, 2414 (2012b)
Wunner, G., Roesner, W., Herold, H., Ruder, H.: Astron. Astrophys. **149**, 102 (1985)
York, D.G., Adelman, J., Anderson, J.E. Jr., Anderson, S.F., Annis, J., Bahcall, N.A., Bakken, J.A., Barkhouser, R., Bastian, S., Berman, E., Boroski, W.N., Bracker, S., Briegel, C., Briggs, J.W., Brinkmann, J., Brunner, R., Burles, S., Carey, L., Carr, M.A., Castander, F.J., Chen, B., Colestock, P.L., Connolly,

A.J., Crocker, J.H., Csabai, I., Czarapata, P.C., Davis, J.E., Doi, M., Dombeck, T., Eisenstein, D., Ellman, N., Elms, B.R., Evans, M.L., Fan, X., Federwitz, G.R., Fiscelli, L., Friedman, S., Frieman, J.A., Fukugita, M., Gillespie, B., Gunn, J.E., Gurbani, V.K., de Haas, E., Haldeman, M., Harris, F.H., Hayes, J., Heckman, T.M., Hennessy, G.S., Hindsley, R.B., Holm, S., Holmgren, D.J., Huang, C.-h., Hull, C., Husby, D., Ichikawa, S.-I., Ichikawa, T., Ivezić, Ž., Kent, S., Kim, R.S.J., Kinney, E., Klaene, M., Kleinman, A.N., Kleinman, S., Knapp, G.R., Korienek, J., Kron, R.G., Kunszt, P.Z., Lamb, D.Q., Lee, B., Leger, R.F., Limmongkol, S., Lindenmeyer, C., Long, D.C., Loomis, C., Loveday, J., Lucinio, R., Lupton, R.H., MacKinnon, B., Mannery, E.J., Mantsch, P.M., Margon, B., McGehee, P., McKay, T.A., Meiksin, A., Merelli, A., Monet, D.G., Munn, J.A., Narayanan, V.K., Nash, T., Neilsen, E., Neswold, R., Newberg, H.J., Nichol, R.C., Nicinski, T., Nonino, M., Okada, N., Okamura, S., Ostriker, J.P., Owen, R., Pauls, A.G., Peoples, J., Peterson, R.L., Petravick, D., Pier, J.R., Pope, A., Pordes, R., Prosapio, A., Rechenmacher, R., Quinn, T.R., Richards, G.T., Richmond, M.W., Rivetta, C.H., Rockosi, C.M., Ruthmansdorfer, K., Sandford, D., Schlegel, D.J., Schneider, D.P., Sekiguchi, M., Sergey, G., Shimasaku, K., Siegmund, W.A., Smee, S., Smith, J.A., Snedden, S., Stone, R., Stoughton, C., Strauss, M.A., Stubbs, C., SubbaRao, M., Szalay, A.S., Szapudi, I., Szokoly, G.P., Thakar, A.R., Tremonti, C., Tucker, D.L., Uomoto, A., Vanden Berk, D., Vogeley, M.S., Waddell, P., Wang, S.-i., Watanabe, M., Weinberg, D.H., Yanny, B., Yasuda, N. (SDSS Collaboration): Astron. J. **120**, 1579 (2000)

Yuasa, T., Makishima, K., Nakazawa, K.: Astrophys. J. **753**, 129 (2012)

Zhang, E., Robinson, E.L., Stiening, R.F., Horne, K.: Astrophys. J. **454**, 447 (1995)

Zharikov, S.V., Tovmassian, G.H., Echevarría, J.: Astron. Astrophys. **390**, 23 (2002)

Zuckerman, B., Koester, D., Reid, I.N., Hünsch, M.: Astrophys. J. **596**, 477 (2003)

Zuckerman, B., Koester, D., Dufour, P., Melis, C., Klein, B., Jura, M.: Astrophys. J. **739**, 101 (2011)

Part III
Neutron Stars

DOI 10.1007/978-1-4939-3550-5_6
Reprinted from *Space Science Reviews* Journal, DOI 10.1007/s11214-014-0102-2

Neutron Stars—Thermal Emitters

Alexander Y. Potekhin · Andrea De Luca · José A. Pons

Received: 3 July 2014 / Accepted: 11 September 2014 / Published online: 10 October 2014

Abstract Confronting theoretical models with observations of thermal radiation emitted by neutron stars is one of the most important ways to understand the properties of both, superdense matter in the interiors of the neutron stars and dense magnetized plasmas in their outer layers. Here we review the theory of thermal emission from the surface layers of strongly magnetized neutron stars, and the main properties of the observational data. In particular, we focus on the nearby sources for which a clear thermal component has been detected, without being contaminated by other emission processes (magnetosphere, accretion, nebulae). We also discuss the applications of the modern theoretical models of the formation of spectra of strongly magnetized neutron stars to the observed thermally emitting objects.

Keywords Neutron stars · Magnetic fields · Thermal emission · Stellar atmospheres

1 Introduction

One of the first expectations of neutron-star (NS) astrophysics, dating back to the epoch when such sources where theoretically proposed, is the possibility to detect thermal radiation

A.Y. Potekhin (✉)
Ioffe Institute, Politekhnicheskaya 26, 194021 Saint Petersburg, Russia
e-mail: palex@astro.ioffe.ru

A.Y. Potekhin
Central Astronomical Observatory at Pulkovo, Pulkovskoe Shosse 65, 196140 Saint Petersburg, Russia

A. De Luca
INAF—Istituto di Astrofisica Spaziale e Fisica Cosmica Milano, via E. Bassini 15, 20133 Milano, Italy
e-mail: deluca@iasf-milano.inaf.it

A. De Luca
INFN—Istituto Nazionale di Fisica Nucleare, sezione di Pavia, via A. Bassi 6, 27100 Pavia, Italy

J.A. Pons
Departament de Física Aplicada, Universitat d'Alacant, Ap. Correus 99, 03080 Alacant, Spain
e-mail: jose.pons@ua.es

from their hot (millions of K) surfaces (Zwicky 1938). Observation of the first sources of cosmic X-rays in the Early Sixties suggested that NSs could indeed be detected, which added impetus to theoretical work as well as to modeling of expected NS observational properties. The crucial point, attracting a lot of interest in the astrophysical community, was that the study of NS cooling could constrain the physical properties of superdense matter in the interiors of the NSs (Bahcall and Wolf 1965), under conditions that cannot be studied in terrestrial laboratories. Discovery of NSs as radio pulsars (Hewish et al. 1968) further boosted NS cooling studies as well as observational efforts. However, the first detection of thermal radiation from the surface of INSs (Cheng and Helfand 1983; Brinkmann and Ögelman 1987) had to wait the launch of focusing X-ray telescopes HEAO2/*Einstein* (1978–1981) and *EXOSAT* (1983–1986), which allowed for a dramatic leap forward in sensitivity to faint point X-ray sources.

Our understanding of the physics of NSs dramatically improved in the last decades. Thanks to multiwavelength observations (with an important role played by X-ray observations), we have discovered with much surprise that not all Isolated NSs (INSs) behave as radio pulsars, but that there exist a rich diversity of INS classes, including Rotation-Powered Pulsars, Rotating Radio Transients, Anomalous X-ray Pulsars, Soft Gamma Repeaters, Central Compact Objects in supernova remnants, and the Magnificent Seven thermally emitting NSs (see the next section). It is now commonly accepted that magnetic fields plays an important role in shaping the electromagnetic emission properties of INSs, e.g. mediating the conversion of their rotational energy into radiation (as in rotation-powered sources), or directly acting as the energy reservoir for most of the INS luminosity (as in magnetars).

One of the main challenges is to disentangle different emission components, overlapping in the X-ray energy range. After discriminating thermal emission, a detailed study of the thermal spectra can yield precious information about the NS surface temperature distribution, the properties of dense magnetized plasmas in their envelopes and atmospheres, as well as set constraints on the equation of state of the ultradense matter in the NS cores.

Besides INSs, some NSs with observed thermal spectra also reside in binary systems. In low-mass X-ray binary systems (LMXBs), a NS accretes matter from a less massive star (a Main Sequence star or a white dwarf), alternating periods of intense accretion and periods of quiescence. When accretion stops and the residual heat diffuses out from the crust, X-ray radiation comes from the heated NS surface (Brown et al. 1998). During the last decade, such quiescent sources (qLMXBs) yield ever increasing amount of valuable information on the NSs. Their spectra are successfully interpreted with models of NS atmospheres (see Potekhin 2014, for a discussion and references). Another class of NSs in binaries with thermal spectra are X-ray bursters—accreting NSs in close binary systems, which produce X-ray bursts with intervals from hours to days (see, e.g. Strohmayer and Bildsten 2006, for a review). During intervals between the bursts, a burster's atmosphere does not essentially differ from an atmosphere of a cooling NS. In such periods, the bulk of the observed X-ray radiation arises from transformation of gravitational energy of the accreting matter into thermal energy. Some of the bursts (so-called long bursts, which last over a minute) occur during periods when the accretion rate is low enough for the luminosity not to exceed a few percent of the Eddington limit. In this cases, a thermal atmospheric spectrum can be observed (e.g., Suleimanov et al. 2011).

As we will see, the number of known NSs with an unambiguously identified thermal component in the spectrum is not large, but it steadily increases. Some of them can be understood with models of nonmagnetic atmospheres, whereas others are believed to be endowed with strong magnetic fields, which must be taken into account. After the seminal work of Romani (1987), the nonmagnetic neutron-star atmospheres have been studied in many works (see Zavlin 2009 for a review). Databases of neutron-star hydrogen atmosphere model spectra have been published (Zavlin et al. 1996; Gänsicke et al. 2002; Heinke et al. 2006),[1] and a computer code for their calculation has been released (Haakonsen et al. 2012). A database of carbon atmosphere model spectra has been also published recently (Suleimanov et al. 2014).[2] Model spectra were calculated for neutron-star atmospheres composed of different chemical elements from H to Fe (e.g., Rajagopal and Romani 1996; Pons et al. 2002; Heinke et al. 2006; Ho and Heinke 2009) and mixtures of different elements (Gänsicke et al. 2002; Pons et al. 2002). In general, the thermal spectra of NSs in binaries (e.g., bursters and qLMXBs mentioned above) are interpreted with theoretical models without magnetic fields. Thermal components of spectra of several millisecond pulsars (e.g., PSR J0437-4715, Bogdanov 2013 and some thermally emitting INSs (e.g., the source in Cassiopeia A, Ho and Heinke 2009) have also been interpreted with the nonmagnetic atmosphere models. In this paper, however, we will not consider the nonmagnetic models (see Potekhin 2014 for a more general review, which includes a discussion of both the nonmagnetic and magnetic cases), but instead we will focus on models of thermal spectra of the INSs that are significantly affected by strong magnetic fields.

The paper is organized as follows. In Sect. 2 we give account of the different classes of INSs. In Sect. 3 we list INSs with confirmed thermal emission and give a summary of their main characteristics. In Sect. 4 we consider the definitions and concepts that are important for the theory of formation of thermal spectra of NSs with strong magnetic fields. The latter theory is described in Sect. 5. Section 6 is devoted to the theory of a magnetized condensed surface as an alternative to gaseous NS atmospheres. In Sect. 7 we describe transformation of the local spectra into the spectra seen by a distant observer. Examples of interpretation of observed spectra with the use of theoretical models of partially ionized, strongly magnetized NS atmospheres are considered in Sect. 8. In Sect. 9, we give brief conclusions. In the Appendix we briefly describe the effects of thermal motion of atoms in a strong magnetic field on the atomic quantum-mechanical characteristics and on the ionization equilibrium of plasmas, that underlie calculations of the opacities for the strongly magnetized NS atmospheres.

2 The Families of Isolated Neutron Stars

A short account of the main properties of the different classes of INSs, with a focus on their emission in the X-ray range, is useful to set the context for the observational panorama of thermal emitters.

- *Rotation-Powered Pulsars (RPPs).* This is the class of INSs with the largest observational database, more than 2200 sources being known,[3] mostly from radio surveys. A large population of radio-silent RPPs is also emerging, thanks to gamma-ray observations by the Fermi mission (see Abdo et al. 2013, and references therein).

  Rotation of a magnetized INS induces electric fields which accelerate particles in the magnetosphere surrounding the star, initiating electromagnetic cascades. This mechanism produces synchrotron and curvature radiation along a peculiar beaming pattern, ultimately

[1] Models NSA, NSAGRAV, and NSATMOS in the database *XSPEC* (Arnaud 1996).

[2] Model CARBATM in the database *XSPEC* (Arnaud 1996).

[3] See e.g. the ATNF pulsar database at http://www.atnf.csiro.au/research/pulsar/psrcat/ (Manchester et al. 2005)

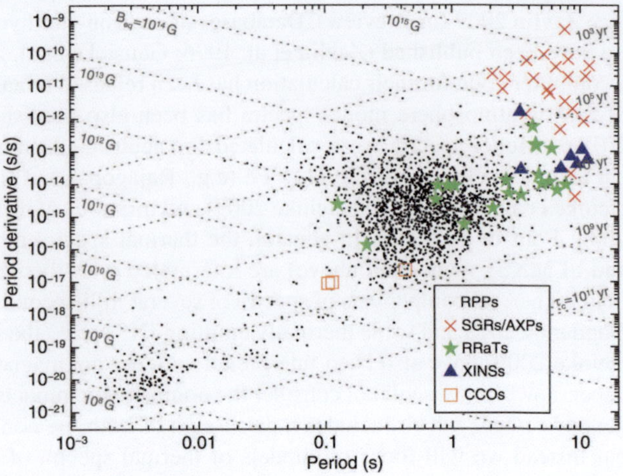

Fig. 1 $P-\dot{P}$ diagram for the ~ 2300 currently known neutron stars with measured P and $\dot{P}$. Different classes are marked with *different symbols*. *Lines* showing constant $\dot{E}_{\rm rot}$ as well as constant inferred surface dipolar magnetic field are superimposed

related to the magnetic field configuration, at the expense of the dissipation of the NS rotational energy (see Harding and Lai 2006, for a review). Because of such rotation-powered emission (see also Gwinn et al. 2014, this issue), RPPs are typically observed as pulsating sources from radio wavelengths[4] to very high energy gamma-rays—the spin-down luminosity $\dot{E}_{\rm rot} = 4\pi^2 I \dot{P} P^{-3}$ (where $I \sim 10^{45}$ g cm^2 is the NS moment of inertia) being indeed much larger than the inferred electromagnetic luminosities. However, as a matter of fact, most of the spin-down luminosity is carried away by a relativistic particle wind which, interacting with the surrounding medium, powers bright pulsar wind nebulae, seen as diffuse structures in the radio, X-ray, and gamma-ray energy ranges.

In this picture, assuming the observed $\dot{E}_{\rm rot}$ to be due to magnetic dipole radiation yields an estimate of the surface dipole field at the magnetic equator of the star $B_{\rm eq} = 3.2 \times 10^{19} (P\dot{P})^{1/2}$ G, as well as an estimate of the NS age $\tau_{\rm c} = P/(2\dot{P})$, which is called characteristic age. The first estimate assumes $\sqrt{I_{45}}/(R_6^3 \sin\alpha) = 1$, where $I_{45} = I/10^{45}$ g cm^2, R_6 is the stellar radius R in units of 10^6 cm, and α is the angle between the magnetic and rotational axes. More realistic calculations give similar relations between B and $P\dot{P}$ (Spitkovsky 2006; Beskin et al. 2013). The second estimate assumes a birth period much shorter than the current period P, as well as a non-variable magnetic field. As can be seen in Fig. 1, the bulk of RPPs have P in the 0.1–1 s range and $B_{\rm eq} \sim 10^{12}$ G. The separate subclass of Millisecond Pulsars (MSPs), accounting for ~ 10 % of the RPP sample, is supposed to include very old NSs with a different evolutionary history, having experienced accretion in a long-lived binary system which eventually spun-up the NSs to $P \sim$ a few ms and quenched their magnetic field to 10^8–10^9 G. For this reason MSPs are also called recycled pulsars (Bisnovatyi-Kogan 2006, and references therein).

More than 120 RPPs have been detected in the soft X-ray energy range (see, e.g., Marelli et al. 2011; Abdo et al. 2013), and several of them display thermal emission from their hot surfaces in addition to the rotation-powered emission. While a thermal component related to the NS cooling is apparent in a few sources, an additional component with a higher temperature and a smaller emitting area can also be observed (De Luca et al. 2005), probably related to re-heating of the magnetic polar caps by bombardment of magnetospheric particles. We note that RPPs with the largest inferred magnetic fields

[4] Indeed, their discovery as radio pulsars was the first observational evidence for the existence of NSs

($B_{eq} \sim 10^{13} - 10^{14}$ G) are often considered as a separate, "High-B" class, (HB, see Ng and Kaspi 2011 for a review). HB RPPs show some evidence for a larger thermal luminosity when compared to RPPs of similar ages and can be the link between magnetars and RPPs, with magnetic field decay playing an important role for their thermal evolution (Aguilera et al. 2008; Pons et al. 2009). At least in one case (PSR J1846-0258, Gavriil et al. 2008), a peculiar time variability, reminiscent of the behavior of the anomalous X-ray pulsars (see below), has been seen in a HB RPP.

- *Rotating Radio Transients (RRaTs).* Discovered as sources of repeated, bright, short (~ 1 ms) radio bursts occurring at integral multiples of an underlying periodicity, RRaTs were initially considered as a new class of INSs (McLaughlin et al. 2006). High time resolution radio surveys are unveiling new RRaTs (about 70 are currently known—see Burke-Spolaor 2014 for a recent review).

  It has been observed that RRaTs, as a class, have a larger magnetic field with respect to the bulk of RPPs, which prompted speculations about possible relationships of such sources with magnetars (or with HB RPPs, or with the Magnificent Seven NSs considered below). Interestingly enough, the most active RRaT (PSR J1819-1458)—the only source of its class being detected in soft X-rays—has an X-ray phenomenology (as well as a position in the $P - \dot{P}$ plane) fully similar to the one of the Magnificent Seven, with a pulsed thermal-like emission and a broad spectral absorption feature superimposed to the continuum (Miller et al. 2011). In any case, it is currently believed that RRaTs are not substantially physically different from RPPs—which also show time variability of the radio pulses—RRaTs have just a somewhat extreme temporal behavior.

- *Soft Gamma Repeaters (SGRs)* and *Anomalous X-ray Pulsars (AXPs).* About two dozen such sources are known (Olausen and Kaspi 2014). Originally thought to form two different classes of objects, they were later shown to form a single SGR/AXP class (Gavriil et al. 2002). Detailed reviews of their phenomenology and physics are included in this issue, see also Mereghetti (2008, 2013), Turolla and Esposito (2013).

  The period of SGR/AXPs (in the 2–12 s range) is much longer than the one of the bulk of RPPs; their period derivative (showing peculiar time variability) is also very large, corresponding to dipole magnetic fields of the order of $B \sim 10^{14} - 10^{15}$ G (see Fig. 1).

  These objects show a very complex high energy phenomenology. This includes a persistent, pulsed soft X-ray component (often with a transient behavior), dominated by thermal-like emission, typically with a luminosity largely exceeding the spin-down luminosity. Moreover, SGR/AXPs display a spectacular flaring emission component (see Rea and Esposito 2011, for a review), best seen from a few keV to a few hundreds keV. They are typically radio silent, although very bright radio pulses have been observed from a few sources.

  SGR/AXPs are generally believed to be magnetars (see the review by Olausen and Kaspi 2014), i.e. NSs ultimately powered by the decay of their magnetic field (Duncan and Thompson 1992).[5] The magnetic field of magnetars is supposed to be much more intense than for other classes of INSs and to have a peculiar topology.

- *Central Compact Objects (CCOs).* This class includes a dozen of point sources, lying close to the center of young (0.3–10 kyr) Supernova Remnants, discovered in the soft

[5] There are alternative hypotheses about the nature of SGR/AXPs, assuming their braking by an accretion disk (e.g., Trümper et al. 2013, and references therein) or a "magnetic slab" (Bisnovatyi-Kogan and Ikhsanov 2014), as well as the models of drift waves in the magnetosphere of a neutron star with $B \sim 10^{12}$ G (Malov 2010) or rapidly rotating massive ($M > M_{\odot}$) white dwarfs with $B \sim 10^{8}$ G (Boshkaev et al. 2013).

X-ray band and supposed to be young INSs (De Luca 2008; Gotthelf et al. 2013). They have no counterparts at any other wavelength, nor associated diffuse nebulae, nor any manifestation of rotation-powered magnetospheric activity. Such sources only display steady, thermal-like X-ray emission, typically with a two-temperature blackbody spectrum and very small emitting areas. Absorption features superimposed to the continuum have been observed in two sources.

Measurement of a tiny period derivative for the only three pulsating CCOs point to very low surface dipole fields ($B_{\rm eq} \sim 10^{11}$ G, Gotthelf et al. 2013). This is consistent with an interpretation of the absorption lines as electron cyclotron features. In this picture, it is not clear what is the origin of the apparent, relatively large thermal luminosities and large temperature anisotropies for these NSs. The spin-down luminosity is too low to explain the hot spot emission, while X-ray timing, as well as deep optical/infrared observations, rule out the existence of companion stars as well as accretion from debris discs (Halpern and Gotthelf 2010; De Luca et al. 2011). It has been suggested that a strong crustal toroidal magnetic field could channel heat from the star interior, producing hot spots on the surfaces, while not influencing the NS spin-down (see, e.g., Shabaltas and Lai 2012).

As another puzzle, the inferred, relatively high birth rate of CCOs clashes with the apparently underpopulated region of the $P - \dot{P}$ diagram in which they reside (Kaspi 2010; Halpern and Gotthelf 2010—see Fig. 1). It has been proposed that the weak observed dipole field could result from prompt accretion of 10^{-4}–10^{-3} $M_\odot$ fallback material soon after the SN explosion. The buried field would re-emerge later, on a 10^3–10^5 yr time scale (depending on the amount of accreted material), turning a CCO into a RPP, or even into a magnetar (Viganò and Pons 2012).

- *The Magnificent Seven, or X-ray INSs (XINSs)*. These seven INSs were discovered in the *ROSAT All Sky Survey*. Reviews on their properties are given by Haberl (2007), Turolla (2009), Kaplan and van Kerkwijk (2009), to whom the reader is referred for more details and references.

  The XINSs have spinning periods in the same range as SGR/AXPs, but the period derivative points to a dipole field ~ 1 order of magnitude lower for the XINSs (but still well above the one of the bulk of RPPs—see Fig. 1).

  These sources display a thermal-like spectrum, from the optical to the soft X-ray range.[6] Their thermal luminosity is somewhat larger than expected from conventional cooling of NSs at the same characteristic age. Broad (often multiple) spectral distortions compatible with absorption features are seen, superimposed on the thermal continuum. The nature of these features is unclear. Suggested explanations include, for example, atomic or proton-cyclotron features (e.g., Haberl 2007), photoionization in a relatively dense cloud in the vicinity of the NS (Hambaryan et al. 2009), or a result of a complex strongly inhomogeneous distribution of temperature over the surface (Viganò et al. 2014). In one case, the observed spectral shape has been tentatively explained as originated in a thin partially-ionized hydrogen atmosphere above a condensed iron surface (see Sect. 8.2).

  All XINSs have an optical/UV counterpart, with a flux exceeding by a factor 5–50 (around 5000 Å) the expected value, based on the extrapolation of the soft X-ray spectrum

[6]For some of them, pulsed radio emission has been also detected at very low frequency 111 MHz at the Pushchino Radio Astronomy Observatory—see Malofeev et al. (2007), Teplykh et al. (2011). Such detections are very intriguing and await confirmation from other observations at similar frequencies, e.g., by the LOFAR telescope.

(Kaplan et al. 2011). Such optical/UV excesses show power-law spectra, but the observed slopes differ from source to source and are generally not consistent with the Rayleigh-Jeans tail of a blackbody spectrum. The origin of the optical/UV excess is not understood (atmospheric effects? magnetospheric activity? evidence for an emitting region larger and cooler than the one seen in X-rays?).

All XINSs are steady emitters, with the exception of RX J0720.4-3125, whose time variability has not yet been understood (see Hohle et al. 2012, possible explanations range from magnetar-like activity to a peculiar accretion episode).

The XINSs are rather close (from 120 pc to a few hundred pc) to the Solar system. These distance estimates are based on direct annual parallax measurements or on comparison of the observed photoelectric absorption to models for the 3D distribution of local interstellar medium, and they are consistent with the large observed proper motions. Back-projection of the NS space trajectories allowed to infer a kinematic age for a few sources, based on possible association with clusters of massive stars (see e.g. Tetzlaff et al. 2011). Interestingly enough, the kinematic ages are systematically smaller than characteristic ages τ_c, being thus more consistent with the relatively high thermal luminosities.

It has been suggested that the magnetic field of the XINSs could have decayed since their birth, affecting their rotational and thermal evolution (Popov et al. 2010). The relationship of the XINSs to other classes of INSs is in any case still poorly understood (aged magnetars? HB RPPs with unfavorable radio beaming? extreme RRaTs?).

Unifying the apparent diversity of the classes of INSs in a coherent physical scenario is a major goal in the astrophysics of INSs. A study in this perspective was performed by Viganò et al. (2013), who modeled the coupled evolution of temperature and magnetic field (driving the rotational evolution) of INSs and proved the possibility to explain the overall properties of the classes of SGR/AXPs, HB RPPs, and of the XINSs by varying the initial magnetic field, mass and envelope composition in a unique parent population of INSs.

3 Isolated Neutron Stars with Thermal Spectra

Viganò et al. (2013) gathered and thoroughly re-analyzed all the best available data on isolated, thermally emitting NSs were in a consistent way. We refer the interested reader to that reference for more details. The data sample of 40 sources was compared to theoretical models of the magneto-thermal evolution of NSs, in an attempt to explain the phenomenological diversity of SGR/AXPs, high-B radio-pulsars, and isolated nearby NSs by only varying their initial magnetic field, mass and envelope composition. The cooling theory of NSs and several issues characteristic of magnetars are also in other sections of this issue. In this chapter, we focus on the thermal emission from sources not cataloged as magnetar candidates (SGR/AXPs). The sample of selected sources (see Table 1) includes:

- Eight CCOs, including the very young NS in Cassiopeia A, and the only three CCOs with measured values of P and $\dot{P}$. We have ignored the other CCOs candidates, since they have spectral information with poor statistics and/or a very uncertain age of the associated SNR.
- 13 rotation powered pulsars, including the Vela pulsar and the so-called Three Musketeers (PSR B0656, PSR B1055 and the γ-ray-loud Geminga; De Luca et al. 2005). We have excluded most of the young pulsars, many of which are associated with pulsar wind nebulae, since in those cases data are compatible with non-thermal emission powered by the rotational energy loss, which is orders of magnitude larger than their X-ray luminosity

Table 1 Emission properties of the thermally emitting neutron stars. T_{bb} and R_{bb} are the temperature and radius inferred by the `bbodyrad` model. T_{nsa} is the temperature inferred by the `nsa` model with acceptable associated radius R_{nsa}, also indicated. T_{cool} is either the lower temperature for models including 2 BB, compatible with emission from the entire surface, or the upper limit for cases showing emission from a small spot $R_{bb} \sim$ a few km. In the latter case, L_{cool} is the associated upper limit to the "hidden" thermal luminosity. Data are from Viganò et al. (2013) as well as from www.neutronstarcooling.info. All radii, temperatures and luminosities are the values as measured by a distant observer

Source	T_{bb} [10^6 K]	R_{bb} [km]	$T_{nsa/rcs}$ [10^6 K]	R_{nsa} [km]	Best fit model	T_{cool} [10^6 K]	$\log(L_{cool})$ [erg s^{-1}]
CXOU J185238.6+004020	5.1	0.9	3.37	3.0	BB/nsa	<1.1	<33.1
1E 1207.4-5209**	2.2	9.6	1.7	7.4	BB*/nsa*	<0.7	<32.2
RX J0822-4300	4.6	1.7	2.37	6.4	BB/nsa	<1.0	<32.9
CXO J232327.9+584842	5.2	1.7	3.34	2.7	BB/nsa	<1.3	<33.3
1WGA J1713.4-3949	4.8	0.4	–	–	BB+PL	–	–
CXOU J085201.4-461753	4.7	0.28	3.1	1.2	BB/nsa	–	–
XMMU J172054.5-372652	4.9	2.6	–	–	BB+PL	–	–
XMMU J173203.3-344518	5.7	–	2.6^C	13^C	C atm	–	–
PSR J0538+2817	1.9	2.6	–	–	BB+PL	<0.6	<31.9
PSR B1055-52	2.2	0.3	–	–	2BB+PL	0.8	–
PSR J0633+1746	1.6	0.1	–	–	2BB+PL	0.49	–
PSR B1706-44	1.7	3.3	–	–	BB+PL	<0.7	<32.2
PSR B0833-45	1.4	5.0	0.93	9.4	(BB/nsa)+PL	<0.5	<31.5
PSR B0656+14	1.2	2.4	–	–	2BB+PL	0.6	–
PSR B2334+61	1.9	1.1	1.0	7.9	BB/nsa	<0.6	<31.9
PSR J1740+1000	2.0	0.4	0.6	10.3	2BB/nsa	0.9	–
PSR J1741-2054	0.7	12	–	–	BB	–	–
PSR J1357-6429	1.6	2.0	0.74	10.0	(BB/nsa)+PL	–	–
PSR J0726-2612	1.0	4.6	–	–	BB	<0.46	<31.5
PSR J1119-6127**	3.1	1.5	–	–	BB	<1.4	<32.9
PSR J1819-1458	1.5	12.3	–	–	BB	–	–
PSR J1718-3718	2.2	2.0	–	–	BB	<1.0	<32.9
RX J0420.0-5022	0.6	3.4	–	–	BB	–	–
RX J1856.5-3754**	0.73	4.1	–	–	BB	–	–
RX J2143.0+0654	1.24	2.3	–	–	BB	–	–
RX J0720.4-3125	0.97	5.7	–	–	BB	–	–
RX J0806.4-4123	1.17	1.2	0.63	8.2	BB*/nsa*	–	–
RX J1308.6+2127**	1.09	5.0	–	–	BB*	–	–
RX J1605.3+3249	1.15	0.9	0.49	7.0	BB*/nsa*	–	–

*Absorption line(s) `gabs` included in the fit

**See discussion of a more elaborated analysis in Sect. 8

^{C}Fits to a carbon atmosphere and assuming $d = 3.2$ kpc

(i.e. RX J0007.0+7303 in SNR CTA1; Caraveo et al. 2010). We also exclude several old pulsars (Zavlin and Pavlov 2004) with thermal emission from a tiny hot spot (a few tens of m^2), since the temperature of the small hot spots is probably unrelated to the cooling history of the NS. Our list also includes four high-B radio-pulsars magnetic fields $B_{eq} \sim 10^{13} - 10^{14}$ G and good quality spectra. We have excluded the AXP-like pulsar PSR J1846-0258 since during quiescence its X-ray emission does not show a significant thermal component (Ng et al. 2008; Livingstone et al. 2011), and it is orders of magnitude smaller than its rotational energy loss.

- We have included the only RRaT detected so far in X-ray (PSR J1819-1458).
- The Magnificent Seven (XINSs). All of them have good spectra, and most of them have well determined timing properties and good distance determinations (sometimes with direct parallax measurements).

We summarize the main properties of thermal emitters in Table 1 (taken from Viganò et al. 2013 and extended). All the data presented here with links to abundant references can also be found in the website http://www.neutronstarcooling.info/.

Although both, luminosities and temperatures can be obtained by spectral analysis, it is usually difficult to determine them accurately. The luminosity is always subject to the uncertainty in the distance measurement, while the inferred effective temperature depends on the choice of the emission model (blackbody vs. atmosphere models, composition, condensed surface, etc.), and it carries large theoretical uncertainties in the case of strong magnetic fields. We often find that more than one model can fit equally well the data, without any clear, physically motivated preference for one of them. Photoelectric absorption from interstellar medium further constitutes a source of error in temperature measurements, since the value of the hydrogen column density $N_{\rm H}$ is correlated to the temperature value obtained in spectral fits. Different choices for the absorption model and the metal abundances can also yield different results for the temperature. In addition, in the very common case of the presence of inhomogeneous surface temperature distributions, only an approximation with two or three regions at different temperatures is usually employed. Moreover, in the case of data with few photons and/or strong absorption features, the temperature is poorly constrained by the fit, adding a large statistical error to the systematic one.

4 Atoms and Matter in Strong Magnetic Fields

A general review of the physics of matter in strong magnetic fields is given by D. Lai (2014) in this issue. Here we consider only the concepts that are crucial for the theory of formation of thermal spectra of NSs with strong magnetic fields.

4.1 Landau Quantization

It is convenient to express the magnetic field by its strength in atomic units, γ, or in relativistic units, b:

$$\gamma = B/B_0 = 425.44 B_{12}, \qquad b = \hbar\omega_{\rm c}/\big(m_{\rm e}c^2\big) = B/B_{\rm QED} = B_{12}/44.14. \tag{1}$$

Here, $B_0 = m_{\rm e}^2 c e^3/\hbar^3$ is the atomic unit of magnetic field, $B_{12} \equiv B/10^{12}$ G, $\omega_{\rm c} = eB/m_{\rm e}c$ is the electron cyclotron frequency, and $B_{\rm QED} = m_{\rm e}^2 c^3/(e\hbar) = B_0/\alpha_{\rm f}^2$ is the critical (Schwinger) field, above which specific effects of quantum electrodynamics (QED) become pronounced. We call a magnetic field *strong*, if $\gamma \gg 1$, and *superstrong*, if $b \gtrsim 1$.

The motion of charged particles in a magnetic field $\boldsymbol{B}$ is quantized in discrete Landau levels, whereas the longitudinal (parallel to $\boldsymbol{B}$) momentum of the particle can change continuously. In the nonrelativistic theory, the threshold excitation energy of the Nth Landau level is $N\hbar\omega_{\rm c}$ ($N = 0, 1, 2, \ldots$). In the relativistic theory, it is $E_N = m_{\rm e}c^2$ $(\sqrt{1+2bN} - 1)$. The wave functions that describe an electron in a magnetic field (Sokolov and Ternov 1986) have a characteristic transverse scale of the order of the "magnetic length" $a_{\rm m} = (\hbar c/eB)^{1/2} =$

$a_B/\sqrt{\gamma}$, where a_B is the Bohr radius. The momentum projection on the magnetic field remains a good quantum number, therefore these projections have the usual Maxwellian distribution at thermodynamic equilibrium. For transverse motion, however, we have the discrete Boltzmann distribution over N.

In practice, Landau quantization becomes important when the electron cyclotron energy $\hbar\omega_c$ is at least comparable to both the electron Fermi energy ϵ_F and temperature T (in energy units, i.e., 10^6 K = 86.17 eV). If $\hbar\omega_c$ is appreciably larger than both these energies, then most electrons reside on the ground Landau level in thermodynamic equilibrium, and the field is called *strongly quantizing*. It is the case, if conditions $\rho < \rho_B$ and $\zeta_e \gg 1$ are fulfilled simultaneously, where

$$\rho_B = \frac{m_i}{\pi^2\sqrt{2}a_m^3 Z} = 7045\frac{A}{Z}B_{12}^{3/2}\ \mathrm{g\,cm^{-3}}, \qquad \zeta_e = \frac{\hbar\omega_c}{T} = 134.34\frac{B_{12}}{T_6}, \tag{2}$$

$m_i = Am_u$ is the ion mass, m_u is the unified atomic mass unit, and T_6 is temperature in units of 10^6 K. In NS atmospheres, these conditions are satisfied, as a rule, at $B \gtrsim 10^{11}$ G. In the opposite limit $\zeta_e \ll 1$, the field can be considered as *nonquantizing*. In the magnetospheres, which have lower densities, electrons can condensate on the lowest Landau level even at $B \sim 10^8$ G because of the violation of the LTE conditions (e.g., Mészáros 1992), but this is not the case in the photospheres (the atmospheric layers whose thermal state is determined by the radiative flux and where the observed spectrum is mainly formed; (see Potekhin 2014)).

For ions, the parameter ζ_e is replaced by

$$\zeta_i = \hbar\omega_{ci}/T = 0.0737(Z/A)B_{12}/T_6. \tag{3}$$

Here, $\omega_{ci} = ZeB/(m_i c)$ is the ion cyclotron frequency, Ze is the ion charge, and $\hbar\omega_{ci} = 6.35(Z/A)B_{12}$ eV is the ion cyclotron energy. In magnetar atmospheres, where $B_{12} \gtrsim 100$ and $T_6 \lesssim 10$, the parameter ζ_i is not small, therefore the quantization of the ion motion should be taken into account. A parameter analogous to ρ_B is unimportant for ions, because they are nondegenerate in NS envelopes.

4.2 Bound Species in Strong Magnetic Fields

As first noticed by Cohen et al. (1970), atoms with bound states should be much more abundant at $\gamma \gg 1$ than at $\gamma \lesssim 1$ in a NS atmosphere at the same temperature. This difference is caused by the magnetically-induced increase of binding energies (and decrease of size) of atoms in so-called *tightly-bound states*, which are characterized by electron-charge concentration at short distances to the nucleus. Therefore it is important to consider the bound states and bound-bound and bound-free transitions in a strong magnetic field even for light-element atmospheres, which would be almost fully ionized at $T \sim 10^5$ K in the nonmagnetic case.

Most studies of atoms in strong magnetic fields have considered an atom with an infinitely heavy (fixed in space) nucleus (see, e.g., Garstang 1977, Ruder et al. 1994, for reviews). This model is rather crude, but it is a convenient first approximation. In this section we review this model. An outline of more accurate treatments, which take the effects of finite atomic mass into account, is given in the Appendix.

4.2.1 One-Electron Atoms and Ions

The H atom in a magnetic field is well studied. At $B > 10^9$ G, its only electron resides at the ground Landau level $N = 0$. Since N is fixed, the quantum state is determined by two other quantum numbers: $s = 0, 1, 2, \ldots$, which corresponds to the electron angular-momentum projection on the magnetic-field direction, $-\hbar s$, and $\nu = 0, 1, 2, \ldots$, which corresponds mainly to the motion along $\boldsymbol{B}$. The tightly-bound states all have $\nu = 0$. Their binding energies logarithmically increase with increasing γ (asymptotically as $\sim \ln^2 \gamma$ Ry, where 1 Ry = 13.6057 eV is the Rydberg constant in energy units). Non-zero values of ν correspond to loosely-bound states, whose binding energies are confined within 1 Ry.

Accurate calculations of the properties of the bound states of a non-moving H atom in a strong magnetic field have been performed in many works (see Ruder et al. 1994, for a review). The B-dependences of binding energies are well approximated by analytical functions (Potekhin 2014). Continuum wave functions and photoionization cross-sections have also been calculated (Potekhin et al. 1997).

In the approximation of an infinite nuclear mass, the energy of any one-electron ion is related to the H-atom energy as $E(Z_\mathrm{n}, B) = Z_\mathrm{n}^2 E(1, B/Z_\mathrm{n}^2)$ (Surmelian and O'Connel 1974). Analogous similarity relations exist also for the cross sections of radiative transitions (Wunner et al. 1982). However, all these relations are violated by motion across the magnetic field. Even for an atom at rest, the account of the finite nuclear mass can be important at $s \neq 0$ (see the Appendix).

4.2.2 Many-Electron Atoms and Ions

According to the Thomas-Fermi model, a typical size of an atom with a large nuclear charge Z_n is proportional to $\gamma^{-2/5}$ in the interval $Z_\mathrm{n}^{4/3} \ll \gamma \ll Z_\mathrm{n}^3$ (Kadomtsev 1970), but this model breaks down at $\gamma \gtrsim Z_\mathrm{n}^3$ (Lieb et al. 1992). In particular, it cannot describe the difference of the transverse and longitudinal atomic sizes, which becomes huge in such strong fields. In this case, however, a good starting approximation is the so called adiabatic approximation, which presents each electron orbital as a product of the Landau function that describes free electron motion transverse to the field (Sokolov and Ternov 1986) and a function describing a one-dimensional motion along $\boldsymbol{B}$ in an effective potential, similar to a truncated one-dimensional Coulomb potential (Haines and Roberts 1969). At $\gamma \gg Z_\mathrm{n}^3$, all electron shells of the atom are strongly compressed in the directions transverse to the field. In the ground state, atomic sizes along and transverse to $\boldsymbol{B}$ can be estimated as (Kadomtsev and Kudryavtsev 1971)

$$l_\perp \approx \sqrt{2Z_\mathrm{n} - 1}\, a_\mathrm{m}, \qquad l_z \approx \frac{Z_\mathrm{n}^{-1} a_\mathrm{B}}{\ln[\sqrt{\gamma}/(Z_\mathrm{n}\sqrt{2Z_\mathrm{n} - 1})]}. \tag{4}$$

In this case, the binding energy of the ground state increases with increasing $\boldsymbol{B}$ asymptotically as $E^{(0)} \sim -Z_\mathrm{n}\hbar^2/(m_\mathrm{e} l_z^2)$ (Kadomtsev and Kudryavtsev 1971).

Thomas-Fermi results are useful as an order-of-magnitude estimate. More accurate calculations of binding energies and oscillator strengths of many-electron atoms were performed with different methods. For atoms with small nuclear charge numbers Z_n, such as helium, a sufficiently accurate and practical method is the Hartree-Fock method with trial orbitals in the adiabatic approximation (Ruder et al. 1994; Miller and Neuhauser 1991; Medin et al. 2008). But the condition of applicability of the adiabatic approximation $\gamma \gg Z_\mathrm{n}^3$ is too restrictive for larger Z_n. It is overcome in the mesh Hartree-Fock method, where each orbital is numerically determined as a function of the longitudinal (z) and radial coordinates (e.g.,

Ivanov and Schmelcher 2000, and references therein). This method, however, is computationally expensive. Mori and Hailey (2002) proposed a "hybrid" method, where corrections to the adiabatic Hartree approximation due to electron exchange and admixture of higher Landau levels are treated as perturbations. The latter method proved to be practical for modeling NS atmospheres containing atoms and ions of elements with $2 < Z_{\rm n} \lesssim 10$, because it can provide an acceptable accuracy at moderate computational expenses.

4.2.3 Molecules and Molecular Ions

Best studied molecules and molecular ions are diatomic systems, especially the H_2^+ ion (Kappes and Schmelcher 1996, and references therein) and the H_2 molecule (Schmelcher et al. 2001, and references therein). Lai (2001) obtained approximate expressions for binding energies of low-lying levels of the H_2 molecule at $\gamma \gtrsim 10^3$. These energies increase approximately at the same rate $\propto (\ln\gamma)^2$ as the binding energies of tightly-bound states of the atom. In such strong fields, the ground state of the H_2 molecule is the state where the spins of both electrons are counter-aligned to $\boldsymbol{B}$ and the molecular axis is parallel to $\boldsymbol{B}$, unlike the weak fields where the ground state is ${}^1\Sigma_g$.

In moderately strong fields (with $\gamma \sim 1$–10), the behavior of the molecular terms is complicated. If the H_2 molecule is oriented along $\boldsymbol{B}$, then its states ${}^1\Sigma_g$ and ${}^3\Pi_u$ are metastable at $0.18 < \gamma < 12.3$ and decay into ${}^3\Sigma_u$, which is unbound (Detmer et al. 1998). It turns out, however, that the orientation along $\boldsymbol{B}$ is not optimal in this case. For example, the lowest energy is provided by the orientation of the molecule in the triplet state at 90° to $\boldsymbol{B}$ at $\gamma = 1$, and 37° at $\gamma = 10$ (Kubo 2007).

Strong magnetic fields stabilize the He_2 molecule and its ions He_2^+, He_2^{2+}, and He_2^{3+}, which do not exist in the absence of the field. Mori and Heyl (2007) have performed the most complete study of their binding energies in NS atmospheres. The ions HeH^{++}, H_3^{++}, and other exotic molecular ions, which become stable in strong magnetic fields, were also considered (see Turbiner 2007, Turbiner et al. 2010, and references therein). An evaluation of the ionization equilibrium shows that, at densities, temperatures, and magnetic fields characteristic of NSs, the abundance of such ions (as well as that of H_2^+ ions considered by Khersonskii 1987) is too small to affect the thermal spectrum.

There are very few results on molecules composed of atoms heavier than He. In particular, Medin and Lai (2006a) applied the density-functional method to calculations of binding energies of various molecules from H_n to Fe_n with n from 1 through 8 at B from 10^{12} G to 2×10^{15} G. The earlier studies of heavy molecules in strong magnetic fields are discussed in the review by Lai (2001).

4.3 Atmosphere Thermodynamics

According to the Bohr-van Leeuwen theorem, the magnetic field does not affect thermodynamics of classical charged particles. The situation differs in quantum mechanics. The importance of the quantum effects depends on the parameters $\zeta_{\rm e}$ and $\zeta_{\rm i}$ (Eqs. (2), (3)).

Studies of thermodynamics of magnetic NS atmospheres, as a rule, are based on the decomposition of the Helmholtz free energy

$$F = F_{\rm id}^{(\rm e)} + F_{\rm id}^{(\rm i)} + F_{\rm int} + F_{\rm ex}, \tag{5}$$

where $F_{\rm id}^{(\rm e)}$ and $F_{\rm id}^{(\rm i)}$ describe the ideal electron and ion gases, $F_{\rm int}$ includes internal degrees of freedom for bound states, and $F_{\rm ex}$ is a nonideal component due to interactions between plasma particles. All the necessary thermodynamic functions are then expressed through derivatives of F over ρ and T.

4.3.1 Equation of State

The free energy of N_i nondegenerate nonrelativistic ions is given by

$$\frac{F_\mathrm{id}^{(\mathrm{i})}}{N_\mathrm{i}T} = \ln\left(2\pi\frac{n_\mathrm{i}\lambda_\mathrm{i}a_\mathrm{m}^2}{Z}\right) + \ln\left(1-\mathrm{e}^{-\zeta_\mathrm{i}}\right) - 1 + \frac{\zeta_\mathrm{i}}{2} + \ln\left(\frac{\sinh[g_\mathrm{i}\zeta_\mathrm{i}(2s_\mathrm{i}+1)/4]}{\sinh(g_\mathrm{i}\zeta_\mathrm{i}/4)}\right), \tag{6}$$

where $\lambda_\mathrm{i} = [2\pi\hbar^2/(m_\mathrm{i}T)]^{1/2}$ is the thermal de Broglie wavelength for the ions, s_i is the spin number, and g_i is the spin-related g-factor (for instance, $s_\mathrm{i} = 1/2$ and $g_\mathrm{i} = 5.5857$ for the proton). All the terms in (6) have clear physical meanings. At $\zeta_\mathrm{i} \to 0$, the first and second terms give together $\ln(n_\mathrm{i}\lambda_\mathrm{i}^3)$, which corresponds to the three-dimensional Boltzmann gas. The first term corresponds to the one-dimensional Boltzmann gas model at $\zeta_\mathrm{i} \gg 1$. The second-last term in (6) gives the energy $\hbar\omega_\mathrm{ci}/2$ of zero-point oscillations of every ion transverse to the magnetic field. Finally, the last term represents the energy of magnetic moments in a magnetic field.

The ideal electron-gas part of the free energy $F_\mathrm{id}^{(\mathrm{e})}$ can be expressed through the Fermi-Dirac integrals (see Potekhin and Chabrier 2013, for explicit expressions). In a strongly quantizing magnetic field, the electron Fermi momentum equals $p_\mathrm{F} = 2\pi^2 a_\mathrm{m}^2\hbar n_\mathrm{e}$, where $n_\mathrm{e} = n_\mathrm{i}Z$ is the electron number density. Therefore, with increasing n_e at a fixed B, the degenerate electrons begin to fill the first Landau level when n_e reaches $n_B = (\pi^2\sqrt{2}a_\mathrm{m}^3)^{-1}$. This value just corresponds to the density ρ_B in Eq. (2). The ratio of the Fermi momentum p_F in the strongly quantizing field to its nonmagnetic value $\hbar(3\pi^2 n_\mathrm{e})^{1/3}$ equals $[4\rho^2/(3\rho_B^2)]^{1/3}$. Therefore, the Fermi energy at a given density $\rho < \sqrt{3/4}\rho_B$ becomes smaller with increasing B, which means that a strongly quantizing magnetic field relieves the electron-gas degeneracy. For this reason, strongly magnetized NS atmospheres remain mostly nondegenerate, despite their densities are orders of magnitude higher than the nonmagnetic atmosphere densities. For the nondegenerate electron gas, $F_\mathrm{id}^{(\mathrm{e})}$ takes the form of Eq. (6) (with the obvious replacements of n_i, λ_i, ζ_i, g_i, s_i, and Z by n_e, λ_e, ζ_e, 2, $\frac{1}{2}$, and 1, respectively).

The nonideal free-energy part F_ex contains the Coulomb and exchange contributions of the electrons and the ions, and the electron-ion polarization energy, and also interactions of ions and electrons with atoms and molecules. In turn, the interaction between the ions is described differently depending on the phase state of matter. The terms that constitute F_ex depend on magnetic field only if it quantizes the motion of these interacting particles. Here we will not discuss these terms but address an interested reader to Potekhin and Chabrier (2013) and references therein. This nonideality is usually negligible in the NS atmospheres, but it determines the formation of a condensed surface, which will be considered in Sect. 6.

4.4 Ionization Equilibrium

For atmosphere simulations, it is necessary to determine the fractions of different bound states, because they affect the spectral features that are caused by bound-bound and bound-free transitions. The solution to this problem is laborious and ambiguous. The principal difficulty in the chemical model of plasmas, namely the necessity to distinguish the bound and free electrons and "attribute" the bound electrons to certain nuclei, becomes especially acute at high densities, where the atomic sizes cannot be anymore neglected with respect to their distances. Current approaches to the solution of this problem are based, as a rule, on the concept of the so-called occupation probabilities of quantum states.

Let us consider a quantum state κ of an ion lacking j electrons, with binding energy $E_{j,\kappa}$ and quantum statistical weight $g_{j,\kappa}$. An occupation probability $w_{j,\kappa}$ is an additional statistical weight of this quantum state, caused by interactions with surrounding plasma (in general, this weight is not necessarily less than unity, therefore it is not quite a probability). As first noted by Fermi (1924), occupation probabilities $w_{j,\kappa}$ cannot be arbitrary but should be consistent with F_{ex}. Minimizing F with account of the Landau quantization leads to a system of ionization-equilibrium equations for $n_j \equiv \sum_\kappa n_{j,\kappa}$ (e.g., Rajagopal et al. 1997)

$$\frac{n_j}{n_{j+1}} = \frac{n_\mathrm{e}\lambda_\mathrm{e}^3}{2} \frac{\sinh\zeta_j}{\zeta_j} \frac{\zeta_{j+1}}{\sinh\zeta_{j+1}} \frac{\tanh\zeta_\mathrm{e}}{\zeta_\mathrm{e}} \frac{\mathcal{Z}_{\text{int},j}}{\mathcal{Z}_{\text{int},j+1}} \exp\left(\frac{E_{j,\text{ion}}}{T}\right), \tag{7}$$

where $\mathcal{Z}_{\text{int},j} = \sum_\kappa g_{j,\kappa} w_{j,\kappa} \exp[(E_{j,\kappa} - E_{j,\text{gr.st}})/(T)]$ is internal partition function for the jth ion type, $E_{j,\text{gr.st}}$ is its ground-state binding energy, $E_{j,\text{ion}} = E_{j,\text{gr.st}} - E_{j+1,\text{gr.st}}$ is its ionization energy, and ζ_j is the magnetic quantization parameter (3). Equation (7) differs from the usual Saha equation, first, by the terms with ζ_e and ζ_j, representing partition functions for distributions of free electrons and ions over the Landau levels, and second, by the occupation probabilities $w_{j,\kappa}$ in the expressions for the partition functions $\mathcal{Z}_{\text{int},j}$. Here, $w_{j,\kappa}$ are the thermodynamic occupation probabilities, which determine the complete destruction of an atom with increasing pressure. They should not be confused with the optical occupation probabilities, which determine dissolution of spectral lines because of the Stark shifts in plasma microfields (see Potekhin 1996 for discussion and references).

Equation (7) was applied to modeling partially ionized atmospheres of NSs, composed of iron, oxygen, and neon (Rajagopal et al. 1997; Mori and Hailey 2006; Mori and Ho 2007). The effects related to the finite nuclear masses (the Appendix) were either ignored or treated as a small perturbation. A more accurate treatment, which rigorously takes these effects into account, is outlined in Appendix A.2.

5 Formation of Spectra in Strongly Magnetized Atmospheres

5.1 Radiative Transfer in Normal Modes

Propagation of electromagnetic waves in magnetized plasmas was studied in many works, the book by Ginzburg (1970) being the most complete of them. At radiation frequency ω much larger than the electron plasma frequency $\omega_{\text{pe}} = (4\pi e^2 n_\mathrm{e}/m_\mathrm{e}^*)^{1/2}$, where $m_\mathrm{e}^* \equiv m_\mathrm{e}\sqrt{1 + p_\mathrm{F}^2/(m_\mathrm{e}c)^2}$, the waves propagate in the form of two polarization modes, extraordinary (hereafter denoted by subscript $j = 1$) and ordinary ($j = 2$). They have different polarization vectors $\boldsymbol{e}_j$ and different absorption and scattering coefficients, which depend on the angle θ_B between $\boldsymbol{B}$ and the wave vector $\boldsymbol{k}$. The modes interact with each another through scattering. Gnedin and Pavlov (1974) formulated the radiative transfer problem in terms of these modes. They showed that in strongly magnetized NS atmospheres a strong Faraday depolarization occurs, except for narrow frequency ranges near resonances. Therefore, it is sufficient to consider specific intensities of the two normal modes instead of the four components of the Stokes vector. The radiative transfer equation for these specific intensities is (Kaminker et al. 1982)

$$\cos\theta_k \frac{\mathrm{d}I_{\omega,j}(\hat{\mathbf{k}})}{\mathrm{d}y_{\text{col}}} = \varkappa_{\omega,j}(\hat{\mathbf{k}}) I_{\omega,j}(\hat{\mathbf{k}}) - \varkappa_{\omega,j}^{\mathrm{a}}(\hat{\mathbf{k}}) \frac{\mathcal{B}_{\omega,T}}{2} - \sum_{j'=1}^{2} \int_{(4\pi)} \varkappa_{\omega,j'j}^{\mathrm{s}}(\hat{\mathbf{k}}', \hat{\mathbf{k}}) I_{\omega,j'}(\hat{\mathbf{k}}')\mathrm{d}\hat{\mathbf{k}}', \tag{8}$$

where $\hat{\mathbf{k}} = \boldsymbol{k}/|k|$ is the unit vector along the wave vector $\boldsymbol{k}$, $y_{\text{col}} = \int_r^\infty (1 + z_g)\rho(r)\mathrm{d}r$ is the column density, and the factor $(1 + z_g)$ is the relativistic scale change in the gravitational field, z_g being the surface redshift (see Sect. 7). Here, $I_{\omega,j}$ denotes the specific intensity of the polarization mode j per unit circular frequency (if I_ν is the specific intensity per unit frequency, then $I_\omega = I_\nu/(2\pi)$; see Zheleznyakov 1996), $\mathcal{B}_{\omega,T} = (\hbar\omega^3/4\pi^3c^2)(e^{\hbar\omega/T} - 1)^{-1}$ is the specific intensity of nonpolarized blackbody radiation, and

$$\varkappa_{\omega,j}(\hat{\mathbf{k}}) \equiv \varkappa^{\mathrm{a}}_{\omega,j}(\hat{\mathbf{k}}) + \sum_{j'=1}^{2}\int_{(4\pi)} \varkappa^{\mathrm{s}}_{\omega,j'j}(\hat{\mathbf{k}}', \hat{\mathbf{k}})\mathrm{d}\hat{\mathbf{k}}', \tag{9}$$

The dependence of the absorption and scattering opacities $\varkappa^{\mathrm{a}}$, $\varkappa^{\mathrm{s}}$ on ray directions $(\hat{\mathbf{k}}, \hat{\mathbf{k}}')$ is affected by $\boldsymbol{B}$. Therefore, the emission of a magnetized atmosphere, unlike the nonmagnetic case, depends not only on the angle θ_k between the ray and the normal direction to the stellar surface, θ_{n}, but also on the angle θ_{n} between $\boldsymbol{B}$ and the normal, and the angle φ_k between the projections of $\boldsymbol{B}$ and $\boldsymbol{k}$ onto the surface.

In the plane-parallel limit, and assuming that the magnetic field is constant in the thin photospheric layer, the equations for hydrostatic and energy balance are the same as in the absence of magnetic field (see, e.g., Suleimanov et al. 2012b):

$$\frac{\mathrm{d}P}{\mathrm{d}y_{\mathrm{col}}} = g - g_{\mathrm{rad}}, \qquad g_{\mathrm{rad}} \approx \frac{2\pi}{c}\int_0^\infty \mathrm{d}\omega\varkappa_\omega \int_0^\pi \cos\theta_k I_\omega(\hat{\mathbf{k}}) \sin\theta_k \mathrm{d}\theta_k, \tag{10}$$

$$\int_0^\infty \mathrm{d}\omega \int_{(4\pi)} I_\omega(\hat{\mathbf{k}}) \cos\theta_k \mathrm{d}\hat{\mathbf{k}} = F_{\mathrm{ph}}, \tag{11}$$

where $I_\omega = \sum_{j=1}^{2} I_{\omega,j}$ is the total specific intensity.

The diffusion equation for the normal modes was derived by Kaminker et al. (1982). For the plane-parallel atmosphere it reads (Zavlin 2009)

$$\frac{\mathrm{d}}{\mathrm{d}y_{\mathrm{col}}} D_{\omega,j} \frac{\mathrm{d}}{\mathrm{d}y_{\mathrm{col}}} J_{\omega,j} = \bar{\varkappa}^{\mathrm{a}}_{\omega,j}\left[J_{\omega,j} - \frac{\mathcal{B}_{\omega,T}}{2}\right] + \bar{\varkappa}^{\mathrm{s}}_{\omega,12}[J_{\omega,j} - J_{\omega,3-j}]. \tag{12}$$

Here,

$$\bar{\varkappa}^{\mathrm{a}}_{\omega,j} = \frac{1}{4\pi}\int_{(4\pi)} \varkappa^{\mathrm{a}}_{\omega,12}\mathrm{d}\hat{\mathbf{k}}, \qquad \bar{\varkappa}^{\mathrm{s}}_{\omega,j} = \frac{1}{4\pi}\int_{(4\pi)} \mathrm{d}\hat{\mathbf{k}}' \int_{(4\pi)} \mathrm{d}\hat{\mathbf{k}}\varkappa^{\mathrm{s}}_{\omega,12}(\hat{\mathbf{k}}', \hat{\mathbf{k}}), \tag{13}$$

$$J_{\omega,j} = \frac{1}{4\pi}\int_{(4\pi)} I_{\omega,j}(\hat{\mathbf{k}})\mathrm{d}\hat{\mathbf{k}}, \qquad D_{\omega,j} = \frac{1}{3\varkappa^{\mathrm{eff}}_{\omega,j}} = \frac{\cos^2\theta_{\mathrm{n}}}{3\varkappa^{\parallel}_{\omega,j}} + \frac{\sin^2\theta_{\mathrm{n}}}{3\varkappa^{\perp}_{\omega,j}}, \tag{14}$$

$$\left\{\begin{matrix} (\varkappa_j^{\parallel})^{-1} \\ (\varkappa_j^{\perp})^{-1} \end{matrix}\right\} = \frac{3}{4}\int_0^\pi \left\{\begin{matrix} 2\cos^2\theta_B \\ \sin^2\theta_B \end{matrix}\right\} \frac{\sin\theta_B \mathrm{d}\theta_B}{\varkappa_j(\theta_B)}. \tag{15}$$

The effective opacity for nonpolarized radiation is $\varkappa^{\mathrm{eff}} = 2/(3D_{\omega,1} + 3D_{\omega,2})$. The diffusion approximation (12) serves as a starting point in an iterative method (Shibanov and Zavlin 1995), which allows one to solve the system (8) more accurately.

5.2 Plasma Polarizability

In Cartesian coordinates with the z-axis along $\boldsymbol{B}$, the plasma dielectric tensor is (Ginzburg 1970)

$$\boldsymbol{\varepsilon} = \mathbf{I} + 4\pi\chi = \begin{pmatrix} \varepsilon_\perp & \mathrm{i}\varepsilon_\wedge & 0 \\ -\mathrm{i}\varepsilon_\wedge & \varepsilon_\perp & 0 \\ 0 & 0 & \varepsilon_\parallel \end{pmatrix}, \tag{16}$$

where $\mathbf{I}$ is the unit tensor, $\boldsymbol{\chi} = \boldsymbol{\chi}^{\mathrm{H}} + \mathrm{i}\boldsymbol{\chi}^{\mathrm{A}}$ is the complex polarizability tensor of plasma, $\boldsymbol{\chi}^{\mathrm{H}}$ and $\boldsymbol{\chi}^{\mathrm{A}}$ are its Hermitian and anti-Hermitian parts, respectively. This tensor becomes diagonal in the cyclic (or rotating) coordinates with unit vectors $\hat{\mathbf{e}}_{\pm 1} = (\hat{\mathbf{e}}_x \pm \mathrm{i}\hat{\mathbf{e}}_y)/\sqrt{2}$, $\hat{\mathbf{e}}_0 = \hat{\mathbf{e}}_z$. Under the assumption that the electrons and ions lose their regular velocity, acquired in an electromagnetic wave, by collisions with an effective frequency ν_{eff} independent of the velocities, then the cyclic components of the polarizability tensor are (Ginzburg 1970, Sect. 10)

$$\chi_\alpha = -\frac{1}{4\pi} \frac{\omega_{\mathrm{pe}}^2}{(\omega + \alpha\omega_{\mathrm{c}})(\omega - \alpha\omega_{\mathrm{ci}}) + \mathrm{i}\omega\nu_{\mathrm{eff}}} \tag{17}$$

($\alpha = 0, \pm 1$). A more rigorous kinetic theory leads to results which cannot be described by Eq. (17) with the same frequency ν_{eff} for the Hermitian and anti-Hermitian components χ_α^{H} and χ_α^{A} (Ginzburg 1970, Sect. 6).

The anti-Hermitian part of the polarizability tensor determines the opacities: $\varkappa_\alpha(\omega) = 4\pi\omega\chi_\alpha^{\mathrm{A}}(\omega)/(\rho c)$. Then the Kramers-Kronig relation gives (Bulik and Pavlov 1996; Potekhin et al. 2004)

$$\chi_\alpha^{\mathrm{H}}(\omega) = \frac{c\rho}{4\pi^2\omega} \left\{ \int_0^\omega \left[\varkappa_\alpha(\omega + \omega') - \varkappa_\alpha(\omega - \omega')\right] \frac{\mathrm{d}\omega'}{\omega'} \right.$$
$$\left. + \int_{2\omega}^\infty \frac{\varkappa_\alpha(\omega')}{\omega' - \omega} \mathrm{d}\omega' - \int_0^\infty \frac{\varkappa_{-\alpha}(\omega')}{\omega' + \omega} \mathrm{d}\omega' \right\}. \tag{18}$$

Thus we can calculate the polarizability tensor $\boldsymbol{\chi}$ from the opacities $\varkappa_\alpha(\omega)$.

5.3 Vacuum Polarization

In certain ranges of density ρ and frequency ω, normal-mode properties are dramatically affected by a specific QED effect called vacuum polarization. The influence of the vacuum polarization on the NS emission was studied in detail by Pavlov and Gnedin (1984). If the vacuum polarization is weak, then it can be linearly added to the plasma polarization. Then the complex dielectric tensor can be written as $\boldsymbol{\varepsilon}' = \mathbf{I} + 4\pi\chi + 4\pi\chi^{\mathrm{vac}}$, where $\boldsymbol{\chi}^{\mathrm{vac}} = (4\pi)^{-1}\,\mathrm{diag}(\bar{a}, \bar{a}, \bar{a} + \bar{q})$ is the vacuum polarizability tensor, and diag(...) denotes the diagonal matrix. The magnetic susceptibility of vacuum is determined by expression $\boldsymbol{\mu}^{-1} = \mathbf{I} + \mathrm{diag}(\bar{a}, \bar{a}, \bar{a} + \bar{m})$. Adler (1971) obtained the vacuum polarizability coefficients $\bar{a}$, $\bar{q}$, and $\bar{m}$ that enter these equations in an explicit form at $b \ll 1$, Heyl and Hernquist (1997) expressed them in terms of special functions in the limits of $b \ll 1$ and $b \gg 1$. Kohri and Yamada (2002) presented their numerical calculations. Finally, Potekhin et al. (2004) described them by simple analytic expressions.

$$\bar{a} = -\frac{2\alpha_{\mathrm{f}}}{9\pi} \ln\left(1 + \frac{b^2}{5} \frac{1 + 0.25487 b^{3/4}}{1 + 0.75 b^{5/4}}\right), \qquad \bar{q} = \frac{7\alpha_{\mathrm{f}}}{45\pi} b^2 \frac{1 + 1.2b}{1 + 1.33b + 0.56b^2},$$
$$\bar{m} = -\frac{\alpha_{\mathrm{f}}}{3\pi} \frac{b^2}{3.75 + 2.7 b^{5/4} + b^2}. \tag{19}$$

The coefficients $\bar{a}$, $\bar{q}$, and $\bar{m}$ are not small at $B \gtrsim 10^{16}$ G. In this case, the vacuum refraction coefficients substantially differ from unity, and the vacuum that surrounds a NS acts as a birefringent lens, which distorts and additionally polarizes thermal radiation (Heyl and Shaviv 2002; van Adelsberg and Perna 2009). At weaker B, the vacuum polarization results in a resonance, which manifests in a mutual conversion of normal modes, which will be considered in Sect. 5.6.

5.4 Polarization Vectors of the Normal Modes

Shafranov (1967) obtained the polarization vectors $\boldsymbol{e}_j$ for fully ionized plasmas. Ho and Lai (2003) presented their convenient expressions in terms of the coefficients $\varepsilon_\perp$, $\varepsilon_\|$, $\varepsilon_\wedge$, $\bar{a}$, $\bar{q}$, and $\bar{m}$, including the contributions of electrons, ions, and vacuum polarization. In the Cartesian coordinate system (xyz) with the z-axis along the wave vector $\boldsymbol{k}$ and with $\boldsymbol{B}$ in the plane x–z, one has

$$\boldsymbol{e}_j = \begin{pmatrix} e_{j,x} \\ e_{j,y} \\ e_{j,z} \end{pmatrix} = \frac{1}{\sqrt{1+K_j^2+K_{z,j}^2}} \begin{pmatrix} \mathrm{i}K_j \\ 1 \\ \mathrm{i}K_{z,j} \end{pmatrix}, \tag{20}$$

where

$$K_j = \beta\left\{1+(-1)^j\left[1+\frac{1}{\beta^2}+\frac{\bar{m}}{1+\bar{a}}\frac{\sin^2\theta_B}{\beta^2}\right]^{1/2}\right\}, \tag{21}$$

$$K_{z,j} = -\frac{(\varepsilon'_\perp-\varepsilon'_\|)K_j\cos\theta_B+\varepsilon_\wedge}{\varepsilon'_\perp\sin^2\theta_B+\varepsilon'_\|\cos^2\theta_B}\sin\theta_B, \tag{22}$$

$$\beta = \frac{\varepsilon'_\|-\varepsilon'_\perp+\varepsilon_\wedge^2/\varepsilon'_\perp+\varepsilon'_\|\bar{m}/(1+\bar{a})}{2\varepsilon_\wedge}\frac{\varepsilon'_\perp}{\varepsilon'_\|}\frac{\sin^2\theta_B}{\cos\theta_B}, \tag{23}$$

$\varepsilon'_\perp = \varepsilon_\perp + \bar{a}$, and $\varepsilon'_\| = \varepsilon_\| + \bar{a} + \bar{q}$.

5.5 Opacities

In the approximation of isotropic scattering, at a given frequency ω, the opacities can be presented in the form (e.g., Kaminker et al. 1982)

$$\varkappa_j^{\mathrm{a}} = \sum_{\alpha=-1}^{1} |e_{j,\alpha}(\theta_B)|^2\frac{\sigma_\alpha^{\mathrm{a}}}{m_\mathrm{i}}, \qquad \varkappa_{jj'}^{\mathrm{s}} = \frac{3}{4}\sum_{\alpha=-1}^{1}|e_{j,\alpha}(\theta_B)|^2\frac{\sigma_\alpha^{\mathrm{s}}}{m_\mathrm{i}}\int_0^\pi \left|e_{j',\alpha}(\theta'_B)\right|^2\sin\theta'_B\,\mathrm{d}\theta'_B, \tag{24}$$

where $\sigma_\alpha^{\mathrm{a,s}}$ are the absorption and scattering cross sections for the three basic polarizations. These include contributions of photon interaction with free electrons or ions (free-free transitions) as well as with bound states of atoms and ions (bound-bound and bound-free transitions).

Figure 2 shows basic opacities ($\sigma_\alpha^{\mathrm{a}}/m_\mathrm{i}$ in Eq. (24)) at $B = 3\times10^{13}$ G, $\rho = 1\ \mathrm{g\,cm^{-3}}$ and $T = 3.16\times10^5$ K (the left panel) and corresponding normal-mode absorption opacities $\varkappa_j^{\mathrm{a}}$ for $\theta_B = 10^\circ$. One can clearly distinguish the features reflecting the peaks at the ion cyclotron frequency and the resonant atomic frequencies, and the line crossings related to the behavior of the plasma polarizability as function of frequency. For comparison, we show also opacities for the fully ionized plasma model under the same conditions. They miss the features related to the atomic resonances, and their values is underestimated by orders of magnitude in a wide frequency range. In the remaining of this subsection we extend our discussion about the form of the different contributions to the opacity.

5.5.1 Scattering

Scattering cross-sections in NS atmospheres are well known (Ventura 1979; Kaminker et al. 1982; Mészáros 1992). For $\alpha = -1$, the photon-electron scattering has a resonance at ω_c.

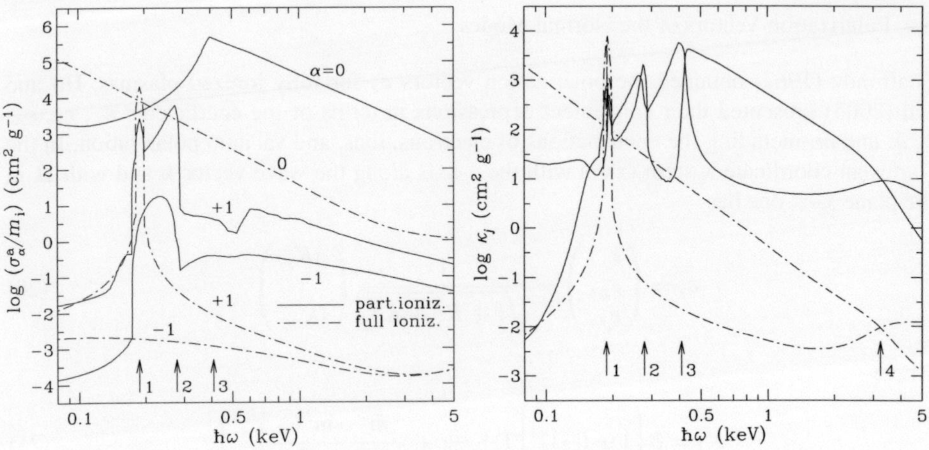

Fig. 2 Logarithm of spectral opacities of a H plasma at $B = 3 \times 10^{13}$ G, $T = 3.16 \times 10^5$ K, $\rho = 1\ \mathrm{g\,cm^{-3}}$. *Solid curves*: partially ionized plasma model; *dot-dashed curves*: fully-ionized plasma model. *Left panel*: basic opacities for $\alpha = 0, \pm 1$. *Right panel*: opacities for two normal modes $j = 1, 2$ propagating at the angle $\theta_B = 10°$ to the field lines; *the lower* (*upper*) *curve* of each type corresponds to the extraordinary (ordinary) wave. *The arrows* indicate the features at resonant frequencies: 1—the ion cyclotron resonance $\omega = \omega_{ci}$; 2—energy threshold for a transition between the lowest two levels $\hbar\omega = |E^{(0)}_{0,0} - E^{(0)}_{1,0}|$; 3—the ground-state binding energy $\hbar\omega = |E^{(0)}_{0,0}|$; 4 (at *the right panel*)—the vacuum resonance

Outside a narrow (about the Doppler width) frequency interval around ω_c, the cross sections for the basic polarizations $\alpha = 0, \pm 1$ are

$$\sigma^{s,e}_\alpha = \frac{\omega^2}{(\omega + \alpha\omega_c)^2 + \nu^2_{e,\alpha}} \sigma_T, \tag{25}$$

where $\sigma_T = (8\pi/3)(e^2/m_e c^2)^2$ is the nonmagnetic Thomson cross section, and $\nu_{e,\alpha}$ are effective damping factors (see below).

The photon-ion scattering cross section looks analogously,

$$\sigma^{s,i}_\alpha = \left(\frac{m_e}{m_i}\right)^2 \frac{\omega^2 Z^4}{(\omega - \alpha\omega_{ci})^2 + \nu^2_{i,\alpha}} \sigma_T. \tag{26}$$

Unlike the nonmagnetic case, in superstrong fields one cannot neglect the scattering on ions, because of the resonance at ω_{ci}. In each case, the damping factor $\nu_{e,\alpha}$ or $\nu_{i,\alpha}$ is equal to half of the total rate of spontaneous and collisional decay of the state with energy $\hbar\omega$ (see discussion in Potekhin and Lai 2007). The spontaneous decay rates are

$$2\nu^s_e = \frac{4}{3}\frac{e^2}{m_e c^3}\omega^2, \qquad 2\nu^s_i = \frac{4}{3}\frac{(Ze)^2}{m_i c^3}\omega^2. \tag{27}$$

As shown by Potekhin and Chabrier (2003) for a proton-electron plasma, the damping factors including the scattering and free-free processes can be approximately written as

$$\nu_{e,\alpha} = \nu^s_e + \nu^{ff}_\alpha(\omega_c), \qquad \nu_{e,\alpha} = \nu^s_e + (m_e/m_i)\nu^{ff}_\alpha(\omega_{ci}), \tag{28}$$

where $\nu^{ff}_\alpha(\omega)$ is the effective free-free frequency given by Eq. (31).

5.5.2 Absorption

Without magnetic field, absorption or emission of a photon by a free electron is impossible without involvement of another particle, which would accept the difference between the values of the total momentum of the electron and the photon before and after the absorption. In a quantizing magnetic field, a photon can be absorbed or emitted by a free electron in a transition between Landau levels. In the nonrelativistic or dipole approximation, such transitions occur between the neighboring levels at the frequency ω_c. In the relativistic theory, the multipole expansion leads to an appearance of cyclotron harmonics (Zheleznyakov 1996). Absorption cross-sections at these harmonics were derived by Pavlov et al. (1980) in the Born approximation without allowance for the magnetic quantization of electron motion, and represented in a compact form by Suleimanov et al. (2012a).

The quantization of electron motion leads to the appearance of cyclotron harmonics in the nonrelativistic theory as well. Pavlov and Panov (1976) derived photon absorption cross-sections for an electron which moves in a magnetic field and interacts with a nonmoving point charge. This model is applicable at $\omega \gg \omega_{ci}$. In the superstrong field of magnetars, the latter condition is unacceptable. A more accurate treatment of absorption of a photon by the system of a finite-mass ion and an electron yields (Potekhin and Chabrier 2003; Potekhin 2010)

$$\sigma_\alpha^{ff}(\omega) = \frac{4\pi e^2}{m_e c} \frac{\omega^2 \nu_\alpha^{ff}(\omega)}{(\omega + \alpha\omega_c)^2(\omega - \alpha\omega_{ci})^2 + \omega^2 \tilde{\nu}_\alpha^2(\omega)}, \tag{29}$$

where ν_α^{ff} is an effective photoabsorption collision frequency, and $\tilde{\nu}_\alpha$ is a damping factor. In the electron-proton plasma, taking into account the scattering and free-free absorption, we have (Potekhin and Chabrier 2003)

$$\tilde{\nu}_\alpha = (1 + \alpha\omega_c/\omega)\nu_{i,\alpha}(\omega) + (1 - \alpha\omega_{ci}/\omega)\nu_{e,\alpha}(\omega) + \nu_\alpha^{ff}(\omega). \tag{30}$$

We see from (29) that σ_{-1}^{ff} and σ_{+1}^{ff} have a resonance at the frequencies ω_c and ω_{ci}, respectively. The effective free-free absorption frequency can be written as

$$\nu_\alpha^{ff}(\omega) = \frac{4}{3}\sqrt{\frac{2\pi}{m_e T}} \frac{n_e e^4}{\hbar\omega} \Lambda_\alpha^{ff}(\omega), \tag{31}$$

where $\Lambda_\alpha^{ff}(\omega)$ is a dimensionless Coulomb logarithm ($\Lambda_\alpha^{ff} = (\pi/\sqrt{3}) g_\alpha^{ff}$, where g_α^{ff} is a Gaunt factor). Without the magnetic field, Λ_α^{ff} is a smooth function of ω. In a quantizing magnetic field, however, it has peaks at the multiples of both the electron and ion cyclotron frequencies for all polarizations α. However, these two types of peaks are different. Unlike the electron cyclotron harmonics, the ion cyclotron harmonics are so weak that they can be safely neglected in the NS atmospheres (see Potekhin 2010).

5.5.3 Bound-Bound and Bound-Free Transitions

The calculation of the cross section of photons with bound states of atoms and ions is very complex. It implies averaging of the cross sections of photon and atom absorption over all values of $K_\perp$. Since the distribution over $K_\perp$ is continuous for the atoms and discrete for the ions, such averaging for atoms reduces to an integration over $K_\perp$, analogous to Eq. (45), whereas for ions it implies summation with an appropriate statistical weight. To date, such calculation has been realized for atoms of hydrogen (Potekhin and Chabrier 2003, 2004) and helium (Mori and Heyl 2007). In the Appendix, we briefly discuss different issues related to bound states and their interaction with photons with the account of atomic thermal motion in strong magnetic fields.

5.6 Spectra of Magnetic Atmospheres

Shibanov et al. (1992) calculated spectra from strongly magnetized NS atmospheres, which was assumed to be fully ionized. Later they created a database of model spectra (Pavlov et al. 1995) and included it in the *XSPEC* package (Arnaud 1996) under the name NSA. They have shown that the spectra of magnetic hydrogen and helium atmospheres are softer than the respective nonmagnetic spectra, but harder than the blackbody spectrum with the same temperature. In addition to the spectral energy distribution, these authors have also studied the polar diagram and polarization of the outgoing emission, which proved to be quite nontrivial because of redistribution of energy between the normal modes. The thermal radiation of a magnetized atmosphere is strongly polarized, and the polarization sharply changes at the cyclotron resonance with increasing frequency. At contrast to the isotropic blackbody radiation, radiation of a magnetic atmosphere consists of a narrow ($< 5°$) pencil beam along the magnetic field and a broad fan beam with typical angles $\sim 20°$–$60°$ (Zavlin et al. 1995; see also van Adelsberg and Lai 2006). These calculations have thus fully confirmed the early analysis by Gnedin and Sunyaev (1974). Later, analogous calculations were performed by other research groups (Zane et al. 2001; Ho and Lai 2003; van Adelsberg and Lai 2006). They paid special attention to manifestations of the ion cyclotron resonance in observed spectra, which was suggested by some SGR/AXP data. Lai and Ho (2002) showed that vacuum polarization leads to a conversion of the normal modes: a photon related to one mode transforms, with certain probability, into a photon of the other mode while crossing a surface with a certain critical density, depending of the photon energy as

$$\rho = 0.00964(A/Z)(\hbar\omega/\mathrm{keV})^2 B_{12}^2/f_B^2\ \mathrm{g\,cm^{-3}}, \tag{32}$$

where $f_B^2 = \alpha_f b^2/[15\pi(\bar{q}+\bar{m})]$, while $\bar{q}$ and $\bar{m}$ are the vacuum-polarization coefficients (Sect. 5.3); $f_B \approx 1$ at $B \lesssim 10^{14}$ G. The energy $\hbar\omega$ in Eq. (32) corresponds to the line crossing in Fig. 2, indicated by arrow labelled '4'. It follows from Eq. (32) that for $B \sim 10^{14}$ G this energy coincides with the ion cyclotron energy at the density where the atmosphere is optically thin for the extraordinary mode, but optically thick for the ordinary mode. Under such conditions, the mode conversion strongly suppresses the ion cyclotron feature in the emission spectrum.

For the first computations of partially ionized atmospheres of NSs with magnetic fields $B \sim 10^{12}$–10^{13} G (Miller 1992; Rajagopal et al. 1997), the properties of atoms in magnetic fields were calculated by the adiabatic Hartree-Fock method (Miller and Neuhauser 1991). The atomic motion was either ignored (Miller 1992) or treated approximately by the perturbation theory (Rajagopal et al. 1997). Later, Potekhin et al. (2004) constructed a strongly magnetized hydrogen atmosphere model beyond the framework of the adiabatic approximation, including partial ionization and effects of the atomic motion. The calculated spectra revealed a narrow absorption line at the proton cyclotron energy and some features related to atomic transitions. As well as in the fully ionized plasma model, the intensity has a maximum at higher energies relative to the maximum of the Planck function, but at lower energies relative to the nonmagnetic H atmosphere model. Therefore, the model of a fully-ionized atmosphere with a strong magnetic field can yield a realistic temperature, but does not reproduce the spectral features caused by atomic transitions.

6 A Condensed Surface as an Alternative to Gaseous Atmospheres

Ruderman (1971) suggested that a strong magnetic field can stabilize polymer chains directed along the field lines, and that the dipole-dipole attraction of these chains may result in a condensed phase. Later works confirmed this conjecture, but the binding and sublimation energies turned out to be smaller than Ruderman assumed (see Medin and Lai 2006b, and references therein).

From the thermodynamics point of view, the magnetic condensation is nothing but the plasma phase transition caused by the strong electrostatic attraction between the ionized plasma particles. This attraction gives a negative contribution to pressure $P_{\rm ex}$, which, at low temperatures, is not counterbalanced until the electrons become degenerate with increasing density. In the absence of a magnetic field, such phase transitions were studied theoretically since 1930s (see Ebeling and Norman 2003, for a review). In this case, the temperature of the outer layers of a NS $T \gtrsim (10^5\text{–}10^6)$ K exceeds the critical temperature $T_{\rm crit}$ for the plasma phase transition. However, we have seen in Sect. 4.3.1 that a quantizing magnetic field lifts electron degeneracy. As a result, $T_{\rm crit}$ increases with increasing B, which may enable such phase transition.

Lai (2001) estimated the condensed-surface density as

$$\rho_{\rm s} \approx 561\eta A Z^{-3/5} B_{12}^{6/5} \ {\rm g\,cm^{-3}}, \tag{33}$$

where η is an unknown factor of the order of unity. In the ion-sphere model (Salpeter 1961), the electrons are replaced by a uniform negative background, and the potential energy per ion is estimated as the electrostatic energy of the ionic interaction with the negative background contained in the sphere of radius $a_{\rm i} = (4\pi n_{\rm i}/3)^{-1/3}$. By equating $|P_{\rm ex}|$ to the pressure of degenerate electrons, one obtains Eq. (33) with $\eta = 1$. This estimate disregards the effects of ion correlations, electron-gas polarizability, and bound state formation. Applying different versions of the Thomas-Fermi method to the treatment of the electron polarization, one gets quite different results: for example, the zero-temperature Thomas-Fermi results for Fe at 10^{10} G $\leqslant B \leqslant 10^{13}$ G (Rögnvaldsson et al. 1993) can be described by Eq. (33) with $\eta \approx 0.2 + 0.01/B_{12}^{0.56}$, and in a finite-temperature Thomas-Fermi model (Thorolfsson et al. 1998) there is no phase transition at all.

Medin and Lai (2006b, 2007) estimated the condensation energy by the density functional method and calculated the equilibrium density of a saturated vapor of the atoms and polymer chains of helium, carbon, and iron above the respective condensed surfaces at $1 \lesssim B_{12} \leqslant 10^3$. By equating this density to $\rho_{\rm s}$, they found $T_{\rm crit}$ at several B values. Unlike previous authors, Medin and Lai (2006b, 2007) have considered the electron band structure in the condensed phase. Their results for $\rho_{\rm s}$ can be described by Eq. (33) with $\eta = 0.517 + 0.24/B_{12}^{1/5} \pm 0.011$ for carbon and $\eta = 0.55 \pm 0.11$ for iron, and the critical temperature can be evaluated as $T_{\rm crit} \sim 5 \times 10^4 Z^{1/4} B_{12}^{3/4}$ K (Potekhin and Chabrier 2013).

When magnetic field increases from 10^{12} G to 10^{15} G, the cohesive energy, calculated by Medin and Lai (2007) for the condensed surface, varies monotonically from 0.07 keV to 5 keV for helium, from 0.05 keV to 20 keV for carbon, and from 0.6 keV to 70 keV for iron. The power-law interpolation between these limits roughly agrees with numerical results. The electron work function changes in the same B range from 100 eV to (600 ± 50) eV. With the calculated energy values, Medin and Lai (2007) determined the conditions of electron and ion emission in the vacuum gap above the polar cap of a pulsar and the conditions of gap formation, and calculated the pulsar "death lines".

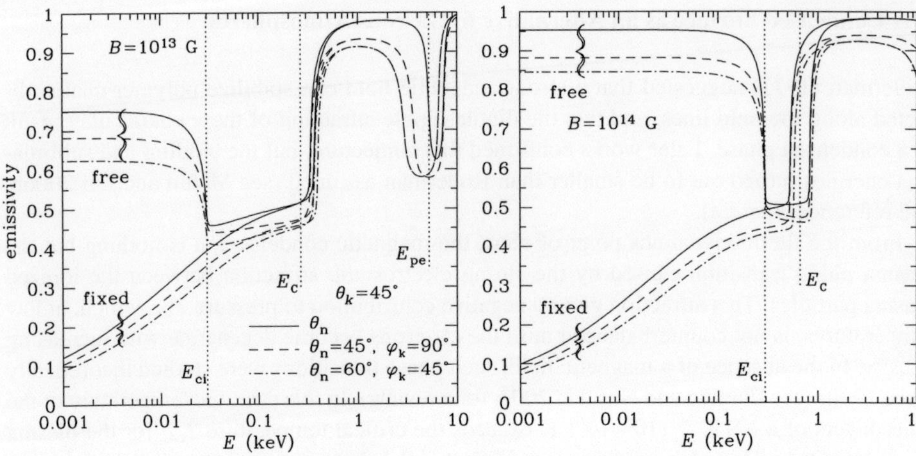

Fig. 3 Dimensionless emissivity of a condensed iron surface at $B = 10^{13}$ G (*left panel*) and 10^{14} G (*right panel*), averaged over polarizations, is shown as a function of energy of a photon emitted at the angle $\theta_k = 45°$, for different magnetic-field inclination angles $\theta_{\rm n}$ and azimuthal angles φ_k. For each parameter set, two curves are obtained in the models of free and fixed ions. Vertical dotted lines mark positions of the characteristic energies: the ion cyclotron energy $E_{\rm ci} = \hbar\omega_{\rm ci}$, the electron plasma energy $E_{\rm pe} = \hbar\omega_{\rm pe}$, and the hybrid energy $E_{\rm C}$. The groups of curves marked "*free*" and "*fixed*" correspond to the models of free and fixed ions, mentioned in the text

6.1 Radiation from a Naked Neutron Star

For NSs with a liquid or solid condensed surface, the formation of thermal spectra depends on its reflectivity, first calculated by Itoh (1975) and Lenzen and Trümper (1978) under simplifying assumptions. Detailed calculations of the reflection properties of a strongly magnetized metallic surface were presented by Brinkmann (1980) and revisited in several more recent papers (Turolla et al. 2004; van Adelsberg et al. 2005; Pérez-Azorín et al. 2005; Potekhin et al. 2012). The authors determined the normal-mode polarization vectors in the medium under the surface, expressed the complex refraction coefficients as functions of the angles θ_k and φ_k that determine the direction of a reflected ray, and expanded the complex electric amplitudes of the incident, reflected, and transmitted waves over the respective basic polarization vectors. The coefficients of these expansions, which are found from Maxwell boundary conditions, determine the surface reflectivity for each incident-wave polarization, $r_{\omega,j}$. Then the total dimensionless emissivity equals $1 - \frac{1}{2}(r_{\omega,1} + r_{\omega,2})$.

The early works assumed that the ions are firmly fixed in the crystalline lattice. More recent works (van Adelsberg et al. 2005; Pérez-Azorín et al. 2005; Potekhin et al. 2012) consider not only this model, but also the opposite limit of free ions. The real reflectivity of the surface lies probably between the limits given by these two extreme models, but this problem is not yet definitely solved.

Figure 3 shows examples of the emissivity, normalized to the blackbody intensity, as a function of the photon energy $E = \hbar\omega$, in both the free- and fixed-ion limits, for the wave-vector inclination angle $\theta_{\rm k} = 45°$, $B = 10^{13}$ G and 10^{14} G, and different values of the magnetic-field inclination $\theta_{\rm n}$ and azimuthal angles φ_k. The characteristic energies $E_{\rm ci} = \hbar\omega_{\rm ci}$, $E_{\rm pe} = \hbar\omega_{\rm pe}$, and $E_{\rm C} = E_{\rm ci} + E_{\rm pe}^2/\hbar\omega_{\rm c}$ are marked. The spectral features near these energies are explained in van Adelsberg et al. (2005). For instance, the emissivity suppression at $E_{\rm ci} \lesssim E \lesssim E_{\rm C}$ is due to the strong damping of one of the two normal modes in the plasma in this energy range. There is a resonant absorption, depending on the directions of the incident wave and the magnetic field, near $E_{\rm pe}$. The local flux density of radiation from a condensed surface is equal to the Planck function $\mathcal{B}_{\omega,T}$ multiplied by the normalized emissivity.

In Fig. 3, the emissivity is averaged over polarizations. But $r_{\omega,1} \neq r_{\omega,2}$, hence the thermal emission of a condensed surface is polarized, the polarization depending in a nontrivial way on the frequency ω and angles $\theta_{\rm n}$, θ_k, and φ_k. For example, the degree of linear polarization can reach tens percent near the frequencies $\omega_{\rm ci}$ and $\omega_{\rm pe}$, which makes promising the polarization diagnostics of NSs with condensed surfaces. The intensity and the polarization degree can be evaluated using analytical approximations for a condensed iron surface at $B = 10^{12}$–10^{14} G (Potekhin et al. 2012).

6.2 Thin and Layered Atmospheres

Motch et al. (2003) suggested that some NSs can possess a hydrogen atmosphere of a finite thickness above the solid iron surface. If the optical depth of such atmosphere is small for some wavelengths and large for other ones, this should lead to a peculiar spectrum, different from the spectra of thick atmospheres. Ho et al. (2007) and Suleimanov et al. (2009, 2010b) calculated such spectra using simplified boundary conditions for the radiative transfer equation at the inner boundary of the atmosphere. More accurate boundary conditions (Potekhin et al. 2012) take into account that an extraordinary or ordinary wave, falling from the atmosphere on the surface, gives rise to reflected waves of both normal polarizations, whose intensities add to the respective intensities of the waves emitted by the condensed surface.

In general, local spectra of radiation emitted by thin hydrogen atmospheres over a condensed surface reveal a narrow absorption line corresponding to the proton cyclotron resonance in the atmosphere, features related to atomic transitions broadened by motion effects (the Appendix), and a kink corresponding to the ion cyclotron energy of the substrate ions. Some of these features may be absent, depending on the atmosphere thickness and magnetic field strength. At high energies, the spectrum is determined by the condensed-surface emission, which is softer than the spectrum of the thick hydrogen atmosphere.

The origin of the thin H atmospheres remains hazy. Ho et al. (2007) discussed three possible scenarios. First, it is the accretion from the interstellar medium. But its rate should be very low, in order to accumulate the hydrogen mass $4\pi R^2 y_{\rm col} \sim 10^{-20} M_\odot$ in $\sim 10^6$ years. Another scenario assumes diffusive nuclear burning of a hydrogen layer accreted soon after the formation of the NS (Chiu and Salpeter 1964; Chang and Bildsten 2003). But this process is too fast at the early cooling epoch, when the star is relatively hot, and would have rapidly consumed all the hydrogen on the surface (Chiu and Salpeter 1964; Chang and Bildsten 2004). The third possibility is a self-regulating mechanism, driven by nuclear spallation in collisions with ultrarelativistic particles at the regions of open field lines (Jones 1978). An estimate of the penetration depth of the magnetospheric accelerated particles indicates that this process could create a hydrogen layer of the necessary thickness $y_{\rm col} \sim 1$ g cm^{-2} (Ho et al. 2007).

It is natural to consider also an atmosphere having a helium layer beneath the hydrogen layer. Indeed, all three scenarios assume that a hydrogen-helium mixture appears originally at the surface, and the strong gravity quickly separates these two elements. Such "sandwich atmosphere" was considered by Suleimanov et al. (2009), where the authors showed that its spectrum can have two or three absorption lines in the range $E \sim (0.2$–$1)$ keV at $B \sim 10^{14}$ G.

7 Modeling Observed Spectra

In order to apply the local spectra models described in Sects. 5 and 6 to observations, one has to calculate a synthetic spectrum, which would be observed at a large distance D from the star. Such calculation should include the effects of General Relativity, which are significant.

The photon frequency, which equals ω in the local inertial reference frame, undergoes a redshift to a smaller frequency ω_∞ in the remote observer's reference frame. Therefore a thermal spectrum with effective temperature $T_{\rm eff}$, measured by the remote observer, corresponds to a lower effective temperature

$$T_{\rm eff}^\infty = T_{\rm eff}/(1+z_g), \tag{34}$$

where $z_g \equiv \omega/\omega_\infty - 1 = (1-x_{\rm g})^{-1/2} - 1$ is the redshift parameter and the compactness parameter $x_g = 2GM/c^2R$ of a typical NS lies between $1/5$ and $1/2$. Here and hereafter the symbol ∞ indicates that the quantity is measured at a large distance from the star and can differ from its value near the surface.

Along with the radius R that is determined by the equatorial length $2\pi R$ in the local reference frame, one often considers an *apparent radius* for a remote observer,

$$R_\infty = R(1+z_g). \tag{35}$$

With decreasing R, z_g increases so that the apparent radius has a minimum, $\min R_\infty \approx$ 12–14 km (Haensel et al. 2007, Chap. 6).

The apparent photon luminosity $L_{\rm ph}^\infty$ and the luminosity in the stellar reference frame $L_{\rm ph}$ are determined by the Stefan-Boltzmann law

$$L_{\rm ph}^\infty = 4\pi\sigma_{\rm SB}R_\infty^2\left(T_{\rm eff}^\infty\right)^4, \qquad L_{\rm ph} = 4\pi\sigma_{\rm SB}R^2T_{\rm eff}^4 \tag{36}$$

with $\sigma_{\rm SB} = \pi^2/(60\hbar^3c^2)$ and $T_{\rm eff}^\infty$ in energy units. According to (34)–(35),

$$L_{\rm ph}^\infty = (1-x_{\rm g})L_{\rm ph} = L_{\rm ph}/(1+z_g)^2. \tag{37}$$

In the absence of spherical symmetry, it is convenient to define a local effective surface temperature $T_{\rm s}$ by the relation $F_{\rm ph}(\theta,\varphi) = \sigma_{\rm SB}T_{\rm s}^4$, where $F_{\rm ph}$ is the local radial flux density at the surface point, determined by the polar angle (θ) and azimuth (φ) in the spherical coordinate system. Then

$$L_{\rm ph} = \int_0^\pi \sin\theta\,{\rm d}\theta \int_0^{2\pi} {\rm d}\varphi\, F_{\rm ph}(\theta,\varphi). \tag{38}$$

The same relation connects the apparent luminosity $L_{\rm ph}^\infty$ (37) with the apparent flux $F_{\rm ph}^\infty = \sigma_{\rm SB}(T_{\rm s}^\infty)^4$ in the remote system, in accord with the relation $T_{\rm s}^\infty = T_{\rm s}/(1+z_{\rm g})$ analogous to (34).

The expressions (34), (35) and (37) agree with the GR concepts of light ray bending and time dilation near a massive body. If the angle between the wave vector $\boldsymbol{k}$ and the normal to the surface $\mathbf{n}$ at the emission point is θ_k, then the observer receives a photon whose wave vector makes an angle $\theta > \theta_k$ with $\mathbf{n}$. The rigorous theory of the influence of the light bending near a star on its observed spectrum has been developed by Pechenick et al. (1983) and cast in a convenient form by Page (1995) and Pavlov and Zavlin (2000). Beloborodov (2002) suggested the simple approximation

$$\cos\theta_k = x_{\rm g} + (1-x_{\rm g})\cos\theta, \tag{39}$$

Fig. 4 Integral spectra of a hydrogen atmosphere of a neutron star with $M = 1.4M_\odot$, $R = 12$ km, and with different effective temperatures $T_{\rm eff}$ ($\log T_{\rm eff}$ (K) from 5.5 to 6.8 with step 0.1). *The dashed* and *solid lines* represent the model with a dipole field of strength $B_{\rm p} = 10^{13}$ G at the pole and oriented along and across the line of sight, respectively. For comparison, *the dotted curve* shows the model with a constant field $B = 10^{13}$ G, normal to the surface

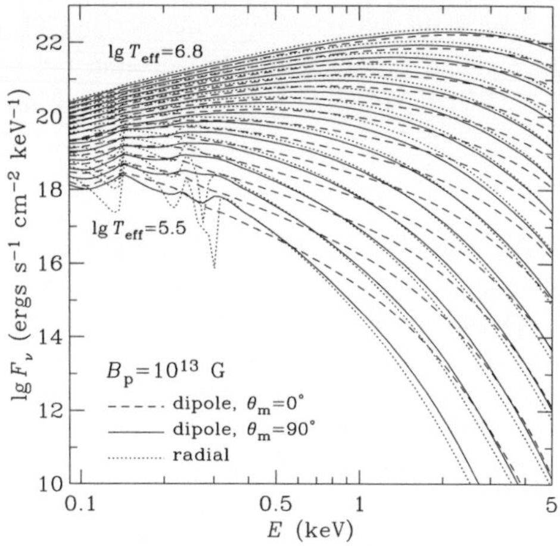

which is applicable at $x_{\rm g} < 0.5$ with an error within a few percent. At $\cos\theta_k < x_{\rm g}$, Eq. (39) gives $\theta > \pi/2$, which allows the observer to look behind the NS horizon.

Magnetic fields and temperatures of NSs vary from one surface point to another. In order to reproduce the radiation spectrum that comes to an observer, one can use the equations derived by Poutanen and Gierliński (2003) (see also Poutanen and Beloborodov 2006). In the particular case of a slowly rotating spherical star they give the following expression for spectral flux density related to the circular frequency $\omega_\infty = \omega\sqrt{1 - r_g/R}$ at a large distance D from the star:

$$F_{\omega_\infty} = (1 - x_g)^{3/2} \frac{R^2}{D^2} \int I_\omega(\boldsymbol{k}; \theta, \varphi) \cos\theta_k \sin\theta \mathrm{d}\theta \mathrm{d}\varphi, \tag{40}$$

where the integration is performed under the condition $\cos\theta_k > 0$. The problem is complicated, because the surface distributions of the magnetic field and the temperature are not known in advance. A conventional fiducial model is the relativistic dipole (Ginzburg and Ozernoi 1965), while the temperature distribution, consistent with the magnetic-field distribution, is found from calculations of heat transport in NS envelopes considered by Page et al. (2014) in this issue.

For the model of partially ionized H atmospheres described in Sect. 5, synthetic spectra were calculated by Ho et al. (2008). Examples are shown in Fig. 4. We see that the spectral features are strongly smeared by the averaging over the surface, and the spectrum depends on the magnetic axis orientation $\theta_{\rm m}$. When the star rotates, the latter dependence leads to the phase dependence of the spectra.

Such spectra of partially ionized, strongly magnetized NS atmospheres composed of hydrogen have been included in the package *XSPEC* (Arnaud 1996) under the names NSMAX and NSMAXG (see Ho 2014 and references therein).

8 Theory Versus Observations

Theoretical models of nonmagnetic atmospheres are successfully applied to analyses of spectra of many NSs with relatively weak magnetic fields $B \lesssim 10^9$ G. However, the nonmagnetic models are inadequate for the strongly magnetized NSs ($B \gg 10^9$ G). The theoretical framework for modeling the atmospheres of such stars is described above. As argued in Sect. 4, models of strongly magnetized NS atmospheres must take the bound species and their radiative transitions into account. Let us consider a few examples where models of magnetized, partially ionized atmospheres have been used to study their thermal radiation.

8.1 RX J1856.5-3754

In the case of RX J1856.5-3754, it is surprising the absence of absorption features and it is necessary to consider the spectrum in the X-ray and optical ranges simultaneously. The X-ray and optical spectra correspond to substantially different effective temperatures if fitted separately with blackbodies. To solve this problem, Ho et al. (2007) involved the model of a thin atmosphere described in Sect. 6.2. The measured spectrum of RX J1856.5-3754 was fitted in the entire range from X-rays to optical within observational errorbars. The best agreement between the theoretical and observed spectra has been achieved at the atmosphere column density $y_{\text{col}} = 1.2\ \text{g cm}^{-2}$, $B \sim (3\text{–}4) \times 10^{12}$ G, $T_{\text{eff}}^{\infty} = (4.34 \pm 0.03) \times 10^5$ K, $z_g = 0.25 \pm 0.05$, and $R_\infty = 17.2^{+0.5}_{-0.1} D_{140}$ km. Here, the errors are given at the 1σ significance level, and $D_{140} \equiv D/(140\ \text{pc})$. Note that a fit of the observed X-ray spectrum with the Planck function yields a 70 % higher temperature and a 3.5 times smaller radius of the emitting surface. Such huge difference exposes the importance of a correct physical interpretation of an observed spectrum for evaluation of NS parameters.

With the aid of expressions (34)–(35), we obtain from these estimates $T_{\text{eff}} = (5.4 \pm 1.1) \times 10^5$ K, $R = 13.8^{+0.9}_{-0.6} D_{140}$ km, and $M = 1.68^{+0.22}_{-0.15} D_{140} M_\odot$. Ho et al. (2007) adopted the value $D = 140$ pc from Kaplan et al. (2002). Using a recent update of the distance estimate, $D = 123^{+11}_{-15}$ pc (Walter et al. 2010), one obtains $R = 12.1^{+1.3}_{-1.6}$ km and $M = 1.48^{+0.16}_{-0.19} M_\odot$. Nevertheless, the given interpretation of the spectrum is somewhat questionable, since it does not agree with the magnetic-field estimate $B \approx 1.5 \times 10^{13}$ G, obtained for this star from timing analysis (van Kerkwijk and Kaplan 2008).

Using the same thin-atmosphere model, Ho (2007) analyzed the light curve of RX J1856.5-3754 and obtained constraints on the geometry of rotational and magnetic axes. It turned out that the light curve can be explained if one of these angles is small ($< 6°$), while the other angle lies between 20° and 45°. In this case, the radio emission around the magnetic poles does not cross the line of sight. As noted by Ho (2007), this may explain the non-detection of this star as a radio pulsar (Kondratiev et al. 2009).

8.2 RX J1308.6+2127

Hambaryan et al. (2011) analyzed the X-ray spectrum of the X-ray source RX J1308.6+2127 (RBS 1223), which reveals a more complex structure than RX J1856.5 -3754. It can be described by a wide absorption line centered around $\hbar\omega = 0.3$ keV, superposed on the Planck spectrum, with the line parameters depending on the stellar rotation phase. Using all 2003–2007 *XMM-Newton* observations of this star, the authors obtained a set of X-ray spectra for different rotation phases. They tried to interpret these spectra with different models, assuming magnetic fields $B \sim 10^{13}\text{–}10^{14}$ G, different atmosphere compositions, possible presence of a condensed surface and a finite atmosphere.

As a result, the authors managed to describe the observed spectrum and its rotational phase dependence with the use of the model of an iron surface covered by a partially ionized hydrogen atmosphere with $y_{\text{col}} \sim 1\text{–}10\ \text{g cm}^{-2}$, with mutually consistent asymmetric bipolar distributions of the magnetic field and the temperature, with the polar values $B_{\text{p1}} = B_{\text{p2}} = (0.86 \pm 0.02) \times 10^{14}$ G, $T_{\text{p1}} = 1.22^{+0.02}_{-0.05}$ MK, and $T_{\text{p2}} = 1.15 \pm 0.04$ MK. The magnetic field and temperature proved to be rather smoothly distributed over the surface. When compared to the theoretical model of Pérez-Azorín et al. (2006), such smooth distribution implies that the crust does not contain a superstrong toroidal magnetic field. The effective temperature is $T_{\mathrm{eff}} \approx 0.7$ MK. The gravitational redshift is estimated to be $z_g = 0.16^{+0.03}_{-0.01}$, which converts into $(M/M_{\odot})/R_6 = 0.87^{+0.13}_{0.05}$ and suggests a stiff EOS of the NS matter.

Note that the paper by Hambaryan et al. (2011) preceded that by Potekhin et al. (2012), where the treatment of the condensed surface and thin atmosphere was improved (Sect. 6). An analysis of the spectrum of RX J1308.6+2127 with the use of the improved results remains to be done in the future.

8.3 1E 1207.4-5209

The discovery of absorption lines in the spectrum of CCO 1E 1207.4-5209 at energies $E \sim 0.7N$ keV ($N = 1, 2, \ldots$) immediately entrained the natural assumption that they are caused by cyclotron harmonics (Bignami et al. 2003). As shown in Potekhin (2010), such harmonics can be only electronic, as the ion harmonics are unobservable. Therefore, this interpretation implies $B \approx 7 \times 10^{10}$ G. Mori et al. (2005) argued that only the first and second lines in the spectrum of 1E 1207.4-5209 are statistically significant, but some authors take also the third and fourth lines into account. This hypothesis was developed by Suleimanov et al. (2010a, 2012a), who considered two types of the electron cyclotron harmonics: the quantum oscillations of the Coulomb logarithm and the relativistic thermal harmonics (Sect. 5.5.2). An analogous explanation of the shape of the spectrum may possibly be applied also to CCO PSR J0821-4300 (Gotthelf et al. 2013).

Mori and Hailey (2006) have critically analyzed the earlier hypotheses about the origin of the absorption lines in the spectrum of 1E 1207.4-5209 and suggested their own explanation. They analyzed and rejected such interpretations as the lines of molecular hydrogen ions, helium ions, and also as the cyclotron lines and their harmonics. One of the arguments against the latter interpretation is that the fundamental cyclotron line should have much larger depth in the atmosphere spectrum than actually observed. Another argument is that the cyclotron lines and harmonics have small widths at a fixed B, therefore their observed width in the integral spectrum is determined by the B distribution. Thus the width of all lines should be the same, but observations do not confirm it. These arguments of Mori and Hailey (2006) were neglected by Suleimanov et al. (2010a, 2012a). Both groups of authors studied the cyclotron harmonics in spectra of fully ionized plasmas. This approach is indeed justified for the CCOs, because the impact of bound states on the spectra is small at $B \lesssim 10^{11}$ G and $T \gtrsim 10^6$ K (Potekhin, Chabrier, and Ho, in preparation).

As an alternative, Mori and Hailey (2006) and Mori and Ho (2007) suggested models of atmospheres composed of mid-Z elements. The authors found that an oxygen atmosphere with magnetic field $B = 10^{12}$ G provides a spectrum similar to the observed one. However, the constraint $B < 3.3 \times 10^{11}$ G obtained by Halpern and Gotthelf (2010) disagrees with this model, but rather favors the electron-cyclotron interpretation of the lines.

Unlike the cases of RX J1856.5-3754 and RX J1308.6+2127 that were considered above, there is no published results of a detailed fitting of the observed spectrum of 1E 1207.4-5209 with a theoretical model. Thus all suggested explanations of the spectrum of this object remain hypothetical.

8.4 Rotation Powered Pulsars

8.4.1 PSR J1119-6127

Ng et al. (2012) applied the partially ionized, strongly magnetized hydrogen atmosphere model (Ho et al. 2008) to interpretation of observations of pulsar J1119-6127, for which the estimate based on spindown gives an atypically high field $B = 4 \times 10^{13}$ G. In the X-ray range, it emits pulsed radiation, which has apparently mostly thermal nature. At fixed $D = 8.4$ kpc and $R = 13$, the bolometric flux gives an estimate of the mean effective temperature $T_{\rm eff} \approx 1.1$ MK. It was difficult to explain, however, the large pulsed fraction (48 ± 12 %) by the thermal emission. Ng et al. (2012) managed to reproduce the X-ray light curve of this pulsar assuming that one of its magnetic poles is surrounded by a heated area, which occupies 1/3 of the surface, is covered by hydrogen and heated to 1.5 MK, while the temperature of the opposite polar cap is below 0.9 MK.

8.4.2 PSR B0943+10

Storch et al. (2014) applied a similar analysis to interpretation of observations of pulsar B0943+10, which shows correlated radio and X-ray mode switches. The authors have taken $B = 2 \times 10^{12}$ G inferred from the pulsar spindown, assumed $M = 1.2M_\odot$ and $R = 12$ km, and modeled the emitting area as a hot spot covered by a partially ionized hydrogen atmosphere. They found that an atmosphere with $T_{\rm eff} \approx (1.4\text{–}1.5)$ MK and emission radius $R_{\rm em} \approx 85$ m matches the radio-quiet X-ray spectrum, whereas previous blackbody fits gave $T_{\rm bb} = 3$ MK and $R_{\rm bb} \approx 20\text{–}30$ m. The authors showed that the large X-ray pulse fraction observed during the radio quiet phase can be explained by including the beaming effect of a magnetic atmosphere, while remaining consistent with the dipole field geometry constrained by radio observations.

8.4.3 PSR J0357+3205

A middle-aged radio-quiet gamma-ray pulsar J0357+3205 was discovered in gamma-rays with Fermi and later in X-rays with Chandra and XMM-Newton observatories. It produces an unusual thermally-emitting pulsar wind nebula observed in X-rays. Kirichenko et al. (2014) fitted the spectrum of this pulsar with several different multicomponent models. In the physically realistic case where the incomplete ionization of the atmosphere was taken into account, they used the NSMAX model (Ho et al. 2008) for the thermal spectral component and a power-law model for the nonthermal one and fixed $M = 1.4M_\odot$ and $B = 10^{12}$ G. They obtained an acceptable fit ($\chi^2 = 1.05/244$) with a very loose constraint on the radius, $R_\infty = 8^{+12}_{-5}(D/500\ \text{pc})$ km.

9 Conclusions

We have considered the main features of neutron-star atmospheres and radiating surfaces and outlined the current state of the theory of the formation of their spectra. The interpretation of observations enters a qualitatively new phase, free from the assumptions of a blackbody spectrum or the "canonical model" of neutron stars. Spectral features, compatible with absorption lines in some cases, have been discovered in thermal spectra of strongly magnetized neutron stars. On the agenda is their detailed theoretical description, which may provide information on the surface composition, temperature and magnetic field distributions. However, in order to disentangle these parameters, a number of problems related to the theory of magnetic atmospheres and radiating surfaces still have to be solved.

Acknowledgements The work of A.P. has been partly supported by the RFBR (grant 14-02-00868) and by the Program "Leading Scientific Schools of RF" (grant NSh 294.2014.2).

Appendix: The Effects of Finite Atomic Masses

In this Appendix, we give a brief account of the effects of motion of atomic nuclei in strong magnetic fields on the quantum-mechanical characteristics of bound species and the ionization equilibrium of partially ionized plasmas (for a more detailed review, see Potekhin 2014)

A.1 The Finite-Mass Effects on Properties of Atoms

An atomic nucleus of finite mass, as any charged particle, undergoes oscillations in the plane (xy) perpendicular to $\boldsymbol{B}$, which are quantized in the ion Landau levels. In an atom or a molecule, these oscillations are entangled with the electron motion. Therefore the longitudinal projections of the orbital moments of the electrons and the nucleus are not conserved separately. Different atomic quantum numbers correspond to different oscillation energies of the atomic nucleus, multiple of its cyclotron energy $\hbar\omega_{\rm ci}$. As a result, the energy of every level gets an addition, which is non-negligible if the magnetic parameter γ is not small compared to the nucleus-to-electron mass ratio $m_{\rm i}/m_{\rm e}$.

For the hydrogen atom and hydrogenlike ions, a conserved quantity is $\hbar s$, which corresponds to the difference of longitudinal projections of orbital moments of the atomic nucleus and the electron, and the sum $N+s$ plays role of a nuclear Landau number, N being the electron Landau number. For the bound states in strong magnetic fields, $N=0$, therefore the nuclear oscillatory addition to the energy equals $s\hbar\omega_{\rm ci}$. Thus the binding energy of a hydrogen atom at rest is

$$E_{s\nu} = E_{s\nu}^{(0)} - \hbar\omega_{\rm ci} s, \tag{41}$$

where $E_{s\nu}^{(0)}$ is the binding energy in the approximation of non-moving nucleus. It follows that the values of s are limited for the bound states. In particular, all bound states have $s=0$ at $B > 6\times 10^{13}$ G.

The account of the finite nuclear mass is more complicated for multielectron atoms. Al-Hujaj and Schmelcher (2003) have shown that the contribution of the nuclear motion to the binding energy of a non-moving atom equals $\hbar\omega_{\rm ci} S(1+\delta(\gamma))$, where $(-S)$ is the total magnetic quantum number of the atom, and $|\delta(\gamma)| \ll 1$.

The astrophysical simulations assume finite temperatures, hence thermal motion of particles. The theory of motion of a system of point charges in a constant magnetic field was reviewed by Johnson et al. (1983). The canonical momentum $\boldsymbol{P}$ is not conserved in a magnetic field. A relevant conserved quantity is pseudomomentum

$$\boldsymbol{K} = \boldsymbol{P} + \frac{1}{2c}\boldsymbol{B} \times \sum_i q_i \boldsymbol{r}_i. \tag{42}$$

If the system is electrically neutral as a whole, then all Cartesian components of $\boldsymbol{K}$ can be determined simultaneously (i.e., their quantum-mechanical operators commute with each other). For a charged system (an ion), one can determine K^2 simultaneously with either K_x or K_y, but K_x and K_y do not commute. The specific effects related to collective motion of a system of charged particles are especially important in NS atmospheres at $\gamma \gg 1$. In particular, so called decentered states may become populated, where an electron is localized mostly in a "magnetic well" aside from the Coulomb center.

For a hydrogen atom, $\boldsymbol{K} = \boldsymbol{P} - (e/2c)\boldsymbol{B} \times \boldsymbol{r}$, where $\boldsymbol{r}$ connects the proton and the electron. Early studies of this particular case were done by Gor'kov and Dzyaloshinskiĭ (1968),

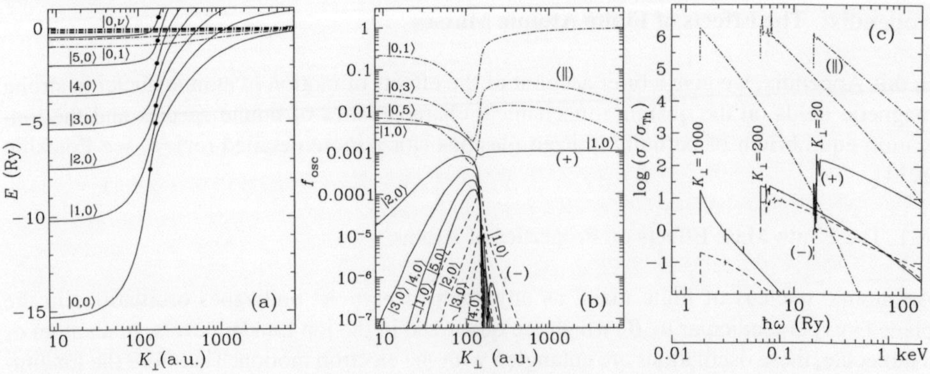

Fig. 5 (**a**) energies, (**b**) oscillator strengths, and (**c**) photoionization cross-sections for a hydrogen atom moving across magnetic field $B = 2.35 \times 10^{12}$ G. Energies of states $|s, 0\rangle$ (*solid curves*) and $|0, \nu\rangle$ (*dot-dashed curves*) are shown as functions of the transverse pseudomomentum $K_\perp$ (in atomic units). The heavy dots on the solid curves are the inflection points at $K_\perp = K_c$. The $K_\perp$-dependence of oscillator strengths (**b**) is shown for transitions from the ground state to the states $|s, 0\rangle$ under influence of radiation with polarization $\alpha = +1$ (*solid curves*) and $\alpha = -1$ (*dashed curves*), and also for transitions into states $|0, \nu\rangle$ for $\alpha = 0$ (*dot–dashed curves*). Cross sections of photoionization (**c**) under the influence of radiation with $\alpha = +1$ (*solid curves*), $\alpha = -1$ (*dashed curves*), and $\alpha = 0$ (*dot-dashed curves*) are shown for the ground state as functions of the photon energy in Ry (*the upper x-scale*) and keV (*the lower x-scale*) at $K_\perp = 20$ a.u. (*the right curve*), $K_\perp = 200$ a.u. (*the middle curve*), and $K_\perp = 1000$ a.u. (*the left curve* of every type)

Burkova et al. (1976), Ipatova et al. (1984). Numerical calculations of the energy spectrum of the hydrogen atom with an accurate treatment of the effects of motion across a strong magnetic field were performed by Vincke et al. (1992) and Potekhin (1994). Bound-bound radiative transitions of a moving H atom in a plasma were studied by Pavlov and Potekhin (1995), and bound-free transitions by Potekhin and Pavlov (1997).

Figure 5 shows the energies, oscillator strengths, and photoionization cross-sections of a hydrogen atom moving in a magnetic field with $\gamma = 1000$. The reference point is taken to be the sum of the zero-point oscillation energies of free electron and proton, $(\hbar\omega_c + \hbar\omega_{ci})/2$.Therefore the negative energies in Fig. 5 a correspond to bound states ($E_{s\nu} = -E > 0$). At small transverse pseudomomenta $K_\perp$, the energies of low levels in Fig. 5a exceed the binding energy of the field-free hydrogen atom (1 Ry) by an order of magnitude. However, the binding energy decreases with increasing $K_\perp$, and it can become negative for the states with $s \neq 0$ due to the term $\hbar\omega_{ci}s$ in Eq. (41). Such states are metastable. In essence, they are continuum resonances. Note that the transverse atomic velocity equals $\partial E/\partial \boldsymbol{K}$, therefore it attains a maximum at the inflection points ($K_\perp = K_c$) on the curves in Fig. 5a and decreases with further increase of $K_\perp$, while the average electron-proton distance continues to increase. At $K_\perp > K_c$ the atom goes into the decentered state, where the electron and proton are localized near their guiding centers, separated by distance $r_* = (a_B^2/\hbar)K_\perp/\gamma$.

Figure 5b shows oscillator strengths for the main transitions from the ground state to excited discrete levels. Since the atomic wave-functions are symmetric with respect to the z-inversion for the states with even ν, and antisymmetric for odd ν, only the transitions that change the parity of ν are allowed for the polarization along the field ($\alpha = 0$), and only those preserving the parity for the orthogonal polarizations ($\alpha = \pm 1$). For the atom at rest, in the dipole approximation, due to the conservation of the z-projection of the total angular momentum of the system, absorption of a photon with polarization $\alpha = 0, \pm 1$ results in the change of s by α. This selection rule for a non-moving atom manifests itself in vanishing oscillator strengths at $K_\perp \to 0$ for $s \neq \alpha$. In an appropriate coordinate system (Burkova et al. 1976; Potekhin 1994), the symmetry is restored at $K_\perp \to \infty$, therefore the transition with $s = \alpha$ is the only one that survives also in the limit of large pseudomomenta. But in the intermediate region of $K_\perp$, where the transverse atomic velocity is not small, the cylindrical symmetry is broken, so that transitions to other levels are allowed. Thus the corresponding oscillator strengths in Fig. 5b have maxima at $K_\perp \approx K_c$. Analytical approximations for these oscillator strengths, as well as for the dependences of the binding energies $E_{s\nu}(K_\perp)$, are given in Potekhin (1998).

Figure 5c shows photoionization cross-sections for hydrogen in the ground state as functions of photon energy at three values of $K_\perp$. The leftward shift of the ionization threshold with increasing $K_\perp$ corresponds to the decrease of the binding energy that is shown in Fig. 5a, while the peaks and dips on the curves are caused by resonances at transitions to metastable states $|s, \nu; K\rangle$ with positive energies (see Potekhin and Pavlov 1997, for a detailed discussion).

Quantum-mechanical calculations of the characteristics of the He^+ ion that moves in a strong magnetic field are performed by Bezchastnov et al. (1998), Pavlov and Bezchastnov (2005). The basic difference from the case of a neutral atom is that the ion motion is restricted by the field in the transverse plane, therefore the values of K^2 are quantized (Johnson et al. 1983). Clearly, the similarity relations for the ions with nonmoving nuclei (Sect. 4.2.1) do not hold anymore.

Currently there is no detailed calculation of binding energies, oscillator strengths, and photoionization cross-sections for atoms and ions other than H and He^+, arbitrarily moving in a strong magnetic field. For such species one usually neglects the decentered states and uses a perturbation theory with respect to $K_\perp$ (e.g., Mori and Hailey 2002, Medin et al. 2008). This approximation can be sufficient for simulations of relatively cool atmospheres of moderately magnetized NSs. A condition of applicability of the perturbation theory for an atom with mass $m_a = A m_u$ requires $T/E^{(0)} \ll m_a/(\gamma m_e) \approx 4A/B_{12}$ (Potekhin 2014). If $B \lesssim 10^{13}$ G and $T \lesssim 10^6$ K, it is satisfied for low-lying levels of carbon and heavier atoms.

A.2 The Finite-Mass Effects on the Ionization Equilibrium and Thermodynamics

Since quantum-mechanical characteristics of an atom in a strong magnetic field depend on the transverse pseudomomentum $K_\perp$, the atomic distribution over $K_\perp$ cannot be written in a closed form, and only the distribution over longitudinal momenta K_z remains Maxwellian. The first complete account of these effects has been taken in Potekhin et al. (1999) for hydrogen atmospheres. Let $p_{s\nu}(K_\perp)\mathrm{d}^2K_\perp$ be the probability of finding a hydrogen atom in the state $|s, \nu\rangle$ in the element $\mathrm{d}^2K_\perp$ near $\boldsymbol{K}_\perp$ in the plane of transverse pseudomomenta. Then the number of atoms in the element d^3K of the pseudomomentum space equals

$$\mathrm{d}N(\boldsymbol{K}) = N_{s\nu}\frac{\lambda_a}{2\pi\hbar}\exp\left(-\frac{K_z^2}{2m_aT}\right)p_{s\nu}(K_\perp)\mathrm{d}^3K, \tag{43}$$

where m_a is the mass of the atom, $\lambda_a = [2\pi\hbar^2/(m_aT)]^{1/2}$ is its thermal wavelength, and $N_{s\nu} = \int \mathrm{d}N_{s\nu}(\boldsymbol{K})$ is the total number of atoms with given discrete quantum numbers. The distribution $N_{s\nu}p_{s\nu}(K_\perp)$ is not known in advance, but should be calculated in a self-consistent way by minimization of the free energy including the nonideal terms. It is convenient to define deviations from the Maxwell distribution with the use of generalized occupation probabilities $w_{s\nu}(K_\perp)$. Then the atomic contribution to the free energy equals (Potekhin et al. 1999)

$$F_{\mathrm{id}}^{\mathrm{at}}+F_{\mathrm{int}}^{\mathrm{at}}=T\sum_{s\nu}N_{s\nu}\int\ln\left[n_{s\nu}\lambda_{\mathrm{a}}^{3}\frac{w_{s\nu}(K_{\perp})}{\exp(1)\mathcal{Z}_{s\nu}}\right]p_{s\nu}(K_{\perp})\mathrm{d}^{2}K_{\perp}, \tag{44}$$

where

$$\mathcal{Z}_{s\nu}=\frac{\lambda_{\mathrm{a}}^{2}}{(2\pi\hbar^{2})}\int_{0}^{\infty}w_{s\nu}(K_{\perp})\mathrm{e}^{E_{s\nu}(K_{\perp})/T}K_{\perp}\mathrm{d}K_{\perp}. \tag{45}$$

The nonideal part of the free energy that describes atom-atom and atom-ion interactions and is responsible for the pressure ionization has been calculated by Potekhin et al. (1999) with the use of the hard-sphere model. The plasma model included also hydrogen molecules H_2 and chains H_n, which become stable in the strong magnetic fields. For this purpose, approximate formulae of Lai (2001) have been used, which do not take full account of the motion effects, therefore the results of Potekhin et al. (1999) are reliable only when the molecular fraction is small.

This hydrogen-plasma model underlies thermodynamic calculations of hydrogen atmospheres of NSs with strong and superstrong magnetic fields (Potekhin and Chabrier 2003, 2004). Mori and Heyl (2007) applied the same approach with slight modifications to strongly magnetized helium plasmas. One of the modifications was the use of the plasma microfield distribution from Potekhin et al. (2002) for calculation of the $K_\perp$-dependent occupation probabilities. Mori and Heyl considered atomic and molecular helium states of different ionization degrees. Their treatment included rotovibrational molecular levels and the dependence of binding energies on orientation of the molecular axis relative to $\boldsymbol{B}$. The $K_\perp$-dependence of the energy, $E(K_\perp)$, was described by an analytical fit, based on an extrapolation of adiabatic calculations at small $K_\perp$. The effects of motion of atomic and molecular ions were not considered.

References

A.A. Abdo, M. Ajello, A. Allafort et al., Astrophys. J. Suppl. Ser. **208**, 17 (2013)
S.L. Adler, Ann. Phys. (NY) **67**, 599 (1971)
D.N. Aguilera, J.A. Pons, J.A. Miralles, Astron. Astrophys. **486**, 255 (2008)
O.-A. Al-Hujaj, P. Schmelcher, Phys. Rev. A **67**, 023403 (2003)
K.A. Arnaud, in *Astronomical Data Analysis Software and Systems V*, ed. by G. Jacoby, J. Barnes. ASP Conf. Ser., vol. 101 (1996), p. 17. http://starchild.gsfc.nasa.gov/xanadu/xspec/
J.N. Bahcall, R.A. Wolf, Astrophys. J. **142**, 1254 (1965)
A.M. Beloborodov, Astrophys. J. **566**, L85 (2002)
V.S. Beskin, Ya.N. Istomin, A.A. Philippov, Phys. Usp. **56**, 164 (2013)
V.G. Bezchastnov, G.G. Pavlov, J. Ventura, Phys. Rev. A **58**, 180 (1998)
G.F. Bignami, P.A. Caraveo, A. De Luca, S. Mereghetti, Nature **423**, 725 (2003)
G.S. Bisnovatyi-Kogan, Phys. Usp. **49**, 53 (2006)
G.S. Bisnovatyi-Kogan, N.R. Ikhsanov, Astron. Rep. **58**, 217 (2014)
S. Bogdanov, Astrophys. J. **762**, 96 (2013)
K. Boshkaev, L. Izzo, J.A. Rueda, R. Ruffini, Astron. Astrophys. **555**, A151 (2013)
W. Brinkmann, Astron. Astrophys. **82**, 352 (1980)
W. Brinkmann, H. Ögelman, Astron. Astrophys. **182**, 71 (1987)
E.F. Brown, L. Bildsten, R.E. Rutledge, Astrophys. J. **504**, L95 (1998)
T. Bulik, G.G. Pavlov, Astrophys. J. **469**, 373 (1996)
S. Burke-Spolaor, in *Proceedings of IAUS 291 "Neutron Stars and Pulsars: Challenges and Opportunities after 80 years"*, ed. by J. van Leeuwen (Cambridge University Press, Cambridge, 2014), p. 95
L.A. Burkova, I.E. Dzyaloshinskiĭ, G.P. Drukarev, B.S. Monozon, Sov. Phys. JETP **44**, 276 (1976)
P.A. Caraveo, A. De Luca, M. Marelli et al., Astrophys. J. **725**, L6 (2010)

P. Chang, L. Bildsten, Astrophys. J. **585**, 464 (2003)
P. Chang, L. Bildsten, Astrophys. J. **605**, 830 (2004)
A.F. Cheng, D.J. Helfand, Astrophys. J. **271**, 271 (1983)
H.-Y. Chiu, E.E. Salpeter, Phys. Rev. Lett. **12**, 413 (1964)
R. Cohen, J. Lodenquai, M. Ruderman, Phys. Rev. Lett. **25**, 467 (1970)
A. De Luca, in *40 Years of Pulsars: Millisecond Pulsars, Magnetars and More*, ed. by C. Bassa, Z. Wang, A. Cumming, V.M. Kaspi. AIP Conf. Proc., vol. 983 (2008), p. 311
A. De Luca, P. Caraveo, S. Mereghetti, M. Negroni, G.F. Bignami, Astrophys. J. **623**, 1051 (2005)
A. De Luca, R.P. Mignani, A. Sartori et al., Astron. Astrophys. **525**, 106 (2011)
T. Detmer, P. Schmelcher, L.S. Cederbaum, Phys. Rev. A **57**, 1767 (1998)
R.C. Duncan, C. Thompson, Astrophys. J. **392**, L9 (1992)
W. Ebeling, G. Norman, J. Stat. Phys. **110**, 861 (2003)
E. Fermi, Z. Phys. **26**, 54 (1924)
B.T. Gänsicke, T.M. Braje, R.M. Romani, Astron. Astrophys. **386**, 1001 (2002)
R.H. Garstang, Rep. Prog. Phys. **40**, 105 (1977)
F.P. Gavriil, V.M. Kaspi, P.M. Woods, Nature **419**, 142 (2002)
F.P. Gavriil, M.E. Gonzalez, E.V. Gotthelf, V.M. Kaspi, M.A. Livingstone, P.M. Woods, Science **319**, 1802 (2008)
V.L. Ginzburg, *The Propagation of Electromagnetic Waves in Plasmas*, 2nd edn. (Pergamon, London, 1970)
V.L. Ginzburg, L.M. Ozernoi, Sov. Phys. JETP **20**, 689 (1965)
Yu.N. Gnedin, G.G. Pavlov, Sov. Phys. JETP **38**, 903 (1974)
Yu.N. Gnedin, R.A. Sunyaev, Astron. Astrophys. **36**, 379 (1974)
L.P. Gor'kov, I.E. Dzyaloshinskiĭ, Sov. Phys. JETP **26**, 449 (1968)
E.V. Gotthelf, J.P. Halpern, J. Alford, Astrophys. J. **765**, 58 (2013)
C.R. Gwinn et al., Space Sci. Rev. (2014, this issue)
C.B. Haakonsen, M.L. Turner, N.A. Tacik, R.E. Rutledge, Astrophys. J. **749**, 52 (2012). https://github.com/McPHAC/
F. Haberl, Astrophys. Space Sci. **308**, 181 (2007)
P. Haensel, A.Y. Potekhin, D.G. Yakovlev, *Neutron Stars 1: Equation of State and Structure* (Springer, New York, 2007)
L.K. Haines, D.H. Roberts, Am. J. Phys. **37**, 1145 (1969)
J.P. Halpern, E.V. Gotthelf, Astrophys. J. **709**, 436 (2010)
V. Hambaryan, R. Neuhäuser, F. Haberl, M.M. Hohle, A.D. Schwope, Astron. Astrophys. **497**, L9 (2009)
V. Hambaryan, V. Suleimanov, A.D. Schwope et al., Astron. Astrophys. **534**, A74 (2011)
A.K. Harding, D. Lai, Rep. Prog. Phys. **69**, 2631 (2006)
C.O. Heinke, G.B. Rybicki, R. Narayan, J.E. Grindlay, Astrophys. J. **644**, 1090 (2006)
A. Hewish, S.J. Bell, J.D.H. Pilkington, P.F. Scott, R.A. Collins, Nature **217**, 709 (1968)
J.S. Heyl, L. Hernquist, Phys. Rev. D **55**, 2449 (1997)
J.S. Heyl, N.J. Shaviv, Phys. Rev. D **66**, 023002 (2002)
W.C.G. Ho, Mon. Not. R. Astron. Soc. **380**, 71 (2007)
W.C.G. Ho, in *Magnetic Fields Throughout Stellar Evolution (Proc. IAU Symp. 302)*, ed. by M. Jardine, P. Petit, H.C. Spruit (Cambridge University Press, Cambridge, 2014), p. 435
W.C.G. Ho, C.O. Heinke, Nature **462**, 71 (2009)
W.C.G. Ho, D. Lai, Mon. Not. R. Astron. Soc. **338**, 233 (2003)
W.C.G. Ho, D.L. Kaplan, P. Chang, M. van Adelsberg, A.Y. Potekhin, Mon. Not. R. Astron. Soc. **375**, 821 (2007)
W.C.G. Ho, A.Y. Potekhin, G. Chabrier, Astrophys. J. Suppl. Ser. **178**, 102 (2008)
M.M. Hohle, F. Haberl, J. Vink et al., Mon. Not. R. Astron. Soc. **423**, 1194 (2012)
I.P. Ipatova, A.Yu. Maslov, A.V. Subashiev, Sov. Phys. JETP **60**, 1037 (1984)
N. Itoh, Mon. Not. R. Astron. Soc. **173**, 1P (1975)
M.V. Ivanov, P. Schmelcher, Phys. Rev. A **61**, 022505 (2000)
P. Jones, Mon. Not. R. Astron. Soc. **184**, 807 (1978)
B.R. Johnson, J.O. Hirschfelder, K.H. Yang, Rev. Mod. Phys. **55**, 109 (1983)
B.B. Kadomtsev, Sov. Phys. JETP **31**, 945 (1970)
B.B. Kadomtsev, V.S. Kudryavtsev, JETP Lett. **13**, 42 (1971)
A.D. Kaminker, G.G. Pavlov, Yu.A. Shibanov, Astrophys. Space Sci. **86**, 249 (1982)
D.L. Kaplan, M.H. van Kerkwijk, Astrophys. J. **705**, 798 (2009)
D.L. Kaplan, M.H. van Kerkwijk, J. Anderson, Astrophys. J. **571**, 447 (2002)
D.L. Kaplan, A. Kamble, M.H. van Kerkwijk, W.C.G. Ho, Astrophys. J. **736**, 117 (2011)
U. Kappes, P. Schmelcher, Phys. Rev. A **54**, 1313 (1996)
V.M. Kaspi, Proc. Natl. Acad. Sci. USA **107**, 7147 (2010)

V.K. Khersonskii, Sov. Astron. **31**, 646 (1987)
A. Kirichenko, A. Danilenko, Yu. Shibanov et al., Astron. Astrophys. **564**, A81 (2014)
K. Kohri, S. Yamada, Phys. Rev. D **65**, 043006 (2002)
V.I. Kondratiev, M.A. McLaughlin, M. Burgay et al., Astrophys. J. **702**, 692 (2009). Erratum: Astrophys. J. **708**, 910 (2010)
A. Kubo, J. Phys. Chem. A **111**, 5572 (2007)
D. Lai, Rev. Mod. Phys. **73**, 629 (2001)
D. Lai, Space Sci. Rev. (2014, this issue)
D. Lai, W.C.G. Ho, Astrophys. J. **566**, 373 (2002)
R. Lenzen, J. Trümper, Nature **271**, 216 (1978)
E.H. Lieb, J.P. Solovej, J. Yngvason, Phys. Rev. Lett. **69**, 749 (1992)
M.A. Livingstone, C.-Y. Ng, V.M. Kaspi, F.P. Gavriil, E.V. Gotthelf, Astrophys. J. **730**, 66 (2011)
V.M. Malofeev, O.I. Malov, D.A. Teplykh, Astrophys. Space Sci. **308**, 211 (2007)
I.F. Malov, Astron. Rep. **54**, 925 (2010)
R.N. Manchester, G.B. Hobbs, A. Teoh, M. Hobbs, Astron. J. **129**, 1993 (2005)
M. Marelli, A. De Luca, P.A. Caraveo, Astrophys. J. **733**, 82 (2011)
M.A. McLaughlin, A.G. Lyne, D.R. Lorimer et al., Nature **439**, 817 (2006)
Z. Medin, D. Lai, Phys. Rev. A **74**, 062507 (2006a)
Z. Medin, D. Lai, Phys. Rev. A **74**, 062508 (2006b)
Z. Medin, D. Lai, Mon. Not. R. Astron. Soc. **382**, 1833 (2007)
Z. Medin, D. Lai, A.Y. Potekhin, Mon. Not. R. Astron. Soc. **383**, 161 (2008)
S. Mereghetti, Astron. Astrophys. Rev. **15**, 225 (2008)
S. Mereghetti, Braz. J. Phys. **43**, 35 (2013)
P. Mészáros, *High-Energy Radiation from Magnetized Neutron Stars* (University of Chicago Press, Chicago, 1992)
J.J. Miller, M.A. McLaughlin, N. Rea et al., Astrophys. J. **776**, 104 (2011)
M.C. Miller, Mon. Not. R. Astron. Soc. **255**, 129 (1992)
M.C. Miller, D. Neuhauser, Mon. Not. R. Astron. Soc. **253**, 107 (1991)
K. Mori, C.J. Hailey, Astrophys. J. **564**, 914 (2002)
K. Mori, C.J. Hailey, Astrophys. J. **648**, 1139 (2006)
K. Mori, J. Heyl, Mon. Not. R. Astron. Soc. **376**, 895 (2007)
K. Mori, W.C.G. Ho, Mon. Not. R. Astron. Soc. **377**, 905 (2007)
K. Mori, J.C. Chonko, C.J. Hailey, Astrophys. J. **631**, 1082 (2005)
C. Motch, V.E. Zavlin, F. Haberl, Astron. Astrophys. **408**, 323 (2003)
C.-Y. Ng, P.O. Slane, B.M. Gaensler, J.P. Hughes, Astrophys. J. **686**, 508 (2008)
C.-Y. Ng, V.M. Kaspi, in *Astrophysics of Neutron Stars 2010: A Conference in Honor of M. Ali Alpar*, ed. by E. Göğüş, T. Belloni, Ü. Ertan. AIP Conf. Proc., vol. 1379 (2011), p. 60
C.-Y. Ng, V.M. Kaspi, W.C.G. Ho et al., Astrophys. J. **761**, 65 (2012)
S.A. Olausen, V.M. Kaspi, Astrophys. J. Suppl. Ser. **212**, 6 (2014)
D. Page, Astrophys. J. **442**, 273 (1995)
D. Page, J.A. Pons, A.Y. Potekhin, Space Sci. Rev. (2014, this issue)
G.G. Pavlov, V.G. Bezchastnov, Astrophys. J. **635**, L61 (2005)
G.G. Pavlov, Yu.N. Gnedin, Sov. Sci. Rev., E, Astrophys. Space Phys. Rev. **3**, 197 (1984)
G.G. Pavlov, A.N. Panov, Sov. Phys. JETP **44**, 300 (1976)
G.G. Pavlov, A.Y. Potekhin, Astrophys. J. **450**, 883 (1995)
G.G. Pavlov, V.E. Zavlin, Astrophys. J. **529**, 1011 (2000)
G.G. Pavlov, Yu.A. Shibanov, D.G. Yakovlev, Astrophys. Space Sci. **73**, 33 (1980)
G.G. Pavlov, Yu.A. Shibanov, V.E. Zavlin, R.D. Meyer, in *The Lives of the Neutron Stars*, ed. by M.A. Alpar, Ü. Kiziloğlu, J. van Paradijs. NATO ASI Series C, vol. 450 (Kluwer, Dordrecht, 1995), p. 71
K.R. Pechenick, C. Ftaclas, J.M. Cohen, Astrophys. J. **274**, 846 (1983)
J.F. Pérez-Azorín, J.A. Miralles, J.A. Pons, Mon. Not. R. Astron. Soc. **433**, 275 (2005)
J.F. Pérez-Azorín, J.A. Miralles, J.A. Pons, Astron. Astrophys. **451**, 1009 (2006)
J.A. Pons, F.M. Walter, J.M. Lattimer et al., Astrophys. J. **564**, 981 (2002)
J.A. Pons, J.A. Miralles, U. Geppert, Astron. Astrophys. **496**, 207 (2009)
S.B. Popov, J.A. Pons, J.A. Miralles, P.A. Boldin, B. Posselt, Mon. Not. R. Astron. Soc. **401**, 2675 (2010)
A.Y. Potekhin, J. Phys. B, At. Mol. Opt. Phys. **27**, 1073 (1994)
A.Y. Potekhin, Phys. Plasmas **3**, 4156 (1996)
A.Y. Potekhin, J. Phys. B, At. Mol. Opt. Phys. **31**, 49 (1998)
A.Y. Potekhin, Astron. Astrophys. **518**, A24 (2010)
A.Y. Potekhin, Phys. Usp. **57**(8), (2014, in press). arXiv:1403.0074. doi:10.3367/UFNe.0184.201408a.0793
A.Y. Potekhin, G. Chabrier, Astrophys. J. **585**, 955 (2003)

A.Y. Potekhin, G. Chabrier, Astrophys. J. **600**, 317 (2004)
A.Y. Potekhin, G. Chabrier, Astron. Astrophys. **550**, A43 (2013)
A.Y. Potekhin, D. Lai, Mon. Not. R. Astron. Soc. **376**, 793 (2007)
A.Y. Potekhin, G.G. Pavlov, Astrophys. J. **483**, 414 (1997)
A.Y. Potekhin, G.G. Pavlov, J. Ventura, Astron. Astrophys. **317**, 618 (1997)
A.Y. Potekhin, G. Chabrier, Yu.A. Shibanov, Phys. Rev. E **60**, 2193 (1999). Erratum: Phys. Rev. E **63**, 019901 (2000)
A.Y. Potekhin, D. Gilles, G. Chabrier, Phys. Rev. E **65**, 036412 (2002)
A.Y. Potekhin, D. Lai, G. Chabrier, W.C.G. Ho, Astrophys. J. **612**, 1034 (2004)
A.Y. Potekhin, V.F. Suleimanov, M. van Adelsberg, K. Werner, Astron. Astrophys. **546**, A121 (2012)
J. Poutanen, A.M. Beloborodov, Mon. Not. R. Astron. Soc. **373**, 836 (2006)
J. Poutanen, M. Gierliński, Mon. Not. R. Astron. Soc. **343**, 1301 (2003)
M. Rajagopal, R. Romani, Astrophys. J. **461**, 327 (1996)
M. Rajagopal, R. Romani, M.C. Miller, Astrophys. J. **479**, 347 (1997)
N. Rea, P. Esposito, in *High-Energy Emission from Pulsars and Their Systems*, ed. by N. Rea, D.F. Torres (Springer, Berlin, 2011), p. 247
Ö.E. Rögnvaldsson, I. Fushiki, E.H. Gudmundsson, C.J. Pethick, J. Yngvason, Astrophys. J. **416**, 276 (1993)
R.W. Romani, Astrophys. J. **313**, 718 (1987)
H. Ruder, G. Wunner, H. Herold, F. Geyer, *Atoms in Strong Magnetic Fields* (Springer, Berlin, 1994)
M.A. Ruderman, Phys. Rev. Lett. **27**, 1306 (1971)
E.E. Salpeter, Astrophys. J. **134**, 669 (1961)
P. Schmelcher, T. Detmer, L.S. Cederbaum, Phys. Rev. A **64**, 023410 (2001)
N. Shabaltas, D. Lai, Astrophys. J. **748**, 148 (2012)
V.D. Shafranov, in *Reviews of Plasma Physics*, vol. 3, ed. by M.A. Leontovich (Consultants Bureau, New York, 1967), p. 1
Yu.A. Shibanov, V.E. Zavlin, Astron. Lett. **21**, 3 (1995)
Yu.A. Shibanov, G.G. Pavlov, V.E. Zavlin, J. Ventura, Astron. Astrophys. **266**, 313 (1992)
A.A. Sokolov, I.M. Ternov, *Radiation from Relativistic Electrons* (Am. Inst. of Phys., New York, 1986)
A. Spitkovsky, Astrophys. J. **648**, 51 (2006)
N.I. Storch, W.C.G. Ho, D. Lai, S. Bogdanov, C.O. Heinke, Astrophys. J. **789**, L27 (2014)
T. Strohmayer, L. Bildsten, in *Compact Stellar X-Ray Sources*, ed. by W.H.G. Lewin, M. van der Klis (Cambridge University Press, Cambridge, 2006), p. 113
V. Suleimanov, A.Y. Potekhin, K. Werner, Astron. Astrophys. **500**, 891 (2009)
V.F. Suleimanov, G.G. Pavlov, K. Werner, Astrophys. J. **714**, 630 (2010a)
V. Suleimanov, V. Hambaryan, A.Y. Potekhin et al., Astron. Astrophys. **522**, A111 (2010b)
V. Suleimanov, J. Poutanen, M. Revnivtsev, K. Werner, Astrophys. J. **742**, 122 (2011)
V.F. Suleimanov, G.G. Pavlov, K. Werner, Astrophys. J. **751**, 15 (2012a)
V. Suleimanov, J. Poutanen, K. Werner, Astron. Astrophys. **545**, A120 (2012b)
V.F. Suleimanov, D. Klochkov, G.G. Pavlov, K. Werner, Astrophys. J. Suppl. Ser. **210**, 13 (2014)
G.L. Surmelian, R.F. O'Connel, Astrophys. J. **190**, 741 (1974). Erratum: Astrophys. J. **204**, 311 (1976)
D. Teplykh, A. Rodin, V. Malofeev, S. Logvinenko, in *Radio Pulsars: An Astrophysical Key to Unlock the Secrets of the Universe*, ed. by M. Burgay, N. D'Amico, P. Esposito, A. Pellizzoni, A. Possenti. AIP Conf. Proc., vol. 1357 (2011), p. 201
N. Tetzlaff, T. Eisenbeiss, R. Neuhäuser, M.M. Hohle, Mon. Not. R. Astron. Soc. **417**, 617 (2011)
A. Thorolfsson, Ö.E. Rögnvaldsson, J. Yngvason, E.H. Gudmundsson, Astrophys. J. **502**, 847 (1998)
J. Trümper, K. Denneri, N.D. Kylafis, Ü. Ertan, A. Zezas, Astrophys. J. **764**, 49 (2013)
A.V. Turbiner, Astrophys. Space Sci. **308**, 267 (2007)
A.V. Turbiner, J.C. López Vieyra, N.L. Guevara, Phys. Rev. A **81**, 042503 (2010)
R. Turolla, in *Neutron Stars and Pulsars*, ed. by W. Becker (Springer, Berlin, 2009), p. 141
R. Turolla, P. Esposito, Int. J. Mod. Phys. **22**, 1330024 (2013)
R. Turolla, S. Zane, J.J. Drake, Astrophys. J. **603**, 265 (2004)
M. van Adelsberg, D. Lai, Mon. Not. R. Astron. Soc. **373**, 1495 (2006)
M. van Adelsberg, R. Perna, Mon. Not. R. Astron. Soc. **399**, 1523 (2009)
M. van Adelsberg, D. Lai, A.Y. Potekhin, P. Arras, Astrophys. J. **628**, 902 (2005)
M.H. van Kerkwijk, D.L. Kaplan, Astrophys. J. **673**, L163 (2008)
J. Ventura, Phys. Rev. D **19**, 1684 (1979)
D. Viganò, J.A. Pons, Mon. Not. R. Astron. Soc. **425**, 248 (2012)
D. Viganò, N. Rea, J.A. Pons, D.N. Aguilera, J.A. Miralles, Mon. Not. R. Astron. Soc. **434**, 123 (2013)
D. Viganò, R. Perna, N. Rea, J.A. Pons, Mon. Not. R. Astron. Soc. **443**, 31 (2014)
M. Vincke, M. Le Dourneuf, D. Baye, J. Phys. B **25**, 2787 (1992)
F.M. Walter, T. Eisenbeiß, J.M. Lattimer et al., Astrophys. J. **724**, 669 (2010)

G. Wunner, H. Ryder, W. Schmitt, H. Herold, Mon. Not. R. Astron. Soc. **198**, 769 (1982)
V.E. Zavlin, in *Neutron Stars and Pulsars*, ed. by W. Becker (Springer, New York, 2009), p. 181
S. Zane, R. Turolla, L. Stella, A. Treves, Astrophys. J. **560**, 384 (2001)
V.E. Zavlin, G.G. Pavlov, Yu.A. Shibanov, J. Ventura, Astron. Astrophys. **297**, 441 (1995)
V.E. Zavlin, G.G. Pavlov, Astrophys. J. **616**, 452 (2004)
V.E. Zavlin, G.G. Pavlov, Yu.A. Shibanov, Astron. Astrophys. **315**, 141 (1996)
V.V. Zheleznyakov, *Radiation in Astrophysical Plasmas* (Kluwer, Dordrecht, 1996)
F. Zwicky, Astrophys. J. **88**, 522 (1938)

DOI 10.1007/978-1-4939-3550-5_7
Reprinted from *Space Science Reviews* Journal, DOI 10.1007/s11214-015-0173-8

Radio Pulsars

V.S. Beskin[1,2] · S.V. Chernov[3] · C.R. Gwinn[4] · A.A. Tchekhovskoy[5]

Received: 6 March 2015 / Accepted: 6 June 2015 / Published online: 19 June 2015

Abstract Almost 50 years after radio pulsars were discovered in 1967, our understanding of these objects remains incomplete. On the one hand, within a few years it became clear that neutron star rotation gives rise to the extremely stable sequence of radio pulses, that the kinetic energy of rotation provides the reservoir of energy, and that electromagnetic fields are the braking mechanism. On the other hand, no consensus regarding the mechanism of coherent radio emission or the conversion of electromagnetic energy to particle energy yet exists. In this review, we report on three aspects of pulsar structure that have seen recent progress: the self-consistent theory of the magnetosphere of an oblique magnetic rotator; the location, geometry, and optics of radio emission; and evolution of the angle between spin and magnetic axes. These allow us to take the next step in understanding the physical nature of the pulsar activity.

Keywords Neutron stars · Pulsars

✉ V.S. Beskin
beskin@td.lpi.ru

S.V. Chernov
chernov@td.lpi.ru

C.R. Gwinn
cgwinn@physics.ucsb.edu

A.A. Tchekhovskoy
atchekho@berkeley.edu

1 Lebedev Physical Institute, Russian Academy of Sciences, Leninskii prosp. 53, Moscow, 119991, Russia

2 Moscow Institute of Physics and Technology, Institutskii per. 9, Dolgoprudnyi, Moscow Region, 141700, Russia

3 Lebedev Physical Institute, Russian Academy of Sciences, Profsoyuznaya st. 84/32, Moscow, 117997, Russia

4 Physics Dept., University of California, Santa Barbara, Broida Hall, Santa Barbara, CA, 93106, United States

5 Department of Astronomy, University of California, Berkeley, CA, 94720, United States

1 Introduction

Radio pulsars are the archetypal observed neutron stars. Their discovery at the end of the 1960s (Hewish et al. 1968) was definitely one of the major astrophysical events of the 20th century. Their discovery confirmed the theoretical prediction of neutron stars in the 1930s (Landau 1932; Baade and Zwicky 1934). Neutron stars have mass M of about 1.2–2.0 $M_\odot$, near the Chandrasekhar mass limit 1.4 $M_\odot$; and radius R of only 10–15 km. They result from the collapse of typical massive stars in the final stage of their evolution (Shapiro and Teukolsky 1985); or from white dwarfs, when accretion from a companion star pushes them over the Chandrasekhar limit (Whelan and Iben 1973; Bailyn and Grindlay 1990; Nomoto and Kondo 1991; Schwab et al. 2015). These formation mechanisms provide the simplest explanation for both the observed short spin periods P to as small as $P = 1.39$ ms, and superstrong magnetic fields with $B_0 \sim 10^{12}$ G.

Most radio pulsars are solitary. Of the more than 2400 pulsars known by the end of 2014, only about 230 were members of binary systems.[1] Even in binary systems, mass transfer from the companion star to the neutron star is negligible. The radio luminosities of pulsars are low relative to the sensitivities of even the largest radiotelescopes, so that our catalog of pulsars is not complete even to a distance of a kpc. Because the Milky Way is an order of magnitude larger, we can observe only a small fraction of "active" pulsars. Because the duration of the active life of pulsars is small, the total number of extinguished pulsars in our Galaxy must be about 10^8–10^9 (Manchester et al. 2005).

2 Theoretical Overview

2.1 Early Pulsar Paradigm—Vacuum Dipole

The basic physical processes determining the observed activity of radio pulsars were understood almost immediately after their discovery (Pacini 1967; Gold 1968). In particular, it quickly became clear that the highly-regular pulsed radio emission that gives rise to their name is related to the rotation of neutron stars. Furthermore, it was evident that radio pulsars are powered by the rotational energy of the neutron star, and the mechanism of energy release is related to their superstrong magnetic fields, with $B_0 \sim 10^{12}$ G. The Larmor formula for energy loss form a magnetic dipole provides an estimate of energy losses (Landau and Lifshitz 1989):

$$W_{\rm tot} = -I_{\rm r}\Omega\dot{\Omega} \tag{1}$$

$$\approx \frac{1}{6}\frac{B_0^2\Omega^4R^6}{c^3}\sin^2\chi \tag{2}$$

where $I_{\rm r} \sim MR^2$ is the moment of inertia of the neutron star, χ is the angle between the magnetic dipole axis and the spin axis, and $\Omega = 2\pi/P$ is the angular velocity of neutron star rotation. Finally, the strength of the magnetic field at the polar cap is B_0.

For most pulsars, energy losses range from 10^{31}–10^{34} erg s^{-1} and can reach 10^{38}–10^{39} erg s^{-1} for very young, fast pulsars, such as the Crab and Vela pulsars. These energy losses correspond to the observed spin-down rate $\mathrm{d}P/\mathrm{d}t \sim 10^{-15}$, or to the spin-down time $\tau_{\rm D} = P/2\dot{P} \sim$ 1–10 Myr.

[1] See ATNF catalog: http://www.atnf.csiro.au/people/pulsar/psrcat/.

After the measurement of the rotational slow-down $\dot{P}$ of the Crab pulsar (Richards and Comella 1969), it was quickly realized that:

- the rate of the energy loss of the rotating neutron star $W_{\rm tot} \approx 5 \times 10^{38}$ erg s^{-1} (1) coincides with the power required to illuminate the Crab Nebula (Gold 1969), and
- the dynamical age of the radio pulsar $\tau_{\rm D} = P/2\dot{P} \approx 1000$ years coincides with the explosion of the historical supernova AD 1054 that brought the Crab Nebula into existence (Comella et al. 1969).

These associations cemented the identification of pulsars as rotating neutron stars. In contrast to these phenomena, radio emission amounts to only 10^{-4}–10^{-6} of total energy losses. For most pulsars this corresponds to 10^{26}–10^{28} erg s^{-1}, 5–7 orders of magnitude less than the luminosity of the Sun.

2.2 Electron-Positron Generation

Goldreich and Julian (1969) showed shortly after the discovery of pulsars that a pulsar's rotating magnetic field will acquire a corotating charge density that opposes induced electric fields and $\mathbf{J} \times \mathbf{B}$ forces. As Sturrock (1971) quickly realized, individual photons can generate electron-positron pairs when they cross lines of the magnetic field, by the process $\gamma + B \to e^+ + e^- + B$. The photon energy must exceed the threshold $2m_{\rm e}c^2$. The probability per-unit-length for conversion of a photon with energy $\mathcal{E}_\gamma$ far above this threshold propagating at an angle of θ to the magnetic field $\mathbf{B}$ is (Berestetsky et al. 1982)

$$w = \frac{3\sqrt{3}}{16\sqrt{2}} \frac{e^3 B \sin\theta}{\hbar m_{\rm e} c^3} \exp\left(-\frac{8}{3}\frac{B_\hbar}{B\sin\theta}\frac{m_{\rm e}c^2}{\mathcal{E}_\gamma}\right). \tag{3}$$

Here, the characteristic value $B_\hbar = m_{\rm e}^2 c^3/e\hbar \approx 4.4 \times 10^{13}$ G is the magnetic field for which the energy gap between two Landau levels reaches the rest energy of an electron: $\hbar\omega_B = m_{\rm e}c^2$. As gamma-quanta are radiated by particles moving along the curved magnetic field lines, one can evaluate the photon free path as (Ruderman and Sutherland 1975)

$$l_\gamma \approx \frac{8}{3\Lambda} R_{\rm c} \frac{B_\hbar}{B} \frac{m_{\rm e}c^2}{\mathcal{E}_\gamma}. \tag{4}$$

Here $R_{\rm c}$ is the curvature radius, and $\Lambda \approx 20$ is the logarithmic factor. As $l_\gamma \ll R$ for high enough photon energy, the vacuum magnetosphere of a neutron star with magnetic field $B_0 \sim 10^{12}$ G is unstable to the generation of charged particles.

In the very strong magnetic field of the neutron star, charged particles can move only along magnetic field lines. Therefore two substantially different regions must develop in the pulsar magnetosphere: regions of open and closed magnetic field lines (see Figs. 1, 2). Closed magnetic field lines do not intersect the light cylinder, where co-rotation speed equals that of light, at radius $R_{\rm L} = c/\Omega$ ($\sim 10^{10}$ cm for ordinary pulsars). Particles on these field lines turn out to be captured. Open field lines intersect the light cylinder, and particles on these field lines can travel to infinity. Consequently, plasma must be continuously regenerated near the magnetic poles of a neutron star (see Fig. 3).

In addition to the primary plasma generated by individual photons and the magnetic field, as discussed above, a secondary plasma forms from the longitudinal electric field (which accelerates particles up to energies high enough to radiate hard γ-quanta), as first indicated

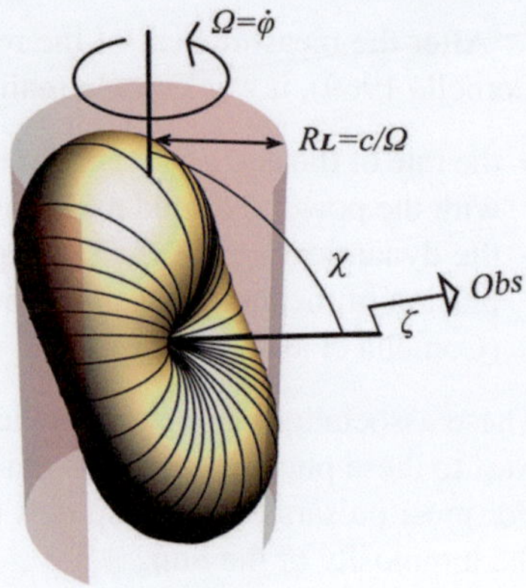

Fig. 1 Vacuum dipole model of a pulsar, showing the region of closed magnetic field lines and the light cylinder. The angle between the rotation and magnetic poles is χ, often called the "inclination angle"; and the minimum angle between the line of sight and the magnetic axis is ζ, sometimes called the "impact angle". Sometimes χ is denoted by α, and ζ by β. See Lyne and Manchester (1988)

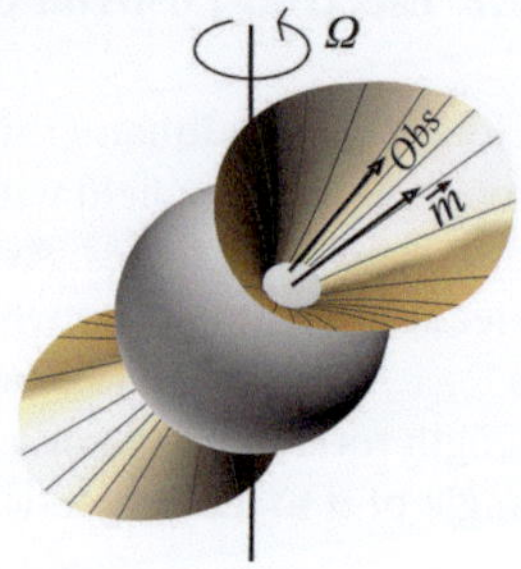

Fig. 2 Intersection of the cone of the last open field lines with the surface of the neutron star, showing the polar cap. The vector **m** is the axis of the magnetic dipole field, and Obs points toward the observer

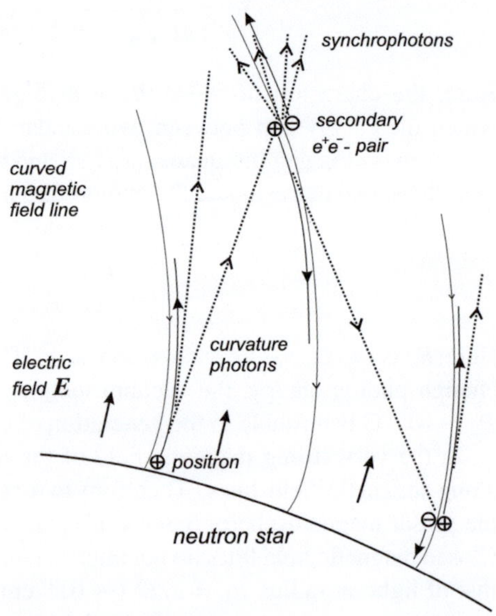

Fig. 3 Structure of the particle generation region. The primary particles are accelerated along the curved magnetic field lines and begin to radiate hard γ-rays. These curvature photons (*dotted lines*) propagating in the curved magnetic field reach the particle generation threshold and create electron-positron pairs. Taken from Beskin (1999)

by Sturrock (1971) and then studied in more detail by Ruderman and Sutherland (1975), as well as by Eidman's group (Al'ber et al. 1975). The continuous escape of particles along the open field lines leads to formation of a strong electric field along the magnetic field. This longitudinal electric field forms in the vicinity of the magnetic poles. The secondary plasma-

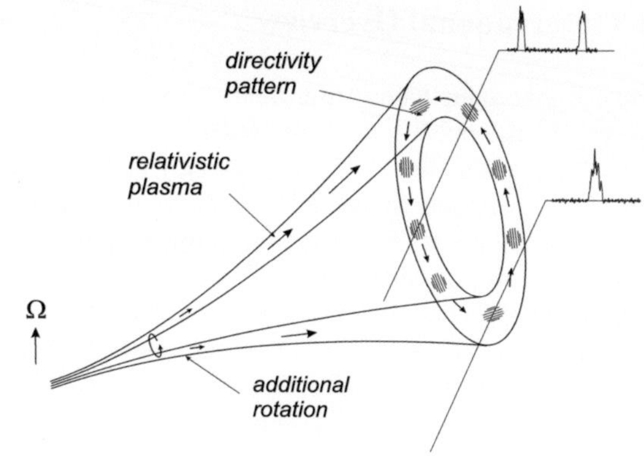

Fig. 4 The hollow cone model. If the intensity of the radio emission is directly connected with the outflowing plasma density, radio emission must decrease near the magnetic pole. Consequently, we expect a double profile when the line of sight passes near the magnetic pole, and a single profile when it passes further away. Taken from Beskin (1999)

generation condition determines its height. Another model, based on the assumption of free particle escape from the neutron star surface, was first studied by Arons' group (Fawley et al. 1977; Scharlemann et al. 1978; Arons and Scharlemann 1979), and recently in more detail by Istomin and Sobyanin (2011a,b), Medin and Lai (2010), Timokhin (2010), Chen and Beloborodov (2013) and Timokhin and Arons (2013).

2.3 Hollow-Cone Model

The hollow-cone model (Radhakrishnan and Cooke 1969; Dyks et al. 2004) explains the basic observed properties of radio emission in the context of the above particle generation processes, without reference to a microphysical model for that emission. This model, already proposed at the end of the 1960s, perfectly accounts for the basic geometric properties of the radio emission. This model proposes that outflowing plasma launches radio emission tangent to open magnetic field lines at a particular altitude above the surface of the neutron star. The characteristic frequency of radiation may depend on altitude: the "radius-to-frequency mapping" (Ruderman and Sutherland 1975). Plasma density and geometry of open field lines define a "directivity pattern". The observed average pulse is a cut across this directivity pattern.

Secondary particle generation is impossible in a nearly rectilinear magnetic field because, first, little curvature radiation is emitted; and second, photons emitted by relativistic particles propagate at small angles to the magnetic field. Therefore, as shown in Fig. 4, in the central region of the open magnetic field lines, a decrease in secondary plasma density is expected. If we make the rather reasonable assumption that radio emission is less when the outflowing plasma density is less, the intensity of radio emission must decrease in the center of the region of open field lines, corresponding to the center of the directivity pattern. Therefore, if without going into details,[2] we should expect a single (one-hump) mean profile in pulsars in which the line of sight intersects the directivity pattern far from its center and the double (two-hump) profile for the central passage. This is precisely as observed in reality (Lyne and Graham-Smith 1998).

[2]Actually, the mean profiles have a rather complex structure, see e.g., Rankin (1983, 1990), Lyne and Graham-Smith (1998).

3 Observational Overview

Pulsars take their name from their remarkably stable periodic emission. The rotation frequency of the pulse train is the angular velocity of the neutron star. Folding the observed pulse train at this fundamental frequency yields an average pulse profile. In most cases this profile is extremely stable, both in form, and in arrival phase at the rotational frequency. This stability allows for precision timing of pulsars, with remarkable applications in structure and evolution of stellar systems containing pulsars, and in tests of special and general relativity (Camenzind 2007). The stability of the mean profile suggests that rotation carries the line of sight through a beam of emitted radiation locked to the surface of the neutron star, and that relatively permanent features of the neutron star and its co-rotating magnetosphere determine the shape of that beam.

However, pulsar emission shows a remarkable degree of variability on all timescales, extending from nanoseconds to months or years. The stable pulse profiles that characterize that stability appear only after 100 or more pulses are added together, for pulsars strong enough to detect variability of emission. Indeed, Popov et al. (2006) suggest that individual micropulses are the "atoms" of pulsar emission. For those who seek to understand emission processes of pulsars, as well as those who merely wish to exploit pulse stability for other scientific goals, pulse variability can provide crucial insights.

Observations of pulsars occupy a multi-dimensional space. The fundamental observables include intensity and polarization of electromagnetic radiation, as function of time in pulse phase and over many pulses. These fundamental observables show both deterministic and random properties, with random properties in particular showing variations over all timescales. At radio wavelengths, pulsar spectra are nearly power-law, but comparisons of pulse shape and structure among wavelength ranges yields great insight into emission geometry and processes (see, for example, Shearer et al. 2003; Lommen et al. 2007; Harding et al. 2008; Strader et al. 2013).

Among the important quantities derived from observations of radio emission from pulsars are the spindown rate, polarization as a function of pulse phase, size of emission region, and evolution of angle between the spin and magnetic dipole.

4 Magnetosphere of an Oblique Magnetic Rotator

A pulsar represents an elegant problem in electrodynamics: a rotating, conducting sphere with a dipole magnetic field (Beskin and Zheltoukhov 2014). This simple picture is complicated by the necessity of a corotating charge distribution, the roles of open and closed field lines, and energy transport by an outflowing wind (Goldreich and Julian 1969). As we summarize in this section, models have progressed from analytic studies of aligned rotators with simple magnetic field configuration and massless charges to self-consistent models including oblique magnetic fields, realistic particle masses, and a range of length scales.

4.1 Current Losses

If the pair creation process is sufficiently effective, magnetic dipole radiation will not carry energy away from the rotating neutron star, because the plasma that fills the magnetosphere fully screens any low-frequency radiation from the neutron star (Beskin et al. 1983, 1993; Mestel et al. 1999). However, in this case, electric currents extract rotational energy from the neutron star, through the Ampère force of one current on another. The currents in question

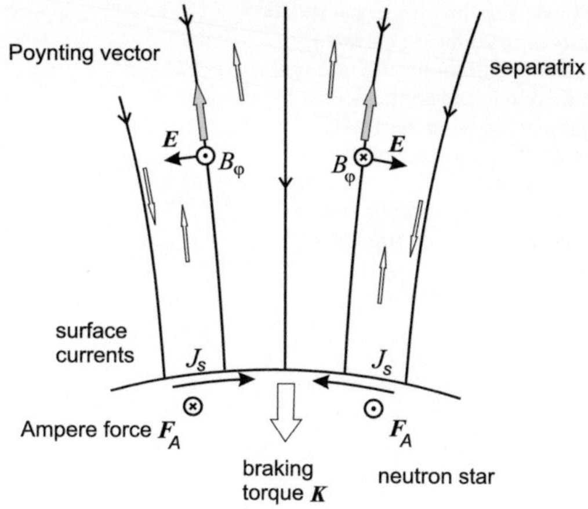

Fig. 5 Schematic view of the axisymmetric polar cap showing magnetospheric current density (*open arrows*), surface currents, Ampére force on surface currents, and braking torque. Here only the symmetric current i_s is present. Taken from Beskin (1999)

are those along magnetic field lines in the magnetosphere and across the pulsar's polar cap, acting together with those responsible for the magnetic field of the neutron star. Just as in the case of magnetic dipole radiation, energy release from the rotating neutron star is related to the electromagnetic energy flux given by the Poynting vector, and the total energy losses can be again estimated using the Larmor formula (2).

The braking torque **K** of the Ampère force results in the following time evolution of the angular velocity Ω and the inclination angle χ:

$$I_r\dot{\Omega} = K_\parallel \cos\chi + K_\perp \sin\chi, \tag{5}$$

$$I_r\Omega\dot{\chi} = K_\perp \cos\chi - K_\parallel \sin\chi, \tag{6}$$

where two components of the torque **K** parallel and perpendicular to the magnetic dipole **m** can be written in the form (Beskin et al. 1993)

$$K_\parallel = -c_\parallel \frac{B_0^2\Omega^3R^6}{c^3} i_s, \tag{7}$$

$$K_\perp = -c_\perp \frac{B_0^2\Omega^3R^6}{c^3}\left(\frac{\Omega R}{c}\right) i_a. \tag{8}$$

Here the coefficients $c_\parallel$ and $c_\perp$ are factors of the order of unity dependent on the profile of the longitudinal current and the form of the polar cap.

The scalar current from the polar cap i has been divided into symmetric and antisymmetric contributions, i_s and i_a, depending upon whether the direction of the current is the same in the north and south parts of the polar cap, or opposite. For an axisymmetric rotating neutron star ($\chi = 0$, Fig. 5), we have $i_a = 0$ and $i_s = 1$ (Goldreich and Julian 1969). Conversely, for the orthogonal rotator (Fig. 6) we have $i_s = 0$ and $i_a = 1$. Here we apply normalization to the Goldreich-Julian current, $I_{GJ} = \pi R_0^2 j_{GJ}$, where $R_0 \approx R(\Omega R/c)^{1/2}$ is the polar cap radius, and $j_{GJ} = \langle|\Omega \cdot \mathbf{B}|\rangle/2\pi$ (with scalar product) is the mean current density within the polar cap. Note that for $i_s \approx i_a \approx 1$, (7) and (8) imply that:

$$K_\perp \sim \left(\frac{\Omega R}{c}\right) K_\parallel. \tag{9}$$

Therefore, $K_\perp \ll K_\parallel$. We will use these expressions in the following sections.

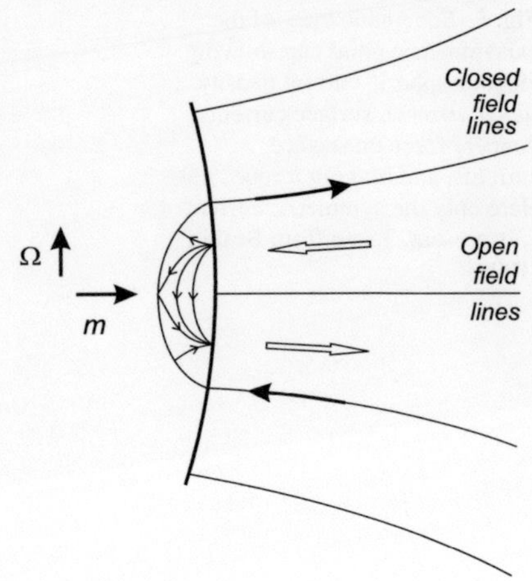

Fig. 6 For the orthogonal rotator only antisymmetric current i_a (i.e., the current having different direction in the north and south parts of the polar cap) takes place. The structure of the surface currents within the polar cap and along the separatrix is also shown. Taken from Beskin et al. (2013)

If we suppose that, in reality, the longitudinal current j is determined by the local charge density $\rho_{GJ} = -\boldsymbol{\Omega} \cdot \mathbf{B}/2\pi c$, and note that ρ_{GJ} is proportional to $\cos\chi$ in the vicinity of the polar cap, one can write down

$$i_s = i_s^A \cos\chi, \tag{10}$$

$$i_a = i_a^A \sin\chi. \tag{11}$$

Consequently, the relations (5)–(6) can be rewritten in the form

$$I_r \dot{\Omega} = K_\parallel^A + \left[K_\perp^A - K_\parallel^A\right] \sin^2\chi, \tag{12}$$

$$I_r \Omega \dot{\chi} = \left[K_\perp^A - K_\parallel^A\right] \sin\chi \cos\chi. \tag{13}$$

As we see, both expressions contain the factor $[K_\perp^A - K_\parallel^A]$. This implies that the sign of $\dot{\chi}$ is given by the χ-dependence of the energy losses (Philippov et al. 2014a). In other words, the inclination angle χ will evolve to 90° (to counter-alignment) if the total energy losses decrease for larger inclinational angles, and to co-alignment if they increase with inclination angle.

Because the plasma filling the pulsar magnetosphere is secondary (in other words, it is produced by the primary particles accelerated by the longitudinal electric field), at any point out to the light cylinder, the energy density of the electromagnetic field must be much larger than the energy density of the magnetospheric plasma. For the same reason, energy transport is given by the Poynting vector (see Fig. 5).

4.2 Split-Monopole Model

The remarkable analytical solution found by Michel (1973) serves to illustrate the transport of energy by Poynting flux. In the force-free approximation, when massless charged particles move radially with the velocity of light, and with the Goldreich-Julian current density $j_{GJ} = \rho_{GJ} c$, a split-monopole magnetic field is the exact solution to Maxwell equation, both inside and beyond the light cylinder (see Fig. 7). The monopolar magnetic field is split so that the magnetic flux converges in the southern hemisphere and diverges in the northern one. In this solution, Ampère forces from longitudinal currents along magnetic field lines, and from corotation currents from rotating charge density, are fully compensated.

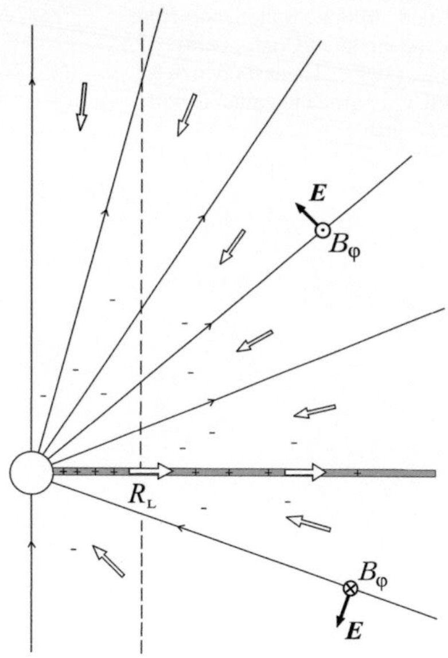

Fig. 7 The Michel split-monopole solution, in which electric field E_θ has magnitude equal to the toroidal magnetic field B_φ. This solution requires a conducting current sheet, to close electric currents (*outlined arrows*). Taken from Beskin (1999)

In the Michel split-monopole solution, the electric field **E** has only a θ-component, and is equal in magnitude to the toroidal component of the magnetic field:

$$B_\varphi = E_\theta = -B_0\left(\frac{\Omega R}{c}\right)\frac{R}{r}\sin\theta. \tag{14}$$

At distances larger than the light cylinder radius, this magnetic field becomes larger than the poloidal magnetic field $B_\mathrm{p} = B_0(R/r)^2$. On the other hand, in this solution the total magnetic field remains larger than the electric field everywhere, so that the fields form electromagnetic waves only at infinity.

Because magnetic flux converges in the lower hemisphere and diverges in the upper one in the split-monopole solution, a current sheet must lie in the equatorial plane (see Figs. 7, 8). This sheet closes the longitudinal electric currents elsewhere in the magnetosphere. This structure of the magnetic field and current sheet has been confirmed numerically (Contopoulos et al. 1999; Ogura and Kojima 2003; Gruzinov 2005; Komissarov 2006; McKinney 2006; Timokhin 2006).

Using Eq. (14), one easily finds that the Poynting vector $\mathbf{S} = (c/4\pi)\mathbf{E}\times\mathbf{B}$ is:

$$S(\theta) = \frac{B_0^2 c}{4\pi}\left(\frac{\Omega R}{c}\right)^2\frac{R^2}{r^2}\sin^2\theta. \tag{15}$$

This implies that the energy flux is concentrated near the equatorial plane. This θ-dependence of the energy flux is used by many authors (Bogovalov and Khangoulyan 2002; Komissarov and Lyubarsky 2003). On the other hand, at large distances $r \gg R_\mathrm{L}$, Ingraham (1973) and Michel (1974) found another asymptotically radial solution, with $E_\theta(\theta) = B_\varphi(\theta)$, resulting in a radial Poynting vector with arbitrary θ-dependence. In all of these solutions, the relation

$$S(\theta) \propto B_r^2(\theta)\sin^2\theta \tag{16}$$

is valid.

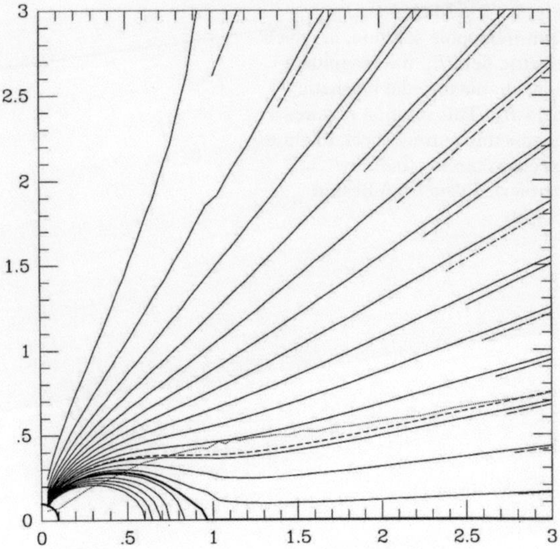

Fig. 8 Magnetospheric structure in the model of Contopoulos et al. (1999). The last open field line is assumed to coincide with the equator

Bogovalov (1999) generalized the split-monopole model, showing that in the force-free approximation the "inclined split monopole field" is a solution of the problem as well. In this solution,

$$B_\varphi = E_\theta = -B_0\left(\frac{\Omega R}{c}\right)\frac{R}{r}\sin\theta\,\mathrm{sign}\,\Theta \tag{17}$$

and $B_\mathrm{p} = B_0(R/r)^2\,\mathrm{sign}\,\Theta$, where

$$\Theta = \sin\chi\,\sin\theta\,\sin(\varphi - \Omega t + \Omega r/c) + \cos\chi\,\cos\theta. \tag{18}$$

In this case, within the cones $\theta < \pi/2 - \chi$, $\pi - \theta < \pi/2 - \chi$ around the rotation axes, the electromagnetic field is not time dependent; whereas in the equatorial region, the electromagnetic fields change the sign at the instant $\Theta = 0$. In other words, the condition $\Theta = 0$ defines the location of the current sheet. We stress that the expression (18) for the shape of the current sheet remains true for the other radial asymptotic solutions, with $E_\theta(\theta) = B_\varphi(\theta)$ but arbitrary θ-dependence, as well (Arzamasskiy et al. 2015a). Numerical simulations obtained recently for the oblique force-free rotator confirm this conclusion as well (Spitkovsky 2006; Kalapotharakos and Contopoulos 2009; Kalapotharakos et al. 2012b; Tchekhovskoy et al. 2013; Philippov et al. 2014a).

4.3 Magnetohydrodynamic Models

As was already stressed, recently numerical simulations have become possible that can simulate the structure of plasma-filled magnetospheres from first principles. Contopoulos et al. (1999) found an iterative way to do this and obtained the first solution for an *aligned* force-free pulsar magnetosphere that extended out to infinity (see Fig. 8). Their results were subsequently verified by other groups within force-free and magnetohydrodynamic (MHD) approximations (e.g., Gruzinov 2005; Timokhin 2006; McKinney 2006; Komissarov 2006; Parfrey et al. 2012; Ruiz et al. 2014) as well as using particle-in-cell (PIC) approach (Philippov and Spitkovsky 2014; Chen and Beloborodov 2014; Cerutti et al. 2014; Belyaev 2014).

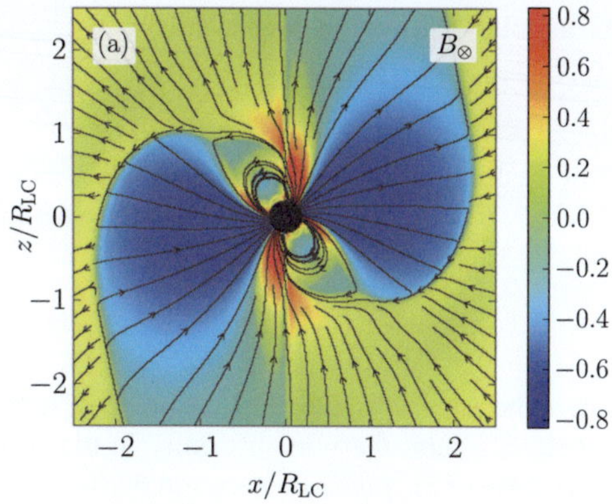

Fig. 9 Slice through the $\mathbf{m}-\boldsymbol{\Omega}$ plane of a relativistic MHD simulation of an oblique pulsar magnetosphere (obliquity angle $\chi = 60°$) taken after 3 rotations. *Solid lines* show field lines as traced in the image plane. Into-the—plane magnetic field component, $B_\otimes$, is shown with *color* (*red*—into plane, *blue*—out of plane). Taken from Tchekhovskoy et al. (2013)

Spitkovsky (2006) carried out the first 3D, oblique pulsar magnetosphere simulations. Using the force-free approximation, he found that pulsar spindown luminosity increases with increasing obliquity angle, χ, which is the angle between the rotational and magnetic axes. The spindown obtained in such force-free and MHD models is well-described by

$$W_{\rm tot} = W_{\rm aligned}\big(1 + \sin^2 \chi\big), \tag{19}$$

where $W_{\rm aligned} = m^2\Omega^4/c^3$ is the spindown luminosity of an aligned plasma-filled pulsar magnetosphere, and $m = B_0 R^3/2$ is the magnetic dipole moment of the pulsar. More recently, these results were confirmed using time-dependent 3D force-free (Kalapotharakos and Contopoulos 2009; Pétri 2012; Kalapotharakos et al. 2012a), MHD (Tchekhovskoy et al. 2013), and PIC (Philippov et al. 2014b) studies. Figure 9 shows a vertical slice through the results of a 3D MHD simulation of an oblique pulsar magnetosphere with obliquity angle $\chi = 60°$. One can clearly see the closed zone that extends out to the light cylinder located at $|x| = R_{\rm LC}$, beyond which starts a warped magnetospheric current sheet, across which all field components undergo a jump. The structure of this current sheet is presently poorly understood, in particular it is not known if in the perfect conductivity limit the magnitude of the magnetic field in the sheet vanishes.

What causes this increase of spindown luminosity with the increase of pulsar obliquity? It turns out that there are two factors: (i) an increase in the amount of open magnetic flux, which accounts for about 40 % of the increase, and (ii) redistribution of open magnetic flux toward the equatorial plane of the pulsar magnetosphere, which accounts for the remaining 60 %.

In fact, the spindown trend (19) can be reproduced via a simple toy model. Suppose that the magnetic field that reaches the light cylinder in an oblique rotator, with inclination angle χ, retains the dipolar structure at $r = r_0 \gg R_{\rm LC}$,

$$B_r = \frac{B_0 r_0^2}{r^2} \sin\theta_{\rm m}, \tag{20}$$

where $\theta_{\rm m}$ is the magnetic colatitude, or the angle away from the magnetic axis,

$$\theta_{\rm m} = \arccos(\sin\chi \sin\theta \cos\varphi + \cos\theta \cos\chi). \tag{21}$$

How would the pulsar spindown change if we kept the total open magnetic flux, $\Phi_{\rm open} = \pi r_0^2 B_0$, fixed, and inclined the pulsar, i.e., increased χ? To find this out, let us first compute the angular distribution of φ-averaged B_r^2:

$$\langle B_r^2 \rangle_\varphi = B_0^2 \big(0.5 \sin^2 \chi \sin^2 \theta + \cos^2 \theta \cos^2 \chi \big). \tag{22}$$

Now, making use of the fact that $B_\varphi \approx E_\theta = -B_r \Omega r \sin\theta / c$ and the radial Poynting flux is $S_r = c E_\theta B_\varphi / 4\pi \approx (B_r \Omega r \sin\theta)^2 / 4\pi c$, we obtain (Tchekhovskoy et al. 2015):

$$W_{\rm toy}(\chi) = \iint S_r \, {\rm d}\omega = \frac{\Omega^2 \Phi_{\rm open}^2}{7.5\pi^2 c} \big(1 + \sin^2 \chi \big), \tag{23}$$

where the integral is over, e.g., a sphere of radius r_0. Clearly, if the total magnetic flux $\Phi_{\rm open}$ is held constant, the nonuniformity in the surface distribution of magnetic flux causes an *enhancement in spindown losses* at higher inclination angles, consistent with the numerical simulations (see Eq. (19)). In the simulations, we find that the magnetic flux itself is an increasing function of χ, $\Phi_{\rm open} \propto (1 + 0.2 \sin^2 \chi)$,[3] so the two effects—of the non-uniformity of the open magnetic flux and the change in the amount of open magnetic flux—have very similar inclination-dependences.

In reality, the structure of the plasma-filled magnetosphere is of course more complex than given by Eq. (20), but the qualitative effect is the same: the inclination of the magnetic axis relative to the rotational axis leads to the shift of the peak of $|B_r|$ away from the axis and toward the equatorial plane and an increase in the spindown luminosity (Tchekhovskoy et al. 2015).

More recently, PIC models have been developed and, in those cases when the magnetospheric polar cascade is efficiently operating and is able to fill the magnetosphere with abundant plasma, are in agreement in the amount of spindown and large-scale dissipation as in MHD simulations (see e.g. Philippov and Spitkovsky 2014; Chen and Beloborodov 2014; Cerutti et al. 2014; Belyaev 2014; Philippov et al. 2014b). Interestingly, if a mechanism of pair formation operates only near the surface of the star, aligned pulsar magnetospheres in PIC simulations do not reach a force-free state (Chen and Beloborodov 2014). In fact, PIC simulations, into which simplified physics of the polar cascade was included, show the development of the polar cascade and of a force-free—like magnetosphere only for high inclinations, $\alpha > 40^\circ$ (Philippov et al. 2014b).

5 Observations: Energy Loss from Pulsars

5.1 Spindown

In principle, the time rate of change of pulse period $\dot{P}$ is easy to measure. Because individual pulses can be numbered, period and period derivative are among the fundamental parameters of a timing model. Period derivative is easily associated with the loss of rotational kinetic energy via electromagnetic radiation and particle wind. The Larmor formula for magnetic dipole radiation then directly associates energy loss with the magnetic moment of the neutron star. This provides a characteristic scale.

Pulsars with periods longer than a fraction of a second show timing noise: random variations of pulse arrival time that change slowly with time (Helfand et al. 1980). These variations are most extreme for the young Crab and Vela pulsars (Boynton et al. 1972; Lyne and

[3] This relation coincides exactly with one obtained by Beskin et al. (1993) analytically.

Graham-Smith 1998; Scott et al. 2003; Dodson et al. 2007). Among millisecond pulsars, B1937+214 shows timing noise, but other millisecond pulsars may not (Kaspi et al. 1994; Cognard et al. 1995).

Several pulsars show clear variations in spindown rate associated with changes in pulse properties. The radio pulsars B1931+24, J1832+0029, and J1841-0500 intermittently switch between an "on" radio-loud state in which they appear as ordinary radio pulsars, and an "off" state in which no radio emission is detected. The spin-down rate is higher in the "on" state than the "off" state, by a factor of $f_{\mathrm{on}\to\mathrm{off}} = \dot{P}_{\mathrm{on}}/\dot{P}_{\mathrm{off}} \approx 1.5$ for B1931+24 (Kramer et al. 2006) and J1832+0029 (Lyne 2009), and $f_{\mathrm{on}\to\mathrm{off}} \approx 2.5$ for J1841-0500 (Camilo et al. 2012). The gamma-ray pulsar J2021+4026 Allafort et al. (2013) displays two states with intensities different by 20 % and with distinct pulse profiles, each associated with a different spindown rate: $f_{\mathrm{on}\to\mathrm{off}} = 1.04$. Pulsar B0919+06 shows quasiperiodic variations between two states with different spindown rates and different pulse profiles (Perera et al. 2015). Lyne et al. (2010) propose that the phenomenon of intermittency is quite general: they find that timing noise for six pulsars can be expressed as the superposition of two states, characterized by distinct pulse profiles and spindown rates, with rather rapid changes between states. From these discussions it is clear that magnetospheric structure affects spindown, as one would suspect from theoretical considerations discussed in Sect. 2.3 above.

Kramer et al. (2006) (see also Beskin and Nokhrina 2007) proposed that two distinct magnetospheric states lead to the observed difference in spindown rates. They associated the "off" state with a magnetosphere depleted of charge, and the "on" state with magnetospheric currents sufficient to produce the observed change in spindown. Li et al. (2012a) observe that the simplest model for the "on" state is the force-free magnetosphere (Spitkovsky 2006), which exhibits spindown rates at least three times that of a vacuum dipole. They suggest a modified picture where the "on" state is the force-free magnetosphere, and the "off" state has no charge on open field lines, but carries the Goldreich-Julian charge on closed field lines. This leads to ratios $f_{\mathrm{on}\to\mathrm{off}} = 1.2$ to 2.9 for inclination angles of $\chi > 30^\circ$. Smaller inclinations lead to larger $f_{\mathrm{on}\to\mathrm{off}}$.

5.2 σ-Problem

Thus, we see that all analytical and numerical force-free models of the pulsar magnetosphere demonstrate the existence of an almost-radial highly-magnetized wind, flowing outward from the pulsar magnetosphere. On the other hand, observations show that most energy far from the neutron star must be carried by relativistic particles (Kennel and Coroniti 1984a,b). For example, the analysis of the emission from the Crab Nebula in the shock region located at a distance of $\sim 10^{17}$ cm from the pulsar in the region of interaction of the pulsar wind with the supernova remnant definitely shows that the total flux W_{em} of the electromagnetic energy in this region is no more than $\sim 10^{-3}$ of the particle energy flux W_{part}. Thus, in the asymptotically far region of pulsar models, the Poynting flux must be completely converted into an outgoing particle flux before reaching the reverse shock at distances of ~ 0.1 pc. Axisymmetric numerical models of jets from radio pulsars are constructed exactly under this assumption (Kirk et al. 2009, and references therein).

The transformation from Poynting flux to particles apparently occurs much closer to the neutron star, at distances comparable to the size of the light cylinder. This is evidenced by the detection of variable optical emission from companions in some close binary systems involving radio pulsars (Fruchter et al. 1988, 1990; Kulkarni et al. 1988; Ryba and Taylor 1991; Stappers et al. 1996; Roberts 2011; Pallanca et al. 2012; Romani et al. 2012; Kaplan et al. 2013; Breton et al. 2013). This variable optical emission with a period equal exactly to the orbital period of the binary can be naturally related to the heating of the companion's surface facing the radio pulsar. It was found that the energy reradiated by the companion star almost matches the total energy emitted by the radio pulsar into the corresponding solid angle. Clearly, this fact cannot be understood either in the magnetic-dipole radiation model or by assuming a Poynting-dominated strongly-magnetized outflow, since the transformation coefficient of a low-frequency electromagnetic wave cannot be close to unity. Only if a significant fraction of the energy is related to the relativistic particle flux can the heating of the star's surface be effective enough. Moreover, eclipses of the double-pulsar system show effects of the particle wind from one object impinging upon (Lyne et al. 2004; Jenet and Ransom 2004; McLaughlin et al. 2004; Lyutikov 2004; Demorest et al. 2004). Therefore, the so-called σ-problem—the question as to how the energy can be converted from electromagnetic fields to particles in the pulsar wind—remains one of great unsolved problems of modern astrophysics. We note that the σ-problem appears to be rather general and in addition to neutron-star powered outflows it applies to black-hole powered, collimated outflows known as astrophysical jets, such as in the magnetically-arrested disk (MAD) scenario (Tchekhovskoy et al. 2011; Tchekhovskoy and McKinney 2012; Zamaninasab et al. 2014; Zdziarski et al. 2014; Ghisellini et al. 2014; Tchekhovskoy 2015). Theoretical models suggest that the jets accelerate roughly up to the equipartition between the magnetic and kinetic energies, beyond which the acceleration slows down dramatically, locking in a substantial fraction of energy in the magnetic form (Tchekhovskoy et al. 2009; Komissarov et al. 2009; Lyubarsky 2010; however, see Tchekhovskoy et al. 2010).

6 Theory: Polarization and Refraction of Radio Emission

6.1 Polarization

Pulsar emission is usually highly linearly polarized, with a small fraction of circular polarization. Like the mean profile, the profiles in polarization states are stable and are characteristic of the pulsar. This long-term stability of the mean properties indicates that the pulse arises as a cut through a radiation cone, with properties that are set by stable properties of the underlying neutron star. It is widely assumed that the polarization is determined by magnetic fields in or above the emission region. Those magnetic fields, in turn, are anchored in the solid crust of the neutron-star (Manchester 1995).

Rapid swings of the position angle of linear polarization through the pulse, first observed in the Vela pulsar by Radhakrishnan et al. (1969), suggest that a vector fixed in the frame of the rotating star influences the direction of linear polarization, a geometric inference known as the rotating-vector model. Radhakrishnan and Cooke (1969) proposed that this vector is the magnetic pole of the pulsar's nearly-dipolar magnetic field; this physical interpretation is known as the magnetic-pole model.

Radiotelescopes can measure polarization properties of individual pulses for a number of strong pulsars. Such studies indicate the presence of orthogonal modes, with polarization differing by 90°, and intensities varying from pulse to pulse (Manchester et al. 1975; Backer et al. 1976; Cordes et al. 1978; Backer and Rankin 1980; Stinebring et al. 1984a,b; McKinnon 2003). For most pulsars, present radiotelescopes can determine only average polarization properties; nevertheless the presence of two competing orthogonal modes can explain the observed departures from the characteristic pattern, for most of these weaker pulsars. Thus, the rotating-vector model, with two orthogonal linearly-polarized modes, successfully describes the characteristic swing of the angle of polarization with pulse phase

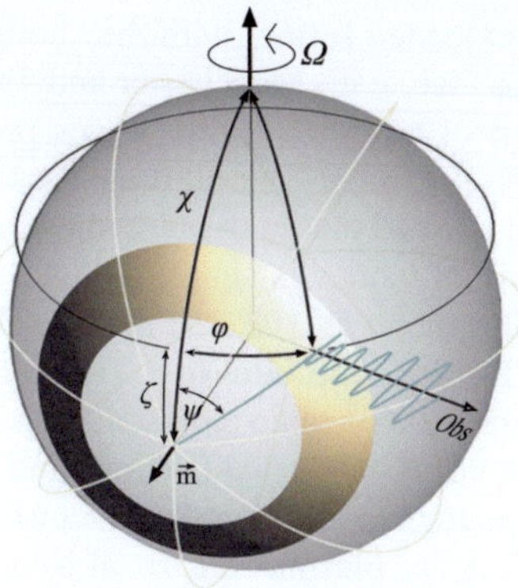

Fig. 10 Geometry for the rotating vector model. The emission surface is shown as a golden band. As the pulsar rotates, the line of sight to the observer traverses a circle of constant latitude, producing an observed pulse each time the emission surface crosses the line of sight. The dipolar magnetic field, projected onto the radius of the emission surface, radiates from the magnetic pole **m**. The instantaneous polarization of the observed radiation is parallel to the field lines, at position angle ψ at the observer

for most pulsars, across a wide range of pulsar parameters and despite observational selection effects (Rankin 1983, 1986; Lyne and Manchester 1988). The two modes are usually interpreted as the X-mode, with wave electric field perpendicular to stellar magnetic field ($E_W \perp B_0$); and the O-mode, with a component of electric field parallel to stellar field ($E_W \| B_0$). Both modes appear to be present, at some level, for all radio pulsars.

Work to model the polarization properties of pulsars in more detail, including the circular-polarized profile, have led to mapping of the polarization properties on the Poincaré sphere describing the Stokes parameters (McKinnon 2009; Chung and Melatos 2011a,b). These show a rich variety of patterns, with greater modulation of polarization being indicative of more complex patterns. Analysis of these patterns suggest emission, or refractive scattering within the pulsar's light cylinder. For some pulsars, the emission, or reprocessing region is inferred to lie at altitudes of 10 to 40 % of the light-cylinder radius.

6.2 Rotating Vector Model

The standard relation for the rotating vector model describes variation of the position angle of polarization ψ in the mean profile, under the assumption that the hollow-cone model is valid. In other words, it assumes that all absorption is absent, and that the magnetic field is dipolar in the emission region, precisely where the polarization is determined. This relation takes the form:

$$\psi = \arctan\left(\frac{\sin\chi \sin\phi}{\sin\chi \cos\zeta \cos\phi - \sin\zeta \cos\chi}\right). \tag{24}$$

Here, once again, χ is the inclination angle of the magnetic dipole to the rotation axis, ζ is the angle between the rotation axis and the direction toward the observer, and ϕ is the phase of the pulse. Figure 10 illustrates the geometry. Equation (24) has been used for many years in estimating the pulsar inclination angle, which is a very important parameter for determining the structure of the magnetosphere. Aberration and retardation effects (Blaskiewicz et al. 1991) have been included in only some studies (Mitra and Li 2004; Krzeszowski et al. 2009).

The rotating vector model, extended further with the "hollow cone" model, is based on the following three assumptions (see, e.g., Manchester and Taylor 1977): the formation of polarization occurs at the point of emission; radio waves propagate along straight lines; and cyclotron absorption can be neglected. But all these assumptions turn out to be incorrect. Barnard and Arons (1986) showed that in the innermost regions of the magnetosphere, the refraction of one of the normal modes is significant. After publication of the work of Mikhailovskii's group (Mikhailovskii et al. 1982), it became clear that cyclotron absorption can significantly affect the radio emission intensity. The influence of the magnetosphere plasma on variation of the polarization of radio emission propagating in the internal regions of the magnetosphere also must not be neglected (Petrova and Lyubarskii 2000).

The "limiting polarization" is the most important effect of magnetospheric propagation. Radio emission in the region of dense plasma consists of a superposition of normal modes: in particular, the principal axes of the polarization ellipse must remain aligned with the magnetic-field direction in the picture plane. Polarization in the vacuum region is independent of magnetic field. Hence, between the two lies a transition layer, past which the polarization is no longer affected by the magnetospheric plasma. For typical parameters of the pulsar magnetosphere, the formation of polarization occurs not at the emission point but at a distance of about $0.1R_{\rm L}$ from it (Cheng and Ruderman 1979; Barnard 1986). Taking this effect into account should also explain the observed fraction of circular polarization of the order of (5–10) %. Therefore, a consistent theory of radio wave propagation in the magnetosphere is required for a quantitative comparison of theoretical results on radio emission with observational data.

6.3 Propagation Effects

At present, the theory of radio wave propagation in the magnetosphere of a pulsar can be considered to provide the necessary precision (Petrova 2006; Wang et al. 2010; Beskin and Philippov 2011, 2012; Kravtsov and Orlov 1980). Four normal modes exist in the magnetosphere (Beskin et al. 1993; Lyne and Graham-Smith 1998). Two of them are plasma modes and two are electromagnetic, which are capable of departing from the magnetosphere. An extraordinary wave (the X-mode) with the polarization perpendicular to the magnetic field in the picture plane propagates along a straight line, while an ordinary wave (the O-mode) undergoes refraction and deviates from the magnetic axis. An important point here is that for typical magnetosphere parameters, refraction occurs at distances up to $0.1R_{\rm L}$, i.e., it can be considered separately from the cyclotron absorption and the limiting polarization. As shown in Fig. 11, the pulsar B0329+54 shows both X- and O-modes, with the O-mode displaying deviations from the rotating-vector model because of refraction (Edwards and Stappers 2004).

Based on the Kravtsov and Orlov (1980) method, Beskin and Philippov (2012) have used such a theory of wave propagation in a realistic pulsar magnetosphere, taking corrections to the dipole magnetosphere into account (based on the results obtained by numerical simulation in Spitkovsky 2006), together with the drift of plasma particles in crossed electric and magnetic fields, and a realistic particle distribution function. The theory developed allows dealing with an arbitrary profile of the spatial plasma distribution, which may differ from the one in the hollow-cone model, because precisely the inhomogeneous plasma distribution leads to the characteristic 'patchy' directivity pattern (Rankin 1990).

The main result consists in the prediction of a correlation between the sign of the circular polarization (the Stokes parameter V) and the sign of the derivative of the change in the polarization of the position angle, ψ, along the profile, $\mathrm{d}\psi/\mathrm{d}\phi$, where ϕ is the phase of

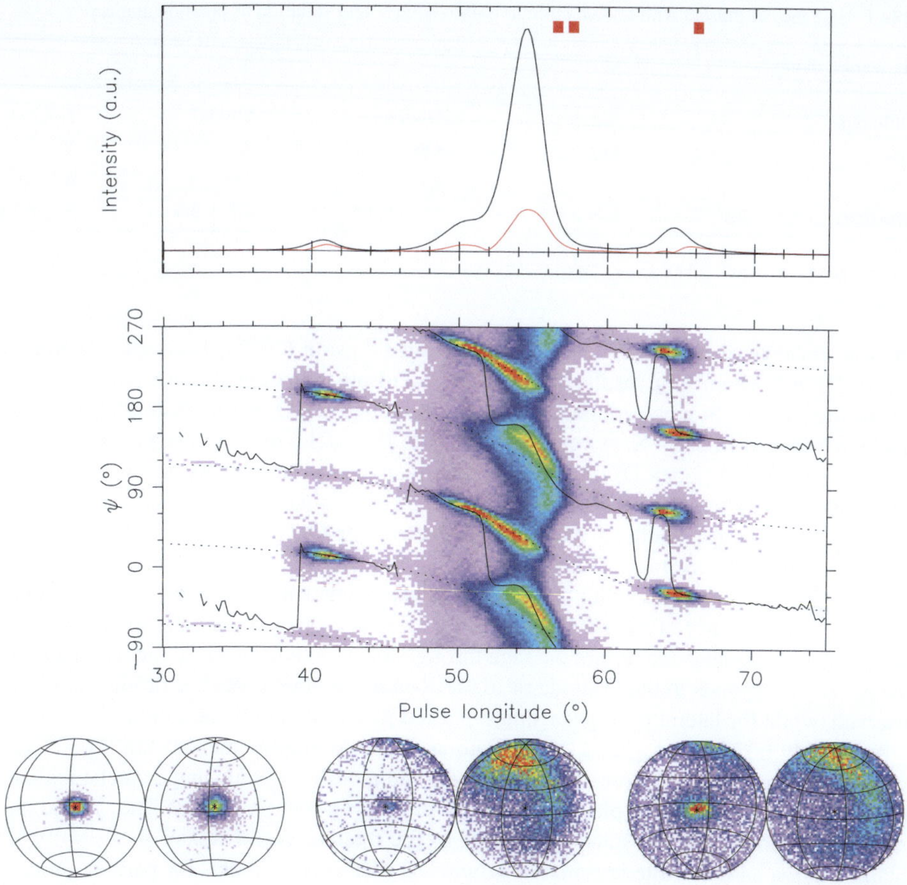

Fig. 11 Polarization properties of the strong pulsar B0329+54. *Upper panel*: average pulse profile, showing both total (*black*) and polarized (*red*) intensity plotted with pulse phase. *Middle panel*: histogram of angle of linear polarization ψ plotted with pulse phase. *Dotted curves* show polarization for the rotating-vector model. *Red* shows most common values, ranging through less-common *green, blue, and violet*, to *white* for not observed. One linear polarization (presumably X) tracks the rotating-vector model well, while the other (presumably O) shows large variations in polarization and timing. *Lower panels*: polarization parameters on the Poincaré sphere, for the 3 intervals of pulse phase indicated by *red boxes* at in the *upper left panel*. Each pair of disks shows 2 hemispheres. Linear polarization lies on the equator; pure right circular polarization is at the upper pole, pure left circular at the lower. This image from Edwards and Stappers (2004) is reproduced by courtesy of B.W. Stappers

the radio pulse. For the ordinary mode, these signs must be opposite to each other, while for the extraordinary mode, they must coincide. Figure 11 shows this pattern as well. As was noted, refraction of the ordinary wave leads to a deviation of beams from the rotation axis, and therefore the ordinary wave pattern should be broader than for the extraordinary wave. In the case of the ordinary mode, double radio emission profiles should mainly be observed, while single profiles should be observed in the case of the narrower extraordinary mode (Beskin et al. 1993).

As was shown by Andrianov and Beskin (2010), observations fully confirm the prediction of correlation between signs of V and $\mathrm{d}\psi/\mathrm{d}\phi$. The analysis used over 70 pulsars with well-traced variation of the position angle and the sign of the circular polarization V, cho-

Table 1 Statistics of pulsars with known circular polarization V and variation of position angle ψ

Polarization mode	O		X	
Profile type	Single	Double	Single	Double
Class	O_S	O_D	X_S	X_D
Number of pulsars	6	23	45	6
Normalized pulse width[a]	6.8 ± 3.1	10.7 ± 4.5	6.5 ± 2.9	5.3 ± 3.0

[a]Normalized pulse width given as: $\sqrt{P}W_{50}$ ($\mathrm{s}^{1/2}$ deg)

sen from reviews of pulse profiles Weltevrede and Johnston (2008), Hankins and Rankin (2010). Table 1 presents the results of the analysis. Pulsars with opposite signs of the derivative $\mathrm{d}\psi/\mathrm{d}\phi$ and the Stokes parameter V were placed in class O, while those with identical signs were placed in class X. As can be seen from the Table, most of the pulsars exhibiting a double-peaked (index D) profile indeed correspond to the ordinary wave, while most of the pulsars with single-peaked profiles (index S) correspond to the extraordinary wave. Moreover, the average width of the radiation pattern for O_D pulsars is indeed about two times larger than the average width of the radiation pattern for X_S pulsars. For the pulse width, the analysis used the width at the 50 % intensity level W_{50}, normalized to the pulsar period P. The existence of a certain number of pulsars of classes O_D and X_S should not give rise to surprise, because for central passage through the directivity pattern, independently of whether it corresponds to the O-mode or to the X-mode, a double-peaked profile should be observed, while for lateral passage, a single-peaked profile should be observed.

Accurately taking propagation effects into account, Andrianov and Beskin (2010), Beskin and Philippov (2011) showed that such a variation of the position angle can be realized only under conditions of low plasma density or high mean particle energy. They found significant deviations from the standard relation of the rotating vector model (Eq. (24)) were obtained in the case of quite reasonable parameters that satisfy models for particle production: for example, a multiplicity $n_e/n_{GJ} \sim 10^4$ and an average Lorentz factor $\gamma \sim 50$.

7 Observations: Polarization and Pulsar Size

7.1 Pulsar Emission Region Size and Shift

The radio emission regions of pulsars lie within the light cylinder, and so have angular sizes from Earth of nanoarcseconds or less. Resolving such an angle at radio wavelengths requires an instrument with an aperture approaching an AU, beyond the capabilities of even the longest VLBI baselines (Kardashev et al. 2013). However, radio-wave scattering by the dilute, turbulent interstellar plasma yields the required angular resolution and offer some of the information provided by a lens of that aperture.

Interstellar scattering affects almost all astrophysical sources at decimeter wavelengths, and for many at shorter wavelengths. For most radio pulsar observations, scattering is "strong" in the sense that paths contributing to the electric field measured at the observer differ in length by many wavelengths. Hence, these paths behave like a corrupt lens (Gwinn et al. 1998). The angular extent on the sky of these paths, θ, delineates the "scattering disk". The scale of variation of the impulse-response function at the observer, S_{ISS}, is the diffractive spot size of that aperture, $\sim \lambda/\theta$. (Here, the subscript "ISS" indicates "interstellar scatter-

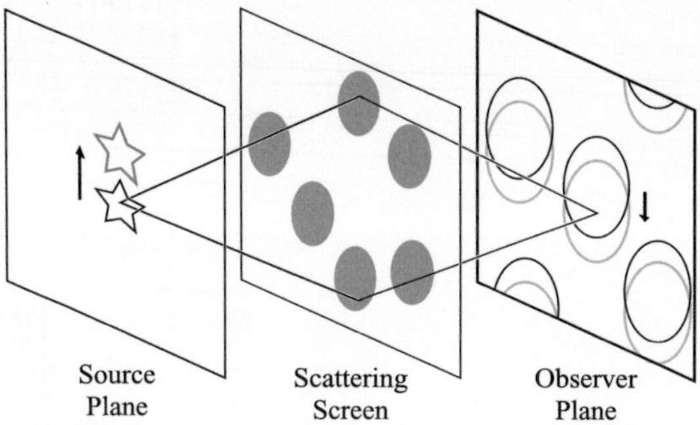

Fig. 12 Geometry for studies of pulsar structure by scattering. Radiation emitted by the pulsar at left travels to the scattering screen, where fluctuations in plasma density change its phase. The radiation arrives at the plane of the observer at right from along many paths, with different phases and amplitudes. Interference among paths produces a random diffraction pattern in the plane of the observer. When the source shifts to a different position, because of either proper motion or a rotational shift in the location of the pulsar emission region, the phases of the paths are modified, to produce a reflex motion of the diffraction pattern in the plane of the observer

ing".) The scattering produces a random diffraction pattern in the observer plane with lateral scale S_{ISS}. The effective resolution limit of the corrupt lens, at the source, is $M S_{\mathrm{ISS}}$, where the magnification factor $M = D/R$ is equal to the distance D of the scattering material from the observer, divided by its distance R from the source. Interstellar scattering does not remove information from the pulsar signal; rather, it adds a great deal of information about the paths taken. The challenge facing the observer is to extract the spatial information about the source from the scattered pulsar signal.

Studies of the sizes of pulsar emission regions using interstellar scattering fall into two categories. One category relies upon the fact that if the emission point of the pulsar shifts across the pulse, the random image in the plane of the observer will undergo a reflex shift, as illustrated in Fig. 12. Proper motion causes a similar shift, but over time spans of many pulses. Correlation of the scintillation spectrum across pulse phases with later or earlier times yields the shift of the emission point (Backer 1975; Cordes et al. 1983; Wolszczan and Cordes 1987; Smirnova et al. 1996; Gupta et al. 1999; Pen et al. 2014).

A second category invokes the decreased modulation for scintillation of an extended source. ("Stars twinkle, planets do not.") The depth of modulation reveals the size of the emission region (Cohen et al. 1966; Readhead and Hewish 1972; Hewish et al. 1974; Gwinn et al. 2012b; Johnson et al. 2012). More precisely, source size affects the distribution of flux density for a scintillating source. In strong scattering many different paths, with lengths differing by many radians of phase, contribute to the electric field measured at a point in the observer plane. The observer implicitly sums over these paths, so that the observed phase and amplitude have the character of a random walk. The optics of this effect are similar to those of the reflex shift: different parts of the source produce shifted, incoherent diffraction patterns at the observer, who sums over them. Thus, finite source size affects the distribution of intensity at one antenna, or that of correlated flux density between the two antennas of an interferometer, principally by shifting the lowest and highest intensities toward the central part of the distribution (Scheuer 1968; Gwinn 2001; Johnson and Gwinn 2012, 2013). For realistic observations, the contributions of background noise, and of the noiselike statistics of the source itself, must be taken into account (Gwinn et al. 2011, 2012a; Johnson and Gwinn 2012, 2013).

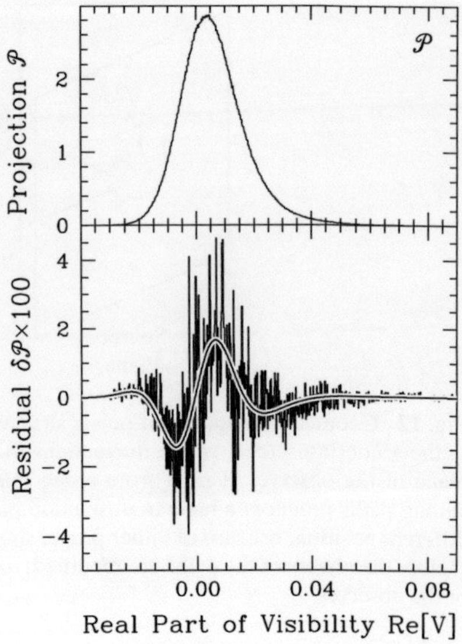

Fig. 13 Observed distribution of visibility projected onto the real axis $\mathcal{P}$, in bins along the real axis. *Upper*: observed distribution. *Lower*: residuals to best-fitting model with zero size for the Vela pulsar at $\lambda = 18$ cm. Curve shows difference of finite- and zero-size models. After Gwinn et al. (2012b)

7.2 Observations

7.2.1 *Size of the Vela Pulsar's Radio Emission Region: $\lambda = 18$ cm*

The fundamental observable of interferometry is visibility, the product of electric fields at a pair of antennas (Thompson et al. 2001). Because electric fields of all astrophysical sources are noiselike, this product must be averaged over some range of time and frequency. For a scintillating source, this averaging must be less than the scales of variation of the scintillation pattern with time and frequency, to preserve the variation of visibility from scintillation (Gwinn et al. 2000).

For a scintillating point source, in the absence of noise, the distribution of interferometric visibility is sharply peaked at the origin (Gwinn 2001). The effect of a small but finite emission size is to soften the sharp peak, shift it from the origin, and narrow the distribution. As compared with a point-source model, the finite-size distribution without noise peaks at larger real part, but has lower probability density at large and small visibility, for the same average flux density (or equivalently, the same mean visibility).

Noise broadens the distribution of visibility. Although noise blurs the distributions and their projections, the difference of point-source and finite-size distributions persists, with a characteristic W-shaped signature, as Fig. 13 shows. To compare with pulsar observations, we must also incorporate the effects of intrinsic variability. Rapid variability modifies the noise statistics, while variability over longer times broadens the distribution (Gwinn et al. 2011, 2012a). Consequences of these effects differ from those of emission size.

Because finite size narrows the distribution of visibility, and noise broadens it, the difference of best-fitting models with finite size and zero size has a characteristic W-shaped signature. Figure 13 shows one example of a fit for a range early in the pulse. The characteristic W-shaped residual is evident, indicating the presence of a finite emission size. A model including one additional parameter, for finite size for the pulsar's emission region, matches this residual accurately with significance exceeding 40σ. The inferred size of the emission region is 420 km. From fits to gates as a function of pulse phase, we find that the size of the pulsar emission region is large at the beginning of the pulse, declines to near zero size near

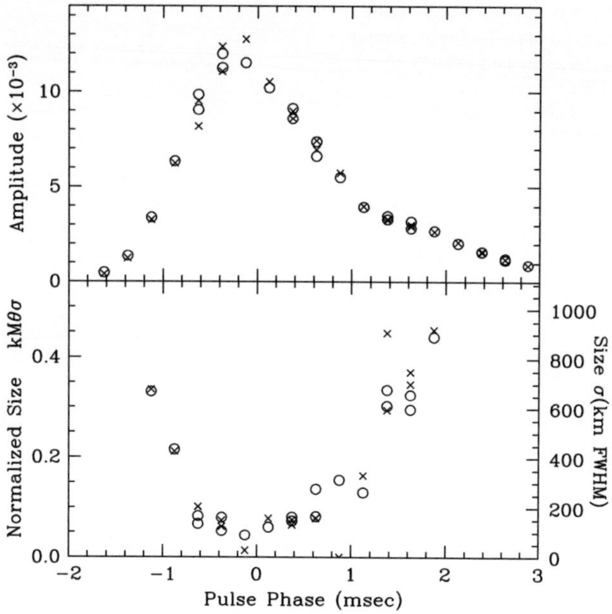

Fig. 14 Best-fitting amplitude (*top panel*) and source size ($kM\theta\sigma$) (*lower panel*) plotted with pulse gate, for 4 gates in 6 spectral ranges. The model for the emission region assumes a circular Gaussian distribution of emission. After Gwinn et al. (2012b)

the middle of the pulse, and then increases again to nearly 1000 km at the end of the pulse. The quoted sizes indicate the full width at half maximum of an equivalent circular Gaussian model.

Theoretical models of pulsar emission typically take their starting point in the geometrical models described above. Hakobyan and Beskin (2014) made theoretical calculations of the images of pulsars as a function of pulse phase, using generic expressions for the emission altitude and beam shape. They include effects of refraction by the magnetospheric plasma, and investigate emission heights up to 100× the radius of the neutron star. They find a characteristic U-shaped curve of the form seen in Fig. 14. This form results from the greater curvature of field lines further from the magnetic pole, and the consequently greater set of loci that can emit in a given direction. Interestingly, they find that the size of the emission region is much larger than its shift over the course of a pulse. Yuen and Melrose (2014) investigate a similar model, and find that the shift of the emission region over a pulse is indeed small. They suggest from geometrical arguments that emission arises at altitudes of more than 10 % of the light-cylinder radius. Lyutikov et al. (1999) comes to similar conclusions based on emission physics.

7.2.2 Size of the Vela Pulsar's Emission Region at $\lambda = 40$ *cm from Nyquist-Limited Statistics*

The unique nature of pulsar emission allows an elegant solution to determination of the distribution of intensity of a variable, scintillating source: the formation of spectra that contain all single-pulse power. Such spectra require a Fourier transform of a data stream that spans the entire pulse, including any scatter-broadening. Without any averaging at all (that is, at the Nyquist limit of the data stream), such spectra show the influence of finite source size. Johnson and Gwinn (2012) calculated the distribution of Nyquist-sampled spectra, for scintillating sources with and without effects of size, including the effects of averaging

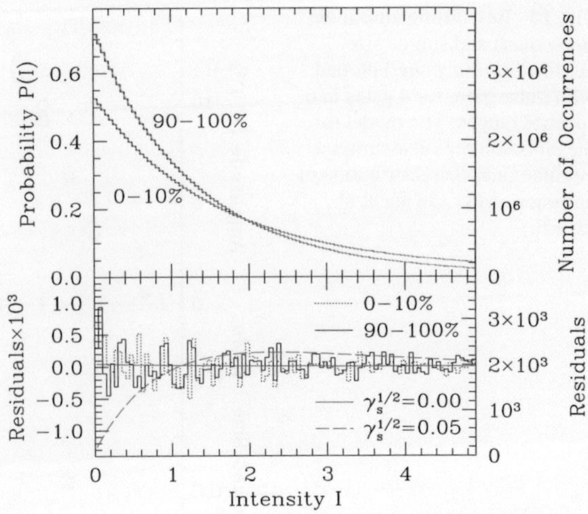

Fig. 15 Observed and model PDFs of intensity for the subsets of pulses in the top and bottom decile by pulse intensity. Theoretical residual curves are completely determined by a single parameter, the source size. Plotted results correspond to a point source and a source that extends over 20 km; the latter is clearly inconsistent with the observed statistics. From Johnson et al. (2012)

and temporal decorrelation. With knowledge of background noise from off-pulse spectra, the intensities of individual pulses, and the scintillation timescale, these statistics provide a measure of source size. A great strength of this technique is that it can measure size for individual pulses, or narrow classes of pulses.

Johnson et al. (2012) used the Nyquist-sampled technique to find the size of the Vela pulsar at 40-cm wavelength, using baseband recording of the pulsar's electric field, at the Green Bank Telescope. They found that the size was consistent with a pointlike source in all cases. The observational upper limit depended upon the set or subset of pulses analyzed. Figure 15 shows a typical example, the distribution of intensity for the brightest 10 % of pulses, and for the weakest 10 %. The distributions are normalized to the mean intensity in both cases, so differences arise from the difference in signal-to-noise ratio. The size is expressed in terms of the characteristic scales of interstellar scattering by the parameter $\gamma_s = (2\pi M\theta\sigma/\lambda)^2$. Here, θ is the angular broadening by interstellar scatter, and σ is the size of a model Gaussian distribution of intensity at the source, both expressed as standard deviation. The observing wavelength is λ, and M is the ratio of the distance of the observer from the scattering screen, to that of the pulsar from the screen. As the figure shows, both distributions are clearly inconsistent with a size as large as $\sigma_c = 20$ km, corresponding to a full-width at half-maximum of 47 km of an assumed circular Gaussian emission region. Figure 16 shows the size of the pulsar as measured in individual pulses, with a range of signal-to-noise ratios. From an fit to their full sample of pulses, they obtained a 3σ upper limit of $\sigma_c < 4$ km (FWHM < 9 km). These sizes are comparable to the size of the neutron star, and suggest a very concentrated emission region. Theory would suggest that the shift of the emission region is still smaller (Hakobyan and Beskin 2014; Yuen and Melrose 2014).

At face value, our results for the size of the emission region of the Vela pulsar at wavelengths of 18 and 40 cm appear inconsistent (Gwinn et al. 2012b; Johnson et al. 2012). How can the size of the emission region change by an order of magnitude, with a change of only $2\times$ in observing wavelength? Longer wavelengths are thought to arise at higher altitudes, apparently exacerbating the discrepancy (Sturrock 1971; Ruderman and Sutherland 1975). Repeated observations have confirmed the observational results.

Refraction of a emergent double-peaked component at $\lambda = 18$ cm may be responsible. At $\lambda = 40$ cm the pulse profile contains a single "core" component, but at $\lambda = 18$ cm an additional, double, "cone" component appears (Komesaroff et al. 1974; Kern et al. 2000;

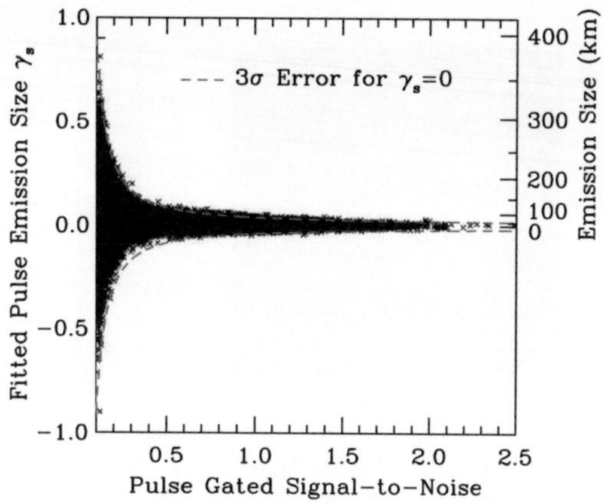

Fig. 16 Inferred emission sizes of individual pulses at $\lambda = 40$ cm. The size is displayed as a function of the single-pulse signal-to-noise ratios S; both linear polarizations are plotted. Because S determines the standard error for each measurement to excellent accuracy, we omit error bars and instead show the expected $\pm 3\sigma$ errors about $\gamma_S = 0$. We do not obtain a statistically significant detection of emission size for any pulse. From Johnson et al. (2012)

Johnson et al. 2012). As discussed in Sect. 6.3, a double-peaked component indicates the presence of the O-mode, and the effects of refraction; whereas a single-peaked component indicates the X-mode and no refraction, and consequently a smaller size. Magnetospheric refraction might be stronger at the shorter wavelength (Arons and Barnard 1986; Barnard and Arons 1986; Lyutikov and Parikh 2000; Hirano and Gwinn 2001; Hakobyan and Beskin 2014). This matches the observed pattern.

7.2.3 Femtoarcsecond Astrometry of Pulsar B0834+06

Pen et al. (2014) extended the comparison of pulsar scattering patterns at different pulse phases. They model the scattering as the interference of a set of points at the screen, the "speckles". Their method isolates the wavefields of each pair of interfering speckles. Interference of each pair of speckles acts as a 2-slit interferometer to cause the pulsar intensity observed at Earth to vary with a specific timescale and bandwidth. A shift of the position of the emission region of the pulsar, over the course of a pulse, causes a reflex shift of the interference pattern from each pair. Pen et al. inverted the very-long baseline interferometry observations of this pulsar by Brisken et al. (2010) to infer the structure of the speckles at the scattering screen, as shown in Fig. 17. They use a holographic technique to partially descatter the data (see also Walker et al. 2008). Their technique corrects for the phase of each speckle relative to its neighbors, and so effectively concentrates the power and boosts the signal-to-noise ratio. The application of this technique to PSR 0834+06 yields an astrometric determination of the phase shift across the pulse profile equivalent to an angular resolution of 150 picoarcseconds, or 10 km at the distance of the pulsar. This remarkable accuracy is comparable to the shift in position of the pulsar due to proper motion, over a single pulse. In particular, they found that the velocity of the radio image in the picture plane is about 1000 $\mathrm{km\,s^{-1}}$, in good agreement with theoretical prediction (Hakobyan and Beskin 2014).

8 Magnetic Axis Alignment

8.1 Theoretical Predictions for Motion of the Magnetic Axis

As was shown above, resulting from the MHD theory of the neutron star magnetosphere, the magnetospheric torque acting on the surface of a neutron star gives the positive factor

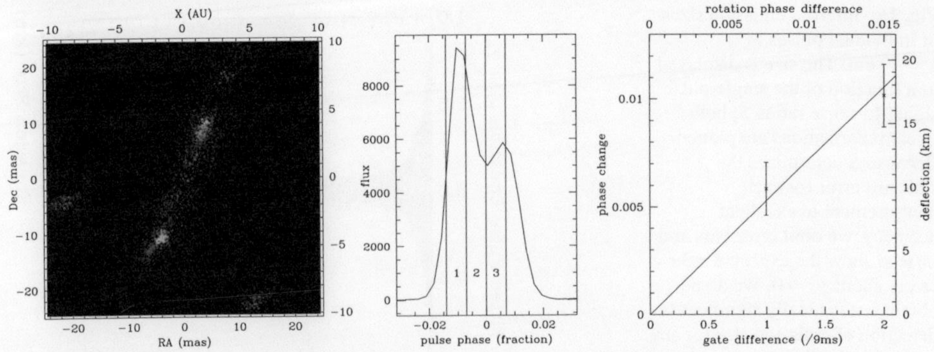

Fig. 17 *Left panel*: inferred image of the speckles that scatter pulsar B0834+06. Color is proportional to time delay, in a periodic hue map. This image becomes the celestial interferometer aperture, for imaging pulsar B0834+06. *Center*: pulse profile, with 3 bins indicated. *Right*: inferred shift of emission region with pulse phase. This image from Pen et al. (2014) is reproduced courtesy of U.-L. Pen

$[K_{\perp}^{A} - K_{\parallel}^{A}]$ in (12)–(13) corresponding to the alignment evolution of the inclination angle. On the other hand, according to (9), this implies that the antisymmetric current i_{a} is to be large enough. E.g., for orthogonal rotator the longitudinal current j is to be 10^3–10^4 times larger than the local Goldreich-Julian one $j_{\mathrm{GJ}}^{\mathrm{loc}} = |\boldsymbol{\Omega}\mathbf{B}|/2\pi$. Recent simulations of pair production in the inner gap (Timokhin and Arons 2013) suggest that the microphysics of the cascade near the polar cap can support the large currents ($j \gg j_{\mathrm{GJ}}^{\mathrm{loc}}$) required by the global magnetospheric structure (it could be accompanied by an efficient heating of the polar cap). Similar results are obtained also by global, 3D PIC simulations of pulsar magnetospheres (Philippov et al. 2014b). In fact, force-free, MHD, and PIC simulations all find that even though for an orthogonal pulsar j_{GJ} essentially vanishes due to the midplane symmetry, the magnitude of the current flowing out along the open magnetic field lines is very similar to that of the aligned pulsar. The results of force-free and MHD simulations tell us that (Philippov et al. 2014a),

$$K_{\parallel} = -K_{\mathrm{aligned}} \cos\chi, \tag{25}$$

$$K_{\perp} = -2K_{\mathrm{aligned}} \sin\chi, \tag{26}$$

where $K_{\mathrm{aligned}} = W_{\mathrm{aligned}}/\Omega = \mu^2\Omega^3/c^3$ is the spindown torque of an aligned rotator. Therefore,

$$I_{\mathrm{r}}\dot{\Omega} \approx -K_{\mathrm{aligned}}\big(1 + \sin^2\chi\big), \tag{27}$$

$$I_{\mathrm{r}}\Omega\,\dot{\chi} \approx -K_{\mathrm{aligned}} \sin\chi \cos\chi. \tag{28}$$

Thus, the force-free and MHD simulation results suggest that pulsars tend to become *aligned* with time. This is not surprising in the context of previous discussion: pulsars tend to evolve toward the lowest luminosity state, e.g., toward the aligned state (see (19)). Vacuum pulsars become aligned exponentially fast, even before they have a chance to spin down substantially, and generically end up with a period that is a few times their birth period. If most pulsars were born as millisecond rotators, this presents a problem in pulsar population synthesis studies, as this would imply that most pulsars would have millisecond periods, yet we observe many pulsars with periods of $\sim$ second. In contrast, plasma-filled pulsars come into alignment much slower, as a power-law in time, $\chi \propto t^{-1/2}$, so both the spindown and alignment proceed at a similar rate (Philippov et al. 2014a).

On the other hand, if there is some restriction of the value of the longitudinal current flowing through the polar cap (no numerical simulation has such a restriction), the situation can be different. Such an alternative model in which both symmetric and antisymmetric currents correspond to the local Goldreich-Julian value was considered by Beskin et al. (1983, 1993). They calculated the torque associated with the Ampére force arising from the interaction of the neutron star poloidal field with the surface currents (these currents close the longitudinal currents flowing in the region of open magnetosphere). One can prove by straightforward but cumbersome calculation that the two approaches are identical. This is a crucial assumption because pulsar spindown luminosity is proportional to the magnetospheric current squared.

For $i_a \approx i_a \approx 1$ Eqs. (12)–(13) can be rewritten as

$$I_r \dot{\Omega} \approx K_{\parallel}^{A} \cos^2 \chi, \tag{29}$$

$$I_r \Omega \dot{\chi} \approx K_{\parallel}^{A} \sin \chi \cos \chi; \tag{30}$$

for orthogonal rotator $\cos \chi < (\Omega R/c)^{1/2}$ we have

$$I_r \dot{\Omega} \approx \left(\frac{\Omega R}{c}\right) K_{\parallel}^{A}. \tag{31}$$

As for $\cos \chi > (\Omega R/c)^{1/2}$ evolutionary equations (29)–(30) have an integral

$$\Omega \sin \chi = \text{const}, \tag{32}$$

this model predicts the evolution of the inclination angle toward an orthogonal configuration. Thus, two theoretical models of the neutron star evolution give approximately identical predictions for the period derivative $\dot{P}$, but opposite ones for the evolution of the inclination angle χ.

8.2 Observational Constraints on Evolution of Inclination Angle

Measurement of the rate of change of position angle with pulse phase at the center of the pulse $\mathrm{d}\psi/\mathrm{d}\phi$ yields only a measure of the minimum angle between the line of sight and the magnetic axis, ζ, as inspection of Eq. (24) shows. The angle ζ is sometimes called the "impact angle" (see Fig. 10). Consequently estimates of the inclination angle χ are indirect, and observational tests of the theories for evolution of χ in the previous section are difficult.

As inspection of Eq. (10) shows, measurement of the rate of change of position angle with pulse phase at the center of the pulse $\mathrm{d}\psi/\mathrm{d}\phi$ yields only a measure of the minimum angle between the line of sight and the magnetic axis, ζ, sometimes called the "impact angle" (see Fig. 1). Consequently estimates of the inclination angle χ are indirect, and observational tests of the theories for evolution of χ in the previous section are difficult.

In a careful study, Tauris and Manchester (1998) compared the inclination angles χ and rotation period P for nearly 100 pulsars. They found that χ decreases as P increases, over this sample of the pulsar population. They made the straightforward assumption that the beam from the pulsar is round, as the hollow-cone model discussed in Sect. 2.3 and Fig. 1 suggest.[4] They used beam radii as a function of pulse period derived by Gould (1994) from

[4]Narayan and Vivekanand (1983) suggest that pulsar beams are, instead, elongated; and that their elongation decreases as the pulsar ages.

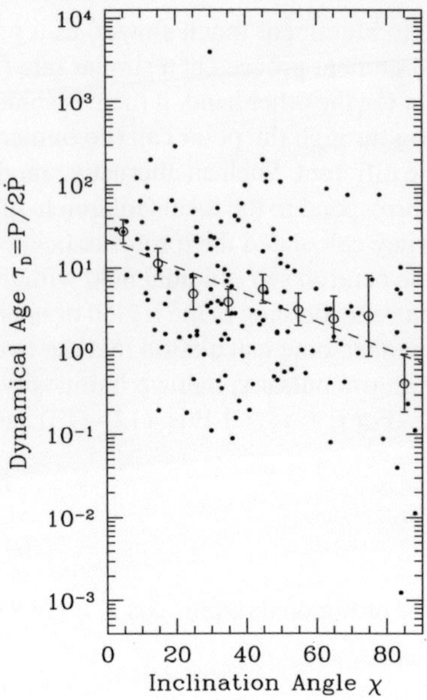

Fig. 18 Dependence of the mean inclination angle χ as a function of the pulsar dynamical age τ_D. Statistically this angle decreases with P and, hence, with the age τ_D. After Tauris and Manchester (1998), using data of Rankin (1993a) and Manchester et al. (2005)

comparisons among pulsars with similar periods but different impact angles ζ, and from pulsars with an interpulse (assumed to be nearly orthogonal: $\chi = \pi/2$) by Rankin (1990). From the observed pulse width, Tauris and Manchester (1998) then inferred the angular separation of the line of sight and the rotation axis, and so the inclination angle χ. Figure 18 illustrates their results, using data from Rankin (1993a) and Manchester et al. (2005). Weltevrede and Johnston (2008) reached similar conclusions, by comparing the sample of pulsars with interpulses with the full population.

As Fig. 18 shows, observations reveal average statistical inclination angles $\langle\chi\rangle$ indisputably decrease as the period P of pulsars increases and its derivative $\dot{P}$ decreases. Therefore, the average inclination angle decreases as the dynamic age increases. Correspondingly, pulsars with longer periods exhibit relatively larger pulse widths $W_r = W_r^{(0)}/\sin\chi$, where $W_r^{(0)}$ is the width of the directivity pattern (Rankin 1990; Gould 1994; Young et al. 2010). These results definitely speak in favor of the alignment mechanism. On the other hand, recently Lyne et al. (2013) on the analysis of the 45 years observations of the Crab pulsar concluded that its inclination angle increases with time. However, the effects of stellar nonsphericity, leading to free precession, can account for this seemingly odd behavior (Arzamasskiy et al. 2015b).

The average inclination angle $\langle\chi\rangle_{\tau_D}$ for a given range of ages τ_D can decrease, even if the inclination angles of individual pulsars *increases* with time, in accord with Eq. (32). This is a consequence of the dependence of the magnetospheric charge density on the inclination angle χ. For example, in the picture of Ruderman and Sutherland (1975), radio emission results from a secondary electron-position cascade, initiated by pair-production from curvature-radiation photons. The acceleration of primary electrons within a gap produced these curvature photons. The Goldreich-Julian charge density $\rho_{GJ} \approx \Omega B \cos\chi/(2\pi c)$ sets the accelerating potential across the gap. As a pulsar ages, ρ_{GJ} and the accelerating potential decline with Ω. The area of the polar cap also decreases as fewer field lines penetrate the light cylinder, and those remaining within the polar cap are less curved. As Eq. (4) shows,

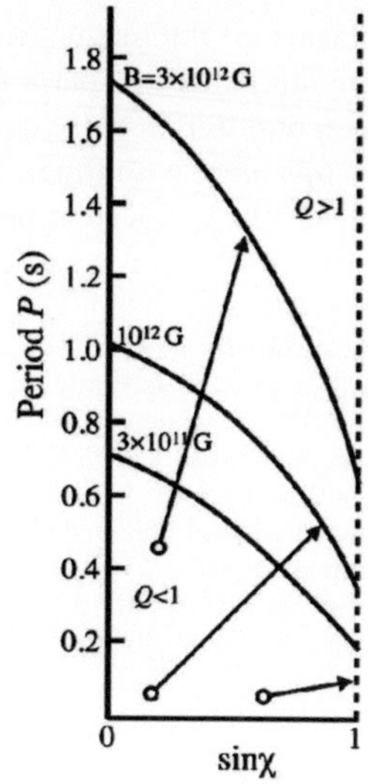

Fig. 19 Pulsar extinction line in a $P - \sin \chi$ diagram for different magnetic fields. Arrows show the evolution tracks of individual pulsars in the model of the current losses (29)–(30) (Beskin et al. 1993)

the mean free path to pair production increases; the gap becomes wider. When the gap width is comparable to the polar-cap radius, the cascade, and radio emission, terminate.

Because the charge density depends on $\cos \chi$ as well as B_0 and Ω, death comes to pulsars with different inclinations, but the same magnetic field, at different spin periods P. Evaluation of the Ruderman and Sutherland model yields $P_{max} \propto B_0^{8/9} (\cos \chi)^{2/3}$. Indeed, as can be seen from Fig. 19, for given values of the pulsar period P and the magnetic field B_0, the production of particles is suppressed precisely at inclinations χ close to $\pi/2$, where the magnetic dipole is nearly orthogonal. Therefore, neutron stars above and to the right of the extinction lines in Fig. 19 do not appear as radio pulsars.

Because of this dependence of the pulsar extinction line on χ, the average inclination angles of the observed populations can decrease as the dynamic age increases. A detailed analysis, carried out in Beskin et al. (1984) (see also Beskin and Nokhrina 2004; Beskin and Eliseeva 2005) on the basis of a kinetic equation describing the distribution of pulsars confirms this picture quantitatively.

Clearly, in any case, that the inclination angle χ is a key hidden parameter: without taking it into account, it is impossible to construct a consistent theory of the evolution of radio pulsars. Eliseeva et al. (2006) include this possibility in their work, suggesting a possible direction for further improvements in models for the evolution of neutron stars (Lipunov et al. 1996; Story et al. 2007; Popov and Prokhorov 2007; Gullón et al. 2014).

9 Summary

Pulsars provide elegant, although not simple, laboratories for fundamental electromagnetic processes at high energies. The basic picture of their structures, involving strong magnetic fields and rapid rotation, generation of electron-positron pairs and an energetic wind that carries away the rotational kinetic energy of the pulsar, became clear not long after their discovery a half-century ago. Recent work has begun to uncover the detailed structures of their magnetospheres; the location, size, and optics of their radio emission regions; and evolution of their spins. Some important questions not far from solution include the effect of plasma on the magnetosphere and the possible existence of multiple states; conversion of Poynting flux to a particle wind (the σ-problem); the location, size, and properties of emission of different pulse components; and whether rotation and magnetic axes tend to co-align or mis-align with time. Further insightful theory, careful statistical studies and targeted observations will lead to deeper understanding, as pulsars continue their role as the archetypal observable neutron stars.

Acknowledgements The authors wish to thank the International Space Science Institute for its hospitality. V. Beskin and S. Chernov were supported by Russian Foundation for Basic Research (project N 14-02-00831) C.R. Gwinn acknowledges support of the US National Science Foundation (AST-1008865). A. Tchekhovskoy was supported by NASA through Einstein Postdoctoral Fellowship grant number PF3-140131 awarded by the Chandra X-ray Center, which is operated by the Smithsonian Astrophysical Observatory for NASA under contract NAS8-03060, and NASA via High-End Computing (HEC) Program through the NASA Advanced Supercomputing (NAS) Division at Ames Research Center that provided access to the Pleiades supercomputer, as well as NSF through an XSEDE computational time allocation TG-AST100040 on NICS Kraken, Nautilus, TACC Stampede, Maverick, and Ranch.

References

Ya.I. Al'ber, Z.N. Krotova, V.Ya. Eidman, Astrophysics **11**, 189 (1975)
A. Allafort, L. Baldini, J. Ballet et al., Astrophys. J. **777**, L2 (2013)
A.S. Andrianov, V.S. Beskin, Astron. Lett. **36**, 248 (2010)
J. Arons, J.J. Barnard, Astrophys. J. **302**, 120 (1986)
J. Arons, E.T. Scharlemann, Astrophys. J. **231**, 854 (1979)
L. Arzamasskiy, A. Philippov, A. Tchekhovskoy, Evolution of non-spherical pulsars with plasma-filled magnetospheres, arXiv:1504.06626 (2015a)
L. Arzamasskiy, V. Beskin, V. Prokofev, On the primary beam deceleration in the pulsar wind, arXiv:1505.03864 (2015b)
W. Baade, F. Zwicky, Proc. Nat. Acad. Sci. **20**, 254 (1934)
D.C. Backer, Astron. Astrophys. **43**, 395 (1975)
D.C. Backer, J.M. Rankin, Astrophys. J. Suppl. Ser. **42**, 143 (1980)
D.C. Backer, J.M. Rankin, D.B. Campbell, Nature **263**, 202 (1976)
C.D. Bailyn, J.E. Grindlay, Astrophys. J. **353**, 159 (1990)
J.J. Barnard, Astrophys. J. **303**, 280 (1986)
J.J. Barnard, J. Arons, Astrophys. J. **302**, 138 (1986)
M.A. Belyaev, arXiv:1412.2819 (2014)
V.B. Berestetsky, E.M. Lifshits, L.P. Pitaevsky, *Relativistic Quantum Theory* (Pergamon, Oxford, 1982)
V.S. Beskin, Sov. Phys. Usp. **41**, 1071 (1999)
V.S. Beskin, S.A. Eliseeva, Astron. Lett. **31**, 263 (2005)
V.S. Beskin, E.E. Nokhrina, Astron. Lett. **30**, 685 (2004)
V.S. Beskin, E.E. Nokhrina, Astrophys. Space Sci. **308**, 569 (2007)
V.S. Beskin, A.A. Philippov, arXiv:1101.5733 (2011)
V.S. Beskin, A.A. Philippov, Mon. Not. R. Astron. Soc. **425**, 814 (2012)
V.S. Beskin, A.A. Zheltoukhov, Phys. Usp. **57**, 799 (2014)
V.S. Beskin, A.V. Gurevich, Ya.N. Istomin, Sov. Phys. JETP **58**, 235 (1983)
V.S. Beskin, A.V. Gurevich, Ya.N. Istomin, Astrophys. Space Sci. **102**, 301 (1984)
V.S. Beskin, A.V. Gurevich, Ya.N. Istomin, *Physics of the Pulsar Magnetosphere* (Cambridge University Press, Cambridge, 1993)
V.S. Beskin, Y.N. Istomin, A.A. Philippov, Phys. Usp. **56**, 164 (2013)
M. Blaskiewicz, J.M. Cordes, I. Wasserman, Astrophys. J. **370**, 643 (1991)
S.V. Bogovalov, Astron. Astrophys. **349**, 1017 (1999)
S.V. Bogovalov, D.V. Khangoulyan, Astron. Lett. **28**, 373 (2002)
P.E. Boynton, E.J. Groth, D.P. Hutchinson et al., Astrophys. J. **175**, 217 (1972)
R.P. Breton, M.H. van Kerkwijk, M.S.E. Roberts et al., Astrophys. J. **769**, 108 (2013)
W.F. Brisken, J.-P. Macquart, J.J. Gao et al., Astrophys. J. **708**, 232 (2010)
M. Camenzind, *Compact Objects in Astrophysics: White Dwarfs, Neutron Stars and Black Holes* (Springer, Berlin, 2007)

F. Camilo, S.M. Ransom, S. Chatterjee et al., Astrophys. J. **746**, 63 (2012)
B. Cerutti, A. Philippov, K. Parfrey, A. Spitkovsky, arXiv:1410.3757 (2014)
A.Y. Chen, A.M. Beloborodov, Astrophys. J. **762**, 9 (2013)
A.Y. Chen, A.M. Beloborodov, Astrophys. J. **795**, L22 (2014)
A.F. Cheng, M.A. Ruderman, Astrophys. J. **229**, 348 (1979)
C.T.Y. Chung, A. Melatos, Mon. Not. R. Astron. Soc. **411**, 2471 (2011a)
C.T.Y. Chung, A. Melatos, Mon. Not. R. Astron. Soc. **415**, 1703 (2011b)
I. Cognard, G. Bourgois, J.-F. Lestrade et al., Astron. Astrophys. **296**, 169 (1995)
M.H. Cohen, E.J. Gundermann, H.E. Hardebeck et al., Science **153**, 745 (1966)
J.M. Comella, H.D. Craft, R.V.E. Lovelace, J.M. Sutton, Nature **221**, 453 (1969)
I. Contopoulos, D. Kazanas, C. Fendt, Astrophys. J. **511**, 351 (1999)
J.M. Cordes, J.M. Rankin, D.C. Backer, Astrophys. J. **223**, 961 (1978)
J.M. Cordes, V. Boriakoff, J.M. Weisberg, Astrophys. J. **268**, 370 (1983)
P. Demorest, R. Ramachandran, D.C. Backer et al., Astrophys. J. **615**, L137 (2004)
R. Dodson, D. Lewis, P. McCulloch, Astrophys. Space Sci. **308**, 585 (2007)
J. Dyks, B. Rudak, A.K. Harding, Astrophys. J. **607**, 939 (2004)
R.T. Edwards, B.W. Stappers, Astron. Astrophys. **421**, 681 (2004)
S.A. Eliseeva, S.B. Popov, V.S. Beskin, arxiv:astro-ph/0611320 (2006)
W.M. Fawley, J. Arons, E.T. Scharlemann, Astrophys. J. **217**, 227 (1977)
A.S. Fruchter, D.R. Stinebring, J.H. Taylor, Nature **333**, 237 (1988)
A.S. Fruchter, G. Berman, G. Bower et al., Astrophys. J. **351**, 642 (1990)
G. Ghisellini, F. Tavecchio, L. Maraschi et al., Nature **515**, 376 (2014)
T. Gold, Nature **218**, 731 (1968)
T. Gold, Nature **221**, 25 (1969)
P. Goldreich, W.H. Julian, Astrophys. J. **157**, 869 (1969)
D.M. Gould, Ph.D. thesis, University of Manchester (1994)
A. Gruzinov, Phys. Rev. Lett. **94**, 021101 (2005)
M. Gullón, J. Miralles, D. Viganò, J. Pons, Mon. Not. R. Astron. Soc. **443**, 1891 (2014)
Y. Gupta, N.D.R. Bhat, A.P. Rao, Astrophys. J. **520**, 173 (1999)
C.R. Gwinn, Astrophys. J. **554**, 1197 (2001)
C.R. Gwinn, M.C. Britton, J.E. Reynolds et al., Astrophys. J. **505**, 928 (1998)
C.R. Gwinn, M.C. Britton, J.E. Reynolds et al., Astrophys. J. **531**, 902 (2000)
C.R. Gwinn, M.D. Johnson, T.V. Smirnova, D.R. Stinebring, Astrophys. J. **733**, 52 (2011)
C.R. Gwinn, M.D. Johnson, J.E. Reynolds et al., Astrophys. J. **758**, 6 (2012a)
C.R. Gwinn, M.D. Johnson, J.E. Reynolds et al., Astrophys. J. **758**, 7 (2012b)
H.L. Hakobyan, V.S. Beskin, Astron. Rep. **58**, 889 (2014)
T.H. Hankins, J.M. Rankin, Astron. J. **139**, 168 (2010)
A.K. Harding, J.V. Stern, J. Dyks, M. Frackowiak, Astrophys. J. **680**, 1378 (2008)
D.J. Helfand, J.H. Taylor, P.R. Backus, J.M. Cordes, Astrophys. J. **237**, 206 (1980)
A. Hewish, S.J. Bell, J.D. Pilkington et al., Nature **217**, 709 (1968)
A. Hewish, A.C.S. Readhead, P.J. Duffett-Smith, Nature **252**, 657 (1974)
C. Hirano, C.R. Gwinn, Astrophys. J. **553**, 358 (2001)
R. Ingraham, Astrophys. J. **186**, 625 (1973)
Y.N. Istomin, D.N. Sobyanin, JETP Lett. **113**, 592 (2011a)
Y.N. Istomin, D.N. Sobyanin, JETP Lett. **113**, 605 (2011b)
F.A. Jenet, S.M. Ransom, Nature **428**, 919 (2004)
M.D. Johnson, C.R. Gwinn, Astrophys. J. **755**, 179 (2012)
M.D. Johnson, C.R. Gwinn, Astrophys. J. **768**, 170 (2013)
M.D. Johnson, C.R. Gwinn, P. Demorest, Astrophys. J. **758**, 8 (2012)
C. Kalapotharakos, I. Contopoulos, Astron. Astrophys. **496**, 495 (2009)
C. Kalapotharakos, I. Contopoulos, D. Kazanas, Mon. Not. R. Astron. Soc. **420**, 2793 (2012a)
C. Kalapotharakos, D. Kazanas, A. Harding, I. Contopoulos, Astrophys. J. **749**, 2 (2012b)
D.L. Kaplan, V.B. Bhalerao, M.H. van Kerkwijk et al., Astrophys. J. **765**, 158 (2013)
N.S. Kardashev, V.V. Khartov, V.V. Abramov et al., Astron. Rep. **57**, 153 (2013)
V.M. Kaspi, J.H. Taylor, M.F. Ryba, Astrophys. J. **428**, 713 (1994)
C.F. Kennel, F.V. Coroniti, Astrophys. J. **283**, 694 (1984a)
C.F. Kennel, F.V. Coroniti, Astrophys. J. **283**, 710 (1984b)
J.S. Kern, T.H. Hankins, J.M. Rankin, in *IAU Colloq. 177: Pulsar Astronomy—2000 and Beyond*, vol. 202, ed. by M. Kramer, N. Wex, N. Wielebinski (2000), p. 257
J.G. Kirk, Y. Lyubarsky, J. Petri, Astrophys. Space Sci. Libr. **357**, 421 (2009)
M.M. Komesaroff, P.M. McCulloch, J.M. Rankin, Nature **252**, 210 (1974)

S.S. Komissarov, Mon. Not. R. Astron. Soc. **367**, 19 (2006)
S.S. Komissarov, Yu.E. Lyubarsky, Mon. Not. R. Astron. Soc. **344**, L93 (2003)
S.S. Komissarov, N. Vlahakis, A. Königl, M.V. Barkov, Mon. Not. R. Astron. Soc. **394**, 1182 (2009)
M. Kramer, A.G. Lyne, J.T. O'Brien et al., Science **312**, 549 (2006)
Yu.A. Kravtsov, Yu.I. Orlov, *Geometricheskaya Optika Neodnorodnykh Sred (Geometrical Optics of Inhomogeneous Media)* (Moscow, Nauka, 1980). Translated into English (Springer, Berlin, 1990)
K. Krzeszowski, D. Mitra, Y. Gupta et al., Mon. Not. R. Astron. Soc. **393**, 1617 (2009)
S.R. Kulkarni, S. Djorgovski, A.S. Fruchter, Nature **334**, 504 (1988)
L.D. Landau, Phys. Z. Sowjetunion **1**, 271 (1932)
L.D. Landau, E.M. Lifshitz, *The Classical Theory of Fields* (Pergamon Press, Oxford, 1989)
J. Li, A. Spitkovsky, A. Tchekhovskoy, Astrophys. J. **746**, 60 (2012a)
J. Li, A. Spitkovsky, A. Tchekhovskoy, Astrophys. J. **746**, L24 (2012b)
V.M. Lipunov, K.A. Postnov, M.E. Prokhorov, Astron. Astrophys. **310**, 489 (1996)
A. Lommen, J. Donovan, C. Gwinn et al., Astrophys. J. **657**, 436 (2007)
A.G. Lyne, Astrophys. Space Sci. Libr. **357**, 67 (2009)
A.G. Lyne, F. Graham-Smith, *Pulsar Astronomy* (Cambridge University Press, Cambridge, 1998)
A.G. Lyne, R.N. Manchester, Mon. Not. R. Astron. Soc. **234**, 477 (1988)
A.G. Lyne, M. Burgay, M. Kramer et al., Science **303**, 1153 (2004)
A. Lyne, G. Hobbs, M. Kramer et al., Science **329**, 408 (2010)
A. Lyne, F. Graham-Smith, P. Weltevrede et al., Science **342**, 598 (2013)
Y.E. Lyubarsky, Mon. Not. R. Astron. Soc. **402**, 353 (2010)
M. Lyutikov, Mon. Not. R. Astron. Soc. **353**, 1095 (2004)
M. Lyutikov, A. Parikh, Astrophys. J. **541**, 1016 (2000)
M. Lyutikov, R.D. Blandford, G. Machabeli, Mon. Not. R. Astron. Soc. **305**, 338 (1999)
R.N. Manchester, J. Astrophys. Astron. **16**, 107 (1995)
R.N. Manchester, J.H. Taylor, *Pulsars* (Freeman, New York, 1977)
R.N. Manchester, J.H. Taylor, G.R. Huguenin, Astrophys. J. **196**, 83 (1975)
R.N. Manchester, G.B. Hobbs, A. Teoh, M. Hobbs, Astron. J. **129**, 1993 (2005)
J.C. McKinney, Mon. Not. R. Astron. Soc. **368**, L30 (2006)
M.M. McKinnon, Astrophys. J. **590**, 1026 (2003)
M.M. McKinnon, Astrophys. J. **692**, 459 (2009)
M.A. McLaughlin, M. Kramer, A.G. Lyne et al., Astrophys. J. **613**, L57 (2004)
Z. Medin, D. Lai, Mon. Not. R. Astron. Soc. **406**, 1379 (2010)
L. Mestel, P. Panagi, S. Shibata, Mon. Not. R. Astron. Soc. **309**, 388 (1999)
F.C. Michel, Astrophys. J. **180**, L133 (1973)
F.C. Michel, Astrophys. J. **187**, 585 (1974)
A.B. Mikhailovskii et al., Sov. Astron. Lett. **8**, 369 (1982)
D. Mitra, X.H. Li, Astron. Astrophys. **421**, 15–228 (2004)
R. Narayan, M. Vivekanand, Astron. Astrophys. **122**, 45 (1983)
K. Nomoto, Y. Kondo, Astrophys. J. **367**, L19 (1991)
J. Ogura, Y. Kojima, Prog. Theor. Phys. **109**, 619 (2003)
F. Pacini, Nature **216**, 567 (1967)
C. Pallanca, R.P. Mignani, E. Dalessandro et al., Astrophys. J. **755**, 180 (2012)
K. Parfrey, A.M. Beloborodov, L. Hui, Mon. Not. R. Astron. Soc. **423**, 1416 (2012)
U.-L. Pen, J.-P. Macquart, A.T. Deller, W. Brisken, Mon. Not. R. Astron. Soc. **440**, L36 (2014)
B.B.P. Perera, B.W. Stappers, P. Weltevrede et al., Mon. Not. R. Astron. Soc. **446**, 1380 (2015)
J. Pétri, Mon. Not. R. Astron. Soc. **424**, 605 (2012)
S. Petrova, Mon. Not. R. Astron. Soc. **368**, 1764 (2006)
S. Petrova, Yu. Lyubarskii, Astron. Astrophys. **355**, 1168 (2000)
A.A. Philippov, A. Spitkovsky, Astrophys. J. **785**, L33 (2014)
A. Philippov, A. Tchekhovskoy, J.G. Li, Mon. Not. R. Astron. Soc. **441**, 1879 (2014a)
A.A. Philippov, A. Spitkovsky, B. Cerutti, arXiv:1412.0673 (2014b)
S.B. Popov, M.E. Prokhorov, Phys. Usp. **50**, 1123 (2007)
M.V. Popov, V.A. Soglasnov, V.I. Kondratiev et al., Astron. Rep. **50**, 55 (2006)
V. Radhakrishnan, D.J. Cooke, Astrophys. J. Lett. **3**, 225 (1969)
V. Radhakrishnan, D.J. Cooke, M.M. Komesaroff, D. Morris, Nature **221**, 443 (1969)
J.M. Rankin, Astrophys. J. **274**, 333 (1983)
J.M. Rankin, Astrophys. J. **301**, 901 (1986)
J.M. Rankin, Astrophys. J. **352**, 247 (1990)
J.M. Rankin, Astrophys. J. Suppl. Ser. **85**, 145 (1993a)
J.M. Rankin, Astrophys. J. **405**, 285 (1993b)

A.C.S. Readhead, A. Hewish, Nature **236**, 440 (1972)
D.W. Richards, J.M. Comella, Nature **222**, 551 (1969)
M.S.E. Roberts, *American Institute of Physics Conference Series*, vol. 1357 (2011), p. 127
R.W. Romani, A.V. Filippenko, J.M. Silverman et al., Astrophys. J. **760**, L36 (2012)
M.A. Ruderman, P.G. Sutherland, Astrophys. J. **196**, 51 (1975)
M. Ruiz, V. Paschalidis, S.L. Shapiro, Phys. Rev. D **89**, 084045 (2014)
M.F. Ryba, J.H. Taylor, Astrophys. J. **380**, 557 (1991)
E.T. Scharlemann, W.M. Fawley, J. Arons, Astrophys. J. **222**, 297 (1978)
P.A.G. Scheuer, Nature **218**, 920 (1968)
J. Schwab, E. Quataert, L. Bildsten, Therma runaway during the evolution of ONeMg cores twoards accretion-induced collapse, arXiv:1504.05194 (2015)
D.M. Scott, M.H. Finger, C.A. Wilson, Mon. Not. R. Astron. Soc. **344**, 412 (2003)
S.L. Shapiro, S.A. Teukolsky, *Black Holes, White Dwarfs, and Neutron Stars* (Wiley-Interscience, New York, 1985)
A. Shearer, B. Stappers, P. O'Connor et al., Science **301**, 493 (2003)
T.V. Smirnova, V.I. Shishov, V.M. Malofeev, Astrophys. J. **462**, 289 (1996)
A. Spitkovsky, Astrophys. J. **648**, L51 (2006)
B.W. Stappers, M.S. Bessell, M. Bailes, Astrophys. J. **473**, L119 (1996)
D.R. Stinebring, J.M. Cordes, J.M. Rankin et al., Astrophys. J. Suppl. Ser. **55**, 247 (1984a)
D.R. Stinebring, J.M. Cordes, J.M. Weisberg et al., Astrophys. J. Suppl. Ser. **55**, 279 (1984b)
S.A. Story, P.L. Gonthier, A. Harding, Astrophys. J. **671**, 713 (2007)
M.J. Strader, M.D. Johnson, B.A. Mazin et al., Astrophys. J. **779**, L12 (2013)
P.A. Sturrock, Astrophys. J. **164**, 529 (1971)
T.M. Tauris, R.N. Manchester, Mon. Not. R. Astron. Soc. **298**, 625 (1998)
A. Tchekhovskoy, Astrophys. Space Sci. Libr. **414**, 45 (2015), doi:10.1007/978-3-319-10356-3_3
A. Tchekhovskoy, J.C. McKinney, Mon. Not. R. Astron. Soc. **423**, L55 (2012)
A. Tchekhovskoy, J.C. McKinney, R. Narayan, Astrophys. J. **699**, 1789 (2009)
A. Tchekhovskoy, R. Narayan, J.C. McKinney, New Astron. **15**, 749 (2010)
A. Tchekhovskoy, R. Narayan, J.C. McKinney, Mon. Not. R. Astron. Soc. **418**, L79 (2011)
A. Tchekhovskoy, A. Spitkovsky, J.G. Li, Mon. Not. R. Astron. Soc. **435**, L1 (2013)
A. Tchekhovskoy, A. Philippov, A. Spitkovsky, Three-dimensional analytical description of magnetised winds from oblique pulsars, arXiv:1503.01467 (2015)
A.R. Thompson, J.M. Moran, G.W. Swenson Jr., *Interferometry and Synthesis in Radio Astronomy*, 2nd edn. (Wiley, New York, 2001)
A.N. Timokhin, Mon. Not. R. Astron. Soc. **368**, 1055 (2006)
A.N. Timokhin, Mon. Not. R. Astron. Soc. **408**, 2092 (2010)
A.N. Timokhin, J. Arons, Mon. Not. R. Astron. Soc. **429**, 20 (2013)
M.A. Walker, L.V.E. Koopmans, D.R. Stinebring, W. van Straten, Mon. Not. R. Astron. Soc. **388**, 1214 (2008)
C. Wang, D. Lai, J. Han, Mon. Not. R. Astron. Soc. **403**, 569 (2010)
P. Weltevrede, S. Johnston, Mon. Not. R. Astron. Soc. **391**, 1210 (2008)
J. Whelan, I. Iben, Astrophys. J. **186**, 1007 (1973)
A. Wolszczan, J.M. Cordes, Astrophys. J. **320**, L35 (1987)
M.D.T. Young et al., Mon. Not. R. Astron. Soc. **402**, 1317 (2010)
R. Yuen, D.B. Melrose, Publ. Astron. Soc. Aust. **31**, 39 (2014)
M. Zamaninasab, E. Clausen-Brown, T. Savolainen, A. Tchekhovskoy, Nature **510**, 126 (2014)
A.A. Zdziarski, M. Sikora, P. Pjanka, A. Tchekhovskoy, arXiv:1410.7310 (2014)

DOI 10.1007/978-1-4939-3550-5_8
Reprinted from *Space Science Reviews* Journal, DOI 10.1007/s11214-015-0180-9

Neutron Stars—Cooling and Transport

Alexander Y. Potekhin[1,2,3] · **José A. Pons**[4] · **Dany Page**[5]

Received: 6 March 2015 / Accepted: 1 July 2015 / Published online: 18 July 2015

Abstract Observations of thermal radiation from neutron stars can potentially provide information about the states of supranuclear matter in the interiors of these stars with the aid of the theory of neutron-star thermal evolution. We review the basics of this theory for isolated neutron stars with strong magnetic fields, including most relevant thermodynamic and kinetic properties in the stellar core, crust, and blanketing envelopes.

Keywords Neutron stars · Magnetic fields · Dense matter · Thermal emission · Heat transport

1 Introduction

The first works on neutron star cooling and thermal emission (Stabler 1960; Tsuruta 1964; Chiu and Salpeter 1964; Morton 1964; Bahcall and Wolf 1965a, 1965b) appeared at the epoch of the discoveries of X-ray sources outside the Solar System in the rocket and balloon experiments (Giacconi et al. 1962; Bowyer et al. 1964a; Bowyer et al. 1964b). The authors estimated cooling rates and surface temperatures in order to answer the question, whether a

✉ A.Y. Potekhin
palex@astro.ioffe.ru

J.A. Pons
jose.pons@ua.es

D. Page
page@astro.unam.mx

1 Ioffe Institute, Politekhnicheskaya 26, 194021 Saint Petersburg, Russia

2 Central Astronomical Observatory at Pulkovo, Pulkovskoe Shosse 65, 196140 Saint Petersburg, Russia

3 St. Petersburg State Polytechnical University, Polyteknicheskaya 29, 195251 Saint Petersburg, Russia

4 Departament de Física Aplicada, Universitat d'Alacant, Ap. Correus 99, 03080 Alacant, Spain

5 Instituto de Astronomía, Universidad Nacional Autónoma de México, México, D.F. 04510, Mexico

neutron star can be detected in this kind of experiments. However, the first attempts failed to prove the relation between neutron stars and newly discovered X-ray sources. In particular, Bowyer et al. (1964b) measured the size of the X-ray source in the Crab Nebula from observations during a lunar occultation on July 7, 1964. Their result, $\sim 10^{13}$ km, indicated that the source was much larger than a neutron star should be. Ironically, there was a neutron star there, the famous Crab pulsar, but it was hidden within a compact plerion pulsar nebula. Kardashev (1964) and later Pacini (1967) conjectured that the Crab Nebula could be powered by the neutron-star rotational energy, which was transferred to the nebula via the magnetic field, but this model remained a hypothesis. Curiously, the Crab pulsar was observed as a scintillating radio source since 1962 (Hewish and Okoye 1968), but the nature of this source remained unclear. Sandage et al. (1966) identified Sco X-1, the first detected and the brightest cosmic X-ray source, as an optical object of 13th magnitude. Shklovsky (1967) analyzed these observations and concluded that the X-ray radiation of Sco X-1 originated from the accretion of matter onto a neutron star from its companion. Later this conjecture was proved to be true (de Freitas Pacheco et al. 1977), but at the time it was refuted (Cameron 1967). Because of these early confusions, the first generally accepted evidence of neutron stars was provided only by the discovery of pulsars (Hewish et al. 1968) after a successful competition of the theoretical interpretation of pulsars as rotating neutron stars (Gold 1968) with numerous alternative hypotheses (see, e.g., the review by Ginzburg 1971).

The foundation of the rigorous cooling theory was laid by Tsuruta (1964) and Tsuruta and Cameron (1966), who formulated the main elements of the theory: the relation between the internal and surface temperatures of a neutron star, the neutrino and photon cooling stages, etc. After the discovery of neutron stars, a search for their soft X-ray thermal emission has become a topical challenge, which stimulated the development of the cooling theory. The first decade of this development was reviewed by Tsuruta (1979) and Nomoto and Tsuruta (1981a).

Thorne (1977) presented the complete set of equations describing the mechanical and thermal structure and evolution of a spherically symmetric star at hydrostatic equilibrium in the framework of General Relativity (GR). The GR effects on the thermal evolution of neutron stars were first included into the cooling calculations by Glen and Sutherland (1980), Nomoto and Tsuruta (1981b), Van Riper and Lamb (1981). A generally relativistic cooling code for a spherically symmetric non-barotropic star was written by Richardson et al. (1982). Nomoto and Tsuruta (1986, 1987) studied neutron star cooling using an updated physics input and discussed the role of different physical factors for thermal relaxation of different models of neutron stars. Tsuruta (1986) provided a comprehensive review of the neutron star cooling theory with a comparison of the results of different research groups obtained by mid-1980s.

The early studies of the neutron-star cooling were mostly focused on the standard scenario where the neutrino emission from the stellar core was produced mainly by the modified Urca (Murca) processes, which compete with neutrino emission via plasmon decay, nucleon bremsstrahlung, etc. The enhanced (accelerated) cooling due to the direct Urca (Durca) processes was believed possible only if the core contains a pion condensate or a quark plasma (e.g., Tsuruta 1979; Glen and Sutherland 1980; Van Riper and Lamb 1981; Richardson et al. 1982). By the end of 1980s a new cooling agent, kaon condensate, was introduced (Brown et al. 1988; Page and Baron 1990). The studies of the enhanced cooling were intensified after the discovery by Lattimer et al. (1991) that the Durca process is allowed in the neutron star core with the standard nuclear composition for some realistic equations of state (EoS) without "exotic" models. The standard and Durca-enhanced neutron star cooling scenarios were compared in a number of numerical simulations starting from Page and Applegate (1992), who also noticed that nucleon superfluidity becomes the strongest cooling regulator in the Durca-allowed stellar kernels. This result triggered a flow of papers on the cooling of superfluid neutron stars.

The progress in the theoretical studies of the neutron-star thermal evolution was influenced in the 1980s and 1990s by the spectacular progress of the X-ray astronomy, notably due to the space observatories *Einstein* (1978–1981), *EXOSAT* (1983–1986), and *ROSAT* (1990–1998). *ROSAT* was the first to reliably detect X-ray thermal radiation from isolated neutron stars. This theoretical and observational progress was reviewed by Tsuruta (1998), Yakovlev and Pethick (2004), Page et al. (2004).

In the 21st century, the data collected by X-ray observatories *Chandra* and *XMM-Newton* give a new impetus to the development of the cooling theory. Some new theoretical results on the cooling of neutron stars and relation of the theory to observations were reviewed by Yakovlev et al. (2008), Page (2009), Tsuruta (2009). Recently, 2D simulations of the fully coupled thermal and magnetic field evolution have been possible (Pons et al. 2014; Viganò et al. 2013), mostly motivated by the increasing number of observed magnetars and high magnetic field pulsars.

The theory of thermal evolution of neutron stars has different aspects associated with rotation, accretion, etc. In this review, we will mostly focus on the physics that determines thermal structure and evolution of slowly rotating non-accreting neutron stars, whose thermal emission can be substantially affected by strong magnetic fields. The processes of formation of thermal spectra in the outermost layers of such stars are explicitly excluded from this paper but considered in the companion review (Potekhin et al. 2015, hereafter Paper I). We will pay a special attention to the effects of strong magnetic fields on the thermal structure and heat conduction in the crust and heat-blanketing envelopes of neutron stars.

2 The Essential Physics of Neutron Star Cooling

In this section we briefly present the essential physical ingredients needed to build a model of a cooling neutron star regardless its magnetic field. The effects of strong magnetic fields will be discussed in subsequent sections, starting from Sect. 4.

2.1 Structure and Composition of a Neutron Star

A neutron star is born hot ($\approx 10^{11}$ K) and lepton-rich, but only a few days after its birth, its temperature drops to a few $\times 10^9$ K. Thus, the Fermi energy ε_F of all particles is much higher than the kinetic thermal energy in most of the star volume, except in the thin outermost layers (a few meters thick), which does not affect the mechanical and thermal structure of the rest of the star. Therefore, a good approximation is to describe the state of matter as cold nuclear matter in beta equilibrium, resulting in an effectively barotropic EoS. The mechanical structure of the star is decoupled from its thermal structure and can be calculated only once and kept fixed during the thermal evolution simulations.

To a very good approximation, the mechanical structure can be assumed to be spherical. Appreciable deviations from the spherical symmetry can be caused by ultra-strong magnetic fields ($B \gtrsim 10^{17}$ G) or by rotation with ultra-short periods (less than a few milliseconds), but we will not consider such extreme cases. Then the space-time is described by the Schwarzschild metric (e.g., Misner et al. 1973)

$$\mathrm{d}s^2 = -\mathrm{e}^{2\Phi(r)} c^2 \mathrm{d}t^2 + \mathrm{e}^{2\Lambda(r)} \mathrm{d}r^2 + r^2 \big(\mathrm{d}\theta^2 + \sin^2\theta \,\mathrm{d}\varphi^2\big), \tag{1}$$

where (r, θ, φ) are the standard spherical coordinates, $e^{2\Lambda(r)} = 1 - 2GM_r/c^2 r$, and $\Phi(r)$ is determined by equation

$$\mathrm{d}\Phi(r)/\mathrm{d}P(r) = -\left[P(r) + \rho(r)c^2\right]^{-1} \tag{2}$$

with the boundary condition $e^{2\Phi(R)} = 1 - r_g/R$ at the stellar radius R. Here, $r_g = 2GM/c^2 = 2.95(M/M_\odot)$ km is the Schwarzschild radius, $M \equiv M_R$ is the stellar mass, $M_r = 4\pi \int_0^r \rho(r) r^2 \mathrm{d}r$ is the mass inside a sphere of radius r, G is the gravitational constant, c is the speed of light, P is the pressure, and ρ is the gravitational mass density.

The mechanical structure of a spherically symmetric star is described by the Tolman-Oppenheimer-Volkoff equation

$$\frac{\mathrm{d}P}{\mathrm{d}r} = -\frac{GM_r\rho}{r^2}\left(1 + \frac{P}{\rho c^2}\right)\left(1 + \frac{4\pi r^3 P}{M_r c^2}\right)\left(1 - \frac{2GM_r}{rc^2}\right)^{-1/2}, \tag{3}$$

where r is the radial coordinate measured from the stellar center. In order to determine the stellar mechanical structure, Eq. (3) should be supplemented by an EoS, which depends on a microscopic physical model (Sect. 2.4). Several qualitatively different regions can be distinguished in a neutron star, from the center to the surface: the inner and outer core, the mantle, the inner and outer crust, the ocean, and the atmosphere (e.g., Haensel et al. 2007).

The outer core of a neutron star has mass density $0.5\rho_0 \lesssim \rho \lesssim 2\rho_0$, where $\rho_0 = 2.8 \times 10^{14}\ \mathrm{g\,cm^{-3}}$ is the nuclear saturation density (the typical density of a heavy atomic nucleus). It is usually several kilometers thick and contains most of the stellar mass. The outer core is mostly composed of neutrons with an admixture of the protons and leptons—electrons and muons ($npe\mu$ matter).

The inner core, which can exist in rather massive neutron stars, $M \gtrsim 1.5\ M_\odot$, occupies the central part with $\rho \gtrsim 2\rho_0$. It is defined as the region where the composition is uncertain, but probably more rich than simply neutrons and protons. Its composition and properties are not well known because the results of their calculation strongly depend on details on the theoretical model of collective fundamental interactions. Some of the proposed models envision the following hypothetical options:

1. hyperonization of matter—the appearance of various hyperons (first of all, Λ- and Σ^--hyperons—$npe\mu\Lambda\Sigma$ matter);
2. pion condensation—formation of a Bose condensate of collective interactions with the properties of π-mesons;
3. kaon condensation—formation of a similar condensate of K-mesons;
4. deconfinement—phase transition to quark matter.

The last three options are often called *exotic* (Haensel et al. 2007, Chap. 7). In this paper we will not consider the exotic matter in any detail.

In the stellar *crust and ocean* the matter is less extraordinary: it contains electrons, nuclei composed of protons and neutrons, and, in the inner crust, quasi-free neutrons. Nevertheless, this region is also under extreme conditions (density, temperature, magnetic field) that cannot be reproduced in the laboratory. In the crust, which is normally $\sim$1 km thick, the nuclei are arranged into a crystalline lattice, and in the ocean with a typical depth from a few to $\sim$100 meters (depending on temperature) they form a liquid (see Sect. 2.4.1).

With increasing density, nuclei become progressively neutron-rich due to the beta-captures that are favored by the increase of pressure of the degenerate electrons. Neutrons start to drip out of nuclei at density $\rho_{\mathrm{drip}} = 4.3 \times 10^{11}\ \mathrm{g\,cm^{-3}}$. Thus at $\rho > \rho_{\mathrm{drip}}$ neutron-rich nuclei are embedded in the sea of quasi-free neutrons.

At the bottom of the crust, the nuclei may take rodlike and platelike shapes to compose so called *pasta phases* of nuclear matter (Pethick and Ravenhall 1995). Then they form a *mantle* with anisotropic kinetic properties (Pethick and Potekhin 1996). Thermodynamic stability of the pasta phase state and, therefore, the existence of the mantle depends on the model of nuclear interactions. Lorenz et al. (1993) demonstrated stability of the pasta phases at $\rho \gtrsim 10^{14}\ \mathrm{g\,cm^{-3}}$ for the FPS EoS model of Pandharipande and Ravenhall (1989), but they were not found to be stable in modern EoS models SLy (Douchin and Haensel 2001) and BSk (Pearson et al. 2012).

The strong gravity drives the rapid separation of chemical elements in the crust and the ocean. Estimates of characteristic sedimentation time range from seconds to months, depending on local conditions and composition (see, e.g., Eq. (20) in Potekhin 2014), which is a very short timescale compared to the stellar age. Therefore the envelopes are thought to be made of chemically pure layers, which are separated by narrow transition bands of diffusive mixing (De Blasio 2000; Chang et al. 2010).

2.2 Thermal Evolution Equations

The multidimensional heat transport and thermal evolution equations in a locally flat reference frame read (e.g., Aguilera et al. 2008; Pons et al. 2014; Viganò et al. 2013)

$$c_\mathrm{v}\,\mathrm{e}^{\Phi}\frac{\partial T}{\partial t} + \nabla\cdot\left(\mathrm{e}^{2\Phi}\boldsymbol{F}\right) = \mathrm{e}^{2\Phi}(H - Q_\nu), \quad \boldsymbol{F} = -\mathrm{e}^{-\Phi}\hat{\kappa}\cdot\nabla\left(\mathrm{e}^{\Phi}T\right), \tag{4}$$

where $\boldsymbol{F}$ is the heat flux density, H is the heating power per unit volume, c_v is specific heat (Sects. 2.5, 3.2, and 4.2), Q_ν is neutrino emissivity (Sects. 2.6, 3.3, 4.3), $\hat{\kappa}$ is the thermal conductivity tensor (Sects. 2.7, 3.4, and 4.4), and $\nabla = (\mathrm{e}^{-\Lambda(r)}\partial/\partial r, r^{-1}\partial/\partial\theta, (r\sin\theta)^{-1}\partial/\partial\varphi)$ in compliance with Eq. (1). The inner boundary condition to the system of equations (4) is $\boldsymbol{F} = 0$ at $r = 0$. The outer boundary condition is determined by the properties of a *heat-blanketing envelope*, which serves as a mediator of the internal heat into the outgoing thermal radiation. It will be considered in Sect. 5. Solutions to the thermal evolution equations and their implications are briefly reviewed in Sect. 6.

For weak magnetic fields, we can assume that the temperature gradients are essentially radial, and that in most of the star volume (inner crust and core) the conductivity tensor is simply a scalar quantity times the identity matrix. In this limit, corrections for deviations from the 1D approximation have little effect on the total luminosity. However, for strong fields and neutron stars with locally intense internal heating sources, such as magnetars, a more accurate description, beyond the 1D approximation, must be considered.

2D calculations of thermal structure and evolution of strongly magnetized neutron stars have been done by several groups (Geppert et al. 2004, 2006; Pérez-Azorin et al. 2006; Aguilera et al. 2008; Kaminker et al. 2012, 2014). In some of these works (Geppert et al. 2006; Pérez-Azorin et al. 2006; Aguilera et al. 2008), neutron-star models with superstrong ($B \sim 10^{15}$–10^{16} G) toroidal magnetic fields in the crust were considered, in addition to the less strong ($B \sim 10^{12}$–10^{14} G) poloidal component that penetrates from the crust into the magnetosphere. The latter models help to explain the strongly non-uniform distribution of the effective temperature over the neutron-star surface and the possible energy source for magnetars outbursts (Pons and Perna 2011; Pons and Rea 2012). Only recently (Viganò et al. 2013), the fully coupled evolution of temperature and magnetic field has been studied with detailed numerical simulations, which allow one to follow the long-term evolution of magnetars and their connection with other neutron star classes. Some results of such calculations will be discussed in Sect. 6.

2.3 Basic Plasma Parameters

In this section we introduce several basic parameters of Coulomb plasmas that are used below. To be concrete, we start with electrons and ions (including bare atomic nuclei). When other charged particles are present, their respective parameters are defined analogously, with the obvious replacements of particle mass, charge, number density, etc.

Since the major constituents of the neutron-star matter are mostly degenerate, an important parameter is the Fermi energy, which (without the rest energy) equals

$$\varepsilon_{\mathrm{F}} = c\sqrt{(mc)^2 + (p_{\mathrm{F}})^2} - mc^2, \tag{5}$$

where m is the particle mass, and p_{F} is the Fermi momentum. For instance, for the Fermi gas in the absence of a quantizing magnetic field, $p_{\mathrm{F}} = \hbar(3\pi^2 n)^{1/3}$, where n is the number density, and $\hbar$ is the reduced Planck constant. It is convenient to use the dimensionless density parameter related to the Fermi momentum of electrons, $x_{\mathrm{r}} = p_{\mathrm{F}e}/m_e c$, where m_e is the electron mass. In the outer core and the envelopes, as long as the baryons are non-relativistic, $x_{\mathrm{r}} \approx (\rho_6 Y_e)^{1/3}$, where Y_e is the number of electrons per baryon and $\rho_6 \equiv \rho/10^6$ g cm^{-3}.

Thermal de Broglie wavelengths of free ions and electrons are usually defined as $\lambda_{\mathrm{i}} = \sqrt{2\pi\hbar^2/m_{\mathrm{i}}T}$ and $\lambda_e = \sqrt{2\pi\hbar^2/m_e T}$, where $m_{\mathrm{i}} = Am_{\mathrm{u}}$ is the ion mass, and m_{u} is the unified atomic mass unit. Here and hereafter, we use T in energy units and suppress the Boltzmann constant (i.e., 10^6 K $= 86.17$ eV). The quantum effects on ion motion are important either at $\lambda_{\mathrm{i}} \gtrsim a_{\mathrm{i}}$ or at $T \ll T_{\mathrm{p}}$, where $T_{\mathrm{p}} = \hbar\omega_{\mathrm{p}}$ is the ion plasma temperature, and $\omega_{\mathrm{p}} = (4\pi e^2 n_{\mathrm{i}} Z^2/m_{\mathrm{i}})^{1/2}$ is the ion plasma frequency. Debye temperature of a crystal Θ_{D} is closely related to the plasma temperature. In the harmonic approximation for the Coulomb crystal, $\Theta_{\mathrm{D}} \approx 0.45\, T_{\mathrm{p}}$ (Carr 1961).

The Coulomb plasmas are called strongly coupled if the parameter $\Gamma_{\mathrm{C}} = (Ze)^2/a_{\mathrm{i}}T$, which estimates the electrostatic to thermal energy ratio, is large. Here, $a_{\mathrm{i}} \equiv (\frac{4}{3}\pi n_{\mathrm{i}})^{-1/3}$ is the ion sphere, or Wigner-Seitz cell, radius, and n_{i} is the ion number density. If the plasma only consists of electrons and non-relativistic ions of one kind, which is typical for neutron-star envelopes, then

$$T_{\mathrm{p}} = 7.832(Z/A)\sqrt{\rho_6} \times 10^6\ \mathrm{K}, \qquad \Gamma_{\mathrm{C}} = 22.747 Z^{5/3}(\rho Y_e)^{1/3}/T_6. \tag{6}$$

Analogously, $T_{\mathrm{p},e} = \hbar(4\pi e^2 n_e/m_e)^{1/2} = 3.34\sqrt{\rho_6 Z/A} \times 10^8$ K is the electron plasma temperature. Other plasma parameters, which become important in a strong magnetic field, will be considered in Sect. 4.1.

2.4 Equation of State

2.4.1 *Equation of State for the Outer Crust and the Ocean*

The composition of the outer crust and the ocean of a neutron star is particularly simple: their basic constituents are electrons and nuclei, which, to a good accuracy, can be treated as pointlike. The EoS of such electron-ion plasmas is well known (see, e.g., Haensel et al. 2007, Chap. 2, and references therein).

The ions thermodynamic state will go from liquid to solid as the star cools, and in the solid state from a classical to a quantum crystal. It is generally assumed that the ions form a crystalline solid and not an amorphous one. This assumption is confirmed by molecular dynamics numerical simulations (Hughto et al. 2011) and corroborated by the analysis of observations of neutron-star crust cooling after an accretion episode (see Sect. 6.2).

The simplest model of the electron-ion plasmas is the one component plasma (OCP) model, which considers Coulomb interactions of identical pointlike ions and replaces the degenerate electron gas by a static uniform charge-compensating background. The OCP has been studied analytically and numerically in many papers (see Haensel et al. 2007, Chap. 2, for references). In the classical regime ($T \gg T_\mathrm{p}$) its thermodynamic functions depend on a single parameter Γ_C. At $\Gamma_\mathrm{C} \ll 1$ the ions form a Debye-Hückel gas, with increasing Γ_C the gas gradually becomes a liquid, and with further increase of Γ_C the OCP liquid freezes. An analysis of Monte Carlo simulations of the OCP shows that its ground state is crystalline when $\Gamma_\mathrm{C} > \Gamma_\mathrm{m} = 175$ (Potekhin and Chabrier 2000). However, supercooling cannot be excluded up to a value $\Gamma_\mathrm{C} \simeq 250$. Indeed, Monte Carlo simulations of freezing of classical OCP (DeWitt et al. 1993) indicate that, as a rule, the ions do not freeze at the equilibrium melting temperature T_m but form a supercooled fluid and freeze at lower T (depending on initial conditions and other parameters). This happens because the phase transition is really tiny.

At $T \lesssim T_\mathrm{p}$, the quantum effects on ion motion become significant. Then thermodynamic functions depend not only on Γ_C, but also on r_s. The quantum effects are especially important for the solid neutron star crust at high densities, although they can also be significant in the deep layers of the ocean composed of light elements (for instance, they prevent solidification of H and He plasmas). The free energy per unit volume of an OCP crystal can be written as

$$\frac{F_\mathrm{lat}}{V} = n_\mathrm{i} C_0 \frac{(Ze)^2}{a_\mathrm{i}} + \frac{3}{2} n_\mathrm{i} u_1 T_\mathrm{p} + \frac{F_\mathrm{th}}{V} + \frac{F_\mathrm{ah}}{V}, \tag{7}$$

where the first term is the classical static-lattice energy, $C_0 \approx -0.9$ is the Madelung constant, and the next two terms describe thermodynamics of the phonon gas in the harmonic approximation (e.g., Kittel 1963): the second term accounts for zero-point quantum vibrations, and $F_\mathrm{th}/V = 3n_\mathrm{i} T \langle \ln[1 - \exp(-\hbar\omega_{\boldsymbol{k}\alpha}/T)] \rangle_\mathrm{ph}$ is the thermal contribution. Here, $u_1 = \langle \omega_{\boldsymbol{k}\alpha} \rangle_\mathrm{ph} / \omega_\mathrm{p} \approx 0.5$ is the reduced first moment of phonon frequencies $\omega_{\boldsymbol{k}\alpha}$, and $\langle \ldots \rangle_\mathrm{ph}$ denotes the averaging over phonon polarizations α and wave vectors $\boldsymbol{k}$ in the first Brillouin zone. The last term in Eq. (7) arises from anharmonic corrections, which have only been studied in detail in the classical regime ($T \gg T_\mathrm{p}$; e.g., Farouki and Hamaguchi 1993 and references therein). An analytical extrapolation of F_ah for any T was proposed in Potekhin and Chabrier (2010).

For mixtures of various ion species, the simplest evaluation of the thermodynamic functions is given by the average of their values for pure substances, weighed with their number fractions, which is called the linear mixing rule (Hansen et al. 1977). The linear mixing rule is accurate within a few percent, if the electrons are strongly degenerate and $\Gamma_\mathrm{C} > 1$ for each of the ion species in the mixture. However, this accuracy may be insufficient for such subtle phenomena as melting/freezing or phase separation in the Coulomb plasmas. Corrections to the linear mixing rule were obtained by Potekhin et al. (2009). Medin and Cumming (2010) used these results to construct a semianalytical model for prediction of the composition and phase state of multicomponent mixtures. Hughto et al. (2012) confirmed the qualitative validity of this model by molecular dynamics simulations.

The OCP model is a reasonable first approximation, but in reality the electrons do not form a uniform background: they interact with each other and with the ions, which gives rise to exchange-correlation and polarization corrections, respectively. The polarization corrections are appreciable even for strongly degenerate plasmas. For instance, they can substantially shift the melting transition away from $\Gamma_\mathrm{C} = 175$ (Potekhin and Chabrier 2013). In the outer envelopes of a neutron star, the electron degeneracy weakens, and one should take the T-dependence of their EoS into account. Analytical fits for all above-mentioned contributions to the EoS of electron-ion plasmas were presented by Potekhin and Chabrier (2010, 2013). Their Fortran implementation is publicly available at http://www.ioffe.ru/astro/EIP/.

An essential input for calculating the EoS is the chemical composition of the plasma. The ground state of the matter in the outer crust can be found following the method of Baym et al. (1971). The procedure, based on the minimization of the Gibbs free energy per nucleon, is described in detail in Haensel et al. (2007). The structure of the crust is completely determined by the experimental nuclear data up to a density of the order $\rho \sim 6 \times 10^{10}\ \mathrm{g\,cm^{-3}}$. At higher densities the nuclei are so neutron rich that they have not yet been experimentally studied, and the composition of these dense layers is model dependent. However, this model dependence is not very significant in the models based on modern nuclear physics data (Haensel and Pichon 1994; Rüster et al. 2006; Pearson et al. 2011).

While a newly born neutron star is made of hot matter in nuclear equilibrium, its subsequent evolution can lead to the formation of regions where the matter is out of nuclear equilibrium. This is the case of an old accreting neutron star. Burning of the helium layer near the surface is followed by electron captures and beta decays in deeper layers. The growing layer of the processed accreted matter pushes down and eventually replaces the original catalyzed (ground-state) crust. These processes were studied by several authors (see Haensel and Zdunik 1990, 2008, and references therein).

2.4.2 *Equation of State for the Inner Crust and the Core*

The pressure in the inner crust of a neutron star is largely created by degenerate neutrons. However, the electrons and nuclei may give an important contribution to the heat capacity (see Sect. 2.5). In the core, there are contributions from neutrons, protons, electrons and muons (and other particles in the exotic models, which we do not consider here). Different theoretical EoSs of the neutron fluid and $npe\mu$ matter have been proposed, based on different methods of theoretical physics: the Brueckner-Bethe-Goldstone theory, the Green's function method, variational methods, the relativistic mean field theory, and the density functional method (see Haensel et al. 2007, Chap. 5, for review). The model of Akmal et al. (1998) (APR) has been often cited as the most advanced one for the core. It was derived from the variational principle of quantum mechanics, under which an energy minimum for the trial wave function was sought. The trial function was constructed by applying the linear combination of operators describing admissible symmetry transformations in the coordinate, spin, and isospin spaces to the Slater determinant consisting of wave functions for free nucleons. The APR EoS exists in several variants, which differ in the effective potentials of nucleon-nucleon interaction and in relativistic boost corrections. The potentials borrowed from earlier publications were optimized so as to most accurately reproduce the results of nuclear physics experiments.

Many theoretical neutron-star EoSs in the literature consist of crust and core segments obtained using different physical models. The crust-core interface there has no physical meaning, and both segments are joined using an ad hoc matching procedure. This generally leads to thermodynamic inconsistencies. The EoS models that avoid this problem by describing the core and the crust in frames of the same physical model are called *unified*. Examples of the unified EoSs are the FPS (Pandharipande and Ravenhall 1989; Lorenz et al. 1993), SLy (Douchin and Haensel 2001), and BSk (Goriely et al. 2010; Pearson et al. 2011, 2012) EoS families. All of them are based on effective Skyrme-like energy density functionals. In particular, the BSk21 model is based on a generalized Skyrme

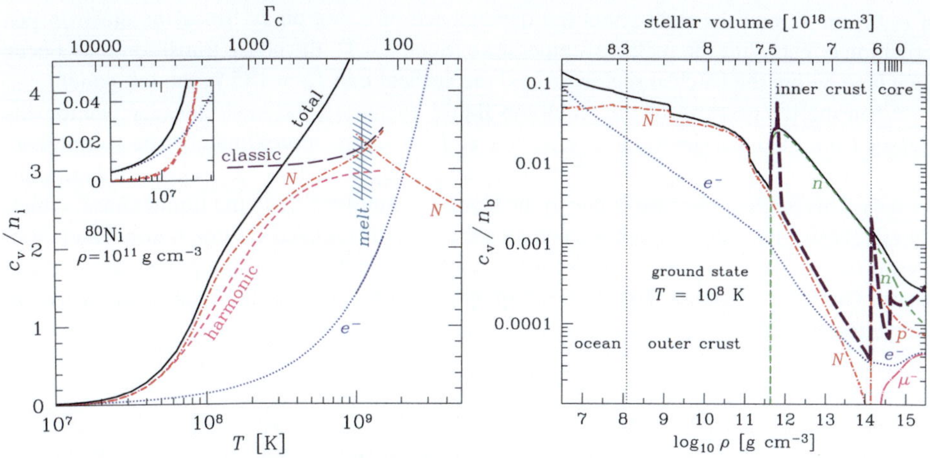

Fig. 1 *Left panel*: Heat capacity per ion versus T (*bottom axis*) and Γ_C (*top axis*) for ^{80}Ni at $\rho = 10^{11}$ g cm^{-3}. The *solid line* displays the total normalized heat capacity c_V/n_i; the *long-dashed line* shows this quantity for a classical Coulomb lattice of ions, including harmonic and anharmonic terms; the *short-dashed line* is the harmonic-lattice approximation in the solid phase; the *dot-dashed line* is the same plus anharmonic and electron polarization corrections in the solid phase. The *dotted line* is the electron Fermi gas contribution. The *vertical line* is the OCP melting point $\Gamma_C = 175$, and the hatched band shows the range $\Gamma_C = 150$–200, where melting is expected to occur in realistic conditions. The *inset* illustrates the competition between the electron and ion contributions t low T. *Right panel*: Heat capacity per baryon as function of mass density from the ocean throughout the crust and core of a neutron star at $T = 10^8$ K. The *solid line* displays the total c_V/n_b, and the *other lines* show its constituents due to the electrons (e^-), neutrons (n) in the inner crust and core, nuclei (N), including electrostatic terms in the ocean and crust but neglecting the neutron entrainment effects in the inner crust (Sect. 2.5.2), protons (p) and muons (μ^-) in the core, assuming that the nucleons are non-superfluid. For comparison, the *thick long dashes* display an example of the total c_V/n_b in the inner crust and core in the case of superfluid nucleons (Sect. 3.2). The *top axis* shows the volume contained inside a sphere with given ρ for a 1.4 $M_\odot$ neutron star. The stellar structure and composition correspond to the BSk21 EoS model

functional that most successfully satisfies various experimental restrictions along with a number of astrophysical requirements (see the discussion in Potekhin et al. 2013).

2.5 Specific Heat

2.5.1 Specific Heat of Electron-Ion Plasmas

The two components that largely dominate the specific heat c_V in the crust are the electron gas and the ions. In the neutron-star crust and core, the electrons form an ultra-relativistic highly degenerate Fermi gas, and their contribution in the heat capacity per unit volume is simply given by

$$c_{V,e} = \frac{p_{Fe}^2}{3\hbar^3 c} T = n_e \frac{\pi^2}{p_{Fe} c} T. \tag{8}$$

In the ocean, where the density is lower, approximation (8) may not work. Then it is advisable to use accurate approximations, cited in Sect. 2.4.1.

In Fig. 1 we show the temperature and density dependences of the normalized heat capacity of the ground-state matter in a neutron star. The left panel illustrates the dependence of c_v/n_i on T, and the right panel the dependence of c_v/n_b on ρ. Since the electron polarization effects shift the melting temperature (Sect. 2.4.1), the phase transition may occur anywhere within the hatched region around the vertical line $\Gamma_C = 175$ in the left panel.

When the temperature of the Coulomb liquid decreases, the heat capacity per ion increases from the ideal-gas value $c_{v,i}/n_i = \frac{3}{2}$ at $T \gg T_m$ to, approximately, the simple harmonic lattice value $c_{v,i}/n_i = 3$ at $T \lesssim T_m$ (the Dulong-Petit law for a classical harmonic crystal). This gradual increase is due to the Coulomb non-ideality in the liquid phase, which effectively smears a phase transition between the strongly coupled Coulomb liquid and OCP crystal (see Baiko et al. 1998). With further cooling, quantum effects suppress the heat capacity. Once the crystal is deep into the quantum regime its specific heat is given by the Debye result

$$c_{v,i}^{(D)} = n_i \frac{12\pi^4}{5} \left(\frac{T}{\Theta_D} \right)^3 . \tag{9}$$

The calculations of Baiko et al. (2001b) show that the Dulong-Petit law applies at temperatures down to $T \simeq T_p$, while the Debye value of Eq. (9) is attained when $T \lesssim 0.1\ T_p$. The same authors present a simple analytical approximation for the heat capacity of a harmonic Coulomb crystal, accurate to a few parts in 10^5 at any T.

However, the harmonic OCP model is an idealization. The anharmonic corrections and electron polarization corrections (Sect. 2.4.1) can amount up to (10–20) % of $c_{v,i}$. Because of the anharmonic effects, $c_{v,i}/n_i$ is not equal to 3 exactly, but is $\sim$10 % larger at $T = T_m$. If the above-mentioned supercooling takes place in stellar matter, various fluid elements solidify at different T below T_m, and the average heat capacity, as function of temperature, can contain a bump, associated with latent heat releases (see Sect. 2.4.6 of Haensel et al. 2007 for a discussion).

The right panel of Fig. 1 shows the density dependence of the total heat capacity, normalized per one nucleon, c_v/n_b, throughout the neutron star from the ocean to the core, and partial contributions to c_v/n_b. Different particle fractions are adopted from the BSk21 model (Goriely et al. 2010; Pearson et al. 2011, 2012), as fitted by Potekhin et al. (2013). Here, we have mostly neglected the effects of nucleon superfluidity to be discussed in Sect. 3. The importance of these effects is demonstrated, however, by the heavy long-dashed line, which displays the total normalized heat capacity suppressed by nucleon superfluidity (see Sect. 3.2).

2.5.2 *Specific Heat of Neutrons*

In the inner crust, besides electrons and nuclei, there are also neutrons. In a thin layer at densities ρ just above the neutron drip point ρ_{drip}, the dripped neutrons are not paired (non-superfluid) and largely dominate c_v. Heat capacity of strongly degenerate non-superfluid neutrons can be accurately evaluated using the above-referenced analytical fits, but since the neutrons are strongly degenerate almost everywhere in the neutron star, the simpler Sommerfeld result for Fermi gases at $T \ll \varepsilon_F$ is usually applicable,

$$c_{v,x} \sim \frac{\pi^2}{2} \frac{n_x T}{\varepsilon_{F,x}}, \tag{10}$$

where x stands for the fermion type (x $= n, p, e, \mu$). For neutrons at ρ only slightly above ρ_{drip}, however, the latter formula is inaccurate because $\varepsilon_{F,n}$ is not sufficiently large. For this

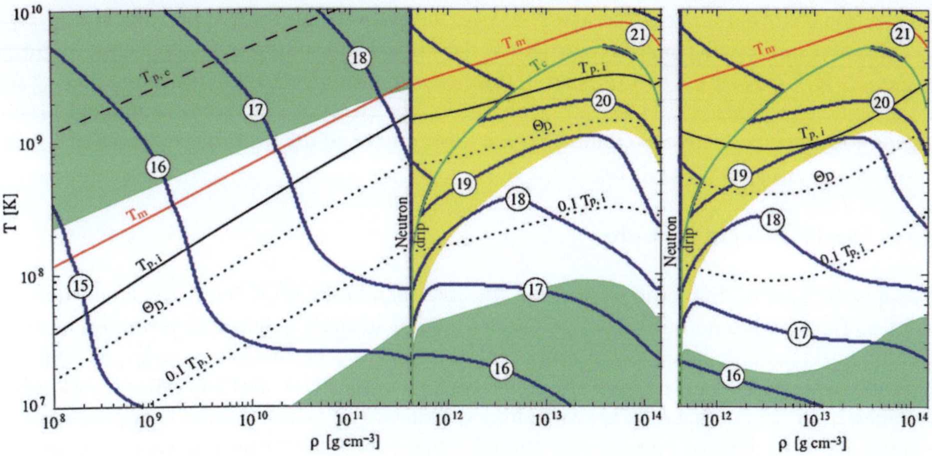

Fig. 2 Iso-contour lines of c_v in the crust, labeled by the value of $\log_{10}(c_\mathrm{v}/\mathrm{erg\,cm^{-3}\,K^{-1}})$. Also shown are the melting curve T_m and the critical temperature for neutron $^1\mathrm{S}_0$ superfluidity, T_c, the electron and ion plasma temperatures, $T_{\mathrm{p},e}$ and T_p respectively, the Debye temperature, $\Theta_\mathrm{D} \simeq 0.45\, T_\mathrm{p}$, that marks the transition from classical to quantum solid and $0.1\, T_\mathrm{p}$ below which the wholly quantum crystal regime is realized. The outer crust chemical composition is from Haensel and Pichon (1994) and inner crust from Negele and Vautherin (1973) with the neutron drip point at $\rho_\mathrm{drip} = 4.3 \times 10^{11}\ \mathrm{g\,cm^{-3}}$. The electron contribution dominates in the two *dark-shadowed* (*green*) *regions* at high T and ρ below ρ_drip and at low T and high ρ, while neutrons dominate in the *light-shadowed* (*yellow*) *region* at high T and ρ above ρ_drip, and ions dominate in the intermediate regime. The *right panel* only displays the inner crust but assuming that about 80 % of the dripped neutrons are entrained, illustrating the resulting increase in $c_{\mathrm{v,i}}$, mainly due to the strong reduction of T_p and Θ_D, significantly extending the regime where $c_{\mathrm{v,i}}$ dominates over $c_{\mathrm{v},e}$

reason, Pastore et al. (2015) proposed an interpolation between Eq. (10) and the ideal-gas limit $c_\mathrm{v} = \frac{3}{2}n$,

$$c_{\mathrm{v,x}} \approx \frac{3}{2} n_\mathrm{x} \left(1 - \mathrm{e}^{-T/T_\mathrm{cl}}\right), \qquad T_\mathrm{cl} = \frac{3\varepsilon_\mathrm{F,x}}{\pi^2}. \tag{11}$$

They also showed that corrections due to the coupling to phonons (e.g., Sect. 1.4.4 in Baym and Pethick 1991) turn out to be unimportant for $c_{\mathrm{v},n}$. Approximation (11) is accurate within 17 % for non-relativistic Fermi gases at any density. For a relativistic Fermi gas, we can preserve this accuracy by using Eq. (5) for $\varepsilon_\mathrm{F,x}$ and multiplying both T_cl and prefactor $\frac{3}{2}$ by the ratio $(mc^2 + 10T)/(mc^2 + 5T)$.

With further density increase, the neutrons become superfluid (Sect. 3), and then their contribution to c_v nearly vanishes. However, even in a superfluid state, the neutrons have a dramatic effect on c_v. Indeed, Flowers and Itoh (1976) noticed that since free neutrons move in a periodic potential created by lattice of atomic nuclei, their energy spectrum should have a band structure, which can affect kinetic and neutrino emission phenomena involving the free neutrons. Chamel (2005) calculated the band structure of these neutrons, in much the same way as electron band structure is calculated in solid state physics. The effect of this band structure is that a large fraction of the dripped neutrons are "locked" to the nuclei, i.e., the thermal motion of the nuclei entrains a significant part of the dripped neutrons resulting in a strongly increased ion effective mass $m_\mathrm{eff,i}$. This increase $m_\mathrm{i} \to m_\mathrm{eff,i}$ significantly increases $c_\mathrm{v,i}$ in the quantum regime since $c_\mathrm{v,i}^{(\mathrm{D})} \propto T_\mathrm{p}^{-1} \propto m_\mathrm{i}^{3/2}$ (Chamel et al. 2013).

The overall "landscape" of crustal specific heat is illustrated in Fig. 2. For highly degenerate electrons $c_{\mathrm{v},e} \propto T$, while for ions $c_\mathrm{v,i}$ decreases as T^3 according to Eq. (9), therefore the electron contribution dominates at $T \ll \Theta_D$, and the ion contribution prevails at $T \gtrsim \Theta_D$ (cf. the inset in the left panel of Fig. 1). On the other hand, in the non-degenerate regime $c_{v,e}/c_{v,i} \sim Z$, therefore the contribution of the electrons dominates again for $Z > 1$ at $T \gtrsim \varepsilon_F$ in the liquid phase (also cf. the left panel of Fig. 1). The effect of dripped neutron band structure on low-level collective excitations in the inner crust and the resulting increase of $c_{v,i}$ is illustrated in the right panel of Fig. 2.

2.5.3 Specific Heat of the Core

The specific heat is simpler to evaluate in the core than in the crust but it has larger uncertainties. The core is a homogeneous quantum liquid of strongly degenerate fermions, and its specific heat is simply taken as the sum of its components contribution: $c_v = \sum_x c_{v,x}$ where x stands for neutrons (n), protons (p), electrons (e), muons (μ), and any other component as hyperons or quarks that may appear at high densities. For each fermionic component, one can use Eq. (10), but for baryons one should replace the bare fermion mass m_x by an effective mass m_x^*, which encapsulates most effects of interactions. In principle, m_x^* should be calculated from the same microphysical interaction as employed for the EoS; cf. Sect. 2.6.3. For leptons (e and μ), interactions have a negligible effect on m_x^* and the bare fermion mass value can be used. The nucleon heat capacity in the core is strongly affected by pairing (superfluidity) effects, as discussed in Sect. 3.2.

2.6 Neutrino Emissivity

The neutrino luminosity of a neutron star is, in most cases, strongly dominated by the core contribution, simply because the core comprises a lion's share of the total mass. The crust contribution can, however, prevail in the case of strong superfluidity in the core, which suppresses the neutrino emissivities. Crust neutrino emission is also essential during the early thermal relaxation phase (the first few decades of the life of the star), or the crust relaxation after energetic transient events (e.g., strong bursts of accretion in X-ray binaries and flares in magnetars).

Yakovlev et al. (2001) reviewed the main neutrino emission mechanisms in neutron star crusts and cores and collected fitting formulae for the neutrino emissivity in each reaction as a function of density and temperature. The summary of the most important processes is given in Table 1. The last column of this table contains references to the analytical fitting formulae that can be directly employed to calculate the relevant emission rates. These processes are briefly described below.

2.6.1 Neutrino Emission in the Crust

There is a variety of neutrino processes acting in the crust. In a non-magnetized crust the most important ones are the *plasmon decay* process and the *electron-ion bremsstrahlung* process (see Table 1). The *pair annihilation* process can be also important if the crust is sufficiently hot.

The total emissivity from the sum of these processes is illustrated in the left panel of Fig. 3. The first thing to notice is the enormous range of values of Q_ν covered in the ρ–T range displayed in this figure, which spans 26 orders of magnitude. This is a direct consequence of the strong T dependence of the neutrino processes. The pair annihilation process is efficient only at low densities and very high temperatures, but when $T \ll T_{F,e}$ very few positrons are present and the process is strongly suppressed. In the whole range of this plot,

Table 1 Main neutrino emission processes[a]

No.	Process/control function	Symbolic notation[b]	Formulae for Q_ν and/or R
In the crust			
1	Plasmon decay	$\Gamma \to \nu + \bar{\nu}$	Eqs. (15)–(32) of [1]
2	Electron-nucleus bremsstrahlung	$e^- + N \to e^- + N + \nu + \bar{\nu}$	Eqs. (6), (16)–(21) of [2]
3	Electron-positron annihilation	$e^- + e^+ \to \nu + \bar{\nu}$	Eq. (22) of [3]
4[c]	Electron synchrotron	$e^- \xrightarrow{B} e^- + \nu + \bar{\nu}$	Eq. (48)–(57) of [3]
In the core			
1[d]	Direct Urca (Durca)	$n \to p + e^- + \bar{\nu}_e$, $p + e^- \to n + \nu_e$	Eq. (120) of [3]
	Magnetic modification[c]	$R_B^{(\mathrm{D})}$	Eqs. (247)–(250) of [3]
	Reduction factors[e]	$R_x^{(\mathrm{D})}$	Eqs. (199), (202)–(206) of [3]
2	Modified Urca (Murca) (neutron branch)	$n + n \to n + p + e^- + \bar{\nu}_e$, $n + p + e^- \to n + n + \nu_e$	Eq. (140) of [3]
	Reduction factors[e]	$R_x^{(\mathrm{M}n)}$	Appendix of [4]
3	Murca (proton branch)	$p + n \to p + p + e^- + \bar{\nu}_e$, $p + p + e^- \to p + n + \nu_e$	Eq. (142) of [3], corrected at $3p_{\mathrm{F}p} > p_{\mathrm{F}n} + p_{\mathrm{F}e}$ as per [4]
	Reduction factors[e]	$R_x^{(\mathrm{M}p)}$	Appendix (and Eq. (25)) of [4]
4	Baryon-baryon bremsstrahlung	$\begin{cases} n + n \to n + n + \nu + \bar{\nu} \\ n + p \to n + p + \nu + \bar{\nu} \\ p + p \to p + p + \nu + \bar{\nu} \end{cases}$	Eq. (165) of [3] Eq. (166) of [3] Eq. (167) of [3]
	Reduction factors[e]	$\begin{cases} R_x^{(nn)} \\ R_x^{(np)} \\ R_x^{(pp)} \end{cases}$	Eqs. (221), (222), (228) of [3] and Eq. (60) of [4] Eq. (220), (229) of [3] and Eq. (54) of [4] Eq. (221) of [3]
5[e]	Cooper pairing of baryons	$\begin{cases} n + n \to [nn] + \nu + \bar{\nu} \\ p + p \to [pp] + \nu + \bar{\nu} \end{cases}$	Eqs. (236), (241) of [3], corrected as per [5] (Sect. 3.3)
6[c,e]	Electron-fluxoid bremsstrahlung	$e^- + f \to e^- + f + \nu + \bar{\nu}$	Eqs. (253), (263), (266)–(268) of [3]

Notes:

[a]References: [1] Kantor and Gusakov (2007); [2] Ofengeim et al. (2014); [3] Yakovlev et al. (2001); [4] Gusakov (2002); [5] Leinson, Leinson (2009, 2010)

[b]Γ means a plasmon, e^- an electron, e^+ a positron, ν a neutrino, $\bar{\nu}$ an antineutrino (in general, of any flavor, but ν_e or $\bar{\nu}_e$ stands for the electron neutrino or antineutrino, respectively), p a proton, n a neutron, $[pp]$ and $[nn]$ their paired states, N stands for an atomic nucleus, and f for a proton fluxoid. At densities where muons are present, they participate in the Urca and bremsstrahlung processes fully analogous to the processes 1, 2, 3, 6 in the core (see details in Ref. [1]). R with subscripts/superscripts signifies a control function (correction factor) due to superfluidity or magnetic field. Subscript x in R_x substitutes for different superfluidity types (proton or neutron, singlet or triplet); B indicates magnetic field

[c]The effect of strong magnetic field (see Sect. 4.3)

[d]At densities beyond the Durca threshold (see Sect. 2.6.2)

[e]The effect of superfluidity (see Sect. 3.3)

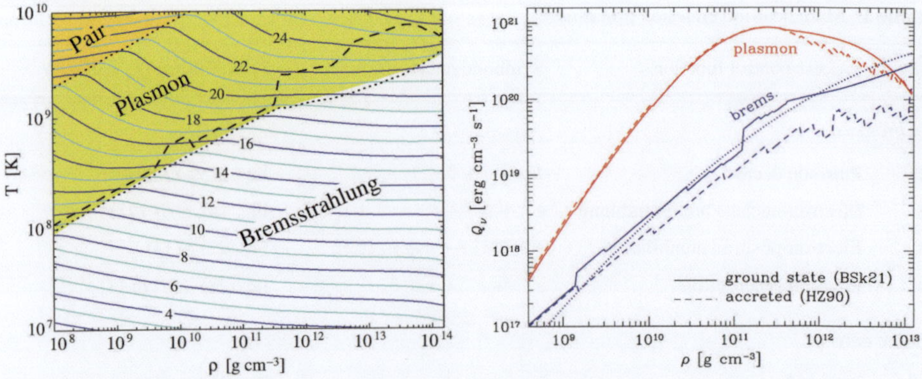

Fig. 3 Neutrino emissivity Q_ν in a non-magnetized crust from the pair annihilation, plasmon decay, and electron-ion bremsstrahlung processes. *Left panel*: The *contour lines* are labeled by the value of $\log_{10}[Q_\nu/(\mathrm{erg\,cm^{-3}\,s^{-1}})]$. Regions where the pair, plasma, and bremsstrahlung processes dominate are indicated: the boundaries happen to be quite well described by the two *dotted lines* that show $\frac{5}{3}T_{\mathrm{p},e}$ and $\frac{1}{13}T_{\mathrm{p},e}$. (Also indicated is the ion melting curve, *dashed line*.) *Right panel*: Density dependences of Q_ν for the ground-state nuclear matter (*solid lines*) and for the accreted crust (*dashed lines*) at $T = 10^9$ K. The *dotted line* represents an older fit to the bremsstrahlung process (see text for detail)

$T_{\mathrm{F,e}} \gg 10^{10}$ K but pair annihilation still dominates at low ρ and high T. In the opposite high-ρ and low-T regime the dominant process is electron-ion bremsstrahlung, for which $Q_\nu^{(\mathrm{brems})} \propto T^8$. At intermediate T and ρ the plasmon decay process is most important and, when it strongly dominates, its emissivity behaves as $Q_\nu^{(\mathrm{pl})} \propto T^4$.

The right panel of Fig. 3 illustrates the density dependence of $Q_\nu^{(\mathrm{pl})}$ and $Q_\nu^{(\mathrm{brems})}$ in either ground-state or accreted crust of a neutron star with $T = 10^9$ K. Pair annihilation is negligible in this case. $Q_\nu^{(\mathrm{pl})}$ is calculated according to Yakovlev et al. (2001) and $Q_\nu^{(\mathrm{brems})}$ according to Ofengeim et al. (2014). For comparison, an older fit to $Q_\nu^{(\mathrm{brems})}$ (Kaminker et al. 1999) is plotted by the dotted line. The ground-state composition and the nuclear size are described by the BSk21 model (Goriely et al. 2010; Pearson et al. 2012, as fitted by Potekhin et al. 2013). The accreted composition is taken from Haensel and Zdunik (1990); in this case the approximation by Itoh and Kohyama (1983) is used for the nuclear size.

The band structure of the energy spectrum of neutrons in the inner crust, which was mentioned in Sect. 2.5.1, should reduce the neutrino reactions of the bremsstrahlung type and initiate an additional neutrino emission due to direct inter-band transitions of the neutrons, in analogy with Cooper pairing of neutrons discussed in Sect. 3.3. These effects have been mentioned by Yakovlev et al. (2001), but remain unexplored.

Electron and positron captures and decays by atomic nuclei (beta processes), which accompany cooling of matter and non-equilibrium nuclear reactions, produce neutrino emission. A pair of consecutive beta capture and decay reactions is a nuclear Urca process. Urca processes involving electrons were put forward by Gamow and Schoenberg (1941), while those involving positrons were introduced by Pinaev (1964). In the neutron star crust, the appropriate neutrino luminosity depends on cooling rate and should be especially strong at $T \sim (2\text{–}4) \times 10^9$ K when the main fraction of free neutrons is captured by nuclei. However, there are other efficient neutrino reactions open at such temperatures, which make the neutrino emission due to beta processes insignificant (Yakovlev et al. 2001). On the other hand, heating produced by non-equilibrium nuclear reactions (the deep crustal heating, Haensel and Zdunik 1990) that accompany accretion mentioned in Sect. 2.4.1, may be more important than non-equilibrium neutrino cooling.

There are a number of other neutrino-emission processes (Yakovlev et al. 2001), which are less efficient than those listed in Table 1. In the inner crust with dripped neutrons, $n-n$ bremsstrahlung is very efficient but it is suppressed by pairing and, hence, only acts in the layers where $T > T_{cn}$, where T_{cn} is the neutron pairing critical temperature (Sect. 3.1). This process operates in a wide range of densities and temperatures, and the density dependence of its emissivity is generally smooth. Neutrino emission from the formation and breaking of Cooper pairs makes a significant contribution, much stronger than the bremsstrahlung, but is confined to a restricted region of ρ and T (Sect. 3.3). In the presence of a very strong magnetic field, some of the above-mentioned processes are modified, and new channels for neutrino emission may open (Sect. 4).

2.6.2 Neutrino Emission in the Core

Yakovlev et al. (2001) discussed a wealth of neutrino reactions which may be important in the $npe\mu\Lambda\Sigma^-$ matter in a neutron star core, which include

1. 8 direct Urca (Durca) processes of the electron or muon production and capture by baryons (baryon direct Urca processes),
2. 32 modified Urca (Murca) processes, also associated with the electron or muon production and capture by baryons (baryon Murca processes),
3. 12 processes of neutrino-pair emission in strong baryon-baryon collisions (baryon bremsstrahlung),
4. 4 Murca processes associated with muon decay and production by electrons (lepton Murca process),
5. 7 processes of neutrino pair emission in Coulomb collisions (Coulomb bremsstrahlung).

In this paper we basically restrict ourselves to the $npe\mu$ matter. We refer the reader to the review by Yakovlev et al. (2001) for the more general case, as well as for a discussion of other exotic models (such as the pion or kaon condensates). It appears that the reactions that proceed in the $npe\mu$ matter are often sufficient for the neutron-star cooling, even when the appearance of the Σ^- and Λ hyperons is allowed. The reason is that these hyperons can appear at high densities only, where the baryon Durca processes are likely to be allowed and dominate, for realistic EoSs.

The Durca cycle consists of the beta decay and electron capture processes (see Table 1). They are threshold reactions open at sufficiently high densities, and not for every EoS model. For the degenerate nucleons they are only possible if the proton fraction exceeds a certain threshold. In the npe matter (without muons) this threshold is $\approx$11 %, which follows readily from the energy and momentum conservation combined with the condition of electric charge neutrality of matter. Indeed, for strongly degenerate fermions the Pauli blocking implies that the reaction is possible only if the energies of the reacting particles are close to their respective Fermi energies. Then the momentum conservation assumes the inequality $p_{\mathrm{Fn}} \leqslant p_{\mathrm{Fe}} + p_{\mathrm{Fp}}$, that is $n_n^{1/3} \leqslant n_e^{1/3} + n_p^{1/3}$. For the npe matter $n_e = n_p$ because of the charge neutrality, therefore $n_n \leqslant 8n_p$, or $n_p \geqslant n_{\mathrm{b}}/9$, where n_{b} is the total baryon number density. The presence of muons can increase this threshold by several percent. If $p_{\mathrm{F}\mu} \geqslant p_{\mathrm{Fn}} - p_{\mathrm{Fp}}$, then the muon Durca process adds to the electron Durca.

If allowed, the Durca processes produce a rapid (enhanced) cooling of neutron stars. If they are forbidden, the main reactions are those of the baryon Murca and bremsstrahlung processes which produce a slow (standard) cooling. The Murca process is a second order process, in which a bystander neutron or proton participates to allow momentum conservation (see Table 1). Since this process involves five degenerate fermions, instead of three for the Durca process, its efficiency is reduced, simply by phase space limitation, by a factor of order $(T/\varepsilon_F)^2$, which gives an overall temperature-dependence T^8 instead of T^6. This reduction, for typical conditions in the neutron-star core, amounts to 6 orders of magnitude. It is certainly the dominant process for not too high densities in absence of pairing, and is the essence of the "standard cooling scenario". However, in presence of superfluidity, neutrino emission by the formation of Cooper pairs (Sect. 3.3) can dominate over the Murca process.

Other neutrino reactions in the core involve neutrino-pair bremsstrahlung in Coulomb collisions lepton modified Urca processes, electron-positron annihilation, etc. All of them are not significant under the typical conditions in the non-exotic core. For instance, the plasmon decay process that is efficient in the neutron star crust (Sect. 2.6.1) is exponentially suppressed in the core, because the electron plasmon energy in the core ($\sim\hbar\omega_p \sim 10$ MeV) is much larger than the thermal energy.

In a strong magnetic field penetrating into the core, some of the above-mentioned processes can be modified, and new channels for neutrino emission may open (see Sect. 4).

2.6.3 Remarks on In-Medium Effects

Neutrino emissivity Q_ν may be strongly modified by in-medium (collective) effects at the high densities of neutron stars (see Voskresensky 2001, for a review). For instance, these effects may result in renormalization of electroweak interaction parameters. Moreover, the in-medium effects may open new channels for neutrino emission. Voskresensky and Senatorov (1986) found that the direct and modified Urca processes appreciably exceed the estimates obtained neglecting the collective effects, provided the density is sufficiently large. On the other hand, the in-medium effects suppress the *nn* bremsstrahlung in the neutron-star core by a factor of 10–20 (Blaschke et al. 1995). According to the study by Schaab et al. (1997), the medium effects on the emissivity of the Murca process cause a more rapid cooling than obtained for the standard scenario and result in a strong density dependence, which gives a smooth crossover from the standard to the enhanced cooling scenario (see Sect. 6.1) for increasing star masses.

The problem of calculation of the in-medium effects in the neutron star matter is complicated. Various theoretical approaches were used to solve it, results of different techniques being different typically by a factor of a few (see, e.g., Blaschke et al. 1995, and references therein). The renormalization of the electroweak coupling is usually taken into account in an approximate manner by replacing the bare baryon masses m_B with effective ones, m_B^* (e.g., Yakovlev et al. 2001). The values of these effective masses should be taken from microscopic theories.

The in-medium effects are also important for the Cooper-pairing neutrino emission mechanism related to baryon superfluidity, as discussed in Sect. 3.3 below, for heat capacity (Sect. 2.5.3), and for baryon heat conduction (Sect. 2.7.3) in the core of a neutron star.

2.7 Thermal Conductivity

The most important heat carriers in the crust and ocean of the star are the electrons. In the atmosphere, the heat is carried mainly by photons. In general, the two mechanisms work in parallel, hence $\kappa = \kappa_r + \kappa_e$, where κ_r and κ_e denote the radiative (r) and electron (e) components of the thermal conductivity κ. The radiative transfer is considered in Paper I. In this paper we will pay most attention to the electron heat conduction mechanism. Both the electron and photon heat conduction are affected by strong magnetic fields. We will consider these effects in Sect. 4.

Table 2 Main contributions to thermal conductivity

No.	Conduction type and regime	References[a]
1[b]	Photon conduction	Eqs. (14)–(20) of [1]
	– plasma cutoff correction	Sect. 3.3 of [2]
	– magnetic field modifications[c]	Eqs. (21)–(23) of [1]
2	Electron conduction in the ocean and the crust:	See Appendix A
	– Electron-ion/electron-phonon scattering	[3] (theory), [4] (public code)
	– the effects of magnetic fields[d]	[5] (theory), [4] (public code)
	– the effects of finite nuclear sizes in the inner crust	[6] (theory), [4] (public code)
	– Electron scattering on impurities in the crust	See Appendix A.4
	– Electron-electron scattering:	
	– strongly degenerate electrons	Eqs. (10), (21)–(23) of [7]
	– arbitrary degeneracy	See Appendix A.3
3	Baryon conduction in the core	Eqs. (7), (12), (21), (28)–(30) of [8]
	– Effects of superfluidity[e]	Eqs. (45)–(48), (50)–(53) of [8]
4	Lepton conduction in the core	Eqs. (4)–(6), (16), (17), (33)–(37) of [9]
	– Effects of superfluidity[e]	Eqs. (45), (54)–(61), (84)–(92)[f] of [9]

Notes:

[a]References: [1] Potekhin and Yakovlev (2001); [2] Potekhin et al. (2003); [3] Potekhin et al. (1999); [4] http://www.ioffe.ru/astro/conduct/; [5] Potekhin, Potekhin (1996, 1999); [6] Gnedin et al. (2001); [7] Shternin and Yakovlev (2006); [8] Baiko et al. (2001a); [9] Shternin and Yakovlev (2007)

[b]For fully ionized atmospheres only. For partially ionized atmospheres, see references in Potekhin (2014)

[c]See Sect. 4.4.1

[d]See Sect. 4.4.2

[e]See Sect. 3.4

[f]The power index 2 should be suppressed at its first occurrences in the third and fourth lines of Eq. (92) of Ref. [9]

The elementary theory in which the effective collision rate ν of heat carriers with effective mass m^* and number density n does not depend on their velocity, gives (Ziman 1960)

$$\kappa = anT/m^*\nu, \tag{12}$$

where a is a numerical coefficient: $a = 3/2$ for a non-degenerate gas, and $a = \pi^2/3$ for strongly degenerate particles. (We remind that we use energy units for T; otherwise a should be multiplied by the squared Boltzmann constant.)

The most important heat carriers and respective scattering processes that control the thermal conductivity κ are listed in Table 2, and briefly discussed below. The last column of the table contains references to either analytical fitting formulae or publicly available computer codes for the evaluation of κ. Figure 4 illustrates the magnitude of κ_e and characteristic temperatures in the crust.

2.7.1 Heat Conduction in the Outer Envelopes

Electron heat conduction is the most important process in the neutron star envelopes that determines thermal luminosity of neutron stars. In this case, $m^* = m_e\sqrt{1+x_r^2}$ in Eq. (12), and $\nu = \nu_e$ is mostly determined by electron-ion (ei) and electron-electron (ee) Coulomb collisions. In the crystalline phase, the electron-ion scattering takes the form of scattering on phonons (collective ion excitations). The Matthiessen rule (e.g., Ziman 1960) assumes that effective frequencies of different collisions simply add up, i.e., $\nu_e = \nu_{\rm ei} + \nu_{ee}$. This is strictly valid for extremely degenerate electrons (Hubbard and Lampe 1969). In general case it remains a good estimate, because $\nu_{ei} + \nu_{ee} \le \nu_e \le \nu_{\rm ei} + \nu_{ee} + \delta\nu$, where $\delta\nu \ll \min(\nu_{\rm ei}, \nu_{ee})$ (Ziman 1960). The relative importance of the different types of collisions and practical formulae for evaluation of ν_e can be different, depending on the composition and phase state of the plasma (see Appendix A).

Chugunov and Haensel (2007) considered an alternative heat transport by the plasma ions (phonons in the solid OCP), which works in parallel with the transport by the electrons. The ion (phonon) heat conduction is usually unimportant in neutron stars. Although the ion thermal conductivity can be larger than the electron conductivity across the strong magnetic field, the multidimensional modeling shows that in such cases the heat is mainly transported by the electrons non-radially (i.e., not straight across the field lines; see Sect. 6).

2.7.2 *Heat Conduction in the Inner Crust*

The inner crust of a neutron star is characterized by the presence of free neutrons. This has two important consequences. First, heat transport by neutrons can compete with the transport by the electrons and phonons. Second, electron-neutron scattering adds to the other electron scattering mechanisms considered above and in Appendix A.

The thermal conductivity by neutrons, κ_n, was studied in several papers (e.g., Flowers and Itoh 1976; Bisnovatyi-Kogan and Romanova 1982). A general expression for κ_n in non-superfluid matter is given by Eq. (12) with $n = n_n$, the number density of neutrons, $m^* = m_n^*$, the neutron effective mass modified by medium effects, and $\nu_n = \nu_{ni} + \nu_{nn}$. The neutron-neutron collision frequency, ν_{nn}, can be calculated in the same manner as in uniform matter of a neutron-star core (Sect. 2.7.3). However, for strongly degenerate neutrons these collisions are much less efficient than the neutron-ion ones. Therefore, one can set $\nu_n \approx \nu_{ni}$, at least for order-of-magnitude estimates. For the scattering of the neutrons by uncorrelated nuclei, $\nu_{ni} = n_{\rm i} {\rm v}_{{\rm F}n} S_{n{\rm i}}$, where ${\rm v}_{{\rm F}n} = p_{{\rm F}n}/\sqrt{c^2 + (p_{{\rm F}n}/m_n^*)^2}$ is the neutron Fermi velocity and $S_{n{\rm i}}$ is the transport cross section. For a crude estimate at sufficiently low neutron energies in the neutron star crust one can set (e.g., Bisnovatyi-Kogan and Romanova 1982) $S_{n{\rm i}} = \pi R_n^2$, where R_n is the neutron radius of an atomic nucleus (fitted, e.g., in Potekhin et al. 2013). Estimated in this way, κ_n is negligible, being at least two orders of magnitude smaller than κ_e in the entire inner crust at $T \lesssim 10^9$ K. However, ν_{ni} can be strongly affected by ion-ion correlations and by superfluidity (Sect. 3.4).

In addition, the electron conduction in the inner crust is affected by the size of a nucleus, which becomes non-negligible compared to the mean distance between the nuclei, so that the approximation of pointlike scatterers is not applicable anymore. Then one should take into account the form factor, which depends on the size and shape of the charge distribution in a nucleus. A finite charge distribution reduces ν_{ei} with respect to the model of a pointlike charge, thereby increasing the conductivity (Gnedin et al. 2001). The effect mainly depends on the ratio of the root mean square charge radius of a nucleus $R_{\rm ch}$ to the Wigner-Seitz cell radius $a_{\rm i}$. Gnedin et al. (2001) presented fitting formulae for the dependences of the thermal and electrical conductivities on the parameter $x_{\rm nuc} = \sqrt{5/3}\, R_{\rm ch}/a_{\rm i}$. The latter parameter has been fitted as function of density for modern BSk models of nuclear matter (Potekhin et al. 2013) and for some other models (Appendix B in Haensel et al. 2007).

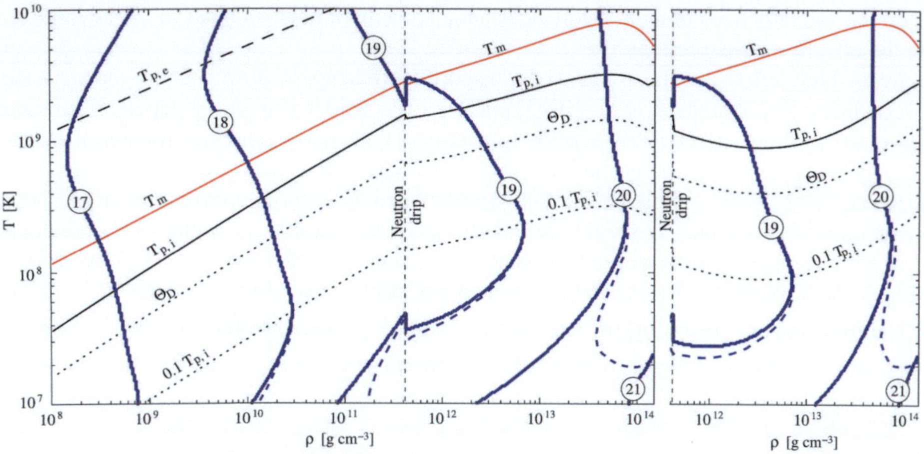

Fig. 4 Iso-contour lines of the electron thermal conductivity κ_e in the crust, labeled by the value of $\log_{10}(\kappa_e/\text{erg s}^{-1}\,\text{cm}^{-1}\,\text{K}^{-1})$, using the results of Gnedin et al. (2001). Also shown are the melting curve T_m, the electron and ion plasma temperatures, $T_{\text{p},e}$ and T_p respectively, the Debye temperature, $\Theta_\text{D} \simeq 0.45T_\text{p}$, that marks the transition from classical to quantum solid and $0.1T_\text{p}$ below which the wholly quantum crystal regime is realized. The crust composition is the same as in Fig. 2. The *right panel* only displays the inner crust but assuming that about 80 % of the dripped neutrons are entrained: the strong reduction of T_p and Θ_D pushes the onset of the wholly quantum regime to lower T. The *dashed contour* lines illustrate the reduction of κ_e from impurity scattering, assuming an impurity parameter $Q_\text{imp} = 1$

2.7.3 Heat Conduction in the Core

The first detailed studies of the kinetic coefficients in neutron star cores were performed by Flowers and Itoh (1979), who constructed the exact solution of the multicomponent system of transport equations in the *npe* matter. But since the proton fraction is small and the electron-neutron interaction is weak, the kinetic coefficients can be split in two almost independent parts—the neutron kinetic coefficients mediated by nucleon-nucleon collisions and electron kinetic coefficients mediated by the collisions between charged particles; the proton kinetic coefficients are small. In the non-superfluid $npe\mu$ matter, the neutrons are the main heat carriers at $T \lesssim 10^8$ K, while the heat transport by leptons e^- and μ^- is competitive at $T \gtrsim 10^9$ K (Shternin and Yakovlev 2007; Shternin et al. 2013).

Baryon Heat Conduction Flowers and Itoh (1979) based their calculations on the free nucleon scattering amplitudes, neglecting the Fermi-liquid effects and nucleon many-body effects. Their results were later reconsidered by Baiko et al. (2001a).

The thermal conductivity is written in the form analogous to Eq. (12):

$$\kappa_n = \frac{\pi^2 T n_e \tau_n}{3m_n^*}, \qquad \kappa_p = \frac{\pi^2 T n_p \tau_p}{3m_p^*}, \tag{13}$$

where the effective relaxation times τ_n and τ_p are provided by solution of the system of algebraic equations (e.g., Shternin et al. 2013)

$$\sum_{j=n,p} \nu_{ij}\tau_j = 1, \qquad \nu_{ij} = \frac{64 m_i^* m_j^{*2} T^2}{5 m_N^2 \hbar^3} S_{ij}\ (i, j = n, p), \tag{14}$$

where ν_{ij} are effective collision frequencies, m_N is the bare nucleon mass in vacuo, and S_{ij} are the effective cross-sections.

Many-body effects in the context of transport coefficients of pure neutron matter were first addressed by Wambach et al. (1993) and later reconsidered in many papers. There are two kinds of these effects: the three-body part of the effective potential for the nucleon-nucleon interactions and the in-medium effects (cf. Sect. 2.6.3) that affect nucleon-nucleon scattering cross-sections. Baiko et al. (2001a) calculated S_{ij} in the approximation of pairwise interactions between nucleons with appropriate effective masses, using the Bonn potential model for the elastic nucleon-nucleon scattering (Machleidt and Holinde 1987) with and without the in-medium effects. They presented the results in the form $S_{ij} = S_{ij}^{(0)} K_{ij}$, where $S_{ij}^{(0)}$ corresponds to scattering of bare particles, and K_{ij} describes the in-medium effects. They also constructed a simple analytical fits to their results for $S_{ij}^{(0)}$ and K_{ij} (referenced in Table 2).

Shternin et al. (2013) studied the many-body effects on the kinetic coefficients of nucleons in the $npe\mu$ matter in beta equilibrium using the Brueckner-Hartree-Fock (BHF) method. According to this study, the three-body forces suppress the thermal conductivity. This suppression is small at low densities but increases to a factor of $\sim$4 at the baryon number density of $n_b = 0.6\ \mathrm{fm}^{-3}$. However, the use of the effective masses partly grasps this difference. For this reason it proves to be sufficient to multiply the conductivities obtained in the effective-mass approximation (Baiko et al. 2001a) by a factor of 0.6 to reproduce the BHF thermal conductivity (Shternin et al. 2013) with an accuracy of several percent in the entire density range of interest.

Lepton Heat Conduction The up-to-date electron and muon contributions to thermal conductivities of neutron star cores were calculated by Shternin and Yakovlev (2007). Their treatment included the Landau damping of electromagnetic interactions owing to the exchange of transverse plasmons. This effect was studied by Heiselberg and Pethick (1993) for a degenerate quark plasma, but was neglected in the previous studies of the lepton heat conductivities in the $npe\mu$ matter (e.g., Flowers and Itoh 1981; Gnedin and Yakovlev 1995).

The electron and muon thermal conductivities are additive, $\kappa_{e\mu} = \kappa_e + \kappa_\mu$, and can be written in the familiar form of Eq. (12):

$$\kappa_e = \frac{\pi^2 T n_e \tau_e}{3m_e^*}, \qquad \kappa_\mu = \frac{\pi^2 T n_\mu \tau_\mu}{3m_\mu^*}, \tag{15}$$

where κ_e and κ_μ are the partial thermal conductivities of electrons and muons, respectively; n_e and n_μ are number densities of these particles, m_e^* and m_μ^* are their dynamical masses at the Fermi surfaces, determined by their chemical potentials. In neutron star cores at beta equilibrium these chemical potentials are equal, therefore $m_e^* = m_\mu^*$. The effective electron and muon relaxation times can be written as (Gnedin and Yakovlev 1995)

$$\tau_e = \frac{\nu_\mu - \nu'_{e\mu}}{\nu_e \nu_\mu - \nu'_{e\mu}\nu'_{\mu e}}, \qquad \tau_\mu = \frac{\nu_e - \nu'_{\mu e}}{\nu_e \nu_\mu - \nu'_{e\mu}\nu'_{\mu e}}, \tag{16}$$

where $\nu_e = \sum_i \nu_{ei} = \nu_{ee} + \nu_{e\mu} + \nu_{ep}$ and $\nu_\mu = \sum_i \nu_{\mu i} = \nu_{\mu\mu} + \nu_{\mu e} + \nu_{\mu p}$ are the total effective collision frequencies of electrons and muons with all charged particles $i = e, \mu, p$; ν_{ei} and $\nu_{\mu i}$ are partial effective collision frequencies, while $\nu'_{e\mu}$ and $\nu'_{\mu e}$ are additional effective collision frequencies, which couple heat transport of the electrons and muons. All these collision frequencies can be expressed as multidimensional integrals over momenta of colliding particles. Shternin and Yakovlev (2007) calculated these integrals in the weak-screening approximation and described the results by simple analytical formulae (referenced in Table 2). In the case of strongly degenerate ultra-relativistic leptons, which is typical for neutron star cores, the latter authors obtained a much simpler expression, which can be written as

$$\kappa_{e,\mu} \approx 20.8c(p_{\mathrm{F}e,\mu}/\hbar)^2. \tag{17}$$

3 Superfluidity and Superconductivity

Soon after the development of the BCS theory (Bardeen et al. 1957), which explains superconductivity by Cooper pairing of fermions (Cooper 1956), Bohr et al. (1958) argued that the same phenomenon of pairing is occurring inside nuclei (later this suggestion was confirmed experimentally). Migdal (1960) extended the idea to the interior of neutron stars. Ginzburg and Kirzhnits (1965) formulated a number of important propositions concerning neutron superfluidity in the interior of neutron stars, the formation of Feynman-Onsager vortices, a critical superfluidity temperature ($T_\mathrm{c} \lesssim 10^{10}$ K) and its dependence on the density ($\rho \sim 10^{13}$–10^{15} g cm^{-3}), and discussed the influence of neutron superfluidity on heat capacity and therefore on the thermal evolution of a neutron star. Baym et al. (1969) and Ginzburg (1970) analyzed the consequences of neutron superfluidity and proton superconductivity: rotation of the superfluid component in the form of quantized vortices and splitting of the internal stellar magnetic field into fluxoids (Sect. 4.3.4). Later many different authors considered various types of pairing of nucleons, hyperons, or quarks using different model potentials.

Although we will not consider exotic models of neutron star cores, let us mention that superfluidity is possible in these models as well. For instance, Takatsuka and Tamagaki (1995) reviewed calculations of neutron and proton superfluid gaps in pion condensed matter. Some authors have discussed superfluidity in quark matter (e.g., Stejner et al. 2009). If hyperons are present, they can also be in a superfluid state (Balberg and Barnea 1998). For a detailed recent review of superfluidity in the interiors of neutron stars, see Page et al. (2014).

3.1 Pairing Types and Critical Temperatures

The Cooper pairing appears as a result of the attraction of particles with the anti-parallel momenta,which is expected to occur, at low enough temperature, in any degenerate system of fermions in which there is an attractive interaction between particles whose momenta $\boldsymbol{p}$ lie close to the Fermi surface (Cooper 1956). The strength of the interaction determines the critical temperature T_c at which the pairing phase transition will occur. In a normal system the particle energy ε varies smoothly when the momentum crosses the Fermi surface, while in the presence of pairing a discontinuity develops, with a forbidden energy zone having a minimum width of $2\Delta_\mathrm{pair}$ at $p = p_\mathrm{F}$, which can be regarded as the binding energy of a Cooper pair.

The BCS equations that describe symmetric nuclear matter in atomic nuclei and asymmetric neutron-rich matter in neutron stars have much in common but have also some differences. For instance, pairing in atomic nuclei takes place in the singlet state of a nucleon pair. In this case, the energy gap is isotropic, that is independent of the orientation of nucleon momenta. On the other hand, one can expect triplet-state pairing in the

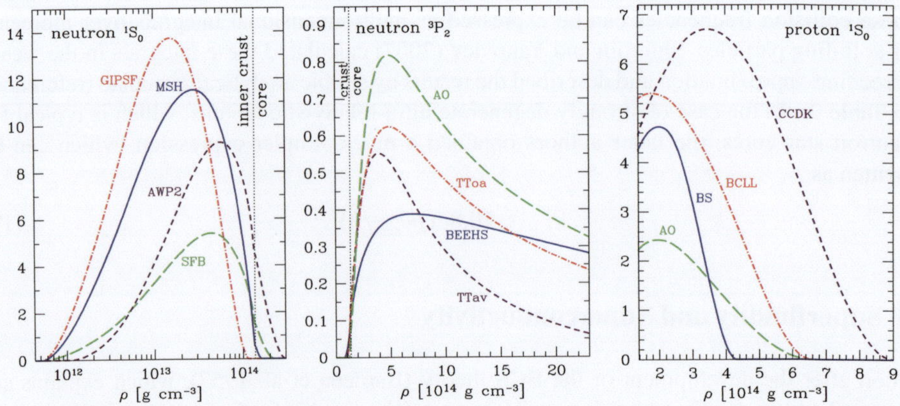

Fig. 5 Critical temperatures of singlet neutron (*left panel*), triplet neutron (*middle panel*), and singlet proton (*right panel*) superfluidities in the inner crust and core of a neutron star, as functions of gravitational mass density ρ, for different superfluidity models, as marked near the curves (see Ho et al. 2015): AO (Amundsen and Østgaard 1985a, 1985b), AWP2 (Ainsworth et al. 1989), BCLL (Baldo et al. 1992), BEEHS (Baldo et al. 1998), BS (Baldo and Schulze 2007), CCDK (Chen et al. 1993), GIPSF (Gandolfi et al. 2008), MSH (Margueron et al. 2008), SFB (Schwenk et al. 2003), TTav and TToa (Takatsuka and Tamagaki 2004)

neutron-star matter, which leads to anisotropic gap. Singlet-state neutron superfluidity develops in the inner neutron star crust and disappears in the core, where an effective neutron-neutron singlet-state attraction transforms into repulsion. Triplet-state neutron superfluidity appears in the neutron-star core. Protons in the core can undergo the singlet-state pairing.

The triplet pair states may have different projections m_J of the total pair momentum onto the quantization axis: $|m_J| = 0$, 1, and 2. The actual (energetically favorable) state may be a superposition of states with different m_J. Owing to uncertainties of microscopic theories this state is still unknown; it depends possibly on density and temperature. In simulations of neutron star cooling, one usually considers the triplet-state pairing with $|m_J| = 0$ and 2, since their effects on the heat capacity and neutrino luminosity are qualitatively different (e.g., Yakovlev et al. 1999b, 2001).

The critical temperature T_c is very sensitive to the strength of the repulsive core of the nucleon-nucleon interaction. It is related to the superfluid energy gap by $T_c = 0.5669\Delta_{\rm pair}$ for the singlet gap (e.g., Lifshitz and Pitaevskiĭ 2002, Sect. 40). For the triplet gap, the situation is more complicated, because the gap is anisotropic (e.g., Amundsen and Østgaard 1985b; Baldo et al. 1992; Yakovlev et al. 1999b). Examples of the dependence of T_c on gravitational mass density in the crust and core of a neutron star are shown in Fig. 5. Here, we employed the gap parametrization of Kaminker et al. (2001) with the parameter values and notations for different models of superfluidity according to Ho et al. (2015) together with the ρ-dependences of free-nucleon number densities n_n and n_p from the fits (Potekhin et al. 2013) for the BSk21 model of crust and core composition. Figure 5 demonstrates a large scatter of theoretical predictions, but also general features. We see that the $^1S_0$ superfluidity of neutrons occurs mostly in the inner crust and the $^3P_2$ superfluidity mostly in the core. The critical temperatures of neutrons in the triplet states, $T_{cn}(^3P_2)$, and protons, $T_{cp}(^1S_0)$, have usually a maximum at a supranuclear density $\rho > \rho_0$. Typical magnitudes of T_c vary from one model to another within a factor of a few. Neutron $^3P_2$ superfluidity has, in general, much lower T_c than $^1S_0$ pairing of neutrons in the inner crust and protons in the core.

3.2 Superfluid Effects on Heat Capacity

Once a component x of the neutron star matter becomes superfluid, its specific heat $c_{\mathrm{v,x}}$ is strongly altered. When T reaches $T_{\mathrm{c,x}}$, the critical temperature for the pairing phase transition, $c_{\mathrm{v,x}}$ jumps by a factor $\gtrsim 2$. However, as T continues to decrease, the heat capacity becomes progressively suppressed. At $T \ll T_{\mathrm{c}}$ the energy gap in the nucleon spectrum strongly reduces the heat capacity even compared to its value in the absence of pairing. These effects are implemented in numerical calculations through "control functions" $R_{\mathrm{c}}(T/T_{\mathrm{c,x}})$ as

$$c_{\mathrm{v,x}} = R_{\mathrm{c}} c_{\mathrm{v,x}}^{(0)}, \tag{18}$$

where $c_{\mathrm{v,x}}^{(0)}$ denotes the value in the normal phase, Eq. (10). The control function depends on the type of pairing. This dependence was studied by Levenfish and Yakovlev (1994). Analytical fitting formulae for R_{c} in the $npe\mu$ matter for the main types of superfluidity listed above are given by Eq. (18) of Yakovlev et al. (1999b).[1]

Three examples of the control functions, calculated according to Yakovlev et al. (1999b) (with the correction mentioned in footnote 1), are shown in the left panel of Fig. 6. One can notice that $c_{\mathrm{v,x}}$ nearly vanishes when T drops below $\sim 0.1 T_{\mathrm{c,x}}$. Therefore, in the case of extensive pairing of baryons, the heat capacity of the core can be reduced to its leptonic part. This would result in a drastic reduction of the total specific heat, as already demonstrated by the heavy long-dashed line in Fig. 1, where we adopted MSH, TToa (assuming $m_J = 0$), and BS superfluidity models for neutrons in the crust and core, and protons in the core, respectively, according to the notations in the caption to Fig. 5.

Another example of the distribution of c_{v} among the various core constituents is shown in the right panel of Fig. 6. Here, we have adopted SFB, BEEHS (with $m_J = 0$), and BCLL pairing gaps. The behavior of c_{v} as function of ρ proves to be qualitatively similar for different sets of superfluid gap models. In all cases this behavior strongly differs from that for unpaired nucleons, which is shown by thin lines for comparison.

3.3 Superfluid Effects on Neutrino Emission

The enormous impact of pairing on the cooling comes directly from the appearance of the energy gap Δ_{pair} at the Fermi surface which leads to a suppression of all processes involving single particle excitations of the paired species. When $T \ll T_{\mathrm{c}}$ the suppression is of the order of $\mathrm{e}^{-\Delta_{\mathrm{pair}}/T}$ and hence dramatic. Its exact value depends on the details of the phase space involved in each specific process. In numerical calculations it is introduced as a control function. As well as for the heat capacity, for the neutrino emissivity one writes

$$Q_\nu = R_{\mathrm{(pairingtype)}}^{\mathrm{(processtype)}} Q_\nu^{(0)}, \tag{19}$$

where $Q_\nu^{(0)}$ relates to the same process in the absence of pairing. These control functions (reduction factors) are available in the form of analytical fits, referenced in Table 1.

The superfluidity not only reduces the emissivity of the usual neutrino reactions but also initiates a specific "pair breaking and formation" (PBF) neutrino emission mechanism. The superfluid or superconducting condensate is in thermal equilibrium with the single particle

[1] In the latter paper, an accidental minus sign in front of the term $(0.2846\mathrm{v})^2$ in the denominator of the fitting formula for R_{c} in the case of "type C" ($^3\mathrm{P}_2$, $|m_J| = 2$) superfluidity must be replaced by the plus sign (Yakovlev, personal communication).

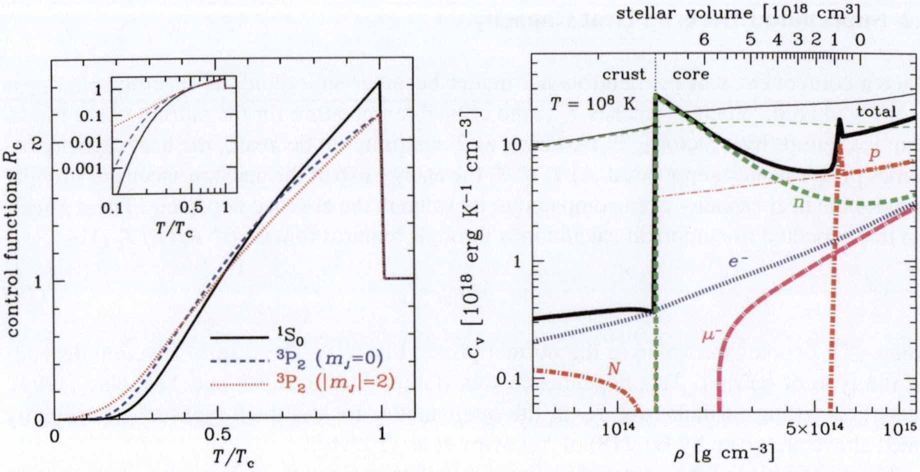

Fig. 6 *Left panel*: Specific heat control functions for the $^1S_0$, $^3P_2$ ($m_J = 0$), and $^3P_2$ ($|m_J| = 2$) types of pairing listed in Sect. 3.1. The *inset* displays the same functions on a logarithmic scale. *Right panel*: Total and partial specific heats near the bottom of the crust and in the core of a neutron star at $T = 10^8$ K as functions of density. The *solid lines* show the total c_V, and the *other lines* show the contributions of electrons (e^-), neutrons (n), nuclei (N, in the crust), muons and protons (μ^- and p, in the core). *Thin lines* show results of a calculation with nucleons assumed to be unpaired, and *thick lines* take pairing into account. The *top axis* shows the volume contained inside a sphere with given ρ for a 1.4 $M_\odot$ neutron star. The stellar structure and composition are adopted from the BSk21 model

("broken pairs") excitations and there is continuous formation and breaking of Cooper pairs. The formation of a Cooper pair liberates energy which can be taken away by a ν–$\bar{\nu}$ pair (Flowers et al. 1976; Voskresensky and Senatorov 1987). This effect is most pronounced near the Fermi surface. When T falls below T_c, the neutrino emissivity produced by the Cooper pairing sharply increases. The PBF mechanism is sensitive to the model adopted for calculating the superfluid gaps in the baryon spectra: it is more important for lower gaps (weaker superfluid). Its emissivity is a sharp function of density and temperature. The main neutrino energy release takes place in the temperature interval between $\sim T_c/5$ and T_c. The control functions and the intensity of the Cooper-pair neutrino emission are available as analytical fits collected by Yakovlev et al. (2001) (see references therein for the original derivations), as indicated in Table 1 above.

Voskresensky and Senatorov (1987) noticed that the PBF mechanism is sensitive to the in-medium renormalization of the nucleon weak-interaction vertex due to strong interactions (cf. Sect. 2.6.3). Later this effect has been reexamined in many papers for different types of baryon pairing—see Leinson, Leinson (2009, 2010) for modern results and a critical analysis of previous works. The net result is that the collective effects virtually kill down the PBF emission for the singlet pairing of nucleons, but leave this mechanism viable for the triplet pairing. Quantitatively, PBF emissivity estimated without in-medium effects (Yakovlev et al. 1999a) has to be multiplied by a small factor of $(p_F/m^*c)^2$ in the $^1S_0$ case, but by a moderate factor of $\approx$0.19 in the $^3P_2$ case. This result lies at the basis of the "minimal cooling scenario" and the explanation of the observed fast cooling of the neutron star in the Cassiopeia A supernova remnant (see Sect. 6).

Superconductivity of protons may also induce another type of neutrino emission, electron-fluxoid scattering, in the presence of a strong magnetic field. It will be addressed in Sect. 4.3.

3.4 Superfluid Effects on Heat Conduction

The effects of nucleon superfluidity on the heat transport in neutron stars were discussed qualitatively by Flowers and Itoh (1976, 1981). The thermal conductivity of electrons and muons was reconsidered by Gnedin and Yakovlev (1995) and later by Shternin and Yakovlev (2007), who obtained accurate analytical expressions valid for a wide class of models of superfluid and non-superfluid matter. Baiko et al. (2001a) reanalyzed the thermal conduction by neutrons, utilizing some new developments in the nucleon–nucleon interaction theory. The latter authors showed that the low-temperature behavior of the nucleon thermal conductivity is very sensitive to the relation between critical temperatures of neutrons and protons.

The lepton heat conduction in the core can also be affected by proton superconductivity, because superconductivity modifies the transverse polarization function and screening functions in neutron-star matter. These effects were studied by Shternin and Yakovlev (2007). These authors, as well as Baiko et al. (2001a), managed to describe the effects of superfluidity by analytical functions, which facilitate their inclusion in simulations of neutron-star thermal evolution (see Table 2).

In the presence of neutron superfluidity, there may be another channel of heat transport, the so-called convective counterflow of the normal component of matter with respect to the superfluid one. This mechanism is known to be quite effective in superfluid helium (e.g., Tilley and Tilley 1990), but in the context of neutron stars the situation is unclear and has not been studied in sufficient detail.

Heat can also be transported through the neutron star crust by collective modes of superfluid neutron matter, called superfluid phonons (Aguilera et al. 2009). At $\rho \approx 10^{12}$–$10^{14}\ \mathrm{g\,cm^{-3}}$ the conductivity due to superfluid phonons was estimated to be significantly larger than that due to lattice phonons and comparable to electron conductivity when $T \approx 10^8$ K. The authors found that this mode of heat conduction could limit the anisotropy of temperature distribution at the surface of highly magnetized neutron stars. However, new studies of the low-energy collective excitations in the inner crust of the neutron star (Chamel 2012; Chamel et al. 2013), including neutron band structure effects, show that there is a strong mixing between the Bogoliubov-Anderson bosons of the neutron superfluid and the longitudinal crystal lattice phonons. In addition, the speed of the transverse shear mode is greatly reduced as a large fraction of superfluid neutrons are entrained by nuclei. This results in an increased specific heat of the inner crust, but also in a decrease of the electron thermal conductivity. On the other hand, the entrainment of the unbound neutrons decreases the density of conduction neutrons, i.e., neutrons that are effectively free. The density of the conduction neutrons can be much smaller than the total density of unbound neutrons (Chamel 2012), which results in a decrease of the neutron thermal conductivity.

4 The Effects of Strong Magnetic Fields

4.1 Magnetic-Field Parameters

Convenient dimensionless parameters that characterize the magnetic field in a plasma are the ratios of the electron cyclotron energy $\hbar\omega_\mathrm{c}$ to the atomic unit of energy, electron rest energy, and temperature:

$$\gamma = \frac{\hbar^3 B}{m_e^2 c e^3} = \frac{B}{B_0}, \qquad b = \frac{\hbar e B}{m_e^2 c^3} = \frac{B}{B_\mathrm{QED}}, \qquad \zeta_e = \frac{\hbar\omega_\mathrm{c}}{T} = 134.34\frac{B_{12}}{T_6}. \tag{20}$$

Here, $\omega_c = eB/m_e c$ is the electron cyclotron frequency, $B_0 = 2.3505 \times 10^9$ G is the atomic unit of magnetic field, $B_{QED} = 4.414 \times 10^{13}$ G is the critical field in Quantum Electrodynamics (Schwinger 1988), and $B_{12} \equiv B/10^{12}$ G.

Motion of charged particles in a magnetic field is quantized in discrete Landau levels. In the non-relativistic theory, the energy of an electron in a magnetic field equals $N\hbar\omega_c + m_e p_z^2/2$, where p_z is the momentum component along $\boldsymbol{B}$, $N = n_L + \frac{1}{2} \mp \frac{1}{2}$ characterizes a Landau level, the term $\mp\frac{1}{2}$ is the spin projection on the field, and n_L is the non-negative integer Landau number related to the quantization of the kinetic motion transverse to the field. In the relativistic theory (e.g., Sokolov and Ternov 1986), the kinetic energy ε of an electron at the Landau level N depends on its longitudinal momentum p_z as

$$\varepsilon_N(p_z) = c\left(m_e^2 c^2 + 2\hbar\omega_c m_e N + p_z^2\right)^{1/2} - m_e c^2. \tag{21}$$

The levels $N \geqslant 1$ are double-degenerate with respect to the spin projection s. Their splitting $\delta\varepsilon$ due to the anomalous magnetic moment of the electron is negligible, because it is much smaller than $\hbar\omega_c$ (e.g., Schwinger 1988; Suh and Mathews 2001):

$$\delta\varepsilon \approx \frac{\alpha_f}{2\pi} \times \begin{cases} \hbar\omega_c \text{ at } b \ll 1, \\ m_e c^2[\ln b - 1.584]^2 \text{ at } b \gg 1, \end{cases} \tag{22}$$

where α_f is the fine structure constant.

The Landau quantization becomes important when the electron cyclotron energy $\hbar\omega_c$ is at least comparable to both the electron Fermi energy ε_F and temperature T. If $\hbar\omega_c$ is appreciably larger than both ε_F and T, then the electrons reside on the ground Landau level, and the field is called *strongly quantizing*. The condition $\hbar\omega_c > T$ is equivalent to $\zeta_e > 1$. The condition $\hbar\omega_c > \varepsilon_{F,e}$ translates into $\rho < \rho_B$, where

$$\rho_B \approx 7045 Y_e^{-1} B_{12}^{3/2} \text{ g cm}^{-3}. \tag{23}$$

In the opposite limit, where either $\zeta_e \ll 1$ or $\rho \gg \rho_B$, the field can be considered as *non-quantizing*.

For the ions, the cyclotron energy is $\hbar\omega_{ci} = Z(m_e/m_i)\hbar\omega_c$, and the Landau quantization is important when the parameter

$$\zeta_i = \hbar\omega_{ci}/T = 0.0737(Z/A)B_{12}/T_6 \tag{24}$$

is not small. The energy spectrum of an ion essentially differs from Eq. (21) because of the non-negligible anomalous magnetic moments. In the non-relativistic theory, the energy of an ion equals $\varepsilon = (n_L + \frac{1}{2})\hbar\omega_{ci} + m_i p_z^2/2 + \frac{1}{4} g_i \zeta_i s_i$, where n_L is the ion Landau number, p_z is the longitudinal momentum, g_i is the g-factor ($g_i = 2$ in the Dirac theory, but, e.g., $g_i = 5.5857$ for the protons), and s_i is the integer quantum number corresponding to the spin projection on $\boldsymbol{B}$ in units of $\hbar/2$. If the ions are relativistic, the situation is much more complicated. For baryons with spin $\frac{1}{2}$ (e.g., protons) the energy spectrum was derived by Broderick et al. (2000).

4.2 Magnetic Field Effects on the Equation of State and Heat Capacity

4.2.1 Magnetized Core

A magnetic field can affect the thermodynamics of the Coulomb plasmas, if the Landau quantization is important, i.e., under the conditions that are quantified in Sect. 4.1. In particular, Eq. (23) can be recast into

$$B \gtrsim (3.8 \times 10^{19}\ \mathrm{G})\big(Y_e n_\mathrm{b}/\mathrm{fm}^{-3}\big)^{2/3}. \tag{25}$$

We have $n_\mathrm{b} \sim 0.1\ \mathrm{fm}^{-3}$ near the crust-core interface, and Y_e is typically several percent throughout the core. Therefore, the electron component of pressure in the core might be affected by the fields $B \gtrsim 10^{18}$ G.

One can easily generalize Eq. (25) for other fermions (μ-mesons, nucleons) in the ideal-gas model. In this case, Y_e should be replaced by the number of given particles per baryon, and the right-hand side should be multiplied by $m_\mu/m_e = 206.77$ for muons and $\sim 10^3$ (of the order of nucleon-to-electron mass and electron-to-nucleon magnetic moment ratios) for protons and neutrons. Accordingly, the partial pressures of muons and nucleons in the core cannot be affected by any realistic ($B \lesssim$ a few $\times\ 10^{18}$ G) magnetic field.

Broderick et al. (2000) developed elaborated models of matter in ultra-magnetized cores of neutron stars. They considered not only the ideal $npe\mu$ gas, but also interacting matter in the framework of the relativistic mean field (RMF) model. The magnetic field affects their EoS at $B \gtrsim 10^{18}$ G. As follows both from the estimates based on the virial theorem (Lai and Shapiro 1991) and from numerical hydrodynamic simulations (e.g., Frieben and Rezzolla 2012), and references therein), this field is close to the upper limit on B for dynamically stable stellar configurations. The effect is even smaller when the magnetization of matter is included consistently in the EoS (Chatterjee et al. 2015). Therefore, it is unlikely that a magnetic modification of the EoS could be important in the cores of neutron stars.

4.2.2 Magnetized Crust and Ocean

At $B \gtrsim 10^{16}$ G, nuclear shell energies become comparable with the proton cyclotron energy. Thus the interaction of nucleon magnetic moments and proton orbital moments with magnetic field may cause appreciable modifications of nuclear shell energies. These modifications and their consequences for magnetars were studied by Kondratyev et al. (2001), who found large changes in the nuclear magic numbers under the influence of such magnetic fields. This effect may alter significantly the equilibrium chemical composition of a magnetar crust.

Muzikar et al. (1980) calculated the triplet-state neutron pairing in magnetized neutron-star cores. According to these calculations, magnetic fields $B \gtrsim 10^{16}$ G make the superfluidity with nodes at the Fermi surface energetically preferable to the usual superfluidity without nodes. Accordingly, the superfluid reduction factors for the heat capacity and neutrino emissivity (the control functions) may be different in ultra-strong fields.

Chamel et al. (2012) studied the impact of superstrong magnetic fields on the composition and EoS of the neutron star crust. In particular, they found that the neutron-drip pressure increases almost linearly by 40 % from its zero-field value in the interval $10^{16}\ \mathrm{G} < B < 5 \times 10^{16}$ G. With further increase of the field strength, the drip pressure becomes directly proportional to B.

Thus the ultra-strong fields $B \gtrsim 10^{16}$ G can affect various aspects of the physics of the inner crust in quite non-trivial way. Hereafter we will consider only fields $B \lesssim 10^{16}$ G. They can be quantizing in the outer crust of a neutron star, but not in the inner crust or the core. Analytical fitting formulae for the thermodynamic functions of the electron-ion plasmas in such

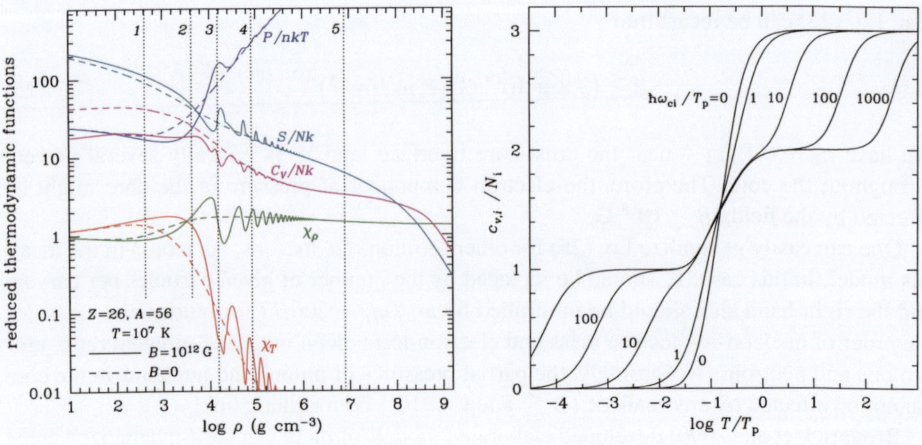

Fig. 7 *Left panel*: Normalized pressure $P/n_{\mathrm{i}}T$; entropy S and heat capacity C_V per one ion, and logarithmic derivatives of pressure over density and temperature, χ_ρ and χ_T, for a fully-ionized non-magnetic (*dashed lines*) and magnetized ($B = 10^{12}$ G, *solid lines*) iron plasma at $T = 10^7$ K. The *vertical dotted lines* mark the densities at which the electron Fermi temperature equals T without (*1*) or with (*2*) the magnetic field, $\rho = \rho_B$ (*3*), $\Gamma_{\mathrm{C}} = \Gamma_{\mathrm{m}}$ (*4*), and $T_{\mathrm{p}} = T$ (*5*). (Figure 6 from Potekhin and Chabrier 2013, reproduced with the permission of ©ESO.) *Right panel*: Normalized thermal phonon contribution to the reduced heat capacity as a function of $\log_{10}(T/T_{\mathrm{p}})$ at different values of the ratio $\hbar\omega_{\mathrm{ci}}/T_{\mathrm{p}}$, marked near the curves

fields, as well as a computer code that implements these fits,[2] were published by Potekhin and Chabrier (2013). Such fields affect the electron part of thermodynamic functions in the outer envelopes only, as illustrated in the left panel of Fig. 7 in the case of fully-ionized iron at $T = 10^7$ K and $B = 10^{12}$ G (for illustration, the density range is extended to $\rho \lesssim 10^5$ neglecting the bound states that can be important in this ρ–T domain). We plot the principal thermodynamic quantities normalized per one ion as functions of density. For comparison we also show them in the absence of quantizing magnetic field. The vertical dotted lines marked by numbers separate different characteristic domains, consecutively entered with increasing density: onset of electron degeneracy at $B = 0$ and at $B = 10^{12}$ G, population of excited Landau levels ($\rho = \rho_B$), melting point with formation of a classical Coulomb crystal ($T_{\mathrm{m}} = T$), and onset of the quantum effects in the crystal ($T_{\mathrm{p}} = T$). The gradually decreasing oscillations correspond to consecutive filling of the electron Landau levels. The magnetic field $B = 10^{12}$ G does not affect the ion contributions at this T.

The contributions of ions to the thermodynamic functions are affected by the magnetic field if the parameter ζ_{i}, defined by Eq. (24), is large. This may occur in a superstrong field of a magnetar. The right panel of Fig. 7 illustrates the effects of a superstrong field on $c_{\mathrm{v,i}}$ for the model of a harmonic Coulomb crystal (Baiko 2009). Here we plot the thermal phonon contribution to the heat capacity of the bcc Coulomb lattice calculated as the derivative $c_{\mathrm{v}} = T\partial S/\partial T$ of the fit to the phonon entropy S given by Eq. (77) of Potekhin and Chabrier (2013). This approximation is more accurate for the heat capacity than the alternative approximation that provides exact fulfillment of the Maxwell relations (Eq. (80) of the same paper). The three steps on the curves in the right panel of Fig. 7 correspond to contributions of three branches of the phonon spectrum, which are affected differently by the quantizing magnetic field.

[2] Also available at http://www.ioffe.ru/astro/EIP/.

4.3 Magnetic Field Effects on Neutrino Emission

4.3.1 Magnetic Durca Process

We have mentioned in Sect. 2.6.2 that the Durca reaction is the most efficient neutrino emission process, but it can only operate above a certain threshold density in the central parts of the cores of sufficiently massive neutron stars. Leinson and Pérez (1998) noted that a superstrong magnetic field can substantially weaken this requirement. An accurate study of this effect was performed by Baiko and Yakovlev (1999). They showed that the border between the open and closed Durca regimes is smeared out over some B-dependent scale and described this smearing by simple formulae. In practice this effect should be very important for neutron stars with $B \gtrsim 10^{16}$ G. At less extreme fields (10^{14} G $\lesssim B \lesssim 10^{16}$ G) it is important for neutron stars whose mass happens to be close (within a few percent) to the Durca threshold mass. Baiko and Yakovlev (1999) also showed that a strong magnetic field has a non-trivial effect (oscillations of the reaction rate) in the permitted domain of the Durca reaction, but the latter effect, albeit interesting, appears to be unimportant.

4.3.2 Pair Annihilation

The e^-e^+ pair annihilation process in strong magnetic fields was studied by Kaminker et al. (1992) and Kaminker and Yakovlev (1994). In a hot, non-degenerate plasma ($T \gtrsim 10^{10}$ K) only ultra-strong magnetic fields $B \gtrsim 10^{16}$ G can significantly affect the neutrino emissivity. Such fields can be quantizing in the $\rho - T$ domain where the pair emission dominates (see Fig. 3). They amplify $Q_{\rm pair}$ by increasing the number densities of electrons and positrons via very strong quantization of their motion. Lower fields may also influence $Q_{\rm pair}$ but less significantly. A field $B \sim 10^{14}$ G may quantize the motion of positrons at $T \lesssim 10^9$ K and increase the positron number density. In this way the presence of a strong magnetic fields greatly enhances $Q_{\rm pair}$ in a not too hot plasma. However, this enhancement usually takes place where the pair annihilation emissivity is much lower than the contribution from other neutrino reactions, and therefore it is unimportant for studies of neutron-star thermal evolution.

4.3.3 Synchrotron Radiation

A relativistic electron propagating in the magnetic field can emit neutrinos because of its rotation around the magnetic field lines. This process is quite analogous to the usual synchrotron emission of photons. The calculation of the corresponding neutrino emissivity, $Q_{\rm syn}$, is similar to that of the pair annihilation process. It was studied, e.g., by Kaminker et al. (1992), Vidaurre et al. (2013), and Bezchastnov et al. (1997). In Fig. 8 we show the plot of $Q_{\rm syn}$ on the $\rho - T$ plane for two field strengths typical for magnetars, $B = 10^{14}$ G and 10^{15} G. It is clear from this plot that the synchrotron process can be dominant in the crust of magnetars in a large temperature range.

4.3.4 Electron-Fluxoid Scattering

The internal stellar magnetic field can be confined in the crust or be distributed over the entire star. In the latter case, a transition to a superconducting state in the course of stellar cooling is accompanied by a dramatic change in the spatial structure of the magnetic field.

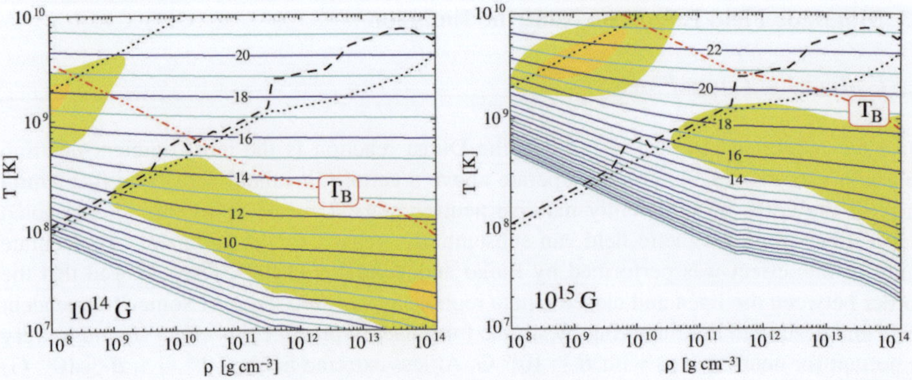

Fig. 8 Neutrino emissivity in a magnetized crust from the synchrotron processes for two, uniform, magnetic field strengths of 10^{14} G (*left panel*) and 10^{15} G (*right panel*). The contour lines are labeled by the value of $\log_{10}[Q_\nu/(\mathrm{erg\,cm^{-3}\,s^{-1}})]$. Regions where this process dominates over the ones shown in Fig. 3 are *lightly shadowed* (in *yellow*) and regions where it dominates by more than a factor of 10 are *darkly shadowed* (in *orange*). The two *dotted lines* show the dominance transitions between the three processes presented in Fig. 3. (Also indicated is the ion melting curve, *dashed line*.)

Initially homogeneous field splits into an ensemble of Abrikosov fluxoids—quantized magnetic flux tubes, which contain a superstrong magnetic field, embedded in the field-free superconducting medium. Neutrino synchrotron radiation is then modified and may be treated as neutrino pair emission due to scattering of electrons on the fluxoids. This mechanism was studied by Kaminker et al. (1997), who obtained an analytical fit to the corresponding neutrino emissivity (referenced in Table 1). The concentration of the field within the fluxoids amplifies the neutrino emissivity, compared to the usual synchrotron regime, when temperatures drops below the critical one for the protons, T_{cp}. As long as T is not much lower than T_{cp}, the Cooper pairing mechanism remains much more powerful (unless $B \gg 10^{16}$ G, in which case the electron-fluxoid scattering may be more powerful at any T). At $T \ll T_{cp}$, the electron-fluxoid scattering becomes the dominant neutrino emission mechanism for the neutron stars with strong and superstrong fields ($B \gtrsim 10^{12}$ G).

4.4 Magnetic Field Effects on Heat Conduction

4.4.1 Photon Heat Conduction

The thermal conductivity κ is related to the opacity $\varkappa$ by the equation

$$\kappa = \frac{16\sigma_{\mathrm{SB}} T^3}{3\rho\varkappa}, \tag{26}$$

where σ_{SB} is the Stefan-Boltzmann constant. The spectral radiative opacities for two normal polarization modes in strongly magnetized neutron-star photospheres are reviewed in Paper I. These opacities, $\varkappa_{\omega,j}(\theta_B)$, where $j = 1, 2$ marks the extraordinary and ordinary polarization modes, depend on the angle θ_B between the wave vector and magnetic field. In the diffusion approximation, they combine into the effective opacities for the transport along ($\varkappa_{\omega,j}^{\|}$) and across ($\varkappa_{\omega,j}^{\perp}$) magnetic field according to

$$\begin{Bmatrix} (\varkappa_{\omega,j}^{\|})^{-1} \\ (\varkappa_{\omega,j}^{\perp})^{-1} \end{Bmatrix} = \frac{3}{4}\int_0^{\pi} \begin{Bmatrix} 2\cos^2\theta_B \\ \sin^2\theta_B \end{Bmatrix} \frac{\sin\theta_B\,\mathrm{d}\theta_B}{\varkappa_{\omega,j}(\theta_B)}. \tag{27}$$

The effective opacity for energy transport at angle θ to $\boldsymbol{B}$ in each polarization mode is given by $1/\bar{\varkappa}_j = \cos^2\theta/\bar{\varkappa}_j^{\|} + \sin^2\theta/\bar{\varkappa}_j^{\perp}$, where $\bar{\varkappa}$ is the Rosseland mean of $\varkappa_\omega$,

$$\frac{1}{\bar{\varkappa}_j} \equiv \int_0^\infty \frac{u(z)}{\varkappa_{\omega,j}}\mathrm{d}z, \quad u(z) = \frac{15}{4\pi^4}\frac{z^4 \mathrm{e}^z}{(\mathrm{e}^z-1)^2}, \; z = \frac{\hbar\omega}{T}. \tag{28}$$

For fully ionized plasmas, the radiative opacities are contributed from the free-free absorption and Thomson scattering. Silant'ev and Yakovlev (1980) studied the Rosseland opacities for a non-polarized radiation in magnetized fully ionized plasmas

$$\varkappa_{\mathrm{r}}^{\|,\perp} = \left[1/\bar{\varkappa}_1^{\|,\perp} + 1/\bar{\varkappa}_2^{\|,\perp}\right]^{-1} \tag{29}$$

using the Born approximation for the free-free contribution. Potekhin and Yakovlev (2001) obtained simple analytical fits for $\varkappa_{\mathrm{r}}^{\|,\perp}$, including a correction to the Born approximation, as functions of ρ, T, Z, A, and the magnetic-field parameter ζ_e defined by Eq. (20). Asymptotically, $\varkappa_{\mathrm{r}} \propto \zeta_e^{-2}$ at $\zeta_e \to \infty$.

At finite but large ζ_e, the radiative opacities of fully ionized matter are strongly reduced. The reduction is $\sim$10 times stronger for the Thomson scattering than for the free-free absorption. In deep, strongly magnetized photospheric layers the Thomson scattering dominates only if $T_6 \gtrsim 10\rho^{2/7} \gtrsim 10 B_{12}^{2/7}$; otherwise the free-free absorption prevails (Potekhin and Yakovlev 2001).

The outermost envelopes of neutron stars can be incompletely ionized in the cases of large Z or B. The presence of bound species can strongly affect the radiative opacities and the spectrum of emitted radiation, as discussed in Paper I. However, the layers that are responsible for the heat flux from the interior of the neutron star to the surface, as a rule, lie at sufficiently large densities, where the plasma is fully ionized by pressure. Therefore the bound species are usually unimportant for the effective surface temperature of a neutron star.

4.4.2 *Electron Heat Conduction*

A non-quantizing magnetic field does not affect thermodynamic functions of the plasma. However, it does affect the electron heat conduction, if the *Hall magnetization parameter*

$$\omega_{\mathrm{g}}\tau \approx 1760 \frac{B_{12}}{\sqrt{1+x_{\mathrm{r}}^2}} \frac{\tau}{10^{-16}\ \mathrm{s}} \tag{30}$$

is not small. Here, $\omega_{\mathrm{g}} = \omega_{\mathrm{c}}/\sqrt{1+x_{\mathrm{r}}^2}$ is the electron gyrofrequency, and τ is the effective relaxation time. In a degenerate Coulomb plasma with a non-quantizing magnetic field, the main contribution is given by the electron-ion scattering according to Appendix A.3. This regime has been studied by Yakovlev and Urpin (1980).

Electric and thermal currents induced in a magnetized plasma under the effect of an electric field $\boldsymbol{E}$, a weak gradient $\nabla\mu$ of the electron chemical potential, and a weak temperature gradient ∇T can be decomposed into conduction and magnetization components (e.g., Hernquist 1984). The latter ones relate to surface effects and must be subtracted. Let $\boldsymbol{j}_e$ and $\boldsymbol{j}_T$ be the conduction components of the electric and thermal current densities. They can be written as

$$\boldsymbol{j}_e = \hat{\sigma}\cdot\boldsymbol{E}^* - \hat{\alpha}\cdot\nabla T, \qquad \boldsymbol{j}_T = \hat{\beta}\cdot\boldsymbol{E}^* - \hat{\tilde{\kappa}}\cdot\nabla T, \tag{31}$$

where $\boldsymbol{E}^* = \boldsymbol{E} + \nabla\mu/e$ is the electrochemical field. The symbols $\hat{\sigma}$, $\hat{\alpha}$, $\hat{\beta}$, and $\hat{\tilde{\kappa}}$ denote second-rank tensors ($\hat{\sigma}$ is the conductivity tensor) which reduce to scalars at $B = 0$. Equations (31) can be rewritten as

$$\boldsymbol{E}^* = \hat{R} \cdot \boldsymbol{j}_e - \hat{Q} \cdot \nabla T, \qquad \boldsymbol{j}_T = -T\hat{Q} \cdot \boldsymbol{j}_e - \hat{\kappa} \cdot \nabla T, \tag{32}$$

where $\hat{R} = \hat{\sigma}^{-1}$, $\hat{Q} = -\hat{R} \cdot \hat{\alpha}$, and $\hat{\kappa} = \hat{\tilde{\kappa}} + T\hat{\alpha} \cdot \hat{Q}$ are the tensors of specific resistance, thermopower, and thermal conductivity, respectively.

Electron heat and charge transport controlled by electron-ion collisions in quantizing magnetic fields of neutron stars was studied by Kaminker and Yakovlev (1981), Yakovlev (1984), Hernquist (1984), Potekhin (1996, 1999). The components of tensors $\hat{\sigma}$, $\hat{\alpha}$, and $\hat{\tilde{\kappa}}$ can be expressed as (Potekhin 1999)

$$\begin{Bmatrix} \sigma_{ij} \\ \alpha_{ij} \\ \tilde{\kappa}_{ij} \end{Bmatrix} = \int_0^\infty \begin{Bmatrix} e^2 \\ e(\mu - \varepsilon)/T \\ (\mu - \varepsilon)^2/T \end{Bmatrix} \frac{\mathcal{N}_B(\varepsilon)}{\varepsilon/c^2} \tau_{ij}(\varepsilon) \left[-\frac{\partial}{\partial \varepsilon} \frac{1}{\mathrm{e}^{(\varepsilon-\mu)/T} + 1} \right] \mathrm{d}\varepsilon, \tag{33}$$

where

$$\mathcal{N}_B(\varepsilon) = \frac{1}{2\pi^2 a_\mathrm{m}^2 \hbar} \sum_{N=0}^{N_\mathrm{max}} (2 - \delta_{N,0}) |p_z|, \tag{34}$$

N_max is the maximum Landau number for a given electron energy ε, and $|p_z|$ depends on ε and N according to Eq. (21). In a non-quantizing magnetic field, i.e., at $N_\mathrm{max} \gg 1$, the sum can be replaced by the integral, which gives $\mathcal{N}_B(\varepsilon) = (p/\hbar)^3/3\pi^2$, where p is the momentum that corresponds to the energy ε. The functions $\tau_{ij}(\varepsilon)$ play role of relaxation times for the components of tensors $\hat{\sigma}$, $\hat{\alpha}$, and $\hat{\tilde{\kappa}}$, determined by electron scattering. In general, they differ from the mean free time $\tau_\mathrm{ei}(\varepsilon) = 1/\nu_\mathrm{ei}(\varepsilon)$ between scattering events for an electron with energy ε. Because of the symmetry properties of the tensors $\hat{\sigma}$, $\hat{\alpha}$, and $\hat{\tilde{\kappa}}$, in the coordinate frame with z axis directed along $\boldsymbol{B}$, there are only three different non-zero components of τ_{ij}: τ_{zz} related to longitudinal currents, $\tau_{xx} = \tau_{yy}$ related to transverse currents, and $\tau_{xy} = -\tau_{yx}$ related to the Hall currents.

In a quantizing magnetic field, there are two different effective mean-free times $\tau_\parallel(\varepsilon)$ and $\tau_\perp(\varepsilon)$, corresponding to electron transport parallel and perpendicular to $\boldsymbol{B}$. In this case, the classical expressions (e.g., Yakovlev and Urpin 1980) are recovered:

$$\tau_{zz} = \tau_\parallel, \qquad \tau_{xx} = \frac{\tau_\perp}{1 + (\omega_\mathrm{g}\tau_\perp)^2}, \qquad \tau_{yx} = \frac{\omega_\mathrm{g}\tau_\perp^2}{1 + (\omega_\mathrm{g}\tau_\perp)^2}. \tag{35}$$

It is convenient to keep using Appendix A.3 for $\tau_\parallel$ and $\tau_\perp$, but with different Coulomb logarithms $\Lambda_\parallel(\varepsilon)$ and $\Lambda_\perp(\varepsilon)$. Potekhin (1999) calculated these Coulomb logarithms and fitted them by analytic expressions. Their Fortran implementation is available at http://www.ioffe.ru/astro/conduct/. In the limit of non-quantizing magnetic field, $\tau_\parallel = \tau_\perp = \tau_\mathrm{ei}(\varepsilon)$ is given by Appendix A.3 with $\varepsilon = \mu$.

When the electrons are strongly degenerate, the derivative in the square brackets in Eq. (33) is sharply peaked. Then Eq. (33) gives

$$\sigma_{ij} \approx \frac{e^2 c^2 n_e}{\mu} \tau_{ij}(\mu), \qquad \kappa_{ij} \approx \tilde{\kappa}_{ij} \approx \frac{\pi^2 T}{3e^2} \sigma_{ij}. \tag{36}$$

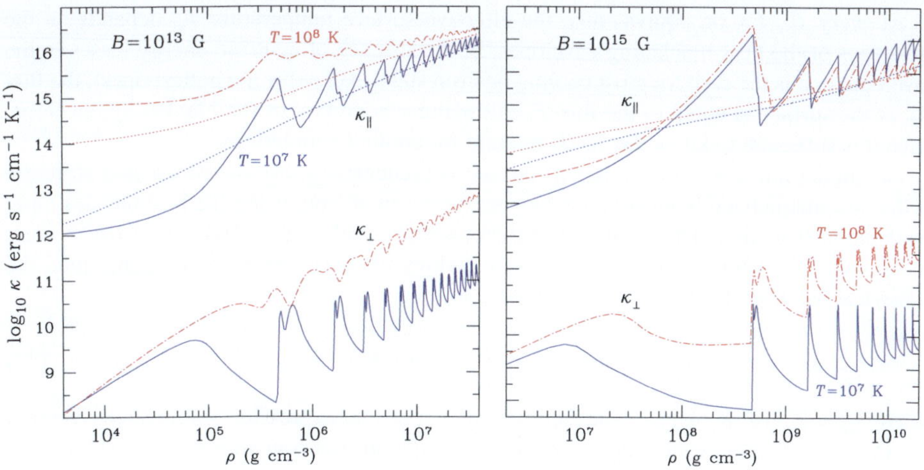

Fig. 9 Electron thermal conductivities along (*upper curves*) and across (*lower curves*) magnetic field $B = 10^{13}$ G (*left panel*) and 10^{15} G (*right panel*) as functions of mass density at temperatures $T = 10^7$ K (*solid lines*) and 10^8 K (*dot-dashed lines*). For comparison, the non-magnetic thermal conductivities are shown by *dotted lines*

The latter relation is the Wiedemann-Franz law generalized to the magnetic case. On the other hand, Eq. (33) satisfactorily describes the conductivities in general, including the opposite case of weakly degenerate electrons.

Figure 9 illustrates the ρ-dependence of the thermal conductivities along ($\kappa_\parallel$) and across ($\kappa_\perp$) the magnetic field. The first, most significant peak at $\kappa_\parallel$ is related to the filling of the first Landau level by the electrons at $\rho \sim \rho_B$. The other peaks correspond to consecutive filling of higher Landau levels.

5 Thermal Structure of Neutron Stars

5.1 Blanketing Envelopes

The very different thermal relaxation timescales of the envelope and the crust of a neutron stars makes computationally expensive to perform cooling simulations in a numerical grid that comprises both regions. Radiative equilibrium is established in the low-density region much faster than the crust evolves, so that the envelope reaches a stationary state on shorter timescales. Thus, the usual approach is to use results of stationary envelope models to obtain a relation between the photon flux F_{ph} radiated from the surface and the flux F_{b} and temperature T_{b} at the crust/envelope boundary, $\rho = \rho_{\mathrm{b}}$. This relation supplements the evolution equations for the interior [Eq. (4)] as an outer boundary condition.

The boundary density ρ_{b} is chosen as a trade-off between two requirements: first, that the thermal relaxation time of the layer with $\rho < \rho_{\mathrm{b}}$ is short compared to the characteristic variability time of the studied thermal radiation, which favors smaller ρ_{b}, and second, that T does not strongly vary at $\rho > \rho_{\mathrm{b}}$, which favors larger ρ_{b}. For weakly magnetized, isolated cooling neutron stars, ρ_{b} is usually set at 10^{10} g cm^{-3} (Gudmundsson et al. 1983), but in general it varies from 10^8 g cm^{-3} for neutron stars with rapid variations of thermal emission (Shternin et al. 2007) to ρ_{drip} for relatively hot and strongly magnetized neutron stars (Potekhin et al. 2003).

At every $T_{\rm b}$, $F_{\rm ph}$ or, equivalently, the effective surface temperature $T_{\rm s}$, depends on the properties of the heat-blanketing envelopes. In the absence of neutrino energy losses in the envelope (that is the case for most cooling neutron stars, except for the hottest ones), the flux $F_{\rm ph}$ at the surface is equal to the flux $F_{\rm b}$ at the inner boundary of the blanketing envelope. Then it is sufficient to know the $T_{\rm b}$–$T_{\rm s}$ relation for cooling simulations.

Gudmundsson et al. (1983) carried out a comprehensive study of the thermal structure of the non-magnetized blanketing envelopes composed of iron, using the best physics input available at that time. They considered the envelopes with $\log_{10} T_{\rm s}[{\rm K}] \geq 5.25$ (there were no reliable calculation of the thermal conductivities for lower temperatures) and fitted the numerical solutions by a remarkably simple formula

$$T_{\rm b} = 1.288 \times 10^8 \left(T_{\rm s6}^4 / g_{14}\right)^{0.455} \text{ K}, \tag{37}$$

where $T_{\rm s6} = T_{\rm s}/10^6$ K. An analytical derivation of a similar expression was given by Ventura and Potekhin (2001). A more accurate but less simple fit was constructed by Potekhin et al. (1997).

The $T_{\rm b}$–$T_{\rm s}$ relation is mainly regulated by the thermal conductivity in the "sensitivity strip" (Gudmundsson et al. 1983) that plays the role of a "bottleneck" for the heat leakage. Its position lies around the line where $\kappa_{\rm r} = \kappa_e$ (as a rule, around $\rho \sim 10^5$–10^7 g cm^{-3} for $B = 0$) and depends on the stellar structure, the boundary temperature $T_{\rm b}$, the magnetic field $\boldsymbol{B}$ in the vicinity of the given surface point, and the chemical composition of the envelope. Since the magnetic field hampers heat transport across $\boldsymbol{B}$, the depth of the sensitivity strip can be different at different places of a star with a strong magnetic field: it lies deeper at the places where the magnetic field lines are parallel to the surface (Ventura and Potekhin 2001).

The blanketing envelopes are more transparent to the heat flux, if they are composed of light chemical elements. This effect was studied in detail by Potekhin et al. (1997) for non-magnetic envelopes and by Potekhin et al. (2003) for strongly magnetized envelopes. The effect is related to the Z-dependence of the collision frequencies ν_{ei}. The higher is Z, the larger is ν_{ei} and the lower is the conductivity. A temperature variation by a factor of 30 can change the thermal conductivity of iron plasma less than altering the chemical composition from Fe to He at a fixed T. This effect has important consequences for the relationship between the surface and internal temperatures of neutron stars. For example, combined effects of strong magnetic fields and light-element composition simplify the interpretation of magnetars: these effects allow one to interpret observations assuming less extreme (therefore, more realistic) heating in the crust (Kaminker et al. 2009; Pons et al. 2014; Viganò et al. 2013).

The envelope is thin (its depth $z_{\rm b} \sim 100$ m, if $\rho_{\rm b} = 10^{10}$ g cm^{-3}) and contains a tiny fraction of the neutron-star mass ($\sim 10^{-7}$, if $\rho_{\rm b} = 10^{10}$ g cm^{-3}). Therefore one can neglect the variation of the gravitational acceleration in this layer. Neglecting also the non-uniformity of the energy flux through the envelope due to the neutrino emission (which is small, if the neutron star is not too hot, as we discuss below) and the variation of the temperature $T_{\rm s}$ over the surface (which varies on larger length scales than $z_{\rm b}$), one can obtain, instead of Eq. (4), the much simpler *thermal structure equation* (Gudmundsson et al. 1983; Van Riper 1988)

$$\frac{d \ln T}{d \ln P} = \frac{3}{16} \frac{P \varkappa}{g} \frac{T_{\rm s}^4}{T^4}, \tag{38}$$

where $\varkappa$ is the total opacity, related to the conductivity κ via Eq. (26).

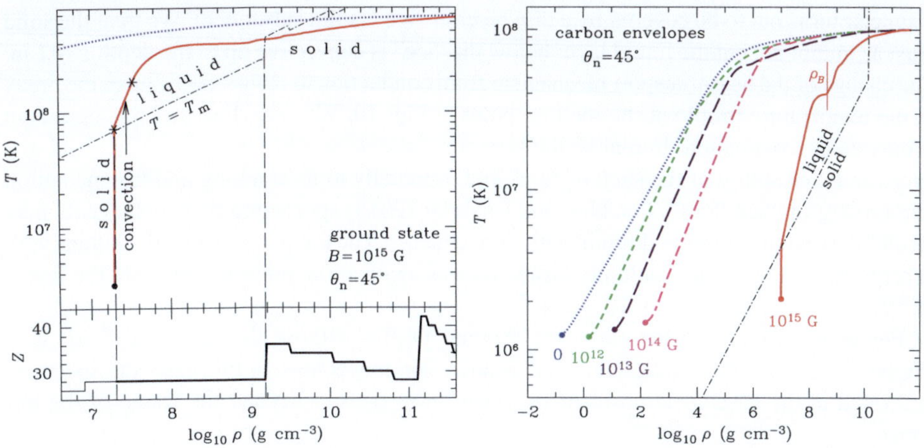

Fig. 10 Thermal structure of blanketing envelopes with different magnetic fields. *Left panel*: temperature profile (*solid line* in the left top panel) for an envelope with ground-state composition, with Z values shown in the bottom left panel, for a neutron star with surface gravity $g = 1.6 \times 10^{14}$ cm s^{-2}, magnetic field $B = 10^{15}$ G directed at 45° to the surface, and internal temperature $T_b = 6.7 \times 10^8$ K, which yields the effective surface temperature $T_s = 3.16 \times 10^6$ K. The *dot-dashed line* is the melting line. The *asterisks* confine the part of the profile where heat is carried by convection. For comparison, the non-magnetic profile with the same T_b is shown by *dotted line*. *Right panel*: temperature profiles for carbon blanketing envelopes for a neutron star with $g = 1.4 \times 10^{14}$ cm s^{-2}, $T_b = 10^8$ K, and magnetic fields $B = 0$ (*dotted line*), 10^{12} G (*short dashes*), 10^{13} G (*long dashes*), 10^{14} G (alternating *short* and *long dashes*), and 10^{15} G (*solid line*). The *vertical segment of the dot-dashed melting line* corresponds to the maximum density for carbon, according to the thermonuclear stability criterion of Potekhin and Chabrier (2012), where carbon gives way to heavier chemical elements, which form a crystal. The *heavy dots mark* the position of the radiative surface, where T equals the effective surface temperature T_s

The assumption of a constant flux, however, breaks down for magnetars, most of which have atypically high effective temperatures. In this case one should solve the complete set of equations, taking neutrino emission and heat sources into account. The neutrino emission from the crust limits the effective surface temperature of a cooling neutron star (Potekhin et al. 2007; cf. Fig. 11 below). This very same effect is what limits the maximum flux in the few days of a magnetar outburst (Pons and Rea 2012). In addition, for magnetars one must go beyond the plane-parallel approximation (see Sect. 5.3).

5.2 The Effects of Strong Magnetic Fields

As seen from Eqs. (30) and (A.3), the Hall magnetization parameter is large in the outer neutron-star envelope at $B \gtrsim 10^{11}$ G. Moreover, the magnetic field can be strongly quantizing in the outermost part of the envelope. In this case the magnetic field can greatly affect the heat conduction and the thermal structure.

Figure 10 shows examples of the temperature profiles in the envelopes. The left panel is a recast of Fig. 8 from Potekhin and Chabrier (2013). Here we show a profile of an ultra-magnetized neutron star, with $B = 10^{15}$ G, and with relatively high surface temperature, $\log_{10} T_s(K) = 6.5$, which is similar to the values evaluated for some magnetars. In this case, thermal photons are radiated from a solid surface, with high mass density $\rho = 2 \times 10^7$ g cm^{-3} just below the surface. The temperature quickly grows at the solid surface and reaches the melting point at the depth $z \approx 7$ cm. Thus, at the given conditions, the liquid ocean of a magnetar turns out to be covered by a thin layer of "ice" (solid substance). We treat the solid crust as immobile, but the liquid layer below the "ice" is convective up to the depth $z \sim 1$ m. The change of the heat-transport mechanism from conduction to convection causes the break of the temperature profile at the melting point in Fig. 10. We underline that this treatment is only an approximation. In reality, the superadiabatic growth of temperature can lead to a hydrostatic instability of the shell of "ice" and eventually to its cracking and fragmentation into turning-up "ice floes". Potekhin and Chabrier (2013) speculated that such events may result in variations of thermal luminosity of magnetars. The temperature profile flattens with density increase, and the Coulomb plasma freezes again at the interface between the layers of ^{66}Ni and ^{86}Kr at $\rho = 1.5 \times 10^9$ g cm^{-3}.

For comparison, we also show the thermal profile without the magnetic field. It is smooth. There is neither magnetic condensation nor convection. In this case, the spectrum is formed in the gaseous atmosphere at much lower density beyond the frame of the left panel.

In the right panel of Fig. 10 we compare temperature profiles for a neutron star with internal temperature 10^8 K and heat blanketing envelopes made of carbon, endowed with different magnetic fields. For the field strengths up to 10^{14} G, the radiation is formed in the gaseous atmosphere, whose density gradually becomes larger with increasing magnetic fields, due to the reduction of the effective opacities discussed in Sect. 4.4.1. The temperature profiles are rather smooth. The blanketing envelopes are liquid at this temperature. At the largest field strength $B = 10^{15}$ G, however, the situation is qualitatively different. As well as in the case of the hotter ultra-magnetized ground-state envelope in the left panel, the heat is radiated from the condensed solid surface. Below the surface, at density 10^7 g cm^{-3}, the temperature quickly grows, which causes melting of the Coulomb crystal with formation of a Coulomb liquid beneath the solid surface. With further density increase, the profile suffers a break at $\rho_B \approx 4.5 \times 10^8$ g cm^{-3} [Eq. (23)], where the electrons start to populate the first excited Landau level, which is associated with the peak of the thermal conductivity around ρ_B (cf. Fig. 9).

As we have seen in Sect. 4.4, the conduction is strongly anisotropic in these conditions. Therefore the effective local surface temperature T_s is non-uniform and depends on the magnetic field geometry. Figure 11 shows examples of the relations between T_s and T_b deep in the crust for the magnetic fields $B = 10^{12}$ G and 10^{15} G perpendicular and parallel to the radial direction. The relations obtained in the 1D approximation (Potekhin et al. 2007) with and without allowance for neutrino emission are plotted by the solid and dotted lines, respectively. We see that at $T_b \lesssim 10^8$ K the neutrino emission does not affect T_s. At higher $T_b \gtrsim 10^9$ K, in contrast, this emission is crucial: if $Q_\nu = 0$, then T_s continues to grow up with increasing T_b, whereas with realistic Q_ν the surface temperature tends to a constant limit, which depends on $\boldsymbol{B}$. In most cases this limit is reached when $T_b \sim 10^9$ K.

Since the distribution of T_s over the neutron-star surface is non-uniform in strong magnetic fields, it is convenient to introduce the overall effective temperature of the star, $T_{\rm eff}$, defined by

$$4\pi\sigma_{\rm SB} R^2 T_{\rm eff}^4 = L_{\rm ph} = \int F_{\rm ph}\, {\rm d}\Sigma = \sigma_{\rm SB} \int T_{\rm s}^4\, {\rm d}\Sigma, \qquad (39)$$

where $F_{\rm ph}$ is the local flux density and ${\rm d}\Sigma$ is the surface element. The quantities T_s, $T_{\rm eff}$, and $L_{\rm ph}$ refer to a local reference frame at the neutron-star surface. The redshifted ("apparent")

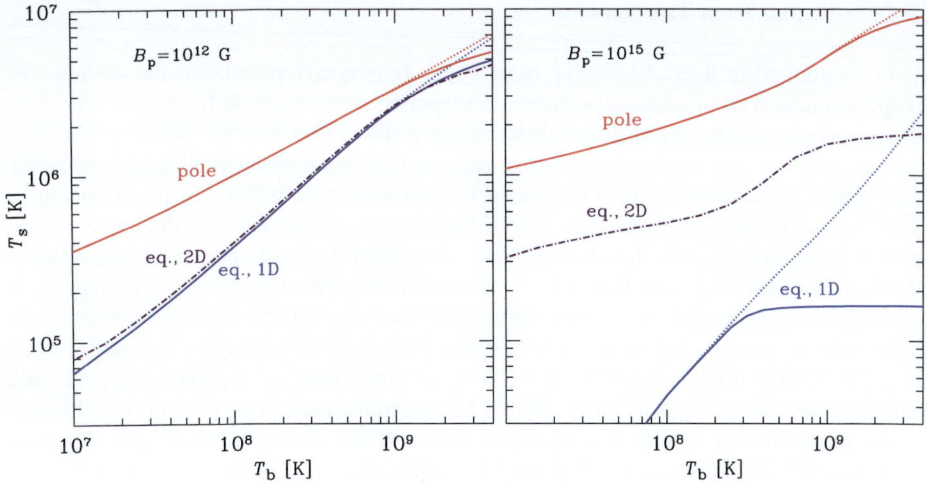

Fig. 11 Local effective surface temperature T_s as function of the temperature T_b at the bottom of a non-accreted heat blanketing envelope with $\rho_b = 10^{10}$ g cm^{-3} for a neutron star with mass $M = 1.4\ M_\odot$, radius $R = 12.6$ km, and the dipole magnetic field with polar strength $B_p = 10^{12}$ G (*left panel*) and 10^{15} G (*right panel*). *Solid lines*—1D calculation with allowance for neutrino emission from the crust, *dotted lines*—neutrino emission is neglected. The *upper dotted* or *solid curve* shows T_s at the magnetic pole, and the *lower curve* curve shows T_s at the equator. The *dot-dashed curve* shows the result of a full 2D calculation for T_s at the magnetic equator

quantities as detected by a distant observer are (Thorne 1977):

$$R^\infty = R/\sqrt{1 - r_g/R}, \qquad T_{\rm eff}^\infty = T_{\rm eff}\sqrt{1 - r_g/R}, \qquad L_{\rm ph}^\infty = (1 - r_g/R)L_{\rm ph}. \qquad (40)$$

The effects of quantizing magnetic fields on the thermal structure of neutron-star envelopes were first studied by Hernquist (1985) and somewhat later by Van Riper (1988) and Schaaf (1990), using the 1D approximation. Van Riper (1988) considered a neutron star with a constant radial magnetic field. In this model, the quantum enhancement of conductivity at ρ near ρ_B, seen in Fig. 9, results in an overall enhancement of the neutron-star photon luminosity $L_{\rm ph}$ at a fixed T_b. Consequently, Van Riper (1991) found a strong effect of the magnetic field $B \sim 10^{13}$ G on the neutron-star cooling. However, Shibanov and Yakovlev (1996) showed that, for the dipole field distribution, the effects of suppression of the heat conduction across $\boldsymbol{B}$ at the loci of nearly tangential field can compensate or even overpower the effect of the conductivity increase near the normal direction of the field lines. This conclusion confirmed the earlier conjectures of Hernquist (1985) and Schaaf (1990). In the 2000s, detailed studies of the T_b–T_s relation in strong magnetic fields were performed for iron envelopes (Potekhin and Yakovlev 2001) and accreted envelopes composed of light elements (Potekhin et al. 2003), as well as for the large-scale (dipole) and small-scale (stochastic) surface magnetic fields (Potekhin et al. 2005). These studies confirmed the conclusions of Shibanov and Yakovlev (1996), but showed that in superstrong fields $B \gtrsim 10^{14}$ G the quantum enhancement of the conductivity and the corresponding increase of T_s at the places where $\boldsymbol{B}$ is nearly radial overpowers the decrease in the regions of nearly tangential field lines, so that $T_{\rm eff}$ at a given T_b increases. However, this may not be the case in the configurations where the field is nearly tangential over a significant portion of the stellar surface as, e.g., in the case of a superstrong toroidal field (Pérez-Azorin et al. 2006; Page et al. 2007).

5.3 Non-Radial Heat Transport

As we mentioned in Sect. 5.1, in the case where $\boldsymbol{B}$ is nearly parallel to the surface, the 1D approximation fails, because the heat is transported along the field lines from the hotter surface regions outside the considered patch of the surface. Therefore the 1D approximation overestimates the heat blanketing effect in regions with nearly tangential magnetic fields. For a dipole field geometry it is an equatorial region, whose width can be estimated as $\sim 10\ \%$ of the radius (Potekhin et al. 2007). Since these regions are also the coldest ones, their contribution to the total flux is negligible. Then the 1D approximation well reproduces the integrated observed flux. However, it is not the case for magnetars, which may have a complex field geometry. A 2D treatment shows that the 1D approach is reliable in the regions where magnetic field lines make a substantial angle to the surface (Kaminker et al. 2012, 2014), but it predicts too low surface temperatures when the tangential magnetic field dominates (see Sect. 3 in Pons et al. 2014). Therefore for magnetars one must go beyond the 1D approximation. Complex field configurations which lack cylindrical symmetry may require the full 3D treatment, which has not been done yet.

In Fig. 11 we show results of 2D calculations in the dipole field geometry, compared with the 1D results. In this case, we see a substantial increase of T_s at the magnetic equator. This effect is especially pronounced for the superstrong field on the right panel. In Appendix B we give an analytical approximation to the T_b–T_s relation in the case of a strong magnetic field, including the effects of neutrino emission from the crust.

6 Thermal Evolution of Neutron Stars

6.1 Cooling Scenarios

Several tens of seconds after birth, the protoneutron-neutron star has lost its excess lepton content, it has finished its residual contraction and becomes transparent to neutrino emission (Burrows and Lattimer 1986; Pons et al. 1999; Roberts 2012). Soon after that, the temperature distribution in the highly conductive stellar core reaches equilibrium, which is preserved thereafter throughout the star lifetime (except during short periods after catastrophic phase transitions in the core postulated by certain hypothetical models).

In the initial cooling stages, the stellar crust is hotter than the core, which is rapidly cooled down by the copious neutrino emission. The cooling wave reaches the surface within 10–100 years; thereafter, the star cools down in the quasistationary regime. Since all currently observed neutron stars are at least several centuries old, they should be in the state of quasistationary cooling, except during transient events with significant energy release in the crust or the ocean discussed below.

Cooling in the quasistationary regime goes through two major stages. The first, *neutrino cooling* stage lasts $\sim 10^5$ years. During this period, the core cools mostly via neutrino emission. The second, *photon cooling* stage begins when the low temperature of the core makes the neutrino energy losses smaller than the losses due to electromagnetic radiation from the surface (see, e.g., Yakovlev and Pethick 2004, and references therein). This occurs at the age of $\approx 10^5$ years, depending on the particular stellar model and local conditions.

A theoretical cooling curve of an isolated neutron star, which shows the photon luminosity of the star L_{ph} or its temperature as a function of age t, depends on the stellar mass M, on the model of superdense matter in the core, which in particular, determines the intensity of neutrino emission and the EoS (and hence the stellar radius R), and on the properties of the envelopes. The latter include the thermal conductivity, which determines L_{ph} at a given internal stellar temperature, the neutrino luminosity Q_ν in the stellar crust, and the intensity of heating sources H. For highly magnetized neutron stars, the cooling curve also depends on the magnetic field $\boldsymbol{B}$ (on both its strength and configuration), since it affects the microphysics (conductivities, EoS, specific heat, etc.). Therefore, in general, the thermal evolution equations (4) should be supplemented by the equations that describe evolution of the magnetic field and electric currents in the star, which leads to the thermomagnetic evolution scenarios (see the review by Mereghetti et al. 2015 and references therein).

By comparing theoretical cooling curves with the observed L_{ph} and t of isolated neutron stars, one can eventually place bounds on the theoretical models of superdense matter. At contrast, most neutron stars in binary systems have an additional source of energy (accretion) and an additional source of X-ray radiation (accretion disk), often much more powerful than the surface thermal emission. For this reason, they cannot be used to test cooling models.

The theoretical cooling scenarios are currently divided into two main classes: "minimal cooling" and "enhanced cooling". The enhanced cooling implies fast neutrino emission processes, such as Durca reactions (Sect. 2.6.2), whereas the minimal cooling does not include such processes, but may be enhanced at the epoch of the onset of the baryon superfluidity via the PBF neutrino emission mechanism (Sect. 3.3), which helps to explain the variety of the observed surface temperatures of cooling neutron stars (Gusakov et al. 2004; Page et al. 2004). A spectacular example is the neutron star CXO J232327.9+584842 in the Cassiopeia A supernova remnant, dubbed Cas A NS, which shows an unexpectedly appreciable temperature decline during several years (Heinke and Ho 2010; Elshamouty et al. 2013) (but see Posselt et al. 2013 for tentative alternative interpretations of the observations). This decline can be comfortably explained by the PBF emission (Page et al. 2011; Shternin et al. 2011; see also Ho et al. 2015 for a recent analysis including modern observational data).

6.2 Heating Mechanisms

In the course of their evolution, some regions of neutron stars may not only cool but also heat up by different mechanisms. For instance, the polar cap of a pulsar can be heated by a stream of electrons or positrons moving along open field lines from the magnetosphere. The temperature of "hot spots" produced by this additional heat deposited onto the stellar surface may be much higher than the average temperature of the star (e.g., Greenstein and Hartke 1983). Non-uniform heating processes occur also during accretion episodes (e.g., Inogamov and Sunyaev 2010). The hot polar caps emit much more intense X-rays than the remaining surface; as a result, such neutron stars become X-ray pulsars. Pulsed X-ray radiation is also observed from thermonuclear explosions of accreted matter at the surface of a rotating neutron star (see, e.g., review by Strohmayer and Bildsten 2006).

On the other hand, a neutron star may also be heated from inside, for example due to dissipation of a strong magnetic field (e.g., Miralles et al. 1998; Urpin and Konenkov 2008; Pons et al. 2014). It has been suggested that dissipation of superstrong magnetic fields may be responsible for the high effective temperatures of magnetars (Thompson 2001; see Mereghetti et al. 2015, for a review). Noticeable liberation of energy in the crust may also occur during starquakes (Haensel et al. 1990; Franco et al. 2004; Perna and Pons 2011). Two heating processes related to the secular spin-down of the star have also been proposed: vortex creep, the dissipative motion of superfluid vortices through the neutron star crust (Alpar et al. 1984), and rotochemical heating, the energy released by non-equilibrium beta decays due to the slow contraction of the neutron star as its centrifugal force decreases (Reisenegger

1995). Gonzalez and Reisenegger (2010) performed a comparative study of several heating mechanisms and found that the rotochemical heating and vortex creep can be most important for classical and millisecond pulsars. Both processes, albeit model-dependent, can keep millisecond pulsars at a surface temperature $T_{\text{eff}} \sim 10^5$ K.

Another class of neutron stars undergoing heating episodes are quasipermanent transients, i.e., those soft X-ray transients (SXTs) whose active and quiescent periods last a few years or longer. During high-state accretion episodes, compression of the crust under the weight of newly accreted matter results in deep crustal heating, driven by exothermic nuclear transformations (Haensel and Zdunik 1990, 2008). For a given neutron star model, one can calculate the heating curve, that is the dependence of the equilibrium accretion-free T_{eff} on the accretion rate averaged over a large period of time. There is a close correspondence between the theory of thermal states of transiently accreting neutron stars and the theory of neutron star cooling (Yakovlev et al. 2003). Comparing the heating curves with a measured equilibrium T_{eff} value, one can constrain parameters connected to properties of dense matter (Yakovlev et al. 2004; Levenfish and Haensel 2007; Ho 2011). Wijnands et al. (2013) discuss prospects of application of such analysis to various classes of X-ray transients. The SXTs that have recently turned into quiescence allow one to probe the state of the neutron-star crust with the observed decline of T_{eff}. Brown et al. (1998) suggested that during this decline the radiation is fed by the heat that was deposited in the crust in the preceding active period. Such cooling is independent of the details of the star structure and composition and therefore its analysis directly yields information on the physics of the crust. Observations of several sources can be interpreted in terms of this hypothesis and yield constraints to the heat conductivity in the neutron-star crust, as, e.g., for KS 1731–260 (Shternin et al. 2007; Brown and Cumming 2009), XTE J1701–462 (Fridriksson et al. 2011; Page and Reddy 2013), EXO 0748–676 (Turlione et al. 2015). The conductivity proves to be rather high, which means that the crust has a regular crystalline (not amorphous) structure. On the other hand, there are similar objects which display variations of thermal flux that do not conform to the thermal-relaxation scenario, which may be caused by a residual slow accretion on the neutron star in quiescence (Rutledge et al. 2002; Coti Zelati et al. 2014; Turlione et al. 2015).

6.3 Thermal Luminosities of Isolated Neutron Stars

The inferred effective temperature depends on the choice of the emission model (blackbody vs. atmosphere models, composition, condensed surface, etc.—see Paper I), which typically results in variation of T_{eff} by a factor $\approx$2–3, and it has even larger theoretical uncertainties in the case of strong magnetic fields. In addition, photoelectric absorption in the interstellar medium further constitutes a source of error in temperature measurements, since the value of the hydrogen column density N_{H} is correlated to the temperature value obtained in spectral fits. Different choices for the absorption model and the metal abundances can also yield different results for the temperature. Last, in the case of data with few photons and/or strong absorption features, the temperature is poorly constrained by the fit, adding a large statistical error to the systematic one.

Because of these uncertainties, the luminosity may often be a better choice to compare data and theoretical cooling models. Since it is an integrated quantity, it averages effects of anisotropy and the choice of spectral model. The main uncertainty on the luminosity is often due to the poorly known distance to the source. In many cases, the distance is known within an error of a few, resulting in up to one order of magnitude of uncertainty in the luminosity. In addition, the interstellar absorption acts predominantly in the energy band in which most of the middle age neutron stars emit ($E \lesssim 1$ keV). Clearly, hottest (magnetars) or closest (XINSs) sources are easier to detect (see Viganò et al. 2013 and Paper I). Similarly to the case of the temperature, the choice of different models of absorption and chemical abundances can yield additional systematic errors on the luminosity. However, for the worst cases, the relative error is about 30 %, making it usually a secondary source of error compared with the distance.

In Table 3 we summarize the properties of cooling neutron stars,[3] and in Fig. 12 we compare the observational data to theoretical cooling curves, from Viganò et al. (2013). Here, the theoretical results are computed by a finite difference method for 2D (axisymmetric) stellar configurations, using the SLy EoS model (Douchin and Haensel 2001) at $\rho > \rho_{\rm drip}$ and the BPS EoS (Baym et al. 1971) at $\rho < \rho_{\rm drip}$. The high Durca threshold of the SLy EoS has been artificially lowered for illustrative purpose to $\rho = 10^{15}$ g cm^{-3}, corresponding to the central density of a star with $M = 1.42\ M_\odot$ (see Viganò 2013 for details). For superfluid gap energies, the phenomenological model of Ho et al. (2012) has been adopted. The other microphysics input is the same as in Sects. 2–4.

In the upper panel of Fig. 12 we show cooling curves for non-magnetized neutron stars with masses ranging between 1.10 and 1.76 $M_\odot$ (lines from top to bottom). After $\approx$100 yr, low mass stars ($M \lesssim 1.4\ M_\odot$) are brighter than high mass stars. For the high-mass family, $M \gtrsim 1.4\ M_\odot$, the Durca processes in the central part of the star result in fast cooling before one hundred years. Within the low-mass family, cooling curves are similar at early ages (<100 yr). The differences at $t \sim 10^2$–10^3 yr are due to the delayed transition of neutrons in the core to a superfluid state, which activates the PBF neutrino emission. After the effect of the transition to a superfluid core is finished, at $t \gtrsim 10^3$ yr, cooling curves for low-mass neutron stars tend to converge again, following the same curve independently of the mass.

We see that luminosities of some objects in the upper panel of Fig. 12 are systematically above the theoretical cooling curves. For the CCOs this discrepancy can be eliminated by considering accreted (more heat-transparent) blanketing envelopes, as the lowest dashed line in the lower panel of Fig. 12 demonstrates. However, the high-B objects still remain systematically hotter than what the theory can explain at $B = 0$. This provides strong evidence in favor of the scenario in which magnetic field decay powers their larger luminosity. In the lower panel we compare the observational data to theoretical cooling curves for different values of the initial magnetic field up to 3×10^{15} G. The most relevant effect of the inclusion of the magnetic field is that it allows to explain objects with high luminosities. Magnetic fields $B \gtrsim 10^{14}$ G are strong enough to noticeably heat up the crust and power the observed X-ray radiation. Another important difference is that the cooling timescale for strongly magnetized objects is several times larger than for the weakly magnetized neutron stars.

7 Conclusions

We have considered the basic physical ingredients needed for theoretical modeling neutron-star thermal evolution and briefly reviewed some recent results on cooling of magnetized neutron stars. The physics behind such thermal evolution is extremely rich. Clearly, we could not consider it in depth in a single review paper. However, the information that we have given, together with the references to the formulae and online resources elsewhere, should

[3] A regularly updated online catalog can be found at http://www.neutronstarcooling.info, with abundant links to references for each source.

Table 3 Cooling neutron stars. t_c is the characteristic age, t_k is the kinematic age, and f_X is the unabsorbed flux in the 1–10 keV band. The range of luminosities L includes both statistical and distance errors; for strongly absorbed sources (i.e., most magnetars) a minimum arbitrary factor of 50 % uncertainty is assumed to account for systematical model-dependent uncertainties. Data have been taken from Viganò et al. (2013) (see references therein and the online catalog in http://www.neutronstarcooling.info)

Source	$\log_{10}(t_c)$ [yr]	$\log_{10}(t_k)$ [yr]	$\log_{10}(f_X)$ [$\mathrm{erg\,cm^{-2}\,s^{-1}}$]	d [kpc]	$\log_{10}(L)$ [erg/s]
CXOU J185238.6+004020	8.3	3.7–3.9	−12.3	7.1	33.5–33.7
1E 1207.4–5209	8.5	3.4–4.3	−11.8	$2.1^{+1.8}_{-0.8}$	33.0–34.0
RX J0822–4300	8.3	3.5–3.6	−11.3	2.2 ± 0.3	33.5–33.7
CXO J232327.9+584842	–	2.5	−11.8	$3.4^{+0.3}_{-0.1}$	33.4–33.6
PSR J0538+2817	5.8	≈4.6	−12.1	1.3 ± 0.2	32.7–32.9
PSR B1055–52	5.7	–	−13.4	0.73 ± 0.15	32.2–32.6
PSR J0633+1746	5.5	–	−12.5	$0.25^{+0.22}_{-0.08}$	31.6–32.5
PSR B1706–44	4.2	–	−12.1	$2.6^{+0.5}_{-0.6}$	31.7–32.1
PSR B0833–45	4.1	3.7–4.2	−10.5	0.28 ± 0.02	32.1–32.3
PSR B0656+14	5.0	≈4.9	−12.6	0.28 ± 0.03	32.7–32.8
PSR B2334+61	4.6	≈4.0	−14.0	$3.1^{+0.2}_{-2.4}$	30.7–32.1
PSR J1740+1000	3.1	–	−13.8	1.4	32.1–32.2
PSR J1741–2054	5.6	–	−12.5	0.8	30.4–31.4
PSR J0726–2612	5.3	–	−14.0	1.0	32.1–32.5
PSR J1119–6127	3.2	3.6–3.9	−13.0	8.4 ± 0.4	33.1–33.4
PSR J1819–1458	5.1	–	−12.6	3.6	33.6–33.9
PSR J1718–3718	4.5	–	−13.2	$4.5^{+5.5}_{-0.0}$	32.8–33.5
RX J0420.0–5022	6.3	–	−17.8	0.34	30.9–31.0
RX J1856.5–3754	6.6	5.5–5.7	−14.4	0.12 ± 0.01	31.5–31.7
RX J2143.0+0654	6.6	–	−13.1	0.43	31.8–31.9
RX J0720.4–3125	6.3	5.8–6.0	−13.3	$0.29^{+0.03}_{-0.02}$	32.2–32.4
RX J0806.4–4123	6.5	–	−13.4	0.25	31.2–31.4
RX J1308.6+2127	6.2	5.9–6.1	−12.1	0.50	32.1–32.2
RX J1605.3+3249	–	5.7–6.7	−13.0	0.35 ± 0.05	30.9–31.0
1E 2259+586	5.4	4.0–4.3	−10.3	3.2 ± 0.2	35.0–35.4
4U 0142+614	4.8	–	−9.8	3.6 ± 0.5	35.4–35.8
CXO J164710.2–455216	5.2	–	−12.2	$4.0^{+1.5}_{-1.0}$	33.1–33.6
XTE J1810–197	4.1	–	−11.7	3.6 ± 0.5	34.0–34.4
1E 1547.0–5408	2.8	–	−11.5	4.5 ± 0.5	34.3–34.7
1E 1048.1–5937	3.7	–	−10.8	2.7 ± 1.0	33.8–34.5
CXOU J010043.1–721	3.8	–	−12.5	60.6 ± 3.8	35.2–35.5
1RXS J170849.0–400910	4.0	–	−10.4	3.8 ± 0.5	34.8–35.1
CXOU J171405.7–381031	3.0	≈3.7	−11.4	13.2 ± 0.2	34.9–35.2
1E 1841–045	3.7	2.7–3.0	−10.4	$9.6^{+0.6}_{-1.4}$	35.2–35.5
SGR 0501+4516	4.2	≈4	−11.3	$1.5^{+1.0}_{-0.5}$	33.2–34.0
SGR 1627–41	3.3	≈3.7	−11.6	11.0 ± 0.2	34.4–34.8
SGR 0526–66	3.5	≈3.7	−12.0	49.7 ± 1.5	35.4–35.8
SGR 1900+14	3.0	3.6–3.9	−11.1	12.5 ± 1.7	35.0–35.4
SGR 1806–20	2.6	2.8–3.0	−10.6	$13.0^{+4.0}_{-3.0}$	35.1–35.5
SGR 0418+5729*	7.6	–	−14.0	2.0	30.7–31.1
Swift J1822.3–1606*	6.2	–	−11.5	1.6 ± 0.3	32.9–33.2

Notes: *The source has been recently discovered in outburst and it could have not yet reached the quiescence level

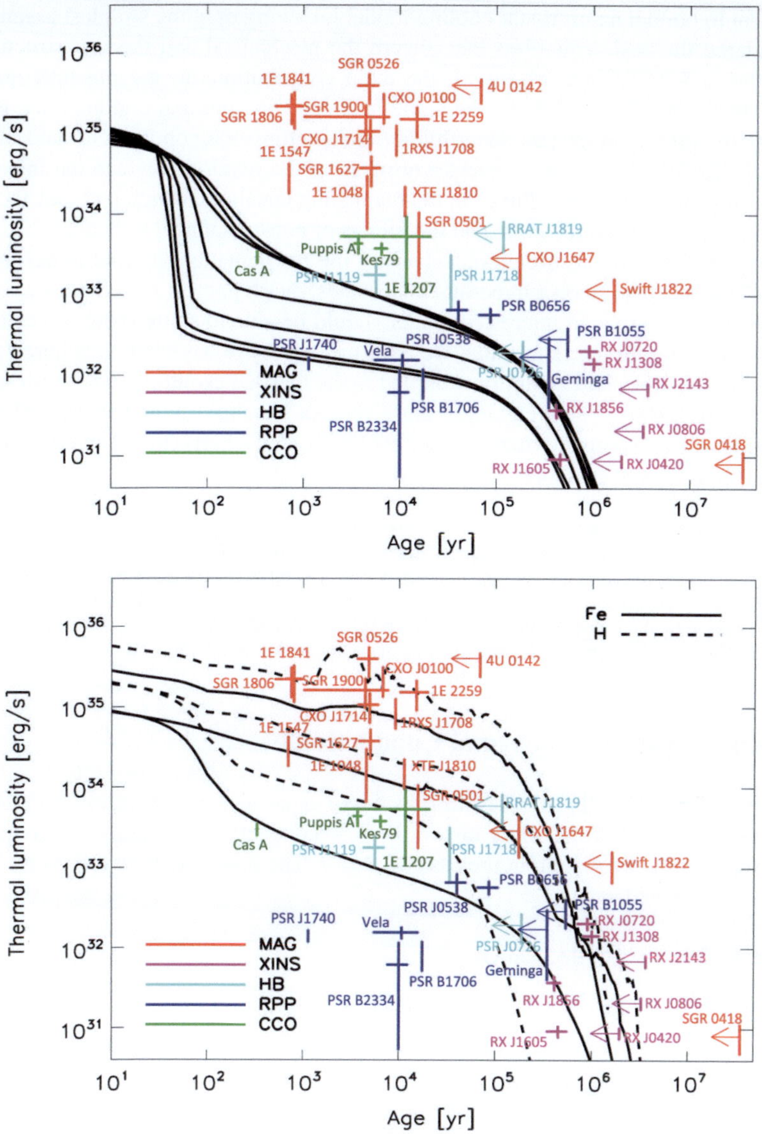

Fig. 12 Comparison between observational data and theoretical cooling curves (from Viganò et al. 2013). The observational estimates of (*errorbars*) or constraints on (*arrows*) the age and thermal luminosity correspond to Viganò et al. (2013) and Table 3. The abbreviations in the legend mark different classes of neutron stars with measured thermal radiation (MAG—magnetar candidates, XINS—X-ray isolated neutron stars, HB—high-B radio pulsars, RPP—rotation powered pulsars, CCO—central compact objects; see Paper I). *Upper panel*: non-magnetic neutron stars with iron envelopes, with $M = (1.10, 1.25, 1.32, 1.40, 1.48, 1.60, 1.70, 1.76)\ M_{\odot}$ (lines from top to bottom). *Lower panel*: a neutron star with $M = 1.4\ M_{\odot}$ and $R = 11.6$ km, and three different cases with initial magnetic field at the pole $B = 0$, 3×10^{14} G, and 3×10^{15} G. The magnetic field topology is that of Model A in Viganò et al. (2013) (crustal confined). We show results for iron envelopes (*solid lines*) and hydrogen envelopes (*dashed lines*)

be sufficient to build a neutron-star cooling model involving only the simplest assumptions. We considered the basic equations that govern the mechanical and thermal structure of a neutron star and its thermal evolution, the main contributions to the physical quantities that enter these equations—namely, EoS and heat capacity, thermal conductivity, neutrino emissivity, the effects of baryon superfluidity and proton superconductivity and of strong magnetic fields. In addition, we present a novel fit to the relation between the internal and external temperatures and heat fluxes in the blanketing envelope, which includes the effects of neutrino emission from the crust and the effects of non-radial heat transport.

In this paper we have restricted ourselves by the $npe\mu$ matter, without either hyperons or "exotic" models that involve hyperon condensates, quark phases, mixed phases, or phase transitions. We hope that an interested reader should be able to study these issues in depth following the literature references that we have provided. We have not considered also the equations of magnetic-field evolution, coupled to the thermal evolution, which is especially important in magnetars. These equations are given, for instance, in the paper by Mereghetti et al. (2015) in this volume, where origin, evolution, and observational manifestations of magnetars are reviewed in depth.

Acknowledgements The authors acknowledge hospitality of organizers and useful discussions with participants at the ISSI Workshop "The Strongest Magnetic Fields in the Universe" (Bern, Switzerland, 3–7 February 2014), where this joint review was initiated. A.P. is grateful to D.G. Yakovlev for useful discussions. The work of A.P. on the effects of strong magnetic fields on blanketing envelopes (Sect. 5.2 and Appendix B) has been supported by the Russian Science Foundation (grant 14-12-00316).

Conflict of Interest: The authors declare that they have no conflict of interest.

Appendix A: Electron Thermal Conductivities

In this Appendix, we briefly overview the physics of electron heat conduction in the neutron-star envelopes, which is the most important heat conduction mechanism as regards the neutron-star thermal evolution, in the case of $B = 0$. The magnetic field effects on the heat conduction are considered in Sect. 4.4.

A.1 Weakly Degenerate Electron Gas

In the case of non-degenerate and non-relativistic electrons (Spitzer and Härm 1953; Braginskiĭ 1958; Spitzer 1962), the effective energy-averaged electron-ion collision frequency is

$$\nu_{ei} = \frac{4}{3}\sqrt{\frac{2\pi}{m_e}}\frac{Z^2 e^4}{T^{3/2}} n_\mathrm{i} \Lambda_{ei}, \tag{A.1}$$

where Λ_{ei} is the Coulomb logarithm. In the considered case Λ_{ei} is a slowly varying function of density and temperature. Its precise value depends on the approximations used to solve the Boltzmann equation, but its order of magnitude is given by the elementary theory, where the Coulomb collision integral is truncated at small and large impact parameters of the electrons. Then $\Lambda_{ei} \sim \ln(r_\mathrm{max}/r_\mathrm{min})$, where r_max and r_min are the maximum and minimum electron impact parameters. The parameter r_max can be set equal to the Debye screening length, $r_\mathrm{max}^{-2} = 4\pi(n_e + Z^2 n_\mathrm{i})e^2/T$. The second parameter can be estimated as $r_\mathrm{min} = \max(\lambda_e, r_\mathrm{cl})$, where λ_e (defined in Sect. 2.3) limits r_min in the high-temperature regime (where the Born approximation holds), and $r_\mathrm{cl} = Ze^2/T$ is the classical closest-approach distance of a thermal electron, which limits r_min in the low-temperature, quasiclassical regime.

A similar effective frequency

$$\nu_{ee} = \frac{8}{3}\sqrt{\frac{\pi}{m_e}}\frac{e^4}{T^{3/2}}n_e\Lambda_{ee} \tag{A.2}$$

characterizes the efficiency of the ee collisions. If $\Lambda_{ee} \sim \Lambda_{ei}$, then $\nu_{ei}/\nu_{ee} \sim Z$, therefore for large Z the ei collisions are much more efficient than the ee collisions.

A.2 Strongly Degenerate Electron Gas

A.2.1 Electron-Ion Scattering

The thermal conductivity of strongly degenerate electrons in a fully ionized plasma is given by Eq. (12) with $a = \pi^2/3$. In order to determine the effective collision frequency that enters this equation, we use the Matthiessen rule $\nu = \nu_{ei} + \nu_{ee}$.

The effective electron-ion collision frequency can be written in the form (Lee 1950; Yakovlev and Urpin 1980)

$$\nu_{ei} = \frac{4Zm_e^* e^4 \Lambda_{ei}}{3\pi\hbar^3} = \frac{Z\Lambda_{ei}\sqrt{1+x_r^2}}{5.7\times10^{-17}\text{ s}}. \tag{A.3}$$

Lee (1950) gave an estimate of the Coulomb logarithm $\Lambda_{ei} = \ln(r_{max}/r_{min})$, with the minimum impact parameter $r_{min} = \hbar/2p_F$ and the maximum impact parameter $r_{max} = a_i$. Yakovlev and Urpin (1980) calculated the conductivities for relativistic degenerate electrons, neglecting electron screening, and obtained a more accurate estimate $r_{max} \approx 0.4a_i$ in the liquid regime. In the solid regime, where the electrons scatter on phonons (collective ion excitations), Yakovlev and Urpin (1980) obtained different approximations for the two distinct cases, $\Theta_D < T < T_m$ and $T < \Theta_D$.

Potekhin et al. (1999) derived a unified treatment of the electron conductivities in the Coulomb liquid and solid and described both regimes by Appendix A.3. Then qualitatively, by order of magnitude, $\Lambda_{ei} \sim 1$ in the ion liquid, and $\Lambda_{ei} \sim T/T_m$ in the Coulomb solid with a melting temperature T_m. The effects of multiphonon scattering, electron screening, and non-Born corrections, have been taken into account, and the Coulomb logarithms in both liquid and solid phases have been fitted by a single analytical formula. A Fortran code and a table of thermal conductivities, based on this formalism, are available online.[4]

At the conditions typical for the envelopes of neutron stars, the electron-phonon scattering proceeds mainly via the Umklapp processes, where the wave vector corresponding to the change of electron momentum lies outside the first Brillouin zone. Raikh and Yakovlev (1982) noticed that if $T \lesssim T_U = T_p Z^{1/3}\alpha_f\sqrt{1+x_r^2}/3x_r$, then the Umklapp processes occur less often ("freeze out"). Then the scattering rate decreases. Raikh and Yakovlev (1982) assumed an extremely strong (exponential) decrease. This implied that at $T < T_U$ the conductivity would be in practice determined by impurities and structure defects of the lattice, rather than by the electron-phonon scattering (Gnedin et al. 2001). However, Chugunov (2012) showed that distortion of electron wave functions due to interaction with the Coulomb lattice destroys this picture and strongly slows down the increase of the conductivity. As a result, the conductivities in neutron star envelopes can be treated neglecting the "freezing-out" of the Umklapp processes.

[4]http://www.ioffe.ru/astro/conduct/.

A.2.2 Electron-Electron Scattering

Although the electron-ion scattering is usually most important for degenerate plasmas, the electron-electron scattering still can be non-negligible for relatively light elements ($Z \lesssim 10$) (Lampe 1968). The expression of ν_{ee} for the relativistic degenerate electrons at $T \ll T_{\mathrm{p}}$ was obtained by Flowers and Itoh (1976). Urpin and Yakovlev (1980) extended it to higher temperatures, where $T_{\mathrm{p}} \lesssim T \ll \varepsilon_{\mathrm{F}}$.

Shternin and Yakovlev (2006) reconsidered the problem including the Landau damping of transverse plasmons, neglected by the previous authors. This effect is due to the difference of the components of the polarizability tensor, responsible for screening the charge-charge and current-current interactions: the transverse current-current interactions undergo "dynamical screening." Shternin and Yakovlev (2006) showed that the Landau damping of transverse plasmons strongly increases ν_{ee} in the domain of $x_{\mathrm{r}} \gtrsim 1$ and $T \ll T_{\mathrm{p}}$ and presented a new fit to ν_{ee} (also implemented in the code referenced in footnote 4).

A.3 The Case of Intermediate Degeneracy

In the case where the electron gas is partially degenerate, that is $T \sim \varepsilon_{\mathrm{F}}$, the thermal and electrical conductivities determined by the electron-ion scattering are satisfactorily evaluated by the thermal averaging procedure [Eq. (33) in Sect. 4.4.2]. For conductivities determined by the electron-electron collisions, there is no such averaging procedure, but we can use an interpolation between the two limiting cases,

$$\nu_{ee} = \nu_{ee}^{\mathrm{deg}} \frac{1 + 625(T/\varepsilon_{\mathrm{F}})^2}{1 + 25T/\varepsilon_{\mathrm{F}} + 271(T/\varepsilon_{\mathrm{F}})^{5/2}}. \tag{A.4}$$

A satisfactory accuracy of this interpolation has been verified by Cassisi et al. (2007).

A.4 Impurities and Mixtures

If the plasma in an envelope is not a pure substance of a single chemical element, then the effective collision frequency ν_{ei} should be modified. The required modification can be different, depending on the state of the plasma and on the amount of impurities. For example, Flowers and Itoh (1976), Yakovlev and Urpin (1980), and Itoh and Kohyama (1993) considered electron scattering by charged impurities in a Coulomb crystal. If the fraction of impurities is small and they are randomly distributed, then electron-impurity scattering can be treated as scattering by charge fluctuations, controlled by the impurity parameter $Q = \langle (Z - \langle Z \rangle)^2 \rangle$, where $\langle Z \rangle \equiv \sum_j Y_j Z_j$, $Y_j = n_j / \sum_j n_j$ is the number fraction of ions of the jth kind, and Z_j is their charge number. Then, using the Matthiessen rule, one can obtain ν_{ei} as a sum of the terms corresponding to the electron-phonon scattering in a homogeneous lattice and to the electron scattering by charge fluctuations. The effective relaxation time for the latter term is given by Appendix A.3 with $Z\Lambda_{ei}$ replaced by $\sum_j Y_j (Z_j - \langle Z \rangle)^2 \Lambda_j / \langle Z \rangle$, where the Coulomb logarithm Λ_j depends generally on j. Neglecting the differences between the Coulomb logarithms, one can thus simply replace Z by $Q/\langle Z \rangle$ in Appendix A.3 to estimate the conductivity due to electron scattering by charged impurities.

An alternative approach is relevant when there is no dominant ion species which forms a crystal (e.g., in a liquid, a gas, or a glassy alloy). In this case, one can use Appendix A.3 with $Z^2 n_{\mathrm{i}} \Lambda_{ei}$ replaced by $\sum_j Z_j^2 n_j \Lambda_j$. An approximation to Λ_j based on the plasma "additivity rule" has been suggested by Potekhin et al. (1999). Neglecting the differences between the

Coulomb logarithms, one arrives at Appendix A.3 with Z replaced by $\sqrt{\langle Z^2\rangle}$. If tabulated conductivities κ_j for pure substances are used, then the best agreement with calculations based on the "additivity rule" is usually given by the estimate

$$\kappa \approx \frac{\sum_j Y_j Z_j \kappa_j}{\sum_j Y_j Z_j} \equiv \frac{\langle \kappa Z\rangle}{\langle Z\rangle}. \tag{A.5}$$

Appendix B: Temperature Relations for Envelopes of Neutron Stars with Magnetic Fields

Here we present an analytical fit to the temperature distribution over a surface of a neutron star with a non-accreted envelope and a dipole magnetic field. We have chosen $\rho_b = 10^{10}\ \mathrm{g\,cm^{-3}}$ and used the BSk21 EoS (Pearson et al. 2012) in the parametrized form (Potekhin et al. 2013). The numerical data have been produced with the 2D code of Viganò et al. (2013) for 5 values of internal temperature T_b from 10^7 K to 10^9 K, 5 values of the magnetic field at the pole B_p from 10^{11} G to 10^{15} G, and 20 values of magnetic colatitude θ at the surface of the neutron star from 0 to $\pi/2$. The use of the 2D code corrects the temperature distribution near the magnetic equator, because the non-radial heat flow increases the equatorial T_s as compared to the 1D model that was employed previously (see Fig. 11 in Sect. 5.2). These data have been supplemented with more detailed calculations at the magnetic pole ($\theta = 0$) using the 1D code of Potekhin et al. (2007) for 36 values of T_b from $10^{6.5}$ K to 10^{10} K and 9 values of B_p from 10^{11} G to 10^{15} G. An important difference from the old results is the inclusion of the neutrino emission from the crust, which is especially important for the magnetars (see Sect. 5). Because of the 2D treatment and the allowance for neutrino emission, the new fit supersedes the previous one (Potekhin et al. 2003), whenever $B > 10^{12}$ G or $T_b \gtrsim 10^8$ K. We stress that its use is restricted by non-accreted (i.e., composed of heavy chemical elements) envelopes in the range of $10^{6.5}\ \mathrm{K} \lesssim T_b \lesssim 10^{10}$ K and $B_p \lesssim 10^{15}$ G, which is covered by the underlying numerical data. For envelopes with $B \lesssim 10^{12}$ G (either non-accreted or accreted), the previous fit can be used, however the surface temperature T_s (but not the flux at the inner boundary, F_b—see item 4 below) should be limited for hot stars according to Eq. (B.4) below.

The fit consists of 3 stages: (1) an expression for the surface temperature at the magnetic pole, T_p, as function of T_b, g, and B_p; (2) an expression for the *ratio* of the polar to the equatorial surface temperatures, T_p/T_{eq}; (3) an expression for the dependence of T_s on the magnetic colatitude θ. Since the thermal conductivities for quantizing magnetic fields (Sect. 4.4.2) are known for the electron-ion but not electron-electron collision mechanism, we multiplied T_s by a correction factor, obtained numerically from a comparison of the results of thermal-structure calculations with and without the *ee* collisions at $B = 0$. At the end of this Appendix we suggest a recipe for relating the flux F_b at the bottom of the heat-blanketing envelope to temperature T_s and thereby to T_b.

1. At the magnetic pole, the effective surface temperature, neglecting neutrino emission from the crust, is approximately given by the expression

$$T_p^{(0)} = \left[g_{14}\left(T_1^4 + (1 + 0.15\sqrt{B_{12}})T_0^4\right)\right]^{1/4} \times 10^6\ \mathrm{K}, \tag{B.1}$$

where

$$T_0 = \left(15.7 T_9^{3/2} + 1.36 T_9\right)^{0.3796}, \qquad T_1 = 1.13 B_{12}^{0.119} T_9^a,$$
$$a = 0.337/(1 + 0.02\sqrt{B_{12}}), \tag{B.2}$$

$T_9 = T_b/10^9$ K, and $B_{12} = B_p/10^{12}$ G. The limiting temperature, at which $T_p(T_b)$ levels off due to the neutrino emission from the crust is approximately given by

$$T_p^{(max)} = \left(5.2 g_{14}^{0.65} + 0.093\sqrt{g_{14} B_{12}}\right) \times 10^6 \text{ K}. \tag{B.3}$$

The corrected surface temperature at the pole, which takes this limit into account, is reproduced by the expression

$$T_p = T_p^{(0)}\left[1 + \left(T_p^{(0)}/T_p^{(max)}\right)^4\right]^{-1/4} \tag{B.4}$$

2. The ratio of the polar to equatorial surface temperatures can be roughly evaluated as

$$\frac{T_p}{T_{eq}} = 1 + \frac{(1230 T_9)^{3.35} B_{12}\sqrt{1 + 2B_{12}^2}}{(B_{12} + 450 T_9 + 119 B_{12} T_9)^4} + \frac{0.0066 B_{12}^{5/2}}{T_9^{1/2} + 0.00258 B_{12}^{5/2}}. \tag{B.5}$$

The numerically calculated T_p/T_{eq} ratio has a complex dependence on T_b and B at $B > 10^{13}$ G. In order to keep our fitting formulae relatively simple, we do not reproduce these oscillations, but instead force the ratio (B.5) to converge to some average value at $B \gg 10^{13}$ G. The numerical data oscillate in a complicated manner around this average, with deviations reaching up to 35 %. For smaller fields, $B \lesssim 3 \times 10^{12}$ G, Eq. (B.5) reproduces the numerical data with typical errors of several percent (up to 10 %). Note that these significant deviations affect only nearly tangential field case, viz. the equatorial region, which is substantially colder than the rest of the surface. Therefore its contribution to the observed flux is usually not very important.

3. Finally, the dependence of the surface temperature on the magnetic colatitude θ is approximately described by the expression

$$\frac{T_s(\theta) - T_{eq}}{T_p - T_{eq}} = \frac{(1 + a_1 + a_2)\cos^2\theta}{1 + a_1\cos\theta + a_2\cos^2\theta},$$

$$\text{where } a_1 = \frac{a_2 T_9^{1/2}}{3}, \quad a_2 = \frac{10 B_{12}}{T_9^{1/2} + 0.1 B_{12} T_9^{-1/4}}. \tag{B.6}$$

4. Note that the outer boundary condition to the thermal evolution equations (4) involves the relation between the heat flux density F_b through the boundary at $\rho = \rho_b$ and the temperature T_b at this boundary. In the absence of the neutrino emission from the crust, this boundary condition is directly provided by the T_b–T_s relation, because in this case (in the plane-parallel approximation) $F_b = \sigma_{SB} T_s^4$. It is not so if a significant part of the energy is carried from the outer crust by neutrinos. In this case we suggest to evaluate the flux through the boundary by the relation $F_b = \sigma_{SB} T_*^4$, where T_* is given by the above approximations for T_s, but without the correction (B.4).

References

D.N. Aguilera, J.A. Pons, J.A. Miralles, Astron. Astrophys. **486**, 255 (2008)
D.N. Aguilera, V. Cirigliano, J.A. Pons, S. Reddy, R. Sharma, Phys. Rev. Lett. **102**, 091101 (2009)
A. Akmal, V.R. Pandharipande, D.G. Ravenhall, Phys. Rev. C **58**, 1804 (1998)
T.L. Ainsworth, J. Wambach, D. Pines, Phys. Lett. B **222**, 173 (1989)
M.A. Alpar, D. Pines, P.W. Anderson, J. Shaham, Astrophys. J. **276**, 325 (1984)

L. Amundsen, E. Østgaard, Nucl. Phys. A **437**, 487 (1985a)
L. Amundsen, E. Østgaard, Nucl. Phys. A **442**, 163 (1985b)
J.N. Bahcall, R.A. Wolf, Astrophys. J. **142**, 1254 (1965a)
J.N. Bahcall, R.A. Wolf, Phys. Rev. **140**, B1452 (1965b)
D.A. Baiko, Phys. Rev. E **80**, 046405 (2009)
D.A. Baiko, D.G. Yakovlev, Astron. Astrophys. **342**, 192 (1999)
D.A. Baiko, A.D. Kaminker, A.Y. Potekhin, D.G. Yakovlev, Phys. Rev. Lett. **81**, 5556 (1998)
D.A. Baiko, P. Haensel, D.G. Yakovlev, Astron. Astrophys. **374**, 151 (2001a)
D.A. Baiko, A.Y. Potekhin, D.G. Yakovlev, Phys. Rev. E **64**, 057402 (2001b)
S. Balberg, N. Barnea, Phys. Rev. C **57**, 409 (1998)
M. Baldo, H.-J. Schulze. PRC750258022007
M. Baldo, J. Cugnon, A. Lejeune, U. Lombardo, Nucl. Phys. A **536**, 349 (1992)
M. Baldo, Ø. Elgarøy, L. Engvik, M. Hjorth-Jensen, H.-J. Schulze, Phys. Rev. C **58**, 1921 (1998)
J. Bardeen, L.N. Cooper, J.R. Schrieffer, Phys. Rev. **108**, 1175 (1957)
G. Baym, C. Pethick, *Landau Fermi-Liquid Theory: Concepts and Applications* (Wiley, New York, 1991)
G. Baym, C. Pethick, D. Pines, Nature **224**, 673 (1969)
G. Baym, C. Pethick, P. Sutherland, Astrophys. J. **170**, 299 (1971)
V.G. Bezchastnov, P. Haensel, A.D. Kaminker, D.G. Yakovlev, Astron. Astrophys. **328**, 409 (1997)
G.S. Bisnovatyi-Kogan, M.M. Romanova, Sov. Phys. JETP **56**, 243 (1982)
D. Blaschke, G. Röpke, H. Schulz, A.D. Sedrakian, D.N. Voskresensky, Mon. Not. R. Astron. Soc. **273**, 596 (1995)
A. Bohr, B.R. Mottelson, D. Pines, Phys. Rev. **110**, 936 (1958)
S. Bowyer, E.T. Byram, T.A. Chubb, H. Friedman, Nature **201**, 1307 (1964a)
S. Bowyer, E.T. Byram, T.A. Chubb, H. Friedman, Science **146**, 912 (1964b)
S.I. Braginskiĭ, Sov. Phys. JETP **6**, 358 (1958)
A. Broderick, M. Prakash, J.M. Lattimer, Astrophys. J. **537**, 351 (2000)
E.F. Brown, A. Cumming, Astrophys. J. **698**, 1020 (2009)
G.E. Brown, K. Kudobera, D. Page, P.M. Pizzochero, Phys. Rev. D **37**, 2042 (1988)
E.F. Brown, L. Bildsten, R.E. Rutledge, Astrophys. J. **504**, L95 (1998)
A. Burrows, J.M. Lattimer, Astrophys. J. **307**, 178 (1986)
A.G.W. Cameron, Nature **215**, 464 (1967)
W.J. Carr, Phys. Rev. **122**, 1437 (1961)
S. Cassisi, A.Y. Potekhin, A. Pietrinferni, M. Catelan, M. Salaris, Astrophys. J. **661**, 1094 (2007)
N. Chamel, Nucl. Phys. A **747**, 109 (2005)
N. Chamel, Phys. Rev. C **85**, 035801 (2012)
N. Chamel, R.L. Pavlov, L.M. Mihailov et al., Phys. Rev. C **86**, 055804 (2012)
N. Chamel, D. Page, S.K. Reddy, Phys. Rev. C **87**, 035803 (2013)
P. Chang, L. Bildsten, P. Arras, Astrophys. J. **723**, 719 (2010)
D. Chatterjee, T. Elghozi, J. Novak, M. Oertel, Mon. Not. R. Astron. Soc. **447**, 3785 (2015)
J.M.C. Chen, J.W. Clark, R.D. Davé, V.V. Khodel, Nucl. Phys. A **555**, 59 (1993)
H.-Y. Chiu, E.E. Salpeter, Phys. Rev. Lett. **12**, 413 (1964)
A.I. Chugunov, Astron. Lett. **38**, 25 (2012)
A.I. Chugunov, P. Haensel, Mon. Not. R. Astron. Soc. **381**, 1143 (2007)
L.N. Cooper, Phys. Rev. **104**, 1189 (1956)
F. Coti Zelati, S. Campana, P. D'Avanzo, A. Melandri, Mon. Not. R. Astron. Soc. **438**, 2634 (2014)
F.V. De Blasio, Astron. Astrophys. **353**, 1129 (2000)
J.A. de Freitas Pacheco, J.E. Steiner, A.D. Neto, Astron. Astrophys. **55**, 111 (1977)
H.E. DeWitt, W.L. Slattery, J. Yang, in *Strongly Coupled Plasma Physics*, ed. by H.M. Van Horn, S. Ichimaru (Univ. Rochester, Rochester, 1993), p. 425
F. Douchin, P. Haensel, Astron. Astrophys. **380**, 151 (2001)
K.G. Elshamouty, C.O. Heinke, G.R. Sivakoff et al., Astrophys. J. **777**, 22 (2013)
R.T. Farouki, S. Hamaguchi, Phys. Rev. E **47**, 4330 (1993)
E. Flowers, N. Itoh, Astrophys. J. **206**, 218 (1976)
E. Flowers, N. Itoh, Astrophys. J. **230**, 847 (1979)
E. Flowers, N. Itoh, Astrophys. J. **250**, 750 (1981)
E. Flowers, M. Ruderman, P. Sutherland, Astrophys. J. **205**, 541 (1976)
L.M. Franco, L. Bennett, R.I. Epstein, Astrophys. J. **543**, 987 (2004)
J.K. Fridriksson, J. Homan, R. Wijnands et al., Astrophys. J. **736**, 162 (2011)
J. Frieben, L. Rezzolla, Mon. Not. R. Astron. Soc. **427**, 3406 (2012)
G. Gamow, M. Schoenberg, Phys. Rev. **59**, 539 (1941)
S. Gandolfi, A.Yu. Illarionov, F. Pederiva, K.E. Schmidt, S. Fantoni, Phys. Rev. C **80**, 045802 (2008)

U. Geppert, M. Küker, D. Page, Astron. Astrophys. **426**, 267 (2004)
U. Geppert, M. Küker, D. Page, Astron. Astrophys. **457**, 937 (2006)
R. Giacconi, H. Gursky, F.R. Paolini, B.B. Rossi, Phys. Rev. Lett. **9**, 439 (1962)
V.L. Ginzburg, Sov. Phys. Usp. **12**, 241 (1970)
V.L. Ginzburg, Sov. Phys. Usp. **14**, 83 (1971)
V.L. Ginzburg, D.A. Kirzhnits, Sov. Phys. JETP **20**, 1346 (1965)
G. Glen, P. Sutherland, Astrophys. J. **239**, 671 (1980)
O.Y. Gnedin, D.G. Yakovlev, Nucl. Phys. A **582**, 697 (1995)
O.Y. Gnedin, D.G. Yakovlev, A.Y. Potekhin, Mon. Not. R. Astron. Soc. **324**, 725 (2001)
T. Gold, Nature **218**, 731 (1968)
D. Gonzalez, A. Reisenegger, Astron. Astrophys. **522**, A16 (2010)
S. Goriely, N. Chamel, J.M. Pearson, Phys. Rev. C **82**, 035804 (2010)
G. Greenstein, G.J. Hartke, Astrophys. J. **271**, 283 (1983)
E.Y. Gudmundsson, C.J. Pethick, R.I. Epstein, Astrophys. J. **272**, 286 (1983)
M.E. Gusakov, Astron. Astrophys. **389**, 702 (2002)
M.E. Gusakov, A.D. Kaminker, D.G. Yakovlev, O.Y. Gnedin, Astron. Astrophys. **423**, 1063 (2004)
P. Haensel, B. Pichon, Astron. Astrophys. **283**, 313 (1994)
P. Haensel, J.L. Zdunik, Astron. Astrophys. **227**, 431 (1990)
P. Haensel, J.L. Zdunik, Astron. Astrophys. **480**, 459 (2008)
P. Haensel, A. Denissov, S. Popov, Astron. Astrophys. **240**, 78 (1990)
P. Haensel, A.Y. Potekhin, D.G. Yakovlev, *Neutron Stars 1: Equation of State and Structure* (Springer, New York, 2007)
J.P. Hansen, G.M. Torrie, P. Vieillefosse, Phys. Rev. A **16**, 2153 (1977)
C.O. Heinke, W.C.G. Ho, Astrophys. J. **719**, L167 (2010)
H. Heiselberg, C.J. Pethick, Phys. Rev. D **48**, 2916 (1993)
L. Hernquist, Astrophys. J. Suppl. Ser. **56**, 325 (1984)
L. Hernquist, Mon. Not. R. Astron. Soc. **213**, 313 (1985)
A. Hewish, S.E. Okoye, Nature **207**, 59 (1968)
A. Hewish, S.J. Bell, J.D.H. Pilkington, P.F. Scott, R.F. Collins, Nature **217**, 709 (1968)
W.C.G. Ho, Mon. Not. R. Astron. Soc. **418**, L99 (2011)
W.C.G. Ho, K. Glampedakis, N. Andersson, Mon. Not. R. Astron. Soc. **422**, 2632 (2012)
W.C.G. Ho, K.G. Elshamouty, C.O. Heinke, A.Y. Potekhin, Phys. Rev. C **91**, 015806 (2015)
W. Hubbard, M. Lampe, Astrophys. J. Suppl. Ser. **18**, 297 (1969)
J. Hughto, A.S. Schneider, C.J. Horowitz, D.K. Berry, Phys. Rev. E **84**, 016401 (2011)
J. Hughto, C.J. Horowitz, A.S. Schneider et al., Phys. Rev. E **86**, 066413 (2012)
N.A. Inogamov, R.A. Sunyaev, Astron. Lett. **36**, 848 (2010)
N. Itoh, Y. Kohyama, Astrophys. J. **275**, 858 (1983)
N. Itoh, Y. Kohyama, Astrophys. J. **404**, 268 (1993). Erratum: Astrophys. J. **420**, 943 (1994)
A.D. Kaminker, D.G. Yakovlev, Theor. Math. Phys. **49**, 1012 (1981)
A.D. Kaminker, D.G. Yakovlev, Astron. Rep. **38**, 809 (1994)
A.D. Kaminker, K.P. Levenfish, D.G. Yakovlev, P. Amsterdamski, P. Haensel, Phys. Rev. D **46**, 3256 (1992)
A.D. Kaminker, D.G. Yakovlev, P. Haensel, Astron. Astrophys. **325**, 391 (1997)
A.D. Kaminker, C.J. Pethick, A.Y. Potekhin, V. Thorsson, D.G. Yakovlev, Astron. Astrophys. **343**, 1009 (1999)
A.D. Kaminker, P. Haensel, D.G. Yakovlev, Astron. Astrophys. **373**, L17 (2001)
A.D. Kaminker, A.Y. Potekhin, D.G. Yakovlev, G. Chabrier, Mon. Not. R. Astron. Soc. **395**, 2257 (2009)
A.D. Kaminker, A.A. Kaurov, A.Y. Potekhin, D.G. Yakovlev, in *Electromagnetic Radiation from Pulsars and Magnetars*, ed. by W. Lewandowski, J. Kijak, A. Slowikowska, O. Maron. Astron. Soc. Pacific Conf. Ser., vol. 466 (2012), p. 237
A.D. Kaminker, A.A. Kaurov, A.Y. Potekhin, D.G. Yakovlev, Mon. Not. R. Astron. Soc. **442**, 3484 (2014)
E.M. Kantor, E.M. Gusakov, Mon. Not. R. Astron. Soc. **381**, 1702 (2007)
N.S. Kardashev, Astron. Ž. **41**, 807 (1964)
C. Kittel, *Quantum Theory of Solids* (Wiley, New York, 1963)
V.N. Kondratyev, T. Maruyama, S. Chiba, Astrophys. J. **546**, 1137 (2001)
D. Lai, S.L. Shapiro, Astrophys. J. **383**, 745 (1991)
M. Lampe, Phys. Rev. **174**, 276 (1968)
J.M. Lattimer, C.J. Pethick, M. Prakash, P. Haensel, Phys. Rev. Lett. **66**, 2701 (1991)
T.D. Lee, Astrophys. J. **111**, 625 (1950)
L.B. Leinson, Phys. Rev. C **79**, 045502 (2009)
L.B. Leinson, Phys. Rev. C **81**, 025501 (2010)
L.B. Leinson, A. Pérez, J. High Energy Phys. **1998**(9), 020 (1998)

K.P. Levenfish, P. Haensel, Astrophys. Space Sci. **308**, 457 (2007)
K.P. Levenfish, D.G. Yakovlev, Astron. Rep. **38**, 247 (1994)
E.M. Lifshitz, L.P. Pitaevskiĭ, *Statistical Physics, Part 2: Theory of the Condensed State* (Butterworth-Heinemann, Oxford, 2002)
C.P. Lorenz, D.G. Ravenhall, C.J. Pethick, Phys. Rev. Lett. **70**, 379 (1993)
R. Machleidt, K. Holinde, Ch. Elster, Phys. Rep. **149**, 1 (1987)
J. Margueron, H. Sagawa, K. Hagino, Phys. Rev. C **77**, 054309 (2008)
Z. Medin, A. Cumming, Phys. Rev. E **81**, 036107 (2010)
S. Mereghetti, J.A. Pons, A. Melatos, Space Sci. Rev. (2015, in press), doi:10.1007/s11214-015-0146-y
A.B. Migdal, Sov. Phys. JETP **10**, 176 (1960)
J.A. Miralles, V. Urpin, D. Konenkov, Astrophys. J. **503**, 368 (1998)
C.W. Misner, K.S. Thorne, J.A. Wheeler, *Gravitation* (Freeman, San Francisco, 1973)
P. Muzikar, J.A. Sauls, J.W. Serene, Phys. Rev. D **21**, 1494 (1980)
D.C. Morton, Nature **201**, 1308 (1964)
J.W. Negele, D. Vautherin, Nucl. Phys. A **207**, 298 (1973)
K. Nomoto, S. Tsuruta, Space Sci. Rev. **30**, 123 (1981a)
K. Nomoto, S. Tsuruta, Astrophys. J. **250**, L19 (1981b)
K. Nomoto, S. Tsuruta, Astrophys. J. **305**, L19 (1986)
K. Nomoto, S. Tsuruta, Astrophys. J. **312**, 711 (1987)
D.D. Ofengeim, A.D. Kaminker, D.G. Yakovlev, Europhys. Lett. **108**, 31002 (2014)
F. Pacini, Nature **216**, 567 (1967)
D. Page, in *Neutron Stars and Pulsars*, ed. by W. Becker. Astrophys. Space Sci. Library, vol. 357 (2009), p. 247
D. Page, J.H. Applegate, Astrophys. J. **394**, L17 (1992)
D. Page, E. Baron, Astrophys. J. **354**, L17 (1990). Erratum: Astrophys. J. **382**, L111–L112 (1991)
D. Page, S. Reddy, Phys. Rev. Lett. **111**, 241102 (2013)
D. Page, J.M. Lattimer, M. Prakash, A.W. Steiner, Astrophys. J. Suppl. Ser. **155**, 623 (2004)
D. Page, U. Geppert, M. Küker, Astrophys. Space Sci. **308**, 403 (2007)
D. Page, M. Prakash, J.M. Lattimer, A.W. Steiner, Phys. Rev. Lett. **106**, 081101 (2011)
D. Page, J.M. Lattimer, M. Prakash, A.W. Steiner, in *Novel Superfluids*, ed. by K.H. Bennemann, J.B. Ketterson (Oxford University Press, Oxford, 2014), p. 550
V.R. Pandharipande, D.G. Ravenhall, in *Nuclear Matter and Heavy Ion Collisions*, ed. by M. Soyeur, H. Flocard, B. Tamain, M. Porneuf. NATO ADS Ser., vol. B205 (Reidel, Dordrecht, 1989), p. 103
A. Pastore, N. Chamel, J. Margueron, Mon. Not. R. Astron. Soc. **448**, 1887 (2015)
J.M. Pearson, S. Goriely, N. Chamel, Phys. Rev. C **83**, 065810 (2011)
J.M. Pearson, N. Chamel, S. Goriely, C. Ducoin, Phys. Rev. C **85**, 065803 (2012)
J.F. Pérez-Azorin, J.A. Miralles, J.A. Pons, Astron. Astrophys. **451**, 1009 (2006)
R. Perna, J.A. Pons, Astrophys. J. **727**, L51 (2011)
C.J. Pethick, A.Y. Potekhin, Phys. Lett. B **427**, 7 (1996)
C.J. Pethick, D.G. Ravenhall, Annu. Rev. Astron. Astrophys. **45**, 429 (1995)
V.S. Pinaev, Sov. Phys. JETP **18**, 377 (1964)
J.A. Pons, R. Perna, Astrophys. J. **741**, 123 (2011)
J.A. Pons, N. Rea, Astrophys. J. **750**, L6 (2012)
J.A. Pons, S. Reddy, M. Prakash, J.M. Lattimer, J.A. Miralles, Astrophys. J. **513**, 780 (1999)
J.A. Pons, J.A. Miralles, U. Geppert, Astron. Astrophys. **496**, 207 (2014)
B. Posselt, G.G. Pavlov, V. Suleimanov, O. Kargaltsev, Astrophys. J. **779**, 186 (2013)
A.Y. Potekhin, Astron. Astrophys. **306**, 999 (1996). Erratum: Astron. Astrophys. **327**, 441 (1997)
A.Y. Potekhin, Astron. Astrophys. **351**, 787 (1999)
A.Y. Potekhin, Phys. Usp. **57**, 735 (2014)
A.Y. Potekhin, G. Chabrier, Phys. Rev. E **62**, 8554 (2000)
A.Y. Potekhin, G. Chabrier, Contrib. Plasma Phys. **50**, 82 (2010)
A.Y. Potekhin, G. Chabrier, Astron. Astrophys. **538**, A115 (2012)
A.Y. Potekhin, G. Chabrier, Astron. Astrophys. **550**, A43 (2013)
A.Y. Potekhin, D.G. Yakovlev, Astron. Astrophys. **374**, 213 (2001)
A.Y. Potekhin, G. Chabrier, D.G. Yakovlev, Astron. Astrophys. **323**, 415 (1997)
A.Y. Potekhin, D.A. Baiko, P. Haensel, D.G. Yakovlev, Astron. Astrophys. **346**, 345 (1999)
A.Y. Potekhin, D.G. Yakovlev, G. Chabrier, O.Y. Gnedin, Astrophys. J. **594**, 404 (2003)
A.Y. Potekhin, V.A. Urpin, G. Chabrier, Astron. Astrophys. **443**, 1025 (2005)
A.Y. Potekhin, G. Chabrier, D.G. Yakovlev, Astrophys. Space Sci. **308**, 353 (2007)
A.Y. Potekhin, G. Chabrier, A.I. Chugunov, H.E. DeWitt, F.J. Rogers, Phys. Rev. E **80**, 047401 (2009)
A.Y. Potekhin, A.F. Fantina, N. Chamel, J.M. Pearson, S. Goriely, Astron. Astrophys. **560**, A48 (2013)

A.Y. Potekhin, A. De Luca, J.A. Pons, Space Sci. Rev. (2015, in press), doi:10.1007/s11214-014-0102-2 (Paper I)
M.E. Raikh, D.G. Yakovlev, Astrophys. Space Sci. **87**, 193 (1982)
A. Reisenegger, Astrophys. J. **442**, 749 (1995)
M.B. Richardson, H.M. Van Horn, K.F. Ratcliff, R.C. Malone, Astrophys. J. **255**, 624 (1982)
L.F. Roberts, Astrophys. J. **755**, 126 (2012)
S.B. Rüster, M. Hempel, J. Schaffner-Bielich, Phys. Rev. C **73**, 035804 (2006)
R.E. Rutledge, L. Bildsten, E.F. Brown, G.G. Pavlov, V.E. Zavlin, Astrophys. J. **577**, 346 (2002)
A. Sandage, P. Osmer, R. Giacconi et al., Astrophys. J. **146**, 316 (1966)
Ch. Schaab, D. Voskresensky, A.D. Sedrakian, F. Weber, M.K. Weigel, Astron. Astrophys. **321**, 591 (1997)
M.E. Schaaf, Astron. Astrophys. **227**, 61 (1990)
A. Schwenk, B. Friman, G.E. Brown, Nucl. Phys. A **713**, 191 (2003)
J. Schwinger, *Particles, Sources, and Fields* (Addison-Wesley, Redwood City, 1988)
Yu.A. Shibanov, D.G. Yakovlev, Astron. Astrophys. **309**, 171 (1996)
I.S. Shklovsky, Astrophys. J. **148**, L1 (1967)
P.S. Shternin, D.G. Yakovlev, Phys. Rev. D **74**, 043004 (2006)
P.S. Shternin, D.G. Yakovlev, Phys. Rev. D **75**, 103004 (2007)
P.S. Shternin, D.G. Yakovlev, P. Haensel, A.Y. Potekhin, Mon. Not. R. Astron. Soc. **382**, L43 (2007)
P.S. Shternin, D.G. Yakovlev, C.O. Heinke, W.C.G. Ho, D.J. Patnaude, Mon. Not. R. Astron. Soc. **412**, L108 (2011)
P.S. Shternin, M. Baldo, P. Haensel, Phys. Rev. C **88**, 065803 (2013)
N.A. Silant'ev, D.G. Yakovlev, Astrophys. Space Sci. **71**, 45 (1980)
A.A. Sokolov, I.M. Ternov, *Radiation from Relativistic Electrons* (Am. Inst. Phys., New York, 1986)
L. Spitzer Jr., *Physics of Fully Ionized Gases* (Wiley, New York, 1962)
L. Spitzer Jr., R. Härm, Phys. Rev. **89**, 977 (1953)
R. Stabler, Ph.D. thesis, Cornell Univ., Ithaca, NY (1960)
M. Stejner, F. Weber, J. Madsen, Astrophys. J. **694**, 1019 (2009)
T. Strohmayer, L. Bildsten, in *Compact Stellar X-Ray Sources*, ed. by W.H.G. Lewin, M. van der Klis (Cambridge University Press, Cambridge, 2006), p. 113
I.-S. Suh, G.J. Mathews, Astrophys. J. **546**, 1126 (2001)
T. Takatsuka, R. Tamagaki, Prog. Theor. Phys. **94**, 457 (1995)
T. Takatsuka, R. Tamagaki, Prog. Theor. Phys. **112**, 37 (2004)
C. Thompson, in *The Neutron Star—Black Hole Connection*, ed. by C. Kouveliotou, J. Ventura, E. Van den Heuvel (Kluwer Academic, Dordrecht, 2001), p. 369
K.S. Thorne, Astrophys. J. **212**, 825 (1977)
D.R. Tilley, J. Tilley, *Superfluidity and Superconductivity* (IOP Publishing, Bristol, 1990)
S. Tsuruta, Neutron star models. Ph.D. thesis, Columbia Univ., New York (1964)
S. Tsuruta, Phys. Rep. **56**, 237 (1979)
S. Tsuruta, Comments Astrophys. **11**, 151 (1986)
S. Tsuruta, Phys. Rep. **292**, 1 (1998)
S. Tsuruta, in *Neutron Stars and Pulsars*, ed. by W. Becker. Astrophys. Space Sci. Library, vol. 357 (2009), p. 289
S. Tsuruta, A.G.W. Cameron, Can. J. Phys. **44**, 1863 (1966)
A. Turlione, D. Aguilera, J. Pons, Astron. Astrophys. **577**, A5 (2015)
V. Urpin, D. Konenkov, Astron. Astrophys. **483**, 223 (2008)
V.A. Urpin, D.G. Yakovlev, Sov. Astron. **24**, 126 (1980)
K.A. Van Riper, Astrophys. J. **329**, 339 (1988)
K.A. Van Riper, Astrophys. J. Suppl. Ser. **75**, 449 (1991)
K.A. Van Riper, F. Lamb, Astrophys. J. **244**, L13 (1981)
J. Ventura, A.Y. Potekhin, in *The Neutron Star—Black Hole Connection*, ed. by C. Kouveliotou, J. Ventura, E. Van den Heuvel. Proceedings of the NATO ASI, Ser. C, vol. 567 (Kluwer Academic, Dordrecht, 2001), p. 393
A. Vidaurre, A. Perez, H. Sivak, J. Bernabeu, J.M. Ibanez, Astrophys. J. **448**, 264 (2013)
D. Viganò, Magnetic Fields in Neutron Stars. Ph.D. thesis, Univ. of Alicante, Alicante (2013)
D. Viganò, N. Rea, J.A. Pons, D.N. Aguilera, J.A. Miralles, Mon. Not. R. Astron. Soc. **434**, 123 (2013)
D.N. Voskresensky, in *Physics of Neutron Star Interiors*, ed. by D. Blaschke, N.K. Glendenning, A. Sedrakian. Lecture Notes in Physics, vol. 578 (2001), p. 467
D.N. Voskresensky, A.V. Senatorov, Sov. Phys. JETP **63**, 885 (1986)
D.N. Voskresensky, A.V. Senatorov, Sov. J. Nucl. Phys. **45**, 411 (1987)
J. Wambach, T.L. Ainsworth, D. Pines, Nucl. Phys. A **555**, 128 (1993)
R. Wijnands, N. Degenaar, D. Page, Mon. Not. R. Astron. Soc. **432**, 2366 (2013)

D.G. Yakovlev, Astrophys. Space Sci. **98**, 37 (1984)
D.G. Yakovlev, C.J. Pethick, Annu. Rev. Astron. Astrophys. **42**, 169 (2004)
D.G. Yakovlev, V.A. Urpin, Sov. Astron. **24**, 303 (1980)
D.G. Yakovlev, A.D. Kaminker, K.P. Levenfish, Astron. Astrophys. **343**, 650 (1999a)
D.G. Yakovlev, K.P. Levenfish, Yu.A. Shibanov, Phys. Usp. **42**, 737 (1999b)
D.G. Yakovlev, A.D. Kaminker, O.Y. Gnedin, P. Haensel, Phys. Rep. **354**, 1 (2001)
D.G. Yakovlev, K.P. Levenfish, P. Haensel, Astron. Astrophys. **407**, 265 (2003)
D.G. Yakovlev, K.P. Levenfish, A.Y. Potekhin, O.Y. Gnedin, G. Chabrier, Astron. Astrophys. **417**, 169 (2004)
D.G. Yakovlev, O.Y. Gnedin, A.D. Kaminker, A.Y. Potekhin, in *40 Years of Pulsars: Millisecond Pulsars, Magnetars and More*, ed. by C. Bassa, Z. Wang, A. Cumming, V.M. Kaspi. AIP Conf. Proc., vol. 983 (2008), p. 379
J.M. Ziman, *Electrons and Phonons* (Oxford University Press, Oxford, 1960)

DOI 10.1007/978-1-4939-3550-5_9
Reprinted from *Space Science Reviews* Journal, DOI 10.1007/s11214-014-0123-x

Magnetic Fields of Neutron Stars in X-Ray Binaries

Mikhail Revnivtsev · Sandro Mereghetti

Received: 27 June 2014 / Accepted: 20 November 2014 / Published online: 5 December 2014

Abstract A substantial fraction of the known neutron stars resides in X-ray binaries—systems in which one compact object accretes matter from a companion star. Neutron stars in X-ray binaries have magnetic fields among the highest found in the Universe, spanning at least the range from $\sim 10^8$ to several 10^{13} G. The magnetospheres around these neutron stars have a strong influence on the accretion process, which powers most of their emission. The magnetic field intensity and geometry, are among the main factors responsible for the large variety of spectral and timing properties observed in the X-ray energy range, making these objects unique laboratories to study the matter behavior and the radiation processes in magnetic fields unaccessible on Earth. In this paper we review the main observational aspects related to the presence of magnetic fields in neutron star X-ray binaries and some methods that are used to estimate their strength.

Keywords Neutron stars · Magnetic field · X-ray binaries

1 Introduction

X-ray binaries (XRBs) were the first astrophysical objects in which relativistic compact stars (neutron stars (NSs) and black holes) were detected. In 1962 the pioneering observations of the sky in the X-ray energy range led to the discovery of the bright source Sco X-1 (Giacconi et al. 1962), which is now known to contain an accreting NS. However, at that time the origin of the X-ray emission and the nature of the Sco X-1 source were completely unknown. Thus, the first ones to recognize the existence of NSs were radio astronomers, who discovered the radio pulsars a few years later (Hewish et al. 1968; Gold 1968). Measurements of the

M. Revnivtsev (✉)
Space Research Institute, Russian Academy of Sciences, Profsoyuznaya 84/32, Moscow 117997, Russia
e-mail: revnivtsev@iki.rssi.ru

S. Mereghetti
INAF, IASF-Milano, v. Bassini 15, 20133 Milano, Italy
e-mail: sandro@iasf-milano.inaf.it

spin-down of radio pulsars, interpreted as a loss of rotational energy due to the emission of magneto-dipole radiation, allowed the first estimates of their magnetic fields (e.g. Gunn and Ostriker 1969). These were in agreement with earlier predictions based on the assumption that the magnetic flux of stars collapsing into NS is conserved, giving rise to NS magnetic fields as high as $B \sim 10^{12}$–10^{16} G (Ginzburg 1964; Woltjer 1964).

The observations obtained by UHURU, the first satellite for X-ray astronomy, showed that many XRBs contain NSs with magnetic fields powerful enough to overcome the enormous gravitational drag of the NS and disrupt the flow of accreting matter, thus creating anisotropic emission patterns. These allowed rotating accreting NSs to be detected as regularly pulsating X-ray sources through a lighthouse effect (Giacconi et al. 1971; Schreier et al. 1972; Tananbaum et al. 1972; Lamb et al. 1973; Davidson and Ostriker 1973).

Since the time of these early discoveries, significant advances in observational astrophysics have provided an extremely rich dataset on NSs. Differences in their intrinsic properties (ages, magnetic fields, spin-periods, etc.), as well as in their environment (e.g., isolated or in binaries, density of interstellar medium, location in globular clusters, etc.), give rise to a variety of manifestations that can be studied in the whole electromagnetic spectrum. In this paper we concentrate on the NSs whose emission is mainly powered by accretion of matter provided by a companion star. The accreting matter acts as a probe in the regions of our interest, in particular around magnetic NSs. Observations in X-rays (from ~1 keV to a few hundred keV) are one of the most informative ways to explore the properties of NSs because this is the energy range where the NS surface and the matter in its immediate vicinity emit a significant part of their bolometric luminosity.

In Sect. 2 we introduce a few basic equations relevant for the physics of accretion and give a brief overview of the main properties of XRBs containing NSs. In the following sections we focus on a few specific aspects closely related to the presence of magnetic fields: the observation of cyclotron lines (Sect. 3), and the interaction between the accretion flow and the NS magnetosphere (Sects. 4, 5, and 6).

2 Accretion-Powered Neutron Star X-ray Binaries

X-ray binaries are powered by accretion onto a compact object which gravitationally captures part of the mass lost by its companion star. The early evidence that many of the observed properties of XRBs depend on the type of companion star led to their classification into the two main classes of high mass X-ray binaries (HMXBs) and low mass X-ray binaries (LMXBs), based on the mass of the companion star. Extensive reviews of this vast subject can be found, e.g., in Lewin and van der Klis (2010). Progress in observations led to the discovery of a large number of XRBs in our Galaxy and in the Magellanic Clouds (see, e.g., Liu et al. 2006, 2007, for recent catalogs of XRBs). At the moment, the most sensitive surveys of the whole Galactic plane obtained in hard X-rays have detected about two hundred sources, nearly equally divided between HMXBs and LMXBs (Krivonos et al. 2010; Bird et al. 2010; Baumgartner et al. 2013).

Before giving an overview of the main properties of XRBs containing NSs, we remind a few basic concepts and definitions that will be used in the following sections (see, e.g., Frank et al. 1992 for a more extensive discussion). We denote with M and R the NS mass and radius, and assume in all the numerical formulae the values $M = 1.4 M_{\odot}$ and $R = 12$ km. The NS moment of inertia is $I \approx 0.4 M R^2$, and its angular velocity is $\omega = 2\pi\nu = 2\pi/P$. The magnetic dipole moment is $\mu = BR^3/2$, where B is the field strength at the NS magnetic poles.

The accreting matter, on its way toward the compact object, releases a large amount of gravitational energy, heats up and emits mostly in the X-ray energy range. In the case of an accreting NS, the accretion luminosity is given by:

$$L_X \approx \frac{GM\dot{M}}{R} \approx 1.5 \times 10^{20} \dot{M} \text{ erg s}^{-1}, \tag{1}$$

where $\dot{M}$ is the mass accretion rate in g s^{-1}. This corresponds to a $\sim$17 % efficiency of conversion of the rest mass energy of the accreting matter.

An upper limit to the accretion luminosity is given by the Eddington luminosity, L_{Edd}, at which the radiation force equals the gravitational attraction, thus stopping the accretion flow:

$$L_{\text{Edd}} = \frac{4\pi GMm_{\text{p}}c}{\sigma_{\text{T}}} \approx 1.3 \times 10^{38} \frac{M}{M_\odot} \text{ erg s}^{-1} \tag{2}$$

(c is the velocity of light, m_{p} is the proton mass, and σ_{T} is the Thompson cross section). The typical luminosities observed in bright XRBs are of the order of $\sim 10^{36}$–10^{38} erg s^{-1}. They imply accretion rates of $\sim 10^{16}$–10^{18} g s^{-1} (corresponding to $\sim 10^{-10}$–$10^{-8} M_\odot \text{ yr}^{-1}$). Much lower luminosities, down to $\sim 10^{31}$ erg s^{-1}, are seen in the quiescent states of transient sources (note that other processes besides accretion might be at work to power these low luminosities).

To discuss the effects produced by the NS magnetic fields on the flow of accreting matter, hence on the X-ray emission properties, it is useful to consider a few characteristic radii around the compact object. Their relative positions, which depend mainly on ω, μ, and density of the inflowing mass, determine the appearance of the NS (see, e.g., Lipunov et al. 1992; Campana et al. 1998; Bozzo et al. 2008).

The accretion radius, r_{a}, defines the region in which the matter is gravitationally captured by the compact object. In the case of a NS accreting from a stellar wind with velocity v_{w}, it is given by:

$$r_{\text{a}} = 2GM/\left(v_{\text{w}}^2 + v_{\text{orb}}^2\right), \tag{3}$$

where v_{orb} is the orbital velocity of the NS (usually negligible compared to the wind velocity v_{w}).

The magnetic energy density, $B^2/8\pi$, around a magnetized NS rapidly rises towards the compact object (as $\propto r^{-6}$ in the case of a dipole and even faster in the case of higher multipole fields). A simple estimate of the distance from the NS where the magnetic pressure becomes equal to the ram pressure of the accretion flow gives the magnetospheric radius (see, e.g., Pringle and Rees 1972; Lamb et al. 1973; Kluźniak and Rappaport 2007):

$$r_{\text{m}} = \xi \left(\frac{\mu^4}{GM\dot{M}^2} \right)^{1/7}, \tag{4}$$

where a dipolar magnetic field has been assumed. The parameter ξ depends on the configuration of the flow: it is $\sim$1 for a spherical inflow (Pringle and Rees 1972), while it can be smaller, $\xi \sim 0.4$, in the case of a Keplerian disk (Ghosh and Lamb 1979). Within r_{m} the motion of the matter is strongly influenced by the geometry of the magnetic field, which rotates rigidly with the star.

Two other relevant radii are the corotation radius, $r_{\text{co}} = (GM/\omega^2)^{1/3}$, and the light-cylinder radius, $r_{\text{lc}} = c/\omega$, where the linear velocity of the rigidly corotating magnetosphere equals, respectively, the Keplerian velocity and the speed of light.

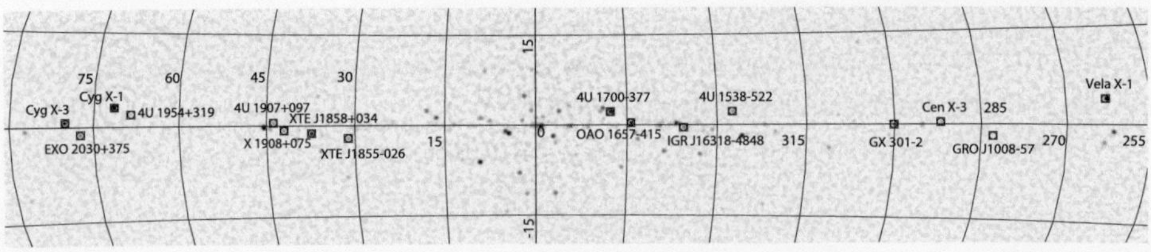

Fig. 1 Image of the central region of the Galactic plane (in Galactic coordinates), obtained by the INTEGRAL observatory in the energy band 17–60 keV. The positions of some of the brightest HMXBs in our Galaxy are indicated. The vast majority of them contain accreting magnetic NSs

2.1 High Mass X-ray Binaries

HMXBs are systems in which the mass donor is typically an OB star with mass greater than 6–7 $M_\odot$. Known HMXBs have orbital periods from 0.2 days (Cyg X-3, with a Wolf-Rayet companion) to almost one year ($P_{\rm orb} \sim 262$ days of SAX J2239+6116) and spin periods ranging from 0.7 s (SMC X-1) up to ~14 ks (1H 1249-637). The detection of X-ray pulsations in most of these systems was one of the first indications for the presence of magnetized NSs, with fields sufficiently strong ($\sim 10^{11-13}$ G) to channel the accretion flow onto the magnetic poles and produce beamed X-ray emission. These magnetic fields are of the same order of those found in young radio pulsars, consistent with the evidence that HMXBs have ages smaller than few tens of Myrs, as implied by the presence of massive stars in these systems.

Extensive surveys of the Galaxy (e.g. Lutovinov et al. 2013) have revealed several tens of persistently bright HMXBs with luminosity $L_{\rm X} \sim 10^{36}$–10^{37} erg s^{-1} (positions of some brightest sources on the sky are shown in Fig. 1). Many of them are located in the spiral arms, associated to regions of recent star formation. Persistent HMXBs typically have OB supergiant companion stars, characterized by strong stellar winds with velocity $v_{\rm w}$ of the order of a thousand km s^{-1} and mass loss rates $\dot{M}_{\rm w} \sim 10^{-7}$–$10^{-5} M_\odot$ yr^{-1}. Such winds provide the matter which is accreted by the NS. Several HMXBs with OB companions are strongly absorbed in the soft X-ray energy range, due to the presence of dense circumstellar material, and could be discovered only recently thanks to observations in the hard X-ray range (Revnivtsev et al. 2003; Walter et al. 2003; Kuulkers 2005).

SFXTs (Supergiant Fast X-ray Transients) have OB-type supergiant companions, like those of the persistent HMXBs, but they are characterized by strong X-ray outbursts of short duration spanning a very large dynamic range—up to 3 or 5 orders of magnitude from quiescence to the peak of their outbursts (see, e.g., Sguera et al. 2006; Sidoli 2012). SFXTs spend most of the time at very low luminosities, and sporadically emit outbursts with duration of the order of few hours or even less (see e.g. review in Paizis and Sidoli 2014), although sometimes superimposed on longer periods of activity. Due to the rarity and short duration of their outbursts, SFXTs escaped detections until recently. Thanks to the extensive coverage of the Galactic plane with the INTEGRAL observatory, this class has now grown up to a dozen of sources (Romano et al. 2014). The mechanism at the basis of SFXT outbursts is not yet understood and several models have been proposed (e.g. in't Zand 2005; Sidoli et al. 2007; Walter and Zurita Heras 2007; Grebenev and Sunyaev 2007; Ducci et al. 2009).

The largest population of HMXBs consists of NSs with Be type companions (e.g. Reig 2011). Most of them are transient systems, which spend a large fraction of time in a low luminosity (or off-) state. Their outbursts are related to the presence of the dense equatorial outflows in the winds of Be stars, which are responsible for the H_α emission lines seen in the optical spectra of these stars. The transient behavior of these XRBs occurs because the NS acquires a higher accretion rate when it crosses the Be equatorial disk and/or because the disk ejection itself is subject to long term variability.

Most NSs in HMXBs accrete matter gravitationally captured from the stellar winds of massive companions which underfill their Roche-lobe (e.g., Davidson and Ostriker 1973). More rarely, the accretion flow is significantly modified by a Roche-lobe influence (e.g. Lamers et al. 1976).

The strong fields of NSs channel the accreting matter onto the magnetic poles, forming an accretion column (e.g. Basko and Sunyaev 1976; Lyubarskii and Sunyaev 1988; Mushtukov et al. 2014). Most of the kinetic energy of the matter falling onto the NS surface should be released as radiation of the accretion column, with a luminosity of the order of $\sim 0.2\dot{M}c^2$ (Eq. (1)). The free fall velocity near the NS surface is $\sim 0.6c$, which corresponds to effective temperatures larger than 10^{12} K (>100 MeV). The observed emission, however, is more concentrated in the 1–20 keV energy band and the spectrum is significantly different from that of a simple blackbody emission—the spectrum is typically a power law ($dN/dE \propto E^{-\Gamma}$) with photon index $\Gamma \sim 1$ up to energies 10–30 keV and an exponential cutoff at higher energies (see, e.g., White et al. 1983; Nagase 1989; Filippova et al. 2005; Caballero and Wilms 2012). The emission pattern is strongly anisotropic, giving rise to easily detectable X-ray pulsations due to the rotation of the NSs. In most sources the pulse profiles show significant variations as a function of energy, luminosity level, orbital phase, and time.

The details of the spectral formation in HMXBs are not yet fully understood, in spite of very serious efforts in this direction (Nagel 1981; Lyubarskii and Sunyaev 1982; Meszaros and Nagel 1985; Burnard et al. 1991; Shibanov et al. 1992; Klein et al. 1996; Becker and Wolff 2005, 2007). Modelling the spectra of the accretion column on NS surfaces of X-ray pulsars is complicated by: a) the presence of powerful radiation pressure, which determines the dynamics of the settling flow, b) the strong magnetic field, which modifies all scattering cross sections and thus the radiation transfer, and c) the presence of fast bulk motion of the infalling matter, which contributes to the Compton upscattering of the outgoing radiation. All these complications should be solved self-consistently.

2.2 Low Mass X-ray Binaries

The companions of NSs in LMXBs are typically late type dwarf stars with mass below 1–2 $M_\odot$. Most LMXBs have orbital periods shorter than one day. The companion stars fill their Roche-lobe and accretion proceeds with the formation of a disk around the compact object. LMXB can have ages up to a few Gyrs, as reflected by their association with the old stellar population in the Galactic bulge and by their presence in globular clusters. The existence of dynamically important magnetic fields in LMXBs, able to channel the accretion flow onto the NS magnetic poles, has been under discussion for a long time because regular pulsations were originally seen only in a few peculiar members of this class, like Her X-1 ($P = 1.24$ s), 4U 1626–67 ($P = 7.7$ s) and GX 1+4 ($P = 130$ s) (see Liu et al. 2007 and ref. therein). Extensive searches for pulsations in other bright LMXBs were unsuccessful (see, e.g., Mereghetti and Grindlay 1987; Vaughan et al. 1994) and the presence of NSs, rather than accreting black holes, in these systems could only be established through the observation of type I X-ray bursts—thermonuclear explosions in the accreted matter which accumulates onto the NS surface (see, e.g., reviews in Lewin et al. 1993; Cumming 2004).

Thanks to the large effective area provided by the instruments on the RXTE satellite, X-ray pulsations were finally discovered in LMXBs (Wijnands and van der Klis 1998). At the moment we know 15 systems of this class in which the NSs create persistent or intermittent

X-ray pulsations with periods of a few milliseconds (see, e.g., Patruno and Watts 2012) and several sources which show millisecond pulsations during type I bursts (e.g. Strohmayer 2001). These findings show that the NSs in LMXBs have spin periods much shorter than those in HMXBs. Their fast rotation is due to the spin-up torque applied on the NS by the rapidly rotating Keplerian accretion disk, which extends close to the star surface. This implies magnetic fields $B \sim 10^{8-10}$ G, smaller than those found in HMXBs, which could be a consequence of the old age of the NSs in LMXBs, possibly coupled to the field decay induced by the accretion process (e.g. Taam and van den Heuvel 1986; Harding and Lai 2006). Such low magnetic fields are at the basis of the well established recycling scenario, which explains the millisecond radio pulsars as old NSs spun-up by the accretion process in LMXBs (Bisnovatyi-Kogan and Komberg 1976; Bhattacharya and van den Heuvel 1991; Srinivasan 2010).

The weak magnetic field of these NSs is also at the basis of the different X-ray spectral properties of LMXBs and HMXBs. In general, at energies below ~20 keV the spectra of LMXBs are much softer than those of HMXBs. This is due to the fact that a large fraction of the observed X-rays originates in optically thick accretion disks, which in LMXBs extend close the compact object where they reach temperatures of only a few keV.

3 Resonant Cyclotron Features in the X-Ray Spectra of Accreting NSs

The energy spectra of several accreting magnetic NSs show the presence of absorption features at energies of tens of keV. Discovered more than thirty years ago (Truemper et al. 1978; Wheaton et al. 1979), they were already expected by theoreticians (Gnedin and Sunyaev 1974). These features were immediately interpreted as due to cyclotron resonant scattering of the outgoing radiation by electrons in a strong magnetic field and provided the first direct evidence for magnetic fields of the order of 10^{12-13} G in the line-forming regions. The line energies are determined by the separation of the quantized Landau energy levels of electrons, which, in a magnetic field of strength B, is given by:

$$E_c = \hbar \frac{eB}{m_e c} = 11.6 \frac{B}{10^{12}\ \mathrm{G}}\ \mathrm{keV} \tag{5}$$

Due to the gravitational redshift z at the position where the scattering occurs, the lines in the observer's frame of reference appear at $E_c/(1+z)$.

Over the last decades, cyclotron absorption features have been detected from more than twenty sources (see Table 1). In a few sources one or more harmonics of the fundamental line are also seen (Santangelo et al. 1999; Tsygankov et al. 2006). One of the best examples is presented in Fig. 2. Note that the line energies given in Table 1 are only approximate values, since many sources show line energy variations as a function of the spin phase, time, and/or flux. In addition, the exact values of the line parameters derived from the spectral fits depend on the spectral model used to describe the continuum. Despite these uncertainties, it is clear that the observed line energies imply magnetic fields in the range $\sim 10^{12}$–10^{13} G, both for NSs with Be and supergiant (SG) companions. Note that Her X-1 and GX 1+4, the only LMXB showing cyclotron lines, are not representative members of the LMXB population: they resulted from different evolutionary histories and have magnetic fields similar to those of HMXBs. Among the most recent discoveries, it is worth noting that the first cyclotron line discovered in a member of the SFXT class indicates a "standard" field value (Bhalerao et al. 2014). This does not support the models invoking strong magnetic fields to explain these sources (e.g. Bozzo et al. 2008).

Table 1 XRBs with cyclotron resonance features (updated from Caballero and Wilms 2012). The 5th column gives the number of harmonics. The + and − symbols in the 6th column indicate a positive or negative correlation between E_c and L_X (tentative reports are indicated by a question mark); the = symbol indicates that E_c has been seen to stay constant during luminosity changes

Name	Type	E_{cyc} keV	$B/(1+z)$ 10^{12} G	Num. harm.	E_c–L_X correl.	References
Swift J1626.6–5156	Be	10	0.9	1	+	DeCesar et al. (2013)
4U 0115+63	Be	11	1	4	−?	Wheaton et al. (1979), Müller et al. (2013b)
KS 1947+300	Be	12	1			Fürst et al. (2014a)
IGR J17544–2619	SFXT	17	1.5			Bhalerao et al. (2014)
4U 1907+09	SG	18	1.5	1		Makishima and Mihara (1992), Cusumano et al. (1998)
4U 1538–52	SG	23	2		+	Clark et al. (1990), Hemphill et al. (2014)
Vela X-1	SG	25	2.1	1	+	Kendziorra et al. (1992), Fürst et al. (2014b)
V 0332+53	Be	26	2.2	2	−	Makishima et al. (1990), Tsygankov et al. (2010)
X Per	Be	29?	2.5			Coburn et al. (2001), Doroshenko et al. (2012b)
Cen X-3	SG	30	2.6			Santangelo et al. (1998), Suchy et al. (2008)
Cep X-4	Be	31	2.7			Mihara et al. (1991), McBride et al. (2007)
RX J0520.5–6932	Be	32	2.8			Tendulkar et al. (2014)
RX J0440.9+4431	Be	32	2.8			Tsygankov et al. (2012)
IGR J16493–4348	SG	32	2.8			D'Aì et al. (2011)
MXB 0656–072	Be	33	2.8			Heindl et al. (2003), McBride et al. (2006)
GX 301–2	SG	35	3			Makishima and Mihara (1992), Kreykenbohm et al. (2004)
XTE J1946+274	Be	36	3.1		+?	Heindl et al. (2001), Müller et al. (2012)
4U 1626–67	LMXB	37	3.2		=	Orlandini et al. (1998), Camero-Arranz et al. (2012b)
Her X-1	LMXB	41	3.5		+	Truemper et al. (1978), Staubert et al. (2007)
A 0535+26	Be	46	4	1	=	Kendziorra et al. (1992), Müller et al. (2013a)
GX 304–1	Be	51	4.4		+	Yamamoto et al. (2011), Klochkov et al. (2012)
1A 1118–61	Be	55	4.7			Doroshenko et al. (2010b), Maitra et al. (2012)
GRO J1008–57	Be	78	6.7			Shrader et al. (1999), Bellm et al. (2014)

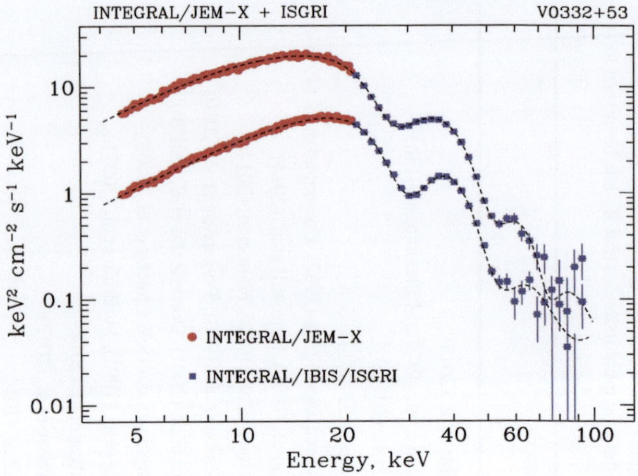

Fig. 2 Energy spectrum of the accretion powered pulsar V0332+53 at two intensity levels measured with the INTEGRAL observatory. At least two cyclotron absorption lines are clearly visible (from Tsygankov et al. 2006)

Variations of the line properties as a function of the source luminosity are particularly interesting since, in principle, they offer the possibility to study how the geometry of the accretion region changes at different mass accretion rates. In the simplest interpretation, the observed variations of line energy, E_c, as a function of the luminosity, L_X, trace the height of the X-ray emitting shock in the accretion column. At luminosities above the local critical value ($L_X \gtrsim 10^{37}$ erg s^{-1}), as the mass accretion rate increases the radiation pressure moves the shock farther from the NS surface, where the magnetic field is weaker. Hence an anticorrelation between E_c and L_X is expected (Basko and Sunyaev 1976; Burnard et al. 1991). This behavior has been observed in V0332+53 (Tsygankov et al. 2006) and in 4U 0115+63 (Nakajima et al. 2006). However, the analysis of another outburst from the latter source indicates that the reported anticorrelation might be an artifact caused by an improper modelling of the underlying variations of the continuum spectrum (Müller et al. 2013b). A positive correlation between E_c and L_X has instead been seen in Her X-1 (Staubert et al. 2007) and in a few other sources (see Table 1). Their different behavior has been attributed to the lower luminosity of these sources, in which gas pressure and Coulomb scattering become important in shaping the geometry and emission pattern of the accretion column (Becker et al. 2012).

There is general agreement that the observed lines are due to resonant electron scattering in a strong magnetic field and models based on the transmission of radiation through a highly magnetized medium in the accretion column can account for most of the observations (see e.g. Schönherr et al. 2007; Nishimura 2008). However, the simple picture outlined above predicts a luminosity-dependence of E_c at high accretion rates stronger than the observed one (typically E_c varies by less than 20–30 % when the luminosity changes by an order of magnitude). A recent model to overcome these problems is presented by Nishimura (2014), while a different explanation is based on the idea that the cyclotron lines are formed as a result of reflection of the accretion column X-ray emission from magnetized matter on the NS surface (Poutanen et al. 2013).

4 Interaction Between NS Magnetosphere and Accretion Flow

Although the details of the interaction between the rotating magnetosphere and the accretion flow are quite complicated and not yet fully understood (see, e.g., Uzdensky 2004; Lai

2014), it is generally assumed that accretion onto the NS surface can occur if the magnetospheric radius (Eq. (4)) is smaller than the corotation radius. If, instead, $r_m > r_{co}$, the centrifugal barrier should stop the flow and inhibit continuous accretion (Illarionov and Sunyaev 1975; Stella et al. 1986). In this case, non-stationary accretion might still proceed as the matter piles up around the intrinsically unstable magnetospheric boundary. Depending on its specific angular momentum, the captured plasma may form a quasi-spherical shell (Pringle and Rees 1972; Ikhsanov 2007; Shakura et al. 2012) or reside in a so-called "dead accretion disk" (Sunyaev and Shakura 1977; Baan 1979; D'Angelo and Spruit 2010, 2012).

Accretion can proceed in two ways: either quasi-spherically or via a geometrically-thin disk. The latter scenario typically occurs when the mass donor star fills its Roche-lobe, as it happens in most LMXBs. In the case of wind-accreting HMXBs, quasi-spherical accretion can occur if the angular momentum of the mass captured within the accretion radius is sufficiently small. This can be evaluated by considering the circularization radius, i.e. the minimum distance from the NS which the accreting matter can reach without loss of angular momentum. The circularization radius, r_{circ}, can be roughly estimated by equating the specific angular momentum of the captured stellar wind matter to that on a Keplerian orbit (see, e.g., Frank et al. 1992):

$$j \sim \frac{\pi r_a^2}{2P_{orb}} \sim \sqrt{GMr_{circ}} \tag{6}$$

This leads to:

$$r_{circ} \sim \frac{4\pi^2 G^3 M^3}{P_{orb}^2 v_w^8} \tag{7}$$

$$r_{circ} \sim 3.3 \times 10^6 \left(\frac{1\ \mathrm{day}}{P_{orb}}\right)^2 \left(\frac{1000\ \mathrm{km\,s^{-1}}}{v_w}\right)^8 \mathrm{cm} \tag{8}$$

For typical values of NS magnetic fields, mass accretion rates, and stellar wind velocities of young OB-type stars, it turns out that, for wide binaries ($P_{orb} >$ a few days), a disk cannot form because $r_{circ} < r_m$. In this case the accreting matter should settle onto the rotating magnetosphere in some kind of a quasi-spherical flow.

The resulting interaction is a long standing problem and still a matter of debate (see, e.g., Arons and Lea 1976; Lamb et al. 1977; Davies et al. 1979; Illarionov and Kompaneets 1990; Ziolkowski 1985; Bozzo et al. 2008; Shakura et al. 2012, 2013). Unsettled questions on details of the magnetosphere–flow interaction sometimes lead to controversial suggestions about the presence of ultrastrong magnetic fields in pulsars with long spin periods (see, e.g. Finger et al. 2010; Doroshenko et al. 2010a; Reig et al. 2012; Ho et al. 2014). The root of this problem is in the assumptions on the values of the torques exerted by the accreting matter settling to the NS magnetosphere.

Contrary to the disk accretion case, where the torques are relatively well understood and the predicted scalings confirmed by observations (see Sect. 4), the issue of the net torque acting on the NS is much less clear in the case of quasi-spherical accretion (see discussion of this problem in Shakura et al. 2012, 2013, Postnov et al. 2014).

In the cases of accretion from a companion star in tight binaries (where Roche-lobe overflow occurs), from the equatorial disks around Be stars, and from low mass companions, the specific angular momentum of the gravitationally-captured matter can be sufficiently large to form a disk before reaching the NS magnetosphere.

Some early works (e.g. Ghosh and Lamb 1978, 1979) assumed that the NS magnetic field can penetrate the disk over a large range of radii, forming a relatively wide transition zone.

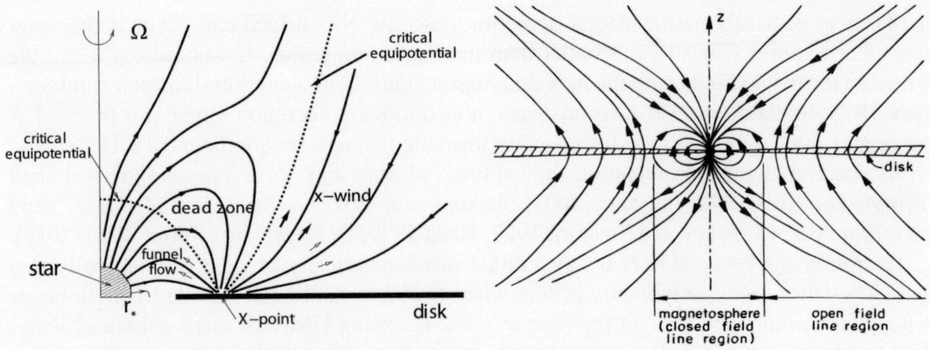

Fig. 3 Models of magnetosphere-accretion disk interaction of Shu et al. (1994) (*left panel*) and Lovelace et al. (1995) (*right panel*). Both models imply that the accretion disk matter is frozen into the NS magnetic field lines only over a relatively small transition region

It was later shown that this scenario requires an unphysically high magnetic diffusivity in the accretion disk (see, e.g., Wang 1995; Lesur and Longaretti 2009). Calculations assuming a more realistic magnetic diffusivity indicate that the NS magnetic field can penetrate the disk only over a quite small range of radii (Campbell 1992; Shu et al. 1994; Lovelace et al. 1995), comparable to the thermal disk scale height $h \sim c_s/\omega_K$, where c_s is the sound speed in the disk and ω_K is the Keplerian frequency. In these models the inner part of the flow is frozen into the closed field lines magnetosphere, and the outer part of the flow contains open field lines leading to the possibility of magnetically driven outflows (see Fig. 3).

The thickness of this transition region influences the structure of the accretion column onto the NS. If the area of the footprint of the accretion flow is small, the release of energy brought by the flow can be locally super-Eddington and can create a radiative shock which stops the infalling plasma. Despite significant theoretical efforts (e.g. Basko and Sunyaev 1976; Ghosh and Lamb 1978, 1979; Campbell 1992; Shu et al. 1994; Lovelace et al. 1995) the width of the transition region remains uncertain, and until recently could not be estimated observationally. A method to estimate the thickness of the transition region based on measurements of the cooling time of matter in the accretion column has been proposed for magnetic white dwarfs (Semena and Revnivtsev 2012; Semena et al. 2014). The inferred plasma penetration depth at the boundary of the magnetosphere is in general agreement with earlier theoretical (Shu et al. 1994; Lovelace et al. 1995). Extending these results to accreting magnetic NSs indicates that the footprint of their accretion columns might occupy an area as small as $<10^{-6}$ of the total NS surface.

4.1 Spin-Period Variations

Matter that moves toward the NS carries angular momentum, which can be added to that of the NS. If the accretion occurs via a Keplerian disk, the spin-up torque acting on the NS can be estimated as $K_{\rm su} \approx \dot{M}\sqrt{GMr_{\rm m}}$ (Pringle and Rees 1972). On the other hand, the rotating magnetosphere can lose angular momentum via different mechanisms. For example, it can transfer angular momentum via large scale magnetic field threading the external regions of the accretion disk, which have a slower rotation (e.g. Ghosh and Lamb 1979; Lovelace et al. 1995; Wang 1995), or it can lose angular momentum by some matter outflow (e.g. Illarionov and Kompaneets 1990). The spin-down torque $K_{\rm sd}$ acting on the NS depends on how the accretion in the binary occurs (e.g. via Keplerian disk or via quasi-spherical flow) and is still a matter of debate (see e.g. Davidson and Ostriker 1973; Kundt 1976; Lipunov 1981; Bisnovatyi-Kogan 1991; Wang 1995; Li and Wang 1996; Kluźniak and Rappaport 2007; Shakura et al. 2012). Leaving aside these complications, we can write the evolution of the spin period of NS accreting from a disk in a generalized form:

$$I\frac{d\omega}{dt} = k_{\rm m}\dot{M}\sqrt{GMr_{\rm m}} - K_{\rm sd} \tag{9}$$

where $k_{\rm m}$ is a numerical constant of the order of unity.

If the spin-up torque is significantly higher than the spin-down torque (e.g. during episodes of high mass accretion rate), the dependence of the spin-up rate on the source luminosity/mass accretion rate can be expressed as follows:

$$\dot{\nu} \sim \frac{\sqrt{GMr_{\rm m}}}{2\pi 0.4MR^2}\dot{M} \sim \frac{G^{3/7}\mu^{2/7}\dot{M}^{6/7}}{2\pi 0.4M^{4/7}R^2} \tag{10}$$

$$\dot{\nu} \sim 3.5 \times 10^{-13}\mu_{30}^{2/7}\left(\frac{\dot{M}}{10^{16}\ {\rm g\,s^{-1}}}\right)^{6/7}\ {\rm s^{-2}} \tag{11}$$

$$\dot{\nu} \sim 2.5 \times 10^{-13}\mu_{30}^{2/7}\left(\frac{L_{\rm X}}{10^{36}\ {\rm erg\,s^{-1}}}\right)^{6/7}\ {\rm s^{-2}} \tag{12}$$

The time scale $t_{\rm su} = \nu/\dot{\nu}$ for spinning-up the accreting NS is

$$t_{\rm su} \sim 3 \times 10^5 \nu^{4/3}\left(\frac{10^{37}{\rm erg\,s^{-1}}}{L_{\rm X}}\right)\left(\frac{r_{\rm co}}{r_{\rm m}}\right)^{1/2}\ {\rm yr} \tag{13}$$

For typical mass accretion rates occurring in high-luminosity sources (10^{37}–10^{38} erg s^{-1}), this time scale is short and spin-up should be clearly observable. Indeed, the spin-up of Cen X-3 was already revealed since its first observations (see Bildsten et al. 1997 for a compilation of spin-period time histories of several accretion-powered pulsars).

To illustrate the dependence of spin-up rate on X-ray luminosity outlined in Eq. (10), let us consider the case of A0535+262, one of the brightest HMXB transients. In this system the NS accretes from its Be-type companion via a thin disk. The mass accretion is not continuous, but it occurs via rare violent outbursts, during which the accretion rate changes by more than three orders of magnitude, resulting in luminosities from (0.5–1) $\times$ 10^{34} to $\sim$$10^{38}$ erg s^{-1}. The dependence of its spin-up rate on the X-ray luminosity gives us a direct probe of the model of disk–magnetosphere interaction and of the properties of the magnetosphere. When A0535+262 is at high luminosity, the accretion disk squeezes the magnetosphere to sizes much smaller than the corotation radius, thus reducing the flow–magnetosphere interaction to the simple scenario outlined above. Figure 4 shows how the observed spin-up rate depends on the X-ray flux, as observed during the bright phase of an outburst which occurred in 1996 (Bildsten et al. 1997). The dotted line indicates the relation $\dot{\nu} \propto \dot{M}^{6/7}$, deduced from the simple model described above. The observed points closely follow the model prediction and yield a NS surface magnetic field $B \sim 10^{12}$ G, which roughly agrees with the value derived from studies of cyclotron absorption lines.

A complete description of the spin-up/spin-down behavior of accreting NSs is much more complicated than the simple picture outlined above, owing to the interaction between the accretion flow and the magnetosphere, which is particularly complex when the NS is rotating close to the equilibrium period, i.e. when $r_{\rm m} \approx r_{\rm co}$, and in the propeller stage, when $r_{\rm m} > r_{\rm co}$. These difficulties are illustrated, e.g., by the case of KS 1947+300, for which the

Fig. 4 Dependence of NS spin-up rate on X-ray flux (i.e. mass accretion rate). The *dotted line* indicates the prediction of the simple model of interaction of a dipole magnetosphere with a Keplerian disk (from Bildsten et al. 1997)

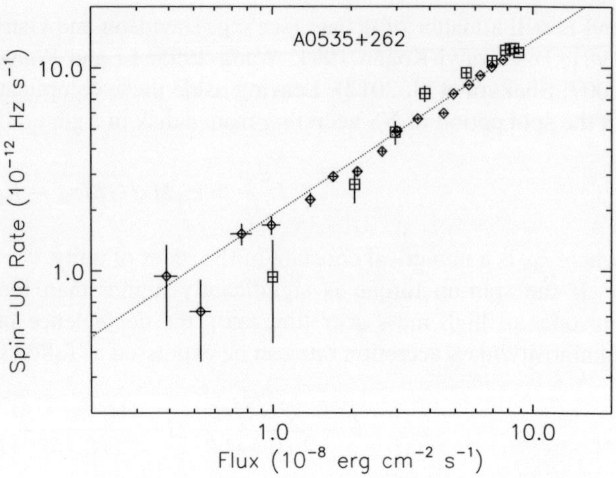

magnetic field derived from the $\dot{\nu}$–L_X correlation (Tsygankov and Lutovinov 2005) is more than one order of magnitude higher than that indicated by the cyclotron line detected in this source.

The existence of counteracting torques (Eq. (9)) implies that: (1) there should be time periods of torque reversal, and (2) NSs should have some equilibrium spin periods. Indeed, observations do demonstrate time periods with zero time derivative of the NS spin value in some disk-accreting systems (e.g. Parmar et al. 1989; Wilson et al. 2002; Baykal et al. 2002). Alternating periods of spin-up and spin-down are common in wind-accreting systems, where the angular momentum of the gravitationally-captured matter is subject to large fluctuations of both signs (see, e.g., the spin history of the wind accreting NS in Vela X-1 in the upper panel of Fig. 5), but they have also been observed in NSs which are known to accrete through a disk (an example is shown in the bottom panel of Fig. 5). A comprehensive review of observational facts about spin-up and spin-down histories can be found in, e.g., Bildsten et al. (1997).

In these cases the observed luminosity/mass accretion rate values were used to make an estimate of the NS magnetic field, adopting some prescription of the spin-down torque K_{sd}. Application of this approach to A0535+262 provides a NS magnetic field value in agreement with that known from position of cyclotron absorption line Doroshenko et al. (2014). The problem of torque reversals in wind-accreting (without Keplerian accretion disk) binaries has been recently re-examined in the context of the quasi-spherical subsonic accretion regime by Shakura et al. (2012). Application of this model to Vela X-1 and GX 301–2 allowed these authors to estimate magnetic fields consistent with those derived from the cyclotron lines.

5 Aperiodic X-Ray Flux Variations as a Tool to Measure NS Magnetospheres

The availability of high-quality data on accreting X-ray binaries has made it possible to investigate the NS magnetospheres using new methods based on the study of their aperiodic X-ray variability. Aperiodic variability in the X-ray flux of accreting sources was discovered at the dawn of X-ray astronomy (e.g., Oda et al. 1971) and since then it has been extensively studied, mostly in the frequency domain, through the use of Fourier analysis techniques

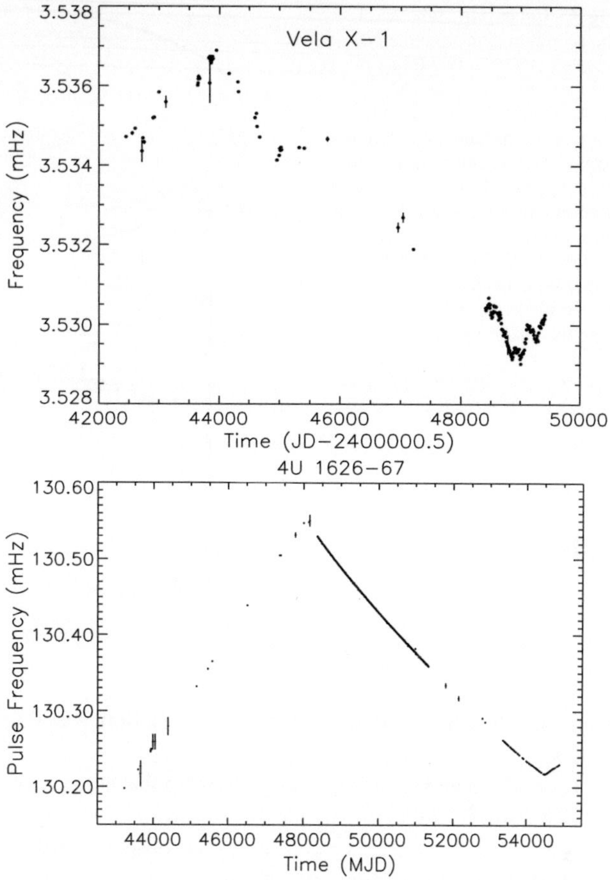

Fig. 5 Examples of spin-up/spin-down transitions of the wind accreting magnetic NS in Vela X-1 (*top panel*, from Bildsten et al. 1997) and the disk accreting magnetic NS in 4U 1626–67 (*bottom panel*, from Camero-Arranz et al. 2010)

(van der Klis 1989). The noise power spectrum in accreting sources has been shown to follow a power law in broad range of frequencies.

It is now widely accepted that this variability originates from the modulation of the instantaneous values of the mass accretion rate at different distances from the central object and their subsequent multiplicative superposition (so called model of propagating fluctuations; Lyubarskii 1997; Churazov et al. 2001; Arévalo and Uttley 2006; Revnivtsev 2008). In the framework of this model, the varying luminosity originates in the central parts of the accretion flow (i.e. the accretion column in the case of X-ray pulsars), but the modulations of the mass accretion rate are inserted into the flow at different (including large) distances from the central object, as a result of the stochastic nature of the viscosity in accretion disks (see, e.g., Balbus and Hawley 1991; Armitage and Reynolds 2003; Hirose et al. 2006).

The fastest variations, originating closer to the compact object, modulate the accretion rate in the disk, which is incoming to these regions from the outer disk regions. This model predicts that, if the accretion disk is truncated at some distance from the compact object, the broad band power spectrum of its luminosity variations should show a break (or steepening) resulting from the lack of variability at the highest frequencies (see Fig. 6). The break frequency, f_b, should correspond to the time scale at which the inner part of the accretion flow modulates the mass accretion rate. As the truncation radius moves inward, f_b should

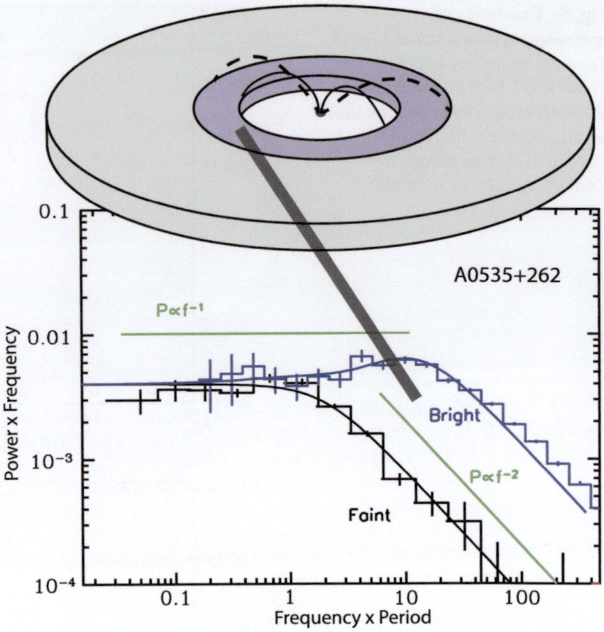

Fig. 6 *Upper part*: scheme of the accretion disk around a NS magnetosphere at two levels of the mass accretion rate. At high accretion rate the inner radius of the disk shrinks and an additional part of the flow with the fastest noise appears. *Lower part*: power spectra (pulse variations removed) of the accreting X-ray pulsar A0535+262 at different X-ray luminosities. The *thick line* relates the inner part of the disk with the fastest variability, not present during low accretion rate periods, when the magnetosphere pushes the disk further away from the NS

increase. Observational support of this prediction was demonstrated by Revnivtsev et al. (2009).

If the characteristic frequency f_0 of the noise at the magnetospheric boundary $r_{\rm m}$ is proportional to the frequency of the Keplerian rotation $\nu_{\rm K}$ of matter at the inner edge of the accretion disk $r_{\rm in} \approx r_{\rm m}$, we can relate the observed break frequency to the instantaneous value of the mass accretion rate $\dot{M}$:

$$2\pi \nu_{\rm K} = (GM)^{1/2} r_{\rm m}^{-3/2} \tag{14}$$

and the break frequency will follow the dependence:

$$f_{\rm b} \propto f_0 \propto \nu_{\rm K}(r_{\rm in}) \propto (GM)^{5/7} \mu^{-6/7} \dot{M}^{3/7} \tag{15}$$

This relation was observed in transiently accreting pulsars (see e.g. Revnivtsev et al. 2009; Tsygankov et al. 2012; Doroshenko et al. 2014, and Fig. 7).

In addition to the broad band noise, accreting X-ray pulsars sometimes show quasi-periodic oscillations (QPO) of their fluxes (see e.g. Shinoda et al. 1990; Finger et al. 1996; Camero-Arranz et al. 2012a). It was found that the QPO frequency varies during the outbursts of the transient A0535+262 and its value strongly depends on the X-ray flux. It is remarkable that the dependence of the QPO frequency on the accretion rate is very similar to that of the frequency of the break in the power spectrum of the source flux variability, which hints on their common origin. It was proposed that the QPO might originate at the inner boundary of the accretion disk, truncated by the NS magnetosphere at the beat frequency between the Keplerian frequency at disk inner edge and the neutron star rotation frequency (Finger et al. 1996).

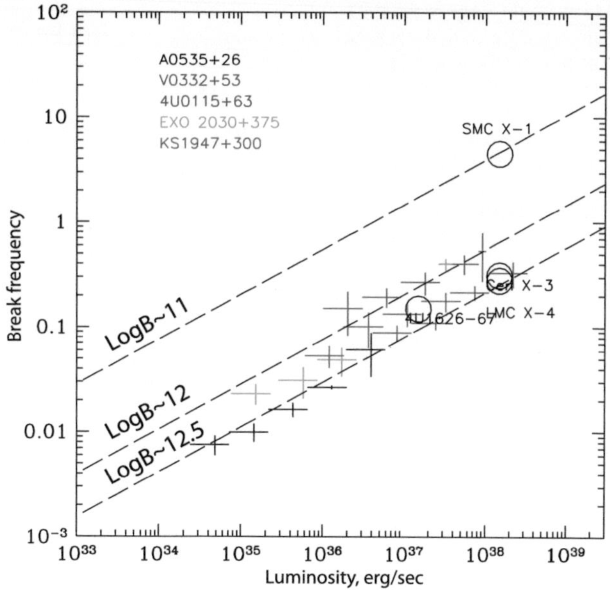

Fig. 7 Luminosity dependence of the break frequency in power spectra of disk accreting X-ray pulsars. *Dashed lines* denote approximate positions of breaks predicted by a simple model of Keplerian disk–dipole magnetosphere interaction with different strengths of NS surface magnetic field. The exact position of these lines should be treated with caution due to uncertainties of NS parameters and properties of the magnetosphere

6 Magnetic Propeller Effect

Due to the increase of magnetospheric radius with the decrease of the mass accretion rate (Eq. (4)), at some stage of transiently accreting NSs (e.g. in Be HMXBs during outbursts or in LMXB transients) the magnetospheric radius can reach the corotation radius and stop the direct accretion regime. This is the so called "propeller effect". The limiting mass accretion rate $\dot{M}$, and the corresponding luminosity, can be estimated by the condition $r_m \sim r_{co}$:

$$\dot{M} \sim \xi^{7/2} \frac{(2\pi)^{7/3} \mu^2}{G^{5/3} M^{5/3} P^{7/3}} \tag{16}$$

$$L_X \sim \xi^{7/2} 2.3 \times 10^{33} \mu_{30}^2 \left(\frac{100\ \mathrm{s}}{P}\right)^{7/3} \ \mathrm{erg\,s^{-1}} \tag{17}$$

Below we consider several consequences of the propeller effect which can be probed with observations.

6.1 Luminosity Drops in HMXBs and LMXBs

When the magnetospheric radius starts to exceed the corotation radius, one should expect to see an abrupt drop of the source luminosity or a switch to unsteady accretion with large luminosity swings. Indeed, at this stage the matter gains angular momentum from the rotating magnetosphere and can not move directly toward the NS. Relatively small variations of the incoming mass rate can lead to large variations of the X-ray luminosity.

Such drops were indeed observed during the decreasing phases of the outbursts of several Be binaries, like, e.g. V0332+53 (Stella et al. 1986), 4U0115+63 (Campana et al. 2001), Vela X-1 (Doroshenko et al. 2011), 4U1907+09 (Doroshenko et al. 2012a). This effect is not easy to catch because it occurs in a short time interval, when the source is crossing the above luminosity limit. An additional complication comes from the fact that when XRBs

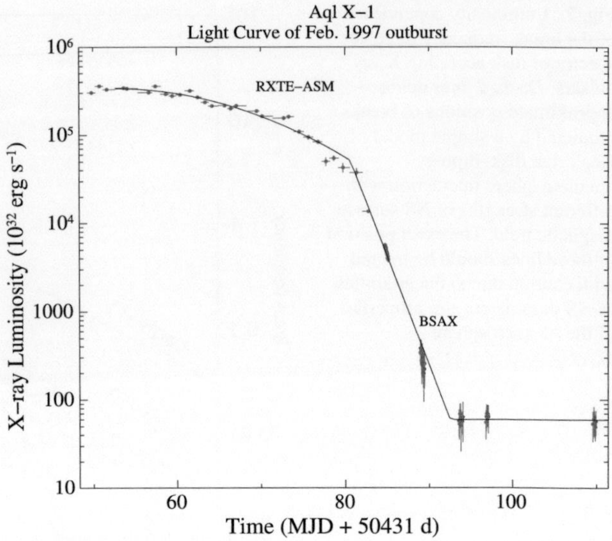

Fig. 8 Light curve of the 1997 outburst of the LMXB transient Aql X-1. After a relatively smooth decline, an abrupt drop of X-ray luminosity can be seen around $L_X \sim 5 \times 10^{36}$ erg s^{-1}. This behavior can be explained by the propeller effect of a rotating magnetosphere with $B \sim (1\text{–}4) \times 10^8$ G at the NS surface (from Campana et al. 2014)

reach low luminosity levels ($L_X < 10^{33}$–10^{34} erg s^{-1}), the contribution provided by X-ray emission of the companion star may become non-negligible.

It is possible that the pure propeller effect, causing the complete ejection of gravitationally captured matter, is unlikely to be realized unless the compact magnetized rotator is really fast (as in the case of the rapidly rotating magnetized white dwarf in the binary AE Aqr, see, e.g., Wynn et al. 1997). In the case of slow rotators some residual accretion might still be possible from a quasi-spherical reservoir above the magnetosphere (see e.g. Shakura et al. 2012) or from so-called "dead-disks" (Illarionov and Sunyaev 1975; D'Angelo and Spruit 2010, 2012).

If the mass supply rate at the magnetospheric boundary persists for some time, one might expect large amplitude quasi-cyclic variations of X-ray luminosity, due to intermittent matter penetration (Baan 1979; D'Angelo and Spruit 2010, 2012). Such variations were observed in several cases: A0535+262 (Caballero et al. 2008; Postnov et al. 2008), EXO 2030+375 (Klochkov et al. 2011).

The influence of the propeller effect has also been observed in transiently accreting LMXBs (Fig. 8), where abrupt drops of X-ray luminosity were seen in the decaying parts of the outbursts (see, e.g., Cui 1997; Gilfanov et al. 1998; Asai et al. 2013; Campana et al. 2014), leading to estimated magnetic fields of a few 10^8 G. Note, that in the case of fast pulsars in LMXBs the inner edge of the disk is much closer to the NS surface than in HMXBs. Thus the luminosity variations associated to the propeller effect are smaller and more difficult to detect. Also considering that the determination of r_m is subject to larger uncertainties in these systems, the above estimates should be taken with some caution (see, e.g., discussion in Patruno and Watts 2012).

6.2 Influence of the Propeller Effect on XRBs Populations

The presence of rotating NS magnetospheres should prevent persistent accretion at low rates (see, e.g., discussion of this point in Lipunov 1982, Stella et al. 1986, Shtykovskiy and Gilfanov 2005; numerical simulation predict similar effect in LMXB populations, see, e.g., Kuranov et al. 2014). In principle, this prediction could be tested with statistical analysis of well defined samples of XRBs, which should reveal a lack of objects in some regions of parameters (e.g. luminosity, spin-period, magnetic field, orbital period). However, such studies are complicated by observational biases and by large uncertainties on many parameters.

Attempts to obtain a well defined sample of X-ray binaries with the help of sky surveys can be traced back to the first X-ray sky surveys (see e.g. Matilsky et al. 1973). However, it took a long time to measure the relevant physical parameters of these binaries (orbital period, distance, etc…), which are needed for quantitative comparisons with the predictions of population synthesis models.

The simplest comparison between observations and population synthesis can be done for the distribution of X-ray luminosities. This is more difficult do for the sources in our Galaxy, due to the uncertainties on their distances, but it is much easier for sources in the Magellanic Clouds. Shtykovskiy and Gilfanov (2005) have shown that the flattening below $L_X < 10^{35}$ erg s^{-1} of the luminosity function of LMC X-ray sources might indicate an influence of the propeller effect. More solid confirmation of this effect is required on much higher statistics of sources at such low luminosities.

Population studies can also be used to test some of the models proposed to explain the SFXTs invoking the role of the rotating magnetosphere in the inhibition of accretion (e.g. Grebenev and Sunyaev 2007; Bozzo et al. 2008). In these models the matter gradually accumulates above the magnetospheric boundary and occasionally finds its way to the NS surface, giving rise to the bright flares which characterize these sources.

A simple picture of stellar wind accretion predicts the existence of an "allowed" region in the P_{orb}–L_X parameter space, determined by the minimum orbital period of a NS with a massive companion not filling its Roche lobe, and the minimum mass accretion rate which can be supplied by such a massive star (Bhadkamkar and Ghosh 2012; Lutovinov et al. 2013). This is based on the fact that the accretion rate onto a NSs at distance a from a companion of mass M_2 depends almost only on the mass loss rate of the secondary $\dot{M}_2$: $L_X \propto a^{-2} f(M_2)$. For OB stars with wind velocities $v_w \sim 1000$ km s^{-1}, the mass loss rate $\dot{M}_2 \propto M_2^{\alpha}$, with $\alpha \sim 2.76$. Therefore, for any fixed separation a (i.e. approximately fixed P_{orb}), the NSs with the smallest mass companions should have the lowest X-ray luminosities. This lower boundary on L_X for persistently accreting sources should have a functional form $L_X \propto a^{-2} \propto P_{orb}^{-4/3}$. If an X-ray luminosity below this limit is observed, some additional mechanism must be invoked to reduce the NS mass accretion rate.

This prediction can be tested using the currently best available flux-limited sample of accreting NSs in HMXBs presented in Lutovinov et al. (2013). This survey is complete over the whole Galaxy down to luminosities (0.5–1) $\times 10^{35}$ erg s^{-1} in the energy range 17–60 keV. As shown in Fig. 9 the persistent NSs with OB giant companions in our Galaxy populate the "allowed" area in the P_{orb}–L_X diagram. On the other hand, a few sources belonging to the class of supergiant fast X-ray transients lie below the L_X lower boundary. According to above mentioned arguments, this suggests that mass accretion onto the NS surface is inhibited (or strongly reduced) as a result of magnetospheric/centrifugal inhibition of accretion.

7 Conclusions

Since the time of their discovery in the late 60s, XRBs have been used as natural laboratories for studies of matter in conditions of extreme density and magnetic fields. A substantial fraction of them contains accreting NSs endowed with magnetic fields sufficiently strong to significantly affect their observed properties. The presence of periodic X-ray pulsations

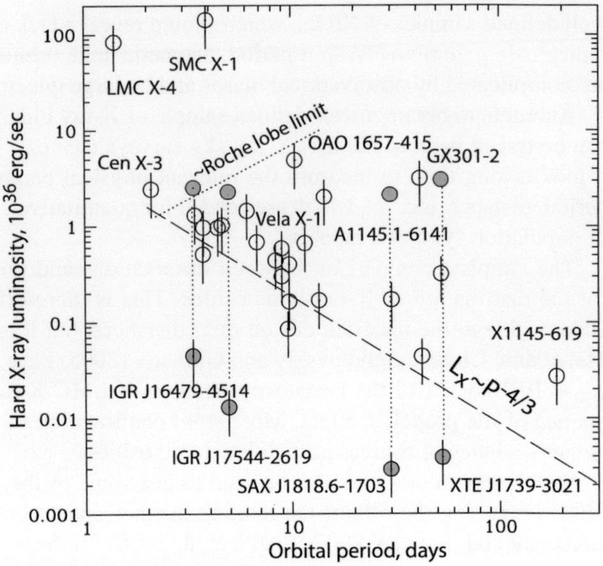

Fig. 9 Hard X-ray luminosity versus orbital period of persistent HMXBs of our Galaxy. Sources below the *dotted line* accrete from stellar wind, sources above this line accrete via Roche lobe overflow. The *dashed line* denotes the approximate lower limit of *persistent* X-ray luminosity which can be provided by accretion from minimally outflowing massive stars (i.e. from young stars with masses around $M_2 \sim 7\text{–}8 M_\odot$). *Filled circles* connected with *vertical dotted lines* denote positions of supergiant fast X-ray transients at their average (*lower circle*) and maximum (*upper circle*) luminosity state. Adapted from Lutovinov et al. (2013)

provided immediate evidence for magnetic fields of the order $\sim 10^{12}$ G in the HMXBs, similar to those deduced in a completely independent way from the spin-down rate of young radio pulsars. At the same time, the rarity of pulsations in LMXBs pointed to much smaller magnetic fields for the NSs contained in these, generally older, systems.

The detection of cyclotron resonance features gives the most direct way to estimate the magnetic field in XRBs. The wealth of good observational data on XRB cyclotron lines is now posing challenges to the theory developed for the spectral formation in accreting magnetized NSs, despite the high level of complexity now reached by these models. However, despite some difficulties in the detailed modelization, it is well established that the observed line energies correspond to magnetic fields in the range $\sim 10^{12}\text{–}10^{13}$ G.

Other ways to estimate the magnetic field in XRBs are less direct than cyclotron line measurements (and often more model-dependent), but have the advantage that they can be applied over a larger range of magnetic field intensities. The study of spin-up and/or spin-down rates in connection with the source luminosity gives the possibility to estimate the large scale magnetic field of the NS, because the exchange of angular momentum between the NS and the surrounding matter is strongly mediated by the magnetic field. The results obtained by these analysis are rather well established in the case of disk accretion during the bright outbursts of transient XRBs, spanning a large range of accretion rate values. In the case of (quasi-)spherical matter flows and/or more complex regimes (e.g. propeller, intermittent accretion, dead accretion disks, etc...), the results are subject to larger uncertainties. Other successful techniques to estimates NS magnetic field strengths rely on the observation of sudden luminosity drops due to the onset of the propeller effect and on the study of rapid aperiodic variability. The first method proved particularly valuable in the case of LMXB transients, where fields of the order of a few 10^8 G were derived.

In general, there is a reasonably good agreement between the magnetic fields estimated with alternative methods and those obtained from cyclotron-line measurements. However, detailed comparisons have been possible only in a limited number of cases and they often rely on some poorly known quantities (e.g., the mass accretion rate, the geometry of the magnetosphere, the magnetization of the accreting plasma). These uncertainties, coupled to some poorly justified assumptions, can explain some discrepant magnetic field estimates reported in the literature. It must also be remembered that the distribution of field values derived from cyclotron lines is affected by the current instrument capabilities, which make it difficult to detect narrow spectral features at energies above a few tens of keV and below ~1 keV. Improvements in this field will be obtained by future missions with a better energy resolution in the soft X-ray range and higher sensitivity at hard X-rays.

Acknowledgements We thank all the staff of the International Space Science Institute and the organizers of the stimulating Workshop "The Strongest Magnetic Fields in the Universe". M.R. acknowledges the support by grant RNF 14-12-01287. S.M. has been supported through financial contribution from the agreement ASI-INAF I/037/12/0.

References

P. Arévalo, P. Uttley, Mon. Not. R. Astron. Soc. **367**, 801 (2006)
P.J. Armitage, C.S. Reynolds, Mon. Not. R. Astron. Soc. **341**, 1041 (2003)
J. Arons, S.M. Lea, Astrophys. J. **207**, 914 (1976)
K. Asai, M. Matsuoka, T. Mihara et al., Astrophys. J. **773**, 117 (2013)
W.A. Baan, Astrophys. J. **227**, 987 (1979)
S.A. Balbus, J.F. Hawley, Astrophys. J. **376**, 214 (1991)
M.M. Basko, R.A. Sunyaev, Mon. Not. R. Astron. Soc. **175**, 395 (1976)
W.H. Baumgartner, J. Tueller, C.B. Markwardt et al., Astrophys. J. Suppl. Ser. **207**, 19 (2013)
A. Baykal, M.J. Stark, J.H. Swank, Astrophys. J. **569**, 903 (2002)
P.A. Becker, M.T. Wolff, Astrophys. J. Lett. **621**, L45 (2005)
P.A. Becker, M.T. Wolff, Astrophys. J. **654**, 435 (2007)
P.A. Becker, D. Klochkov, G. Schönherr et al., Astron. Astrophys. **544**, A123 (2012)
E.C. Bellm, F. Fürst, K. Pottschmidt et al., Astrophys. J. **792**, 108 (2014)
H. Bhadkamkar, P. Ghosh, Astrophys. J. **746**, 22 (2012)
V. Bhalerao, P. Romano, J. Tomsick et al., arXiv:1407.0112 (2014)
D. Bhattacharya, E.P.J. van den Heuvel, Phys. Rep. **203**, 1 (1991)
L. Bildsten, D. Chakrabarty, J. Chiu et al., Astrophys. J. Suppl. Ser. **113**, 367 (1997)
A.J. Bird, A. Bazzano, L. Bassani et al., Astrophys. J. Suppl. Ser. **186**, 1 (2010)
G.S. Bisnovatyi-Kogan, Astron. Astrophys. **245**, 528 (1991)
G.S. Bisnovatyi-Kogan, B.V. Komberg, Sov. Astron. Lett. **2**, 130 (1976)
E. Bozzo, M. Falanga, L. Stella, Astrophys. J. **683**, 1031 (2008)
D.J. Burnard, J. Arons, R.I. Klein, Astrophys. J. **367**, 575 (1991)
I. Caballero, J. Wilms, Mem. Soc. Astron. Ital. **83**, 230 (2012)
I. Caballero, A. Santangelo, P. Kretschmar et al., Astron. Astrophys. **480**, L17 (2008)
A. Camero-Arranz, M.H. Finger, N.R. Ikhsanov, C.A. Wilson-Hodge, E. Beklen, Astrophys. J. **708**, 1500 (2010)
A. Camero-Arranz, M.H. Finger, C.A. Wilson-Hodge et al., Astrophys. J. **754**, 20 (2012a)
A. Camero-Arranz, K. Pottschmidt, M.H. Finger et al., Astron. Astrophys. **546**, A40 (2012b)
S. Campana, M. Colpi, S. Mereghetti, L. Stella, M. Tavani, Astron. Astrophys. Rev. **8**, 279 (1998)
S. Campana, F. Gastaldello, L. Stella et al., Astrophys. J. **561**, 924 (2001)
S. Campana, F. Brivio, N. Degenaar et al., Mon. Not. R. Astron. Soc. **441**, 1984 (2014)
C.G. Campbell, Geophys. Astrophys. Fluid Dyn. **63**, 179 (1992)
E. Churazov, M. Gilfanov, M. Revnivtsev, Mon. Not. R. Astron. Soc. **321**, 759 (2001)
G.W. Clark, J.W. Woo, F. Nagase, K. Makishima, T. Sakao, Astrophys. J. **353**, 274 (1990)
W. Coburn, W.A. Heindl, D.E. Gruber et al., Astrophys. J. **552**, 738 (2001)
W. Cui, Astrophys. J. Lett. **482**, L163 (1997)
A. Cumming, Nucl. Phys. B, Proc. Suppl. **132**, 435 (2004)
G. Cusumano, T. di Salvo, L. Burderi et al., Astron. Astrophys. **338**, L79 (1998)
A. D'Aì, G. Cusumano, V. La Parola et al., Astron. Astrophys. **532**, A73 (2011)
C.R. D'Angelo, H.C. Spruit, Mon. Not. R. Astron. Soc. **406**, 1208 (2010)
C.R. D'Angelo, H.C. Spruit, Mon. Not. R. Astron. Soc. **420**, 416 (2012)
K. Davidson, J.P. Ostriker, Astrophys. J. **179**, 585 (1973)
R.E. Davies, A.C. Fabian, J.E. Pringle, Mon. Not. R. Astron. Soc. **186**, 779 (1979)

M.E. DeCesar, P.T. Boyd, K. Pottschmidt et al., Astrophys. J. **762**, 61 (2013)
V. Doroshenko, A. Santangelo, V. Suleimanov et al., Astron. Astrophys. **515**, A10 (2010a)
V. Doroshenko, S. Suchy, A. Santangelo et al., Astron. Astrophys. **515**, L1 (2010b)
V. Doroshenko, A. Santangelo, V. Suleimanov, Astron. Astrophys. **529**, A52 (2011)
V. Doroshenko, A. Santangelo, L. Ducci, D. Klochkov, Astron. Astrophys. **548**, A19 (2012a)
V. Doroshenko, A. Santangelo, I. Kreykenbohm, R. Doroshenko, Astron. Astrophys. **540**, L1 (2012b)
V. Doroshenko, A. Santangelo, R. Doroshenko et al., Astron. Astrophys. **561**, A96 (2014)
L. Ducci, L. Sidoli, S. Mereghetti, A. Paizis, P. Romano, Mon. Not. R. Astron. Soc. **398**, 2152 (2009)
E.V. Filippova, S.S. Tsygankov, A.A. Lutovinov, R.A. Sunyaev, Astron. Lett. **31**, 729 (2005)
M.H. Finger, R.B. Wilson, B.A. Harmon, Astrophys. J. **459**, 288 (1996)
M.H. Finger, N.R. Ikhsanov, C.A. Wilson-Hodge, S.K. Patel, Astrophys. J. **709**, 1249 (2010)
J. Frank, A. King, D. Raine, *Accretion Power in Astrophysics* (Cambridge University Press, Cambridge, 1992)
F. Fürst, K. Pottschmidt, J. Wilms et al., Astrophys. J. Lett. **784**, L40 (2014a)
F. Fürst, K. Pottschmidt, J. Wilms et al., Astrophys. J. **780**, 133 (2014b)
P. Ghosh, F.K. Lamb, Astrophys. J. Lett. **223**, L83 (1978)
P. Ghosh, F.K. Lamb, Astrophys. J. **232**, 259 (1979)
R. Giacconi, H. Gursky, F.R. Paolini, B.B. Rossi, Phys. Rev. Lett. **9**, 439 (1962)
R. Giacconi, H. Gursky, E. Kellogg, E. Schreier, H. Tananbaum, Astrophys. J. Lett. **167**, L67 (1971)
M. Gilfanov, M. Revnivtsev, R. Sunyaev, E. Churazov, Astron. Astrophys. **338**, L83 (1998)
V.L. Ginzburg, Sov. Phys. Dokl. **9**, 329 (1964)
I.N. Gnedin, R.A. Sunyaev, Astron. Astrophys. **36**, 379 (1974)
T. Gold, Nature **218**, 731 (1968)
S.A. Grebenev, R.A. Sunyaev, Astron. Lett. **33**, 149 (2007)
J.E. Gunn, J.P. Ostriker, Nature **221**, 454 (1969)
A.K. Harding, D. Lai, Rep. Prog. Phys. **69**, 2631 (2006)
W.A. Heindl, W. Coburn, D.E. Gruber et al., Astrophys. J. Lett. **563**, L35 (2001)
W. Heindl, W. Coburn, I. Kreykenbohm, J. Wilms, The Astronomer's Telegram 200, 1 (2003)
P.B. Hemphill, R.E. Rothschild, A. Markowitz et al., Astrophys. J. **792**, 14 (2014)
A. Hewish, S.J. Bell, J.D.H. Pilkington, P.F. Scott, R.A. Collins, Nature **217**, 709 (1968)
S. Hirose, J.H. Krolik, J.M. Stone, Astrophys. J. **640**, 901 (2006)
W.C.G. Ho, H. Klus, M.J. Coe, N. Andersson, Mon. Not. R. Astron. Soc. **437**, 3664 (2014)
N.R. Ikhsanov, Mon. Not. R. Astron. Soc. **375**, 698 (2007)
A.F. Illarionov, D.A. Kompaneets, Mon. Not. R. Astron. Soc. **247**, 219 (1990)
A.F. Illarionov, R.A. Sunyaev, Astron. Astrophys. **39**, 185 (1975)
J.J.M. in't Zand, Astron. Astrophys. **441**, L1 (2005)
E. Kendziorra, B. Mony, P. Kretschmar et al., in *NASA Conference Publication*, ed. by C.R. Shrader, N. Gehrels, B. Dennis, vol. 3137 (1992), p. 217
R.I. Klein, J. Arons, G. Jernigan, J.J.-L. Hsu, Astrophys. J. Lett. **457**, L85 (1996)
D. Klochkov, C. Ferrigno, A. Santangelo et al., Astron. Astrophys. **536**, L8 (2011)
D. Klochkov, V. Doroshenko, A. Santangelo et al., Astron. Astrophys. **542**, L28 (2012)
W. Kluźniak, S. Rappaport, Astrophys. J. Lett. **671**, 1990 (2007)
I. Kreykenbohm, J. Wilms, W. Coburn et al., Astron. Astrophys. **427**, 975 (2004)
R. Krivonos, S. Tsygankov, M. Revnivtsev et al., Astron. Astrophys. **523**, A61 (2010)
W. Kundt, Phys. Lett. A **57**, 195 (1976)
A.G. Kuranov, K.A. Postnov, M.G. Revnivtsev, Astron. Lett. **40**, 29 (2014)
E. Kuulkers, in *Interacting Binaries: Accretion, Evolution, and Outcomes*, ed. by L. Burderi, L.A. Antonelli, F. D'Antona, T. di Salvo, G.L. Israel, L. Piersanti, A. Tornambè, O. Straniero. American Institute of Physics Conference Series, vol. 797 (2005), pp. 402–409
D. Lai, EPJ Web Conf. **64**, 1001 (2014)
F.K. Lamb, C.J. Pethick, D. Pines, Astrophys. J. **184**, 271 (1973)
F.K. Lamb, A.C. Fabian, J.E. Pringle, D.Q. Lamb, Astrophys. J. **217**, 197 (1977)
H.J.G.L.M. Lamers, E.P.J. van den Heuvel, J.A. Petterson, Astron. Astrophys. **49**, 327 (1976)
G. Lesur, P.-Y. Longaretti, Astron. Astrophys. **504**, 309 (2009)
W. Lewin, M. van der Klis, *Compact Stellar X-Ray Sources* (Cambridge University Press, Cambridge, 2010)
W.H.G. Lewin, J. van Paradijs, R.E. Taam, Space Sci. Rev. **62**, 223 (1993)
X.-D. Li, Z.-R. Wang, Astron. Astrophys. **307**, L5 (1996)
V.M. Lipunov, Sov. Astron. **25**, 375 (1981)
V.M. Lipunov, Astrophys. Space Sci. **85**, 451 (1982)
V.M. Lipunov, G. Börner, R.S. Wadhwa, *Astrophysics of Neutron Stars* (1992)
Q.Z. Liu, J. van Paradijs, E.P.J. van den Heuvel, Astron. Astrophys. **455**, 1165 (2006)

Q.Z. Liu, J. van Paradijs, E.P.J. van den Heuvel, Astron. Astrophys. **469**, 807 (2007)
R.V.E. Lovelace, M.M. Romanova, G.S. Bisnovatyi-Kogan, Mon. Not. R. Astron. Soc. **275**, 244 (1995)
A.A. Lutovinov, M.G. Revnivtsev, S.S. Tsygankov, R.A. Krivonos, Mon. Not. R. Astron. Soc. **431**, 327 (2013)
Y.E. Lyubarskii, Mon. Not. R. Astron. Soc. **292**, 679 (1997)
Y.E. Lyubarskii, R.A. Sunyaev, Sov. Astron. Lett. **8**, 330 (1982)
Y.E. Lyubarskii, R.A. Sunyaev, Sov. Astron. Lett. **14**, 390 (1988)
C. Maitra, B. Paul, S. Naik, Mon. Not. R. Astron. Soc. **420**, 2307 (2012)
K. Makishima, T. Mihara, in *Frontiers Science Series*, ed. by Y. Tanaka, K. Koyama (1992), p. 23
K. Makishima, T. Mihara, M. Ishida et al., Astrophys. J. Lett. **365**, L59 (1990)
T. Matilsky, H. Gursky, E. Kellogg et al., Astrophys. J. **181**, 753 (1973)
V.A. McBride, J. Wilms, M.J. Coe et al., Astron. Astrophys. **451**, 267 (2006)
V.A. McBride, J. Wilms, I. Kreykenbohm et al., Astron. Astrophys. **470**, 1065 (2007)
S. Mereghetti, J.E. Grindlay, Astrophys. J. **312**, 727 (1987)
P. Meszaros, W. Nagel, Astrophys. J. **298**, 147 (1985)
T. Mihara, K. Makishima, S. Kamijo et al., Astrophys. J. Lett. **379**, L61 (1991)
S. Müller, M. Kühnel, I. Caballero et al., Astron. Astrophys. **546**, A125 (2012)
S. Müller, C. Ferrigno, M. Kühnel et al., Astron. Astrophys. **551**, A6 (2013b)
D. Müller, D. Klochkov, I. Caballero, A. Santangelo, Astron. Astrophys. **552**, A81 (2013a)
A.A. Mushtukov, V.F. Suleimanov, S.S. Tsygankov, J. Poutanen, arXiv:1409.6457 (2014)
F. Nagase, Publ. Astron. Soc. Jpn. **41**, 1 (1989)
W. Nagel, Astrophys. J. **251**, 288 (1981)
M. Nakajima, T. Mihara, K. Makishima, H. Niko, Astrophys. J. **646**, 1125 (2006)
O. Nishimura, Astrophys. J. **672**, 1127 (2008)
O. Nishimura, Astrophys. J. **781**, 30 (2014)
M. Oda, P. Gorenstein, H. Gursky et al., Astrophys. J. Lett. **166**, L1 (1971)
M. Orlandini, D. Dal Fiume, F. Frontera et al., Astrophys. J. Lett. **500**, L163 (1998)
A. Paizis, L. Sidoli, Mon. Not. R. Astron. Soc. **439**, 3439 (2014)
A.N. Parmar, N.E. White, L. Stella, C. Izzo, P. Ferri, Astrophys. J. **338**, 359 (1989)
A. Patruno, A.L. Watts, arXiv:1206.2727 (2012). A. Patruno, Astrophys. J. Lett. **753**, L12 (2012). doi:10.1088/2041-8205/753/1/L12
K. Postnov, R. Staubert, A. Santangelo et al., Astron. Astrophys. **480**, L21 (2008)
K.A. Postnov, N.I. Shakura, A.Y. Kochetkova, L. Hjalmarsdotter, EPJ Web Conf. **64**, 2002 (2014)
J. Poutanen, A.A. Mushtukov, V.F. Suleimanov et al., Astrophys. J. **777**, 115 (2013)
J.E. Pringle, M.J. Rees, Astron. Astrophys. **21**, 1 (1972)
P. Reig, Astrophys. Space Sci. **332**, 1 (2011)
P. Reig, J.M. Torrejón, P. Blay, Mon. Not. R. Astron. Soc. **425**, 595 (2012)
M. Revnivtsev, in *American Institute of Physics Conference Series*, ed. by M. Axelsson. vol. 1054 (2008), pp. 143–153
M.G. Revnivtsev, S.Y. Sazonov, M.R. Gilfanov, R.A. Sunyaev, Astron. Lett. **29**, 587 (2003)
M. Revnivtsev, E. Churazov, K. Postnov, S. Tsygankov, Astron. Astrophys. **507**, 1211 (2009)
P. Romano, H.A. Krimm, D.M. Palmer et al., Astron. Astrophys. **562**, A2 (2014)
A. Santangelo, S. del Sordo, A. Segreto et al., Astron. Astrophys. **340**, L55 (1998)
A. Santangelo, A. Segreto, S. Giarrusso et al., Astrophys. J. Lett. **523**, L85 (1999)
G. Schönherr, J. Wilms, P. Kretschmar et al., Astron. Astrophys. **472**, 353 (2007)
E. Schreier, R. Levinson, H. Gursky et al., Astrophys. J. Lett. **172**, L79 (1972)
A.N. Semena, M.G. Revnivtsev, Astron. Lett. **38**, 321 (2012)
A. Semena, M. Revnivtsev, A. Buckley et al., Mon. Not. R. Astron. Soc. **442**(2), 1123 (2014)
V. Sguera, A. Bazzano, A.J. Bird et al., Astrophys. J. **646**, 452 (2006)
N. Shakura, K. Postnov, A. Kochetkova, L. Hjalmarsdotter, Mon. Not. R. Astron. Soc. **420**, 216 (2012)
N.I. Shakura, K.A. Postnov, A.Y. Kochetkova, L. Hjalmarsdotter, Phys. Usp. **56**, 321 (2013)
I.A. Shibanov, V.E. Zavlin, G.G. Pavlov, J. Ventura, Astron. Astrophys. **266**, 313 (1992)
K. Shinoda, T. Kii, K. Mitsuda et al., Publ. Astron. Soc. Jpn. **42**, L27 (1990)
C.R. Shrader, F.K. Sutaria, K.P. Singh, D.J. Macomb, Astrophys. J. **512**, 920 (1999)
P. Shtykovskiy, M. Gilfanov, Astron. Astrophys. **431**, 597 (2005)
F. Shu, J. Najita, E. Ostriker et al., Astrophys. J. **429**, 781 (1994)
L. Sidoli, in *Proceedings of "An INTEGRAL View of the High-Energy Sky (the First 10 Years)"—9th INTEGRAL Workshop and Celebration of the 10th Anniversary of the Launch (INTEGRAL 2012) 15–19 October* (Bibliotheque Nationale de France, Paris, 2012). Published online at http://pos.sissa.it/cgi-bin/reader/conf.cgi?confid=176,id.11
L. Sidoli, P. Romano, S. Mereghetti et al., Astron. Astrophys. **476**, 1307 (2007)

G. Srinivasan, New Astron. Rev. **54**, 93 (2010)
R. Staubert, N.I. Shakura, K. Postnov et al., Astron. Astrophys. **465**, L25 (2007)
L. Stella, N.E. White, R. Rosner, Astrophys. J. **308**, 669 (1986)
T.E. Strohmayer, Adv. Space Res. **28**, 511 (2001)
S. Suchy, K. Pottschmidt, J. Wilms et al., Astrophys. J. **675**, 1487 (2008)
R.A. Sunyaev, N.I. Shakura, Sov. Astron. Lett. **3**, 114 (1977)
R.E. Taam, E.P.J. van den Heuvel, Astrophys. J. **305**, 235 (1986)
H. Tananbaum, H. Gursky, E.M. Kellogg et al., Astrophys. J. Lett. **174**, L143 (1972)
S.P. Tendulkar, F. Fürst, K. Pottschmidt et al., Astrophys. J. **795**, 154 (2014). doi:10.1088/0004-637X/795/2/154
J. Truemper, W. Pietsch, C. Reppin et al., Astrophys. J. Lett. **219**, L105 (1978)
S.S. Tsygankov, A.A. Lutovinov, Astron. Lett. **31**, 88 (2005)
S.S. Tsygankov, A.A. Lutovinov, E.M. Churazov, R.A. Sunyaev, Mon. Not. R. Astron. Soc. **371**, 19 (2006)
S.S. Tsygankov, A.A. Lutovinov, A.V. Serber, Mon. Not. R. Astron. Soc. **401**, 1628 (2010)
S.S. Tsygankov, R.A. Krivonos, A.A. Lutovinov, Mon. Not. R. Astron. Soc. **421**, 2407 (2012)
D.A. Uzdensky, Astrophys. Space Sci. **292**, 573 (2004)
M. van der Klis, in *Timing Neutron Stars*, ed. by H. Ögelman, E.P.J. van den Heuvel (1989), p. 27
B.A. Vaughan, M. van der Klis, K.S. Wood et al., Astrophys. J. **435**, 362 (1994)
R. Walter, J. Zurita Heras, Astron. Astrophys. **476**, 335 (2007)
R. Walter, J. Rodriguez, L. Foschini et al., Astron. Astrophys. **411**, L427 (2003)
Y.-M. Wang, Astrophys. J. Lett. **449**, L153 (1995)
W.A. Wheaton, J.P. Doty, F.A. Primini et al., Nature **282**, 240 (1979)
N.E. White, J.H. Swank, S.S. Holt, Astrophys. J. **270**, 711 (1983)
R. Wijnands, M. van der Klis, Nature **394**, 344 (1998)
C.A. Wilson, M.H. Finger, M.J. Coe, S. Laycock, J. Fabregat, Astrophys. J. **570**, 287 (2002)
L. Woltjer, Astrophys. J. **140**, 1309 (1964)
G.A. Wynn, A.R. King, K. Horne, Mon. Not. R. Astron. Soc. **286**, 436 (1997)
T. Yamamoto, M. Sugizaki, T. Mihara et al., Publ. Astron. Soc. Jpn. **63**, 751 (2011)
J. Ziolkowski, Acta Astron. **35**, 185 (1985)

DOI 10.1007/978-1-4939-3550-5_10
Reprinted from *Space Science Reviews* Journal, DOI 10.1007/s11214-015-0146-y

Magnetars: Properties, Origin and Evolution

Sandro Mereghetti[1] · José A. Pons[2] · Andrew Melatos[3]

Received: 16 December 2014 / Accepted: 17 March 2015 / Published online: 25 March 2015

Abstract Magnetars are neutron stars in which a strong magnetic field is the main energy source. About two dozens of magnetars, plus several candidates, are currently known in our Galaxy and in the Magellanic Clouds. They appear as highly variable X-ray sources and, in some cases, also as radio and/or optical pulsars. Their spin periods (2–12 s) and spin-down rates ($\sim 10^{-13}$–10^{-10} s s^{-1}) indicate external dipole fields of $\sim 10^{13-15}$ G, and there is evidence that even stronger magnetic fields are present inside the star and in non-dipolar magnetospheric components. Here we review the observed properties of the persistent emission from magnetars, discuss the main models proposed to explain the origin of their magnetic field and present recent developments in the study of their evolution and connection with other classes of neutron stars.

Keywords Neutron star · Magnetic field · Magnetar

1 Introduction

Magnetars are neutron stars in which the main source of energy is provided by a strong magnetic field, instead of rotation, accretion, nuclear reactions, or cooling. While the bulk of rotation-powered (radio) pulsars have fields in the range $B \sim 10^{11}$–10^{13} G, the external magnetic field of magnetars is typically 10^{13}–10^{15} G and it is likely that their internal field is even stronger. However, the distributions of field intensities for magnetars and "normal"

✉ S. Mereghetti
sandro@iasf-milano.inaf.it

J.A. Pons
jose.pons@ua.es

A. Melatos
amelatos@unimelb.edu.au

[1] INAF IASF-Milano, v. Bassini 15, 20133 Milano, Italy

[2] Departament de Física Aplicada, Universitat d'Alacant, Ap. Correus 99, 03080 Alacant, Spain

[3] School of Physics, University of Melbourne, Parkville, VIC 3010, Australia

neutron stars overlap: there is not a discriminating B threshold between these two classes. Indeed, the presence of a strong dipole field (typically estimated from the star spin period and spin-down rate) is not a sufficient (nor a necessary) condition to trigger "magnetar-like" activity. The latter is in fact mainly related to the presence of a significant toroidal component of the internal field, able to produce magnetospheric twists.

Magnetars are the most variable sources among the different classes of isolated neutron stars: their characterizing property is the emission, in the X-ray and soft γ-ray range, of powerful short bursts which often reach super-Eddington luminosities. More rarely, they also emit intermediate and giant flares, the latter involving the release of up to about 10^{46} erg in less than half a second. Magnetars also show pulsed X-ray emission with typical luminosity of $\sim 10^{35}$ erg s^{-1} in persistent sources, and ranging from $\sim 10^{32}$ to 10^{36} erg s^{-1} in transient ones. The pulsations, caused by the neutron star rotation, have periods of a few seconds which are secularly increasing on timescales from one thousand to several million years ($\dot{P} \sim 10^{-13}$–10^{-10} s s^{-1}).

Due to this variety of phenomena, most of the sources that are now believed to be magnetars were initially classified in different ways and only later recognized as members of the same class of astrophysical objects. Bursts from magnetars had been observed since the end of the 1970s (Mazets et al. 1979a,b). They were initially classified as a sub-class of γ-ray bursts, with the peculiarity of a softer spectrum and of coming repeatedly from the same sky directions (Norris et al. 1991). They were thus named soft γ-ray repeaters (SGRs). A secure identification with astrophysical objects known at other wavelengths was unfeasible with the large positional uncertainties available at that time, but their possible association with supernova remnants[1] suggested a neutron star nature. Other sources that are now believed to be magnetars were discovered as persistent pulsars in the soft X-ray range (<10 keV) and thought to be X-ray binaries powered by accretion, as most of the bright X-ray sources known at that time. It was later pointed out that their narrow period distribution, long term spin-down, soft X-ray spectrum and faint optical counterparts were at variance with the properties of pulsars in massive binaries (Mereghetti and Stella 1995). This led to their denomination as anomalous X-ray pulsars (AXPs).

We now believe that SGRs and AXPs are a single class of objects. In fact, when the persistent X-ray counterparts of SGRs were identified (Murakami et al. 1994; Rothschild et al. 1994; Hurley et al. 1999; Woods et al. 1999b), it was found that they are pulsating sources very similar to the AXPs (Kouveliotou et al. 1998, 1999; Esposito et al. 2009; Kulkarni et al. 2003), and SGR-like bursts were detected from several sources originally classified as AXPs (Gavriil et al. 2002; Kaspi et al. 2003; Woods et al. 2005). About two dozens of AXPs/SGRs are currently known in our Galaxy (plus one in each of the Magellanic Clouds).[2] Most of them show X-ray pulsations and have been seen to emit bursts. For extensive reviews of the AXPs/SGRs observations and of the main models proposed to explain them see Woods and Thompson (2006), Mereghetti (2008, 2011a), Turolla and Esposito (2013).

We believe that the most successful explanation of the AXPs/SGRs is provided by the magnetar model (Thompson and Duncan 1995, 1996), according to which they are neutron stars powered by a strong magnetic field. Alternative interpretations based on isolated neutron stars accreting from fall-back disks formed after the supernova explosion (see, e.g., Alpar 2001; Trümper et al. 2010) require some additional process, besides accretion, in order to explain the powerful bursts and flares observed from these sources.

[1] Ironically, it is now known that the nebulae associated to two of the three first discovered SGRs are not supernova remnants.

[2] An updated list is maintained at http://www.physics.mcgill.ca/~pulsar/magnetar/main.html (Olausen and Kaspi 2014).

In fact, the suggestion that SGRs are neutron stars powered by magnetic energy was first proposed to interpret the exceptional properties[3] of the giant flare emitted by SGR 0526–66 on March 5, 1979 (Paczynski 1992; Duncan and Thompson 1992). In the following years, the original magnetar model has been considerably developed and expanded and it provides now the best explanation for the rich diversity of AXPs and SGRs phenomenology (see, e.g., Beloborodov 2011). One essential feature of the magnetar model is the presence of significant twists in the magnetosphere (Thompson et al. 2002; Beloborodov and Thompson 2007), resulting in a structure quite different from that of the simple dipolar geometry assumed for normal radio pulsars and in magnetospheric currents with a charge density much larger than the classical Goldreich-Julian value. Bursts and flares can be explained by sudden releases of energy in the star interior leading to fractures in the crust (Thompson and Duncan 1995), by field reconnection events in the magnetosphere analogous to those occurring in the Sun (Lyutikov 2006), or by pair plasma fireballs produced by discontinuities in the propagation of fast MHD waves in the magnetosphere (Heyl and Hernquist 2005).

In Sect. 2 we describe the properties of the so called "persistent" emission[4] of magnetars. In the two following sections we review the scenarios that have been advanced for the formation of magnetars (Sect. 3) and for their evolution (Sect. 4).

2 Properties of the Persistent Emission

The main manifestations of magnetars occur in the X-ray energy range. All known confirmed magnetars show pulsations in the soft X-ray band (<10 keV) and many of them have also been detected in hard X-rays, up to $\sim$100–200 keV. About half of the known magnetars have repeatedly been observed at nearly constant X-ray luminosities of $\sim 10^{34}$–10^{35} erg s^{-1}, with only moderate variability (a factor of a few) on long timescales (see, e.g., Mereghetti 2011a). Much larger variability is seen in the transient magnetars, which reach the luminosity level of the "constant" magnetars only during outbursts lasting weeks/months and spend the remaining time at a much fainter, quiescent level, $\sim 10^{32}$ erg s^{-1} or less (see Rea and Esposito 2011, and references therein).

The outbursts of transient magnetars are often associated with the emission of short bursts or flares (e.g., Woods et al. 2005; Esposito et al. 2008; Mereghetti et al. 2009; Aptekar et al. 2009; van der Horst et al. 2010). Short bursts are also emitted by the "constant" magnetars, but they have never been detected from a transient magnetar in the quiescent luminosity level.

2.1 X-Ray Pulsations

The presence of regular pulsations with secularly increasing period, caused by the slowing down of the neutron star rotational velocity, is one of the distinctive properties of magnetars and provides a very useful diagnostic tool for their study. The 23 currently known magnetars have spin periods in a very narrow range (2–12 s), while their period derivatives span five orders of magnitude. Most of them have $\dot{P}$ in the range 10^{-12}–10^{-10} s s^{-1}, but in recent

[3]The association with the supernova remnant N49 in the Large Magellanic Cloud yielded the distance and energetics of this event.

[4]In the lack of a better nomenclature, we use this adjective somehow improperly also for transient and variable sources, just to distinguish this emission from that of the short bursts and of the intermediate/giant flares.

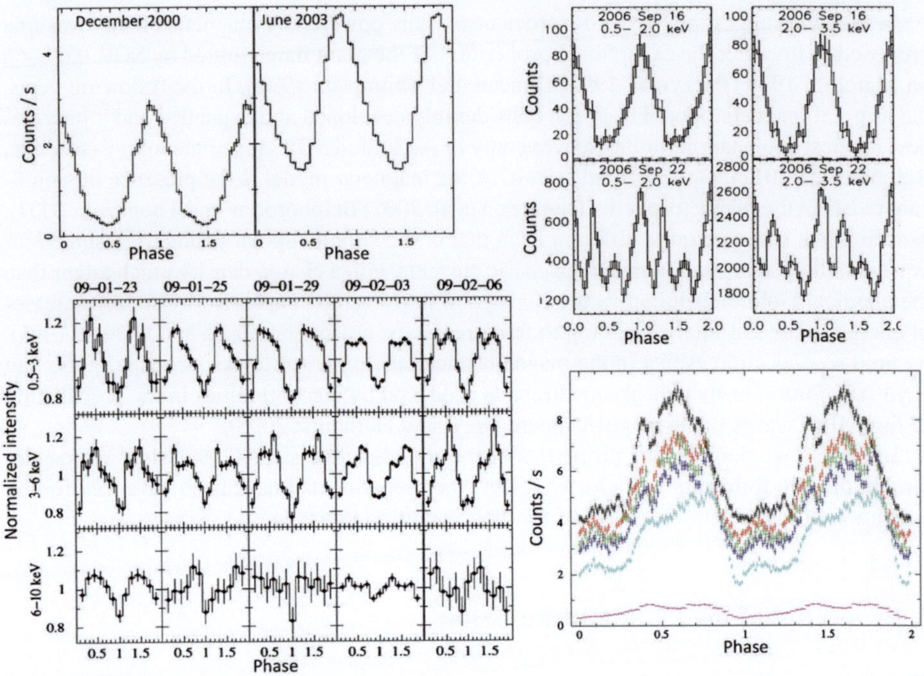

Fig. 1 Examples of X-ray pulse profiles of magnetars (for clarity, two neutron star rotations are shown in each plot). *Top left panel*: the pulsed fraction of the persistent magnetar 1E 1048.1−5937 anticorrelates with the luminosity (Mereghetti et al. 2004): when the flux was about twice that of the normal level, the pulsed fraction was smaller (53 % wrt 89 %); both curves refer to the 0.6–10 keV range and were obtained with the EPIC instrument on *XMM-Newton*. *Top right panel*: pulse profile variations in two energy ranges of the transient magnetar CXOU J1647−4552 (Muno et al. 2007) in quiescence (*upper panels*) and during the outburst (*lower panels*). *Bottom left panel*: evolution of the pulse profiles of 1E 1547.0−5408 during the decaying phase of the January 2009 outburst (Bernardini et al. 2011). *Bottom right panel*: pulse profiles of the transient magnetar SGR 0501+4516 at different luminosity levels (Camero et al. 2014)

years a few "low-$\dot{P}$ magnetars" have been discovered, with spin-down rates as small as 4×10^{-15} s s^{-1}, well in the range of those of rotation-powered pulsars (Rea et al. 2010, 2012, 2013b, 2014; An et al. 2013b; Scholz et al. 2014). The observed distribution of magnetars in the pulsar $P - \dot{P}$ diagram gives information on their evolution and relation with other classes of neutron stars. The lack of observed magnetars with periods longer than 12 s indicates that the spin-down mechanism becomes highly inefficient at large ages and/or that old magnetars become more difficult to detect, for example because their X-ray luminosity decreases and they emit bursts less frequently. The most obvious explanation to account for these effects is magnetic field decay (Colpi et al. 2000; Dall'Osso et al. 2012), as discussed in detail in Sect. 4.

Magnetars display X-ray pulse profiles with a variety of shapes (from simple sinusoids to multipeaked) and spanning a large range of pulsed fractions (from less than 10 % to nearly 100 %). The pulse profiles are energy-dependent (with a tendency toward more complex shapes with increasing energy) and, in many sources, time-variable. Changes in pulse profiles are often connected with bursts/flares and/or glitches, but also long term variations, apparently unrelated to particular events, have been observed. Some examples of pulse profiles are shown in Figs. 1 and 2.

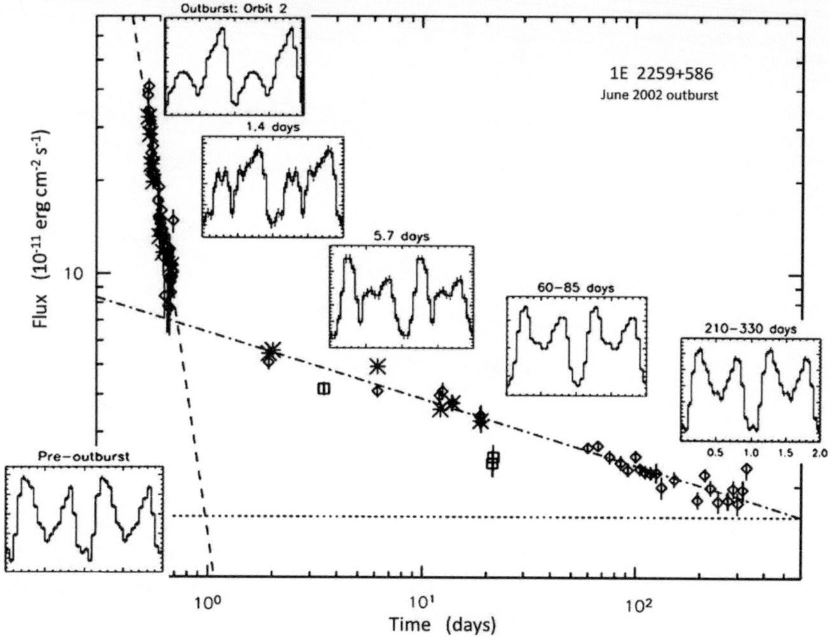

Fig. 2 Evolution of the pulse profile of 1E 2259+586 during the outburst of June 2002 (adapted from Woods et al. 2004). The insets show pulse profiles in the 2–10 keV energy range obtained with the PCA instrument on *RossiXTE*. The inset in the *lower left corner* shows the pre-outburst pulse profile. Only the relative strength of the different peaks can be inferred from these profiles, which are plotted in arbitrary flux units

The spin-down of magnetars is attributed to the angular momentum carried away by (time-variable) magnetized outflows and to dipole radiation losses. In a twisted magnetosphere, the latter effect produces a higher spin-down rate than in pulsars with dipolar field because the twist inflates the poloidal lines and increases the magnetic field at the light cylinder. Variations in the spin-down rate have been detected in practically all the magnetars for which good timing data extending over long periods are available. The $\dot{P}$ variations are generally smaller than $\sim$50 %, but changes as large as a factor of ten over timescales of weeks have sometimes been observed (Gavriil and Kaspi 2004; Dib et al. 2009). The variations in $\dot{P}$ observed in several magnetars are a proof of the dynamic nature of their magnetospheres. Since both the torque and the magnetospheric currents are driven by variations in the twisting of the field lines (Beloborodov 2009), some correlations between the spin-down rate and X-ray emission properties can be expected, and indeed they have been observed (see Sect. 2.6).

Glitches have been observed in many magnetars (Kaspi et al. 2000; Dall'Osso et al. 2003; Woods et al. 2004; Dib et al. 2009). They involve fractional frequency changes $\Delta\nu/\nu \sim 10^{-7}$–$10^{-4}$, similar to those of the strongest glitches of radio pulsars (Dib et al. 2008), but the apparent paucity of magnetars glitches with smaller $\Delta\nu/\nu$ is probably a selection effect. The relation between magnetar glitches and observed changes in the properties of their X-ray radiation (e.g. bursts, flux or pulse profile changes, etc.) is extensively discussed in Dib and Kaspi (2014). While the majority of radiative changes are associated with glitches (or with some other timing anomaly), the converse is not true: several glitches seem to have no consequences on the emitted radiation. Two sources showed episodes in which the spin frequency apparently jumped in a short time interval to a value significantly

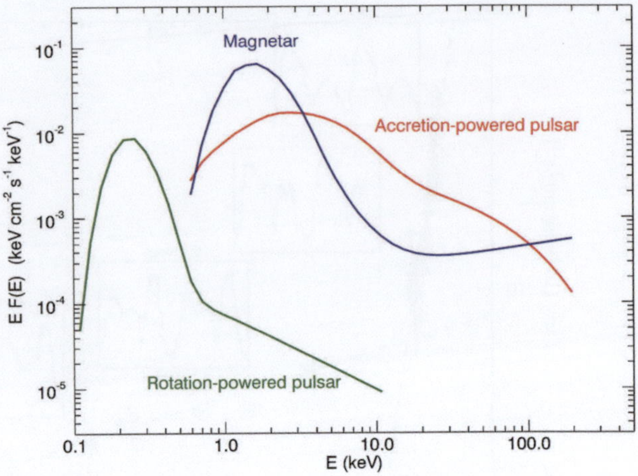

Fig. 3 Comparison of the X-ray spectra of three neutron stars representative of different classes. *Red*: the accretion-powered binary X Persei (adapted from Di Salvo et al. 1998); *green*: the rotation-powered pulsar Geminga (adapted from Jackson and Halpern 2005); *blue*: the magnetically-powered AXP 4U 0142+61 (adapted from Rea et al. 2007)

smaller than that predicted by the observed spin-down rate. These events with negative $\Delta\nu$ have been called antiglitches and cannot be explained with the theories of normal glitches ($\Delta\nu > 0$), which are based on the fact that the neutron star interior rotates faster than the crust and acts as an angular momentum reservoir. However, due to the sparse time sampling of the available data, it is not clear if these events are really occurring on a short timescale, as normal glitches. The first possible antiglitch, with $\Delta\nu/\nu = -10^{-4}$, coincided with the August 1998 giant flare of SGR 1900+14 (Woods et al. 1999a). An analysis of the pulse phases suggests that the frequency variation could have been caused by an increase of more than a factor 100 in the spin-down rate in a few hours after the giant flare (Palmer 2002). However, an alternative possibility requiring only a factor of two increase in the spin-down rate in the three months preceding the giant flare cannot be ruled out. A more convincing case for a magnetar antiglitch, with $\Delta\nu/\nu = -3 \times 10^{-7}$ and connected with a flux increase, has been recently reported for 1E 2259+586 (Archibald et al. 2013). This prompted several interpretations, involving either a sudden event, like the collision with a solid body (Huang and Geng 2014), or a rapid reconfiguration of the magnetosphere with the opening of some of the closed field lines and the emission of an enhanced particle wind (Lyutikov 2013; Tong 2014).

2.2 X-Ray Spectra

In Fig. 3 we compare representative spectra of three different classes of neutron stars: a magnetar (4U 0142+61), a rotation-powered pulsar (Geminga), and an accretion-powered binary at low luminosity (X Persei). If we limit the comparison to the ~2–10 keV range, historically the first one to be explored, the most striking difference between these sources, is the softness of the magnetar spectrum. The first measured spectra of AXPs were in fact well fit by power-laws with photon indexes $\Gamma \sim 3$–4 (Parmar et al. 1998; White et al. 1987; Sugizaki et al. 1997). This characteristic spectral softness gives an immediate diagnostics to recognize magnetar candidates among newly discovered X-ray pulsars. When data of better quality became available, it was found that a better phenomenological fit to the magnetar spectra below 10 keV is provided by a blackbody model with temperature $kT_{BB} \sim 0.5$ keV, plus either a power-law or a second blackbody component (Oosterbroek et al. 1998; White et al. 1996; Patel et al. 2001; Mereghetti et al. 2005b, 2006).

Contrary to the case of rotation-powered neutron stars, for which the thermal and non-thermal components dominate in different energy ranges (see, e.g., Geminga in Fig. 3), the higher temperature of magnetars implies that the blackbody and the power-law contribute in a similar way to the 1–10 keV flux, making more difficult to disentangle and constrain the two components. The requirement of a power-law in the soft X-ray range might simply reflect the inadequacy of a simple blackbody to fit a more complex thermal model, rather than representing a physically distinct process. For this reason, some caution is needed when drawing physical interpretations from some of the correlations between spectral parameters that have been reported in the literature.

Most magnetars are located at low Galactic latitude and thus their spectra are strongly affected by the interstellar absorption, with large column densities $N_H \sim 10^{22}$–10^{23} cm^{-2}. The N_H values required by the blackbody plus power-law fits are often larger than those independently estimated in other ways, suggesting that the power-law cannot be extrapolated to low energy without a cut-off. Good fits are generally obtained with the sum of two-blackbody models, which can be interpreted in terms of regions with different temperatures on the star surface (Halpern and Gotthelf 2005). Thanks to its location in the Small Magellanic Cloud, CXOU J0100−7211 is the magnetar with the lowest interstellar absorption ($N_H \sim 6 \times 10^{20}$ cm^{-2}) and offers the best opportunity to study the X-ray emission at low energy: its spectrum is well fit by the two-blackbody model while the power-law plus blackbody is rejected with high confidence (Tiengo et al. 2008).

On the other hand, a power-law component is certainly present in the hard X-ray range. Several magnetars have been detected up to ~150 keV with large pulsed fractions and spectra flatter than those of accreting X-ray pulsars. As schematically shown in Fig. 3, the latter have exponential cut-offs at a few tens of keV while the spectra of magnetars extend to higher energies. The first studies of the (non-bursting) emission from AXPs/SGRs above ~10 keV were carried out with the INTEGRAL, RXTE and Suzaku satellites. Despite the limited sensitivity and imaging capabilities of the instruments operating in this range, these observations were crucial to demonstrate that the hard X-ray emission represents a non-negligible fraction of the energy output from magnetars (Kuiper et al. 2004, 2006; Mereghetti et al. 2005a; Götz et al. 2006; den Hartog et al. 2008a,b; Enoto et al. 2010b,c). More sensitive observations have been obtained in the last two years with the NuSTAR satellite, thanks to the imaging capability provided by its focusing telescopes covering the 3–79 keV range. These observations allow to carry out spectral and variability analysis on short timescales and to spatially resolve the hard X-ray emission in crowded and/or confused regions (An et al. 2013a; Vogel et al. 2014; Kaspi et al. 2014).

We can summarize the properties of the hard X-ray emission from magnetars as follows:

- The luminosity in the hard component is similar to that observed below 10 keV.
- Fits in the range ~10–200 keV with power-law models give photon index values typically between $\Gamma \sim 1$ and 2 (except in the case of SGR 1900+14, which has $\Gamma \sim 3$, Götz et al. 2006).
- The flux upper limits derived in the MeV region (Kuiper et al. 2006; den Hartog et al. 2006) imply that the spectra cannot extend as power laws to such high energies. Indeed, the data with high statistics show that curved models, like a log-parabolic function, provide better fits than simple power laws (Rea et al. 2007; den Hartog et al. 2008a,b).
- The spectra of the pulsed component are harder than those of the total emission and show phase-dependent variations.
- Hard X-ray emission has been observed also in transient magnetars (1E 1547.0−5408, Enoto et al. 2010a; SGR 0501+4516, Rea et al. 2009; SGR 1729−45, Mori et al. 2013;

Kaspi et al. 2014). In the case of 1E 1547.0−5408 the spectrum hardened as the flux decreased (Kuiper et al. 2012).

- No detections at higher energy have been obtained.[5] The upper limits derived with *Fermi* in the 0.1–10 GeV range (Abdo et al. 2010; Şaşmaz Muş and Göğüş 2010) are incompatible with earlier predictions which assumed emission from the outer magnetospheres of AXPs/SGRs (Cheng and Zhang 2001; Zhang and Cheng 2002). Searches for TeV emission with ground based telescopes gave negative results (Aleksić et al. 2013).

The above description of the magnetar spectra is based on simple phenomenological fits, but in recent years more physically-motivated models to interpret the observed broad band spectra have been developed (Thompson and Beloborodov 2005; Beloborodov 2009). Two main ingredients play an important role in these models: (a) thermal emission from (a part of) the neutron star surface, which is surrounded by a thin atmosphere, and (b) the presence of a magnetosphere with a complex geometry and significant charge density. The magnetosphere affects significantly the emerging spectrum and provides additional emission components due to the presence of accelerated charges. The surface thermal emission results from interior cooling powered by magnetic field dissipation and from external heating caused by backward-flowing charges in the magnetosphere. The presence of a relatively dense plasma in the magnetospheres with a twisted configuration, a distinguishing property of magnetars, has important implications for the emitted spectrum. Resonant cyclotron scattering of the thermal photons can produce hard tails (Thompson et al. 2002).

First steps toward physical modeling taking into account the effects of the strong magnetic field on the thermal emission were done considering radiation transfer in a monodimensional approximation (Lyutikov and Gavriil 2006; Güver et al. 2007) More realistic 3-D computations required a Monte Carlo approach to study the photon propagation in a globally twisted magnetosphere supporting the strong currents that provide a large optical depth to resonant cyclotron scattering (Fernández and Thompson 2007; Nobili et al. 2008a). These models can successfully fit the observed spectra (Rea et al. 2008; Zane et al. 2009), but their validity above a few tens of keV is uncertain, because they are computed for non-relativistic (or only mildly relativistic) particle distributions.[6] Also the assumed geometry is probably oversimplified: the seed photons may have a non-uniform and time-variable temperature distribution over the star surface and the magnetospheric twist may involve only a limited bundle of field lines rather than being global. The modeling of phase-resolved spectra obtained during the evolution of transient magnetar outbursts is a promising approach which can give useful constrains on the models (see, e.g., Albano et al. 2010). Unfortunately, phase-resolved spectra of good statistical quality are not always available, especially in the hard X-ray range.

A model for the hard X-ray spectra of magnetars has been developed by Beloborodov (2013b) who numerically solved the radiative transfer of charged particles flowing in a large twisted magnetic loop. Relativistic electrons ($\gamma \sim 10^3$) are injected by high voltage discharges close to the star surface and flow in the closed loop (rather than along open field lines as in normal pulsars). The resulting emission, strongly beamed along the loop, has a hard power-law shape below the MeV and is significantly suppressed at higher energy. This is in agreement with the observations, but the model parameters depend strongly on the geometry of the loop and orientation of the source (Hascoët et al. 2014). Both phase-averaged

[5] The MeV-GeV source in the region of 1E 2259+586 is well explained as emission from the supernova remnant CTB 109 interacting with molecular clouds (Castro et al. 2012).

[6] The extension to the general case has been derived by Nobili et al. (2008b), but only limited applications have been reported (Zane et al. 2011b).

and phase-resolved spectra of magnetars recently obtained with the *NuSTAR* satellite in the 4–80 keV range have been successfully fit by this model, which, however, does not yet include in a self-consistent way the low energy X-ray emission (An et al. 2013a; Vogel et al. 2014).

2.3 Spectral Features

The detection of cyclotron lines is the most direct way to measure the magnetic field of neutron stars. This method has been successfully applied to accreting neutron stars in X-ray binaries since the beginning of X-ray astronomy (see, e.g., Revnivtsev and Mereghetti 2014). For magnetic fields of $B \sim 10^{14}$–10^{15} G, electron cyclotron lines are in the MeV energy range, where the currently available instruments are not sensitive enough to detect the magnetar emission. On the other hand, the energies of proton cyclotron lines fall in the soft X-ray range, offering, in principle, a direct way to measure the magnetic fields of these objects. This motivated extensive searches for narrow features in the persistent emission from magnetars, which however gave, until recently, only negative results.[7] Some early claims with low statistical significance (Iwasawa et al. 1992; Rea et al. 2003) could not be confirmed with more sensitive observations (Rea et al. 2005), implying either line variability or spurious detection. The best upper limits, obtained with *XMM-Newton* and *Chandra*, yield equivalent widths smaller than a few tens of eV (Tiengo et al. 2005, 2008; Juett et al. 2002; Rea et al. 2009).

Line smearing, caused, e.g., by the superposition of emission from regions of different field strength, is possibly one of the effects that reduce the detectability of cyclotron lines. Phase-resolved spectroscopy could mitigate this problem, but at the cost of a worse sensitivity due to the lower counts statistics of the spectra.

The absorption line recently discovered in the transient magnetar SGR 0418+5729 (Tiengo et al. 2013) shows indeed a strong dependence on the star rotation phase and could be discovered only thanks to the examination of phase-energy images (see Fig. 4). The line energy varies between $\sim$1 and $\gtrsim$5 keV within a small interval of the spin phase. This strong phase-dependence disfavors an explanation in terms of a cyclotron line from electrons, which, given the dipolar field $B_d = 6 \times 10^{12}$ G inferred from the timing parameters $P = 9.1$ s and $\dot{P} = 4 \times 10^{-15}$ s s^{-1} (Rea et al. 2013b), should be at a height of a few stellar radii. A field of 10^{14} G at the star surface was inferred by fitting the X-ray spectrum of SGR 0418+5729 with a magnetic atmosphere model (Güver et al. 2011). As discussed in Tiengo et al. (2013), the phase-dependent absorption line is best interpreted as a cyclotron feature from protons residing in a relatively small magnetic loop with $B \sim (2\text{–}10) \times 10^{14}$ G, much higher than B_d. If this interpretation is correct, this result, besides providing a direct estimate of the magnetic field strength close to the surface of a magnetar, confirms the complex topology of the magnetosphere, in which global and/or localized twists, as predicted by the magnetar model, play an important role.

2.4 Radio Emission

Most magnetars have not been detected in the radio band, despite being located above the death-line in the $P - \dot{P}$ diagram. The few magnetars detected in this band show radio properties very different from those of rotation-powered neutron stars. Radio pulsations were

[7] Several features have been observed during bursts, e.g.: emission lines at $\sim$13–14 keV in 1E 1048.1−5937 (Gavriil et al. 2002; An et al. 2014), in 4U 0142+61 (Gavriil et al. 2011), and in XTE J1810−197 (Woods et al. 2005); an absorption line at 5 keV (and possibly its harmonics) in SGR 1806−20 (Ibrahim et al. 2003); an emission line at 6.4 keV in SGR 1900+14 (Strohmayer and Ibrahim 2000).

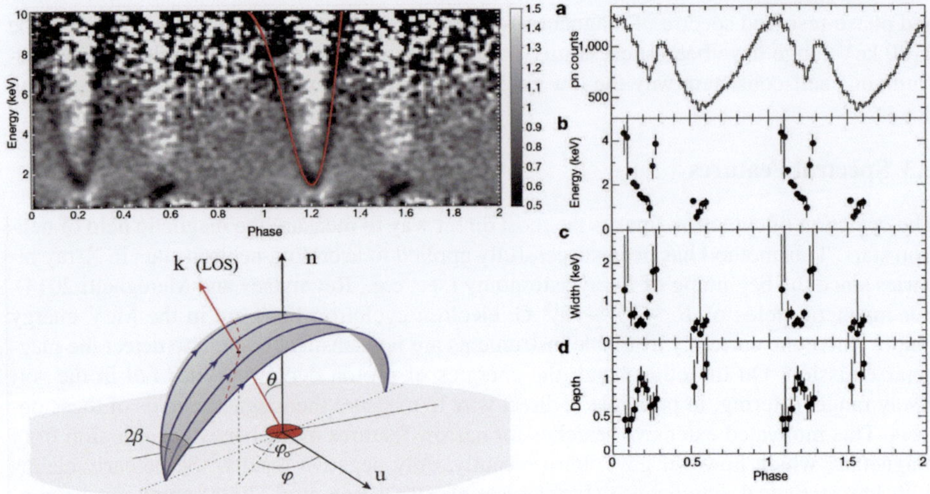

Fig. 4 Phase-dependent line in the spectrum of the transient magnetar SGR 0418+5729 discovered with an *XMM-Newton* observation carried out on 2009 August 12, about two months after the beginning of the outburst (adapted from Tiengo et al. 2013). *Top left*: phase-energy image obtained by binning the source counts into 100 phase bins and 100-eV-wide energy channels and normalizing to the average spectrum and *light curve*. The *red line* indicates (only on one of the two displayed cycles, for clarity) the expected phase-dependence of the line energy in the proton cyclotron model illustrated in the next panel. *Bottom left*: schematic view of the model involving a magnetic loop over the X-ray emitting hot spot. The line of sight (LOS) intercepts the loop at different positions as the star rotates. The magnetic field varies along the loop, causing the observed shift in the line energy. To reproduce the observed feature, the angle between the rotation axis and the normal **n** to the surface at the spot position must be 20° and the LOS must form an angle of 70° with the rotation axis. *Right*: results of the phase-resolved spectroscopy. From top to bottom: 0.3–10 keV pulse profile folded at the spin period of 9.1 s (**a**); line energy (**b**), width (**c**), and depth (**d**) of the cyclotron feature as a function of the spin phase

first detected in two transient magnetars: XTE J1810−197 (Camilo et al. 2006) and 1E 1547.0−5408 (Camilo et al. 2007b). They are characterized by large variability both in flux and pulse profile shape on timescale of days, by a very flat spectrum ($S_\nu \propto \nu^\alpha$, with $\alpha > -0.5$), and high polarization (Camilo et al. 2007c, 2008). The first (and so far the only) magnetar discovered in the radio band, PSR J1622−4950, was reported in 2010 (Levin et al. 2010). Its radio properties are similar to those of the other radio-emitting magnetars. Although no clear signatures of magnetar-like activity have been seen in other wavelengths, its X-ray counterpart decreased in luminosity by a factor over 50 from 2007 to 2011 (Anderson et al. 2012), suggesting that also in this case the radio emission is associated with a transient magnetar. The most recent addition to this small group is the transient magnetar SGR 1729−45 (Mori et al. 2013), which is particularly interesting due to its vicinity to the Galactic center. Its radio dispersion measure (DM $= 1770 \pm 3$ pc cm^{-2}) and Faraday rotation measure (RM $= (-6.696 \pm 0.005) \times 10^4$ rad m^{-2}) are the highest among all known pulsars and indicate a distance very similar to that of the Galactic center (Eatough et al. 2013; Shannon and Johnston 2013). At this distance, the angular separation of 3″ between SGR 1729−45 and the Galactic center black hole Sgr A* corresponds to only ∼0.1 pc and thus SGR 1729−45 has a non-negligible probability of being in a bound orbit with Sgr A* (Rea et al. 2013a). By comparing the DM and RM values of SGR 1729−45 with those of Sgr A*, and considering the density profile of the hot gas seen in X-rays, Eatough et al. (2013) could constrain the magnetic field intensity at the beginning of the accretion flow onto the central black hole to be larger than ~8 mG. If such a field is transported by the accretion flow it can be dynamically important for the accretion process on Sgr A*.

The presence of radio emission gives the possibility to get very accurate positions and to measure proper motion through long baseline radio interferometry. The recently detected proper motion for SGR 1729−45 suggests that this magnetar descends from one of the massive stars in the clockwise-rotating disk around the Galactic center (Bower et al. 2015). Proper motions have been measured also for XTE J1810−197 and 1E 1547.0−5408 (Helfand et al. 2007; Deller et al. 2012) in the radio band, while near IR observations yielded the proper motions of SGR 1806−20 SGR 1900+14 1E 2259+586 and 4U 0142+61 (Tendulkar et al. 2012, 2013). These measurements correspond to transverse velocities of ~100–300 km s^{-1}, not dissimilar from those of rotation powered pulsars (Hobbs et al. 2005).

2.5 Optical and Infrared Emission

The study of magnetars in the optical and infrared is complicated by their intrinsic faintness at these wavelengths and by their location in strongly absorbed and crowded regions of the Galactic plane. Despite these difficulties, counterparts have been found for about one third of the known magnetars, and possible candidates have been suggested for a few other objects. The associations are certain for the three sources showing optical pulsations: 4U 0142+61 (Kern and Martin 2002; Dhillon et al. 2005), 1E 1048.1−5937 (Dhillon et al. 2009), and SGR 0501+4516 (Dhillon et al. 2011), while the other identifications are supported by the detection of long term variability (Israel et al. 2005; Testa et al. 2008; Tam et al. 2004).

The detected counterparts have magnitudes ~23–26 in the optical band and $K \sim 19$–22 in the near infrared (NIR). They are variable, but the relation between the optical and X-ray flux changes is unclear because only few truly simultaneous observations exist and different behaviors have been reported. Correlated variations were seen during the outbursts of XTE J1810−197 (Rea et al. 2004), 1E 2259+586 (Tam et al. 2004), and SGR 0501+4516 (Dhillon et al. 2011), but also cases of apparently uncorrelated or anti-correlated variations were reported (Testa et al. 2008; Camilo et al. 2007a; Durant and van Kerkwijk 2005). The pulse profiles of the three optically pulsed sources show a single broad peak, nearly aligned with the soft X-ray pulse, and pulsed fractions between ~20 % and ~50 %.

In the context of the magnetar scenario a few ideas for the origin of the optical/NIR emission have been put forward, involving non-thermal magnetospheric emission (Eichler et al. 2002; Beloborodov and Thompson 2007; Zane et al. 2011a; Beloborodov 2013a), but a detailed model is still lacking.

2.6 Correlations

Based on the small sample of seven AXPs and SGRs known at that time, Marsden and White (2001) pointed out that the sources with the larger spin-down rate have smaller photon index in their soft X-ray spectra. The long term evolution of the power-law photon index and $\dot{P}$ in SGR 1806−20 indicates that such a correlation between spectral hardness and average spin-down rate holds also for single sources (Mereghetti et al. 2005b). An updated version of the photon index versus spin-down rate plot (top left panel of Fig. 5) confirms the correlation for persistent sources (squares) and for transients in outburst (red triangles), but only for $\dot{\nu} \gtrsim 10^{-14}$ s^{-2} and with some exceptions. The low-$\dot{P}$ sources have spectra harder than what would be expected from the correlation seen at higher spin-down rates, which also becomes less significant if the spectra of transients in quiescence are considered (blue triangles). In Fig. 5 we show how the spectral hardness correlates with other quantities derived from the

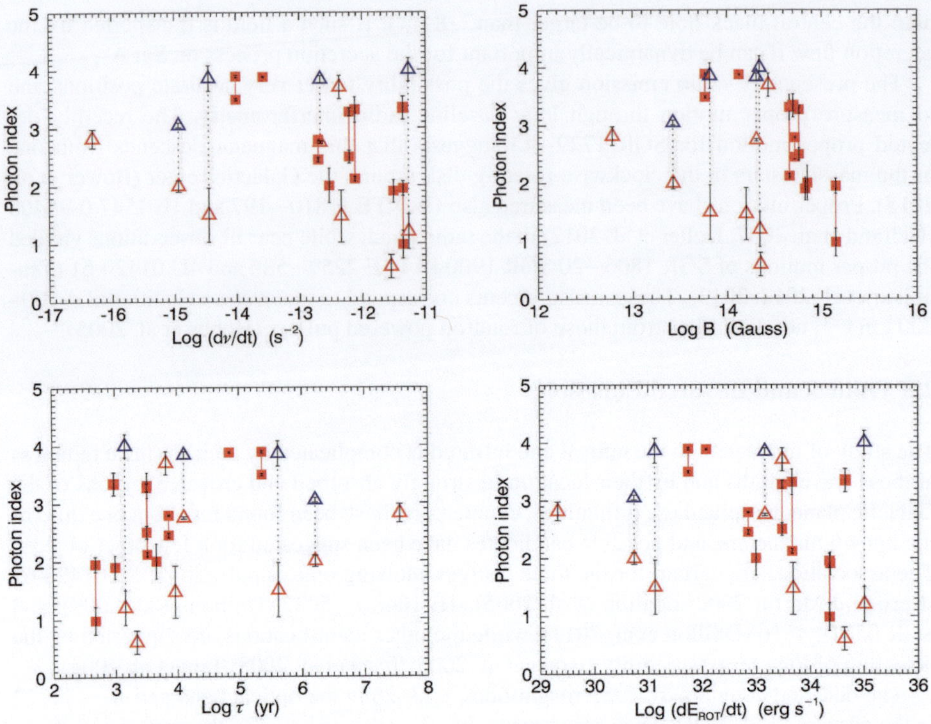

Fig. 5 Dependence of the soft X-ray power-law photon index on different parameters: $\dot{\nu}$ (*top left panel*), B_d (*top right panel*), τ (*bottom left panel*) and $\dot{E}_{rot}$ (*bottom right panel*). *Red squares* indicate persistent sources (maximum and minimum observed values are reported for several sources). *Triangles* indicate transient sources during outburst (*red*) and in during quiescence (*blue*)

timing parameters, such as dipole field $B_d \propto (P\dot{P})^{1/2}$, characteristic age $\tau = P/2\dot{P}$, and spin-down power $\dot{E}_{rot} \propto \dot{P}P^{-3}$. As noted by Kaspi and Boydstun (2010), the best correlation is that with B_d. This is often considered to support the magnetar model: a stronger, and more twisted, field causes a larger spin-down rate as well as stronger magnetospheric currents which harden the spectrum through resonant cyclotron scattering (Baring and Harding 2007). However, the situation is probably more complicated, as also shown by the scatter of the points of Fig. 5, and other parameters might play an important role. As discussed in Sect. 2.2, it is difficult to disentangle the thermal and non-thermal components in the soft X-ray range ($\lesssim$10 keV) and the particular geometry of the twisted bundles of magnetic field lines, not necessarily reflected in the derived B_d values, is the most relevant factor affecting the magnetar emission properties (Beloborodov 2009).

Figure 6 indicates that the blackbody temperatures derived from double component fits to the soft X-ray spectra of magnetars do not show significant correlations with $\dot{P}$ or B_d (Olausen and Kaspi 2014). However, when one compares the average temperature (or thermal luminosity) of magnetars with those of other classes of isolated neutron stars, interesting trends appear: there is a clear correlation between temperature and magnetic field (Aguilera et al. 2008; Olausen et al. 2013) and the magnetars are systematically more luminous than rotation-powered neutron stars of comparable characteristic age (see Sect. 4).

The first observations of persistent emission from magnetars above ~20 keV revealed a difference between the steep spectra of two SGRs with a remarkable record of bursting

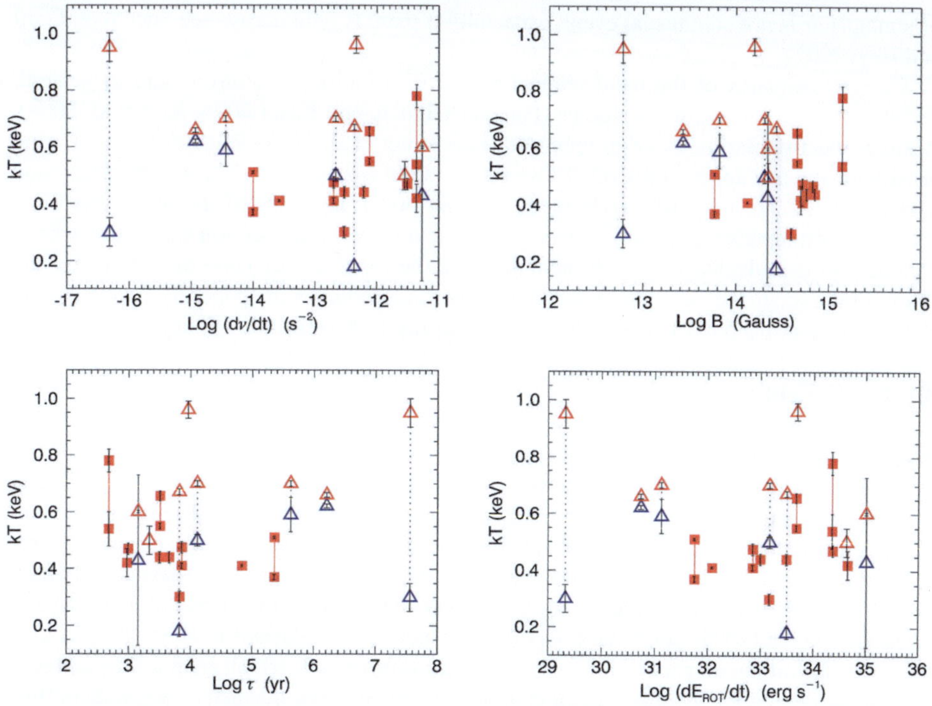

Fig. 6 Dependence of the soft X-ray blackbody temperature on different parameters: $\dot{\nu}$ (*top left panel*), B_d (*top right panel*), τ (*bottom left panel*) and $\dot{E}_{rot}$ (*bottom right panel*). *Red squares* indicate persistent sources (maximum and minimum observed values are reported for several sources). *Triangles* indicate transient sources during outburst (*red*) and in during quiescence (*blue*)

activity (SGR 1806−20 and SGR 1900+14) and those of the much quieter AXPs, which showed instead a significant hardening at high energy (Götz et al. 2006). Enoto et al. (2010c) reported a correlation between the hardness ratio (between the 20–100 keV and 2–10 keV fluxes) and B_d, based on the five persistent magnetars considered by Götz et al. (2006) plus the transient SGR 0501+4516. Similar conclusions were obtained by Kaspi and Boydstun (2010), who examined how the spectral turnover (defined as the difference between the photon indexes of soft and hard X-rays) correlates with spin-down rate and B_d.

3 Magnetar Formation and Origin of Magnetic Field

Scenarios of magnetar formation need to reconcile two somewhat contradictory trends. On the one hand, the strong fields of magnetars argue for special conditions at birth: a highly magnetized progenitor in the fossil field hypothesis (Woltjer 1964), or a rapidly rotating protoneutron star in the shear-driven dynamo hypothesis (Thompson and Duncan 1993). On the other hand, the magnetar birth rate ($\gtrsim 0.3$ century^{-1}; Keane and Kramer 2008) is comparable to rate of the core collapse supernovae (1.9 ± 1.1 century^{-1}), measured by taking a γ-ray census of radioactive ^{26}Al in the Galaxy[8] (Diehl et al. 2006). In this respect, the birth

[8]The estimate from ^{26}Al is subject to systematic uncertainties (included in the quoted error), as the isotopic yield is model dependent, and there is an unknown yield contribution from local spallation processes and

of a magnetar is not an unusual event, especially if there is field decay (see Sect. 4 and Gill and Heyl 2007).

Current estimates of the natal magnetic field distribution of neutron stars in general, based on population synthesis studies (Faucher-Giguère and Kaspi 2006; Kiel et al. 2008), Zeeman spectropolarimetry of progenitors (Landstreet et al. 2007; Wade et al. 2014), and gravitational wave upper limits (Dall'Osso et al. 2009; Mastrano et al. 2011; Mastrano and Melatos 2012), are not sufficiently precise to pin down the shape of the distribution and test whether magnetars populate a second, high-field peak. In this section we present a brief summary of the advantages and disadvantages of the fossil and dynamo hypothesis for the origin of the magnetar magnetic field. The reader is referred to the paper by Ferrario et al. (2015, this issue) and to the in-depth review by Spruit (2009) for more details.

3.1 Fossil Field

Magnetic flux conservation during the collapse of a massive progenitor, e.g. a chemically peculiar star with radius 3 $R_\odot$ and a dipole magnetic field strength $B_0 \sim 10$ kG, produces a natal neutron star magnetic field nominally as large as $B \sim 5 \times 10^{15}$ G, enough to account for all known magnetars (Woltjer 1964; Ruderman 1972). The fossil field scenario is therefore economical as it does not invoke a separate mechanism to produce magnetars, which should naturally derive from the strong-field tail of the progenitor distribution. Ferrario and Wickramasinghe (2006) performed population synthesis calculations to show that the idea accounts also for the magnetization of strong-field white dwarfs, which exhibit similar mass-flux ratios. Population synthesis computations also predict that magnetars originate preferentially from the most massive O stars, consistent with some magnetars being associated with massive star clusters (Muno et al. 2006; Clark et al. 2014; Eikenberry et al. 2004; Vrba et al. 2000) and with their very small scale height on the Galactic plane (Olausen and Kaspi 2014). Therefore magnetars should be more massive than ordinary neutron stars, a claim which cannot be tested observationally at present.

The nominal maximum field $B \sim 5 \times 10^{15}$ G implied by magnetic flux conservation is hard to attain for three reasons. First, only the central ~ 2 % by cross-sectional area of the progenitor collapses to form a protoneutron star, reducing B proportionally. Second, there are too few progenitors with $B_0 \gtrsim 10$ kG to account for the magnetar birth rate inferred observationally (Keane and Kramer 2008; Woods 2008). Third, magnetic core-envelope coupling in the progenitor brakes the core too efficiently to explain the observed neutron star spin distribution (Spruit and Phinney 1998) and leaves no room for the magnetic-dipole braking evolution normally envisaged for magnetars unless supernova kicks play a role. (Core-envelope coupling is an issue for the dynamo scenario too.)

3.2 Protoneutron Star Dynamo

Following flux compression, the magnetic field in a protoneutron star can be amplified further (over ~ 10 s) by dynamo action driven by convection (Thompson and Duncan 1993; Bonanno et al. 2005) or differential rotation (Braithwaite 2006; Moiseenko et al. 2006). The relatively high ($\gtrsim 10$ %) incidence of progenitors with $B_0 \gtrsim 0.25$ kG from spectral class F0 to O4 (Wade et al. 2014) ensures that a seed field is available.

specific star-forming regions. This rate is broadly consistent with extragalactic estimates (Dahlen et al. 2012; Taylor et al. 2014) which are undifferentiated by galactic type, e.g. comparative studies in the Local Volume which are limited statistically by the small number of events (Botticella et al. 2012).

Neutrino-driven convection leads to protoneutron star fields of about 10^{11} G for a neutrino luminosity of $\sim 10^{44}$ erg s^{-1}, which increases $\sim 10^4$-fold by flux conservation when the protoneutron star collapses to form a magnetar (Thompson and Duncan 1993). However, these maximum field values are hard to attain, because dynamos typically operate at $\lesssim 5$ % of equipartition (Cook et al. 2003; Braithwaite 2006) and stratification quenches convection (Spruit 2009).

In contrast, counter-intuitively, a shear-driven dynamo operates more efficiently under stratification (Braithwaite 2006). The poloidal and toroidal field components grow in concert through the action of Tayler and/or magnetorotational instabilities (Braithwaite 2006; Moiseenko et al. 2006), as well as r-mode instabilities (Cheng and Yu 2014); the same instabilities also prevent premature saturation caused by back-reaction stress from the wound-up toroidal field. Differential rotation can also arise from binary mergers (Wickramasinghe et al. 2014), and mean-field magnetohydrodynamics (e.g. the α effect and anisotropic resistivity) and superfluid circulation to assist with amplification (Mastrano et al. 2011; Melatos 2012; Gusakov and Kantor 2013).

We do not expect to see evidence for rapid rotation at birth in the current magnetar population due to magnetic braking. However, a rapidly rotating protomagnetar is expected to power an energetic, relativistic wind for $\sim 10^2$ s after birth, whose energy content is deposited in the supernova remnant. X-ray observations of three supernova remnants associated with magnetars find no evidence for such "over-powering", implying initial spin periods $\gtrsim 5$ ms (Vink and Kuiper 2006; Martin et al. 2014), although this conclusion assumes an idealized Sedov expansion and neglects gravitational radiation which can nullify the over-powering issue (Dall'Osso et al. 2009).

In both dynamo scenarios, magnetic flux tends to escape buoyantly from the dynamo region (Reisenegger and Goldreich 1992; Spruit 2009; Gusakov and Kantor 2013). Indeed, most of the flux would be lost via this process, were it not for helicity conservation, which stabilizes the situation for linked poloidal-toroidal fields under a variety of conditions (Braithwaite and Nordlund 2006; Lander and Jones 2012; Akgün et al. 2013; Ciolfi and Rezzolla 2013; Gourgouliatos and Cumming 2014).

4 Magneto-Thermal Evolution of Magnetars

In neutron stars endowed with strong magnetic fields, the temperature and magnetic field evolution are closely inter-related. On one hand, the dissipation rate of the magnetic field depends on the local value of the electrical resistivity, which is a quantity strongly dependent on temperature. On the other hand, the microphysics ingredients determining the temperature evolution (heat capacity, thermal conductivity, neutrino emission rates) are significantly modified by the presence of a strong field.

A thorough and comprehensive discussion of all the aspects involved in the magneto-thermal evolution of neutron stars can be found in the recent work by Viganò et al. (2013), where the authors present results from simulations including two major novelties extending previous works (Geppert et al. 1999; Page et al. 2000; Pons and Geppert 2007; Aguilera et al. 2008; Pons et al. 2009): the proper treatment of the important Hall term in the induction equation describing the magnetic field evolution (Viganò et al. 2012), and updated microphysics inputs (see Lai 2015, for a review on matter properties in strong fields). In this section, we briefly summarize the equations, the method, and the updated ingredients of the simulations.

4.1 Basic Equations

For our purposes, the small structural deformations induced by rotation and magnetic field can be safely neglected. To include general relativistic effects, we consider the standard static metric

$$ds^2 = -c^2 e^{2\nu(r)} dt^2 + e^{2\lambda(r)} dr^2 + r^2 d\Omega^2, \tag{1}$$

where $e^{2\lambda(r)} = 1 - 2Gm(r)/c^2 r$, $m(r)$ is the enclosed mass within radius r, and $\nu(r)$ is the metric factor accounting for redshift corrections.

The neutron star magneto-thermal evolution is described by the coupled system formed by the energy balance and the Hall induction equations. The first reads:

$$c_v e^{\nu} \frac{\partial T}{\partial t} - \vec{\nabla} \cdot \left[e^{\nu} \hat{\kappa} \cdot \vec{\nabla}(e^{\nu} T)\right] = e^{2\nu}(-\mathcal{Q}_\nu + \mathcal{Q}_h) \tag{2}$$

where c_v is the volumetric heat capacity, $\hat{\kappa}$ is the thermal conductivity tensor, $\mathcal{Q}_\nu$ are the energy losses by neutrino emission per unit volume, and $\mathcal{Q}_h$ is the Joule heating rate per unit volume. This is the first important coupling between the two evolution equations, because $\mathcal{Q}_h = \vec{j}^2/\sigma$, where j is the electrical current determined by the magnetic field geometry and σ the electrical conductivity. The second important effect of the presence of a strong magnetic field in the conduction of heat is the anisotropic conductivity tensor ($\hat{\kappa}$), and the last one is that the magnetic field also affects the rate of neutrino processes, $\mathcal{Q}_\nu$.

In the crust, ions form a Coulomb lattice, while electrons are relativistic, degenerate and can almost freely flow, providing the currents that sustain the magnetic field. The evolution of the magnetic field is governed by the Hall induction equation which, using the same notation as in Pons et al. (2009), has the form:

$$\frac{\partial \vec{B}}{\partial t} = -\vec{\nabla} \times \left[\frac{c^2}{4\pi\sigma} \vec{\nabla} \times (e^{\nu} \vec{B}) + \frac{c}{4\pi e n_e} \left[\vec{\nabla} \times (e^{\nu} \vec{B})\right] \times \vec{B}\right] \tag{3}$$

where the conductivity σ takes into account all the electron processes, which are strongly temperature-dependent, thus resulting in the strong coupling of the magnetic field evolution to the local evolution of temperature. The first term on the right hand side accounts for Ohmic dissipation, while the second term is the Hall term. The previous equation can be cast as

$$\frac{\partial \vec{B}}{\partial t} = -\vec{\nabla} \times \left\{\eta\left(\vec{\nabla} \times (e^{\nu} \vec{B}) + \omega_B \tau_e \left[\vec{\nabla} \times (e^{\nu} \vec{B})\right]\right) \times \vec{B}\right\} \tag{4}$$

where we have introduced the magnetic diffusivity $\eta = \frac{c^2}{4\pi\sigma}$ and the magnetization parameter $\omega_B \tau_e \equiv \frac{\sigma B}{c e n_e}$ (where $\omega_B = eB/m_e^* c$ is the gyration frequency of electrons, with τ_e and m_e^* are the relaxation time and effective mass of electrons). This term regulates whether the evolution is dominated by the diffusive term or by the Hall term. In the regime where $\omega_B \tau_e \gg 1$ (strong magnetic fields, $\gtrsim 10^{14}$ G, and temperatures $\lesssim 5 \times 10^8$ K, see Pons and Geppert 2007; Aguilera et al. 2008; Pons et al. 2009; Viganò et al. 2012 for more details), the Hall term dominates, and the induction equation acquires a hyperbolic character. The Ohmic and Hall timescales vary by orders of magnitude within the crust and during the evolution, depending strongly on density, temperature, and magnetic field intensity and curvature. The main effect of the Hall term is to transfer part of the magnetic energy from large to small scales, as well as between poloidal and toroidal components. In the case of strong toroidal components, it also leads to the formation of discontinuities of the tangential components of the magnetic field, i.e. current sheets, where the dissipation is strongly enhanced (Viganò et al. 2012). This directly affects the thermal evolution through the term $\mathcal{Q}_h$ in Eq. (2).

The previous equations provide a proper description of the physics of the crust, once the microphysical input is provided. In the neutron star core, however, the situation is more complex. The core of neutron stars (or at least a fraction of its volume) is thought to be a type II superconductor (Migdal 1959; Baym et al. 1969). The dynamics of the magnetic field in the core are not clearly understood. Standard Ohmic dissipation is irrelevant due to high conductivity, but other mechanisms such as the interplay between flux-tubes and vortices, magnetic buoyancy, or ambipolar diffusion may operate to expel magnetic flux from the core on timescales comparable to the thermal evolution timescale. The detailed study of these mechanisms is still lacking, and this explains why most previous works considered models with the field confined into the crust or used a very crude approach for the magnetic fields permeating the core (Hollerbach and Rüdiger 2002; Pons and Geppert 2007; Pons et al. 2009; Gourgouliatos and Cumming 2014). Some basic issues as whether or not ambipolar diffusion plays any role at all are still under debate (Glampedakis et al. 2011).

4.2 Results

From detailed numerical simulations solving the above system of equations, one can in principle obtain the local temperature at each point of the neutron star surface and the integrated luminosity, to be compared to observations. But the reality is more complicated. Although both luminosities and temperatures can be obtained by spectral analysis, it is usually difficult to determine them accurately. The luminosity is always subject to the uncertainty in the distance estimate, while the inferred effective temperature depends on the choice of the emission model (blackbody vs. atmosphere models, composition, condensed surface, etc.), and it carries large theoretical uncertainties in the case of strong magnetic fields. It is often found that more than one model can fit equally well the data, without any clear, physically motivated preference for one of them. Photoelectric absorption from interstellar medium further constitutes a source of error in temperature measurements, since the value of the hydrogen column density is covariant with the temperature value obtained in spectral fits. Different choices for the absorption model and the metal abundances can also yield different results for the temperature. In addition, in the very common case of inhomogeneous surface temperature distributions, only an approximation with two or three regions at different temperatures is usually employed. Moreover, in the case of data with few photons and/or strong absorption features, the temperature is poorly constrained by the fit, adding a large statistical error to the systematic one.

With all the previous caveats in mind, the studies of the magneto-thermal evolution of isolated NSs, have explored the influence of their initial magnetic field strength and geometry, their mass, envelope composition, and relevant microphysical parameters such as the impurity content of the innermost part of the crust (the pasta region). The main findings can be summarized as follows (see Figs. 7 and 8):

- *Dependence on the magnetic field strength*: The comparison between a range of theoretical models and the observations (Viganò et al. 2013), has shown that, for the objects born with relatively low fields ($B_p \lesssim 10^{14}$ G), the magnetic field has little effect on the luminosity. Sources of this group, which includes most of the "normal" radio pulsars, have luminosities which are compatible with the predictions of standard cooling models. The bulk of the magnetars, with $B_p \sim$ a few $\times\ 10^{14}$ G (as estimated from their timing parameters), display luminosities generally too high to be compatible with standard cooling alone. The magneto-thermal evolutionary models with $B_p^0 \sim (3–5) \times 10^{14}$ G can account

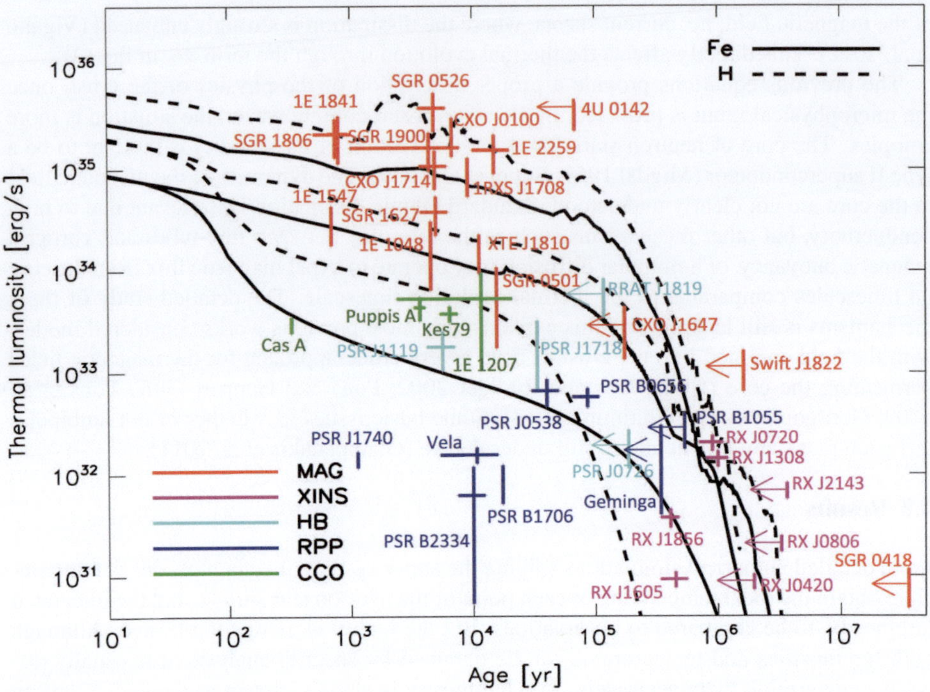

Fig. 7 Thermal X-ray luminosity versus characteristic age for magnetars (MAG) and other classes of isolated neutron stars (from Viganò et al. 2013): X-ray dim isolated neutron stars (XINS), rotation-powered pulsars with high (HB) and normal (RPP) magnetic field, central compact objects in supernova remnants (CCO) (see, e.g., Mereghetti 2011b, for a definition of these classes of sources and their X-ray properties). The *lines* indicate theoretical cooling curves for different compositions of the envelope and three values of the initial magnetic field in the crust (3×10^{15} G, 3×10^{14} G and 0 G, from top to bottom). A neutron star with mass of 1.4 $M_\odot$ and radius 11.6 km has been assumed

for their range of luminosities at the corresponding inferred ages. As these objects evolve and their magnetic fields dissipate, their observational properties (both timing and luminosities) appear compatible with those of the XDINS.[9]

– *Relevance of the Hall term*: The Hall term plays a very important role in the overall magnetic field evolution, strongly enhancing the dissipation of energy over the first $\sim 10^6$ yr of neutron star life, with respect to the purely resistive case. This is due to two main effects: the generation of smaller structures and currents sheets, and the gradual compression of currents and toroidal field towards the crust/core interface. Hence, the rate of field dissipation strongly depends on the resistivity given by the amount of impurities in the innermost region of the crust. A highly impure or amorphous inner crust produces a significant increase in the field dissipation on timescales $\gtrsim 10^5$ yr.
– *Standard vs. fast neutrino cooling scenarios*: For weakly magnetized objects, low mass stars ($M \lesssim 1.4\ M_\odot$) are systematically brighter than high mass stars, because fast neutrino processes are only active above a certain threshold density. This separation is smeared out for highly magnetized stars, for which the overwhelming effect of the magnetic field

[9] X-ray Dim Isolated Neutron Stars, a class of relatively old, purely thermally emitting neutron stars discovered by the *ROSAT* satellite (see, e.g., Turolla 2009, for a review of their properties).

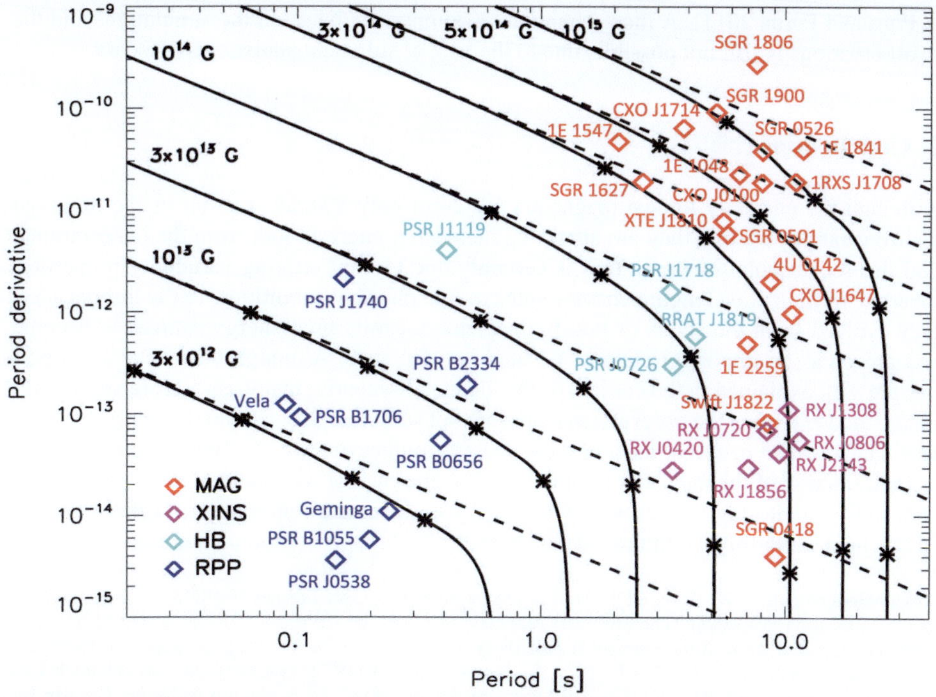

Fig. 8 $P - \dot{P}$ diagram for magnetars and other classes of isolated neutron stars (from Viganò et al. 2013). The *solid lines* indicate theoretical evolutionary tracks for different values of the initial magnetic field. The *points* corresponding to true ages of 10^3, 10^4, 10^5 and 5×10^5 yrs are indicated by the asterisks on each lines. The *dashed lines* represent the evolution without magnetic field decay

makes it hard to distinguish if fast neutrino cooling processes are acting or not, and therefore the mass dependence of the cooling curves is much smaller.

- *Effect of the envelope composition*: As a robust and general trend, light-element envelopes are able to maintain a higher luminosity (up to an order of magnitude) than iron envelopes for a long period of time, $\sim 10^4$ yr, regardless of the magnetic field strength. The most luminous magnetars, with estimated field strengths $\sim 10^{15}$ G from their timing parameters, are barely compatible with the 10^{15} G cooling curve with an iron envelope. However, for the same initial magnetic field, a light envelope is able to account for the luminosity of even the brightest objects.
- *Importance of the initial model*: The initial magnetic field configuration plays a very important role in the observational properties of the NS. If the currents sustaining the magnetic field flow in the core, their dissipation is negligible, comparable with models in which (most of) currents flow in the crust. In particular, the presence of an initial strong dipolar, toroidal field in the crust breaks the symmetry with respect to the equator, resulting in a hemisphere warmer than the other. If most of the initial currents are instead confined into the core, then the reduced heat deposition in the crust results in a much cooler surface compared to the case in which the B field lives in the crust only.
- *Influence in the magnetar outburst rate*: The estimated outburst rate, resulting from breaking of the crust by the strong magnetic stresses, is found to be an increasing function of the initial magnetic field strength and a decreasing function of age (Perna and Pons 2011;

Pons and Perna 2011). A more quantitative comparison between the simulations and the observations is still not possible, due to the lack of sufficient statistics in the data.

5 Conclusions

Although the currently known magnetars represent only a small fraction of the observed neutron star population, they are attracting increasing interest, both from the observational and theoretical point of view. This is certainly due to their striking variability properties, diversity of multiwavelength behavior and extreme physical conditions. In the last decades, they evolved from the status of poorly understood, exotic high-energy sources to become recognized as an important class of isolated neutron stars. Although their general properties are well explained in the context of the magnetar scenario, many aspects are still poorly understood and often the observational data are not sufficient to constrain the model parameters. The transient nature of most of these sources implies that we have now discovered only a small fraction of the magnetar population. The presence of wide field of view instruments constantly monitoring the variable X/γ-ray sky is extremely important to further progress in this field exploiting the improved capabilities of future observational facilities.

Acknowledgements We thank all the staff of the International Space Science Institute and the organizers of the stimulating Workshop "The Strongest Magnetic Fields in the Universe". The work of SM has been partially supported through the agreement ASI-INAF I/037/12/0. JAP acknowledges support of the Spanish national grant AYA 2013-42184-P and of the New Compstar COST action MP1304. AM acknowledges support of an Australian Research Council Discovery Project grant and is grateful to Nicole Darman for assistance with typesetting.

Conflict of interest The authors declare that they have no conflict of interest.

References

A.A. Abdo, M. Ackermann, M. Ajello et al., Astrophys. J. Lett. **725**, L73 (2010)
D.N. Aguilera, J.A. Pons, J.A. Miralles, Astron. Astrophys. **486**, 255 (2008)
T. Akgün, A. Reisenegger, A. Mastrano, P. Marchant, Mon. Not. R. Astron. Soc. **433**, 2445 (2013)
A. Albano, R. Turolla, G.L. Israel et al., Astrophys. J. **722**, 788 (2010)
J. Aleksić, L.A. Antonelli, P. Antoranz et al., Astron. Astrophys. **549**, A23 (2013)
M.A. Alpar, Astrophys. J. **554**, 1245 (2001)
H. An, R. Hascoët, V.M. Kaspi et al., Astrophys. J. **779**, 163 (2013a)
H. An, V.M. Kaspi, R. Archibald, A. Cumming, Astrophys. J. **763**, 82 (2013b)
H. An, V.M. Kaspi, A.M. Beloborodov et al., Astrophys. J. **790**, 60 (2014)
G.E. Anderson, B.M. Gaensler, P.O. Slane et al., Astrophys. J. **751**, 53 (2012)
R.L. Aptekar, T.L. Cline, D.D. Frederiks et al., Astrophys. J. Lett. **698**, L82 (2009)
R.F. Archibald, V.M. Kaspi, C.-Y. Ng et al., Nature **497**, 591 (2013)
M.G. Baring, A.K. Harding, Astrophys. Space Sci. **308**, 109 (2007)
G. Baym, C. Pethick, D. Pines, M. Ruderman, Nature **224**, 872 (1969)
A.M. Beloborodov, Astrophys. J. **703**, 1044 (2009)
A.M. Beloborodov, in *High-Energy Emission from Pulsars and Their Systems*, ed. by D.F. Torres, N. Rea (2011), p. 299
A.M. Beloborodov, Astrophys. J. **777**, 114 (2013a)
A.M. Beloborodov, Astrophys. J. **762**, 13 (2013b)
A.M. Beloborodov, C. Thompson, Astrophys. J. **657**, 967 (2007)
F. Bernardini, G.L. Israel, L. Stella et al., Astron. Astrophys. **529**, A19 (2011)
A. Bonanno, V. Urpin, G. Belvedere, Astron. Astrophys. **440**, 199 (2005)
M.T. Botticella, S.J. Smartt, R.C. Kennicutt et al., Astron. Astrophys. **537**, A132 (2012)
G.C. Bower, A. Deller, P. Demorest et al., Astrophys. J. **798**, 120 (2015)

J. Braithwaite, Astron. Astrophys. **449**, 451 (2006)
J. Braithwaite, Å. Nordlund, Astron. Astrophys. **450**, 1077 (2006)
A. Camero, A. Papitto, N. Rea et al., Mon. Not. R. Astron. Soc. **438**, 3291 (2014)
F. Camilo, S.M. Ransom, J.P. Halpern et al., Nature **442**, 892 (2006)
F. Camilo, I. Cognard, S.M. Ransom et al., Astrophys. J. **663**, 497 (2007a)
F. Camilo, S.M. Ransom, J.P. Halpern, J. Reynolds, Astrophys. J. Lett. **666**, L93 (2007b)
F. Camilo, J. Reynolds, S. Johnston et al., Astrophys. J. Lett. **659**, L37 (2007c)
F. Camilo, J. Reynolds, S. Johnston, J.P. Halpern, S.M. Ransom, Astrophys. J. **679**, 681 (2008)
D. Castro, P. Slane, D.C. Ellison, D.J. Patnaude, Astrophys. J. **756**, 88 (2012)
Q. Cheng, Y.-W. Yu, Astrophys. J. Lett. **786**, L13 (2014)
K.S. Cheng, L. Zhang, Astrophys. J. **562**, 918 (2001)
R. Ciolfi, L. Rezzolla, Mon. Not. R. Astron. Soc. **435**, L43 (2013)
J.S. Clark, B.W. Ritchie, F. Najarro, N. Langer, I. Negueruela, Astron. Astrophys. **565**, A90 (2014)
M. Colpi, U. Geppert, D. Page, Astrophys. J. Lett. **529**, L29 (2000)
J.N. Cook, S.L. Shapiro, B.C. Stephens, Astrophys. J. **599**, 1272 (2003)
T. Dahlen, L.-G. Strolger, A.G. Riess et al., Astrophys. J. **757**, 70 (2012)
S. Dall'Osso, G.L. Israel, L. Stella, A. Possenti, E. Perozzi, Astrophys. J. **599**, 485 (2003)
S. Dall'Osso, S.N. Shore, L. Stella, Mon. Not. R. Astron. Soc. **398**, 1869 (2009)
S. Dall'Osso, J. Granot, T. Piran, Mon. Not. R. Astron. Soc. **422**, 2878 (2012)
A.T. Deller, F. Camilo, J.E. Reynolds, J.P. Halpern, Astrophys. J. Lett. **748**, L1 (2012)
P.R. den Hartog, W. Hermsen, L. Kuiper et al., Astron. Astrophys. **451**, 587 (2006)
P.R. den Hartog, L. Kuiper, W. Hermsen, Astron. Astrophys. **489**, 263 (2008a)
P.R. den Hartog, L. Kuiper, W. Hermsen et al., Astron. Astrophys. **489**, 245 (2008b)
V.S. Dhillon, T.R. Marsh, F. Hulleman et al., Mon. Not. R. Astron. Soc. **363**, 609 (2005)
V.S. Dhillon, T.R. Marsh, S.P. Littlefair et al., Mon. Not. R. Astron. Soc. **394**, L112 (2009)
V.S. Dhillon, T.R. Marsh, S.P. Littlefair et al., Mon. Not. R. Astron. Soc. **416**, L16 (2011)
T. Di Salvo, L. Burderi, N.R. Robba, M. Guainazzi, Astrophys. J. **509**, 897 (1998)
R. Dib, V.M. Kaspi, Astrophys. J. **784**, 37 (2014)
R. Dib, V.M. Kaspi, F.P. Gavriil, Astrophys. J. **673**, 1044 (2008)
R. Dib, V.M. Kaspi, F.P. Gavriil, Astrophys. J. **702**, 614 (2009)
R. Diehl, H. Halloin, K. Kretschmer et al., Nature **439**, 45 (2006)
R.C. Duncan, C. Thompson, Astrophys. J. Lett. **392**, L9 (1992)
M. Durant, M.H. van Kerkwijk, Astrophys. J. **627**, 376 (2005)
R.P. Eatough, H. Falcke, R. Karuppusamy et al., Nature **501**, 391 (2013)
D. Eichler, M. Gedalin, Y. Lyubarsky, Astrophys. J. Lett. **578**, L121 (2002)
S.S. Eikenberry, K. Matthews, J.L. LaVine et al., Astrophys. J. **616**, 506 (2004)
T. Enoto, K. Nakazawa, K. Makishima et al., Publ. Astron. Soc. Jpn. **62**, 475 (2010a)
T. Enoto, K. Nakazawa, K. Makishima et al., Astrophys. J. Lett. **722**, L162 (2010b)
T. Enoto, N. Rea, Y.E. Nakagawa et al., Astrophys. J. **715**, 665 (2010c)
P. Esposito, G.L. Israel, S. Zane et al., Mon. Not. R. Astron. Soc. **390**, L34 (2008)
P. Esposito, A. Tiengo, S. Mereghetti et al., Astrophys. J. Lett. **690**, L105 (2009)
C.A. Faucher-Giguère, V.M. Kaspi, Astrophys. J. **643**, 332 (2006)
R. Fernández, C. Thompson, Astrophys. J. **660**, 615 (2007)
L. Ferrario, D. Wickramasinghe, Mon. Not. R. Astron. Soc. **367**, 1323 (2006)
L. Ferrario et al., Space Sci. Rev. (2015, this issue). doi:10.1007/s11214-015-0138-y
F.P. Gavriil, V.M. Kaspi, Astrophys. J. Lett. **609**, L67 (2004)
F.P. Gavriil, V.M. Kaspi, P.M. Woods, Nature **419**, 142 (2002)
F.P. Gavriil, R. Dib, V.M. Kaspi, Astrophys. J. **736**, 138 (2011)
U. Geppert, D. Page, T. Zannias, Astron. Astrophys. **345**, 847 (1999)
R. Gill, J. Heyl, Mon. Not. R. Astron. Soc. **381**, 52 (2007)
K. Glampedakis, D.I. Jones, L. Samuelsson, Mon. Not. R. Astron. Soc. **413**, 2021 (2011)
D. Götz, S. Mereghetti, A. Tiengo, P. Esposito, Astron. Astrophys. **449**, L31 (2006)
K.N. Gourgouliatos, A. Cumming, Mon. Not. R. Astron. Soc. **438**, 1618 (2014)
M.E. Gusakov, E.M. Kantor, Phys. Rev. D **88**, 101302 (2013)
T. Güver, F. Özel, E. Göğüş, C. Kouveliotou, Astrophys. J. Lett. **667**, L73 (2007)
T. Güver, E. Göğüş, F. Özel, Mon. Not. R. Astron. Soc. **418**, 2773 (2011)
J.P. Halpern, E.V. Gotthelf, Astrophys. J. **618**, 874 (2005)
R. Hascoët, A.M. Beloborodov, P.R. den Hartog, Astrophys. J. Lett. **786**, L1 (2014)
D.J. Helfand, S. Chatterjee, W.F. Brisken et al., Astrophys. J. **662**, 1198 (2007)
J.S. Heyl, L. Hernquist, Astrophys. J. **618**, 463 (2005)
G. Hobbs, D.R. Lorimer, A.G. Lyne, M. Kramer, Mon. Not. R. Astron. Soc. **360**, 974 (2005)

R. Hollerbach, G. Rüdiger, Mon. Not. R. Astron. Soc. **337**, 216 (2002)
Y.F. Huang, J.J. Geng, Astrophys. J. Lett. **782**, L20 (2014)
K. Hurley, P. Li, C. Kouveliotou et al., Astrophys. J. Lett. **510**, L111 (1999)
A.I. Ibrahim, J.H. Swank, W. Parke, Astrophys. J. Lett. **584**, L17 (2003)
G. Israel, S. Covino, R. Mignani et al., Astron. Astrophys. **438**, L1 (2005)
K. Iwasawa, K. Koyama, J.P. Halpern, Publ. Astron. Soc. Jpn. **44**, 9 (1992)
M.S. Jackson, J.P. Halpern, Astrophys. J. **633**, 1114 (2005)
A.M. Juett, H.L. Marshall, D. Chakrabarty, N.S. Schulz, Astrophys. J. Lett. **568**, L31 (2002)
V.M. Kaspi, K. Boydstun, Astrophys. J. Lett. **710**, L115 (2010)
V.M. Kaspi, J.R. Lackey, D. Chakrabarty, Astrophys. J. Lett. **537**, L31 (2000)
V.M. Kaspi, F.P. Gavriil, P.M. Woods et al., Astrophys. J. Lett. **588**, L93 (2003)
V.M. Kaspi, R.F. Archibald, V. Bhalerao et al., Astrophys. J. **786**, 84 (2014)
E.F. Keane, M. Kramer, Mon. Not. R. Astron. Soc. **391**, 2009 (2008)
B. Kern, C. Martin, Nature **417**, 527 (2002)
P.D. Kiel, J.R. Hurley, M. Bailes, J.R. Murray, Mon. Not. R. Astron. Soc. **388**, 393 (2008)
C. Kouveliotou, S. Dieters, T. Strohmayer et al., Nature **393**, 235 (1998)
C. Kouveliotou, T. Strohmayer, K. Hurley et al., Astrophys. J. Lett. **510**, L115 (1999)
L. Kuiper, W. Hermsen, M. Mendez, Astrophys. J. **613**, 1173 (2004)
L. Kuiper, W. Hermsen, P.R. den Hartog, W. Collmar, Astrophys. J. **645**, 556 (2006)
L. Kuiper, W. Hermsen, P.R. den Hartog, J.O. Urama, Astrophys. J. **748**, 133 (2012)
S.R. Kulkarni, D.L. Kaplan, H.L. Marshall et al., Astrophys. J. **585**, 948 (2003)
D. Lai, Space Sci. Rev. **1** (2015). doi:10.1007/s11214-015-0137-z
S.K. Lander, D.I. Jones, Mon. Not. R. Astron. Soc. **424**, 482 (2012)
J.D. Landstreet, S. Bagnulo, V. Andretta et al., Astron. Astrophys. **470**, 685 (2007)
L. Levin, M. Bailes, S. Bates et al., Astrophys. J. Lett. **721**, L33 (2010)
M. Lyutikov, Mon. Not. R. Astron. Soc. **367**, 1594 (2006)
M. Lyutikov, Magnetospheric "anti-glitches" in magnetars. ArXiv e-prints (2013). arXiv:1306.2264
M. Lyutikov, F.P. Gavriil, Mon. Not. R. Astron. Soc. **368**, 690 (2006)
D. Marsden, N.E. White, Astrophys. J. Lett. **551**, L155 (2001)
J. Martin, N. Rea, D.F. Torres, A. Papitto, Mon. Not. R. Astron. Soc. **444**, 2910 (2014)
A. Mastrano, A. Melatos, Mon. Not. R. Astron. Soc. **421**, 760 (2012)
A. Mastrano, A. Melatos, A. Reisenegger, T. Akgün, Mon. Not. R. Astron. Soc. **417**, 2288 (2011)
E.P. Mazets, S.V. Golenetskij, Y.A. Guryan, Sov. Astron. Lett. **5**, 343 (1979a)
E.P. Mazets, S.V. Golentskii, V.N. Ilinskii, R.L. Aptekar, I.A. Guryan, Nature **282**, 587 (1979b)
A. Melatos, Astrophys. J. **761**, 32 (2012)
S. Mereghetti, Astron. Astrophys. Rev. **15**, 225 (2008)
S. Mereghetti, Adv. Space Res. **47**, 1317 (2011a)
S. Mereghetti, in *High-Energy Emission from Pulsars and Their Systems*, ed. by D.F. Torres, N. Rea (2011b), p. 345
S. Mereghetti, L. Stella, Astrophys. J. Lett. **442**, L17 (1995)
S. Mereghetti, A. Tiengo, L. Stella et al., Astrophys. J. **608**, 427 (2004)
S. Mereghetti, D. Götz, I.F. Mirabel, K. Hurley, Astron. Astrophys. **433**, L9 (2005a)
S. Mereghetti, A. Tiengo, P. Esposito et al., Astrophys. J. **628**, 938 (2005b)
S. Mereghetti, P. Esposito, A. Tiengo et al., Astrophys. J. **653**, 1423 (2006)
S. Mereghetti, D. Götz, G. Weidenspointner et al., Astrophys. J. Lett. **696**, L74 (2009)
A.B. Migdal, Nucl. Phys. **13**, 655 (1959)
S.G. Moiseenko, G.S. Bisnovatyi-Kogan, N.V. Ardeljan, Mon. Not. R. Astron. Soc. **370**, 501 (2006)
K. Mori, E.V. Gotthelf, S. Zhang et al., Astrophys. J. Lett. **770**, L23 (2013)
M.P. Muno, J.S. Clark, P.A. Crowther et al., Astrophys. J. Lett. **636**, L41 (2006)
M.P. Muno, B.M. Gaensler, J.S. Clark et al., Mon. Not. R. Astron. Soc. **378**, L44 (2007)
T. Murakami, Y. Tanaka, S.R. Kulkarni et al., Nature **368**, 127 (1994)
L. Nobili, R. Turolla, S. Zane, Mon. Not. R. Astron. Soc. **386**, 1527 (2008a)
L. Nobili, R. Turolla, S. Zane, Mon. Not. R. Astron. Soc. **389**, 989 (2008b)
J.P. Norris, P. Hertz, K.S. Wood, C. Kouveliotou, Astrophys. J. **366**, 240 (1991)
S.A. Olausen, V.M. Kaspi, Astrophys. J. Suppl. Ser. **212**, 6 (2014)
S.A. Olausen, W.W. Zhu, J.K. Vogel et al., Astrophys. J. **764**, 1 (2013)
T. Oosterbroek, A.N. Parmar, S. Mereghetti, G.L. Israel, Astron. Astrophys. **334**, 925 (1998)
B. Paczynski, Acta Astron. **42**, 145 (1992)
D. Page, U. Geppert, T. Zannias, Astron. Astrophys. **360**, 1052 (2000)
D.M. Palmer, Mem. Soc. Astron. Ital. **73**, 578 (2002)
A.N. Parmar, T. Oosterbroek, F. Favata et al., Astron. Astrophys. **330**, 175 (1998)

S.K. Patel, C. Kouveliotou, P.M. Woods et al., Astrophys. J. Lett. **563**, L45 (2001)
R. Perna, J.A. Pons, Astrophys. J. Lett. **727**, L51 (2011)
J.A. Pons, U. Geppert, Astron. Astrophys. **470**, 303 (2007)
J.A. Pons, R. Perna, Astrophys. J. **741**, 123 (2011)
J.A. Pons, J.A. Miralles, U. Geppert, Astron. Astrophys. **496**, 207 (2009)
N. Rea, P. Esposito, in *High-Energy Emission from Pulsars and Their Systems*, ed. by D.F. Torres, N. Rea (2011), p. 247
N. Rea, G.L. Israel, L. Stella et al., Astrophys. J. Lett. **586**, L65 (2003)
N. Rea, V. Testa, G.L. Israel et al., Astron. Astrophys. **425**, L5 (2004)
N. Rea, T. Oosterbroek, S. Zane et al., Mon. Not. R. Astron. Soc. **361**, 710 (2005)
N. Rea, R. Turolla, S. Zane et al., Astrophys. J. Lett. **661**, L65 (2007)
N. Rea, S. Zane, R. Turolla, M. Lyutikov, D. Götz, Astrophys. J. **686**, 1245 (2008)
N. Rea, G.L. Israel, R. Turolla et al., Mon. Not. R. Astron. Soc. **396**, 2419 (2009)
N. Rea, P. Esposito, R. Turolla et al., Science **330**, 944 (2010)
N. Rea, G.L. Israel, P. Esposito et al., Astrophys. J. **754**, 27 (2012)
N. Rea, P. Esposito, J.A. Pons et al., Astrophys. J. Lett. **775**, L34 (2013a)
N. Rea, G.L. Israel, J.A. Pons et al., Astrophys. J. **770**, 65 (2013b)
N. Rea, D. Viganò, G.L. Israel, J.A. Pons, D.F. Torres, Astrophys. J. Lett. **781**, L17 (2014)
A. Reisenegger, P. Goldreich, Astrophys. J. **395**, 240 (1992)
M. Revnivtsev, S. Mereghetti, Space Sci. Rev. (2014). doi:10.1007/s11214-014-0123-x
R.E. Rothschild, S.R. Kulkarni, R.E. Lingenfelter, Nature **368**, 432 (1994)
M. Ruderman, Annu. Rev. Astron. Astrophys. **10**, 427 (1972)
S. Şaşmaz Muş, E. Göğüş, Astrophys. J. **723**, 100 (2010)
P. Scholz, V.M. Kaspi, A. Cumming, Astrophys. J. **786**, 62 (2014)
R.M. Shannon, S. Johnston, Mon. Not. R. Astron. Soc. **435**, L29 (2013)
H.C. Spruit, in *IAU Symposium*, ed. by K.G. Strassmeier, A.G. Kosovichev, J.E. Beckman. IAU Symposium, vol. 259 (2009), pp. 61–74
H. Spruit, E.S. Phinney, Nature **393**, 139 (1998)
T.E. Strohmayer, A.I. Ibrahim, Astrophys. J. Lett. **537**, L111 (2000)
M. Sugizaki, F. Nagase, K. Torii et al., Publ. Astron. Soc. Jpn. **49**, L25 (1997)
C.R. Tam, V.M. Kaspi, M.H. van Kerkwijk, M. Durant, Astrophys. J. Lett. **617**, L53 (2004)
M. Taylor, D. Cinabro, B. Dilday et al., Astrophys. J. **792**, 135 (2014)
S.P. Tendulkar, P.B. Cameron, S.R. Kulkarni, Astrophys. J. **761**, 76 (2012)
S.P. Tendulkar, P.B. Cameron, S.R. Kulkarni, Astrophys. J. **772**, 31 (2013)
V. Testa, N. Rea, R.P. Mignani et al., Astron. Astrophys. **482**, 607 (2008)
C. Thompson, A.M. Beloborodov, Astrophys. J. **634**, 565 (2005)
C. Thompson, R.C. Duncan, Astrophys. J. **408**, 194 (1993)
C. Thompson, R.C. Duncan, Mon. Not. R. Astron. Soc. **275**, 255 (1995)
C. Thompson, R.C. Duncan, Astrophys. J. **473**, 322 (1996)
C. Thompson, M. Lyutikov, S.R. Kulkarni, Astrophys. J. **574**, 332 (2002)
A. Tiengo, S. Mereghetti, R. Turolla et al., Astron. Astrophys. **437**, 997 (2005)
A. Tiengo, P. Esposito, S. Mereghetti, Astrophys. J. Lett. **680**, L133 (2008)
A. Tiengo, P. Esposito, S. Mereghetti et al., Nature **500**, 312 (2013)
H. Tong, Astrophys. J. **784**, 86 (2014)
J.E. Trümper, A. Zezas, Ü. Ertan, N.D. Kylafis, Astron. Astrophys. **518**, A46 (2010)
R. Turolla, in *Astrophysics and Space Science Library*, ed. by W. Becker. Astrophysics and Space Science Library, vol. 357 (2009), p. 141
R. Turolla, P. Esposito, Int. J. Mod. Phys. D **22**, 30024 (2013)
A.J. van der Horst, V. Connaughton, C. Kouveliotou et al., Astrophys. J. Lett. **711**, L1 (2010)
D. Viganò, J.A. Pons, J.A. Miralles, Comput. Phys. Commun. **183**, 2042 (2012)
D. Viganò, N. Rea, J.A. Pons et al., Mon. Not. R. Astron. Soc. **434**, 123 (2013)
J. Vink, L. Kuiper, Mon. Not. R. Astron. Soc. **370**, L14 (2006)
J.K. Vogel, R. Hascoët, V.M. Kaspi et al., Astrophys. J. **789**, 75 (2014)
F.J. Vrba, A.A. Henden, C.B. Luginbuhl et al., Astrophys. J. Lett. **533**, L17 (2000)
G.A. Wade, J. Grunhut, E. Alecian et al., in *IAU Symposium*. IAU Symposium, vol. 302 (2014), pp. 265–269
N.E. White, K.O. Mason, P. Giommi et al., Mon. Not. R. Astron. Soc. **226**, 645 (1987)
N.E. White, L. Angelini, K. Ebisawa, Y. Tanaka, P. Ghosh, Astrophys. J. Lett. **463**, L83 (1996)
D.T. Wickramasinghe, C.A. Tout, L. Ferrario, Mon. Not. R. Astron. Soc. **437**, 675 (2014)
L. Woltjer, Astrophys. J. **140**, 1309 (1964)
P.M. Woods, in *40 Years of Pulsars: Millisecond Pulsars, Magnetars and More*, ed. by C. Bassa, Z. Wang, A. Cumming, V.M. Kaspi. American Institute of Physics Conference Series, vol. 983 (2008), pp. 227–233

P.M. Woods, C. Thompson, Soft gamma repeaters and anomalous X-ray pulsars: Magnetar candidates, in *Compact Stellar X-Ray Sources* (2006), pp. 547–586
P.M. Woods, C. Kouveliotou, J. van Paradijs et al., Astrophys. J. Lett. **524**, L55 (1999a)
P.M. Woods, C. Kouveliotou, J. van Paradijs et al., Astrophys. J. Lett. **519**, L139 (1999b)
P.M. Woods, V.M. Kaspi, C. Thompson et al., Astrophys. J. **605**, 378 (2004)
P.M. Woods, C. Kouveliotou, F.P. Gavriil et al., Astrophys. J. **629**, 985 (2005)
S. Zane, N. Rea, R. Turolla, L. Nobili, Mon. Not. R. Astron. Soc. **398**, 1403 (2009)
S. Zane, L. Nobili, R. Turolla, in *High-Energy Emission from Pulsars and Their Systems*, ed. by D.F. Torres, N. Rea (2011a), p. 329
S. Zane, R. Turolla, L. Nobili, N. Rea, Adv. Space Res. **47**, 1298 (2011b)
L. Zhang, K.S. Cheng, Astrophys. J. **579**, 716 (2002)

Part IV
Accretion, Winds, Jets

DOI 10.1007/978-1-4939-3550-5_11
Reprinted from *Space Science Reviews* Journal, DOI 10.1007/s11214-015-0200-9

Accretion, Outflows, and Winds of Magnetized Stars

Marina M. Romanova[1] · **Stanley P. Owocki**[2]

Received: 18 February 2015 / Accepted: 12 August 2015 / Published online: 7 October 2015

Abstract Many types of stars have strong magnetic fields that can dynamically influence the flow of circumstellar matter. In stars with accretion disks, the stellar magnetic field can truncate the inner disk and determine the paths that matter can take to flow onto the star. These paths are different in stars with different magnetospheres and periods of rotation. External field lines of the magnetosphere may inflate and produce favorable conditions for outflows from the disk-magnetosphere boundary. Outflows can be particularly strong in the propeller regime, wherein a star rotates more rapidly than the inner disk. Outflows may also form at the disk-magnetosphere boundary of slowly rotating stars, if the magnetosphere is compressed by the accreting matter. In isolated, strongly magnetized stars, the magnetic field can influence formation and/or propagation of stellar wind outflows. Winds from low-mass, solar-type stars may be either thermally or magnetically driven, while winds from massive, luminous O and B type stars are radiatively driven. In all of these cases, the magnetic field influences matter flow from the stars and determines many observational properties. In this chapter we review recent studies of accretion, outflows, and winds of magnetized stars with a focus on three main topics: (1) accretion onto magnetized stars; (2) outflows from the disk-magnetosphere boundary; and (3) winds from isolated massive magnetized stars. We show results obtained from global magnetohydrodynamic simulations and, in a number of cases compare global simulations with observations.

Keywords Magnetic field · Accretion disks · Stellar winds · Magnetohydrodynamics

✉ M.M. Romanova
romanova@astro.cornell.edu

S.P. Owocki
owocki@udel.edu

[1] Department of Astronomy and CRSR, Cornell University, Ithaca, NY 14853, USA

[2] Department of Physics & Astronomy, University of Delaware, Newark, DE 19716, USA

1 Introduction

Magnetic fields play important and distinct roles in the dynamics of gas flow around disk-accreting and isolated magnetized stars. In Sects. 1.1 and 1.2, we review different types of disk-accreting magnetized stars, for which the magnetic field determines the processes of accretion and outflow. In Sect. 1.3 we review isolated magnetized stars, where the magnetic field drives or shapes the winds from these stars.

1.1 Accretion onto Magnetized Stars

Different types of *disk-accreting* stars have dynamically important magnetic fields. Some of them are very young and are at the stage of gravitational contraction; others are very old, compact stars. Young ($\lesssim 10^7$ years) low-mass (0.5–$2M_\odot$) stars have significant magnetic fields of a few kG (Donati et al. 2007; Johns-Krull 2007), which is thousands of times stronger than the globally averaged field of the Sun. Some of these stars (called *Classical T Tauri Stars, or CTTSs*) are surrounded by accretion disks, and their fields are strong enough to truncate the disks (see Fig. 1). The truncation (or magnetospheric) radius r_m is determined by the balance between the magnetic stress in the magnetosphere and the matter stress in the disk.[1] For a number of CTTSs, the truncation radii are estimated to be of a few stellar radii. The light curves from such CTTSs show complex patterns of periodic, quasi-periodic, or irregular variability (see Fig. 2; see also Herbst et al. 1994; Stauffer et al. 2014), which are probably connected with complex paths of matter flow around the magnetosphere and rotations of the hot spots that form as a result of the gravitational impact of the free-falling matter to the stellar surface (see, e.g., review by Bouvier et al. 2007a).

Old, compact stars (i.e., white dwarfs, neutron stars) have even stronger magnetic fields, originating from the collapse of a main sequence star under the conservation of magnetic flux. In some *white dwarfs*, called 'polars,' the magnetic field is so strong, ($B \sim 10^7$–10^8 G) that an accretion disk does not form, and matter from the secondary star accretes directly to the large magnetosphere. In white dwarfs known as 'intermediate polars' (IPs), with strong, but less extreme, fields ($B \sim 10^6$ G) an accretion disk forms (e.g., Warner 1995; Hellier 2001). This accretion disk is truncated at large distances from the star, with $r_m \gtrsim 10R_\star$. Such IPs show periodic variability associated with the channelling of disk matter onto the magnetic poles of the star.

Accreting *neutron stars* in binary systems may have very strong magnetic fields, $B \sim 10^{12}$ G, with accretion disks that interact with huge magnetospheres, $r_m \gg R_\star$. Due to their large magnetospheres, neutron stars have relatively long periods of order seconds. There is also a sub-class of neutron stars called accreting millisecond X-ray pulsars (AMXPs, e.g., Wijnands and van der Klis 1998), where the field has decayed to $B \sim 10^7$–10^9 G and the neutron star has spun up due to disk accretion in binary systems (e.g., Bisnovatyi-Kogan and Komberg 1974; Alpar et al. 1982; Belloni et al. 2015). In these stars, as in CTTSs, the disk is truncated at a few stellar radii and the magnetospheric accretion is expected. However, due to the much smaller size of the neutron star, the dynamical time-scale at the truncation radius is only a few milliseconds. In spite of the large difference in size between young CTTSs (about 10^{11} cm) and millisecond-period neutron stars (about 10^6 cm), the physics of the disk-magnetosphere interaction is quite similar: the accretion-induced pulsations are observed in X-ray, which are associated with the rotating hot spots. In addition, high-frequency quasi-periodic oscillations (QPOs) are observed which carry information about processes in the

[1] We further discuss the disk-magnetosphere interaction and the magnetospheric radius in Sect. 2.2.

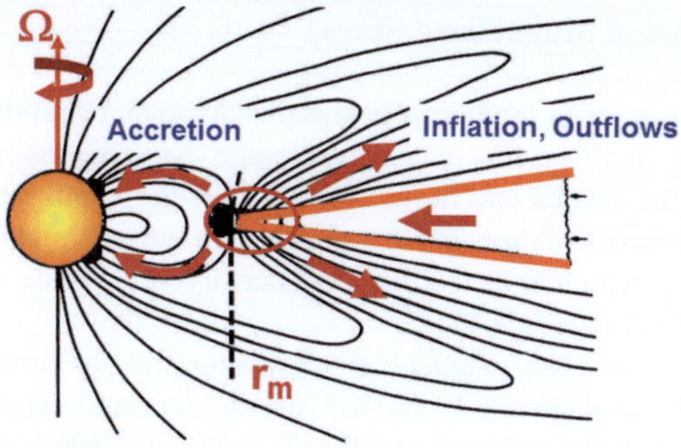

Fig. 1 Sketch shows interaction of a magnetized star with the disk. Matter of the disk is stopped by the magnetosphere. Part of the matter reaches the inner regions of the closed magnetosphere and accretes onto the star in funnel streams. The field lines of the external layers of the magnetosphere can inflate and outflows are possible from the disk-magnetosphere boundary. Based on sketch from Camenzind (1990)

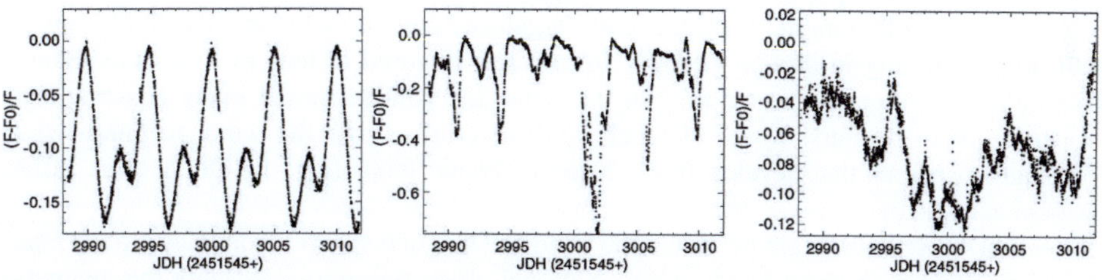

Fig. 2 Different types of light-curves from young stars in NGC 2264 obtained with *CoRoT*: periodic light-curve (*left panel*), AA Tau type light-curve, with characteristic dips (*middle panel*) and irregular light-curve (*right panel*). From Alencar et al. (2010)

inner disk (e.g., Van der Klis 2000; Belloni et al. 2015). Section 2 discusses different aspects of accretion onto magnetized stars.

1.2 Outflows from the Disk-Magnetosphere Boundary

The field lines of the external parts of the magnetosphere may inflate due to the differences in angular velocities of the foot-points of the magnetic loops connecting the star and the disk (Aly and Kuijpers 1990; Camenzind 1990; Shu et al. 1994; Lovelace et al. 1995; Uzdensky et al. 2002). Different signs of episodic inflation are observed in CTTSs (see, e.g., review by Bouvier et al. 2007a). Inflation creates favorable conditions for outflows from the disk-magnetosphere boundary, at which the magnetic field of the star participates in driving outflows (see Fig. 1). Collimated outflows are observed from a number of CTTSs (e.g., Ray et al. 2007). A significant number of CTTSs show signs of outflows in spectral lines, particularly in the near-infrared He I λ10830Å line, for which two distinct components of outflows had been found (e.g., Edwards et al. 2006). Outflows are also observed from accreting compact stars such as accreting white dwarfs in symbiotic binaries (e.g., Sokoloski et al. 2008), as well as from the vicinity of neutron stars (e.g., Fender 2004; Heinz et al. 2007). The origin of these outflows is not yet clear. One of attractive possibilities is that they may originate at the disk-magnetosphere boundary (e.g., Shu et al. 1994; Romanova et al. 2009). In Sect. 3 we discuss different types of outflows from the disk-magnetosphere boundary.

1.3 Winds from Isolated Magnetized Stars

Many stars are isolated, some of them having significant magnetic fields. The magnetic field of the Sun is relatively small—a few Gauss on average. Nonetheless, it plays an important role in the heating of the corona and in the formation of the wind, which is driven by gas pressure. Powerful events of inflation and reconnection of the solar field lines lead to coronal mass ejections (CMEs), which enhance the wind density and induce storms in the Earth's magnetosphere (e.g., Boiko et al. 2012).

In young low-mass stars, the magnetic field is thousands of times stronger than that of the Sun, and much stronger winds and CMEs are expected. These winds can be either thermally or magnetically driven (e.g., Weber and Davis 1967; Lovelace et al. 2008; Vidotto et al. 2009, 2011). In magnetically driven winds, the pressure gradient of the magnetic field determines the acceleration of matter to the wind. This is possible when, for gas pressure P and magnetic field B, the plasma parameter $\beta = 8\pi P/B^2 < 1$ at the base of the wind (Lovelace et al. 2008; Vidotto et al. 2011). In more massive stars ($M > 5M_\odot$) of O, B and A types, powerful winds are driven by the star's radiation pressure. About 10 percent of O and B stars harbor large-scale, organized (often predominantly dipolar) magnetic fields, ranging in dipolar strength from a few hundred to tens of thousands Gauss. This field can influence matter flow in the wind and can determine many observational properties of these stars. It can channel wind material or trap the wind, forming *wind-fed* magnetospheres that develop from closed magnetic loops (e.g., ud-Doula et al. 2013; Owocki et al. 2014).

Section 4 discusses in greater detail the wind-fed magnetospheres from magnetized massive stars along with corresponding observational properties associated with the magnetic field.

The problems of the disk-magnetosphere interaction, outflows and winds from magnetized stars are multidimensional and require global axisymmetric or three-dimensional (3D) numerical simulations. Below, we describe results of recent numerical simulations of accretion and outflows from disk-accreting magnetized stars, wind-fed magnetospheres of isolated massive stars, and show examples of comparisons of models with observations.

2 Magnetospheric Accretion

Magnetospheric accretion is a complex process, in which the inner disk matter interacts with the magnetosphere of the star. The result of such an interaction depends on a number of factors, such as the period of stellar rotation, the structure of the magnetic field of the star, the size of the magnetosphere, the diffusivity at the disk-magnetosphere boundary, properties of the accretion disk, and other factors. We start from a brief overview of studies the magnetospheric accretion.

2.1 Different Studies of the Magnetospheric Accretion

The disk-magnetosphere interaction has been studied in a number of theoretical works (e.g., Pringle and Rees 1972; Lamb et al. 1973; Ghosh and Lamb 1978; Spruit and Taam 1990; Königl 1991; Campbell 1992; Koldoba et al. 2002a). It was predicted that the disk should be truncated by the magnetosphere of the star, and matter should flow above and below the magnetosphere in the funnel flow. Some authors also predicted that the magnetic field lines of the external parts of the magnetosphere may inflate and open due to the differential rotation of the foot-points of the field lines of the external magnetosphere (e.g., Aly and Kuijpers 1990), and outflows are possible along the field lines of the inflated magnetosphere (e.g., Camenzind 1990; Shu et al. 1994; Lovelace et al. 1995).

A number of numerical simulations have been also performed. The process of the disk-magnetosphere interaction is internally multi-dimensional, so that two or three-dimensional simulations are required for investigation of this process. Early axisymmetric simulations (e.g., Stone and Norman 1994; Hayashi et al. 1996; Miller and Stone 1997; Hirose et al. 1997) were relatively brief, only a few Keplerian rotations at the inner disk. Rapid, almost free-fall accretion has been observed due to the initial magnetic braking of the Keplerian disk by the magnetic field of the non-rotating magnetosphere. Strong inflation of the field lines and one-time episode of outflows has been observed. These simulations did show inflation of the field lines and outflows. They also have shown brief events of the magnetospheric accretion; however, the magnetospheric accretion has not been studied due to non-stationary nature of the process. Goodson et al. (1997) has been able to perform longer simulation runs and obtained a few episodes of accretion, inflation and outflows from the disk-magnetosphere boundary. Their simulations confirmed the suggestion of Aly and Kuijpers (1990) about the possibility of the cyclic inflation of the field lines of the external magnetosphere and formation of outflows along the field lines threading the inner disk (Lovelace et al. 1995). However, these authors investigated only the case of a rapidly rotating young star (with period of 1.8 days) in the regime which is close to the propeller regime (e.g., Illarionov and Sunyaev 1975; Lovelace et al. 1999). In subsequent simulations (Goodson et al. 1999; Matt et al. 2002) only rapidly-rotating stars were considered. It is not clear whether these results are general, that is can be applied to any disk-accreting magnetized stars.

Development of quasi-equilibrium initial conditions for the disk-magnetosphere configuration helped to obtain a slow accretion in the disk, and to investigate accretion through the funnel flows (Romanova et al. 2002).[2] Simulations confirmed many aspects of the disk-magnetosphere interaction predicted theoretically, such as the truncation of the accretion disk by the stellar magnetosphere, formation of the funnel flow, and angular momentum flow between the disk and the star. However, in these simulations, the corona above the disk is relatively heavy so that the inflation and outflows were suppressed in most of simulation runs. Simulations with the lower-density corona confirmed accretion through funnel flows and the inflation of the field lines; however, no outflows were observed (Long et al. 2005). Similar simulations were performed by Bessolaz et al. (2008) who also compared the position of the truncation radius of the disk observed in simulations with magnetospheric radii derived from different theoretical studies. Long et al. (2005) studied the angular momentum flow between the disk, star and corona and found conditions for the rotational equilibrium state at which a star neither spins-up, nor spins-down. They found, that in this state, a star rotates somewhat more slowly than the inner disk, because part of angular momentum flows from the star to corona along inflated and partially-inflated field lines. More recently, Zanni and Ferreira (2013) observed many cycles of inflation and reconnection in the external magnetosphere with ejection of matter along the inflated field lines. Simulations show that outflows are strongly enhanced in the propeller regime of accretion (e.g., Romanova et al. 2005b; Ustyugova et al. 2006), or in cases where the accretion disk compresses the magnetic field

[2]In these initial conditions the corona above the disk rotates with the angular velocity of the disk, so that the magnetic field lines do not experience discontinuity at the disk-magnetosphere boundary, and there is no initial magnetic breaking. There is also an initial balance between the gravitation, pressure and centrifugal forces in each point of the simulation region.

of the external magnetosphere (Romanova et al. 2009). These axisymmetric simulations helped to understand many elements of the disk-magnetosphere interaction. They are particularly valuable for investigation of the inflation and outflows in the external magnetosphere, because the coronal density should be low enough for modeling these processes.

The disk-magnetosphere interaction also depends on the properties of an accretion disk. In many studies it is suggested that the outward transport of angular momentum is provided by viscosity, and the viscosity coefficient is proportional to parameter α, where $\alpha < 1$ (Shakura and Sunyaev 1973). This approach has been successfully used in many studies of accretion disks and also has been used in most of modelings of the magnetospheric accretion (e.g., Goodson et al. 1997; Long et al. 2005; Zanni and Ferreira 2013; Romanova et al. 2014). The α-viscosity plays an important role in bringing matter toward the magnetosphere in steady rate. On the other hand, α-viscosity mimics some more complex processes in the disk, such as magnetic turbulence (Shakura and Sunyaev 1973). A significant progress has been made in understanding turbulent disks where the turbulence is driven by the magneto-rotational instability (e.g., Balbus and Hawley 1991, 1998). MRI-driven turbulence has been extensively studied in axisymmetric and local/global 3D MHD simulations (e.g., Hawley et al. 1995; Brandenburg et al. 1995; Stone et al. 1996; Armitage 1998; Hawley 2000; Beckwith et al. 2009; Flock et al. 2011; Simon et al. 2011; McKinney et al. 2014). In most of simulations the central object is non-magnetized, usually a black hole. Accretion onto a magnetized star has been studied in much fewer simulations (Romanova et al. 2011a, 2012). Simulations show that many properties of the magnetospheric accretion are similar in cases of the turbulent and laminar, α-disks, in particular, in cases of the small-scale turbulence.

To model most realistic situation, the disk-magnetosphere interaction should be considered in three dimensions. One of the main issues is that the inner disk matter is expected to penetrate through the magnetosphere (the effective diffusivity) in non-axisymmetric, 3D instabilities, such as the Rayleigh–Taylor and Kelvin–Helmholtz instabilities (e.g., Arons and Lea 1976). Such a penetration will determine the effective "diffusivity" at the disk-magnetosphere boundary and thus the level of interaction between the disk and the star. On the other hand, the dipole magnetic moment may be tilted about the rotational axis of the star, or the magnetic field may be more complex than the dipole field. Therefore, it is important to study the disk-magnetosphere interaction in global 3D simulations. A special "Cubed Sphere" 3D MHD Godunov-type code has been developed for this purpose (Koldoba et al. 2002b) and different 3D simulations were performed for the first time. In this section, we describe mainly the results of the global 3D simulations of the disk-magnetosphere interaction. In particular, we show results of accretion onto a star with a tilted dipole magnetosphere (see Sect. 2.3); accretion through 3D instabilities (see Sect. 2.4); accretion onto stars with relatively large magnetospheres (see Sect. 2.5), and accretion onto stars with complex magnetic fields (see Sect. 2.6). Finally, the 3D MHD models have been developed for CTTS stars with realistic parameters, and results of simulations were compared with observations of these stars (see Sects. 2.7 and 2.8).

2.2 Truncation of the Disk by the Magnetosphere, and Characteristic Radii

If the magnetic field of the star is sufficiently strong, then it truncates the accretion disk at a radius r_m where the magnetic stress in the magnetosphere matches the matter stress in the disk. In the disk, the largest stress is connected with the azimuthal components of the stress tensor: $T_{\phi\phi} = [p + \rho v_\phi^2] + [B^2/8\pi - B_\phi^2/4\pi]$, where ρ, p, B, B_ϕ are the local density, gas pressure, total and azimuthal magnetic field in the disk, respectively. At the innermost

Fig. 3 Slice of the density distribution and selected field lines resulting from 3D MHD simulations of accretion onto a star with a tilted dipole magnetic field. The *dashed line* shows the position of the magnetospheric radius. From Romanova et al. (2004)

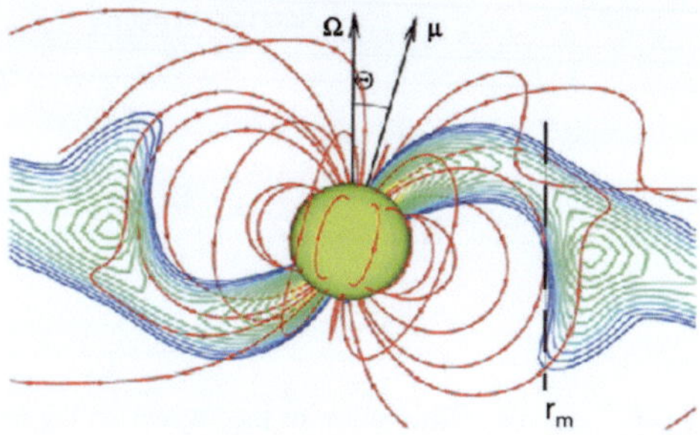

edge of the disk, $B_\phi \ll B$ and the truncation radius can be found from the condition $p + \rho v_\phi^2 = B^2/8\pi$. However, it is better to include the total ram pressure which acts in all three directions and to compare the total matter pressure with the total magnetic pressure

$$p + \rho v^2 = B^2/8\pi, \quad \text{or } \beta_t = 8\pi\left(p + \rho v^2\right)/B^2 = 1, \tag{1}$$

Here, β_t is the generalized plasma parameter, which includes both *thermal and ram pressure* (Romanova et al. 2002). It is analogous to the standard plasma parameter $\beta = 8\pi p/B^2$, but takes into account the ram pressure of the matter flow in the disk. Axisymmetric and global 3D MHD simulations show that the condition $\beta_t = 1$ is useful for finding the magnetospheric radius r_m (e.g., Kulkarni and Romanova 2013). This radius usually corresponds to the *innermost* edge of the disk, where the density drops sharply towards the low-density magnetosphere (in which magnetic pressure is dominant). This formula can be used to find r_m in many situations, including those that are non-stationary. There is also no restriction on the geometry of the magnetic field of the star (see also Sect. 2.6).[3] Fig. 3 shows a snapshot from 3D simulations of a laminar, α-disk. The vertical dashed line shows the magnetospheric radius calculated from the stress balance Eq. (1).

In earlier theoretical studies (e.g., Pringle and Rees 1972; Lamb et al. 1973) the magnetospheric radius was estimated from the balance between the largest components of the stresses, assuming a dipole field from the star and Keplerian orbital speed in the disk; this gives

$$r_m = k\left[\mu^4/\left(\dot{M}^2 G M_\star\right)\right]^{1/7}, \quad k \sim 1, \tag{2}$$

where $\mu = B_\star R_\star^3$ is the magnetic moment of the star with a surface field $B_\star$, $\dot{M}$ is the disk accretion rate, and $M_\star$ and $R_\star$ are the mass and radius of the star, respectively. Axisymmetric disk-accretion simulations by, e.g., Long et al. (2005) found that Eqs. (1) and (2) give similar values of r_m if $k \approx 0.5$ (Long et al. 2005). Bessolaz et al. (2008) compared different approaches for estimating the magnetospheric radius, and found that all give similar values, essentially set by the steep ($B^2 \sim r^{-6}$) decline of the magnetic stress, which dominates over the much more gradual variation of the disk matter stress.

From a series of 3D MHD simulations, Kulkarni and Romanova (2013) fit a scaling for the magnetospheric radius,

$$r_m/R_\star \approx 1.06\left[\mu^4/\left(\dot{M}^2 G M_\star R_\star^7\right)\right]^{1/10}, \tag{3}$$

[3] Sometimes, the condition $\beta = 8\pi p/B^2 = 1$ is used to find the magnetospheric radius (e.g., Bessolaz et al. 2008). This condition, however, yields somewhat a larger radius at which the matter flows from the disk to the funnel stream.

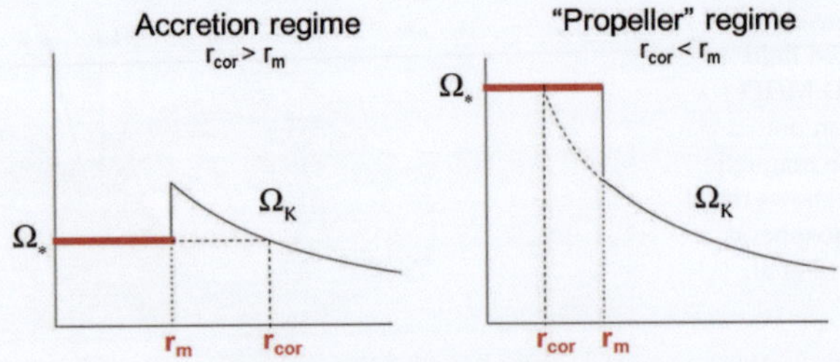

Fig. 4 *Left panel*: Illustration of the "accretion regime" (*left*), for which rigid-body rotation at the stellar rotation frequency $\Omega_\star$ extends to a magnetic radius r_m less than the corotation radius, $r_{\rm cor}$, at which the local Keplerian frequency $\Omega_K = \Omega_\star$; the *sub*-Keplerian rotation of the inner region facilitates accretion of disk material onto the star. *Right panel*: The "propeller regime" (*right*), wherein $r_{\rm cor} > r_m$; the *super*-Keplerian rotation of the inner region now causes disk material to be propelled into an outflow by magneto-centrifugal forces. The condition $r_m = r_{\rm cor}$ thus represents the boundary between these regimes. From Ustyugova et al. (2006)

that is somewhat flatter than the $r_m \sim (\mu^2/\dot{M})^{1/7}$ scaling of Eq. (2). This difference can be attributed to the non-dipole form of the magnetic field, which results from the compression of the magnetosphere by the accretion disk (see Fig. 5, middle panel). Despite this modest difference in dependencies, the formula (2) with coefficient $k \approx 0.5$ still describes the position of the magnetospheric radius with accuracy sufficient in many astrophysical situations. But note that this scaling has only been tested for accretion onto relatively small magnetospheres, $r_m \lesssim 5R_\star$; accretion onto stars with larger magnetospheres has not yet been studied systematically.[4,5]

Another important radius is the *corotation radius*, at which the angular velocity of the star matches the Keplerian angular velocity of the disk: $\Omega_\star = \sqrt{GM_\star/r_{\rm cor}^3}$:

$$r_{\rm cor} = \left[GM_\star/\Omega_\star{}^2\right]^{1/3}. \tag{4}$$

As illustrated in Fig. 4, the condition $r_{\rm cor} = r_m$ characterizes the boundary between "accretion" vs. "propeller" regimes. The left panel demonstrates the case favorable for accretion, in which $r_m < r_{\rm cor}$, and the inner disk rotates more rapidly than the magnetosphere of the star; the accreting matter interacts with the magnetosphere, loses some of its angular momentum, and accretes onto the star. The right panel of Fig. 4 shows the opposite situation, in which $r_{\rm cor} < r_m$ and the matter of the inner disk gains angular momentum; this regime is called the "propeller" regime (e.g., Illarionov and Sunyaev 1975; Alpar and Shaham 1985; Lovelace et al. 1999), in which the inner disk matter can be pushed away by the rapidly-rotating magnetosphere, while some of the matter can be ejected as a wind. This situation will be considered in greater detail in Sect. 3.1.

It is also convenient to characterize the state of the accreting magnetized star with the fastness parameter, ω_s, which is defined as the ratio between the angular velocity of the star,

[4] Simulations of accretion onto stars with larger magnetosphere require much longer computing time. Test simulations of such accretion are described in Sect. 2.5. However, to obtain a formula for magnetospheric radius, multiple simulations are needed.

[5] The compression of the magnetosphere is probably connected with the ram pressure in the radial direction and so may depend on the value of the radial velocity (which is proportional to α-parameter of viscosity). The above described simulations were performed for $\alpha = 0.02$. The possible dependence of the compression on α-parameter should be studied in a separate set of simulations.

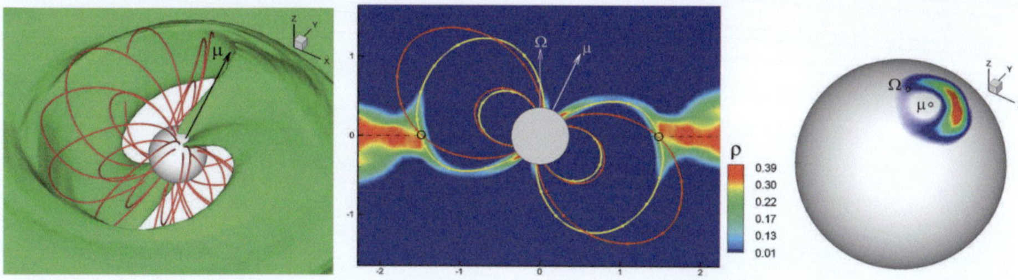

Fig. 5 *Left panel*: A 3D view of the funnel flow from the disk to a magnetized star, where the dipole moment μ is tilted by $\Theta = 20°$ about the rotational axis. One of the density levels is shown in *green*; sample field lines are shown in *red*. *Middle panel*: The slice of density distribution (*color background*) and magnetic field lines (*yellow lines*). *Red lines* show the dipole field lines at $t = 0$. The external field lines are chosen such that both the *yellow* and the *red lines* start at the inner magnetospheric radius r_m, which is marked as a *black circle*. *Right panel*: the energy flux distribution on the surface of the star. *Circles* show the position of the magnetic (μ) and rotational (Ω) axes, respectively. From Kulkarni and Romanova (2013)

$\Omega_\star$ and Keplerian velocity of the inner disk at $r = r_m$:

$$\omega_s = \frac{\Omega_\star}{\Omega_K(r_m)} = \left(\frac{r_m}{r_{\rm cor}}\right)^{3/2}. \tag{5}$$

This parameter efficiently characterizes different states of the accreting magnetized star (e.g., Ghosh 2007). In the following sections we show that many processes at the disk-magnetosphere boundary can be characterized by this parameter.

2.3 Accretion onto Stars with a Tilted Dipole Magnetic Field

The magnetic axis of the dipole moment, μ, can be tilted by an angle, Θ, about the rotational axis, $\Omega_\star$, of the star. The rotational axis of the star is fixed, and it coincides with the rotational axis of the disk. In general, the rotational axis of the star may be also tilted about the rotational axis of the disk, and this leads to secular warping and precession of the inner disk (e.g., Lipunov and Shakura 1980; Lai 1999). In this Chapter, however, we consider only cases in which the rotational axes of the star and disk coincide.

Global 3D MHD simulations of accretion onto stars with different tilt angles of the dipole field, Θ, show that matter flows onto the star via two funnel streams (Romanova et al. 2003, 2004). The matter in the funnel streams is pulled towards the star by the gravitational force. Numerical simulations have shown for the first time the structure and shape of the funnel streams. They have shown that funnel streams are wide in the meridional direction (see left panel of Fig. 5) and narrow in the vertical direction (see slice of the density distribution in the middle panel of Fig. 5). They also show that the low-density matter blankets the whole magnetosphere, while the funnel streams represent the denser parts of the flow (see Fig. 7). The slice in the middle of Fig. 5 shows both the dipole field lines, and the field lines during the funnel accretion.[6] It is still not clear how important such compression will be in stars with much larger magnetospheres, i.e. with $r_m/R_\star > 5$. But we can speculate that the field compression might become less significant in such cases, because only the external parts of the magnetosphere interact with the disk, while the larger part of the magnetosphere is strongly magnetically dominated and so has a nearly dipole structure.

[6]We should note that simulations of other groups also show a strong compression of the magnetosphere (e.g., Bessolaz et al. 2008; Zanni and Ferreira 2013).

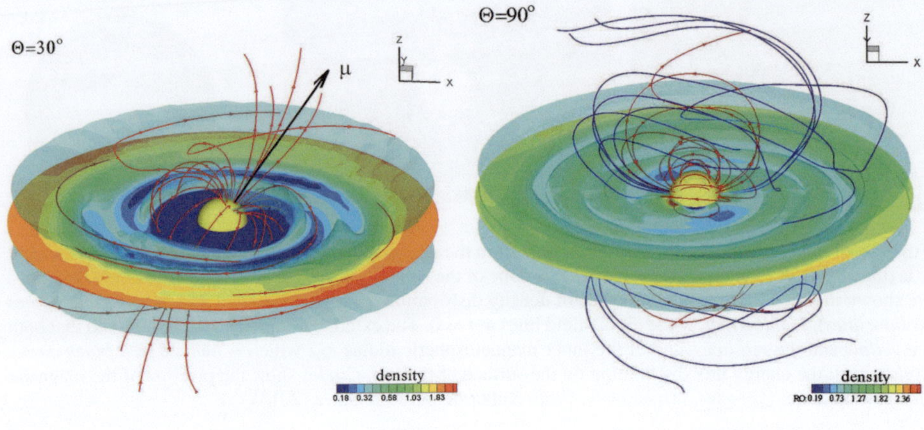

Fig. 6 *Left panel*: Results of 3D MHD simulations of accretion onto a star for which the magnetic axis of the dipole is tilted by $\Theta = 30°$ about the rotational axis (where the rotational axis is aligned with the z-axis). *Right panel*: the dipole axis is tilted by $\Theta = 90°$. The *color background* shows the density distribution in the equatorial plane. The *blue translucent color* shows one of the density levels in 3D. The *lines* show sample magnetic field lines. From Romanova et al. (2005a)

Fig. 7 The two panels compare the configuration of the funnel streams at two density levels: the density in the *left panel* is three times smaller than in the *right panel*. From Romanova et al. (2005a)

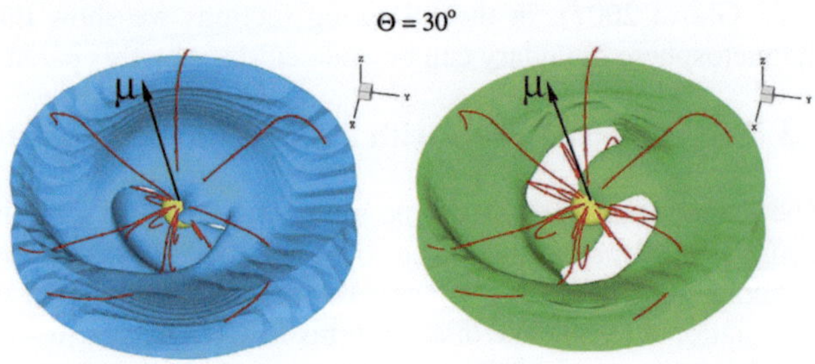

2.3.1 Properties of the Funnel Streams and Hot Spots

The gravitational attraction from the star accelerates the matter in the funnel streams. This, in turn, causes the formation of two hot spots where the matter hits the surface of the star. Numerical simulations helped for the first time to show the likely shape and structure of the hot spots. Results indicate they tend to be crescent-shaped, with densities and temperatures highest near the spot center (see right panel of Fig. 5). This numerical finding led to prediction for the wavelength dependence of the spot radiation, with a different spot size at different wavelengths. For example, in CTTSs, the innermost small region of the spot can radiate in X-rays, while an increasingly larger area radiates in the UV, optical, and IR. In fact, the measurements of the spot sizes in several CTTSs show that the area covered by the hot spots in optical wavelengths is relatively large, about 10 % of the stellar surface (Donati et al. 2007, 2008, 2011). On the other hand, in models of the UV radiation by the accretion shock, the best match with observations is reached if spots cover from less than 1 % up to a few percents of the stellar surface (e.g., Calvet and Gullbring 1998; Gullbring et al. 2000). More generally, observations point to smaller sizes of spots in higher versus lower energy bands. The inhomogeneous structure of the hot spots discovered in

3D simulations (Romanova et al. 2004) can potentially explain these observational results. Before these simulations, hot spots were assumed to have a homogeneous density and temperature distribution. These simulations thus opened a path for understanding the properties of hot spots in different wavebands.

These 3D numerical simulations also have shown for the first time how the shape of the hot spots depends on the tilt of the magnetic axis Θ (Romanova et al. 2004); at small Θ the spots have crescent shape, while as Θ increases to larger values, the spots acquire a bar shape (see also Kulkarni and Romanova 2005). The different shapes predict corresponding differences in the light curves (Romanova et al. 2004), but these also depend on other factors like observer viewing angle. The differences from light curves obtained with the homogeneous, round spots of previous, simplified models is modest, e.g. within $\sim 20\ \%$ in test runs of accreting neutron stars (where relativistic effects were taken into account, Kulkarni and Romanova 2005). While such simple round spots thus provide a reasonable first approximation, a more accurate model of the position and shape of the spot can be important, for example in studies of accreting neutron stars, where the mass and radius (or, their ratio) can be found from relativistically-corrected light-curves (see, e.g., review by Belloni et al. 2015). For convenience in application, an analytic formula has been obtained from multiple 3D simulation runs, with the shape and position of the spots given as a function of the magnetospheric size r_m/R_* and the ratio $r_m/r_{\rm cor}$, which characterizes the level of stellar rotation (Kulkarni and Romanova 2013).

2.3.2 Moving Spots: Phase Shifts and Possible QPOs

Simulations show that the hot spots on the stellar surface are not fixed, but instead move about some preferred position. This motion leads to phase shifts in the light curves (Romanova et al. 2003; Kulkarni and Romanova 2013). In young stars, this results in the phenomenon of "drifting period" which is frequently observed in CTTSs, where the sets of observations are usually brief (only a few stellar rotations). Recent observations by the space instrument *MOST* indicate that the CTTS star TW Hya shows systematic variation of its "period" (Rucinski et al. 2008), which may be connected with the drifting spot on the stellar surface, as found in earlier 3D simulations.

In millisecond pulsars, moving spots may lead to phase shifts in light-curves (Papitto et al. 2007; Patruno et al. 2009) and to timing noise (e.g., Poutanen et al. 2009; Lamb et al. 2009; Ibragimov and Poutanen 2009). Papitto et al. (2007) found a correlation between the pulse phase shifts and the X-ray flux in millisecond pulsar XTE J1814-338 (see also Patruno et al. 2009). The authors argued that the observed phase shifts were due to movements of the hot spot in response to variation in accretion rate. This phenomenon can be explained by the correlation between the hot spot longitude and the location of the magnetospheric radius as shown by Kulkarni and Romanova (2013).

In the case of very small tilts of the dipole moment, $\Theta = (2–5)^\circ$, the funnel streams can be dragged by the inner disk and the spot may systematically rotate faster or slower than the star. This can lead to the phenomenon of the quasi-periodic oscillations (QPOs) in the frequency spectrum and may possibly explain some of QPOs observed in AMXPs. Usually, one or two main high-frequency QPO peaks are observed in AMXPs (see an example in Fig. 8). The observed QPO frequencies show significant temporal variation, but the difference between peaks often corresponds to either the stellar rotation frequency ν_*, or its half-value $\nu_*/2$[7] (e.g., Van der Klis 2000). In many cases, however, the frequency difference

[7]This phenomenon can be possibly explained by the beat-frequency model (e.g., Miller et al. 1998).

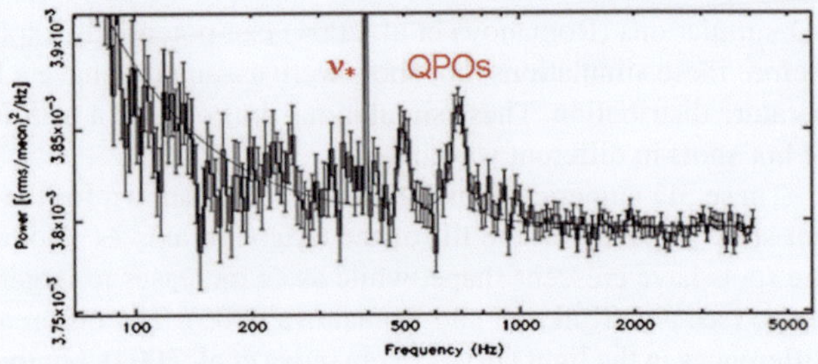

Fig. 8 Power spectral density of the AMXP SAX J1808.4-3658 which shows the frequency of the star, ν_* and two high-frequency QPOs which may carry information about the processes at the disk-magnetosphere boundary. From Wijnands et al. (2003)

is nearly fixed at a value, ~ 300 Hz, that does not correlate with the frequency of the star (Boutloukos et al. 2006; Méndez and Belloni 2007). The origin of these QPOs has not yet been understood. One of QPO peaks may originate at the disk-magnetosphere boundary. The origin of the second peak is less clear, but an interesting, spot-based model has been developed by Bachetti et al. (2010). They performed systematic 3D simulations of accretion onto neutron stars with very small tilts, $\Theta = 2°$, and showed that these rotating spots may possibly explain one of two QPO frequencies in AMXPs, where the difference between QPO frequencies does not vary much in time and is ~ 300 Hz. Bachetti et al. (2010) found that the lower of two of QPO frequencies is connected with the rotation of the funnel stream, while the higher QPO frequency is connected with unstable tongues that rotate with the frequency of the inner disk (see Sect. 2.4). Simulations show that the frequency associated with the rotating funnel streams is about 250–300 Hz lower, and both QPOs move in parallel, which is in accord with observations (Méndez and Belloni 2007). This model, however, is applicable only to stars with small Θ, $\Theta \lesssim 5°$.

2.3.3 Tilted Magnetosphere and Waves in the Disk

Various types of waves can propagate in accretion disks (e.g., Kato et al. 1998; Kato 2004, 2007). A star with a tilted dipole magnetosphere excites bending and density waves through the action of magnetic forces on the inner disk (see, e.g., theoretical studies by Lipunov and Shakura 1980; Lai 1999; Terquem and Papaloizou 2000 and global 3D simulations by Romanova et al. 2013). Global simulations show that, if the angular velocity of the inner disk is close to the angular velocity of the magnetosphere (fastness $\omega_s \sim 1$), a strong bending wave forms at the disk-magnetosphere boundary. Figure 9 shows a snapshot from such a simulation, where a one-armed bending wave forms at the inner edge of the disk.[8] This bending wave may occult or reflect light from the star.

For example, Bouvier et al. (1999) noticed that CTTS AA Tau has dips in the light-curve and, interpreted these as stemming from occultation of the stellar light by a dusty disk-warp (see also Bouvier et al. 2003, 2007b). These dips are quasi-periodic with QPO period of approximately 8.2 days, comparable to the expected rotation period of the star. Doppler tomography observations of AA Tau show that the dominant component of the field is a 2–3 kG dipole field that is tilted at 20° relative to the rotational axis. At this field strength the magnetospheric radius is close to the corotation radius, $r_m \approx r_{\rm cor}$ (Donati et al. 2010). Simulations show that in this situation a large amplitude warp forms and rotates with the frequency of the star, and so could well explain the observed dips (Romanova et al. 2013).

[8] Simulations also show that if a star rotates more slowly than the inner disk, then bending waves are excited at larger distances from the star.

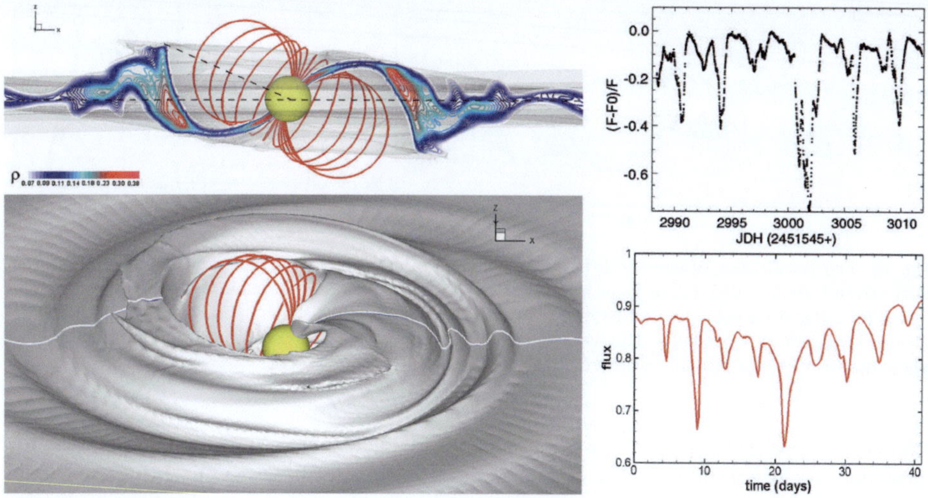

Fig. 9 *Left two panels*: Tilted magnetosphere excites a warp at the disk-magnetosphere boundary. A slice of density distribution (*top panel*) and one of density levels (*bottom panel*) are shown. From Romanova et al. (2013). *Right two panels*: *Top panel*: typical AA Tau type light-curve observed in one of stars in young cluster NGC 2264 (from Alencar et al. 2010). *Bottom panel*: the light-curve obtained from 3D MHD simulations

More recently, Alencar et al. (2010) analyzed the photometric variability of CTTSs in the young cluster NGC 2264 using data obtained by the *CoRoT* satellite; they concluded that AA Tau-like light curves are fairly common, and are present in at least 30–40 % of young stars with inner dusty disks (see also Carpenter et al. 2001; Stauffer et al. 2014, 2015). Figure 9 compares observed and simulated light curves, for a case in which the star is occulted by the dusty warp. Overall, results indicate that the warps can produce dips if the dipole magnetosphere is tilted at sufficiently larger angle, $\Theta \gtrsim 20^\circ$, relative to the rotational axis.

There are also two persistent density waves which form in the inner disk and rotate with the frequency which is lower than the Keplerian frequency of the inner disk. These waves are good candidates for explanation of the one of two QPO frequencies observed in AMXPs. It was frequently suggested that one of QPO frequencies is associated with some inhomogeneities in the inner disk, though the nature of inhomogeneities has not been understood (van der Klis 2006). Numerical simulations show that these inhomogeneities may be associated with persistent density waves at the inner disk (Romanova et al. 2013).

2.3.4 Magnetospheric Accretion from Turbulent Disks

The full dynamical structure of accretion disks is still uncertain. It is commonly inferred that the disks are turbulent and the turbulence can be initiated and supported by the MRI instability. MRI-driven turbulence has been extensively studied in axisymmetric and local/global 3D MHD simulations (e.g., Hawley et al. 1995; Brandenburg et al. 1995; Stone et al. 1996; Armitage 1998; Hawley 2000; Beckwith et al. 2009; Flock et al. 2011; Simon et al. 2011). In most of simulations the central object is non-magnetized, usually a black hole.

Accretion from MRI-driven turbulent disks onto a magnetized star has been investigated numerically in both axisymmetric and 3D simulations (Romanova et al. 2011a, 2012). These

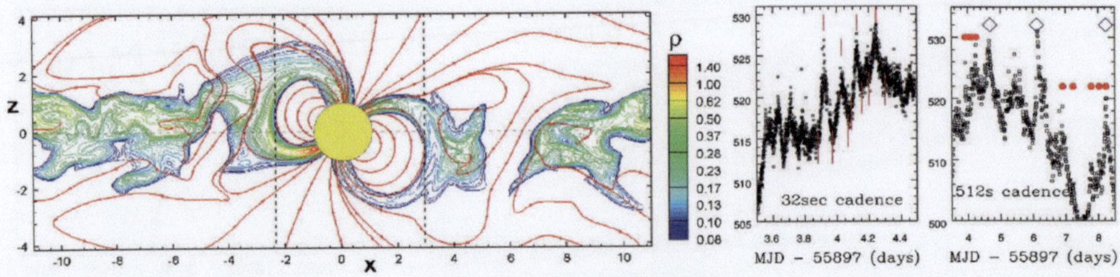

Fig. 10 *Left panel*: slice of density distribution obtained in 3D MHD simulations of matter flow from an MRI-driven turbulent disk (from Romanova et al. 2012). *Right two panels*: example of the irregular light curves for a star from young cluster NGC 2264 where variability on the scale of 1 day (*right panel*) and 0.2 days (*middle panel*) is observed. This variability may be caused, e.g., by accretion from the turbulent disk where the turbulence is present on different scales. From Stauffer et al. (2014)

simulations have shown that the main aspects of magnetospheric accretion are similar: the turbulent disk is truncated by the magnetosphere and matter flows to the star along funnel streams.

It is important to note that the position of the magnetospheric radius does not depend much on the weak magnetic field associated with the MRI-driven turbulence. Its magnetic stress is responsible for the angular momentum transport, but the associated magnetic pressure is smaller than gas pressure, and the ram pressure ρv^2 is much larger than gas or magnetic pressure. This is why the formula in Eq. (1) provides accurately the position of the truncation radius in turbulent disks as well.

The main difference between accretion from the laminar α-disk and turbulent disk is that accretion of the individual cells may lead to variability in the light-curve. The variability time-scale depends on the scale of the turbulent cells $L_{\rm turb}$. If it is small compared with the thickness of the disk, H, then the disk acts as a laminar disk and only small-amplitude variability is expected, as found in axisymmetric simulations (Romanova et al. 2011b).[9] For the largest-scale turbulent cells, $L_{\rm turb} \sim H$, much rarer but higher amplitude flares are expected in light-curves (Romanova et al. 2013). Observations of CTTSs show that many stars have irregular light-curves with variability observed at different time-scales (e.g., Stauffer et al. 2014). Figure 10 (left panel) shows an example of 3D simulations from the disk with a large-scale turbulence, while the left two panels show sample light curves at two different time-scales. Simulations of turbulent accretion with different scales of the turbulence, together with comparisons between modeled and observed light-curves, may help us understand the nature of turbulence in the disk.

2.3.5 *Inflation of the Field Lines*

Three-dimensional simulations show that the field lines of the external magnetosphere inflate due to the disk-magnetosphere interaction. Figure 6 (left panel) shows a snapshot from a simulation in which $\Theta = 30^\circ$. One can see that the field lines of the inner magnetosphere are closed; this is the region in which magnetic energy dominates. Some field lines are trapped in the disk and are azimuthally wrapped by the disk matter. The field lines connecting the polar regions of the star with the disk inflate and wrap around the rotational axis forming a magnetic tower. There is a set of the partially inflated field lines that thread the inner parts

[9] Many axisymmetric simulations also show strong spikes which are connected with very low (only small numerical) diffusivity in the code, and episodic accumulation of matter at the disk-magnetosphere boundary.

of the disk, and transfer the angular momentum between the star and the disk and between the star and corona (see also Zanni and Ferreira 2013). In 3D simulations, a magnetic tower forms, however, most simulation runs show little or no outflow.[10] The inflation of the field lines is observed at different tilts of the dipole magnetosphere, including the extreme case of very large tilt of the magnetosphere, $\Theta = 90°$, where the magnetic axis is located in the equatorial plane of the disk. The right panel of Fig. 6 shows that the magnetic field lines of the external magnetosphere inflate, and also have a tendency to leave the disk and to expand into the corona above and below the disk. Simulations show that the magnetic field lines always wrap around the rotational axis, independent of the tilt angle Θ. However, it is not yet clear whether the outflows represent a common feature of accreting magnetized stars, or whether they are more typical for the case of the rapidly-rotating stars (see Sect. 3.1). Outflows were investigated in greater detail in 2.5D (axisymmetric) simulations, where the density in the corona is lower compared with the 3D simulation cases (see Sect. 3.1).

2.3.6 Stellar Spinup/Spindown from Angular Momentum Gain/Loss

A star may spin up or spin down, depending on the total angular momentum flux onto the star. The net angular momentum flux to or from the surface of the star, $\dot{L}$, is composed of two parts: the angular momentum flux carried by the matter ($\dot{L}_m$) and that carried by the magnetic field ($\dot{L}_f$):

$$\dot{L} = \dot{L}_m + \dot{L}_f, \qquad \dot{L}_m = \int d\mathbf{S} \cdot \rho r v_\phi \mathbf{v_p}, \qquad \dot{L}_f = -\int d\mathbf{S} \cdot r B_\phi \mathbf{B_p}/4\pi, \tag{6}$$

where the fluxes are integrated along the surface of the star; $\mathbf{v_p}$ and $\mathbf{B_p}$ are the poloidal velocity and the poloidal magnetic field.

Simulations show that the disk and the star exchange angular momentum mainly through the funnel streams. Near the disk, the angular momentum in the funnel stream is carried by the matter, and the matter then slightly twists the field lines threading the funnel streams; this twist determines the torque on the star (e.g., Romanova et al. 2002; Long et al. 2005; Zanni and Ferreira 2013). The sign of the term $B_\phi \mathbf{B_p}$ (more precisely, the sign of the B_ϕ component at the stellar surface) determines whether the star spins up or down. At the stellar surface, the angular momentum carried by the field, $\dot{L}_f$ is much larger than that carried by matter, $\dot{L}_f$.

Long et al. (2005) performed a set of 2D simulation runs at different angular velocities of the star, $\Omega_\star$, to find the critical value of the fastness parameter ω_s, corresponding to the rotational equilibrium state (i.e., when $\dot{L} = 0$). He found that, in the rotational equilibrium state $r_{\rm cor}/r_m \approx 1.2$–$1.3$, that is the fastness parameter $\omega_s = 0.76$–0.67. They found that the inner disk rotates more rapidly than the magnetosphere, because the star loses some angular momentum to the corona along the inflated or partially-inflated field lines. This negative torque is compensated by a positive torque from the inner disk. Therefore, the fastness parameter corresponding to the rotational equilibrium state, ω_s, strongly depends on the amount of angular momentum that flows to the inflated or partially-inflated field lines.[11]

[10]This is the main restriction of the current 3D simulations, where the main computing power is used for resolving the low-density closed magnetosphere and the disk, while the density in the corona is relatively high, and matter pressure dominates the magnetic pressure, suppressing magnetic or magneto-centrifugally driven outflows.

[11]This angular momentum depends on the coronal density: it can be larger in case of young stars, which can have strong stellar winds, and smaller in cases of neutron stars.

3D simulations also show that a star can spin up, spins down, or be in a rotational equilibrium state. It is interesting to note that the matter fluxes onto the stellar surface, as well as the torques on the star, are similar across different values of Θ (e.g., Romanova et al. 2003). This is because the magnetospheric radius is approximately the same for different Θ.

In case of steady long-lasting accretion, a magnetized star is expected to be in the rotational equilibrium state. Probably many CTTSs are in this state (see, e.g., review by Bouvier et al. 2007a). In accreting millisecond pulsars, the situation is more complex, because accretion occurs during episodes of accretion outbursts, where the accretion rate increases a few orders of magnitude, then decreases back to small values (see reviews by van der Klis 2006; Belloni et al. 2015). A propeller stage and spinning-down of a neutron star is expected during the rise and decline of the outburst (e.g., Patruno et al. 2009), while the magnetospheric accretion and spinning-up is expected during period of the high accretion rate. These issues are further discussed in Sect. 3.1.

2.4 The Unstable Regime of Accretion due to Rayleigh–Taylor Instability

A new regime of accretion has been found in 3D simulations of stars whose dipole moment is only slightly tilted relative to the rotational axis ($\Theta = 5^\circ$). These simulations were performed in order to better understand AMXPs, for which the tilt of the dipole magnetosphere is expected to be small. The simulations show that, in the unstable accretion regime, the magnetic Rayleigh–Taylor (R–T) instability enables matter to penetrate the magnetosphere via several chaotically-formed "tongues" (Kulkarni and Romanova 2008; Romanova et al. 2008; Bachetti et al. 2010; Kurosawa and Romanova 2013). In earlier studies, it was suggested that the instabilities at the disk-magnetosphere boundary may lead to the mixing of plasma with the field in the external layers of the magnetosphere (Arons and Lea 1976). However, global 3D MHD simulations show that the unstable tongues can deeply penetrate into the magnetosphere. The fact that the magnetic field is not an obstacle for the Rayleigh–Taylor instability has been also shown by Stone and Gardiner (2007a, 2007b) who performed local (in the box) 3D simulations wherein the homogeneous magnetic field is placed at different orientations relative to the boundary between the light and heavy fluids. These simulations and earlier ones by Wang and Robertson (1984, 1985, 1999) show that the Rayleigh–Taylor instability leads to the formation of small-scale waves and filaments that merge to form much larger filaments, which then deeply penetrate into the magnetically-dominated, low-density regions.

The matter in the randomly-forming tongues is accelerated by gravity and is deposited onto the stellar surface as hot spots with irregular shape and random position. The light curves from these spots are often irregular, with several peaks per rotational period. Figure 11 (two left panels) compares 3D views of matter flow in the stable and unstable regimes, as well as the light curves from the hot spots.

Observations of CTTSs show that many CTTSs exhibit irregular, chaotic-looking light curves (Herbst et al. 1994; Alencar et al. 2010; Cody et al. 2014; Stauffer et al. 2014). The right panel of Fig. 11 shows the chaotic-looking light curve of the CTTS TW Hya obtained with the *MOST* satellite (Rucinski et al. 2008). The characteristic time-scale between flares (a few events per rotational period) corresponds to that expected in the unstable regime of accretion.

Figure 12 shows an example of accretion through instabilities. The panels show that matter of the disk pushes the magnetic field lines aside, and penetrates deep onto the magnetosphere, where it is stopped and channeled into the funnel streams. These usually do not live long, less than period of rotation at the inner disk. After depositing matter onto the star,

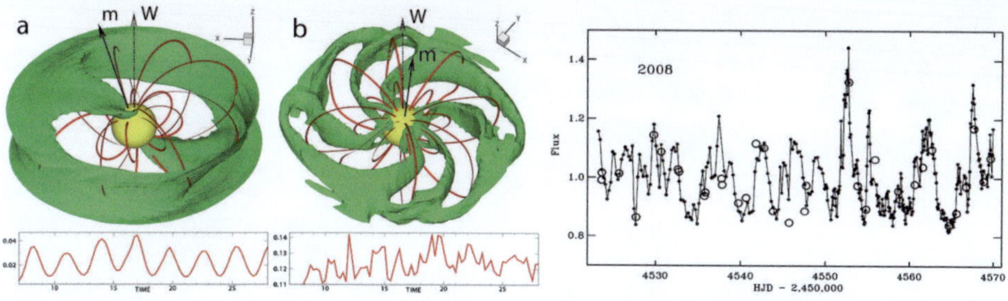

Fig. 11 *Left two panels*: a 3D view of matter flow in the stable (*left*) and unstable (*right*) regime of accretion. *Bottom panels* show the light curves from the hot spots (from Romanova et al. 2008). *Right panel*: The light curve of the CTTS TW Hya obtained by the *MOST* satellite (from Rucinski et al. 2008)

Fig. 12 *Left panel*: 3D view of matter flow in a case where chaotic accretion in multiple tongues dominates. One of the density levels is shown in color, selected magnetic field lines are shown in *red*. *Middle panel*: Same but in the face-on projection. *Right panel*: An equatorial slice of density distribution is shown in color. The figure has been created for the model $\mu 1c2.5\Theta 5a0.02$ calculated by Blinova et al. (2015)

they become weaker and are destroyed by interaction with the magnetosphere. Figure 13 (top panels) shows several consecutive slices of the density distribution. One can see that the number and the position of tongues varies in time. The light-curve calculated from the hot spots (at an observer's angle $i = 45°$) shows somewhat chaotic behavior (see bottom left panel of the same figure). Fourier analysis of this light curve shows several frequencies that are associated with chaotic hot spots, as well as the rotational period of the star. Wavelet analysis shows at which interval the instabilities are stronger.

2.4.1 Theoretical Background for Rayleigh–Taylor Instability

According to theoretical studies (e.g., Chandrasekhar 1961), a magnetic field that is parallel to the boundary between the heavy and light fluids is not an obstacle for the development of R–T unstable modes at the boundary. The sketch in Fig. 14 demonstrates that the situation at the disk-magnetosphere boundary is similar to that considered by Chandrasekhar (1961). In the case of a rotating disk, the inner disk can be unstable if the effective gravitational acceleration is negative: $g_{\rm eff} = g + g_c < 0$, where $g = -GM_\star/r^2$ and $g_c = v_\phi^2/r$ are gravitational and centrifugal accelerations, respectively.

There are a few factors that can suppress the instability, however. Specifically, the radial shear of the angular velocity, $2(r\frac{d\Omega}{dr})^2$, can do so by smearing out the perturbations. Spruit et al. (1995) performed a general analysis of disk stability in the thin disk approximation,

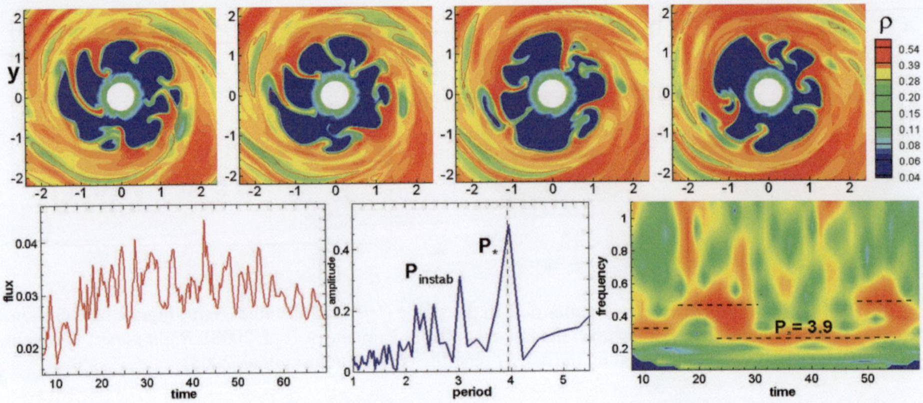

Fig. 13 *Top panels*: Consecutive xy-slices show the density distribution in the chaotic unstable regime in the model with $r_{\rm cor} = 2.5$ at times $t = 16$–18. *Bottom panels*: The light curve from the rotating hot spots calculated at an inclination angle of the rotational axis relative to the observer $i = 45°$ (*bottom left panel*), the Fourier transform obtained from analysis of the light curve (*bottom middle panel*), and the wavelet transform obtained from analysis of the light curve (*bottom right panel*). From Blinova et al. (2015)

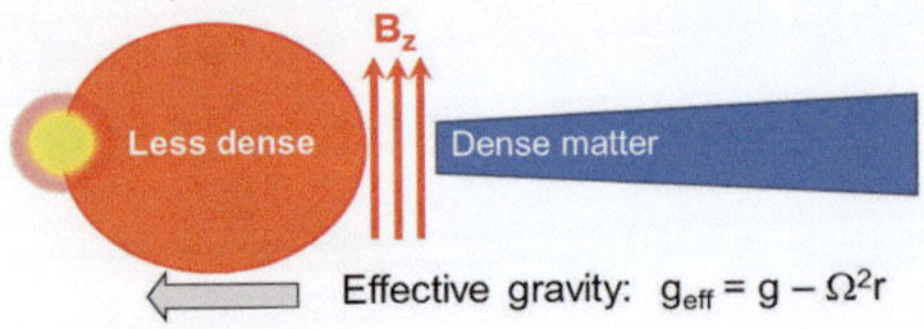

Fig. 14 A sketch of the situation where accretion through R–T instability is favorable. The disk is denser than the magnetosphere, and the effective gravity is directed toward the star; the poloidal dipole field is not an obstacle for the R–T instability

taking the velocity shear into account (see also earlier work by Kaisig et al. 1992). The disk has a surface density Σ and is threaded by a magnetic field with a vertical component, B_z. Their analytical criterion for the development of instability is:

$$\gamma_{B\Sigma}^2 \equiv (-g_{\rm eff})\left|\frac{d}{dr}\ln\frac{\Sigma}{B_z}\right| > 2\Big(r\frac{d\Omega}{dr}\Big)^2 \equiv \gamma_\Omega^2\,. \tag{7}$$

One can see that the sign and value of the effective gravitational acceleration $g_{\rm eff}$ are important in this criterion. Namely, the disk-magnetosphere boundary is unstable if $g_{\rm eff}$ is negative (i.e., when the effective acceleration is directed towards the star). The term $|\frac{d}{dr}\ln\frac{\Sigma}{B_z}|$ characterizes the level of *compression* of the surface density and magnetic field in the disk. Instability occurs if the product of $(-g_{\rm eff})$ and $|\frac{d}{dr}\ln\frac{\Sigma}{B_z}|$ is large enough to overcome the stabilizing effect of the velocity shear, γ_Ω^2. Simulations show that this criterion, developed for a disk, also works well for the disk-magnetosphere boundary (e.g., Kulkarni and Romanova 2008, 2009; Blinova et al. 2015).

2.4.2 The Boundary Between the Stable and Unstable Accretion Regimes

It is important to identify the boundary between the stable and unstable regimes of accretion. Recent studies show that this boundary occurs at a critical ratio of magnetospheric to corotation radii $r_{\rm cor}/r_m = k \approx 1.4$, corresponding to a fastness parameter $\omega_s = (r_m/r_{\rm cor})^{3/2} \approx 0.6$

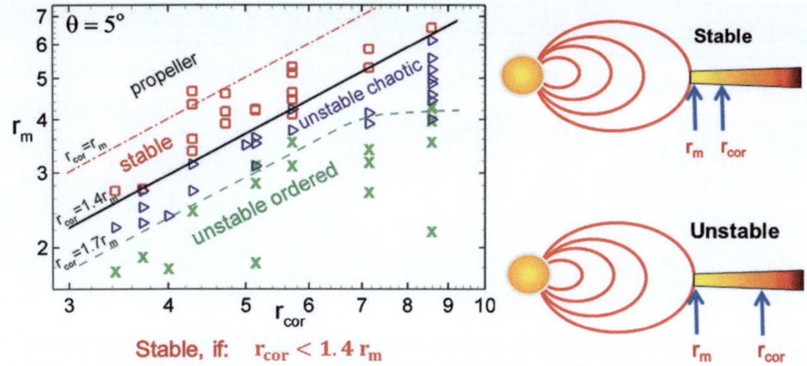

Fig. 15 The boundary between the stable and unstable regimes of accretion strongly depends on the ratio between the magnetospheric and corotation radii (r_m and r_{cor}, respectively). *Left panel*: The result of multiple simulation runs. *Squares*, *triangles*, and *crosses* correspond to the stable, chaotically unstable, and ordered unstable regimes of accretion, respectively (from Blinova et al. 2015). *Right panels*: The positions of the corotation and magnetospheric radii in the cases of the stable and unstable regimes of accretion, respectively

(Blinova et al. 2015). Figure 15 shows the results of multiple simulation runs (symbols). The black solid line shows the boundary between the stable and unstable accretion regimes; accretion is unstable below the line and stable above the line. The instability becomes stronger further away from the line, and it is mildly strong near the line. The red dashed line shows the boundary $r_m = r_{cor}$ ($\omega_s = 1$), above which the propeller regime is expected. These simulations were performed for stars with $\Theta = 5°$ and small viscosity parameter, $\alpha = 0.02$.

Test simulations with larger values of α-parameter of viscosity (i.e., $\alpha = 0.1$), show that the instability is stronger, and the transition to the unstable regime is expected at the larger values of the fastness parameter, $\omega_s > 0.6$ (i.e., closer to the propeller line $\omega_s = 1$), compared with the lower viscosity, $\alpha = 0.02$, case. This can be explained by the fact that, at larger values of α, the compression factor $|\frac{d}{dr} \ln \frac{\Sigma}{B_z}|$ is larger, and thus the instability starts more readily (Blinova et al. 2015).[12] In addition, a course grid resolution suppresses the instability: comparisons of simulations at different grid resolutions show that the instability starts at larger values of ω_s in case of the finer grid, implying that the courser grid effectively acts to suppress the instability (Blinova et al. 2015). This leads to the question whether the boundary between regimes may be located at the line $r_{cor}/r_m \approx 1.0$, if the grid is sufficiently fine? Addressing this will require a systematic set of studies at even higher grid resolution.

An interesting phenomenon has been recently observed in the AMXP SAX J1808.4-3658, where the rotation frequency of the star is known ($\nu_\star = 401$ Hz), and where the two main QPO frequencies vary between values lower and higher than this stellar rotation frequency (see Fig. 8). It has often been suggested that one of frequencies is associated with the frequency of the Keplerian rotation at the inner disk, but clear evidence for this has been lacking. Recent studies (Bult and van der Klis 2015) found that the pulse fraction (which is associated with the magnetospheric accretion onto the surface of the neutron star) strongly decreases when the QPO frequency increases up to the frequency of the star, that is when the inner disk rotates more rapidly than the magnetosphere of the star. At the end of the outburst, when the accretion rate decreases, the QPO frequency decreases, and when it passes the stel-

[12] The effective α-parameter can be estimated from comparisons of observations with models of accretion. For example, Bisnovatyi-Kogan et al. (2014) derived the value $\alpha = 0.1$–0.3 for outburst of accretion in X-ray transient A0535+26/HDE245770.

Fig. 16 *Left panel*: 3D view of matter flow in a case where ordered accretion in two tongues dominates. One of the density levels is shown in color, selected magnetic field lines are shown in *red*. *Middle panel*: Same but in the face-on projection. *Right panel*: An equatorial slice of density distribution is shown in color. From Blinova et al. (2015)

lar frequency, the pulse fraction increases back to large value. This phenomenon provides evidence that this QPO is associated with the inner disk, and that the transition of the inner disk through the point $r_m = r_{\rm cor}$ leads to change in the magnetospheric accretion. One possibility is that accretion becomes unstable and therefore the pulse fraction decreases when the inner disk reaches the radius $r_m = r_{\rm cor}$ and moves closer to the star (van der Klis, privet comm.). If true, then the boundary line between stable and unstable regimes should be at $r_m \approx r_{\rm cor}$. To substantiate this hypothesis, more work is needed in both numerical modeling and observations of different AMXPs.

Modeling of accretion onto stars with larger tilts of the dipole magnetosphere (i.e., $\Theta = 10^\circ, 20^\circ$ and 30°) shows that the instability develops, however more matter accretes in funnel streams above the magnetosphere. The hot spots associated with these streams provide the sinusoidal component of the light curve, while the chaotic hot spots provide the irregular component in the light curve. The amplitude of this irregular component is high for $\Theta = 10^\circ$, but it decreases for models with larger Θ (Blinova et al. 2015).

2.4.3 The Ordered Unstable Regime

Another important regime of accretion has been found in simulations of accretion onto slowly rotating stars, when the condition $r_{\rm cor} \gtrsim 1.7 r_m$ has been satisfied (Blinova et al. 2015). In this regime, one or two ordered tongues form and rotate with the angular velocity of the *inner disk*. Figure 16 shows an example of accretion in this ordered unstable regime. The top panels of Fig. 17 show equatorial slices, and the bottom panels show the light curve associated with rotating hot spots, as well as Fourier and wavelet analysis of this light curve. The period associated with this instability dominates in both the Fourier and wavelet spectra. This boundary between chaotic and this newly found ordered unstable regimes is shown in Fig. 15 as the green dashed line. It corresponds to the fastness parameter $\omega_s \approx 0.45$.

The frequency of oscillations in the ordered unstable regime depends on the accretion rate. When the accretion rate increases, the inner disk moves closer to the star, and the frequency of oscillations increases (and vice versa). This mechanism may be important for understanding the high-frequency oscillations in AMXPs, for which a frequency-luminosity correlation has been observed (e.g., Papitto et al. 2007). Ordered rotation of "tongues" probably reflects an ordered rotation of the density waves in the inner disk; however this issue should be studied further. The QPO radiation may come from the inner disk waves, from the

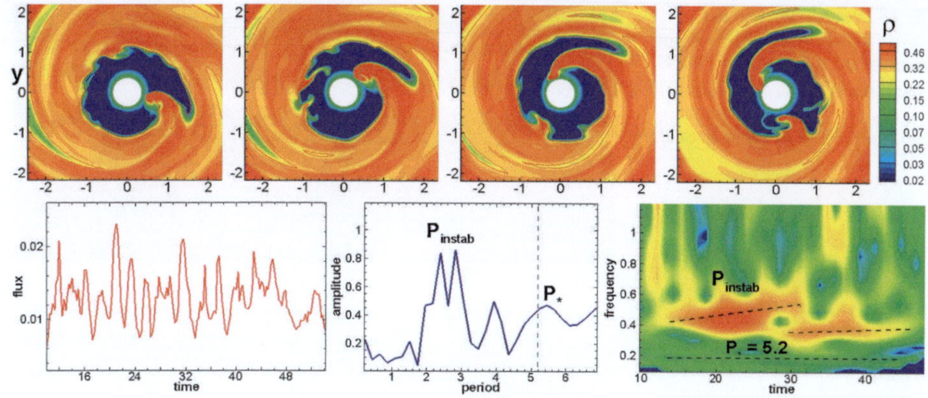

Fig. 17 An example of accretion in the ordered unstable regime. *Top panels*: Consecutive views of xy slices. The *color background* shows the density distribution, and the *lines* show where the kinetic plasma parameter $\beta_t = 1$. *Bottom panels*: The light curve from the rotating hot spots calculated at an inclination angle $i = 45°$ (*bottom left panel*), the Fourier transform obtained from analysis of the light curve (*bottom middle panel*), and the wavelet transform obtained from analysis of the light curve (*bottom right panel*). From Blinova et al. (2015)

ordered tongues, or from the rapidly-rotating hot spots which glide along the surface of the slower-rotating star.

In the chaotic unstable regime, the whole system of chaotic-looking tongues rotates with the angular velocity of the inner disk. This phenomenon could be detected in AMXPs, where the observations cover many rotational periods of the neutron star. This ordered rotation of the system of chaotic spots could induce a QPO with a frequency ν_{QPO} that increases when the inner disk moves inward. During this motion, the unstable tongues become more and more ordered, and the AMXP switches gradually to the ordered regime of unstable accretion. During this process, the quality factor $Q = \nu_{QPO}/\Delta\nu_{QPO}$ increases. In fact observations do indicate that the quality factor of QPOs usually increases with the frequency of QPO (e.g., Barret et al. 2007).

In young stars accretion in the ordered unstable regime is expected during periods of enhanced accretion rate, that is, during EXor and FUor stages of evolution (see, e.g., review by Audard et al. 2014). Any short-period oscillations may be connected with rotation of the hot spots associated with ordered unstable tongues. Recent X-ray observations of the young star V1647 Ori during two accretion outbursts revealed a ~ 1 day period that is very short for a young star, and corresponds almost to the stellar rotation break-up period (Hamaguchi et al. 2012). Numerical models suggest that this type of variability could be connected with the ordered unstable regime of accretion. Future search of the short-term variability in young stars would be helpful for understanding processes during their accretion outburst, like in FUors and EXors.

In the above simulations the unstable regime has been investigated in case of α-disks. In the future, similar studies should be performed with the MRI-driven turbulent disks.

2.5 Accretion onto Stars with Larger Magnetospheres

The simulations described in the above sections are relevant to stars with relatively small magnetospheres, such as CTTSs and AMXPs, where the size of the magnetosphere is only a few times larger than the radius of the star. However, the magnetospheres are larger for

Fig. 18 An XZ-slice in the density distribution in the case of accretion onto a star with a large magnetosphere ($r_m \approx 12R_\star$). The *lines* are sample magnetic field lines. From Romanova et al. (2014)

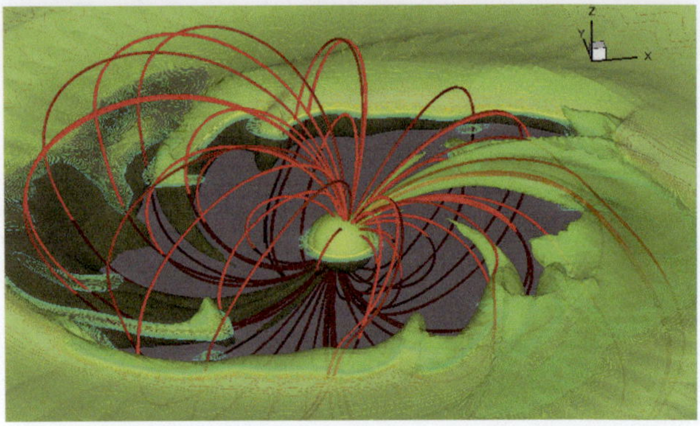

Fig. 19 A 3D view of matter flow onto a star with a large magnetosphere. The *color background* shows the density distribution in the equatorial slice. The *green translucent layer* shows the density distribution in 3D. From Romanova et al. (2014)

intermediate polars, $r_m/R_\star \gtrsim 10$, and much larger for X-ray pulsars. A special set of simulations has been performed to investigate the disk-magnetosphere interaction in systems with larger stellar magnetospheres, where the magnetospheric radius $r_m > 10R_\star$ (Romanova et al. 2014). Much larger values of the stellar magnetic moments were taken for these simulations.

The slice of the density distribution in Fig. 18 shows that the disk matter stops at the distance $r_m \approx 12R_\star$ and flows onto the star in two funnel streams, which are narrow in the vertical direction. Figure 19, on the other hand, shows that the funnel stream is wide in the horizontal direction. The corotation radius $r_{\rm cor} = 20R_\star$ is large compared to the magnetospheric radius, $r_m \approx 12R_\star$, and accretion through R–T instabilities is expected. Simulations show that the matter in the inner disk penetrates through the external layers of the magnetosphere due to the R–T instability, and this instability is observed at the disk-magnetosphere boundary. However, the unstable tongues only penetrate the outer parts of the external magnetosphere, while the matter flows onto the star in two ordered funnel streams. Each funnel stream may split into several streams, because matter flows into the funnel stream from higher-density regions (i.e., from the unstable tongues; see Fig. 19).

2.6 Accretion onto Stars with Complex Magnetic Fields

Measurements of the surface magnetic fields of CTTSs indicate that they have a complex structure (e.g., Johns-Krull et al. 1999; Johns-Krull 2007). Furthermore, measurements of the magnetic fields of nearby low-mass stars using the Zeeman–Doppler technique show that their fields are, also, often complex (e.g., Donati and Collier Cameron 1997; Donati et al. 1999; Jardine et al. 2002). The observed surface magnetic field is often approximated with a set of multipoles whose magnetic moments have different misalignment angles relative to the rotational axis (i.e., different tilts), as well as different phases in the longitudinal direction (e.g., Jardine et al. 2002). The surface magnetic field of a star with a complex magnetic field

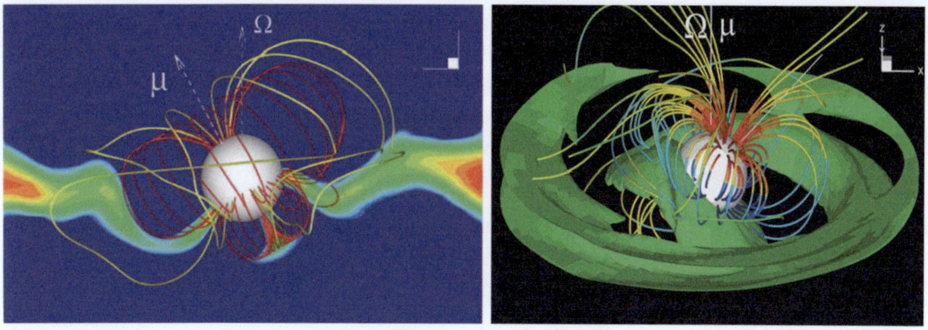

Fig. 20 *Left panel*: the XZ-slice of the density distribution and sample magnetic field lines in the case where the dipole and quadrupole magnetic moments are aligned and both are tilted by an angle $\Theta = 30°$ relative to the rotational axis. *Right panel*: A 3D view of matter flow onto a star for which the quadrupole moment is tilted about the dipole moment by 45°. From Long et al. (2007, 2008)

can be approximated by a superposition of tilted dipole, quadrupole, octupole, and higher order multipoles: $\mathbf{B}_\star = \mathbf{B}_{\rm dip} + \mathbf{B}_{\rm quad} + \mathbf{B}_{\rm oct} + \cdots$.

A star with a quadrupole or octupole magnetic field will also truncate the disk, albeit at smaller distances from the star (compared to that of a pure dipole field). The magnetospheric radius can be derived in analogy with the case of a pure dipole field (see Eq. (2)). For simplicity, we assume that the primary axis of the multipolar field is aligned with the rotation axis, and therefore that the n-th multipolar component of the magnetic field can be written in the form $B_n \sim \mu_n/r^{n+2}$. From the balance of the largest stresses $B^2/8\pi = \rho v_\phi^2$, and taking $v_\phi = v_{\rm Kep}$, we obtain:

$$r_{m,n}^{(0)} = k_n \mu_n^{\frac{4}{4n+3}} \dot{M}^{-\frac{2}{4n+3}} (GM_\star)^{-\frac{1}{4n+3}}, \qquad (8)$$

where $k_n \sim 1$ is a coefficient that can be different for different multipoles. This formula can be applied in the cases of aligned multipoles, and it can also be used for estimates of the magnetospheric radius in the more general case in which the multipole is not aligned with the rotational axis. We should note that the $B_z = 0$ in the disk plane for aligned fields of $2n$-th order multipoles, such as a quadrupole field (Long et al. 2007), and matter could flow directly onto the star in the disk plane. So $r_{m,n}$ does not reflect where the inflowing matter stops, but only the size of the magnetosphere. Gregory (2011) studied accretion onto stars with the complex magnetic fields semi-analytically, suggesting a fixed configuration of the field and showed that complex paths of matter flow are expected.

A number of global 3D MHD simulations have been performed with the goal of better understanding accretion onto stars with complex, non-dipole magnetic fields. Initial simulations were performed to study the superposition of the dipole and quadrupole fields (Long et al. 2007, 2008). These simulations show that matter partially accretes onto the star near the magnetic poles, and partially to the ring associated with the quadrupole component of the field. Figure 20 shows an example of simulations where the dipole and quadrupole moments are aligned (left panel) and misaligned (right panel). In the misaligned case, the magnetic field looks more complex than in the aligned case.

In the case of a superposition of the dipole and octupole fields, usually even a small dipole field determines the magnetospheric accretion, because the octupole field decreases with the distance rapidly ($B_n \sim r^{-5}$). That is why, to demonstrate accretion onto a star with octupolar field, the dipole component has been taken to be very small for demonstration (see Fig. 21; Long et al. 2012). In this case, the matter accretion forms two equatorial belts.

Recent measurements of the surface magnetic field in CTTSs have shown that, in several stars, the dominant components of the field are dipole and octupole, while quadrupole and higher-order components are smaller (Donati et al. 2007, 2008). Using stellar parameters obtained from these observations, the modeling of accretion onto the observed stars was somewhat simplified. Figures 22 and 23 show 3D MHD simulation results for two observed

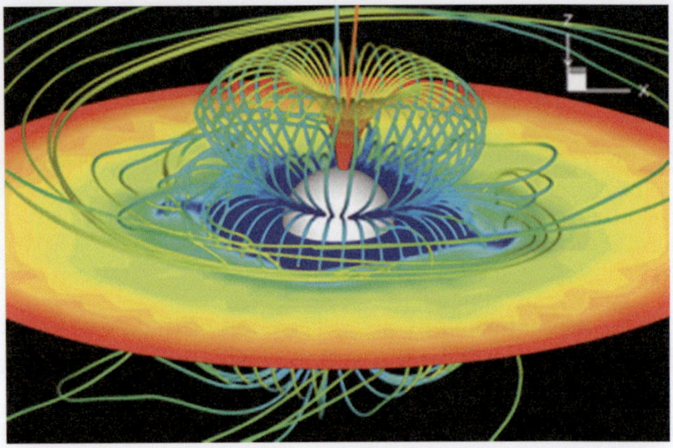

Fig. 21 A 3D view of the magnetic field distribution, as well as a density slice in the equatorial plane, in the case of accretion onto a star with a predominantly octupolar field. The *colors* along the field lines represent different polarities and strengths of the field. The *thick cyan* and *orange lines* represent the rotation and octupole moment axes, respectively (from Long et al. 2012)

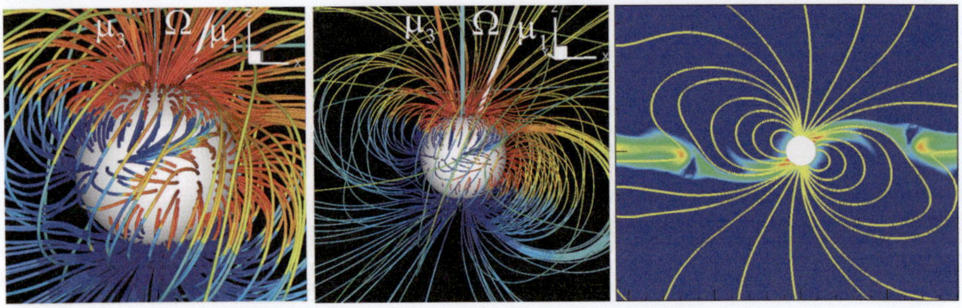

Fig. 22 *Left two panels*: The magnetic field of BP Tau in the 3D MHD model. *Right panel*: The density distribution in the XZ-plane. The *color* of the field lines represents the polarity and strength of the field. The *cyan*, *white*, and *orange lines* represent the rotational axis, the dipole moment, and the octupole moment, respectively. From Long et al. (2011)

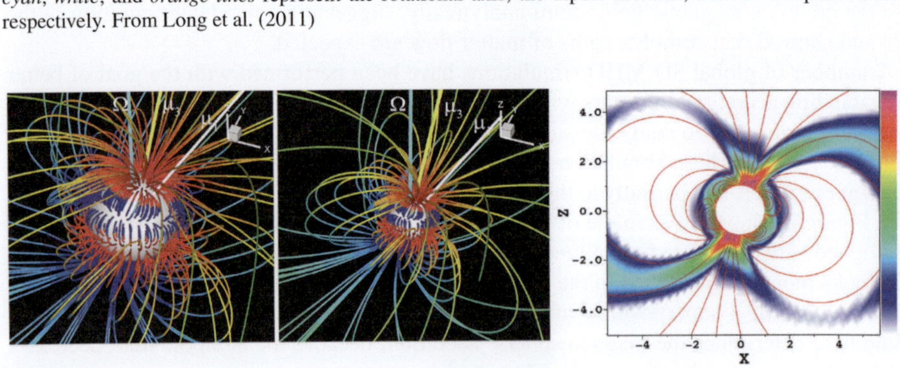

Fig. 23 *Left two panels*: The magnetic field of V2129 Oph in the 3D MHD model (from Romanova et al. 2011b). *Right panel*: The density distribution in the XZ-plane (from Alencar et al. 2012)

CTTSs: BP Tau, where the octupole component is small (Long et al. 2011), and V2129 Oph, where the octupole component is relatively large (Romanova et al. 2011b).

2.7 Modelling Accretion onto the CTTS BP Tau

The surface distribution of the magnetic field of the CTTS BP Tau has been measured with the Doppler spectro-polarimetry technique (Donati et al. 2008). The observed surface magnetic field has been decomposed into spherical harmonics, and it was found that the field is mainly poloidal with only 10 % of the total magnetic energy contained within the toroidal field. The poloidal component can be approximated by dipole ($n = 1$) and octupole ($n = 3$) moments, which each comprise 50 % and 30 % of the magnetic energy, respectively. Other multipoles (up to $n < 10$) have only 10 % of the total magnetic energy. Donati et al. (2008) concluded that the magnetic field of BP Tau is primarily composed of dipole and octupole moments, with $B_{\rm dip} = 1.2$ kG ($\Theta_{\rm dip} \approx 20°$) and $B_{\rm oct} = 1.6$ kG ($\Theta_{\rm oct} \approx 10°$), respectively.

Long et al. (2011) performed global 3D MHD simulations of accretion onto a model star with parameters close to those of BP Tau: $M_\star = 0.7 M_\odot$, $R_\star = 1.95 R_\odot$, and a rotational period of $P_\star = 7.6$ days (corresponding to a corotation radius of $r_{\rm cor} \approx 7.5 R_\star$). In this 3D model, the magnetic field has been approximated using superposed tilted dipole and octupole moments, with polar magnetic field amplitudes of 1.2 kG and 1.6 kG, respectively. Comparison of the dipole and octupole fields (using Eq. (8)) shows that, in BP Tau, the dipole field dominates almost up to the surface of the star.

Figure 22 shows the distribution of the magnetic field near the star (left panel) and at larger distances from the star (middle panel). The right panel shows that matter flows from the disk onto the star in two funnel streams. However, near the stellar surface, the octupole component slightly alters the funnel streams such that matter is deposited closer to the magnetic poles than those originating from a pure dipole. Moreover, the hot spots are rounder relative to the crescent-shaped hot spots associated with the pure dipole case. The shape and position of hot spots obtained in numerical simulations were in good agreement with those obtained from observations (Long et al. 2011).

2.8 Modelling Accretion onto V2129 Oph, and Comparisons of Spectra

In another example, accretion onto a modeled star with parameters close to CTTS V2129 Oph has been investigated (Romanova et al. 2011b). This star has a mass of $M_\star = 1.35 M_\odot$, a radius of $R_\star = 2.1 R_\odot$, and a rotational period of $P_\star \approx 6.5$ days. The magnetic field of this star is dominated by a dipole component of $B_{\rm dip} \approx 0.9$ kG and octupole component of $B_{\rm oct} \approx 2.1$ kG, both of which are tilted at small angles about the rotational axis (Donati et al. 2011).

Figure 23 (left two panels) shows the initial distribution of the magnetic field. The octupole component of the field dominates near the star, while the dipole component dominates at larger distances. The dipole and octupole fields are equal at approximately $1.6 R_\star$. Simulations show that matter flows onto the star in two funnel streams above and below the dipole component of the magnetosphere (see the slice of the density distribution in the right panel of Fig. 23). However, in the region where the octupole field dominates, the streams are redirected by the octupolar field toward higher latitudes on the surface of the star

Three-dimensional simulations provide us with the dynamical characteristics of matter flow around magnetized stars (the distribution of, e.g., density, velocity, temperature). However, for comparisons with observations, it is important to calculate the spectrum of

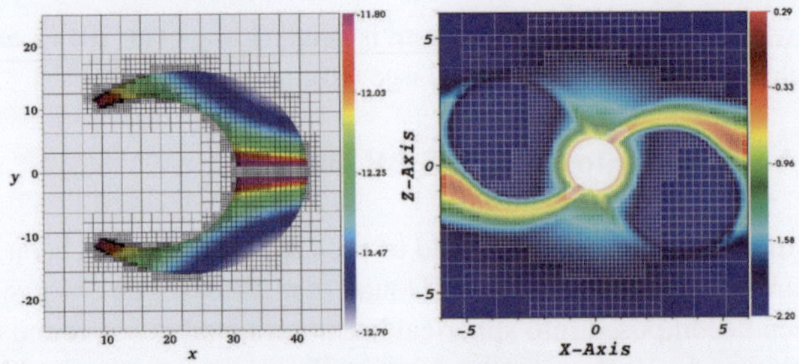

Fig. 24 An example of the restructuring grid used in the radiative transfer code *TORUS* in the case where the funnel stream density and other parameters are determined by the analytical formulae from Hartmann et al. (1994) (*left panel*, from Kurosawa et al. 2004) and from 3D MHD simulations (*right panel*)

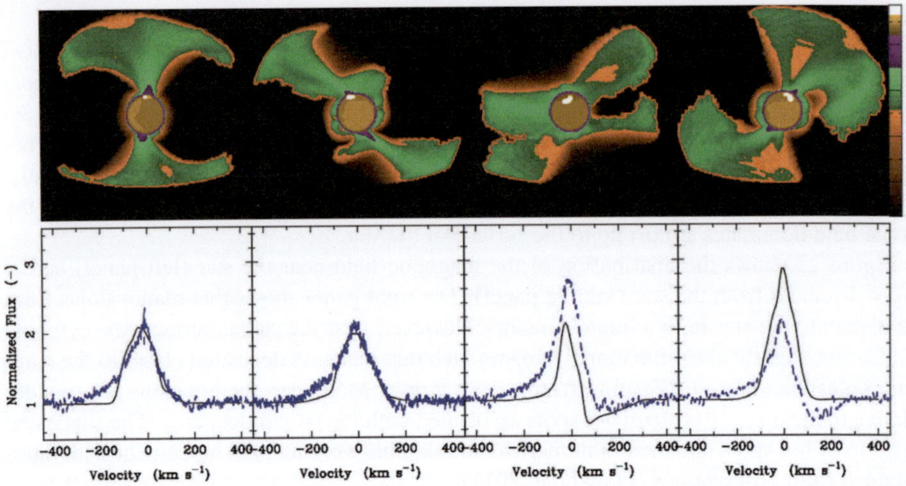

Fig. 25 *Top panels*: The emissivity of the funnel flow calculated in the Hβ spectral line. *Bottom panels*: A comparison of the observed spectrum in Hβ line (*blue line*) with the modeled spectrum (*black line*). From Alencar et al. (2012)

the modeled star. To calculate the radiation from the modeled matter flow and, subsequently, the spectrum in the Hydrogen lines, the 3D Monte Carlo radiative transfer code *TORUS* has been used (e.g., Harries 2000; Kurosawa et al. 2004, 2008). The results of 3D MHD simulations (density, velocity etc. distribution) were projected onto the adaptive mesh refinement grid of the *TORUS* code (see Fig. 24). Figure 25 (top panels) shows the emission of the funnel streams calculated in the Hβ spectral line shown for different phases of stellar rotation. The bottom panel of the same figure compares the modeled and observed spectra of the Hβ line. The plot shows that the observed and modeled spectra are in good agreement (Alencar et al. 2012). This is an exciting example where a 3D MHD model, combined with a 3D stellar radiative transfer model, with realistic parameters has been compared with detailed observations and resulted in a good match. This model shows that global 3D simulations can properly describe realistic matter flow in CTTSs, and that the 3D radiative transfer code *TORUS* can provide realistic spectra. These comparisons of simulations with observations also act as a "proof of concept" of the magnetospheric accretion paradigm, suggested earlier theoretically (e.g., Camenzind 1990;

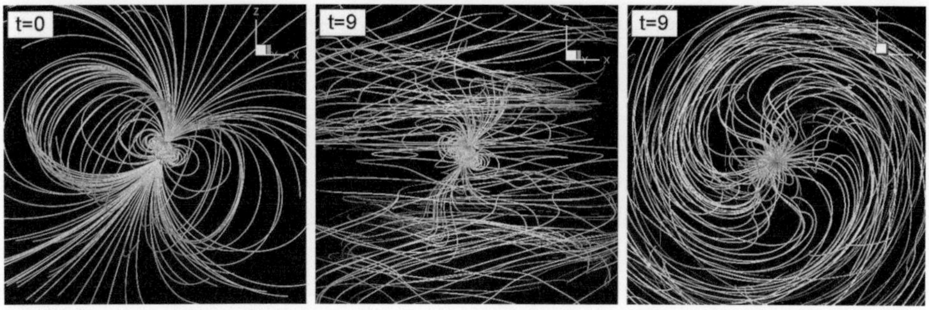

Fig. 26 *Left panel*: The initial configuration of the magnetic field in V2129 Oph on a large scale. *Middle panel*: The magnetic field configuration at $t = 9$. *Right panel*: A top view of the magnetic field configuration. From Romanova et al. (2011b)

Königl 1991). Of course, this is an example of stable accretion, for which the magnetospheric and corotation radii are similar in value, which is typical for the stable accretion case.

Simulations also predict the magnetic configuration of V2129 Oph on a larger scale (see Fig. 26). The disk-magnetosphere interaction leads to the winding and inflation of the external magnetic field lines, which thread the disk at radii $r \gtrsim r_m$. The figure shows that a strong azimuthal component of the field is present, and that magnetic tower structures form about the rotational axis.

2.9 Summary and Future Outlook on Magnetospheric Accretion

As discussed in the above subsections, 3D simulations have shown, for the first time, how matter flows onto stars with a tilted dipole, and onto stars with the complex magnetic field. This has allowed direct modeling of the structure of the funnel stream, and the position and shapes of the hot spots. Specific simulations were performed for CTT stars with realistic parameters, BP Tau and V2129 Oph, and the spectrum has been calculated for V2129 Oph using the 3D radiative transfer code *TORUS*. Comparisons with observations have shown excellent match, providing a strong substantiation of the magnetospheric accretion paradigm.

Numerical simulations also led to discovery of two distinct states, characterized by stable vs. unstable regimes of accretion. The unstable regime may help explain the short-term variability, with characteristic time of a few events per rotational period, that is observed in a significant number of CTTSs.

The above research concentrated on the processes at the disk-magnetosphere boundary and the magnetospheric accretion. Inflation of the field lines is also modeled, but the high coronal density assumed means it is matter-dominated, effectively suppressing possible magneto-centrifugal mechanisms for driving outflows. Future 3D simulations should concentrate on regimes with density low enough to allow magnetically driven outflows. Future simulations can also aim at modeling of stars with larger magnetospheres, in order to better understand accretion onto intermediate polars, as well as X-ray pulsars. Both steps would require more computing power, because larger parts of the simulation region will be covered by both a magnetically dominated corona, and a magnetically dominated, large-scale magnetosphere. Within more computationally tractable 2.5D (axisymmetric) simulations, the next sections describe such investigations of outflows from disk-accreting magnetized stars.

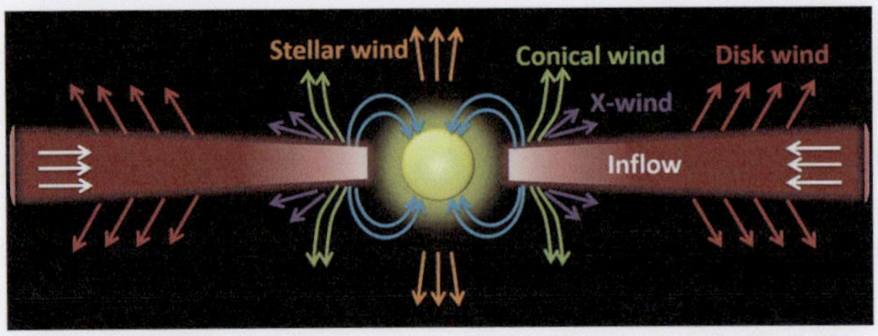

Fig. 27 A sketch illustrating the physical processes in the vicinity of accreting magnetized stars. The inner disk matter may flow onto a star in funnel streams (*blue arrows*), or it may fly away from the disk-magnetosphere boundary either in centrifugally-driven X-winds or magnetically-driven conical winds (*purple* and *green arrows*, respectively). Matter may also flow from the surface of the star in the form of stellar wind, or from the disk in the form of the disk wind (*orange* and *red arrows*, respectively). Sketch by M.L Comins

3 Outflows from the Disk-Magnetosphere Boundary

Different theoretical models have been proposed to explain the winds and jets from accreting stars (see review by Ferreira et al. 2006). Figure 27 demonstrates a few different possibilities. Outflows can be accelerated by a magneto-centrifugal mechanism along the field lines threading the disk at different distances from the star and tilted by $> 30°$ about the z-axis (e.g., Blandford and Payne 1982); or they can originate at the disk-magnetosphere boundary (Shu et al. 1994; Romanova et al. 2009). Stellar winds may also contribute to an outflow if part of the accreting matter is redirected into a stellar wind (Matt and Pudritz 2005, 2008).

In this section, we concentrate on outflows from the disk-magnetosphere boundary. We show results from two scenarios: (1) when the star rotates rapidly in the propeller regime and the rapidly rotating magnetosphere drives outflows; and (2) when the star rotates slowly but the field lines are bunched up at the disk-magnetosphere boundary during episodes of enhanced accretion, in which case the matter flows into a conically shaped wind (e.g., Romanova et al. 2009).

3.1 Propeller-Driven Outflows

The propeller regime, wherein $r_{\rm cor} < r_m$, was studied in 2.5D MHD simulations in the cases of α-disks (Romanova et al. 2005b, 2009; Ustyugova et al. 2006) and turbulent disks (Lii et al. 2014), in which the turbulence is driven by the magneto-rotational instability (MRI, Balbus and Hawley 1991, 1998). We consider these cases in Sects. 3.1.1 and 29.

3.1.1 Propeller Regime Studied with α-Disks

The propeller regime has been initially modelled in terms of α-disks, assuming both axisymmetry and north-south symmetry, using a code with spherical coordinates. In most simulation runs, the viscosity parameter is taken to be $\alpha = 0.1$–0.3, which allows a relatively high rate of matter flow toward the propelling magnetosphere. Magnetic diffusivity is also incorporated, using a formalism analogous to viscous diffusion, now proportional to a diffusivity parameter $\alpha_d < 1$ (e.g., Bisnovatyi-Kogan and Ruzmaikin 1976). Such a diffusivity mimics more complex processes, such as 3D instabilities at the disk-magnetosphere boundary, and is important because it allows penetration of inner disk matter onto field lines of the external magnetosphere. Most simulations (Romanova et al. 2005b, 2009; Ustyugova et al. 2006). take a diffusivity coefficient $\alpha_d = 0.1$; the dependence on α_d has been studied in a special set of simulation runs by Ustyugova et al. (2006).

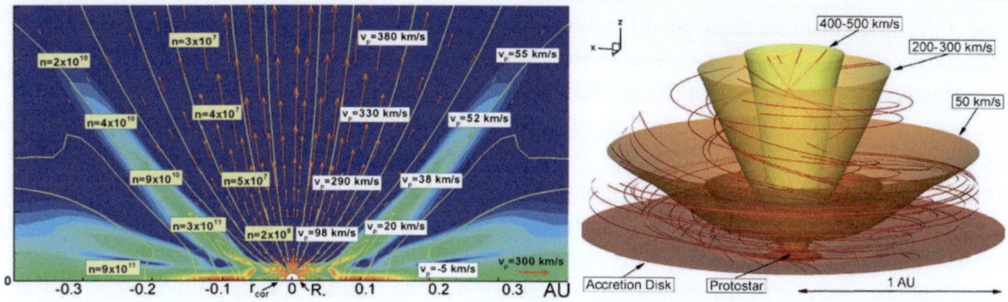

Fig. 28 *Left panel*: The density and velocity distribution in the propeller regime. The numbers correspond to CTTSs (Romanova et al. 2009). *Right panel*: A 3D rendering showing magnetic field lines and density levels corresponding to different velocities

The propeller may be "strong"—where the magnetosphere rotates much more rapidly than the inner disk (fastness parameter $\omega_s \gg 1$)—, or weak—where the angular velocities of the magnetosphere and the disk are comparable ($\omega_s \sim 1$). For a weak propeller, most of matter accretes onto the star, while the angular momentum is transferred outward along the disk by viscosity (e.g., Sunyaev and Shakura 1977; Spruit and Taam 1993). For a strong propeller, significant amounts of matter can be redirected from the inner disk to a direct outflow (Illarionov and Sunyaev 1975; Lovelace et al. 1999). For a case of very strong propeller (with fastness parameter is $\omega_s \approx 4$, and $r_{\rm cor} = 1$; Romanova et al. 2009), Fig. 28 (left panel) shows a simulation snapshot during an outflow episode. Analysis of forces shows that the conical part of outflow is centrifugally driven, while magnetic forces also drive a smaller mass flux into a magnetically-dominated, better-collimated Poynting jet (Lovelace et al. 1991, 2002). Some accretion is also possible, due to the 2D nature of the propeller regime, wherein matter may flow around the centrifugally-rotating magnetosphere.

Propeller-driven outflows are expected at the early stages of stellar evolution, wherein the star still rotates rapidly and retains a strong magnetic field. A strong propeller may redirect most of the disk matter into the outflows. It is possible that the powerful jets from young (Classes 0 and I) stars are connected with such a propeller stage in young star evolution. These stars are usually hidden inside dusty envelopes of forming stars, and it is presently difficult to test this hypothesis. It is also possible that the outflows observed in class II young stars (CTTSs) can also be connected with the propeller regime of accretion. Observations of jets from CTTSs show that they have an "onion-skin" velocity distribution, for which the outflow velocity is higher in the axial regions and decreases away from the axis (e.g., Bacciotti et al. 1999; Dougados et al. 2000). Similar onion-skin structure is seen in simulations of outflows from propelling star (see Fig. 28, right panel), and in simulations by Goodson et al. (1997).

3.1.2 Propeller Regime in Case of Accretion from Turbulent Disks

In the case of MRI-driven turbulent disks, the inward disk accretion can be conveniently provided by this turbulence. However, the MRI-driven turbulence does not provide the diffusivity at the disk-magnetosphere boundary. Some matter penetration is associated with the reconnection events between the stellar field lines and the field lines of the turbulent cells (Romanova et al. 2011b), but this does not provide significant diffusivity. Therefore, in axisymmetric simulations an α-type diffusivity is often added, but only to the region where the disk interacts with the magnetosphere (Lii et al. 2014). Actually, most simulation runs of Lii et al. (2014) assume no explicit diffusivity ($\alpha_d = 0$), so that only the very small numerical diffusivity provides the interaction between the disk and magnetosphere. Such small diffusivity leads to long episodes of the matter accumulation at the disk-magnetosphere boundary, punctuated by rare events of accretion onto the star. Figure 29 shows an example of outflow for the strong propeller case, with $\omega_s \approx 2$ and $r_{\rm cor} = 1.3$. Along with the strong episodic outflows are evident, intervals of matter accumulation and accretion are also evident. Matter accumulates and accretes when a sufficient amount of matter reaches the corotation radius. As discussed in Sect. 2.3.4, the MRI-driven disk behaves as matter-dominated disk. In this propeller case, matter of the disk simply flows above the rapidly rotating magnetosphere of the star. In these propeller simulations, top-bottom symmetry was not assumed, thus allowing matter to flow both above or below the rapidly-rotating magnetosphere, while pushing the magnetosphere in the opposite direction (see also analysis of this phenomenon in Romanova et al. 2011b). The direction of the outflows also changes episodically.

In test simulations with larger diffusivity ($\alpha_d = 0.1$) in the disk-magnetosphere region, the interaction between the disk and the magnetosphere is stronger, and more matter is ejected to the outflows, while smaller amount of matter accretes onto the star (Lii et al. 2014). In both types of simulations of the propeller regime, the disk-magnetosphere interaction has a cyclic character. Initially, the inner disk matter accumulates, then diffusively penetrates the inner parts of the magnetosphere, where it acquires angular momentum; some of the matter is then ejected to a centrifugally-driven wind, while some accretes onto the star. After this, the magnetosphere expands, and the cycle repeats (see also Goodson et al. 1997; Goodson and Winglee 1999). Compared with Goodson et al. (1997), simulations show many more oscillation cycles. This cyclic oscillations are different from cyclic accretion discussed for dead disks, where only weak propellers are considered, and where the excess of angular momentum flows into the dead disk (e.g., Spruit and Taam 1993; D'Angelo and Spruit 2010). There are also similarities: accretion occurs when the inner disk reaches the corotation radius. However, the propeller-driven oscillations are expected on much smaller time-scales compared with the dead disk oscillations, because in the propeller-driven oscillations only the inner part of the disk oscillates, while in dead disks much larger parts of the disk are usually involved.

In cases of both turbulent and α-type disks, the strength of the propeller strongly depends on the fastness parameter ω_s. The estimated value of ω_s is only approximate, because the inner disk strongly oscillates, causing the values of r_m and $\Omega_K(r_m)$ to vary in time. The range of strengths of the propellers varies depending on the fastness parameter. When ω_s is a few times larger than unity, then the propeller is strong and most of the disk matter can be ejected into outflows. When $\omega_s \approx 1$, the propeller is weak and most of the matter accretes onto the star.

The outcome of the propeller regime also strongly depends on the diffusivity at the disk-magnetosphere boundary. The diffusivity should be sufficiently high, so that the matter of the inner disk can interact with the magnetosphere. When the diffusivity is very low, the rapidly-rotating magnetosphere and slowly-rotating disks do not exchange angular momentum, and outflows are not possible. We suggest that, in more realistic 3D simulations, the diffusivity is not very small, because three-dimensional instabilities such as Kelvin–Helmholtz and magnetic interchange instabilities will lead to mixing of the disk matter with the magnetosphere.

3.1.3 Matter Fluxes and Oscillation of the Inner Disk

The amounts of matter flowing onto the star and into the winds varies according to the strength of the propeller. Figure 30 shows an example of these fluxes in the case of a strong

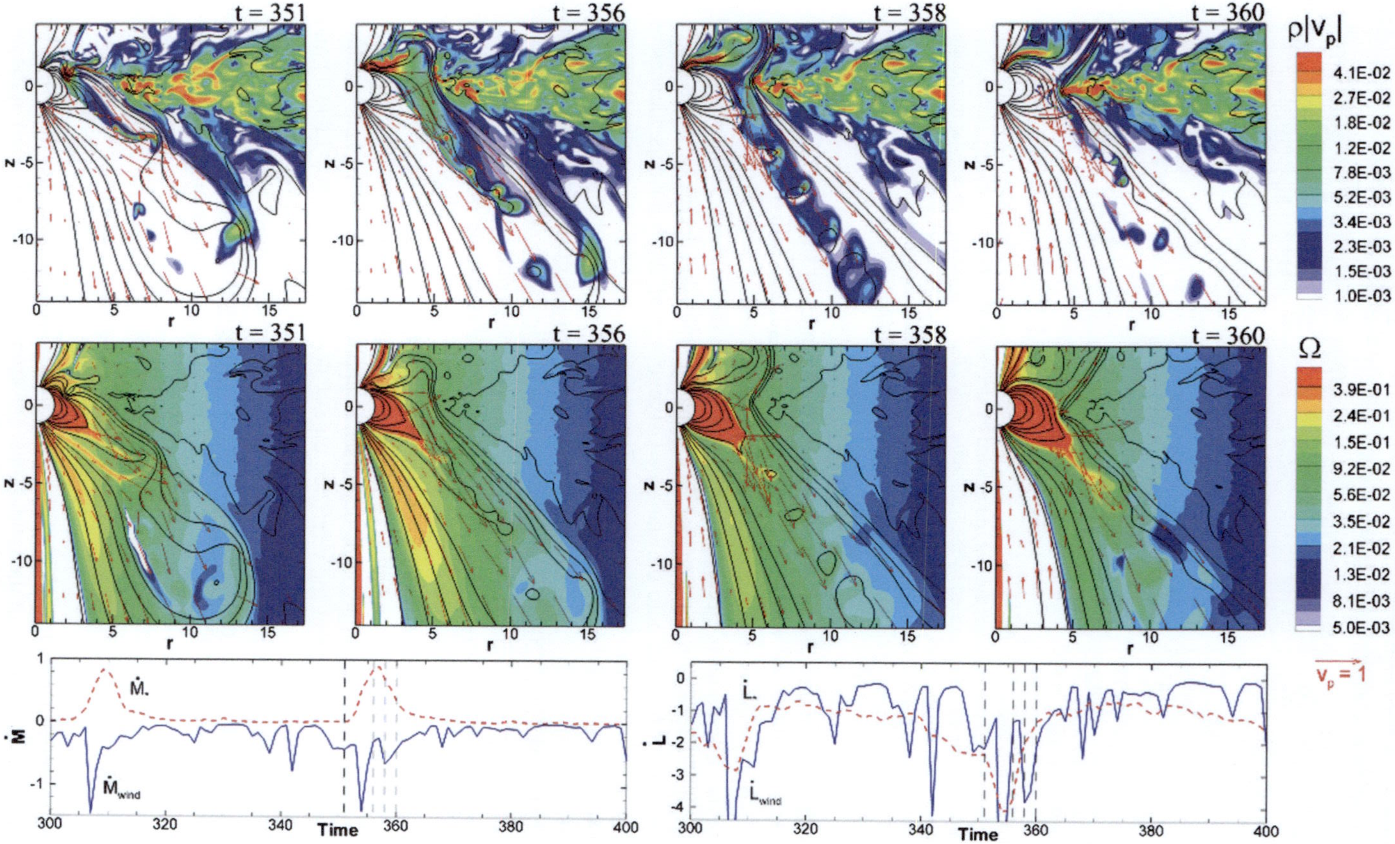

Fig. 29 Axisymmetric simulations of the propeller outflows in the case of accretion from the turbulent disk ($r_{\mathrm{cor}} = 1.3$). The *color background* shows the matter flux (*top row*) and angular velocity (*middle row*). The *bottom row* shows the matter fluxes (*left panel*) and angular momentum fluxes (*right panel*) onto the star and into the wind, respectively. The dimensionless time-scale can be converted to dimensional one by multiplying the dimensional time by P_0 from Table 1. From Lii et al. (2014)

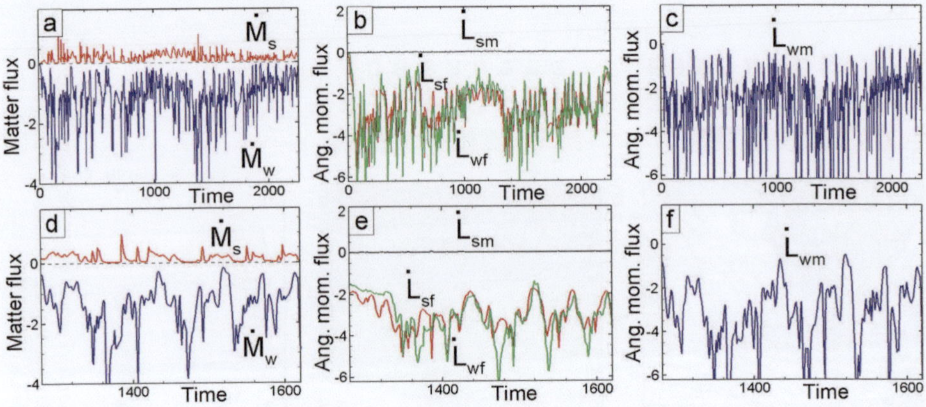

Fig. 30 *Top left panel*: The matter fluxes onto the star, $\dot{M}_s$, and into the wind, $\dot{M}_w$. *Top middle panel*: The angular momentum fluxes carried from the star to the wind by the magnetic field, $\dot{L}_{\rm sf}$, and by matter, $\dot{L}_{\rm sm}$. $\dot{L}_{\rm wf}$ is the angular momentum flux carried by the field to the wind. *Top right panel*: The angular momentum flux carried by the matter to the wind. The *bottom panels* show the same fluxes as the *top panels*, but "zoomed in" to focus on a particular period of time in the simulations. The dimensionless time-scale can be converted to dimensional one by multiplying the dimensionless time by P_0 from Table 1. From Romanova et al. (2009)

propeller, calculated in the case of an α-disk. Figure 30 (panel *a*) shows the matter fluxes onto the star, $\dot{M}_s$, and into the winds, $\dot{M}_w$, integrated over a surface with radius $r = 10$. The matter flux into the wind is much larger than that onto the star, $\dot{M}_w \gg \dot{M}_s$, implying that almost all of the disk matter is ejected from the system into the outflows. If the star rotates more slowly, then the fraction of the matter flux going into the wind decreases, and a larger portion of the matter accretes onto the star (see Ustyugova et al. 2006 for dependences of matter fluxes on $\Omega_\star$, $B_\star$, α, and α_d).

Both fluxes strongly vary and show episodic enhancement of accretion and outflows. The interval between the strongest outbursts increases when the diffusivity coefficient α_d decreases (Ustyugova et al. 2006). The bottom panels of Fig. 30 show the same fluxes as the top panels, but "zoomed in" to focus on a particular time period in the simulation. The time interval between the strongest outbursts in the propeller regime is $\Delta t \approx 50$–70 (in dimensionless units). For protostars and CTTSs ($P_0 = 1.04$ days), this time corresponds to $(\Delta t)_{\rm outb} = 52$–$73$ days. In some young stars (e.g., CTTS HH30/XZ Tau), the outbursts into the jet occur at intervals of a few months. This implies that episodic ejections to the propeller wind may be responsible for some of the outbursts. In application to accreting neutron stars (AMXPs), we take into account the reference time-scale $P_0 = 1.3$ ms (see Table 1) and obtain the dimensional time-scale $(\Delta t)_{\rm outb} \approx 65$–$90$ ms. Therefore, rapid variability on this and smaller time-scales is expected in AMXPs in the propeller regime. This time-scale is larger, if the diffusivity at the disk-magnetosphere boundary is smaller, and vice versa. The longer time scale between bursts is also expected in case of the larger fastness ω_s of the propeller.

3.1.4 Angular Momentum Outflow and Spinning-down of Young Stars

A star in the propeller regime loses its angular momentum along the field lines that originate at the surface of the star. Some of these field lines are strongly inflated and a part of the angular momentum flows into the magnetically-dominated Poynting flux jet (e.g., Ustyugova et al. 2006). The other set of field lines originating on the star connect the star to the disk

Table 1 Reference values for different types of stars. We choose the mass M, radius R_* and equatorial magnetic field B_* of the star. The reference length is R_0, the reference velocity is Keplerian velocity at R_0, the reference time-scale P_0 is the period of rotation at R_0. The reference density ρ_0 is determined at R_0. Reference matter flux $\dot{M}_0$ and angular momentum flux $\dot{L}_0$ are derived from other reference values (see details in Romanova et al. 2009). To apply the simulation results to a particular star one needs to multiply the dimensionless values from the plots by the reference values from this table

	CTTSs	White dwarfs	Neutron stars
M ($M_\odot$)	0.8	1	1.4
R_*	$2R_\odot$	5000 km	10 km
R_0 (cm)	2.8×10^{11}	1.0×10^9	2×10^6
v_0 (cm^{-1})	1.95×10^7	3.6×10^8	9.7×10^9
P_0	1.04 days	17.2 s	1.3 ms
B_* (G)	10^3	10^6	10^9
ρ_0 (g cm^{-3})	4.1×10^{-13}	1.2×10^{-9}	1.7×10^{-6}
$\dot{M}_0$ ($M_\odot$yr^{-1})	2.0×10^{-8}	1.3×10^{-8}	2.0×10^{-9}
$\dot{L}_0$ (erg s^{-1})	3.4×10^{36}	1.6×10^{35}	1.2×10^{33}

and are only partially inflated. A fraction of the angular momentum can also flow out of the star along these field lines (see also Zanni and Ferreira 2013). The middle panels of Fig. 30 show the dimensionless values of the angular momentum fluxes out of the star, calculated at the stellar surface. The angular momentum is primarily carried by the Poynting stresses of the magnetic field (see the red curve for $\dot{L}_{sw}$ in the middle panels), while the direct angular momentum carried by the matter is negligibly small (see the black curve for $\dot{L}_{sm}$, also in the middle panels). Panels *b* and *c* also shows the angular momentum fluxes to the outflows (calculated at the radius $r = 10R_\star$ from the star) carried by the field, $\dot{L}_{\rm wf}$, and by the matter, $\dot{L}_{\rm wm}$. The angular momentum carried by the field to the outflow, $\dot{L}_{\rm wm}$, is approximately equal to the angular momentum out of the surface of the star, $\dot{L}_{sw}$.

In young solar-type stars, the propeller mechanism may be responsible for their spinning down from near critical rotation, which is expected right after their formation, to less than 10 % of this speed, observed in CTTSs (e.g., Romanova et al. 2005b). If a rapidly rotating protostar has a magnetic field strength of a few kG, then it is in the strong propeller regime and can rapidly lose angular momentum. Estimations show that if a protostar rotates initially with period of $P_\star = 1$ day and has a magnetic field of $B = 3 \times 10^3$ G, then it will lose most of its angular momentum to propeller outflows during the time scale of $\tau \approx 3 \times 10^5$ years. If the magnetic field is 1 kG, then this time-scale increases up to $\tau \approx 3 \times 10^6$ years. This is shorter than typical lifetime of CTTSs, which is $\sim 10^7$ years. Therefore, if the magnetic field of protostars is sufficiently strong, then the propeller mechanism may be responsible for the fast spinning-down of these protostars to the slow rotation speeds observed in CTTSs.

3.1.5 Propeller Regime in Transitional Millisecond Pulsars

In accreting neutron stars the propeller regime is expected at the end of the accretion outburst, when the accretion rate decreases, and the truncation radius of the disk can be larger than the corotation radius. For example, in AMXP SAX J1808.4-3658, 1 Hz flaring oscillations have been observed at the end of an outburst (van Straaten et al. 2005). These oscillations may be caused by the inner disk oscillations in the propeller regime (Patruno et al. 2009; Patruno and D'Angelo 2013). For the typical AMXP, the time-scale of spinning down is estimated to be $\tau \approx 2.5 \times 10^7$ years. This time scale is shorter than the expected lifetime of the accreting millisecond pulsars. However, these stars spend relatively little time in the propeller regime.

Recently discovered 'transitional millisecond pulsars', which switch between an accretion stage and a radiopulsar stage (e.g., Papitto et al. 2013; Ferrigno et al. 2014; Linares 2014), support the earlier suggested scenario that millisecond radiopulsars represent recycled pulsars that are re-spun-up by disk accretion in the binary system (Alpar et al. 1982; Bisnovatyi-Kogan and Komberg 1974). Such pulsars are observed as accreting millisecond X-ray pulsars during periods of high accretion rate, and as radiopulsars during periods of low accretion rate. In these stars, the propeller stage is inevitable and is expected between these two regimes. In fact, strong rapid oscillations in X-ray have been observed in the transitional pulsar AMXP IGR J18245-2452, which may be associated with the propeller regime (Ferrigno et al. 2014). Similar oscillations have been observed in another transitional pulsar PSR J1023+0038 (e.g., Patruno et al. 2014). At the same time, enhanced radiation in the radio band has been observed during this stage, which is interpreted as possible outflow from the system. Strong oscillations of the inner disk and outflows were predicted earlier in theoretical and numerical models (e.g., Lovelace et al. 1999; Romanova et al. 2005b; Ustyugova et al. 2006; Lii et al. 2014).

These new discoveries present an opportunity to compare models with observations, both to test and constrain the models and open up a new puzzles. For example, X-ray pulsations were found in PSR J1023+0038 during its very dim state, when the X-ray luminosity was about factor 100 lower than during the main accretion state, when the pulsar is expected to be in the propeller regime (e.g., Archibald et al. 2015; Papitto and Torres 2015). Different explanations were proposed to solve this puzzle (see Archibald et al. 2015). But one key point is that the formal condition for the propeller regime, $r_m > r_{\rm cor}$ only applies to an idealized, one-dimensional picture, wherein vertical centrifugal barrier prevents any accretion onto propelling star. In an axisymmetric, two-dimensional approach, the centrifugal barrier is restricted by the closed magnetosphere of the star. In other parts of the simulation region, matter of the corona rotates more slowly, and therefore disk material can flow above/below this rapidly-rotating magnetosphere and accrete onto the stellar surface. In addition, a small amount of coronal matter may precipitate onto the star along the inflated coronal field lines. Such residual 2D accretion represents an important difference from the idealized 1D picture the propeller regime: Of course, more matter accretes in case of the weaker propeller regime, but even in a strong propeller regime, some matter accretes onto the star (see Fig. 30). This may explain X-ray pulsations of PSR J1023+0038 during very dim state. Detailed comparisons of axisymmetric simulations with observations may explain variability and other features of transitional pulsars at their propeller stage. However, even more advanced, three-dimensional simulations of the propeller regime are required to model this properly. On the other hand, current X-ray telescopes do not yet provide sufficient temporal resolution (about 10 ms) to detect the time variability found in such models.

Similar flaring oscillations have been observed in a few cataclysmic variables, e.g., in AE Aqr (e.g., Mauche 2006) which indicated that these stars may also be in the propeller regime during a part of their life-time. It is often suggested that this CV periods of accretion alternate with periods of ejection (e.g. Wynn et al. 1997).

3.2 Conical Winds from the Disk-Magnetosphere Boundary

Recent numerical simulations reveal a new type of wind that can be important in stars with any rotation rate, if the magnetic flux of the star is compressed by the disk matter at the disk-magnetosphere boundary (see Fig. 31.) This situation is possible, if the inward radial velocity of the disk matter is larger than velocity of the diffusive penetration of the disk matter

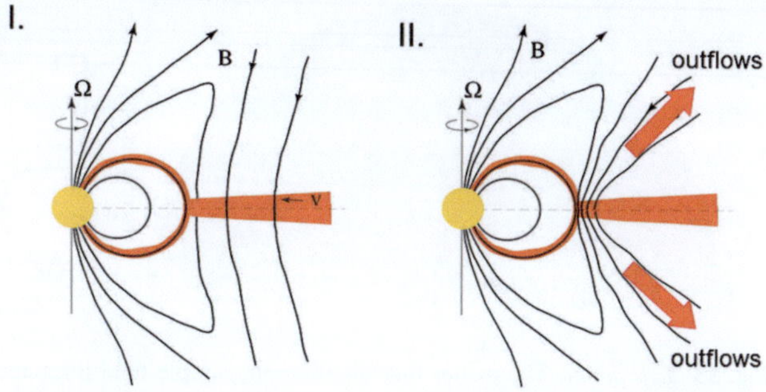

Fig. 31 *Left panel*: An example of accretion in which the velocity of the radial flow of matter towards the star, $v_{\rm rad}$, is comparable to the velocity of the diffusive penetration of the disk matter through the field lines of the outer magnetosphere, $v_{\rm d}$. *Right panel*: An example in which $v_{\rm rad} \gg v_{\rm d}$, the field lines are bunched by the accreting matter, and the situation is favorable for outflows

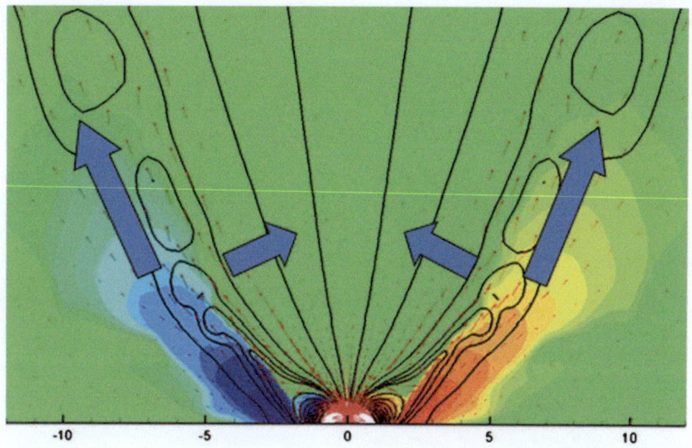

Fig. 32 The *color background* shows the poloidal current, which results from the large azimuthal component of the magnetic field lines above the disk. The magnetic pressure gradient force $F_M \propto -\nabla(r B_\phi)^2$ drives matter into cone-shaped outflows. From Romanova et al. (2009)

through the field lines of the magnetosphere (Romanova et al. 2009; Königl et al. 2011; Lii et al. 2012). Simulations show that if the incoming matter compresses the magnetosphere of the star, the field lines inflate due to the differential rotation between the disk and the star, and conically shaped winds flow out of the inner disk. These winds are driven by the magnetic force, $F_M \propto -\nabla(r B_\phi)^2$ (see Fig. 32), which arises from the wrapping of the field lines above the disk (Lovelace et al. 1991). The wind is also gradually collimated by the magnetic hoop-stress, and it can be strongly collimated when the accretion rate is large (Lii et al. 2012).

Conical winds model differ from earlier proposed X-winds model (Shu et al. 1994) in several respects.[13] Conical winds are expected in stars where the accretion rate strongly increases during the accretion outburst episodes, such as in AMXPs, and in sub-classes of young stars EXors and FUors, where the accretion rate in the disk increases dramatically (see review by Audard et al. 2014). In fact, in FU Ori itself, H_α shows a strong blueshifted absorption, providing direct evidence for outflows. The conical wind model has been applied to FU Ori star and compared with the empirical model based on the spectral analysis of the winds in FU Ori (Calvet et al. 1993). A reasonably good agreement has been found between these models (Königl et al. 2011). Therefore, the conical winds represent an attractive model to explain outflows in stars wherein the accretion rate strongly increases in time. However,

[13](1) In the X-wind model, one of the necessary requirements is the condition $r_m = r_{\rm cor}$, that is the inner disk should rotate with the angular velocity of the magnetosphere. In conical winds model, there is no such restriction: a star may rotate much more slowly than the inner disk, with $r_m \ll r_{\rm cor}$. (2) In the X-wind model matter is driven by the centrifugal force and overall situation is closer to the weak propeller regime; in conical winds model the driving force is mainly the magnetic pressure arising from the winding of the magnetic field lines above the inner parts of the disk.

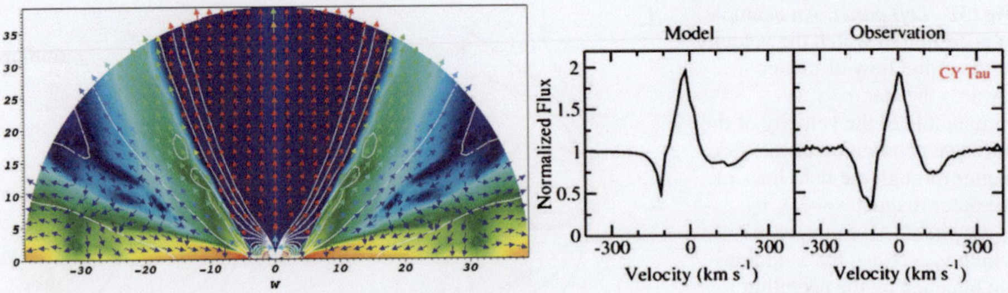

Fig. 33 *Left panel*: The matter flux distribution, sample field lines and homogeneous velocity vectors in a conical wind. *Middle panel*: The spectrum from a conical wind calculated for the He I λ10830Å spectral line using the radiative transfer code *TORUS*. *Right panel*: The spectrum of the He I λ10830Å line observed in the wind from the CTTS CY Tau. From Kurosawa and Romanova (2012)

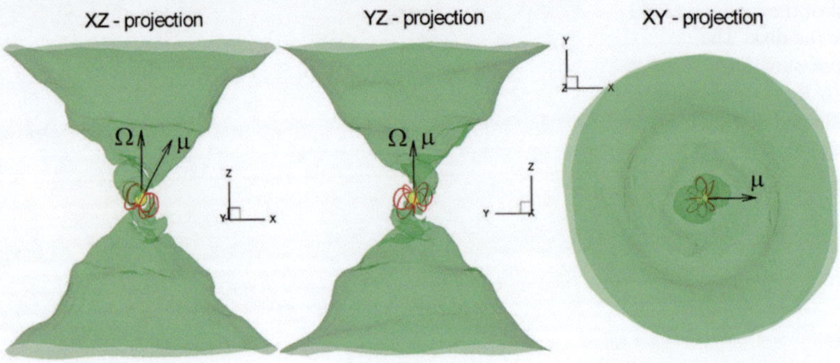

Fig. 34 3D MHD simulations of conical winds from the disk-magnetosphere boundary when the dipole moment of the star is tilted at $\Theta = 30°$ show the wavy structure of the inner wall of the conical shell wind. The *green background* shows one of the density levels, and the *lines* show sample magnetic field lines of the tilted magnetosphere

this model should be tested in 3D simulations to check the role of 3D instabilities at the disk-magnetosphere boundary. The strength of the winds will probably depend on the ratio between the inward penetration of matter through instabilities (as discussed in Sect. 2.4), and outward flow of matter due to the winding of the field lines and action of the magnetic pressure force. Both processes are expected to occur on the time-scale comparable with the Keplerian time-scale. Global 3D simulation are required to further investigate conical winds.

The observational properties of conical winds, applied to CTTSs, were investigated by Kurosawa et al. (2011) and Kurosawa and Romanova (2012). In these studies, the spectra of the H and He lines were calculated using the radiative transfer code *TORUS*. These simulations show that conical winds produce a narrow blue absorption component in the spectrum (see Fig. 33). Such a blue component is frequently observed in the spectra of CTTSs (e.g., Edwards et al. 2006).

Conical winds were also studied in test global 3D simulations where the magnetic axis of the stellar dipolar field has been tilted by the angle $\Theta = 30°$ about the rotational axis (Romanova et al. 2009). These simulations show that, in spite of the tilted magnetosphere, the winds form a conical structure about the rotational axis. However, the interaction of the inner disk with the tilted magnetosphere develops a spiral pattern within the cone-shaped wind (see Fig. 34).

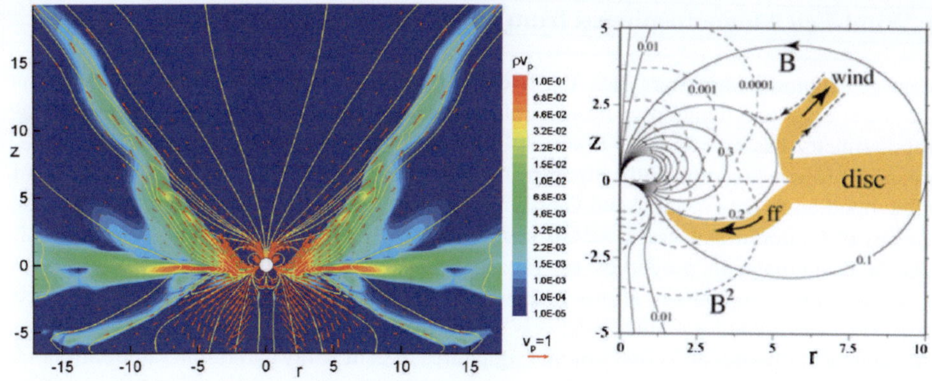

Fig. 35 *Left panel*: a density slice and sample field lines show the result of an axisymmetric simulation of accretion onto a star with superposed dipole and quadrupole magnetic fields. *Right panel*: a sketch of accretion and outflow from such a system (from Lovelace et al. 2010)

3.3 Asymmetric and One-Sided Outflows

In stars with a complex magnetic field, outflows may be asymmetric due to the top-bottom asymmetry of the magnetic field (e.g., Wang et al. 1992). For example, the superposition of an axisymmetric dipole field with a quadrupole field leads to a configuration in which the magnetic flux is larger on one side of the equatorial plane than the other (see Fig. 35, right panel). Axisymmetric simulations of the propeller regime show that stronger outflows are observed on the side where the magnetic flux is larger (see left panel of Fig. 35). In this case, the matter and energy fluxes will be systematically higher in one direction and lower in the other direction. One-sided outflows are observed in a number of young stars (e.g., Bacciotti et al. 1999).

Axisymmetric simulations of the entire region also show that, even in the case of a pure dipole field, outflows are usually one-sided. However, the direction of the outflows switches frequently, and therefore the averaged matter and energy fluxes of the outflows above and below the equatorial plane are expected to be approximately equal in both directions (Lovelace et al. 2010). Recent simulations of the propeller regime in the case of MRI-driven accretion have also shown that the outflows are one-sided, but that the wind switches sides much less frequently (Lii et al. 2014).

3.4 Future Outlook on Outflows from the Disk-Magnetosphere Boundary

In the above subsections, outflows from the disk-magnetosphere boundary were modeled in two scenarios for which matter flows from the disk-magnetosphere boundary: rapidly rotating stars in the propeller regime and slowly-rotating stars (conical winds). Most of the simulations described were performed using an axisymmetric approximation. Future simulations should expand to full three dimensions, so that the disk-magnetosphere interaction through instabilities can be taken into account. Also, outflows at larger scales should be further studied.

4 Wind-Fed Magnetospheres from Magnetized Massive Stars

Massive, luminous, hot stars lack the hydrogen recombination convection zone that induces the magnetic dynamo cycle of cooler, solar-type stars. Nonetheless, modern spectropolarimetry has revealed that about 10 % of O, B and A-type stars harbor large-scale, organized (often predominantly dipolar) magnetic fields ranging in dipolar strength from a few hundred to tens of thousand Gauss. Petit et al. (2013) recently compiled an exhaustive list of 64 confirmed magnetic OB stars with $T_{\rm eff} \gtrsim 16$ kK, along with their physical, rotational and magnetic properties; see Fig. 38 below.

This section summarizes efforts to develop dynamical models for the effects of such large-scale surface fields on the radiatively driven mass outflow from such OB stars. The focus is on the properties and observational signatures (e.g. in X-ray and Balmer line emission) of the resulting *wind-fed* magnetospheres in closed loop regions, and on the stellar rotation *spindown* that results from the angular momentum loss associated with magnetically torqued wind outflow from open field regions. In this way magnetic fields can have a profound effect on the star's rotational evolution, giving rotation periods ranging from weeks to even decades, in strong contrast to the day-timescale periods of non-magnetic massive stars.

The inside-out building of these wind-fed magnetospheres is in some way complementary to the outside-in nature of the accretion-fed magnetospheres discussed in previous sections. But there are also some interesting similarities in the role of the characteristic magnetospheric and corotation radii, which in this case are identified below (see Eqs. (9) and (10) in Sects. 4.1 and 4.2) as the Alfvén radius $R_{\rm A}$ and the Kepler radius $R_{\rm K}$. In particular, the relative sizes of these radii again plays a key role in determining the magnetospheric characteristics, in this case leading to dynamical magnetospheres (DM) when $R_{\rm A} < R_{\rm K}$, and centrifugal magnetospheres (CM) when $R_{\rm A} > R_{\rm K}$. The trapping of high-speed wind outflow also leads to strong shocks and so X-ray emission that can be much harder and stronger than generally seen for cases of non-degenerate stellar accretion.

4.1 Wind Magnetic Confinement Parameter and Alfvén Radius

MHD simulation studies (e.g., ud-Doula and Owocki 2002; ud-Doula et al. 2008) show that the overall net effect of a large-scale, dipole magnetic field in diverting such a hot-star wind can be well characterized by a single *wind magnetic confinement parameter* and its associated *Alfvén radius*,

$$\eta_* \equiv \frac{B_{eq}^2 R_*^2}{\dot{M} v_\infty}; \qquad \frac{R_{\rm A}}{R_*} \approx 0.3 + (\eta_* + 0.25)^{1/4}, \tag{9}$$

where $B_{\rm eq} = B_{\rm p}/2$ is the field strength at the magnetic equatorial surface radius R_*, and $\dot{M}$ and v_∞ are the fiducial mass-loss rate and terminal speed that the star *would have* in the *absence* of any magnetic field. This confinement parameter sets the scaling for the ratio of the magnetic to wind kinetic energy density. For a dipole field, the r^{-6} radial decline of magnetic energy density is much steeper than the r^{-2} decline of the wind's mass and energy density; this means the wind always dominates beyond the Alfvén radius, which scales as $R_{\rm A} \sim \eta_*^{1/4}$ in the limit $\eta_* \gg 1$ of strong confinement.

As shown in Fig. 36, magnetic loops extending above $R_{\rm A}$ are drawn open by the wind, while those with an apex below $R_{\rm A}$ remain closed. Indeed, the trapping of wind upflow from opposite footpoints of closed magnetic loops leads to strong collisions that form X-ray emitting, *magnetically confined wind shocks* (MCWS; Babel and Montmerle 1997a,

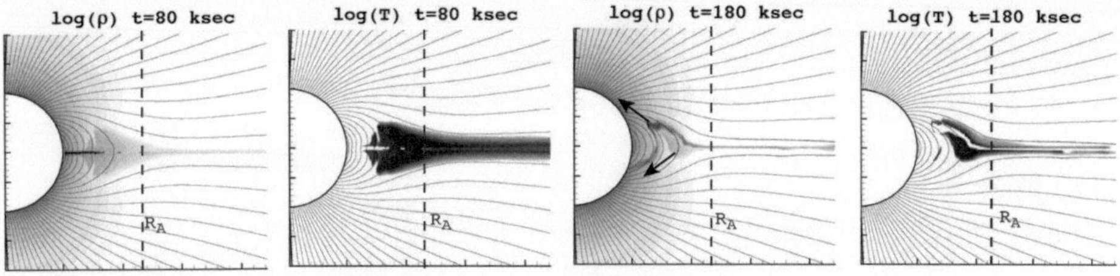

Fig. 36 2D MHD simulation for magnetic channeling and confinement of radiatively driven stellar wind from a non-rotating hot-star with $\eta_* = 15$ (and so $R_A \approx 2.3 R_*$), showing the logarithm of density ρ and temperature T in a meridional plane. At a time 80 ksecs after the initial condition, the magnetic field has channeled wind material into a compressed, hot region about the magnetic equator, much as envisioned in the Magnetically Confined Wind Shock (MCWS) paradigm of Babel and Montmerle (1997a, 1997b). But by a time of 180 ksecs, the cooled equatorial material is falling back toward the star along field lines, in a complex 'snake' pattern. The *darkest areas* of the temperature plots represent gas at $T \sim 10^7$ K, hot enough to produce relatively hard X-ray emission of a few keV. The model reproduces quite well the observed X-ray properties of θ^1 Ori C (Gagné et al. 2005)

1997b). The post-shock temperatures $T \approx 20$ MK are sufficient to produce the moderately hard (~ 2 keV) X-rays observed in the prototypical magnetic O-star θ^1 Ori C (Gagné et al. 2005). As illustrated by the downward arrows in the density plot at a simulation time $t =$ 180 ksec, once this material cools back to near the stellar effective temperature, the high-density trapped material falls back onto the star over a dynamical timescale.

4.2 Orbital Rotation Fraction and Kepler Co-rotation Radius

The dynamical effects of rotation can be analogously parameterized (ud-Doula et al. 2008) in terms of the *orbital rotation fraction*, and its associated *Kepler corotation radius*,

$$W \equiv \frac{V_{\rm rot}}{V_{\rm orb}} = \frac{V_{\rm rot}}{\sqrt{GM_*/R_*}}; \qquad R_{\rm K} = W^{-2/3}\, R_* \tag{10}$$

which depend on the ratio of the star's equatorial rotation speed to the speed to reach orbit near the equatorial surface radius R_*. Insofar as the field within the Alfvén radius is strong enough to maintain *rigid-body rotation*, the Kepler corotation radius $R_{\rm K}$ identifies where the centrifugal force for rigid-body rotation exactly balances the gravity in the equatorial plane. If $R_{\rm A} < R_{\rm K}$, then material trapped in closed loops will again eventually fall back to the surface, forming a *dynamical magnetosphere* (DM). But if $R_{\rm A} > R_{\rm K}$, then wind material located between $R_{\rm K}$ and $R_{\rm A}$ can remain in static equilibrium, forming a *centrifugal magnetosphere* (CM) that is supported against gravity by the magnetically enforced co-rotation. As illustrated in the upper left schematic in Fig. 37, the much longer confinement time allows material in this CM region to build up to a much higher density than in a DM region.

For full 2D MHD simulations in the axisymmetric case of a rotation-axis aligned dipole, the mosaic of color plots in Fig. 37 shows the time vs. height variation of the equatorial mass distribution $\Delta m/\Delta r$ for various combinations of rotation fraction W and wind confinement η_* that respectively increase upward and to the right. This illustrates vividly the DM infall for material trapped below $R_{\rm K}$ and $R_{\rm A}$, vs. the dense accumulation of a CM from confined material near and above $R_{\rm K}$, but below $R_{\rm A}$.

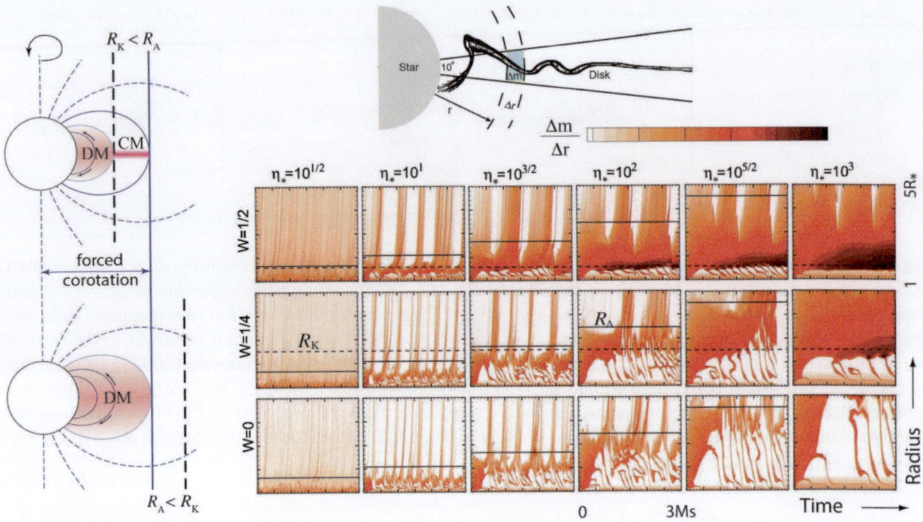

Fig. 37 *Left*: Sketch of the regimes for a dynamical vs. centrifugal magnetosphere (DM vs. CM). The *lower panel* illustrates the case of a slowly rotating star with Kepler radius beyond the Alfvén radius ($R_{\rm K} > R_{\rm A}$); the lack of centrifugal support means that trapped material falls back to the star on a dynamical timescale, forming a DM, with shading illustrating the time-averaged distribution of density. The *upper panel* is for more rapid rotation with $R_{\rm K} < R_{\rm A}$, leading then to a region between these radii where a net outward centrifugal force against gravity is balanced by the magnetic tension of closed loops; this allows material to build up to the much higher density of CM. *Right, Upper*: Contour plot for density at arbitrary snapshot of an isothermal 2D MHD simulation with magnetic confinement parameter $\eta_* = 100$ and critical rotation factor $W = 1/2$. The overlay illustrates the definition of radial mass distribution, $\Delta m/\Delta r$, within 10° of the equator. *Right, Lower*: Density plots for log of $\Delta m/\Delta r$, plotted versus radius (1–5 R_*) and time (0–3 Msec), for a mosaic of 2D MHD models with a wide range of magnetic confinement parameters η_*, and 3 orbital rotation fractions W. The horizontal solid lines indicate the Alfvén radius $R_{\rm A}$ (*solid*) and the horizontal dashed lines show Kepler radius $R_{\rm K}$ (*dashed*)

4.3 Comparison with Observations of Confirmed Magnetic Hot-Stars

For the 64 observationally confirmed magnetic hot-stars ($T_{\rm eff} \gtrsim 16$ kK) compiled by Petit et al. (2013), Fig. 38 plots positions in a log–log plane of $R_{\rm K}$ vs. $R_{\rm A}$. The vertical solid line representing $\eta_* = 1$ separates the domain of non-magnetized or weakly magnetized winds to left, from the domain of stellar magnetospheres to the right. The diagonal line representing $R_{\rm K} = R_{\rm A}$ divides the domain of centrifugal magnetospheres (CM) to the upper right from that for dynamical magnetospheres (DM) to the lower left. Let us now consider how these distinctions in magnetospheric properties organize their observational characteristics.

4.3.1 Balmer-α Line Emission from DM and CM

Figure 39 plots these observed magnetic stars in a diagram comparing the *ratio* $R_{\rm A}/R_{\rm K}$ vs. stellar luminosity, with now the symbol coded to mark the presence (light shading) or absence (black) of magnetospheric Hα emission. The horizontal solid line marks the transition between the CM domain above and the DM domain below, while the vertical dashed line marks the divide between O- and B-type main sequence stars. Note that *all* O-stars show emission, with all but one (Plaskett's star, which has likely been spun-up by mass exchange from its close binary companion; Grunhut et al. 2013.) located among the slow rotators

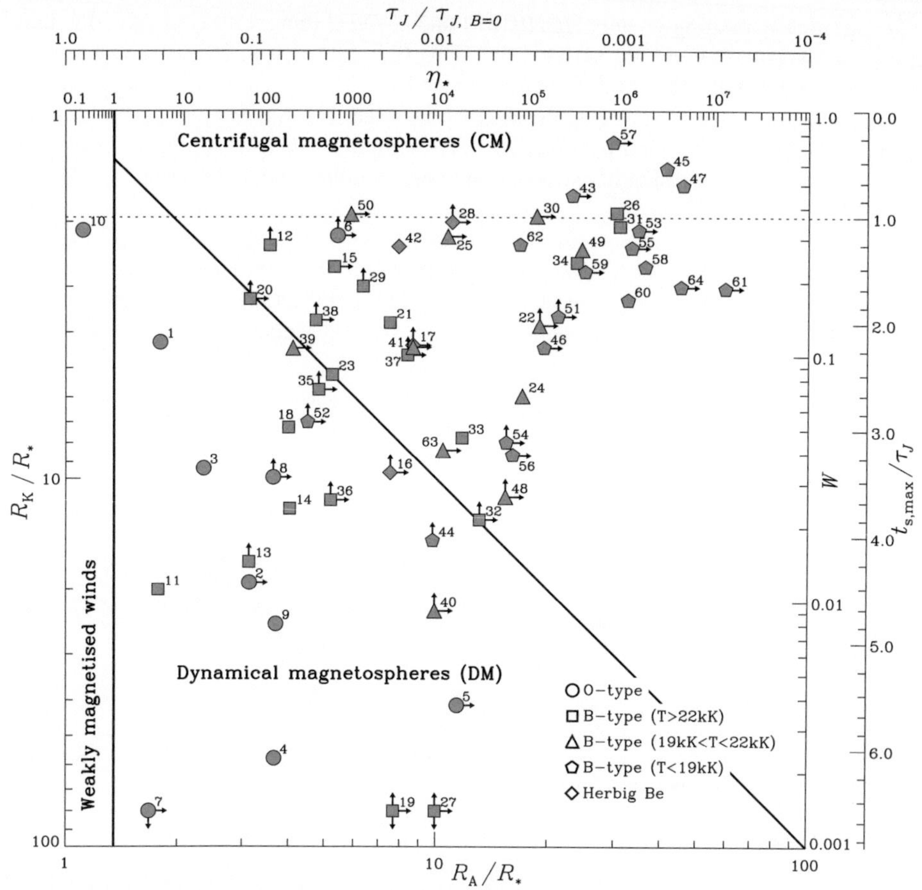

Fig. 38 Classification of 64 observationally confirmed magnetic massive stars in terms of magnetic confinement vs. rotation fraction, characterized here by a log–log plot of Kepler radius R_K increasing downward vs. Alfvén radius R_A increasing to the *right*. The labeled ID numbers are sorted in order of decreasing effective temperature T_{eff}, with stellar identities given in Table 1 of Petit et al. (2013). Stars to the *left* of the *vertical solid line* have only weakly magnetized winds (with $\eta_* < 1$), Star *below* and *left* of the *diagonal solid line* have dynamical magnetospheres (DM) with $R_A < R_K$, while those above and *right of this line* have centrifugal magnetospheres (CM) with $R_A > R_K$. The additional *upper* and *right* axes give respectively the corresponding spindown timescale τ_J, and maximum spindown age $t_{s,max}$, as defined in Section 4.4. Rapidly rotating stars above the *horizontal dotted line* have a maximum spindown age $t_{s,max}$ that is less than one spindown time τ_J

with a DM. By contrast, most B-type stars only show emission if they are well above the $R_A/R_K = 1$ horizontal line, implying a relatively fast rotation and strong confinement that leads to a CM.

The basic explanation for this dichotomy is straightforward. The stronger winds driven by the higher luminosity O-stars can accumulate even within a relatively short dynamical timescale to a sufficient density to give the strong emission in a DM, while the weaker winds of lower luminosity B-stars require the longer confinement and buildup of a CM to reach densities for such emission. This general picture is confirmed by the detailed dynamical models of DM and CM emission that motivated this empirical classification.

For the slowly rotating O-stars HD 191612 and θ^1 Ori C (here with respective ID numbers 4 and 3), both 2D and 3D MHD simulations (Sundqvist et al. 2012; ud-Doula et al. 2013) of the wind-fed DM reproduce quite well the rotational variation of Hα emission. For the 3D simulations of θ^1 Ori C, Fig. 40 shows how wind material trapped in closed loops over the magnetic equator (left panel) leads to circumstellar emission that is strongest during rotational phases corresponding to pole-on views (middle panel). For a pure dipole with the inferred magnetic tilt $\Theta = 45°$, an observer with the inferred inclination $i = 45°$ has perspectives that vary from magnetic pole to equator, leading in the 3D model to the rotational phase variations in Hα equivalent width shown in the right panel (shaded circles). This matches quite well both the modulation and random fluctuation of the observed equivalent width (black dots), though accounting for the asymmetry about minimum will require

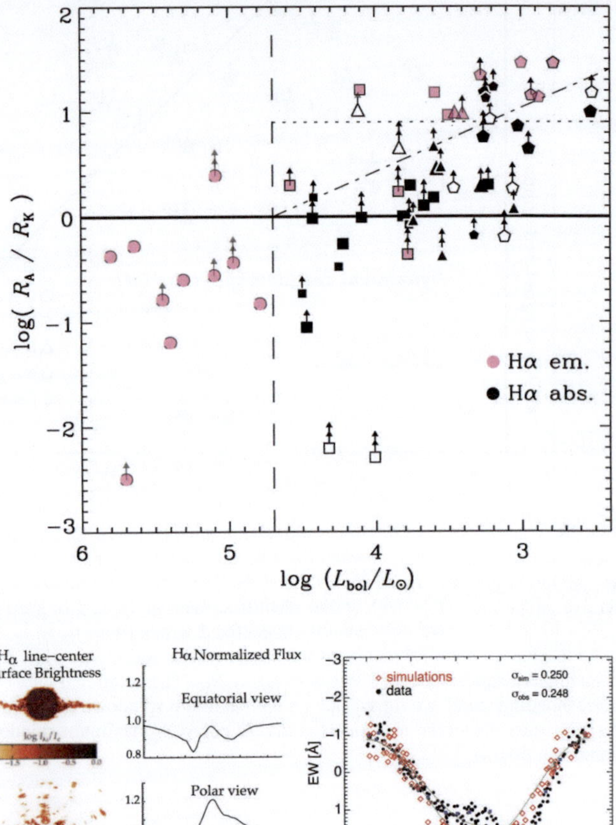

Fig. 39 Location of magnetic massive stars in a log–log plot of R_A/R_K vs. stellar luminosity. The symbol shadings mark the presence (*pink* or *shaded*) or absence (*black*) of magnetospheric Hα emission, with *empty symbols* when no Hα information is available. The *vertical dashed line* represents the luminosity transition between O-type and B-type main sequence stars. The *horizontal dotted line* and the *diagonal dot–dashed line* show division of the CM domain according to potential magnetospheric leakage mechanisms

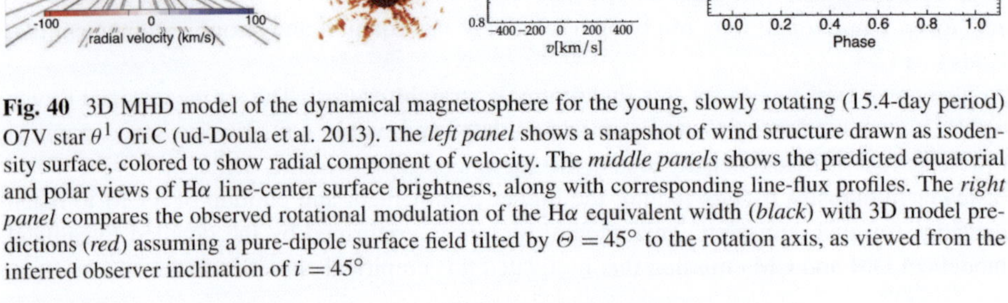

Fig. 40 3D MHD model of the dynamical magnetosphere for the young, slowly rotating (15.4-day period) O7V star θ^1 Ori C (ud-Doula et al. 2013). The *left panel* shows a snapshot of wind structure drawn as isodensity surface, colored to show radial component of velocity. The *middle panels* shows the predicted equatorial and polar views of Hα line-center surface brightness, along with corresponding line-flux profiles. The *right panel* compares the observed rotational modulation of the Hα equivalent width (*black*) with 3D model predictions (*red*) assuming a pure-dipole surface field tilted by $\Theta = 45°$ to the rotation axis, as viewed from the inferred observer inclination of $i = 45°$

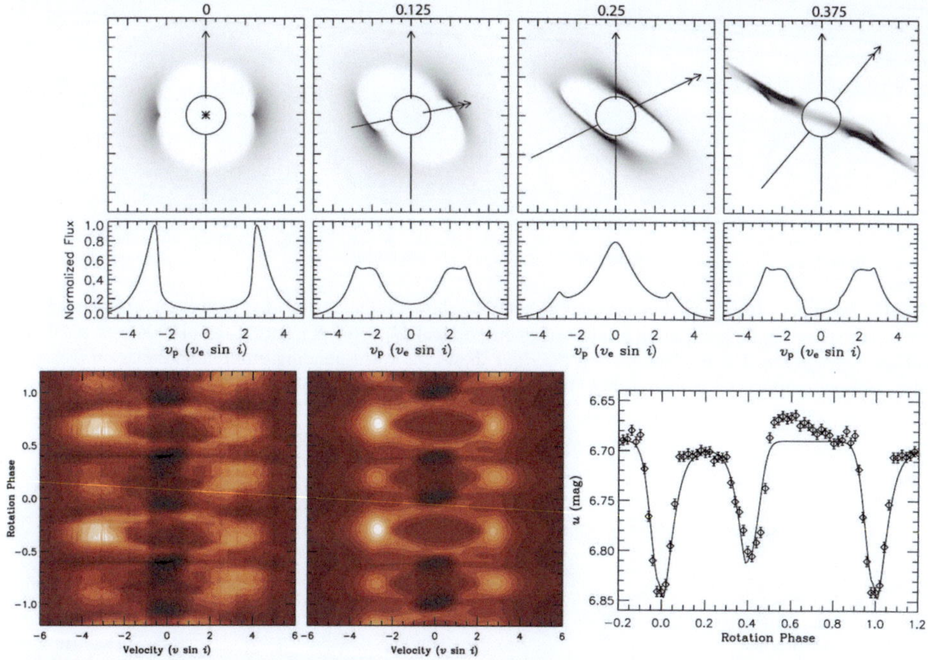

Fig. 41 Observational signatures of the CM in the moderately fast rotator (1.2-day period) B2V star σ Ori E, compared with results from the RRM model (Townsend and Owocki 2005). The *top row* shows surface maps of Hα emission and resulting emission line profiles at the marked rotational phases. The *lower-row* density plots are associated dynamic Hα spectra, showing the variations relative to the photospheric profile over two rotation periods of $\sim 1.2d$; *white* indicates emission, and *black* absorption. The *left panel* is based on echelle observations of the star, while the *central bottom panel* is the prediction from the RRM model. The *lower-right line plot* shows the Strömgren u-band light curve of σ Ori E, revealing the eclipse-like dimmings that occur when its two magnetospheric clouds transit in front of the star. The *solid line* indicates the predictions of an early RRM model

future, more detailed models that include a secondary, higher-order (non-dipole) component of the inferred surface field.

4.3.2 The Rigidly Rotating Magnetosphere (RRM) Model

In modeling the CM of more rapidly rotating, strongly magnetic B-stars like σ Ori E, a key challenge stems from the fact that their wind magnetic confinement parameters are generally of order $\eta_* \sim 10^6$ or more, far beyond the maximum $\eta_* \approx 10^3$ achieved with direct MHD simulations, which are limited by the Courant stability criterion. As an alternative for this *strong-field limit*, Townsend and Owocki (2005) developed a *Rigidly Rotating Magnetosphere* (RRM) model that uses a semi-analytical prescription for the 3D magnetospheric plasma distribution, based on the form and minima of the total gravitational l-plus-centrifugal potential along each separate field line. Townsend et al. (2005) applied this RRM model to synthesize the emission from material trapped in the associated CM of σ Ori E. Figure 41 compares the predicted variation of the dynamic emission spectrum over the 1.2 day rotational period with that obtained from echelle observations of the star. The agreement is again very good, providing strong general support for this RRM model for Hα emission from the CM of σ Ori E.

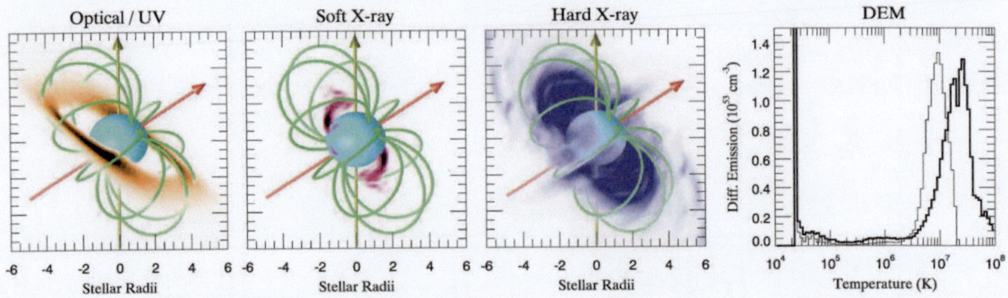

Fig. 42 Snapshots from an RFHD model of σ Ori E, showing the spatial distribution of magnetospheric emission measure in three different temperature bins: optical ($T < 10^6$ K), soft X-ray (10^6 K $< T < 10^7$ K) and hard X-ray ($T > 10^7$ K). The *plot on the right* shows the corresponding differential emission measure, for models with (*thin*) and without (*thick*) thermal conduction

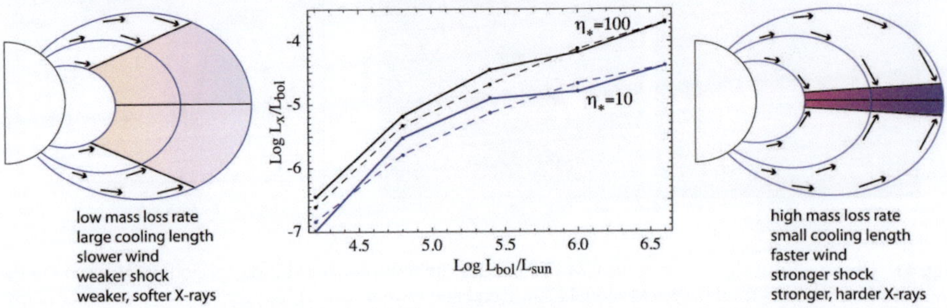

Fig. 43 Scaling of X-ray luminosities L_x (for energies $E_x > 0.3$ keV) with stellar bolometric luminosity L_{bol} (*center*), for X-ray simulations with (*solid*) and without (*dashed*) inverse Compton cooling (from ud–Doula et al. 2014). The *left* and *right schematics* illustrate the effect of "shock retreat" in reducing the strength and hardness of X-rays in lower luminosity stars with lower mass loss rate and thus less efficient radiative cooling

The basic RRM concept has been further developed in a successor *Rigid Field Hydrodynamics* (RFHD) model (Townsend et al. 2007), wherein the time-dependent flow along each individual field line is simulated using a 1D hydro code. By piecing together independent simulations of many different field lines (typically, several *thousand*!), a 3D picture of a star's magnetosphere can be constructed at modest computational cost, leading in turn to predictions for not only Hα but also for X-ray emission (and other wind-related observables) of magnetospheres in the strong-field limit, as shown Fig. 42. A powerful aspect of both the RRM and RFHD models is that, within the strong field limit, they are in principle applicable to arbitrary field topologies, not just the oblique dipole configurations considered so far. Thus, for example, they could be used to model the magnetosphere of HD 37776, which harbors high-order multipoles (Kochukhov et al. 2011).

4.3.3 MHD Scalings for X-Ray Luminosity from MCWS

For the DM cases without dynamically significant rotation, ud-Doula et al. (2014) carried out a MHD simulation parameter study with a full energy equation to compute the X-ray luminosity L_x that results from magnetically confined wind shocks (MCWS). The central panel of Fig. 43 plots the ratio L_x/L_{bol} vs. the bolometric luminosity L_{bol} for models with magnetic confinement $\eta_* = 10$ and 100. The dashed curves assume the post-shock cooling

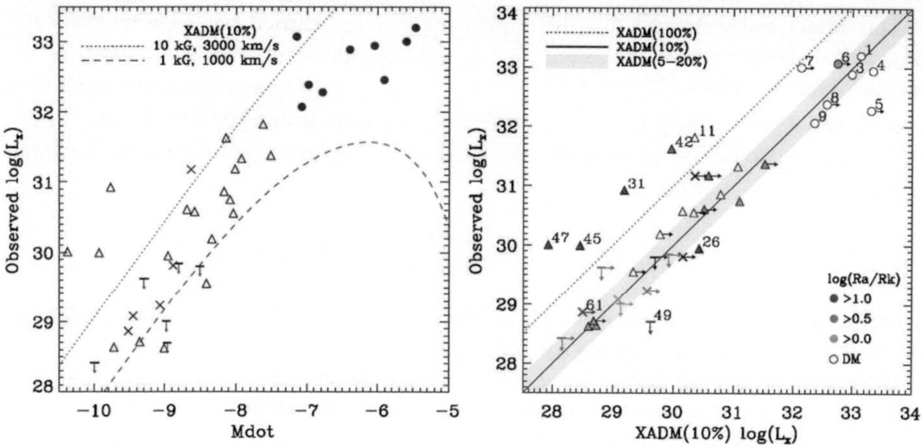

Fig. 44 *Left*: Compilations by Nazé et al. (2014) of X-ray luminosity in the 0.5–10.0 keV band, plotted vs. the log of mass-loss rate, as inferred from stellar parameters using the formula from Vink et al. (2000). *Filled blue dots* correspond to O stars, *black empty triangles* to B stars, and crosses and downward-pointing arrows to faint detections and upper limits on the X-ray luminosity, respectively. The *dotted* and *dashed lines* show associated X-ray luminosities predicted from the X-ray Analytic Dynamical Magnetosphere (XADM) model of ud-Doula et al. (2014), scaled by 10 % efficiency, for the two indicated sets of magnetic field and wind parameters bracketing the parameters of the sample. *Right*: Direct comparison between the observed X-ray luminosity of magnetic stars (as in *left panel*) and the predicted values using the XADM model, assuming the cited overall efficiency factors. The agreement is best for slowly rotating DM stars without centrifugal support (*open symbols*); rapid rotators, with color indicating level of centrifugally supported CMs, tend to have higher L_X. Stars of particular interest are labeled according to their identification number in Table 1 of Nazé et al. (2014)

is purely by radiative emission, while the solid curves account also for the effects of inverse Compton cooling. For the most luminous stars, L_x scales in proportion to the wind mass loss rate, which for line-driven winds follows $\dot{M} \sim L_{bol}^{1.6}$; but at lower L_{bol}, the lower $\dot{M}$ means the radiative cooling becomes inefficient. As illustrated in the left vs. right schematic panels of Fig. 43, the larger cooling layer forces a "shock retreat" back to lower, slower wind outflow, leading to weaker shocks, and so lower, softer X-ray emission.

These MHD simulation results of ud-Doula et al. (2014) were also used to calibrate an 'X-ray Analytic Dynamical Magnetosphere' (XADM) analysis for how the overall X-ray luminosity scales with stellar magnetic field strength and wind mass loss rate. For the subset of relatively slowly rotating magnetic OB stars modeled in these MHD simulations, Fig. 44 shows that, when modified by an overall duty cycle efficiency factor to account for extended intervals of infall of trapped material without much X-ray emission, this XADM scaling gives remarkably close agreement with empirical trends inferred by Nazé et al. (2014).

4.4 Magnetic Wind Braking, Spindown Time, and Spindown Age

Let us now turn to the issue of rotational spindown from magnetic wind braking. In this regard, the case of σ Ori E provides a key testbed, because extended photometric monitoring of the timing of magnetospheric clouds transiting in front of the star (Townsend 2008; see also lower right panel of Fig. 41) has allowed a *direct* measurement of the change in rotation period, yielding a spindown time of 1.34 Myr (Townsend et al. 2010). This is remarkably close to the spindown time *predicted* previously by ud-Doula et al. (2009), based on the same 2D MHD aligned-dipole parameter study used for Fig. 37.

This MHD study showed that the angular momentum carried out by a magnetically torqued stellar wind follows the same simple, split-monopole scaling law derived for the Sun by Weber and Davis (1967), $\dot{J} = \frac{2}{3}\dot{M}\Omega R_A^2$—with, however, the Alfvén radius R_A now given by the *dipole* scaling $R_A \sim \eta_*^{1/4}$, instead the oft-quoted, stronger scaling ($R_A \sim \eta_*^{1/2}$) for a split monopole. This leads to an associated general formula for the rotational braking timescale,

$$\tau_J \equiv \frac{I\Omega}{\dot{J}} = \frac{3}{2} f \tau_{\rm mass} \left(\frac{R_*}{R_A}\right)^2 \approx 0.15 \frac{\tau_{\rm mass}}{\sqrt{\eta_*}}. \qquad (11)$$

Here $\tau_{\rm mass} \equiv M/\dot{M}$ is the stellar mass loss timescale, and $f \approx 0.1$ is a dimensionless measure of the star's moment of inertia $I \equiv f M R_*^2$.

If we assume for simplicity a fixed radius R_* and moment of inertia factor $f \approx 0.1$, as well as a constant angular momentum loss rate $\dot{J}$, then the stellar rotation period P will simply increase exponentially with age t from its initial value, $P(t) = P_o e^{t/\tau_J}$. This can be used to define a star's *spindown age*, t_s, in terms of the spindown time τ_J, and its inferred present-day critical rotation fraction $W = P_{\rm orb}/P$ relative to its initial rotation fraction W_o, $t_s = \tau_J \ln W_o/W$. Taking the initial rotation to be critical, $W_o = 1$, yields a simple upper limit to the spindown age,

$$t_{\rm s,max} = \tau_J \ln(1/W). \qquad (12)$$

If the initial rotation is subcritical, $W_o < 1$, then the actual spindown age is shorter by a time $\Delta t_s = \tau_J \ln W_o$.

In Fig. 38 the upper axis gives the spindown timescale τ_J (normalized by the value in a non-magnetized wind), while the right axis gives the maximum spindown age $t_{\rm s,max}$ (normalized by the spindown time). Stars above the horizontal dotted line have a maximum spindown age that is *less* than a single spindown time. Together with the R_A/R_K vs. luminosity plot in Fig. 39, we can identify some important features and trends:

- All the most rapidly rotating stars are cooler B-type with weak winds, and thus weak braking, despite their strong field. The two most extreme examples (ID 45 and 47) may be very close to critical rotation, and so provide a potential link to Be stars, which have *not* been found to have strong ordered fields, but for which rapid rotation is linked to decretion into an orbiting Keplerian disk.
- The only rapidly rotating O-star is Plaskett's star (ID 6), which has likely been spun up by mass exchange with its close binary companion (Grunhut et al. 2013). Many O-stars have very long rotation period, e.g. 538 days for the field star HD 191612 (ID 4), suggesting substantial main-sequence spindown by wind magnetic braking, with a spindown age comparable to its estimated main-sequence age.
- In contrast, the young Orion cluster star θ^1 Ori C (ID 3) has a moderately slow (15.4-day period) rotation, but is generally thought to be about 1 Myr old (Hillenbrand 1997; Scandariato et al. 2012), much less than its maximum spindown age $t_{\rm s,max} \approx 3\tau_J \approx$ 10 Myr. Thus its zero-age main-sequence (ZAMS) rotation was likely already quite slow, suggesting significant *pre*-main-sequence braking, e.g. by pre-main-sequence (PMS) disk-locking, or through a PMS jet and/or wind, as discussed in Sect. 3.1.2 above.

To reinforce the last point, the recent survey of Herbig Ae/Be stars by Alecian et al. (2013a, 2013b) concludes that magnetic HeAeBe stars have a slower rotation than those without a detected field. Among their sample of non-magnetic stars they further find that those with lower mass evolve toward the ZAMS with a constant angular momentum, whereas higher mass ($> 5M_\odot$) stars show evidence of angular momentum loss during their PMS evolution, most likely as a result of their stronger, radiatively driven mass loss.

4.5 Future Outlook

The above shows there has been substantial progress in our efforts to understand the physical and observational properties of massive-star magnetospheres. But there are still important gaps in this understanding and key limitations to the physical realism of the models developed. The following lists some specific areas for future work:

- *3D MHD of Non-Axisymmetric Cases:* Thus far all MHD simulations, whether run in 2D or 3D, have been restricted to cases with an underlying axial symmetry, assuming a purely dipole field either without dynamically significant rotation, or with rotation that is taken to be aligned with the magnetic dipole axis. Fully 3D simulations are needed for both the many stars with an oblique dipole, as well as cases with more complex, higher-order multi-pole fields.
- *Spindown from oblique dipoles or higher-order multipoles:* An important application of these 3D MHD models will be to analyze the angular momentum loss from oblique dipole fields, as well as from higher-order fields. This will allow determination of generalized spindown scalings for complex fields, and provide the basis for interpreting anticipated future direct measurements of magnetic braking in stars with tilted-dipole or higher multi-pole fields.
- *Non-Ideal MHD and magnetospheric leakage:* In MHD simulations of slowly rotating magnetic stars with a DM, the dynamical infall of material back to the star balances the mass feeding from the stellar wind, yielding an overall mass and density that is in quite good agreement with absorption and emission diagnostics. By contrast, in CM simulations the much longer confinement and mass buildup is limited only by eventual centrifugal breakout of regions beyond the Kepler radius (Townsend and Owocki 2005), and this now leads to an overall predicted CM mass and density that significantly exceeds values inferred by observational diagnostics. To understand better the mass budget of CM's, it will be necessary to investigate additional plasma leakage mechanisms, such as the field line interchange transport that is thought to be key to mass balance of planetary magnetospheres (Kivelson and Southwood 2005). In addition to comparison with emission diagnostics of individual stars, this should aim to derive general scaling laws that can explain the trends for Balmer emission seen in Fig. 39, particularly the boundary between Hα emission and absorption in B-type stars.
- *Rapid rotation and gravity darkening:* To model the rapidly rotating magnetic B-stars with $W = V_{\rm rot}/V_{\rm orb} > 1/2$, there is a need to generalize the lower boundary condition for both MHD and RFHD models to account for stellar oblateness, while also including the effect of gravity darkening for the wind radiative driving. This will also allow a link to Be stars, to constrain upper limits on the dynamical role of (undetected) magnetic fields in their quite distinctively Keplerian (vs. rigid-body) decretion disks. This will also provide a basis for applying such MHD models to PMS disks of HeAeBe stars.

Acknowledgement We thank the organizers of ISSI Workshop "The Strongest Magnetic Fields in the Universe" for excellent meeting and hospitality. Resources supporting this work were provided by the NASA High-End Computing (HEC). MMR acknowledges support by NASA grant NNX14AP30G and NSF grant AST-1211318, and contributions of different collaborators, particularly A.V. Koldoba, G.V. Ustyugova, R. Kurosawa, A.A. Blinova and R.V.E. Lovelace. MMR thanks M. Comins for editing the manuscript. SPO acknowledges support by NASA ATP Grants NNX11AC40G and NNX12AC72G, respectively to University of Delaware and University of Wisconsin, and extensive contributions of collaborators in the MiMeS project, particularly D. Cohen, V. Petit, J. Sundqvist, R. Townsend, A. ud-Doula, and G. Wade.

References

E. Alecian, G.A. Wade, C. Catala et al., Mon. Not. R. Astron. Soc. **429**, 1001 (2013a)
E. Alecian, G.A. Wade, C. Catala et al., Mon. Not. R. Astron. Soc. **429**, 1027 (2013b)
S.H.P. Alencar, P.S. Teixeira, M.M. Guimarães, P.T. McGinnis, J.F. Gameiro et al., Astron. Astrophys. **519**, A88 (2010)
S.H.P. Alencar, J. Bouvier, F.M. Walter, C. Dougados, J.F. Donati, R. Kurosawa, M.M. Romanova et al., Astron. Astrophys. **541**, A116 (2012)
M.A. Alpar, A.F. Cheng, M.A. Ruderman, J. Shaham, Nature **300**, 728 (1982)
M.A. Alpar, J. Shaham, Nature **316**, 239 (1985)
J.J. Aly, J. Kuijpers, Astron. Astrophys. **227**, 473 (1990)
A.M. Archibald, S. Bogdanov, A. Patruno, J.W.T. Hessels, A.T. Deller, C. Bassa et al., Astrophys. J. **807**, 62 (2015)
P. Armitage, Astrophys. J. **501**, L189 (1998)
J. Arons, S.M. Lea, Astrophys. J. **207**, 914 (1976)
M. Audard, P. Ábrahám, M.M. Dunham, J.D. Green, N. Grosso et al., in *Protostars and Planets VI*, ed. by H. Beuther, R. Klessen, C. Dullemond, Th. Henning (University of Arizona Press, Tucson, 2014)
J. Babel, T. Montmerle, Astrophys. J. **485**, L29 (1997a)
J. Babel, T. Montmerle, Astron. Astrophys. **323**, 121 (1997b)
F. Bacciotti, J. Eisloffel, T.P. Ray, Astron. Astrophys. **350**, 917 (1999)
M. Bachetti, M.M. Romanova, A. Kulkarni, L. Burderi, T. di Salvo, Mon. Not. R. Astron. Soc. **403**, 1193 (2010)
S.A. Balbus, J.F. Hawley, Astrophys. J. **376**, 214 (1991)
S.A. Balbus, J.F. Hawley, Rev. Mod. Phys. **70**, 1 (1998)
D. Barret, J.-F. Olive, M.C. Miller, Mon. Not. R. Astron. Soc. **376**, 1139 (2007)
N. Bessolaz, C. Zanni, J. Ferreira, R. Keppens, J. Bouvier, Astron. Astrophys. **478**, 155 (2008)
K. Beckwith, J.F. Hawley, J.H. Krolik, Astrophys. J. **707**, 428 (2009)
G.S. Bisnovatyi-Kogan, B.V. Komberg, Sov. Astron. **18**, 217 (1974)
G.S. Bisnovatyi-Kogan, A.S. Klepnef, F. Giovannelli, *Proc. High Energy Phenomena in Relativistic Outflows*. (HEPRO IV) International Journal of Modern Physics: Conference Series, vol. 28 (2014). arXiv:1408.1924
G.S. Bisnovatyi-Kogan, A.A. Ruzmaikin, Astrophys. Space Sci. **42**, 401 (1976)
R.D. Blandford, D.G. Payne, Mon. Not. R. Astron. Soc. **199**, 883 (1982)
A.A. Blinova, M.M. Romanova, R.V.E. Lovelace, Mon. Not. R. Astron. Soc. (2015, submitted). (arXiv:1501.01948)
S. Boutloukos, M. van der Klis, D. Altamirano et al., Astrophys. J. **653**, 1435 (2006)
J. Bouvier, A. Chelli, S. Allain, L. Carrasco, R. Costero, I. Cruz-Gonzalez, C. Dougados, M. Fernández et al., Astron. Astrophys. **349**, 619 (1999)
J. Bouvier, K.N. Grankin, S.H.P. Alencar, C. Dougados, M. Fernández, G. Basri, C. Batalha, E. Guenther et al., Astron. Astrophys. **409**, 169 (2003)
J. Bouvier, S.H.P. Alencar, T.J. Harries, C.M. Johns-Krull, M.M. Romanova, in *Protostars and Planets V*, ed. by B. Reipurth, D. Jewitt, K. Keil (University of Arizona Press, Tucson, 2007a), p. 479
J. Bouvier, S.H.P. Alencar, T. Boutelier, C. Dougados, Z. Balog, K. Grankin, S.T. Hodgkin, M.A. Ibrahimov, Astron. Astrophys. **463**, 1017 (2007b)
A.I. Boiko, V.N. Melnik, A.A. Konovalenko et al., Adv. Astron. Space Phys. **2**, 76 (2012)
A. Brandenburg, A. Nordlund, R.F. Stein, U. Torkelsson, Astrophys. J. **446**, 741 (1995)
P. Bult, M. van der Klis, Astrophys. J. **798**, L29 (2015)
N. Calvet, L. Hartmann, S.J. Kenyon, Astrophys. J. **402**, 623 (1993)
N. Calvet, E. Gullbring, Astrophys. J. **509**, 802 (1998)
M. Camenzind, Rev. Mod. Astron. **3**, 234 (1990)
C.G. Campbell, Geophys. Astrophys. Fluid Dyn. **63**, 179 (1992)
J.M. Carpenter, L.A. Hillenbrand, M.F. Skrutskie, Astron. J. **121**, 3160 (2001)
S. Chandrasekhar, *Hydrodynamic and Hydromagnetic Stability* (Clarendon, Oxford, 1961), p. 466
T. Chlebowski, F.R. Harnden Jr., S. Sciortino, Astrophys. J. **341**, 427 (1989)
A.M. Cody, J. Stauffer, A. Baglin et al., Astron. J. **147**, 82 (2014). 47 pp.
C.R. D'Angelo, H.C. Spruit, Mon. Not. R. Astron. Soc. **406**, 1208 (2010)
J.-F. Donati, A. Collier Cameron, Mon. Not. R. Astron. Soc. **291**, 1 (1997)
J.-F. Donati, A. Collier Cameron, G.A.J. Hussain, M. Semel, Mon. Not. R. Astron. Soc. **302**, 437 (1999)
J.-F. Donati, M.M. Jardine, S.G. Gregory, P. Petit, J. Bouvier et al., Mon. Not. R. Astron. Soc. **380**, 1297 (2007)
J.-F. Donati, M.M. Jardine, S.G. Gregory, P. Petit et al., Mon. Not. R. Astron. Soc. **386**, 1234 (2008)

J.-F. Donati, M.B. Skelly, J. Bouvier, K.N. Grankin, M.M. Jardine et al., Mon. Not. R. Astron. Soc. **409**, 1347 (2010)
J.-F. Donati, J. Bouvier, F.M. Walter, S.G. Gregory, M.B. Skelly et al., Mon. Not. R. Astron. Soc. **412**, 2454 (2011)
C. Dougados, S. Cabrit, C. Lavalley, F. Ménard, Astron. Astrophys. **357**, L61 (2000)
S. Edwards, W. Fischer, L. Hillenbrand, J. Kwan, Astrophys. J. **646**, 319 (2006)
R.P. Fender, Nature **427**, 222 (2004)
J. Ferreira, C. Dougados, S. Cabrit, Astron. Astrophys. **453**, 785 (2006)
C. Ferrigno, E. Bozzo, A. Papitto, N. Rea, L. Pavan et al., Astron. Astrophys. **567**, A77 (2014). 14 pp.
M. Flock, N. Dzyurkevich, H. Klahr, N.J. Turner, Th. Henning, Astrophys. J. **735**, 122 (2011). 16 pp.
M. Gagné, M.E. Oksala, D.H. Cohen, S.K. Tonnesen, A. ud-Doula, S.P. Owocki, R.H.D. Townsend, J.J. MacFarlane, Astrophys. J. **628**, 986 (2005)
P. Ghosh, in *Rotation and Accretion Powered Pulsars*, ed. by P. Ghosh. World Scientific Series in Astronomy and Astrophysics, vol. 10 (Published by World Scientific Publishing Co., Singapore, 2007)
P. Ghosh, F.K. Lamb, Astrophys. J. **223**, L83 (1978)
A.P. Goodson, R.M. Winglee, K.-H. Böhm, Astrophys. J. **489**, 199 (1997)
A.P. Goodson, R.M. Winglee, Astrophys. J. **524**, 159 (1999)
A.P. Goodson, K.-H. Böhm, R.M. Winglee, Astrophys. J. **524**, 142 (1999)
S.G. Gregory, Am. J. Phys. **79**, 461 (2011)
E. Gullbring, N. Calvet, J. Muzerolle, L. Hartmann, Astrophys. J. **544**, 927 (2000)
J.H. Grunhut, G.A. Wade, M. Leutenegger et al., Mon. Not. R. Astron. Soc. **428**, 1686 (2013)
K. Hamaguchi, N. Grosso, J.H. Kastner et al., Astrophys. J. **754**, 32 (2012)
T.J. Harries, Mon. Not. R. Astron. Soc. **315**, 722 (2000)
L. Hartmann, R. Hewett, N. Calvet, Astrophys. J. **426**, 669 (1994)
M.R. Hayashi, K. Shibata, R. Matsumoto, Astrophys. J. **468**, L37 (1996)
J.F. Hawley, C.F. Gammie, S.A. Balbus, Astrophys. J. **440**, 742 (1995)
J.F. Hawley, Astrophys. J. **528**, 462 (2000)
S. Heinz, N.S. Schulz, W.N. Brandt, D.K. Galloway, Astrophys. J. **663**, L93 (2007)
C. Hellier, *Cataclysmic Variable Stars* (Springer, Berlin, 2001)
W. Herbst, D.K. Herbst, E.J. Grossman, D. Weinstein, Astron. J. **108**, 1906 (1994)
L.A. Hillenbrand, Astron. J. **113**, 1733 (1997)
S. Hirose, Yu. Uchida, K. Shibata, R. Matsumoto, Publ. Astron. Soc. Jpn. **49**, 193 (1997)
A. Ibragimov, J. Poutanen, Mon. Not. R. Astron. Soc. **400**, 429 (2009)
A.F. Illarionov, R.A. Sunyaev, Astron. Astrophys. **39**, 185 (1975)
M. Jardine, K. Wood, A. Collier Cameron, J.-F. Donati, D.H. Mackay, Mon. Not. R. Astron. Soc. **336**, 1364 (2002)
C.M. Johns-Krull, Astrophys. J. **664**, 975 (2007)
C. Johns-Krull, J.A. Valenti, C. Koresko, Astrophys. J. **516**, 900 (1999)
M. Kaisig, T. Tajima, R.V.E. Lovelace, Astrophys. J. **386**, 83 (1992)
M.G. Kivelson, D.J. Southwood, J. Geophys. Res. Space Phys. **110**, 12209 (2005)
S. Kato, J. Fukue, S. Mineshige, *Black-Hole Accretion Disks* (Kyoto University Press, Kyoto, 1998)
S. Kato, Publ. Astron. Soc. Jpn. **56**, 905 (2004)
S. Kato, Publ. Astron. Soc. Jpn. **59**, 451 (2007)
O. Kochukhov, A. Lundin, I. Romanyuk, D. Kudryavtsev, Astrophys. J. **726**, 24 (2011)
A.V. Koldoba, R.V.E. Lovelace, G.V. Ustyugova, M.M. Romanova, Astron. J. **123**, 2019 (2002a)
A.V. Koldoba, M.M. Romanova, G.V. Ustyugova, R.V.E. Lovelace, Astrophys. J. **576**, L53 (2002b)
A. Königl, Astrophys. J. **370**, L39 (1991)
A. Königl, M.M. Romanova, R.V.E. Lovelace, Mon. Not. R. Astron. Soc. **416**, 757 (2011)
A.K. Kulkarni, M.M. Romanova, Astrophys. J. **633**, 349 (2005)
A. Kulkarni, M.M. Romanova, Astrophys. J. **386**, 673 (2008)
A. Kulkarni, M.M. Romanova, Astrophys. J. **398**, 1105 (2009)
A. Kulkarni, M.M. Romanova, Mon. Not. R. Astron. Soc. **433**, 3048 (2013)
R. Kurosawa, T.J. Harries, M.R. Bate, N.H. Symington, Mon. Not. R. Astron. Soc. **351**, 1134 (2004)
R. Kurosawa, M.M. Romanova, T.J. Harries, Mon. Not. R. Astron. Soc. **385**, 1931 (2008)
R. Kurosawa, M.M. Romanova, T.J. Harries, Mon. Not. R. Astron. Soc. **416**, 2623 (2011)
R. Kurosawa, M.M. Romanova, Mon. Not. R. Astron. Soc. **426**, 2901 (2012)
R. Kurosawa, M.M. Romanova, Mon. Not. R. Astron. Soc. **431**, 2673 (2013)
D. Lai, Astrophys. J. **524**, 1030 (1999)
F.K. Lamb, C.J. Pethick, D. Pines, Astrophys. J. **184**, 271 (1973)
F.K. Lamb, S. Boutloukos, S. Van Wassenhove et al., Astrophys. J. **705**, L36 (2009)

P.S. Lii, M.M. Romanova, G.V. Ustyugova, A.V. Koldoba, R.V.E. Lovelace, Mon. Not. R. Astron. Soc. **420**, 2020 (2012)
P.S. Lii, M.M. Romanova, G.V. Ustyugova, A.V. Koldoba, R.V.E. Lovelace, Mon. Not. R. Astron. Soc. **441**, 86 (2014)
M. Linares, Astrophys. J. **795**, 72 (2014)
V.M. Lipunov, N.I. Shakura, Sov. Astron. Lett. **6**, 14 (1980)
M. Long, M.M. Romanova, R.V.E. Lovelace, Astrophys. J. **634**, 1214 (2005)
M. Long, M.M. Romanova, R.V.E. Lovelace, Mon. Not. R. Astron. Soc. **374**, 436 (2007)
M. Long, M.M. Romanova, R.V.E. Lovelace, Mon. Not. R. Astron. Soc. **386**, 1274 (2008)
M. Long, M.M. Romanova, A.K. Kulkarni, J.-F. Donati, Mon. Not. R. Astron. Soc. **413**, 1061 (2011)
M. Long, M.M. Romanova, F.K. Lamb, New Astron. **17**, 232 (2012)
R.V.E. Lovelace, H.L. Berk, J. Contopoulos, Astrophys. J. **379**, 696 (1991)
R.V.E. Lovelace, M.M. Romanova, G.S. Bisnovatyi-Kogan, Mon. Not. R. Astron. Soc. **275**, 244 (1995)
R.V.E. Lovelace, M.M. Romanova, G.S. Bisnovatyi-Kogan, Astrophys. J. **514**, 368 (1999)
R.V.E. Lovelace, H. Li, A.V. Koldoba, G.V. Ustyugova, M.M. Romanova, Astrophys. J. **572**, 445 (2002)
R.V.E. Lovelace, M.M. Romanova, A.W. Barnard, Mon. Not. R. Astron. Soc. **389**, 1233 (2008)
R.V.E. Lovelace, M.M. Romanova, G.V. Ustyugova, A.V. Koldoba, Mon. Not. R. Astron. Soc. **408**, 2083 (2010)
S. Matt, A.P. Goodson, R.M. Winglee, K.-H. Böhm, Astrophys. J. **574**, 232 (2002)
S. Matt, R.E. Pudritz, Astrophys. J. **632**, 135 (2005)
S. Matt, R.E. Pudritz, Astrophys. J. **681**, 391 (2008)
C.W. Mauche, Mon. Not. R. Astron. Soc. **369**, 1983 (2006)
J.C. McKinney, A. Tchekhovskoy, R.D. Blandford, Mon. Not. R. Astron. Soc. **423**, 3083 (2014)
M. Méndez, T. Belloni, Mon. Not. R. Astron. Soc. **381**, 790 (2007)
M.C. Miller, F.K. Lamb, D. Psaltis, Astrophys. J. **508**, 791 (1998)
K.A. Miller, J.M. Stone, Astrophys. J. **489**, 890 (1997)
Y. Nazé, V. Petit, M. Rinbrand et al., Astrophys. J. Suppl. Ser. **215**, 10 (2014)
S.P. Owocki, A. ud-Doula, R.H.D. Townsend et al., Symp. - Int. Astron. Union **302**, 320 (2014)
A. Papitto, T. di Salvo, L. Burderi, M.T. Menna, G. Lavagetto, A. Riggio, Mon. Not. R. Astron. Soc. **375**, 971 (2007)
A. Papitto, C. Ferrigno, E. Bozzo, N. Rea, L. Pavan et al., Nature **501**(7468), 517 (2013)
A. Papitto, D.E. Torres, Astrophys. J. **807**, 33 (2015). 10 pp.
A. Patruno, A.L. Watts, M. Klein-Walt, R. Wijnands, M. van der Klis, Astrophys. J. **707**, 1296 (2009)
A. Patruno, C. D'Angelo, Astrophys. J. **771**, 94 (2013)
T. Belloni, M. Mendez, C.M. Zhang (eds.), *Timing Neutron Stars: Pulsations, Oscillations and Explosions, ASSL* (Springer, Berlin, 2015). Review to appear arXiv:1206.2727
A. Patruno, A.M. Archibald, J.W.T. Hessels, S. Bogdanov, B.W. Stappers, C.G. Bassa et al., Astrophys. J. **781**, L3 (2014). 5 pp.
V. Petit, S.P. Owocki, G.A. Wade, D.H. Cohen, J.O. Sundqvist, M. Gagné, J. Maíz Apellániz, M.E. Oksala, D.A. Bohlender, T. Rivinius, H.F. Henrichs, E. Alecian, R.H.D. Townsend, A. ud-Doula (MiMeS Collaboration), Mon. Not. R. Astron. Soc. **429**, 398 (2013)
J.E. Pringle, M.J. Rees, Astron. Astrophys. **21**, 1 (1972)
J. Poutanen, A. Ibragimov, M. Annala, Astrophys. J. **706**, L129 (2009)
L. Rastätter, K. Schindler, Astrophys. J. **524**, 361 (1999)
T. Ray, C. Dougados, F. Bacciotti, J. Eislöffel, A. Chrysostomou, in *Protostars and Planets V*, ed. by B. Reipurth, D. Jewitt, K. Keil (University of Arizona Press, Tucson, 2007), p. 231
M.M. Romanova, A.K. Kulkarni, R.V.E. Lovelace, Astrophys. J. **673**, L171 (2008)
M.M. Romanova, G.V. Ustyugova, A.V. Koldoba, R.V.E. Lovelace, Astrophys. J. **578**, 420 (2002)
M.M. Romanova, G.V. Ustyugova, A.V. Koldoba, R.V.E. Lovelace, Astrophys. J. **610**, 920 (2004)
M.M. Romanova, R.V.E. Lovelace, G.V. Ustyugova, A.V. Koldoba, in *Protostars and Planets V*. LPI Contribution, vol. 1286 (2005a), p. 8533
M.M. Romanova, G.V. Ustyugova, A.V. Koldoba, R.V.E. Lovelace, Astrophys. J. **635**, 165L (2005b)
M.M. Romanova, G.V. Ustyugova, A.V. Koldoba, R.V.E. Lovelace, Mon. Not. R. Astron. Soc. **399**, 1802 (2009)
M.M. Romanova, G.V. Ustyugova, A.V. Koldoba, R.V.E. Lovelace, Mon. Not. R. Astron. Soc. **416**, 416 (2011a)
M.M. Romanova, G.V. Ustyugova, A.V. Koldoba, R.V.E. Lovelace, Mon. Not. R. Astron. Soc. **421**, 63 (2012)
M.M. Romanova, G.V. Ustyugova, A.V. Koldoba, R.V.E. Lovelace, Mon. Not. R. Astron. Soc. **430**, 699 (2013)
M.M. Romanova, G.V. Ustyugova, A.V. Koldoba, J.V. Wick, R.V.E. Lovelace, Astrophys. J. **595**, 1009 (2003)

M.M. Romanova, M. Long, F.K. Lamb, A.K. Kulkarni, J.-F. Donati, Mon. Not. R. Astron. Soc. **411**, 915 (2011b)
M.M. Romanova, R.V.E. Lovelave, M. Bachetti, A.A. Blinova, A.V. Koldoba et al., in *Physics at the Magnetospheric Boundary*, vol. 64, ed. by E. Bozzo, P. Kretschmar, M. Audard, M. Falanga, C. Ferrigno, EPJ Web of Conferences, Geneva, Switzerland (2014). Id. 05001
S.M. Rucinski et al., Mon. Not. R. Astron. Soc. **391**, 1913 (2008)
G. Scandariato, N. Da Rio, M. Robberto, I. Pagano, K. Stassun, Astron. Astrophys. **545**, A19 (2012)
N.I. Shakura, R.A. Sunyaev, Astron. Astrophys. **24**, 337 (1973)
R.A. Sunyaev, N.I. Shakura, Astron. Ž. **3**, 262 (1977)
F. Shu, J. Najita, E. Ostriker, F. Wilkin, S. Ruden, S. Lizano, Astrophys. J. **429**, 781 (1994)
J.B. Simon, J.F. Hawley, K. Beckwith, Astrophys. J. **730**, 94 (2011)
J.L. Sokoloski, M.P. Rupen, A.J. Mioduszewski, Astrophys. J. **685**, L137 (2008)
H.C. Spruit, R. Stehle, J.C.B. Papaloizou, Mon. Not. R. Astron. Soc. **275**, 1223 (1995)
H.C. Spruit, R.E. Taam, Astrophys. J. **229**, 475 (1990)
H.C. Spruit, R.E. Taam, Astrophys. J. **402**, 393 (1993)
J.M. Stone, J.F. Hawley, S.A. Balbus, C.F. Gammie, Astrophys. J. **463**, 656 (1996)
J. Stauffer, A.M. Cody, A. Baglin et al., Astron. J. **147**, 83 (2014). 34 pp.
J. Stauffer, A.M. Cody, P. McGinnis, L. Rebull, L.A. Lillenbrand et al., Astron. J. **149**, 130 (2015). 30 pp.
J.M. Stone, T.A. Gardiner, Phys. Fluids **19**, 4104 (2007a)
J.M. Stone, T.A. Gardiner, Astrophys. J. **671**, 1726 (2007b)
J.M. Stone, J.M. Norman, Astrophys. J. **433**, 746 (1994)
J.O. Sundqvist, A. ud-Doula, S.P. Owocki, R.H.D. Townsend, I.D. Howarth, G.A. Wade, Mon. Not. R. Astron. Soc. **423**, L21 (2012)
C. Terquem, J.C.B. Papaloizou, Astron. Astrophys. **360**, 1031 (2000)
R.H.D. Townsend, Mon. Not. R. Astron. Soc. **389**, 559 (2008)
R.H.D. Townsend, M.E. Oksala, D.H. Cohen, S.P. Owocki, A. ud-Doula, Astrophys. J. **714**, L318 (2010)
R.H.D. Townsend, S.P. Owocki, Mon. Not. R. Astron. Soc. **357**, 251 (2005)
R.H.D. Townsend, S.P. Owocki, D. Groote, Astrophys. J. **630**, L81 (2005)
R.H.D. Townsend, S.P. Owocki, A. ud-Doula, Mon. Not. R. Astron. Soc. **382**, 139 (2007)
A. ud-Doula, S.P. Owocki, Astrophys. J. **576**, 413 (2002)
A. ud-Doula, S.P. Owocki, R.H.D. Townsend, Mon. Not. R. Astron. Soc. **385**, 97 (2008)
A. ud-Doula, S.P. Owocki, R.H.D. Townsend, Mon. Not. R. Astron. Soc. **392**, 1022 (2009)
A. ud-Doula, J.O. Sundqvist, S.P. Owocki, V. Petit, R.H.D. Townsend, Mon. Not. R. Astron. Soc. **428**, 2723 (2013)
A. ud-Doula, S. Owocki, R. Townsend, V. Petit, D. Cohen, Mon. Not. R. Astron. Soc. **441**, 3600 (2014)
G.V. Ustyugova, A.V. Koldoba, M.M. Romanova, R.V.E. Lovelace, Astrophys. J. **646**, 304 (2006)
D.A. Uzdensky, A. Kñigl, C. Litwin, Astrophys. J. **565**, 1205 (2002)
M. Van der Klis, Annu. Rev. Astron. Astrophys. **38**, 717 (2000)
M. van der Klis, in *Compact Stellar X-Ray Sources*, ed. by W.H.G. Lewin, M. van der Klis (Cambridge University Press, Cambridge, 2006), p. 39
A.A. Vidotto, M. Opher, V. Jatenco-Pereira, T.I. Gombosi, Astrophys. J. **703**, 1734 (2009)
A.A. Vidotto, M. Jardine, M. Opher, J.F. Donati, T.I. Gombosi, Mon. Not. R. Astron. Soc. **412**, 351 (2011)
S. van Straaten, M. van der Klis, R. Wijnands, Astrophys. J. **619**, 455 (2005)
J.S. Vink, A. de Koter, H.J.G.L.M. Lamers, Astron. Astrophys. **362**, 295 (2000)
Y.-M. Wang, J.A. Robertson, Astron. Astrophys. **139**, 93 (1984)
Y.-M. Wang, J.A. Robertson, Astrophys. J. **299**, 85 (1985)
J.C.L. Wang, M.E. Sulkanen, R.V.E. Lovelace, Astrophys. J. **390**, 46 (1992)
B. Warner, *Cataclysmic Variable Stars* (Cambridge University Press, Cambridge, 1995)
E.J. Weber, L. Davis, Astrophys. J. **148**, 217 (1967)
R. Wijnands, M. van der Klis, Nature **394**, 344 (1998)
R. Wijnands, M. van der Klis, J. Homan et al., Nature **424**, 44 (2003)
G.A. Wynn, A.R. King, K. Horne, Mon. Not. R. Astron. Soc. **286**, 436 (1997)
C. Zanni, J. Ferreira, Astron. Astrophys. **550**, A99 (2013)

DOI 10.1007/978-1-4939-3550-5_12
Reprinted from *Space Science Reviews* Journal, DOI 10.1007/s11214-015-0171-x

Pulsar-Wind Nebulae

Recent Progress in Observations and Theory

Oleg Kargaltsev[1] · Benoît Cerutti[2] · Yuri Lyubarsky[3] · Edoardo Striani[4]

Received: 10 March 2015 / Accepted: 1 June 2015 / Published online: 17 June 2015

Abstract In this review we describe recent observational and theoretical developments in our understanding of pulsar winds and pulsar-wind nebulae (PWNe). We put special emphasis on the results from observations of well-characterized PWNe of various types (e.g., torus-jet and bowshock-tail), the most recent MHD modeling efforts, and the status of the flaring Crab PWN puzzle.

Keywords Pulsars: general · ISM: jets and outflows · MHD · Radiation mechanisms: non-thermal · Acceleration of particles

1 Observations of PWNe

1.1 Introduction

Only $\sim 1\ \%$ of the total pulsar spin-down luminosity is emitted as pulsed electromagnetic radiation, the majority of the spin-down luminosity of a pulsar being carried away by a relativistic and highly magnetized pair plasma. These particles are generally believed to be accelerated and randomized in their pitch angle either upstream or at the pulsar wind

✉ O. Kargaltsev
kargaltsev@gwu.edu

B. Cerutti
bcerutti@astro.princeton.edu

Y. Lyubarsky
lyub@bgu.ac.il

E. Striani
edoardo.striani@iaps.inaf.it

1 George Washington University, 105 Corcoran Hall, Washington DC, 20052, USA

2 Department of Astrophysical Sciences, Princeton University, Princeton, NJ 08544, USA

3 Ben-Gurion University, P.O. Box 653, Beer-Sheva 84105, Israel

4 Department of Physics, University of Torino, Via P. Giuria 1, 10125 Turin, Italy

termination shock. The radiation produced but these particles downstream of the termination shock is often seen as a pulsar-wind nebula (PWN).

Most of the recent progress in our understanding of PWNe has been spurred by X-ray and TeV γ-ray observations. The majority of PWNe has been discovered in one of these bands, and many are seen in both (see Kargaltsev et al. 2013b). To study PWN emission, it is important to disentangle the pulsar and pulsar wind contributions either by spatially resolving the nebula from the pulsar or by isolating the PWN component in the spectrum (e.g., the PWN contribution is expected to dominate in TeV). High-resolution images from *Chandra X-ray Observatory* (*Chandra*) revealed two dominant PWN morphologies: torus-jet and bowshock–tail.[1] In addition, a few objects with varying and puzzling morphologies can be seen in *Chandra* images (Kargaltsev and Pavlov 2008). The PWN properties (size, morphology, and spectrum) can be expected to depend on the pulsar parameters (spin-down properties, pulsar velocity, and the angles between the spin, magnetic dipole, and velocity vectors) and on the environment (e.g., ambient pressure, magnetic field, and radiation field). The limited angular resolution of the ground-based TeV arrays (such as H.E.S.S. and VERITAS) does not allow us to detect TeV emission from the same particles that produce bright and compact X-ray nebulae in the vicinity of the pulsar. Instead, TeV images reveal much larger structures filled with the aged particles that may have accumulated over substantial part of pulsar's lifetime (see de Jager and Djannati-Ataï 2009). The TeV emission is usually attributed to inverse Compton (IC) scattering of background optical/IR photons off aged electrons, although, in denser environments, the contribution of neutral pion decay to the γ-rays emission could play an important role (if the pulsar winds indeed contain the so far elusive relativistic protons).

A general overview of PWN physics and X-ray observations was presented by Gaensler and Slane (2006) and Kargaltsev and Pavlov (2008), while PWN theory was recently reviewed by Amato (2014) and Bucciantini (2014). Here we focus on some of the most recent observational results[2] and their implications (Sect. 1), discuss the latest theoretical advances in MHD modeling (Sect. 2), and review the non-MHD scenarios that can explain the puzzling Crab PWN flares (Sect. 3).

1.2 The Crab and Vela PWNe as Prototypes of Young PWNe in SNRs

The Crab and Vela PWNe are often considered to be archetypal representatives of torus-jet PWNe. Since the Vela PWN is a factor of 10–20 older than the Crab PWN, one can look for evolutionary changes by comparing the two.[3] PWNe of this type are usually found around young pulsars whose velocities are smaller than the speed of sound inside their host SNRs. These environments can be characterized by relatively high pressures and temperatures (Bamba et al. 2010). There is also evidence that pulsars powering torus-jet PWNe are likely to have substantial misalignment between the spin and magnetic dipole axes (e.g., Crab and Vela pulsars; Moffett and Hankins 1999 and Johnston et al. 2005, respectively) which may play a pivotal role in formation of this type of morphology.

[1] Such classification is possible only for sufficiently bright and relatively nearby PWNe.

[2] A more detailed overview of the observational properties of population of relic PWNe can be found in Kargaltsev et al. (2013b).

[3] Since the PWN properties and evolution depend on the environment, one should not forget that the progenitor and SNR properties may be quite different for the Crab and Vela pulsars.

1.2.1 Multiwavelength Properties of the Crab

The Crab nebula has been studied with nearly all major telescopes since its discovery (see Hester 2008 and references therein). However, it was not until the *Hubble Space Telescope* (*HST*; Hester et al. 1995) and *Chandra* (Weisskopf et al. 2000) era when the intricate and complex structure of the nebula was revealed (see Fig. 1 for the feature nomenclature introduced by Hester et al. 1995). These observations have also shown that the bright inner part of the nebula is very dynamic, with apparent velocities corresponding to up to $0.5c$ (in projection onto the sky) as measured, e.g., from the shifts in wisp positions (Hester et al. 2002). The changes in the nebula are more complex than simple translational motion (e.g. steady expansion). They include variations in brightness (e.g., Inner Knot; Melatos et al. 2005) and shape (e.g., Sprite; Bietenholz et al. 2004; Hester et al. 2002). The wisp shapes can also be very different and while most of the wisps can be described as a ripple pattern with ripples moving away from the pulsar some of the wisps appear at the same location (e.g., Thin Wisp in Fig. 1). The prominent south-eastern (SE) jet (see Fig. 1) also shows quite remarkable changes in its shape, based on X-ray images (taken over 14-year baseline), which could be explained by either precession of the curved jet or by the motion of kinks along the jet (Weisskopf 2012). Finally, the unexpected detection of γ-ray flares by *Fermi* LAT and *AGILE* (Buehler et al. 2012; Abdo et al. 2011b; Tavani et al. 2011) suggests a significant energy release rate (which can reach 4×10^{36} erg s^{-1}; Buehler et al. 2012) on timescales of hours (Mayer et al. 2013); however, it was not possible so far to pinpoint the location of the flaring region because of the lack of a "smoking gun" at lower frequencies. Consequently, the lack of information about the site of the flare in the PWN has led to a variety of models being suggested (see Sect. 3). Given the lack of contemporaneous variability at lower frequencies, it might be possible that some of the energy released during the process associated with the γ-ray flares will manifest itself as more gradual flux changes at lower frequencies[4] occurring on much longer timescales (e.g., hard X-ray variability reported by Kouzu et al. 2013; Wilson-Hodge et al. 2011).

As a baseline for further comparison a multiwavelength (MW), high-resolution snapshot of the Crab PWN was obtained within a single day (Krassilchtchikov et al. in prep.; K+15 hereafter). Figure 2 shows *Chandra*, *HST* (NIR, optical), and Karl G. Jansky Very Large Array (JVLA) images from this latest MW campaign. These data can also be used to measure the contemporaneous spectra of the prominent PWN features. Both the radio and optical (broad-V band) images reveal prominent filaments of which only some are coincident (which implies different MW spectra). Also, the SE jet ("1" in Fig. 2), which is prominent in the X-ray image, appears to have counterparts in the optical and NIR images but not in the radio image, confirming the earlier findings of Bietenholz et al. (2004). On the other hand, a jet-like structure ("2" in Fig. 2) located to the east of the SE X-ray jet in the JVLA image does not have an X-ray or NIR counterpart and coincides with one of the thermal optical filaments.[5] Finally, in the NIR image, to the west of the X-ray jet there is another bright linear feature ("3" in Fig. 2) extending southward from pulsar. The feature has a radio counterpart, a very faint X-ray counterpart, and does not coincide with any of the optical filaments. The lack of X-ray counterpart (label with numbers on Fig. 1) suggests that emission from this feature must be synchrotron continuum produced by a cooled population

[4]See animation at http://home.gwu.edu/~kargaltsev/Crab.html.

[5]Overall, the large degree of correlation between the radio structure and optical filaments suggests than most of the radio emission is related to the SNR filaments.

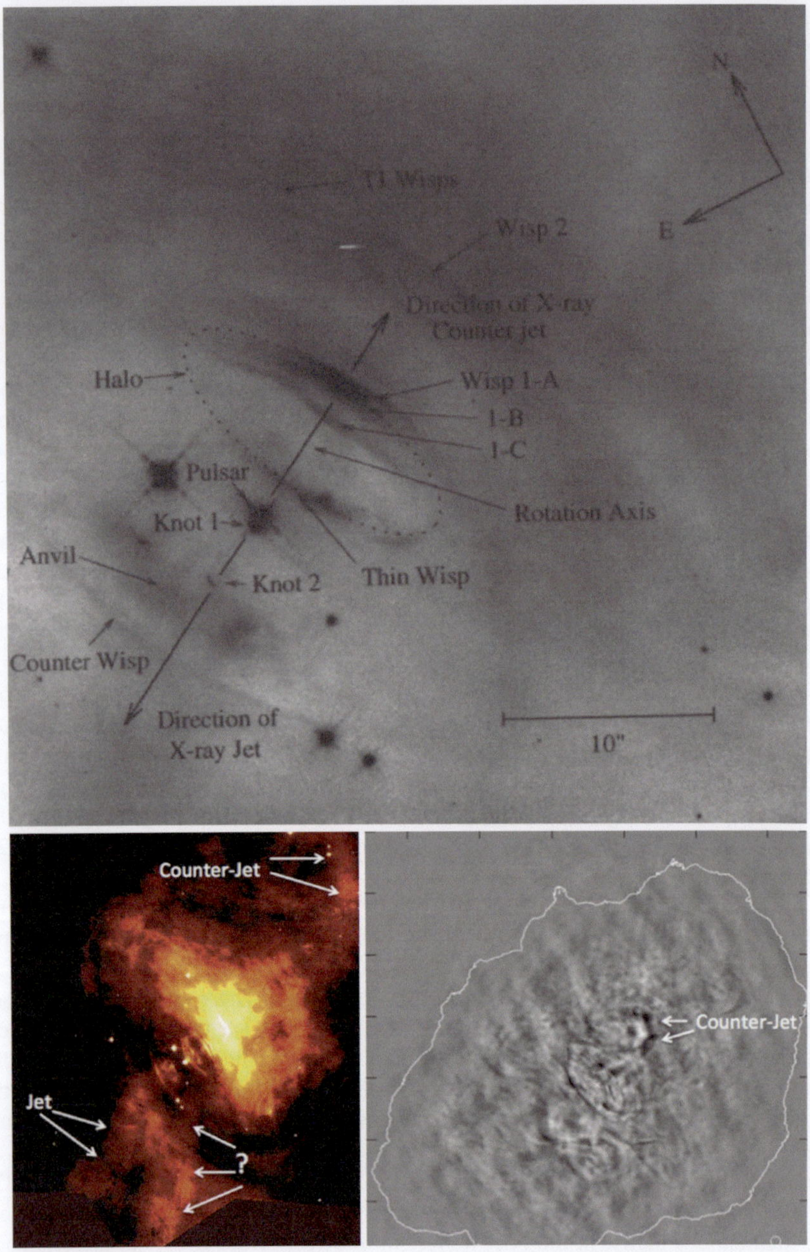

Fig. 1 The *top panel* introduces the most prominent features of the Crab PWN and their "conventional" names (from Hester et al. 1995). The *bottom left panel* shows the 12.5 ks exposure image obtained with *HST* ACS F550M. The image has been produced by a combining series of auxiliary images obtained during the 09/2005–12/2005 polarimetry campaign (Hester 2008). The F550M filter avoids any strong emission lines and provides a relatively unobstructed view of the synchrotron nebula. The "?" mark enigmatic feature discussed in Sect. 1.2.1 (also labeled as "3" in Fig. 2, *top right panel*). Notice that the feature labeled as the counter-jet in the F550M image also appears to show large variability in the difference image (shown in the *bottom right panel*) produced from 2 JVLA observations obtained on 2001 April 16 and 2012 August 26 (from Bietenholz et al. 2014)

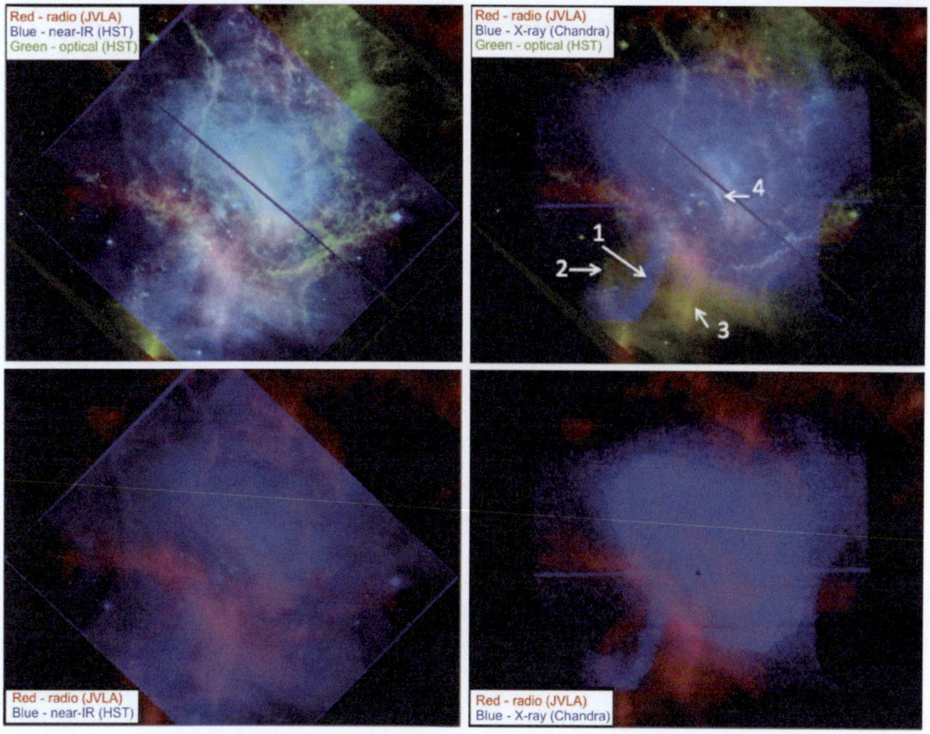

Fig. 2 False color MW images of the Crab PWN (see the *legends* in the panels) based on the observations described and analyzed in K+15. Numbers refer to the PWN features mention in the text

of X-ray particles. It is, however, difficult to explain the presence of this low-energy feature in the conventional axisymmetric paradigm with equatorial and polar outflows, where only one jet and one counter-jet are expected.[6] On the other hand, the "line-free" (F550M) optical image shown in Fig. 1 supports the axial outflow paradigm by revealing better than ever the other side of the axial "backbone" of the torus (dubbed as a counter jet, or NW jet). We note that at the brightness/contrast level chosen in the F550M image (Fig. 1, bottom left panel) the counterpart of the SE X-ray jet is barely seen, while the NW jet and the SE "jet-like" feature (see above) are clearly seen. Similar to the SE X-ray jet, the counter-jet appears to be hardly discernible in the radio images (see e.g., right panel in Fig. 3) but it stands out in the 2-epoch difference image produced by Bietenholz et al. (2014) (bottom right panel in Fig. 1). Finally, the counter-jet is so faint in X-rays (if preset at all) that it cannot be discerned from the torus emission. Therefore, the frequency-dependent differences in brightness between the jet and counter-jet appear to be more complex than those expected for a simple scenario with the frequency-invariant Doppler boost (see e.g., Schweizer et al. 2013).

K+15 found that the location of the bright optical/NIR wisp ("4" in Fig. 2) only approximately coincides with the X-ray ring,[7] and the wisp brightness in the NIR image drops much

[6]However, one can imagine that the jet stayed at one position ("3" in Fig. 2) for a long time and then relatively quickly moved to the other position ("1" in Fig. 2).

[7]The outer edge of the optical/NIR wisp is about $1''$ further away from the pulsar, with the X-ray ring emission trailing behind or possibly being sandwiched between the bright wisp and the fainter wisp.

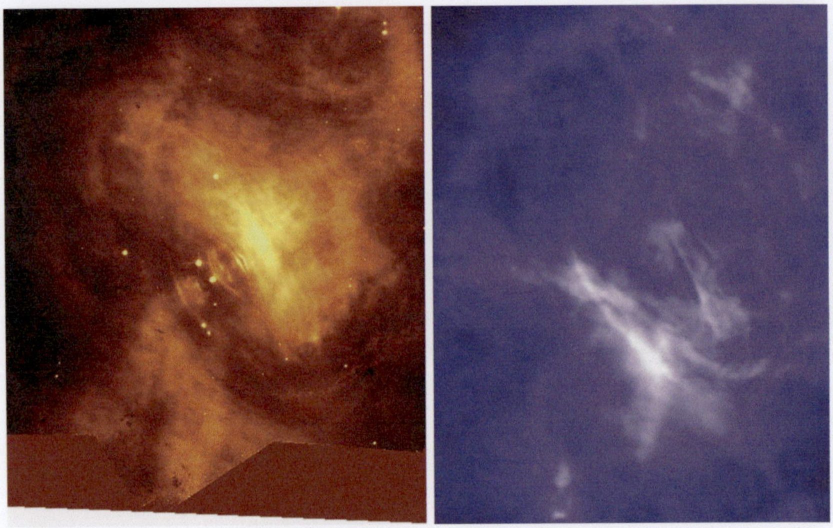

Fig. 3 *Left:* 12.5 ks exposure image of the Crab PWN obtained with *HST* ACS F550M. The image has been produced by combining a series of auxiliary images obtained during the 09/2005–12/2005 polarimetry campaign (Hester 2008). The F550M filter avoids any strong emission lines and provides a relatively unobstructed view of the synchrotron nebula. *Right:* JVLA image of the Crab (from Bietenholz et al. 2014)

faster with the distance from the symmetry axis of the nebula compared to the X-ray ring brightness. This supports the Schweizer et al. (2013) findings (based on earlier optical and X-ray monitoring) who concluded that the X-ray and optical emission must be produced by different populations of particles. Furthermore, according to Schweizer et al. (2013), the fits with the Doppler-boosted tilted ring model require noticeably different (higher) flow velocities for the optical wisps ($\approx 0.9c$) compared to the X-ray wisps, which made Schweizer et al. (2013) question the simple "boosted-ring" model (see, however, Olmi et al. 2015). K+15 also found that for most individual features of the PWN (e.g., wisps) the NIR-optical-FUV spectra are significantly harder than the contemporaneously measured X-ray spectra suggesting cooling breaks whose frequencies are feature-dependent.

The overall morphology of the Crab nebula has been reproduced in the relativistic magnetohydrodynamic simulations with anisotropic energy flux (most recently by Volpi et al. 2008; Camus et al. 2009; Porth et al. 2014). The simulated images display wisp-like features and variable structures along the symmetry axis due to the Doppler boosting of the emission from the oblique termination shock that has been associated with the Crab's Inner Knot (Komissarov and Lyubarsky 2003; see, however critique in Melatos et al. 2005). On the other hand, some challenges to the MHD models still remain (Bühler and Blandford 2014). The predicted bright arc due to the emission from the relativistic post-shock flow originating from the termination shock (Komissarov and Lyubarsky 2003; Del Zanna et al. 2006) is absent in the X-ray images where we instead see the patchy (likely consisting of multiple knots) inner ring which may or may not appear to be brighter on the northwestern side (depending on the observation epoch). In this sense the optical (or NIR) images featuring a bright wisp NW of the pulsar appear to be in a better correspondence with the predictions of the MHD models. The models also predict co-spatial small-scale structures in the optical and X-rays while this is generally not observed (see above). Only parts of the X-ray inner ring are seen in the optical and most (if not all) optical wisps also do not appear to have X-ray counterparts (e.g., the Thin Wisp[8] labeled in the top panel of Fig. 1 is lacking any nearby counterpart while other wisps are offset from the possibly associated X-ray bright features; see Schweizer et al. 2013 and K+15). Therefore, it yet remains to be shown whether more advanced models can fully capture the rich MW structure and variability of the Crab PWN. It seems that a complex injection spectrum is required to achieve this, hinting that there may be multiple acceleration sites throughout the PWN with possibly different acceleration mechanisms. The most recent, advanced 3D models predict somewhat disordered structure of the magnetic field and suggest the need for the in-situ particle acceleration outside the termination shock region (Porth et al. 2014). Diffusion transport may become more important in the case of disordered magnetic field.

1.2.2 Multiwavelength Properties of the Vela PWN

The Vela pulsar is a factor of 20 older than Crab, hence some evolutionary differences are expected. In addition, the differences can be attributed to a different progenitor type, different ISM, and different properties of the pulsar (e.g., magnetic field, angles between the spin and magnetic dipole axis, or the orientation and magnitude of the pulsar velocity). Figure 4 (bottom left) shows a deep *Chandra* ACIS image of the Vela PWN (see also zoomed-in view of the compact nebula in Fig. 5) produced by combining images from the latest observational campaign (comprised of eight 40 ks observations taken with one-week intervals; see Durant et al. 2013). The overall morphology of the bright compact PWN can be described as an axisymmetric double-arc structure with two axial jets having different brightnesses and widths. The bright, compact X-ray PWN is located inside the larger double-lobe radio PWN (Dodson et al. 2003). Interestingly, the radio lobes appear to be filled by fainter X-ray emission which is particularly clearly seen in the ACIS hard band (1–8 keV; see Fig. 4) thus suggesting that the radiating particles have not cooled too much. Therefore, the pure advection model proposed by Kennel and Coroniti (1984a) may need to be augmented with some other transport mechanism (e.g., diffusion) capable of moving energetic particles away from the pulsar more rapidly. We also note that the bright, double-lobed radio nebula is surrounded by a much larger ($\sim 2^\circ$ in diameter) radio-emitting structure called Vela X (see Sect. 1.5.4) which is filled with bright filaments and fainter diffuse continuum (Fig. 4, bottom right).

1.2.3 The Crab and Vela PWNe: Similarities and Differences

The double-arc X-ray morphology of the Vela PWN is quite different from the Crab PWN, with its single ring, and torus (likely comprised of multiple wisps) seen in the ACIS images (see Fig. 4). These differences can hardly be attributed to the larger age of the Vela PWN (because the X-ray emission from compact PWN mostly comes from freshly injected electrons) or different ambient (SNR) medium properties (these could become progressively more important further away from the termination shock). It is also unlikely to be due to the difference in the angles between the pulsar spin axis and the line of sight because these angles are believed to be similar (127° for Vela (Helfand et al. 2001) and $\approx 120^\circ$ for Crab (Weisskopf et al. 2012)). The only other important parameters could be the angle between the spin and magnetic dipole[9] axis (still rather poorly constrained, e.g., $\sim 43^\circ$ for Vela (Johnston et al.

[8] Here we are following nomenclature introduced by Hester et al. (1995), see Fig. 1 (top panel).

[9] In principle, the magnetic field may deviate from the dipolar configuration more for the Vela pulsar than for the Crab pulsar. Indeed, braking indices, n, of the Crab ($n = 2.5$) and Vela ($n = 1.4$) are very different, and the Vela pulsar is much more "glitchy" compared to Crab.

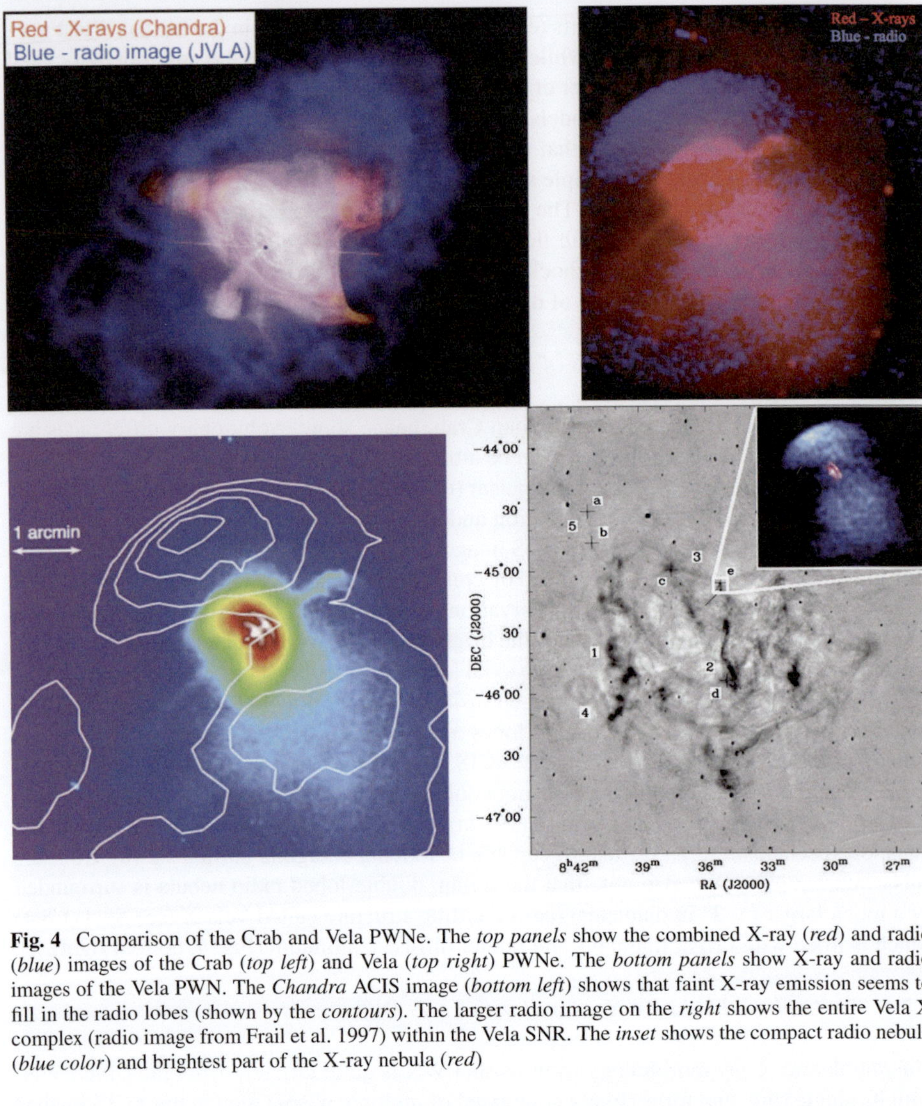

Fig. 4 Comparison of the Crab and Vela PWNe. The *top panels* show the combined X-ray (*red*) and radio (*blue*) images of the Crab (*top left*) and Vela (*top right*) PWNe. The *bottom panels* show X-ray and radio images of the Vela PWN. The *Chandra* ACIS image (*bottom left*) shows that faint X-ray emission seems to fill in the radio lobes (shown by the *contours*). The larger radio image on the *right* shows the entire Vela X complex (radio image from Frail et al. 1997) within the Vela SNR. The *inset* shows the compact radio nebula (*blue color*) and brightest part of the X-ray nebula (*red*)

2005) and $\approx 45°–70°$ for Crab [(Lyne et al. 2013)]) and different pulsar velocities[10] (the projected onto the sky velocities are $60d_{0.3}$ and $\sim 120d_2$ km s^{-1} for Vela and Crab, respectively[11]). These velocities are smaller than the typical sound speed inside the young SNR, and therefore the Mach number $\mathcal{M}$ should be $\lesssim 1$ for both pulsars.[12] Thus the differences in the compact PWN morphologies are more likely to be attributed to the different angles

[10]The parameter that determines to what degree a PWN is affected by the pulsar motion is the ratio of the pulsar velocity to the local ambient sound speed (Mach number $\mathcal{M}$). The medium within the younger Crab SNR is hotter than in the Vela SNR.

[11]Here the distances are scaled as $d_{0.3} = d/(300 \text{ pc})$ and $d_2 = d/2{,}000$ pc.

[12]It is likely that the Mach number is somewhat larger for the Vela PWN where we see some effect of the motion (Pavlov et al. 2003).

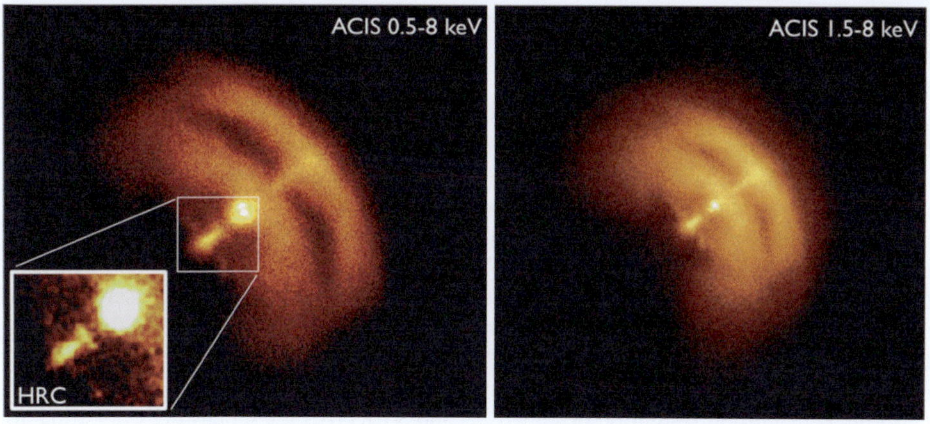

Fig. 5 Deep *Chandra* ACIS and HRC (*inset*) images of the compact Vela PWN (Levenfish et al. 2013)

between the spin and magnetic dipole axis. It is also possible that the different degrees of deviation of the NS magnetic field from that of an ideal, centered dipole have some impact on the PWN. Even for the compact parts of the PWNs some of the differences in X-ray morphologies could still be attributed to the longer synchrotron cooling time for the Vela PWN which is expected to have weaker magnetic field (Pavlov et al. 2003).

We also note that the physical connection between the bright inner jets in the Vela PWN and the fainter large-scale outer jets (see e.g., Fig. 4 bottom left panel) has not been established yet. The bright SE inner jet of the Vela pulsar suddenly appears out of the orthogonal bar-like feature (shock in the polar backflow? See Komissarov and Lyubarsky 2003) at about 5.4″ from the pulsar and then nearly as abruptly fades away at about 10.5″ from the pulsar (see Fig. 5, left panel). In the very deep Vela PWN image the outer jets are visible up to much larger scales ($\simeq$ 2–3′); however, we do not see any smooth transition from the inner axial jets and hence we cannot establish a firm link between the two (except that both structures are extending along the PWN symmetry axis[13]). Note that simulated X-ray images based on MHD models (e.g. Del Zanna et al. 2006; Camus et al. 2009) do not show such structures. In the Crab PWN the dynamical feature called "Sprite" (possibly an analog of the bar feature in the Vela PWN) seems to be the place from which the SE jet originates. However, the Crab's SE jet does not undergo any dramatic transitions in brightness until it bends and terminates at about 1.2′ from the pulsar. Morphologically, this "kinked" jet resembles the outer jets of the Vela PWN rather than the straight and bright inner jets. This leaves no obvious analog to the bight inner jets of the Vela PWN in the Crab PWN. As it has been discussed by Pavlov et al. (2003) and Durant et al. (2013), the fact that the NW outer jet of the Vela PWN is brighter than the SE jet is at odds with the 3D orientation inferred from the arc brightness distribution. The optical image of the Crab in Fig. 1 demonstrates a similar discrepancy assuming that the above mentioned optical "counter-jet" is the actual NW jet.

Unfortunately, the compact Vela PWN has not been detected in the optical despite considerable efforts (Mignani et al. 2003; Moran et al. 2014) hence direct comparison with the Crab PWN is not possible in this band. Zyuzin et al. (2013) recently reported the extended

[13]In the Radhakrishnan and Deshpande (2001) model the arcs are the traces of the particle beams from the two magnetic poles and the inner jets are the Doppler-boosted projections of the beams.

feature seen in the NIR (K_s band) which could be associated with bar at the base of the SE jet or could be analog of the Crab's inner knot. However, this result still requires confirmation.

It is also interesting to contrast the radio morphologies of the Crab and Vela PWNe. The filamentary structure of the Crab resembles that of Vela X, however, the latter has a much larger angular extent ($r \approx 60'$ for Vela vs. $\approx 2.2'$) and it is much more asymmetric. On the other hand, the Vela PWN is a factor of 7 closer (hence, it should appear larger) and a factor of 10–20 older (hence, it had more time to expand). Therefore, the different angular sizes are not surprising. We also note that in the Vela PWN the TeV emission comes from the region of brightest radio filament (Frail et al. 1997; Aharonian et al. 2006b) which is filled with the ejecta (based on the X-ray spectra; LaMassa et al. 2008) thus providing denser target for the putative relativistic hadrons that might be present in the pulsar wind. Therefore, it may turn out that some of the prominent thermal Crab filaments are TeV bright. If confirmed, it could provide evidence for the elusive hadronic component in pulsar winds (Arons 1998). Unfortunately, current resolution of the HESS and VERITAS telescopes does not allow to test this hypothesis (the existing data only suggest that the TeV emission is confined to within $< 1.7'$ from the pulsar; Aharonian et al. 2006d). Although it is plausible that the filamentary radio morphology from the Crab PWN is analogous to that of Vela X, there is no analogy in the Crab for the compact radio PWN found in Vela by Dodson et al. (2003). The ATCA images reveal a double lobe structure, with the lobes being on each side of the X-ray PWN symmetry axis. The radio lobes, extending out to $3'$–$4'$, exceed the size of the X-ray arcs by a factor of 5 but they nonetheless appear to be filled with faint X-ray emission well seen in the harder (1–8 keV) band (see Fig. 4). Even if a similar structure in the Crab would be smaller by a factor of 10–100, it would have been resolved in the JVLA images (unless it is much fainter than the filamentary structure). We also note that no wisp-like structures are seen in the X-ray, optical, or radio images of the Vela PWN.

The X-ray spectral indices of the Crab and Vela PWNe are very different. Figure 6 shows spectral maps for the Crab and Vela PWNe. One can see that the spectra are the hardest (photon indices are the smallest) for the inner ring (in Crab PWN) or for the arcs (in Vela PWN) suggesting that these structures are associated with freshly injected accelerated particles. Interestingly, for the outer arc in Vela PWN the spectrum softens noticeably away from the symmetry axis while this is not the case for the inner arc in the Vela PWN or the inner ring in the Crab PWN.

1.3 Bow Shocks and Tails: PWNe Around Supersonically Moving Pulsars

Pulsar average 3D velocities have been found to be $\sim 400\ \mathrm{km\,s^{-1}}$ for isotropic velocity distribution (see Hobbs et al. 2005). This implies that the majority of pulsars stay within their host SNR environment for a few tens of thousands years although some particularly fast-moving pulsars can leave it earlier. Once the pulsar leaves the SNR,[14] it moves in a very different environment which has a much lower sound speed. For comparison, the sound speed in the middle-aged SNR(such as Vela SNR) can be on the order of a few hundred $\mathrm{km\,s^{-1}}$. The transition between the two very different environments should have a dramatic effect on the PWN of the high-speed pulsar. Once the pulsar is moving in the medium where its speed substantially exceeds the ambient sound speed (i.e. Mach number $\mathcal{M} = v_p/c_s \gg 1$, where v_p and c_s are the pulsar and sound speeds, respectively), the PWN shape should be strongly distorted by the ram pressure of the medium. If initially the wind was isotropic (this is obviously a great oversimplification, see above) the PWN would acquire a cometary

[14] Alternatively, an old SNR can break-up and dissolve.

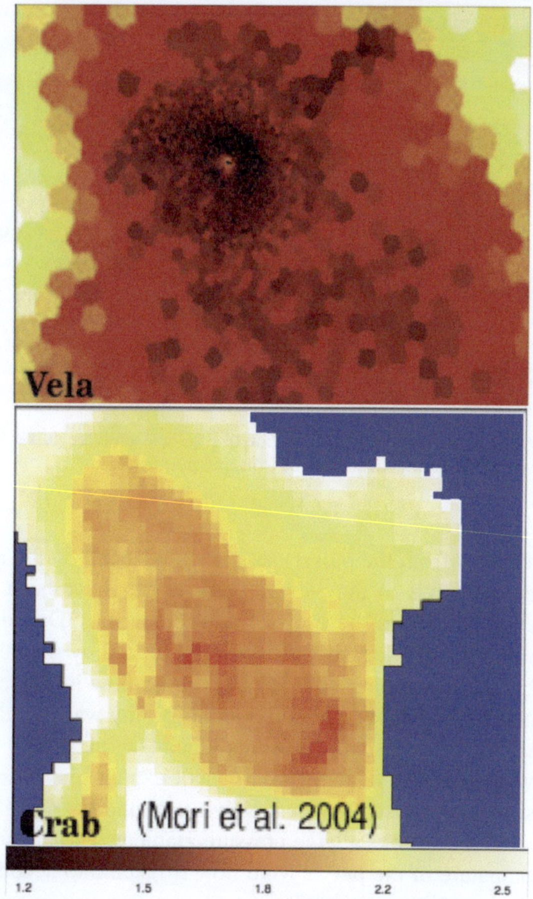

Fig. 6 Photon index maps for Crab (Mori et al. 2004) and Vela PWNe (Kargaltsev et al. 2013a) obtained from *Chandra* ACIS data. The *color bar* at the *bottom* shows photon index values

shape with the pulsar wind being confined to within the surface formed by the contact discontinuity (CD) separating the shocked ambient medium and the shocked pulsar wind (see Fig. 7). Typically, it is assumed that the pulsar wind is shocked in the termination shock (TS) at distances, r_s, substantially smaller than the distance to the contact discontinuity, r_{cd}, even at the apex of the bowshock (see Fig. 7). For very fast moving pulsars r_{cd} may become so small that for some of the X-ray emitting electrons the gyration radius, r_g, would become $\sim r_s \sim r_{cd}$ which may lead to leakage of the electrons from the apex of the bowshock (Bandiera 2008; see also below). Numerical simulations by Bucciantini et al. (2005) indicate high flow speeds ($\gg v_p$) in the shocked pulsar wind outflow behind the moving pulsar, suggesting that an extended pulsar wind tail should form.[15] For realistically anisotropic pulsar winds (with equatorial and polar outflows), in addition to the Mach number, the appearance of the head of the bow shock PWNe and properties of the pulsar tails (to within

[15] The simulations of Bucciantini et al. (2005) do not extend further than $25r_{cd,0}$ (where $r_{cd,0}$ is the scant-off distance at the apex of the bowshock) from the pulsar due to numerical challenges. Also, the model neglects the impact of the magnetic field on the flow dynamics. It is reasonable to expect that the pulsar tail physics may have some similarities with that of leptonic AGN jets for the case when the pulsar spin axis are parallel to its velocity vector (except that magnetic hoop stress may turn out to be larger in the case of pulsar tails). Therefore, some of the AGN jet simulations may be relevant for the pulsar tails.

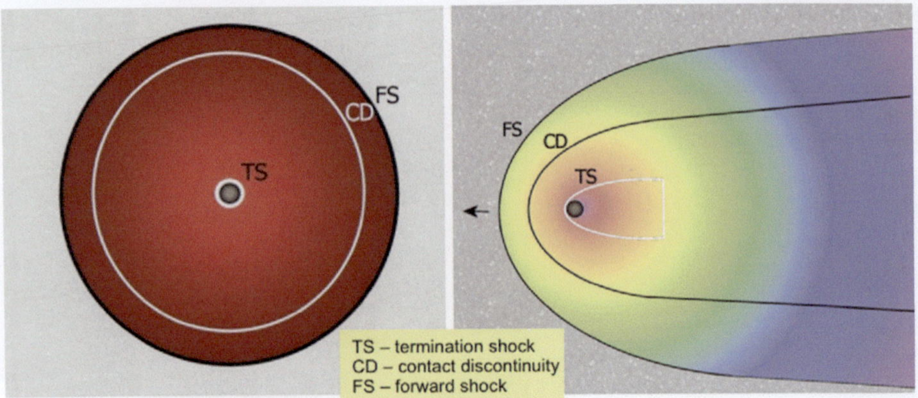

Fig. 7 A schematic representation of PWNe around the pulsar at rest (*left*) and supersonically moving pulsar (*right*) for an idealized case of isotropic pulsar wind

a few r_s from the pulsar) should also depend on the angle between the velocity vector and the spin axis of the pulsar. These effects have been investigated numerically by Vigelius et al. (2007) in the limit of non-relativistic 3D hydrodynamics who found that the bow shock morphology is only weakly affected by the pulsar-wind momentum flux anisotropy but the morphology of the pulsar wind flow in the tail is strongly affected.[16] On the other hand, the ambient medium non-uniformity was found to be greatly affecting the bow shock symmetry and shape. Overall, Vigelius et al. (2007) concluded that "the anisotropy of the wind momentum flux alone cannot explain the observed bow shock morphologies". The simulations also show that Kelvin-Helmholtz (KH) instabilities can develop if the ambient medium exhibits a large pressure gradient. These could be further amplified if the relativistic nature of the pulsar wind flow is taken into account due to the increased velocity shear (Bucciantini et al. 2005). In such situation it is possible that the shocked ambient material can be entrained in the pulsar wind flow altering its structure, dynamics, and emission properties. The entrainment of ambient matter in the pulsar wind is largely an unexplored area (see, however, Lyutikov 2003). Further simulations of bow shock PWNe taking into account all these effects (relativistic velocities, ambient medium non-uniformity and entrainment, pulsar wind anisotropy and dynamical role of the magnetic fields, 3D geometry and instabilities) can provide a realistic picture for comparison with the observations. It may also be possible to make progress by advancing the analytical models of these outflows. Romanova et al. (2005) has constructed a model of a pulsar magnetotail for the axisymmetric case (the pulsar velocity is co-aligned with its spin axis). In this model the pulsar wind remains collimated at large distances from the pulsar forming a magnetotail where an equipartition is reached between the magnetic energy and the relativistic particle energy. The model predictions for the shape of the magnetotail appear to agree with the data in some cases (e.g., PSR J1101–6101; Halpern et al. 2014 and PSR J1747–2958; Gaensler et al. 2004; Yusef-Zadeh and Gaensler 2005) and disagree in others (e.g., PSR J1509–5850; Kargaltsev et al. 2008), possibly, discriminating between the axisymmetric and non-axisymmetric cases.

[16]The simulations by Vigelius et al. (2007) only extended for a few $r_{cd,0}$ from the pulsar, much smaller than the scales of extended tails seen in X-rays.

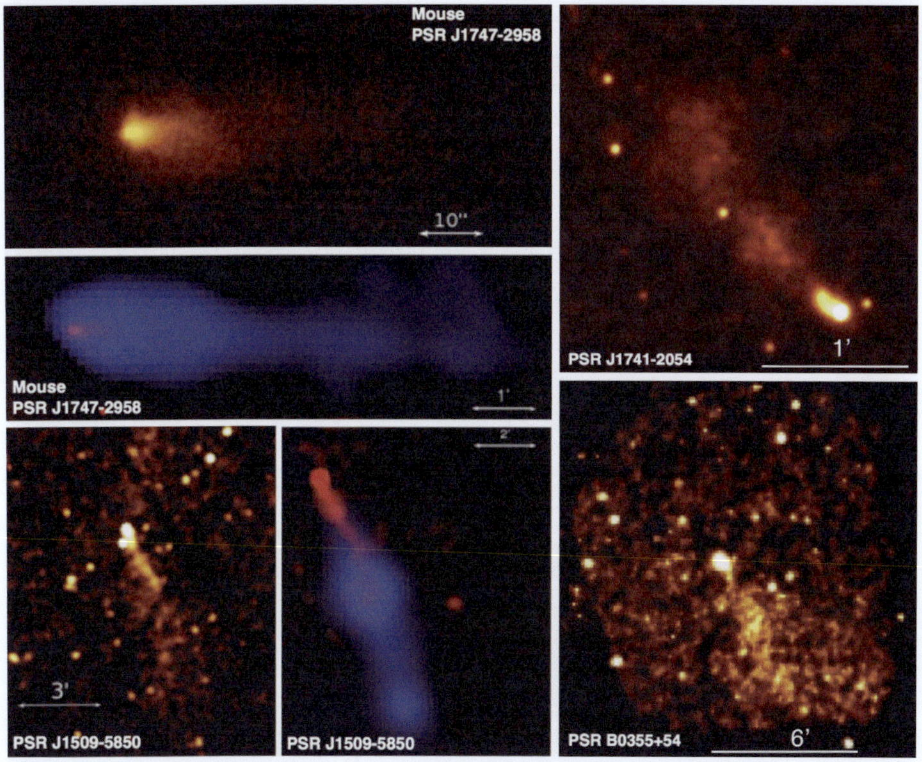

Fig. 8 X-ray (*red*) and radio (*blue*) images of pulsar tails produced from archival *Chandra* and JVLA data. The J1509-5850 radio image is based on the Ng et al. (2010) analysis of Australia Telescope Compact Array observation

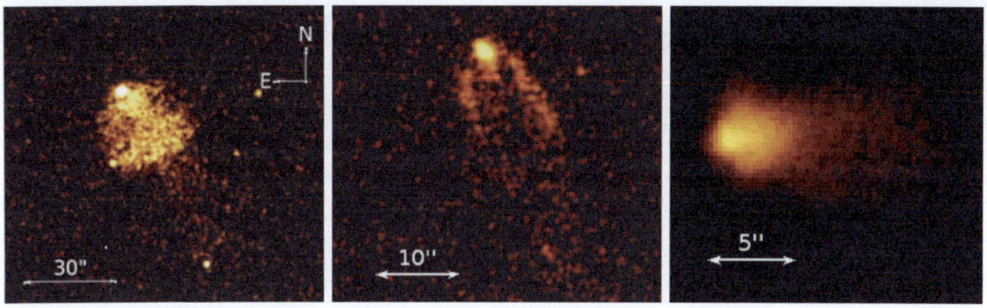

Fig. 9 *Chandra* ACIS images of the head regions of B0355+54, J1509–5058, and Mouse PWNe (*left* to *right*). Notice very different morphologies of B0355+54 and J1509–5058 PWNe. The images are produced from archival *Chandra* data

Chandra and *XMM-Newton* observations provided for the first time X-ray images of PWNe around supersonically moving pulsars (see examples in Figs. 8, 9, 10, and 11). Some of these images display structures (see e.g., Figs. 8 and 9) that are broadly consistent with the theoretical expectations. Indeed, in the X-ray images shown in Fig. 8 one can identify bright PWN heads accompanied by much fainter extended tails (see Fig. 8). Interestingly, the X-ray bright PWN head may or may not be bright in the radio (cf. PWN of PSR J1747–

Fig. 10 570-ks *Chandra* ACIS image of Geminga PWN (Posselt et al. in prep.)

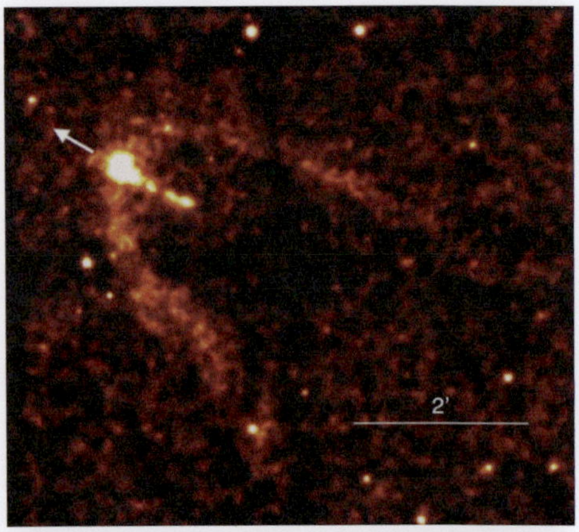

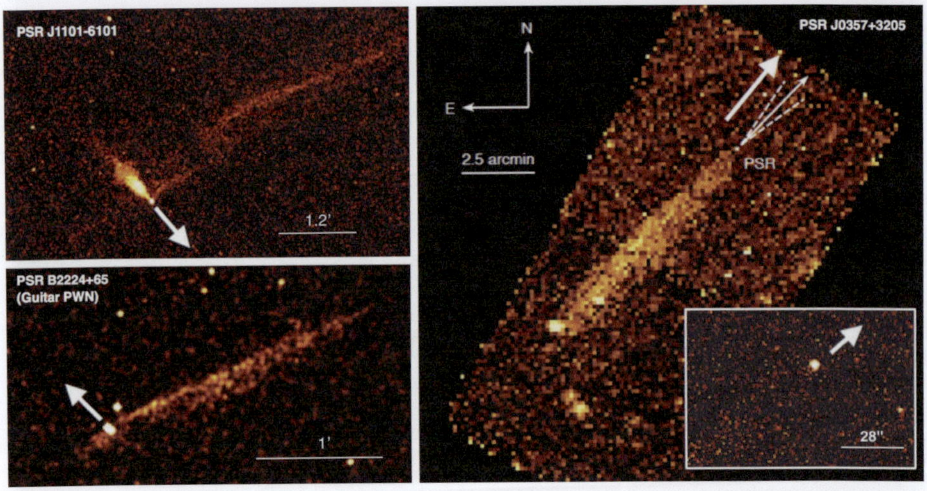

Fig. 11 Puzzling PWN morphologies are seen in the *Chandra* images of three high-speed pulsars. The *arrows* show pulsar proper motion directions. In two cases (*left panels*) the extended structures are orthogonal to the pulsar proper motion. For PSR J0357+3205 (*right panel*) the extended X-ray emission brightens further away from the pulsar while it is very dim in the immediate vicinity of the pulsar (shown in the *inset*). See Pavan et al. (2014); De Luca et al. (2013); Johnson and Wang (2010) for detailed analysis

2958 ("Mouse") and PWN of PSR J1509–5058). Radio polarimetry of two extended tails (Mouse and J1509–5058; Yusef-Zadeh and Gaensler 2005 and Ng et al. 2010) shows that the magnetic field direction is predominantly transverse in the case of the J1509–5058 tail and aligned with the tail in the case of the Mouse tail. This may suggest that the spin axis is more aligned with the velocity vector for J1509–5058 than for the Mouse pulsar (see Fig. 3 from Romanova et al. 2005). Recently, deep, high-resolution observations with *Chandra* revealed interesting structures of the bright bow shock PWN heads. Images of J1509–5058 and B0355+54 PWNe, shown in Fig. 9, exhibit contrasting morphologies. The head of the B0355+54 PWN shows symmetric, filled "mushroom cap" morphology with emission being somewhat brighter near the center than on the sides. On the other hand, the head of the J1509–5058 PWN, comprised of two bent arcs resembling a bow structure (the arcs, however, do not quite connect to the pulsar), is mostly "empty" except for the slight extension just behind the pulsar. This structure remarkably resembles that of the nearby ($d \approx 250$ pc) Geminga PWN (see Fig. 10) as it would be seen at a much larger distance of J1509–5058 ($d \approx 4$ kpc). The bow-shaped X-ray emission can either be associated with the forward shock in the ambient medium (unlikely, because the pulsar velocity must be very high to produce X-rays by heating ISM) or pulsar jets. In the latter scenario the outflows from J1509–5058 and Geminga must be dominated by the luminous jets rather than the equatorial component (cf. Crab and Vela PWNe). This may be difficult to reconcile in the Komissarov and Lyubarsky (2004) model where the jet formation is intimately connected to the diverted by the magnetic field hoops stress equatorial outflow (backflow) which helps to collimate the polar outflow. Furthermore, the recent 3D simulations (see Sects. 2.2 and 2.3) suggest reduced axial compression and weaker jets compared to the 2D simulations. On the other hand, most numerical simulations are designed to reproduce the Crab and Vela structures with a large angle between the NS magnetic dipole and spin axis. If this angle is small, the outflow dynamics could be substantially different. If the side arcs of J1509 are indeed jets, it would also be difficult to explain the ordered helical magnetic field morphology in the extended tail (revealed by the radio polarimetry; Ng et al. 2010) because such a structure would be more natural for the axially symmetric case (Romanova et al. 2005). Thus, although it is plausible that qualitative morphological differences in the appearances of compact PWNe can be attributed to the geometrical factors (i.e. angles between the line-of-sight, velocity vector, spin axis, and dipole axis), these dependencies are yet to be understood.

Particularly interesting and puzzling is the transition region between the bright PWN head and the faint extended tail. For instance, in the B0355+54 PWN ("Mushroom") the drop in the surface brightness at the trailing edge of the Mushroom "cap" is nearly as sharp as at the leading edge (this makes it unlikely to be due to the synchrotron burn-off). In the conventional (isotropic pulsar wind) bow-shock tail models this could be associated with the back surface of the bullet-shaped termination shock (Bucciantini et al. 2005; Gaensler et al. 2004). However, this interpretation does not appear to work for the J1509–5058 PWN which lacks the emission from putative back surface of the termination shock. The transition is also much smoother in the Mouse PWN (see Fig. 9, right panel). The B0355+54 PWN image also shows much fainter, narrow "stem" attached to the Mushroom cap which makes it tempting to associate the bright trailing edge of the Mushroom cap with the equatorial termination shock that has been pushed back by the ram pressure. In this scenario the stem and the brighter middle part of the Mushroom cap would be associated with polar outflow (a jet). However, even in this case the drop in the brightness at the trailing edge of the Mushroom cap may be too abrupt. For instance, in the deep *Chandra* images of the compact Vela PWN one can see the effect of the motion onto the inner ring (commonly associated with the termination shock) with the particles being blown back off the inner ring (see Fig. 5, right panel). We do not observe such a smooth transition behind the Mushroom "cap".

The PWNe behind several very fast-moving pulsars display puzzling morphologies (see Fig. 11). These are the "Lighthouse nebula" with PSR J1101-6101 (Halpern et al. 2014; Pavan et al. 2011; Tomsick et al. 2012), the "Guitar nebula" with PSR B2224+65 (Johnson and Wang 2010; Hui and Becker 2007), and the "Turtle nebula" with PSR J0357+3205 (Marelli et al. 2013). The first two display bizarre extended features *orthogonal* to the pulsar's proper motion directions. The third PWN represents a long and luminous tail in the direction opposite to that of the pulsar's motion; however, close to the pulsar the tail is very faint (undetectable) with no sign of a bright "head" (or compact nebula) near the pulsar (cf. Mouse, B0355+54, or J1509–5058 PWNe). Similar puzzling behavior is seen for PSR J1101-6101 in the Lighthouse PWN. To explain these structures, several hypothesizes have been suggested. Sideways structures in PSRs J1101-6103 and B2224+65 could be pulsar jets (e.g., Johnson and Wang 2010; Pavan et al. 2014), although the one-sidedness and high X-ray efficiencies ($L_X/\dot{E}$) of these structures remain puzzling. The leakage of the wind particles from the apex of the bow shock pushed too close to the termination shock may be an alternative possibility (Bandiera 2008). In the latter case, the morphologies of the sideways features are expected to follow the morphology of the magnetic field in the surrounding ambient medium (which, in these cases, is the ISM well outside the pulsar's host SNRs).

The bowshock PWNe with tails are often found around older pulsars moving fast through the rarefied ISM (perhaps due to the observational bias; see KP08). Unlike torus-jet PWNe, the bowshocks are often seen in H_α. However, H_α bowshocks and bright X-ray tails are rarely seen together (exceptions are PSR J1741-2054, PSRs J2124-3358, and binary B1957+20). The scarcity of such cases suggests that the pre-ionization of the oncoming ISM by the high-energy radiation from the compact PWN and/or pulsar can play an important role. The cases with both H_α and radio emission are also very rare but this may be due to the limited number of objects observed sufficiently deep in the radio. The sizes of extended pulsar tails can reach 7–8 pc in X-rays (e.g., PWN of PSR J1509–5850) and up to ~ 17 pc in radio (e.g., Mouse PWN). These two PWNe, well studied both in X-rays and radio, show remarkably contrasting behavior. The Mouse PWN appears to be the brightest closer to the pulsar *both in radio and X-rays* while the radio emission from the J1509–5850 PWN is lacking near the pulsar and becomes brightest only a few arc minutes away from the pulsar (see Fig. 8). The morphology of the radio PWN of PSR J1101–6101 is similar to that of J1509–5850 radio PWN. Furthermore, the radio polarization measurements indicate that in the Mouse tail the magnetic field direction is predominantly parallel to the tail (Yusef-Zadeh and Gaensler 2005), in the J1509–5850 PWN it is mostly perpendicular to the tail (Ng et al. 2010). The differences may be related to different angles between the pulsar velocity and spin axes in these two PWNe.

PWNe with H_α bow shocks are particularly interesting because the H_α emission allows one to map the structure of the forward shock not only in coordinate space but also in velocity space through the measurements of the Doppler shifts in hydrogen lines across the forward shock (Romani et al. 2010; Brownsberger and Romani 2014). For instance, Romani et al. (2010) performed spectroscopic observations for J1741–2054 and measured the radial velocities up to $\simeq 50\ \mathrm{km\,s^{-1}}$ consistent with the bowshock model implying pulsar speed of ~ 150–$200\ \mathrm{km\,s^{-1}}$ and inclination angle[17] of about 75° (see also Auchettl et al. 2015).

Measurements of spectral line (H_α or/and H_β) fluxes can provide accurate diagnostics of the ambient medium density if other parameters (pulsar velocity $V_{\rm psr}$, distance to the pulsar d, and the stand-off distance r_{cd}) are well constrained. A nearby millisecond PSR J0437–4715 with $\dot{E} = 5.5 \times 10^{33}\ \mathrm{erg\,s^{-1}}$ represents an example of such kind. The pulsar, located at the (parallax) distance of $d = 156$ pc and moving with $V_{\rm psr} = 134\ \mathrm{km\,s^{-1}}$ (for the inclination $i = 53°$ inferred from the bowshock shape), shows a prominent H_α bow shock (Fig. 12) with a stand-off distance $r_s = 2.2 \times 10^{16}$ cm. This implies ISM density $n_H = \dot{E}/4\pi c r_s^2 m_H V_{\rm psr}^2 = 0.1\ \mathrm{cm^{-3}}$. For slow ($V_{\rm psr} < 10^3\ \mathrm{km\,s^{-1}}$) bowshocks the H_α yield (i.e., number of H_α photons per incoming neutral) is $\epsilon_{H_\alpha} = 0.6(v/100\ \mathrm{km\,s^{-1}})^{-1/2}$ according to Heng and McCray (2007) and the H_α flux $f_{H_\alpha} \approx 0.0074(v/134\ \mathrm{km\,s^{-1}})^{1/2}(r_s/2.2 \times 10^{16}\ \mathrm{cm})^2(d/156\ \mathrm{pc})^{-2}(n_H/0.1\ \mathrm{cm^{-3}})\xi_{HI}$ photons $\mathrm{cm^{-2}\,s^{-1}}$ (see e.g., Chatterjee and Cordes 2002; Brownsberger and Romani 2014). The measured H_α flux from the J0437–4715's bow shock apex, $f_{H_\alpha} = 6.7 \times 10^{-3}$ photons $\mathrm{cm^{-2}\,s^{-1}}$ (Brownsberger and Romani 2014), implies that the neutral H fraction $\xi_{HI} \approx 0.9$. This is somewhat surprising because the classical Strömgren radius for the ionizing radiation produced by the pulsar appears

[17] The angle between the line of sight and the pulsar velocity vector.

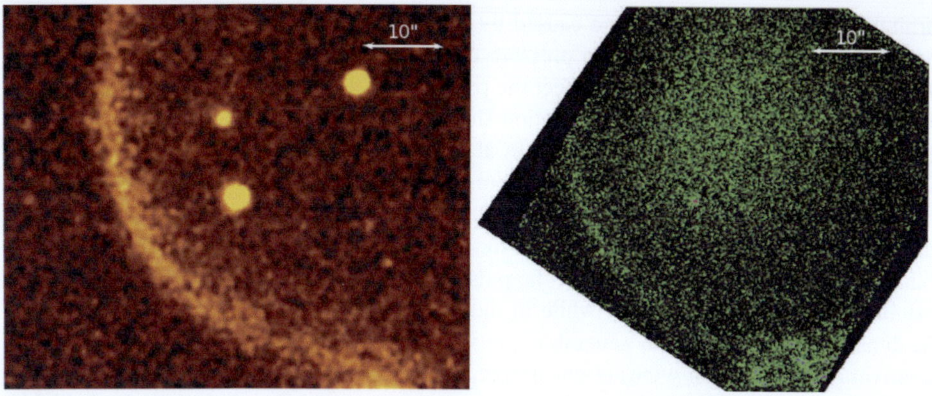

Fig. 12 H_α (*left*) and far-UV (*right*) images of the bow shock around PSR J0437–4715. Faint, amorphous diffuse emission seen around the pulsar in the *right panel* is the instrumental artifact (thermal glow of *HST*/SBC detector) while the structure seen at the *bottom* is a background galaxy. Both *right* and *left panel* images show the same area of sky. (Rangelov et al. in prep.)

to be much larger than the distance to the H_α bowshock apex indicating that the classical formula is probably inapplicable when ionizing radiation is X-rays. PSR J0437–4715 bow shock is the only one from which the far-UV (FUV; see Fig. 12) emission has been detected (other pulsars with H_α bowshocks may simply be too far, so FUV photons are easily absorbed). The measurement of the FUV spectral slope, $\alpha \sim 1.5$ (for $F_\nu \propto \nu^{-\alpha}$), suggests a non-thermal emission mechanism (likely synchrotron) apparently coming from the vicinity of the forward shock (FUV and H-alpha bow-shocks coincide within the measurements errors, $\simeq 0.5''$). This is puzzling because it either implies acceleration of the ambient medium electrons to very high energies or escape of the pulsar wind electrons into the forward shock zone. The latter would imply that these electrons, invisible (in the FUV) inside the bow shock, for some reason become FUV emitters in the forward shock region. No firm detection of X-ray emission has been reported for the pulsar wind of J0437–4715 yet, although an analysis of archival *Chandra* ACIS data indicates that a compact, $r \simeq 2''$–$3''$, asymmetric (most of the emission is ahead of the pulsar) PWN with the luminosity of 3.8×10^{28} erg s^{-1} may be present (Rangelov et al. in prep.).

So far B1951+32, J1509–5850, and J2124–3358 are the only three solitary pulsars where both X-rays and H_α have been detected. However, in J1509–5850, H_α emission is believed to come from the photo-ionized medium rather than from the bow shock (see the discussion in Brownsberger and Romani 2014) and for B1951+32 the surrounding environment is complex due to possible contribution to H_α from the filaments of the host SNR CTB 80. Therefore, it is important to perform deep H-alpha imaging and spectroscopy of nearby pulsars with X-ray bow shocks to study the connection between the pulsar wind and shocked ambient medium regions.

1.4 PWNe in Binaries

Sufficiently energetic non-accreting binary pulsars may be able to power detectable PWNe. In binary systems, in addition to the pulsar parameters, the PWN morphology and appearance would depend on the properties of the binary companion and parameters of the binary orbit. Even if the companion star is lacking any wind itself (e.g., it is a cold white dwarf or old low-B neutron star) the pulsar wind will be affected by the ram pressure due to its own orbital motion if the ram pressure caused by the orbital motion is comparable to or larger than the ambient pressure (or ram pressure due to the motion of the binary as a whole). Thus, binary motion can strongly affect the pulsar wind flow (and morphology of the magnetic field) downstream of the termination shock or even the termination shock itself (see e.g., Bosch-Ramon et al. 2012; Zabalza et al. 2013). If the companion star has a powerful wind the interaction becomes even more complex with the outcome critically dependent on the ratio of momentum fluxes of the two winds $\eta = \dot{E}c^{-1}/\dot{M}_w v_w$ (where $\dot{M}_w$ and v_w are the mass loss rate and the massive star wind velocity), a key parameter in colliding wind binaries (see Dubus 2013). If $\eta \gg 1$, the pulsar wind dominates (e.g., for B1957+20—the original "black widow" system) while in the opposite case the companion star wind will be dynamically dominant. In general, η can vary with the orbital position (1) if the pulsar wind (or the massive star wind) is anisotropic, and the spin axis of the pulsar (or the massive star) do not coincide with the orbital angular momentum vector or (2) if the orbit is highly eccentric.

The most famous example of a pulsar binary system where all these effects play a significant role is LS 2883 with the young energetic pulsar B1259–63 (B1259 hereafter) in an eccentric 3.4-year orbit around a massive O-star. Although direct observations of pulsar wind in B1259 may not be feasible (except perhaps for the VLBI imaging observations; see Moldón et al. 2012), there are indirect ways to learn about the pulsar and stellar wind properties and their interaction. These include multi-wavelength spectrum and flux measurements as a function of orbital phase (Chernyakova et al. 2014), pulsar radio signal variability measurements (Shannon et al. 2014 and references therein), and, a high-resolution X-ray imaging, which recently revealed a dynamic structure associated with the binary (Kargaltsev et al. 2014b). An X-ray emitting cloud was found to be moving away from the binary (see Fig. 13) with the velocity of $\approx 0.07c$ (Pavlov et al. 2015), which, together with the lack of deceleration, implies either a hadronic cloud with very large mass $\gtrsim 10^{27} n (d/2.3\ \mathrm{kpc})^2$ g moving in the O-star wind with density n or a lighter cloud moving in the rapidly expanding, *unshocked* relativistic pulsar wind (implies $\eta \gg 1$ in the polar O-star wind). The former scenario implies that the cloud was ejected from the binary during the 2011 periastron passage, when the pulsar interacted with the excretion disk of the massive O-star. However, the corresponding kinetic energy of the cloud must be very large, $\sim 2 \times 10^{45}$ erg, and it must have been launched via a complex interaction between the pulsar wind and excretion disk of the O-star with the energy source being problematic (pulsar's $\dot{E}$ can only provide $\sim 10^{42}$ erg during the disk passage). The latter scenario is at odds with the common assumption of $\eta \lesssim 1$ for such kind of binaries but it does not require extreme values of mass and energy for the cloud (Pavlov et al. 2015).

We note that there are other systems where pulsars (albeit these may not be as young and energetic) might orbit massive (often Be or O type) stars, and it is plausible that other TeV gamma-ray binaries (e.g., LS 5039, LS I +61 303, and HESS J0632+057) also host pulsars (Dubus 2013; Kargaltsev et al. 2014a and references therein). Indeed, Durant et al. (2011) reported evidence for amorphous arc minute-scale X-ray emission with a hard spectrum around LS 5039 which was interpreted as synchrotron radiation from ultrarelativistic (pulsar wind?) particles escaping from the system.

Among other types of binaries which can shed light on the properties of pulsar winds at smaller distances from the pulsar through the interaction with the companion are the famous double pulsar (see e.g., Arons et al. 2005), recently reported very eccentric binary with PSR J2032+4127 (Lyne et al. 2015), and black widow pulsar B1957+20 (Huang et al. 2012).

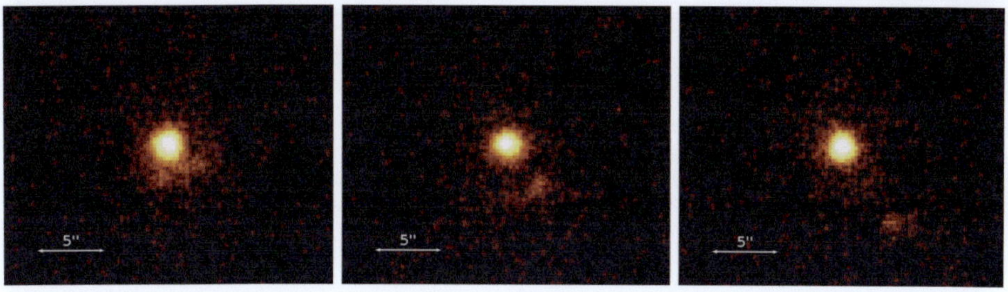

Fig. 13 *Chandra* observations (on MJD 55912, MJD 56431, and MJD 56696 *left* to *right*) of LS 2883/B1259-63 reviling the relativistic, $v = (0.074 \pm 0.006)c$, motion of cloud ejected from the binary. See the corresponding movie at http://home.gwu.edu/~kargaltsev/B1259.html

1.5 PWN Spectra

It is commonly accepted that radio-to-MeV emission in PWNe is the synchrotron emission. This naturally explains the observed high degree of polarization in X-rays, optical, and radio (Moran et al. 2013; Bietenholz et al. 2004; Weisskopf et al. 1978). Multiwavelength observations and modeling show that IC scattering on Cosmic Microwave Background and NIR/IR background starts to dominate PWN spectra in GeV (Acero et al. 2013) and completely dominates synchrotron at TeV energies (Atoyan and Aharonian 1996, Volpi et al. 2008, Olmi et al. 2014, and Gelfand et al. 2009; see the Crab and Vela PWN spectra in Fig. 14 for example). The theoretical models of particle acceleration and non-thermal emission in PWNe predict, in the ideal MHD framework, a cutoff around 150 MeV (see Sect. 3.1). These models well reproduce the observed cutoff in the Crab Nebula spectrum (see left side of Fig. 14). However, the slow acceleration rates of ideal MHD models (as, e.g., diffusive shock acceleration) do not allow fast variability in the nebular emission (see Sect. 3 for Crab flares). The questions about the location and distribution of acceleration sites as well as acceleration mechanism in the pulsar winds and PWNe also remain open (see Sect. 2.4).

If the pulsar wind contains relativistic protons, it is possible that hadronic emission (due to neutral pion decay) contribution can become appreciable in dense environments (Bartko and Bednarek 2008). Multiwavelength emission from bow shock PWNe (including pulsar tails) produced by pulsars moving in a low-density ISM (outside their host SNRs) should be purely leptonic. However, surprisingly few of these objects have been detected in TeV (one of the deepest limits, 0.1 % of $\dot{E}$ in 1–10 TeV, is obtained with VERITAS for the tail of B0355+54; Brett McArthur, private communication). Recent review of the observational X-ray and TeV properties for the population of 91 PWNe can be found in Kargaltsev et al. (2013b).

1.5.1 Spatially Resolved X-Ray Spectra

Deep *Chandra* ACIS observations of a few bright PWNe allow us to create spectral maps with high-spatial resolution (Kargaltsev et al. 2013a). These maps are expected to provide manifestation of synchrotron burn-off (X-ray spectra should become softer farther away from the pulsar) which depends on the strength of magnetic field and bulk flow speed (Kennel and Coroniti 1984b; Wang et al. 2001; Reynolds 2009) but they may also contain signatures of spatially distributed (in-situ) particle acceleration or rapid particle diffusion. These spectral maps demonstrate that the pulsar spectra measured just downstream of the termination shock can differ substantially. From Table 1 one can see that although for the Crab PWN

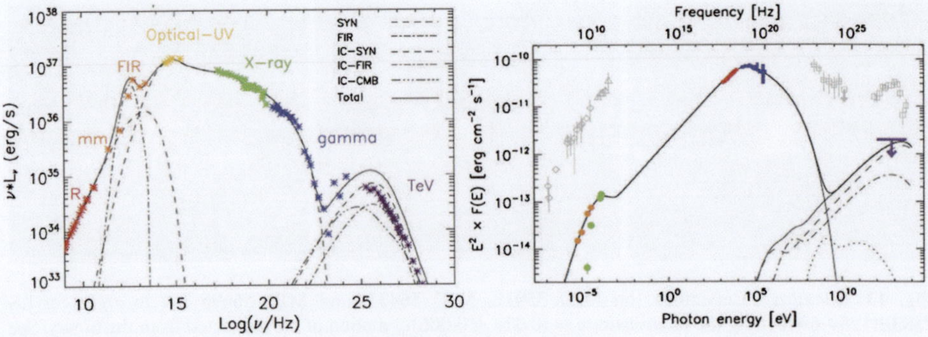

Fig. 14 Multiwavelength spectra of the Crab (*left*; from Volpi et al. 2008) and Vela PWN (*right*; spectrum from $r = 6'$ from pulsar shown in *color* while the relic Vela X plerion spectrum is in *grey*; from Mattana et al. 2011)

Table 1 Photon indices (Γ) and electron SED slopes ($p = 2\Gamma - 1$) measured from *Chandra ACIS* data for the innermost regions in 9 bright PWNe (Kargaltsev et al. 2014c)

PWN	Γ	p
Crab	1.80 ± 0.05	2.6
Vela	1.30 ± 0.05	1.6
3C58	1.9 ± 0.07	2.8
G320.4-1.2	1.4 ± 0.1	1.8
Kes 75	1.9 ± 0.1	2.8
G21.5-0.9	1.40 ± 0.06	1.8
G11.2-0.3	1.5 ± 0.1	2.0
CTB 80	1.7 ± 0.1	2.4
G54.1+0.3	1.50 ± 0.05	2.0

the inferred (assuming synchrotron emission model) slope of the electron SED, $p = 2\Gamma - 1$, is consistent with the $p = 2.1$–2.8 expected from the commonly invoked Fermi acceleration mechanism (Achterberg et al. 2001; Sironi and Spitkovsky 2009a, 2011b), for the Vela PWN the spectrum is much harder ($p \approx 1.6$) suggesting that a different mechanism might be at work (e.g., magnetic reconnection; Sironi and Spitkovsky 2014). At least some of the long pulsar tails (e.g., tail of PSR J1509–5850) tend to show very little evidence of cooling (in terms of spectral softening in X-rays) which either suggests ongoing in-situ acceleration along the tail or extremely fast bulk flow (Klingler et al. in prep.).

1.5.2 X-Ray and TeV Efficiencies of PWNe

The substantial number of PWNe detected in X-rays and TeV allows one to investigate the population properties. Here we will only consider X-ray and TeV radiative PWN efficiencies ($\eta_{\gamma,\mathrm{X}} = L_{\gamma,\mathrm{X}}/\dot{E}$) and refer the readers to Kargaltsev et al. (2013b) for the analysis of other TeV and X-ray properties of PWNe. Figure 15 (based on the information collected in Tables 1–3 in Kargaltsev et al. 2013b, with some updates) shows the X-ray and TeV luminosities of PWNe (and PWN candidates for TeV). Notice a very large spread of X-ray efficiencies and a noticeably smaller spread for the TeV efficiencies. While there is a clear correlation between $\dot{E}$ and η_X there is no noticeable correlation in TeV. Finally, the majority of PWNe that are underluminous in X-rays appears to be around γ-ray loud, radio-emitting pulsars with small magnetic inclination angles (based on Rookyard et al. 2014).

1.5.3 Search for New PWNe in GeV γ-Rays

PWNe are the most numerous source class that emerged from the H.E.S.S. Galactic Plane Survey (Carrigan et al. 2013), and about 100 PWNe or PWNe candidates are known in X-ray. Nevertheless only very few sources are detected in the MeV–GeV range, because this energy range falls between the tail of the synchrotron emission and the rising part of the IC emission and because of the lack of sensitive instrumentation with good angular resolution. Often, the search for PWN at GeV energies has to be carried out in the presence of γ-ray-loud pulsar by looking for the off-pulse emission which could come from the PWN (see Abdo et al. 2013). In Acero et al. (2013), a search for GeV emission from 58 TeV PWNe and unidentified sources was performed, with the requirements of (1) good GeV and TeV spectrum connection and (2) extended emission. A total of thirty sources were detected in the GeV range, for energies above 10 GeV; among them, 11 sources are PWN candidates, and three are reliably identified as new PWNe. These new sources are associated with young (age between 1 and 30 kyr) and powerful pulsars with $\dot{E}$ between 10^{36} and 10^{39} erg s^{-1}. It is interesting to study how the multiwavelength properties of the PWNe evolve with the properties of the host pulsar. It was found in Acero et al. (2013) that there is no correlation between the GeV luminosity and the age and the spin-down luminosity of the pulsars, and the same for the GeV-to-TeV luminosity ratio. On the other hand, the GeV-to-X-ray luminosity ratio appears to increase with age. Even less is known about PWN emission at the MeV energies due to the poor imaging capabilities of the MeV telescopes. The dip is expected to occur at these energies in the spectra of $\sim$10 to $\sim$100 kyr-old PWNe which are most frequently found in X-rays and TeV. Measuring the cut-off energy of the synchrotron spectrum in hard X-rays constrains maximum energies of the electrons in PWNe.

1.5.4 Multiwavelength Spectra

Spatially-resolved spectroscopy is primarily done in X-rays, thanks to the superb resolution of *Chandra*. It remains rather challenging to obtain spatially-resolved spectral measurements at other frequencies. In the optical-NIR this is primarily due to the faintness of the PWN emission (except for the Crab PWN) and contamination by various background sources. In the radio, this is challenging because of the interferometric nature of high-resolution observations and difficulties in combing the requirements of imaging both the large- and small-scale structures. At higher energies (MeV, GeV and TeV), the resolution of the existing telescopes is often insufficient to resolve PWNe, and in those cases when large relic PWNe can be resolved (Aharonian et al. 2006e; Abramowski et al. 2012) the limited signal-to-noise ratio typically precludes spatially resolved studies (see, however, Aharonian et al. 2006a; Van Etten and Romani 2011). Therefore, the MW spectra of most PWNe are, by necessity, spatially-integrated which in many cases introduces bias and systematic uncertainties that are difficult to account for (Van Etten and Romani 2011). Indeed, as one can see from Fig. 14, the MW spectrum of the compact Vela PWN is very different from the MW spectrum of the relic plerion Vela X (which may be a peculiar one in other respects also, see below). The PWN luminosities shown in Fig. 15 are calculated for very different regions in most PWNe and therefore should be treated as such (i.e. cannot be treated as if both X-ray and TeV photons are emitted from the same PWN region). If both TeV and X-ray

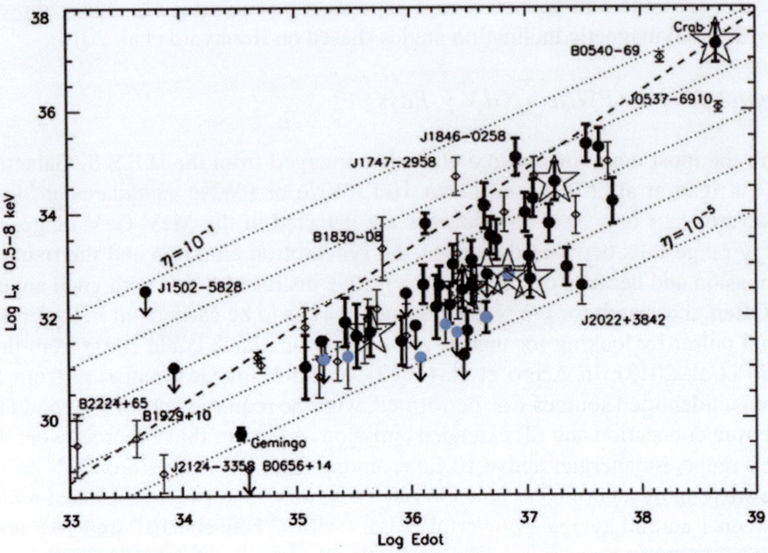

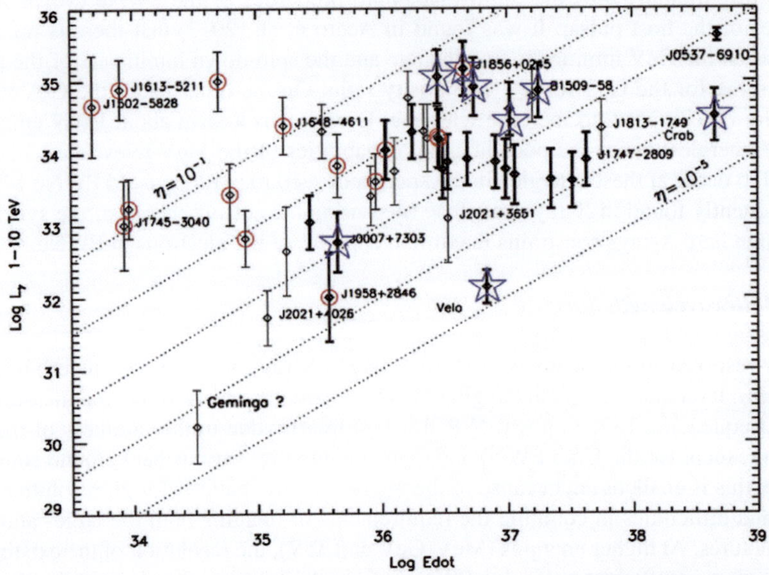

Fig. 15 *Top*: X-ray luminosities of PWNe and PWN candidates vs. pulsar's $\dot{E}$. TeV PWNe and TeV PWN candidates are shown in *red*. The *dotted straight lines* correspond to constant X-ray efficiencies; the upper bound, $\log L_X^{\rm cr} = 1.51 \log \dot{E} - 21.4$, is shown by a *dashed line* (Kargaltsev et al. 2012). The PWNe detected in GeV by *Fermi* are marked by *stars*. *Blue filled circles* are the pulsars with confidently measured small ($< 10^\circ$) magnetic inclination angles from Rookyard et al. (2014). *Bottom*: TeV luminosities of PWNe and PWN candidates vs. pulsar's $\dot{E}$. *Thin error bars* mark questionable associations. The PWNe undetected in X-rays are marked by *circles*. PWNe detected by *Fermi* are marked by *stars*. The *dotted lines* correspond to constant values of the TeV γ-ray efficiency $\eta_\gamma = L_\gamma/\dot{E}$. The detection of TeV emission from Geminga region is, so far, based solely on Milagro result (Abdo et al. 2009) which has not been confirmed by any other observatories (e.g., VERITAS)

efficiencies (luminosities) of PWNe are compared to those derived from a one-zone PWN model, one should be aware of the limitations of this approach. Multi-zone models, taking into account both advection and diffusion, appear to be the next logical step in theoretical development and in preparation for Cherenkov Array Telescope (CTA; see e.g., Fortson and CTA Consortium 2015).

As an example (albeit possibly an unusual one) of a relic PWN one can consider Vela X. The nearby Vela SNR ($\sim 8°$ in diameter) has a large region of non-thermal radio emission surrounding the Vela pulsar (see e.g., Bock et al. 1998). One of the brightest radio filaments in Vela X, positioned at the southwest of the pulsar, was detected in X rays with *ROSAT*, *ASCA*, *Suzaku*, and *XMM-Newton* (Markwardt and Ögelman 1995, 1997; Mori et al. 2008; LaMassa et al. 2008) and, more recently, in GeV γ-rays with *AGILE* (Pellizzoni et al. 2010) and *Fermi* LAT (Abdo et al. 2010b) and very high energies (0.5–70 TeV) with HESS (Aharonian et al. 2006c) and CANGAROO (Enomoto et al. 2006). The bright X-ray and VHE emission regions are positionally coincident (they sometimes referred to as a "cocoon"), and have been commonly dubbed a relic PWN displaced to the south by the unequal pressure of the reverse shock propagating within the SNR. However, subsequent deeper observations with H.E.S.S. (Abramowski et al. 2012) and *Fermi* LAT (Grondin et al. 2013) revealed a fainter extended emission whose morphology appears to correlate with the the double-lobe, large ($\sim 2°$ in extent) structure found at 61 GHz in the WMAP images. This yields strong support to the scenario where two different populations of electrons are needed to reproduce the radio/GeV halo and the X-ray/TeV cocoon, respectively (de Jager et al. 2008).

2 Theory of PWNe

2.1 General Properties of Pulsar Winds and Structure of PWNe

Theoretical studies of PWNe concentrated on interaction of the relativistic pulsar wind with the surrounding plasma. The morphology of nebulae is described in the scope of magnetohydrodynamic (MHD) models, which proved to be very successful. In order to understand the radiation spectra, one has to develop realistic particle acceleration models; this is a much more difficult task. Here we shortly review the recent development of the PWNe theory; for more comprehensive reviews, see, e.g. Kirk et al. (2009); Arons (2012); Bühler and Blandford (2014).

The general features of PWNe are basically dictated by the physics of the pulsar wind:

- The pulsar wind is composed predominantly of electron-positron pairs, may be with some admixture of ions. The pair content of PWNe suggests that the pair density in the wind is rather high, more than enough to ensure that the wind could be described as an MHD flow (de Jager et al. 1996; Bucciantini et al. 2011).
- The wind is strongly magnetized; most of the energy is transferred, at least initially, as Poynting flux.
- The wind is highly anisotropic: the Poynting flux is maximal in the equatorial belt and goes to zero at the axis (Michel 1973; Bogovalov 1999; Tchekhovskoy et al. 2013).
- An obliquely rotating magnetosphere produces variable electromagnetic fields that propagate in the wind as MHD waves; specifically in the equatorial belt, where the magnetic field changes sign every half of period, the so called striped wind is formed.

The PWN is in fact a bubble filled predominantly by relativistic particles and magnetic fields; it is inflated by the pulsar wind that continuously pumps into the surrounding medium the energy in the form of relativistic particles and magnetic fields. Within the nebula, the fields and the particles are roughly in equipartition therefore the main question is how the Poynting flux in the pulsar wind is converted into the energy of particles (the so called σ-problem).

Even though the details of the energy transformation process remain obscure, the general picture is robust: most of the energy is transferred in the wind by alternating electromagnetic field; and the energy is transferred to the plasma when the alternating field dissipates. This conclusion follows from the strong anisotropy of the MHD wind, in which energy is predominantly transferred in the equatorial belt where the striped wind is formed. Of course the fraction of the energy transferred by alternating fields depends on the angle between the magnetic and rotational axes of the pulsar: an aligned rotator does not produce alternating fields at all whereas the energy from a perpendicular rotator is totally transferred by alternating fields. Due to the strong anisotropy of the pulsar wind, most of the energy is transferred by alternating fields even in a moderately oblique case. Komissarov (2013) calculated the fraction of the energy flux due to alternating fields making use of the split monopole model of the pulsar wind (Michel 1973; Bogovalov 1999), in which the Poynting flux is distributed as $\sin^2\theta$, where θ is the polar angle. He found that even if the angle between the rotational and magnetic axes is 45°, as much as 72 % of the total energy is transferred by alternating fields. In real pulsar winds, this fraction is even larger because the Poynting flux in the wind from a rotating dipole magnetosphere is more concentrated to the equatorial belt than in the monopole wind; according to numerical simulations by Tchekhovskoy et al. (2013), the angular distribution of the Poynting flux is close to $\sin^4\theta$ in this case.

Even though the dissipation mechanisms for variable fields in pulsar winds are still debated, one has to stress that the particle Larmor radii within the nebula significantly exceed the wavelength of the waves in the pulsar wind. Therefore these waves could not penetrate into the nebula in any case; even if they survive within the wind, they dissipate at the termination shock front (Lyubarsky 2003b; Pétri and Lyubarsky 2007; Sironi and Spitkovsky 2011a). Therefore the post-shock flow must be the same as it would be if the dissipation has already been fully completed in the wind. The structure of the nebula is determined by the distribution of the total energy and mean magnetic field in the wind. In the equatorial belt, where most of the energy is transferred, the mean field is weak therefore relatively weakly magnetized plasma is injected into the nebula in this region. At higher latitudes, the magnetic field does not change sign and the variable magnetic fields could propagate in the form of fast magnetosonic waves. These waves efficiently decay as a result of non-linear steepening (Lyubarsky 2003a) however, they could not transfer a large fraction of the Poynting flux so that the flow magnetization in this region remains large even after they decay.

Taking into account the above properties of the pulsar wind, the observed morphology of PWNe, and first of all the characteristic disk-jet structure (Weisskopf et al. 2000; Gaensler and Slane 2006), is naturally explained (Lyubarsky 2002). Namely, the disk is formed by the relatively weakly magnetized equatorial flow, which transfers most of the energy. In the Crab, such a disk is clearly seen in the *Chandra* map because the X-ray emitting electrons rapidly loose their energy thus making X-rays a tracer for the freshly injected plasma. In other PWNe, like Vela, the disk may even not be seen at all because the high energy electrons fill a much larger volume. The high latitude flow remains magnetized therefore it is compressed by the magnetic hoop stress to form a jet-like feature at the axis. An important point is that in the highly relativistic, super-sonic (more exactly, super-fast-magnetosonic) wind, the magnetic collimation is inefficient; the "jet" is formed beyond the termination shock where the flow is decelerated. Inasmuch as the pulsar wind is anisotropic, the termination shock is highly non-spherical: it lies much closer to the pulsar in the polar regions than in the equatorial belt; therefore the "jet" appears to originate from the pulsar (see also Bogovalov and Khangoulian 2002).

This general conjecture has been confirmed by axisymmetric MHD simulations (Komissarov and Lyubarsky 2004; Del Zanna et al. 2004, 2006; Volpi et al. 2008; Bucciantini et al. 2011). These simulations were able to explain also nontrivial features of the fine structure, such as a mysterious knot, which is located within $1''$ from the Crab pulsar (Hester et al. 1995). Namely, the knot is a Doppler-beamed emission from the patch of the highly oblique termination shock where the post-shock flow is still relativistic and directed towards the observer (Komissarov and Lyubarsky 2004). Simulations with a better resolution and therefore with a lower numerical viscosity (Camus et al. 2009) revealed bright fine filaments moving away from the termination shock with a good fraction of the speed of light. These are highly reminiscent of the so-called wisps of Crab nebula (Hester et al. 1995).

In spite of all these successes, important basic problems have remained unresolved. Namely, axisymmetric simulations reproduce the observed structure of the nebula only if the wind magnetization at high latitudes was chosen to be relatively small, significantly smaller than one could expect from theoretical considerations. This discrepancy was resolved only recently when fully 3D simulations have been performed (Porth et al. 2014). Let us describe these recent developments in more details.

2.2 The σ-Problem and 3D Simulations of PWNe

The problem of the magnetic to the plasma energy transformation in pulsar winds is generally referred to as the σ-problem because the flow magnetization, defined as the ratio of the Poynting to the plasma energy fluxes, is typically denoted by σ. The pulsar wind starts as a highly Poynting dominated ($\sigma \sim 10^4$–10^6); on the other hand, there is a pervasive belief that one can account for the morphology of PWNe, including the remarkable jet-torus structure, only if just upstream of the termination shock, σ does not exceed 0.01. Such a tremendous drop in the flow magnetization looks so mysterious that the problem was sometimes referred to as the σ-paradox. However, one has to stress that what we really need to consider is the mean field because alternating fields inevitably decay. As it was mentioned above, they transfer most of the energy in the pulsar wind therefore the magnetization due to the mean field is not large, which makes the σ-problem not so severe. Let us discuss the issue in a bit more details.

First of all one has to stress that the strong constraints on the wind magnetization at the termination shock mentioned above were obtained in spherically or axially symmetric models of PWNs (Rees and Gunn 1974; Kennel and Coroniti 1984a; Emmering and Chevalier 1987; Begelman and Li 1992). The reason for the required low value of σ is that in these models, the magnetic field strength grows with radius within the nebula so that the field would exceed the equipartition value and pinch the flow too much if the magnetization at the termination shock is not extremely small. The behavior of the magnetic field could be easily understood if one takes into account that the field in the far zone of the wind is practically azimuthal, and in the axisymmetric flow, the field lines remain coaxial circular loops. The radius of the field line increases when the flow expands, the field strength being determined by the conservation of the magnetic flux within the toroidal magnetic tube. At the termination shock, the flow compresses so that the magnetic field increases three times. The flow within the nebula is subsonic therefore the pressure and the density of the plasma do not change significantly. Therefore the volume of the toroidal magnetic tube remains

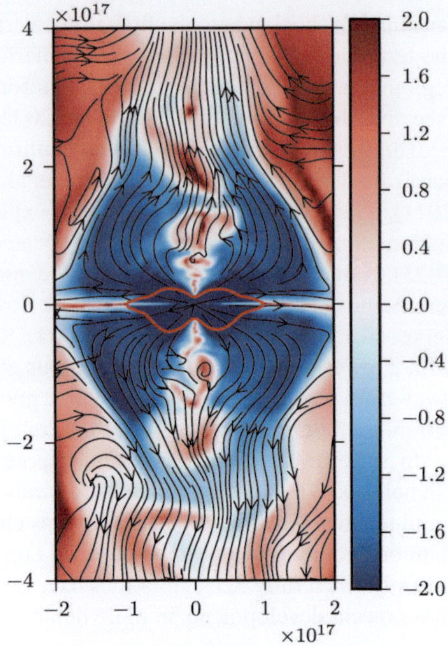

Fig. 16 Formation of the polar outflow in the 3D MHD simulations by Porth et al. (2014). The *colour* images show the distribution of lg β in the yz plane, where β is the ratio of the thermal to the magnetic pressure. The *black lines* show the momentary streamlines and the *red line* the termination shock

roughly constant. In this case, the cross section of the tube decreases when the tube radius increases, which implies an increase in the magnetic field roughly linearly with the radius.

Taking into account that the size of the nebula is about an order of magnitude larger than the radius of the termination shock, one finds that in the axisymmetric flow, the field strength in the main body of the nebula exceeds that in the wind just upstream of the shock $\sim 3 \times 10 = 30$ times whereas σ grows roughly three orders of magnitude. The problem can be alleviated if the kink instability destroys the concentric field structure in the nebula (Begelman 1998). Then the magnetic loops could come apart and one expects that in 3D, the mean field strength is not amplified much by expansion of the flow. In this case, σ just upstream of the termination shock might not need to be so unreasonably small as was found in axisymmetric models.

This idea can be checked only by 3D simulations of plasma flow within the nebula. As the first step, Mizuno et al. (2011) simulated the 3D evolution of a simple cylindrical model of PWNe developed earlier by Begelman and Li (1992). This model describes a static cylindrical configuration with a relativistically hot plasma such that the thermal pressure is balanced by the hoop stress of a purely toroidal magnetic field. The simulations clearly demonstrated that the kink instability does develop in the system and destroys the regular concentric structure of the magnetic field thus relaxing the hoop stress and triggering magnetic dissipation. This proves that 3D effects play crucial role in the evolution of PWNe. However, these simulations do not claim to model PWN, simply because the continuous injection of magnetic flux and energy into PWN by their pulsar winds is not accounted for.

The first realistic 3D simulations of PWNe were performed by Porth et al. (2014). They used qualitatively the same setup as in 2D simulations, namely, the nebula is pumped by a strongly anisotropic pulsar wind with the magnetization determined only by the mean magnetic field as if the alternating component of the field has completely dissipated. The difference was in the magnetization at high latitudes, where the magnetic field does not change sign. In 2D simulations, the observed morphology was reproduced only if the high

Fig. 17 Field lines in the 3D simulations by Porth et al. (2014). The *lines* are *coloured* according to their orientation, sections with dominating azimuthal component being *blue* and those with dominating poloidal component being *red*. The surface of the termination shock is also shown, using the *magenta contour*

latitude σ was chosen to be as small as 0.1 even though according to the pulsar wind theory, it should remain significant, not less than a few. Porth et al. (2014) took $\sigma = 1 \div 3$ at high latitudes, as it should be.

According to the results of their 3D simulations, the azimuthal component of the magnetic field is still dominant in the inner part of the nebula, which is filled mainly with freshly injected plasma. The hoop stress of this field is still capable of producing noticeable axial compression close to the termination shock and driving polar outflows, required to explain the Crab jet, and jets of other PWNe (Fig. 16). However, these are much more moderate than in 2D models. In the main body of the nebula, the highly organized coaxial configuration of magnetic field is largely destroyed by the kink instability (Fig. 17) therefore the global evolution of the PWN in 2D and 3D cases differs radically (Fig. 18). If the high latitude magnetization is large, the 2D models develop extremely strong polar jets, which burst through the supernova shell. In contrast, in the 3D models the z-pinch configuration is destroyed by the kink instability so that the polar outflows are less powerful and eventually lose collimation, as observed.

Important observational constraints are imposed by polarization measurements that reveal a high degree of polarization in the central part of the Crab nebula with the polarization vector parallel to the pulsar axis, which may be considered as direct evidence for the azimuthal field (Hester 2008). The simulations do show that in spite of the strong disruption of the azimuthal magnetic field, the polarization remains substantial, particularly in the inner part of the nebula. The polarization direction on the scale of the torus clearly indicates an azimuthal field because the photon magnetic vector appears to curve around the torus.

The 3D simulations also show a bright knot, which was discovered in the previous 2D simulations (Komissarov and Lyubarsky 2004) and identified with the inner knot observed in the Crab nebula (Hester et al. 1995). This emission comes from the immediate vicinity of the termination shock, it is highly Doppler beamed and originates in the high-speed part of the post-shock flow. A correlation was found between the knot position and the flux, such that brighter states correspond to a smaller offset between the knot and the location of the pulsar, which is in excellent agreement with the recent optical observations of Crab's knot (Moran et al. 2013). The simulated polarization degree and polarization angle in the knot also agree with observations.

The termination shock is found to be unsteady due to an intricate feedback mechanism between the shock and the nebula flow. The inhomogeneities, formed in the post-shock flow

Fig. 18 Dependence of the total pressure distribution, lg p_{tot}, on the imposed symmetry (Porth et al. 2014). The *upper panel* shows the pressure distribution in the xz plane of the 3D simulations and the *lower panel* in the corresponding 2D run. One sees that the strong axial compression observed in 2D simulations is an artifact of the imposed symmetry

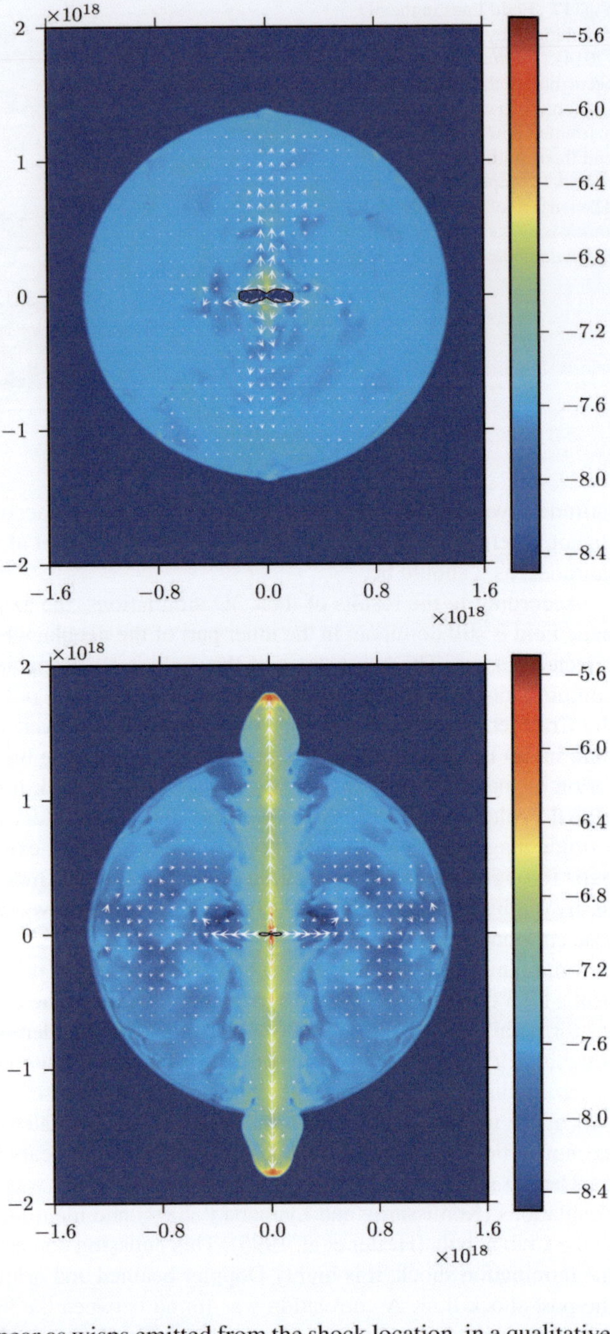

as a result of this variability, appear as wisps emitted from the shock location, in a qualitative agreement with the observations of the Crab nebula.

In the main body of the nebula, the kink instability not only destroys the regular magnetic field structure; the excited turbulence yields efficient magnetic dissipation. In simulations, this occurs at the grid scale via numerical resistivity. To become efficient, it requires creation of ever smaller scale structures in the magnetic field distribution. However, it is important that the processes which drive the development of such structures occurs on scales above the grid scale so that the dissipation rate is determined sufficiently accurately. The simulations with doubled resolution show the same dissipation rate, which suggests that the high degree of dissipation observed in these simulations is not far from being realistic. This also agrees with the observations of synchrotron and inverse-Compton emission of the Crab nebula, which show that the magnetic field is energetically sub-dominant to the population of relativistic electrons by a factor of ~ 30 (Hillas et al. 1998).

The above results suggest that the magnetic dissipation inside PWNe is a key factor of their dynamics. Combined with the magnetic dissipation in the striped zone of the pulsar wind, it allows to reconcile the observations of the Crab nebula with the expected high magnetization of such winds, thus finally resolving the long-lasting σ-problem.

2.3 The Structure of the Highly Magnetized Region in PWNe

Even after decay of the alternating fields, the pulsar wind remains highly magnetized at high latitudes. The energy and momentum flux in this domain of the wind is relatively small therefore the termination shock approaches significantly closer to the pulsar near the axis than at the equator. This region is not quite well resolved in simulations; they just show rapid disruption of the flow by the kink instability. On the other hand, the highly magnetized region is of special interest because the recently discovered strong, short gamma-ray flares from the Crab nebula (Abdo et al. 2011b; Tavani et al. 2011) are generally attributed to a rapid magnetic energy release via, e.g., reconnection (see Sect. 3), which assumes a magnetically dominated region within the nebula. Lyubarsky (2012) developed a simple model clarifying the structure of the high latitude flow.

In a highly magnetized flow, only weak shocks could arise therefore the termination shock at high latitudes is weak; and the postshock flow in this region remains radial and relativistic. Lyubarsky (2012) found a simple analytical solution for such a relativistic postshock flow. According to this solution, the flow initially expands and decelerates but eventually becomes to converge because the magnetic hoop stress is not counterbalanced either by the poloidal field or by the plasma pressure. In the converging flow, magnetic energy is converted into the plasma energy therefore the plasma accelerates and heats. If the flow remained axisymmetric, it would eventually be focused at the axis, the magnetic energy being transferred to the plasma. The focus occurs on the axis of the system at the distance from the pulsar $\sim \theta_0^2 a$, where θ_0 is the opening angle of the highly magnetized part of the wind, a the equatorial radius of the termination shock. This point may be identified with the base of the observed jet.

An important point is that in a converging flow, even small perturbations eventually destroy the regular structure. The reason is that if converging loops are initially shifted one with respect to another by a displacement much less than their radius, the distortion becomes strong when the radius of the loops approaches the initial displacement. One has to conclude that when the axisymmetric flow is focused into a point at the axis, the magnetic loops inevitably come apart close enough to the converging point giving rise to a specific turbulence of shrinking magnetic loops. Hence one can expect that the energy of the highly magnetized part of the pulsar wind is released in a small region close to the converging point; this gives rise to the observed jet. Relativistic turbulent motions in highly magnetized plasma imply $E \approx B$ so that in the energy release region, particles could be efficiently accelerated either via the second order Fermi mechanism or via the magnetic reconnection. Therefore the synchrotron gamma-ray emission in the hundreds MeV band, both persistent and flaring, could come from a small region at the base of the jet.

2.4 The Unsolved Problem: Origin of PWN Spectra

One sees that the overall morphology of PWNe is now well understood in the scope of MHD models. However, our ignorance of the physical processes giving rise to particle acceleration forces us to treat the injection particle spectra in PWNe as free parameters (Del Zanna et al. 2006; Volpi et al. 2008; Olmi et al. 2014), and this freedom in interpreting the data limits the level of scrutiny to which MHD models can be subjected. The radiation spectrum carries information about the particle acceleration processes. Most of the observed radiation (from the radio up to a few hundred MeV) is synchrotron emission, with only the peak in the very high energy gamma-ray band being attributed to the inverse Compton scattering of synchrotron photons off high-energy electrons and positrons. The observed strong polarization in the radio, optical and X-ray bands is a supporting evidence for the synchrotron origin of the nebula emission.

The synchrotron part of the spectrum may be described as a broken power-law. The generic observational feature of PWNe is a flat radio spectrum, $\mathcal{F}_\nu \propto \nu^{-\alpha}$, with α between 0 and 0.3, extending in some cases out to the infrared. At high frequencies, the spectrum softens, and in the X-ray band, $\alpha > 1$. Such an injection spectrum suggests an unusual acceleration process. The observed radio spectrum implies a power-law energy distribution of injected electrons, $N(E) \propto E^{-\kappa}$, with a shallow slope $1 < \kappa < 1.6$. Such an energy distribution is remarkable in that most of the particles are found at the low energy end of the distribution, whereas particles at the upper end of the distribution dominate the energy density of the plasma. Specifically in the Crab Nebula, the observed emission spectrum implies that the particles in the energy range from $E_{\rm min} < 100$ MeV to $E_{\rm break} \sim 1$ TeV are injected into the nebula with a spectral slope $\kappa = 1.6$, so most of the injected energy ($\sim 5 \cdot 10^{38}$ erg/s) is carried by TeV particles, whereas ~ 100 times more particles are found at low energies of less than 100 MeV. This means that the acceleration process somehow transfers most of the total energy of the system to a handful of energetic particles, leaving only a small fraction of the energy for the majority of the particles. This is not what one would normally expect from the conventional first-order Fermi acceleration process, in which the particle flow is randomized at the shock and only a fraction of the upstream kinetic energy is deposited in highly accelerated particles.

Lyubarsky (2003b) proposed that the flat energy distribution is formed in the course of the particle acceleration by driven reconnection of the alternating magnetic field at the pulsar wind termination shock. As a model for such a process, Lyubarsky and Liverts (2008) have performed PIC simulations of driven magnetic reconnection in a pair plasma. Two stripes of opposite magnetic polarity were compressed by means of an external force, which would imitate the effect of a shock. They found that driven magnetic reconnection can produce flat non-thermal particle spectrum, with $\kappa \approx 1$. Realistic 3D PIC simulations of the shock in a striped wind were performed by Sironi and Spitkovsky (2011a). They found that the spectrum of accelerated particles depends on the parameter $\xi = \lambda/\sigma r_l$, where λ is the stripe wavelength, $r_L = mc^2\Gamma/eB$ the Larmor radius corresponding to the upstream Lorentz factor of the flow, Γ, and the upstream magnetic field, B. It turns out that broad particle spectra with flat slopes ($1 < \kappa < 2$) could be formed by the shock-driven reconnection only if the above parameter is not less than a few hundreds. In the opposite case, the spectrum resembles a Maxwellian distribution.

Note that $\sigma\Gamma$ is in fact the Lorentz factor the particles would achieve if the whole spin-down energy were equally distributed between them. Therefore the parameter ξ is in fact equal, to within a numerical factor, to the ratio of the pulsar light cylinder radius to the Larmor radius acquired by the particles when a significant fraction of the Poynting flux is converted to the plasma energy. The latter is generally very large because the magnetic field at the pulsar wind termination shock is weak therefore ξ could hardly exceed unity. The condition for the formation of the non-thermal tail could be achieved only if the pulsar wind is overloaded by pairs; then the energy per particle may be small enough so that the Larmor radius remains small. At present, there is no reason, neither observational not theoretical, to believe in such an necessary extraordinary large pair production in pulsars. Therefore the problem of flat particle energy spectra in PWNe remains unsolved.

3 Implications of the Crab Flares

The rapid variability is now a well-established intrinsic property of the Crab Nebula in the GeV gamma-ray band[18] (Tavani et al. 2011; Abdo et al. 2011b). The flares were not predicted by the models and they generally do not fit in the framework of the classical theory of pulsar wind nebulae and particle acceleration. This unexpected phenomenon is also a challenge for observers, because the Nebula is routinely used as a standard candle for cross-calibrating X-ray and gamma-ray instruments. We explain why the flares are so challenging for the models of the Crab Nebula in Sect. 3.1, and we briefly review some of the current attempts to model the flares in Sect. 3.4 (see also the reviews by Arons 2012; Bühler and Blandford 2014).

3.1 The Puzzling Features of the Flares

With more than 6 years of data, we know that the mysterious engine at the origin of the gamma-ray flares turns on about once or twice a year for about a week.[19] Outside of these spectacular events, identified as the "flares", the > 100 MeV lightcurve remains apparently restless with continuous small variations of the flux (Buehler et al. 2012; Striani et al. 2013), as if the engine never really switches off. The duration of the flares indicates that the emitting region must be surprisingly small compared with the size of the Nebula. For a typically week-long episode, the length-scale of the accelerator is of order $ct_{\rm flare} \sim 10^{16}$ cm, i.e., much smaller than the size of the termination shock radius, which is of order 0.1 pc. The brightest events present intra-flare variability timescales as short as about 8 hrs (see Fig. 19, and Balbo et al. 2011; Buehler et al. 2012; Mayer et al. 2013), which put even more severe constraints on the size of the particle acceleration site. Consequently, a tiny fraction of the Crab Nebula is radiating ~ 10 times more flux than the entire quiescent Nebula in the GeV band. During the April 2011 flare, the gamma-ray flux peaked at about 1 % of the spin-down power of the pulsar ($L_{\rm sd} = 5 \times 10^{38}$ erg/s) (Buehler et al. 2012). This is putting strong constraints on the energetic budget required to power the flares.

The gamma-ray flare spectrum appears at the high-energy end of the quiescent synchrotron spectrum, and extends up to about 1 GeV. The flaring component is usually attributed to synchrotron radiation emitted by 10^{15} eV (or PeV) electron-positron pairs in a $\sim$ milli-Gauss magnetic field. Other radiative processes such as inverse Compton scattering or Bremsstrahlung are far too inefficient to cool down the particles over the duration of the flare.[20] It was already known that the Crab Nebula accelerates particles up to PeV

[18] The X-ray flux of the Crab Nebula is also variable but to a ~ 10 % level over a year-timescale (Wilson-Hodge et al. 2011). This will not be discussed here, because this phenomenon does not appear to be directly connected to the flares.

[19] See http://fermi.gsfc.nasa.gov/ssc/data/access/lat/msl_lc/source/Crab_Pulsar.

[20] The flux of the inverse-Compton component above > 100 GeV remains constant during the flares (H.E.S.S. Collaboration et al. 2014; Aliu et al. 2014).

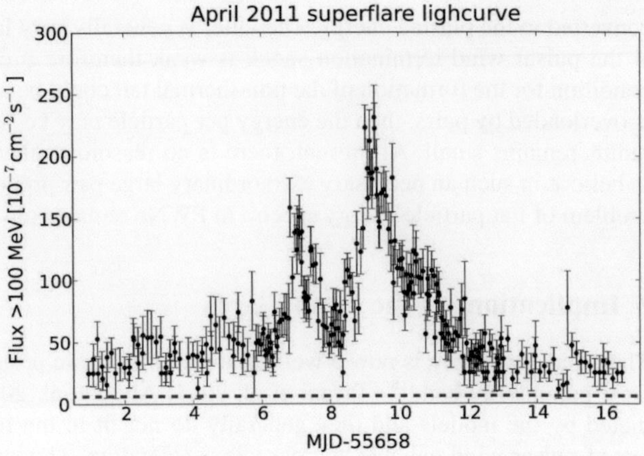

Fig. 19 Gamma-ray lightcurve above 100 MeV of the April 2011 super-flare, measured by the Fermi-LAT (Buehler et al. 2012). The *horizontal dashed blue line* shows the "quiescent" synchrotron flux > 100 MeV

energies (e.g., de Jager and Harding 1992). What is new, however, is the evidence that such particles are accelerated over such a short timescale (the rise and decay timescales of the flares range between 6 hours and few days). In fact, the gyration time of the PeV particles is of order the duration of the flare themselves. Hence, the particles must be accelerated over a sub-Larmor timescale, i.e., the acceleration process must be extremely efficient. Diffuse shock-acceleration is not adequate to explain the flares because it operates only over multiple gyrations of the particles moving back and forth through the shock front. In addition, the inferred flaring particle spectrum can be as hard as $dN/d\gamma \propto \gamma^{-1.6}$ (Buehler et al. 2012) which is inconsistent with the steep power-law (i.e., of index $\lesssim -2$) usually expected in diffuse shock-acceleration (e.g., Blandford and Eichler 1987; Sironi and Spitkovsky 2009a).

3.2 Multiwavelength Follow-up of Crab PWN Flares

During the gamma-ray flares the Crab was observed in radio, infrared, optical, X-ray and TeV energies, but no substantial variation in the flux emission at these wavelengths was measured. After the discovery of the gamma-ray flares, the Chandra X-ray observatory started to observe the Crab approximately every month. Five observations were carried out during the major gamma-ray flare of April 2011. The bright Anvil region (see Fig. 1) and several other regions (that are known to be active) exhibit time variability during the flaring activity time. Nevertheless, despite these hints in the X-ray data, there is no evidence for statistical significant variations associated with the flare (Weisskopf et al. 2013). The near-IR observations performed by Keck's NIRC2 revealed that the inner knot (knot-1) was slightly brighter when compared to previous observations. Indeed knot-1, which is the brightest feature from the Nebula in the near infrared energy band, was reported to show flux variation at this wavelength of the order of $\simeq 35$ %. However this variation is well within the range typically observed from this region. The radio observations performed with VLA did not reveal anything interesting. No other point source, a part from the pulsar, was found. Therefore, no "smoking gun" has been identified from the X-ray, near-IR and radio observations.

In terms of X-ray and optical counterparts of the Crab flares, we can describe two scenarios:

1. A simultaneous brightening of X-rays and optical locations associated with the gamma-ray flares. This scenario would favor "shock-driven" power-law models of particle acceleration;

2. A delayed response of the optical/X-ray emission, with a timescale that depends on the radiative cooling properties of the accelerated particle population. This scenario would favor an extremely efficient acceleration mechanism, likely to saturate the particle energy to a maximum value, and that could be modeled with a quasi-monoenergetic particle distribution;

The absence of a strongly enhanced X-ray and/or optical location in the Nebula in coincidence with the gamma-ray flares tends to exclude the first scenario. However in Weisskopf et al. (2013) the possibility to model the photon spectrum of the flare of April 2011 with a power-law with index $\Gamma \sim 1.3$ connecting gamma rays and X rays was investigated and could not be ruled out.

Because of the poor angular resolution of gamma-ray telescopes, localizing the emission site of the Crab flares is a big challenge. Several regions can be considered as candidate for the acceleration and emitting region. Among them, we can identify 3 particularly interesting possibilities:

1. instabilities in the Anvil at the South-East jet base. Variability in this region was detected by optical and X-ray observations both during the September 2010 and April 2011 gamma-ray flares (Tavani et al. 2011; Weisskopf et al. 2013). MHD simulations of the Crab South-East jet (Mignone et al. 2013) revealed substantial jet deviation and magnetic dissipation. The jet, fed by highly magnetized and relativistic plasma ($\sigma \sim 1$–10, Mignone et al. 2013), could be a region of magnetic reconnection. Kink instability in the jet could trigger magnetic reconnection and consequent particle acceleration.
2. tearing mode and reconnection on the termination shock "ring". This region is known to be highly variable. The three most variable spots during the April 2011 flare are located along the ring. The highly variable wisps are observed to originate from this region. Recent simulations (Porth et al. 2014) recently proposed that efficient magnetic reconnection could take place in the Nebula right after the termination shock;
3. variation in the observed emission from knot-1. Variability up to ~ 20–30 % is known. Its variability is interpreted as a variable Doppler factor δ, but no indication for optical variability has been observed so far. Simulations (Camus et al. 2009; Komissarov et al. 2007) show that at 100 MeV the inner knot is the brightest feature of the Nebula, and that the magnetic field can be up to 10 times larger than the average Nebular magnetic field.

3.3 Proposed Models of the Flares

It is usually expected that synchrotron spectrum observed in astrophysical sources cuts off far below 100 MeV (Blumenthal and Gould 1970; Guilbert et al. 1983). This limit is given by the balance of two antagonist forces acting on the particles: (i) the accelerating electric force, and (ii) the radiation-reaction-force opposite to the particle's motion due to the emission of synchrotron photons. Hence, there is a maximum energy limited by synchrotron losses rather than by the size of the accelerator (e.g., Aharonian et al. 2002; Medvedev 2003). Then, one can show that the maximum (critical) synchrotron photon energy should be

$$\epsilon_{\rm max}^{\rm sync} = \frac{9 m_{\rm e} c^2}{4\alpha_{\rm F}} \left(\frac{E}{B_\perp}\right) \approx 160 \times \left(\frac{E}{B_\perp}\right) \text{MeV}, \tag{1}$$

where $m_{\rm e}$ is the rest mass of the electron, $\alpha_{\rm F} \approx 1/137$ is the fine structure constant, E the electric field, and $B_\perp$ the magnetic field perpendicular to the particle's direction of motion. A particle accelerated above the radiation reaction limit would radiate away most of its energy within a sub-Larmor cycle. In most cases, ideal MHD applies, $E \ll B$,

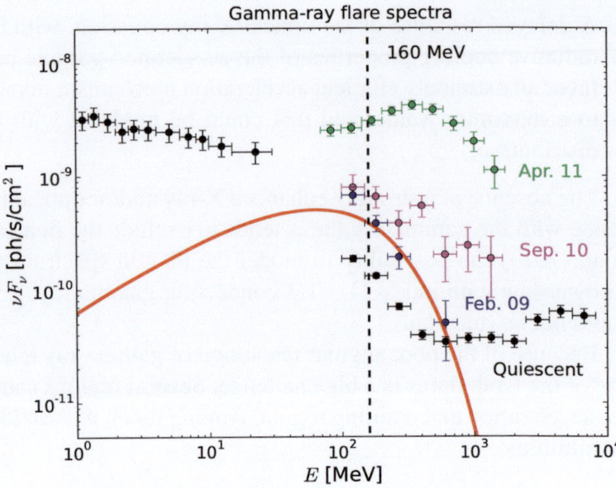

Fig. 20 Measured gamma-ray spectra of the Crab Nebula in the 1 MeV–10 GeV band. *Black data-points* show the "quiescent" spectra of the Nebula. The *green*, *magenta*, and *blue points* are respectively the > 100 MeV spectra during the April 2011, September 2010 and February 2009 flares measured by the *Fermi*-LAT (Bühler and Blandford 2014). The *red solid line* is the spectra obtained with 3D PIC simulations with radiation reaction of a reconnecting current sheet in the Crab Nebula (Cerutti et al. 2014). The model assumes that the Nebula is 2 kpc away

so the synchrotron spectrum should cut $\ll 100$ MeV. This rule of thumb generally applies well to astrophysical sources, and in particular to the quiescent Crab Nebula where the synchrotron spectrum turns over at about the 100 MeV limit (de Jager et al. 1996; Abdo et al. 2010a). However, the flaring emission is systematically extending significantly above 160 MeV, up to about 1 GeV (see Fig. 20). Unless the emitting region moves at highly relativistic speeds ($\gtrsim 0.9\,c$), it implies that $E \gtrsim 5B_\perp$, suggesting that a non-ideal MHD process may be at work. This is, once again, difficult to explain with classical models of particle acceleration.

All of the estimates derived above from observations are quite conservative, in a sense that we ignored the effect of beaming (geometrical or relativistic), spatial and/or temporal inhomogeneities. Every model proposed so far is taking advantage of one or more of these effects to alleviate the tight constraints imposed by the flares. For instance, one feature commonly invoked in models is a strong inhomogeneity of the flaring region, in particular in the magnetic field structure. Bykov et al. (2012) proposed that the flares occur around the equatorial belt where the annihilation of the striped pulsar wind at the shock accelerates particles (Lyubarsky 2003b; Pétri and Lyubarsky 2007; Sironi and Spitkovsky 2011a) and, crucially for this model, generates magnetic turbulence. They find that a concentration of fluctuating magnetic field can generate an intermittent, strongly polarized gamma-ray signal that is most pronounced at the high-energy end of the synchrotron spectrum. In this model, synchrotron photons above 100 MeV can be emitted if the magnetic field varies over a timescale shorter than the synchrotron cooling time of the particles which is determined only on the RMS value of the field. The observed gamma-ray variability would then be given by the statistical properties of the magnetic fluctuations (lifetime, amplitude).

The model by Teraki and Takahara (2013) also relies on a highly inhomogeneous turbulent flow, but in which the coherence length of the magnetic field, λ_B, is extremely short compared with the cooling length of the particles and even compared with the formation length of the synchrotron photons, $\lambda_{\rm sync} = m_{\rm e}c^2/eB \gg \lambda_B$.[21] In this regime, the particles emit the so-called "jitter" radiation rather than the classical synchrotron radiation (see, e.g., Medvedev 2000). The cooling rate of the particles is identical to syn-

[21] This might be challenging to achieve in relativistic shocks (Sironi and Spitkovsky 2009b).

chrotron but the emitted spectra can differ significantly. In particular, the critical photon energy increases by a factor $\epsilon_{\rm jitt}/\epsilon_{\rm sync} \sim \lambda_{\rm sync}/\lambda_B \gg 1$ in the jitter regime, and hence > 100 MeV gamma rays could be emitted even by particles below the radiation reaction limit. In addition, jitter radiation can produce a harder spectrum than synchrotron, specifically, $F_\nu \propto \nu$ instead of $F_\nu \propto \nu^{1/3}$ for a mono-energetic population of particles. However, we note that jitter radiation is not needed to explain the flare spectra. So far, observations are fully consistent with synchrotron radiation (Buehler et al. 2012; Weisskopf et al. 2013), but this could be tested in the future.

A more natural way to explain the Crab flares is to invoke a relativistic bulk motion of the flaring region with a modest Lorentz factor $\Gamma \gtrsim 2$. Indeed, the relativistic motion of the source can boost $\lesssim 100$ MeV synchrotron photons emitted in the co-moving frame by a $\sim \Gamma$ factor above the radiation reaction limit (Yuan et al. 2011; Bednarek and Idec 2011; Komissarov and Lyutikov 2011; Lyutikov et al. 2012; Clausen-Brown and Lyutikov 2012). A Doppler boost would also relax the tight constraints on the size of the region and the duration of the flare in the co-moving frame, and beam the emission in the frame of the observer which would also help at reducing the energetic constraints.[22] Lastly, the Doppler beaming would explain the observed correlation between the gamma-ray flux and the cut-off energy in the spectra during the April 2011 flare (Buehler et al. 2012). Although this simple solution solves many problems at once, there is still no definite evidence that such a relativistic flow exists in the Crab Nebula. Observations show only mildly relativistic flows with proper motion of order half the speed of light (e.g., Hester 2008). However, in principle, highly relativistic flows could emerge in the polar regions of the Nebula because the relativistic shock is oblique and magnetized (weak shock) at high latitudes (e.g., Lyubarsky 2012; Komissarov 2013). In particular, Komissarov and Lyutikov (2011) argued that the flares may originate from the emission from the oblique shock Doppler-boosted towards the observer, that they associate with the well-known bright compact structure near the pulsar (the so-called "inner knot"). Unfortunately, the brightness of the knot does not show any variations correlated with the gamma-ray flares (see Tavani et al. 2011; Lobanov et al. 2011; Weisskopf et al. 2013 and references therein) contrary to what the model predicts.

Alternatively, Uzdensky et al. (2011) proposed that magnetic reconnection could accelerate particles well above the radiation reaction limit, and hence result in the emission of > 100 MeV synchrotron radiation. Indeed, as pointed out by Kirk (2004), inside a reconnection layer the magnetic field is small and even vanishes at its center while the reconnection electric field is maximum, i.e., we are in the situation where $E \gg B$. Thus, in principle, a particle trapped deep inside the reconnection layer could be linearly accelerated by the electric field to arbitrary high energies with little synchrotron losses. The maximum energy of the particle would be limited only by the length of the reconnection layer L, i.e., $\mathcal{E}_{\rm max} \sim eEL = e\beta_{\rm rec} B_0 L$ where $\beta_{\rm rec} = E/B_0$ is the dimensionless reconnection rate, E is the reconnection electric field and B_0 is the reconnecting magnetic field. Using a test-particle approach with prescribed static fields, Contopoulos (2007) and Cerutti et al. (2012a) showed that the high-energy particles are naturally trapped and confined deeply inside the layer, where they follow the relativistic analog of Speiser orbits (Speiser 1965; Uzdensky et al. 2011). This scenario was successfully tested using state-of-the-art particle-in-cell (PIC) simulations of 2D and 3D relativistic reconnection with guide field and, more importantly, with the radiation reaction force (Cerutti et al. 2013, 2014). Furthermore, these studies revealed that a natural outcome of relativistic reconnection is the strong beaming and bunching of the energetic particles (see also Cerutti et al. 2012b). The combination of both

[22] This solution is often proposed to account for the ultra-rapid gamma-ray flares in blazars.

effects results in several bright ultra-rapid synchrotron flares above 100 MeV consistent with the observed intra-flare variability ($\lesssim 6$ hrs) if, by chance, the beam crosses the line of sight of a distant observer. Thanks to beaming, this model can also explain the overall energetics of the flares (Fig. 20), the flux/cut-off energy correlation and, at least qualitatively, the hard particle spectrum.[23]

The reconnection scenario works best in a highly magnetized flow (i.e., $\sigma \gg 1$) which may be hard to find in the nebula, except in the polar regions and in the jets. As predicted by Begelman (1998) and as recently shown by Mizuno et al. (2011); Porth et al. (2014); Mignone et al. (2013), the polar regions and the jets are subject to kink instabilities which results in important magnetic dissipation (see Sect. 2.2), and may ultimately power the Crab flares (Cerutti et al. 2012a; Lyubarsky 2012; Komissarov 2013; Mignone et al. 2013).

3.4 Comparison with GeV Flares in the PSR B1259–63/LS 2883 Binary

The gamma-ray binary PSR B1259−63 that contains a 48-ms pulsar in a 3.4-yrs eccentric orbit around an O star (see Sect. 1.4), is a well-known TeV gamma-ray emitter (Aharonian et al. 2005; H.E.S.S. Collaboration et al. 2013). The very-high energy radiation is often interpreted as inverse Compton scattering of the UV stellar photons on relativistic pairs accelerated near the shock front between the pulsar wind and the stellar wind (e.g., Tavani and Arons 1997; Kirk et al. 1999; Khangulyan et al. 2007). The model predicts a maximum of GeV gamma-ray emission close to periastron which was indeed observed for the first time by the Fermi-LAT during the 2010 periastron passage (Tam et al. 2011; Abdo et al. 2011a). A few weeks after the peak of the periastron emission faded away, and against all expectations, a bright flare appeared in the Fermi data. The flare is about 10-times brighter than the predicted emission at periastron, which represents a gamma-ray luminosity comparable to the pulsar spin-down power. A radiative efficiency close to 100 % is in principle achievable with inverse Compton scattering, but the density of stellar photons is far too low to explain the flux at these phases, unless there are extra sources of radiation close to the pulsar (see, e.g., Khangulyan et al. 2012; Dubus and Cerutti 2013), or significant Doppler boosting of the emission towards the observer (Dubus et al. 2010; Kong et al. 2012). The flare peaks at 300 MeV and is seen only in the GeV band, which suggests that the particle energy distribution must be very narrow. These properties remind us of the Crab-flare events (see Sect. 3.1). Although both flares share similar properties, they have also important differences. The week-long flares in the Crab reach at most 1 % of the Crab pulsar spin-down power, whereas the highest day-average flux reaches nearly 100 % of the pulsar spin-down flux during flaring period associated with the second disk passage. Unlike Crab PWN the flares in B1259–63 appear to be periodic. i.e. they occurred at similar binary phases (close to the second disk passage) during the past two binary cycles (Tam et al. 2015; Caliandro et al. 2015).

While it may be that the underlying nature of flares in both cases is the synchrotron radiation associated with reconnection in the magnetized relativistic plasma, the details should differ. In the Crab PWN, the reconnection can be driven by the growth of instabilities or other random process, while in B1259–63, it can be driven by the magnetic field distortion and compression caused by the pulsar passage through the excretion disk. It is possible that the reconnection happens in the tail of the PWN after the pulsar passage through the disk which could explain the delay between the GeV flare and the peak of X-ray flux (Tam et al. 2015;

[23]Recent studies show that reconnection produces hard particle spectra $dN/d\gamma \propto \gamma^{-1}$, $\gamma^{-1.5}$ for $\sigma \gg 1$ (Sironi and Spitkovsky 2014; Guo et al. 2014; Werner et al. 2014).

Caliandro et al. 2015). An alternative scenario, considered by Khangulyan et al. (2012), where the flare is due to the IC radiation associated with the increase in volume occupied by the unshocked pulsar wind when the excretion disk is strongly perturbed, also remains a possibility. One may be able to differentiate between the two scenarios once the γ-ray variability timescales and the state of the disk during the pulsar passage are better probed by the observations.

Acknowledgements We thank George Pavlov for his valuable comments and discussions. We are grateful to Blagoy Rangelov for creating the merged F550M Crab image from the archival *HST* data and help with the manuscript editing. We acknowledge support from NASA grants AR8-9009X, GO3-14084X, GO3-14057C, G03-14082A, NNX09AC81G, NNX09AC84G, HST-GO-13043.09, and G02-13085C. B.C. acknowledges support from the Lyman Spitzer Jr. Fellowship awarded by the Department of Astrophysical Sciences at Princeton University, and the Max-Planck/Princeton Center for Plasma Physics. Y.L. acknowledges support from Israeli Science Foundation under the grant 719/14.

References

A.A. Abdo, B.T. Allen, T. Aune, D. Berley, C. Chen, G.E. Christopher, T. DeYoung, B.L. Dingus, R.W. Ellsworth, M.M. Gonzalez, J.A. Goodman, E. Hays, C.M. Hoffman, P.H. Hüntemeyer, B.E. Kolterman, J.T. Linnemann, J.E. McEnery, T. Morgan, A.I. Mincer, P. Nemethy, J. Pretz, J.M. Ryan, P.M. Saz Parkinson, A. Shoup, G. Sinnis, A.J. Smith, V. Vasileiou, G.P. Walker, D.A. Williams, G.B. Yodh, Milagro observations of multi-TeV emission from Galactic sources in the Fermi Bright Source List. Astrophys. J. Lett. **700**, 127–131 (2009). doi:10.1088/0004-637X/700/2/L127

A.A. Abdo, M. Ackermann, M. Ajello, W.B. Atwood, M. Axelsson, L. Baldini, J. Ballet, G. Barbiellini, M.G. Baring, D. Bastieri, K. Bechtol, R. Bellazzini, B. Berenji, R.D. Blandford, E.D. Bloom, E. Bonamente, A.W. Borgland, J. Bregeon, A. Brez, M. Brigida, P. Bruel, T.H. Burnett, G.A. Caliandro, R.A. Cameron, F. Camilo, P.A. Caraveo, J.M. Casandjian, C. Cecchi, Ö. Çelik, A. Chekhtman, C.C. Cheung, J. Chiang, S. Ciprini, R. Claus, I. Cognard, J. Cohen-Tanugi, L.R. Cominsky, J. Conrad, C.D. Dermer, A. de Angelis, A. de Luca, F. de Palma, S.W. Digel, E.d.C.e. Silva, P.S. Drell, R. Dubois, D. Dumora, C. Espinoza, C. Farnier, C. Favuzzi, S.J. Fegan, E.C. Ferrara, W.B. Focke, M. Frailis, P.C.C. Freire, Y. Fukazawa, S. Funk, P. Fusco, F. Gargano, D. Gasparrini, N. Gehrels, S. Germani, G. Giavitto, B. Giebels, N. Giglietto, F. Giordano, T. Glanzman, G. Godfrey, I.A. Grenier, M.-H. Grondin, J.E. Grove, L. Guillemot, S. Guiriec, Y. Hanabata, A.K. Harding, M. Hayashida, E. Hays, R.E. Hughes, G. Jóhannesson, A.S. Johnson, R.P. Johnson, T.J. Johnson, W.N. Johnson, S. Johnston, T. Kamae, H. Katagiri, J. Kataoka, N. Kawai, M. Kerr, J. Knödlseder, M.L. Kocian, M. Kramer, F. Kuehn, M. Kuss, J. Lande, L. Latronico, S.-H. Lee, M. Lemoine-Goumard, F. Longo, F. Loparco, B. Lott, M.N. Lovellette, P. Lubrano, A.G. Lyne, A. Makeev, M. Marelli, M.N. Mazziotta, J.E. McEnery, C. Meurer, P.F. Michelson, W. Mitthumsiri, T. Mizuno, A.A. Moiseev, C. Monte, M.E. Monzani, E. Moretti, A. Morselli, I.V. Moskalenko, S. Murgia, T. Nakamori, P.L. Nolan, J.P. Norris, A. Noutsos, E. Nuss, T. Ohsugi, N. Omodei, E. Orlando, J.F. Ormes, M. Ozaki, D. Paneque, J.H. Panetta, D. Parent, V. Pelassa, M. Pepe, M. Pesce-Rollins, M. Pierbattista, F. Piron, T.A. Porter, S. Rainò, R. Rando, P.S. Ray, M. Razzano, A. Reimer, O. Reimer, T. Reposeur, S. Ritz, L.S. Rochester, A.Y. Rodriguez, R.W. Romani, M. Roth, F. Ryde, H.F.-W. Sadrozinski, D. Sanchez, A. Sander, P.M. Saz Parkinson, J.D. Scargle, C. Sgrò, E.J. Siskind, D.A. Smith, P.D. Smith, G. Spandre, P. Spinelli, B.W. Stappers, M.S. Strickman, D.J. Suson, H. Tajima, H. Takahashi, T. Tanaka, J.B. Thayer, J.G. Thayer, G. Theureau, D.J. Thompson, S.E. Thorsett, L. Tibaldo, D.F. Torres, G. Tosti, A. Tramacere, Y. Uchiyama, T.L. Usher, A. Van Etten, V. Vasileiou, N. Vilchez, V. Vitale, A.P. Waite, E. Wallace, P. Wang, K. Watters, P. Weltevrede, B.L. Winer, K.S. Wood, T. Ylinen, M. Ziegler, Fermi Large Area Telescope observations of the Crab pulsar and nebula. Astrophys. J. **708**, 1254–1267 (2010a). doi:10.1088/0004-637X/708/2/1254

A.A. Abdo, M. Ackermann, M. Ajello, A. Allafort, L. Baldini, J. Ballet, G. Barbiellini, D. Bastieri, K. Bechtol, R. Bellazzini, B. Berenji, R.D. Blandford, E.D. Bloom, E. Bonamente, A.W. Borgland, A. Bouvier, J. Bregeon, A. Brez, M. Brigida, P. Bruel, T.H. Burnett, S. Buson, G.A. Caliandro, R.A. Cameron, P.A. Caraveo, S. Carrigan, J.M. Casandjian, C. Cecchi, Ö. Çelik, A. Chekhtman, C.C. Chung, J. Chiang, S. Ciprini, R. Claus, J. Cohen-Tanugi, J. Conrad, A. de Angelis, F. de Palma, M. Dormody, E.d.C.e. Silva, P.S. Drell, R. Dubois, D. Dumora, C. Farnier, C. Favuzzi, S.J. Fegan, W.B. Focke, P. Fortin, M. Frailis, Y. Fukazawa, S. Funk, P. Fusco, F. Gargano, N. Gehrels, S. Germani, G. Giavitto, N. Giglietto, F. Giordano, T. Glanzman, G. Godfrey, I.A. Grenier, M.-H. Grondin, J.E. Grove, L. Guillemot, S. Guiriec, A.K.

Harding, M. Hayashida, E. Hays, D. Horan, R.E. Hughes, M.S. Jackson, G. Jóhannesson, A.S. Johnson, T.J. Johnson, W.N. Johnson, S. Johnston, T. Kamae, H. Katagiri, J. Kataoka, N. Kawai, M. Kerr, J. Knödlseder, M. Kuss, J. Lande, L. Latronico, S.-H. Lee, M. Lemoine-Goumard, M. Llena Garde, F. Longo, F. Loparco, B. Lott, M.N. Lovellette, P. Lubrano, A. Makeev, M. Marelli, M.N. Mazziotta, J.E. McEnery, C. Meurer, P.F. Michelson, W. Mitthumsiri, T. Mizuno, A.A. Moiseev, C. Monte, M.E. Monzani, A. Morselli, I.V. Moskalenko, S. Murgia, T. Nakamori, P.L. Nolan, J.P. Norris, A. Noutsos, E. Nuss, T. Ohsugi, N. Omodei, E. Orlando, J.F. Ormes, M. Ozaki, D. Paneque, J.H. Panetta, D. Parent, V. Pelassa, M. Pepe, M. Pesce-Rollins, M. Pierbattista, F. Piron, T.A. Porter, S. Rainò, R. Rando, P.S. Ray, N. Rea, A. Reimer, O. Reimer, T. Reposeur, S. Ritz, A.Y. Rodriguez, R.W. Romani, M. Roth, F. Ryde, H.F.-W. Sadrozinski, D. Sanchez, A. Sander, P.M. Saz Parkinson, J.D. Scargle, C. Sgrò, E.J. Siskind, D.A. Smith, P.D. Smith, G. Spandre, P. Spinelli, M.S. Strickman, D.J. Suson, H. Tajima, H. Takahashi, T. Takahashi, T. Tanaka, J.B. Thayer, J.G. Thayer, D.J. Thompson, L. Tibaldo, D.F. Torres, G. Tosti, A. Tramacere, Y. Uchiyama, T.L. Usher, A. Van Etten, V. Vasileiou, C. Venter, N. Vilchez, V. Vitale, A.P. Waite, P. Wang, P. Weltevrede, B.L. Winer, K.S. Wood, T. Ylinen, M. Ziegler, Fermi Large Area Telescope observations of the Vela-X pulsar wind nebula. Astrophys. J. **713**, 146–153 (2010b). doi:10.1088/0004-637X/713/1/146

A.A. Abdo, M. Ackermann, M. Ajello, A. Allafort, J. Ballet, G. Barbiellini, D. Bastieri, K. Bechtol, R. Bellazzini, B. Berenji, R.D. Blandford, E. Bonamente, A.W. Borgland, J. Bregeon, M. Brigida, P. Bruel, R. Buehler, S. Buson, G.A. Caliandro, R.A. Cameron, F. Camilo, P.A. Caraveo, C. Cecchi, E. Charles, S. Chaty, A. Chekhtman, M. Chernyakova, C.C. Cheung, J. Chiang, S. Ciprini, R. Claus, J. Cohen-Tanugi, L.R. Cominsky, S. Corbel, S. Cutini, F. D'Ammando, A. de Angelis, P.R. den Hartog, F. de Palma, C.D. Dermer, S.W. Digel, E.d.C.e. Silva, M. Dormody, P.S. Drell, A. Drlica-Wagner, R. Dubois, G. Dubus, D. Dumora, T. Enoto, C.M. Espinoza, C. Favuzzi, S.J. Fegan, E.C. Ferrara, W.B. Focke, P. Fortin, Y. Fukazawa, S. Funk, P. Fusco, F. Gargano, D. Gasparrini, N. Gehrels, S. Germani, N. Giglietto, P. Giommi, F. Giordano, M. Giroletti, T. Glanzman, G. Godfrey, I.A. Grenier, M.-H. Grondin, J.E. Grove, E. Grundstrom, S. Guiriec, C. Gwon, D. Hadasch, A.K. Harding, M. Hayashida, E. Hays, G. Jóhannesson, A.S. Johnson, T.J. Johnson, S. Johnston, T. Kamae, H. Katagiri, J. Kataoka, M. Keith, M. Kerr, J. Knödlseder, M. Kramer, M. Kuss, J. Lande, S.-H. Lee, M. Lemoine-Goumard, F. Longo, F. Loparco, M.N. Lovellette, P. Lubrano, R.N. Manchester, M. Marelli, M.N. Mazziotta, P.F. Michelson, W. Mitthumsiri, T. Mizuno, A.A. Moiseev, C. Monte, M.E. Monzani, A. Morselli, I.V. Moskalenko, S. Murgia, T. Nakamori, M. Naumann-Godo, A. Neronov, P.L. Nolan, J.P. Norris, A. Noutsos, E. Nuss, T. Ohsugi, A. Okumura, N. Omodei, E. Orlando, D. Paneque, D. Parent, M. Pesce-Rollins, M. Pierbattista, F. Piron, T.A. Porter, A. Possenti, S. Rainò, R. Rando, P.S. Ray, M. Razzano, S. Razzaque, A. Reimer, O. Reimer, T. Reposeur, S. Ritz, H.F.-W. Sadrozinski, J.D. Scargle, C. Sgrò, R. Shannon, E.J. Siskind, P.D. Smith, G. Spandre, P. Spinelli, M.S. Strickman, D.J. Suson, H. Takahashi, T. Tanaka, J.G. Thayer, J.B. Thayer, D.J. Thompson, S.E. Thorsett, L. Tibaldo, O. Tibolla, D.F. Torres, G. Tosti, E. Troja, Y. Uchiyama, T.L. Usher, J. Vandenbroucke, V. Vasileiou, G. Vianello, V. Vitale, A.P. Waite, P. Wang, B.L. Winer, M.T. Wolff, D.L. Wood, K.S. Wood, Z. Yang, M. Ziegler, S. Zimmer, Discovery of high-energy gamma-ray emission from the binary system PSR B1259-63/LS 2883 around periastron with Fermi. Astrophys. J. Lett. **736**, 11 (2011a). doi:10.1088/2041-8205/736/1/L11

A.A. Abdo, M. Ackermann, M. Ajello, A. Allafort, L. Baldini, J. Ballet, G. Barbiellini, D. Bastieri, K. Bechtol, R. Bellazzini, B. Berenji, R.D. Blandford, E.D. Bloom, E. Bonamente, A.W. Borgland, A. Bouvier, T.J. Brandt, J. Bregeon, A. Brez, M. Brigida, P. Bruel, R. Buehler, S. Buson, G.A. Caliandro, R.A. Cameron, A. Cannon, P.A. Caraveo, J.M. Casandjian, Ö. Çelik, E. Charles, A. Chekhtman, C.C. Cheung, J. Chiang, S. Ciprini, R. Claus, J. Cohen-Tanugi, L. Costamante, S. Cutini, F. D'Ammando, C.D. Dermer, A. de Angelis, A. de Luca, F. de Palma, S.W. Digel, E. do Couto e Silva, P.S. Drell, A. Drlica-Wagner, R. Dubois, D. Dumora, C. Favuzzi, S.J. Fegan, E.C. Ferrara, W.B. Focke, P. Fortin, M. Frailis, Y. Fukazawa, S. Funk, P. Fusco, F. Gargano, D. Gasparrini, N. Gehrels, S. Germani, N. Giglietto, F. Giordano, M. Giroletti, T. Glanzman, G. Godfrey, I.A. Grenier, M.-H. Grondin, J.E. Grove, S. Guiriec, D. Hadasch, Y. Hanabata, A.K. Harding, K. Hayashi, M. Hayashida, E. Hays, D. Horan, R. Itoh, G. Jóhannesson, A.S. Johnson, T.J. Johnson, D. Khangulyan, T. Kamae, H. Katagiri, J. Kataoka, M. Kerr, J. Knödlseder, M. Kuss, J. Lande, L. Latronico, S.-H. Lee, M. Lemoine-Goumard, F. Longo, F. Loparco, P. Lubrano, G.M. Madejski, A. Makeev, M. Marelli, M.N. Mazziotta, J.E. McEnery, P.F. Michelson, W. Mitthumsiri, T. Mizuno, A.A. Moiseev, C. Monte, M.E. Monzani, A. Morselli, I.V. Moskalenko, S. Murgia, T. Nakamori, M. Naumann-Godo, P.L. Nolan, J.P. Norris, E. Nuss, T. Ohsugi, A. Okumura, N. Omodei, J.F. Ormes, M. Ozaki, D. Paneque, D. Parent, V. Pelassa, M. Pepe, M. Pesce-Rollins, M. Pierbattista, F. Piron, T.A. Porter, S. Rainò, R. Rando, P.S. Ray, M. Razzano, A. Reimer, O. Reimer, T. Reposeur, S. Ritz, R.W. Romani, H.F.-W. Sadrozinski, D. Sanchez, P.M.S. Parkinson, J.D. Scargle, T.L. Schalk, C. Sgrò, E.J. Siskind, P.D. Smith, G. Spandre, P. Spinelli, M.S. Strickman, D.J. Suson, H. Takahashi, T. Takahashi, T. Tanaka, J.B. Thayer, D.J. Thompson, L. Tibaldo, D.F. Torres, G. Tosti, A. Tramacere, E. Troja, Y. Uchiyama, J. Vandenbroucke, V. Vasileiou, G. Vianello, V. Vitale, P. Wang,

K.S. Wood, Z. Yang, M. Ziegler, Gamma-ray flares from the Crab Nebula. Science **331**, 739 (2011b). doi:10.1126/science.1199705

A.A. Abdo, M. Ajello, A. Allafort, L. Baldini, J. Ballet, G. Barbiellini, M.G. Baring, D. Bastieri, A. Belfiore, R. Bellazzini, et al., The second Fermi Large Area Telescope catalog of gamma-ray pulsars. Astrophys. J. Suppl. Ser. **208**, 17 (2013). doi:10.1088/0067-0049/208/2/17

A. Abramowski, F. Acero, F. Aharonian, A.G. Akhperjanian, G. Anton, S. Balenderan, A. Balzer, A. Barnacka, Y. Becherini, J. Becker Tjus, K. Bernlöhr, E. Birsin, J. Biteau, A. Bochow, C. Boisson, J. Bolmont, P. Bordas, J. Brucker, F. Brun, P. Brun, T. Bulik, S. Carrigan, S. Casanova, M. Cerruti, P.M. Chadwick, A. Charbonnier, R.C.G. Chaves, A. Cheesebrough, G. Cologna, J. Conrad, C. Couturier, M. Dalton, M.K. Daniel, I.D. Davids, B. Degrange, C. Deil, P. deWilt, H.J. Dickinson, A. Djannati-Ataï, W. Domainko, L.O.. Drury, F. Dubois, G. Dubus, K. Dutson, J. Dyks, M. Dyrda, K. Egberts, P. Eger, P. Espigat, L. Fallon, C. Farnier, S. Fegan, F. Feinstein, M.V. Fernandes, D. Fernandez, A. Fiasson, G. Fontaine, A. Förster, M. Füßling, M. Gajdus, Y.A. Gallant, T. Garrigoux, H. Gast, B. Giebels, J.F. Glicenstein, B. Glück, D. Göring, M.-H. Grondin, S. Häffner, J.D. Hague, J. Hahn, D. Hampf, J. Harris, S. Heinz, G. Heinzelmann, G. Henri, G. Hermann, A. Hillert, J.A. Hinton, W. Hofmann, P. Hofverberg, M. Holler, D. Horns, A. Jacholkowska, C. Jahn, M. Jamrozy, I. Jung, M.A. Kastendieck, K. Katarzyński, U. Katz, S. Kaufmann, B. Khélifi, D. Klochkov, W. Kluźniak, T. Kneiske, N. Komin, K. Kosack, R. Kossakowski, F. Krayzel, P.P. Krüger, H. Laffon, G. Lamanna, J.-P. Lenain, D. Lennarz, T. Lohse, A. Lopatin, C.-C. Lu, V. Marandon, A. Marcowith, J. Masbou, G. Maurin, N. Maxted, M. Mayer, T.J.L. McComb, M.C. Medina, J. Méhault, U. Menzler, R. Moderski, M. Mohamed, E. Moulin, C.L. Naumann, M. Naumann-Godo, M. de Naurois, D. Nedbal, N. Nguyen, J. Niemiec, S.J. Nolan, S. Ohm, E. de Oña Wilhelmi, B. Opitz, M. Ostrowski, I. Oya, M. Panter, D. Parsons, M. Paz Arribas, N.W. Pekeur, G. Pelletier, J. Perez, P.-O. Petrucci, B. Peyaud, S. Pita, G. Pühlhofer, M. Punch, A. Quirrenbach, M. Raue, A. Reimer, O. Reimer, M. Renaud, R. de los Reyes, F. Rieger, J. Ripken, L. Rob, S. Rosier-Lees, G. Rowell, B. Rudak, C.B. Rulten, V. Sahakian, D.A. Sanchez, A. Santangelo, R. Schlickeiser, A. Schulz, U. Schwanke, S. Schwarzburg, S. Schwemmer, F. Sheidaei, J.L. Skilton, H. Sol, G. Spengler, Ł. Stawarz, R. Steenkamp, C. Stegmann, F. Stinzing, K. Stycz, I. Sushch, A. Szostek, J.-P. Tavernet, R. Terrier, M. Tluczykont, C. Trichard, K. Valerius, C. van Eldik, G. Vasileiadis, C. Venter, A. Viana, P. Vincent, H.J. Völk, F. Volpe, S. Vorobiov, M. Vorster, S.J. Wagner, M. Ward, R. White, A. Wierzcholska, D. Wouters, M. Zacharias, A. Zajczyk, A.A. Zdziarski, A. Zech, H.-S. Zechlin, Probing the extent of the non-thermal emission from the Vela X region at TeV energies with H.E.S.S. Astron. Astrophys. **548**, 38 (2012). doi:10.1051/0004-6361/201219919

F. Acero, M. Ackermann, M. Ajello, A. Allafort, L. Baldini, J. Ballet, G. Barbiellini, D. Bastieri, K. Bechtol, R. Bellazzini, R.D. Blandford, E.D. Bloom, E. Bonamente, E. Bottacini, T.J. Brandt, J. Bregeon, M. Brigida, P. Bruel, R. Buehler, S. Buson, G.A. Caliandro, R.A. Cameron, P.A. Caraveo, C. Cecchi, E. Charles, R.C.G. Chaves, A. Chekhtman, J. Chiang, G. Chiaro, S. Ciprini, R. Claus, J. Cohen-Tanugi, J. Conrad, S. Cutini, M. Dalton, F. D'Ammando, F. de Palma, C.D. Dermer, L. Di Venere, E.d.C.e. Silva, P.S. Drell, A. Drlica-Wagner, L. Falletti, C. Favuzzi, S.J. Fegan, E.C. Ferrara, W.B. Focke, A. Franckowiak, Y. Fukazawa, S. Funk, P. Fusco, F. Gargano, D. Gasparrini, N. Giglietto, F. Giordano, M. Giroletti, T. Glanzman, G. Godfrey, T. Grégoire, I.A. Grenier, M.-H. Grondin, J.E. Grove, S. Guiriec, D. Hadasch, Y. Hanabata, A.K. Harding, M. Hayashida, K. Hayashi, E. Hays, J. Hewitt, A.B. Hill, D. Horan, X. Hou, R.E. Hughes, Y. Inoue, M.S. Jackson, T. Jogler, G. Jóhannesson, A.S. Johnson, T. Kamae, T. Kawano, M. Kerr, J. Knödlseder, M. Kuss, J. Lande, S. Larsson, L. Latronico, M. Lemoine-Goumard, F. Longo, F. Loparco, M.N. Lovellette, P. Lubrano, M. Marelli, F. Massaro, M. Mayer, M.N. Mazziotta, J.E. McEnery, J. Mehault, P.F. Michelson, W. Mitthumsiri, T. Mizuno, C. Monte, M.E. Monzani, A. Morselli, I.V. Moskalenko, S. Murgia, T. Nakamori, R. Nemmen, E. Nuss, T. Ohsugi, A. Okumura, M. Orienti, E. Orlando, J.F. Ormes, D. Paneque, J.H. Panetta, J.S. Perkins, M. Pesce-Rollins, F. Piron, G. Pivato, T.A. Porter, S. Rainò, R. Rando, M. Razzano, A. Reimer, O. Reimer, T. Reposeur, S. Ritz, M. Roth, R. Rousseau, P.M. Saz Parkinson, A. Schulz, C. Sgrò, E.J. Siskind, D.A. Smith, G. Spandre, P. Spinelli, D.J. Suson, H. Takahashi, Y. Takeuchi, J.G. Thayer, J.B. Thayer, D.J. Thompson, L. Tibaldo, O. Tibolla, M. Tinivella, D.F. Torres, G. Tosti, E. Troja, Y. Uchiyama, J. Vandenbroucke, V. Vasileiou, G. Vianello, V. Vitale, M. Werner, B.L. Winer, K.S. Wood, Z. Yang, Constraints on the Galactic population of TeV pulsar wind nebulae using Fermi Large Area Telescope observations. Astrophys. J. **773**, 77 (2013). doi:10.1088/0004-637X/773/1/77

A. Achterberg, Y.A. Gallant, J.G. Kirk, A.W. Guthmann, Particle acceleration by ultrarelativistic shocks: Theory and simulations. Mon. Not. R. Astron. Soc. **328**, 393–408 (2001). doi:10.1046/j.1365-8711.2001.04851.x

F.A. Aharonian, A.A. Belyanin, E.V. Derishev, V.V. Kocharovsky, V.V. Kocharovsky, Constraints on the extremely high-energy cosmic ray accelerators from classical electrodynamics. Phys. Rev. D **66**(2), 023005 (2002). doi:10.1103/PhysRevD.66.023005

F. Aharonian, A.G. Akhperjanian, K.-M. Aye, A.R. Bazer-Bachi, M. Beilicke, W. Benbow, D. Berge, P. Berghaus, K. Bernlöhr, C. Boisson, O. Bolz, I. Braun, F. Breitling, A.M. Brown, J. Bussons Gordo, P.M. Chadwick, L.-M. Chounet, R. Cornils, L. Costamante, B. Degrange, A. Djannati-Ataï, L.C. O'Drury, G. Dubus, D. Emmanoulopoulos, P. Espigat, F. Feinstein, P. Fleury, G. Fontaine, Y. Fuchs, S. Funk, Y.A. Gallant, B. Giebels, S. Gillessen, J.F. Glicenstein, P. Goret, C. Hadjichristidis, M. Hauser, G. Heinzelmann, G. Henri, G. Hermann, J.A. Hinton, W. Hofmann, M. Holleran, D. Horns, O.C. de Jager, S. Johnston, B. Khélifi, J.G. Kirk, N. Komin, A. Konopelko, I.J. Latham, R. Le Gallou, A. Lemière, M. Lemoine-Goumard, N. Leroy, O. Martineau-Huynh, T. Lohse, A. Marcowith, C. Masterson, T.J.L. McComb, M. de Naurois, S.J. Nolan, A. Noutsos, K.J. Orford, J.L. Osborne, M. Ouchrif, M. Panter, G. Pelletier, S. Pita, G. Pühlhofer, M. Punch, B.C. Raubenheimer, M. Raue, J. Raux, S.M. Rayner, I. Redondo, A. Reimer, O. Reimer, J. Ripken, L. Rob, L. Rolland, G. Rowell, V. Sahakian, L. Saugé, S. Schlenker, R. Schlickeiser, C. Schuster, U. Schwanke, M. Siewert, O. Skjæraasen, H. Sol, R. Steenkamp, C. Stegmann, J.-P. Tavernet, R. Terrier, C.G. Théoret, M. Tluczykont, G. Vasileiadis, C. Venter, P. Vincent, H.J. Völk, S.J. Wagner, Discovery of the binary pulsar PSR B1259-63 in very-high-energy gamma rays around periastron with HESS. Astron. Astrophys. **442**, 1–10 (2005). doi:10.1051/0004-6361:20052983

F. Aharonian, A.G. Akhperjanian, A.R. Bazer-Bachi, M. Beilicke, W. Benbow, D. Berge, K. Bernlöhr, C. Boisson, O. Bolz, V. Borrel, I. Braun, A.M. Brown, R. Bühler, I. Büsching, S. Carrigan, P.M. Chadwick, L.-M. Chounet, R. Cornils, L. Costamante, B. Degrange, H.J. Dickinson, A. Djannati-Ataï, L.C. O'Drury, G. Dubus, K. Egberts, D. Emmanoulopoulos, P. Espigat, F. Feinstein, E. Ferrero, A. Fiasson, G. Fontaine, S. Funk, S. Funk, M. Füßling, Y.A. Gallant, B. Giebels, J.F. Glicenstein, P. Goret, C. Hadjichristidis, D. Hauser, M. Hauser, G. Heinzelmann, G. Henri, G. Hermann, J.A. Hinton, A. Hoffmann, W. Hofmann, M. Holleran, D. Horns, A. Jacholkowska, O.C. de Jager, E. Kendziorra, B. Khélifi, N. Komin, A. Konopelko, K. Kosack, I.J. Latham, R. Le Gallou, A. Lemière, M. Lemoine-Goumard, T. Lohse, J.M. Martin, O. Martineau-Huynh, A. Marcowith, C. Masterson, G. Maurin, T.J.L. McComb, E. Moulin, M. de Naurois, D. Nedbal, S.J. Nolan, A. Noutsos, K.J. Orford, J.L. Osborne, M. Ouchrif, M. Panter, G. Pelletier, S. Pita, G. Pühlhofer, M. Punch, B.C. Raubenheimer, M. Raue, S.M. Rayner, A. Reimer, O. Reimer, J. Ripken, L. Rob, L. Rolland, G. Rowell, V. Sahakian, A. Santangelo, L. Saugé, S. Schlenker, R. Schlickeiser, R. Schröder, U. Schwanke, S. Schwarzburg, A. Shalchi, H. Sol, D. Spangler, F. Spanier, R. Steenkamp, C. Stegmann, G. Superina, J.-P. Tavernet, R. Terrier, C.G. Théoret, M. Tluczykont, C. van Eldik, G. Vasileiadis, C. Venter, P. Vincent, H.J. Völk, S.J. Wagner, M. Ward, Energy dependent γ-ray morphology in the pulsar wind nebula HESS J1825-137. Astron. Astrophys. **460**, 365–374 (2006a). doi:10.1051/0004-6361:20065546

F. Aharonian, A.G. Akhperjanian, A.R. Bazer-Bachi, M. Beilicke, W. Benbow, D. Berge, K. Bernlöhr, C. Boisson, O. Bolz, V. Borrel, I. Braun, F. Breitling, A.M. Brown, R. Bühler, I. Büsching, S. Carrigan, P.M. Chadwick, L.-M. Chounet, R. Cornils, L. Costamante, B. Degrange, H.J. Dickinson, A. Djannati-Ataï, L.C. O'Drury, G. Dubus, K. Egberts, D. Emmanoulopoulos, B. Epinat, P. Espigat, F. Feinstein, E. Ferrero, G. Fontaine, S. Funk, S. Funk, Y.A. Gallant, B. Giebels, J.F. Glicenstein, P. Goret, C. Hadjichristidis, D. Hauser, M. Hauser, G. Heinzelmann, G. Henri, G. Hermann, J.A. Hinton, W. Hofmann, M. Holleran, D. Horns, A. Jacholkowska, O.C. de Jager, B. Khélifi, N. Komin, A. Konopelko, I.J. Latham, R. Le Gallou, A. Lemière, M. Lemoine-Goumard, T. Lohse, J.M. Martin, O. Martineau-Huynh, A. Marcowith, C. Masterson, T.J.L. McComb, M. de Naurois, D. Nedbal, S.J. Nolan, A. Noutsos, K.J. Orford, J.L. Osborne, M. Ouchrif, M. Panter, G. Pelletier, S. Pita, G. Pühlhofer, M. Punch, B.C. Raubenheimer, M. Raue, S.M. Rayner, A. Reimer, O. Reimer, J. Ripken, L. Rob, L. Rolland, G. Rowell, V. Sahakian, L. Saugé, S. Schlenker, R. Schlickeiser, U. Schwanke, H. Sol, D. Spangler, F. Spanier, R. Steenkamp, C. Stegmann, G. Superina, J.-P. Tavernet, R. Terrier, C.G. Théoret, M. Tluczykont, C. van Eldik, G. Vasileiadis, C. Venter, P. Vincent, H.J. Völk, S.J. Wagner, M. Ward, First detection of a VHE gamma-ray spectral maximum from a cosmic source: HESS discovery of the Vela X nebula. Astron. Astrophys. **448**, 43–47 (2006b). doi:10.1051/0004-6361:200600014

F. Aharonian, A.G. Akhperjanian, A.R. Bazer-Bachi, M. Beilicke, W. Benbow, D. Berge, K. Bernlöhr, C. Boisson, O. Bolz, V. Borrel, I. Braun, F. Breitling, A.M. Brown, R. Bühler, I. Büsching, S. Carrigan, P.M. Chadwick, L.-M. Chounet, R. Cornils, L. Costamante, B. Degrange, H.J. Dickinson, A. Djannati-Ataï, L.C. O'Drury, G. Dubus, K. Egberts, D. Emmanoulopoulos, B. Epinat, P. Espigat, F. Feinstein, E. Ferrero, G. Fontaine, S. Funk, S. Funk, Y.A. Gallant, B. Giebels, J.F. Glicenstein, P. Goret, C. Hadjichristidis, D. Hauser, M. Hauser, G. Heinzelmann, G. Henri, G. Hermann, J.A. Hinton, W. Hofmann, M. Holleran, D. Horns, A. Jacholkowska, O.C. de Jager, B. Khélifi, N. Komin, A. Konopelko, I.J. Latham, R. Le Gallou, A. Lemière, M. Lemoine-Goumard, T. Lohse, J.M. Martin, O. Martineau-Huynh, A. Marcowith, C. Masterson, T.J.L. McComb, M. de Naurois, D. Nedbal, S.J. Nolan, A. Noutsos, K.J. Orford, J.L. Osborne, M. Ouchrif, M. Panter, G. Pelletier, S. Pita, G. Pühlhofer, M. Punch, B.C. Raubenheimer, M. Raue, S.M. Rayner, A. Reimer, O. Reimer, J. Ripken, L. Rob, L. Rolland, G. Rowell, V. Sahakian, L. Saugé, S. Schlenker, R. Schlickeiser, U. Schwanke, H. Sol, D. Spangler, F. Spanier, R. Steenkamp, C. Stegmann, G. Superina, J.-P. Tavernet, R. Terrier, C.G. Théoret, M. Tluczykont, C. van Eldik, G.

Vasileiadis, C. Venter, P. Vincent, H.J. Völk, S.J. Wagner, M. Ward, First detection of a VHE gamma-ray spectral maximum from a cosmic source: HESS discovery of the Vela X nebula. Astron. Astrophys. **448**, 43–47 (2006c). doi:10.1051/0004-6361:200600014

F. Aharonian, A.G. Akhperjanian, A.R. Bazer-Bachi, M. Beilicke, W. Benbow, D. Berge, K. Bernlöhr, C. Boisson, O. Bolz, V. Borrel, I. Braun, F. Breitling, A.M. Brown, R. Bühler, I. Büsching, S. Carrigan, P.M. Chadwick, L.-M. Chounet, R. Cornils, L. Costamante, B. Degrange, H.J. Dickinson, A. Djannati-Ataï, L.C. O'Drury, G. Dubus, K. Egberts, D. Emmanoulopoulos, P. Espigat, F. Feinstein, E. Ferrero, A. Fiasson, G. Fontaine, S. Funk, S. Funk, Y.A. Gallant, B. Giebels, J.F. Glicenstein, P. Goret, C. Hadjichristidis, D. Hauser, M. Hauser, G. Heinzelmann, G. Henri, G. Hermann, J.A. Hinton, W. Hofmann, M. Holleran, D. Horns, A. Jacholkowska, O.C. de Jager, B. Khélifi, N. Komin, A. Konopelko, K. Kosack, I.J. Latham, R. Le Gallou, A. Lemière, M. Lemoine-Goumard, T. Lohse, J.M. Martin, O. Martineau-Huynh, A. Marcowith, C. Masterson, T.J.L. McComb, M. de Naurois, D. Nedbal, S.J. Nolan, A. Noutsos, K.J. Orford, J.L. Osborne, M. Ouchrif, M. Panter, G. Pelletier, S. Pita, G. Pühlhofer, M. Punch, B.C. Raubenheimer, M. Raue, S.M. Rayner, A. Reimer, O. Reimer, J. Ripken, L. Rob, L. Rolland, G. Rowell, V. Sahakian, L. Saugé, S. Schlenker, R. Schlickeiser, U. Schwanke, H. Sol, D. Spangler, F. Spanier, R. Steenkamp, C. Stegmann, G. Superina, J.-P. Tavernet, R. Terrier, C.G. Théoret, M. Tluczykont, C. van Eldik, G. Vasileiadis, C. Venter, P. Vincent, H.J. Völk, S.J. Wagner, M. Ward, Observations of the Crab Nebula with HESS. Astron. Astrophys. **457**, 899–915 (2006d). doi:10.1051/0004-6361:20065351

F. Aharonian, A.G. Akhperjanian, A.R. Bazer-Bachi, M. Beilicke, W. Benbow, D. Berge, K. Bernlöhr, C. Boisson, O. Bolz, V. Borrel, I. Braun, F. Breitling, A.M. Brown, P.M. Chadwick, L.-M. Chounet, R. Cornils, L. Costamante, B. Degrange, H.J. Dickinson, A. Djannati-Ataï, L.O. Drury, G. Dubus, D. Emmanoulopoulos, P. Espigat, F. Feinstein, G. Fontaine, Y. Fuchs, S. Funk, Y.A. Gallant, B. Giebels, S. Gillessen, J.F. Glicenstein, P. Goret, C. Hadjichristidis, M. Hauser, G. Heinzelmann, G. Henri, G. Hermann, J.A. Hinton, W. Hofmann, M. Holleran, D. Horns, A. Jacholkowska, O.C. de Jager, B. Khélifi, N. Komin, A. Konopelko, I.J. Latham, R. Le Gallou, A. Lemière, M. Lemoine-Goumard, N. Leroy, T. Lohse, J.M. Martin, O. Martineau-Huynh, A. Marcowith, C. Masterson, T.J.L. McComb, M. de Naurois, S.J. Nolan, A. Noutsos, K.J. Orford, J.L. Osborne, M. Ouchrif, M. Panter, G. Pelletier, S. Pita, G. Pühlhofer, M. Punch, B.C. Raubenheimer, M. Raue, J. Raux, S.M. Rayner, A. Reimer, O. Reimer, J. Ripken, L. Rob, L. Rolland, G. Rowell, V. Sahakian, L. Saugé, S. Schlenker, R. Schlickeiser, C. Schuster, U. Schwanke, M. Siewert, H. Sol, D. Spangler, R. Steenkamp, C. Stegmann, J.-P. Tavernet, R. Terrier, C.G. Théoret, M. Tluczykont, G. Vasileiadis, C. Venter, P. Vincent, H.J. Völk, S.J. Wagner, The H.E.S.S. survey of the inner Galaxy in very high energy gamma rays. Astrophys. J. **636**, 777–797 (2006e). doi:10.1086/498013

E. Aliu, S. Archambault, T. Aune, W. Benbow, K. Berger, R. Bird, A. Bouvier, J.H. Buckley, V. Bugaev, K. Byrum, M. Cerruti, X. Chen, L. Ciupik, M.P. Connolly, W. Cui, J. Dumm, M. Errando, A. Falcone, S. Federici, Q. Feng, J.P. Finley, P. Fortin, L. Fortson, A. Furniss, N. Galante, G.H. Gillanders, S. Griffin, S.T. Griffiths, J. Grube, G. Gyuk, D. Hanna, J. Holder, G. Hughes, T.B. Humensky, P. Kaaret, M. Kertzman, Y. Khassen, D. Kieda, F. Krennrich, S. Kumar, M.J. Lang, M. Lyutikov, G. Maier, S. McArthur, A. McCann, K. Meagher, J. Millis, P. Moriarty, R. Mukherjee, A. O'Faoláin de Bhróithe, R.A. Ong, A.N. Otte, N. Park, J.S. Perkins, M. Pohl, A. Popkow, J. Quinn, K. Ragan, J. Rajotte, L.C. Reyes, P.T. Reynolds, G.T. Richards, E. Roache, G.H. Sembroski, F. Sheidaei, A.W. Smith, D. Staszak, I. Telezhinsky, M. Theiling, J.V. Tucci, J. Tyler, A. Varlotta, S.P. Wakely, T.C. Weekes, A. Weinstein, R. Welsing, D.A. Williams, A. Zajczyk, B. Zitzer, A search for enhanced very high energy gamma-ray emission from the 2013 March Crab Nebula flare. Astrophys. J. Lett. **781**, 11 (2014). doi:10.1088/2041-8205/781/1/L11

E. Amato, The theory of pulsar wind nebulae. Int. J. Mod. Phys. Conf. Ser. **28**, 60160 (2014). doi:10.1142/S2010194514601604

J. Arons, On the coupling of rotation powered pulsars to plerionic nebulae. Mem. Soc. Astron. Ital. **69**, 989 (1998)

J. Arons, Pulsar wind nebulae as cosmic pevatrons: A current sheet's tale. Space Sci. Rev. **173**, 341–367 (2012). doi:10.1007/s11214-012-9885-1

J. Arons, D.C. Backer, A. Spitkovsky, V.M. Kaspi, Probing relativistic winds: The case of PSR J0737-3039 A and B, in *Binary Radio Pulsars*, ed. by F.A. Rasio, I.H. Stairs Astronomical Society of the Pacific Conference Series, vol. 328, 2005, p. 95

A.M. Atoyan, F.A. Aharonian, On the mechanisms of gamma radiation in the Crab Nebula. Mon. Not. R. Astron. Soc. **278**, 525–541 (1996)

K. Auchettl, P. Slane, R.W. Romani, B. Posselt, G.G. Pavlov, O. Kargaltsev, C. Ng, T. Temim, M.C. Weisskopf, A. Bykov, D.A. Swartz, X-ray analysis of the proper motion and pulsar wind nebula for PSR J1741-2054. ArXiv e-prints (2015)

M. Balbo, R. Walter, C. Ferrigno, P. Bordas, Twelve-hour spikes from the Crab Pevatron. Astron. Astrophys. **527**, 4 (2011). doi:10.1051/0004-6361/201015980

A. Bamba, K. Mori, S. Shibata, Chandra view of pulsar wind nebula tori. Astrophys. J. **709**, 507–511 (2010). doi:10.1088/0004-637X/709/1/507

R. Bandiera, On the X-ray feature associated with the Guitar nebula. Astron. Astrophys. **490**, 3–6 (2008). doi:10.1051/0004-6361:200810666

H. Bartko, W. Bednarek, γ-Ray emission from PWNe interacting with molecular clouds. Mon. Not. R. Astron. Soc. **385**, 1105–1109 (2008). doi:10.1111/j.1365-2966.2008.12870.x

W. Bednarek, W. Idec, On the variability of the GeV and multi-TeV gamma-ray emission from the Crab Nebula. Mon. Not. R. Astron. Soc. **414**, 2229–2234 (2011). doi:10.1111/j.1365-2966.2011.18539.x

M.C. Begelman, Instability of toroidal magnetic field in jets and plerions. Astrophys. J. **493**, 291–300 (1998). doi:10.1086/305119

M.C. Begelman, Z.-Y. Li, An axisymmetric magnetohydrodynamic model for the Crab pulsar wind bubble. Astrophys. J. **397**, 187–195 (1992). doi:10.1086/171778

M.F. Bietenholz, J.J. Hester, D.A. Frail, N. Bartel, The Crab Nebula's wisps in radio and optical. Astrophys. J. **615**, 794–804 (2004). doi:10.1086/424653

M.F. Bietenholz, Y. Yuan, R. Buehler, A.P. Lobanov, R. Blandford, The variability of the Crab Nebula in radio: No radio counterpart to gamma-ray flares. ArXiv e-prints (2014)

R. Blandford, D. Eichler, Particle acceleration at astrophysical shocks: A theory of cosmic ray origin. Phys. Rep. **154**, 1–75 (1987). doi:10.1016/0370-1573(87)90134-7

G.R. Blumenthal, R.J. Gould, Bremsstrahlung, synchrotron radiation, and Compton scattering of high-energy electrons traversing dilute gases. Rev. Mod. Phys. **42**, 237–271 (1970). doi:10.1103/RevModPhys.42.237

D.C.-J. Bock, D.A. Frail, R.J. Sault, A.J. Green, D.K. Milne, New observations of the Vela X region. Mem. Soc. Astron. Ital. **69**, 919 (1998)

S.V. Bogovalov, On the physics of cold MHD winds from oblique rotators. Astron. Astrophys. **349**, 1017–1026 (1999)

S.V. Bogovalov, D.V. Khangoulian, On the origin of the torus and jet-like structures in the centre of the Crab Nebula. Mon. Not. R. Astron. Soc. **336**, 53–55 (2002). doi:10.1046/j.1365-8711.2002.06027.x

V. Bosch-Ramon, M.V. Barkov, D. Khangulyan, M. Perucho, Simulations of stellar/pulsar-wind interaction along one full orbit. Astron. Astrophys. **544**, 59 (2012). doi:10.1051/0004-6361/201219251

S. Brownsberger, R.W. Romani, A Survey for Hα Pulsar Bow Shocks. Astrophys. J. **784**, 154 (2014). doi:10.1088/0004-637X/784/2/154

N. Bucciantini, Pulsar wind nebulae modeling. Int. J. Mod. Phys. Conf. Ser. **28**, 60162 (2014). doi:10.1142/S2010194514601628

N. Bucciantini, E. Amato, L. Del Zanna, Relativistic MHD simulations of pulsar bow-shock nebulae. Astron. Astrophys. **434**, 189–199 (2005). doi:10.1051/0004-6361:20042205

N. Bucciantini, J. Arons, E. Amato, Modelling spectral evolution of pulsar wind nebulae inside supernova remnants. Mon. Not. R. Astron. Soc. **410**, 381–398 (2011). doi:10.1111/j.1365-2966.2010.17449.x

R. Buehler, J.D. Scargle, R.D. Blandford, L. Baldini, M.G. Baring, A. Belfiore, E. Charles, J. Chiang, F. D'Ammando, C.D. Dermer, S. Funk, J.E. Grove, A.K. Harding, E. Hays, M. Kerr, F. Massaro, M.N. Mazziotta, R.W. Romani, P.M. Saz Parkinson, A.F. Tennant, M.C. Weisskopf, Gamma-ray activity in the Crab Nebula: The exceptional flare of 2011 April. Astrophys. J. **749**, 26 (2012). doi:10.1088/0004-637X/749/1/26

R. Bühler, R. Blandford, The surprising Crab pulsar and its nebula: a review. Rep. Prog. Phys. **77**(6), 066901 (2014). doi:10.1088/0034-4885/77/6/066901

A.M. Bykov, G.G. Pavlov, A.V. Artemyev, Y.A. Uvarov, Twinkling pulsar wind nebulae in the synchrotron cut-off regime and the γ-ray flares in the Crab Nebula. Mon. Not. R. Astron. Soc. **421**, 67–71 (2012). doi:10.1111/j.1745-3933.2011.01208.x

G.A. Caliandro, C.C. Teddy Cheung, J. Li, D.F. Torres, K. Wood, High energy γ-ray emission from PSR B1259$-$63 during 2014 and 2010 periastron passages. ArXiv e-prints (2015)

N.F. Camus, S.S. Komissarov, N. Bucciantini, P.A. Hughes, Observations of 'wisps' in magnetohydrodynamic simulations of the Crab Nebula. Mon. Not. R. Astron. Soc. **400**, 1241–1246 (2009). doi:10.1111/j.1365-2966.2009.15550.x

S. Carrigan, F. Brun, R.C.G. Chaves, C. Deil, A. Donath, H. Gast, V. Marandon, M. Renaud (H.E.S.S. Collaboration), The H.E.S.S. Galactic Plane Survey—maps, source catalog and source population. ArXiv e-prints (2013)

B. Cerutti, D.A. Uzdensky, M.C. Begelman, Extreme particle acceleration in magnetic reconnection layers: Application to the gamma-ray flares in the Crab Nebula. Astrophys. J. **746**, 148 (2012a). doi:10.1088/0004-637X/746/2/148

B. Cerutti, G.R. Werner, D.A. Uzdensky, M.C. Begelman, Beaming and rapid variability of high-energy radiation from relativistic pair plasma reconnection. Astrophys. J. Lett. **754**, 33 (2012b). doi:10.1088/2041-8205/754/2/L33

B. Cerutti, G.R. Werner, D.A. Uzdensky, M.C. Begelman, Simulations of particle acceleration beyond the classical synchrotron burnoff limit in magnetic reconnection: An explanation of the Crab flares. Astrophys. J. **770**, 147 (2013). doi:10.1088/0004-637X/770/2/147

B. Cerutti, G.R. Werner, D.A. Uzdensky, M.C. Begelman, Three-dimensional relativistic pair plasma reconnection with radiative feedback in the Crab Nebula. Astrophys. J. **782**, 104 (2014). doi:10.1088/0004-637X/782/2/104

S. Chatterjee, J.M. Cordes, Bow shocks from neutron stars: Scaling laws and Hubble Space Telescope observations of the Guitar Nebula. Astrophys. J. **575**, 407–418 (2002). doi:10.1086/341139

M. Chernyakova, A.A. Abdo, A. Neronov, M.V. McSwain, J. Moldón, M. Ribó, J.M. Paredes, I. Sushch, M. de Naurois, U. Schwanke, Y. Uchiyama, K. Wood, S. Johnston, S. Chaty, A. Coleiro, D. Malyshev, I. Babyk, Multiwavelength observations of the binary system PSR B1259-63/LS 2883 around the 2010–2011 periastron passage. Mon. Not. R. Astron. Soc. **439**, 432–445 (2014). doi:10.1093/mnras/stu021

E. Clausen-Brown, M. Lyutikov, Crab nebula gamma-ray flares as relativistic reconnection minijets. Mon. Not. R. Astron. Soc. **426**, 1374–1384 (2012). doi:10.1111/j.1365-2966.2012.21349.x

I. Contopoulos, The magnetic field topology in the reconnecting pulsar magnetosphere. Astron. Astrophys. **472**, 219–223 (2007). doi:10.1051/0004-6361:20077167

O.C. de Jager, A.K. Harding, The expected high-energy to ultra-high-energy gamma-ray spectrum of the Crab Nebula. Astrophys. J. **396**, 161–172 (1992). doi:10.1086/171706

O.C. de Jager, A.K. Harding, P.F. Michelson, H.I. Nel, P.L. Nolan, P. Sreekumar, D.J. Thompson, Gamma-ray observations of the Crab Nebula: A study of the synchro-Compton spectrum. Astrophys. J. **457**, 253 (1996). doi:10.1086/176726

O.C. de Jager, P.O. Slane, S. LaMassa, Probing the radio to X-ray connection of the Vela X pulsar wind nebula with Fermi LAT and H.E.S.S. Astrophys. J. Lett. **689**, 125–128 (2008). doi:10.1086/595959

O.C. de Jager, A. Djannati-Ataï, Implications of HESS observations of pulsar wind nebulae, in *Neutron Stars and Pulsars*, ed. by W. Becker Astrophysics and Space Science Library, vol. 357, 2009, p. 451. doi:10.1007/978-3-540-76965-1_17

A. De Luca, R.P. Mignani, M. Marelli, D. Salvetti, N. Sartore, A. Belfiore, P. Saz Parkinson, P.A. Caraveo, G.F. Bignami, PSR J0357+3205: A fast-moving pulsar with a very unusual X-ray trail. Astrophys. J. Lett. **765**, 19 (2013). doi:10.1088/2041-8205/765/1/L19

L. Del Zanna, E. Amato, N. Bucciantini, Axially symmetric relativistic MHD simulations of pulsar wind nebulae in supernova remnants. On the origin of torus and jet-like features. Astron. Astrophys. **421**, 1063–1073 (2004). doi:10.1051/0004-6361:20035936

L. Del Zanna, D. Volpi, E. Amato, N. Bucciantini, Simulated synchrotron emission from pulsar wind nebulae. Astron. Astrophys. **453**, 621–633 (2006). doi:10.1051/0004-6361:20064858

R. Dodson, D. Lewis, D. McConnell, A.A. Deshpande, The radio nebula surrounding the Vela pulsar. Mon. Not. R. Astron. Soc. **343**, 116–124 (2003). doi:10.1046/j.1365-8711.2003.06653.x

G. Dubus, Gamma-ray binaries and related systems. Astron. Astrophys. Rev. **21**, 64 (2013). doi:10.1007/s00159-013-0064-5

G. Dubus, B. Cerutti, G. Henri, Relativistic Doppler-boosted emission in gamma-ray binaries. Astron. Astrophys. **516**, 18 (2010). doi:10.1051/0004-6361/201014023

G. Dubus, B. Cerutti, What caused the GeV flare of PSR B1259-63? Astron. Astrophys. **557**, 127 (2013). doi:10.1051/0004-6361/201321741

M. Durant, O. Kargaltsev, G.G. Pavlov, C. Chang, G.P. Garmire, Extended X-ray emission in the vicinity of the microquasar LS 5039: Pulsar wind nebula? Astrophys. J. **735**, 58 (2011). doi:10.1088/0004-637X/735/1/58

M. Durant, O. Kargaltsev, G.G. Pavlov, J. Kropotina, K. Levenfish, The helical jet of the Vela pulsar. Astrophys. J. **763**, 72 (2013). doi:10.1088/0004-637X/763/2/72

R.T. Emmering, R.A. Chevalier, Shocked relativistic magnetohydrodynamic flows with application to pulsar winds. Astrophys. J. **321**, 334–348 (1987). doi:10.1086/165632

R. Enomoto, K. Tsuchiya, Y. Adachi, S. Kabuki, P.G. Edwards, A. Asahara, G.V. Bicknell, R.W. Clay, Y. Doi, S. Gunji, S. Hara, T. Hara, T. Hattori, S. Hayashi, Y. Higashi, R. Inoue, C. Itoh, F. Kajino, H. Katagiri, A. Kawachi, S. Kawasaki, T. Kifune, R. Kiuchi, K. Konno, L.T. Ksenofontov, H. Kubo, J. Kushida, Y. Matsubara, Y. Mizumoto, M. Mori, H. Muraishi, Y. Muraki, T. Naito, T. Nakamori, D. Nishida, K. Nishijima, M. Ohishi, J.R. Patterson, R.J. Protheroe, Y. Sakamoto, M. Sato, S. Suzuki, T. Suzuki, D.L. Swaby, T. Tanimori, H. Tanimura, G.J. Thornton, S. Watanabe, T. Yamaoka, M. Yamazaki, S. Yanagita, T. Yoshida, T. Yoshikoshi, M. Yuasa, Y. Yukawa, A search for sub-TeV gamma rays from the Vela pulsar region with CANGAROO-III. Astrophys. J. **638**, 397–408 (2006). doi:10.1086/498858

L. Fortson (CTA Consortium), Science with the Cherenkov Telescope Array, in *American Astronomical Society Meeting Abstracts*. American Astronomical Society Meeting Abstracts, vol. 225, 2015, pp. 336–403

D.A. Frail, M.F. Bietenholz, C.B. Markwardt, A radio/X-ray comparison of the Vela X region. Astrophys. J. **475**, 224–230 (1997)

B.M. Gaensler, E. van der Swaluw, F. Camilo, V.M. Kaspi, F.K. Baganoff, F. Yusef-Zadeh, R.N. Manchester, The mouse that soared: High-resolution X-ray imaging of the pulsar-powered bow shock G359.23-0.82. Astrophys. J. **616**, 383–402 (2004). doi:10.1086/424906

B.M. Gaensler, P.O. Slane, The evolution and structure of pulsar wind nebulae. Annu. Rev. Astron. Astrophys. **44**, 17–47 (2006). doi:10.1146/annurev.astro.44.051905.092528

J.D. Gelfand, P.O. Slane, W. Zhang, A dynamical model for the evolution of a pulsar wind nebula inside a nonradiative supernova remnant. Astrophys. J. **703**, 2051–2067 (2009). doi:10.1088/0004-637X/703/2/2051

M.-H. Grondin, R.W. Romani, M. Lemoine-Goumard, L. Guillemot, A.K. Harding, T. Reposeur, The Vela-X pulsar wind nebula revisited with four years of Fermi Large Area Telescope observations. Astrophys. J. **774**, 110 (2013). doi:10.1088/0004-637X/774/2/110

P.W. Guilbert, A.C. Fabian, M.J. Rees, Spectral and variability constraints on compact sources. Mon. Not. R. Astron. Soc. **205**, 593–603 (1983)

F. Guo, H. Li, W. Daughton, Y.-H. Liu, Formation of hard power laws in the energetic particle spectra resulting from relativistic magnetic reconnection. Phys. Rev. Lett. **113**(15), 155005 (2014). doi:10.1103/PhysRevLett.113.155005

H.E.S.S. Collaboration, A. Abramowski, F. Acero, F. Aharonian, A.G. Akhperjanian, G. Anton, S. Balenderan, A. Balzer, A. Barnacka, Y. Becherini, J. Becker Tjus, K. Bernlöhr, E. Birsin, J. Biteau, C. Boisson, J. Bolmont, P. Bordas, J. Brucker, F. Brun, P. Brun, T. Bulik, S. Carrigan, S. Casanova, M. Cerruti, P.M. Chadwick, R.C.G. Chaves, A. Cheesebrough, S. Colafrancesco, G. Cologna, J. Conrad, C. Couturier, M. Dalton, M.K. Daniel, I.D. Davids, B. Degrange, C. Deil, P. deWilt, H.J. Dickinson, A. Djannati-Ataï, W. Domainko, L.O. Drury, G. Dubus, K. Dutson, J. Dyks, M. Dyrda, K. Egberts, P. Eger, P. Espigat, L. Fallon, C. Farnier, S. Fegan, F. Feinstein, M.V. Fernandes, D. Fernandez, A. Fiasson, G. Fontaine, A. Förster, M. Füßling, M. Gajdus, Y.A. Gallant, T. Garrigoux, H. Gast, B. Giebels, J.F. Glicenstein, B. Glück, D. Göring, M.-H. Grondin, M. Grudzińska, S. Häer, J.D. Hague, J. Hahn, D. Hampf, J. Harris, S. Heinz, G. Heinzelmann, G. Henri, G. Hermann, A. Hillert, J.A. Hinton, W. Hofmann, P. Hofverberg, M. Holler, D. Horns, A. Jacholkowska, C. Jahn, M. Jamrozy, I. Jung, M.A. Kastendieck, K. Katarzyński, U. Katz, S. Kaufmann, B. Khélifi, S. Klepser, D. Klochkov, W. Kluźniak, T. Kneiske, D. Kolitzus, N. Komin, K. Kosack, R. Kossakowski, F. Krayzel, P.P. Krüger, H. Lan, G. Lamanna, J. Lefaucheur, M. Lemoine-Goumard, J.-P. Lenain, D. Lennarz, T. Lohse, A. Lopatin, C.-C. Lu, V. Marandon, A. Marcowith, J. Masbou, G. Maurin, N. Maxted, M. Mayer, T.J.L. McComb, M.C. Medina, J. Méhault, U. Menzler, R. Moderski, M. Mohamed, E. Moulin, C.L. Naumann, M. Naumann-Godo, M. de Naurois, D. Nedbal, N. Nguyen, J. Niemiec, S.J. Nolan, L. Oakes, S. Ohm, E. de Oña Wilhelmi, B. Opitz, M. Ostrowski, I. Oya, M. Panter, R.D. Parsons, M. Paz Arribas, N.W. Pekeur, G. Pelletier, J. Perez, P.-O. Petrucci, B. Peyaud, S. Pita, G. Pühlhofer, M. Punch, A. Quirrenbach, S. Raab, M. Raue, A. Reimer, O. Reimer, M. Renaud, R. de los Reyes, F. Rieger, J. Ripken, L. Rob, S. Rosier-Lees, G. Rowell, B. Rudak, C.B. Rulten, V. Sahakian, D.A. Sanchez, A. Santangelo, R. Schlickeiser, A. Schulz, U. Schwanke, S. Schwarzburg, S. Schwemmer, F. Sheidaei, J.L. Skilton, H. Sol, G. Spengler, Ł. Stawarz, R. Steenkamp, C. Stegmann, F. Stinzing, K. Stycz, I. Sushch, A. Szostek, J.-P. Tavernet, R. Terrier, M. Tluczykont, C. Trichard, K. Valerius, C. van Eldik, G. Vasileiadis, C. Venter, A. Viana, P. Vincent, H.J. Völk, F. Volpe, S. Vorobiov, M. Vorster, S.J. Wagner, M. Ward, R. White, A. Wierzcholska, P. Willmann, D. Wouters, M. Zacharias, A. Zajczyk, A.A. Zdziarski, A. Zech, H.-S. Zechlin, H.E.S.S. observations of the binary system PSR B1259-63/LS 2883 around the 2010/2011 periastron passage. Astron. Astrophys. **551**, 94 (2013). doi:10.1051/0004-6361/201220612

H.E.S.S. Collaboration, A. Abramowski, F. Aharonian, F. Ait Benkhali, A.G. Akhperjanian, E. Angüner, G. Anton, S. Balenderan, A. Balzer, A. Barnacka, et al., H.E.S.S. observations of the Crab during its March 2013 GeV gamma-ray flare. Astron. Astrophys. **562**, 4 (2014). doi:10.1051/0004-6361/201323013

J.P. Halpern, J.A. Tomsick, E.V. Gotthelf, F. Camilo, C.-Y. Ng, A. Bodaghee, J. Rodriguez, S. Chaty, F. Rahoui, Discovery of X-ray pulsations from the INTEGRAL source IGR J11014-6103. ArXiv e-prints (2014)

D.J. Helfand, E.V. Gotthelf, J.P. Halpern, Vela pulsar and its synchrotron nebula. Astrophys. J. **556**, 380–391 (2001). doi:10.1086/321533

K. Heng, R. McCray, Balmer-dominated shocks revisited. Astrophys. J. **654**, 923–937 (2007). doi:10.1086/509601

J.J. Hester, The Crab Nebula: An astrophysical chimera. Annu. Rev. Astron. Astrophys. **46**, 127–155 (2008). doi:10.1146/annurev.astro.45.051806.110608

J.J. Hester, P.A. Scowen, R. Sankrit, C.J. Burrows, J.S. Gallagher III, J.A. Holtzman, A. Watson, J.T. Trauger, G.E. Ballester, S. Casertano, J.T. Clarke, D. Crisp, R.W. Evans, R.E. Griffiths, J.G. Hoessel, J. Krist, R. Lynds, J.R. Mould, E.J. O'Neil Jr., K.R. Stapelfeldt, J.A. Westphal, WFPC2 studies of the Crab Nebula. I. HST and ROSAT imaging of the synchrotron nebula. Astrophys. J. **448**, 240 (1995). doi:10.1086/175956

J.J. Hester, K. Mori, D. Burrows, J.S. Gallagher, J.R. Graham, M. Halverson, A. Kader, F.C. Michel, P. Scowen, Hubble Space Telescope and Chandra monitoring of the Crab synchrotron nebula. Astrophys. J. Lett. **577**, 49–52 (2002). doi:10.1086/344132

A.M. Hillas, C.W. Akerlof, S.D. Biller, J.H. Buckley, D.A. Carter-Lewis, M. Catanese, M.F. Cawley, D.J. Fegan, J.P. Finley, J.A. Gaidos, F. Krennrich, R.C. Lamb, M.J. Lang, G. Mohanty, M. Punch, P.T. Reynolds, A.J. Rodgers, H.J. Rose, A.C. Rovero, M.S. Schubnell, G.H. Sembroski, G. Vacanti, T.C. Weekes, M. West, J. Zweerink, The spectrum of teravolt gamma rays from the Crab Nebula. Astrophys. J. **503**, 744–759 (1998). doi:10.1086/306005

G. Hobbs, D.R. Lorimer, A.G. Lyne, M. Kramer, A statistical study of 233 pulsar proper motions. Mon. Not. R. Astron. Soc. **360**, 974–992 (2005). doi:10.1111/j.1365-2966.2005.09087.x

R.H.H. Huang, A.K.H. Kong, J. Takata, C.Y. Hui, L.C.C. Lin, K.S. Cheng, X-ray studies of the Black Widow pulsar PSR B1957+20. Astrophys. J. **760**, 92 (2012). doi:10.1088/0004-637X/760/1/92

C.Y. Hui, W. Becker, X-ray emission properties of the old pulsar PSR B2224+65. Astron. Astrophys. **467**, 1209–1214 (2007). doi:10.1051/0004-6361:20066562

S.P. Johnson, Q.D. Wang, The pulsar B2224+65 and its jets: A two epoch X-ray analysis. Mon. Not. R. Astron. Soc. **408**, 1216–1224 (2010). doi:10.1111/j.1365-2966.2010.17200.x

S. Johnston, G. Hobbs, S. Vigeland, M. Kramer, J.M. Weisberg, A.G. Lyne, Evidence for alignment of the rotation and velocity vectors in pulsars. Mon. Not. R. Astron. Soc. **364**, 1397–1412 (2005). doi:10.1111/j.1365-2966.2005.09669.x

O. Kargaltsev, Z. Misanovic, G.G. Pavlov, J.A. Wong, G.P. Garmire, X-ray observations of parsec-scale tails behind two middle-aged pulsars. Astrophys. J. **684**, 542–557 (2008). doi:10.1086/589145

O. Kargaltsev, G.G. Pavlov, Pulsar wind nebulae in the Chandra era, in *40 Years of Pulsars: Millisecond Pulsars, Magnetars and More*, ed. by C. Bassa, Z. Wang, A. Cumming, V.M. Kaspi American Institute of Physics Conference Series, vol. 983, 2008, pp. 171–185. doi:10.1063/1.2900138

O. Kargaltsev, M. Durant, G.G. Pavlov, G. Garmire, Chandra Pulsar Survey (ChaPS). Astrophys. J. Suppl. Ser. **201**, 37 (2012). doi:10.1088/0067-0049/201/2/37

O. Kargaltsev, G. Pavlov, M. Durant, X-ray observations of pulsar-wind nebulae: Current status and future prospects, in *The Fast and the Furious: Energetic Phenomena in Isolated Neutron Stars, Pulsar Wind Nebulae and Supernova Remnants*, ed. by J.-U. Ness, 2013a, p. 7

O. Kargaltsev, B. Rangelov, G.G. Pavlov, Gamma-ray and X-ray properties of pulsar wind nebulae and unidentified Galactic TeV sources. ArXiv e-prints (2013b)

O. Kargaltsev, A. Bykov, A. Krassilchtchikov, B. Rangelov, G. Pavlov, N. Klingler, J. Kropotina, K. Levenfish, The inner structure of the Vela pulsar-wind nebula, in *40th COSPAR Scientific Assembly*, 2–10 August 2014, Moscow, Russia. COSPAR Meeting, vol. 40, 2014c, p. 1398. Abstract E1.16-36-14

O. Kargaltsev, G.G. Pavlov, M. Durant, I. Volkov, J. Hare, The dynamic X-ray nebula powered by the pulsar B1259-63. Astrophys. J. **784**, 124 (2014b). doi:10.1088/0004-637X/784/2/124

O. Kargaltsev, B. Rangelov, J. Hare, G.G. Pavlov, Chandra imaging of gamma-ray binaries. Astron. Nachr. **335**, 301–306 (2014a). doi:10.1002/asna.201312036

C.F. Kennel, F.V. Coroniti, Confinement of the Crab pulsar's wind by its supernova remnant. Astrophys. J. **283**, 694–709 (1984a). doi:10.1086/162356

C.F. Kennel, F.V. Coroniti, Magnetohydrodynamic model of Crab nebula radiation. Astrophys. J. **283**, 710–730 (1984b). doi:10.1086/162357

D. Khangulyan, S. Hnatic, F. Aharonian, S. Bogovalov, TeV light curve of PSR B1259-63/SS2883. Mon. Not. R. Astron. Soc. **380**, 320–330 (2007). doi:10.1111/j.1365-2966.2007.12075.x

D. Khangulyan, F.A. Aharonian, S.V. Bogovalov, M. Ribó, Post-periastron gamma-ray flare from PSR B1259-63/LS 2883 as a result of comptonization of the cold pulsar wind. Astrophys. J. Lett. **752**, 17 (2012). doi:10.1088/2041-8205/752/1/L17

J.G. Kirk, Particle acceleration in relativistic current sheets. Phys. Rev. Lett. **92**(18), 181101 (2004). doi:10.1103/PhysRevLett.92.181101

J.G. Kirk, L. Ball, O. Skjæraasen, Inverse Compton emission of TeV gamma rays from PSR B1259-63. Astropart. Phys. **10**, 31–45 (1999). doi:10.1016/S0927-6505(98)00041-3

J.G. Kirk, Y. Lyubarsky, J. Petri, The theory of pulsar winds and nebulae, in *Neutron Stars and Pulsars*, ed. by W. Becker Astrophysics and Space Science Library, vol. 357, 2009, p. 421. doi:10.1007/978-3-540-76965-1_16

S.S. Komissarov, Magnetic dissipation in the Crab nebula. Mon. Not. R. Astron. Soc. **428**, 2459–2466 (2013). doi:10.1093/mnras/sts214

S.S. Komissarov, Y.E. Lyubarsky, The origin of peculiar jet-torus structure in the Crab nebula. Mon. Not. R. Astron. Soc. **344**, 93–96 (2003). doi:10.1046/j.1365-8711.2003.07097.x

S.S. Komissarov, Y.E. Lyubarsky, Synchrotron nebulae created by anisotropic magnetized pulsar winds. Mon. Not. R. Astron. Soc. **349**, 779–792 (2004). doi:10.1111/j.1365-2966.2004.07597.x

S.S. Komissarov, M. Barkov, M. Lyutikov, Tearing instability in relativistic magnetically dominated plasmas. Mon. Not. R. Astron. Soc. **374**, 415–426 (2007). doi:10.1111/j.1365-2966.2006.11152.x
S.S. Komissarov, M. Lyutikov, On the origin of variable gamma-ray emission from the Crab nebula. Mon. Not. R. Astron. Soc. **414**, 2017–2028 (2011). doi:10.1111/j.1365-2966.2011.18516.x
S.W. Kong, K.S. Cheng, Y.F. Huang, Modeling the multiwavelength light curves of PSR B1259-63/LS 2883. II. The effects of anisotropic pulsar wind and Doppler boosting. Astrophys. J. **753**, 127 (2012). doi:10.1088/0004-637X/753/2/127
T. Kouzu, M.S. Tashiro, Y. Terada, S. Yamada, A. Bamba, T. Enoto, K. Mori, Y. Fukazawa, K. Makishima, Spectral variation of hard X-ray emission from the Crab Nebula with the Suzaku Hard X-Ray Detector. Publ. Astron. Soc. Jpn. **65**, 74 (2013). doi:10.1093/pasj/65.4.74
S.M. LaMassa, P.O. Slane, O.C. de Jager, Probing the nature of the Vela X cocoon. Astrophys. J. Lett. **689**, 121–124 (2008). doi:10.1086/595958
K.P. Levenfish, A.M. Bykov, M. Durant, O.Y. Kargaltsev, Y.A. Kropotina, G.G. Pavlov, A.M. Krassilchtchikov, Y.A. Uvarov, Finest persistent structures in the Vela PWN. Mem. Soc. Astron. Ital. **84**, 588 (2013)
A.P. Lobanov, D. Horns, T.W.B. Muxlow, VLBI imaging of a flare in the Crab nebula: More than just a spot. Astron. Astrophys. **533**, 10 (2011). doi:10.1051/0004-6361/201117082
A. Lyne, F. Graham-Smith, P. Weltevrede, C. Jordan, B. Stappers, C. Bassa, M. Kramer, Evolution of the magnetic field structure of the Crab pulsar. Science **342**, 598–601 (2013). doi:10.1126/science.1243254
A. Lyne, B. Stappers, M. Keith, P. Ray, M. Kerr, F. Camilo, T. Johnson, The binary nature of PSR J2032+4127. ArXiv e-prints (2015)
Y.E. Lyubarsky, On the structure of the inner Crab Nebula. Mon. Not. R. Astron. Soc. **329**, 34–36 (2002). doi:10.1046/j.1365-8711.2002.05151.x
Y.E. Lyubarsky, Fast magnetosonic waves in pulsar winds. Mon. Not. R. Astron. Soc. **339**, 765–771 (2003a). doi:10.1046/j.1365-8711.2003.06221.x
Y.E. Lyubarsky, The termination shock in a striped pulsar wind. Mon. Not. R. Astron. Soc. **345**, 153–160 (2003b). doi:10.1046/j.1365-8711.2003.06927.x
Y.E. Lyubarsky, Highly magnetized region in pulsar wind nebulae and origin of the Crab gamma-ray flares. Mon. Not. R. Astron. Soc. **427**, 1497–1502 (2012). doi:10.1111/j.1365-2966.2012.22097.x
Y. Lyubarsky, M. Liverts, Particle acceleration in the driven relativistic reconnection. Astrophys. J. **682**, 1436–1442 (2008). doi:10.1086/589640
M. Lyutikov, Mass-loading of pulsar winds. Mon. Not. R. Astron. Soc. **339**, 623–632 (2003). doi:10.1046/j.1365-8711.2003.06141.x
M. Lyutikov, D. Balsara, C. Matthews, Crab GeV flares from the corrugated termination shock. Mon. Not. R. Astron. Soc. **422**, 3118–3129 (2012). doi:10.1111/j.1365-2966.2012.20831.x
M. Marelli, A. De Luca, D. Salvetti, N. Sartore, A. Sartori, P. Caraveo, F. Pizzolato, P.M. Saz Parkinson, A. Belfiore, PSR J0357+3205: The tail of the turtle. Astrophys. J. **765**, 36 (2013). doi:10.1088/0004-637X/765/1/36
C.B. Markwardt, H. Ögelman, An X-ray jet from the Vela pulsar. Nature **375**, 40–42 (1995). doi:10.1038/375040a0
C.B. Markwardt, H.B. Ögelman, The ASCA spectrum of the VELA pulsar jet. Astrophys. J. Lett. **480**, 13–16 (1997). doi:10.1086/310616
F. Mattana, D. Götz, R. Terrier, L. Bouchet, G. Ponti, M. Falanga, M. Renaud, I. Caballero, S. Soldi, J.A. Zurita Heras, S. Schanne, Extended hard X-ray emission from the Vela pulsar wind nebula. Astrophys. J. Lett. **743**, 18 (2011). doi:10.1088/2041-8205/743/1/L18
M. Mayer, R. Buehler, E. Hays, C.C. Cheung, M.S. Dutka, J.E. Grove, M. Kerr, R. Ojha, Rapid gamma-ray flux variability during the 2013 March Crab Nebula flare. Astrophys. J. Lett. **775**, 37 (2013). doi:10.1088/2041-8205/775/2/L37
M.V. Medvedev, Theory of "jitter" radiation from small-scale random magnetic fields and prompt emission from gamma-ray burst shocks. Astrophys. J. **540**, 704–714 (2000). doi:10.1086/309374
M.V. Medvedev, Constraint on electromagnetic acceleration of highest energy cosmic rays. Phys. Rev. E **67**(4), 045401 (2003). doi:10.1103/PhysRevE.67.045401
A. Melatos, D. Scheltus, M.T. Whiting, S.S. Eikenberry, R.W. Romani, F. Rigaut, A. Spitkovsky, J. Arons, D.J.B. Payne, Near-infrared, kilosecond variability of the wisps and jet in the Crab pulsar wind nebula. Astrophys. J. **633**, 931–940 (2005). doi:10.1086/468176
F.C. Michel, Rotating magnetospheres: An exact 3-D solution. Astrophys. J. Lett. **180**, 133 (1973). doi:10.1086/181169
R.P. Mignani, A. De Luca, O. Kargaltsev, G.G. Pavlov, S. Zaggia, P.A. Caraveo, W. Becker, Search for the optical counterpart of the Vela pulsar X-ray nebula. Astrophys. J. **594**, 419–427 (2003). doi:10.1086/376829

A. Mignone, E. Striani, M. Tavani, A. Ferrari, Modelling the kinked jet of the Crab nebula. Mon. Not. R. Astron. Soc. **436**, 1102–1115 (2013). doi:10.1093/mnras/stt1632

Y. Mizuno, Y. Lyubarsky, K.-I. Nishikawa, P.E. Hardee, Three-dimensional relativistic magnetohydrodynamic simulations of current-driven instability. II. Relaxation of pulsar wind nebula. Astrophys. J. **728**, 90 (2011). doi:10.1088/0004-637X/728/2/90

D.A. Moffett, T.H. Hankins, Polarimetric properties of the Crab pulsar between 1.4 and 8.4 GHz. Astrophys. J. **522**, 1046–1052 (1999). doi:10.1086/307654

J. Moldón, M. Ribó, J.M. Paredes, Periodic morphological changes in gamma-ray binaries, in *High Energy Gamma-Ray Astronomy: 5th International Meeting on High Energy Gamma-Ray Astronomy*, ed. by F.A. Aharonian, W. Hofmann, F.M. Rieger American Institute of Physics Conference Series, vol. 1505, 2012, pp. 386–389. doi:10.1063/1.4772278

P. Moran, A. Shearer, R.P. Mignani, A. Słowikowska, A. De Luca, C. Gouiffès, P. Laurent, Optical polarimetry of the inner Crab nebula and pulsar. Mon. Not. R. Astron. Soc. **433**, 2564–2575 (2013). doi:10.1093/mnras/stt931

P. Moran, R.P. Mignani, A. Shearer, HST optical polarimetry of the Vela pulsar and nebula. Mon. Not. R. Astron. Soc. **445**, 835–844 (2014). doi:10.1093/mnras/stu1791

K. Mori, D.N. Burrows, J.J. Hester, G.G. Pavlov, S. Shibata, H. Tsunemi, Spatial variation of the X-ray spectrum of the Crab Nebula. Astrophys. J. **609**, 186–193 (2004). doi:10.1086/421011

K. Mori, O. Kargaltsev, G. Pavlov, M. Yamamoto, S. Shibata, H. Tsunemi, Suzaku mapping observations of the Vela X, in *37th COSPAR Scientific Assembly*. COSPAR Meeting, vol. 37, 2008, p. 2105

C.-Y. Ng, B.M. Gaensler, S. Chatterjee, S. Johnston, Radio polarization observations of G319.9-0.7: A bow-shock nebula with an azimuthal magnetic field powered by pulsar J1509-5850. Astrophys. J. **712**, 596–603 (2010). doi:10.1088/0004-637X/712/1/596

B. Olmi, L. Del Zanna, E. Amato, R. Bandiera, N. Bucciantini, On the magnetohydrodynamic modelling of the Crab nebula radio emission. Mon. Not. R. Astron. Soc. **438**, 1518–1525 (2014). doi:10.1093/mnras/stt2308

B. Olmi, L. Del Zanna, E. Amato, N. Bucciantini, Constraints on particle acceleration sites in the Crab nebula from relativistic magnetohydrodynamic simulations. Mon. Not. R. Astron. Soc. **449**, 3149–3159 (2015). doi:10.1093/mnras/stv498

L. Pavan, E. Bozzo, G. Pühlhofer, C. Ferrigno, M. Balbo, R. Walter, IGR J11014-6103: A newly discovered pulsar wind nebula? Astron. Astrophys. **533**, 74 (2011). doi:10.1051/0004-6361/201117379

L. Pavan, P. Bordas, G. Pühlhofer, M.D. Filipović, A. De Horta, A. O'Brien, M. Balbo, R. Walter, E. Bozzo, C. Ferrigno, E. Crawford, L. Stella, The long helical jet of the Lighthouse nebula, IGR J11014-6103. Astron. Astrophys. **562**, 122 (2014). doi:10.1051/0004-6361/201322588

G.G. Pavlov, M.A. Teter, O. Kargaltsev, D. Sanwal, The variable jet of the Vela pulsar. Astrophys. J. **591**, 1157–1171 (2003). doi:10.1086/375531

G.G. Pavlov, J. Hare, O. Kargaltsev, B. Rangelov, M. Durant, An extended X-ray object ejected from the PSR B1259-63/LS 2883 binary. ArXiv e-prints (2015)

A. Pellizzoni, A. Trois, M. Tavani, M. Pilia, A. Giuliani, G. Pucella, P. Esposito, S. Sabatini, G. Piano, A. Argan, G. Barbiellini, A. Bulgarelli, M. Burgay, P. Caraveo, P.W. Cattaneo, A.W. Chen, V. Cocco, T. Contessi, E. Costa, F. D'Ammando, E. Del Monte, G. De Paris, G. Di Cocco, G. Di Persio, I. Donnarumma, Y. Evangelista, M. Feroci, A. Ferrari, M. Fiorini, F. Fuschino, M. Galli, F. Gianotti, A. Hotan, C. Labanti, I. Lapshov, F. Lazzarotto, P. Lipari, F. Longo, M. Marisaldi, M. Mastropietro, S. Mereghetti, E. Moretti, A. Morselli, L. Pacciani, J. Palfreyman, F. Perotti, P. Picozza, C. Pittori, A. Possenti, M. Prest, M. Rapisarda, A. Rappoldi, E. Rossi, A. Rubini, P. Santolamazza, E. Scalise, P. Soffitta, E. Striani, M. Trifoglio, E. Vallazza, S. Vercellone, F. Verrecchia, V. Vittorini, A. Zambra, D. Zanello, P. Giommi, S. Colafrancesco, A. Antonelli, L. Salotti, N. D'Amico, G.F. Bignami, Detection of gamma-ray emission from the Vela pulsar wind nebula with AGILE. Science **327**, 663 (2010). doi:10.1126/science.1183844

J. Pétri, Y. Lyubarsky, Magnetic reconnection at the termination shock in a striped pulsar wind. Astron. Astrophys. **473**, 683–700 (2007). doi:10.1051/0004-6361:20066981

O. Porth, S.S. Komissarov, R. Keppens, Three-dimensional magnetohydrodynamic simulations of the Crab nebula. Mon. Not. R. Astron. Soc. **438**, 278–306 (2014). doi:10.1093/mnras/stt2176

V. Radhakrishnan, A.A. Deshpande, Vela, its X-ray nebula, and the polarization of pulsar radiation. Astron. Astrophys. **379**, 551–556 (2001). doi:10.1051/0004-6361:20011321

M.J. Rees, J.E. Gunn, The origin of the magnetic field and relativistic particles in the Crab Nebula. Mon. Not. R. Astron. Soc. **167**, 1–12 (1974)

S.P. Reynolds, Synchrotron-loss spectral breaks in pulsar-wind nebulae and extragalactic jets. Astrophys. J. **703**, 662–670 (2009). doi:10.1088/0004-637X/703/1/662

R.W. Romani, M.S. Shaw, F. Camilo, G. Cotter, G.R. Sivakoff, The Balmer-dominated bow shock and wind nebula structure of γ-ray pulsar PSR J1741-2054. Astrophys. J. **724**, 908–914 (2010). doi:10.1088/0004-637X/724/2/908

M.M. Romanova, G.A. Chulsky, R.V.E. Lovelace, Winds, B-fields, and magnetotails of pulsars. Astrophys. J. **630**, 1020–1028 (2005). doi:10.1086/431727
S.C. Rookyard, P. Weltevrede, S. Johnston, Investigating the magnetic inclination angle distribution of γ-ray-loud radio pulsars. ArXiv e-prints (2014)
T. Schweizer, N. Bucciantini, W. Idec, K. Nilsson, A. Tennant, M.C. Weisskopf, R. Zanin, Characterization of the optical and X-ray properties of the north-western wisps in the Crab nebula. Mon. Not. R. Astron. Soc. **433**, 3325–3335 (2013). doi:10.1093/mnras/stt995
R.M. Shannon, S. Johnston, R.N. Manchester, The kinematics and orbital dynamics of the PSR B1259-63/LS 2883 system from 23 yr of pulsar timing. Mon. Not. R. Astron. Soc. **437**, 3255–3264 (2014). doi:10.1093/mnras/stt2123
L. Sironi, A. Spitkovsky, Particle acceleration in relativistic magnetized collisionless pair shocks: Dependence of shock acceleration on magnetic obliquity. Astrophys. J. **698**, 1523–1549 (2009a). doi:10.1088/0004-637X/698/2/1523
L. Sironi, A. Spitkovsky, Synthetic spectra from particle-in-cell simulations of relativistic collisionless shocks. Astrophys. J. Lett. **707**, 92–96 (2009b). doi:10.1088/0004-637X/707/1/L92
L. Sironi, A. Spitkovsky, Acceleration of particles at the termination shock of a relativistic striped wind. Astrophys. J. **741**, 39 (2011a). doi:10.1088/0004-637X/741/1/39
L. Sironi, A. Spitkovsky, Particle acceleration in relativistic magnetized collisionless electron-ion shocks. Astrophys. J. **726**, 75 (2011b). doi:10.1088/0004-637X/726/2/75
L. Sironi, A. Spitkovsky, Relativistic reconnection: An efficient source of non-thermal particles. Astrophys. J. Lett. **783**, 21 (2014). doi:10.1088/2041-8205/783/1/L21
T.W. Speiser, Particle trajectories in model current sheets. 1. Analytical solutions. J. Geophys. Res. **70**, 4219–4226 (1965). doi:10.1029/JZ070i017p04219
E. Striani, M. Tavani, V. Vittorini, I. Donnarumma, A. Giuliani, G. Pucella, A. Argan, A. Bulgarelli, S. Colafrancesco, M. Cardillo, E. Costa, E. Del Monte, A. Ferrari, S. Mereghetti, L. Pacciani, A. Pellizzoni, G. Piano, C. Pittori, M. Rapisarda, S. Sabatini, P. Soffitta, M. Trifoglio, A. Trois, S. Vercellone, F. Verrecchia, Variable gamma-ray emission from the Crab Nebula: Short flares and long "waves". Astrophys. J. **765**, 52 (2013). doi:10.1088/0004-637X/765/1/52
P.H.T. Tam, R.H.H. Huang, J. Takata, C.Y. Hui, A.K.H. Kong, K.S. Cheng, Discovery of GeV γ-ray emission from PSR B1259-63/LS 2883. Astrophys. J. Lett. **736**, 10 (2011). doi:10.1088/2041-8205/736/1/L10
P.H.T. Tam, K.L. Li, J. Takata, A.T. Okazaki, C.Y. Hui, A.K.H. Kong, High-energy observations of PSR B1259-63/LS 2883 through the 2014 periastron passage: Connecting X-rays to the GeV flare. Astrophys. J. Lett. **798**, 26 (2015). doi:10.1088/2041-8205/798/1/L26
M. Tavani, J. Arons, Theory of high-energy emission from the pulsar/Be star system PSR 1259-63. I. Radiation mechanisms and interaction geometry. Astrophys. J. **477**, 439–464 (1997)
M. Tavani, A. Bulgarelli, V. Vittorini, A. Pellizzoni, E. Striani, P. Caraveo, M.C. Weisskopf, A. Tennant, G. Pucella, A. Trois, E. Costa, Y. Evangelista, C. Pittori, F. Verrecchia, E. Del Monte, R. Campana, M. Pilia, A. De Luca, I. Donnarumma, D. Horns, C. Ferrigno, C.O. Heinke, M. Trifoglio, F. Gianotti, S. Vercellone, A. Argan, G. Barbiellini, P.W. Cattaneo, A.W. Chen, T. Contessi, F. D'Ammando, G. DeParis, G. Di Cocco, G. Di Persio, M. Feroci, A. Ferrari, M. Galli, A. Giuliani, M. Giusti, C. Labanti, I. Lapshov, F. Lazzarotto, P. Lipari, F. Longo, F. Fuschino, M. Marisaldi, S. Mereghetti, E. Morelli, E. Moretti, A. Morselli, L. Pacciani, F. Perotti, G. Piano, P. Picozza, M. Prest, M. Rapisarda, A. Rappoldi, A. Rubini, S. Sabatini, P. Soffitta, E. Vallazza, A. Zambra, D. Zanello, F. Lucarelli, P. Santolamazza, P. Giommi, L. Salotti, G.F. Bignami, Discovery of powerful gamma-ray flares from the Crab Nebula. Science **331**, 736 (2011). doi:10.1126/science.1200083
A. Tchekhovskoy, A. Spitkovsky, J.G. Li, Time-dependent 3D magnetohydrodynamic pulsar magnetospheres: Oblique rotators. Mon. Not. R. Astron. Soc. **435**, 1–5 (2013). doi:10.1093/mnrasl/slt076
Y. Teraki, F. Takahara, Jitter radiation model of the Crab gamma-ray flares. Astrophys. J. **763**, 131 (2013). doi:10.1088/0004-637X/763/2/131
J.A. Tomsick, A. Bodaghee, J. Rodriguez, S. Chaty, F. Camilo, F. Fornasini, F. Rahoui, Is IGR J11014-6103 a pulsar with the highest known kick velocity? Astrophys. J. Lett. **750**, 39 (2012). doi:10.1088/2041-8205/750/2/L39
D.A. Uzdensky, B. Cerutti, M.C. Begelman, Reconnection-powered linear accelerator and gamma-ray flares in the Crab Nebula. Astrophys. J. Lett. **737**, 40 (2011). doi:10.1088/2041-8205/737/2/L40
A. Van Etten, R.W. Romani, Multi-zone modeling of the pulsar wind nebula HESS J1825-137. Astrophys. J. **742**, 62 (2011). doi:10.1088/0004-637X/742/2/62
M. Vigelius, A. Melatos, S. Chatterjee, B.M. Gaensler, P. Ghavamian, Three-dimensional hydrodynamic simulations of asymmetric pulsar wind bow shocks. Mon. Not. R. Astron. Soc. **374**, 793–808 (2007). doi:10.1111/j.1365-2966.2006.11193.x
D. Volpi, L. Del Zanna, E. Amato, N. Bucciantini, Non-thermal emission from relativistic MHD simulations of pulsar wind nebulae: From synchrotron to inverse Compton. Astron. Astrophys. **485**, 337–349 (2008). doi:10.1051/0004-6361:200809424

Q.D. Wang, E.V. Gotthelf, Y.-H. Chu, J.R. Dickel, Detection of an X-ray pulsar wind nebula and tail in SNR N157B. Astrophys. J. **559**, 275–281 (2001). doi:10.1086/322392

M. Weisskopf, Observations of the Crab Nebula with the Chandra X-ray observatory, in *Proceedings of "An INTEGRAL View of the High-Energy Sky (the First 10 Years)"—9th INTEGRAL Workshop and Celebration of the 10th Anniversary of the Launch*, INTEGRAL 2012, 15–19 October 2012 (Bibliotheque Nationale de France, Paris, 2012), p. 4. Published online at http://pos.sissa.it/cgi-bin/reader/conf.cgi?confid=176, id. 4

M.C. Weisskopf, E.H. Silver, H.L. Kestenbaum, K.S. Long, R. Novick, A precision measurement of the X-ray polarization of the Crab Nebula without pulsar contamination. Astrophys. J. Lett. **220**, 117–121 (1978). doi:10.1086/182648

M.C. Weisskopf, J.J. Hester, A.F. Tennant, R.F. Elsner, N.S. Schulz, H.L. Marshall, M. Karovska, J.S. Nichols, D.A. Swartz, J.J. Kolodziejczak, S.L. O'Dell, Discovery of spatial and spectral structure in the X-ray emission from the Crab Nebula. Astrophys. J. Lett. **536**, 81–84 (2000). doi:10.1086/312733

M.C. Weisskopf, R.F. Elsner, J.J. Kolodziejczak, S.L. O'Dell, A.F. Tennant, Unraveling the geometry of the Crab Nebula's "inner ring". Astrophys. J. **746**, 41 (2012). doi:10.1088/0004-637X/746/1/41

M.C. Weisskopf, A.F. Tennant, J. Arons, R. Blandford, R. Buehler, P. Caraveo, C.C. Cheung, E. Costa, A. de Luca, C. Ferrigno, H. Fu, S. Funk, M. Habermehl, D. Horns, J.D. Linford, A. Lobanov, C. Max, R. Mignani, S.L. O'Dell, R.W. Romani, E. Striani, M. Tavani, G.B. Taylor, Y. Uchiyama, Y. Yuan, Chandra, Keck, and VLA observations of the Crab Nebula during the 2011-April gamma-ray flare. Astrophys. J. **765**, 56 (2013). doi:10.1088/0004-637X/765/1/56

G.R. Werner, D.A. Uzdensky, B. Cerutti, K. Nalewajko, M.C. Begelman, The extent of power-law energy spectra in collisionless relativistic magnetic reconnection in pair plasmas. ArXiv e-prints (2014)

C.A. Wilson-Hodge, M.L. Cherry, G.L. Case, W.H. Baumgartner, E. Beklen, P. Narayana Bhat, M.S. Briggs, A. Camero-Arranz, V. Chaplin, V. Connaughton, M.H. Finger, N. Gehrels, J. Greiner, K. Jahoda, P. Jenke, R.M. Kippen, C. Kouveliotou, H.A. Krimm, E. Kuulkers, N. Lund, C.A. Meegan, L. Natalucci, W.S. Paciesas, R. Preece, J.C. Rodi, N. Shaposhnikov, G.K. Skinner, D. Swartz, A. von Kienlin, R. Diehl, X.-L. Zhang, When a standard candle flickers. Astrophys. J. Lett. **727**, 40 (2011). doi:10.1088/2041-8205/727/2/L40

Q. Yuan, P.-F. Yin, X.-F. Wu, X.-J. Bi, S. Liu, B. Zhang, A statistical model for the γ-ray variability of the Crab Nebula. Astrophys. J. Lett. **730**, 15 (2011). doi:10.1088/2041-8205/730/2/L15

F. Yusef-Zadeh, B.M. Gaensler, A radio study of the mouse, G359.23-0.82. Adv. Space Res. **35**, 1129–1136 (2005). doi:10.1016/j.asr.2005.03.003

V. Zabalza, V. Bosch-Ramon, F. Aharonian, D. Khangulyan, Unraveling the high-energy emission components of gamma-ray binaries. Astron. Astrophys. **551**, 17 (2013). doi:10.1051/0004-6361/201220589

D. Zyuzin, Y. Shibanov, A. Danilenko, R.E. Mennickent, S. Zharikov, The Vela pulsar and its likely counter-jet in the K_s band. Astrophys. J. **775**, 101 (2013). doi:10.1088/0004-637X/775/2/101

DOI 10.1007/978-1-4939-3550-5_13
Reprinted from *Space Science Reviews* Journal, DOI 10.1007/s11214-015-0174-7

Disks and Jets

Gravity, Rotation and Magnetic Fields

John F. Hawley[1] · **Christian Fendt**[2] · **Martin Hardcastle**[3] · **Elena Nokhrina**[4] · **Alexander Tchekhovskoy**[5]

Received: 26 November 2014 / Accepted: 11 June 2015 / Published online: 19 June 2015

Abstract Magnetic fields are fundamental to the dynamics of both accretion disks and the jets that they often drive. We review the basic physics of these phenomena, the past and current efforts to model them numerically with an emphasis on the jet-disk connection, and the observational constraints on the role of magnetic fields in the jets of active galaxies on all scales.

Keywords Accretion · Accretion disks · Black hole physics · Jets · MHD · Instabilities

1 Introduction

The topic of this workshop was the diverse astrophysical phenomena enabled by the strongest magnetic fields in the Universe. Gamma ray bursts and magnetars provide two

A. Tchekhovskoy is an Einstein Fellow.

✉ J.F. Hawley
jh8h@virginia.edu

C. Fendt
fendt@mpia.de

M. Hardcastle
m.j.hardcastle@herts.ac.uk

E. Nokhrina
nokhrinaelena@gmail.com

A. Tchekhovskoy
atchekho@berkeley.edu

1 Department of Astronomy, University of Virginia, Charlottesville, USA

2 Max Planck Institute for Astronomy, Königstuhl 17, 69117 Heidelberg, Germany

3 School of Physics, Astronomy and Mathematics, U. Hertfordshire, Hatfield, UK

4 Moscow Institute of Physics and Technology, Moscow, Russia

5 Department of Physics and Department of Astronomy, UC Berkeley, Berkeley, CA, USA

examples of extreme phenomena where extraordinarily strong fields are manifest. However, the concept of "strong" is relative. A magnetic field may be weak in the sense that the energy density associated with the field is small compared to other energies, e.g., thermal, rotational, gravitational. Such "weak" magnetic fields can, however, enable phenomena that otherwise would not be possible. In this contribution we consider just such an example, namely accretion disks and the jets they produce. Magnetic fields are fundamental to both accretion and outflows, and while in some specific cases the magnetic fields may be strong in the conventional sense of total energetics, for the most part the fields are likely to be relatively weak by those same conventional metrics. Nevertheless, magnetic fields play the central dynamic role.

Gravity is the dominant force in creating the structure in the universe. When gravity is combined with angular momentum, the result is a disk. Since some angular momentum will inevitably be present, disks are arguably the most basic type of object in the universe. Other structures, such as stars, result from disks that have somehow shed most of their angular momentum. Accretion is also a major astrophysical power source. As Lynden-Bell observed (Lynden-Bell 1969), the gravitational energy available in black hole accretion greatly exceeds that available through nuclear reactions, with efficiencies anywhere from 6–40 % mc^2, versus less than 1 % for nuclear fusion.

Direct observational evidence for magnetic fields in accretion disks is very limited (measurements exist, e.g., for FU Orionis, Donati et al. 2005). The strongest case for magnetic fields is theoretical. We know from observations that disks accrete at rates that require considerable internal stress to transport angular momentum. As we shall discuss, this requires magnetic fields.

Indirect evidence for magnetic fields in accreting systems comes from observations of highly collimated outflows, or jets. Jets are one of the most striking signatures resulting from accretion, and are seemingly ubiquitous; they are associated with young stars, micro-quasars or X-ray binaries (MQs, XRBs), and active galactic nuclei (AGN). The current understanding of jet formation is that outflows are launched by *magnetohydrodynamic* (MHD) processes in the close vicinity of the central object (Blandford and Payne 1982; Pudritz and Norman 1983; Pudritz et al. 2007; Shang et al. 2007) It is furthermore clear that accretion and ejection are related to each other. Jets may be a consequence of accretion, but accretion may, in turn, be a consequence of a jet. Accretion occurs when gas loses some of its angular momentum, and one very efficient way to remove angular momentum from a disk is to eject it vertically into a jet or outflow (Wardle and Königl 1993; Ferreira and Pelletier 1993; Li 1995; Ferreira 1997). Not all details of the physical processes involved are completely understood yet.

In stellar sources, a central stellar magnetic field is surrounded by a disk carrying its own magnetic flux. Such a geometrical setup can be found in young stars, cataclysmic variables, high-mass and low-mass X-ray binaries, and other micro-quasar systems. Systems such as AGN or μ-quasars have a black hole at their center, and a black hole cannot have an intrinsic magnetic field. The surrounding accretion disk, however, can support the magnetic flux needed for jet launching. In this case, the interaction of the black hole itself with the ambient field may launch a highly magnetized Poynting flux-driven axial flow (Blandford and Znajek 1977).

Although jets are found in a wide range of systems, from protostars to AGN, from the observational point of view protostellar and relativistic jets are quite different. For protostars we can measure Doppler-shifted forbidden line emission to obtain gas velocities, densities or temperatures, and these provide a good estimate on the mass flux of the outflow. Evidence concerning the magnetic field, however, is mainly indirect: magnetic flares of large-scale protostellar stellar magnetic fields indicate that protostellar jets originating from an area close to the star, originate in a magnetized environment. As noted above, in a few objects there are direct measurements of the strength or orientation of magnetic fields in the disks (e.g. Donati et al. 2005; Stephens et al. 2014). Concerning the jet magnetic field, direct evidence is less clear. So far an estimate is available for only one source, HH80/81, obtained from observed synchrotron emission (Carrasco-González et al. 2010). The situation is completely reversed for relativistic jets. While there is plenty of information concerning the jet magnetic field structure, and some constraints on field strengths, speeds are harder to measure, there is no direct measure of the mass fluxes involved, and the magnetic properties of the accretion flow are not directly observable. We do not even know for certain whether the matter content of these jets is leptonic or hadronic (Sect. 5).

While protostars and AGN are the most common jet sources known, the typical characteristics of jet systems—the existence of an accretion disk and a strong magnetic field—are also present in other astrophysical sources, namely in cataclysmic variables, high mass or low mass X-ray binaries, and pulsars. Resolved Chandra imaging of the Crab (Hester et al. 2002) and the Vela pulsar (Pavlov et al. 2003) do indeed show elongated, highly time-variable but persistent structures emerging along the rotational axis of these systems. Whether these structures are similar to the jets observed in protostars and AGN or are features intrinsic to the pulsar wind nebula is not yet clear. *Persistent* jets are not widely observed in cataclysmic variables (either in the highly transient dwarf novae or the more stable nova-like objects), although there are reports of unresolved, steep-spectrum radio emission, consistent with jet activity, in both types of system (Körding et al. 2008, 2011). It is not clear why jets are not more commonly observed in these systems, although some jet launching models suggesting explanations exist in the literature (Soker and Lasota 2004). One basic reason may be the lack of axisymmetry in cataclysmic variables, as this is thought to be another major condition to launch a jet for a considerable period of time (Fendt and Cemeljic 1998). Alternatively, the observability of jet-related synchrotron emission depends strongly on the local environment of the jet, and it is possible that the conditions for this emission to be observed are simply not typically met in CVs.

In this workshop proceeding we will review the many fundamental roles played by magnetic fields in accretion and outflows, both from a theoretical and an observational point of view. The principal processes involved in jet formation can be summarized as follows (Blandford and Payne 1982; Pudritz and Norman 1983; Pudritz et al. 2007; Shang et al. 2007),

(1) Jets are powered by gravitational energy released through accretion and by rotational energy of the disk and/or the central star (or black hole). Magnetic flux is provided by the star-disk (or black hole-disk) system, possibly by a disk or stellar dynamo, or by the advection of the interstellar field. The star-disk system also drives an electric current.
(2) Accreting plasma is diverted and launched as a plasma wind (from the stellar or disk surface) coupled to the magnetic field and accelerated magneto-centrifugally.
(3) Inertial forces wind up the poloidal field inducing a toroidal component.
(4) The jet plasma is accelerated magnetically (conversion of Poynting flux).
(5) The toroidal field tension collimates the outflow into a high-speed jet beam.
(6) The plasma velocities subsequently exceed the speed of the magnetosonic waves. The super-fast magnetosonic regime is causally decoupled from the surrounding medium.
(7) Where the outflow meets the ISM, a shock develops, thermalizing the jet energy.

The overall process as outlined above is complex and takes place over a wide range of both length- and time-scales. Due to this complexity, the aspects of the jet problem have to

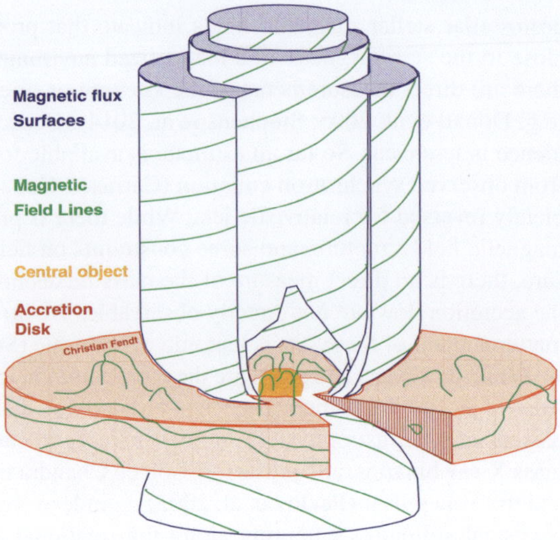

Fig. 1 Model sketch of MHD jet formation, indicating the six generic problems to be solved: launching, acceleration, and collimation; disk structure; magnetic field origin; asymptotic interaction; central spine jet

be tackled independently. One may distinguish six principal topics (Fig. 1), roughly corresponding to the stages in the overall picture described above. These include:

(1) The accretion disk structure and its evolution, including thermal effects and the origin of turbulence.
(2) The origin of the jet magnetic field, possibly through a disk dynamo, a stellar dynamo, or by the advection of ambient magnetic field.
(3) The ejection of disk material into wind, thus the transition from accretion to ejection.
(4) The collimation and acceleration of ejected material into jets.
(5) The propagation of the asymptotic jet, its stability and interaction with ambient medium. A related question is the feedback of jets on star or galaxy formation.
(6) The possible presence and impact of a central spine jet, e.g. a stellar wind or black hole jet, in comparison to a jet originating with the disk. Under what circumstances are disk jets and spine jets present or absent?

In this article we begin by describing the theoretical support for the role of magnetic fields in the accretion disk itself (Sect. 2). We then discuss the modeling of magnetically driven jets (Sect. 3) and their relationship to the disk (Sect. 4). Observational constraints on field properties and their relation to models are discussed in Sect. 5. Our conclusions are presented in Sect. 6.

2 Magnetic Fields and Accretion Disks

We begin with the accretion disk itself, and the fundamental role played by magnetic fields. Throughout the history of disk theory, "shedding the angular momentum" has been easier said than done. The absence of an obvious mechanism to do so was considered a serious objection to Laplace's disk hypothesis for solar system formation, for example. By the time that the basic theory of accretion was laid out by Shakura and Sunyaev (1973) and Lynden-Bell and Pringle (1974) it was accepted that Nature accomplished the transport of angular momentum by some means, though that means was unknown to astronomers. Shakura and Sunyaev introduced the "alpha viscosity" parameterization as a means of sidestepping the uncertainty. In this model the internal stress is proportional to the total pressure, $T_{R\phi} = \alpha P$.

The physical viscosity in the gas within an accretion disk is known to be far too small to account for observed accretion rates. If the disk were turbulent, however, the Reynolds stresses associated with that turbulence could transport angular momentum at the necessary rate. This seemed like a straightforward solution to the problem. Because the viscosity is so small, the Reynolds number (ratio of characteristic velocity times a characteristic length divided by the viscosity) of the gas is correspondingly huge, and in terrestrial contexts high Reynolds number flows are generally turbulent. But gas in Keplerian orbits is stable to perturbations by the Rayleigh criterion, which simply requires that angular momentum increase with radius, $dL/dR > 0$. The positive epicyclic frequency associated with Keplerian orbital flows is strongly stabilizing, and it is not energetically favorable for turbulence to develop and be sustained from the background angular momentum distribution (Balbus et al. 1996; Balbus and Hawley 1998).

The linear stability of Keplerian flows was, of course, recognized early on as a difficulty for the creation of turbulence in disks. However, it was known that in the laboratory simple shear flows display *nonlinear* instability. Balbus et al. (1996) argued that this nonlinear instability represented the boundary between the Rayleigh-stable and Rayleigh-unstable regimes; when the epicyclic frequency goes to zero the linear response vanishes, leaving nonlinear terms as the lowest order effect. Given the importance of the question of the nonlinear stability of hydrodynamic Keplerian flow, several groups have carried out fluid experiments. Ji et al. (2006) examined the stability of a Rayleigh-stable Couette flow and found no evidence of significant turbulence. Paoletti and Lathrop (2011), on the other hand, found a breakdown into turbulence in their experiment. The experiments of Schartman et al. (2012) found no transition to turbulence, and attributed the Paoletti and Lathrop (2011) results to turbulence generated at the end caps of the Couette cylinder, a conclusion further supported by the numerical simulations of Avila (2012). More recent laboratory experiments by Edlund and Ji (2014) are also consistent with stability. While work will no doubt continue on this important issue, at the moment the weight of the theoretical and experimental data lie with the nonlinear hydrodynamic stability of Keplerian flow.

2.1 Basic MRI Physics

It was, of course, recognized early on that magnetic fields could transport angular momentum through Maxwell stresses, $-B_r B_\phi/4\pi$, but unless the magnetic pressure were near the thermal pressure, it was expected that the effective α would be relatively small. As it turns out, magnetic fields render the disk linearly unstable to the *magneto-rotational instability* (Balbus and Hawley 1991, MRI) even when (and, indeed, *only* when) the magnetic field is weak. The MRI can be visualized intuitively in the form of a spring connecting two orbiting masses (Balbus and Hawley 1998), one in a slightly lower orbit than the other. Because angular velocity decreases with R, $\Omega \propto R^{-3/2}$, the inner mass pulls ahead of the outer mass. This causes the spring to stretch; the tension force pulls back on the inner mass, and pulls forward on the outer. The effect of this is to transfer some angular momentum from the inner to the outer mass, but in doing so the inner mass drops to a lower orbit, increasing its angular velocity, increasing the relative separation of the masses and increasing the spring tension force. The process runs away unless the spring is strong enough to force the masses to remain together, which happens only if the tension exceeds the orbital dynamical force. If we replace the spring with a magnetic field, the spring tension is replaced with the magnetic tension associated with the Maxwell stress. In a magnetized disk the linear stability requirement is no longer the Rayleigh criterion, but is instead a requirement that angular *velocity* increase outward, $d\Omega/dR > 0$. A Keplerian disk is unstable.

There are several remarkable properties of the MRI. One is that the question of stability does not depend on the magnetic field strength or orientation. The stability criterion,

$$(\mathbf{k} \cdot \mathbf{v_A})^2 > -\frac{d\Omega^2}{d\ln R}, \tag{1}$$

includes the magnetic field through the Alfv́en speed $\mathbf{v_A}$ but only in combination with a wavenumber vector $\mathbf{k}$. Hence, for any magnetic field, however weak, one can find an unstable wavenumber so long as Ω decreases with R.

The stability criterion for a Keplerian disk is

$$\frac{2\pi}{\lambda_{\rm MRI}} v_A > \sqrt{3}\Omega, \tag{2}$$

which indicates that the unstable wavelengths, $\lambda_{\rm MRI}$, are those that are longer than the distance an Alfvén wave travels in one orbit. The growth rate of the MRI is proportional to $k \cdot v_A$, peaking at a value of $\frac{3}{4}\Omega$ for Keplerian disks for a wavelength (Balbus and Hawley 1998)

$$\lambda_{\rm MRI} = \frac{4}{\sqrt{15}} \frac{2\pi v_A}{\Omega}. \tag{3}$$

As a practical matter, for an increasingly weak field, $\lambda_{\rm MRI}$ should eventually fall to a scale small enough that ideal MHD no longer applies. When the MRI wavelengths are comparable to the resistive scale, for example, the field diffuses through the gas faster than the instability can grow, resulting in stabilization. At the other extreme, namely the strong field limit, the shortest unstable wavelength must fit into the disk. When $\lambda_{\rm MRI}$ is comparable to the scale height of the disk, H, then $v_A \sim c_s$, where c_s is the sound speed. This is a strong field indeed, and this condition is often expressed in terms of the plasma β parameter, which is the ratio of the thermal to magnetic pressure, as $\beta \leq 1$. Strong fields do not necessarily mean a stable, quiet disk, however. First, when $\beta < 1$ the fields are dynamically important and can transport angular momentum directly, possibly through a wind or a jet (see, for example, Lesur et al. 2013). Second, the stability condition on a toroidal field requires the Alfvén speed to be comparable to the *orbital* speed, rather than the sound speed; confining fields of this strength within a disk would seem to be problematic.

To conclude, disks for which the gas has sufficient conductivity as to be magnetized will be linearly unstable to the MRI. The action of the MRI is to transfer angular momentum outward through the disk, precisely what is needed for accretion.

2.2 MRI-Driven Turbulence

The linear analysis establishes that Keplerian orbits are unstable in the presence of a weak magnetic field. For any further understanding of the properties of an unstable disk we must turn to numerical simulations. Accretion simulations can be global or local. Global simulations model the whole disk, or at least a region of significant radial extent. Local simulations, on the other hand, attempt to focus on the properties of the gas within a small region of the disk. For accretion disks, local simulations use an approximation known as the *shearing box* (Hawley et al. 1995). The shearing box domain is assumed to be centered at some radial location R that has an orbital frequency $\Omega(R)$. The size of the box is taken to be much less than R so that one can assume a local Cartesian geometry. The tidal gravitational and

Coriolis forces are retained. Radial boundary conditions are established by assuming that the box is surrounded by identical boxes sliding past at the appropriate shear rate. Through use of the shearing box approximation one can study the details of MRI-driven turbulence on scales that are much smaller than the scale height of the disk or the radial distance from the central star.

Extensive shearing box simulations have led to the following general conclusions:

- The MRI leads to MHD turbulence characterized by an anisotropic stress tensor: the $R\phi$ magnetic stress component, $T_{R\phi}$, is large, which leads to significant radial transport of net angular momentum (Hawley et al. 1995).
- A net Reynolds stress is also present in the turbulence, and it too transports angular momentum, but the Maxwell stress is consistently larger than the Reynolds stress by a factor of 3–4. If the magnetic fields are removed, the turbulence quickly dies out (Hawley et al. 1995).
- Time and space variations in the turbulence can be large. The turbulence is chaotic (Winters et al. 2003). Characterizing the stress in terms of an α parameter makes sense only in a space- and time-averaged sense.
- The stress is proportional to the magnetic pressure, $T_{R\phi} = \alpha_{\rm mag} P_{\rm mag}$, with an $\alpha_{\rm mag} = 0.4$–0.5. This implies that the traditional Shakura-Sunyaev value is $\alpha \sim 1/\beta$, where β is determined by $P_{\rm mag}$, the total magnetic pressure in the turbulent flow.
- The magnetic energy typically saturates at a value $\beta \sim 10$–100. This is somewhat dependent on the character of the background magnetic field. Simulations with net vertical field tend to saturate at stronger field values.
- The local model is valid in so far as the energy released by the stresses is thermalized promptly, on an eddy turnover time-scale of order Ω^{-1} (Simon et al. 2009). The rapid thermalization of the energy is required for α-disk theory to be applicable (Balbus and Papaloizou 1990).
- The MRI acts like an MHD dynamo in that it sustains positive magnetic field energies in the face of considerable dissipation (Hawley et al. 1996). Further, stratified shearing box simulations exhibit a behavior that can be modeled as an α–Ω dynamo (Brandenburg et al. 1995; Gressel 2010; Guan and Gammie 2011). This is, however, a small-scale dynamo; shearing box simulations have not provided any evidence for the existence of a dynamo capable of producing a large scale field.
- In simulations that include resistivity and viscosity, the amplitude of the MRI-driven turbulent fluctuations and the resulting angular momentum transport are a function of magnetic Prandtl number. Turbulence is not sustained for $P_{\rm M} < 1$ (Lesur and Longaretti 2007; Simon and Hawley 2009), i.e., when the resistivity is greater than the viscosity.

The local simulations tell us a considerable amount about the MRI and the resulting turbulence, but they cannot address questions of a global nature, such as the net accretion rate into the central star, losses due to winds and outflows, and the mechanisms by which jets might be generated. We know, however, that disks will be magnetized, and this implies the possibility of jet generation from the disk. In the next section we discuss the modeling of such jets.

3 Magnetic Jets from Disks—Theory and Simulations

For jet theory and simulation researchers typically consider only a subset of the issues listed in Sect. 1 and employ a variety of simplifications. An early example of a jet simulation that focused on the propagation of the jet and interaction with the surrounding medium is given by Norman et al. (1982) who carried out some of the first high-resolution simulations of hydrodynamic axisymmetric cylindrical jet flows. Some of the earliest simulations of questions related to jet launching, collimation in relation to disk properties were given by Uchida and Shibata (1984) (see also Shibata and Uchida 1985; Uchida and Shibata 1985).

Since those early efforts, considerable work has been done on MHD jet modeling. One may distinguish i) between steady-state models and time-dependent numerical simulations, and also ii) between simulations considering the jet formation only from a fixed-in-time disk surface and simulations considering also the launching process, thus taking into account disk and jet evolution together.

Clearly, many jet properties depend on the mass loading of the jet, which can only be inferred from a treatment of the accretion-ejection process. While numerical simulations of the accretion-ejection structure potentially provide the time-evolution of the launching process, a number of constraints that were found, had been discovered already by previous steady state models (see below).

3.1 Magneto-centrifugal Disk Winds

Steady-state modeling of magneto-centrifugally launched disk winds have mostly followed the self-similar Blandford and Payne (1982) approach (e.g. Contopoulos and Lovelace 1994a; Contopoulos 1994b). Some fully two-dimensional models have been proposed (Pelletier and Pudritz 1992; Li 1993), including some that take into account the central stellar dipole (Fendt et al. 1995). Further, some numerical solutions have been proposed by e.g., Wardle and Königl (1993); Königl et al. (2010); Salmeron et al. (2011) in weakly ionized accretion disks that are threaded by a large-scale magnetic field as a wind-driving accretion disk. They have studied the effects of different regimes for ambipolar diffusion or Hall and Ohm diffusivity dominance in these disk. Self-similar steady-state models have also been applied to the jet launching domain (Wardle and Königl 1993; Ferreira and Pelletier 1993; Li 1995; Ferreira 1997; Casse and Ferreira 2000) connecting the collimating outflow with the accretion disk structure. In particular, the fact that large scale magnetic fields need to be close to equipartition in order to launch jets via the Blandford-Payne mechanism is now well accepted, but was first established with self-similar studies (Ferreira and Pelletier 1993; Li 1995; Ferreira 1997). A similar comment can be made also concerning the turbulent magnetic diffusivity required in accretion disks.

In addition to the steady-state approach, the magneto-centrifugal jet formation mechanism has been the subject of a number of time-dependent numerical studies. In particular, Ustyugova et al. (1995) and Ouyed and Pudritz (1997) demonstrated the feasibility of the MHD self-collimation property of jets. Among these works, some studies have investigated artificial collimation (Ustyugova et al. 1999), a more consistent disk boundary condition (Krasnopolsky et al. 1999; Anderson et al. 2005), the effect of magnetic diffusivity on collimation (Fendt and Cemeljic 2002), the impact of the disk magnetization profile on collimation (Fendt 2006; Pudritz et al. 2006), or the impact of reconnection flares on the mass flux in jets from a two-component magnetic field consisting of a stellar dipole superposed on a disk magnetic field (Fendt 2009).

In the context of core-collapse gamma-ray bursts, Tchekhovskoy et al. (2008) studied the acceleration of magnetically-dominated jets confined by an external medium and demonstrated that jets gradually accelerate under the action of magnetic forces to Lorentz factors $\Gamma \gtrsim 1000$ as they travel from the compact object to the stellar surface. However, Komissarov et al. (2009) pointed out an important problem: such confined, magnetized jets were too tightly collimated for their Lorentz factors to be consistent with observations. Does this mean that gamma-ray burst jets are not magnetically powered? It turns out that magnetized jets are still viable: as they exit the star and become deconfined, they experience an additional, substantial burst of acceleration that brings their properties into agreement with observations (Tchekhovskoy et al. 2010a).

In the context of stellar jets, Ramsey and Clarke (2011) studied the large-scale jet formation process spanning the whole range from the disk surface out to scales of more than 1000 AU. Further extensions of this model approach have included the implementation of radiation pressure by line forces as applied to jets in AGN (Proga et al. 2000; Proga and Kallman 2004) or massive young stars (Vaidya et al. 2011), which may well affect acceleration and collimation of the jet material. Simulations of relativistic MHD jet formation were presented by Porth and Fendt (2010) finding collimated jets just as for the non-relativistic case (above). Applying relativistic polarized synchrotron radiative transport to these MHD simulation data through postprocessing yields mock observations of small-scale AGN jets (Porth et al. 2011).

The stability of the jet formation site has been studied using 3D simulations for the non-relativistic (Ouyed et al. 2003) and the relativistic case (Porth 2013). Self-stabilization of the jet formation mechanism seems to be enforced by the magnetic "backbone" of the jet, the very inner highly magnetized axial jet region. These studies complement simulations investigating the stability of jet propagation on the asymptotic scales. Here, usually a collimated jet is injected into an ambient gas distribution, either for the non-relativistic (Stone and Norman 1992; Todo et al. 1993; Stone and Norman 1994; Stone and Hardee 2000) or the relativistic (Mignone et al. 2010; Keppens et al. 2008) case.

In the aforementioned studies, the jet-launching accretion disk is taken into account as a boundary condition, *prescribing* a certain mass flux or magnetic flux profile in the outflow. This may be a reasonable setup in order to investigate jet formation, i.e. the acceleration and collimation process of a jet. However, such simulations cannot tell the efficiency of mass loading or angular momentum loss from disk to jet, or cannot determine which kind of disks launch jets and under which circumstances.

It is therefore essential to extend the jet formation setup and include the launching process in the simulations—that is, to simulate the accretion-ejection transition. Clearly, this approach is computationally much more expensive. The typical time scales for the jet and disk region differ substantially; disk physics operates on the Keplerian time scale (which increases as $R^{3/2}$), and on the even-longer viscous and the diffusive time scales. The jet follows a much faster dynamical time scale. As this approach is also limited by spatial and time resolution, jet launching simulations to date have employed a rather simple disk model—namely, an α-prescription for the disk turbulent magnetic diffusivity and viscosity, and without considering radiative effects.

Numerical simulations of the launching of MHD jets from accretion disks have been presented by Kudoh et al. (1998); Kato et al. (2002) and Casse and Keppens (2002); Casse and Keppens (2004). These simulations treat the ejection of a collimated outflow out of an evolving disk through which a magnetic field is threaded. To prevent the field from accreting itself, the disk is resistive, allowing the field to slip through. Zanni et al. (2007) further developed this approach with emphasis on how resistive effects modify the dynamical evolution. An additional central stellar wind was considered by Meliani et al. (2006). Further studies have concerned the effects of the absolute field strength or the field geometry, in particular investigating field strengths around and below equipartition (Kuwabara et al. 2005; Tzeferacos et al. 2009; Murphy et al. 2010). These latter simulations follow several hundreds of (inner) disk orbital periods, providing sufficient time evolution to also reach a (quasi) steady state for the fast jet flow.

Parameter studies have been able to disentangle the effects of magnetic field strength (magnetization) and magnetic diffusivity (strength, scale height) on mass loading and jet speed (Sheikhnezami et al. 2012). Tzeferacos et al. (2013) considered how entropy affects the launching process, in particular how disk heating and cooling influence the launching process. They find that heating at the disk surface enhances the mass load, as predicted in the steady state modeling by Casse and Ferreira (2000).

Most jets and outflows are observed as *bipolar* streams. Very often, observations show asymmetrical jets and counterjets. For protostellar jets one exception is HH 212, which shows an almost perfectly symmetric bipolar structure (Zinnecker et al. 1998). For relativistic jets Doppler beaming may play a role for the observed jet asymmetries, and, in fact, many well-behaved jets can be modeled on the assumption that Doppler beaming dominates the apparent asymmetry (Laing and Bridle 2014). However, environmental asymmetries must be also considered. It is therefore interesting to investigate the evolution of both hemispheres of a *global* jet-disk system in order to see whether and how a global asymmetry in the large-scale outflow can be governed by the disk evolution. Fendt and Sheikhnezami (2013) have been able to trigger jet asymmetries by disturbing the hemispheric symmetry of the jet-launching accretion disk and find mass flux or jet velocity differences between jet and counter jet of up to 20 %.

von Rekowski et al. (2003) and von Rekowski and Brandenburg (2004) investigated the origin of the magnetic field driving the jet by including a mean-field disk dynamo in the star or in the disk. Asymmetric ejections of stellar wind components were found from offset multi-pole stellar magnetospheres (Lovelace et al. 2010). The most recent work considers the time ($> 5 \times 10^5$ dynamical time steps) evolution on a large spherical grid (> 2000 inner disk radii), including the action of a mean-field disk dynamo that builds up the jet magnetic field. A variable dynamo action may cause the time-dependent ejection of jet material (Stepanovs and Fendt 2014; Stepanovs et al. 2014).

The main jet launching processes seem to be well understood, in the sense that a large number of independent simulation studies give consistent results. Several issues are not yet resolved, however. These include: (i) the origin of jet knots, and (ii) the coupling of the small scale disk physics to the global jet outflow. In the end there is a good chance that the answers to both questions are interrelated. A full numerical simulation that addresses these questions would require substantial effort, including full 3D, high resolution, and more complete physics, in particular the treatment of thermal effects. First steps in this direction have already been taken.

Concerning the origin of jet knots, it is still unclear whether knots are signatures of internal shocks of jet material launched episodically with different speed, external shocks of jet material with the ambient medium, or re-collimation shocks. One observation supporting the intrinsic origin of jet knots is that of the jet of HH 212, which shows an almost perfectly symmetric bipolar structure with an identical knot separation for jet and counter jet (Zinnecker et al. 1998). Such a structure can only be generated by a mechanism intrinsic to the jet source.

The knot separation in protostellar jets typically corresponds to time scales of $\tau_{\rm kin} \simeq$ 10–100 yrs. By contrast, a typical time scale for the jet launching area would be about 10–20 days, that is the Keplerian period close to an inner disk radius of 0.1 AU. This time scale increases up to one year if the jet launching radius is larger, say up to 3–5 AU (Frank et al. 2014). However, outflows launched from such large radii would probably not achieve the high velocities of 300–500 $\mathrm{km\,s^{-1}}$ observed for jets. Thus, the mechanism responsible for the jet knots must be intrinsic to the disk, and triggered by a physical processes on a rather long time scale. Candidates are disk thermal (FU Orionis) or accretion instabilities, a mid-term variation of the jet-launching magnetic field, or MRI-active/dead disks.

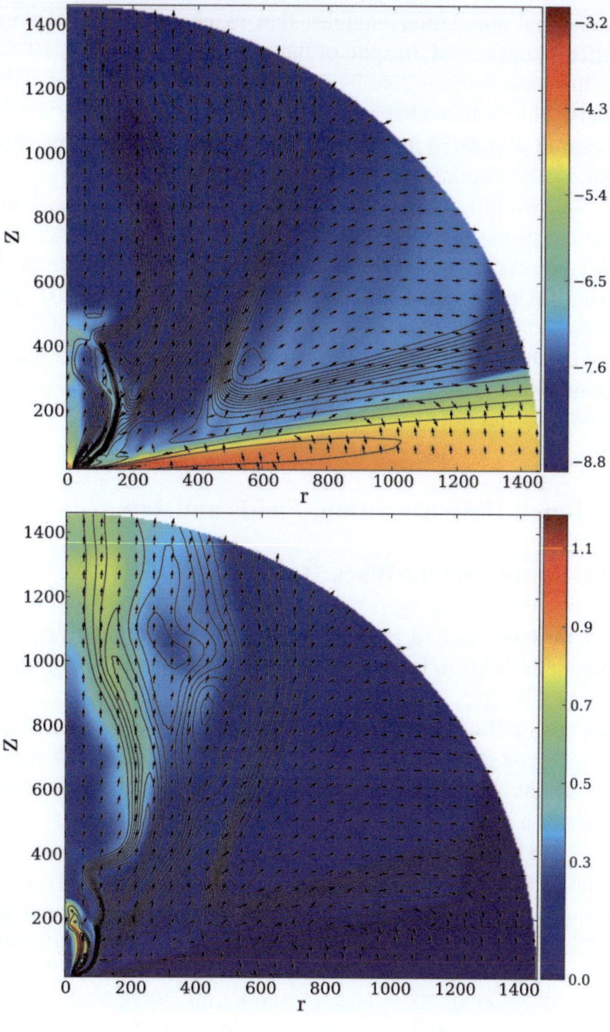

Fig. 2 Time evolution of the inflow-outflow structure for a dynamo-generated disk magnetic field. Shown is the density (*top*, *colors*, in logarithmic scale), and poloidal velocity (*bottom*, *colors*, linear scale), the poloidal magnetic flux (*thin black lines*), and the poloidal velocity vectors. The dynamical time step is about 10^5, while the dynamo time scale (switch on/off) is 10^3. Two ejections of material launched at $\Delta t = 2000$ are more easily seen in the velocity figure (*below*), than in the *top* figure, which shows a slightly later time when the preceding knot has left the grid already. Length units is the inner disk radius of ~ 0.1 AU. Figure taken from Stepanovs and Fendt (2014)

The simulations of Stepanovs et al. (2014) provide an example of episodic knot ejections. They apply an α–Ω mean-field disk dynamo to generate the jet launching magnetic field. In contrary, all previous simulations of this kind start with a strong initial magnetic field distribution. The episodic ejections are triggered by switching on and off the dynamo term every 2000 dynamical time steps (Fig. 2).

The idea of a connection between the small scale disk physics and the global disk outflow is motivated by the fact that jet launching relies on both the existence of a large scale magnetic field and the existence of a (turbulent) magnetic diffusivity, which allows for accretion through the field lines, for angular momentum transfer, and, essentially, also for the mass loading onto the jet magnetic field. The disk turbulence is a result of the magnetorotational instability (MRI, see above) and thus of the small scale physics.

First results combining a mixed numerical-analytical treatment indicate that large scale MRI modes may produce magnetically driven outflows, however, these flows seem to be 3D-unstable (Lesur et al. 2013), and thus may be a transient effect. On the other hand, local numerical simulations suggest that magnetocentrifugal winds can be launched when the MRI is suppressed, in spite of the low magnetization $\sim 10^{-5}$ (Bai and Stone 2013a, 2013b). In this case, outflows are launched from the disk surface region where the plasma β is about unity. This is also what Murphy et al. (2010) have observed in global launching simulations. Lesur et al. (2013) have found that the MRI near equipartition does not lead to turbulence, but can be responsible for jet launching.

The overall goal is still to answer the question: what kind of disks launch jets and what kind of disks do not? It is clear from statistical arguments for both protostars and extragalactic sources that jets are a relatively short-lived phenomenon. The typical jet propagation time scale for protostellar jets is about $\tau_{\rm dyn} \equiv L_{\rm jet}/V_{\rm jet} \simeq 10^4$ yrs, while protostellar disks live some 10^6 yrs. A similar argument can be raised for AGN jets based on the fact that there are far more radio-quiet AGN (quasars) than radio-loud AGN (jet sources), although at least some types of radio-loud AGN appear to be almost continuously in an 'on' state. So far, modeling has not provided an answer to this question.

4 Black Hole Accretion, Fields and Jets

4.1 Simulating the Black Hole Jet

As we have seen, a plethora of models and simulations have demonstrated the ability of magnetic fields to launch, accelerate and collimate jet outflows. But simulations designed to study jets have, as discussed in the previous section, typically used one or another specific set of initial or boundary conditions that pre-suppose a disk structure, a mass-energy flux, a magnetic field configuration, or some combination of these. An alternative approach is to focus not on the jet, but on the accretion disk itself and see when and under what circumstances a jet might *develop* from the flow.

Most global simulations begin with an orbiting axisymmetric hydrostatic equilibrium (torus) a few tens of gravitational radii from the black hole. The initial magnetic field is entirely contained within the matter, so that there is no net magnetic flux and no magnetic field on either the outer boundary or the event horizon. A favorite configuration consists of large concentric dipolar loops (Hawley 2000; McKinney and Gammie 2004; De Villiers et al. 2005; Hawley and Krolik 2006; McKinney 2006). The development of the MRI subsequently leads to the establishment of outflows from the disk. The wind generation seems to be largely an outcome of thermodynamics. In simulations that do not include cooling, the combination of the heat released from accretion and the buildup of magnetic pressure lead to pressure-driven outflows. In the absence of a large scale vertical field through the disk, however, this wind is neither collimated nor unbound.

Global simulations have nevertheless produced jets that are consistent with the Blandford-Znajek mechanism, through the creation of a substantial axial field that threads the black hole. For example, in a model that begins with a torus containing dipolar field loops, differential rotation rapidly generates toroidal field. This field increases the total pressure within the torus, driving the its inner edge inward. As the field is stretched radially, the Maxwell stress transports angular momentum outward, further enhancing the inflow. Subsequent evolution carries the inner edge of the disk, and the magnetic field within, down to the black hole. The field expands rapidly into the nearly empty *funnel* region along the black hole axis. Gas drains off the field lines and into the hole leaving behind a low β axial field. Frame dragging by the rotating hole powers an outgoing Poynting flux. The power of the Poynting flux jet is determined by the spin of the black hole and the strength of the magnetic field (Blandford and Znajek 1977), as we discuss below. The simulated jet's power matches well with the predictions of the Blandford-Znajek model (McKinney and Gammie 2004; McKinney 2005; Tchekhovskoy et al. 2011).

The matter content within the Blandford-Znajek jet is very small; the angular momentum in the disk material is too great to enter the axial funnel. In the simulations the jet density is typically set by the value assigned to the numerical vacuum. How these jets become mass-loaded in Nature is still something of an open issue. Since the boost factor of the jet is determined by the mass loading, Γ is typically large, e.g., $\Gamma \sim 10$ (McKinney 2005). No particular weight can be given to any value of Γ found in simulations, although it is safe to say that the feasibility of values as large as those observed (Sect. 5) has been demonstrated.

Although there is no appreciable matter within the funnel, some of the field lines that lie just outside the funnel pass through the accretion flow, and those field lines can accelerate matter outward in a collimating flow. This "funnel wall jet" was observed in early pseudo-Newtonian simulations (Hawley and Balbus 2002), and is a regular feature of fully relativistic jet simulations, e.g., Hawley and Krolik (2006).

Because the field lines in the jet rotate, the jet carries angular momentum as well as energy away from the black hole. The angular momentum flux can be large, comparable to that brought down by accretion (Gammie et al. 2004; Hawley and Krolik 2006; Beckwith et al. 2008). One potentially important consequence of this was noted by Gammie et al. (2004) who pointed out that electromagnetic angular momentum flux in the jet rises so steeply with black hole spin that it may limit a/M to $\simeq 0.93$ (see also McKinney and Gammie 2004; Hawley and Krolik 2006; Beckwith et al. 2008) or even down to $\lesssim 0.1$, as we discuss below (Tchekhovskoy et al. 2012; Tchekhovskoy 2015). The possibility that magnetic fields could play such an important role in black hole physics is certainly in keeping with the spirit of this workshop!

For a magnetic field strength B at the event horizon, jet power is (roughly) given by the magnetic energy density, B^2, times jet cross-section at its base, r_g^2 (where $r_g = GM/c^2$ is the black hole gravitational radius and M its mass), times the speed $v \sim c$ at which the field moves out (Blandford and Znajek 1977; Tchekhovskoy et al. 2010b),

$$P_{\text{jet}} \sim (a/M)^2 B^2 r_g^2 c \sim \left(a^2 c/M^2 r_g^2\right) \Phi^2 = k\Phi^2, \tag{4}$$

where the effect of black hole rotation is in the $(a/M)^2$ pre-factor, and the magnetic field is expressed in terms of the flux $\Phi \sim B r_g^2$. Since in the course of an observation M and a usually do not change appreciably, the factor $k = a^2 c/M^2 r_g^2$ is a constant, and all changes in jet power are due to changes in the magnetic flux, Φ.

What is the possible range of Φ and the corresponding range of P_{jet}? The lower limit is clear: zero magnetic flux leads to zero power. But what sets the maximum value of Φ? In the simulations of Beckwith et al. (2009) the black hole field strength is comparable to the (mostly gas) pressure in the inner disk, $\beta \sim 1$, suggesting that, in general, equipartition holds. This further suggests that even higher black hole magnetic field strengths are possible for more energetic accretion flows. Tchekhovskoy et al. (2011) carried out numerical simulations of black hole disk-jet systems designed specifically to address the question of maximum field strength. The simulation begins with an initial thick, hot, large scale torus (as shown in Fig. 3(a, b)). Because of the large torus size, it contained a particularly large amount of magnetic flux, much larger than in previous work. Figure 3(a, b) shows the initial condition, compared with initial conditions typically used in previous simulations (Fig. 3(c)). Using these initial conditions Tchekhovskoy et al. (2011) found that magnetic flux on the black hole saturated at an equipartition point, one where the magnetic pressure was sufficient to halt accretion against the inward pull of gravity. In other words, the

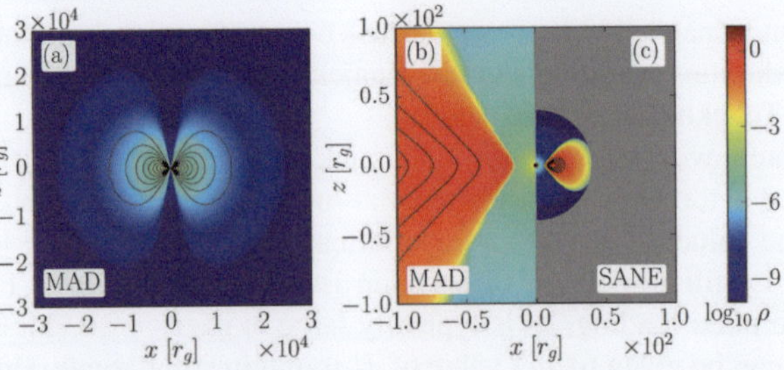

Fig. 3 Panels (**a**) and (**b**) show a meridional slice of density with *color* (*red* shows high, *blue* low density, see *color bar*) and magnetic field lines with *black lines* for the initial conditions of the simulations of Tchekhovskoy et al. (2011) The initial magnetic field is fully contained within the torus, and have a sufficient amount of magnetic flux to saturate the black hole with magnetic flux and to lead to the development of a magnetically-arrested disk (MAD). Panel (**c**) contrasts this initial condition with a typical initial condition in earlier simulations where the torus extent is smaller, and the magnetic flux is reduced. This is labelled "SANE" ICs, for "standard and normal evolution" (Narayan et al. 2012). Figure adapted from Tchekhovskoy (2015)

dynamically-important magnetic flux obstructed the accretion and led to what is dubbed a *magnetically arrested disk*, or a "MAD"; in such a condition no further flux could be carried down to the hole. In contrast, models where magnetic field falls short of saturating the black hole were referred to as "SANE" accretion models, for "standard and normal evolution" by Narayan et al. (2012). In the MAD simulation Φ and $P_{\rm jet}$ are greatly increased, as we discuss below (Tchekhovskoy et al. 2011, 2012; McKinney et al. 2012; Tchekhovskoy 2015).

How do such strong magnetic fields manage to stay in force-balance with the disk? As in previous disk simulations, e.g., Beckwith et al. (2009), the black hole field strength is comparable to the total pressure in the disk. In MADs, magnetic and gas pressures in the disk are comparable, $\beta \sim 1$, but the much stronger black hole magnetic field in the MAD case compresses the inner disk substantially, so the inner regions of MAD disks are actually over-pressured compared to previous simulations by a factor of ~ 5–10.

Figure 4 shows the structure of the simulated disk-jet system. An accretion disk around the central black hole leads to a pair of jets that extend out to much larger distances than the black hole horizon radius and are collimated into small opening angles by outflows from the disk (not shown). Jet magnetic field is predominantly toroidal, which reflects the fact that the jets are produced by the rotation of the black hole, which twists the magnetic field lines into tightly-wound helices.

Even though the accretion disk and the jets turn out to be highly time-variable, in a time-average sense the structure of the flow is remarkably simple. The poloidal magnetic field lines are shown with black solid lines in Fig. 4(b). The group of field lines highlighted in green connects to the black hole and makes up the twin polar jets. As discussed above, these field lines have little to no gas attached to them; disk gas cannot cross magnetic field lines and has too much angular momentum to enter the funnel. The large Poynting flux and very small inertia yields highly relativistic velocities. The field lines highlighted in blue connect to the disk body and make up the magnetic field bundle that produces the slow, heavy disk wind, whose power is much smaller than $P_{\rm jet}$ for rapidly spinning black holes.

The jet power can be characterized by the *outflow energy efficiency* η as

$$\eta = \frac{P_{\rm outflow}}{\dot{M}c^2}. \tag{5}$$

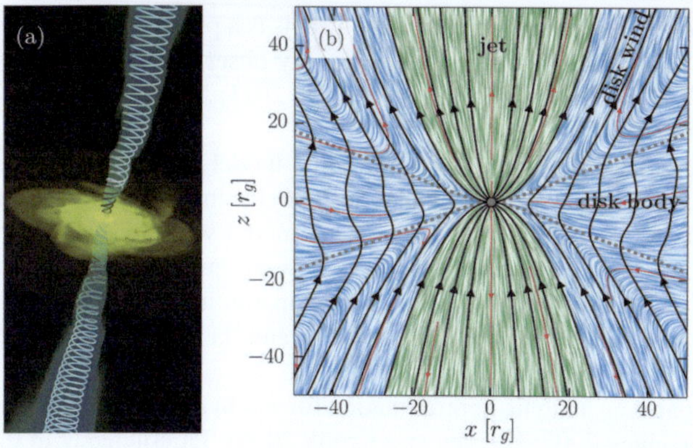

Fig. 4 Panel (**a**): A 3D rendering of a MAD disk-jet simulation. Dynamically-important magnetic fields are twisted by the rotation of a black hole (too small to be seen in the image) at the center of an accretion disk. The azimuthal magnetic field component clearly dominates the jet structure. Density is shown with *color*: disk body is shown with *yellow* and jets with *cyan-blue color*; jet magnetic field lines are *cyan bands*. The image size is approximately $300r_g \times 800r_g$. Panel (**b**): Vertical slice through a MAD disk-jet simulation averaged in time and azimuth. Ordered, dynamically-important magnetic fields remove the angular momentum from the accreting gas even as they obstruct its infall onto a rapidly spinning black hole ($a/M = 0.99$). *Gray filled circle* shows the black hole, *black solid lines* show poloidal magnetic field lines, and *gray dashed lines* indicate density scale height of the accretion flow, which becomes strongly compressed vertically by black hole magnetic field near the event horizon. The symmetry of the time-average magnetic flux surfaces is broken, due to long-term fluctuations in the accretion flow. Streamlines of velocity are depicted both as *thin red lines* and with colored "iron filings", which are better at indicating the fine details of the flow structure. The flow pattern is a standard hourglass shape: equatorial *disk inflow* at low latitudes, which turns around and forms a *disk wind outflow* (labeled as "disk body" and "disk wind", respectively, and highlighted in *blue*), and twin polar *jets* at high latitudes (labeled as "jet" and highlighted in *green*). Figure taken from Tchekhovskoy (2015)

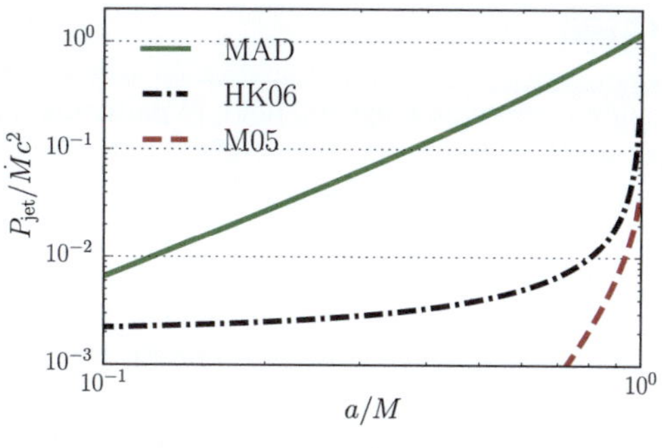

Fig. 5 Comparison of jet energy efficiency obtained in MAD simulations, $P_{\rm jet}/\dot{M}c^2$, which is shown with *green solid line*, with previously reported approximations of simulated jet power: Hawley and Krolik (2006), which is shown with *black dash-dotted line* (HK06), and McKinney (2005), which is shown with *brown dashed line* (M05), plotted over the range $0 \le a/M \le 0.99$. MADs can produce much more powerful jets because they have the maximum possible amount of magnetic flux threading the black hole

For the MAD simulation, the accretion rate $\dot{M}$ fluctuates considerably, but does maintain a well-defined average value. In the MAD flow, for a rapidly spinning black hole, $a/M = 0.99$, the total outflow power, $P_{\rm outflow}$, which is the sum of the black hole powered jet, $P_{\rm jet}$ and the disk powered wind, exceeds the accretion power, $\eta > 1$, i.e., *net* energy is extracted from the black hole (Tchekhovskoy et al. 2011). This is a promising result since high values of jet power are consistent with or are preferred by observations (Rawlings and Saunders 1991; Ghisellini et al. 2010, 2014; Fernandes et al. 2011; McNamara et al. 2011; Nemmen and Tchekhovskoy 2014).

The MAD flow hearkens back to the ion-torus model of Rees et al. (1982) where the role of the thick torus is to confine and anchor the fields that extract rotational energy from the hole, and accretion *per se* is near zero. This model was put originally forward as an explanation for systems where high jet power is seen, but very little luminosity emerges from the central engine. In the MAD flow, accretion continues at a non-zero rate regulated by the magnetic interchange instability in the disk and the strength of the central magnetic field.

Figure 5 compares an analytic approximation for $P_{\rm jet}$ in MAD simulations to some earlier simulations (McKinney 2005; Hawley and Krolik 2006). Jet efficiency in the MAD case is $\propto a^2$, suggesting the ratio of magnetic flux squared to mass accretion rate is roughly constant (Tchekhovskoy et al. 2012; Tchekhovskoy 2015). If MADs achieve the maximum possible amount of magnetic flux threading the black hole then they have achieved the maximum possible jet power for a given mass accretion rate (see Fig. 5). The other models show a rapid rise as a/M approaches 1. Hawley and Krolik (2006) attribute this to greater local field amplification for the extreme Kerr holes; the field has "room to grow," and is a factor of a few smaller than in the MAD simulations.

Efficiencies $\eta > 1$ can lead to spindown of the black hole. For example, in an AGN accreting at 10 % Eddington luminosity, the central black hole is spun down to near-zero spin, $a/M \lesssim 0.1$, in $\tau \simeq 3 \times 10^8$ years (Tchekhovskoy et al. 2012; Tchekhovskoy 2015). This value is interesting astrophysically because it is comparable to a characteristic quasar lifetime (Sect. 5). Over this time period, jets can extract a substantial fraction of the central black hole spin energy and deposit it into the ambient medium. The central galaxy in the cluster MS0735.6+7421 may be one such example (McNamara et al. 2009).

4.2 Origin of the Jet Field

Jets appear to require some large-scale magnetic field, in particular a net vertical field connected either to the disk itself or the central star or black hole. This raises the question of the origin of that field. In the jet-forming global simulations, the initial condition consists of dipole loops embedded within initial orbiting gas, typically a torus of some thickness. In essence, the subsequent evolution (which is a consequence of the initial conditions) transfers the dipole field down to the black hole. Other initial field configurations, such as quadrupolar, toroidal, smaller scale loops, have been found to be far less effective in creating jets (Beckwith et al. 2008).

Simulations that produce jets rely on favorable initial conditions. What happens in Nature? At least two possibilities exist. One is that the required field is generated *in situ* by the disk itself through a dynamo process, as a natural consequence of accretion. Some example dynamo models were discussed above (e.g., Fig. 2). The other possibility is that the required field is carried into the central disk/star, from the companion star in a binary system and from the interstellar medium in an AGN. The advantage of advection is that a relatively weak net field at large radius could be significantly amplified simply by the geometric factor in going from large to small radius. For example, magnetic fields at the edge of the sphere of influence of a supermassive black hole, i.e., $r \sim 100$ pc, are plausibly $B \sim \mu$G. Suppose these fields maintain their coherence over roughly a similar length scale. The magnetic flux contained in patch of field this size, $\Phi_{\rm patch} \approx 10^{35}$ G cm^2, is much larger than the flux necessary to saturate a black hole with magnetic field, $\Phi_{\rm MAD} \approx 10^{33.5}$ G cm^2 (for a black hole of mass $M = 10^9 M_\odot$ accreting at 10 % of Eddington luminosity, Narayan et al. 2003; Tchekhovskoy 2015).

Under what general circumstances can a net field be advected in by the accretion flow? The question is traditionally framed in the terms used by van Ballegooijen (1989), namely that accretion rate is set by the viscosity ν and the magnetic diffusion by the resistivity η. Whether or not a field can be advected inward depends on the magnetic Prandtl number $P_m = \nu/\eta$. If P_m exceeds 1 then field advection will not occur (see also Lubow et al. 1994). In accretion disks the viscosity and resistivity are not the physical values, but instead are effective values resulting from the turbulence (Guan and Gammie 2009; Lesur and Longaretti 2009; Fromang and Stone 2009).

In global simulations, flux in the initial torus (e.g. Fig. 5) finds its way onto the black hole, despite starting from a large radius. This suggests that geometrically-thick disks are capable of transporting large-scale magnetic flux inward, at least from the distances of $r \sim 100 r_g$. In standard α-disk theory, infall rates are proportional to $(h/r)^2$. If h/r is close to unity the infall time becomes relatively rapid, and the inflow rate can exceed the diffusion rate. Thick, radiatively inefficient flows, then, seem to have the best chance of transporting flux inward.

The question of whether a turbulent, radiatively efficient disk could transport flux inward remains an interesting one. Beckwith et al. (2009) studied the question through a global simulation of an initial orbiting torus with an $h/r \sim 0.1$ embedded in a vertical magnetic field. They found that there was no net flux transported through the disk itself; the turbulence transported angular momentum and mass, but not net flux. Net flux was nevertheless brought down to the black hole by a coronal inflow surrounding the accretion disk. Initially the field lies along cylinders of constant rotation, but at high altitude B^2/ρ is much larger than deep inside the disk, and the field is subject to a finite-amplitude version of the MRI. The sign relation between the radial field component and the toroidal field component is the usual one, so the Maxwell stress has the right sign to transport angular momentum outward. The angular momentum flux due to the Maxwell stress is large compared to the fluid angular momentum density. Fluid elements at high altitude, both above and below the equator, are therefore driven inward quickly, carrying half-loops of field with them. When one loop approaches the central black hole from above and outward while another approaches the disk from below and outward, their local field directions are opposed, and they can reconnect. Reconnection changes the field topology, creating a closed loop at larger radius and an open field line at small radius. The open field line, the one carrying the net flux, is located where the reconnection occurred, which is at a radius considerably smaller than the initial radius of the field line. Thus, reconnection causes the flux to move inward in large jumps. All of this takes place faster than the mean mass inflow rate within the disk at the vicinity of the flux line's initial position.

In a similar vein, Suzuki and Inutsuka (2014) carried out a series of simulations with a relatively thick disk (sound speed ~ 0.1 of the orbital speed) with an initial weak magnetic field. In these simulations net vertical flux was carried inward by a rapidly infalling layer near the disks surface. All these simulations transport net flux through a mechanism that is governed by large-scale torques and resulting rapid inflows, rather than by small-scale turbulence where the turbulent velocities exceed the mean inward drift velocity associated with accretion. It appears plausible, therefore, that net inward flux transport can occur for certain types of accretion flows.

Testing jet-generation models, including quantitative predictions for the magnetic flux close to the black hole, fundamentally depends on observations that can give insights into the field properties both at the jet base and further out. We discuss those in the following section.

5 Observational Constraints on Jet Magnetic Fields

5.1 Introduction: Basic Jet Properties

In this section we ask what the observational constraints are on the field strengths and configurations in jets, and whether these observational results can be connected to jet-launching models. (For a more detailed review of these issues see Pudritz et al. 2012; parts of that work are summarized in this section.) We focus on relativistic AGN jets, since these are by far the best-studied of the systems producing synchrotron radiation, and begin by setting out some basic properties of the jets in radio-loud AGN.

AGN jets are fast outflows carrying kinetic powers that are estimated to be between $\sim 10^{42}$ and $\sim 10^{47}$ erg s^{-1} (e.g., Rawlings and Saunders 1991). Bulk speeds are known to be highly relativistic on small scales ($\Gamma \sim 10$–30, and perhaps higher in some objects: Lister et al. 2009); on kpc scales some jets have decelerated to sub-relativistic speeds (Laing and Bridle 2002) while in more powerful systems jets are at least mildly relativistic out to hundred-kpc scales (Mullin and Hardcastle 2009) and may retain a highly relativistic spine with $\Gamma \sim 10$ out to those scales, though direct observational evidence for this remains debatable (e.g., Tavecchio et al. 2000; Hardcastle 2006; Marshall et al. 2011; Konar and Hardcastle 2013).

Jet composition remains an open question. Observations of synchrotron radiation require relativistic leptons and magnetic fields to be present, but *direct* observational constraints on the presence of a hadronic component are hard to obtain. Some indirect constraints suggest that the bulk of the required energy for jets can be carried by the radiating leptons without any need for additional components on scales from pc to kpc (e.g., Wardle et al. 1998; Croston et al. 2005; Wykes et al. 2013), it is certain that all jets must entrain material from stellar winds in the host galaxy (Bicknell 1994) so their composition may change with distance along the jet, and there is some evidence from pressure balance arguments that this is indeed the case (Croston and Hardcastle 2014).

One important observational constraint on jet launching mechanisms comes from the fact that they appear to be able to originate in AGN with a wide range of different properties. The vast majority of radio-loud AGN are so-called 'low-excitation radio galaxies' which show no evidence at any wavelength for a standard radiatively efficient accretion disk (Hardcastle et al. 2009; Best and Heckman 2012). However, the existence of powerful radio-loud quasars with $L/L_{\rm Edd} \sim 1$ in the optical shows that radiatively inefficient accretion is not directly connected to radio jet activity. In fact it appears that radio-loud AGN have a very wide range of $L/L_{\rm Edd}$, even when jet power is taken into account (Mingo et al. 2014).

Finally, it is important to note that observations of synchrotron radiation require a particle acceleration mechanism. There is indirect or direct evidence for local particle acceleration on all scales of radio-loud AGN up to scales of tens of kpc. Shocks are almost certainly implicated—some of the best-understood sites of particle acceleration, the hotspots of powerful double radio galaxies, are clearly physically associated with the jet-termination shock and show properties consistent with simple first-order Fermi acceleration (Meisenheimer et al. 1997). However, other mechanisms may well be necessary to explain diffuse particle acceleration distributed on scales of many kpc (e.g., Hardcastle et al. 2007). When interpreting observations, it is necessary to bear in mind that the particles being observed at a given location may owe their energy spectrum to acceleration in some region that is spatially quite distinct.

Because jets can be persistent (ages of the order 10^8 years have been estimated for some sources based on spectral or dynamical age estimates) they produce large-scale, long-lasting structures consisting of material that has passed up the relativistic jet (Scheuer 1974), generally called lobes or plumes. The physical conditions in these structures are to some extent determined by the jet and so they give important clues to aspects of the jets that are hard to investigate directly. Because the observational techniques used are rather different, it is usual to discuss the large-scale lobes, the kpc-scale jets and the pc-scale jets separately, a convention that we follow in the remainder of this section.

5.2 Observational Tools

The combination of relativistic leptons and magnetic field gives rise to synchrotron emission, while relativistic leptons and a photon field give us inverse-Compton emission. These are the basic tools available to estimate the magnetic field properties in jets and their products.

Optically thin synchrotron radiation cannot on its own give a measurement of magnetic field strength. For simplicity, let us represent the electron energy spectrum as a power law with index p, so that $N(E) = N_0 E^{-p}$. Then the synchrotron emissivity in the optically thin regime is given by

$$J(\nu) = C N_0 \nu^{-\frac{(p-1)}{2}} B^{\frac{(p+1)}{2}} \tag{6}$$

where C is a constant depending only weakly on p. Equation (6) implies that the same observed emissivity can be produced by any combination of the number density of electrons (scaling as N_0) and the strength of the field B. However, total intensity emission does give important information about the magnitude of any spatial *variation* in field strength (combined with electron density, energy spectrum etc) given the strong dependence on B of emissivity (Eq. (6)). The direction of observed polarization in optically thin regions gives us an (emission-weighted, line-of-sight integrated, projected) estimate of the local direction of magnetic field, which is our best probe of the vector properties of the field, while fractional polarization tells us about its ordering: for a uniform field the fractional polarization for power-law electron energy index p is $\Pi = (p+1)/(p+\frac{7}{3})$.

Faraday rotation can complicate the interpretation of polarization. For the simple case of an external Faraday screen, the rotation angle ϕ is given by

$$\phi = \frac{c^2}{\nu^2} K \int_0^s n_{\rm th} \mathbf{B}.\mathbf{ds} \tag{7}$$

where K is a constant, $n_{\rm th}$ is the number density of thermal electrons, and the integral is along the line of sight to the source. If the value of ϕ varies within the resolution element, depolarization rather than simple rotation with $\phi \propto \lambda^2$ will be observed. In some cases, therefore polarized intensity, and in particular the dependence of fractional polarization or position angle on frequency, actually tells us more about the foreground magnetoionic medium (in the host or the Milky Way) than it does about the source itself. The effects of internal and external Faraday-active media on polarization are well understood in some simple analytic cases (Burn 1966; Jones and O'Dell 1977; Cioffi and Jones 1980).

Optically thick synchrotron radiation is an essential tool for field strength measurements on the pc scale; the turnover frequency for synchrotron self-absorption is sensitively dependent on the magnetic field strength, with the absorption coefficient going as $B^{(p+2)/2}$. However, as the turnover also depends on the normalization of the electron energy spectrum, this method depends on good information about the structure of the components where the turnover is observed.

The combination of inverse-Compton and synchrotron emission gives an excellent constraint on the mean field strength, if both processes can be measured from the same region; inverse-Compton emissivity depends only on the number density of electrons (for a known photon field, often the CMB) so that the field strength may be directly estimated from the synchrotron emissivity. Again, it is necessary to know the geometry of the emitting region accurately,

Finally, if all else fails, it is traditional to resort to the assumptions of equipartition,

$$\frac{B^2}{8\pi} = \int EN(E)\mathrm{d}E \tag{8}$$

or minimum energy (Burbidge 1956)

$$U_{\mathrm{tot}} = \frac{B^2}{8\pi} + \int EN(E)\mathrm{d}E: \quad \frac{\mathrm{d}U_{\mathrm{tot}}}{\mathrm{d}B} = 0 \tag{9}$$

which, putting in B-dependent estimates of N_0 from Eq. (6), give fairly similar magnetic field values for a given volume emissivity. Until recently there has been little observational justification for the equipartition assumption (though see below) and it should still be applied with caution.

5.3 Observations: Large-Scale Components

In the large-scale lobes and plumes of radio galaxies the gold standard for field strength measurements is the inverse-Compton technique. Large numbers of powerful (FRII: Fanaroff and Riley 1974) radio galaxies now have global lobe field strength measurements (Kataoka and Stawarz 2005; Croston et al. 2005) using X-ray inverse-Compton measurements made with *XMM* and *Chandra*. These imply field strengths close to, but somewhat below, the equipartition values: a typical field strength in a lobe is of the order 10 μG. Fewer such measurements exist for the lower-power FRI radio galaxies, in part because of their typically brighter thermal environments but also because some part of their internal pressure is probably provided by thermal material (Hardcastle and Croston 2010). However, a magnetic field strength around 1 μG, again close to the equipartition value, has been estimated for the nearby FRI radio galaxy Centaurus A based on *Fermi* γ-ray detections of its large-scale lobes (Abdo et al. 2010).

Inverse-Compton observations also provide important evidence for the spatial intermittency of the magnetic field: the inverse-Compton surface brightness (which, for scattering of the CMB, depends only on the electron energy density and the line-of-sight depth) is seen to be much more uniform than the synchrotron surface brightness (Hardcastle and Croston 2005). This is as expected if the lobes are turbulent with an energetically sub-dominant magnetic field, and similar effects are seen in numerical MHD modeling of lobes (Tregillis et al. 2004; Hardcastle and Krause 2014).

Turning to field direction, the apparent field directions inferred from polarization measurements in the lobes are almost always perpendicular to the jet far away from the jet termination, with mean fractional polarization in the range 20–40 % at high frequencies. This is consistent with the idea that a toroidal component of the field dominates on large scales, as would be expected if an initially disordered field expands into the lobes, but compression of a disordered field can give similar results (Laing 1980).

5.4 Observations: kpc-Scale Jets and Termination Features

Inverse-Compton measurements of field strengths in jets are difficult because of the existence of *in situ* particle acceleration, which can give rise to high-energy (e.g. X-ray) synchrotron emission. Arguments for high bulk speeds in kpc-scale jets, discussed above, assume that the X-ray emission from these jets is synchrotron, but there is a degeneracy between the magnetic field strength and the (a priori unknown) jet speed and angle to the line of sight,, so that these observations are not very useful for measurements. In low-power jets, where the dynamics are better understood, TeV γ-ray observations place some constraints on magnetic field strength (Hardcastle and Croston 2011), implying $B > 20$ μG to avoid over-producing the observed TeV emission. These models depend on a good knowledge of the photon field, in this case the starlight at the centre of the elliptical host galaxy.

The magnetic field direction is much better understood. In powerful jets the field direction is almost universally along the jet (although it is not clear whether the synchrotron emission from the jet is a reliable tracer of the jet itself—it may simply show a boundary layer). In low-power jets, observations of polarization are essential to models that break the degeneracy between jet speed and angle to the line of sight (Laing and Bridle 2002) and when this is done it is found that the field tends to evolve from a predominantly longitudinal configuration on the smallest scale to a predominantly toroidal one after jet deceleration: no significant radial component is required by the models (Laing and Bridle 2014). Helical field models on these scales for these well-studied low-power jets are ruled out by the lack of transverse polarization asymmetry.

Powerful jets terminate in compact (1–10 kpc scale), bright features known as hotspots, which are thought to trace the jet termination shock. As the downstream material should be subsonic, and so at most mildly relativistic, hotspots are easier to understand than the jets themselves. Their brightness means that they are good sources of inverse-Compton emission via the synchrotron self-Compton process, producing X-ray emission (Harris et al. 1994). Inverse-Compton modeling implies field strengths of the order 100 μG in hotspots of powerful sources (Hardcastle et al. 2004; Kataoka and Stawarz 2005). Magnetic field structures in hotspots, as revealed by high-resolution polarization imaging, are complex, but there is a general tendency for the projected field direction to be perpendicular to the jet (Leahy et al. 1997).

5.5 Observations: pc-Scale Jets

Inverse-Compton methods are very uncertain on small scales—high-energy emission from the jet is often hard to distinguish from emission from other components of AGN. In objects where this is not the case, such as blazars, the jet geometry, bulk speed and angle to the line of sight are likely to be poorly known. Most methods for estimating the magnetic field strength on parsec scales rely on inferences from very long baseline interferometry (VLBI) in the radio, and hence on synchrotron emission. Typically, the parsec-scale radio morphology of a bright AGN manifests a one-sided jet structure due to Doppler boosting, which enhances the emission of the approaching jet. As noted above, Lorentz factors on these scales can be high, $\Gamma \sim 10$–30: constraints come from observations of apparent motions (Cohen et al. 2007; Lister et al. 2013) and from radio variability (Jorstad et al. 2005; Hovatta et al. 2009; Savolainen et al. 2010).

The apparent base of the jet, which is often the brightest and least-resolved part, is commonly called the *core*. As noted above, synchrotron self-absorption, which depends on both the particle number density N_e and the magnetic field B, will affect the appearance of the jet; the core is thought to represent the jet region where the optical depth is equal to unity. As the number density and magnetic field depend on distance from the jet base, the position of the apparent core is expected to depend on observing frequency. This is the basis of the *core-shift* method for estimating magnetic field strengths (Lobanov 1998; O'Sullivan and Gabuzda 2009b), evaluating the relation between core flux at given frequency and mass of black hole (Heinz and Sunyaev 2003), jet composition (Hirotani 2005), and flow magnetization (Nokhrina et al. 2015). This method necessarily assumes that the jet is well described by the simple, homogeneous conical jet models of Blandford and Königl (1979). The latter assumption is supported by multi-frequency observations by Sokolovsky et al. (2011).

As an alternative to the core-shift method, individual components of the jet can be fitted with self-absorption models (e.g. Savolainen et al. 2008); this method is less model-dependent but has larger random errors as a result of the uncertainties on component size, geometry and Doppler factor. Magnetic field strengths derived by either method are of the order 1 G on pc scales. Recently, Zdziarski et al. (2015) have proposed a "hybrid" method of doing core-shift analysis which does not rely on the equipartition assumption, again obtaining similar magnetic field strength values.

On pc scales, the inferred jet magnetic field direction is normally either parallel or perpendicular to the jet direction (e.g. Lister and Homan 2005), with jets in more powerful objects tending to have parallel fields. Misaligned polarization that is definitively not the result of Faraday rotation is rare. The interpretation of these field structures varies: they are often attributed to internal jet dynamics, e.g. shocks in the case of transverse apparent field direction, but could equally well be the result of an underlying helical magnetic field geometry with varying pitch angle (Lyutikov et al. 2005). (As noted above, helical fields are expected in most jet generation models due to the toroidal component introduced by field winding: see Sects. 1, 4.2.) As pointed out by Blandford (1993), the helical-field model would be supported by observations of transverse gradients of Faraday rotation across the jet, provided that there are thermal electrons in the jet or in its immediate environment to provide the required magnetoionic medium (Eq. (7)). Detection of such gradients is observationally difficult, since the jets are poorly resolved, but they have been observed in a number of objects (e.g., Asada et al. 2002; O'Sullivan and Gabuzda 2009a; Croke et al. 2010). While their interpretation is still controversial (Taylor and Zavala 2010) these gradients provide at least some direct evidence for helical field structures in the inner parts of the jets.

5.6 Core-Shifts and Jet Magnetization

Two important MHD parameters for describing relativistic flows are the Michel magnetization parameter σ_{M} and the multiplicity parameter λ. The first one tells us how strongly the flow is magnetized at its origin, and it determines the maximum possible bulk Lorentz factor of the flow. The second one is the dimensionless multiplicity parameter $\lambda = N_{\mathrm{e}}/N_{\mathrm{GJ}}$, which is defined as the ratio of the number density N_{e} to the Goldreich-Julian (GJ) number density $N_{\mathrm{GJ}} = \Omega B/2\pi ce$—the minimum number density needed for screening the longitudinal electric field. These two parameters are connected with total jet power P_{jet} by equation (Beskin 2010)

$$\sigma_{\mathrm{M}} \approx \frac{1}{\lambda}\left(\frac{P_{\mathrm{jet}}}{P_{\mathrm{A}}}\right)^{1/2}. \tag{10}$$

Here $P_{\mathrm{A}} = m_{\mathrm{e}}^2 c^5/e^2 \approx 10^{17}$ erg/s is the minimum energy loss rate of a central engine which can accelerate particles to relativistic energies.

There are two theoretical models of plasma production in a jet. In the first one, pairs are produced by two-photon collisions with photons with sufficient energy produced by the inner parts of the accretion disk (Blandford and Znajek 1977). In this case we expect $\lambda \sim 10^{10}$–10^{13}, and Michel magnetization parameter $\sigma_M \sim 10$–10^3. The second model of the pair production is a cascade process in non-zero electric field in a region with the zero GJ plasma density due to general relativity effects (Beskin et al. 1992; Hirotani and Okamoto 1998). This model gives $\lambda \sim 10^2$–10^3, and magnetization $\sigma_M \sim 10^{10}$–10^{13}. So in both scenarios the flow is strongly magnetized at its base, but this large difference in the estimates for the magnetization parameter σ_M leads to two different pictures of the flow structure in jets. Indeed, as has been shown by several authors (Beskin and Nokhrina 2006; Tchekhovskoy et al. 2009; Komissarov et al. 2009), for well-collimated magnetically dominated MHD jets the Lorentz factors of the particle bulk motion follows the relation

$$\Gamma \approx r_\perp / R_L, \tag{11}$$

where $r_\perp$ is the distance from the jet axis, and $R_L = c/\Omega$ is the light cylinder radius. The relation (11) holds until the flow reaches the equipartition regime—the Poynting flux is approximately equal to the particle kinetic energy flux. Further acceleration is ineffective. For ordinary jets $r_\perp/R_L \sim 10^4$–10^5. As a result, using the universal asymptotic solution (11), one can find that values $\sigma_M \sim 100$ correspond to the saturation regime when there is approximately equipartition between the Poynting flux P_{em} and the particle kinetic energy flux P_{part}. On the other hand, for $\sigma_M \sim 10^{12}$ the jet remains magnetically dominated ($P_{part} \ll P_{em}$). Thus, the determination of the Michel magnetization parameter σ_M is a key point in the analysis of the internal structure of relativistic jets.

The core-shift measurements described above (Gould 1979; Marscher 1983; Sokolovsky et al. 2011; Pushkarev et al. 2012) might provide an observational way to probe λ and σ_M. The ideal MHD flow of plasma with exactly the drift velocity in crossed electric and magnetic fields does not emit synchrotron radiation. However, some internal dissipative processes, for example internal shocks and magnetic reconnection, can produce particles with a non-thermal spectrum that can account for the observed emission. By measuring the core-shift discussed in the previous subsection, one can obtain the combination of magnetic field magnitude B_{cs} and electron number density N_{cs} in the region with optical depth equal to unity. On the other hand, B_{cs} and N_{cs} can be related through the flow magnetization $\sigma = \sigma_M/\Gamma - 1$ (Tchekhovskoy et al. 2009) assuming that the flow is in the saturation regime, i.e. $\sigma_{cs} \approx \sigma_M/2 \approx \Gamma_{cs}$ (Beskin and Nokhrina 2006; Tchekhovskoy et al. 2009; Komissarov et al. 2009). Combining the latter with the definitions of λ and σ_M, one can obtain both λ and σ_M as a function of core-shift and total jet power P_{jet}. The latter can be estimated by correlating the radio jet luminosity to total power needed to form the cavities in surrounding gas (Cavagnolo et al. 2010). This analysis leads to estimates of $\lambda \sim 10^{13}$ and $\sigma \sim 10$ (Nokhrina et al. 2015). The measurements of core-shift by Pushkarev et al. (2012) thus allow the order of Michel magnetization at the jet base to be estimated, and it can be concluded that the MHD flow on the observable scales are in the saturation regime, i.e. there is no effective plasma acceleration by the MHD mechanism downstream. Moreover, the order of σ_M correlates well with the estimated Lorentz factors of jet flows discussed above.

Recently, it has become possible to use the core-shift to measure the magnetic flux threading the jets and central black holes in a rather large sample of radio-loud active galactic nuclei (Zamaninasab et al. 2014; Zdziarski et al. 2015). Dynamically-important black hole magnetic fields were inferred, indicating that the central black holes of most if not all radio-loud AGN are in a MAD state (Sect. 4.2). MADs have also been inferred in tidal disruption events (Tchekhovskoy et al. 2014) and core-collapse gamma-ray bursts (Tchekhovskoy and

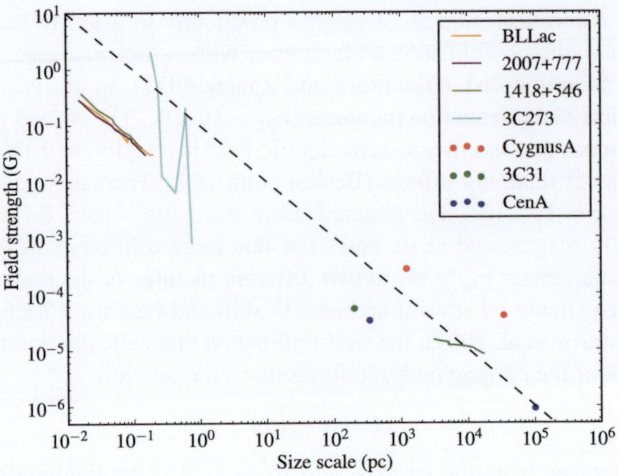

Fig. 6 Magnetic field strengths on different size scales illustrated using a few well-studied objects on pc and kpc scales. The size scale plotted here is the transverse radius of the component in which the field is measured, not the distance from the nucleus (though in general the two scale). Data are taken from O'Sullivan and Gabuzda (2009b), Savolainen et al. (2008), Hardcastle et al. (2004), Hardcastle and Croston (2010, 2011), Croston and Hardcastle (2014), and Abdo et al. (2010). The *solid line* shows an arbitrarily normalized line of $B \propto r^{-1}$—we would expect the normalization of this line to depend on the jet power so it is not surprising that all points do not lie on the line

Giannios 2014). This suggests that MADs are perhaps not rare or unusual as their name might imply, but possibly quite the opposite.

5.7 Observational Summary

To summarize, we now have relatively good information about the magnetic field strengths on kpc scales, and excellent information about the vector properties of the field, which, however, are probably not strongly connected to the field properties at jet generation. On pc scales and below, magnetic field strengths are not as well constrained, but such constraints from, e.g., core-shift observations as we have are consistent with the expectations from models, while polarization observations are certainly consistent with the presence of a helical field structure on small scales.

The field strength constraints obtained for some of the objects discussed above are summarized in Fig. 6, which shows a broad consistency with an r^{-1} scaling over the observable ~ 7 orders of magnitude in physical scale. Fields of the order 10^4 G, and thus magnetic fluxes $\sim 10^{33}$ G cm^2 would be implied by an extrapolation to scales comparable to the gravitational radius of the central black hole. In principle, magnetic fluxes at this level could be dynamically important (plasma $\beta \sim 1$) and so could affect the flow of material in the disk as discussed in Sect. 4.2.

6 Conclusions

Accretion disks and the jets they produce are fundamentally magnetic systems. A great deal of progress has been made in the theory of these systems, and in obtaining their properties from observation. We began by considering how the entire complex system is broken into different components for theoretical analysis. Where then do we stand with those components?

First, accretion disks are turbulent, and that turbulence is almost certainly created by the instability to the MRI. This MRI-induced turbulence produces internal stress that can transport angular momentum at the rates required.

Although it seems clear that disks are magnetic, jets appear to require a more organized magnetic field than is produced by MRI turbulence. The origin of the jet magnetic field remains uncertain. A disk dynamo remains a possibility, and dynamo action has been seen in some simulations, but the field produced is not the type required for jet launching. In the case of stellar jets, the central star may also be the source of the required field through its own internal dynamo. The possibility remains that the required field could be carried in from large radii; however, MRI turbulence does not seem to accomplish that on its own. More direct infall, taking place outside the turbulent disk proper, or in a hot, thick accretion flow with relatively large infall velocities, could work if the resulting field is able to reconnect as needed to establish the required topology on the central star or black hole, or through the central accretion disk.

The general model of jets as magnetized, accelerated and collimated disk winds is well established today by observations and theoretical investigations. While for young stars the disk wind seems to play the major role in jet launching (Blandford-Payne), the jets from AGN and other relativistic sources are plausibly dominated by processes energetically supported directly by the spinning black hole (Blandford-Znajek, Blandford and Znajek 1977), and observations, at least on the smallest scales, show good consistency with the expectations of this type of model, as discussed in Sect. 5. There are still many unknowns. For example, the matter content of relativistic jets is as yet unclear, and in many cases their propagation speed is poorly constrained; there is no direct measurement of the magnetic field of jets from stars or of the accretion disk.

Global MHD simulations are able to treat the global jet launching that is the accretion-ejection process. Usually an initial large-scale magnetic field has been assumed so far in these models, only a few simulations were able to consider mean-field disk dynamo. How such global, axisymmetric mean-field models match the disk microphysics and the respective local disk simulations, is still an open question. The magnetized disks considered in jet launching simulations use a rather simple model for the (turbulent) magnetic diffusivity or viscosity and also neglect radiation and radiative transport processes. However, all these processes seem to be essential for the disk physics.

Future progress will clearly come from more sophisticated simulation models, considering the physical effects mentioned above. The codes and computational resources may not yet be ready for such an ambitious goal, as the treatment of the additional physics will require new numerical methods, and also a higher numerical resolution, together with a fully 3D treatment.

Simulations have clearly demonstrated the viability of the Blandford-Znajek type jet powered by a rotating black hole. Those simulations further suggest that a dipole field is required, in addition to the rotating hole, and that the field strength is set by equipartition with the surrounding pressure (gas, ram, magnetic) associated with the accretion disk. The issue of the presence of such jets comes back to the origin of the axial black hole field, a problem that remains uncertain, as discussed above.

Finally, we have seen that observations, particularly of AGN jets, are starting to provide direct tests of jet-launching models through constraints on the bulk speed, particle content, kinetic power, and magnetic field strength and configuration of jets. Much progress on at least some of these issues can be expected from the improvement in observational capabilities at arcsecond resolutions (and thus generally kpc scales) to be provided by next-generation radio facilities such as the upgraded JVLA and LOFAR, and in the future the SKA. However, the most direct constraints remain those provided by long-baseline interferometers such as the VLBA and EVN, and continued support of these is vital if we are to be able to make direct tests of jet models in the future.

Acknowledgements The authors would like to thank the ISSI team for their support and hospitality, and for the organization of a great workshop and fruitful discussions during the week in Bern. This work was also supported by NSF grant AST0908869 (JFH) and by funding from the UK STFC (MJH). AT was supported by NASA through Einstein Postdoctoral Fellowship grant number PF3-140115 awarded by the *Chandra* X-ray Center, which is operated by the Smithsonian Astrophysical Observatory for NASA under contract NAS8-03060, and NASA via High-End Computing (HEC) Program through the NASA Advanced Supercomputing (NAS) Division at Ames Research Center that provided access to the Pleiades supercomputer, as well as NSF through an XSEDE computational time allocation TG-AST100040 on NICS Kraken, Nautilus, TACC Stampede, Maverick, and Ranch. AT used Enthought Canopy Python distribution to generate the figures for this work. JFH wishes to thank Julian Krolik for useful discussions.

References

A.A. Abdo et al., Science **328**, 725 (2010)
J.M. Anderson, Z.-Y. Li, R. Krasnopolsky, R.D. Blandford, Astrophys. J. **630**, 945 (2005)
K. Asada et al., Proc. Astron. Soc. Jpn. **54**, L39 (2002)
M. Avila, Phys. Rev. Lett. **108**, 124501 (2012)
X.-N. Bai, J.M. Stone, Astrophys. J. **767**, 30 (2013a)
X.-N. Bai, J.M. Stone, Astrophys. J. **769**, 76 (2013b)
S.A. Balbus, J.F. Hawley, Astrophys. J. **376**, 214 (1991)
S.A. Balbus, J.F. Hawley, Rev. Mod. Phys. **70**, 1 (1998)
S.A. Balbus, J.C.B. Papaloizou, Astrophys. J. **521**, 650 (1990)
S.A. Balbus, J.F. Hawley, J.M. Stone, Astrophys. J. **467**, 76 (1996)
K.R.C. Beckwith, J.F. Hawley, J.H. Krolik, Astrophys. J. **678**, 1180 (2008)
K.R.C. Beckwith, J.F. Hawley, J.H. Krolik, Astrophys. J. **707**, 428 (2009)
V.S. Beskin, Phys. Usp. **53**, 1199 (2010)
V.S. Beskin, E.E. Nokhrina, Mon. Not. R. Astron. Soc. **367**, 375 (2006)
V.S. Beskin, Ya.N. Istomin, V.I. Pariev, Sov. Astron. **36**, 642 (1992)
P.N. Best, T.M. Heckman, Mon. Not. R. Astron. Soc. **421**, 1569 (2012)
G.V. Bicknell, Astrophys. J. **422**, 542 (1994)
R.D. Blandford, Astrophys. Space Sci. Libr. **103**, 15 (1993)
R.D. Blandford, A. Königl, Astrophys. J. **232**, 34 (1979)
R.D. Blandford, D.G. Payne, Mon. Not. R. Astron. Soc. **199**, 883 (1982)
R.D. Blandford, R.L. Znajek, Mon. Not. R. Astron. Soc. **179**, 433 (1977)
A. Brandenburg, A. Nordlund, R.F. Stein, U. Torkelsson, Astrophys. J. **446**, 741 (1995)
G. Burbidge, Astrophys. J. **124**, 416 (1956)
B.J. Burn, Mon. Not. R. Astron. Soc. **133**, 67 (1966)
C. Carrasco-González, L.F. Rodríguez, G. Anglada, J. Martí, J.M. Torrelles, M. Osorio, Science **179**, 433 (2010)
F. Casse, J. Ferreira, Astron. Astrophys. **361**, 1178 (2000)
F. Casse, R. Keppens, Astrophys. J. **581**, 988 (2002)
F. Casse, R. Keppens, Astrophys. J. **601**, 90 (2004)
K.W. Cavagnolo, B.R. McNamara, P.E.J. Nulsen et al., Astrophys. J. **720**, 1066 (2010)
D.F. Cioffi, T.W. Jones, Astron. J. **85**, 368 (1980)
M.H. Cohen et al., Astrophys. J. **658**, 232 (2007)
J. Contopoulos, R.V.E. Lovelace, Astrophys. J. **429**, 139 (1994a)
J. Contopoulos, Astrophys. J. **432**, 508 (1994b)
S.M. Croke, S.P. O'Sullivan, D.C. Gabuzda, Mon. Not. R. Astron. Soc. **402**, 259 (2010)
J.H. Croston, M.J. Hardcastle, Mon. Not. R. Astron. Soc. **438**, 3310 (2014)
J.H. Croston et al., Astrophys. J. **626**, 733 (2005)
J.-P. De Villiers, J.F. Hawley, J.H. Krolik, S. Hirose, Astrophys. J. **620**, 878 (2005)
J.-F. Donati, F. Paletou, J. Bouvier, J. Ferreira, Nature **438**, 466 (2005)

E.M. Edlund, H. Ji, Phys. Rev. E **89**, 021004 (2014)
B.L. Fanaroff, J.M. Riley, Mon. Not. R. Astron. Soc. **167**, 31P (1974)
Ch. Fendt, Astrophys. J. **651**, 272 (2006)
Ch. Fendt, Astrophys. J. **692**, 346 (2009)
Ch. Fendt, M. Cemeljic, Astron. Astrophys. **334**, 750 (1998)
Ch. Fendt, M. Cemeljic, Astron. Astrophys. **395**, 1045 (2002)
Ch. Fendt, S. Sheikhnezami, Astrophys. J. **774**, 12 (2013)
Ch. Fendt, M. Camenzind, S. Appl, Astron. Astrophys. **300**, 791 (1995)
C.A.C. Fernandes, M.J. Jarvis, S. Rawlings et al., Mon. Not. R. Astron. Soc. **411**, 1909 (2011)
J. Ferreira, Astron. Astrophys. **319**, 340 (1997)
J. Ferreira, G. Pelletier, Astron. Astrophys. **276**, 625 (1993)
A. Frank et al., in *Protostars and Planets VI*, ed. by H. Beuther, R.S. Klessen, C.P. Dullemond, T. Henning (University of Arizona Press, Tucson, 2014)
S. Fromang, J.M. Stone, Astron. Astrophys. **507**, 19 (2009)
C.F. Gammie, S.L. Shapiro, J.C. McKinney, Astrophys. J. **602**, 312 (2004)
G. Ghisellini, F. Tavecchio, L. Foschini et al., Mon. Not. R. Astron. Soc. **402**, 497 (2010)
G. Ghisellini, F. Tavecchio, L. Maraschi, A. Celotti, T. Sbarrato, Nature **515**, 376 (2014)
R.J. Gould, Astron. Astrophys. **76**, 306 (1979)
O. Gressel, Mon. Not. R. Astron. Soc. **405**, 41 (2010)
X. Guan, C.F. Gammie, Astrophys. J. **697**, 1901 (2009)
X. Guan, C.F. Gammie, Astrophys. J. **728**, 130 (2011)
M.J. Hardcastle, Mon. Not. R. Astron. Soc. **366**, 1465 (2006)
M.J. Hardcastle, J.H. Croston, Mon. Not. R. Astron. Soc. **363**, 649 (2005)
M.J. Hardcastle, J.H. Croston, Mon. Not. R. Astron. Soc. **404**, 2018 (2010)
M.J. Hardcastle, J.H. Croston, Mon. Not. R. Astron. Soc. **415**, 133 (2011)
M.J. Hardcastle, M.G. Krause, Mon. Not. R. Astron. Soc. **443**, 1482 (2014)
M.J. Hardcastle, D.E. Harris, D.M. Worrall, M. Birkinshaw, Astrophys. J. **612**, 729 (2004)
M.J. Hardcastle, J.H. Croston, R.P. Kraft, Astrophys. J. **669**, 893 (2007)
M.J. Hardcastle, D.A. Evans, J.H. Croston, Mon. Not. R. Astron. Soc. **396**, 1929 (2009)
D.E. Harris, C.L. Carilli, R.A. Perley, Nature **367**, 713 (1994)
J.F. Hawley, Astrophys. J. **528**, 462 (2000)
J.F. Hawley, S.A. Balbus, Astrophys. J. **573**, 738 (2002)
J.F. Hawley, J.H. Krolik, Astrophys. J. **641**, 103 (2006)
J.F. Hawley, C.F. Gammie, S.A. Balbus, Astrophys. J. **440**, 742 (1995)
J.F. Hawley, C.F. Gammie, S.A. Balbus, Astrophys. J. **464**, 690 (1996)
S. Heinz, R. Sunyaev, Mon. Not. R. Astron. Soc. **343**, L59–L64 (2003)
J.J. Hester, K. Mori, D. Burrows, J.S. Gallagher, J.R. Graham, M. Halverson, A. Kader, F.C. Michel, P. Scowen, Astrophys. J. **577**, L49 (2002)
K. Hirotani, Astrophys. J. **619**, 73 (2005)
K. Hirotani, I. Okamoto, Astrophys. J. **497**, 563 (1998)
T. Hovatta, E. Valtaoja, M. Tornikoski, A. Lähteenmäki, Astron. Astrophys. **498**, 723 (2009)
H. Ji, M. Burin, E. Schartman, J. Goodman, Nature **444**, 343 (2006)
T.W. Jones, S.L. O'Dell, Astrophys. J. **214**, 522 (1977)
S.G. Jorstad et al., Astron. J. **130**, 1418 (2005)
J. Kataoka, L. Stawarz, Astrophys. J. **622**, 797 (2005)
S.X. Kato, T. Kudoh, K. Shibata, Astrophys. J. **565**, 1035 (2002)
R. Keppens, Z. Meliani, B. van der Holst, F. Casse, Astron. Astrophys. **486**, 663 (2008)
S.S. Komissarov, N. Vlahakis, A. Königl, M.V. Barkov, Mon. Not. R. Astron. Soc. **394**, 1182 (2009)
C. Konar, M.J. Hardcastle, Mon. Not. R. Astron. Soc. **395**, 457 (2013)
A. Königl, R. Salmeron, M. Wardle, Mon. Not. R. Astron. Soc. **401**, 479 (2010)
E. Körding, M. Rupen, C. Knigge, R. Fender, V. Dhawan, M. Templeton, T. Muxlow, Science **320**, 1318 (2008). 2008
E. Körding, C. Knigge, T. Tzioumis, R. Fender, Mon. Not. R. Astron. Soc. **418**, L129 (2011)
R. Krasnopolsky, Z.-Y. Li, R.D. Blandford, Astrophys. J. **526**, 631 (1999)
T. Kudoh, R. Matsumoto, K. Shibata, Astrophys. J. **508**, 186 (1998)
T. Kuwabara, K. Shibata, T. Kudoh, R. Matsumoto, Astrophys. J. **621**, 921 (2005)
R.A. Laing, Mon. Not. R. Astron. Soc. **193**, 439 (1980)
R.A. Laing, A.H. Bridle, Mon. Not. R. Astron. Soc. **336**, 328 (2002)
R.A. Laing, A.H. Bridle, Mon. Not. R. Astron. Soc. **437**, 3405 (2014)
J.P. Leahy et al., Mon. Not. R. Astron. Soc. **291**, 20 (1997)
G. Lesur, P.-Y. Longaretti, Mon. Not. R. Astron. Soc. **378**, 1471 (2007)

G. Lesur, P.-Y. Longaretti, Astron. Astrophys. **504**, 309 (2009)
G. Lesur, J. Ferreira, G.I. Ogilvie, Astron. Astrophys. **550**, A61 (2013)
Z.-Y. Li, Astrophys. J. **415**, 118 (1993)
Z.-Y. Li, Astrophys. J. **444**, 848 (1995)
M.L. Lister, D.C. Homan, Astron. J. **130**, 1389 (2005)
M.L. Lister et al., Astron. J. **138**, 1874 (2009)
M.L. Lister et al., Astron. J. **146**, 120 (2013)
A.P. Lobanov, Astron. Astrophys. **330**, 79 (1998)
R.V.E. Lovelace, M.M. Romanova, G.V. Ustyugova, A.V. Koldoba, Mon. Not. R. Astron. Soc. **408**, 2083 (2010)
S.H. Lubow, J.C.B. Papaloizou, J.E. Pringle, Mon. Not. R. Astron. Soc. **267**, 235 (1994)
D. Lynden-Bell, Nature **223**, 690 (1969)
D. Lynden-Bell, J.E. Pringle, Mon. Not. R. Astron. Soc. **168**, 603 (1974)
M. Lyutikov, V.I. Pariev, D.C. Gabuzda, Mon. Not. R. Astron. Soc. **360**, 869 (2005)
A.P. Marscher, Astrophys. J. **264**, 296 (1983)
H.L. Marshall et al., Astrophys. J. Suppl. Ser. **193**, 15 (2011)
J.C. McKinney, Astrophys. J. **630**, L5 (2005)
J.C. McKinney, Mon. Not. R. Astron. Soc. **368**, 1561 (2006)
J.C. McKinney, C.F. Gammie, Astrophys. J. **611**, 977 (2004)
J.C. McKinney, A. Tchekhovskoy, R.D. Blandford, Mon. Not. R. Astron. Soc. **423**, 3083 (2012)
B.R. McNamara, F. Kazemzadeh, D.A. Rafferty et al., Astrophys. J. **698**, 594 (2009)
B.R. McNamara, M. Rohanizadegan, P.E.J. Nulsen, Astrophys. J. **727**, 39 (2011)
K. Meisenheimer, M.G. Yates, H.-J. Röser, Astron. Astrophys. **325**, 57 (1997)
Z. Meliani, F. Casse, C. Sauty, Astron. Astrophys. **460**, 1 (2006)
A. Mignone, P. Rossi, G. Bodo, A. Ferrari, S. Massaglia, Mon. Not. R. Astron. Soc. **402**, 7 (2010)
B. Mingo et al., Mon. Not. R. Astron. Soc. **440**, 269 (2014)
L.M. Mullin, M.J. Hardcastle, Mon. Not. R. Astron. Soc. **398**, 1989 (2009)
G.C. Murphy, J. Ferreira, C. Zanni, Astron. Astrophys. **512**, 82 (2010)
R. Narayan, I.V. Igumenshchev, M.A. Abramowicz, Proc. Astron. Soc. Jpn. **55**, L69 (2003)
R. Narayan, A. Sädowski, R.F. Penna, A.K. Kulkarni, Mon. Not. R. Astron. Soc. **426**, 3241 (2012)
R.S. Nemmen, A. Tchekhovskoy, arXiv:1406.7420 (2014)
E.E. Nokhrina, V.S. Beskin, Y.Y. Kovalev, A.A. Zheltoukhov, Mon. Not. R. Astron. Soc. **447**, 2726 (2015)
M.L. Norman, K.-H.A. Winkler, L.L. Smarr, M.D. Smith, Astron. Astrophys. **113**, 285 (1982)
S.P. O'Sullivan, D.C. Gabuzda, Mon. Not. R. Astron. Soc. **393**, 429 (2009a)
S.P. O'Sullivan, D.C. Gabuzda, Mon. Not. R. Astron. Soc. **400**, 26 (2009b)
R. Ouyed, R.E. Pudritz, Astrophys. J. **482**, 710 (1997)
R. Ouyed, D.A. Clarke, R.E. Pudritz, Astrophys. J. **582**, 292 (2003)
M.S. Paoletti, D.P. Lathrop, Phys. Rev. Lett. **106**, 024501 (2011)
G.G. Pavlov, M.A. Teter, O. Kargaltsev, D. Sanwal, Astrophys. J. **591**, 1157 (2003)
G. Pelletier, R.E. Pudritz, Astrophys. J. **394**, 117 (1992)
O. Porth, Mon. Not. R. Astron. Soc. **429**, 2482 (2013)
O. Porth, Ch. Fendt, Astrophys. J. **709**, 1100 (2010)
O. Porth, Ch. Fendt, Z. Meliani, B. Vaidya, Astrophys. J. **709**, 1100 (2011)
D. Proga, T.R. Kallman, Astrophys. J. **616**, 688 (2004)
D. Proga, J.M. Stone, T.R. Kallman, Astrophys. J. **543**, 686 (2000)
R.E. Pudritz, C.A. Norman, Astrophys. J. **274**, 677 (1983)
R.E. Pudritz, C. Rogers, R. Ouyed, Mon. Not. R. Astron. Soc. **365**, 1131 (2006)
R.E. Pudritz et al., in *Protostars and Planets V*, ed. by B. Reipurth, D. Jewitt, K. Keil (University of Arizona Press, Tucson, 2007)
R.E. Pudritz, M.J. Hardcastle, D.C. Gabuzda, Space Sci. Rev. **169**, 27 (2012)
A.B. Pushkarev, T. Hovatta, Y.Y. Kovalev et al., Astron. Astrophys. **545**, A113 (2012)
J.P. Ramsey, D.A. Clarke, Astrophys. J. **728**, 11 (2011)
S. Rawlings, R. Saunders, Nature **349**, 138 (1991)
M.J. Rees, M.C. Begeman, R.D. Blandford, E.S. Phinney, Nature **295**, 17 (1982)
R. Salmeron, A. Königl, M. Wardle, Mon. Not. R. Astron. Soc. **412**, 1163 (2011)
T. Savolainen, K. Wiik, E. Valtaoja, M. Tornikoski, in *Extragalactic Jets: Theory and Observation from Radio to Gamma Ray*, ed. by T.A. Rector, D.S. De Young (2008), p. 451
T. Savolainen, D.C. Homan, T. Hovatta, M. Kadler, Y.Y. Kovalev, M.L. Lister, E. Ros, J.A. Zensus, Astron. Astrophys. **512**, A24 (2010)
E. Schartman, H. Ji, M. Burin, J. Goodman, Astron. Astrophys. **543**, A94 (2012)
P.A.G. Scheuer, Mon. Not. R. Astron. Soc. **166**, 513 (1974)

N.I. Shakura, R.A. Sunyaev, Astron. Astrophys. **24**, 337 (1973)
H. Shang, Z.-Y. Li, N. Hirano, in *Protostars and Planets V*, ed. by B. Reipurth, D. Jewitt, K. Keil (University of Arizona Press, Tucson, 2007)
S. Sheikhnezami, Ch. Fendt, O. Porth, B. Vaidya, J. Ghanbari, Astrophys. J. **757**, 65 (2012)
K. Shibata, Y. Uchida, Proc. Astron. Soc. Jpn. **37**, 31 (1985)
J.B. Simon, J.F. Hawley, Astrophys. J. **707**, 833 (2009)
J.B. Simon, J.F. Hawley, K. Beckwith, Astrophys. J. **690**, 974 (2009)
N. Soker, J.P. Lasota, Astron. Astrophys. **422**, 1039 (2004)
K.V. Sokolovsky et al., Astron. Astrophys. **532**, A38 (2011)
D. Stepanovs, Ch. Fendt, Astrophys. J. **793**, 31 (2014)
D. Stepanovs, Ch. Fendt, S. Sheikhnezami, Astrophys. J. **796**, 29 (2014)
I.W. Stephens, L.W. Looney, W. Kwon, M. Fernández-López, A.M. Hughest, L.G. Mundy, R.M. Crutcher, Z.-Y. Li, R. Rao, Nature **514**, 597 (2014)
J.M. Stone, P.E. Hardee, Astrophys. J. **540**, 192 (2000)
J.M. Stone, M.L. Norman, Astrophys. J. **389**, 297 (1992)
J.M. Stone, M.L. Norman, Astrophys. J. **433**, 746 (1994)
T.K. Suzuki, S.-I. Inutsuka, Astrophys. J. **784**, 121 (2014)
F. Tavecchio, L. Maraschi, R.M. Sambruna, C.M. Urry, Astrophys. J. **544**, L23 (2000)
G.B. Taylor, R. Zavala, Astrophys. J. **722**, L183 (2010)
A. Tchekhovskoy, Launching of active galactic nuclei jets, in *The Formation and Disruption of Black Hole Jets*, ed. by I. Contopoulos, D. Gabuzda, N. Kylafis. Astrophysics and Space Science Library, vol. 414 (Springer, Berlin, 2015), pp. 45–82, Chapter 3. doi:10.1007/978-3-319-10356-3_3
A. Tchekhovskoy, D. Giannios, Mon. Not. R. Astron. Soc. **447**, 327 (2014)
A. Tchekhovskoy, J.C. McKinney, R. Narayan, Mon. Not. R. Astron. Soc. **388**, 551 (2008)
A. Tchekhovskoy, J.C. McKinney, R. Narayan, Astrophys. J. **699**, 1789 (2009)
A. Tchekhovskoy, R. Narayan, J.C. McKinney, New Astron. **15**, 749 (2010a)
A. Tchekhovskoy, R. Narayan, J.C. McKinney, Astrophys. J. **711**, 50 (2010b)
A. Tchekhovskoy, R. Narayan, J.C. McKinney, Mon. Not. R. Astron. Soc. **418**, L79 (2011)
A. Tchekhovskoy, J.C. McKinney, R. Narayan, J. Phys. Conf. Ser. **372**, 012040 (2012). arXiv:1202.2864
A. Tchekhovskoy, B.D. Metzger, D. Giannios, L.Z. Kelley, Mon. Not. R. Astron. Soc. **437**, 2744 (2014)
Y. Todo, Y. Uchida, T. Sato, R. Rosner, Astrophys. J. **403**, 164 (1993)
I.L. Tregillis, T.W. Jones, D. Ryu, Astrophys. J. **601**, 778 (2004)
P. Tzeferacos, A. Ferrari, A. Mignone, C. Zanni, G. Bodo, S. Massaglia, Mon. Not. R. Astron. Soc. **400**, 820 (2009)
P. Tzeferacos, A. Ferrari, A. Mignone, C. Zanni, G. Bodo, S. Massaglia, Mon. Not. R. Astron. Soc. **428**, 3151 (2013)
Y. Uchida, K. Shibata, Proc. Astron. Soc. Jpn. **36**, 105 (1984)
Y. Uchida, K. Shibata, Proc. Astron. Soc. Jpn. **37**, 515 (1985)
G.V. Ustyugova, A.V. Koldoba, M.M. Romanova, V.M. Chechetkin, R.V.E. Lovelace, Astrophys. J. **439**, 39 (1995)
G.V. Ustyugova, A.V. Koldoba, M.M. Romanova, V.M. Chechetkin, R.V.E. Lovelace, Astrophys. J. **516**, 221 (1999)
B. Vaidya, Ch. Fendt, H. Beuther, O. Porth, Astrophys. J. **742**, 56 (2011)
A.A. van-Ballegooijen, in *Accretion Disks and Magnetic Fields in Astrophysics*, ed. by G. Belvedere (1989), p. 99
B. von Rekowski, A. Brandenburg, Astron. Astrophys. **420**, 17 (2004)
B. von Rekowski, A. Brandenburg, W. Dobler, Astron. Astrophys. **398**, 825 (2003)
M. Wardle, A. Königl, Astrophys. J. **410**, 218 (1993)
J.F.C. Wardle, D.C. Homan, R. Ojha, D.H. Roberts, Nature **395**, 457 (1998)
W.F. Winters, S.A. Balbus, J.F. Hawley, Mon. Not. R. Astron. Soc. **340**, 519 (2003)
S. Wykes et al., Astron. Astrophys. **558**, A19 (2013)
M. Zamaninasab, E. Clausen-Brown, T. Savolainen, A. Tchekhovskoy, Nature **510**, 126 (2014)
C. Zanni, A. Ferrari, R. Rosner, G. Bodo, S. Massaglia, Astron. Astrophys. **469**, 811 (2007)
A.A. Zdziarski, M. Sikora, P. Pjanka, A. Tchekhovskoy, Mon. Not. R. Astron. Soc. **451**, 927 (2015)
H. Zinnecker, M.J. McCaughrean, J.T. Rayner, Nature **394**, 862 (1998)

Part V
Physical Processes

DOI 10.1007/978-1-4939-3550-5_14
Reprinted from *Space Science Reviews* Journal, DOI 10.1007/s11214-015-0191-6

Gamma-Ray Bursts as Sources of Strong Magnetic Fields

Jonathan Granot[1] · Tsvi Piran[2] · Omer Bromberg[3] · Judith L. Racusin[4] · Frédéric Daigne[5]

Received: 7 July 2015 / Accepted: 21 July 2015 / Published online: 4 August 2015

Abstract Gamma-Ray Bursts (GRBs) are the strongest explosions in the Universe, which due to their extreme character likely involve some of the strongest magnetic fields in nature. This review discusses the possible roles of magnetic fields in GRBs, from their central engines, through the launching, acceleration and collimation of their ultra-relativistic jets, to the dissipation and particle acceleration that power their γ-ray emission, and the powerful blast wave they drive into the surrounding medium that generates their long-lived afterglow emission. An emphasis is put on particular areas in which there have been interesting developments in recent years.

Keywords Gamma-ray bursts · Magnetic fields · MHD · Neutron stars · Jets · Radiation mechanisms: non-thermal

✉ J. Granot
granot@openu.ac.il

T. Piran
tsvi@phys.huji.ac.il

O. Bromberg
omerb@astro.princeton.edu

J.L. Racusin
judith.racusin@nasa.gov

F. Daigne
daigne@iap.fr

1 Department of Natural Sciences, The Open University of Israel, 1 University Road, P.O. Box 808, Ra'anana 4353701, Israel

2 Racah Institute of Physics, The Hebrew University of Jerusalem, Jerusalem 91904, Israel

3 Department of Astrophysical Sciences, Princeton University, 4 Ivy Ln., Princeton, NJ 08544, USA

4 Astrophysics Science Division, NASA Goddard Space Flight Center, Greenbelt, MD, USA

5 Institut d'Astrophysique de Paris, UMR 7095 Université Pierre et Marie Curie, CNRS, 98 bis, boulevard Arago, 75014 Paris, France

1 Introduction

Gamma-ray Bursts (GRBs) are among the most extreme objects in the Universe. They are the most luminous cosmic explosions, and therefore serve as beacons at the edge of the visible Universe that can be used as cosmic probes. GRBs provide short timescale insight into end-stage stellar evolution, and serve as probes of extreme physics such as strong gravity, very large densities and magnetic fields, extremely energetic particles, and relativistic bulk motions. They are also promising sources of high-energy neutrinos and gravitational waves.

GRBs can be roughly divided into two main sub-classes: (i) Long-duration ($\gtrsim 2$ s) soft-spectrum bursts that are found in star-forming regions and are associated with broad-lined Type Ic supernovae, implying a massive star progenitor, which is most likely low-metalicity and rapidly rotating near this cataclysmic end of its life, and lives in a gas-rich environment not far from where it was formed. In order to produce a GRB, the central engine must drive a strong relativistic jet that bores its way through the stellar envelope and produces the GRB well outside of the progenitor star; (ii) Short-duration ($\lesssim 2$ s) hard-spectrum bursts that are thought to arise from the merger of a binary neutron star system (or a neutron star and a stellar-mass black hole) that emits gravitational waves as it inspirals and coalesces, producing a central engine driven jet. Such systems live in low density environments, possibly with a prior supernova kick that pushed them into the outskirts of their host galaxies. A third sub-class, whose importance was realized only recently (Soderberg et al. 2006; Campana et al. 2006; Liang et al. 2007; Virgili et al. 2009; Bromberg et al. 2011, 2012; Nakar and Sari 2012), involves low-luminosity GRBs, whose overall isotropic equivalent radiated energy is $E_{\gamma,\mathrm{iso}} \lesssim 10^{49}$ erg. They also typically have a smooth, single-peaked light curve, and their νF_ν spectrum typically peaks at a lower than average photon energy (usually $E_\mathrm{p} \lesssim 100\,\mathrm{keV}$). While observed rarely, because of their low luminosity, they are the most numerous group in nature (in terms of their rate per unit volume). They most likely do not arise from the same emission mechanism as regular long GRBs (e.g., Bromberg et al. 2012; Nakar and Sari 2012).

The phenomenology of GRBs is generally separated into two observational phases: the prompt emission and the afterglow. These two phases are traditionally differentiated largely based upon instrumental measurement methods, but they do seem to also be physically distinct—they arise from different emission mechanisms and occur at different distances from the central source. However, the dividing lines between the prompt emission and the afterglow have blurred in recent years. In the standard Fireball model (e.g., Piran 1999, 2004; Kumar and Zhang 2015), the prompt emission (i.e. the burst of γ-rays) is due to dissipative internal shocks within the outflow, while the long-lived broadband afterglow is the result of the jet driving a strong relativistic forward shock into the surrounding medium as it decelerates and transfers its energy to the shocked external medium. Unless the outflow is highly magnetized when it is decelerated by the external medium, this deceleration can occur through a strong reverse shock that results in a bright optical flash, which is also sometimes detected (e.g., Akerlof et al. 1999; Sari and Piran 1999; Mundell et al. 2007b; Racusin et al. 2008; Vestrand et al. 2014) just after the onset of the prompt emission, and decays largely independently of the forward shock emission.

The field of GRBs is relatively young, with several revolutions in our understanding of these objects thanks to new observations over the last two decades. The role of magnetic fields in GRBs is relevant to many topics in this field. They affect the properties of the compact object (neutron star and/or black hole) that powers the central engine, and how it launches the jet. Magnetic fields may also play an important role in the acceleration and collimation of the relativistic jets in GRBs, as well as in their composition. They can contribute to the energy dissipation and particle acceleration that powers the prompt GRB emission, and may play a key role in its emission mechanism. A strong magnetic field can suppress the reverse shock and its emission. During the afterglow, the amplification of the weak magnetic field in the external medium by the afterglow shock and its subsequent behavior in the shocked external medium downstream of the shock play a key role in the particle acceleration by the shock and in the shaping of the afterglow emission. In fact, it seems hard to find any important part of GRB physics where magnetic fields might be safely ignored.

In this review, we explore the evidence for extreme magnetic fields in GRBs, and how magnetic fields are intertwined with our understanding of the mechanisms that produce the relativistic jets that power these objects. As we cannot cover here all of the relevant topics in detail, we have instead chosen to focus on specific topics in which there has been recent progress (see e.g. Piran 2005, for an earlier review). First, in Sect. 2 a brief overview is given on the progenitors of both long and short GRBs, with the main thrust being devoted to the possible role of millisecond magnetars—newly born, very rapidly rotating and highly magnetized neutron stars—a topic that has recently received a lot of attention in the literature and in the GRB community. Next, Sect. 3 discusses the dynamics of GRB jets. It starts with long GRB jets as they bore their way out of their massive star progenitors, and then moves on to discuss more generally the possible role of magnetic fields in the acceleration and collimation of GRB jets, both in steady-state and highly time-variable outflows, as well as in the interaction of the jet with the external medium and the reverse shock. In Sect. 4 we discuss the role of magnetic fields in the dissipation and radiation that power the prompt γ-ray emission, and what GRB observations can tell us about the conditions within the emitting region. Finally, Sect. 5 is devoted to the role of magnetic fields in the afterglow. It focuses on their effects on the afterglow and reverse shock emission and their polarization, and how this can teach us about the magnetic field structure in the GRB outflow and its amplification in the afterglow shock as well as its structure and possible decay further downstream of this shock. Our conclusions are discussed in Sect. 6.

2 GRB Progenitors, Central Engine, and the Role of Magnetars

Long duration GRBs are associated with Type Ic supernovae (e.g., Woosley and Bloom 2006), which directly relate them to the death of massive stars stripped of their hydrogen and helium. This supports the popular Collapsar model (Woosley 1993; MacFadyen and Woosley 1999) in which a central engine lunches a relativistic jet that penetrates the stellar envelope and powers the GRB. Typically, within the Collapsar models the central engine is considered to be a an accreting newly-formed stellar-mass black hole at the center of the progenitor star. The most popular model for short duration GRBs features the merger of two neutron stars in a tight binary system (Eichler et al. 1989), which may again form a black hole surrounded by an accretion disk as they coalesce. Therefore, in both long and short duration GRBs, despite their very different progenitors, the central engine that is formed during the explosion and launches the relativistic jets might still be similar in nature—accretion onto a newly formed black hole.

An attractive alternative possibility that has gained popularity in recent years is that GRB central engines may involve magnetars (Usov 1992; Duncan and Thompson 1992; Bucciantini et al. 2008; Dall'Osso et al. 2009)—highly magnetized neutron stars with surface magnetic fields of order $B \sim 10^{15}$ G, which in this case are newly born and very rapidly rotating, with ~ 1 ms periods (and hence dubbed millisecond magnetars). In this model the main energy source is the neutron star's rotational energy, and a very strong magnetic field is needed for a rapid extraction of this rotational energy and to channel it into a relativistic outflow. A rapidly rotating neutron star may arise in the collapse of a rotating stellar core and the magnetic field can be amplified in this collapse (e.g., Duncan and Thompson 1992). As the magnetar's energy is naturally extracted in the form of a Poynting flux (though this flux is initially not significantly collimated) it naturally leads to a magnetically dominated outflow. Collimation of the outflow into a narrow jet may, however, be facilitated by the interaction of the outgoing strong magnetohydrodynamic (MHD) wind with the progenitor star's envelope (e.g., Bucciantini et al. 2007, 2008, 2009; Bromberg et al. 2014).

After the first ~ 10–100 s or so from the neutron star formation, the neutrino-driven winds subside and the baryon loading on the MHD wind significantly decreases. As a result the initial wind magnetization parameter σ_0 significantly increases and becomes $\gg 1$ (the magnetization parameter σ is the Poynting-to-matter energy flux ratio, or proper enthalpy density ratio). At this stage, the neutron star spin-down and its associate luminosity are approximately given by the magnetic dipole in vacuum formula (which also approximately holds in the force-free regime; Spitkovsky 2006),

$$L(t) = \frac{B^2 R^6 \Omega_0^4/(6c^3)}{[1 + 2B^2 R^6 \Omega_0^2 t/(6Ic^3)]^2} = \frac{E_0}{t_0(1+t/t_0)^2} \approx L_0 \times \begin{cases} 1 & (t < t_0), \\ (t/t_0)^{-2} & (t > t_0), \end{cases} \quad (1)$$

where I is the neutron star's moment of inertia, R is its radius, B is the surface dipole magnetic field at the pole, Ω_0 is the initial angular velocity, $E_0 \approx \frac{1}{2} I \Omega_0^2$ is the initial rotational energy, $L_0 = E_0/t_0$ is the initial spin-down luminosity and

$$t_0 = \frac{3Ic^3}{B^2 R^6 \Omega_0^2} \approx 2 \times 10^3 \left(\frac{B}{10^{15}\ \mathrm{G}}\right)^{-2} \left(\frac{P}{1\ \mathrm{ms}}\right)^2 \mathrm{s} \quad (2)$$

is the initial spin-down time (using typical values of $R \approx 10$ km, $I \approx 10^{45}$ g cm^2). This spindown luminosity initially (at $t < t_0$) has a plateau at L_0, and then (at $t > t_0$) falls off as t^{-2}. Both L_0 and t_0 can be tuned with the proper choice of the initial angular velocity Ω_0 and the magnetic field B. With a choice of $B \sim 10^{15.5-16}$ G one can arrange t_0 fit the prompt duration in which case the magnetar is invoked to power the prompt GRB. With a lower magnetic field of order $10^{14.5-15}$ G, t_0 is of order several thousand seconds, comparable to the duration of the plateau phase in some X-ray afterglows.

The magnetar model gained a lot of popularity with the discovery by *Swift* of plateaus in the X-ray afterglow light curves of many GRBs (Nousek et al. 2006; Zhang et al. 2006), whose shape resembles the overall shape of Magnetar's spindown luminosity (Troja et al. 2007; Dall'Osso et al. 2009, 2011; Rowlinson et al. 2013, 2014). Somewhat surprisingly, even though this tentative evidence[1] for magnetar-like activity was obtained for the afterglow phase, it was interpreted in the community as evidence for a magnetar operating as the main source of energy for the prompt emission as well. Both interpretations face some difficulties.

The magnetic field needed to produce the prompt emission is larger by about one order of magnitude than the one observed even in the strongest magnetars. This may not be that

[1] These plateaus have several alternative explanation, which are at least as compelling as the magnetar explanation, such as promptly ejected slow material that gradually catches up with the afterglow shock (Nousek et al. 2006; Granot and Kumar 2006), time-varying afterglow shock microphysical parameters (Granot et al. 2006), viewing angle effects (Eichler and Granot 2006), or a two-component jet (Peng et al. 2005; Granot et al. 2006).

puzzling as there is ample evidence of magnetic field decay in magnetars (Dall'Osso et al. 2012), and the observed magnetars are typically a few thousand years old. It is possible that the magnetic fields of newborn magnetars are large enough.

A more serious problem concerns the energy budget. The rotational energy of a typical neutron star, even when rotating at breakup velocity, is at best marginally sufficient to power the most powerful GRBs (Cenko et al. 2010). This is especially so if we also take into account the efficiency of converting this rotational energy into the prompt flux of γ-rays. Of course, magnetars could still power less energetic GRBs. However, this would require one to invoke two kinds of central engines, as a different energy source would be needed to power the most energetic GRBs.

Even more perplexing is the situation concerning the longer duration plateaus in the afterglow light curves. Here, the needed values of magnetic fields are indeed typical for those arising in the observed magnetars, and the overall energy budget is reasonable as well. However, another question arises: if a low magnetic field magnetar has powered the afterglow plateau, then what has powered the prompt GRB? Can a magnetar fire twice? The simple answer, according to Eq. (1), is no. This is as long as the magnetic field remains constant during the slowdown time scale. However, one can come up with a fine-tuned model in which the magnetic field decays on a timescale shorter than t_0. In this case the duration of the magnetar activity is not determined by t_0, but by the magnetic field decay time. Once the magnetic field has decayed, a second slower magnetar phase appears with a new t_0. Overall such a model requires a the magnetic field that is extremely large initially, leading to the prompt emission and then it decreases, just at the right time (and before all the rotational energy is exhausted) to a lower level in which the weaker magnetar powers (using the remaining rotational energy) the afterglow plateau.

An alternative option is as follows: in the first $\sim$ 10–100 s or so after the formation of the neutron star, the strong neutrino-driven winds cause a large baryon loading on the MHD wind that prevents the formation of a very high initial magnetization ($\sigma_0 \lesssim$ 10–100). Therefore, during at least part of this time the spindown luminosity can significantly deviate from the form Eq. (1)—the formula for a magnetic dipole in vacuum—and may in fact be significantly higher, and closer to the result for a magnetic monopole in vacuum, since most of the magnetic field lines are opened by the strong baryon loading (e.g., Metzger et al. 2011). This can increase the spindown luminosity by a factor of $\sim (R_L/R)^2 \sim 10^{1.5}$, where $R_L = c/\Omega_0$ is the initial value of the light-cylinder radius. However, as in the early magnetic field decay scenario mentioned above, also this solution would require fine tuning in order to extract just the right amount of rotational energy over just the right timescale. Moreover, in this case the bulk of the large amount of energy that is released on the timescale of the prompt GRB is given to relatively low-σ_0 baryon-rich material, which could not attain sufficiently large asymptotic Lorentz factors that are needed to power a GRB.

Another possible solution to this problem was suggested recently (Rezzolla and Kumar 2015; Ciolfi and Siegel 2015)—the "time reversal model", which postulates magnetar activity for the plateau but an accretion disk for the prompt phase. According to this model, first a magnetar with $t_0 \sim 10^4$ s is born and launches a fast MHD wind whose interaction with slower matter that was ejected earlier produces the afterglow plateau. In this scenario the magnetar is a supramassive neutron star, i.e. supported against gravitational collapse by its very fast rotation, so once it spins down significantly it collapses to a black hole, and an accretion disk that forms during this collapse powers the prompt emission.[2] In spite of this

[2]This scenario is rather similar to the "supranova model" that was suggested much earlier (Vietri and Stella 1998, 1999).

reversed time sequence the plateau is observed after the prompt emission because it involved the interaction of the winds and this phase introduces a time delay (in the observer frame). While it is appealing, this model requires the formation of a disk during the collapse of the supramassive neutron star. However, Margalit et al. (2015) have recently argued that this is impossible.

A different possible solution is if the prompt GRB is powered by the energy in a strong initial differential rotation (Kluźniak and Ruderman 1998). The strong differential rotation winds-up strong toroidal magnetic field loops, which are buoyantly pushed out of the neutron star surface and power the prompt GRB. This lasts until they exhaust all of the differential rotation energy on the timescale of the prompt GRB emission of long duration GRBs. The rotational energy of the remaining uniformly rotating neutron star could then power the plateaus on its longer magnetic dipole spindown time t_0. This might possibly work for long duration GRBs that are not too energetic (as the energy in differential rotation is somewhat lower than the total rotational energy).

Recently, it was suggested that millisecond magnetars might also be at work in short duration GRBs (e.g., Fan and Xu 2006; Rosswog 2007; Metzger et al. 2008; Rowlinson et al. 2010, 2013; Bucciantini et al. 2012; Fan et al. 2013a; Gompertz et al. 2014). Newly formed millisecond magnetars were suggested to produce the extended emission seen on a timescale of $\sim 10^2$ s in some short GRBs (e.g. Metzger et al. 2008; Bucciantini et al. 2012). In this picture the initial short GRB spike may arise from short-lived accretion following the merger. The extended emission is driven by the spindown power that is released over $\sim 10^2$ s, and takes several seconds to break out of the surrounding mildly relativistic material that is ejected quasi-isotropically during the merger. In this scenario, however, the jets that power the short GRB itself (the initial hard spike) are launched within the first second or so after the formation of a newly-born millisecond magnetar, when the neutrino-driven wind is very vigorous, the magnetization is low ($\sigma_0 \lesssim 1$), and the baryon loading is very significant near the star and the inner accretion disc where the jet may be launched. It is therefore unclear whether in this case the jets could eventually reach a high enough Lorentz factor to produce the GRB. Moreover, in a binary merger the neutron star is formed extremely rapidly rotating (near breakup), and its rotational energy of a few $\times 10^{52}$ erg is eventually injected into the afterglow shock. This should naturally produce a bright afterglow emission while the observed afterglows of short GRBs (either with or without an extended emission) are typically much dimmer than those of long GRBs. While this might in part be attributed to a lower external density on average, this cannot fully account for the dimmer afterglows of short GRBs essentially over the entire broad-band spectrum, from radio to GeV energies.

3 Jet Propagation and Dynamics

The question of the jet composition is still a major issue in our understanding of GRBs. It affects the location of the emission site, the mechanism of the emission and the particle acceleration. There are two main possibilities that are commonly discussed in the literature: a hydrodynamic jet and a Poynting flux dominated jet (for the jet dynamics discussed here we do not make the distinction between baryonic and $e^{\pm}$ pairs particle content). The main advantage of a hydrodynamic jet is fast and robust acceleration, which allows the jet to reach very high Lorentz factors relatively close to the central source. Magnetic acceleration, on the other hand, is slower and less robust. However, hydromagnetic jet launching implies dynamically strong magnetic fields near the central source, which can naturally avoid excessive baryonic loading into the central part of the jet, and thus allow it to reach large asymptotic Lorentz factors far from the source. The required very small baryon loading is hard to naturally achieve in a purely hydrodynamic jet. Here we focus on the possible role of magnetic fields in the jet dynamics and propagation, while keeping in mind these two main options for the jet composition.

3.1 Jet Propagation in the Stellar Envelope

In order to produce a regular GRB, a collapsar jet needs to successfully break out of its progenitor star. After it breaks out, the jet can accelerate freely and eventually generate the observed γ-ray photons far from the star in a region where they can escape (see Sect. 4). Before it emerges from the stellar surface, the jet propagates inside the star by pushing the stellar material in front of it, forming a bow shock ahead of the jet. The stellar material that crosses this shock is heated and forms a cocoon around the jet, which in turn applies pressure on the jet and collimates it. The collimated jet propagates at a different velocity than a freely expanding jet. It continuously injects energy into the cocoon through a slower moving *head* that forms at the front of the jet. The head dissipates the jet's energy and channels it into the cocoon. Therefore, the continuous propagation of the jet through the star depends on the supply of fresh energy from the source. If the engine stops injecting energy, the head will essentially stop propagating once the information about the energy cutoff will reach it, and the jet will fail. The *breakout time*, $t_{\rm b}$, is defined as the time of the engine shutoff for which the information about the shutoff reaches the jet's head when it is at the edge of the star. If the engine stops working at a time $t_e < t_{\rm b}$, the head will "feel" this cutoff while it is inside the star and will stop propagating. In this case the jet will not break out and it will not produce a regular GRB.[3] Since the information travels outwards at very close to the speed of light, the breakout time is related to the time at which the jet's head reaches the outer edge of the star through

$$t_{\rm b} = \int_0^{R_*} \frac{dz}{\beta_{\rm h}(z)c} - \frac{R_*}{c} \equiv \frac{R_*}{c}\frac{1-\bar{\beta}_{\rm h}}{\bar{\beta}_{\rm h}}, \qquad (3)$$

where $\beta_{\rm h}(z)c$ is the instantaneous velocity of the jet's head at a distance z from the central source (along its symmetry axis, using cylindrical coordinates), and $\bar{\beta}_{\rm h}c$ is its average velocity.

Simple analytic solutions to Eq. (3) can be obtained in two limits: (i) a non-relativistic limit, characterized by a proper speed (in units of c) $u_{\rm h} = \Gamma_{\rm h}\beta_{\rm h} \ll 1$ (where $\Gamma_{\rm h}$ is the head's Lorentz factor) in which $t_{\rm b} \simeq R_*/\bar{\beta}_{\rm h}c$, and (ii) the relativistic limit, characterized by $u_{\rm h} \gg 1$, in which $t_{\rm b} \simeq R_*/2\bar{\Gamma}_{\rm h}^2 c$. The transition between these two limits occurs when $t_{\rm b} \simeq R_*/c$, which according to Eq. (3) corresponds to $\bar{\beta}_{\rm h} \simeq 1/2$. The jet's head is initially sub-relativistic, but it accelerates in the steep density profile inside the star ($\xi = -d\log\rho/d\log r > 2$). Therefore, if the jet becomes relativistic at some radius, $R_{\rm rel}$, where $u_{\rm h} \simeq 1$, then it will remain so until it will break out.

In order to calculate the breakout time ones needs a proper model for the propagation of the jet in the star. Such a model needs to consider the evolution of the jet and the cocoon self-consistently, as they affect one another. The propagation velocity of the head is determined by its cross section, which is set by the collimation of the jet. The head's velocity, in turn, controls the energy injection into the cocoon, which determines the collimating pressure. The dynamics of this system can be described in a relatively simple way in two extreme

[3] A failed jet produces, most likely, a low-luminosity GRB when a shock wave generated by the dissipated energy breaks out from the seller envelope.

cases of a purely hydrodynamic jet and a purely electromagnetic (Poynting flux dominated) jet.

3.1.1 The Breakout Time of Collapsar Jets

Close to the injection point the jet's internal pressure, p_j, is much larger than the cocoon's pressure, p_c. Therefore, initially the jet material expands freely until the collimation point where the jet's pressure equals the cocoon's pressure, $p_j = p_c$. Above this point the jet is collimated by the cocoon's pressure, and its behavior depends on its magnetization.

In the hydrodynamic case, the collimation of the jet leads to the formation of a collimation shock at the base of the jet (e.g., Bromberg and Levinson 2007). Above this point the jet maintains a roughly cylindrical shape due to a relatively uniform pressure in the cocoon (e.g., Zhang et al. 2003; Matzner 2003; Bromberg et al. 2011). The jet material remains relativistic with a roughly constant Lorentz factor $\Gamma_j \sim 1/\theta_0$, where θ_0 is the jet injection (or initial) half-opening angle. At the head of the jet the relativistic jet material decelerates abruptly through a strong reverse shock. Since the jet is roughly cylindrical upstream of the reverse shock, the width of the head—its cylindrical radius r_j and its corresponding cross-section $\Sigma_j = \pi r_j^2$, are set by the width and cross-section of the jet at the collimation point, which are shown to be

$$\Sigma_j = \pi r_j^2 \simeq \frac{L_j \theta_0^2}{4cp_c}, \tag{4}$$

where L_j is the (one sided) jet luminosity. The velocity of the head was shown (Matzner 2003; Bromberg et al. 2011) to follow

$$\beta_h = \frac{\beta_j}{1 + \tilde{L}^{-1/2}}, \tag{5}$$

where the dimensionless parameter

$$\tilde{L} = \frac{\rho_j h_j \Gamma_j^2}{\rho_a} \simeq \frac{L_j}{\Sigma_j \rho_a c^3}, \tag{6}$$

represents the ratio between the energy density of the jet ($L_j/\Sigma_j c$) and the rest-mass energy density of the surrounding medium ($\rho_a c^2$) at the location of the head. Here $h_j = 1 + 4p_j/\rho_j c^2$ is the dimensionless specific enthalpy of the jet material just upstream of the termination shock at the base of its head.

Bromberg et al. (2011) have obtained approximate analytic expressions for the propagation velocity of a hydrodynamic jet, and demonstrated that for typical stellar and jet properties, the jets head propagates at a velocity that is at most mildly relativistic. Therefore, in this case the solution to Eq. (3) is in the transition region between the relativistic and the non-relativistic limits, and can be approximated following (Bromberg et al. 2015). In order to obtain a useful analytic solution, they approximated the exact integration (shown by the *dashed blue line* in Fig. 1) by:

$$t_{b,hyd} \simeq 6.5 R_{*,4R_\odot} \left[\left(\frac{L_j}{L_{rel}} \right)^{-2/3} + \left(\frac{L_j}{L_{rel}} \right)^{-2/5} \right]^{1/2} \text{ s}, \tag{7}$$

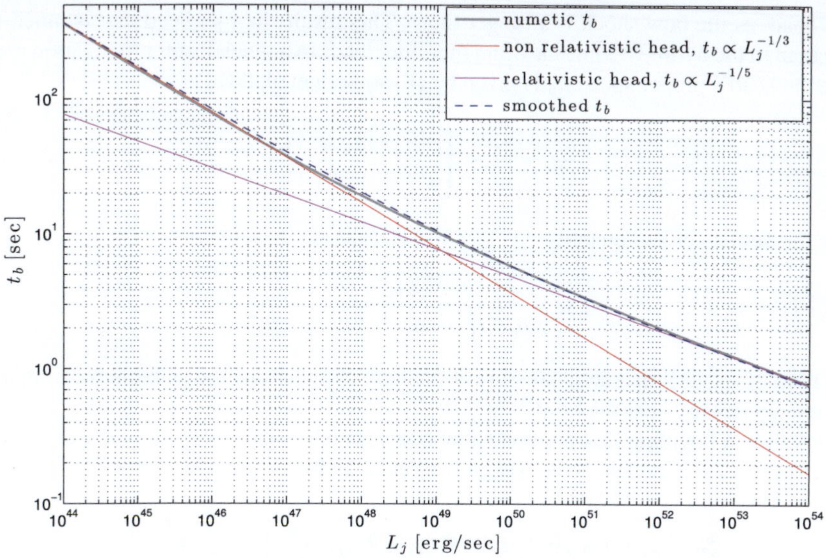

Fig. 1 The breakout time, $t_{\rm b}$, as a function of $L_{\rm j}$ calculated for a jet with an opening angle $\theta_j = 7°$, and a star with a mass $M_* = 15 M_\odot$, radius $R_* = 4 R_\odot$ and a power-law density profile $\rho \propto r^{-2.5}$. The *gray solid curve* tracks the exact integration of Eq. (3), the *red and magenta lines* show the analytic approximation for the non relativistic and the relativistic cases respectively. The *dashed blue line* follows the smoothed analytic solution for $t_{\rm b}$ from Eq. (7) (this figure is taken from Bromberg et al. 2015)

where $L_{\rm rel}$ is the transition luminosity between a non-relativistic and a relativistic case:

$$L_{\rm rel} \simeq 1.6 \times 10^{49}\ R_{*,4R_\odot}^{-1} M_{*,15M_\odot} \theta_{0.84}^4 \left(\frac{3-\xi}{0.5}\right)^{7/5} \left(\frac{5-\xi}{2.5}\right)^{4/5} \left(\frac{7-\xi}{4.5}\right)^{15/2} \text{erg}\,\text{s}^{-1}. \quad (8)$$

As canonical parameters we have used here a stellar mass of $M_* = 15 M_\odot$, a stellar radius $R_* = 4 R_\odot$ and we assume a power-law density profile: $\rho_* \propto r^{-\xi}$ with $\xi = 2.5$. Hereafter we measure masses and radii in units of solar mass and solar radius respectively and use the subscript '$*$' to denote properties of the progenitor star. For all other quantities we use the dimensionless form $A_x \equiv A/10^x$ measured in c.g.s units. For a typical collapsar (one sided) jet luminosity of $L_j \sim 2 \times 10^{49}$ erg s^{-1}, and injection angle of $\theta_j = 7°$ the corresponding breakout time is $t_{\rm b}(L_{\rm rel}) \simeq 9$ s.

In a Poynting flux dominated jet the situation is different. The cocoon's pressure is typically strong enough to collimate the jet before it looses causal contact with the axis. In this case the poloidal magnetic field is comparable to the toroidal field in the comoving frame of the flow, and shocks are inhibited. This leads to a smooth transition of the jet material from a free expansion state, close to the engine, to a collimated state. The jet material remains in a strong causal contact also above the collimation point. Therefore as it approaches the head, it does not shock. Moreover, it can be shown (Bromberg et al. 2014) that under these conditions the jet's proper velocity u_j is approximately equal to the ratio of its cylindrical radius $r_{\rm j}$ and the light cylinder radius $r_{\rm L}$: $u_j \simeq r_{\rm j}/r_{\rm L}$, and the same also holds at the jet's head, $u_{\rm h} \simeq r_{\rm h}/r_{\rm L}$. Therefore, the jet material gradually decelerates and becomes narrower as it approaches the head until at the head its velocity matches that of the shocked stellar material just behind the front tip of the bow shock. This deceleration and narrowing of the jet towards its head is assisted by the fact that the cocoon's pressure becomes larger closer to the head, as the bow shock is stronger there. This results in a jet head that is much narrower than in the hydrodynamic case and therefore leads to a much faster propagation speed, where the head's proper speed u_h is given by (Bromberg et al. 2014):

$$u_\mathrm{h} \sim \frac{r_\mathrm{h}}{r_\mathrm{L}} \sim \begin{cases} a^{1/5} & (u_\mathrm{h} \ll 1), \\ a^{1/6} & (u_\mathrm{h} \gg 1), \end{cases} \tag{9}$$

where the dimensionless quantity

$$a \equiv \frac{L_\mathrm{j}}{\pi \rho_a c^3 r_\mathrm{L}^2} = \frac{p_\mathrm{L}}{\rho_a c^2} \approx 1.2 \frac{L_{50}}{\rho_4 r_{L7}^2}, \tag{10}$$

is the ratio of the jet's magnetic pressure at the light cylinder and the ambient medium's rest mass energy density near the head.[4]

Therefore, a Poynting flux dominated jet becomes relativistic at a radius R_rel deep inside the star, even with a modest power (Bromberg et al. 2014):

$$\frac{R_\mathrm{rel}}{R_*} \simeq 1.4 \times 10^{-2} \left[L_{49.3}^{-1} M_{*,15M_\odot} R_{*,4R_\odot}^{-3} r_{L,7}^2 \left(\frac{3-\xi}{0.5} \right) \right]^{1/\xi}. \tag{11}$$

This implies that here only the relativistic asymptotic solution ($u_\mathrm{h} \approx a^{1/6}$) is relevant. The corresponding breakout time is (Bromberg et al. 2015):

$$t_\mathrm{b,mag} \simeq 0.8 L_{49.3}^{-1/3} M_{*,15M_\odot}^{1/3} r_{L,7}^{2/3} \left(\frac{0.5}{3-\xi} \right)^{2/3} \mathrm{s}. \tag{12}$$

This time is much shorter than the breakout time of a hydrodynamic jet with a similar luminosity.

3.1.2 *Observational Evidence for the Jet Breakout Time*

After the jet emerges from the stellar envelope it dissipates its energy at a large distance and produces the GRB. On average, the overall behavior of the prompt emission does not vary significantly during the burst (the second half of the prompt emission is rather similar to the first one). This suggests that the prompt emission arises at a more or less constant radius and not in a propagating single shell. A single shell would have expanded by a factor of ~ 10–100 during the duration of a burst and it is unlikely to maintain constant conditions as it emits the prompt γ-ray emission over such a wide range of radii. This implies, in turn, that the GRB activity follows the central engine's activity (Sari and Piran 1997), and that the GRB lasts as long as the central engine is active. Therefore, within the Collapsar model, the observed GRB duration (usually denoted by T_{90}, which measures the time over which the central 90 % of the prompt photon counts are detected) is the difference between the engine operation time, t_e, and the breakout time, t_b, namely $T_{90} = t_e - t_\mathrm{b}$ (not accounting for the cosmological time dilation here).

The breakout time essentially serves as a threshold time: a regular GRB is formed only if $t_e > t_\mathrm{b}$. Bromberg et al. (2012) have shown that in such a case one would expect a plateau in the duration distribution of GRBs, dN_GRB/dT_{90}, at durations that are shorter than t_b. The

[4] This analysis does not account for 3D effects that can slow down the head's propagation speed (Bromberg and Tchekhovskoy 2015).

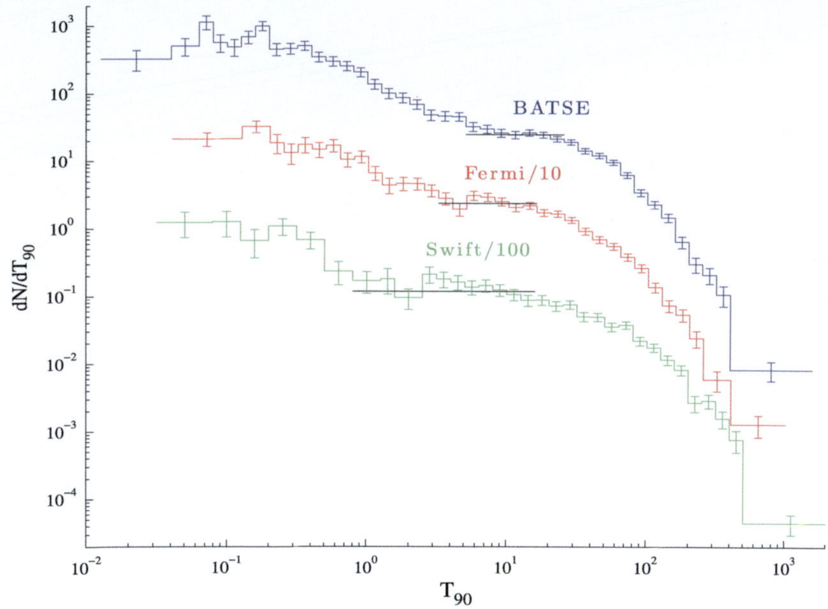

Fig. 2 The duration distribution, $dN_{\rm GRB}/dT_{90}$ of BATSE (*blue*), *Fermi* (*red*) and *Swift* (*green*) GRBs. The *different curves* are shifted in order to avoid overlap. The data bins are evenly spaced in logarithmic scale with $\Delta\log(T_{90}) = 0.1$. Bins with less than 5 events are combined with their neighbors in order to achieve statistical significance. The *black horizontal lines* mark the bins that fit a plateau at a confidence interval up to 2σ (this figure is taken from Bromberg et al. 2015)

logic behind this is as follow. At the time when the jet's head breaches the edge of the star, it is already disconnected from the engine and cannot transmit information backward to the engine. In other words, the engine cannot "tell" when the jet breaks out of the star and we do not expect that t_e and $t_{\rm b}$ will be strongly related to each other. In fact, for a given $t_{\rm b}$ we expect to have a distribution of engine activity times, where some are shorter ($t_e < t_{\rm b}$) and some are longer ($t_e > t_{\rm b}$) than $t_{\rm b}$. In this case the probability of observing a GRB with duration T_{90} is equal to the probability that the engine will work for a time $t_e = T_{90} + t_{\rm b}$: $P_{\rm GRB}(T_{90}) \equiv P_e(t_e = T_{90} + t_{\rm b})$. This probability has a simple description in two limits:

$$P_{\rm GRB}(T_{90}) \approx \begin{cases} P_e(t_{\rm b}) & (T_{90} \ll t_{\rm b}), \\ P_e(T_{90}) & (T_{90} \gg t_{\rm b}). \end{cases} \tag{13}$$

Now, if there is a dominant population of GRBs with a typical $t_{\rm b}$, then at short durations $P_{\rm GRB}(T_{90}) \to P_e(t_{\rm b}) = \text{const}$, we expect to get a plateau at durations $T_{90} \ll t_{\rm b}$.

Figure 2 depicts the duration distribution, $dN_{\rm GRB}/dT_{90}$, of BATSE[5] (2100 GRBs), *Fermi*-GBM[6] (1310 GRBs) and *Swift*[7] (800 GRBs). To fit a plateau in each data set (Bromberg et al. 2015) looked for the maximal number of bins that are consistent with a plateau at a

[5] http://gammaray.msfc.nasa.gov/batse/grb/catalog/current/ from April 21, 1991 until August 17, 2000.

[6] http://heasarc.gsfc.nasa.gov/W3Browse/fermi/fermigbrst.html, from July 17, 2008 until February 14, 2014.

[7] http://swift.gsfc.nasa.gov/archive/grb_table/, from December 17, 2004 until February 14, 2014.

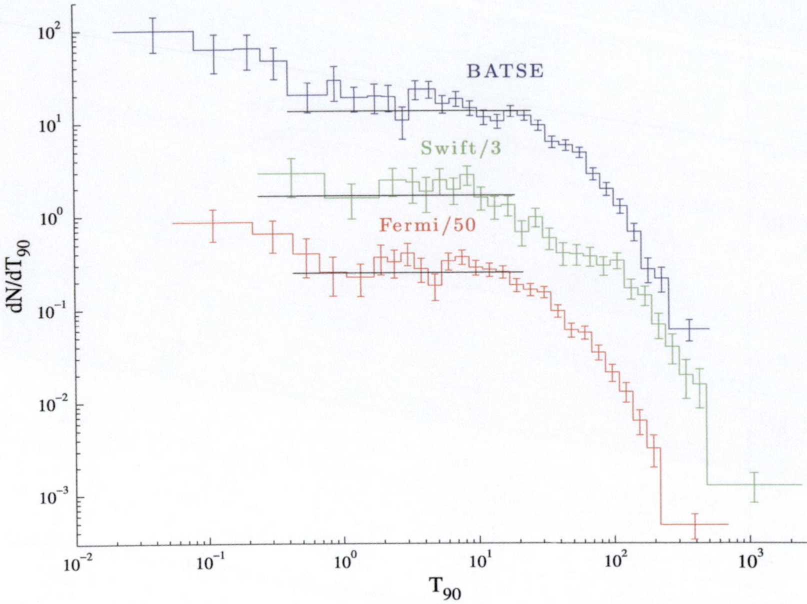

Fig. 3 The duration distribution, $dN_{\mathrm{GRB}}/dT_{90}$ of the soft GRBs. The analysis is the same as in Fig. 2, only the data from each satellite contains only events that are softer than the median hardness of the long GRBs with durations $T_{90} > 20$ s. For BATSE, this corresponds to GRBs having a hardness ratio $HR_{32} < 2.6$, for *Fermi* the GRBs have a power law spectral index < -1.5, and for *Swift* the GRBs have a power law spectral index < -1.7. The analysis here updates the analysis in Bromberg et al. (2013) using a more complete recent sample (this figure is taken from Bromberg et al. 2015)

confidence level ≤ 95 % (2σ).[8] The extent of the best fitted plateaus is 5–25 s in the BATSE data (7.19/4 χ^2/d.o.f.), 2.5–17 s in the *Fermi*-GBM data (10/5 χ^2/d.o.f.), and 1–20 s in the *Swift* data (15.85/9 χ^2/d.o.f.). Bromberg et al. (2015) accounted for three free parameters in the fit: the height of the plateau and the two opposite ends of the plateau line. The differences between the maximal durations of the plateaus can be mostly attributed to the different sensitivity and triggering algorithms of the different detectors.

At short durations, the plateau is concealed by the increasing number of non-Collapsar ("short") GRBs having a typical duration of less than a few seconds (Bromberg et al. 2013). As non-Collapsars have on-average harder spectra than Collapsars (e.g. Kouveliotou et al. 1993), the relative number of non-Collapsars can be reduced by choosing a hardness threshold (for each sample) and selecting only the events that are softer than this threshold. This should lead to a less prominent "bump" at short duration. If the plateau is indeed an intrinsic property of the (softer) Collapsars duration distribution, it should extend to shorter durations in a softer subsample. To examine this effect they selected in each sample all the events that are softer than the median hardness of long GRBs ($T_{90} > 20$ s) in the sample (see Bromberg et al. 2013, for further details). Figure 3 shows the duration distribution of the soft GRB subsamples. The plateaus indeed extend to much shorter durations than in the complete samples, supporting our hypothesis. The extent of the best fitted plateaus is 0.4–25 s in the

[8]The confidence level is defined here as $\int_0^{\chi^2} P(x,\nu)dx$, where $P(\chi^2,\nu)$ is the probability density function of χ^2 with ν degrees of freedom (Press et al. 1992).

BATSE data (20.75/12 χ^2/d.o.f.), 0.4–17 s in the *Fermi*-GBM data (8.7/10 χ^2/d.o.f.) and 0.2–20 s in the *Swift* data (9.04/8 χ^2/d.o.f.).

Taking a median redshift of $z \simeq 2$ for *Swift* GRBs and $z \simeq 1$ for *Fermi*-GBM and BATSE bursts, Bromberg et al. (2015) find that in the GRBs' rest frame, these plateaus extend up to intrinsic durations of 7–12 s, consistent with the results obtained by Bromberg et al. (2012). Note that the actual t_b may be somewhat longer than the duration that marks the end of the plateau, but it cannot be shorter. We use the duration interval of 7–12 s as our best estimate for the typical t_b.

3.1.3 Implications: The Jet Composition at Early Times

Equations (7), (8) and (12) use parameter values inferred from typical GRB jets, after accounting for the jet opening angle (e.g., Bloom et al. 2003). From these equations it can be seen that a breakout time of 7–12 s is consistent with that expected for a hydrodynamic jet from a WR star with a radius of $\sim (3\text{–}6) R_\odot$. On the other hand, these breakout times are too long for typical parameters expected for a Poynting dominated jet. To account for these breakout times, the light cylinder of the compact object at the base of the jet should be of the order of $r_L \simeq (2.5\text{–}5) \times 10^8$ cm, corresponding to an angular frequency of $\Omega_m \simeq 60\text{–}120$ rad/s at the base of the jet. Such a frequency is too low to allow the engine to power a typical GRB jet (Bromberg et al. 2015).

The conclusion arising from this analysis is that during most of its propagation within the star the jet has a low magnetization and it propagates as a hydrodynamic jet (unless 3D effects significantly increase $t_{b,mag}$). This result leads to some interesting implications for the properties of long GRB central engines and the conditions at the base of the jets. One possibility is that the jet is launched hydrodynamically at the source. The most probable process for that is neutrino–anti-neutrino annihilation above the rotational axis of the central engine (e.g. Eichler et al. 1989; Levinson and Eichler 1993). This scenario can work only if the accretion rate is $\gtrsim 0.1\ M_\odot\ \mathrm{s}^{-1}$, so that neutrino emission is large enough to power the observed jets (Kawanaka et al. 2013; Levinson and Globus 2013). The high accretion rate must be sustained throughout the entire duration of the GRB, which can last from tens to hundreds of seconds. Though a duration of $\lesssim 30$ s seems to be consistent with such a model (e.g. Lindner et al. 2010), its seems unlikely to be capable of powering longer duration GRBs.

A second possibility is that the jet is launched Poynting flux dominated but it dissipates most of its magnetic energy close to the source, and it then propagates as a hydrodynamic jet. An appealing process for such efficient dissipation is the kink instability (Lyubarskij 1992; Eichler 1993; Spruit et al. 1997; Begelman 1998; Lyubarskii 1999; Giannios and Spruit 2006). Bromberg et al. (2014), however, have shown via analytic considerations, that collapsar jets are less likely to be disrupted by the kink instability. Thus a different process, possibly internal to the jet, may be needed to dissipate the jet energy. A definite answer will be obtained only via 3D numerical simulations, which are underway. In one such work, Bromberg and Tchekhovskoy (2015) show that indeed kink instability is unlikely to disrupt a typical collapsar jet. Nevertheless, kink modes can grow internally in the jet and lead to efficient dissipation of the magnetic energy via reconnection of the magnetic field lines without compromising the jets's integrity. The outcome of such dissipation is a jet with an equipartition between thermal and magnetic energy, which propagates more or less like a hydrodynamic jet.

A third possibility is that the jet changes its character with time. Our conclusion concerning the jet composition applies only to the initial phase, while its head is still within the stellar envelope. This phase, which lasts ~ 10 s, must be predominantly hydrodynamic. Once the jet has breached the star it can be Poynting flux dominated. This would require a more complicated central engine that switches from one mode to another. While this seems contrived, remarkably, some magnetar models suggest such a possibility (Metzger et al. 2011). One can also imagine accretion disk models that initially cool via neutrinos and later on as the accretion rate decreases, become Poynting flux dominated (Kawanaka et al. 2013). However, all such models require some degree of coincidence as the central engine does not receive any feedback from the propagating jet and there is no a priori reason that the transition from one composition to the other would take place just at the right stage.

3.2 Jet Steady State Acceleration

Magnetic acceleration and thermal acceleration are the two main competing mechanisms for the acceleration of GRB jets or outflows. Thus, the acceleration mechanism is tightly related to the outflow composition and in particular its degree of magnetization, which is both highly uncertain and of great interest. In other sources of relativistic jets or outflows, there are currently better constraints on the composition. Pulsar winds are almost certainly Poynting flux dominated near the central source. This most likely also holds for active galactic nuclei (AGN) and tidal disruption events, as in these sources the central accreting black hole is supermassive, and therefore even close to it the Thompson optical depth, τ_T, may not be high enough for thermal acceleration by radiation pressure (the main competition to magnetic acceleration) to work efficiently (e.g., Ghisellini 2012). In GRBs or micro-quasars, however, thermal acceleration could also work (since $\tau_T \gg 1$ is possible, or even likely close enough to the source), and the dominant acceleration mechanism is less clear.

First, let us consider the thermal acceleration of a steady, axisymmetric, and unmagnetized flow that is initially relativistically hot with $p \gg \rho c^2$. Let the jet cross section be $\Sigma \propto r^2$ where r is its cylindrical radius. The relativistic equation of state implies $p \propto \rho^{4/3}$, while mass and energy conservation read $\Gamma \rho c \Sigma = \text{const}$ and $\Gamma^2(\rho c^2 + 4p)c\Sigma = \text{const}$, respectively (where we have assumed a relativistic velocity, $\beta = v/c \approx 1$). The ratio of the two last expressions gives the Bernoulli equation—the total energy per unit rest energy (which is conserved without any significant energy losses from the system), $(1 + 4p/\rho c^2)\Gamma = \text{const}$. As long as the flow is relativistically hot ($p \gg \rho c^2$) it accelerates as $\Gamma \propto \rho/p \propto \rho^{-1/3} \propto \Sigma^{1/2} \propto r$. This reproduces the familiar result for a spherical or conical flow for which $\Gamma \propto r \propto z$, i.e. the Lorentz factor grows linearly with the distance z from the central source. Therefore, thermal acceleration is relatively fast, efficient and robust.

Let us now do a similar simple analysis for a cold and initially highly-magnetized flow, with $\sigma_0 = B_0^2/4\pi\rho_0 c^2 \gg 1$ (e.g., Komissarov 2011). Let the flow be steady, axisymmetric, and ideal MHD (i.e. without magnetic dissipation). Let us consider the flow between two magnetic flux surfaces defined by r and $r + \delta r$ (which are both functions of z). Flux freezing (ideal MHD) implies that the poloidal and tangential magnetic field components scale as $B_p \propto 1/r\delta r$ and $B_\phi \propto 1/\delta r$, respectively, in the lab frame. Therefore, the tangential field component rapidly dominates far from the source, so that $B \approx B_\phi \approx \Gamma B'$ where B' in the magnetic field in the comoving frame of the outflowing plasma (in which the electric field vanishes). Altogether this gives $B = \Gamma B' \propto 1/\delta r$. Mass and energy conservation read $\Gamma \rho c \Sigma = \text{const}$ and $\Gamma^2(\rho c^2 + B'^2/4\pi)c\Sigma = \text{const}$, respectively, where $\Sigma \propto r\delta r$. Their ratio implies a total energy per unit rest energy of $(1+\sigma)\Gamma = (1+\sigma_0)\Gamma_0 = \Gamma_{\max}$ where $\sigma = B'^2/4\pi\rho c^2 \propto r/\Gamma\delta r$ is the magnetization parameter. Therefore, this results in the following Lorentz factor evolution:

$$\frac{\Gamma}{\Gamma_0} = 1 + \sigma_0\left(1 - \frac{\delta r_0}{r_0}\frac{r}{\delta r}\right), \qquad \frac{\Gamma}{\Gamma_{\max}} = 1 - \left(1 - \frac{\Gamma_0}{\Gamma_{\max}}\right)\frac{\delta r_0}{r_0}\frac{r}{\delta r}. \tag{14}$$

This immediately implies that for a conical (or spherical) flow, in which $\delta r \propto r$ and $\delta r/r = \delta r_0/r_0$, the Lorentz factor essentially remains constant, $\Gamma \approx$ const, and the flow hardly accelerates. This result can be understood by simple energy considerations. As long as there is no expansion along the direction of motion, the volume of a fluid element scales as $\propto r^2$ while its magnetic energy density scales as $\propto B^2 \propto r^{-2}$, implying a constant magnetic energy and no conversion into kinetic energy.

More generally, Eq. (14) implies that in order for the flow to accelerate, $r/\delta r$ must decrease, i.e. streamlines must diverge faster than conical. For power-law streamlines, $z = z_0(r/r_0)^\alpha = z_0[(r + \delta r)/(r_0 + \delta r_0)]^\alpha$, one has $r/\delta r = r_0/\delta r_0$ so there is still no acceleration. If one allows the power law index to vary with $r_0 = r(z_0)$, i.e. $\alpha = \alpha(r_0)$, then one finds $\delta r/r = (\delta r_0/r_0)[1 - r_0\alpha'(r_0)\alpha^{-2}(r_0)\ln(z/z_0)]$, and the condition for acceleration becomes $\alpha' = d\alpha/dr_0 < 0$. Altogether one can see that such steady-state, axisymmetric ideal MHD acceleration is quite delicate and requires a very particular configuration of the magnetic field lines. Satisfying this requirement is not trivial, and in particular it requires lateral causal contact across the jet.

A key open question regarding outflows that start out highly magnetized near the central source is how they convert most of their initial electromagnetic energy to other forms, namely bulk kinetic energy or the energy in the random motions of the particles that also produce the radiation we observe from these sources. It is suggested by observations of relevant sources, such as AGN, GRBs or pulsar wind nebulae that the outflow magnetization is rather low at large distances from the source. This is the essence of the well-known σ problem—how to transform from $\sigma \gg 1$ near the source to $\sigma \ll 1$ very far from the source.

It has been shown early on that a highly magnetized steady spherical flow accelerates only up to an asymptotic Lorentz factor $\Gamma_\infty \sim \sigma_0^{1/3}$, and magnetization $\sigma_\infty \sim \sigma_0^{2/3}$ (Goldreich and Julian 1970) where $\sigma_0 \gg 1$ is the initial value of the magnetization parameter σ, implying that most of the energy remains in electromagnetic form (a Poynting flux dominated flow). This is valid for any such *unconfined* flow, i.e. where the external pressure is effectively negligible. A sufficiently large external pressure can help collimate and accelerate the flow. It has been found (Lyubarsky 2009, 2010b; Komissarov et al. 2009) that for a power law external pressure profile, $p_{\rm ext} \propto z^{-\kappa}$, the collimation and acceleration can proceed in two distinct regimes.

For $\kappa > 2$, the *weak confinement* regime, the external pressure drops fast enough such that the flow loses lateral causal contact while it is still highly magnetized, and from that point on it becomes conical and essentially stops accelerating. This collimation-induced acceleration increases Γ_∞ and decrease σ_∞ by up to a factor of $\sim \theta_{\rm j}^{-2/3}$ compared to the unconfined (quasi-spherical) case, where $\theta_{\rm j}$ is the asymptotic jet half-opening angle. This arises because lateral causal contact in the jet is maintained as long as $\theta_{\rm j}$ does not exceed the Mach angle, $\theta_{\rm j} \lesssim \theta_{\rm M} \sim \sigma^{1/2}/\Gamma$, where energy conservation implies $\sigma\Gamma \sim \sigma_0$ (for $\sigma_0 \gg 1$ and $\Gamma_0 \sim 1$) as long as the flow remains highly magnetized ($\sigma \gg 1$).

For $\kappa \le 2$, the *strong confinement* regime, the external pressure drops slowly enough that the jet maintains lateral causal contact throughout its collimation-induced acceleration process. In this case about half of the initial magnetic energy is converted into kinetic energy and the flow becomes only mildly magnetized, $\sigma_\infty \sim 1$, while the Lorentz factor approaches its maximal possible value, $\Gamma_\infty \sim \sigma_0$. In this regime the collimation and acceleration proceed as $\Gamma \propto r \propto z^{\kappa/4}$ and the jet remains narrow, $\Gamma_\infty\theta_{\rm j} \sim 1$.

The main problem with this picture, however, is that even under the most favorable conditions the asymptotic magnetization is $\sigma_\infty \ge 1$, which does not allow efficient energy dissipation in internal shocks within the outflow (Lyubarsky 2009, 2010b; Komissarov et al. 2009). It has been found (Tchekhovskoy et al. 2010; Komissarov et al. 2010) that a sudden drop in the external pressure, as may occur when a GRB jet exits its progenitor star, can result in a sudden additional acceleration that can lead to $\Gamma_\infty\theta_j \gg 1$ as inferred in GRBs, but still with $\sigma_\infty \geq 1$.

These important limitations of the "standard" steady, axisymmetric and non-dissipative (or ideal MHD) acceleration have, on the one hand, led to the suggestion that the jets might remain Poynting flux dominated at large distances from the source and the observed emission is the result of magnetic reconnection events rather than internal shocks (Blandford 2002; Lyutikov and Blandford 2003; Lyutikov 2006). On the other hand, other models suggested increasing the acceleration efficiency by relaxing one of the standard assumptions, such as axi-symmetry—leading to non-axisymmetric instabilities that randomize the magnetic field orientation (Heinz and Begelman 2000). Since a highly tangled magnetic field effectively behaves like a relativistic fluid (with an adiabatic index of 4/3) this leads to efficient acceleration, similar to thermal acceleration of relativistic outflows. What is more, both the kink instability mentioned above (Drenkhahn and Spruit 2002), as well as other instabilities (such as the Kruskal-Schwarzschild instability in a striped wind; Lyubarsky 2010a) can lead to magnetic reconnection, i.e. gradual magnetic dissipation, which in turn enhances the acceleration due to the conversion of magnetic to thermal energy, where the thermal pressure efficiently accelerates the outflow.

3.3 Impulsive Magnetic Acceleration

Replacing the usual steady-state assumption by strong time-dependence is a natural alternative. This impulsive regime was sparsely studied, and mainly in the non-relativistic case (Contopoulos 1995). Recently, a new impulsive magnetic acceleration mechanism was found that operates in the relativistic case (Granot et al. 2011), which can be much more efficient than magnetic acceleration in steady flows, and can lead to low magnetizations, $\sigma \ll 1$, thus enabling efficient dissipation in internal shocks. This qualitatively different behavior of impulsive outflows can be very relevant for GRBs, as well as for other relativistic jet sources such as tidal disruptions or flares in AGN or micro-quasars, or even giant flares in soft gamma repeaters (SGRs, thought to be magnetars), which also triggered renewed interest in this topic (e.g., Levinson 2010; Lyutikov 2011; Granot 2012a,b; Komissarov 2012).

Figure 4 (*left panel*) shows results for an impulsive magnetic acceleration test case: a cold, initially uniform plasma shell (of with width l_0, rest mass density ρ_0 and magnetic field B_0), highly magnetized ($\sigma_0 = B_0^2/4\pi\rho_0 c^2 \gg 1$) and at rest, with a conducting "wall" at its back and vacuum in front of it. A strong, self-similar rarefaction wave forms at its front (vacuum interface) and propagates towards its back, reaching the wall at $t = t_0 \approx l_0/c$. By this time the shell's energy-weighted mean Lorentz factor and magnetization are $\langle\Gamma\rangle \sim \sigma_0^{1/3}$ and $\langle\sigma\rangle \sim \sigma_0^{2/3}$. At $t > t_0$ the shell detaches from the wall, keeps an almost constant width ($l \approx 2l_0$) and accelerates as $\langle\Gamma\rangle \sim \sigma_0/\langle\sigma\rangle \sim (\sigma_0 t/t_0)^{1/3}$ up to the coasting time $t_c = \sigma_0^2 t_0$. At $t > t_c$ the shell coasts at $\langle\Gamma\rangle \sim \sigma_0$, its width grows ($l/2l_0 \sim t/t_c$) and its magnetization rapidly decreases ($\langle\sigma\rangle \sim t_c/t$), leading to complete conversion of magnetic to kinetic energy that allows strong internal shocks to form that can lead to large radiative efficiencies.

3.4 Interaction with the External Medium and the Reverse Shock

Let us now consider the evolution of a similar shell in spherical geometry that propagates into an external medium with a power-law density profile, $\rho_{ext} = AR^{-k}$, following Granot (2012a). The main results are shown in the *right panel* of Fig. 4. The initial shell magnetization σ_0 and density $\rho_0 \propto 1/\sigma_0$ are allowed to vary while keeping fixed the values of

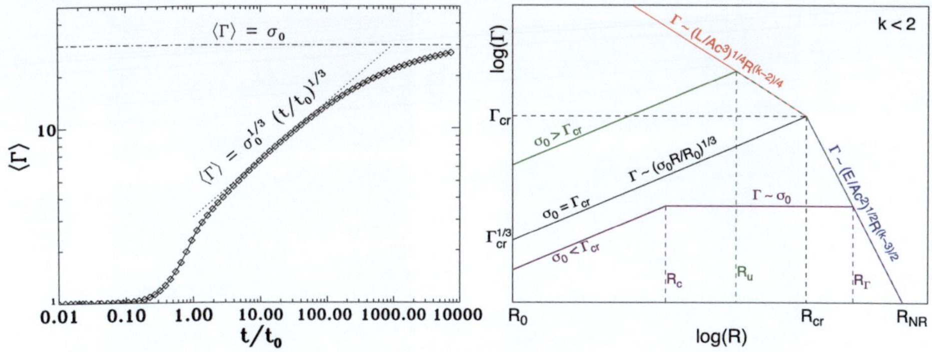

Fig. 4 *Left*: test case for impulsive magnetic acceleration: the energy-weighted mean Lorentz factor $\langle\Gamma\rangle$ of a finite cold shell of plasma initially uniform (with width l_0, rest mass density ρ_0 and magnetic field B_0), highly magnetized ($\sigma_0 = B_0^2/4\pi\rho_0 c^2 \gg 1$; $\sigma_0 = 30$ was used here) and at rest, whose back leans against a conducting "wall" while its front faces vacuum, versus the time t in units of the shell's initial fast magnetosonic crossing time $t_0 \approx l_0/c$. The analytic expectations (*dotted* and *dashed-dotted lines*) and the results of numerical simulations (*diamond symbols* joined by a *solid line*) are in very good agreement. (This figure is taken from Granot et al. 2011.) *Right*: evolution of the typical (or energy-weighted average) Lorentz factor Γ with the distance $R \approx ct$ from the central source, for a finite shell similar to that described in the *left panel*, but for a spherical shell propagating into an external medium with a power-law density profile, $\rho_{\rm ext} = AR^{-k}$ (this figure is taken from Granot 2012a)

the initial time or length scale ($t_0 \approx R_0/c$ or R_0), energy ($E \sim Lt_0 \approx LR_0/c$ or power L), and external density ($k < 2$ in this figure, and A or $\rho_{\rm ext}(R_0) = AR_0^{-k}$), which imply fixed $\Gamma_{\rm cr} \sim (f_0\sigma_0)^{1/(8-2k)}$ where $f_0 = \rho_0/\rho_{\rm ext}(R_0)$ and $R_{\rm cr} \sim R_0\Gamma_{\rm cr}^2$. Shown are the two dynamical regimes most relevant for GRBs. The purple line shows regime I ($1 < \sigma_0 < \Gamma_{\rm cr}$ or a sufficiently low external density) where the shell initially expands as if into vacuum (as described in the *left panel*) and only after becoming kinetically dominated and expanding radially is it significantly decelerated by the external medium through a strong relativistic reverse shock, that can produce a bright emission that peaks on a timescale larger than the duration of the prompt GRB emission (the familiar low-σ "thin shell"; Sari and Piran 1995). Eventually, most of the energy is transfered to the shocked external medium and the flow approaches the Blandford and McKee (1976) self-similar solution.

In regime II ($1 < \Gamma_{\rm cr} < \sigma_0 < \Gamma_{\rm cr}^{3(4-k)/2}$), depicted by the green line in the *right panel* of Fig. 4, the shell is significantly affected by the external medium while it is still Poynting dominated (at $R > R_u \sim R_0(f_0\sigma_0^{-1/3})^{3/(10-3k)}$), thus suppressing the reverse shock (which is either non-existent or very weak). The shell remains highly magnetized and gradually transfers its energy to the shocked external medium through pdV work across the contact discontinuity up to $R_{\rm cr}$, after which the flow approaches the Blandford-McKee solution. In this regime no significant reverse shock emission is expected, and the onset of the afterglow (i.e. the peak of the emission from the shocked external medium) is expected to be on a timescale comparable to the prompt GRB duration (i.e. a high-σ "thick shell").

In addition, there are other regimes not shown in this figure, but all of the regimes are mapped in the relevant parameter space in Fig. 5. In regime III ($1 < \Gamma_{\rm cr}^{3(4-k)/2} < \sigma_0$) the external density is high enough that there is no impulsive acceleration stage where $\langle\Gamma\rangle \propto R^{1/3}$, and instead $\langle\Gamma\rangle \sim \sigma_0/\langle\sigma\rangle \propto R^{(k-2)/4}$ at $R_0 < R < R_{\rm cr} \sim R_{\rm dec}$, and then approaches the Blandford-McKee solution (its observational signatures are expected to be similar to regime II). In regime IV ($\Gamma_{\rm cr} < 1$) the external density is so high that the flow remains Newtonian all along (as might happen while the GRB jet is propagating inside a

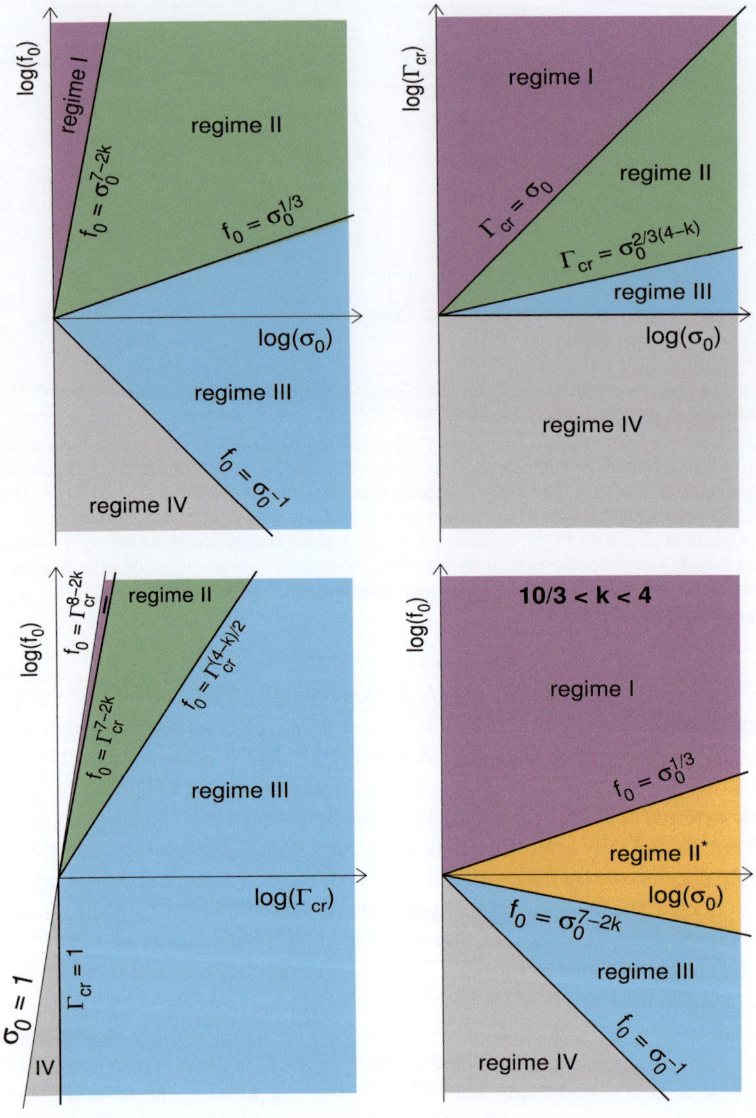

Fig. 5 Phase space diagrams of the different dynamical regimes: in the f_0–σ_0 plane for $k < 10/3$ (*top left panel*), Γ_{cr}–σ_0 plane for $k < 10/3$ (*top right panel*), f_0–Γ_{cr} plane for $k < 10/3$ (*bottom left panel*), and in the f_0–σ_0 plane for $10/3 < k < 4$ (*bottom right panel*). Each regime is labeled and denoted by a different color, and the borders between the different regimes are indicated by labeled *thick black lines* (this figure is taken from Granot 2012a)

massive star progenitor). There is also an "exotic" regime II* that exists only in a highly stratified external medium ($10/3 < k < 4$).

Under realistic conditions, GRB variability times are in practice typically large enough that the flow should first undergo quasi-steady collimation-induced acceleration that saturates, and only later the impulsive acceleration kicks in and operates until the flow becomes kinetically dominated (see Fig. 6). Moreover, one typically expects the outflow from the cen-

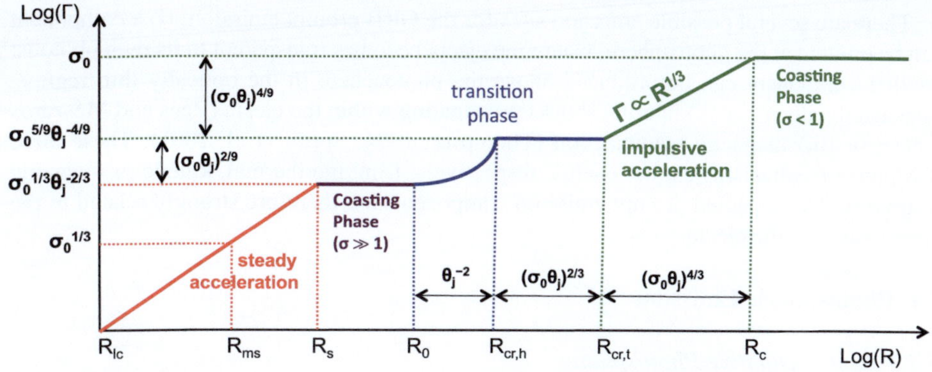

Fig. 6 An illustration of the expected transition from (quasi-) steady collimation-induced acceleration near the central source (*in red*) to impulsive acceleration further away from the source (*in green*). The energy weighted mean Lorentz factor Γ is shown against the distance R from the central source, and a few critical radii and Lorentz factors are indicated following the notations of Granot et al. (2011) and Granot (2012a)

tral source to consist of many sub-shells rather than a single continuous shell. The effects of such multiple sub-shells in the outflow can be important, and the collisions between them may provide efficient energy dissipation that can power the GRB emission (Granot 2012b; Komissarov 2012). They may also allow a low-σ "thick shell", i.e. a strong relativistic reverse shock peaking on a timescale comparable to the prompt GRB emission, which is not possible for a single shell. For a long-lived source (e.g. AGN) with initial sub-shell widths l_0 and separations $l_{\rm gap}$, each sub-shell can expand by a factor of $1 + l_{\rm gap}/l_0$. Its magnetic energy decreases by the same factor (where $\sigma_\infty \sim l_0/l_{\rm gap}$), and may be converted to kinetic or internal energy, or radiation. For a finite source activity time, the merged shell can still expand further and convert more magnetic energy into other forms (even without interaction with an eternal medium). Important related points that warrant further study are the transition from quasi-steady collimation-induced acceleration to impulsive acceleration, both in a single shell and in multiple sub-shells, as well as the dissipation in the interaction between sub-shells and its effect on the outflow acceleration and the resulting emission, such as a possible photospheric spectral component.

4 Dissipation and Prompt Emission

As discussed above, GRBs must be associated with relativistic outflows ejected by a stellar mass compact source, with a bulk Lorentz factor $\Gamma \gtrsim 100$ in order to avoid the *compactness problem* (Baring and Harding 1997; Lithwick and Sari 2001; Granot et al. 2008; Hascoët et al. 2012b). This also naturally explains the afterglow through the deceleration of the ejecta by the external medium, whereas the observed fast prompt variability implies that the prompt emission must be produced by internal dissipation within the ejecta (Sari and Piran 1997). Therefore, the analysis of the GRB prompt emission may provide valuable information on magnetic fields within an ultra-relativistic jet. It can put unique constraints on the state of the jet at the end of the acceleration phase, and more specifically on the geometry of the magnetic field and the magnetization at a large distance to the central source, where the γ-ray emission is produced. This is, however, a difficult task as it requires a full understanding of the nature of the dissipative mechanisms and of the radiative processes at work.

There are several possible emission sites for the GRB prompt emission: (i) a component can be emitted at the photosphere, where the ejecta becomes transparent to its own photons; another component can be produced above the photosphere in the optically thin regime, associated with either (ii) internal shocks propagating within the ejecta (Rees and Meszaros 1994); or (iii) magnetic reconnection (Thompson 1994; Spruit et al. 2001). These three dissipative mechanisms extract energy, respectively, from the thermal, kinetic or magnetic reservoirs. The expected prompt emission components are therefore strongly related to the composition of the ejecta.

4.1 Photospheric Emission

4.1.1 Non-dissipative Photospheres

When internal dissipation below the photosphere is negligible, the expected thermal emission at the photosphere is well understood with precise predictions (Paczynski 1986; Goodman 1986; Shemi and Piran 1990; Meszaros et al. 1993). Only few theoretical uncertainties remain, mainly related to the lateral structure of the jet (see e.g. Lundman et al. 2013; Deng and Zhang 2014). Assuming that the photosphere is above the saturation radius, the photospheric radius is given by (Mészáros and Rees 2000; Mészáros et al. 2002; Daigne and Mochkovitch 2002; Hascoët et al. 2013)

$$R_{\rm ph} \simeq \frac{\kappa \dot{E}}{8\pi c^3 \Gamma^3 (1+\sigma)} \simeq 3\times 10^{13} \frac{\kappa_{0.2} \dot{E}_{53}}{\Gamma_2^3 (1+\sigma)}\ {\rm cm}, \tag{15}$$

where $\kappa_{0.2}$ is the matter opacity in units of $0.2\ {\rm cm^2\,g^{-1}}$, $\Gamma = 100\Gamma_2$ is its bulk Lorentz factor, $\dot{E} = 10^{53}\dot{E}_{53}\ {\rm erg\,s^{-1}}$ is the isotropic equivalent jet power, and σ is the magnetization parameter at the end of the acceleration, so that $\dot{E}/(1+\sigma)$ is the isotropic equivalent kinetic power. The observed photospheric luminosity and temperature are

$$L_{\rm ph} \simeq \epsilon_{\rm th} \dot{E} \left(\frac{R_{\rm ph}}{\Gamma R_0}\right)^{-2/3} \quad \text{and} \quad T_{\rm ph} \simeq \frac{T_0}{1+z}\left(\frac{R_{\rm ph}}{\Gamma R_0}\right)^{-2/3}, \tag{16}$$

where the initial temperature equals $T_0 = (\epsilon_{\rm th}\dot{E}/4\pi\sigma R_0^2)^{1/4}$. Here $\epsilon_{\rm th}$ is the thermal fraction of the jet power at the base of the flow, located at the initial radius R_0, and $1-\epsilon_{\rm th}$ is therefore the initial magnetic fraction. In the case of a passive magnetic field carried by the outflow without contributing to the acceleration, the initial thermal fraction $\epsilon_{\rm th}$ and the magnetization at the photosphere σ are related by $\sigma_{\rm passive} = \sigma_0 = (1-\epsilon_{\rm th})/\epsilon_{\rm th}$. An efficient magnetic acceleration leads to $\sigma < \sigma_{\rm passive}$ (Spruit et al. 2001). The predicted spectrum is quasi-thermal, with an exponential cutoff at high-energy and a power law at low-energy with a photon-index of $\alpha \simeq 0.4$, which differs from the $\alpha = 1$ slope of the Raleigh-Jeans spectrum due to the peculiar geometric shape of a relativistic photosphere (Goodman 1986; Beloborodov 2011).

4.1.2 Dissipative Photospheres

If dissipation occurs below the photosphere, the emitted spectrum can be significantly different than the previous case: a high-energy tail can be produced by comptonization due to the presence of relativistic electrons (Thompson 1994; Mészáros and Rees 2000; Rees and Mészáros 2005; Giannios and Spruit 2007; Beloborodov 2010), and the low-energy slope can be modified by synchrotron radiation (Pe'er et al. 2006; Vurm et al. 2011). The resulting observed spectrum may now appear non-thermal, with several components. This scenario is more uncertain than the previous one. The nature of the sub-photospheric dissipative mechanism must be identified, with several candidates: early internal shocks, gradual magnetic reconnection (Thompson 1994; Giannios and Spruit 2007), neutron-proton collisions (Beloborodov 2010), etc. An important relevant issue is related to the photon production efficiency and thermalization deep within the ejecta (Vurm et al. 2013).

4.2 Non-thermal Emission in the Optically Thin Regime

4.2.1 Electron Acceleration and Synchrotron Radiation

Non-thermal emission can be produced above the photosphere if some internal dissipation processes can lead to efficient electron acceleration. In this case, two natural candidates for the dominant radiative process are the synchrotron radiation and the inverse Compton scatterings of synchrotron photons by relativistic electrons (SSC). However, the measurement of the prompt γ-ray spectrum over a broad spectral range (keV-GeV) in a few bursts by *Fermi*/GBM+LAT can rule out the possibility of SSC being dominant in the soft γ-ray range, as it would lead either to a strong synchrotron peak at lower energy, or a strong second inverse Compton peak at higher energy, which are not observed (Bošnjak et al. 2009; Piran et al. 2009). Therefore, the discussion is focussed on the synchrotron radiation of relativistic electrons, with several possible dissipation mechanisms responsible for the acceleration of electrons.

4.2.2 Internal Shocks

If the magnetization at a large distance to the central source is sufficiently low, strong internal shocks are expected to form and propagate within the ejecta due to the variability of the ejected outflow. A large range of radii is expected,

$$R_{\mathrm{is}} \simeq 2\Gamma^2 c(t_{\mathrm{var}} \to t_{\mathrm{GRB}}) \simeq \Gamma_2^2\big(6\times 10^{12} t_{\mathrm{var},-2} \to 6\times 10^{15} t_{\mathrm{GRB},1}\big)\ \mathrm{cm}, \tag{17}$$

where $t_{\mathrm{var},-2}$ is the shortest timescale of variability, in units of 10^{-2} s and $t_{\mathrm{GRB},1}$ is the total duration of the relativistic ejection, in units of 10 s. The shocks are expected to be mildly relativistic, except for a very large amplitude of variation of the initial Lorentz factor. The dynamics of the internal shocks phase has been studied in detail (Kobayashi et al. 1997; Daigne and Mochkovitch 1998, 2000), from a simple ballistic approximation to a full hydrodynamical code, and is well understood. Up to $f_{\mathrm{d}} \simeq 40\ \%$ of the kinetic energy can be dissipated (for a low magnetization outflow, $\sigma \ll 1$), depending again on the initial distribution of the Lorentz factor.

On the other hand, large uncertainties on the emission remain, due to the poor understanding of the microphysics of mildly relativistic shocks (for a recent review of relativistic collisionless shocks see Sironi et al. 2015). It is usually parametrized by assuming that a fraction ϵ_{B} of the internal energy is injected into an amplified random magnetic field at the shock, whose structure is not known, and a fraction ϵ_{e} is injected into a fraction ζ of electrons, which are therefore accelerated into a non-thermal distribution with slope p ($dN_e/d\gamma_e \propto \gamma_e^{-p}$). To reach the soft γ-ray domain by synchrotron radiation, the fraction of accelerated electrons must be low, $\zeta \lesssim 10^{-2}$ (Daigne and Mochkovitch 1998; Bošnjak et al. 2009; Daigne et al. 2011; Beniamini and Piran 2013). On the other hand, values of ϵ_{e} close to equipartition ($\epsilon_e = 1/3$) are required to explain the huge luminosities of GRBs. If the radiative efficiency is high, a fraction $f_{IS} \simeq f_d \epsilon_e \simeq 0.01$–$0.1$ of the initial kinetic power can be converted into radiation in internal shocks. With such assumptions, the non-thermal emission in the comoving frame of the shocked regions can be computed with a detailed radiative model including all the relevant processes, namely synchrotron radiation and slef-absorption, inverse Compton scatterings and photon-photon annihilation. The contributions of each internal shock can then be added with an integration over equal-arrival time surface of photons to the observer in the source frame to produce synthetic light curves and spectra that can be directly compared to observations. The predicted spectrum shows several components, a strong synchrotron peak in the soft γ-ray range and a weaker inverse Compton peak at higher energy (Bošnjak et al. 2009).

4.2.3 Magnetic Reconnection

If the magnetization at large distances remains high ($\sigma \gtrsim 1$), then internal shocks are either significantly suppressed or in some cases cannot form altogether (Mimica and Aloy 2010; Narayan et al. 2011). In such cases, electrons may be accelerated predominantly in magnetic reconnection sites. This scenario is even more uncertain (less understood) than the two previous ones (photosphere, internal shocks), but it is under extensive investigation (for a recent review on relativistic magnetic reconnection see Kagan et al. 2015). Some authors considered a gradual reconnection starting at a small radius and extending up to $R_{rec} \sim 10^{13}$ cm (Drenkhahn and Spruit 2002; Giannios 2008). In this case, most of the dissipation occurs below the photosphere, corresponding to the dissipative photosphere scenario discussed above. If, on the other hand, reconnection remains inefficient below the photosphere, it could occur at larger radii in the optically thin regime. In the simulations by McKinney and Uzdensky (2012), a catastrophic dissipation of the magnetic field occurs at $R_{rec} \simeq 10^{13}$–10^{14} cm when reconnection enters a rapid collisionless mode. Zhang and Yan (2011) proposed another scenario where reconnection is triggered by internal shocks, the so-called ICMART model. The typical radius may be as large as $R_{rec} \simeq 10^{15}$ cm. The microphysics in the reconnection sites is also uncertain. One expects many electron acceleration sites, which may move relativistically in the outflow's rest frame. The non-thermal electron distribution may be somewhat harder than in shock acceleration (see e.g. Sironi and Spitkovsky 2014; Kagan et al. 2015), and the acceleration process slower, which can lead to a different shape for the synchrotron emission. Contrary to internal shocks, detailed calculations of the light curves and spectra based on a detailed radiative model coupled to a dynamical simulation are not yet available.

4.3 Magnetic Field in Emission Sites

The models discussed above have very different implications for the magnetic field:

- **Case 1**: most of the prompt emission is due to a dissipative photosphere. Then, the magnetic field must be large enough at the photosphere to produce synchrotron radiation and affect the low-energy spectrum. If this magnetic field is generated by the dissipation process (e.g. shocks; Sironi et al. 2015), it is most probably random. Otherwise, an ordered field must be present. The corresponding initial magnetization must either be low (otherwise the photospheric emission is weak), or high with very efficient reconnection below the photosphere, which then leads to a possible candidate for the sub-photospheric dissipation process.
- **Case 2**: the prompt emission is mostly non-thermal, from an optically thin region:

- For internal shocks to be the dominant dissipation process, the magnetization at large distances from the source must be low. A random field is generated locally at the shock front, where the electrons are accelerated. However, the magnetic field felt by the radiating electrons must be considered far behind the shock front (as the radiative cooling length is much larger than the plasma skin depth) where its strength and structure are not well known.
- If reconnection dominates the dissipation then σ must be large far from the source. The ordered field is destroyed at the reconnection sites, but if electrons have enough time to migrate from their acceleration site before radiating, their emission may still be mostly in the large-scale ordered field.

Observations of the GRB prompt emission, discussed in the next subsection, can put strong constraints on these various emission models.

4.4 Constraints from the Observed Prompt Soft γ-Ray Emission

4.4.1 Light Curves

All the scenarios discussed above can reproduce the observed variable light curves. There are, however, important differences:

- (Dissipative) **photosphere**: the emission radius is low (see Eq. (15)). Therefore the curvature effect, i.e. the spreading of photon arrival times from different angles with respect to the line of sight over an angular timescale $t_\theta = R/2\Gamma^2 c$ of a flash of photons emitted at the same time and radius, is negligible: the observed light curve directly traces the activity of the central engine.
- **Internal shocks**: the light curves trace the source activity (Kobayashi et al. 1997; Daigne and Mochkovitch 1998), but two effects now affect the observed pulse shapes: the curvature effect (due to a larger radius) dominates the pulse decay (Genet and Granot 2009; Willingale et al. 2010), and the radial or hydrodynamic timescale due to shock propagation, $t_r = \Delta R/2\Gamma^2 c$, dominates the pulse rise and overall shape (Daigne and Mochkovitch 1998, 2003; Bošnjak and Daigne 2014).
- **Reconnection**: again, the light curve traces the source activity, with new effects due to relativistic bulk motion in the local jet's frame. Relativistic motions of emitting plasma in the jet's frame cause rapid variability (that should show up as a distinct component in the Fourier power spectrum), while a slower envelope may arise from their combined effect (Zhang and Zhang 2014) or from slower emitting plasma. This can be tested by characterizing the observed variability. Analysis of GRB light curves shows a continuum of timescales (see e.g. Beloborodov et al. 2000; Guidorzi et al. 2012), which does not support the reconnection model of Zhang and Zhang (2014) (see however Gao et al. 2012). A possible concern appears if the emission is produced by many relativistically moving emitters: the predicted pulse shape may be too symmetric compared to observations (Lazar et al. 2009). However, both concerns (the power spectrum and pulse shapes) may be solved if the reconnection occurs in relatively ordered thin layers located between anti-parallel regions in the outflow (with a geometry of thin quasi-spherical shells) and the relativistic motions in the jet's frame are limited to these layers (Beniamini and Granot 2015). Such a model may also account for many of the correlations that are observed in the prompt emission.

4.4.2 Polarization

Measuring the polarization in the γ-ray domain remains challenging. A very large degree of polarization was claimed by Coburn and Boggs (2003), but it was later refuted by others (Rutledge and Fox 2004; Wigger et al. 2004) as not being statistically significant. Only a few later measurements (by INTEGRAL, GAP) are available, however with a low or moderate significance (McGlynn et al. 2007; Götz et al. 2009; Yonetoku et al. 2011, 2012; Götz et al. 2014). Such measurements (if reliable) can put constraints on the magnetic field geometry in the emission sites (e.g., Granot and Königl 2003; Granot 2003; Lyutikov et al. 2003; Nakar et al. 2003). Current observations seem to favor synchrotron radiation in an ordered field with patches, which would favor emission in the optically thin regime above the photosphere. It is however not trivial to justify a highly ordered field in the internal shocks model (a turbulent field at the shock is required for particle acceleration; The structure of the field on intermediate scales between the plasma and the dynamical scales is less known), or in the reconnection model (in principle the structured field is destroyed by reconnection, but the remaining field can still possess significant structure and electrons may also radiate somewhat outside of these localized reconnection regions). Either way, more definitive polarization observations are needed before strong conclusions can be drawn.

4.4.3 Spectrum

The prompt soft γ-ray spectrum is usually fitted by a phenomenological model introduced by Band et al. (1993), which consists of two power laws with low- and high-energy photon indices of α and β, smoothly connected at the peak energy $E_{\rm p}$. This eliminates non-dissipative photospheres for the dominant contribution to the emission, which puts a strong constraint on the initial magnetization (low $\epsilon_{\rm th}$, Daigne and Mochkovitch 2002), favoring magnetic acceleration of the outflow. The remaining scenarios for the prompt GRB emission are either a dissipative photosphere, or a combination of a weak photospheric emission and a non-thermal component due to shocks or reconnection. The discussion is then focussed on the general shape of the spectrum, and the low-energy photon index α, which is observed to be close to $\alpha \simeq -1$ (Preece et al. 2000; Kaneko et al. 2006; Nava et al. 2011; Gruber et al. 2014).

- **Dissipative photospheres**: the value of α can be reproduced by adjusting the magnetization, which controls the synchrotron emission at low energies. The theoretical instantaneous spectral peak is narrower than the observed time-integrated spectral peak, but the comparison should be made using a theoretical time-integrated spectrum, which should broaden it.
- **Internal shocks**: to reproduce the high luminosities and the short timescale variability of GRBs, the radiating electrons must be in the fast cooling regime (Cohen et al. 1997; Sari et al. 1998; Ghisellini et al. 2000), i.e. their radiative timescale must be shorter than the dynamical timescale that governs the adiabatic cooling for the spherical expansion. This leads to a predicted photon index $\alpha \leq -3/2$, in contradiction with observations (the so-called "synchrotron line-of-death", Preece et al. 1998). Another potential problem is that the resulting spectrum is too broad around the peak. The two problems are naturally connected. Several possibilities have been discussed to solve this issue: (i) inverse Compton scatterings in the Klein-Nishina regime affect the cooling of electrons, leading to photon indices $\alpha \lesssim -1$ (Derishev et al. 2001; Wang et al. 2009; Nakar et al. 2009; Daigne et al. 2011). This puts a constraint on the strength of the magnetic field, which should be small,

with $\epsilon_B \lesssim 10^{-3}$ (Daigne et al. 2011; Barniol Duran et al. 2012); (ii) in the marginally fast cooling regime (Daigne et al. 2011; Beniamini and Piran 2013), where the radiative timescale is close to the dynamical timescale but still below, the electron radiative efficiency can remain large enough ($\gtrsim 50$ %) to explain the observed luminosities, but the synchrotron spectrum is strongly affected: the intermediate region of the spectrum below the peak with a photon index $-3/2$ disappears and the slope $\alpha = -2/3$ usually associated with the inefficient slow cooling regime is measured. This regime is also favored by weak magnetic fields; (iii) in the fast cooling regime, electrons radiate on timescales which are long compared to the plasma scale at the shock front, but small compared to the dynamical timescale. Then they experience a magnetic field that is not necessarily the same as the turbulent field just behind the shock (see the structure of the magnetic field in the simulations, e.g. Keshet et al. 2009). If the field is decaying on this intermediate scale, it will affect the synchrotron spectrum and can lead to a hard spectrum, with $-1 \lesssim \alpha \lesssim -2/3$ (Derishev 2007; Lemoine 2013; Uhm and Zhang 2014; Zhao et al. 2014).
- **Reconnection**: the slow electron heating in the turbulent field can lead to hard synchrotron spectra with $\alpha \simeq -1$ (Uhm and Zhang 2014). It is unclear if the expected hard power-law index $p \lesssim 1.5$ of the non-thermal electron distribution (e.g., Sironi and Spitkovsky 2014) can be identified in the observed spectrum. A potential issue is that the presence of many emitting regions that move relativistically in random directions in the jet's fame might lead to much broader spectra than observed. This may be alleviated, however, in models where these regions move predominantly along the thin reconnection layer that is located between regions of oppositely-directed magnetic field in the flow, normal to the jet's bulk motion (Beniamini and Granot 2015).

Recently the description of observed GRB spectra in the soft γ-ray range has been greatly improved by *Fermi*/GBM observations. An important result is the identification of significant deviations from the Band spectrum, which seem to be related to the presence of a weak thermal component[9] below the dominant non-thermal one (see e.g. Ryde et al. 2010, 2011; Guiriec et al. 2011; Burgess et al. 2011, 2014; Axelsson et al. 2012; Guiriec et al. 2013, 2015), as illustrated in Fig. 7. A natural explanation is to associate the quasi-thermal weak component to a (non-dissipative) photosphere and the Band component to synchrotron radiation from electrons accelerated either in shocks or in reconnection: Fig. 8 shows an example of a synthetic burst with these two contributions in the case of internal shocks. The weakness of the photospheric emission puts interesting constraints on the initial magnetization of the outflow (Daigne and Mochkovitch 2002; Hascoët et al. 2013), favoring an efficient magnetic acceleration, with a large range of initial magnetization in the GRB population, $\epsilon_{th} \lesssim 0.01$ ($\sigma_0 \gtrsim 100$) in most cases where no detection is made and $\epsilon_{th} \simeq 0.01$–$0.1$ ($\sigma_0 \simeq 10$–100) in less frequent cases like GRB 100724B (Hascoët et al. 2013). GRB 090902B with $\epsilon_{th} \simeq 0.3$–1 ($\sigma_0 \lesssim 2.3$) remains an exception within long GRBs, and the short GRB 120323A appears as an intermediate case between GRBs 100724B and 090902B with $\epsilon_{th} \simeq 0.1$–0.5 ($\sigma_0 \simeq 1$–9) (Guiriec et al. 2013). The fact that the photospheric emission seems brighter in the only case of detection in a short GRB (GRB 120323A, Guiriec et al. 2013) may indicate a different acceleration mechanism. If this interpretation is correct, these recent detections rule out purely thermal acceleration (standard fireball) at least in long GRBs.

4.5 Constraints on a Poynting Flux Dominated Outflow

Several authors (Lyutikov 2006; Giannios and Spruit 2006; Zhang and Yan 2011) have proposed that GRB jets are Poynting flux dominated all the way up to the emission region.

[9] Or possibly even a dominant photospheric component in the case of GRB 090902B.

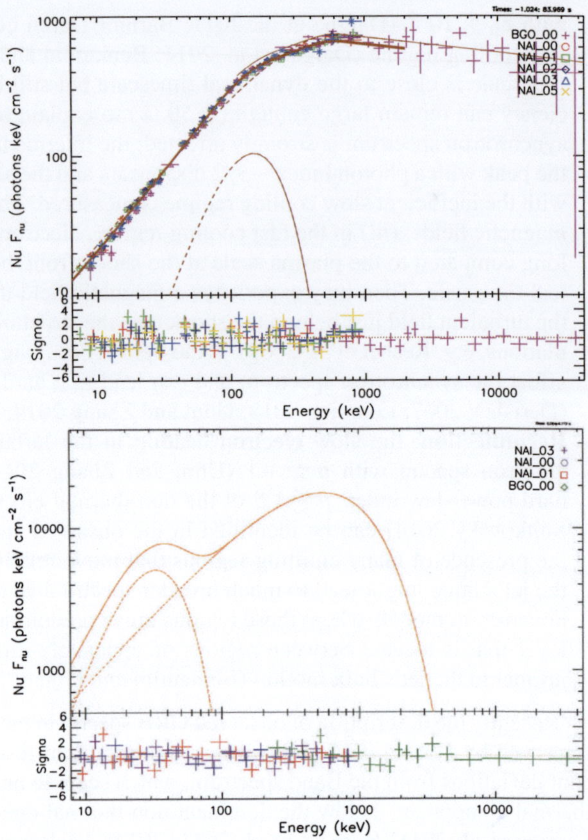

Fig. 7 Two examples of quasi-thermal components detected in GRB prompt spectra. *Top*: a weak quasi-thermal component at $T \simeq 38$ keV in the long GRB 100724B (from Guiriec et al. 2011). *Bottom*: a quasi-thermal component at $T \simeq 12$ keV in the short GRB 120323A (from Guiriec et al. 2013)

The prompt γ-rays arise, in this case, from a process that converts this magnetic energy to radiation. Obviously, this cannot take place directly and one has to invoke some sort of magnetic dissipation (e.g. reconnection) that converts the magnetic energy to accelerated electrons (or electron-positron pairs) that emit the observed γ-rays. Particular support for this idea came with the claim of strong polarization in the prompt emission by Coburn and Boggs (2003), which was later refuted (Rutledge and Fox 2004; Wigger et al. 2004). Such polarization could arise if the magnetic field is ordered and this will arise naturally if the magnetic field is dominant (Granot and Königl 2003; Granot 2003; Lyutikov et al. 2003).

However, the efficiency of the synchrotron emission process poses serious constraints on models in which the emission region is Poynting flux dominated (Beniamini and Piran 2014). Consider a Poynting flux dominated outflow and an observed (isotropic equivalent) γ-ray luminosity L_γ. This luminosity immediately sets a lower limit on the strength of the magnetic field B in the rest frame[10] of the central source, $L_\gamma < R_{\rm em}^2 B^2 c$, where $R_{\rm em}$ is the emission radius. Accelerated electrons effectively emit synchrotron radiation. The critical issue here is that synchrotron emission is too efficient. The accelerated electrons cool so rapidly in a strong magnetic field that their lower bands (X-rays and optical) synchrotron emission would produce a signal that is much stronger than the observed emission in these bands.

[10]The magnetic field in the jet's frame is $B' = B/\Gamma$, where Γ is its bulk Lorentz factor.

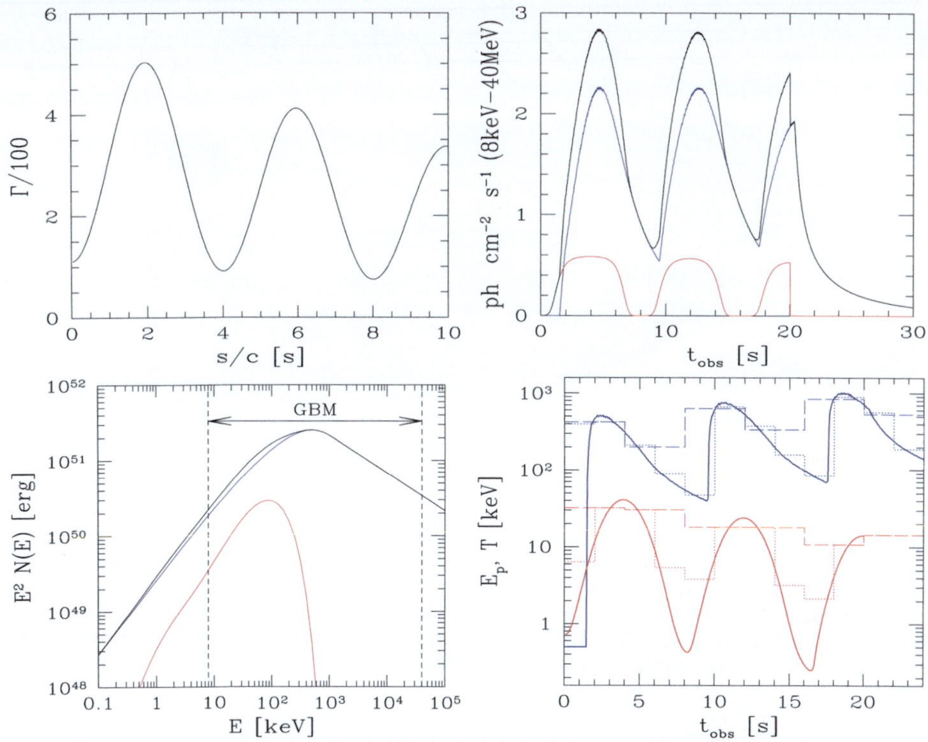

Fig. 8 An example of a synthetic GRB with the contribution from the photosphere and internal shocks computed self-consistently. The photospheric emission is plotted *in red*, the non-thermal emission from internal shocks *in blue*, the total *in black*. *Top-left*: initial distribution of the outflow Lorentz factor at the end of the acceleration phase. *Top-right*: light-curves in the GBM energy range. *Bottom-left*: spectrum. *Bottom-right*: spectral evolution (observed peak energy of the non-thermal component and temperature of the photosphere). The *dashed and dotted lines* show the expected result when integrating the spectrum over different timescales. The parameters are $\dot{E} = 10^{53}$ erg/s, $\epsilon_{th} = 0.03$ (high initial magnetization, $\sigma_0 = 32.3$), $\sigma = 0.1$ (low magnetization at large distance), $R_0 = 3 \times 10^7$ cm (all panels are taken from Hascoët et al. 2013)

The observed prompt upper limits in the optical or the X-rays set strong constraints on the conditions within the emitting region. First, if the observed γ-rays are due to some other (non-synchrotron) emission process then this process must be extremely efficient and its cooling time should be significantly shorter than the relevant synchrotron cooling time (see e.g. Fig. 9).

Alternatively, if the observed prompt γ-ray emission is synchrotron then there must be a rapid reaccelerating process that keeps the electrons with the right Lorentz factor so that they would not cool too much and emit strongly in lower energy bands, in particular in soft X-rays (Ghisellini and Celotti 1999; Kumar and McMahon 2008; Fan 2010). This requires strong fine tuning as the Lorentz factor range in which the electrons must be kept is rather narrow (a factor of ~ 3–10). Multi-zone configurations in which the electrons escape the emitting region before cooling and over-producing X-ray or optical emission are also a possibility (e.g. as in the ICMRAT model Zhang and Yan 2011). Beniamini and Piran (2014) considered several such two-zone toy models (in which electrons are accelerated in one region and emit in the other) but proper conditions could not be found in any of them.

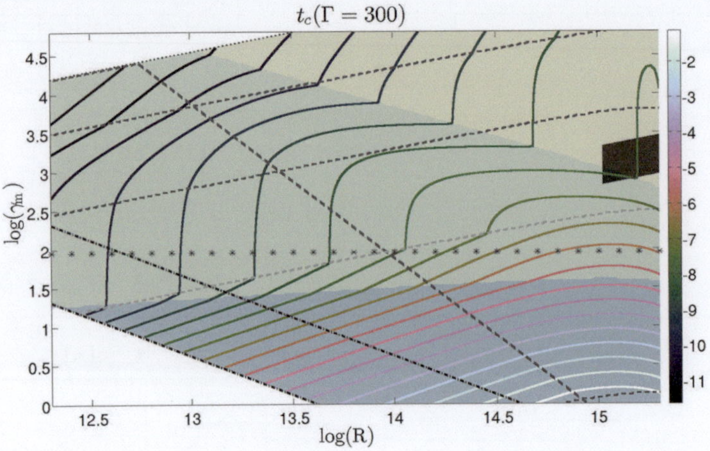

Fig. 9 If the dominant γ-ray emission mechanism is not synchrotron, then in order for it to be able to tap a significant fraction of the electrons' energy its cooling time, t_c, must be shorter than that due to synchrotron radiation, $t_{c,syn}$, which is depicted here by the *contour lines*. The observational constraints $F_{\nu,syn,opt} < 1$ mJy, $F_{\nu,syn,X\text{-}ray} < 1$ mJy and $\nu_{syn,LAT} F_{\nu,syn,LAT} < 10^{-7}$ erg s^{-1} cm^{-2} further constrain t_c. Beyond the corresponding lines, t_c should be significantly shorter than $t_{c,syn}$ in order for the synchrotron not to overproduce the upper limits on the optical, X-ray or GeV fluxes. Within the *black region* the synchrotron emission produces the observed prompt γ-rays. The conditions $\tau_T < 1, 10$ (*dot-dashed lines*; a Thompson optical depth that is not too large) define general limits on the parameter space (see Beniamini and Piran 2014 for more details)

These considerations pose severe constraints on prompt emission models that involve Poynting flux dominated outflows. Any emission model in such a regime should satisfy these constraints. Lacking a model that satisfies all these constraints, it is likely that if the outflow is initially Poynting flux dominated then the magnetic energy is dissipated before the emitting region, where it must be subdominant.

4.6 Spectral Diversity—Spectral Evolution

The peak energy E_p varies a lot from one GRB to another, from a few to tens of keV (X-ray Flashes, X-ray Rich GRBs; Sakamoto et al. 2005) to over 10 MeV (Axelsson et al. 2012). An important property is that short GRBs are harder with larger peak energies (Kouveliotou et al. 1993; Guiriec et al. 2010). Spectral evolution is also found: when time-resolved spectroscopy is possible, the GRB spectrum is always found to strongly evolve during the prompt phase (see e.g. Lu et al. 2012, Burgess et al. 2014, Guiriec et al. 2015 for recent analyses of *Fermi* GRBs, or Preece et al. 2014 for a very bright case where the spectral evolution can be studied in great detail); E_p typically varies over more than a factor of 30 within an individual pulse. Spectral and temporal properties appear correlated within GRB pulses: hardness following the intensity, pulses being narrower and peaking earlier at higher energies, etc.

Both the spectral diversity between different GRBs and the spectral evolution within individual GRBs are hard to reproduce by models. Reconnection models are barely developed enough to allow discussion of these observations. In dissipative photospheres, variations in E_p are related to changes in the properties of the outflow ejection leading to a change in the location of the photosphere (Pe'er 2008; Beloborodov 2013; Deng and Zhang 2014). A potential issue is to explain how the dissipative process adjusts to always remain located just below the photosphere (unless it always occurs over a wide range of radii, in which case it should also occur above the photosphere, so this would not be a pure photospheric model). In internal shocks, the spectral evolution is reproduced qualitatively (Daigne and Mochkovitch 1998, 2003; Asano and Mészáros 2011; Bošnjak and Daigne 2014), and can even be reproduced quantitatively with some constraints on microphysics parameters (Bošnjak and Daigne 2014), which may indicate non-universal values in mildly relativistic shocks, as suggested for instance by (Bykov and Meszaros 1996, see also Bykov et al. 2012). The spectral diversity is also naturally explained by variations in the lifetime and variability of the central engine (Barraud et al. 2005). The hardness-duration relation is well reproduced (Daigne and Mochkovitch 1998; Bošnjak and Daigne 2014).

4.7 Constraints at Other Wavelengths

The discussion above was centered on observations in the soft γ-ray range, where the prompt emission is observed in most GRBs. We discuss here briefly some additional constraints coming from observations of the prompt emission at other wavelengths.

4.7.1 The End of the Prompt Emission: The X-Ray Early Steep Decay

Swift/XRT discovered in most GRB X-ray afterglows an early steep decay at the end of the prompt phase, before recovering a plateau and/or a standard afterglow decay (Nousek et al. 2006; Zhang et al. 2006; O'Brien et al. 2006). A natural explanation is provided by the high-latitude tail of the prompt emission, once the on-axis emission has stopped (Kumar and Panaitescu 2000; Genet and Granot 2009). It can reproduce the observed temporal decay and spectral evolution (Liang et al. 2006; Willingale et al. 2010). This puts a strong constraint on the emission radius at the end of the prompt phase, which may be fulfilled by internal shocks, and also possibly by reconnection models if the radius is large enough (Hascoët et al. 2012a). On the other hand, it is incompatible with photospheric models, which must instead explain the rapid decay phase by a universal behavior of the central engine when it is switching off.

4.7.2 Prompt GeV Emission

Fermi/LAT detects GeV emission in some GRBs (Ackermann et al. 2013). As detection requires enough photons in its energy range (tens of MeV to $\gtrsim 300$ GeV), it detects mainly very bright GRBs, in terms of both their GeV fluence and their total fluence (and thus also in terms of $E_{\gamma,\mathrm{iso}}$). For the same reason, LAT detects a smaller fraction of short GRBs compared to soft γ-ray instruments, since their fluence is typically much smaller than that of long GRBs. Bright enough LAT GRBs show a distinct high-energy spectral component, usually fitted by a power law (e.g., Abdo et al. 2009; Ackermann et al. 2010, 2013). The observed variability in the prompt LAT light curve indicates an internal origin. It is followed by a long-lasting emission (with a power law in time and energy) that likely originates from the deceleration phase or early afterglow. In dissipative photospheric models, it is hard to produce GeV photons due to strong $\gamma\gamma$ annihilation. However, additional processes such as later scatterings of prompt photons by the external medium can explain this GeV emission (see e.g. Beloborodov et al. 2014). In reconnection models, spectral models cannot make such predictions yet. In internal shocks, such multi-component spectra are expected (Guetta and Granot 2003; Bošnjak et al. 2009; Asano and Mészáros 2012; Bošnjak and Daigne 2014); the fact that the GeV component is usually weaker than the soft γ-ray component constrains the strength of the magnetic field, implying that it must be weak, $\epsilon_\mathrm{B} \lesssim 10^{-2}$ (Daigne et al. 2011; Bošnjak and Daigne 2014).

4.8 Prompt Emission Summary

The dissipation mechanism and radiative processes responsible for the prompt GRB emission are still not well understood due to the complex physics involved, both on large and micro-scales. The lack of strong thermal components in GRB spectra suggests a high initial magnetization in GRB outflows, while prompt GRB observational constraints imply a low magnetization in the emission region. Put together, this strongly suggests either very efficient conversion of magnetic to kinetic energy, which leaves a low magnetization in the emission region, and allows for efficient internal shocks (Granot et al. 2011; Granot 2012b), or strong magnetic reconnection that converts magnetic energy to thermal energy and accelerates particles, and yet somehow leaves a low enough magnetization where these particle radiate most of their energy.

A weak thermal emission can be produced at the photosphere, followed by a dominant non-thermal emission at larger radii. Depending on the efficiency of the acceleration and the resulting magnetization at large distances, the dissipation leading to the emission can occur either in shocks or in magnetic reconnection. In both cases, the dominant radiative process should be synchrotron emission. Only in the first case (internal shocks), detailed simulations coupling a dynamical calculation with a detailed radiative model are available. To have a good agreement between the observed spectrum and the predicted one, detailed modeling is needed, where the strength and structure of the magnetic field play a crucial role: moderately efficient inverse Compton scatterings in the Klein-Nishina regime are needed, which requires a weak field, and a decay of the magnetic field far from the shock front is also probably required.

An alternative is to explain the whole soft γ-ray emission by a dissipative photosphere. Its nature, however, must then be elucidated. The magnetic field could again play a vital role, via the magnetic reconnection below the photosphere, which is a natural candidate. Further progress can come from more observations over a broad spectral range with time-resolved spectroscopy, additional and firmer polarization measurements, and improvements in the modeling of the expected spectrum and spectral evolution in each model. Unfortunately, it remains limited by the current knowledge of the microphysics (structure of the magnetic field, particle acceleration) in mildly relativistic shocks and magnetic reconnection (Sironi et al. 2015; Kagan et al. 2015).

5 Magnetic Fields in the Afterglow

Eventually, the GRB outflow is decelerated by the external medium. It drives a strong relativistic blast wave—the afterglow (or external forward) shock—into the surrounding medium. It transfers most of its energy to the shocked external medium (via pdV work across the contact discontinuity that separates them) at a distance $R_{\rm dec}$ from the central source—the deceleration radius. Radiation from $R_{\rm dec}$ reaches the observer at the deceleration time, $T_{\rm dec}$. At $R > R_{\rm dec}$ the original outflow composition no longer affects the dynamics (or emission) of the afterglow shock. However, the outflow magnetization can greatly affect the reverse shock (or external reverse shock, as it is formed due to the interaction with the external medium), whose strength and emission can be greatly suppressed if the outflow is highly magnetized, $\sigma(R_{\rm dec}) \gtrsim 1$.

5.1 The Afterglow Emission

The dominant emission mechanism in the afterglow is thought to be synchrotron radiation, which is produced by relativistic electrons accelerated at the afterglow shock that gyrate in the magnetic fields within the shocked external medium.[11] Such a synchrotron origin of the afterglow emission is strongly supported by the detection of linear polarization at the level of $\sim$ 1–3 % in several optical or NIR afterglows (see Sect. 5.2), and by the shape of the broadband spectrum, which consists of several power law segments that smoothly join at some typical break frequencies. Figure 10 shows the possible resulting afterglow spectra. Broadband (radio to γ-ray) afterglows fit such synchrotron spectra far better than the prompt emission. The broad and mostly featureless smoothly broken power-law shapes of afterglow spectra evolve and fade more slowly over time, and have characteristic frequencies that vary as a power law with time, roughly according to the theoretically expected power-law indices (Sari et al. 1998; Granot and Sari 2002). Synchrotron self-Compton (SSC)—the inverse-Compton scattering of the synchrotron photons to (much) higher energies by the same population of relativistic electrons that emits the synchrotron photons—can sometimes dominate the afterglow flux in the X-rays (Sari and Esin 2001; Harrison et al. 2001), and may affect the synchrotron emission by increasing the electron cooling.

Relativistic collisionless shock physics (e.g., how they amplify the magnetic field and accelerate a non-thermal population of relativistic particles) are still not well understood from first principles (e.g. Sironi et al. 2015). Thus, simple assumptions are usually made that conveniently parameterize our ignorance. The electrons are assumed to be (instantly) shock-accelerated into a power-law distribution of energies, $dN/d\gamma_e \propto \gamma_e{}^{-p}$ for $\gamma_e > \gamma_m$, and then cool both adiabatically and due to radiative losses.[12] The relativistic electrons are assumed to hold a fraction ϵ_e of the internal energy immediately behind the shock, while the magnetic field is assumed to hold a fraction ϵ_B of the internal energy everywhere in the shocked region. Both the temporal and spectral indices depend on the power law index p of the electron energy distribution. The temporal index (i.e. the rate of flux decay) also depends on the circumburst density profile, which is parameterized in Fig. 10 as a power law of index k with the distance R from the central source, $\rho_{\rm ext} \propto R^{-k}$, with $k = 0$ and $k = 2$, respectively, corresponding to an ISM and a stellar wind—WIND. The temporal index can also be affected by other factors, such as energy losses or injection into the afterglow shock, the afterglow jet angular structure and the viewing angle relative to the jet symmetry axis, or time evolution of the shock microphysics parameters ϵ_e and/or ϵ_B.

5.2 Polarization: Afterglow and Reverse Shock

The detection of linear polarization of a few percent in the optical and NIR *afterglow* of several GRBs (see Covino et al. 2004, and references therein) was considered as a confirmation that synchrotron radiation is the dominant afterglow emission mechanism. The synchrotron emission from a fluid element with a locally uniform magnetic field is linearly polarized in the direction perpendicular to the projection of the magnetic field onto the plane normal to the wave vector. Since the source moves relativistically, one must account for aberration of light when calculating the observed local direction of polarization. Figure 11 shows the predicted local polarization map from emission by an ultra-relativistic expanding shell, for two

[11] In an alternative scenario, the afterglow emission is dominated at early times by the contribution of a long-lived reverse shock (Uhm and Beloborodov 2007; Genet et al. 2007), which allows to reproduce more easily the observed diversity and variability, such as X-ray plateaus (Uhm et al. 2012; Hascoët et al. 2014) or X-ray flares (Hascoet et al. 2015), though in this scenario a transition to forward shock dominance is expected at late times but not observed.

[12] It is usually also further assumed that practically all of the electrons take part in this acceleration process and form such a non-thermal (power-law) distribution, leaving no thermal component (which is not at all clear or justified; e.g. Eichler and Waxman 2005).

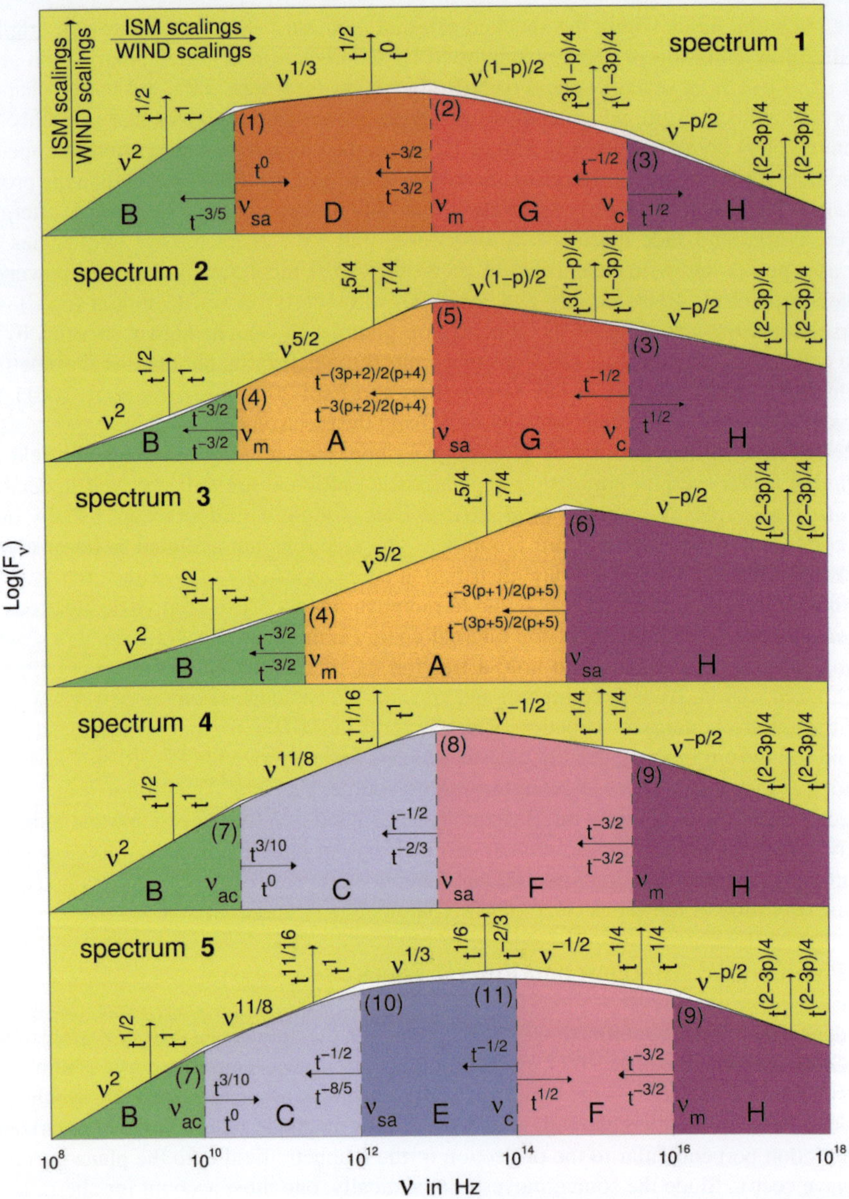

Fig. 10 The afterglow synchrotron spectrum, calculated for the Blandford and McKee (1976) spherical self-similar solution, under standard assumptions, using the accurate form of the synchrotron spectral emissivity and integration over the emission from the whole volume of shocked material behind the forward (afterglow) shock (for details see Granot and Sari 2002, from which this figure is taken). The different panels show the five possible broad band spectra of the afterglow synchrotron emission, each corresponding to a different ordering of the spectral break frequencies. Each spectrum consists of several power-law segments (PLSs; each shown with a different color and labeled by a different letter A–H) that smoothly join at the break frequencies (numbered 1–11). The broken power-law spectrum, which consists of the asymptotic PLSs that abruptly join at the break frequencies (and is widely used in the literature), is shown for comparison. Most PLSs appear in more than one of the five different broad band spectra. Indicated next to the arrows are the temporal scaling of the break frequencies and the flux density at the different PLSs, for a uniform (ISM; $k=0$) and stellar wind (WIND; $k=2$) external density profile

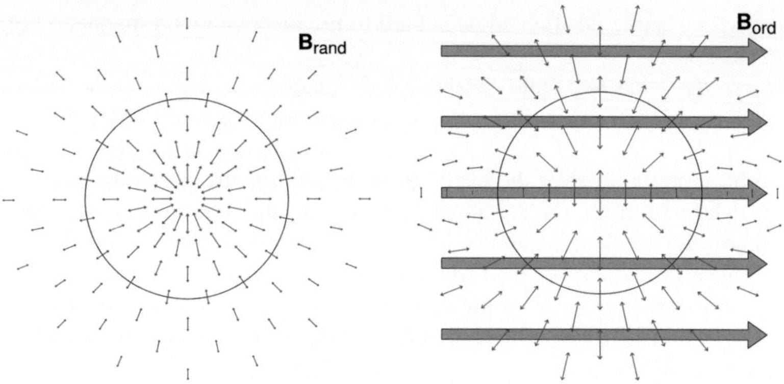

Fig. 11 The predicted polarization map for synchrotron emission from a thin spherical ultra-relativistic shell expanding with a Lorentz factor $\Gamma \gg 1$. The *double-sided arrows* show the direction of the linear polarization (the wave electric vector), while their length depends monotonically on the polarized intensity (in a non-trivial way, for display purposes). The circle indicates an angle of $1/\Gamma$ around the line of sight to the central source, and contains the region responsible for most of the observed flux. *Left*: for a magnetic field that is random within the plane of the shell (normal to the radial direction), for which the polarization direction always points at the center of the image, where the polarization vanishes (due to symmetry consideration). *Right*: for an ordered magnetic field within the plane of the shell that is coherent over angular scales $\gg 1/\Gamma$ (Granot and Königl 2003). In this case the direction of the ordered magnetic field clearly breaks the symmetry around the center of the image, resulting in a large net polarization. For simplicity, the map is for a constant emission radius, rather than for a constant photon arrival time

different magnetic field structures: a magnetic field that is random within the plane normal to the radial direction (*left panel*) as could be expected from a shock-produced field (e.g., Medvedev and Loeb 1999), and an ordered magnetic field normal to the radial direction (*right panel*; as could be expected in the prompt or reverse-shock emission for a magnetic field coherent on angular scales $\gg 1/\Gamma$ that is advected from the central source).

The afterglow image is almost always unresolved, so we can only measure the (weighted) average polarization over the whole image. Therefore, a shock produced magnetic field that is symmetric about the shock normal will procure no net polarization for a spherical flow (as in this case the polarization pattern across the image is symmetric around its center, and the polarization averages out to zero when summed over the whole image). For a shock-produced magnetic field, one thus needs to break this symmetry of the emission to produce net polarization. A simple and natural way of doing this is considering a jet, or narrowly collimated outflow (e.g., Sari 1999; Ghisellini and Lazzati 1999). In this picture a jet geometry together with a line of sight that is not along the jet symmetry axis (but still within the jet aperture, in order to see the prompt GRB) is needed to break the symmetry of the afterglow image around our line of sight. Other models for afterglow polarization include a magnetic field that is coherent over patches of a size comparable to that of causally connected regions (Gruzinov and Waxman 1999), polarization that is induced by microlensing (Loeb and Perna 1998) or by scintillations in the radio (Medvedev and Loeb 1999), a small ordered magnetic field component originating from the circumburst medium (Granot and Königl 2003), clumps in the external medium (Granot and Königl 2003), or a very inhomogeneous jet angular structure—a "patchy shell" with "hot spots" (Granot and Königl 2003; Nakar and Oren 2004). The many possible causes of polarization, and the degeneracy with other factors makes it difficult to robustly determine the magnetic field structure in the emitting region from afterglow polarization measurements. Nonetheless, a high degree of linear polarization with a stable position angle is hard to produce without a magnetic field that is ordered on large scales.

The *reverse shock* has two main observational signatures: a sharply-peak "optical flash" (e.g., Akerlof et al. 1999) on a timescale comparable to the prompt GRB T_{90}, and a "radio flare" (e.g., Kulkarni et al. 1999; Frail et al. 2000; Berger et al. 2003) that peaks on a timescale of a day or so after the GRB. In such cases, if the relatively bright observed emission is indeed from the reverse shock, this implies that the outflow was not strongly magnetized near the deceleration epoch, $\sigma(R_{\rm dec}) \ll 1$. Moreover, the polarization properties of the synchrotron emission from the reverse shock provide a powerful and unique probe for the magnetic field structure in the original outflow. Early optical polarization measurements from $T \gtrsim T_{\rm dec}$ have finally been obtained in the last eight years or so (Mundell et al. 2007a, 2013; Steele et al. 2009).

On the one hand, there is a strict upper limit on the degree of linear polarization from GRB 060418 of $P < 8$ % (2-σ) at $T = 203$ s after the GRB trigger, while the deceleration time suggested by the early optical light curve of this GRB is $T_{\rm dec} = 153 \pm 10$ s (Molinari et al. 2007) and its prompt emission lasted only $T_{90} = 52 \pm 1$ s. The fact that $T_{\rm dec}/T_{90} \approx 3$ suggest a "thin shell" in this case, which is consistent with a moderate magnetization ($\sigma(R_{\rm dec}) \ll 1$) that allows a strong reverse shock with bright emission. However, the polarization is fairly low near $T_{\rm dec} = 153 \pm 10$ s (at $T = 203$ s), which suggests that either the reverse shock emission even near its peak is for some reason greatly sub-dominant compared to the (very weakly polarized) forward shock emission, or more likely that in this GRB there is hardly any ordered magnetic field in the ejecta on angular scales $\gtrsim 1/\Gamma$ that cover most of the visible region.

On the other hand, GRB 090102 had a prompt duration of $T_{90} = 27$ s and an optical linear polarization of $P = 10.2 \pm 1.3$ % in a 60 s exposure starting at $T = 161$ s after the trigger time (Steele et al. 2009). Its optical light curve shows a power-law decay $F_\nu \propto t^{-\alpha}$ with $\alpha = 1.50 \pm 0.06$ from $T \sim 40$ s to $T \sim 1000$ s and then flattens to $\alpha = 0.97 \pm 0.03$ (Gendre et al. 2010). This suggests a deceleration time $T_{\rm dec} \lesssim 40$ s, well before the polarization measurement. However, the optical emission may be dominated by the reverse shock up to the break time of ~ 1000 s, and in particular during the polarization measurement. In the latter case this might possibly explain the measured polarization as arising from an ordered magnetic field component in the ejecta, though a purely ordered field on the scale of the whole emitting region (angular scale $\gtrsim 1/\Gamma$) would produce a significantly larger polarization of several tens of percent (Granot and Königl 2003; Granot 2003; Lyutikov et al. 2003; Nakar et al. 2003), which would suggest either a smaller magnetic field coherence length or a dominant contribution from the much less polarized external forward shock emission.

Finally, GRB 120308A that lasted $T_{90} \sim 100$ s (between $T \approx -30$ s and $T \approx 70$ s post-trigger) showed an optical linear polarization of $P = 28 \pm 4$ % in an exposure between $T = 240$ s and 323 s (Mundell et al. 2013), which gradually decreased to $P = 16^{+5}_{-4}$ % over the next ten minutes, while keeping an approximately constant position angle (to within an accuracy of about 15°). The optical light curve peaked at around $T \sim 300$ s, during the time bin in which the largest polarization was measured, and subsequently decayed, with a possible transition from reverse to forward shock domination of the optical emission around ~ 1000 s. This strongly suggests the presence of a large-scale ordered magnetic field in the original GRB ejecta.

Observations of radio flares at roughly a day after the GRB have so far produced no detection of polarization. However, these observations have enabled to set strict upper limits on a possible linear or circular polarization (Granot and Taylor 2005). The strictest limits are for GRB 991216, for which the 3-σ upper limits on the linear and circular polarization are $P_{\rm lin} < 7\ \%$ and $P_{\rm circ} < 9\ \%$, respectively. These limits provide interesting constraints on existing GRB models (Granot and Taylor 2005), and in particular are hard to reconcile with a predominantly ordered toroidal magnetic field in the GRB outflow together with a "structured" jet, where the energy per solid angle drops as the inverse square of the angle from the jet axis, as is expected in some models in which the outflow is Poynting flux dominated.

Recently, the detection of circular polarization was reported in the optical afterglow of GRB 121024A by Wiersema et al. (2014). In particular, they measured a circular polarization of $P_{\rm circ} = 0.61 \pm 0.13\ \%$ at $T = 0.15$ days after the GRB. The linear polarization during that time was $P_{\rm lin} \sim 4\ \%$ implying a circular to linear polarization ratio of $P_{\rm circ}/P_{\rm lin} \sim 0.15$. A very recent detailed study that examined different assumptions for the magnetic field configuration, jet geometry and electron pitch-angle distribution (Nava et al. 2015) concluded that such a relatively high $P_{\rm circ}/P_{\rm lin}$ ratio cannot be produced by synchrotron emission from the afterglow (i.e. forward external) shock, which suggests an alternative origin.

5.3 Maximum Synchrotron Photon Energy

Since the afterglow emission from the shocked external medium is independent of the outflow composition, it can much more "cleanly" probe the physics of relativistic collisionless shocks, and serve as a testbed for how the very weak upstream magnetic fields in the pristine surrounding medium are amplified in the shock, and how the particles are accelerated in this shock and radiate in the downstream magnetic field.

A recent challenge to the standard synchrotron afterglow scenario was raised by the exceptional GRB 130427A (Ackermann et al. 2014; Perley et al. 2014; Kouveliotou et al. 2013; Maselli et al. 2014). This was a very energetic GRB (with an isotropic equivalent γ-ray energy of $E_{\gamma,\rm iso} = 1.4 \times 10^{54}$ erg) and it occured relatively nearby (at redshift $z = 0.34$). Therefore, it was extremely bright and in particular, it was detected by *Fermi*-LAT (100 MeV–100 GeV) for nearly a day, including a 95 GeV photon several minutes after the burst and a 32 GeV photon after 9 hours (Ackermann et al. 2014). Altogether, this GRB has a large number of high-energy photons that clearly violate (Ackermann et al. 2014) the maximum synchrotron photon energy limit,[13]

$$E_{\rm syn,max} \sim \frac{\Gamma}{(1+z)}\frac{m_e c^2}{\alpha} = 3.5\left(\frac{2}{1+z}\right)\frac{\Gamma}{100}\ \text{GeV}, \tag{18}$$

which is obtained by equating the electron acceleration time to its synchrotron cooling time, assuming that it is accelerated and radiates its synchrotron emission in same magnetic field strength. This has been argued in order to rule out an afterglow synchrotron origin of the late-time high-energy LAT photons, and in particular motivated suggestions for an origin in a distinct SSC spectral component (e.g. Ackermann et al. 2014; Tam et al. 2013; Liu et al. 2013; Fan et al. 2013b).

However, Kouveliotou et al. (2013) have shown that the optical to GeV spectrum is consistent with a single spectral component that very nicely matches the expectation for synchrotron afterglow emission (Granot and Sari 2002, see Fig. 10). Figure 12 shows their spectral fit, and demonstrates that there is hardly any room for a distinct (SSC) spectral component to dominate the observed flux at the highest LAT energies (above several GeV

[13]The exact numerical coefficient depends on the exact assumptions, and in particular on whether the acceleration time is assumed to be a fraction of or a complete Larmor gyration time, which is in any case a very fast acceleration, and arguably even unrealistically so. Here $\alpha \approx 1/137$ is the fine structure constant.

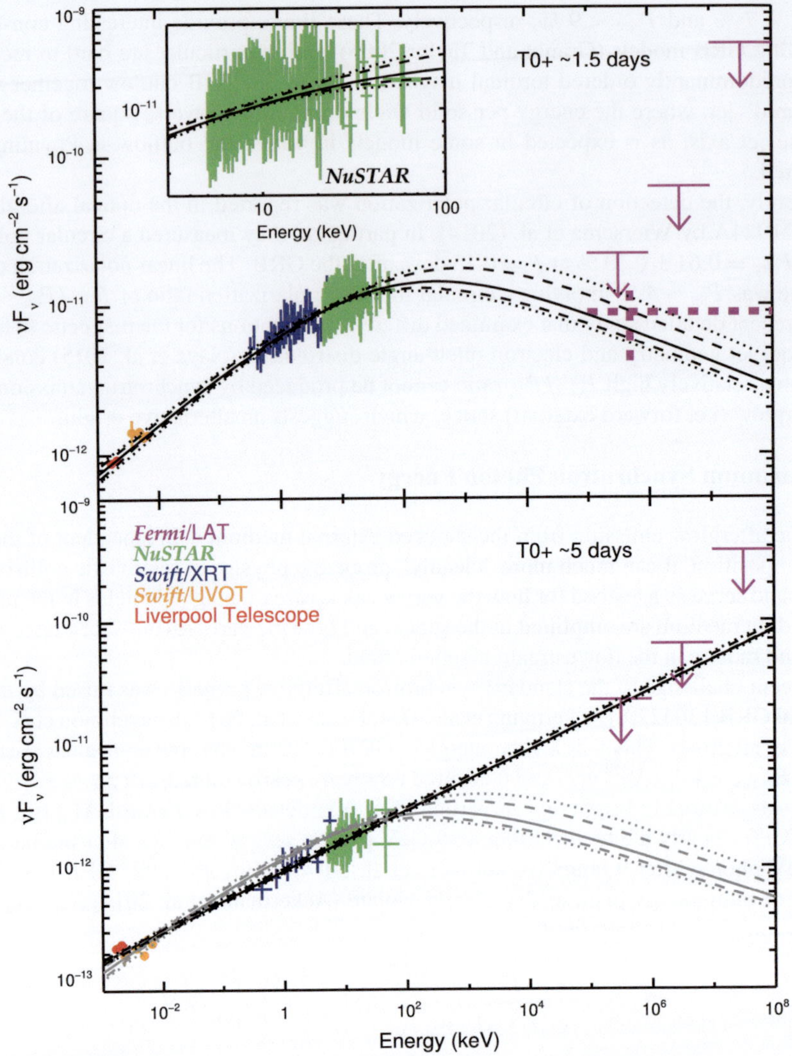

Fig. 12 The optical to GeV spectrum of GRB 130427A fit with the afterglow synchrotron model of Granot and Sari (2002). Broadband SEDs are shown during the first (*top-panel*) and the second (*bottom-panel*) NuSTAR epochs. The *Fermi*/LAT upper-limits are shown as arrows and the extrapolation of the LAT flux light curve is shown as a dashed magenta cross (only during the first epoch). The second epoch (*bottom-panel*) is fit with a power law (*black lines*); the fit to the first epoch is scaled down and superposed on the second epoch data for comparison (*in gray*) (this figure is taken from Kouveliotou et al. 2013)

or so, as is needed to avoid violating $E_{\rm syn,max}$). This conclusion is strengthened by strict upper limits on the > 100 GeV flux measured by VERITAS at three different epochs near the first NuSTAR observation (Aliu et al. 2014).

Therefore, this comprises very compelling evidence for a genuine violation of the $E_{\rm syn,max}$ in this case, which is much harder to circumvent compared with previous *Fermi*/LAT GRBs (Abdo et al. 2009; Piran and Nakar 2010; Atwood et al. 2013). Thus, one should start to seriously consider how this limit can indeed be violated. A possible solution may lie in relaxing the assumption of a uniform magnetic field and instead allowing for a lower magnetic field acceleration region and a higher magnetic field synchrotron radiation region (e.g., Lyutikov 2010; Kumar et al. 2012). Such a situation might arise for diffusive shock acceleration (Fermi Type I) if the tangled shock-amplified magnetic field decays on a short length-scale behind the shock front (where most of the high-energy radiation is emitted), while the highest energy electrons are accelerated in the lower magnetic field further downstream (Kumar et al. 2012). In fact, such a scenario has recently been suggested (Lemoine 2013; Lemoine et al. 2013), and also shown to significantly alleviate the previously very high γ-ray radiative efficiencies inferred for most *Swift* GRBs (Beniamini et al. 2015).

6 Conclusions

In this review we have demonstrated that magnetic fields clearly play a vital role in GRBs, practically in every important aspect of this phenomenon. Here we briefly summarize some of our main conclusions, and stress both what was learned so far, as well as what still needs to be carefully studied. Following the main text, the discussion proceeds more or less in order of increasing distance from the central source.

Magnetic fields most likely play a crucial role in the launching of GRB jets. Moreover, hydromagnetic jet launching implies dynamically strong magnetic fields near the central source, which can naturally help avoid an excessive baryonic loading into the jet. Such a low baryon loading is essential in order for the jet to be able to reach sufficiently large Lorentz factors ($\Gamma \gtrsim 100$–500) that are inferred from prompt GRB observations. The jet launching definitely requires many further detailed studies, also (semi-) analytic, but mainly numerical studies. The latter are, however, involved as they require general-relativistic magnetohydrodynamics (GRMHD) codes, coupled with neutrino, plasma and radiation physics. Therefore, is it likely to take many years before such studies will provide definitive answers.

Millisecond-magnetar models for the GRB central engine, discussed in Sect. 2, face serious challenges that still must be overcome. Models where the GRB arises from the delayed collapse of a supramassive millisecond magnetar (such as the "time reversal model"; Rezzolla and Kumar 2015; Ciolfi and Siegel 2015) face serious difficulty as this requires the formation of a disk during this collapse, which was recently argued to not be possible (Margalit et al. 2015). Producing both the prompt GRB emission and the X-ray plateau observed by *Swift* in long GRBs is challenging for millisecond-magnetar models. Models invoking millisecond magnetars in short GRBs also face many problems, such as how to produce the short GRB, or hide the huge amount of rotational energy (a few $\times 10^{52}$ erg) that is injected into the afterglow shock, while short GRB afterglows are very dim. More detailed studies of the relevant physics, both analytic and numerical, are definitely needed in order to produce more robust and realistic predictions that could be tested in more detail against the relevant observations. A relatively simple example is that a highly-relativistic pulsar-like magnetar MHD wind is invoked to explain the X-ray plateaus observed by *Swift*. However, most studies consider only the forward shock emission and ignore the emission from the MHD wind itself (see, however, Dai 2004), which may involve magnetic reconnection and the resulting particle acceleration at the termination shock of the pulsar wind, similar to pulsar wind nebulae (see, e.g., Lyubarsky and Kirk 2001; Lyubarsky 2003; Sironi and Spitkovsky 2014). The birth process of a millisecond magnetar, either in the core-collapse of a massive star or in a binary neutron star merger, and how it evolves in these messy environments, is obviously very involved but there is definitely a lot of room for improvement and new studies on such systems.

The GRB outflow composition, and in particular its degree of magnetization is an important open question. On the one hand, in Sect. 3.1 we have shown very compelling evidence based on the theoretical jet propagation time in the progenitor star of a long GRB and the observed GRB duration distribution, that the jet appears to have modest magnetization ($\sigma \lesssim 1$) throughout most of the time it takes for it to bore its way out of the star. Moreover, in Sect. 4.5 we argued based on the prompt GRB observations (both γ-ray detections and upper limits in optical and soft X-ray) that the emission region has a low magnetization. On the other hand, hydromagnetic jet launching is much more promising than a pure thermal one, and can naturally help avoid excessive baryon loading, suggesting a high magnetization near the source ($\sigma_0 \gg 1$). A similar conclusion is also strongly suggested from the fact that quasi-thermal components in the prompt GRB spectrum are typically highly sub-dominant (see Sect. 4). Taking these two lines of evidence together, it appears that the magnetization significantly decreases as the jet propagates from the source out to large distances. This can occur through two main channels (or some combination of the two): (i) very efficient conversion of magnetic to kinetic energy that leaves a low magnetization in the emission region and allows for efficient internal shocks, through first quasi-steady and then impulsive magnetic acceleration (see Fig. 6; Granot et al. 2011; Granot 2012b), or (ii) strong magnetic reconnection that converts magnetic energy into thermal energy and accelerates particles, and yet somehow leaves a sufficiently low magnetization where these particles radiate most of their energy.

Magnetic fields can also play a very important role in the energy dissipation that leads to the GRB emission. On the one hand, sufficiently low magnetization ($\sigma < 1$ or even $\sigma \lesssim 0.1$) is needed for efficient energy dissipation in internal shocks. On the other hand, while a large magnetization ($\sigma > 1$) effectively suppresses internal shocks, or the reverse shock, it can lead to significant dissipation through magnetic reconnection under appropriate conditions (e.g. if the source ejects outflow with a magnetic field that occasionally changes its polarity, or through certain instabilities).

Another vital role of magnetic fields in GRBs is in particle acceleration. Within the outflow the magnetization can be high, allowing efficient magnetic reconnection, which can directly convert a good fraction of the dissipated magnetic energy into the random motions of the particles that it accelerates in this process. The exact magnetic field strength and geometry throughout the emission region also greatly affect the radiation of the accelerated particles, and therefore their observable signatures. In particular, magnetic fields are vital to the synchrotron emission, which dominates in the afterglow and also likely plays a key role in the prompt GRB emission.

Both optically thin internal shocks as well as the external forward and reverse shocks are collisionless, and mediated by electromagnetic fields (through collective plasma interactions). Moreover, while the outflow itself typically has a large enough magnetic field for efficient synchrotron radiation, the external medium has a very low magnetization that must be significantly amplified at the afterglow shock front in order to produce the observed afterglow emission. Therefore, magnetic field amplification in relativistic collisionless shocks and its possible decay behind the shock are vital for understanding the afterglow physics and interpreting afterglow observations. Moreover they can strongly affect the particle acceleration in the afterglow shock, and the resulting afterglow emission. Such physics might hold the key to unravel a puzzle (see Sect. 5.3) arising from observations of the very bright and relatively nearby long GRB 130427A, which show an apparent violation of the maximum synchrotron photon energy limit, $E_{\rm syn,max}$.

The different roles of magnetic fields in GRBs are numerous and diverse. Many of them are only starting to be understood, while new roles are still being occasionally discovered.

Since magnetic fields appear in almost all aspects of GRB physics, future studies of their properties and effects are likely to greatly improve our understanding of GRBs, and lead to fundamental progress.

Acknowledgements The authors acknowledge support from the ISF grant 719/14 (JG), as well as from the I-CORE Program—ISF grant 1829/12, the ISA grant 3-10417, and an ISF-CNSF grant (TP).

References

A.A. Abdo et al., Fermi observations of GRB 090902B: A distinct spectral component in the prompt and delayed emission. Astrophys. J. Lett. **706**, 138–144 (2009). doi:10.1088/0004-637X/706/1/L138

M. Ackermann et al., Fermi observations of GRB 090510: A short-hard gamma-ray burst with an additional, hard power-law component from 10 keV TO GeV energies. Astrophys. J. **716**, 1178–1190 (2010). doi:10.1088/0004-637X/716/2/1178

M. Ackermann et al., The first Fermi-LAT gamma-ray burst catalog. Astrophys. J. Suppl. Ser. **209**, 11 (2013). doi:10.1088/0067-0049/209/1/11

M. Ackermann et al., Fermi-LAT observations of the gamma-ray burst GRB 130427A. Science **343**, 42–47 (2014). doi:10.1126/science.1242353

C. Akerlof et al., Observation of contemporaneous optical radiation from a γ-ray burst. Nature **398**, 400–402 (1999). doi:10.1038/18837

E. Aliu et al., Constraints on very high energy emission from GRB 130427A. Astrophys. J. Lett. **795**, 3 (2014). doi:10.1088/2041-8205/795/1/L3

K. Asano, P. Mészáros, Spectral-temporal simulations of internal dissipation models of gamma-ray bursts. Astrophys. J. **739**, 103 (2011). doi:10.1088/0004-637X/739/2/103

K. Asano, P. Mészáros, Delayed onset of high-energy emissions in leptonic and hadronic models of gamma-ray bursts. Astrophys. J. **757**, 115 (2012). doi:10.1088/0004-637X/757/2/115

W.B. Atwood et al., New Fermi-LAT event reconstruction reveals more high-energy gamma rays from gamma-ray bursts. Astrophys. J. **774**, 76 (2013). doi:10.1088/0004-637X/774/1/76

M. Axelsson et al., GRB110721A: An extreme peak energy and signatures of the photosphere. Astrophys. J. Lett. **757**, 31 (2012). doi:10.1088/2041-8205/757/2/L31

D. Band et al., BATSE observations of gamma-ray burst spectra. I—Spectral diversity. Astrophys. J. **413**, 281–292 (1993). doi:10.1086/172995

M.G. Baring, A.K. Harding, The escape of high-energy photons from gamma-ray bursts. Astrophys. J. **491**, 663–686 (1997)

R. Barniol Duran, Ž. Bošnjak, P. Kumar, Inverse-Compton cooling in Klein-Nishina regime and gamma-ray burst prompt spectrum. Mon. Not. R. Astron. Soc. **424**, 3192–3200 (2012). doi:10.1111/j.1365-2966.2012.21533.x

C. Barraud, F. Daigne, R. Mochkovitch, J.L. Atteia, On the nature of X-ray flashes. Astron. Astrophys. **440**, 809–817 (2005). doi:10.1051/0004-6361:20041572

M.C. Begelman, Instability of toroidal magnetic field in jets and plerions. Astrophys. J. **493**, 291 (1998). doi:10.1086/305119

A.M. Beloborodov, Collisional mechanism for gamma-ray burst emission. Mon. Not. R. Astron. Soc. **407**, 1033–1047 (2010). doi:10.1111/j.1365-2966.2010.16770.x

A.M. Beloborodov, Radiative transfer in ultrarelativistic outflows. Astrophys. J. **737**, 68 (2011). doi:10.1088/0004-637X/737/2/68

A.M. Beloborodov, Regulation of the spectral peak in gamma-ray bursts. Astrophys. J. **764**, 157 (2013). doi:10.1088/0004-637X/764/2/157

A.M. Beloborodov, B.E. Stern, R. Svensson, Power density spectra of gamma-ray bursts. Astrophys. J. **535**, 158–166 (2000). doi:10.1086/308836

A.M. Beloborodov, R. Hascoët, I. Vurm, On the origin of GeV emission in gamma-ray bursts. Astrophys. J. **788**, 36 (2014). doi:10.1088/0004-637X/788/1/36

P. Beniamini, J. Granot (2015, in preparation)

P. Beniamini, T. Piran, Constraints on the synchrotron emission mechanism in gamma-ray bursts. Astrophys. J. **769**, 69 (2013). doi:10.1088/0004-637X/769/1/69

P. Beniamini, T. Piran, The emission mechanism in magnetically dominated gamma-ray burst outflows. Mon. Not. R. Astron. Soc. **445**, 3892–3907 (2014). doi:10.1093/mnras/stu2032

P. Beniamini, L. Nava, R. Barniol Duran, T. Piran, Energies of GRB blast waves and prompt efficiencies as implied by self-consistent modeling of X-ray and LAT afterglows. ArXiv e-prints (2015)

E. Berger, A.M. Soderberg, D.A. Frail, S.R. Kulkarni, A radio flare from GRB 020405: Evidence for a uniform medium around a massive stellar progenitor. Astrophys. J. Lett. **587**, 5–8 (2003). doi:10.1086/375158

R.D. Blandford, To the lighthouse, in *Lighthouses of the Universe: The Most Luminous Celestial Objects and Their Use for Cosmology*, ed. by M. Gilfanov, R. Sunyeav, E. Churazov (2002), p. 381. doi:10.1007/10856495_59

R.D. Blandford, C.F. McKee, Fluid dynamics of relativistic blast waves. Phys. Fluids **19**, 1130–1138 (1976). doi:10.1063/1.861619

J.S. Bloom, D.A. Frail, S.R. Kulkarni, Gamma-ray burst energetics and the gamma-ray burst Hubble diagram: Promises and limitations. Astrophys. J. **594**, 674–683 (2003). doi:10.1086/377125

Ž. Bošnjak, F. Daigne, Spectral evolution in gamma-ray bursts: Predictions of the internal shock model and comparison to observations. Astron. Astrophys. **568**, 45 (2014). doi:10.1051/0004-6361/201322341

Ž. Bošnjak, F. Daigne, G. Dubus, Prompt high-energy emission from gamma-ray bursts in the internal shock model. Astron. Astrophys. **498**, 677–703 (2009). doi:10.1051/0004-6361/200811375

O. Bromberg, A. Levinson, Hydrodynamic collimation of relativistic outflows: Semianalytic solutions and application to gamma-ray bursts. Astrophys. J. **671**, 678–688 (2007). doi:10.1086/522668

O. Bromberg, A. Tchekhovskoy (2015, in preparation)

O. Bromberg, E. Nakar, T. Piran, R. Sari, The propagation of relativistic jets in external media. Astrophys. J. **740**, 100 (2011). doi:10.1088/0004-637X/740/2/100

O. Bromberg, E. Nakar, T. Piran, R. Sari, An observational imprint of the collapsar model of long gamma-ray bursts. Astrophys. J. **749**, 110 (2012). doi:10.1088/0004-637X/749/2/110

O. Bromberg, E. Nakar, T. Piran, R. Sari, Short versus long and collapsars versus non-collapsars: A quantitative classification of gamma-ray bursts. Astrophys. J. **764**, 179 (2013). doi:10.1088/0004-637X/764/2/179

O. Bromberg, J. Granot, Y. Lyubarsky, T. Piran, The dynamics of a highly magnetized jet propagating inside a star. Mon. Not. R. Astron. Soc. **443**, 1532–1548 (2014). doi:10.1093/mnras/stu995

O. Bromberg, J. Granot, T. Piran, On the composition of GRBs' Collapsar jets. Mon. Not. R. Astron. Soc. **450**, 1077–1084 (2015). doi:10.1093/mnras/stv226

N. Bucciantini, E. Quataert, J. Arons, B.D. Metzger, T.A. Thompson, Magnetar-driven bubbles and the origin of collimated outflows in gamma-ray bursts. Mon. Not. R. Astron. Soc. **380**, 1541–1553 (2007). doi:10.1111/j.1365-2966.2007.12164.x

N. Bucciantini, E. Quataert, J. Arons, B.D. Metzger, T.A. Thompson, Relativistic jets and long-duration gamma-ray bursts from the birth of magnetars. Mon. Not. R. Astron. Soc. **383**, 25–29 (2008). doi:10.1111/j.1745-3933.2007.00403.x

N. Bucciantini, E. Quataert, B.D. Metzger, T.A. Thompson, J. Arons, L. Del Zanna, Magnetized relativistic jets and long-duration GRBs from magnetar spin-down during core-collapse supernovae. Mon. Not. R. Astron. Soc. **396**, 2038–2050 (2009). doi:10.1111/j.1365-2966.2009.14940.x

N. Bucciantini, B.D. Metzger, T.A. Thompson, E. Quataert, Short gamma-ray bursts with extended emission from magnetar birth: Jet formation and collimation. Mon. Not. R. Astron. Soc. **419**, 1537–1545 (2012). doi:10.1111/j.1365-2966.2011.19810.x

J.M. Burgess et al., Constraints on the synchrotron shock model for the Fermi GRB 090820A observed by gamma-ray burst monitor. Astrophys. J. **741**, 24 (2011). doi:10.1088/0004-637X/741/1/24

J.M. Burgess et al., Time-resolved analysis of Fermi gamma-ray bursts with fast- and slow-cooled synchrotron photon models. Astrophys. J. **784**, 17 (2014). doi:10.1088/0004-637X/784/1/17

A.M. Bykov, P. Meszaros, Electron acceleration and efficiency in nonthermal gamma-ray sources. Astrophys. J. Lett. **461**, 37 (1996). doi:10.1086/309999

A. Bykov, N. Gehrels, H. Krawczynski, M. Lemoine, G. Pelletier, M. Pohl, Particle acceleration in relativistic outflows. Space Sci. Rev. **173**, 309–339 (2012). doi:10.1007/s11214-012-9896-y

S. Campana et al., The association of GRB 060218 with a supernova and the evolution of the shock wave. Nature **442**, 1008–1010 (2006). doi:10.1038/nature04892

S.B. Cenko et al., The collimation and energetics of the brightest swift gamma-ray bursts. Astrophys. J. **711**, 641–654 (2010). doi:10.1088/0004-637X/711/2/641

R. Ciolfi, D.M. Siegel, Short gamma-ray bursts in the "Time-reversal" scenario. Astrophys. J. Lett. **798**, 36 (2015). doi:10.1088/2041-8205/798/2/L36

W. Coburn, S.E. Boggs, Polarization of the prompt γ-ray emission from the γ-ray burst of 6 December 2002. Nature **423**, 415–417 (2003). doi:10.1038/nature01612

E. Cohen, J.I. Katz, T. Piran, R. Sari, R.D. Preece, D.L. Band, Possible evidence for relativistic shocks in gamma-ray bursts. Astrophys. J. **488**, 330–337 (1997)

J. Contopoulos, A simple type of magnetically driven jets: An astrophysical plasma gun. Astrophys. J. **450**, 616 (1995). doi:10.1086/176170

S. Covino, G. Ghisellini, D. Lazzati, D. Malesani, Polarization of gamma-ray burst optical and near-infrared afterglows, in *Gamma-Ray Bursts in the Afterglow Era*, ed. by M. Feroci, F. Frontera, N. Masetti, L. Piro. Astronomical Society of the Pacific Conference Series, vol. 312 (2004), p. 169

Z.G. Dai, Relativistic wind bubbles and afterglow signatures. Astrophys. J. **606**, 1000–1005 (2004). doi:10.1086/383019

F. Daigne, R. Mochkovitch, Gamma-ray bursts from internal shocks in a relativistic wind: Temporal and spectral properties. Mon. Not. R. Astron. Soc. **296**, 275–286 (1998). doi:10.1046/j.1365-8711.1998.01305.x

F. Daigne, R. Mochkovitch, Gamma-ray bursts from internal shocks in a relativistic wind: A hydrodynamical study. Astron. Astrophys. **358**, 1157–1166 (2000)

F. Daigne, R. Mochkovitch, The expected thermal precursors of gamma-ray bursts in the internal shock model. Mon. Not. R. Astron. Soc. **336**, 1271–1280 (2002). doi:10.1046/j.1365-8711.2002.05875.x

F. Daigne, R. Mochkovitch, The physics of pulses in gamma-ray bursts: Emission processes, temporal profiles and time-lags. Mon. Not. R. Astron. Soc. **342**, 587–592 (2003). doi:10.1046/j.1365-8711.2003.06575.x

F. Daigne, Ž. Bošnjak, G. Dubus, Reconciling observed gamma-ray burst prompt spectra with synchrotron radiation? Astron. Astrophys. **526**, 110 (2011). doi:10.1051/0004-6361/201015457

S. Dall'Osso, S.N. Shore, L. Stella, Early evolution of newly born magnetars with a strong toroidal field. Mon. Not. R. Astron. Soc. **398**, 1869–1885 (2009). doi:10.1111/j.1365-2966.2008.14054.x

S. Dall'Osso, G. Stratta, D. Guetta, S. Covino, G. De Cesare, L. Stella, Gamma-ray bursts afterglows with energy injection from a spinning down neutron star. Astron. Astrophys. **526**, 121 (2011). doi:10.1051/0004-6361/201014168

S. Dall'Osso, J. Granot, T. Piran, Magnetic field decay in neutron stars: From soft gamma repeaters to 'weak-field magnetars'. Mon. Not. R. Astron. Soc. **422**, 2878–2903 (2012). doi:10.1111/j.1365-2966.2012.20612.x

W. Deng, B. Zhang, Low energy spectral index and E_p evolution of quasi-thermal photosphere emission of gamma-ray bursts. Astrophys. J. **785**, 112 (2014). doi:10.1088/0004-637X/785/2/112

E.V. Derishev, Synchrotron emission in the fast cooling regime: Which spectra can be explained? Astrophys. Space Sci. **309**, 157–161 (2007). doi:10.1007/s10509-007-9421-z

E.V. Derishev, V.V. Kocharovsky, V.V. Kocharovsky, Physical parameters and emission mechanism in gamma-ray bursts. Astron. Astrophys. **372**, 1071–1077 (2001). doi:10.1051/0004-6361:20010586

G. Drenkhahn, H.C. Spruit, Efficient acceleration and radiation in Poynting flux powered GRB outflows. Astron. Astrophys. **391**, 1141–1153 (2002). doi:10.1051/0004-6361:20020839

R.C. Duncan, C. Thompson, Formation of very strongly magnetized neutron stars—Implications for gamma-ray bursts. Astrophys. J. Lett. **392**, 9–13 (1992). doi:10.1086/186413

D. Eichler, Magnetic confinement of jets. Astrophys. J. **419**, 111 (1993). doi:10.1086/173464

D. Eichler, J. Granot, The case for anisotropic afterglow efficiency within gamma-ray burst jets. Astrophys. J. Lett. **641**, 5–8 (2006). doi:10.1086/503667

D. Eichler, E. Waxman, The efficiency of electron acceleration in collisionless shocks and gamma-ray burst energetics. Astrophys. J. **627**, 861–867 (2005). doi:10.1086/430596

D. Eichler, M. Livio, T. Piran, D.N. Schramm, Nucleosynthesis, neutrino bursts and gamma-rays from coalescing neutron stars. Nature **340**, 126–128 (1989). doi:10.1038/340126a0

Y.-Z. Fan, The spectrum of γ-ray burst: A clue. Mon. Not. R. Astron. Soc. **403**, 483–490 (2010). doi:10.1111/j.1365-2966.2009.16134.x

Y.-Z. Fan, D. Xu, The X-ray afterglow flat segment in short GRB 051221A: Energy injection from a millisecond magnetar? Mon. Not. R. Astron. Soc. **372**, 19–22 (2006). doi:10.1111/j.1745-3933.2006.00217.x

Y.-Z. Fan et al., A supramassive magnetar central engine for GRB 130603B. Astrophys. J. Lett. **779**, 25 (2013a). doi:10.1088/2041-8205/779/2/L25

Y.-Z. Fan et al., High-energy emission of GRB 130427A: Evidence for inverse Compton radiation. Astrophys. J. **776**, 95 (2013b). doi:10.1088/0004-637X/776/2/95

D.A. Frail et al., The enigmatic radio afterglow of GRB 991216. Astrophys. J. Lett. **538**, 129–132 (2000). doi:10.1086/312807

H. Gao, B.-B. Zhang, B. Zhang, Stepwise filter correlation method and evidence of superposed variability components in gamma-ray burst prompt emission light curves. Astrophys. J. **748**, 134 (2012). doi:10.1088/0004-637X/748/2/134

B. Gendre et al., Testing gamma-ray burst models with the afterglow of GRB 090102. Mon. Not. R. Astron. Soc. **405**, 2372–2380 (2010). doi:10.1111/j.1365-2966.2010.16601.x

F. Genet, J. Granot, Realistic analytic model for the prompt and high-latitude emission in GRBs. Mon. Not. R. Astron. Soc. **399**, 1328–1346 (2009). doi:10.1111/j.1365-2966.2009.15355.x

F. Genet, F. Daigne, R. Mochkovitch, Can the early X-ray afterglow of gamma-ray bursts be explained by a contribution from the reverse shock? Mon. Not. R. Astron. Soc. **381**, 732–740 (2007). doi:10.1111/j.1365-2966.2007.12243.x

G. Ghisellini, Jetted active galactic nuclei. Int. J. Mod. Phys. Conf. Ser. **8**, 1–12 (2012). doi:10.1142/S2010194512004345

G. Ghisellini, A. Celotti, Quasi-thermal comptonization and gamma-ray bursts. Astrophys. J. Lett. **511**, 93–96 (1999). doi:10.1086/311845

G. Ghisellini, D. Lazzati, Polarization light curves and position angle variation of beamed gamma-ray bursts. Mon. Not. R. Astron. Soc. **309**, 7–11 (1999). doi:10.1046/j.1365-8711.1999.03025.x

G. Ghisellini, A. Celotti, D. Lazzati, Constraints on the emission mechanisms of gamma-ray bursts. Mon. Not. R. Astron. Soc. **313**, 1–5 (2000). doi:10.1046/j.1365-8711.2000.03354.x

D. Giannios, Prompt GRB emission from gradual energy dissipation. Astron. Astrophys. **480**, 305–312 (2008). doi:10.1051/0004-6361:20079085

D. Giannios, H.C. Spruit, The role of kink instability in Poynting-flux dominated jets. Astron. Astrophys. **450**, 887–898 (2006). doi:10.1051/0004-6361:20054107

D. Giannios, H.C. Spruit, Spectral and timing properties of a dissipative γ-ray burst photosphere. Astron. Astrophys. **469**, 1–9 (2007). doi:10.1051/0004-6361:20066739

P. Goldreich, W.H. Julian, Stellar winds. Astrophys. J. **160**, 971 (1970). doi:10.1086/150486

B.P. Gompertz, P.T. O'Brien, G.A. Wynn, Magnetar powered GRBs: Explaining the extended emission and X-ray plateau of short GRB light curves. Mon. Not. R. Astron. Soc. **438**, 240–250 (2014). doi:10.1093/mnras/stt2165

J. Goodman, Are gamma-ray bursts optically thick? Astrophys. J. Lett. **308**, 47–50 (1986). doi:10.1086/184741

D. Götz, P. Laurent, F. Lebrun, F. Daigne, Ž. Bošnjak, Variable polarization measured in the prompt emission of GRB 041219A using IBIS on board INTEGRAL. Astrophys. J. Lett. **695**, 208–212 (2009). doi:10.1088/0004-637X/695/2/L208

D. Götz et al., GRB 140206A: The most distant polarized gamma-ray burst. Mon. Not. R. Astron. Soc. **444**, 2776–2782 (2014). doi:10.1093/mnras/stu1634

J. Granot, The most probable cause for the high gamma-ray polarization in GRB 021206. Astrophys. J. Lett. **596**, 17–21 (2003). doi:10.1086/379110

J. Granot, Interaction of a highly magnetized impulsive relativistic flow with an external medium. Mon. Not. R. Astron. Soc. **421**, 2442–2466 (2012a). doi:10.1111/j.1365-2966.2012.20473.x

J. Granot, The effects of sub-shells in highly magnetized relativistic flows. Mon. Not. R. Astron. Soc. **421**, 2467–2477 (2012b). doi:10.1111/j.1365-2966.2012.20474.x

J. Granot, A. Königl, Linear polarization in gamma-ray bursts: The case for an ordered magnetic field. Astrophys. J. Lett. **594**, 83–87 (2003). doi:10.1086/378733

J. Granot, P. Kumar, Distribution of gamma-ray burst ejecta energy with Lorentz factor. Mon. Not. R. Astron. Soc. **366**, 13–16 (2006). doi:10.1111/j.1745-3933.2005.00121.x

J. Granot, R. Sari, The shape of spectral breaks in gamma-ray burst afterglows. Astrophys. J. **568**, 820–829 (2002). doi:10.1086/338966

J. Granot, G.B. Taylor, Radio flares and the magnetic field structure in gamma-ray burst outflows. Astrophys. J. **625**, 263–270 (2005). doi:10.1086/429536

J. Granot, A. Königl, T. Piran, Implications of the early X-ray afterglow light curves of Swift gamma-ray bursts. Mon. Not. R. Astron. Soc. **370**, 1946–1960 (2006). doi:10.1111/j.1365-2966.2006.10621.x

J. Granot, J. Cohen-Tanugi, E. do Couto e Silva, Opacity buildup in impulsive relativistic sources. Astrophys. J. **677**, 92–126 (2008). doi:10.1086/526414

J. Granot, S.S. Komissarov, A. Spitkovsky, Impulsive acceleration of strongly magnetized relativistic flows. Mon. Not. R. Astron. Soc. **411**, 1323–1353 (2011). doi:10.1111/j.1365-2966.2010.17770.x

D. Gruber et al., The Fermi GBM gamma-ray burst spectral catalog: Four years of data. Astrophys. J. Suppl. Ser. **211**, 12 (2014). doi:10.1088/0067-0049/211/1/12

A. Gruzinov, E. Waxman, Gamma-ray burst afterglow: Polarization and analytic light curves. Astrophys. J. **511**, 852–861 (1999). doi:10.1086/306720

D. Guetta, J. Granot, High-energy emission from the prompt gamma-ray burst. Astrophys. J. **585**, 885–889 (2003). doi:10.1086/346221

C. Guidorzi et al., Average power density spectrum of Swift long gamma-ray bursts in the observer and in the source-rest frames. Mon. Not. R. Astron. Soc. **422**, 1785–1803 (2012). doi:10.1111/j.1365-2966.2012.20758.x

S. Guiriec et al., Time-resolved spectroscopy of the three brightest and hardest short gamma-ray bursts observed with the Fermi gamma-ray burst monitor. Astrophys. J. **725**, 225–241 (2010). doi:10.1088/0004-637X/725/1/225

S. Guiriec et al., Detection of a thermal spectral component in the prompt emission of GRB 100724B. Astrophys. J. Lett. **727**, 33 (2011). doi:10.1088/2041-8205/727/2/L33

S. Guiriec et al., Evidence for a photospheric component in the prompt emission of the short GRB 120323A and its effects on the GRB hardness-luminosity relation. Astrophys. J. **770**, 32 (2013). doi:10.1088/0004-637X/770/1/32

S. Guiriec et al., Towards a better understanding of the GRB phenomenon: A new model for GRB prompt emission and its effects on the new non-thermal $L_{\mathrm{i}}^{\mathrm{NT}}$–$E_{\mathrm{peak,i}}^{\mathrm{rest,NT}}$ relation. ArXiv e-prints (2015)

F.A. Harrison et al., Broadband observations of the afterglow of GRB 000926: Observing the effect of inverse Compton scattering. Astrophys. J. **559**, 123–130 (2001). doi:10.1086/322368

R. Hascoët, F. Daigne, R. Mochkovitch, Accounting for the XRT early steep decay in models of the prompt gamma-ray burst emission. Astron. Astrophys. **542**, 29 (2012a). doi:10.1051/0004-6361/201219339

R. Hascoët, F. Daigne, R. Mochkovitch, V. Vennin, Do Fermi large area telescope observations imply very large Lorentz factors in gamma-ray burst outflows? Mon. Not. R. Astron. Soc. **421**, 525–545 (2012b). doi:10.1111/j.1365-2966.2011.20332.x

R. Hascoët, F. Daigne, R. Mochkovitch, Prompt thermal emission in gamma-ray bursts. Astron. Astrophys. **551**, 124 (2013). doi:10.1051/0004-6361/201220023

R. Hascoët, F. Daigne, R. Mochkovitch, The prompt-early afterglow connection in gamma-ray bursts: Implications for the early afterglow physics. Mon. Not. R. Astron. Soc. **442**, 20–27 (2014). doi:10.1093/mnras/stu750

R. Hascoet, A.M. Beloborodov, F. Daigne, R. Mochkovitch, X-ray flares from dense shells formed in gamma-ray burst explosions. ArXiv e-prints (2015)

S. Heinz, M.C. Begelman, Jet acceleration by tangled magnetic fields. Astrophys. J. **535**, 104–117 (2000). doi:10.1086/308820

D. Kagan, L. Sironi, B. Cerutti, D. Giannios, Relativistic magnetic reconnection in pair plasmas and its astrophysical applications. Space Sci. Rev. (2015). doi:10.1007/s11214-014-0132-9

Y. Kaneko, R.D. Preece, M.S. Briggs, W.S. Paciesas, C.A. Meegan, D.L. Band, The complete spectral catalog of bright BATSE gamma-ray bursts. Astrophys. J. Suppl. Ser. **166**, 298–340 (2006). doi:10.1086/505911

N. Kawanaka, T. Piran, J.H. Krolik, Jet luminosity from neutrino-dominated accretion flows in gamma-ray bursts. Astrophys. J. **766**, 31 (2013). doi:10.1088/0004-637X/766/1/31

U. Keshet, B. Katz, A. Spitkovsky, E. Waxman, Magnetic field evolution in relativistic unmagnetized collisionless shocks. Astrophys. J. Lett. **693**, 127–130 (2009). doi:10.1088/0004-637X/693/2/L127

W. Kluźniak, M. Ruderman, The central engine of gamma-ray bursters. Astrophys. J. Lett. **505**, 113–117 (1998). doi:10.1086/311622

S. Kobayashi, T. Piran, R. Sari, Can internal shocks produce the variability in gamma-ray bursts? Astrophys. J. **490**, 92 (1997). doi:10.1086/512791

S.S. Komissarov, Magnetic acceleration of relativistic jets. Mem. Soc. Astron. Ital. **82**, 95 (2011)

S.S. Komissarov, Shock dissipation in magnetically dominated impulsive flows. Mon. Not. R. Astron. Soc. **422**, 326–346 (2012). doi:10.1111/j.1365-2966.2012.20609.x

S.S. Komissarov, N. Vlahakis, A. Königl, M.V. Barkov, Magnetic acceleration of ultrarelativistic jets in gamma-ray burst sources. Mon. Not. R. Astron. Soc. **394**, 1182–1212 (2009). doi:10.1111/j.1365-2966.2009.14410.x

S.S. Komissarov, N. Vlahakis, A. Königl, Rarefaction acceleration of ultrarelativistic magnetized jets in gamma-ray burst sources. Mon. Not. R. Astron. Soc. **407**, 17–28 (2010). doi:10.1111/j.1365-2966.2010.16779.x

C. Kouveliotou, C.A. Meegan, G.J. Fishman, N.P. Bhat, M.S. Briggs, T.M. Koshut, W.S. Paciesas, G.N. Pendleton, Identification of two classes of gamma-ray bursts. Astrophys. J. Lett. **413**, 101–104 (1993). doi:10.1086/186969

C. Kouveliotou et al., NuSTAR observations of GRB 130427A establish a single component synchrotron afterglow origin for the late optical to multi-GeV emission. Astrophys. J. Lett. **779**, 1 (2013). doi:10.1088/2041-8205/779/1/L1

S.R. Kulkarni et al., Discovery of a radio flare from GRB 990123. Astrophys. J. Lett. **522**, 97–100 (1999). doi:10.1086/312227

P. Kumar, E. McMahon, A general scheme for modelling γ-ray burst prompt emission. Mon. Not. R. Astron. Soc. **384**, 33–63 (2008). doi:10.1111/j.1365-2966.2007.12621.x

P. Kumar, A. Panaitescu, Afterglow emission from naked gamma-ray bursts. Astrophys. J. Lett. **541**, 51–54 (2000). doi:10.1086/312905

P. Kumar, B. Zhang, The physics of gamma-ray bursts & relativistic jets. Phys. Rep. **561**, 1–109 (2015). doi:10.1016/j.physrep.2014.09.008

P. Kumar, R.A. Hernández, Ž. Bošnjak, R. Barniol Duran, Maximum synchrotron frequency for shock-accelerated particles. Mon. Not. R. Astron. Soc. **427**, 40–44 (2012). doi:10.1111/j.1745-3933.2012.01341.x

A. Lazar, E. Nakar, T. Piran, Gamma-ray burst light curves in the relativistic turbulence and relativistic subjet models. Astrophys. J. Lett. **695**, 10–14 (2009). doi:10.1088/0004-637X/695/1/L10

M. Lemoine, Synchrotron signature of a relativistic blast wave with decaying microturbulence. Mon. Not. R. Astron. Soc. **428**, 845–866 (2013). doi:10.1093/mnras/sts081

M. Lemoine, Z. Li, X.-Y. Wang, On the magnetization of gamma-ray burst blast waves. Mon. Not. R. Astron. Soc. **435**, 3009–3016 (2013). doi:10.1093/mnras/stt1494

A. Levinson, Interaction of a magnetized shell with an ambient medium: Limits on impulsive magnetic acceleration. Astrophys. J. **720**, 1490–1499 (2010). doi:10.1088/0004-637X/720/2/1490

A. Levinson, D. Eichler, Baryon purity in cosmological gamma-ray bursts as a manifestation of event horizons. Astrophys. J. **418**, 386 (1993). doi:10.1086/173397

A. Levinson, N. Globus, Ultra-relativistic, neutrino-driven flows in gamma-ray bursts: A double transonic flow solution in a Schwarzschild spacetime. Astrophys. J. **770**, 159 (2013). doi:10.1088/0004-637X/770/2/159

E.W. Liang et al., Testing the curvature effect and internal origin of gamma-ray burst prompt emissions and X-ray flares with swift data. Astrophys. J. **646**, 351–357 (2006). doi:10.1086/504684

E. Liang, B. Zhang, F. Virgili, Z.G. Dai, Low-luminosity gamma-ray bursts as a unique population: Luminosity function, local rate, and beaming factor. Astrophys. J. **662**, 1111–1118 (2007). doi:10.1086/517959

C.C. Lindner, M. Milosavljević, S.M. Couch, P. Kumar, Collapsar accretion and the gamma-ray burst X-ray light curve. Astrophys. J. **713**, 800–815 (2010). doi:10.1088/0004-637X/713/2/800

Y. Lithwick, R. Sari, Lower limits on Lorentz factors in gamma-ray bursts. Astrophys. J. **555**, 540–545 (2001). doi:10.1086/321455

R.-Y. Liu, X.-Y. Wang, X.-F. Wu, Interpretation of the unprecedentedly long-lived high-energy emission of GRB 130427A. Astrophys. J. Lett. **773**, 20 (2013). doi:10.1088/2041-8205/773/2/L20

A. Loeb, R. Perna, Microlensing of gamma-ray burst afterglows. Astrophys. J. **495**, 597–603 (1998). doi:10.1086/305337

R.-J. Lu et al., A comprehensive analysis of Fermi gamma-ray burst data. II. E $_p$ evolution patterns and implications for the observed spectrum-luminosity relations. Astrophys. J. **756**, 112 (2012). doi:10.1088/0004-637X/756/2/112

C. Lundman, A. Pe'er, F. Ryde, A theory of photospheric emission from relativistic, collimated outflows. Mon. Not. R. Astron. Soc. **428**, 2430–2442 (2013). doi:10.1093/mnras/sts219

Y.E. Lyubarskii, Kink instability of relativistic force-free jets. Mon. Not. R. Astron. Soc. **308**, 1006–1010 (1999). doi:10.1046/j.1365-8711.1999.02763.x

Y.E. Lyubarskij, Energy release in strongly magnetized relativistic winds. Sov. Astron. Lett. **18**, 356 (1992)

Y.E. Lyubarsky, The termination shock in a striped pulsar wind. Mon. Not. R. Astron. Soc. **345**, 153–160 (2003). doi:10.1046/j.1365-8711.2003.06927.x

Y. Lyubarsky, Asymptotic structure of Poynting-dominated jets. Astrophys. J. **698**, 1570–1589 (2009). doi:10.1088/0004-637X/698/2/1570

Y.E. Lyubarsky, A new mechanism for dissipation of alternating fields in Poynting-dominated outflows. Astrophys. J. Lett. **725**, 234–238 (2010a). doi:10.1088/2041-8205/725/2/L234

Y.E. Lyubarsky, Transformation of the Poynting flux into kinetic energy in relativistic jets. Mon. Not. R. Astron. Soc. **402**, 353–361 (2010b). doi:10.1111/j.1365-2966.2009.15877.x

Y. Lyubarsky, J.G. Kirk, Reconnection in a striped pulsar wind. Astrophys. J. **547**, 437–448 (2001). doi:10.1086/318354

M. Lyutikov, The electromagnetic model of gamma-ray bursts. New J. Phys. **8**, 119 (2006). doi:10.1088/1367-2630/8/7/119

M. Lyutikov, A high-sigma model of pulsar wind nebulae. Mon. Not. R. Astron. Soc. **405**, 1809–1815 (2010). doi:10.1111/j.1365-2966.2010.16553.x

M. Lyutikov, Dynamics of strongly magnetized ejecta in gamma-ray bursts. Mon. Not. R. Astron. Soc. **411**, 422–426 (2011). doi:10.1111/j.1365-2966.2010.17696.x

M. Lyutikov, R. Blandford, Gamma ray bursts as electromagnetic outflows. ArXiv Astrophysics e-prints (2003)

M. Lyutikov, V.I. Pariev, R.D. Blandford, Polarization of prompt gamma-ray burst emission: Evidence for electromagnetically dominated outflow. Astrophys. J. **597**, 998–1009 (2003). doi:10.1086/378497

A.I. MacFadyen, S.E. Woosley, Collapsars: Gamma-ray bursts and explosions in "Failed Supernovae". Astrophys. J. **524**, 262–289 (1999). doi:10.1086/307790

B. Margalit, B.D. Metzger, A.M. Beloborodov, Does the collapse of a supramassive neutron star leave a debris disk? ArXiv e-prints (2015)

A. Maselli et al., GRB 130427A: A nearby ordinary monster. Science **343**, 48–51 (2014). doi:10.1126/science.1242279

C.D. Matzner, Supernova hosts for gamma-ray burst jets: Dynamical constraints. Mon. Not. R. Astron. Soc. **345**, 575–589 (2003). doi:10.1046/j.1365-8711.2003.06969.x

S. McGlynn et al., Polarisation studies of the prompt gamma-ray emission from GRB 041219a using the spectrometer aboard INTEGRAL. Astron. Astrophys. **466**, 895–904 (2007). doi:10.1051/0004-6361:20066179

J.C. McKinney, D.A. Uzdensky, A reconnection switch to trigger gamma-ray burst jet dissipation. Mon. Not. R. Astron. Soc. **419**, 573–607 (2012). doi:10.1111/j.1365-2966.2011.19721.x

M.V. Medvedev, A. Loeb, Generation of magnetic fields in the relativistic shock of gamma-ray burst sources. Astrophys. J. **526**, 697–706 (1999). doi:10.1086/308038

P. Mészáros, M.J. Rees, Steep slopes and preferred breaks in gamma-ray burst spectra: The role of photospheres and comptonization. Astrophys. J. **530**, 292–298 (2000). doi:10.1086/308371

P. Meszaros, P. Laguna, M.J. Rees, Gasdynamics of relativistically expanding gamma-ray burst sources—Kinematics, energetics, magnetic fields, and efficiency. Astrophys. J. **415**, 181–190 (1993). doi:10.1086/173154

P. Mészáros, E. Ramirez-Ruiz, M.J. Rees, B. Zhang, X-ray-rich gamma-ray bursts, photospheres, and variability. Astrophys. J. **578**, 812–817 (2002). doi:10.1086/342611

B.D. Metzger, E. Quataert, T.A. Thompson, Short-duration gamma-ray bursts with extended emission from protomagnetar spin-down. Mon. Not. R. Astron. Soc. **385**, 1455–1460 (2008). doi:10.1111/j.1365-2966.2008.12923.x

B.D. Metzger, D. Giannios, T.A. Thompson, N. Bucciantini, E. Quataert, The protomagnetar model for gamma-ray bursts. Mon. Not. R. Astron. Soc. **413**, 2031–2056 (2011). doi:10.1111/j.1365-2966.2011.18280.x

P. Mimica, M.A. Aloy, On the dynamic efficiency of internal shocks in magnetized relativistic outflows. Mon. Not. R. Astron. Soc. **401**, 525–532 (2010). doi:10.1111/j.1365-2966.2009.15669.x

E. Molinari et al., REM observations of GRB 060418 and GRB 060607A: The onset of the afterglow and the initial fireball Lorentz factor determination. Astron. Astrophys. **469**, 13–16 (2007). doi:10.1051/0004-6361:20077388

C.G. Mundell et al., Early optical polarization of a gamma-ray burst afterglow. Science **315**, 1822 (2007a). doi:10.1126/science.1138484

C.G. Mundell et al., The remarkable afterglow of GRB 061007: Implications for optical flashes and GRB fireballs. Astrophys. J. **660**, 489–495 (2007b). doi:10.1086/512605

C.G. Mundell et al., Highly polarized light from stable ordered magnetic fields in GRB 120308A. Nature **504**, 119–121 (2013). doi:10.1038/nature12814

E. Nakar, Y. Oren, Polarization and light-curve variability: The "Patchy-Shell" model. Astrophys. J. Lett. **602**, 97–100 (2004). doi:10.1086/382729

E. Nakar, R. Sari, Relativistic shock breakouts—A variety of gamma-ray flares: From low-luminosity gamma-ray bursts to type Ia supernovae. Astrophys. J. **747**, 88 (2012). doi:10.1088/0004-637X/747/2/88

E. Nakar, T. Piran, E. Waxman, Implications of the bold gamma-ray polarization of GRB 021206. J. Cosmol. Astropart. Phys. **10**, 5 (2003). doi:10.1088/1475-7516/2003/10/005

E. Nakar, S. Ando, R. Sari, Klein-Nishina effects on optically thin synchrotron and synchrotron self-Compton spectrum. Astrophys. J. **703**, 675–691 (2009). doi:10.1088/0004-637X/703/1/675

R. Narayan, P. Kumar, A. Tchekhovskoy, Constraints on cold magnetized shocks in gamma-ray bursts. Mon. Not. R. Astron. Soc. **416**, 2193–2201 (2011). doi:10.1111/j.1365-2966.2011.19197.x

L. Nava, G. Ghirlanda, G. Ghisellini, A. Celotti, Spectral properties of 438 GRBs detected by Fermi/GBM. Astron. Astrophys. **530**, 21 (2011). doi:10.1051/0004-6361/201016270

L. Nava, E. Nakar, T. Piran, Linear and circular polarization in GRB afterglows. ArXiv e-prints (2015)

J.A. Nousek, C. Kouveliotou, D. Grupe, K.L. Page, J. Granot, E. Ramirez-Ruiz et al., Evidence for a canonical gamma-ray burst afterglow light curve in the swift XRT data. Astrophys. J. **642**, 389–400 (2006). doi:10.1086/500724

P.T. O'Brien et al., The early X-ray emission from GRBs. Astrophys. J. **647**, 1213–1237 (2006). doi:10.1086/505457

B. Paczynski, Gamma-ray bursters at cosmological distances. Astrophys. J. Lett. **308**, 43–46 (1986). doi:10.1086/184740

A. Pe'er, Temporal evolution of thermal emission from relativistically expanding plasma. Astrophys. J. **682**, 463–473 (2008). doi:10.1086/588136

A. Pe'er, P. Mészáros, M.J. Rees, The observable effects of a photospheric component on GRB and XRF prompt emission spectrum. Astrophys. J. **642**, 995–1003 (2006). doi:10.1086/501424

F. Peng, A. Königl, J. Granot, Two-component jet models of gamma-ray burst sources. Astrophys. J. **626**, 966–977 (2005). doi:10.1086/430045

D.A. Perley et al., The afterglow of GRB 130427A from 1 to 10^{16} GHz. Astrophys. J. **781**, 37 (2014). doi:10.1088/0004-637X/781/1/37

T. Piran, Gamma-ray bursts and the fireball model. Phys. Rep. **314**, 575–667 (1999). doi:10.1016/S0370-1573(98)00127-6

T. Piran, The physics of gamma-ray bursts. Rev. Mod. Phys. **76**, 1143–1210 (2004). doi:10.1103/RevModPhys.76.1143

T. Piran, Magnetic fields in gamma-ray bursts: A short overview, in *Magnetic Fields in the Universe: From Laboratory and Stars to Primordial Structures*, ed. by E.M. de Gouveia dal Pino, G. Lugones, A. Lazarian. American Institute of Physics Conference Series, vol. 784 (2005), pp. 164–174. doi:10.1063/1.2077181

T. Piran, E. Nakar, On the external shock synchrotron model for gamma-ray bursts' GeV emission. Astrophys. J. Lett. **718**, 63–67 (2010). doi:10.1088/2041-8205/718/2/L63

T. Piran, R. Sari, Y.-C. Zou, Observational limits on inverse Compton processes in gamma-ray bursts. Mon. Not. R. Astron. Soc. **393**, 1107–1113 (2009). doi:10.1111/j.1365-2966.2008.14198.x

R.D. Preece, M.S. Briggs, R.S. Mallozzi, G.N. Pendleton, W.S. Paciesas, D.L. Band, The synchrotron shock model confronts a "Line of Death" in the BATSE gamma-ray burst data. Astrophys. J. Lett. **506**, 23–26 (1998). doi:10.1086/311644

R.D. Preece, M.S. Briggs, R.S. Mallozzi, G.N. Pendleton, W.S. Paciesas, D.L. Band, The BATSE gamma-ray burst spectral catalog. I. High time resolution spectroscopy of bright bursts using high energy resolution data. Astrophys. J. Suppl. Ser. **126**, 19–36 (2000). doi:10.1086/313289

R. Preece et al., The first pulse of the extremely bright GRB 130427A: A test lab for synchrotron shocks. Science **343**, 51–54 (2014). doi:10.1126/science.1242302

W.H. Press, S.A. Teukolsky, W.T. Vetterling, B.P. Flannery, *Numerical Recipes in C. The Art of Scientific Computing* (1992)

J.L. Racusin, S.V. Karpov, M. Sokolowski, J. Granot et al., Broadband observations of the naked-eye γ-ray burst GRB080319B. Nature **455**, 183–188 (2008). doi:10.1038/nature07270

M.J. Rees, P. Meszaros, Unsteady outflow models for cosmological gamma-ray bursts. Astrophys. J. Lett. **430**, 93–96 (1994). doi:10.1086/187446

M.J. Rees, P. Mészáros, Dissipative photosphere models of gamma-ray bursts and X-ray flashes. Astrophys. J. **628**, 847–852 (2005). doi:10.1086/430818

L. Rezzolla, P. Kumar, A novel paradigm for short gamma-ray bursts with extended X-ray emission. Astrophys. J. **802**, 95 (2015). doi:10.1088/0004-637X/802/2/95

S. Rosswog, Last moments in the life of a compact binary system: Gravitational waves, gamma-ray bursts and magnetar formation. Rev. Mex. Astron. Astrofís. **27**, 57–79 (2007)

A. Rowlinson et al., The unusual X-ray emission of the short Swift GRB 090515: Evidence for the formation of a magnetar? Mon. Not. R. Astron. Soc. **409**, 531–540 (2010). doi:10.1111/j.1365-2966.2010.17354.x

A. Rowlinson, P.T. O'Brien, B.D. Metzger, N.R. Tanvir, A.J. Levan, Signatures of magnetar central engines in short GRB light curves. Mon. Not. R. Astron. Soc. **430**, 1061–1087 (2013). doi:10.1093/mnras/sts683

A. Rowlinson, B.P. Gompertz, M. Dainotti, P.T. O'Brien, R.A.M.J. Wijers, A.J. van der Horst, Constraining properties of GRB magnetar central engines using the observed plateau luminosity and duration correlation. Mon. Not. R. Astron. Soc. **443**, 1779–1787 (2014). doi:10.1093/mnras/stu1277

R.E. Rutledge, D.B. Fox, Re-analysis of polarization in the γ-ray flux of GRB 021206. Mon. Not. R. Astron. Soc. **350**, 1288–1300 (2004). doi:10.1111/j.1365-2966.2004.07665.x

F. Ryde et al., Identification and properties of the photospheric emission in GRB090902B. Astrophys. J. Lett. **709**, 172–177 (2010). doi:10.1088/2041-8205/709/2/L172

F. Ryde et al., Observational evidence of dissipative photospheres in gamma-ray bursts. Mon. Not. R. Astron. Soc. **415**, 3693–3705 (2011). doi:10.1111/j.1365-2966.2011.18985.x

T. Sakamoto et al., Global characteristics of X-ray flashes and X-ray-rich gamma-ray bursts observed by HETE-2. Astrophys. J. **629**, 311–327 (2005). doi:10.1086/431235

R. Sari, Linear polarization and proper motion in the afterglow of beamed gamma-ray bursts. Astrophys. J. Lett. **524**, 43–46 (1999). doi:10.1086/312294

R. Sari, A.A. Esin, On the synchrotron self-Compton emission from relativistic shocks and its implications for gamma-ray burst afterglows. Astrophys. J. **548**, 787–799 (2001). doi:10.1086/319003

R. Sari, T. Piran, Hydrodynamic timescales and temporal structure of gamma-ray bursts. Astrophys. J. Lett. **455**, 143 (1995). doi:10.1086/309835

R. Sari, T. Piran, Variability in gamma-ray bursts: A clue. Astrophys. J. **485**, 270 (1997). doi:10.1086/304428

R. Sari, T. Piran, GRB 990123: The optical flash and the fireball model. Astrophys. J. Lett. **517**, 109–112 (1999). doi:10.1086/312039

R. Sari, T. Piran, R. Narayan, Spectra and light curves of gamma-ray burst afterglows. Astrophys. J. Lett. **497**, 17–20 (1998). doi:10.1086/311269

A. Shemi, T. Piran, The appearance of cosmic fireballs. Astrophys. J. Lett. **365**, 55–58 (1990). doi:10.1086/185887

L. Sironi, A. Spitkovsky, Relativistic reconnection: An efficient source of non-thermal particles. Astrophys. J. Lett. **783**, 21 (2014). doi:10.1088/2041-8205/783/1/L21

L. Sironi, U. Keshet, M. Lemoine, Relativistic shocks: Particle acceleration and magnetization. Space Sci. Rev. (2015). doi:10.1007/s11214-015-0102-2
A.M. Soderberg et al., Relativistic ejecta from X-ray flash XRF 060218 and the rate of cosmic explosions. Nature **442**, 1014–1017 (2006). doi:10.1038/nature05087
A. Spitkovsky, Time-dependent force-free pulsar magnetospheres: Axisymmetric and oblique rotators. Astrophys. J. Lett. **648**, 51–54 (2006). doi:10.1086/507518
H.C. Spruit, T. Foglizzo, R. Stehle, Collimation of magnetically driven jets from accretion discs. Mon. Not. R. Astron. Soc. **288**, 333–342 (1997)
H.C. Spruit, F. Daigne, G. Drenkhahn, Large scale magnetic fields and their dissipation in GRB fireballs. Astron. Astrophys. **369**, 694–705 (2001). doi:10.1051/0004-6361:20010131
I.A. Steele, C.G. Mundell, R.J. Smith, S. Kobayashi, C. Guidorzi, Ten per cent polarized optical emission from GRB090102. Nature **462**, 767–769 (2009). doi:10.1038/nature08590
P.-H.T. Tam, Q.-W. Tang, S.-J. Hou, R.-Y. Liu, X.-Y. Wang, Discovery of an extra hard spectral component in the high-energy afterglow emission of GRB 130427A. Astrophys. J. Lett. **771**, 13 (2013). doi:10.1088/2041-8205/771/1/L13
A. Tchekhovskoy, R. Narayan, J.C. McKinney, Magnetohydrodynamic simulations of gamma-ray burst jets: Beyond the progenitor star. New Astron. **15**, 749–754 (2010). doi:10.1016/j.newast.2010.03.001
C. Thompson, A model of gamma-ray bursts. Mon. Not. R. Astron. Soc. **270**, 480 (1994)
E. Troja et al., Swift observations of GRB 070110: An extraordinary X-ray afterglow powered by the central engine. Astrophys. J. **665**, 599–607 (2007). doi:10.1086/519450
Z.L. Uhm, A.M. Beloborodov, On the mechanism of gamma-ray burst afterglows. Astrophys. J. Lett. **665**, 93–96 (2007). doi:10.1086/519837
Z.L. Uhm, B. Zhang, Fast-cooling synchrotron radiation in a decaying magnetic field and γ-ray burst emission mechanism. Nat. Phys. **10**, 351–356 (2014). doi:10.1038/nphys2932
Z.L. Uhm, B. Zhang, R. Hascoët, F. Daigne, R. Mochkovitch, I.H. Park, Dynamics and afterglow light curves of gamma-ray burst blast waves with a long-lived reverse shock. Astrophys. J. **761**, 147 (2012). doi:10.1088/0004-637X/761/2/147
V.V. Usov, Millisecond pulsars with extremely strong magnetic fields as a cosmological source of gamma-ray bursts. Nature **357**, 472–474 (1992). doi:10.1038/357472a0
W.T. Vestrand et al., The bright optical flash and afterglow from the gamma-ray burst GRB 130427A. Science **343**, 38–41 (2014). doi:10.1126/science.1242316
M. Vietri, L. Stella, A gamma-ray burst model with small baryon contamination. Astrophys. J. Lett. **507**, 45–48 (1998). doi:10.1086/311674
M. Vietri, L. Stella, Supranova events from spun-up neutron stars: An explosion in search of an observation. Astrophys. J. Lett. **527**, 43–46 (1999). doi:10.1086/312386
F.J. Virgili, E.-W. Liang, B. Zhang, Low-luminosity gamma-ray bursts as a distinct GRB population: A firmer case from multiple criteria constraints. Mon. Not. R. Astron. Soc. **392**, 91–103 (2009). doi:10.1111/j.1365-2966.2008.14063.x
I. Vurm, A.M. Beloborodov, J. Poutanen, Gamma-ray bursts from magnetized collisionally heated jets. Astrophys. J. **738**, 77 (2011). doi:10.1088/0004-637X/738/1/77
I. Vurm, Y. Lyubarsky, T. Piran, On thermalization in gamma-ray burst jets and the peak energies of photospheric spectra. Astrophys. J. **764**, 143 (2013). doi:10.1088/0004-637X/764/2/143
X.-Y. Wang, Z. Li, Z.-G. Dai, P. Mészáros, GRB 080916C: On the radiation origin of the prompt emission from keV/MeV TO GeV. Astrophys. J. Lett. **698**, 98–102 (2009). doi:10.1088/0004-637X/698/2/L98
K. Wiersema et al., Circular polarization in the optical afterglow of GRB 121024A. Nature **509**, 201–204 (2014). doi:10.1038/nature13237
C. Wigger, W. Hajdas, K. Arzner, M. Güdel, A. Zehnder, Gamma-ray burst polarization: Limits from RHESSI measurements. Astrophys. J. **613**, 1088–1100 (2004). doi:10.1086/423163
R. Willingale, F. Genet, J. Granot, P.T. O'Brien, The spectral-temporal properties of the prompt pulses and rapid decay phase of gamma-ray bursts. Mon. Not. R. Astron. Soc. **403**, 1296–1316 (2010). doi:10.1111/j.1365-2966.2009.16187.x
S.E. Woosley, Gamma-ray bursts from stellar mass accretion disks around black holes. Astrophys. J. **405**, 273–277 (1993). doi:10.1086/172359
S.E. Woosley, J.S. Bloom, The supernova gamma-ray burst connection. Annu. Rev. Astron. Astrophys. **44**, 507–556 (2006). doi:10.1146/annurev.astro.43.072103.150558
D. Yonetoku et al., Detection of gamma-ray polarization in prompt emission of GRB 100826A. Astrophys. J. Lett. **743**, 30 (2011). doi:10.1088/2041-8205/743/2/L30
D. Yonetoku et al., Magnetic structures in gamma-ray burst jets probed by gamma-ray polarization. Astrophys. J. Lett. **758**, 1 (2012). doi:10.1088/2041-8205/758/1/L1
B. Zhang, H. Yan, The internal-collision-induced magnetic reconnection and turbulence (ICMART) model of gamma-ray bursts. Astrophys. J. **726**, 90 (2011). doi:10.1088/0004-637X/726/2/90

B. Zhang, B. Zhang, Gamma-ray burst prompt emission light curves and power density spectra in the ICMART model. Astrophys. J. **782**, 92 (2014). doi:10.1088/0004-637X/782/2/92

W. Zhang, S.E. Woosley, A.I. MacFadyen, Relativistic jets in collapsars. Astrophys. J. **586**, 356–371 (2003). doi:10.1086/367609

B. Zhang et al., Physical processes shaping gamma-ray burst X-ray afterglow light curves: Theoretical implications from the swift X-ray telescope observations. Astrophys. J. **642**, 354–370 (2006). doi:10.1086/500723

X. Zhao, Z. Li, X. Liu, B.-b. Zhang, J. Bai, P. Mészáros, Gamma-ray burst spectrum with decaying magnetic field. Astrophys. J. **780**, 12 (2014). doi:10.1088/0004-637X/780/1/12

DOI 10.1007/978-1-4939-3550-5_15
Reprinted from *Space Science Reviews* Journal, DOI 10.1007/s11214-015-0181-8

Relativistic Shocks: Particle Acceleration and Magnetization

L. Sironi[1] · U. Keshet[2] · M. Lemoine[3]

Received: 5 June 2015 / Accepted: 2 July 2015 / Published online: 2 September 2015

Abstract We review the physics of relativistic shocks, which are often invoked as the sources of non-thermal particles in pulsar wind nebulae (PWNe), gamma-ray bursts (GRBs), and active galactic nuclei (AGN) jets, and as possible sources of ultra-high energy cosmic-rays. We focus on particle acceleration and magnetic field generation, and describe the recent progress in the field driven by theory advances and by the rapid development of particle-in-cell (PIC) simulations. In weakly magnetized or quasi parallel-shocks (i.e. where the magnetic field is nearly aligned with the flow), particle acceleration is efficient. The accelerated particles stream ahead of the shock, where they generate strong magnetic waves which in turn scatter the particles back and forth across the shock, mediating their acceleration. In contrast, in strongly magnetized quasi-perpendicular shocks, the efficiencies of both particle acceleration and magnetic field generation are suppressed. Particle acceleration, when efficient, modifies the turbulence around the shock on a long time scale, and the accelerated particles have a characteristic energy spectral index of $s_\gamma \simeq 2.2$ in the ultra-relativistic limit. We discuss how this novel understanding of particle acceleration and magnetic field generation in relativistic shocks can be applied to high-energy astrophysical phenomena, with an emphasis on PWNe and GRB afterglows.

Keywords Acceleration of particles · Galaxies: active · Gamma rays: bursts · Magnetic fields · Pulsars: general · Radiation mechanisms: non-thermal · Relativistic processes

All authors contributed equally to this review.

✉ L. Sironi
lsironi@cfa.harvard.edu

U. Keshet
ukeshet@bgu.ac.il

M. Lemoine
lemoine@iap.fr

1 Harvard-Smithsonian Center for Astrophysics, Cambridge, MA 02138, USA

2 Physics Department, Ben-Gurion University of the Negev, Be'er-Sheva 84105, Israel

3 Institut d'Astrophysique de Paris, CNRS-UPMC, 98 bis boulevard Arago, 75014 Paris, France

1 Introduction

In pulsar wind nebulae (PWNe), gamma-ray bursts (GRBs), and jets from active galactic nuclei (AGNs), signatures of non-thermal processes are revealed by power-law radiation spectra spanning an extremely wide range of wavelengths, from radio to X-rays, and beyond. Yet, it is still a mystery how the emitting particles can be accelerated up to ultra-relativistic energies and how the strong magnetic fields are generated, as required in order to explain the observations. In most models, non-thermal particles and near-equipartition fields are thought to be produced at relativistic shock fronts, but the details of the mechanisms of particle acceleration and magnetic field generation are still not well understood.

Particle acceleration in shocks is usually attributed to the Fermi process, where particles are energized by bouncing back and forth across the shock. Despite its importance, the Fermi process is still not understood from first principles. The highly nonlinear coupling between accelerated particles and magnetic turbulence—which is generated by the particles, and at the same time governs their acceleration—is extremely hard to incorporate in analytic models. Only in recent years, thanks to major breakthroughs on analytical and numerical grounds, has our understanding of the Fermi process in relativistic shocks significantly advanced. This is the subject of the present review.

Relativistic shocks pose some unique challenges with respect to their non-relativistic counterparts. For example, the distribution of accelerated particles can no longer be approximated as isotropic if the shock is relativistic. In a relativistic shock, the electric and magnetic fields significantly mix as one switches between upstream and downstream frames of reference. And unlike non-relativistic shocks, where some aspects of the theory can be tested by direct spacecraft measurements, relativistic shocks are only constrained by remote observations. For recent reviews of relativistic shocks, see Bykov and Treumann (2011), Bykov et al. (2012).

This chapter is organized as follows. First, we review recent analytical advances on the theory of particle acceleration in relativistic shocks, arguing that the accelerated particle spectrum and its power-law slope in the ultra-relativistic limit, $s_\gamma \equiv -d\log N/d\log\gamma \simeq 2.2$ (where γ is the particle Lorentz factor), are fairly robust (Sect. 2). Here, we assume *a priori* that some magnetic turbulence exists on both sides of the shock, such that the Fermi process can operate. Next, we describe the plasma instabilities that are thought to be most relevant for generating this turbulence (Sect. 3), stressing the parameter regime where the so-called Weibel (or "filamentation") instability—which is often thought to mediate the Fermi process in weakly magnetized relativistic shocks—can grow. Then, we summarize recent findings from particle-in-cell (PIC) simulations of relativistic shocks, where the non-linear coupling between particles and magnetic waves can be captured from first principles (Sect. 4). Finally, we describe the astrophysical implications of these results for the acceleration of ultra high energy cosmic rays (UHECRs) and for the radiative signatures of PWNe and GRB afterglows (Sect. 5; for a review of PWNe, see Kargaltsev et al. 2015 in the present volume; for a review of GRBs, see Granot et al. 2015 in the present volume). We briefly conclude in Sect. 6.

2 Particle Acceleration in Relativistic Shocks

Diffusive (Fermi) acceleration of charged particles in collisionless shocks is believed to be responsible for the production of non-thermal distributions of energetic particles in many astronomical systems (Blandford and Eichler 1987; Malkov and Drury 2001, but see, e.g.

Arons and Tavani 1994 for a discussion of alternative shock acceleration processes). The Fermi acceleration process in shocks is still not understood from first principles: particle scattering in collisionless shocks is due to electromagnetic waves formed around the shock, but no present analytical formalism self-consistently calculates the generation of these waves, the scattering and acceleration of particles, and the backreaction of these particles on the waves and on the shock itself.

The theory of particle acceleration was first developed mainly by evolving the particle distribution under some Ansatz for the scattering mechanism (e.g. diffusion in pitch angle), within the "test particle" approximation, where modifications of wave and shock properties due to the high energy particles are neglected. This phenomenological approach proved successful in explaining the spectrum of relativistic particle distributions inferred from observations, although a more careful approach is needed to account for the energy fraction deposited in each particle species (electrons, positrons, protons, and possibly heavier ions), and to test the Ansatz of the scattering prescription.

For *non-relativistic* shocks, the linear theory of diffusive particle acceleration, first developed in 1977 (Krymskii 1977; Axford et al. 1978; Bell 1978; Blandford and Ostriker 1978), yields a power-law distribution $d^3N/d^3p \propto p^{-s_p}$ of particle momenta p, with a spectral index

$$s_p = s_\gamma + 2 = 3\beta_u/(\beta_u - \beta_d). \tag{1}$$

Here, β is the fluid velocity normalized to the speed of light c in the frame of the shock, which is assumed planar and infinite, and subscripts u (d) denote the upstream (downstream) plasma. For strong shocks in an ideal gas of adiabatic index $\Gamma = 5/3$, this implies $s_p = 4$ (i.e. $s_\gamma = 2$; constant energy per logarithmic energy interval, since $p^2 d^3N/d^3p \propto p^{-2}$), in agreement with observations.

The lack of a characteristic momentum scale, under the above assumptions, implies that the spectrum remains a power-law in the relativistic case, as verified numerically (e.g., Bednarz and Ostrowski 1998; Achterberg et al. 2001). The particle drift downstream of the shock implies that more particles are moving downstream than upstream; this anisotropy is of order of β_u when measured in the downstream frame (Keshet and Waxman 2005). Thus, while particle anisotropy is negligible for non-relativistic shocks, the distribution becomes highly anisotropic in the relativistic case, even when measured in the more isotropic downstream frame. Consequently, one must simultaneously determine the spectrum and the angular distribution of the particles, which is the main difficulty underlying the analysis of test particle acceleration when the shock is relativistic.

Observations of GRB afterglows led to the conclusion that highly relativistic collisionless shocks produce a power-law distribution of high energy particles with $s_p = 4.2 \pm 0.2$ (Waxman 1997; Freedman and Waxman 2001; Berger et al. 2003). This triggered a numerical investigation of particle acceleration in such shocks, showing that s_p indeed approaches the value of 4.2 for large shock Lorentz factors ($\gamma_u \equiv (1-\beta_u^2)^{-1/2} \gg 1$), in agreement with GRB observations, provided that particle scattering is sufficiently isotropic.

The spectral index s_p was calculated under the test particle approximation for a wide range of shock velocities, various equations of state, and different scattering prescriptions. This was achieved by approximately matching numerical eigenfunctions of the transport equation between upstream and downstream (Kirk and Schneider 1987; Heavens and Drury 1988; Kirk et al. 2000), by Monte Carlo simulations (Bednarz and Ostrowski 1998; Achterberg et al. 2001; Ellison and Double 2002; Lemoine and Pelletier 2003; Niemiec and Ostrowski 2004; Lemoine and Revenu 2006; Ellison et al. 2013), by expanding the distribution parallel to the shock front (Keshet and Waxman 2005), and by solving for moments of the angular distribution (Keshet 2006).

These studies have assumed rest frame diffusion in pitch angle or in the angle between particle velocity and shock normal. These two assumptions yield similar spectra in the limit of ultra-relativistic shocks (Ostrowski and Bednarz 2002). As discussed later in this review, one expects these assumptions to hold at relativistic shocks. However, some scenarios involve the conversion of the accelerated species into a neutral state and then back—e.g. proton to neutron and then back to proton via photo-hadronic interactions (Derishev et al. 2003) or electron to photon and then back to electron through Compton and pair production interactions (Stern and Poutanen 2008)—in which case the particle may have time to suffer a large angle deflection upstream of the shock, leading to large energy gains and generically hard spectra (Bednarz and Ostrowski 1998; Meli and Quenby 2003; Blasi and Vietri 2005).

For isotropic, small-angle scattering in the fluid frame, expanding the particle distribution about the shock grazing angle (Keshet and Waxman 2005) leads to a generalization of the non-relativistic Eq. (1) that reads

$$s_p = \left(3\beta_u - 2\beta_u\beta_d^2 + \beta_d^3\right)/(\beta_u - \beta_d), \tag{2}$$

in agreement with numerical studies (Kirk et al. 2000; Achterberg et al. 2001) over the entire range of β_u and β_d. In particular, in the ultra-relativistic shock limit, the spectral index becomes

$$s_p(\beta_u \to 1, \beta_d \to 1/3) = 38/9 = 4.222\ldots \tag{3}$$

The spectrum is shown in Fig. 1 for different equations of state, as a function of the shock four-velocity $\gamma_u\beta_u$.

The above analyses assumed that the waves scattering the particles move, on average, with the bulk fluid velocity. More accurately, one should replace β by the mean velocity of the waves that are scattering the particles. In the shock precursor (see Sect. 3.1), the scattering waves are expected to be slower than the incoming flow, leading to a softer spectrum (smaller β_u in Eq. (2)).

Small-angle scattering can be parameterized by the angular diffusion function $\mathcal{D} \equiv \langle(\Delta\theta)^2/\Delta t\rangle$, where θ is the angle of the particle velocity, taken here with respect to the shock normal, and angular brackets denote an ensemble average. The function $\mathcal{D} = \mathcal{D}(\theta, p, z)$ should be specified on both sides of the shock, and in general depends on θ, on the particle momentum p, and on its distance z from the shock front.

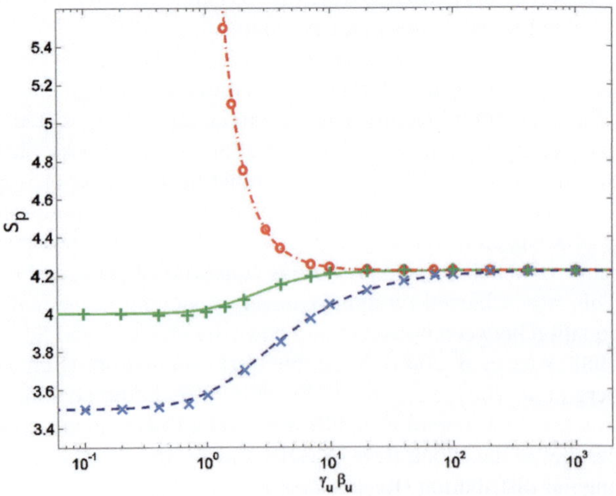

Fig. 1 Spectral index according to Eq. (2) (Keshet and Waxman 2005, *curves*) and to a numerical eigenfunction method (Kirk et al. 2000, *symbols*), as a function of $\gamma_u\beta_u$, for three different types of shocks (Kirk and Duffy 1999): a strong shock with the Jüttner/Synge equation of state (*solid curve* and *crosses*), a strong shock with fixed adiabatic index $\Gamma = 4/3$ (*dashed curve* and *x-marks*), and for a relativistic gas where $\beta_u\beta_d = 1/3$ (*dash-dotted curve* and *circles*)

For scattering off waves with a small coherence length $\lambda \ll r_L$, where $r_L = (pc/eB)$ is the Larmor radius, roughly $(r_L/\lambda)^2$ uncorrelated scattering events are needed in order to produce an appreciable deflection, so $\mathcal{D} \sim r_L^2 c/\lambda \propto p^2$ (Achterberg et al. 2001; Pelletier et al. 2009). Here, B is the magnetic field, and e is the electron's charge. Simulations (Sironi et al. 2013) confirm this scaling at early times; some implications are discussed in Sect. 4. The precise dependence of $\mathcal{D}$ upon z is not well known. It is thought that $\mathcal{D}$ slowly and monotonically declines away from the shock, as the energy in self-generated fields decreases. However, the extents of the upstream precursor and downstream magnetized region are not well constrained observationally, and in general are numerically unaccessible in the foreseeable future.

For an evolved magnetic configuration, it is natural to assume that the diffusion function is approximately separable in the form $\mathcal{D} = D(\theta)D_2(p,z)$. Here, D_2 (which may be approximately separable as well, but see Katz et al. 2007) can be eliminated from the transport equation by rescaling z, such that the spectrum depends only on the angular part $D(\theta)$.

The spectrum is typically more sensitive to the downstream diffusion function D_d than it is to the upstream D_u. In general, an enhanced D_d along (opposite to) the flow yields a softer (harder) spectrum; the trend is roughly reversed for D_u (Keshet 2006). Thus, the spectrum may deviate significantly from that of isotropic diffusion, in particular in the ultra-relativistic limit (Kirk et al. 2000; Keshet 2006). However, the spectral slope s is not sensitive to localized (in θ) deviations of D from isotropy, at angles perpendicular to the flow (Keshet 2006). For roughly forward-backward symmetric scattering in the downstream frame, as suggested by PIC simulations, s is approximately given by its isotropic diffusion value in Eq. (2) (Keshet et al., in preparation).

Particle acceleration is thought to be efficient, at least in weakly magnetized or quasi-parallel shocks, as discussed below. Thus, the relativistic particles are expected not only to generate waves, but also to slow down and heat the bulk plasma (Blandford and Eichler 1987). As particles with higher energies are expected to diffuse farther upstream and slow the plasma, lower-energy particles are effectively accelerated by a slower upstream. Consequently, if the scattering waves are assumed to move with the bulk plasma, the spectrum would no longer be a power-law. However, this effect may be significant only for mildly relativistic shocks, with Lorentz factors below $\gamma_u \sim 3$ (Ellison and Double 2002; Ellison et al. 2013).

To understand the energy, composition, and additional features of the accelerated particles, such as the acceleration time and energy cutoffs, one must not only analyze the scattering of these particles (for example, by deriving $\mathcal{D}$), but also address the injection problem, namely the process by which a small fraction of particles becomes subject to significant acceleration. Such effects were investigated using Monte-Carlo techniques (Bednarz and Ostrowski 1998; Ellison et al. 2013), in the so-called "thermal leakage" model, where fast particles belonging to the downstream Maxwellian are assumed to be able to cross the shock into the upstream. More self-consistent results on particle injection based on PIC simulations are presented in Sect. 4. To uncover the physics behind the injection and acceleration processes, we next review the generation of electromagnetic waves in relativistic shocks.

3 Plasma Instabilities in Relativistic Shocks

3.1 The Shock Precursor

The collisionless shock transition is associated with the build-up of some electromagnetic barrier, which is able to slow down and nearly isotropize the incoming unshocked plasma. In media of substantial magnetization,[1] $\sigma \gtrsim 10^{-2}$, this barrier can result from the compression of the background magnetic field (as a result of the Lorentz transformation to the frame of a relativistic shock, the most generic configuration is that of a quasi-perpendicular field), while at lower magnetizations, it is understood to arise from the generation of intense micro-turbulence in the shock "precursor", as explained hereafter and illustrated in Fig. 2.

At high magnetization, the gyration of the ambient particles in the background compressed magnetic field can trigger a synchrotron maser instability, which sends precursor electromagnetic waves into the upstream (Langdon et al. 1988; Hoshino and Arons 1991; Hoshino et al. 1992; Gallant et al. 1992). As incoming electrons and positrons interact with these waves, they undergo heating (Hoshino 2008; Sironi and Spitkovsky 2011b), but acceleration seemingly remains inefficient (Sect. 4.1).

At magnetizations $\sigma \lesssim 10^{-2}$, the interpenetration of the incoming background plasma and the supra-thermal particles, which have been reflected on the shock front or which are undergoing Fermi cycles around the shock, leads to anisotropic micro-instabilities over an extended region in front of the shock, called the "precursor" here. These instabilities then build up a magnetic barrier, up to a level[2] $\epsilon_B \sim 10^{-2}$–10^{-1}, sufficient to deflect strongly the incoming particles and thus mediate the shock transition. This picture, first envisioned by Moiseev and Sagdeev (1963), has been recently demonstrated in ab initio PIC simulations

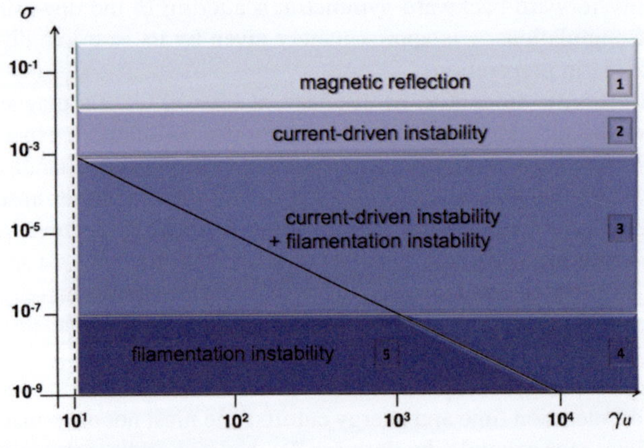

Fig. 2 Phase diagram of relativistic collisionless shocks in the plane (γ_u, σ); this figure assumes $\gamma_u > 10$ and $\xi_{cr} = 0.1$, where the parameter $\xi_{cr} = e_{cr}/[\gamma_u(\gamma_u - 1)n'mc^2]$ characterizes the energy density of supra-thermal particles (e_{cr}) relative to the incoming energy flux, as measured in the shock rest frame. In region 1, the shock transition is initiated by magnetic reflection in the compressed background field, while in regions 2–5, the magnetic barrier is associated to the growth of micro-instabilities, as indicated. The *solid diagonal line* indicates values of σ and γ_u above which the filamentation instability would not have time to grow, in the absence of deceleration resulting from the compensation of the perpendicular current of the supra-thermal particles gyrating in the background field. See Sect. 3.1 and Lemoine et al. (2014b) for a detailed discussion

[1]The magnetization is defined as $\sigma = B^2/[4\pi\gamma_u(\gamma_u - 1)n'mc^2]$ in terms of B, the large-scale background magnetic field measured in the shock front rest frame, and n', the proper upstream particle density. The mass m is m_p for an electron-proton shock, and m_e for an electron-positron shock, i.e. it corresponds to the mass of the particles which carry the bulk of the inertia. For a perpendicular shock, in which the background magnetic field in the shock frame is perpendicular to the flow, the magnetization can also be written as $\sigma = (u_A/c)^2$, with u_A the Alfvén four-velocity of the upstream plasma.

[2]The parameter ϵ_B denotes the magnetization of the turbulence, $\epsilon_B = \delta B^2/[4\pi\gamma_u(\gamma_u - 1)n'mc^2]$, where δB is the fluctuating magnetic field.

(Spitkovsky 2005; Kato 2007; Spitkovsky 2008a). The generation of micro-turbulence in the shock precursor is thus a key ingredient in the formation of the shock and in the development of the Fermi process, as anticipated analytically (Lemoine et al. 2006) and from Monte Carlo simulations (Niemiec et al. 2006), and demonstrated by PIC simulations (Spitkovsky 2008b; Sironi and Spitkovsky 2009a, 2011b), see hereafter.

As seen in the background plasma (upstream) rest frame, the supra-thermal particles form a highly focused beam, with an opening angle $\sim 1/\gamma_u$ and a mean Lorentz factor $\overline{\gamma}_{|u} \sim \gamma_u^2$. In contrast, boosting back to the shock frame, this supra-thermal particle distribution is now open over $\sim\pi/2$, with a mean Lorentz factor $\overline{\gamma}_{|\mathrm{sh}} \gtrsim \gamma_u$, while the incoming plasma is highly focused, with a mean Lorentz factor γ_u. A host of micro-instabilities can in principle develop in such anisotropic configurations, see the general discussion by Bret (2009). However, in the deep relativistic regime, the restricted length scale of the precursor imposes a strong selection of potential instabilities, since a background plasma volume element remains subject to the growth of instabilities only while it crosses the precursor. In the shock rest frame, this time scale is $t_{\times,B} \simeq \omega_\mathrm{c}^{-1}$ in the presence of a quasi-perpendicular background field[3] (a common field geometry in relativistic flows), or $t_{\times,\delta B} \simeq \gamma_u \epsilon_B^{-1} \omega_\mathrm{p}^{-1} (\omega_\mathrm{p}\lambda_{\delta B}/c)^{-1}$ if the scattering is dominated by short scale turbulence of magnetization ϵ_B and coherence length $\lambda_{\delta B}$ (assuming that the waves are purely magnetic in the rest frame of the background plasma), see e.g. Milosavljević and Nakar (2006a), Lemoine and Pelletier (2010) and Plotnikov et al. (2013). This small length scale implies that only the fastest modes can grow, which limits the discussion to a few salient instabilities.

Before proceeding further, one should stress that the above estimates for $t_\times$ do not account for the influence of particles accelerated to higher energies, which can propagate farther into the upstream plasma and thus seed instabilities with smaller growth rate and on larger spatial scales. While such particles do not carry the bulk of the energy if the spectral index $s_\gamma > 2$, it is anticipated that they should nevertheless influence the structure of the precursor, see in particular Milosavljević and Nakar (2006a), Katz et al. (2007), Medvedev and Zakutnyaya (2009), Pelletier et al. (2009) and Reville and Bell (2014) for general analytical discussions, as well as Keshet et al. (2009) for an explicit numerical demonstration of their potential influence. Similarly, the above estimates do not make a distinction between electron-positron and electron-ion shocks; in particular, it is understood that ω_c and ω_p refer to the species which carries the bulk of the energy (i.e. ions for electron-ion shocks). PIC simulations have demonstrated that in electron-ion shocks, electrons are heated in the precursor to nearly equipartition with the ions, meaning that in the shock transition their relativistic inertia becomes comparable to that of ions (e.g. Sironi et al. 2013); hence one does not expect a strong difference between the physics of electron-positron and electron-ion shocks from the point of view of micro-instabilities, and unless otherwise noted, this difference will be omitted in the following. The microphysics of electron heating in the precursor nevertheless remains an important open question, see Gedalin et al. (2008, 2012), Plotnikov et al. (2013) and Kumar et al. (2015) for recent discussions of this issue; indeed, the average Lorentz factor of electrons at the shock transition directly impacts the peak frequency of the synchrotron radiation of relativistic blast waves.

In the context of relativistic weakly magnetized shocks, the most celebrated instability is the Weibel-like filamentation mode, which derives from the anisotropy of the particle distribution function. In a simplified two-stream picture, as envisaged for the precursor of relativistic collisionless shocks, particles of opposite charges are deflected by the Lorentz

[3] $\omega_\mathrm{c} \equiv eB_{|u}/mc$ represents the upstream frame cyclotron frequency (and $B_{|u}$ is the magnetic field in the upstream frame) while $\omega_\mathrm{p} \equiv (4\pi n' e^2/m)^{1/2}$ denotes the plasma frequency.

force of transverse magnetic field fluctuations into current filaments of alternate polarity (e.g. Gruzinov and Waxman 1999; Medvedev and Loeb 1999; Brainerd 2000; Wiersma and Achterberg 2004; Lyubarsky and Eichler 2006; Achterberg and Wiersma 2007; Achterberg et al. 2007; Lemoine and Pelletier 2010, 2011; Bret et al. 2010; Rabinak et al. 2011; Nakar et al. 2011; Shaisultanov et al. 2012). The current carried by the particles then positively feeds the magnetic fluctuations, leading to fast growth, even in the absence of a net large-scale magnetic field. In the rest frame of the background plasma, this instability grows in the linear regime as fast as[4] $\Im\omega \simeq \xi_{\rm cr}^{1/2}\omega_{\rm p}$, with maximum growth on scales of the order of $c/\omega_{\rm p}$; in the filament rest frame, this instability is mostly of magnetic nature, i.e. $\Re\omega \sim 0$. Several branches of this instability have been discussed in the literature, in particular the "oblique mode", which involves a resonance with electrostatic modes. Even though this latter mode grows slightly faster than the fully transverse filamentation mode, it suffers from Landau damping once the electrons are heated to relativistic temperatures, while the transverse filamentation mode appears relatively insensitive to temperature effects. Thus, at a first order approximation, the transverse filamentation mode indeed appears to dominate the precursor at very low magnetizations. Its non-linear evolution, however, remains an open question; analytical estimates suggest that it should saturate at values $\epsilon_B \ll 10^{-2}$ via trapping of the particles (Wiersma and Achterberg 2004; Lyubarsky and Eichler 2006; Achterberg and Wiersma 2007; Achterberg et al. 2007), while PIC simulations see a continuous growth of magnetic energy density even when the non-linear filamentary structures have been formed (e.g. Keshet et al. 2009; Sironi et al. 2013). Whether additional instabilities such as a kinking of the filaments contribute in the non-linear phase thus remains debated, see for instance Milosavljević and Nakar (2006b).

At moderate magnetization levels, another fast instability can be triggered by the perpendicular current (transverse to both the magnetic field and the shock normal) seeded in the precursor by the supra-thermal particles during their gyration around the background field (Lemoine et al. 2014b,c). The compensation of this current by the background plasma on its entry into the precursor leads to a deceleration of the flow, which modifies somewhat the effective timescale available for the growth of plasma instabilities, and destabilizes the modes of the background plasma. The growth rate for this instability can be as large as $\Im\omega \sim \omega_{\rm p}$, indicating that it can compete with the Weibel filamentation mode at moderate magnetizations. If the supra-thermal particle beam carries a net charge (in the shock rest frame), or a net transverse current, other similar instabilities are to be expected (e.g. Pelletier et al. 2009; Casse et al. 2013; Reville and Bell 2014). The phase space study of Lemoine et al. (2014b) concludes that the filamentation mode likely dominates at magnetization levels $\sigma \lesssim 10^{-7}$, while this perpendicular current-driven instability dominates at $10^{-3} \lesssim \sigma \lesssim 10^{-2}$; in between, both instabilities combine to form a complex precursor structure. Interestingly, these results do not seem to depend on the shock Lorentz factor, in good agreement with PIC simulations (Sironi and Spitkovsky 2009a, 2011b; Sironi et al. 2013).

Finally, one should mention the particular case of quasi-parallel (subluminal) configurations: there, a fraction of the particles can in principle escape to infinity along the magnetic field and seed other, larger scale, instabilities. One prime candidate is the relativistic generalization of the Bell streaming instability (e.g. Milosavljević and Nakar 2006a; Reville et al. 2006), which is triggered by a net longitudinal current of supra-thermal particles; this instability has indeed been observed in PIC simulations (Sironi and Spitkovsky 2011b). Of course, such a parallel configuration remains a special case in the deep relativistic regime. In

[4]The maximal growth rate of the Weibel instability is related to the plasma frequency of the beam of supra-thermal particles, $\omega_{\rm pb}$, though $\Im\omega \simeq \omega_{\rm pb}$, with $\omega_{\rm pb} \simeq \xi_{\rm cr}^{1/2}\omega_{\rm p}$.

mildly relativistic shock waves, with $\gamma_u \beta_u \sim 1$, locally parallel configurations become more frequent, hence one could expect such instabilities to play a key role in seeding large scale turbulence.

3.2 Downstream Magnetized Turbulence

How the magnetized turbulence evolves downstream of the shock is an important question, with direct connections to observations. The previous discussion suggests that the coherence length of the fields generated in Weibel-like instabilities should be comparable to the plasma skin-depth, $c/\omega_{\rm p}$. However, magnetic power on such scales is expected to decay rapidly through collisionless phase mixing (Gruzinov 2001), while modeling of GRB afterglow observations rather indicates that magnetic fields persist over scales $\sim 10^7$–$10^9\, c/\omega_{\rm p}$ downstream (Gruzinov and Waxman 1999).

In a relativistic plasma, small-scale turbulence is dissipated at a damping rate[5] $\Im\omega \simeq -k^3c^3/\omega_{\rm p}^2$ (Chang et al. 2008; Lemoine 2015) as a function of the wavenumber k, indicating that small scales are erased early on. Larger modes can survive longer; power on scales exceeding the Larmor radius of the bulk plasma decays on long, $\Im\omega \propto k^2$ MHD scales (Keshet et al. 2009). It is not clear at present whether the small-scale turbulence manages to evolve to larger scales through inverse cascade effects (e.g. Medvedev et al. 2005; Katz et al. 2007; Zrake 2014), whether it is dissipated but at a rate which allows to match the observations (Lemoine 2013; Lemoine et al. 2013), or whether a large-scale field is seeded in the downstream plasma by some external instabilities (e.g. Sironi and Goodman 2007; Couch et al. 2008; Levinson 2009).

PIC simulations have quantified the generation of upstream current filaments by pinching instabilities (e.g. Silva et al. 2003; Frederiksen et al. 2004; Jaroschek et al. 2005; Spitkovsky 2005, 2008a; Chang et al. 2008), and resolved the formation of shocks in two- and three-dimensional (2D and 3D) pair plasma (Spitkovsky 2005; Kato 2007; Chang et al. 2008; Keshet et al. 2009; Sironi and Spitkovsky 2009a, 2011b; Haugbølle 2011; Sironi et al. 2013) and ion-electron plasma (Spitkovsky 2008a; Martins et al. 2009; Sironi et al. 2013). These simulations revealed a rapid decay of the magnetic field downstream at early times (Gruzinov 2001; Chang et al. 2008). Yet, a slow evolution of the plasma configuration takes place on $>10^3/\omega_{\rm p}$ timescales, involving a gradual increase in the scale of the magnetic structures, and consequently their slower dissipation downstream (Keshet et al. 2009).

This long-term evolution is driven entirely by the high-energy particles accelerated in the shock; it is seen both upstream (e.g. in the precursor) and downstream, both of which become magnetized at increasingly large distances from the shock, and with an increasingly flat magnetic power-spectrum downstream (on large scales, where the power initially declines with increasing wavelength; Keshet et al. 2009). A flatter magnetic power spectrum at the shock implies a larger fraction of the magnetic energy stored in long-wavelength modes, which may survive farther from the shock. Indeed, the index of a power-law spectrum of magnetic fluctuations directly controls how fast the magnetic energy density, integrated over wavenumbers, decays behind the shock (Chang et al. 2008; Lemoine 2015); the scale-free limit corresponds to a flat magnetic power spectrum (Katz et al. 2007).

Properly capturing the backreaction of high energy particles requires large simulation boxes and large particle numbers, to guarantee that the largest scale fields and the highest energy particles are included. The largest available simulations at the present, with length L

[5] The shock crossing conditions imply that the relativistic plasma frequency of the shocked downstream plasma is roughly the same as the plasma frequency of the upstream plasma; no distinction will be made here between these quantities.

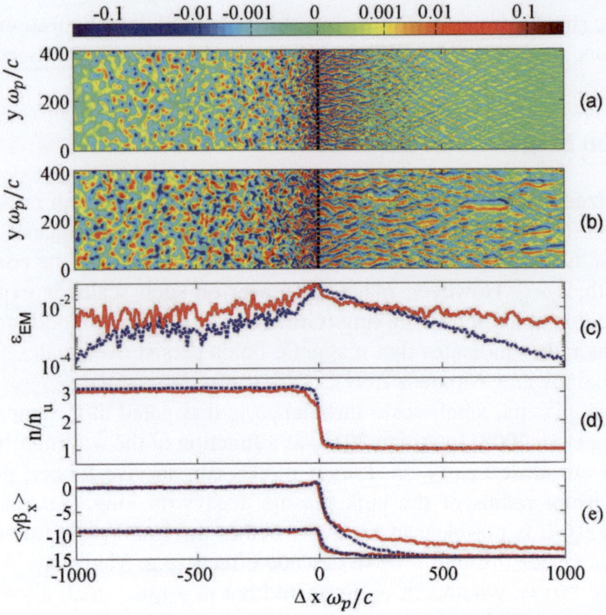

Fig. 3 Pair plasma evolution within $1000c/\omega_{\rm p}$ of the shock, from Keshet et al. (2009). The simulation is performed in the downstream frame, and the upstream flow moves with a Lorentz factor $\gamma_r = 15$ (so, γ_r is the relative Lorentz factor between the upstream and downstream regions). The normalized transverse magnetic field $\mathrm{sign}(B)\epsilon_B$ (color scale stretched in proportion to $\epsilon_B^{1/4}$ to highlight weak features) is shown at (**a**) early ($t_1 = 2250\omega_{\rm p}^{-1}$), and (**b**) late ($t_2 = 11925\omega_{\rm p}^{-1}$) times. Here $\Delta x \equiv x - x_{\rm sh}$ is the distance from the shock, with $x_{\rm sh}$ (*dashed vertical line*) defined as the location of median density between far upstream and far downstream. Also shown are the transverse averages (at t_1, *dashed blue*, and t_2, *solid red*) of (**c**) the electromagnetic energy $\epsilon_{\rm EM} \equiv [(B^2 + E^2)/8\pi]/[(\gamma_r - 1)\gamma_r n' mc^2]$ (with E the electric field amplitude in the downstream frame, included in the definition of $\epsilon_{\rm EM}$ because in the simulation frame the induced electric field in the upstream medium is $E \sim B$) normalized to the upstream kinetic energy, (**d**) density normalized to the far upstream density $n_u = \gamma_r n'$, and (**e**) particle momentum $\gamma\beta$ (with β the velocity in c units) in the x-direction averaged over all particles (higher $\langle\gamma\beta_x\rangle$) and over downstream-headed particles only

and time T scales of $(L\omega_{\rm p}/c)^2(T\omega_{\rm p}) \lesssim 10^{11}$, show no sign of convergence at $T \gtrsim 10^4 c/\omega_{\rm p}$ (Keshet et al. 2009; Sironi et al. 2013). This is illustrated in Fig. 3 for a pair-plasma shock in 2D.

For magnetized shocks, the situation is different, as we describe below (Sironi et al. 2013). At strong magnetizations, and for the quasi-perpendicular field geometry most relevant for relativistic flows, particle acceleration is suppressed, and the shock quickly reaches a steady state. At low (but nonzero) quasi-perpendicular magnetization, the shock evolves at early times similarly to the case of unmagnetized shocks (i.e. $\sigma = 0$). Particle acceleration proceeds to higher and higher energies, and modes of longer and longer wavelength appear. However, the maximum particle energy stops evolving once it reaches a threshold $\gamma_{\rm sat} \propto \sigma^{-1/4}$ (Sironi et al. 2013), and at that point the overall shock structure approaches a steady state (Sironi et al. 2013).[6]

[6]This conclusion regarding the saturation of the maximum particle Lorentz factor at $\gamma_{\rm sat}$ has been tested in electron-positron shocks having $\sigma = 10^{-4}$–10^{-3} by Sironi et al. (2013), with the largest PIC study available to date. We caution that further nonlinear evolution, beyond the timespan covered by current PIC simulations, might be present in shocks with lower magnetization.

4 PIC Simulations of Relativistic Shocks

Only in the last few years, thanks to important advances in numerical algorithms and computer capabilities, plasma simulations have been able to tackle the problem of particle acceleration in relativistic shocks from first principles. In the following, we describe the major advances driven by large-scale PIC simulations in our understanding of particle acceleration in relativistic shocks. PIC codes can model astrophysical plasmas in the most fundamental way (Birdsall and Langdon 1991; Buneman 1993; Spitkovsky 2005), as a collection of charged macro-particles that are moved by the Lorentz force. The currents deposited by the macro-particles on the computational grid are then used to solve for the electromagnetic fields via Maxwell's equations. The loop is closed self-consistently by extrapolating the fields to the macro-particle locations, where the Lorentz force is computed.

Full PIC simulations can capture, from first principles, the acceleration physics of both electrons and ions. However, such simulations must resolve the electron plasma skin depth $c/\omega_{\rm pe}$, which is typically much smaller than astrophysical scales. Hence, most simulations can only cover limited time and length scales, and usually with low dimensionality (1D or 2D instead of 3D) and small ion-to-electron mass ratios (the ion skin depth $c/\omega_{\rm pi}$ is a factor of $\sqrt{m_i/m_e}$ larger than the electron skin depth $c/\omega_{\rm pe}$). The results discussed below pertain to simulation durations of order $\sim 10^3$–$10^4 \omega_{\rm pe}$ in electron-positron shocks and $\sim 10^3 \omega_{\rm pi}$ in electron-ion shocks (but with reduced mass ratios), so a careful extrapolation is needed to bridge these microscopic scales with the macroscopic scales of astrophysical interest. Yet, as we review below, PIC simulations provide invaluable insight into the physics of particle injection and acceleration in astrophysical sources.

The structure of relativistic shocks and the efficiency of particle acceleration depend on the conditions of the upstream flow, such as bulk velocity, magnetic field strength and field orientation. PIC simulations have shown that the shock physics and the efficiency of particle acceleration are insensitive to the shock Lorentz factor (modulo an overall shift in the energy scale), in the regime $\gamma_r \gg 1$ of ultra-relativistic flows (e.g. Sironi et al. 2013). Here, γ_r is the relative Lorentz factor between the upstream and downstream regions. Below, we only discuss results for shocks where the upstream Lorentz factor with respect to the downstream frame is $\gamma_r \gtrsim 5$, neglecting the trans- and non-relativistic regimes that are outside the scope of this review. We discuss the physics of both electron-positron shocks and electron-ion shocks (up to realistic mass ratios), neglecting the case of electron-positron-ion shocks presented by e.g. Hoshino et al. (1992), Amato and Arons (2006), Stockem et al. (2012), which might be relevant for PWNe. As found by Sironi and Spitkovsky (2009a, 2011b), Sironi et al. (2013), for highly relativistic flows, the main parameter that controls the shock physics is the magnetization σ. Below, we distinguish between shocks propagating into strongly magnetized media ($\sigma \gtrsim 10^{-3}$) and weakly magnetized or unmagnetized shocks ($\sigma \lesssim 10^{-3}$).

4.1 Particle Acceleration in Strongly Magnetized Shocks

For high magnetizations ($\sigma \gtrsim 10^{-3}$ in electron-positron flows, or $\sigma \gtrsim 3 \times 10^{-5}$ in electron-ion flows), the shock structure and acceleration properties depend critically on the inclination angle θ between the upstream field and the shock direction of propagation (Sironi and Spitkovsky 2009a, 2011b). If the magnetic obliquity is larger than a critical angle $\theta_{\rm crit}$, charged particles would need to move along the field faster than the speed of light in order to outrun the shock ("superluminal" configurations). In Fig. 4, we show how the critical angle

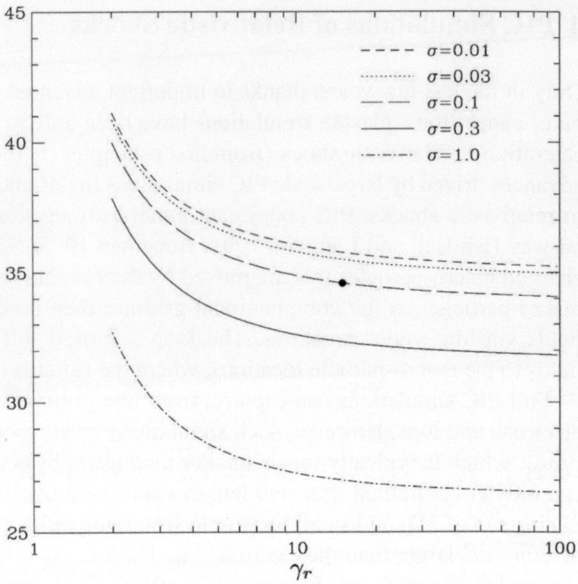

Fig. 4 Critical obliquity angle θ_{crit} (measured in the downstream frame) that separates subluminal and superluminal configurations (Sironi and Spitkovsky 2009a), as a function of the flow Lorentz factor γ_r and the magnetization σ, as indicated in the label. The *filled black circle* indicates our reference case with $\gamma_r = 15$ and $\sigma = 0.1$

θ_{crit} (as measured in the downstream frame) depends on the flow velocity and magnetization. In the limit of $\sigma \ll 1$ and $\gamma_r \gg 1$, the critical obliquity approaches the value $\theta_{crit} \simeq 34°$.

Only "subluminal" shocks ($\theta \lesssim \theta_{crit}$) are efficient particle accelerators (Sironi and Spitkovsky 2009a, 2011b; Sironi et al. 2013), in agreement with the analytical findings of Begelman and Kirk (1990). As illustrated in Fig. 5, a stream of shock-accelerated particles propagates ahead of the shock (panel (c)), and their counter-streaming with the incoming flow generates magnetic turbulence in the upstream region (panel (b)). In turn, such waves govern the acceleration process, by providing the turbulence required for the Fermi mechanism. In the particular case of Fig. 5—a relativistic shock with $\gamma_r = 15$, $\sigma = 0.1$ and $\theta = 15°$ propagating into an electron-ion plasma—the upstream turbulence is dominated by Bell-like modes (Reville et al. 2006; Lemoine and Pelletier 2010, 2011). The downstream particle spectrum in subluminal shocks shows a pronounced non-thermal tail of shock-accelerated particles with a power-law index $2 \lesssim s_\gamma \lesssim 3$ (panel (d)). The tail contains $\sim$5% of the particles and $\sim$20% of the flow energy at time $2250\omega_{pi}^{-1}$; both values appear to be time-converged, within the timespan covered by the simulations.

In contrast, superluminal shocks ($\theta \gtrsim \theta_{crit}$) show negligible particle acceleration (Gallant et al. 1992; Hoshino 2008; Sironi and Spitkovsky 2009a, 2011b; Sironi et al. 2013). Here, due to the lack of significant self-generated turbulence, charged particles are forced to slide along the background field lines, whose orientation prohibits repeated crossings of the shock. This inhibits the Fermi process, and in fact the particle distribution behind superluminal shocks is purely thermal. The same conclusion holds for both electron-positron and electron-ion flows. In electron-ion shocks, the incoming electrons are heated up to the ion energy, due to powerful electromagnetic waves emitted by the shock into the upstream medium, as a result of the synchrotron maser instability [studied analytically by Lyubarsky (2006), and with 1D PIC simulations by e.g. Langdon et al. (1988), Gallant et al. (1992), Hoshino et al. (1992), Hoshino (2008)]. Yet, such heating is not powerful enough to permit an efficient injection of electrons into the Fermi acceleration process at superluminal electron-ion shocks.

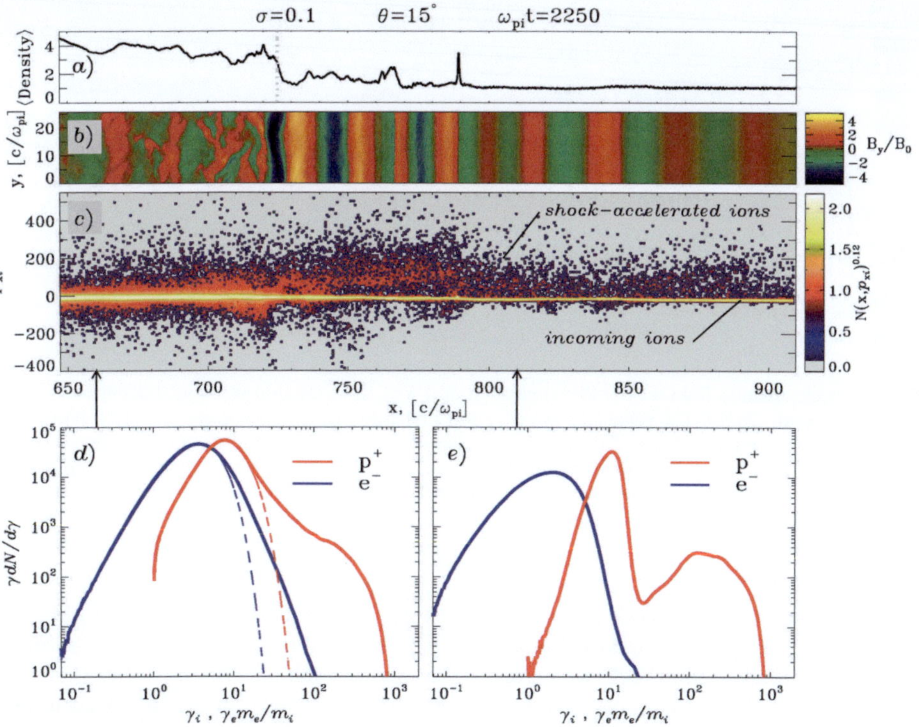

Fig. 5 Structure of an electron-ion subluminal shock with $\gamma_r = 15$, $\sigma = 0.1$ and $\theta = 15°$, from Sironi and Spitkovsky (2011b). The simulation is performed in the downstream frame. The shock front is located at $x \sim 725\,c/\omega_{\rm pi}$ (*vertical dotted red line* in panel (**a**)), and it separates the upstream region (to its right) from the compressed downstream region (to its left). A stream of shock-accelerated ions propagates ahead of the shock (see the diffuse cloud in the momentum space $x - p_{xi}$ of panel (**c**) to the right of the shock, at $x \gtrsim 725\,c/\omega_{\rm pi}$). Their interaction with the upstream flow (narrow beam to the right of the shock in panel (**c**)) generates magnetic turbulence ahead of the shock (see the transverse waves in panel (**b**), to the right of the shock). In turn, such waves govern the process of particle acceleration. The particle energy spectrum (*red* for ions, *blue* for electrons) is shown in panels (**d**) and (**e**) at two different locations, as indicated by the *arrows* at the bottom of panel (**c**). The particle spectrum behind the shock (*solid lines* in panel (**d**)) is not compatible with a simple thermal distribution (*dashed lines*), showing a clear non-thermal tail of high-energy particles, most notably for ions. The highest energy ions can propagate ahead of the shock, where they populate a bump in the upstream ion spectrum (*red line* in panel (**e**))

If magnetized superluminal shocks are responsible for producing the radiating particles in astrophysical relativistic sources, the strong electron heating observed in electron-ion shocks implies that the putative power-law tail in the electron spectrum should start from energies higher than the ion bulk kinetic energy. For models of GRBs and AGN jets that require a power-law distribution extending down to lower energies, the presence of such shocks would suggest that electron-positron pairs may be a major component of the flow.

4.2 Particle Acceleration in Weakly Magnetized and Unmagnetized Shocks

Weakly magnetized shocks ($\sigma \lesssim 10^{-3}$ in electron-positron flows, $\sigma \lesssim 3 \times 10^{-5}$ in electron-ion flows) are governed by electromagnetic plasma instabilities (see Sect. 3.1), that generate magnetic fields stronger than the background field. Such shocks do accelerate particles

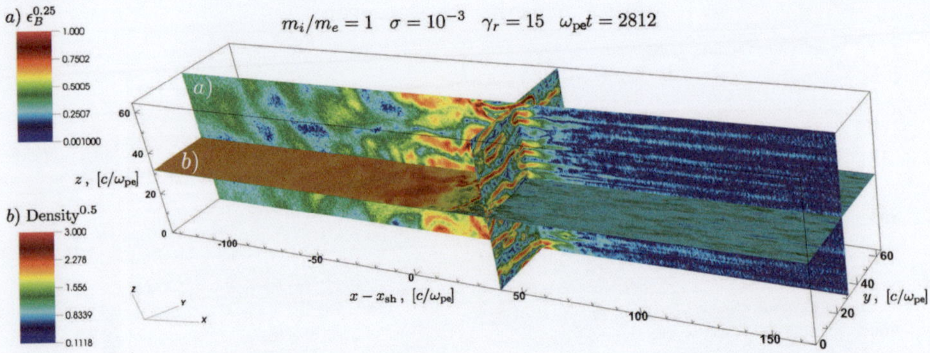

Fig. 6 Shock structure from the 3D PIC simulation of a $\sigma = 10^{-3}$ electron-positron shock with $\gamma_r = 15$, from Sironi et al. (2013). The simulation is performed in the downstream frame and the shock propagates along $+\hat{x}$. We show the xy slice of the particle number density (normalized to the upstream density), and the xz and yz slices of the magnetic energy fraction ϵ_B. A stream of shock-accelerated particles propagates ahead of the shock, and their counter-streaming motion with respect to the incoming flow generates magnetic turbulence in the upstream via electromagnetic micro-instabilities. In turn, such waves provide the scattering required for particle acceleration

self-consistently, regardless of the magnetic obliquity angle (Spitkovsky 2008a,b; Martins et al. 2009; Haugbølle 2011; Sironi et al. 2013). The stream of shock-accelerated particles propagates ahead of the shock, triggering the Weibel instability. The instability generates filamentary magnetic structures in the upstream region, as shown in Fig. 6, which in turn scatter the particles back and forth across the shock, mediating Fermi acceleration.

The accelerated particles in weakly magnetized shocks populate in the downstream region a power-law tail $dN/d\gamma \propto \gamma^{-s_\gamma}$ with a slope $s_\gamma \sim 2.5$, that contains $\sim$3% of the particles and $\sim$10% of the flow energy.[7] In electron-ion shocks, the acceleration process proceeds similarly for the two species, since the electrons enter the shock nearly in equipartition with the ions, as a result of strong pre-heating in the self-generated Weibel turbulence (Spitkovsky 2008a; Martins et al. 2009; Sironi et al. 2013). In both electron-positron and electron-ion shocks, the maximum energy of the accelerated particles scales in time as $\gamma_{\max} \propto t^{1/2}$ (Sironi et al. 2013), as shown in Fig. 7. More precisely, the maximum particle Lorentz factor in the downstream frame scales as

$$\gamma_{\max} \simeq 0.5\, \gamma_r (\omega_{pe} t)^{1/2} \tag{4}$$

$$\gamma_{\max,i} \sim \frac{\gamma_{\max,e} m_e}{m_i} \simeq 0.25\, \gamma_r (\omega_{pi} t)^{1/2} \tag{5}$$

in electron-positron and in electron-ion shocks, respectively (Sironi et al. 2013). This scaling is shallower than the so-called (and commonly assumed) Bohm limit $\gamma_{\max} \propto t$, and it naturally results from the small-scale nature of the Weibel turbulence generated in the shock layer (see Fig. 6).

The increase of the maximum particle energy over time proceeds up to a saturation Lorentz factor (once again, measured in the downstream frame) that is constrained by the magnetization σ of the upstream flow according to

[7] These values are nearly independent of the flow composition and magnetization, in the regime of weakly magnetized shocks. Also, they are measured at time $\sim 10^4\, \omega_{pe}^{-1}$ in electron-positron shocks and at $\sim 10^3\, \omega_{pi}^{-1}$ in electron-ion shocks, but they appear remarkably constant over time, within the timespan covered by the simulations.

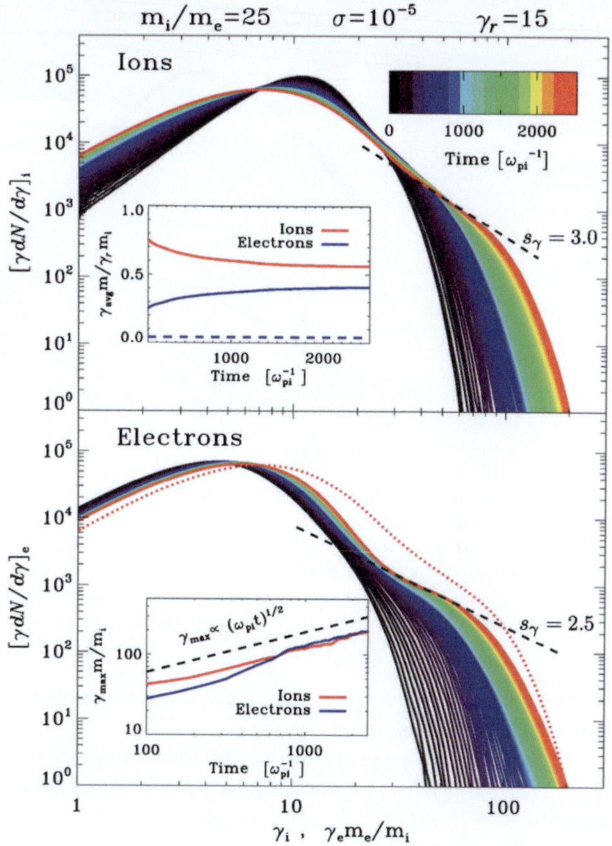

Fig. 7 Temporal evolution of the downstream particle spectrum, from the 2D simulation of a $\gamma_r = 15$ electron-ion ($m_i/m_e = 25$) shock propagating into a flow with magnetization $\sigma = 10^{-5}$, from Sironi et al. (2013). The evolution of the shock is followed from its birth (*black curve*) up to $\omega_{\rm pi}t = 2500$ (*red curve*). In the *top panel* we show the ion spectrum and in the *bottom panel* the electron spectrum. The non-thermal tails approach at late times a power law with a slope $s_\gamma = 3.0$ for ions and $s_\gamma = 2.5$ for electrons (*black dashed lines* in the two panels). In the *bottom panel*, we overplot the ion spectrum at $\omega_{\rm pi}t = 2500$ with a *red dotted line*, showing that ions and electrons are nearly in equipartition. *Inset* of the top panel: mean downstream ion (*red*) and electron (*blue*) energy, in units of the bulk energy of an upstream particle. The *dashed blue line* shows the electron energy at injection. *Inset* of the bottom panel: temporal evolution of the maximum Lorentz factor of ions (*red*) and electrons (*blue*), scaling as $\propto (\omega_{\rm pi}t)^{1/2}$ at late times (*black dashed line*)

$$\gamma_{\rm sat} \simeq 4\,\gamma_r\,\sigma^{-1/4} \tag{6}$$

$$\gamma_{{\rm sat},i} \sim \frac{\gamma_{{\rm sat},e} m_e}{m_i} \simeq 2\,\gamma_r\,\sigma^{-1/4} \tag{7}$$

in electron-positron and electron-ion shocks, respectively. The saturation of the maximum particle energy is shown in Fig. 8 for a shock with $\sigma = 10^{-3}$. Further energization is prevented by the fact that the self-generated turbulence is confined within a region of thickness $L_{B,{\rm sat}} \propto \sigma^{-1/2}$ around the shock (Sironi et al. 2013).

5 Astrophysical Implications

5.1 Acceleration of Ultra-High Energy Cosmic Rays

Relativistic shock waves have long been considered as prime candidates for the acceleration of cosmic rays to the highest energies observed, $E \sim 10^{20}$ eV. Indeed, a naive extrapolation of the acceleration time scale in the sub-relativistic regime ($t_{\rm acc} \sim t_{\rm scatt}/\beta_u^2$, with $t_{\rm scatt}$ the scattering timescale) suggests that relativistic shocks (i.e. $\beta_u \sim 1$) accelerate particles on shorter time scales than non-relativistic shocks (i.e. $\beta_u \ll 1$), at a given $t_{\rm scatt}$. For given radiative loss and escape time scales, this implies that relativistic shocks would be accelerating particles to much higher energies than non-relativistic shocks. However, the situation

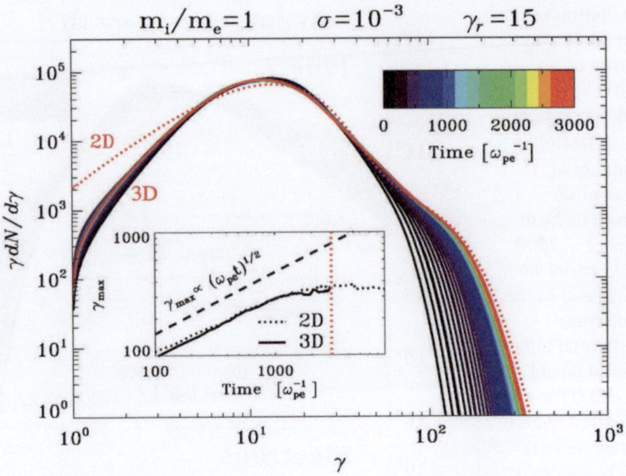

Fig. 8 Time evolution of the downstream particle spectrum from the 3D PIC simulation of a $\sigma = 10^{-3}$ electron-positron shock with $\gamma_r = 15$, from Sironi et al. (2013). The evolution of the shock is followed from its birth (*black curve*) up to $\omega_{pe}t = 3000$ (*red curve*). We overplot the spectrum at $\omega_{pe}t = 3000$ from a 2D simulation with the same parameters (*red dotted line*), showing excellent agreement at high energies. The *inset* shows that the maximum particle Lorentz factor grows as $\gamma_{max} \propto t^{1/2}$, before saturating at $\gamma_{sat} \propto \sigma^{-1/4}$. The results are consistent between 2D (*dotted*) and 3D (*solid*)

is more complex than it appears; in particular, in relativistic shock waves, t_{scatt} may be much larger than usually assumed.

As mentioned repeatedly in the previous paragraphs, particle acceleration in the relativistic regime $\gamma_u\beta_u \gg 1$ around a steady planar shock wave, operates only if intense micro-turbulence has been excited in the shock precursor, as demonstrated analytically (Lemoine et al. 2006), by Monte Carlo simulations (Niemiec et al. 2006) and by PIC simulations (Sironi et al. 2013); consequences for the acceleration of particles to ultra-high energies have been discussed in several papers, e.g. by Pelletier et al. (2008), Lemoine and Waxman (2009), Lemoine (2011), Eichler and Pohl (2011), Bykov et al. (2012), Sironi et al. (2013) or more recently by Reville and Bell (2014).

Scattering in small-scale turbulence leads to a downstream residence time $t_{scatt} \sim r_L^2/(\lambda_{\delta B}c)$, with r_L the Larmor radius of the particle and $\lambda_{\delta B}$ the coherence length scale of the turbulence. This implies that the (shock frame) acceleration timescale t_{acc} grows quadratically with the energy, which fits well the result seen in PIC simulations that the maximum energy grows as the square root of time. In other words, as the particle energy grows, the acceleration timescale departs more and more from the Bohm estimate, which is generally used to compute the maximum energy. Comparing for instance the acceleration timescale, which is at least equal to the above downstream residence time, with the dynamical timescale r/γ_u in the shock rest frame (r denoting the radius of the shock front in the upstream rest frame), one finds a maximum energy $E_{max} \lesssim e\delta Br(\gamma_u\lambda_{\delta B}/r)^{1/2}$, with δB the strength of the turbulent field expressed in the shock frame; the above maximal energy has been written in the upstream (observer) frame. The factor in the brackets generally takes very small values, because $\lambda_{\delta B} \sim c/\omega_p$ while r is a macroscopic length scale; this maximal energy is thus far below the so-called Hillas estimate $e\,\delta Br$, which corresponds to a Bohm estimate for t_{scatt}.

Another way to phrase the problem is as follows (see Lemoine and Waxman 2009 for a discussion): assume that the acceleration timescale is written $t_{acc} = \mathcal{A} r_L/c$, and derive the maximum energy by comparing t_{acc} with $t_{dyn} = r/(\gamma\beta c)$ as above for a jet moving at velocity β towards the observer. Then one finds that acceleration of particles of charge Z to $10^{20} E_{20}$ eV requires that the isotropic equivalent magnetic luminosity of the object exceeds: $L_B \gtrsim 10^{45}\, Z^{-2} E_{20}^2 \mathcal{A}^2 \gamma^2$ erg/s, a very large number indeed, all the more so if $\mathcal{A} \gg 1$. For acceleration at ultra-relativistic shock waves, $\mathcal{A}$ is much larger than unity (while the Bohm estimate corresponds to $\mathcal{A} \sim 1$), with typical values $\mathcal{A} \sim E/(\gamma_u m_p c^2)$.

In summary, particle acceleration at ultra-relativistic shock waves does not appear fast enough to produce particles of ultra-high energies. In particular, when the above arguments are applied to the case of the external shock of a GRB, the maximal energy is found to be of the order of 10^{16} eV (Plotnikov et al. 2013; Sironi et al. 2013; Reville and Bell 2014).

It is important however to note three caveats in the above arguments. One is that as $\gamma_u \beta_u \to 1$, i.e. for mildly relativistic shock waves, the nature of the turbulence remains unknown and one cannot exclude that scattering would be closer to a Bohm estimate. Two facts support such a speculation: (1) the precursor increases in size as γ_u diminishes, which suggests that MHD-scale instabilities could arise and excite large scale turbulence; and (2) the obliquity becomes less of a problem for mildly relativistic shock waves, suggesting that large scale turbulence could possibly lead to acceleration in this regime. A second caveat is the fact that PWNe are very efficient particle accelerators, even though one would expect the opposite in the absence of reconnection or other dissipative processes, due to the large magnetization of the flow (Sect. 5.3). More precisely, synchrotron photons are observed with energies as high as 100 MeV, which means that pairs are accelerated up to the radiation-reaction limit, i.e. with an acceleration time scale close to the theoretical Bohm scaling. Such empirical evidence suggests that ions could also be accelerated to very high energies, if ions are indeed injected along with pairs in the wind. In the Crab Nebula, such a maximal energy would be limited by the confinement in the nebular turbulence to values of the order of 10^{17} eV (for $Z = 1$); more powerful nebulae, associated with young pulsars born with a few millisecond periods, could however confine (and potentially accelerate) protons up to the highest energies (Lemoine et al. 2014a). Finally, as the nonlinear evolution of weakly magnetized or parallel shocks over long timescales is not yet understood, some of the above estimates, pertaining e.g. to the diffusive properties and extent of the magnetic field, may be altered on macroscopic times.

5.2 Radiative Signatures of Relativistic Blast Waves

In line with the previous discussion, one can compute the maximal energy for electrons and derive the maximal synchrotron photon energy. Using an acceleration time scale $t_{\rm acc} \simeq r_L^2/(\lambda_{\delta B} c)$ and comparing to synchrotron losses in the self-generated micro-turbulence, characterized by its magnetization ϵ_B, one derives a maximum synchrotron photon energy of the order of a few GeV in the early phase of GRB afterglows, i.e. during the first hundreds of seconds (Kirk and Reville 2010; Plotnikov et al. 2013; Lemoine 2013; Wang et al. 2013; Sironi et al. 2013). Let us stress that in the latter study, this estimate has been derived from PIC simulations with a self-consistent measurement of the acceleration time scale in the self-consistent magnetic field. The synchrotron radiation of electrons accelerated at the external ultra-relativistic shock of GRBs can thus produce the bulk of the long-lasting $>$100 MeV emission detected by the Fermi satellite (e.g. Kumar and Barniol Duran 2009; Ackermann 2010; De Pasquale 2010; Ghisellini et al. 2010). The photons that have been observed with energies in excess of $\gtrsim$10 GeV probably result from inverse Compton interactions (Wang et al. 2013). Interestingly, the recent GRB130427A has revealed a long-lasting emission with a possible break in the spectrum at an energy of a GeV, characteristic of a turn-over between the synchrotron and the synchrotron self-Compton components (Tam et al. 2013), in good qualitative agreement with the above arguments.

Other potential radiative signatures of the shock microphysics come from the small-scale nature of the turbulence and its long-term evolution in the blast. As discussed in Sect. 3.2, one notably expects this turbulence to relax through collisionless damping on hundreds of c/ω_p (Chang et al. 2008; Keshet et al. 2009; Lemoine 2015) while the electrons typically cool on much longer length scales. In GRB external blast waves, the shocked region is typically 7–9 orders of magnitude larger than c/ω_p in size, which leaves room for a substantial evolution of ϵ_B, even if it decreases as a mild power-law in distance from the shock, as suggested by the above studies. Since the electron cooling length depends on the inverse of the electron Lorentz factor, particles of different initial Lorentz factors emit their energy in regions of different magnetic field strength, leading to a non-standard synchrotron spectrum (Rossi and Rees 2003; Derishev 2007; Lemoine 2013), which could in principle be used as a tomograph of the evolution of the micro-turbulence downstream of the shock. Interestingly, in this picture the decay index of the turbulence is related to the long-wavelength content of the power spectrum of magnetic fluctuations at the shock front, which is unknown so far, as it is known to be modified by the acceleration of higher energy particles (Keshet et al. 2009). Finally, it is interesting to note that the recent broad-band analysis of GRB afterglows seen from the radio up to GeV energies has indeed revealed spectral signatures of a decaying magnetic field (Lemoine et al. 2013), with a decay law scaling with distance from the shock roughly as $\Delta x^{-0.5}$ (Δx being the proper distance to the shock in the downstream frame).

As discussed in Sect. 3.2, there are alternative possibilities however; it has been suggested for instance that the turbulence could evolve in a self-similar way as a function of distance to the shock, maintaining a uniform ϵ_B thanks to an inverse cascade process (Katz et al. 2007). It is also possible that external sources seed the blast with a large scale long-lived turbulence, e.g. through a Rayleigh-Taylor instability at the contact discontinuity (Levinson 2009) or through small scale dynamos following the interaction of the shock front with external inhomogeneities (Sironi and Goodman 2007; Couch et al. 2008). Hopefully, future high accuracy observational data will provide diagnostics which can be confronted with numerical simulations.

The possibility that the small scale nature of the turbulence gives rise to diffusive (or jitter) synchrotron radiation rather than conventional synchrotron radiation has also attracted attention (e.g. Medvedev 2000, 2006; Fleishman 2006; Mao and Wang 2011; Medvedev et al. 2011; Kelner et al. 2013). In particular, jitter radiation has been proposed as a solution for the fact that GRBs prompt spectra below the peak frequency are not always compatible with the predictions of synchrotron emission (the so-called "line of death" puzzle, see Preece 1998). In the jitter regime, particles are deflected by less than $1/\gamma$ (γ is the electron Lorentz factor) as they cross a wavelength $\lambda_{\delta B}$, implying that coherence of the emission is maintained over several coherence cells of the turbulence. This regime thus takes place whenever the wiggler parameter $a \equiv e\delta B\lambda_{\delta B}/mc^2 \ll 1$, while the standard synchrotron approximation becomes valid in the opposite limit. However, it is easy to verify that in the vicinity of the shock $a \sim \overline{\gamma}_{|\rm sh}$, with $\overline{\gamma}_{|\rm sh}$ the average Lorentz factor of the supra-thermal electrons in the shock rest frame, suggesting that jitter signatures must be weak.

The absence of jitter radiation in relativistic shocks has been demonstrated from first principles by computing the radiation from particles in PIC simulations (Sironi and Spitkovsky 2009b), which produce spectra entirely consistent with synchrotron radiation in the fields generated by the Weibel instability (Fig. 9). The so-called "jitter" regime is recovered only by artificially reducing the strength of the fields, such that the parameter a becomes much smaller than unity. So, if the GRB prompt emission results from relativistic unmagnetized shocks, it seems that resorting to the jitter regime is not a viable solution for the "line of

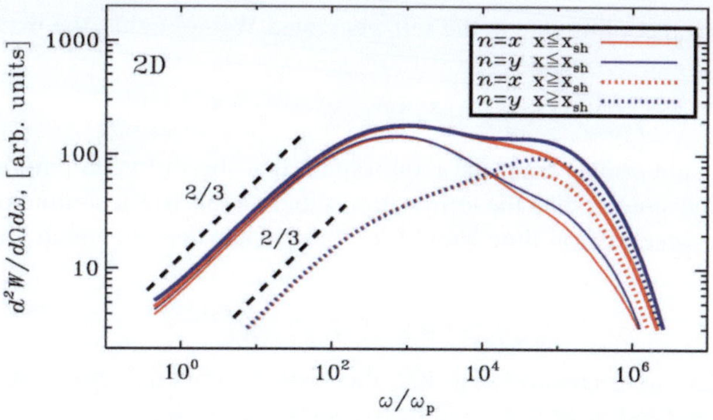

Fig. 9 *Ab initio* photon spectrum (*thick solid lines*) from the 2D PIC simulation of an unmagnetized (i.e. $\sigma = 0$) pair shock. *Red lines* are for head-on emission ($\hat{n} = \hat{x}$, along the shock direction of propagation), *blue lines* for edge-on emission ($\hat{n} = \hat{y}$, along the shock front). The slope at low frequencies is 2/3 (*black long-dashed lines*), proving that the spectra are consistent with synchrotron radiation from a 2D particle distribution (in 3D, the predicted slope of 1/3 is obtained). By separating the relative contribution of downstream ($x \leq x_{sh}$; *thin solid lines*) and upstream ($x \geq x_{sh}$; *dotted lines*) particles, one sees that upstream particles contribute significantly to the total emission (*thick solid lines*), especially at high frequencies. Frequencies are in units of the plasma frequency ω_p

death" puzzle. At frequencies above the peak, the synthetic spectra from PIC simulations show, somewhat unexpectedly, that the contribution of the upstream medium to the total emission is not negligible (Fig. 9), yet it is omitted in most models. This causes the radiation spectrum to be flatter than the corresponding downstream spectrum, thus partly masking the contribution of downstream thermal particles.

5.3 Radiative Signatures of Pulsar Wind Nebulae

The spectrum of PWNe consists of two components, where the low energy component, most likely dominated by synchrotron, shows a cutoff at a few tens of MeV. The fact that synchrotron emission reaches these energies, despite the rapid synchrotron cooling, implies that particle acceleration in the nebula is an extremely fast process (de Jager and Harding 1992), which challenges our understanding of particle acceleration in relativistic shocks.

Around the equatorial plane of obliquely-rotating pulsars, the wind consists of toroidal stripes of opposite magnetic polarity, separated by current sheets of hot plasma. It is still a subject of active research whether the alternating stripes will dissipate their energy into particle heat ahead of the termination shock, or whether the wind remains dominated by Poynting flux till the termination shock (Lyubarsky and Kirk 2001; Kirk and Skjæraasen 2003; Sironi and Spitkovsky 2011a). If the stripes are dissipated far ahead of the termination shock, the upstream flow is weakly magnetized and the pulsar wind reaches a terminal Lorentz factor (in the frame of the nebula) $\gamma_r \sim L_{sd}/m_e c^2 \dot{N} \simeq 3.7 \times 10^4 L_{sd,38.5} \dot{N}_{40}^{-1}$, where $L_{sd} \equiv 3 \times 10^{38} L_{sd,38.5}$ ergs s^{-1} is the spin-down luminosity, parameterized according to the Crab (the Crab Nebula is the prototype of PWNe), and $\dot{N} = 10^{40} \dot{N}_{40}$ s^{-1} is the particle flux entering the nebula, including the radio-emitting electrons (Bucciantini et al. 2011).

For electron-positron flows, as appropriate for pulsar winds, the maximum particle Lorentz factor in the downstream frame increases with time as $\gamma_{max} \sim 0.5\, \gamma_r\, (\omega_p t)^{1/2}$ (see Sect. 4). The plasma frequency ω_p can be computed from the number density ahead of the termination shock, which is $n_{TS} = \dot{N}/(4\pi R_{TS}^2 c)$, assuming an isotropic particle flux. Here, $R_{TS} \equiv 3 \times 10^{17} R_{TS,17.5}$ cm is the termination shock radius. Balancing the acceleration rate with the synchrotron cooling rate in the self-generated Weibel fields, the maximum electron Lorentz factor is

$$\gamma_{\rm sync,e} \simeq 3.5 \times 10^8 L_{\rm sd,38.5}^{1/6} \dot{N}_{40}^{-1/3} \epsilon_{B,-2.5}^{-1/3} R_{\rm TS,17.5}^{1/3}. \tag{8}$$

A stronger constraint comes from the requirement that the diffusion length of the highest energy electrons be smaller than the termination shock radius (i.e. a confinement constraint). Alternatively, the acceleration time should be shorter than $R_{\rm TS}/c$, which yields the critical limit

$$\gamma_{\rm conf,e} \simeq 1.9 \times 10^7 L_{\rm sd,38.5}^{3/4} \dot{N}_{40}^{-1/2}, \tag{9}$$

which is generally more constraining than the cooling-limited Lorentz factor $\gamma_{\rm sync,e}$. The corresponding synchrotron photons will have energies

$$h\nu_{\rm conf,e} \simeq 0.17\, L_{\rm sd,38.5}^{2} \dot{N}_{40}^{-1} \epsilon_{B,-2.5}^{1/2} R_{\rm TS,17.5}^{-1}\ {\rm keV} \tag{10}$$

which are apparently too small to explain the X-ray spectrum of the Crab, extending to energies beyond a few tens of MeV. We conclude that Fermi acceleration at the termination shock of PWNe is not a likely candidate for producing X-ray photons via the synchrotron process, and valid alternatives should be investigated.

One possibility—magnetic dissipation of the striped pulsar wind in and around the shock front itself—has been extensively studied, with the conclusion that particle acceleration along extended X-lines formed by tearing of the current sheets may contribute to the flat particle distribution (with spectral index $s_\gamma \simeq 1.5$) required to explain the far infrared and radio spectra of PWNe (e.g., Lyubarsky 2003; Sironi and Spitkovsky 2011a, 2012).[8] Indeed, hard particle spectra are found to be a generic by-product of magnetic reconnection in the relativistic regime appropriate for pulsar winds (Sironi and Spitkovsky 2014; Sironi et al. 2015, see also Kagan et al. 2015 in the present volume). However, further acceleration to gamma-ray emitting energies by the Fermi process cannot occur in the shock that terminates the pulsar wind, if particle scattering depends only on the self-generated turbulence.

Yet, the steady-state hard X-ray and gamma-ray spectra of PWNe do look like the consequences of Fermi acceleration—particle distributions with $s_\gamma \simeq 2.4$ are implied by the observations. In this regard, we argue that the wind termination shock might form in a macroscopically turbulent medium, with the outer scale of the turbulence driven by the large-scale shear flows in the nebula (Komissarov and Lyubarsky 2004; Del Zanna et al. 2004; Camus et al. 2009). If the large-scale motions drive a turbulent cascade to shorter wavelengths, back-scattering of the particles in this downstream turbulence, along with upstream reflection by the transverse magnetic field of the wind, might sustain Fermi acceleration to higher energies.

Another "external" influence of reconnection on the shock structure, that might lead to particle acceleration to higher energies, may be connected to the accelerator behind the recently discovered gamma-ray flares in the Crab Nebula (Abdo 2011). Runaway acceleration of electrons and positrons at reconnection X-lines, a linear accelerator, may inject energetic beams into the shock, with the mean energy per particle approaching the whole open field line voltage, $\gtrsim 10^{16}$ V in the Crab (Arons 2012), as required to explain the Crab GeV flares. This high-energy population can drive cyclotron turbulence when gyrating in the shock-compressed fields, and resonant absorption of the cyclotron harmonics can accelerate the electron-positron pairs in a broad spectrum, with maximum energy again comparable to the whole open field line voltage (Hoshino et al. 1992; Amato and Arons 2006).

[8] Yet, as described by Sironi and Spitkovsky (2011a), this would imply an extremely high plasma flux in the pulsar wind, well above the existing estimates.

6 Conclusions

There has been significant progress in our understanding of relativistic shocks in recent years, thanks to both analytical work and numerical simulations. The highly nonlinear problem of particle acceleration and magnetic field generation in shocks—with the accelerated particles generating the turbulence that in turn mediates their acceleration—is being tackled from first principles, assessing the parameter regime where particle acceleration in relativistic shocks is efficient. In this chapter, we have described the basic analytical formalism of test particle acceleration in relativistic shocks, leading to the "universal" energy slope $s_\gamma \simeq 2.2$ in the ultra-relativistic limit; we have unveiled the most relevant plasma instabilities that mediate injection and acceleration in relativistic shocks; and we have summarized recent results of large-scale PIC simulations concerning the efficiency and rate of particle acceleration in relativistic shocks, and the long-term evolution of the self-generated magnetic turbulence. Our novel understanding of particle acceleration and magnetic field generation in relativistic shocks has profound implications for the modeling of relativistic astrophysical sources, most importantly PWNe, GRBs, and AGN jets.

Acknowledgements We gratefully thank Boaz Katz, Guy Pelletier, Anatoly Spitkovsky and Eli Waxman for their collaboration on many of the issues discussed here. U.K. is supported by the European Union Seventh Framework Programme (FP7/2007–2013) under grant agreement No. 293975, by an IAEC-UPBC joint research foundation grant, and by an ISF-UGC grant. M.L. acknowledges support by the ANR-14-CE33-0019 MACH project.

References

A.A. Abdo, Gamma-ray flares from the Crab Nebula. Science **331**, 739 (2011). doi:10.1126/science.1199705

A. Achterberg, J. Wiersma, The Weibel instability in relativistic plasmas. I. Linear theory. Astron. Astrophys. **475**, 1–18 (2007). doi:10.1051/0004-6361:20065365

A. Achterberg, Y.A. Gallant, J.G. Kirk, A.W. Guthmann, Particle acceleration by ultrarelativistic shocks: theory and simulations. Mon. Not. R. Astron. Soc. **328**, 393–408 (2001). doi:10.1046/j.1365-8711.2001.04851.x

A. Achterberg, J. Wiersma, C.A. Norman, The Weibel instability in relativistic plasmas. II. Nonlinear theory and stabilization mechanism. Astron. Astrophys. **475**, 19–36 (2007). doi:10.1051/0004-6361:20065366

M. Ackermann, Fermi observations of GRB 090510: a short-hard gamma-ray burst with an additional, hard power-law component from 10 keV TO GeV energies. Astrophys. J. **716**, 1178–1190 (2010). doi:10.1088/0004-637X/716/2/1178

E. Amato, J. Arons, Heating and nonthermal particle acceleration in relativistic, transverse magnetosonic shock waves in proton-electron-positron plasmas. Astrophys. J. **653**, 325–338 (2006). doi:10.1086/508050

J. Arons, Pulsar wind nebulae as cosmic pevatrons: a current sheet's tale. Space Sci. Rev. **173**, 341–367 (2012). doi:10.1007/s11214-012-9885-1

J. Arons, M. Tavani, Relativistic particle acceleration in plerions. Astrophys. J. Suppl. Ser. **90**, 797–806 (1994). doi:10.1086/191905

W.I. Axford, E. Leer, G. Skadron, Acceleration of cosmic rays by shock waves, in *Cosmophysics*, ed. by V.A. Dergachev, G.E. Kocharov (1978), pp. 125–134

J. Bednarz, M. Ostrowski, Energy spectra of cosmic rays accelerated at ultrarelativistic shock waves. Phys. Rev. Lett. **80**, 3911–3914 (1998). doi:10.1103/PhysRevLett.80.3911

M.C. Begelman, J.G. Kirk, Shock-drift particle acceleration in superluminal shocks—a model for hot spots in extragalactic radio sources. Astrophys. J. **353**, 66–80 (1990). doi:10.1086/168590

A.R. Bell, The acceleration of cosmic rays in shock fronts, I. Mon. Not. R. Astron. Soc. **182**, 147–156 (1978)

E. Berger, S.R. Kulkarni, D.A. Frail, A standard kinetic energy reservoir in gamma-ray burst afterglows. Astrophys. J. **590**, 379–385 (2003). doi:10.1086/374892

C.K. Birdsall, A.B. Langdon, *Plasma Physics via Computer Simulation* (CRC Press, Boca Raton, 1991)

R. Blandford, D. Eichler, Particle acceleration at astrophysical shocks—a theory of cosmic-ray origin. Phys. Rep. **154**, 1 (1987). doi:10.1016/0370-1573(87)90134-7

R.D. Blandford, J.P. Ostriker, Particle acceleration by astrophysical shocks. Astrophys. J. Lett. **221**, 29–32 (1978). doi:10.1086/182658
P. Blasi, M. Vietri, On particle acceleration around shocks. II. A fully general method for arbitrary shock velocities and scattering media. Astrophys. J. **626**, 877–886 (2005). doi:10.1086/430164
J.J. Brainerd, A plasma instability theory of gamma-ray burst emission. Astrophys. J. **538**, 628–637 (2000). doi:10.1086/309136
A. Bret, Weibel, two-stream, filamentation, oblique, Bell, Buneman... which one grows faster? Astrophys. J. **699**, 990–1003 (2009). doi:10.1088/0004-637X/699/2/990
A. Bret, L. Gremillet, D. Bénisti, Exact relativistic kinetic theory of the full unstable spectrum of an electron-beam-plasma system with Maxwell-Jüttner distribution functions. Phys. Rev. E **81**(3), 036402 (2010). doi:10.1103/PhysRevE.81.036402
N. Bucciantini, J. Arons, E. Amato, Modelling spectral evolution of pulsar wind nebulae inside supernova remnants. Mon. Not. R. Astron. Soc. **410**, 381–398 (2011). doi:10.1111/j.1365-2966.2010.17449.x
O. Buneman, in *Computer Space Plasma Physics* (Terra Scientific, Tokyo, 1993), p. 67
A.M. Bykov, R.A. Treumann, Fundamentals of collisionless shocks for astrophysical application. 2. Relativistic shocks. Astron. Astrophys. Rev. **19**, 42 (2011). doi:10.1007/s00159-011-0042-8
A. Bykov, N. Gehrels, H. Krawczynski, M. Lemoine, G. Pelletier, M. Pohl, Particle acceleration in relativistic outflows. Space Sci. Rev. **173**, 309–339 (2012). doi:10.1007/s11214-012-9896-y
N.F. Camus, S.S. Komissarov, N. Bucciantini, P.A. Hughes, Observations of 'wisps' in magnetohydrodynamic simulations of the Crab Nebula. Mon. Not. R. Astron. Soc. **400**, 1241–1246 (2009). doi:10.1111/j.1365-2966.2009.15550.x
F. Casse, A. Marcowith, R. Keppens, Non-resonant magnetohydrodynamics streaming instability near magnetized relativistic shocks. Mon. Not. R. Astron. Soc. **433**, 940–951 (2013). doi:10.1093/mnras/stt772
P. Chang, A. Spitkovsky, J. Arons, Long-term evolution of magnetic turbulence in relativistic collisionless shocks: electron-positron plasmas. Astrophys. J. **674**, 378–387 (2008). doi:10.1086/524764
S.M. Couch, M. Milosavljević, E. Nakar, Shock vorticity generation from accelerated ion streaming in the precursor of ultrarelativistic gamma-ray burst external shocks. Astrophys. J. **688**, 462–469 (2008). doi:10.1086/592194
O.C. de Jager, A.K. Harding, The expected high-energy to ultra-high-energy gamma-ray spectrum of the Crab Nebula. Astrophys. J. **396**, 161–172 (1992). doi:10.1086/171706
M. De Pasquale, Swift and Fermi observations of the early afterglow of the short gamma-ray burst 090510. Astrophys. J. Lett. **709**, 146–151 (2010). doi:10.1088/2041-8205/709/2/L146
L. Del Zanna, E. Amato, N. Bucciantini, Axially symmetric relativistic MHD simulations of pulsar wind nebulae in supernova remnants: on the origin of torus and jet-like features. Astron. Astrophys. **421**, 1063–1073 (2004). doi:10.1051/0004-6361:20035936
E.V. Derishev, Synchrotron emission in the fast cooling regime: which spectra can be explained? Astrophys. Space Sci. **309**, 157–161 (2007). doi:10.1007/s10509-007-9421-z
E.V. Derishev, F.A. Aharonian, V.V. Kocharovsky, V.V. Kocharovsky, Particle acceleration through multiple conversions from a charged into a neutral state and back. Phys. Rev. D **68**(4), 043003 (2003). doi:10.1103/PhysRevD.68.043003
D. Eichler, M. Pohl, Can ultrahigh-energy cosmic rays come from gamma-ray bursts? Cosmic rays below the ankle and galactic gamma-ray bursts. Astrophys. J. Lett. **738**, 21 (2011). doi:10.1088/2041-8205/738/2/L21
D.C. Ellison, G.P. Double, Nonlinear particle acceleration in relativistic shocks. Astropart. Phys. **18**, 213–228 (2002). doi:10.1016/S0927-6505(02)00142-1
D.C. Ellison, D.C. Warren, A.M. Bykov, Monte Carlo simulations of nonlinear particle acceleration in parallel trans-relativistic shocks. Astrophys. J. **776**, 46 (2013). doi:10.1088/0004-637X/776/1/46
G.D. Fleishman, Diffusive synchrotron radiation from relativistic shocks in gamma-ray burst sources. Astrophys. J. **638**, 348–353 (2006). doi:10.1086/498732
J.T. Frederiksen, C.B. Hededal, T. Haugbølle, Å. Nordlund, Magnetic field generation in collisionless shocks: pattern growth and transport. Astrophys. J. Lett. **608**, 13–16 (2004). doi:10.1086/421262
D.L. Freedman, E. Waxman, On the energy of gamma-ray bursts. Astrophys. J. **547**, 922–928 (2001). doi:10.1086/318386
Y.A. Gallant, M. Hoshino, A.B. Langdon, J. Arons, C.E. Max, Relativistic, perpendicular shocks in electron-positron plasmas. Astrophys. J. **391**, 73–101 (1992). doi:10.1086/171326
M. Gedalin, M.A. Balikhin, D. Eichler, Efficient electron heating in relativistic shocks and gamma-ray-burst afterglow. Phys. Rev. E **77**(2), 026403 (2008). doi:10.1103/PhysRevE.77.026403
M. Gedalin, E. Smolik, A. Spitkovsky, M. Balikhin, Electron heating by filamentary instability. Europhys. Lett. **97**, 35002 (2012). doi:10.1209/0295-5075/97/35002
G. Ghisellini, G. Ghirlanda, L. Nava, A. Celotti, GeV emission from gamma-ray bursts: a radiative fireball? Mon. Not. R. Astron. Soc. **403**, 926–937 (2010). doi:10.1111/j.1365-2966.2009.16171.x

J. Granot, T. Piran, O. Bromberg, J.L. Racusin, R. Daigne, Gamma-ray bursts as sources of strong magnetic fields. Space Sci Rev. (2015). doi:10.1007/s11214-015-0191-6 (this issue), SSSI-54 Chapter 14.

A. Gruzinov, Gamma-ray burst phenomenology, shock dynamo, and the first magnetic fields. Astrophys. J. Lett. **563**, 15–18 (2001). doi:10.1086/324223

A. Gruzinov, E. Waxman, Gamma-ray burst afterglow: polarization and analytic light curves. Astrophys. J. **511**, 852–861 (1999). doi:10.1086/306720

T. Haugbølle, Three-dimensional modeling of relativistic collisionless ion-electron shocks. Astrophys. J. Lett. **739**, 42 (2011). doi:10.1088/2041-8205/739/2/L42

A.F. Heavens, L.O. Drury, Relativistic shocks and particle acceleration. Mon. Not. R. Astron. Soc. **235**, 997–1009 (1988)

M. Hoshino, Wakefield acceleration by radiation pressure in relativistic shock waves. Astrophys. J. **672**, 940–956 (2008). doi:10.1086/523665

M. Hoshino, J. Arons, Preferential positron heating and acceleration by synchrotron maser instabilities in relativistic positron-electron-proton plasmas. Phys. Fluids, B Plasma Phys. **3**, 818–833 (1991). doi:10.1063/1.859877

M. Hoshino, J. Arons, Y.A. Gallant, A.B. Langdon, Relativistic magnetosonic shock waves in synchrotron sources—shock structure and nonthermal acceleration of positrons. Astrophys. J. **390**, 454–479 (1992). doi:10.1086/171296

C.H. Jaroschek, H. Lesch, R.A. Treumann, Ultrarelativistic plasma shell collisions in γ-ray burst sources: dimensional effects on the final steady state magnetic field. Astrophys. J. **618**, 822–831 (2005). doi:10.1086/426066

D. Kagan, L. Sironi, B. Cerutti, D. Giannios, Relativistic magnetic reconnection in pair plasmas and its astrophysical applications. Space. Sci Rev. (2015). doi:10.1007/s11214-014-0132-9

O. Kargaltsev, B. Cerutti, Y. Lyubarsky, E. Striani, Pulsar-wind nebulae. Space. Sci Rev. (2015). doi:10.1007/s11214-015-0171-x

T.N. Kato, Relativistic collisionless shocks in unmagnetized electron-positron plasmas. Astrophys. J. **668**, 974–979 (2007). doi:10.1086/521297

B. Katz, U. Keshet, E. Waxman, Self-similar collisionless shocks. Astrophys. J. **655**, 375–390 (2007). doi:10.1086/509115

S.R. Kelner, F.A. Aharonian, D. Khangulyan, On the jitter radiation. Astrophys. J. **774**, 61 (2013). doi:10.1088/0004-637X/774/1/61

U. Keshet, Analytical study of diffusive relativistic shock acceleration. Phys. Rev. Lett. **97**(22), 221104 (2006). doi:10.1103/PhysRevLett.97.221104

U. Keshet, E. Waxman, Energy spectrum of particles accelerated in relativistic collisionless shocks. Phys. Rev. Lett. **94**(11), 111102 (2005). doi:10.1103/PhysRevLett.94.111102

U. Keshet, B. Katz, A. Spitkovsky, E. Waxman, Magnetic field evolution in relativistic unmagnetized collisionless shocks. Astrophys. J. Lett. **693**, 127–130 (2009). doi:10.1088/0004-637X/693/2/L127

J.G. Kirk, P. Duffy, Topical review: particle acceleration and relativistic shocks. J. Phys. G, Nucl. Part. Phys. **25**, 163 (1999). doi:10.1088/0954-3899/25/8/201

J.G. Kirk, B. Reville, Radiative signatures of relativistic shocks. Astrophys. J. Lett. **710**, 16–20 (2010). doi:10.1088/2041-8205/710/1/L16

J.G. Kirk, P. Schneider, On the acceleration of charged particles at relativistic shock fronts. Astrophys. J. **315**, 425–433 (1987). doi:10.1086/165147

J.G. Kirk, O. Skjæraasen, Dissipation in Poynting-flux-dominated flows: the σ-problem of the Crab Pulsar wind. Astrophys. J. **591**, 366–379 (2003). doi:10.1086/375215

J.G. Kirk, A.W. Guthmann, Y.A. Gallant, A. Achterberg, Particle acceleration at ultrarelativistic shocks: an eigenfunction method. Astrophys. J. **542**, 235–242 (2000). doi:10.1086/309533

S.S. Komissarov, Y.E. Lyubarsky, Synchrotron nebulae created by anisotropic magnetized pulsar winds. Mon. Not. R. Astron. Soc. **349**, 779–792 (2004). doi:10.1111/j.1365-2966.2004.07597.x

G.F. Krymskii, A regular mechanism for the acceleration of charged particles on the front of a shock wave. Dokl. Akad. Nauk SSSR **234**, 1306–1308 (1977)

P. Kumar, R. Barniol Duran, On the generation of high-energy photons detected by the Fermi satellite from gamma-ray bursts. Mon. Not. R. Astron. Soc. **400**, 75–79 (2009). doi:10.1111/j.1745-3933.2009.00766.x

R. Kumar, D. Eichler, M. Gedalin, Electron heating in a relativistic, Weibel-unstable plasma. arXiv:1501.05466 [astro-ph] (2015)

A.B. Langdon, J. Arons, C.E. Max, Structure of relativistic magnetosonic shocks in electron-positron plasmas. Phys. Rev. Lett. **61**, 779–782 (1988). doi:10.1103/PhysRevLett.61.779

M. Lemoine, Acceleration to ultra-high energies, in *American Institute of Physics Conference Proc.*, ed. by H. Sagawa, Y. Kawasaki, T. Sako, M. Takeda, Y. Tsunesada. American Institute of Physics Conference Series, vol. 1367 (2011), pp. 70–75. doi:10.1063/1.3628718

M. Lemoine, Synchrotron signature of a relativistic blast wave with decaying microturbulence. Mon. Not. R. Astron. Soc. **428**, 845–866 (2013). doi:10.1093/mnras/sts081

M. Lemoine, Nonlinear collisionless damping of Weibel turbulence in relativistic blast waves. J. Plasma Phys. **81**, 45101 (2015). doi:10.1017/S0022377814000920

M. Lemoine, G. Pelletier, Particle transport in tangled magnetic fields and Fermi acceleration at relativistic shocks. Astrophys. J. Lett. **589**, 73–76 (2003). doi:10.1086/376353

M. Lemoine, G. Pelletier, On electromagnetic instabilities at ultra-relativistic shock waves. Mon. Not. R. Astron. Soc. **402**, 321–334 (2010). doi:10.1111/j.1365-2966.2009.15869.x

M. Lemoine, G. Pelletier, Dispersion and thermal effects on electromagnetic instabilities in the precursor of relativistic shocks. Mon. Not. R. Astron. Soc. **417**, 1148–1161 (2011). doi:10.1111/j.1365-2966.2011.19331.x

M. Lemoine, B. Revenu, Relativistic Fermi acceleration with shock compressed turbulence. Mon. Not. R. Astron. Soc. **366**, 635–644 (2006). doi:10.1111/j.1365-2966.2005.09912.x

M. Lemoine, E. Waxman, Anisotropy vs chemical composition at ultra-high energies. J. Cosmol. Astropart. Phys. **11**, 9 (2009). doi:10.1088/1475-7516/2009/11/009

M. Lemoine, G. Pelletier, B. Revenu, On the efficiency of Fermi acceleration at relativistic shocks. Astrophys. J. Lett. **645**, 129–132 (2006). doi:10.1086/506322

M. Lemoine, Z. Li, X.-Y. Wang, On the magnetization of gamma-ray burst blast waves. Mon. Not. R. Astron. Soc. **435**, 3009–3016 (2013). doi:10.1093/mnras/stt1494

M. Lemoine, K. Kotera, J. Pétri, On ultra-high energy cosmic ray acceleration at the termination shock of young pulsar winds. arXiv:1409.0159 [astro-ph] (2014a)

M. Lemoine, G. Pelletier, L. Gremillet, I. Plotnikov, A fast current-driven instability in relativistic collisionless shocks. Europhys. Lett. **106**, 55001 (2014b). doi:10.1209/0295-5075/106/55001

M. Lemoine, G. Pelletier, L. Gremillet, I. Plotnikov, Current-driven filamentation upstream of magnetized relativistic collisionless shocks. Mon. Not. R. Astron. Soc. **440**, 1365–1378 (2014c). doi:10.1093/mnras/stu213

A. Levinson, Convective instability of a relativistic ejecta decelerated by a surrounding medium: an origin of magnetic fields in gamma-ray bursts? Astrophys. J. Lett. **705**, 213–216 (2009). doi:10.1088/0004-637X/705/2/L213

Y.E. Lyubarsky, The termination shock in a striped pulsar wind. Mon. Not. R. Astron. Soc. **345**, 153–160 (2003). doi:10.1046/j.1365-8711.2003.06927.x

Y. Lyubarsky, Electron-ion coupling upstream of relativistic collisionless shocks. Astrophys. J. **652**, 1297–1305 (2006). doi:10.1086/508606

Y. Lyubarsky, D. Eichler, Are gamma-ray burst shocks mediated by the Weibel instability? Astrophys. J. **647**, 1250–1254 (2006). doi:10.1086/505523

Y. Lyubarsky, J.G. Kirk, Reconnection in a striped pulsar wind. Astrophys. J. **547**, 437–448 (2001). doi:10.1086/318354

M.A. Malkov, L.O. Drury, Nonlinear theory of diffusive acceleration of particles by shock waves. Rep. Prog. Phys. **64**, 429–481 (2001). doi:10.1088/0034-4885/64/4/201

J. Mao, J. Wang, Gamma-ray burst prompt emission: jitter radiation in stochastic magnetic field revisited. Astrophys. J. **731**, 26 (2011). doi:10.1088/0004-637X/731/1/26

S.F. Martins, R.A. Fonseca, L.O. Silva, W.B. Mori, Ion dynamics and acceleration in relativistic shocks. Astrophys. J. Lett. **695**, 189–193 (2009). doi:10.1088/0004-637X/695/2/L189

M.V. Medvedev, Theory of "jitter" radiation from small-scale random magnetic fields and prompt emission from gamma-ray burst shocks. Astrophys. J. **540**, 704–714 (2000). doi:10.1086/309374

M.V. Medvedev, The theory of spectral evolution of the gamma-ray burst prompt emission. Astrophys. J. **637**, 869–872 (2006). doi:10.1086/498697

M.V. Medvedev, A. Loeb, Generation of magnetic fields in the relativistic shock of gamma-ray burst sources. Astrophys. J. **526**, 697–706 (1999). doi:10.1086/308038

M.V. Medvedev, O.V. Zakutnyaya, Magnetic fields and cosmic rays in GRBs: a self-similar collisionless foreshock. Astrophys. J. **696**, 2269–2274 (2009). doi:10.1088/0004-637X/696/2/2269

M.V. Medvedev, M. Fiore, R.A. Fonseca, L.O. Silva, W.B. Mori, Long-time evolution of magnetic fields in relativistic gamma-ray burst shocks. Astrophys. J. Lett. **618**, 75–78 (2005). doi:10.1086/427921

M.V. Medvedev, J.T. Frederiksen, T. Haugbølle, Å. Nordlund, Radiation signatures of sub-Larmor scale magnetic fields. Astrophys. J. **737**, 55 (2011). doi:10.1088/0004-637X/737/2/55

A. Meli, J.J. Quenby, Particle acceleration in ultra-relativistic oblique shock waves. Astropart. Phys. **19**, 649–666 (2003). doi:10.1016/S0927-6505(02)00257-8

M. Milosavljević, E. Nakar, The cosmic-ray precursor of relativistic collisionless shocks: a missing link in gamma-ray burst afterglows. Astrophys. J. **651**, 979–984 (2006a). doi:10.1086/507975

M. Milosavljević, E. Nakar, Weibel filament decay and thermalization in collisionless shocks and gamma-ray burst afterglows. Astrophys. J. **641**, 978–983 (2006b). doi:10.1086/500654

S.S. Moiseev, R.Z. Sagdeev, Collisionless shock waves in a plasma in a weak magnetic field. J. Nucl. Energy **5**, 43–47 (1963). doi:10.1088/0368-3281/5/1/309
E. Nakar, A. Bret, M. Milosavljević, Two-stream-like instability in dilute hot relativistic beams and astrophysical relativistic shocks. Astrophys. J. **738**, 93 (2011). doi:10.1088/0004-637X/738/1/93
J. Niemiec, M. Ostrowski, Cosmic-ray acceleration at relativistic shock waves with a "realistic" magnetic field structure. Astrophys. J. **610**, 851–867 (2004). doi:10.1086/421730
J. Niemiec, M. Ostrowski, M. Pohl, Cosmic-ray acceleration at ultrarelativistic shock waves: effects of downstream short-wave turbulence. Astrophys. J. **650**, 1020–1027 (2006). doi:10.1086/506901
M. Ostrowski, J. Bednarz, Comment on the first-order Fermi acceleration at ultra-relativistic shocks. Astron. Astrophys. **394**, 1141–1144 (2002). doi:10.1051/0004-6361:20021173
G. Pelletier, M. Lemoine, A. Marcowith, Fermi acceleration at relativistic shocks, in *American Institute of Physics Conference Proc.*, ed. by F.A. Aharonian, W. Hofmann, F. Rieger. American Institute of Physics Conference Series, vol. 1085 (2008), pp. 61–70. doi:10.1063/1.3076750
G. Pelletier, M. Lemoine, A. Marcowith, On Fermi acceleration and magnetohydrodynamic instabilities at ultra-relativistic magnetized shock waves. Mon. Not. R. Astron. Soc. **393**, 587–597 (2009). doi:10.1111/j.1365-2966.2008.14219.x
I. Plotnikov, G. Pelletier, M. Lemoine, Particle transport and heating in the microturbulent precursor of relativistic shocks. Mon. Not. R. Astron. Soc. **430**, 1280–1293 (2013). doi:10.1093/mnras/sts696
R.D. Preece, The synchrotron shock model confronts a "line of death" in the BATSE gamma-ray burst data. Astrophys. J. Lett. **506**, 23–26 (1998). doi:10.1086/311644
I. Rabinak, B. Katz, E. Waxman, Long-wavelength unstable modes in the far upstream of relativistic collisionless shocks. Astrophys. J. **736**, 157 (2011). doi:10.1088/0004-637X/736/2/157
B. Reville, A.R. Bell, On the maximum energy of shock-accelerated cosmic rays at ultra-relativistic shocks. Mon. Not. R. Astron. Soc. **439**, 2050–2059 (2014). doi:10.1093/mnras/stu088
B. Reville, J.G. Kirk, P. Duffy, A current-driven instability in parallel, relativistic shocks. Plasma Phys. Control. Fusion **48**, 1741–1747 (2006). doi:10.1088/0741-3335/48/12/004
E. Rossi, M.J. Rees, Gamma-ray burst afterglow emission with a decaying magnetic field. Mon. Not. R. Astron. Soc. **339**, 881–886 (2003). doi:10.1046/j.1365-8711.2003.06242.x
R. Shaisultanov, Y. Lyubarsky, D. Eichler, Stream instabilities in relativistically hot plasma. Astrophys. J. **744**, 182 (2012). doi:10.1088/0004-637X/744/2/182
L.O. Silva, R.A. Fonseca, J.W. Tonge, J.M. Dawson, W.B. Mori, M.V. Medvedev, Interpenetrating plasma shells: near-equipartition magnetic field generation and nonthermal particle acceleration. Astrophys. J. Lett. **596**, 121–124 (2003). doi:10.1086/379156
L. Sironi, J. Goodman, Production of magnetic energy by macroscopic turbulence in GRB afterglows. Astrophys. J. **671**, 1858–1867 (2007). doi:10.1086/523636
L. Sironi, A. Spitkovsky, Particle acceleration in relativistic magnetized collisionless pair shocks: dependence of shock acceleration on magnetic obliquity. Astrophys. J. **698**, 1523–1549 (2009a). doi:10.1088/0004-637X/698/2/1523
L. Sironi, A. Spitkovsky, Synthetic spectra from particle-in-cell simulations of relativistic collisionless shocks. Astrophys. J. Lett. **707**, 92–96 (2009b). doi:10.1088/0004-637X/707/1/L92
L. Sironi, A. Spitkovsky, Acceleration of particles at the termination shock of a relativistic striped wind. Astrophys. J. **741**, 39 (2011a). doi:10.1088/0004-637X/741/1/39
L. Sironi, A. Spitkovsky, Particle acceleration in relativistic magnetized collisionless electron-ion shocks. Astrophys. J. **726**, 75 (2011b). doi:10.1088/0004-637X/726/2/75
L. Sironi, A. Spitkovsky, Particle-in-cell simulations of shock-driven reconnection in relativistic striped winds. Comput. Sci. Discov. **5**(1), 014014 (2012). doi:10.1088/1749-4699/5/1/014014
L. Sironi, A. Spitkovsky, Relativistic reconnection: an efficient source of non-thermal particles. Astrophys. J. Lett. **783**, 21 (2014). doi:10.1088/2041-8205/783/1/L21
L. Sironi, A. Spitkovsky, J. Arons, The maximum energy of accelerated particles in relativistic collisionless shocks. Astrophys. J. **771**, 54 (2013). doi:10.1088/0004-637X/771/1/54
L. Sironi, M. Petropoulou, D. Giannios, Relativistic jets shine through shocks or magnetic reconnection? Mon. Not. R. Astron. Soc. **450**, 183–191 (2015). doi:10.1093/mnras/stv641
A. Spitkovsky, Simulations of relativistic collisionless shocks: shock structure and particle acceleration, in *Astrophysical Sources of High Energy Particles and Radiation*, ed. by T. Bulik, B. Rudak, G. Madejski. AIP Conf. Ser., vol. 801 (2005), p. 345. doi:10.1063/1.2141897
A. Spitkovsky, On the structure of relativistic collisionless shocks in electron-ion plasmas. Astrophys. J. Lett. **673**, 39–42 (2008a). doi:10.1086/527374
A. Spitkovsky, Particle acceleration in relativistic collisionless shocks: Fermi process at last? Astrophys. J. Lett. **682**, 5–8 (2008b). doi:10.1086/590248

B.E. Stern, J. Poutanen, Radiation from relativistic jets in blazars and the efficient dissipation of their bulk energy via photon breeding. Mon. Not. R. Astron. Soc. **383**, 1695–1712 (2008). doi:10.1111/j.1365-2966.2007.12706.x

A. Stockem, F. Fiúza, R.A. Fonseca, L.O. Silva, Acceleration in perpendicular relativistic shocks for plasmas consisting of leptons and hadrons. Astrophys. J. **755**, 68 (2012). doi:10.1088/0004-637X/755/1/68

P.-H.T. Tam, Q.-W. Tang, S.-J. Hou, R.-Y. Liu, X.-Y. Wang, Discovery of an extra hard spectral component in the high-energy afterglow emission of GRB 130427A. Astrophys. J. Lett. **771**, 13 (2013). doi:10.1088/2041-8205/771/1/L13

X.-Y. Wang, R.-Y. Liu, M. Lemoine, On the origin of >10 GeV photons in gamma-ray burst afterglows. Astrophys. J. Lett. **771**, 33 (2013). doi:10.1088/2041-8205/771/2/L33

E. Waxman, Gamma-ray–burst afterglow: supporting the cosmological fireball model, constraining parameters, and making predictions. Astrophys. J. Lett. **485**, 5–8 (1997). doi:10.1086/310809

J. Wiersma, A. Achterberg, Magnetic field generation in relativistic shocks: an early end of the exponential Weibel instability in electron-proton plasmas. Astron. Astrophys. **428**, 365–371 (2004). doi:10.1051/0004-6361:20041882

J. Zrake, Inverse cascade of nonhelical magnetic turbulence in a relativistic fluid. Astrophys. J. Lett. **794**, 26 (2014). doi:10.1088/2041-8205/794/2/L26

DOI 10.1007/978-1-4939-3550-5_16
Reprinted from *Space Science Reviews* Journal, DOI 10.1007/s11214-014-0132-9

Relativistic Magnetic Reconnection in Pair Plasmas and Its Astrophysical Applications

D. Kagan[1,2] · L. Sironi[3] · B. Cerutti[4] · D. Giannios[5]

Received: 21 July 2014 / Accepted: 27 December 2014 / Published online: 13 January 2015

Abstract This review discusses the physics of magnetic reconnection—a process in which the magnetic field topology changes and magnetic energy is converted to kinetic energy—in pair plasmas in the relativistic regime. We focus on recent progress in the field driven by theory advances and the maturity of particle-in-cell codes. This work shows that fragmentation instabilities at the current sheet can play a critical role in setting the reconnection speed and affect the resulting particle acceleration, anisotropy, bulk flows, and radiation. Then, we discuss how this novel understanding of relativistic reconnection can be applied to high-energy astrophysical phenomena, with an emphasis on pulsars, pulsar wind nebulae, and active galactic nucleus jets.

Keywords Acceleration of particles · Galaxies: active · Instabilities · Magnetic reconnection · Pulsars: general · Radiation mechanisms: non-thermal · Relativistic processes

All authors contributed equally to this review.

✉ D. Kagan
daniel.kagan@astro.huji.ac.il

L. Sironi
lsironi@cfa.harvard.edu

B. Cerutti
bcerutti@astro.princeton.edu

D. Giannios
dgiannio@purdue.edu

1 Racah Institute of Physics, The Hebrew University of Jerusalem, Jerusalem 91904, Israel

2 Raymond and Beverly Sackler School of Physics and Astronomy, Tel Aviv University, Tel Aviv 69978, Israel

3 Harvard-Smithsonian Center for Astrophysics, Cambridge, MA 02138, USA

4 Department of Astrophysical Sciences, Princeton University, Princeton, NJ 08544, USA

5 Department of Physics and Astronomy, Purdue University, 525 Northwestern Avenue, West Lafayette, IN 47907, USA

1 Introduction

Magnetic reconnection is a common phenomenon in which the topology of magnetic field lines is changed and magnetic energy is converted to kinetic energy. Interpretations of space plasma measurements (e.g., Chen et al. 2008; Øieroset et al. 2011) and astronomical observations suggest that reconnection occurs in many places in the Universe. Because the length scale of magnetic fields in astrophysical plasmas is extremely large, of order the size of astrophysical sources, while low plasma resistivity means that the characteristic scale of dissipation is very small, magnetic field lines are typically "frozen" into the astrophysical plasma, inhibiting dissipation. The topological change in the field lines produced by reconnection can break flux freezing and facilitate dissipative energy conversion.

In this review, we focus on reconnection in pair plasmas in the relativistic regime, in which the magnetic energy before the fields reconnect is significantly greater than the total enthalpy of the particles, so that the particles become relativistic when they enter the reconnection region. This condition is precisely stated as

$$\sigma \equiv \frac{B^2}{4\pi mnc^2 w_n} > 1 \tag{1}$$

where B is the magnetic field, n is the total particle number density including all species, and w_n is the enthalpy per particle (assumed to be the same for both species), given by $w_n = \gamma + P/(mnc^2)$, where γ is the mean particle Lorentz factor and P is the particle pressure.[1]

Relativistic reconnection may be of importance in astrophysical magnetically dominated systems such as Pulsar Wind Nebulae (PWN), as well as relativistic jets in Active Galactic Nuclei (AGN) or Gamma Ray Bursts (GRB) which may be magnetically dominated. The observed radiation from such systems is typically highly energetic and nonthermal. Because shock acceleration of particles through the Fermi process is likely to be inefficient in magnetically dominated flows (Sironi and Spitkovsky 2011b; Sironi et al. 2013), it is expected that reconnection is responsible for the acceleration of high energy particles and the production of radiation in these magnetically dominated systems. The role of relativistic reconnection in particle acceleration and radiation is a primary subject of this review.

This paper is organised as follows. In the remainder of Sect. 1, we review simple models of relativistic reconnection (Sect. 1.1) and discuss the physics of particle acceleration in relativistic reconnection (Sect. 1.2). In Sect. 2, we discuss simulations of relativistic reconnection and the resulting particle acceleration, anisotropies, and bulk flows. In Sect. 3, we explore the application of relativistic reconnection in astrophysical systems; this section includes predictions of the radiation spectrum resulting from reconnection in those systems. Finally, in Sect. 4 we present our conclusions.

1.1 Models of Reconnection

We now discuss models of reconnection in detail. Whenever regions of opposite magnetic polarity are present, Maxwell's equations imply that there will be a current sheet in between. In this current layer, magnetic field lines can diffuse across the plasma to reconnect at one or more X-lines. During reconnection, magnetized plasma approaches the central plane of the

[1] If $w_n \gg 1$ but $\sigma < 1$, the plasma is initially relativistic but reconnection is typically weak, so the relativistic reconnection discussed in this review typically fulfills condition (1).

current layer with an asymptotic inflow velocity $v_{\rm in}$, which is also known as the reconnection velocity. After passing the X-line, plasma is expelled from the vicinity of the X-line to either side at the outflow velocity $v_{\rm out}$, which is typically assumed to equal the characteristic speed of magnetic disturbances in plasma, the Alfvén velocity $v_{\rm A}$. In the relativistic regime, $v_{\rm A} = c\sqrt{\sigma/(1+\sigma)} \sim c$. The dimensionless reconnection rate is usually defined as $r_{\rm rec} \equiv v_{\rm in}/v_{\rm out}$.

Outside of the current sheet, non-ideal effects are negligible and the magnetohydrodynamic (MHD) condition

$$\mathbf{E} + \frac{1}{c}\langle \mathbf{v} \rangle \times \mathbf{B} = 0, \tag{2}$$

holds, where $\mathbf{E}$ is the electric field, $\mathbf{B}$ is the magnetic field, and $\langle \mathbf{v} \rangle$ is the mean particle velocity.

In a steady-state configuration which is quasi-two dimensional and does not vary strongly perpendicular to the plane of reconnection, the electric field throughout the reconnection region $\mathbf{E}_{\rm rec}$ is uniform and can be found by applying the condition (2) outside the current sheet, giving

$$\mathbf{E}_{\rm rec} = -\frac{1}{c}(\mathbf{v}_{\rm in} \times \mathbf{B}_0), \tag{3}$$

where $\mathbf{B}_0$ is the reversing magnetic field outside the current sheet. Because there is no velocity flow inside the current sheet, the electric field there is sustained by some non-ideal effect which is responsible for dissipation. The reconnection rate $r_{\rm rec}$ may be related to the electric field by the equation

$$r_{\rm rec} \equiv \frac{v_{\rm in}}{v_{\rm out}} = \frac{E_{\rm rec}}{(v_{\rm A}/c)B_0}. \tag{4}$$

1.1.1 Sweet-Parker Resistive and Kinetic Relativistic Reconnection

Defining δ and L to be the thickness and length of the current sheet, the conservation of mass from the reconnection inflow to the outflow in an incompressible plasma requires

$$\frac{\delta}{L} = r_{\rm rec} = \frac{v_{\rm in}}{v_{\rm out}} \sim \frac{v_{\rm in}}{v_{\rm A}}. \tag{5}$$

This equation is not always applicable to relativistic reconnection due to the possible presence of relativistic bulk flows which violate the incompressibility assumption, but it does apply in the simple steady-state models we discuss in this section.

In the Sweet-Parker resistive model of reconnection, L is taken to be the macroscopic length scale of the magnetic field, while the thickness δ is determined by the dissipation rate that can be sustained by resistivity. The dimensionless parameter that determines the importance of collisional resistivity is the Lundquist number $S \equiv v_{\rm A}L/\eta$, where η is the magnetic diffusivity produced by resistivity. Lyubarsky (2005) has shown that the reconnection rate for relativistic Sweet-Parker resistive reconnection is

$$r_{\rm rec} = \frac{\delta}{L} \sim \frac{1}{\sqrt{S}}, \tag{6}$$

which is identical to the result for non-relativistic Sweet-Parker resistive reconnection. Since the Lundquist number S is very large in astrophysical plasmas (depending on the application, $S \sim 10^{20}$ may be a typical value), Sweet-Parker reconnection is extremely slow. On the other hand, solar flares are believed to be powered by magnetic reconnection requiring that $v_{in}/v_A \sim 0.1$!

Since the collisional resistivity is often extremely small in magnetically dominated astrophysical plasmas, kinetic effects resulting from individual particle motions are likely to be more important than resistivity in many systems.

The characteristic frequency of kinetic effects is the plasma oscillation frequency ω_p, given by

$$\omega_p = \sqrt{\frac{4\pi n q^2}{w_n m}}, \tag{7}$$

where q is the charge of the particles. Kinetic effects become important on spatial scales smaller than the corresponding inertial length c/ω_p (also known as "skin depth"). Comisso and Asenjo (2014) have shown that when kinetic effects are important, the reconnection rate in the relativistic case is given by

$$r_{rec} = \frac{c}{\omega_p L}. \tag{8}$$

Because c/ω_p is small compared to the macroscopic scale L of the field lines, steady-state Sweet-Parker kinetic reconnection is still relatively slow.

1.1.2 Fast Reconnection and the Tearing and Plasmoid Instabilities

There have been many attempts to identify effects that would result in current sheets with smaller aspect ratios L/δ, to allow for faster reconnection. The most basic of these models is the Petschek mechanism (Petschek 1964), which assumes that oblique slow shocks are present around a central X-point, and they effectively limit the length of the reconnection region. Simulations in the non-relativistic regime have found that this configuration is unstable unless an anomalous localised resistivity is present in the center of the reconnection layer, i.e., at the X-line (Uzdensky and Kulsrud 2000). If the aspect ratio of the reconnection region is larger than ~100, oblique slow shocks can form at the end of the reconnection exhausts (Liu et al. 2012; Higashimori and Hoshino 2012), but it is uncertain whether these shocks are analogous to those in the Petschek model. Despite the difficulty in confirming the viability of this mechanism, the name "Petschek reconnection" is often used to describe fast reconnection because kinetic effects can produce an effective anomalous resistivity. Below, we occasionally use the relativistic "Petschek" model derived by Lyubarsky (2005) to parameterise the properties of fast reconnection in the relativistic regime.

Most other models of fast reconnection focus on the effects of instabilities in the current layer. In any current sheet, the oppositely oriented fields constitute a source of free energy. An important instability that draws on this energy is the tearing instability, which at the same time mediates and is mediated by reconnection. The tearing instability produces an alternating series of narrow X-lines where reconnection can occur, separated by large flux ropes. In turn, steady reconnection equilibria contain thin current sheets, which themselves can be unstable to the tearing instability. The nonideal effect that violates flux freezing to produce reconnection at these X-lines may be collisional resistivity, or it may arise from kinetic effects, so the tearing instability, like reconnection, can take both resistive and kinetic forms. The growth rate of the tearing instability depends strongly on the width of the current sheet. For fast growth, the sheet width must be comparable to those associated with resistive or kinetic reconnection (Biskamp 2000; Pétri and Kirk 2007).

A Sweet-Parker resistive current sheet is thin enough that a resistive instability of the Sweet-Parker current sheet, called the plasmoid instability, may break the sheet into X-lines and magnetic islands, thus lowering its aspect ratio L/δ and leading to relatively fast reconnection rates $r_{\mathrm{rec}} \sim 0.01$ even at high Lundquist numbers, for which the unperturbed Sweet-Parker reconnection would be extremely slow (Loureiro et al. 2007; Samtaney et al. 2009; Huang and Bhattacharjee 2010). However, the corresponding reconnection rate in the relativistic case is significantly lower, $r_{\mathrm{rec}} \sim 0.0001$ (Zanotti and Dumbser 2011). In a long kinetic current sheet whose width is comparable to the skin depth, the kinetic tearing instability can grow quickly and break up the current sheet into X-lines and flux ropes, which can result in fast reconnection at $r_{\mathrm{rec}} \sim 0.1$ (e.g., Birn and Hesse 2001). A phase diagram of reconnection has been proposed uniting Sweet-Parker and plasmoid configurations for resistive and kinetic reconnection, with the transition from resistive to kinetic reconnection occurring when the Sweet-Parker sheet width approaches the skin depth, and the transition from Sweet-Parker to plasmoid configurations occurring as the aspect ratio of the reconnection region increases (Ji and Daughton 2011; Comisso and Asenjo 2014). The transition between resistive and kinetic regimes has been proposed as a possible explanation of observed variability in reconnection sites (Goodman and Uzdensky 2008) and the onset of fast reconnection far from the central engine in a Poynting flux model of GRBs (McKinney and Uzdensky 2012). In this review, we focus on the study of kinetic relativistic reconnection and the particle acceleration and radiation that can be produced by such reconnection, because kinetic effects will often dominate in relativistic magnetically dominated astrophysical plasmas, which are typically nearly collisionless.

1.2 Particle Acceleration and Radiation in Reconnection

As discussed earlier, it is thought that magnetic reconnection is likely to be responsible for the acceleration of particles in systems that are magnetically dominated. As particles cross the current sheet at the X-line, they are forced to return into the current sheet by the reversing magnetic field, following Speiser orbits (Speiser 1965). Particles following such orbits can be accelerated in the direction perpendicular to the plane of reconnection (e.g. Zenitani and Hoshino 2001) by the reconnection electric field. Other acceleration mechanisms, in both X-lines and flux ropes, have been found in kinetic simulations; for a review of these mechanisms see Oka et al. (2010), as well as the discussion in Sect. 2.3.

The energy gain per unit time for a charged particle accelerated electromagnetically is given in general by

$$\frac{dW}{dt} = q\mathbf{E} \cdot \mathbf{v} \sim qEc. \tag{9}$$

Particles accelerated in relativistic magnetically dominated systems are typically thought to radiate via the synchrotron mechanism, which tends to place a fundamental constraint on the maximum energy of electromagnetically accelerated particles. The total synchrotron power emitted by a particle is given approximately by

$$\frac{dW}{dt} \sim \frac{2q^4B^2\gamma^2}{3m^2c^3}. \tag{10}$$

In regions where the MHD condition (2) holds, $E \leq B$ and setting $E = B$ allows the derivation of a maximum γ for charged particles, which corresponds to a maximum radiation frequency referred to as the synchrotron burnoff limit. However, during reconnection particles experiencing extreme acceleration at the X-line can spend most of their time deep in the reconnection layer where $E > B$ (Uzdensky et al. 2011). Thus, they are able to evade this restriction and produce radiation beyond the burnoff limit, as we demonstrate in Sect. 3.1.

2 Particle-in-Cell Simulations of Reconnection

2.1 Numerical Setup

2.1.1 Numerical Techniques

The most common method for simulating the kinetic dynamics of a reconnecting plasma involves the use of a particle-in-cell (PIC) code that evolves the discretized equations of electrodynamics—Maxwell's equations and the Lorentz force law. See Birdsall and Langdon (1991) for a detailed discussion of this method. PIC codes can model astrophysical plasmas from first principles, as a collection of charged macro-particles that are moved by integration of the Lorentz force. Each macroparticle represents many physical particles. Currents associated with the macro-particles are deposited on a grid on which Maxwell's equations is discretized. Electromagnetic fields are then advanced via Maxwell's equations, with particle currents as the source term. Finally, the updated fields are extrapolated to the particle locations and used for the computation of the Lorentz force, so the loop is closed self-consistently. So long as current deposition is the only effect of the macro-particles on the field quantities, charge conservation is ensured. This approach is capable of treating all effects present in collisionless plasmas, including particle acceleration to high energies. To ensure that kinetic effects are resolved in the simulation, it is necessary that the grid spacing be much smaller than the skin depth c/ω_{p}, and that the timestep be much smaller than the corresponding timescale ω_{p}^{-1}. To ensure that the momentum space distribution is adequately sampled, keep particle noise at a low level, and reduce the effects of unphysical collisions due to the relatively small number of particles in a Debye sphere, it is necessary that there be several particles per cell for each particle species.

2.1.2 The Harris Current Sheet

The starting equilibrium of most reconnection simulations is the Harris current sheet, which is an exact 1D equilibrium of plasma physics (Harris 1962). It is characterised by the field profile

$$\mathbf{B} = B_0 \tanh \frac{y}{\delta}\, \hat{\mathbf{x}} + \kappa B_0 \hat{\mathbf{z}}, \tag{11}$$

where δ is the half-thickness of the current sheet, which must be of the same order as c/ω_{p} for fast reconnection to occur. The quantity κ sets the relative strength of a uniform "guide" field (orthogonal to the reconnection plane) which may be present in realistic reconnection configurations. For most of the discussion below, we will assume $\kappa = 0$ for the sake of simplicity.

The particles within the current sheet in the Harris equilibrium are initialised in a drifting Maxwell-Juttner thermal distribution in which positively and negatively charged particles have equal and opposite bulk velocities $\boldsymbol{\beta}_+ = -\boldsymbol{\beta}_- = \boldsymbol{\beta}$ (in units of the speed of light) and drifting Lorentz factors of $\gamma_d = 1/\sqrt{1 - \boldsymbol{\beta}^2}$.

The density profile of the Harris current sheet including both electrons and positrons in the simulation frame is

$$n = n_0 \operatorname{sech}^2 \frac{y}{\delta}. \tag{12}$$

Pressure equilibrium requires that $B_0^2 = 8\pi n_0 T_0$, where T_0 is the temperature of the particles (in units of mc^2) in the current sheet including the Boltzmann constant in the simulation frame. Ampère's Law requires that

$$\boldsymbol{\beta}_+ = -\boldsymbol{\beta}_- = B_0/(4\pi n_0 q \delta)(-\hat{\mathbf{z}}). \tag{13}$$

This simple configuration is unstable to the tearing instability and is useful for studying reconnection. An additional uniform background population of particles with rest-frame density n_b and no drift velocity is typically added to the current sheet population. Thus, the total density in the simulation frame of all particles in the middle of the current sheet is $n_0 + n_b$, whereas the total density in the background plasma away from the current sheet is n_b. Using the expression for pressure equilibrium above allows us to express the value of σ far from the current sheet as

$$\sigma = \frac{2 n_0 T_0}{n_b w_{n,b}}, \tag{14}$$

where $w_{n,b}$ is the mean enthalpy of the particles in the background plasma. Note that the value of $n_0 T_0$ is a Lorentz invariant. This equilibrium can be modified while retaining the same value of σ_b by increasing the temperature T_0 and decreasing the value of n_0/n_b to produce an equilibrium with less density contrast but a difference in temperature between the populations; for a detailed discussion, see Melzani et al. (2013). This modification is used in the simulations in this paper.

While the Harris sheet is the most common initial condition for studying reconnection, it should be mentioned that there are other possibilities. Reconnection can be initialised using a force-free current sheet (Guo et al. 2014), and dynamical scenarios such as X-point collapse (Graf von der Pahlen and Tsiklauri 2014). Finally, fully three dimensional configurations (Pontin 2011, and references therein) are likely to be the most realistic starting points for simulation, but only a few PIC simulations have used such configurations (Baumann and Nordlund 2012; Olshevsky et al. 2013).

2.2 Structure of the Reconnection Layer

We now present the structure and the dynamics of the reconnection layer, discussing the results of 2D and 3D PIC simulations. We concentrate on the case of an electron-positron plasma, which has been most widely explored in the literature, both in 2D (Zenitani and Hoshino 2001, 2005, 2007; Zenitani and Hesse 2008; Jaroschek et al. 2004, 2008; Bessho and Bhattacharjee 2005, 2007, 2012; Daughton and Karimabadi 2007; Lyubarsky and Liverts 2008; Cerutti et al. 2012b, 2013; Werner et al. 2014) and in 3D (Zenitani and Hoshino 2008; Yin et al. 2008; Liu et al. 2011; Sironi and Spitkovsky 2011a, 2012; Kagan et al. 2013; Cerutti et al. 2014b; Sironi and Spitkovsky 2014; Guo et al. 2014). The physics of relativistic electron-proton reconnection, yet still at an early stage of investigation, shows remarkable similarities with electron-positron reconnection (Melzani et al. 2014).

As described above, the reconnection layer is set up in Harris equilibrium, with the magnetic field reversing at $y = 0$. For the sake of simplicity, we discuss here the case of anti-parallel fields, without a guide field component. The strength of the alternating fields is parameterized by the magnetization σ defined in Eq. (1). Here, we assume that the background plasma far from the current sheet is cold, so $w_n \sim 1$ and $\sigma = B_0^2/4\pi m n_b c^2$.

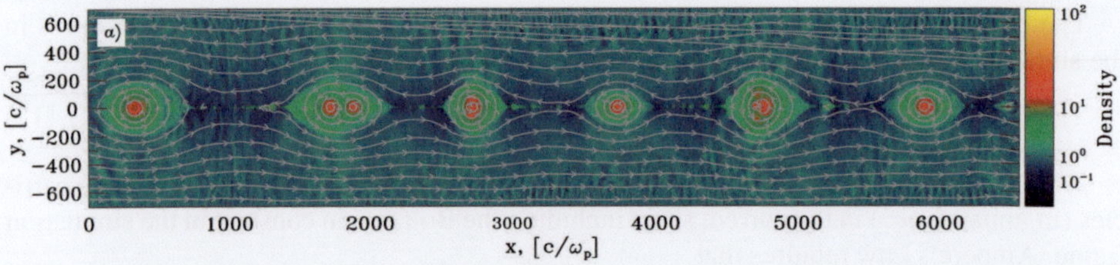

Fig. 1 Structure of the particle density in the reconnection layer at $\omega_{\mathrm{p}}t = 3000$, from a 2D simulation of $\sigma = 10$ reconnection presented in Sironi and Spitkovsky (2014)

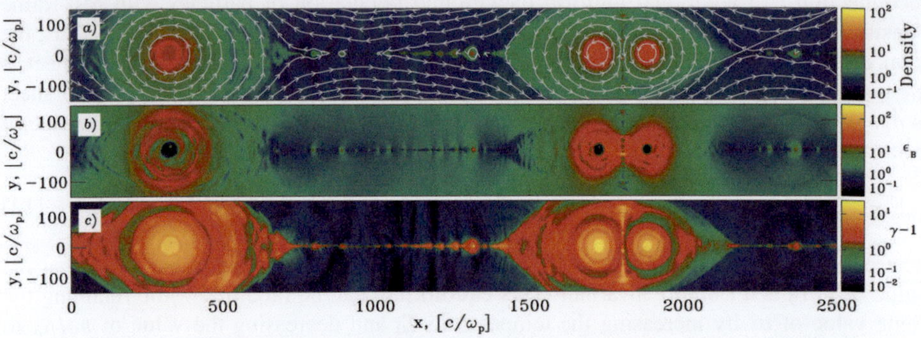

Fig. 2 Structure of the reconnection layer at $\omega_{\mathrm{p}}t = 3000$, from a 2D simulation of $\sigma = 10$ reconnection discussed in Sironi and Spitkovsky (2014). This figure is a zoom-in at $0 \le x \le 2500\, c/\omega_{\mathrm{p}}$ of Fig. 1. We present (**a**) particle density, in units of the number density far from the current sheet (with overplotted magnetic field lines), (**b**) magnetic energy fraction $\epsilon_B = B^2/8\pi m n_b c^2$ and (**c**) mean kinetic energy per particle

As a result of the tearing instability, the reconnection layer fragments into a series of magnetic islands (or flux tubes), separated by X-points. Over time, the islands coalesce and grow to larger scales (Daughton and Karimabadi 2007 have described a similar evolution in non-relativistic reconnection). The structure of the reconnection region at late times is presented in Fig. 1, from a large-scale 2D simulation in a $\sigma = 10$ pair plasma presented in Sironi and Spitkovsky (2014). By zooming into the region $0 \lesssim x \lesssim 2500\, c/\omega_{\mathrm{p}}$ (here, the inertial length c/ω_{p} is measured taking the density far from the current sheet), we see that each X-line is further fragmented into a number of smaller islands. This is a result of the secondary tearing mode (or "plasmoid instability") discussed by Uzdensky et al. (2010). The secondary islands lie at $700\, c/\omega_{\mathrm{p}} \lesssim x \lesssim 1400\, c/\omega_{\mathrm{p}}$ in Fig. 2. They are overdense (Fig. 2a), filled with hot particles (Fig. 2c) and confined by strong fields (Fig. 2b). In between each pair of secondary islands, a secondary X-point mediates the transfer of energy from the fields to the particles. As shown in the next section, efficient particle acceleration occurs at the X-points.

The reconnection rate is $r_{\mathrm{rec}} \equiv v_{\mathrm{in}}/v_{\mathrm{out}} \sim v_{\mathrm{in}}/c \simeq 0.08$ for $\sigma = 10$, nearly constant at late times. The reconnection rate depends on the plasma magnetization. In the case of vanishing guide field, Sironi and Spitkovsky (2014) quote that the reconnection rate in 2D increases from $r_{\mathrm{rec}} \simeq 0.03$ for $\sigma = 1$ to $r_{\mathrm{rec}} \simeq 0.12$ for $\sigma = 30$, and it is nearly independent of σ for larger magnetizations, in agreement with the analytical model by Lyubarsky (2005).

After entering the current sheet, the flow is advected towards the large magnetic islands by the tension force of the reconnected magnetic field (in Fig. 2a–c, the major islands are

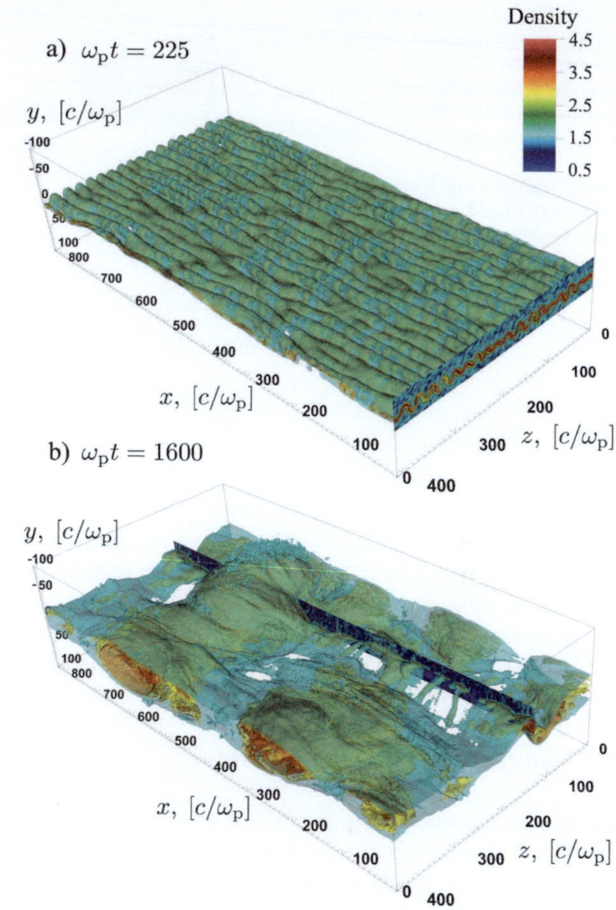

Fig. 3 Structure of the particle density at two different times: (**a**) $\omega_p t = 250$ and (**b**) $\omega_p t = 1600$. The plot refers to a 3D simulation of $\sigma = 10$ reconnection without a guide field, presented in Sironi and Spitkovsky (2014). The 2D slices in the top and bottom panels (at $x = 0$ and $z = 130\,c/\omega_p$, respectively) show the particle number density in that plane

at $200\,c/\omega_p \lesssim x \lesssim 500\,c/\omega_p$ and $1600\,c/\omega_p \lesssim x \lesssim 1900\,c/\omega_p$). Pushed by the ram pressure of the reconnection outflows, the major islands move along the layer, merging with neighboring islands. A merger event in indeed seen at $x \sim 1800\,c/\omega_p$ in Fig. 2. The current sheet formed between the two merging islands is unstable to the tearing mode, and it breaks into a series of secondary islands along the y direction (orthogonal to the primary current sheet).

The evolution of 3D reconnection at late times parallels closely the 2D physics described above, even in the absence of a guide field.[2] As shown in Fig. 3a, the early phases of evolution are governed by the so-called drift-kink (DK) mode (Zenitani and Hoshino 2008; Cerutti et al. 2014b; Sironi and Spitkovsky 2014). The DK instability corrugates the current sheet in the z direction, broadening the layer and inhibiting the growth of the tearing mode at early times. However, at later times the evolution is controlled by the tearing instability (Sironi and Spitkovsky 2014), that produces in the xy plane a series of magnetic islands (or rather, flux tubes), in analogy to the 2D physics. The reconnection layer at late times is

[2]The presence of a strong guide field orthogonal to the reconnecting plane guarantees that the 3D physics will resemble the 2D results, see Guo et al. (2014).

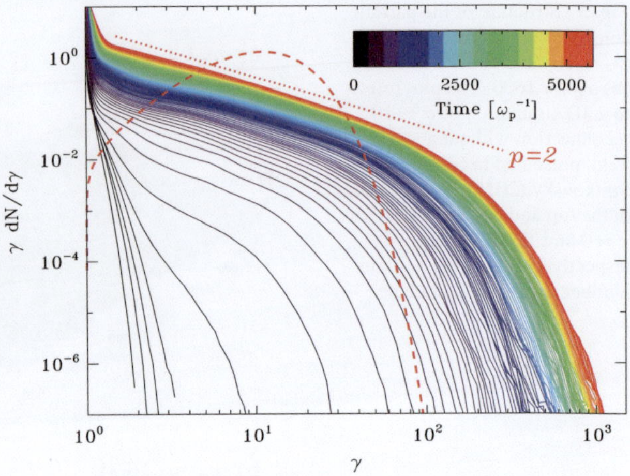

Fig. 4 Temporal evolution of the particle energy spectrum, from a 2D simulation of $\sigma = 10$ reconnection by Sironi and Spitkovsky (2014). The spectrum at late times resembles a power-law with slope $p = 2$ (*dotted red line*), and it clearly deviates from a Maxwellian with mean energy $(\sigma + 1)\, mc^2$ (*dashed red line*, which assumes complete field dissipation)

organized into a few major islands (see the overdense plasmoids in Fig. 3b), separated by underdense regions (transparent in Fig. 3b) where field dissipation by reconnection is most efficient. In short, at late times the 3D physics parallels closely the 2D evolution presented above (yet, with a smaller reconnection rate, $r_{\rm rec} \simeq 0.02$ in 3D versus $r_{\rm rec} \simeq 0.08$ in 2D). As discussed in the next section, this has important implications for the acceleration performance of relativistic reconnection in 3D.

2.3 Non-thermal Particle Acceleration

Relativistic reconnection is an efficient source of non-thermal particles. In Fig. 4 we present the time evolution of the particle energy spectrum, from a 2D simulation of reconnection with $\sigma = 10$ performed by Sironi and Spitkovsky (2014). A generic by-product of relativistic reconnection is the generation of a broad non-thermal spectrum extending to ultra-relativistic energies. For $\sigma = 10$, the spectrum at $\gamma \gtrsim 1.5$ can be fitted with a power-law of slope $p \equiv -d \log N / d \log \gamma \sim 2$ (dotted red line). The spectrum clearly departs from a Maxwellian distribution with mean energy $(\sigma + 1)\, mc^2$ (red dashed line, which assumes complete field dissipation). As shown in Fig. 5, the power-law slope depends on the flow magnetization, being harder for higher σ ($p \sim 1.5$ for $\sigma = 50$, compare solid and dotted green lines). The slope is steeper for lower magnetizations ($p \sim 4$ for $\sigma = 1$, solid and dotted black lines), approaching the result of non-relativistic reconnection, yielding poor acceleration efficiencies (Drake et al. 2010). In the limit $\sigma \gg 1$, Guo et al. (2014) and Werner et al. (2014) have confirmed the trend described above, arguing that the non-thermal slope asymptotes to $p \simeq 1$ for highly magnetized flows.

For magnetizations $\sigma \gtrsim 10$ that yield $p \lesssim 2$, the increase in maximum energy over time is expected to terminate, since the mean energy per particle cannot exceed $(\sigma + 1)\, mc^2$. For a power-law of index $1 < p < 2$ starting from $\gamma_{\rm min} = 1$, the maximum Lorentz factor should saturate at $\gamma_{\rm max} \sim [(\sigma + 1)(2 - p)/(p - 1)]^{1/(2-p)}$. For $\sigma \lesssim 10$ (where $p \gtrsim 2$), the increase in maximum energy does not stop, but it slows down at late times.

In short, 2D simulations of relativistic reconnection produce hard populations of non-thermal particles. However, the structure of X-points in 3D is different from 2D, as emphasized in the previous section. In particular, the DK mode is expected to result in heating, not in particle acceleration (Zenitani and Hoshino 2007). Figure 6 presents the temporal evo-

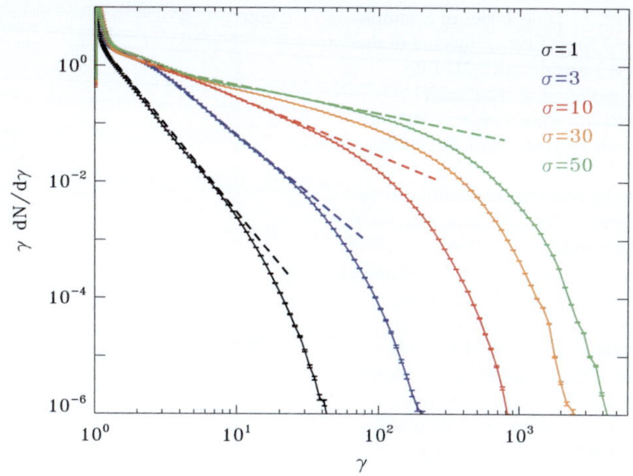

Fig. 5 Dependence of the spectrum on the magnetization, as indicated in the legend. The *dotted lines* refer to power-law slopes of −4, −3, −2 and −1.5 (from *black* to *green*)

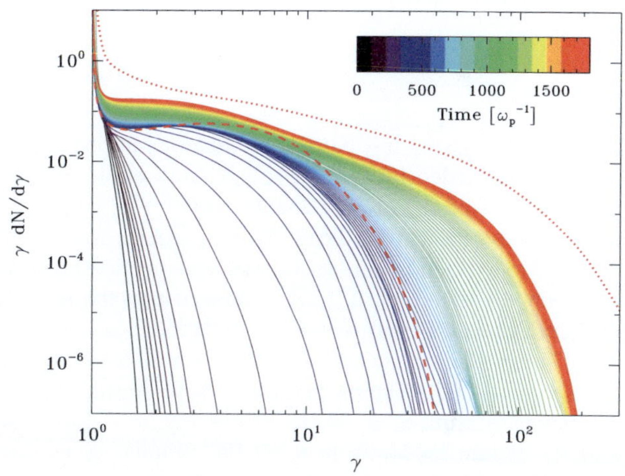

Fig. 6 Temporal evolution of the particle energy spectrum, from a 3D simulation of $\sigma = 10$ reconnection by Sironi and Spitkovsky (2014). The spectra from two 2D simulations with in-plane (out-of-plane, respectively) anti-parallel fields are shown with *red dotted* (*dashed*, respectively) *lines*

lution of the particle spectrum in a 3D simulation with $\sigma = 10$, by Sironi and Spitkovsky (2014). The spectrum at early times is quasi-thermal (black to blue lines in Fig. 6), and it resembles the distribution resulting from the DK mode (the red dashed line shows the spectrum from a 2D simulation with out-of-plane anti-parallel fields, to target the contribution of the DK mode). As discussed above, the DK mode is the fastest to grow, but the sheet evolution at late times is controlled by the tearing instability, in analogy to the 2D physics with in-plane fields. In fact, the spectrum at late times (cyan to red lines in Fig. 6) presents a pronounced high-energy power-law. The power-law slope is $p \sim 2.3$, close to the $p \sim 2$ index of 2D simulations with in-plane fields. With respect to the 2D spectrum (dotted red line in Fig. 6), the normalization and the upper energy cutoff of the 3D spectrum are smaller, due to the lower reconnection rate ($r_{\rm rec} \simeq 0.02$ in 3D versus $r_{\rm rec} \simeq 0.08$ in 2D), so that fewer particles enter the current sheet per unit time, where they get accelerated by a weaker electric field $E_{\rm rec} \sim r_{\rm rec} B_0$.

The mechanism of particle acceleration at X-points has been the subject of various investigations, with analytical (Larrabee et al. 2003; Bessho and Bhattacharjee 2012) or numer-

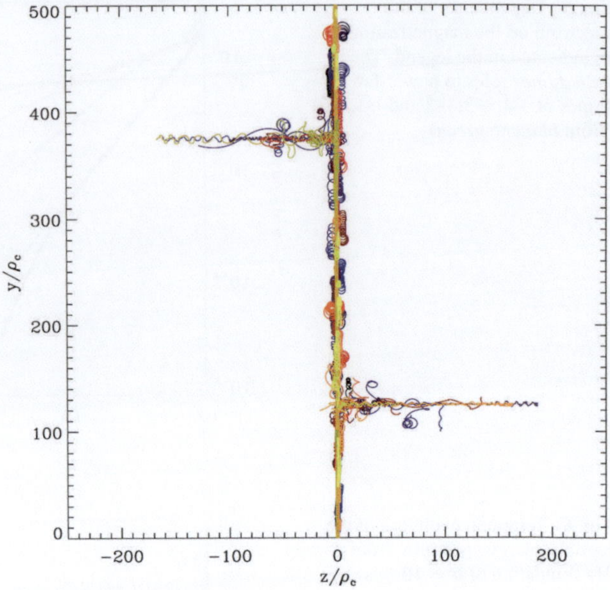

Fig. 7 Trajectories of a sample of 150 particles projected in the (yz)-plane from a 2D PIC simulation of relativistic reconnection with $\sigma = 10$, and without guide field. Each orbit are drawn with a different color to increase the readability of this figure. The simulation starts with two anti-parallel Harris sheets of temperature $kT = mc^2$ located at $y/\rho_c \sim 125$ and 375, where $\rho_c = mc^2/eB_0$. Particles are accelerated along the z-axis within the current layers where the electric field is maximum, and they follow special orbits known as relativistic Speiser orbits. The further the particle gets along the z-axis, the more energetic the particle will become

ical methods.[3] Using test particle simulations in prescribed electromagnetic fields, Nodes et al. (2003), Uzdensky et al. (2011), Cerutti et al. (2012a) found that reconnection naturally produces beams of high-energy particles aligned with the reconnection electric field present within the current layer. These particles follow relativistic Speiser orbits as they are moving back and forth across the reconnection layer. For a steady Sweet-Parker configuration, Uzdensky et al. (2011) showed that the meandering width of the Speiser orbit decreases as the energy of the particle increases, i.e., the most energetic particles, with larger Lorentz factor, are also the most focused along the electric field (see also Kirk 2004; Contopoulos 2007). The properties of these special orbits are also well captured by PIC simulations (Cerutti et al. 2012a, 2013). Figure 7 shows the trajectory of a sample of 150 particles chosen randomly in a 2D PIC simulation with $\sigma = 10$. The particle orbits are projected in the plane perpendicular to the reconnecting field, i.e., here in the (yz)-plane (reconnection happens in the xy-plane). Away from the two layers (located at $y/\rho_c \sim 125$ and 375, with $\rho_c = mc^2/eB_0$), the particles are well magnetized: they gyrate along the field lines and remain at $z = 0$. In contrast, the particles that enter the layer are efficiently boosted along the direction of the electric field (the z-axis) and follow relativistic Speiser orbits. The further the particle gets along the z-direction, the more energetic it will be.

The trajectories of a sample of particles extracted from a 2D simulation with $\sigma = 10$ (in Fig. 8, from Sironi and Spitkovsky 2014) also illustrate the mechanism for the formation

[3]Particle acceleration in magnetic islands (as opposed to X-lines or X-points) is also widely discussed in the literature, both in non-relativistic reconnection (e.g., Drake et al. 2006; Oka et al. 2010)—where the particles are adiabatic, and they bounce several times between the two edges of an island—and relativistic reconnection (Liu et al. 2011; Guo et al. 2014), where the energy gain might come just from a single bounce. However, the inflowing particles interact at first with the X-points, where they get energy from the dissipating fields. It is this first acceleration episode (that we describe below) which will establish the spectral slope and strongly affect the future history of the inflowing particles. In fact, particles accelerated to high energies at the X-point are likely to experience further acceleration via reflection off of moving magnetic disturbances (e.g., in contracting islands or in between two merging islands), which might eventually dominate the overall energy gain.

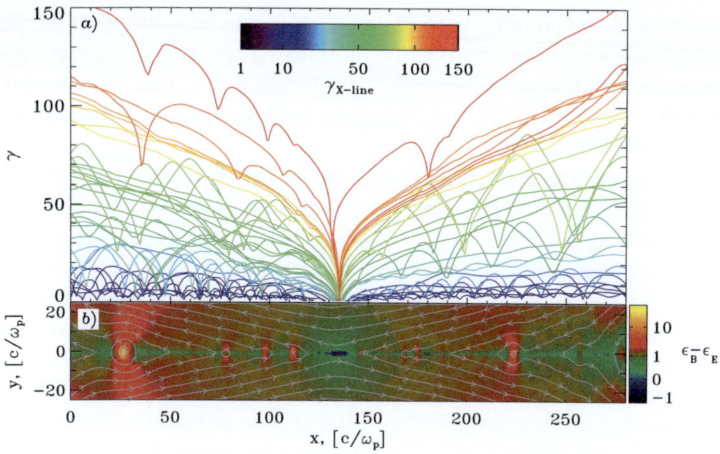

Fig. 8 (**a**) Energy evolution of a sample of selected particles interacting with a major X-point, as a function of the location x along the current sheet. Colors are scaled with $\gamma_{\text{X-line}}$, the Lorentz factor attained at the outflow boundary of the X-line (at $x = 0$ or $280\,c/\omega_{\text{p}}$, depending on the particle). (**b**) $\epsilon_B - \epsilon_E$ at the time when the particles interact with the X-point (here, $\epsilon_E = E^2/8\pi m n_b c^2$ is the electric energy fraction)

of the power-law tail in the particle spectrum. At the X-point located at $x \sim 135\,c/\omega_{\text{p}}$ the magnetic energy is smaller than the electric energy (blue region in Fig. 8b), so the particles become unmagnetized and they get accelerated along z by the reconnection electric field. The final energy of the particles—the color in Fig. 8a indicates the Lorentz factor measured at the outflow boundary of the X-line—directly correlates with the location at the moment of interaction with the current sheet (as argued in the analytical models by Larrabee et al. 2003; Bessho and Bhattacharjee 2012). Particles interacting closer to the center of the X-point (darkest blue in Fig. 8b) are less prone to be advected away along x by the reconnected magnetic field, so they can stay longer in the acceleration region and reach higher Lorentz factors (orange and red lines in Fig. 8a). In other words, energetic particles turn slowly into the reconnected field (B_y in Fig. 8), because the Larmor radius is proportional to γ, so that they spend even more time at the X-point than particles with lower energies. This is an argument originally proposed by Zenitani and Hoshino (2001), that may also explain the power-law nature of the spectrum (along with the impact parameter of the particles in the current sheet). Indeed, a broad power-law distribution is then established, as a result of the different energy histories of particles interacting at different distances from the X-point.

We point out that the most energetic particles (red and orange curves in Fig. 8) are slowly turning around the reconnected magnetic field B_y, and still have a positive $q\mathbf{E} \cdot \mathbf{v}$, so that they gain energy even outside of the blue region (where $|\mathbf{E}| > |\mathbf{B}|$). On the other hand, the green and blue particles experience also the electric fields surrounding the secondary islands, which explains the oscillations in their energy curves.

2.4 Particle Anisotropy and Bulk Motions

It is now well established that relativistic reconnection is an efficient source of non-thermal particle acceleration (see previous section). In usual astrophysical environments, these energetic particles would emit non-thermal radiation via, e.g., synchrotron or inverse Compton scattering. Due to relativistic aberrations, the radiation emitted by highly relativistic particles (with $\gamma \gg 1$) is beamed within a cone of semi-aperture angle $\sim 1/\gamma \ll 1$ along the direction of motion of the emitting particle. As a result, any anisotropy in the particle distribution results in an anisotropic distribution of radiation which is of critical importance in astronomy because the observer probes only one direction at a time. The overall energetic budget or even the shape of the particle spectrum inferred from observations could differ significantly from the isotropically averaged quantities.

Figure 9 presents the angular distribution of the particle 4-velocity vectors as a function of the particle energy, from a 2D PIC simulation with $\sigma \approx 10$ and with no guide field as first reported by Cerutti et al. (2012b). The low-energy particles ($\gamma \sim 1$, top panel) present little anisotropy because these particles have not been accelerated at X-points. At higher energies ($\gamma \gtrsim \sigma$, middle and bottom panel), the particles exhibit clear sign of anisotropy with two beams pointing roughly towards the $\pm x$-directions, i.e., along the reconnection exhausts. Hence, the beams are not necessarily pointing along the reconnection electric field because the tension of the reconnected field lines pushes the particles away from the X-points in the form of a reconnection outflow towards the magnetic islands (see Fig. 2a, and top panel in Fig. 11). Nonetheless, the direction of the beam of energetic particles is not static: it wiggles rapidly within the (xz)-plane (along the horizontal axis in Fig. 9), which results in rapid flares of energetic radiation when the beam crosses the line of sight of a distant observer (Cerutti et al. 2012b). This result has interesting application to astrophysical flares, and in particular to the recently discovered >100 MeV gamma-ray flares discovered in the Crab Nebula (Cerutti et al. 2013, 2014b) (see Sect. 3.1). The Crab flare case is quite extreme in the sense that the particles emitting >100 MeV synchrotron radiation should be accelerated and radiating over a sub-Larmor timescale, so the highest energy radiation should keep the imprint of the particle anisotropy (regardless of the acceleration process), while the low-energy radiation should be more isotropic.

The pronounced anisotropy discussed above lasts for some limited amount of time. Indeed, when the high-energy particles reach the magnetic islands, they isotropize quickly in the strong fields shown in Fig. 2c and they do not contribute to the beamed emission. Since most of the particles at late times are contained in the major islands, it is not surprising that the long-term momentum spectra show little signs of anisotropy, as shown in Fig. 10. Even the residual difference between the momentum spectra along $+z$ and $-z$ (red solid and dashed lines, respectively) diminishes at later times (the 2D momentum spectra at $\omega_{\rm p}t = 1800$ were similar to the 3D results in Fig. 10, showing that the anisotropy decays over time).

It is important to stress that this beaming mechanism is strongly energy-dependent. It should be distinguished from the Doppler boosting due to a relativistic bulk motion in the flow which beams all the particles and radiation by the same factor. In fact, relativistic reconnection produces also relativistic bulk flows as anticipated by Lyubarsky (2005), and constitutes the cornerstone of the fast-variability models for blazar jets by Giannios et al. (2009) (see Sect. 3.2). Figure 11 shows the three components of the fluid velocity vector normalized by the speed of light, $\boldsymbol{\beta} = \mathbf{v}/c$, for the same simulation as in Fig. 9 (where $\sigma = 10$ and with no guide field) and at the same stage. The x-component presents the characteristic signature of a dipolar relativistic flow at every X-point where $\beta_x \approx \pm 0.5$, which corresponds to the reconnection outflow accelerated by the tension of the newly reconnected field lines (i.e., $v_{\rm out}/c$ defined in Sect. 1.1). The y-component shows the inflow of particles from the upstream towards the X-point that feeds the reconnection process with fresh plasma (i.e., $v_{\rm in}/c$ in Sect. 1.1). This motion is due to the $\mathbf{E}_z \times \mathbf{B}_x$ drift velocity, and is about $\beta_y \approx \pm 0.3$ in this particular simulation. The z-component is related to the electric current carried by counter-streaming electrons and positrons around the X-points. The corresponding fluid velocity is about $\beta_z = \beta_+ = -0.6$ for the positrons and $\beta_z = \beta_- = +0.6$ for the electrons,

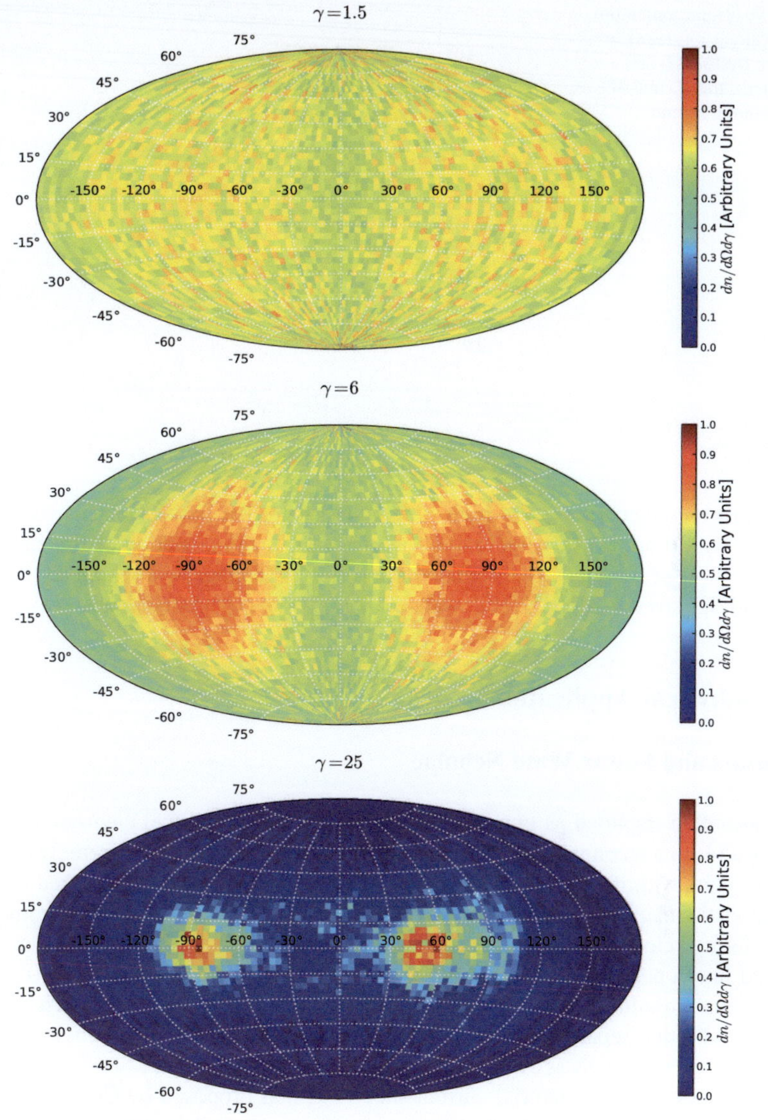

Fig. 9 Angular distribution of the particle 4-velocity vectors **u**, $dn/d\Omega d\gamma$ (contour plot), in three energy bins: $\gamma = \mathbf{1.5 \pm 0.1}$ (*top*), $\gamma = \mathbf{6 \pm 0.3}$ (*middle*), and $\gamma = \mathbf{25 \pm 1.2}$ (*bottom*). In this projection (Aitoff), each direction is given by the latitude angle ($\sin\phi = u_y/|u|$ with $-90^\circ < \phi < +90^\circ$, *vertical axis*) and the longitude angle ($\cos\lambda = u_z/\sqrt{u_x^2 + u_z^2}$ with $-180^\circ < \lambda < +180^\circ$, *horizontal axis*). The precise geometry of the simulation is shown in Fig. 11. These results were obtained from a 2D PIC simulation with $\sigma = 10$ with no guide field (see also Cerutti et al. 2012b, 2013, and Cerutti et al. 2014b in 3D)

but the net velocity is close to zero if both fluids are combined. Overall, the bulk Lorentz factor is only close to unity in this simulation (see Fig. 12), which demonstrates that the anisotropic particle distributions is not related to the relativistic Doppler beaming. This being said, according to Lyubarsky (2005), the bulk Lorentz factor of the outflow in relativistic Petscheck-like reconnection should scale as $\gamma_{\rm out} \sim \sqrt{\sigma}$. Indeed, it is hard to envision a sce-

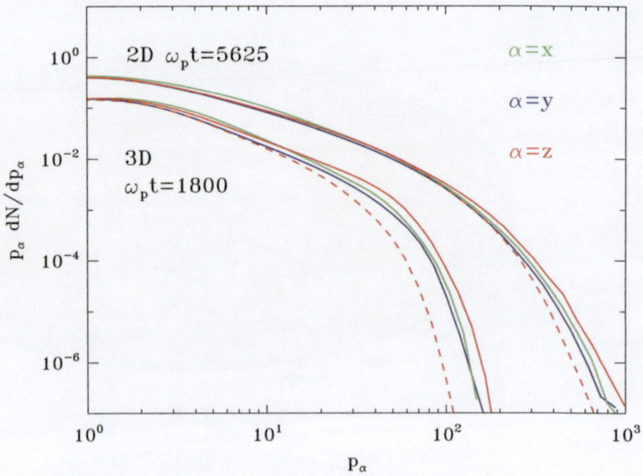

Fig. 10 Positron momentum spectrum along x (*green*), y (*blue*), $+z$ (*red solid*) and $-z$ (*red dashed*), for 2D and 3D, as indicated in the legend

nario of fast reconnection (in the high σ regime) where the outflowing material is not in relativistic bulk motion. PIC simulation runs that follow the evolution of the current sheet on a longer time scale typically find that the $\gamma_{\rm out} \sim \sqrt{\sigma}$ scaling works in the high-σ regime (Sironi and Spitkovsky 2014; K. Nalewajko 2013, private communication).

3 Astrophysical Applications

3.1 Pulsars and Pulsar Wind Nebulae

Pulsars are often regarded as one of the most suitable astrophysical environment for relativistic pair plasma reconnection. These objects are known to generate extremely magnetized plasma of pairs within their co-rotating magnetosphere. The plasma is released in the form of a relativistic magnetized wind beyond the light-cylinder surface, which is defined where the co-rotating velocity with the star equals the speed of light. In the wind region, the magnetic field lines open up and become mostly toroidal due to the fast rotation of the neutron star. This configuration naturally results in the formation of an equatorial current sheet (or "striped wind") that separates the two magnetic polarities. This is the relativistic analog of the well-known ballerina's skirt shaped heliospheric current sheet.

Reconnection in the equatorial current sheet was first proposed by Coroniti (1990) and Michel (1994) as a remedy to the "sigma-problem", i.e., to explain the transition between a Poynting-flux dominated flow formed close to the neutron star ($\sigma \gg 1$) to the observed low-σ pulsar wind nebulae. However, Lyubarsky and Kirk (2001) noticed that the dissipation of the current sheet would be followed by the acceleration of the wind. In the Crab pulsar, the wind would reach the termination shock before reconnection could proceed, unless the pulsar injects pairs at a higher rate than usually expected (Kirk and Skjæraasen 2003). As an alternative to the classical magnetospheric models (e.g., polar-cap, outer-gap, slot-gap), Lyubarskii (1996) suggested that reconnection in the striped wind could also explain the high-energy gamma-ray emission observed in pulsars (Kirk et al. 2002; Pétri 2012; Arka and Dubus 2013; Uzdensky and Spitkovsky 2014).

If, however, reconnection is inefficient in the wind zone, the striped wind is forced to dissipate at the termination shock (Lyubarsky 2003). Using particle-in-cell simulations, Pétri and Lyubarsky (2007) in 1D and Sironi and Spitkovsky (2011a) in 2D and 3D showed that

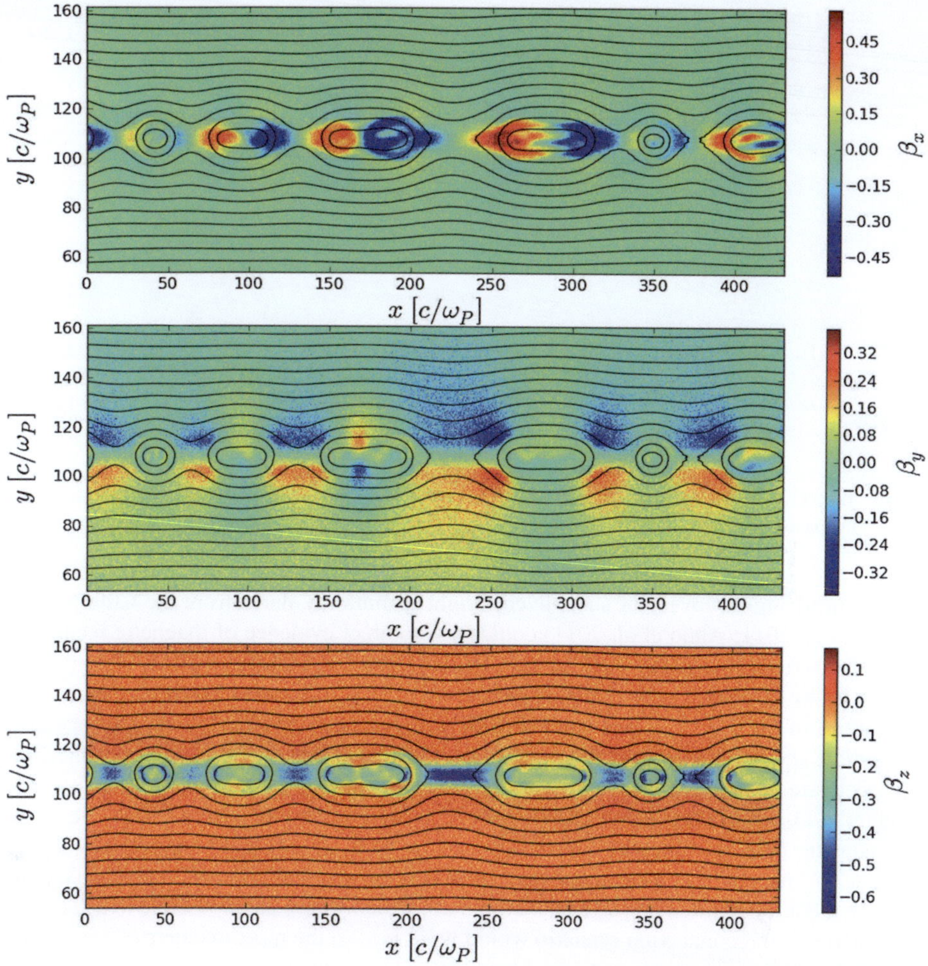

Fig. 11 Positron fluid velocity $\boldsymbol{\beta} = \mathbf{v}/c$ in the x- (*top*), y- (*middle*), and z-directions (*bottom*), for the same simulation as in Fig. 9 ($\sigma = 10$, no guide field). The *black solid lines* show the magnetic field lines. The electron fluid velocity maps are identical, except that $\beta_{z,\text{electrons}} = -\beta_{z,\text{positrons}}$

shock-driven reconnection is able to annihilate the magnetic structure and efficiently accelerates particles regardless of the wind properties for large magnetizations. Whether the dissipation occurs in the wind or at the termination shock, it solves only partially the sigma-problem because the striped wind covers only a fraction of the solid angle set by the inclination angle between the rotation axis and the magnetic axis. Hence, the wind and the nebula should remain magnetically dominated at high latitudes (except for an orthogonal rotator). But, as we know from observations, pulsar wind nebulae are particle kinetic energy dominated flows, so there must be an extra mechanism to dissipate the remaining Poynting flux. Lyubarskij (1992) and Begelman (1998) argued that pulsar wind nebulae should be subject to non-axisymmetric kink-like instabilities. Their hypothesis was recently corroborated by 3D relativistic MHD simulations by Mizuno et al. (2011) and Porth et al. (2013, 2014). The dissipation of the magnetic energy could be done during the non-linear development of these instabilities via non-ideal MHD effects such as magnetic reconnection.

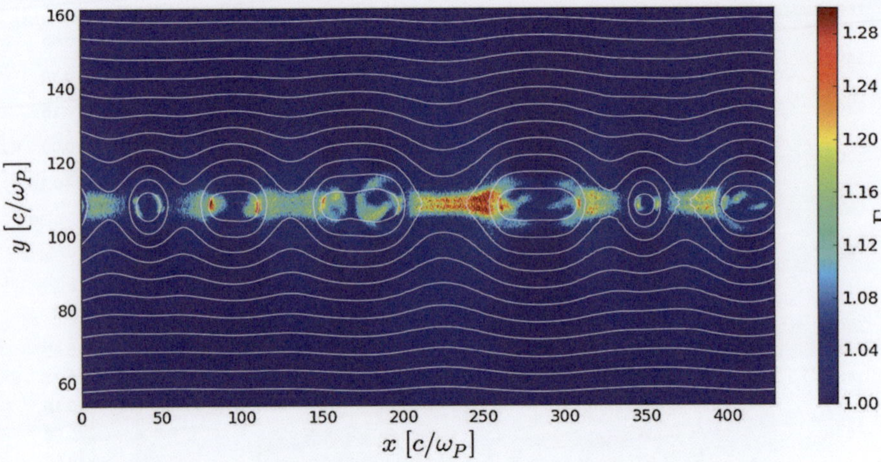

Fig. 12 Total Lorentz factor of the positron fluid, $\Gamma = 1/\sqrt{1-\beta^2}$, computed from Fig. 11. The *white solid lines* are magnetic field lines

The surprising discovery of short-lived, bright gamma-ray flares from the Crab Nebula (Tavani et al. 2011; Abdo et al. 2011) could be the direct evidence of magnetic reconnection in the Nebula (Uzdensky et al. 2011; Clausen-Brown and Lyutikov 2012; Cerutti et al. 2014a). Using 2D and 3D PIC simulations, Cerutti et al. (2013, 2014b) showed that most of the features of the flares can be explained with relativistic reconnection (timescale, energetics, particle and photon spectra). In particular, these studies demonstrated that reconnection can accelerate particles above the synchrotron radiation burn-off limit (Guilbert et al. 1983; de Jager et al. 1996) deep inside the reconnection layer where the electric field overcome the magnetic field (see Fig. 13), as anticipated by Kirk (2004) and Contopoulos (2007) (Sect. 1.2). This result is crucial because it can explain the emission of >100 MeV synchrotron radiation emitted during every Crab flare, which would be impossible to achieve in ideal MHD. The reconnection scenario would work best in the most magnetized regions of the nebula, i.e., near the poles and possibly in the jets (Cerutti et al. 2012a; Lyubarsky 2012; Komissarov 2013; Mignone et al. 2013). Unfortunately, the current gamma-ray telescopes do not have the angular resolution to pin down the precise location of the flares within the Nebula.

3.2 Jets from Active Galactic Nuclei

Jets from active-galactic nuclei have been monitored for decades at practically all accessible electromagnetic wavelengths resulting in a very rich phenomenology (Urry and Padovani 1995). When the jet is pointing close to our line of sight, it is referred to as a "blazar". Recent observational progress in the blazar field has been immense. In particular, Cherenkov telescopes can now detect minute timescale variability in an increasing number of blazars (Aharonian et al. 2007). These novel results strongly constrain the hydrodynamical models for the jet emission.

A broader consensus has emerged regarding the qualitative nature of the "central engine". The energy source in this view is a spinning black hole or the inner accretion disk threaded by a strong magnetic field (see, e.g., Blandford and Payne 1982). This field transfers rotation energy outward as a Poynting flux. While part of the magnetic energy is used for the

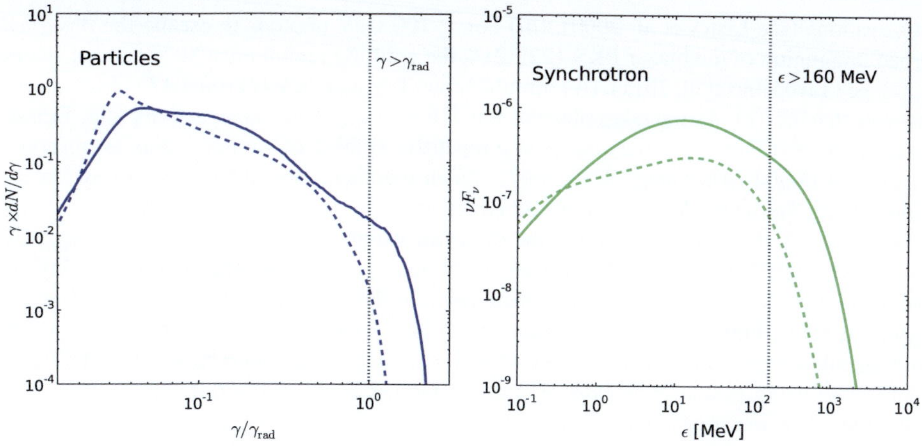

Fig. 13 Isotropically-averaged particle spectrum ($\gamma dN/d\gamma$, *left panel*) and synchrotron radiation energy distribution (νF_ν, *right panel*) in a 2D (*solid line*) and 3D (*dashed line*) PIC simulations of relativistic reconnection, including the effect of the radiation reaction force on the particles. The *vertical dotted lines* show the radiation-reaction limited energy of a particle if $E = B_0$ ($\gamma = \gamma_{\rm rad}$, *left*), and the corresponding maximum synchrotron photon energy ($\epsilon = 160$ MeV independent of E and B_0, *right*). Figure adapted from Cerutti et al. (2014a)

bulk acceleration of the jet, much of the energy remains in the magnetic field (Lyubarsky 2010) and is available to power the jet emission through dissipation by instabilities and magnetic reconnection (Giannios et al. 2009). In this picture, the jet is expected to be magnetically dominated in the emitting region, i.e., one deals with relativistic reconnection. In the following we show that applying our current understanding of relativistic reconnection to the physical conditions expected is blazar jets, reconnection can account for the extreme energetics and timescales inferred by blazar observations (for a similar approach to the modeling of the emission from gamma-ray bursts see Giannios and Spruit 2005; Lyutikov 2006; Zhang and Yan 2011). The possibility that ultra-high-energy cosmic ray acceleration takes place at the current sheets of the reconnection regions of powerful jets is investigated in Giannios (2010).

The magnetic reconnection model for blazar emission: Blazar emission varies on timescales typically ranging from hours to years and is thought to reflect, in part, variations of the gas properties in the black-hole vicinity.[4] The recently discovered ultra-fast TeV flares from several blazars[5] (see, e.g., Aharonian et al. 2007; Albert et al. 2007) are strongly challenging the models for the blazar emission (Ghisellini and Tavecchio 2008; Giannios et al. 2009). This rare but generic blazar activity has several very revealing properties. (i) Fast flares have $\sim$10 minute variability timescale, i.e., a factor $\sim$100 shorter than the light-crossing time of the size of the black hole, pointing to extremely compact emitting regions. (ii) The emitting material must move with $\Gamma_{\rm em} \gtrsim 50$–100 for the TeV radiation to avoid annihilation by soft radiation fields at the source (Begelman et al. 2008; Finke et al. 2008); these values of $\Gamma_{\rm em}$ are much larger than the bulk jet motion $\Gamma_j \sim 10$ typically inferred in blazars from radio

[4]Several hours is the event-horizon light-crossing time of a billion solar-mass black hole—mass typically inferred for the central engine in blazars: $t_{\rm cross} = 2GM_{\rm BH}/c^3 \simeq 10^4 M_9$ s.

[5]Including Mrk 421, Mrk 501, PKS 2155-304, PKS 1222-216, and BL Lac.

observations (see Lister et al. 2009). (iii) For $\gtrsim 100$ GeV photons to escape the *observed* broad line region of the blazar PKS 1222-216, the emitting region must be located at scales $\gtrsim 0.5$ pc (Tavecchio et al. 2011). (iv) Simultaneous TeV and GeV (*Fermi-LAT*) observations indicate that the TeV flaring takes place on top of longer day-long blazar activity (e.g. Tanaka et al. 2011). (v) Fast flares may come in a repetitive fashion of similar events as observed in PKS 2155-304 (Aharonian et al. 2007). Taken together, these inferences are extremely constraining for the models for the blazar emission.

Giannios et al. (2009) argued that the ultra-fast variability must be generated internally in the jet by MHD instabilities. In strongly magnetized jets, the reconnection process injects energetic particles in compact, fast moving regions. These regions are natural emitters of powerful flares. Furthermore, the emitting material is expected to be faster than the jet on average allowing for TeVs to escape the source. For a jet moving with bulk $\Gamma_j \sim 10$–20 and a plasmoid being ejected with bulk $\gamma_{\rm out} \simeq \sqrt{\sigma}$ (as measured in the rest frame of the jet), the emitting region moves with $\Gamma_{\rm em} \simeq 2\Gamma_j \gamma_{\rm out}$ (in the frame of the host galaxy). For $\sigma \sim$ several, one can easily account for the required $\Gamma_{\rm em} \gtrsim 50$. Applications of the model to fit spectra of specific sources are reported in Giannios et al. (2010), Nalewajko et al. (2011).

The Giannios et al. (2009) model is based on a simplified picture for the reconnection geometry adopting a steady state reconnection model. As pointed out by Narayan and Piran (2012) steady reconnection cannot account for the fastest evolving blazar flares because the variability timescale is limited by the reconnection speed $\beta_{\rm in} < 1$. However, assuming steady reconnection is over-simplistic. Solar and Earth magnetosphere observations and recent advances in theory and numerical simulations (see previous sections) have revealed that reconnection is an inherently time-dependent, highly dynamic, process (see, e.g., Lin et al. 2005; Park et al. 2006; Karlický and Kliem 2010). These time-dependent aspects of reconnection are crucial in understanding the fastest timescales involved in blazar flaring. For the physical conditions prevailing in jets, the reconnection current sheets are expected to suffer from tearing instabilities that lead to their fragmentation to a large number of plasmoids (Loureiro et al. 2007; Bhattacharjee et al. 2009). The plasmoids grow rapidly through mergers before leaving the reconnection region. Occasionally plasmoids undergo significant growth to a scale of order of that of the reconnection region, forming "monster" plasmoids (Uzdensky et al. 2010; see Fig. 14; left panel). The relativistic motion of the plasmoids in the rest frame of the jet results in additional beaming of their emission (i.e., beyond that induced by the jet motion). When the layer's orientation is such that plasmoids beam their emission towards the observer, powerful and fast evolving flares emerge. *Here we focus on the characteristic observed timescales and luminosities resulting from plasmoids that form in the reconnection region.* For simplicity, we assume that the dissipated energy is efficiently converted into radiation.[6]

Giannios (2013) demonstrated that a broad range of blazar phenomenology can be qualitatively understood in the context of plasmoid-dominated reconnection. The virtue of the model is that it can be applied to all blazar sources with observed fast flaring for similar adopted parameters. The model favors pc-scale dissipation for the origin of the fast flaring and provides theoretical motivation for such dissipation distance. Another interesting aspect of the model is that a sequence of fast flares is expected to have similar timescale set by the size of the reconnection layer as observed in PKS 2155. This work has demonstrated that the tight energetic, emitter Lorentz factor, and timescale constraints (i)–(v) are satisfied in the

[6]In practice the blazar emission is likely to result of ultrarelativistic electrons cooling via synchrotron radiation and Compton scattering. As discussed in previous sections, relativistic reconnection is an effective means of accelerating particles to such extreme energies.

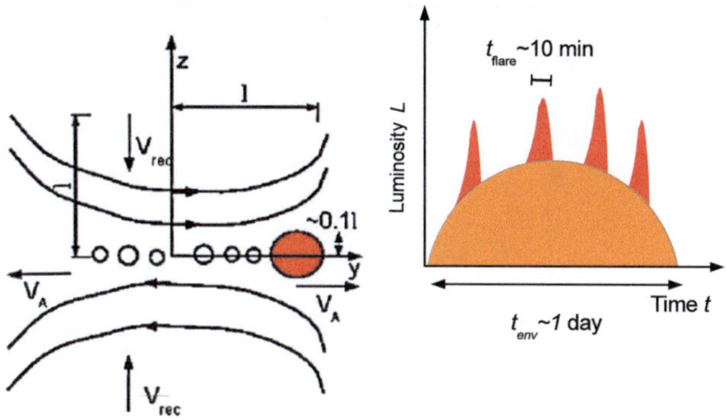

Fig. 14 *Left Panel*: Schematic representation of the geometry of reconnection process shown in a frame comoving with the jet. Magnetic field lines of opposite polarity annihilate at the x–y plane with speed $v_{\rm rec} = \beta_{\rm in} c$. The reconnection layer fragments to a large number of plasmoids. Regularly, plasmoids undergo multiple mergers resulting in a "monster" plasmoid (*shaded blob*). *Right Panel*: Sketch of the emission from plasmoid-dominated reconnection. The reconnection proceeds on a global timescale $t_{\rm rec} = l/\beta_{\rm in} c$, powering $\sim$1 day long flares (or envelope emission). Regularly, plasmoids grow to become "monster" plasmoids (*shaded blob*) giving rise to powerful, fast-evolving flares of duration $t_{\rm flare} \sim 10$ minutes. Several fast flares are expected from a single reconnection event

reconnection model. *More importantly, the basic assumptions of the Giannios 2013 analysis on the properties of the reconnection layer have been fully verified by PIC simulations since then (see previous sections).*

In the following of this section we make a plausibility argument for the model: we estimate the characteristic observed timescales and luminosities resulting from plasmoids that form in the reconnection region (for full derivations see Giannios 2013). To this end, we consider a blob (or plasmoid) emerging from the reconnection layer moving with the Alfvén speed of the reconnection upstream, i.e., with a corresponding bulk Lorentz factor $\gamma_{\rm out} \simeq \sqrt{\sigma}$ (measured in the jet rest frame) and of size $R''_p = f l'$, where l' is the characteristic scale of the reconnection region and f is a dimensionless parameter of the order of 0.1, as expected for the largest, "monster" plasmoids (Uzdensky et al. 2011); hereafter, primed (double primed) quantities are measured in the rest frame of the jet (emitting blob).[7] The observed characteristic variability time for the plasmoid emission is $t_{\rm v} \simeq R''_p/\delta_p c$, where δ_p is the Doppler boost of the plasmoid radiation towards the observer. For a central engine in which the magnetic field varies on a dynamical time $\sim R_{\rm Sch}/c$, the characteristic scale of the reconnection region can be estimated to be $l' \simeq \Gamma_j R_{\rm Sch}$ resulting in

$$t_{\rm v} = \frac{f \Gamma_j R_{\rm Sch}}{\delta_p c} = 400 f_{-1} \Gamma_{j,20} M_9 \delta_{p,50}^{-1} \ {\rm s}, \tag{15}$$

where $\delta_p = 50\delta_{p,50}$, $f = 0.1 f_{-1}$, $\Gamma_j = 20\Gamma_{j,20}$. $f \sim 0.1$ describes the largest plasmoids expected in the layer (Uzdensky et al. 2010). Flaring on several minute timescale is therefore expected in this picture.

[7]We assume that the plasmoid instability operates across the whole length of the current sheet, as opposed to a situation where central, very compact, dissipation region forms and is surrounded by extended magnetic separatrices (the slow shocks in Petscheck model) across which most of the plasma flows. In the latter case, the monster plasmoids may be smaller.

Consider a jet emerging from a supermassive black hole with (isotropic equivalent) power $L_{\rm iso}$, opening angle θ_j and Lorentz factor Γ_j. We also assume that $\theta_j \Gamma_j = 0.2$ as indicated by observations (Pushkarev et al. 2009). The typical bulk Lorentz factor of gamma-ray active blazars is $\Gamma_j \sim 10$–20 (Savolainen et al. 2010; Piner et al. 2012). The energy density at the dissipation, or "blazar", zone is

$$U'_j = \frac{L_{\rm iso}}{4\pi(\theta_j R_{\rm diss})^2 \delta_j^4 c}. \quad (16)$$

The dissipation distance $R_{\rm diss}$ is estimated requiring that the reconnection proceeds within the expansion time of the jet ($R_{\rm diss}/\Gamma_j c \sim l'/\epsilon c$).

Pressure balance across the reconnection layer requires the energy density of the plasmoid to be similar to that of the jet $U''_p \sim U'_j$. Assuming efficient conversion of dissipated energy into radiation, the rest-frame luminosity of the plasmoid is thus $L_{p,\rm obs} = \delta_p^4 L'' = \delta_p^4 U''_p 4\pi R''^2_p c$. Putting everything together, the observed luminosity of the plasmoid is (Giannios 2013)

$$L_{p,\rm obs} = 10^{47} \frac{\beta^2_{{\rm in},-1} f^2_{-1} \delta^4_{p,50} L_{{\rm iso},48}}{\delta^4_{j,20}} \text{ erg/s}. \quad (17)$$

The Doppler factor of the plasmoid δ_p depends on several parameters. It is related to Γ_j, $\gamma_{\rm out}$, the angle of the plasmoid with respect to the jet motion and the observer's angle of sight. For typical situations where the reconnection layer is at a large $\theta \sim \pi/2$ angle with respect to the jet propagation (as seen in the jet rest) and fairly aligned with the observer (giving powerful flares) $\delta_p \sim \Gamma_j \gamma_{\rm out}$. One can see (see Eq. (17)) that powerful flares on a timescale of $\sim$10 min is possible even with very modest relativistic motions within the jet $\gamma_{\rm out} \sim 2$.

Ejection of multiple monster plasmoids: During a reconnection event multiple monster plasmoids are expected to form. 2D simulations (Loureiro et al. 2012) indicate that monster plasmoids form every few Alfvén times $t_{\rm A}$ or at a rate of $\sim 0.3 t_{\rm A}^{-1}$. It appears likely that 2D simulations underestimate the rate of formation of monster plasmoids. The actual rate may be higher when the 3D structure of the layer is considered (Sironi and Spitkovsky 2014). If monster plasmoids emerge at a rate $\sim(0.3$–$3)t_{\rm A}^{-1}$, some $(3$–$30)/\beta_{{\rm in},-1}$ plasmoids are expected from a single reconnection layer powering multiple flares. A sketch of such pattern is shown in Fig. 14.

The "envelope emission" from the reconnection region: The bulk motion of a monster plasmoid is expected to be similar to the speed of other structures (e.g. smaller plasmoids) leaving the reconnection region. When the plasmoid emission is beamed towards the observer (powering a fast flare), the overall emission from the current layer is also beamed by a similar factor. The emission from the layer forms a slower-evolving "envelope". In the following we estimate the timescale and luminosity of the emission from the reconnection layer.

At the dissipation distance $R_{\rm diss}$, the reconnection proceeds within the expansion time of the jet ($R_{\rm diss}/\Gamma_j c \sim l'/\beta_{\rm in} c$) which is observed to last for $t_{\rm exp,obs} \simeq R_{\rm diss}/\Gamma_j^2 c$. Therefore, $t_{\rm exp,obs}$ corresponds to the observed duration of the envelope emission which is simply (using also Eq. (15)):

$$t_{\rm env} = \frac{R_{\rm diss}}{\Gamma_j^2 c} = 10^5 \frac{M_9}{\beta_{{\rm in},-1}} \text{ s}. \quad (18)$$

The duration of the envelope emission is ~days. Such timescale is characteristic of blazar flares.

The (lab frame) energy available to power the envelope emission is $E_{\rm env} = U_j 2l'^3/\Gamma_j$, where $U_j = \Gamma_j^2 U_j'$ is the energy density of the jet and $2l'^3/\Gamma_j$ accounts for (lab frame) volume of the reconnection region that powers each minijet (see Fig. 14). The emitted luminosity of the reconnection region is $E_{\rm env}/t_{\rm env}$. It can be converted into *observed* luminosity by accounting for beaming factor of the emission $\sim\delta_p^2$:

$$L_{\rm env,obs} \simeq 2\Gamma_j^2 \delta_p^2 l'^2 U_j' \beta_{\rm in} c = 3 \times 10^{46} \frac{\Gamma_{j,20}^2 \delta_{p,50}^2 \beta_{\rm in,-1}^3 L_{\rm iso,48}}{\delta_{j,20}^4} \text{ erg/s}. \tag{19}$$

The envelope emission is quite bright. Dividing Eqs. (17) and (19), one arrives to a fairly simple expression for the ratio of the plasmoid to envelope luminosities $L_p/L_{\rm env} \sim 3 f_{-1}^2 \delta_{p,50}^2/(\Gamma_{j,20}^2 \beta_{\rm in,-1})$. The luminosity contrast depends only on the Lorentz factor of the minijet in the rest frame of the jet $\gamma_p \simeq \delta_p/\Gamma_j$, the size of the plasmoid parametrized by f, and the reconnection sped $\beta_{\rm in}$. The observed luminosity ratio is of order unity constraining $\delta_{p,50}/\Gamma_{j,20} \sim 1$ for $\beta_{\rm in} \sim f \sim 0.1$. The ratio $\delta_{p,50}/\Gamma_{j,20}$ is determined by the reconnection-induced bulk motions in the jet and points to $\gamma_{\rm out} \sim 2$ or, equivalently, moderately magnetized jet with $\sigma \sim$ several.

Most of the current numerical work on relativistic reconnection (and this review so far) has focused on the case of electron-positron plasmas. The composition of the jet flow is still an open question but an electron-proton jet is a strong possibility. Electron-ion reconnection is more challenging, on a numerical level, than electron-positron reconnection, since the computation has to resolve the small scales of electrons, yet the system evolves on the longer ion timescales. However, the physics of relativistic electron-proton reconnection, yet still at an early stage of investigation, shows remarkable similarities with electron-positron reconnection (e.g., Melzani et al. 2014). A detailed investigation of relativistic reconnection in the case of unequal mass charges is of paramount importance to obtaining predictions for the acceleration of electrons and cosmic rays in blazar jets.

4 Conclusion

There has been significant progress in our understanding of relativistic reconnection in recent years, thanks to both analytical works and numerical simulations. One important outcome is that plasma instabilities in current sheets play a crucial role in the dynamics of reconnection. In particular, the tearing instability which fragments the current sheet, leads to fast reconnection and efficient non-thermal particle acceleration. Particle-in-cell simulations are now large enough to unambiguously identify broad, hard power laws in the particle energy distributions (in the high-magnetization limit). The power-law index is typically harder than the universal ~ -2 index expected in shock acceleration. These impressive developments were also motivated by puzzling observations of high-energy phenomena in the Universe, especially flaring gamma-ray sources. Ultra-rapid gamma-ray flares discovered in the Crab Nebula and in several AGN jets are too fast and too bright to be explained by conventional models. Particle beaming and relativistic bulk motions associated with relativistic reconnection can alleviate these difficulties. We expect fast new developments in this field, with more applications to astrophysical objects.

Acknowledgements We thank the referees for useful comments that helped to improve the manuscript. L.S. is supported by NASA through Einstein Postdoctoral Fellowship grant number PF1-120090 awarded by the Chandra X-ray Center, which is operated by the Smithsonian Astrophysical Observatory for NASA under contract NAS8-03060. B.C. acknowledges support from the Lyman Spitzer Jr. Fellowship awarded by the Department of Astrophysical Sciences at Princeton University, and the Max-Planck/Princeton Center for Plasma Physics. D.G. acknowledges support from the NASA grant NNX13AP13G.

References

A.A. Abdo, M. Ackermann, M. Ajello, A. Allafort, L. Baldini, J. Ballet, G. Barbiellini, D. Bastieri, K. Bechtol, R. Bellazzini, B. Berenji, R.D. Blandford, E.D. Bloom, E. Bonamente, A.W. Borgland, A. Bouvier, T.J. Brandt, J. Bregeon, A. Brez, M. Brigida, P. Bruel, R. Buehler, S. Buson, G.A. Caliandro, R.A. Cameron, A. Cannon, P.A. Caraveo, J.M. Casandjian, Ö. Çelik, E. Charles, A. Chekhtman, C.C. Cheung, J. Chiang, S. Ciprini, R. Claus, J. Cohen-Tanugi, L. Costamante, S. Cutini, F. D'Ammando, C.D. Dermer, A. de Angelis, A. de Luca, F. de Palma, S.W. Digel, E. do Couto e Silva, P.S. Drell, A. Drlica-Wagner, R. Dubois, D. Dumora, C. Favuzzi, S.J. Fegan, E.C. Ferrara, W.B. Focke, P. Fortin, M. Frailis, Y. Fukazawa, S. Funk, P. Fusco, F. Gargano, D. Gasparrini, N. Gehrels, S. Germani, N. Giglietto, F. Giordano, M. Giroletti, T. Glanzman, G. Godfrey, I.A. Grenier, M.-H. Grondin, J.E. Grove, S. Guiriec, D. Hadasch, Y. Hanabata, A.K. Harding, K. Hayashi, M. Hayashida, E. Hays, D. Horan, R. Itoh, G. Jóhannesson, A.S. Johnson, T.J. Johnson, D. Khangulyan, T. Kamae, H. Katagiri, J. Kataoka, M. Kerr, J. Knödlseder, M. Kuss, J. Lande, L. Latronico, S.-H. Lee, M. Lemoine-Goumard, F. Longo, F. Loparco, P. Lubrano, G.M. Madejski, A. Makeev, M. Marelli, M.N. Mazziotta, J.E. McEnery, P.F. Michelson, W. Mitthumsiri, T. Mizuno, A.A. Moiseev, C. Monte, M.E. Monzani, A. Morselli, I.V. Moskalenko, S. Murgia, T. Nakamori, M. Naumann-Godo, P.L. Nolan, J.P. Norris, E. Nuss, T. Ohsugi, A. Okumura, N. Omodei, J.F. Ormes, M. Ozaki, D. Paneque, D. Parent, V. Pelassa, M. Pepe, M. Pesce-Rollins, M. Pierbattista, F. Piron, T.A. Porter, S. Rainò, R. Rando, P.S. Ray, M. Razzano, A. Reimer, O. Reimer, T. Reposeur, S. Ritz, R.W. Romani, H.F.-W. Sadrozinski, D. Sanchez, P.M.S. Parkinson, J.D. Scargle, T.L. Schalk, C. Sgrò, E.J. Siskind, P.D. Smith, G. Spandre, P. Spinelli, M.S. Strickman, D.J. Suson, H. Takahashi, T. Takahashi, T. Tanaka, J.B. Thayer, D.J. Thompson, L. Tibaldo, D.F. Torres, G. Tosti, A. Tramacere, E. Troja, Y. Uchiyama, J. Vandenbroucke, V. Vasileiou, G. Vianello, V. Vitale, P. Wang, K.S. Wood, Z. Yang, M. Ziegler, Gamma-ray flares from the Crab Nebula. Science **331**, 739 (2011). doi:10.1126/science.1199705

F. Aharonian, A.G. Akhperjanian, A.R. Bazer-Bachi, B. Behera, M. Beilicke, W. Benbow, D. Berge, K. Bernlöhr, C. Boisson, O. Bolz, V. Borrel, T. Boutelier, I. Braun, E. Brion, A.M. Brown, R. Bühler, I. Büsching, T. Bulik, S. Carrigan, P.M. Chadwick, A.C. Clapson, L.-M. Chounet, G. Coignet, R. Cornils, L. Costamante, B. Degrange, H.J. Dickinson, A. Djannati-Ataï, W. Domainko, L.O. Drury, G. Dubus, J. Dyks, K. Egberts, D. Emmanoulopoulos, P. Espigat, C. Farnier, F. Feinstein, A. Fiasson, A. Förster, G. Fontaine, S. Funk, S. Funk, M. Füßling, Y.A. Gallant, B. Giebels, J.F. Glicenstein, B. Glück, P. Goret, C. Hadjichristidis, D. Hauser, M. Hauser, G. Heinzelmann, G. Henri, G. Hermann, J.A. Hinton, A. Hoffmann, W. Hofmann, M. Holleran, S. Hoppe, D. Horns, A. Jacholkowska, O.C. de Jager, E. Kendziorra, M. Kerschhaggl, B. Khélifi, N. Komin, K. Kosack, G. Lamanna, I.J. Latham, R. Le Gallou, A. Lemière, M. Lemoine-Goumard, J.-P. Lenain, T. Lohse, J.M. Martin, O. Martineau-Huynh, A. Marcowith, C. Masterson, G. Maurin, T.J.L. McComb, R. Moderski, E. Moulin, M. de Naurois, D. Nedbal, S.J. Nolan, J.-P. Olive, K.J. Orford, J.L. Osborne, M. Ostrowski, M. Panter, G. Pedaletti, G. Pelletier, P.-O. Petrucci, S. Pita, G. Pühlhofer, M. Punch, S. Ranchon, B.C. Raubenheimer, M. Raue, S.M. Rayner, M. Renaud, J. Ripken, L. Rob, L. Rolland, S. Rosier-Lees, G. Rowell, B. Rudak, J. Ruppel, V. Sahakian, A. Santangelo, L. Saugé, S. Schlenker, R. Schlickeiser, R. Schröder, U. Schwanke, S. Schwarzburg, S. Schwemmer, A. Shalchi, H. Sol, D. Spangler, Ł. Stawarz, R. Steenkamp, C. Stegmann, G. Superina, P.H. Tam, J.-P. Tavernet, R. Terrier, C. van Eldik, G. Vasileiadis, C. Venter, J.P. Vialle, P. Vincent, M. Vivier, H.J. Völk, F. Volpe, S.J. Wagner, M. Ward, A.A. Zdziarski, An exceptional very high energy gamma-ray flare of PKS 2155-304. Astrophys. J. Lett. **664**, 71–74 (2007). doi:10.1086/520635

J. Albert, E. Aliu, H. Anderhub, P. Antoranz, A. Armada, C. Baixeras, J.A. Barrio, H. Bartko, D. Bastieri, J.K. Becker, W. Bednarek, K. Berger, C. Bigongiari, A. Biland, R.K. Bock, P. Bordas, V. Bosch-Ramon, T. Bretz, I. Britvitch, M. Camara, E. Carmona, A. Chilingarian, J.A. Coarasa, S. Commichau, J.L. Contreras, J. Cortina, M.T. Costado, V. Curtef, V. Danielyan, F. Dazzi, A. De Angelis, C. Delgado, R. de los Reyes, B. De Lotto, E. Domingo-Santamaría, D. Dorner, M. Doro, M. Errando, M. Fagiolini, D. Ferenc, E. Fernández, R. Firpo, J. Flix, M.V. Fonseca, L. Font, M. Fuchs, N. Galante, R.J. García-López, M. Garczarczyk, M. Gaug, M. Giller, F. Goebel, D. Hakobyan, M. Hayashida, T. Hengstebeck, A. Herrero, D. Höhne, J. Hose, D. Hrupec, C.C. Hsu, P. Jacon, T. Jogler, R. Kosyra, D. Kranich, R. Kritzer, A.

Laille, E. Lindfors, S. Lombardi, F. Longo, J. López, M. López, E. Lorenz, P. Majumdar, G. Maneva, K. Mannheim, O. Mansutti, M. Mariotti, M. Martínez, D. Mazin, C. Merck, M. Meucci, M. Meyer, J.M. Miranda, R. Mirzoyan, S. Mizobuchi, A. Moralejo, D. Nieto, K. Nilsson, J. Ninkovic, E. Oña-Wilhelmi, N. Otte, I. Oya, D. Paneque, M. Panniello, R. Paoletti, J.M. Paredes, M. Pasanen, D. Pascoli, F. Pauss, R. Pegna, M. Persic, L. Peruzzo, A. Piccioli, E. Prandini, N. Puchades, A. Raymers, W. Rhode, M. Ribó, J. Rico, M. Rissi, A. Robert, S. Rügamer, A. Saggion, T. Saito, A. Sánchez, P. Sartori, V. Scalzotto, V. Scapin, R. Schmitt, T. Schweizer, M. Shayduk, K. Shinozaki, S.N. Shore, N. Sidro, A. Sillanpää, D. Sobczynska, A. Stamerra, L.S. Stark, L. Takalo, F. Tavecchio, P. Temnikov, D. Tescaro, M. Teshima, D.F. Torres, N. Turini, H. Vankov, V. Vitale, R.M. Wagner, T. Wibig, W. Wittek, F. Zandanel, R. Zanin, J. Zapatero, Variable very high energy γ-ray emission from Markarian 501. Astrophys. J. **669**, 862–883 (2007). doi:10.1086/521382

I. Arka, G. Dubus, Pulsed high-energy γ-rays from thermal populations in the current sheets of pulsar winds. Astron. Astrophys. **550**, 101 (2013). doi:10.1051/0004-6361/201220110

G. Baumann, Å. Nordlund, Particle-in-cell simulation of electron acceleration in solar coronal jets. Astrophys. J. Lett. **759**, 9 (2012). doi:10.1088/2041-8205/759/1/L9

M.C. Begelman, Instability of toroidal magnetic field in jets and plerions. Astrophys. J. **493**, 291–300 (1998). doi:10.1086/305119

M.C. Begelman, A.C. Fabian, M.J. Rees, Implications of very rapid TeV variability in blazars. Mon. Not. R. Astron. Soc. **384**, 19–23 (2008). doi:10.1111/j.1745-3933.2007.00413.x

N. Bessho, A. Bhattacharjee, Collisionless reconnection in an electron-positron plasma. Phys. Rev. Lett. **95**(24), 245001 (2005). doi:10.1103/PhysRevLett.95.245001

N. Bessho, A. Bhattacharjee, Fast collisionless reconnection in electron-positron plasmas. Phys. Plasmas **14**(5), 056503 (2007). doi:10.1063/1.2714020

N. Bessho, A. Bhattacharjee, Fast magnetic reconnection and particle acceleration in relativistic low-density electron-positron plasmas without guide field. Astrophys. J. **750**, 129 (2012). doi:10.1088/0004-637X/750/2/129

A. Bhattacharjee, Y.-M. Huang, H. Yang, B. Rogers, Fast reconnection in high-Lundquist-number plasmas due to the plasmoid instability. Phys. Plasmas **16**(11), 112102 (2009). doi:10.1063/1.3264103

C.K. Birdsall, A.B. Langdon, *Plasma Physics via Computer Simulation* (1991)

J. Birn, M. Hesse, Geospace Environment Modeling (GEM) magnetic reconnection challenge: resistive tearing, anisotropic pressure and Hall effects. J. Geophys. Res. **106**, 3737–3750 (2001). doi:10.1029/1999JA001001

D. Biskamp, *Magnetic Reconnection in Plasmas* (2000)

R.D. Blandford, D.G. Payne, Hydromagnetic flows from accretion discs and the production of radio jets. Mon. Not. R. Astron. Soc. **199**, 883–903 (1982)

B. Cerutti, D.A. Uzdensky, M.C. Begelman, Extreme particle acceleration in magnetic reconnection layers: application to the gamma-ray flares in the Crab Nebula. Astrophys. J. **746**, 148 (2012a). doi:10.1088/0004-637X/746/2/148

B. Cerutti, G.R. Werner, D.A. Uzdensky, M.C. Begelman, Beaming and rapid variability of high-energy radiation from relativistic pair plasma reconnection. Astrophys. J. Lett. **754**, 33 (2012b). doi:10.1088/2041-8205/754/2/L33

B. Cerutti, G.R. Werner, D.A. Uzdensky, M.C. Begelman, Simulations of particle acceleration beyond the classical synchrotron burnoff limit in magnetic reconnection: an explanation of the Crab Flares. Astrophys. J. **770**, 147 (2013). doi:10.1088/0004-637X/770/2/147

B. Cerutti, G.R. Werner, D.A. Uzdensky, M.C. Begelman, Gamma-ray flares in the Crab Nebula: a case of relativistic reconnection?. Phys. Plasmas **21**(5), 056501 (2014a). doi:10.1063/1.4872024

B. Cerutti, G.R. Werner, D.A. Uzdensky, M.C. Begelman, Three-dimensional relativistic pair plasma reconnection with radiative feedback in the Crab Nebula. Astrophys. J. **782**, 104 (2014b). doi:10.1088/0004-637X/782/2/104

L.-J. Chen, A. Bhattacharjee, P.A. Puhl-Quinn, H. Yang, N. Bessho, S. Imada, S. Mühlbachler, P.W. Daly, B. Lefebvre, Y. Khotyaintsev, A. Vaivads, A. Fazakerley, E. Georgescu, Observation of energetic electrons within magnetic islands. Nat. Phys. **4**, 19–23 (2008). doi:10.1038/nphys777

E. Clausen-Brown, M. Lyutikov, Crab Nebula gamma-ray flares as relativistic reconnection minijets. Mon. Not. R. Astron. Soc. **426**, 1374–1384 (2012). doi:10.1111/j.1365-2966.2012.21349.x

L. Comisso, F.A. Asenjo, Thermal-inertial effects on magnetic reconnection in relativistic pair plasmas. Phys. Rev. Lett. **113**(4), 045001 (2014). doi:10.1103/PhysRevLett.113.045001

I. Contopoulos, The magnetic field topology in the reconnecting pulsar magnetosphere. Astron. Astrophys. **472**, 219–223 (2007). doi:10.1051/0004-6361:20077167

F.V. Coroniti, Magnetically striped relativistic magnetohydrodynamic winds—the Crab Nebula revisited. Astrophys. J. **349**, 538–545 (1990). doi:10.1086/168340

W. Daughton, H. Karimabadi, Collisionless magnetic reconnection in large-scale electron-positron plasmas. Phys. Plasmas **14**(7), 072303 (2007). doi:10.1063/1.2749494

O.C. de Jager, A.K. Harding, P.F. Michelson, H.I. Nel, P.L. Nolan, P. Sreekumar, D.J. Thompson, Gamma-ray observations of the Crab Nebula: a study of the synchro-Compton spectrum. Astrophys. J. **457**, 253 (1996). doi:10.1086/176726

J.F. Drake, M. Swisdak, H. Che, M.A. Shay, Electron acceleration from contracting magnetic islands during reconnection. Nature **443**, 553–556 (2006). doi:10.1038/nature05116

J.F. Drake, M. Opher, M. Swisdak, J.N. Chamoun, A magnetic reconnection mechanism for the generation of anomalous cosmic rays. Astrophys. J. **709**, 963–974 (2010). doi:10.1088/0004-637X/709/2/963

J.D. Finke, C.D. Dermer, M. Böttcher, Synchrotron self-Compton analysis of TeV X-ray-selected BL Lacertae Objects. Astrophys. J. **686**, 181–194 (2008). doi:10.1086/590900

G. Ghisellini, F. Tavecchio, Rapid variability in TeV blazars: the case of PKS2155-304. Mon. Not. R. Astron. Soc. **386**, 28–32 (2008). doi:10.1111/j.1745-3933.2008.00454.x

D. Giannios, UHECRs from magnetic reconnection in relativistic jets. Mon. Not. R. Astron. Soc. **408**, 46–50 (2010). doi:10.1111/j.1745-3933.2010.00925.x

D. Giannios, Reconnection-driven plasmoids in blazars: fast flares on a slow envelope. Mon. Not. R. Astron. Soc. **431**, 355–363 (2013). doi:10.1093/mnras/stt167

D. Giannios, H.C. Spruit, Spectra of Poynting-flux powered GRB outflows. Astron. Astrophys. **430**, 1–7 (2005). doi:10.1051/0004-6361:20047033

D. Giannios, D.A. Uzdensky, M.C. Begelman, Fast TeV variability in blazars: jets in a jet. Mon. Not. R. Astron. Soc. **395**, 29–33 (2009). doi:10.1111/j.1745-3933.2009.00635.x

D. Giannios, D.A. Uzdensky, M.C. Begelman, Fast TeV variability from misaligned minijets in the jet of M87. Mon. Not. R. Astron. Soc. **402**, 1649–1656 (2010). doi:10.1111/j.1365-2966.2009.16045.x

J. Goodman, D. Uzdensky, Reconnection in marginally collisionless accretion disk coronae. Astrophys. J. **688**, 555–558 (2008). doi:10.1086/592345

J. Graf von der Pahlen, D. Tsiklauri, The effect of guide-field and boundary conditions on collisionless magnetic reconnection in a stressed X-point collapse. Phys. Plasmas **21**(1), 012901 (2014). doi:10.1063/1.4861258

P.W. Guilbert, A.C. Fabian, M.J. Rees, Spectral and variability constraints on compact sources. Mon. Not. R. Astron. Soc. **205**, 593–603 (1983)

F. Guo, H. Li, W. Daughton, Y.-H. Liu, Formation of hard power-laws in the energetic particle spectra resulting from relativistic magnetic reconnection. ArXiv e-prints, 2014

E.G. Harris, On a plasma sheath separating regions of oppositely directed magnetic field. Nuovo Cimento **23**, 115–121 (1962)

K. Higashimori, M. Hoshino, The relation between ion temperature anisotropy and formation of slow shocks in collisionless magnetic reconnection. J. Geophys. Res. Space Sci. **117**, 1220 (2012). doi:10.1029/2011JA016817

Y.-M. Huang, A. Bhattacharjee, Scaling laws of resistive magnetohydrodynamic reconnection in the high-Lundquist-number, plasmoid-unstable regime. Phys. Plasmas **17**(6), 062104 (2010). doi:10.1063/1.3420208

C.H. Jaroschek, H. Lesch, R.A. Treumann, Relativistic kinetic reconnection as the possible source mechanism for high variability and flat spectra in extragalactic radio sources. Astrophys. J. Lett. **605**, 9–12 (2004). doi:10.1086/420767

C.H. Jaroschek, M. Hoshino, H. Lesch, R.A. Treumann, Stochastic particle acceleration by the forced interaction of relativistic current sheets. Adv. Space Res. **41**, 481–490 (2008). doi:10.1016/j.asr.2007.07.001

H. Ji, W. Daughton, Phase diagram for magnetic reconnection in heliophysical, astrophysical, and laboratory plasmas. Phys. Plasmas **18**(11), 111207 (2011). doi:10.1063/1.3647505

D. Kagan, M. Milosavljević, A. Spitkovsky, A flux rope network and particle acceleration in three-dimensional relativistic magnetic reconnection. Astrophys. J. **774**, 41 (2013). doi:10.1088/0004-637X/774/1/41

M. Karlický, B. Kliem, Reconnection of a kinking flux rope triggering the ejection of a microwave and hard X-ray source I. Observations and interpretation. Sol. Phys. **266**, 71–89 (2010). doi:10.1007/s11207-010-9606-4

J.G. Kirk, Particle acceleration in relativistic current sheets. Phys. Rev. Lett. **92**(18), 181101 (2004). doi:10.1103/PhysRevLett.92.181101

J.G. Kirk, O. Skjæraasen, Dissipation in Poynting-flux-dominated flows: the σ-problem of the Crab Pulsar wind. Astrophys. J. **591**, 366–379 (2003). doi:10.1086/375215

J.G. Kirk, O. Skjæraasen, Y.A. Gallant, Pulsed radiation from neutron star winds. Astron. Astrophys. **388**, 29–32 (2002). doi:10.1051/0004-6361:20020599

S.S. Komissarov, Magnetic dissipation in the Crab Nebula. Mon. Not. R. Astron. Soc. **428**, 2459–2466 (2013). doi:10.1093/mnras/sts214

D.A. Larrabee, R.V.E. Lovelace, M.M. Romanova, Lepton acceleration by relativistic collisionless magnetic reconnection. Astrophys. J. **586**, 72–78 (2003). doi:10.1086/367640

J. Lin, Y.-K. Ko, L. Sui, J.C. Raymond, G.A. Stenborg, Y. Jiang, S. Zhao, S. Mancuso, Direct observations of the magnetic reconnection site of an eruption on 2003 November 18. Astrophys. J. **622**, 1251–1264 (2005). doi:10.1086/428110

M.L. Lister, M.H. Cohen, D.C. Homan, M. Kadler, K.I. Kellermann, Y.Y. Kovalev, E. Ros, T. Savolainen, J.A. Zensus, MOJAVE: monitoring of jets in active galactic nuclei with VLBA experiments. VI. Kinematics analysis of a complete sample of blazar jets. Astron. J. **138**, 1874–1892 (2009). doi:10.1088/0004-6256/138/6/1874

W. Liu, H. Li, L. Yin, B.J. Albright, K.J. Bowers, E.P. Liang, Particle energization in 3D magnetic reconnection of relativistic pair plasmas. Phys. Plasmas **18**(5), 052105 (2011). doi:10.1063/1.3589304

Y.-H. Liu, J.F. Drake, M. Swisdak, The structure of the magnetic reconnection exhaust boundary. Phys. Plasmas **19**(2), 022110 (2012). doi:10.1063/1.3685755

N.F. Loureiro, A.A. Schekochihin, S.C. Cowley, Instability of current sheets and formation of plasmoid chains. Phys. Plasmas **14**(10), 100703 (2007). doi:10.1063/1.2783986

N.F. Loureiro, R. Samtaney, A.A. Schekochihin, D.A. Uzdensky, Magnetic reconnection and stochastic plasmoid chains in high-Lundquist-number plasmas. Phys. Plasmas **19**(4), 042303 (2012). doi:10.1063/1.3703318

Y.E. Lyubarskij, Energy release in strongly magnetized relativistic winds. Sov. Astron. Lett. **18**, 356 (1992)

Y.E. Lyubarskii, A model for the energetic emission from pulsars. Astron. Astrophys. **311**, 172–178 (1996)

Y.E. Lyubarsky, The termination shock in a striped pulsar wind. Mon. Not. R. Astron. Soc. **345**, 153–160 (2003). doi:10.1046/j.1365-8711.2003.06927.x

Y.E. Lyubarsky, On the relativistic magnetic reconnection. Mon. Not. R. Astron. Soc. **358**, 113–119 (2005). doi:10.1111/j.1365-2966.2005.08767.x

Y.E. Lyubarsky, Transformation of the Poynting flux into kinetic energy in relativistic jets. Mon. Not. R. Astron. Soc. **402**, 353–361 (2010). doi:10.1111/j.1365-2966.2009.15877.x

Y.E. Lyubarsky, Highly magnetized region in pulsar wind nebulae and origin of the Crab gamma-ray flares. Mon. Not. R. Astron. Soc. **427**, 1497–1502 (2012). doi:10.1111/j.1365-2966.2012.22097.x

Y. Lyubarsky, J.G. Kirk, Reconnection in a striped pulsar wind. Astrophys. J. **547**, 437–448 (2001). doi:10.1086/318354

Y. Lyubarsky, M. Liverts, Particle acceleration in the driven relativistic reconnection. Astrophys. J. **682**, 1436–1442 (2008). doi:10.1086/589640

M. Lyutikov, The electromagnetic model of gamma-ray bursts. New J. Phys. **8**, 119 (2006). doi:10.1088/1367-2630/8/7/119

J.C. McKinney, D.A. Uzdensky, A reconnection switch to trigger gamma-ray burst jet dissipation. Mon. Not. R. Astron. Soc. **419**, 573–607 (2012). doi:10.1111/j.1365-2966.2011.19721.x

M. Melzani, C. Winisdoerffer, R. Walder, D. Folini, J.M. Favre, S. Krastanov, P. Messmer, Apar-T: code, validation, and physical interpretation of particle-in-cell results. Astron. Astrophys. **558**, 133 (2013). doi:10.1051/0004-6361/201321557

M. Melzani, R. Walder, D. Folini, C. Winisdoerffer, J.M. Favre, The energetics of relativistic magnetic reconnection: ion-electron repartition and particle distribution hardness. ArXiv e-prints, 2014

F.C. Michel, Magnetic structure of pulsar winds. Astrophys. J. **431**, 397–401 (1994). doi:10.1086/174493

A. Mignone, E. Striani, M. Tavani, A. Ferrari, Modelling the kinked jet of the Crab Nebula. Mon. Not. R. Astron. Soc. **436**, 1102–1115 (2013). doi:10.1093/mnras/stt1632

Y. Mizuno, Y. Lyubarsky, K.-I. Nishikawa, P.E. Hardee, Three-dimensional relativistic magnetohydrodynamic simulations of current-driven instability. II. Relaxation of pulsar wind nebula. Astrophys. J. **728**, 90 (2011). doi:10.1088/0004-637X/728/2/90

K. Nalewajko, D. Giannios, M.C. Begelman, D.A. Uzdensky, M. Sikora, Radiative properties of reconnection-powered minijets in blazars. Mon. Not. R. Astron. Soc. **413**, 333–346 (2011). doi:10.1111/j.1365-2966.2010.18140.x

R. Narayan, T. Piran, Variability in blazars: clues from PKS 2155-304. Mon. Not. R. Astron. Soc. **420**, 604–612 (2012). doi:10.1111/j.1365-2966.2011.20069.x

C. Nodes, G.T. Birk, H. Lesch, R. Schopper, Particle acceleration in three-dimensional tearing configurations. Phys. Plasmas **10**, 835–844 (2003). doi:10.1063/1.1542612

M. Øieroset, T.D. Phan, J.P. Eastwood, M. Fujimoto, W. Daughton, M.A. Shay, V. Angelopoulos, F.S. Mozer, J.P. McFadden, D.E. Larson, K.-H. Glassmeier, Direct evidence for a three-dimensional magnetic flux rope flanked by two active magnetic reconnection X lines at Earth's magnetopause. Phys. Rev. Lett. **107**(16), 165007 (2011). doi:10.1103/PhysRevLett.107.165007

M. Oka, T.-D. Phan, S. Krucker, M. Fujimoto, I. Shinohara, Electron acceleration by multi-island coalescence. Astrophys. J. **714**, 915–926 (2010). doi:10.1088/0004-637X/714/1/915

V. Olshevsky, G. Lapenta, S. Markidis, Energetics of kinetic reconnection in a three-dimensional null-point cluster. Phys. Rev. Lett. **111**, 045002 (2013). doi:10.1103/PhysRevLett.111.045002. http://link.aps.org/doi/10.1103/PhysRevLett.111.045002

H.K. Park, N.C. Luhmann Jr., A.J.H. Donné, I.G.J. Classen, C.W. Domier, E. Mazzucato, T. Munsat, M.J. van de Pol, Z. Xia, Observation of high-field-side crash and heat transfer during sawtooth oscillation in magnetically confined plasmas. Phys. Rev. Lett. **96**(19), 195003 (2006). doi:10.1103/PhysRevLett.96.195003

J. Pétri, High-energy emission from the pulsar striped wind: a synchrotron model for gamma-ray pulsars. Mon. Not. R. Astron. Soc. **424**, 2023–2027 (2012). doi:10.1111/j.1365-2966.2012.21350.x

J. Pétri, J.G. Kirk, Growth rates of the Weibel and tearing mode instabilities in a relativistic pair plasma. Plasma Phys. Control. Fusion **49**, 1885–1896 (2007). doi:10.1088/0741-3335/49/11/009

J. Pétri, Y. Lyubarsky, Magnetic reconnection at the termination shock in a striped pulsar wind. Astron. Astrophys. **473**, 683–700 (2007). doi:10.1051/0004-6361:20066981

H.E. Petschek, Magnetic field annihilation. NASA Spec. Publ. **50**, 425 (1964)

B.G. Piner, A.B. Pushkarev, Y.Y. Kovalev, C.J. Marvin, J.G. Arenson, P. Charlot, A.L. Fey, A. Collioud, P.A. Voitsik, Relativistic jets in the radio reference frame image database. II. Blazar jet accelerations from the first 10 years of data (1994–2003). Astrophys. J. **758**, 84 (2012). doi:10.1088/0004-637X/758/2/84

D.I. Pontin, Three-dimensional magnetic reconnection regimes: a review. Adv. Space Res. **47**, 1508–1522 (2011). doi:10.1016/j.asr.2010.12.022

O. Porth, S.S. Komissarov, R. Keppens, Solution to the sigma problem of pulsar wind nebulae. Mon. Not. R. Astron. Soc. **431**, 48–52 (2013). doi:10.1093/mnrasl/slt006

O. Porth, S.S. Komissarov, R. Keppens, Three-dimensional magnetohydrodynamic simulations of the Crab Nebula. Mon. Not. R. Astron. Soc. **438**, 278–306 (2014). doi:10.1093/mnras/stt2176

A.B. Pushkarev, Y.Y. Kovalev, M.L. Lister, T. Savolainen, Jet opening angles and gamma-ray brightness of AGN. Astron. Astrophys. **507**, 33–36 (2009). doi:10.1051/0004-6361/200913422

R. Samtaney, N.F. Loureiro, D.A. Uzdensky, A.A. Schekochihin, S.C. Cowley, Formation of plasmoid chains in magnetic reconnection. Phys. Rev. Lett. **103**(10), 105004 (2009). doi:10.1103/PhysRevLett.103.105004

T. Savolainen, D.C. Homan, T. Hovatta, M. Kadler, Y.Y. Kovalev, M.L. Lister, E. Ros, J.A. Zensus, Relativistic beaming and gamma-ray brightness of blazars. Astron. Astrophys. **512**, 24 (2010). doi:10.1051/0004-6361/200913740

L. Sironi, A. Spitkovsky, Acceleration of particles at the termination shock of a relativistic striped wind. Astrophys. J. **741**, 39 (2011a). doi:10.1088/0004-637X/741/1/39

L. Sironi, A. Spitkovsky, Particle acceleration in relativistic magnetized collisionless electron-ion shocks. Astrophys. J. **726**, 75 (2011b). doi:10.1088/0004-637X/726/2/75

L. Sironi, A. Spitkovsky, Particle-in-cell simulations of shock-driven reconnection in relativistic striped winds. Comput. Sci. Discov. **5**(1), 014014 (2012). doi:10.1088/1749-4699/5/1/014014

L. Sironi, A. Spitkovsky, Relativistic reconnection: an efficient source of non-thermal particles. Astrophys. J. Lett. **783**, 21 (2014). doi:10.1088/2041-8205/783/1/L21

L. Sironi, A. Spitkovsky, J. Arons, The maximum energy of accelerated particles in relativistic collisionless shocks. Astrophys. J. **771**, 54 (2013). doi:10.1088/0004-637X/771/1/54

T.W. Speiser, Particle trajectories in model current sheets. 1. Analytical solutions. J. Geophys. Res. **70**, 4219–4226 (1965). doi:10.1029/JZ070i017p04219

Y.T. Tanaka, Ł. Stawarz, D.J. Thompson, F. D'Ammando, S.J. Fegan, B. Lott, D.L. Wood, C.C. Cheung, J. Finke, S. Buson, L. Escande, S. Saito, M. Ohno, T. Takahashi, D. Donato, J. Chiang, M. Giroletti, F.K. Schinzel, G. Iafrate, F. Longo, S. Ciprini, Fermi large area telescope detection of bright γ-ray outbursts from the peculiar quasar 4C +21.35. Astrophys. J. **733**, 19 (2011). doi:10.1088/0004-637X/733/1/19

M. Tavani, A. Bulgarelli, V. Vittorini, A. Pellizzoni, E. Striani, P. Caraveo, M.C. Weisskopf, A. Tennant, G. Pucella, A. Trois, E. Costa, Y. Evangelista, C. Pittori, F. Verrecchia, E. Del Monte, R. Campana, M. Pilia, A. De Luca, I. Donnarumma, D. Horns, C. Ferrigno, C.O. Heinke, M. Trifoglio, F. Gianotti, S. Vercellone, A. Argan, G. Barbiellini, P.W. Cattaneo, A.W. Chen, T. Contessi, F. D'Ammando, G. DeParis, G. Di Cocco, G. Di Persio, M. Feroci, A. Ferrari, M. Galli, A. Giuliani, M. Giusti, C. Labanti, I. Lapshov, F. Lazzarotto, P. Lipari, F. Longo, F. Fuschino, M. Marisaldi, S. Mereghetti, E. Morelli, E. Moretti, A. Morselli, L. Pacciani, F. Perotti, G. Piano, P. Picozza, M. Prest, M. Rapisarda, A. Rappoldi, A. Rubini, S. Sabatini, P. Soffitta, E. Vallazza, A. Zambra, D. Zanello, F. Lucarelli, P. Santolamazza, P. Giommi, L. Salotti, G.F. Bignami, Discovery of powerful gamma-ray flares from the Crab Nebula. Science **331**, 736 (2011). doi:10.1126/science.1200083

F. Tavecchio, J. Becerra-Gonzalez, G. Ghisellini, A. Stamerra, G. Bonnoli, L. Foschini, L. Maraschi, On the origin of the γ-ray emission from the flaring blazar PKS 1222+216. Astron. Astrophys. **534**, 86 (2011). doi:10.1051/0004-6361/201117204

C.M. Urry, P. Padovani, Unified schemes for radio-loud active galactic nuclei. Publ. Astron. Soc. Pac. **107**, 803 (1995). doi:10.1086/133630

D.A. Uzdensky, R.M. Kulsrud, Two-dimensional numerical simulation of the resistive reconnection layer. Phys. Plasmas **7**, 4018–4030 (2000). doi:10.1063/1.1308081

D.A. Uzdensky, A. Spitkovsky, Physical conditions in the reconnection layer in pulsar magnetospheres. Astrophys. J. **780**, 3 (2014). doi:10.1088/0004-637X/780/1/3

D.A. Uzdensky, B. Cerutti, M.C. Begelman, Reconnection-powered linear accelerator and gamma-ray flares in the Crab Nebula. Astrophys. J. Lett. **737**, 40 (2011). doi:10.1088/2041-8205/737/2/L40

D.A. Uzdensky, N.F. Loureiro, A.A. Schekochihin, Fast magnetic reconnection in the plasmoid-dominated regime. Phys. Rev. Lett. **105**(23), 235002 (2010). doi:10.1103/PhysRevLett.105.235002

G.R. Werner, D.A. Uzdensky, B. Cerutti, K. Nalewajko, M.C. Begelman, The extent of power-law energy spectra in collisionless relativistic magnetic reconnection in pair plasmas. ArXiv e-prints, 2014

L. Yin, W. Daughton, H. Karimabadi, B.J. Albright, K.J. Bowers, J. Margulies, Three-dimensional dynamics of collisionless magnetic reconnection in large-scale pair plasmas. Phys. Rev. Lett. **101**(12), 125001 (2008). doi:10.1103/PhysRevLett.101.125001

O. Zanotti, M. Dumbser, Numerical simulations of high Lundquist number relativistic magnetic reconnection. Mon. Not. R. Astron. Soc. **418**, 1004–1011 (2011). doi:10.1111/j.1365-2966.2011.19551.x

S. Zenitani, M. Hesse, Self-regulation of the reconnecting current layer in relativistic pair plasma reconnection. Astrophys. J. **684**, 1477–1485 (2008). doi:10.1086/590425

S. Zenitani, M. Hoshino, The generation of nonthermal particles in the relativistic magnetic reconnection of pair plasmas. Astrophys. J. Lett. **562**, 63–66 (2001). doi:10.1086/337972

S. Zenitani, M. Hoshino, Relativistic particle acceleration in a folded current sheet. Astrophys. J. Lett. **618**, 111–114 (2005). doi:10.1086/427873

S. Zenitani, M. Hoshino, Particle acceleration and magnetic dissipation in relativistic current sheet of pair plasmas. Astrophys. J. **670**, 702–726 (2007). doi:10.1086/522226

S. Zenitani, M. Hoshino, The role of the guide field in relativistic pair plasma reconnection. Astrophys. J. **677**, 530–544 (2008). doi:10.1086/528708

B. Zhang, H. Yan, The internal-collision-induced magnetic reconnection and turbulence (ICMART) model of gamma-ray bursts. Astrophys. J. **726**, 90 (2011). doi:10.1088/0004-637X/726/2/90

MIX
Papier aus verantwortungsvollen Quellen
Paper from responsible sources
FSC® C105338

If you have any concerns about our products,
you can contact us on
ProductSafety@springernature.com

In case Publisher is established outside the EU,
the EU authorized representative is:
Springer Nature Customer Service Center GmbH
Europaplatz 3, 69115 Heidelberg, Germany

Printed by Libri Plureos GmbH
in Hamburg, Germany